T0190070

Noise in Nanoscale Semiconductor Devices

Tibor Grasser
Editor

Noise in Nanoscale Semiconductor Devices

Editor
Tibor Grasser
Institute for Microelectronics
Technische Universität Wien
Wien, Austria

ISBN 978-3-030-37502-7 ISBN 978-3-030-37500-3 (eBook)
https://doi.org/10.1007/978-3-030-37500-3

This Springer imprint is published by the registered company Springer Nature Switzerland AG.
The registered company address is: Gewerbestrasse 11, 6330 Cham, Switzerland

Contents

Origins of 1/f Noise in Electronic Materials and Devices: A Historical Perspective

D. M. Fleetwood

1 Introduction

Low-frequency (1/f) noise is observed in a remarkable number of diverse cosmological [1–3], biological [4–7], economic [8–10], and electronic materials and devices [11–48]. It has also been found in music and speech [49–51], dynamics of sandpiles and avalanches [52–54], automotive and computer-network traffic flow [55–57], and statistics of earthquakes, floods, and other natural phenomena [1, 50, 58, 59]. Due to its technological relevance, the noise of electronic materials has received the most attention.

From its discovery in 1925 by Johnson [11] until the late 1960s, it was generally accepted that the low-frequency excess (1/f) noise of electronic materials and devices was caused primarily by defects and impurities [12–17]. The 1/f noise of semiconductor devices was considered to be a surface effect caused by surface charge trapping [14]. This model was generalized and extended to MOSFETs (metal–oxide–semiconductor field-effect transistors) shortly after their introduction into commercial use [15–17, 22, 25]. The discovery of 1/f noise in thin metal films [19] and its similarly-scaled magnitude to noise in semiconductors [18] led Hooge and Vandamme to propose instead that the noise was due to bulk mobility fluctuations caused by lattice (phonon) scattering [21]. These and other models were tested extensively from ~1969 to ~1994 [1, 23, 24, 27, 29–31, 36, 37, 39–42, 48]. After a lot of debate, it is now clear that, for most technologically relevant electronic devices, carrier number fluctuations due to charge trapping predominantly cause the observed noise [29, 31, 38, 48]. For thin metal films, the noise is generally due to mobility fluctuations caused by carrier-defect scattering [24, 28–30, 48]. The

D. M. Fleetwood (✉)
Vanderbilt University, Nashville, TN, USA
e-mail: dan.fleetwood@vanderbilt.edu

T. Grasser (ed.), *Noise in Nanoscale Semiconductor Devices*,
https://doi.org/10.1007/978-3-030-37500-3_1

defect-activation model of Dutta and Horn describes the noise of most metals and semiconductor devices remarkably well [24, 29, 48]. Over the last ~ 25 years, noise measurements have been increasingly used in combination with other experimental, computational, and theoretical methods to obtain information about the defects and impurities that are the underlying cause of the observed fluctuations [29, 31, 35, 37, 39, 43–46, 48].

In this chapter, the principles that underlie the number and (Hooge) mobility fluctuation models of low-frequency noise in electronic devices are reviewed. Evidence is summarized that demonstrates the applicability of the number fluctuation model [14, 17] and the importance of defects and impurities [48] to the noise of typical electronic devices, e.g., MOSFETs (including those based on SiC and two-dimensional materials) and HEMTs (high electron mobility transistors). The Dutta–Horn model of $1/f$ noise [24] is briefly reviewed. This model provides a detailed description of the effective energy distributions of defects that largely determine the low-frequency noise of metal films and microelectronic devices and materials [48]. Examples are provided for a number of diverse material systems that emphasize its general applicability and utility.

2 Number Fluctuations: Application to MOSFETs

A random process with a single characteristic time τ leads to a Lorentzian power spectrum Eq. (1). At high frequency, the voltage-noise power spectral density S_V scales as $\sim 1/f^2$, and at low frequency, the noise is nearly independent of frequency [1, 11, 13]. Bernamont showed in 1937 [12] that, if the noise results from processes with a distribution of characteristic times $D(\tau)$, and if $D(\tau) \sim 1/\tau$ for times $\tau_1 < \tau < \tau_2$, and the pre-factor A is independent of frequency, then the resulting noise,

$$S_V = A \int_{\tau_1}^{\tau_2} \left[D(\tau) \left(1 + \omega^2 \tau^2\right)^{-1} \right] d\,\tau, \tag{1}$$

is proportional to $\sim 1/f$ for $1/\tau_2 < f < 1/\tau_1$ [1, 11–14]. Typically, the observed noise results from a thermally activated random process, for which

$$\tau = \tau_o \exp\left(E/kT\right), \tag{2}$$

where τ_o is a constant, E is the activation energy, k is the Boltzmann constant, and T is the temperature. When $D(E)$ is nearly constant, then $D(\tau)$ is proportional to $\sim 1/\tau$, the above conditions are satisfied, and $1/f$ noise is found [24, 29, 48].

Noise due to pure tunneling processes is also observed, more often at low temperature than at room temperature [60]. A first-order tunneling model was developed by McWhorter in 1957 to describe noise in Ge due to the exchange of carriers between the semiconductor and surface oxide traps [14]. The simplest form of the model, as applied to a large area Si MOSFET with oxide thicker than ~3–5 nm, attributes the noise to tunnel-assisted charge exchange between the channel and defects in the near-interfacial SiO_2 [14–17]. For defects distributed approximately uniformly in space within the near-interfacial oxide and uniformly in energy within the band gap, fluctuations in the density of traps N_t lead to a power-spectral density S_{N_t} given by [14, 17, 61]:

$$S_{N_t}(f, T) = \frac{kT\, D_t(E_f)}{\text{LW}\ \ln(\tau_1/\tau_0)\, f} \tag{3}$$

where $D_t(E_f)$ is the number of traps per unit energy per unit area at the Fermi level E_f (Fig. 1 [62]), and τ_0 and τ_1 are the minimum and maximum tunneling times, respectively [14, 17, 39, 61].

For constant drain current I_d and gate bias V_g, in the linear region of MOS operation, fluctuations in gate voltage δV_g are related to changes in trapped charge density δQ_t via:

$$\delta V_g = \delta Q_t/C_{ox} = q\,\delta N_t/C_{ox} \tag{4}$$

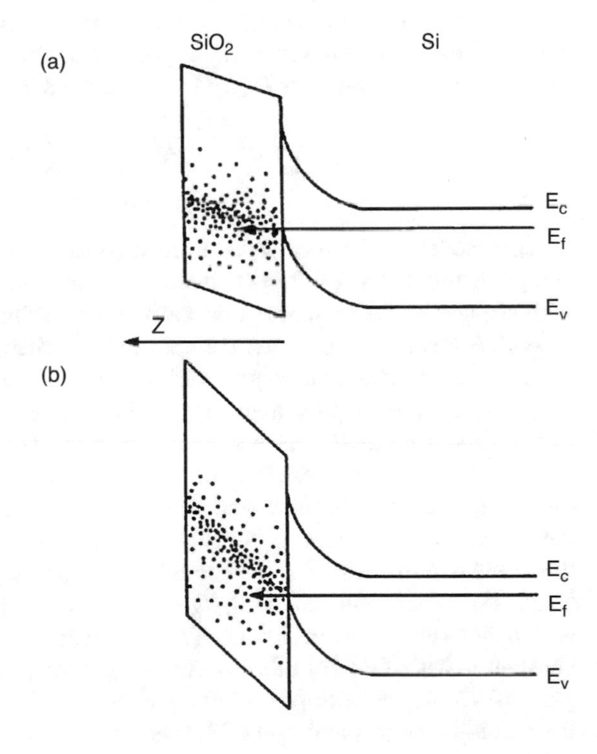

Fig. 1 Energy bands for a *p*MOS Si/SiO_2 transistor for (**a**) lower and (**b**) higher applied electric field. The dots represent trapping sites in SiO_2. (After Surya and Hsiang [62], © 1986, American Institute of Physics, AIP)

where C_{ox} is the gate oxide capacitance per unit area, and q is the magnitude of the electron charge. Fluctuations in gate voltage δV_g lead to fluctuations in drain voltage δV_d [17, 61]:

$$\delta V_d = \left(\partial V_d/\partial V_g\right)\, \delta V_g. \tag{5}$$

For a MOSFET in the linear mode of operation, in terms of δN_t, this reduces to [17, 61]:

$$\delta V_d = \frac{qV_d}{C_{ox}\left(V_g - V_t\right)} \delta N_t \tag{6}$$

where V_t, V_g, and V_d are the threshold, gate, and drain voltages. Thus, drain voltage fluctuations are related to the fluctuations in the trapped charge density via [17, 61]:

$$S_{V_d}\left(f, T\right) = \frac{q^2 {V_d}^2}{{C_{ox}}^2 \left(V_g - V_t\right)^2} S_{N_t}\left(f, T\right) \tag{7}$$

where S_{V_d} is the excess drain-voltage noise power spectral density, which is the quantity measured in a typical noise study.

Combining Eqs. (3) and (7) yields a first-order expression that relates the measured excess drain-voltage noise power spectral density S_{V_d} of a MOSFET to the density of relevant traps $D_t(E_f)$ [17, 25, 39, 48, 61]:

$$S_{V_d} = \frac{q^2}{C_{ox}^2} \frac{V_d^2}{\left(V_g - V_t\right)^2} \frac{k_B \mathrm{T} D_t\left(E_f\right)}{\mathrm{LW} \ln\left(\tau_1/\tau_0\right)} \frac{1}{f}. \tag{8}$$

This model has been extended to include the effects of nonuniform spatial and energy distributions [24, 29, 48] and/or correlated mobility fluctuations associated with changes in charge states of the defects responsible for the noise [33, 63].

Figure 1 shows that, when the gate bias is changed, the Si surface potential changes only slightly, but band bending is more significant in the SiO_2 insulator [61, 62]. Thus, varying the applied gate bias enables one to probe different regions of semiconductor and/or insulator band gaps via 1/*f* noise measurements [25, 38, 48, 61, 62]. In the simplest form of the number fluctuation model, which assumes tunneling is the rate-limiting step that leads to 1/*f* noise, the portion of the SiO_2 defect energy distribution most easily accessible to measurements is that closest to the Si and within a few kT of the Fermi level in energy [13, 14, 17, 38, 48, 64]. When $D_t(E_f)$ is approximately constant, $S_{V_d} \sim (V_g - V_t)^{-2}$, as a direct consequence of the assumptions embodied in Eqs. (3)–(7). However, neither $D_t(E_f)$ nor $D(\tau)$ are often constant in MOSFETs or other electronic devices [24, 29, 35, 48, 65]. Hence, the gate voltage dependence of the noise can be more complex than anticipated under the assumptions of the original McWhorter model [14, 38, 48, 66]. Nonconstant values of $D_t(E_f)$ and/or $D(\tau)$ also lead to deviations of the frequency dependence of the noise from a pure 1/*f* power law [24, 29, 48].

3 Mobility Fluctuations: Hooge's Model

A comparison of published studies of low-frequency noise in a large number of semiconductor devices and materials (with surface oxides) and measurements of noise in thin, continuous gold films (with no surface oxides) [19] led Hooge in 1969 to assert that "1/*f* noise is no surface effect" [18]. The empirical observation of Hooge was that:

$$S_V/{V_d}^2 = S_I/I^2 \approx \alpha_H/N_c f \tag{9}$$

where α_H is a dimensionless parameter and N_c is the number of carriers in the material. Hooge estimated α_H to be of the order of 2×10^{-3}, with about two orders of magnitude variation among the materials considered [18]. When applied to a MOSFET, this expression becomes:

$$S_V/{V_d}^2 \approx \alpha_H\,(q/\mathrm{LW}C_{ox})\,f^{-1}\left(V_g-V_t\right)^{-1}. \tag{10}$$

Hence, the gate voltage dependence of the noise differs from that in the number fluctuation model in Eq. (8).

A further comparison of the 1/*f* noise in Ge and GaAs materials with different ratios of lattice (phonon) and impurity scattering led Hooge and Vandamme to conclude that "lattice scattering causes 1/*f* noise" [21] and impurity scattering does not. Efforts were made to derive a theoretical foundation for this model [67–69]; however, Eqs. (9) and (10) remain mostly empirical [18, 19, 21, 23, 24, 29, 34, 36, 40].

The simplicity of Eqs. (9) and (10) and their utility as a first-order benchmark of noise magnitudes [22–24, 27, 29, 39–41] and deviations of the voltage dependence of the noise of some semiconductor devices from Eq. (8) with constant $D_t(E_f)$ [23, 34, 36, 38, 40, 48, 66, 69] have led the Hooge mobility model to remain popular [40, 47, 69–71]. This is true despite (1) its lack of firm theoretical foundation; (2) internal inconsistencies, e.g., impurity scattering is treated differently in [18, 21], and individual carriers typically are not present in devices long enough (ms to s time scales) to exhibit the fluctuations in mobility needed to account for the noise, as pointed out in the 1980s by Weissman [29, 72, 73]; and (3) compelling evidence that defects and impurities are primarily responsible for the noise of microelectronic devices and materials [24, 29, 38, 48], as discussed below.

4 Noise in Metals: Dutta–Horn Model

The observation of 1/*f* noise in thin gold films, as well as additional studies showing 1/*f* noise in a wealth of unrelated systems [1, 4, 20, 49, 50, 52], led to a burst of experimental and theoretical activity in the 1970s and 1980s attempting to discover and/or develop fundamental theories of the fluctuations and their underlying origins. Studies of the noise of metal films led to great progress [24, 29, 30].

Figure 2 illustrates two basic features of the noise of metals. In Fig. 2a, it is shown that $S_V \sim 1/f^{1.15}$ for 0.002 Hz $< f <$ 100 Hz [74]. The low frequency limit in Fig. 1 is determined by the measuring time, and the high frequency limit is determined by the relative magnitudes of the $1/f$ noise and thermal noise. Figure 2b confirms that the noise magnitude of Pt films and wires increases inversely with decreasing sample volume ($\sim 1/N_A$, where N_A is the number of atoms, which for metals is approximately equal to the number of carriers N_c) [74].

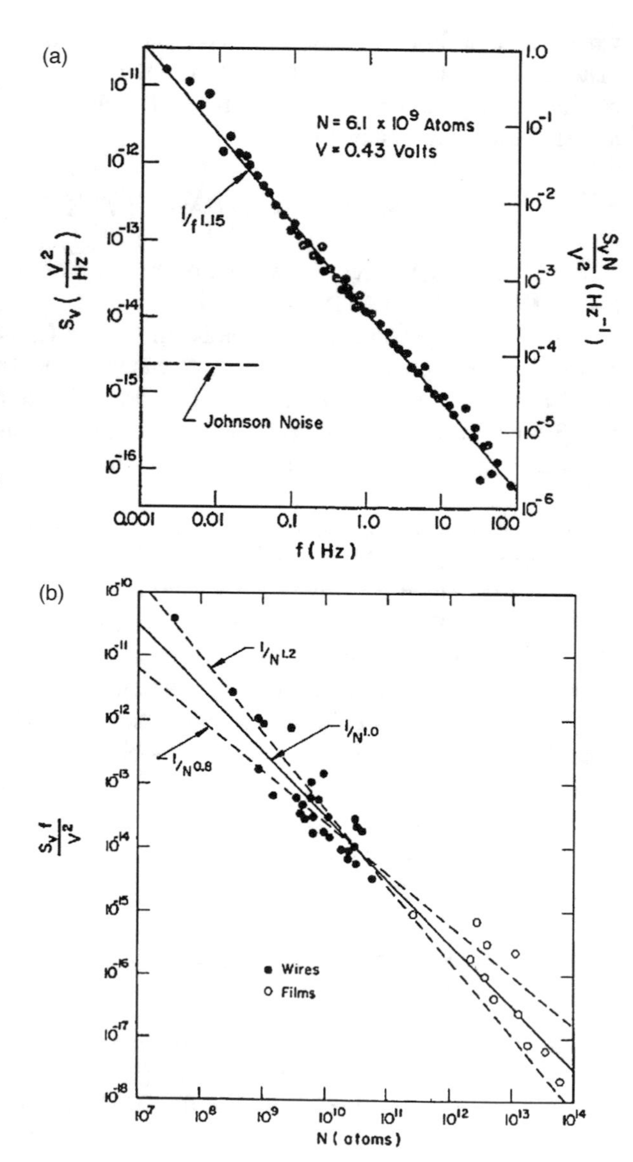

Fig. 2 (**a**) Excess voltage-noise power spectral density S_V (left-hand scale) and normalized noise magnitude $S_V N/V^2$ (right-hand scale; N is the number of atoms) as a function of f for a platinum nanowire. The Johnson (thermal) noise level for this wire is indicated, and subtracted to obtain the excess noise. (**b**) $S_V f/V^2$ at $f = 10$ Hz as a function of N for platinum films and wires. (After Fleetwood and Giordano [74], © 1983, AIP)

Voss and Clarke proposed that the low-frequency noise of thin metal films is caused by thermal fluctuations, and described a series of experimental studies that appeared to provide strong evidence in support of this model [20]. However, follow-up studies failed to confirm many of their findings. For example, in contrast to predictions of the thermal fluctuation model [20], the noise of metal films in general (1) did not scale with the temperature coefficient of resistance of metal films, (2) did not exhibit spatial correlations, and (3) except in special circumstances, did not scale with frequency in a manner consistent with thermal fluctuations [24, 29]. This led to the thermal fluctuation model [20] largely being abandoned in the 1980s [24, 29, 30, 48, 75–79]. However, the studies of Voss and Clarke on the connections between 1/*f* noise and music [49, 50], and later investigations of fractals and low-frequency noise in biological systems by Voss and others have profoundly influenced the field [5, 80].

Of efforts to develop a universal and fundamental theory of 1/*f* noise, the most controversial is the quantum noise theory of Handel [81–84], which attracted a lot of attention from 1980 to 1995, despite significant objections to underlying assumptions [24, 27, 29, 85, 86]. Predictions of the model, e.g., that deviations from pure Poisson statistics in radioactive decay should occur as a result of the same processes that lead 1/*f* noise [87], were not verified, and theoretical objections were unresolved [29, 73, 88], so interest has waned.

A breakthrough in understanding the origin of low-frequency noise in metal films was the demonstration by Eberhard, Dutta, and Horn that the noise is strongly temperature dependent [24, 89]. Importantly, changes in noise magnitude at a given frequency with temperature correlate strongly with changes in the frequency dependence of the noise [24, 90]. Dutta and Horn demonstrated that if the noise is caused by a random thermally activated process having a broad distribution of energies relative to kT, but not necessarily constant $D(E)$, the correlation between the frequency and temperature dependences of the noise can be described via [24, 90]:

$$\alpha\left(\omega, T\right) = 1 - \frac{1}{\ln\left(\omega\tau_0\right)}\left(\frac{\partial \ln S_V(T)}{\partial \ln T} - 1\right). \tag{11}$$

where τ_o is the characteristic time of the process leading to the noise, and $\omega = 2\pi f$. For details of the derivation of Eq. (11), and/or alternative formalisms that lead to similar expression, see [24, 29, 76]. For noise that satisfies Eq. (11), one can infer the shape of the defect-energy distribution $D(E_o)$ from noise measurements versus temperature via:

$$D\left(E_0\right) \propto \frac{\omega}{kT} S_V\left(\omega, T\right) \tag{12}$$

where the defect energy is related to the temperature and frequency through the simple expression [24, 90]:

$$E_o \approx -kT \ln\left(\omega\tau_o\right). \tag{13}$$

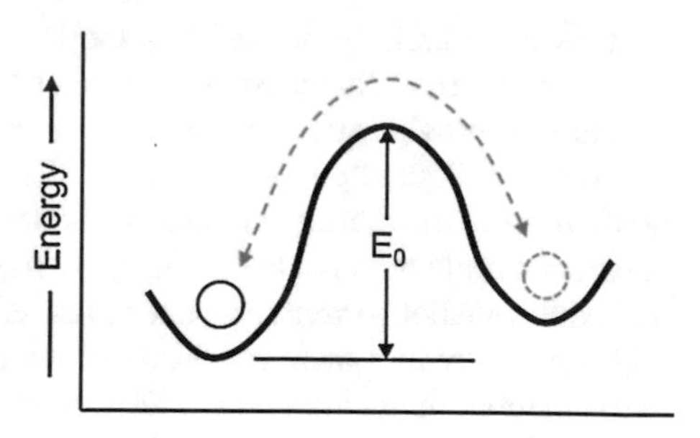

Fig. 3 Schematic illustration of a system with two configurations with different energy levels, charge states, and/or carrier scattering rates. E_o is the energy barrier for the system to move reversibly from one configurational state to another. (After Weissman [29], © 1988, AIP)

If the noise results from thermally activated processes involving two energy levels, E_o is the barrier that the system must overcome to move from one configurational state to another [24, 90] as illustrated in Fig. 3 [24, 29, 91].

Equations (11)–(13) assume: (1) The excess noise is due to the superposition of random, uncorrelated processes having thermally activated characteristic times. (2) The distribution of activation energies varies slowly with respect to kT. (3) The process is characterized by an attempt frequency $f_o = 1/\tau_o$ that is much higher than the frequency at which the noise is measured. (4) The coupling constant between the random processes and the resistance, and hence the integrated noise magnitude over all frequencies, is constant with temperature [24, 90]. For the latter assumption to be satisfied, defects cannot be created or annealed during the noise measurement process [75].

Over the following ~10 years, a large number of studies were performed to evaluate the extent to which the Dutta–Horn model describes $1/f$ noise in metals [24, 29, 30]. For example, Fig. 4a shows the temperature dependence of the noise magnitude of a AuPd nanowire with a 53-nm diameter [75]. Significant decreases in noise and resistivity were observed through heating cycles as a result of defect annealing. Figure 4b demonstrates that the Dutta–Horn model describes accurately the changes in frequency dependence α that occur [48].

Scofield et al. performed a comprehensive comparison of noise magnitudes for a variety of metal films with the fraction of the resistivity caused by defect and/or impurity scattering, finding a strong correlation [76]. Pelz and Clarke extended the work of Martin [92] to develop a local-interference model [28] that provides order-of-magnitude estimates of the noise magnitude of metals with moderate disorder [28, 29, 48, 75, 76]. These results and related studies [24, 28–30, 48, 76] convincingly rule out a significant role for lattice scattering as the origin of the noise of metals. The noise of more highly disordered films and/or metals at cryogenic temperatures can be described by this mechanism and/or universal conductance fluctuation model [93, 94]. Thus, the work of Hooge and co-workers [18, 19, 21, 23, 36] served primarily to (1) provide an easy way to parameterize low-frequency noise, and (2) stimulate a large amount of work that led to a comprehensive understanding of the noise [24, 29, 30, 48].

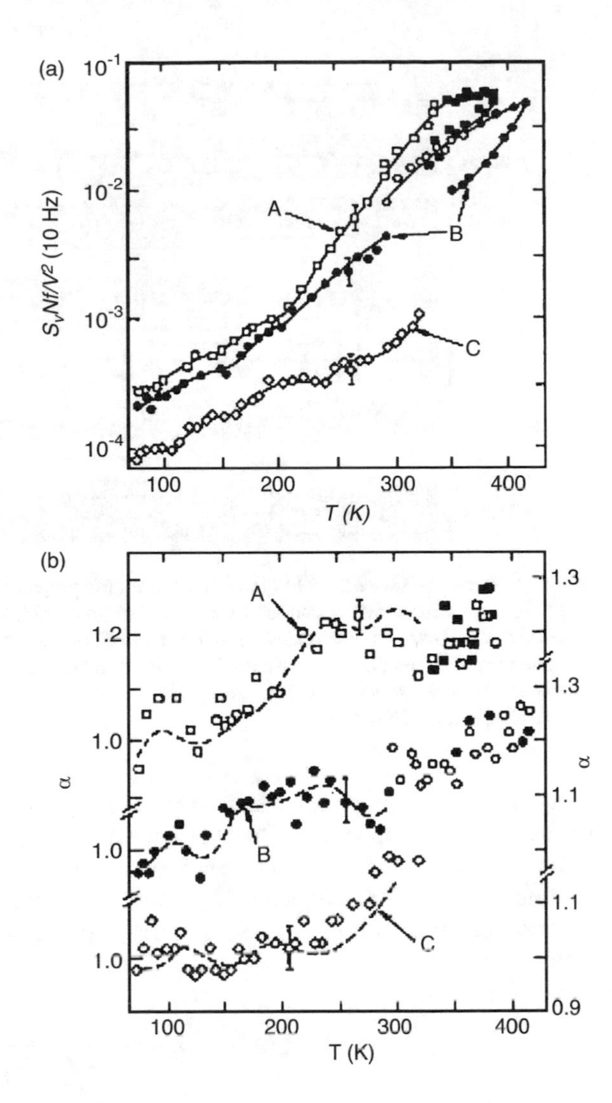

Fig. 4 (**a**) Normalized noise magnitude at $f = 10$ Hz and (**b**) measured and predicted values of $\alpha = -\partial \ln S_V/\partial \ln f$ as functions of temperature during three separate cooling and heating sequences (A-C) for a 53-nm diameter AuPd nanowire. Between B and C, the wire was heated to ~470 K in Ar. (After Fleetwood and Giordano [75], © 1985, AIP)

5 MOS Transistors: Defect Densities and Microstructure

Evidence that thermally activated processes associated with charge trapping are also important to MOS 1/*f* noise is shown in Fig. 5 [26]. At the lowest temperatures, only a single prominent trap is active in these μm-scale transistors, leading to random telegraph noise (RTN). The noise power spectral density in this case is Lorentzian in form [1, 11, 13]. As the temperature is increased, resistance switching rates become faster, and more traps become active. For higher temperatures and/or larger devices, discrete resistance fluctuations are not observed. Instead, 1/*f* noise is found [26]. The electronic properties (capture and emission times, energy, cross section, etc.) of

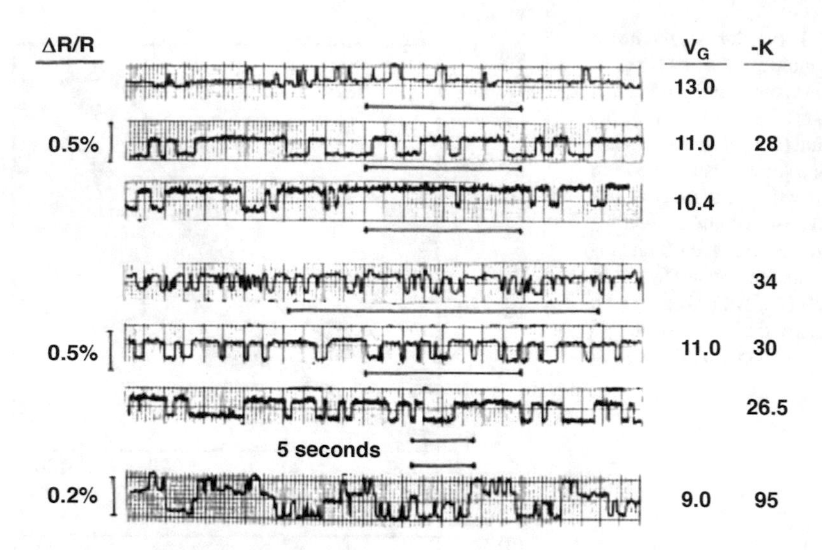

Fig. 5 Discrete resistance switching events (random telegraph noise) as a function of gate voltage and temperature for a *p*MOS transistor with a 65 nm gate oxide and dimensions $L = 1.0$ μm and $W = 0.15$ μm. At higher temperatures and lower values of gate voltage, the signal transitions from a region in which only discrete resistance fluctuations are observed to a region in which 1/*f* noise is observed. For transistors with larger gate area on the same chip, only 1/*f* noise is observed. (After Ralls et al. [26], © 1984, AIP)

a large number of individual defects in MOS devices have characterized [31, 44–46, 95–99], strongly supporting a number fluctuation origin for both the RTN and 1/*f* noise. This is true for both *n* and *p*MOS devices [38, 43–46, 48].

Often it is convenient to explicitly parameterize the gate voltage and frequency dependence of the noise due to carrier number fluctuations via an expression of the form:

$$S_{V_d}\left(f, V_d, V_g\right) = \frac{K}{f^\alpha} \frac{V_d^2}{\left(V_g - V_t\right)^\beta} \tag{14}$$

Equation (14) with $\alpha = -\partial \ln S_V/\partial \ln f = 1$ and $\beta = 1$ is the Hooge model Eq. (10). Number fluctuations with constant $D_t(E_f)$ lead to $\beta = 2$ Eq. (8); deviations from $\alpha = 1$ and $\beta = 2$ are evidence of nonuniform $D_t(E_f)$ [38, 48]. Values of β other than 1 or 2 are often found because nonuniform $D_t(E)$ values are typical in electronic devices [24, 29, 32, 35, 38, 39, 43–46, 48, 62, 100–107]. Variations in $D_t(E)$ occur naturally from process variations during fabrication. High-field stress, aging, exposure to moisture and/or ionizing radiation, etc., can also change $D_t(E_f)$ for a single device, often significantly [35, 48, 100, 101].

Figure 6 shows (a) noise magnitudes and (b) defect energy distributions inferred via the method of Hung et al. [33, 101] for *p*MOS transistors that were (1) not exposed to moisture (control) or irradiated, (2) exposed to moisture but not

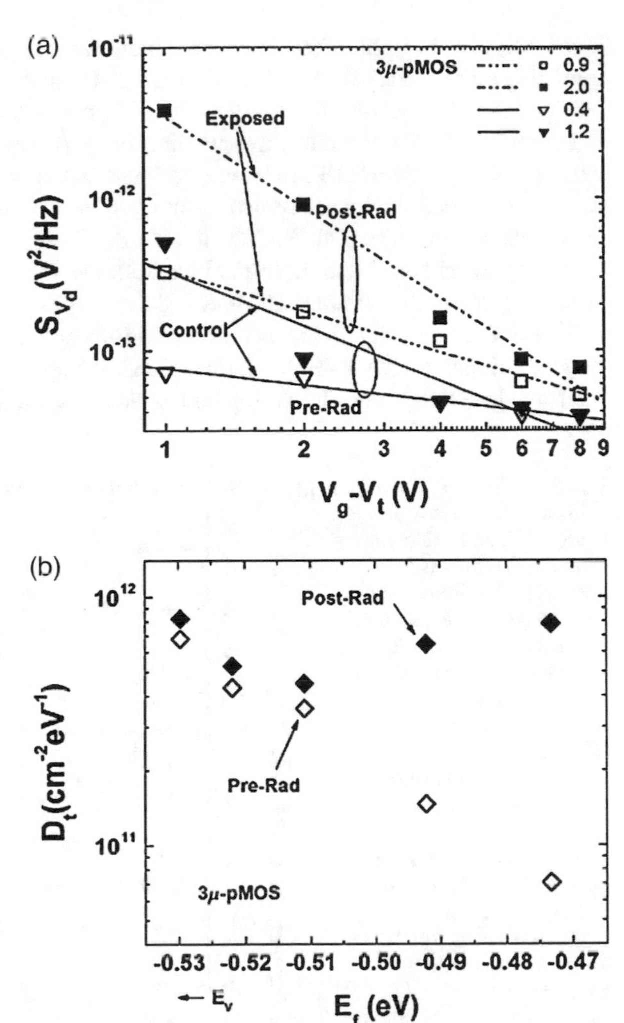

Fig. 6 (**a**) S_{V_d} at ~10 Hz vs. $V_g - V_t$ for packaged, fully processed and passivated, *p*MOS transistors with 37 nm oxides, and dimensions $L = (3.45 \pm 0.10)$ μm and $W = (16.0 \pm 0.5)$ μm. Results are shown for devices with or without exposure to moisture (85% relative humidity at 130 °C for 1 week). Control and moisture-exposed unlidded devices were measured before and after irradiation with 10-keV X-rays to 500 krad(SiO_2) at a dose rate of 31.5 krad(SiO_2)/min at 6 V gate bias. During noise measurements, the drain voltage V_d was held at a constant −100 mV. (**b**) Inferred trap distributions as a function of the Fermi level for these devices and irradiation conditions. (After Francis et al. [66], © 2010, IEEE)

irradiated, (3) irradiated, but not exposed to moisture, or (4) both exposed to moisture and irradiated. For the control *p*MOS device in Fig. 6a before irradiation, $\beta = 0.4$. After irradiation, the average value of β is ~1.2, but the slope is multivalued. For the moisture-exposed *p*MOS device, prior to irradiation, $\beta = 0.9$, and after irradiation, $\beta = 2.0$ [48, 101]. Clearly, these kinds of variations in β are not consistent with the Hooge model as reflected in Eq. (10).

Before irradiation, the inferred defect-energy distribution in Fig. 6b increases toward the valence band edge, a trend often observed in MOS devices [48, 106, 107]. That the effective defect-energy distribution before irradiation or high field stress often increases toward the valence band edge more strongly for *p*MOS devices than toward the conduction band edge for *n*MOS devices [38, 40, 108–110] may result from a relatively larger role of interface traps in *p*MOS noise than *n*MOS noise [48, 111–113]. Interface traps may function more commonly as a trap-assisted tunneling intermediary for hole injection into SiO_2 than electron injection, since the barrier for hole injection (~4.8 eV) is much greater than the barrier for electron injection (~3.1 eV), and tunneling probability is strongly influenced by the energy barrier at the interface [113]. However, and perhaps more likely, these differences in defect energy distributions may also result from the different roles and configuration dynamics of defects near the Si/SiO_2 interface [43–46, 48].

Much experimental and theoretical work shows that O vacancies in SiO_2 play a dominant role in the 1/*f* noise of MOS transistors [35, 39, 43–46, 48, 61, 111, 114–117]. For example, Fig. 7 shows that the low-frequency noise of MOS transistors is proportional to the radiation-induced oxide-trap charge density (but not the interface-trap charge density) during both irradiation and annealing [115]. The noise

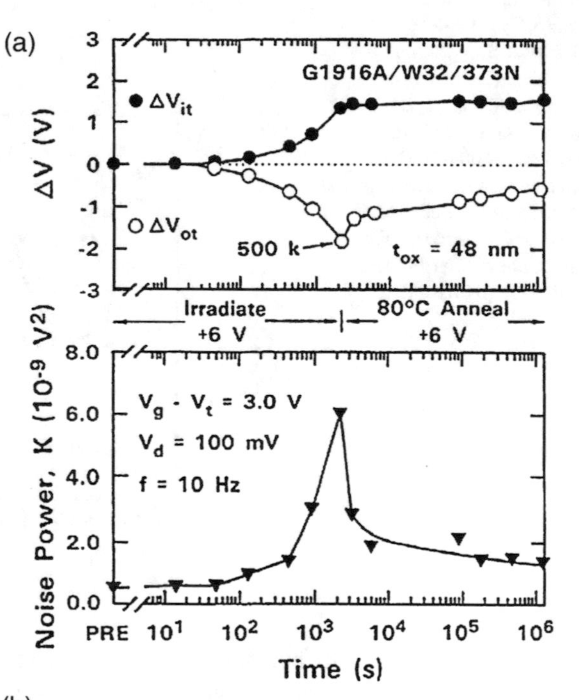

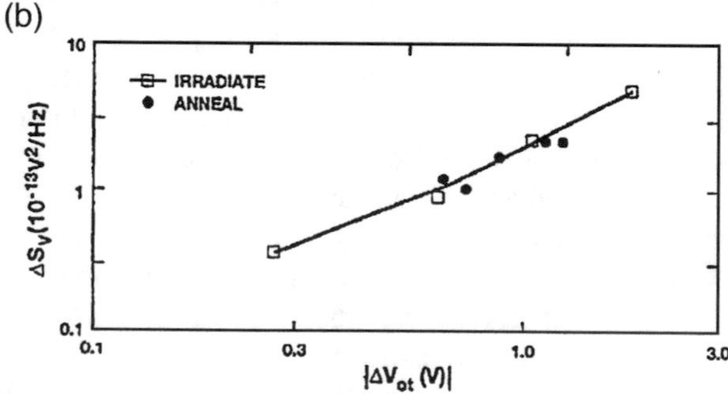

Fig. 7 (**a**) Threshold voltage shifts due to interface-trap charge ΔV_{it} and oxide-trap charge ΔV_{ot} (top) and normalized low-frequency noise magnitude *K* (bottom) as functions of irradiation and annealing time for an *n*MOS transistor with $L = (3.45 \pm 0.10)$ μm and $W = (16.0 \pm 0.5)$ μm irradiated to 500 krad(SiO_2) with Co-60 gamma rays at a dose rate of 13 krad(SiO_2)/min with a gate bias of 6 V. (**b**) Change in low-frequency noise magnitude during irradiation and room-temperature annealing. (After Meisenheimer and Fleetwood [115], © 1990, IEEE)

of these particular devices is not strongly temperature dependent [100], so the choice of room temperature for the noise measurements does not affect the conclusion in this case.

Figure 8 illustrates a strong correlation between radiation-induced oxide-trap charge density and O vacancy density in SiO_2 [117–126], suggesting that both the noise and the radiation-induced-hole trapping are associated with a common defect and/or defect precursor. The noise of unirradiated devices also correlates strongly with postirradiation oxide-trap charge density [61, 114, 117]. Supporting evidence for the importance of O vacancy-related defects, and/or their complexes with hydrogen, to 1/*f* noise in MOS devices from experimental studies and first-principles calculations is summarized in [39, 43–46, 48, 61, 111–117].

In a *p*MOS transistor, capture and emission of a hole from an unrelaxed (dimer) O vacancy near the Si/SiO_2 can lead to noise [43–46, 48]. A schematic representation is shown in Fig. 9 [43–46, 48], which extends the model of Lelis et al. [121]. The time dependence and energetics of the processes illustrated in Fig. 9 were updated to incorporate effects of near-interfacial hydrogen by Grasser et al. [127–129]. In an *n*MOS transistor, the capture and release of electrons at O vacancies in the (strained) near-interfacial SiO_2 layer can lead to noise [43, 48]. O vacancies and hydrogen-related defects also play key roles in the low-frequency noise of MOS devices with high-K gate dielectrics [48, 130–134], as well as MOSFETs based on two-dimensional materials [47, 91, 135].

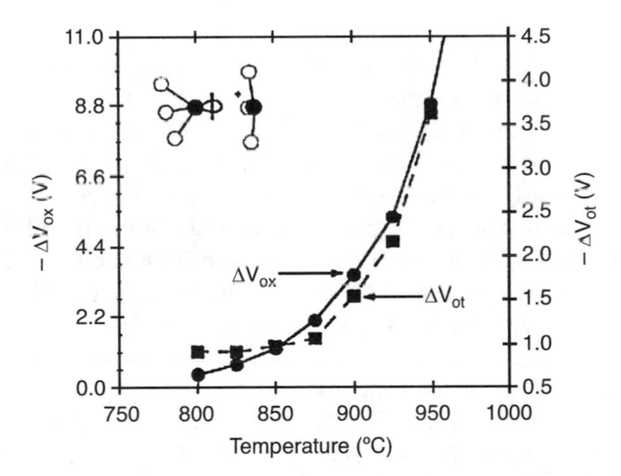

Fig. 8 Comparison of measured ΔV_{ot} (squares, right-hand scale) and predicted ΔV_{ox} based on O vacancy densities in the SiO_2 bulk derived from electron-paramagnetic resonance (EPR) measurements and a model based on Fick's law of diffusion (circles, left-hand scale). ΔV_{ot} data are for 1 Mrad(SiO_2) X-ray irradiation of capacitors with 45 nm oxides at 10 V bias (from [120]). ΔV_{ox} model results (from [123]) assume all vacancies capture a radiation-induced hole, leading to the overprediction. The inset shows a relaxed O vacancy (E_γ'), with the trapped hole on the right, and an unbonded electron on the left, which is EPR active. (After Fleetwood et al. [117], © 1995, Elsevier)

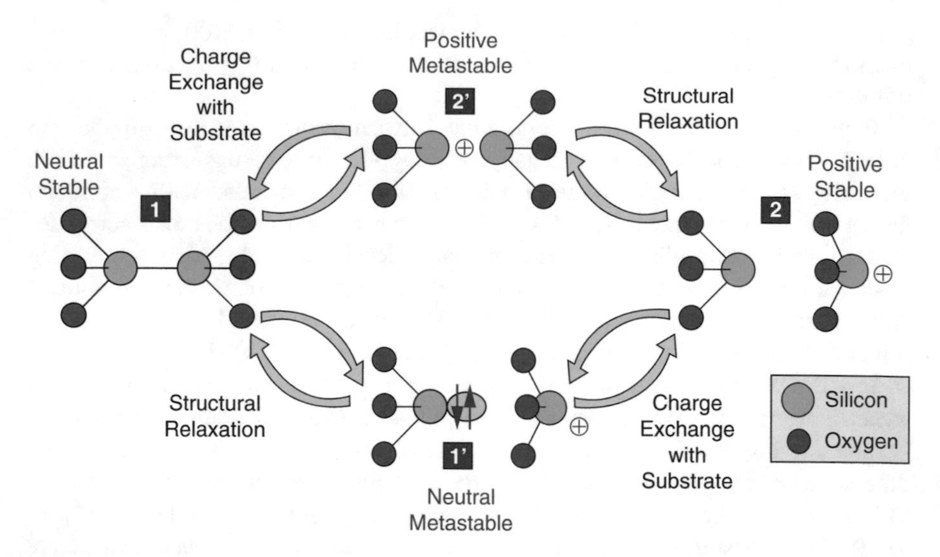

Fig. 9 Schematic illustration of (1) an O vacancy that can capture a hole (2′) and then release it (1) or relax into configuration (2). The further capture of an electron by a trapped hole (2) neutralizes the trapped positive charge, forming a dipole (1′). This defect can reversibly exchange an electron with Si, or relax to reform the initial vacancy structure (1), under suitable bias conditions. (After Grasser et al. [46], © 2012, Elsevier.) More recent versions of this figure in [127–129] are more complex and include hydrogen

The defects that cause the noise in MOS defects typically are border traps, which are near-interfacial oxide traps that exchange charge with the underlying Si on the time scales of measurements [35, 43–46, 48, 136–139]. Effective border-trap densities estimated via current-voltage, capacitance-voltage, and/or charge-pumping measurements typically are in general agreement with estimates of trap densities derived from a simple number fluctuation model of the noise, Eq. (8) [39, 48, 101, 116]. Moreover, the effective energy scale derived from the Dutta–Horn model, Eq. (13), is consistent with energy scales for radiation-induced-hole trapping and emission, derived independently using the thermally stimulated current method [43, 122].

Over the last ~30 years, the Dutta–Horn model has been applied successfully not only to SiO_2-based MOSFETs, but also to a wide variety of additional semiconductor devices and materials [48]. A recent, illustrative example for a SiGe *p*MOSFET with a high-K gate dielectric is shown in Fig. 10, showing excellent agreement between the model and experimental data [134]. For additional examples, see [48].

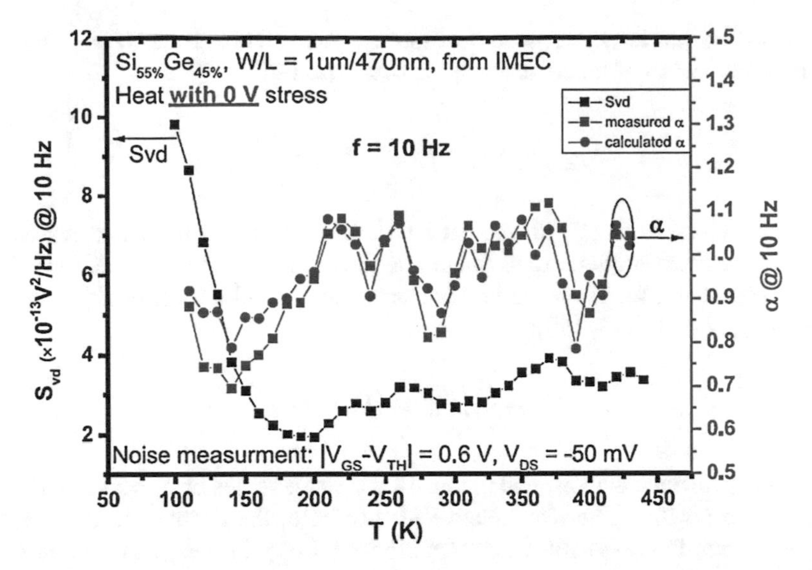

Fig. 10 S_{V_d} at 10 Hz (left axis) and spectral slope $\alpha = -\partial \ln S_V / \partial \ln f$ at $f = 10$ Hz as a function of temperature [right axis: red squares denote measured data, and blue circles denote calculated values from Dutta–Horn analysis via Eq. (11)] for $Si_{0.55}Ge_{0.45}$ *p*MOSFETs fabricated on an *n*-type Si wafer with a 4.0 nm $Si_{0.55}Ge_{0.45}$ layer deposited onto a 2.0 nm Si buffer at imec, Belgium. A Si cap layer was oxidized during processing to form a thin SiO_2 layer. Added to this was ~2 nm HfO_2. (After Duan et al. [134], © IEEE, 2016)

6 GaN/AlGaN HEMTs: Number Fluctuations

Although compelling evidence demonstrates that the low-frequency noise of Si-based MOS devices is dominated by carrier number fluctuations, the Hooge mobility fluctuation model is often used to analyze the low-frequency noise in AlGaN/GaN HEMTs [70, 71, 140–145]. Wang et al. have recently provided strong evidence that the noise is caused by number fluctuations [65], as we now discuss.

In contrast to a MOSFET, for which the entire conducting channel is gated, a HEMT includes a narrow gated region and a wider ungated region to reduce the device switching time and enhance its frequency response (inset of Fig. 11). The channel resistance of a HEMT is therefore the sum of the resistances of the ungated and gated regions R_U and R_G [140–142, 146]:

$$R_{total} = R_G + R_U = \frac{L_{gate} V_{th}}{W q \mu n \left(V_g - V_{th}\right)} + R_U \tag{15}$$

where μ is the channel mobility, n is the areal carrier density in the two-dimensional electron gas (2DEG), L_{gate} is the length of the gated channel region, and W is the channel width. In the Hooge model of low-frequency noise for a HEMT, the noise of the gated region of the channel is [23, 140–142, 146]:

$$S_{R_{\text{total}}} = S_{R_G} + S_{R_U} = S_{R_U} + \frac{\alpha_H R_{\text{ch}}^2}{Nf} \tag{16}$$

For V_g close to threshold, the carrier density is low in the gated region; the noise in the gated portion of the channel is the dominant noise source. The Hooge model expression for the noise in this case reduces to the form of Eq. (10), with $N = nL_{\text{gate}}W$ [140–142, 146]:

$$S_{v_d} = \frac{\alpha_H}{Nf} V_d^2 \propto \left(V_g - V_{\text{th}}\right)^{-1}$$

This dependence is illustrated in Fig. 11 for small values of $v_g = (V_g - V_{\text{th}})^{-1}$.

At more positive gate bias, relative to threshold, the channel electron density increases, and the noise still originates predominantly in the gated portion of the

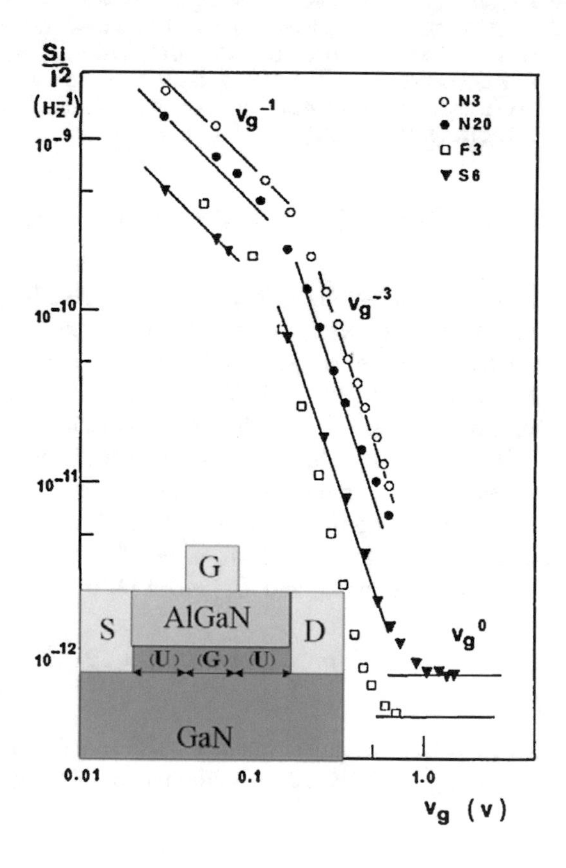

Fig. 11 Relative current noise power spectral density vs. $v_g = (V_g - V_{\text{th}})^{-1}$, showing the approximate voltage dependences assumed in the Hooge mobility model of low-frequency noise in a number of differently processed HEMTs. The inset is a schematic cross section of a GaN HEMT, where the gated (G) and ungated (U) portions of the channel are labeled. (After Peransin et al. [146], © IEEE, 1990 and Wang et al. [65] © IEEE, 2017)

HEMT. Here $R_G << R_U$, R_G is proportional to $(V_g - V_{th})^{-2}$, and the Hooge model can be approximated as [140–142, 146]:

$$\frac{S_{v_d}}{V^2} = \frac{S_{R_{total}}}{R^2_{total}} = \frac{\alpha_H R_G{}^2}{Nf R_U^2} \propto \left(V_g - V_{th}\right)^{-3} \tag{17}$$

This stronger voltage dependence is illustrated in Fig. 11 for intermediate values of $v_g = (V_g - V_{th})^{-1}$.

For more positive values of $v_g = (V_g - V_{th})$, both the resistance and noise are dominated by the ungated portion of the channel, and the noise becomes independent of gate bias (Fig. 11) [23, 140–142, 146]:

$$S_{v_d} = \frac{\alpha_H}{N_U f} V_d^2 \propto \left(V_g - V_{th}\right)^0 \tag{18}$$

where N_U is the number of carriers in the ungated channel.

For small values of $(V_g - V_{th})^{-1}$, where the Hooge model assumes S_{V_d} is proportional to $1/N \sim (V_g - V_{th})^{-1}$, the number fluctuation model (Eqs. 7 and 8) predicts instead that S_{V_d} is proportional to $1/N^2$ [17, 61], and $S_{V_d} \sim (V_g - V_{th})^{-2}$ for constant $D_t(E)$ [17, 25, 48, 61]. In the intermediate voltage region, for which the Hooge model predicts the noise to scale as $(V_g - V_{th})^{-3}$, the number fluctuation model predicts that the noise will scale as $\sim (V_{g_s} - V_{th})^{-4}$ when $D_t(E)$ is approximately constant [65].

Figure 12 shows the gate voltage dependence of S_{V_d} before and after proton irradiation for devices manufactured by Qorvo, Inc., as processed, and after proton irradiation under the semi-ON bias condition [65]. Only small changes in noise are observed for these devices and irradiation conditions. For reference, S_{V_d} is plotted at 10 Hz and 100 Hz as a function of $V_g - V_{th}$. The slope of the gate voltage

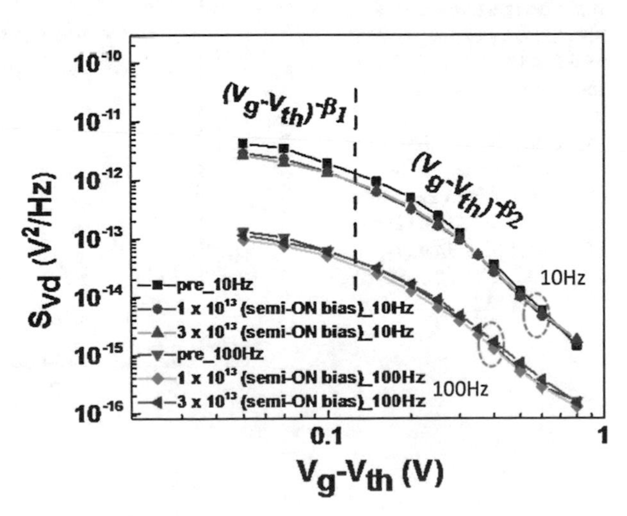

Fig. 12 S_{V_d} at 10 Hz and 100 Hz as a function of $V_g - V_{th}$ for AlGaN/GaN HEMTs manufactured by Qorvo, Inc., before and after proton irradiation with the Vanderbilt Pelletron to fluences up to $3 \times 10^{13}/\text{cm}^2$ that were biased during irradiation under the semi-ON condition. $V_d = 30$ mV during the noise measurements. (After Wang et al. [65], © 2017, IEEE)

dependence in the narrow region very close to V_{th}, β_1, is close to 1.0 before and after proton irradiation. The slope of the gate voltage dependence in the extended region, 0.12 V < $V_g - V_{th}$ < 0.80 V, β_2, is ~ 3.7 before proton irradiation, and ~ 3.5 after proton irradiation.

For noise due to number fluctuations, in cases where the defect-energy distribution is relatively uniform, Ghibaudo et al. showed that $S_{V_d}/V_d^2 = S_I/I^2$ is proportional to $(g_m/I_d)^2$, where g_m is the transconductance [34]. Figure 13 shows this comparison before and after proton irradiation for the devices of Fig. 12 [65]. Generally good agreement is observed, with small deviations from perfect correlation resulting from nonuniformities in the defect-energy distribution.

Figure 14 shows the low-frequency noise measured over a temperature range of 85 K to 450 K for GaN/AlGaN HEMTs similar to those of Figs. 12 and 13. Devices were irradiated with 1.8 MeV protons to a fluence of 10^{13} protons/cm^2 using the Vanderbilt Pelletron facility, under semi-ON bias conditions [147]. A strongly nonuniform defect-energy dependence is observed, consistent with other studies of the low-frequency noise of GaN/AlGaN HEMTs [48, 104, 148, 149]. The ~ 0.2 eV trap visible before irradiation is attributed to an oxygen DX center (O_N) in AlGaN [147, 149], and the ~ 0.7 eV level is likely a N anti-site defect in GaN or a hydrogenated Ga vacancy/O_N complex [147, 150], as shown in Figs. 15 and 16. N vacancy-related defects add to the low-energy peak after irradiation [104, 147–151]. These defect identifications are based on extensive density-functional-theory calculations of defect energies and charge states in GaN/AlGaN HEMT devices [147–153].

Figure 17 shows the noise of an AlGaN/GaN HEMT from the University of California, Santa Barbara, processed differently than the devices of Figs. 12, 13 and 14, and exposed to a series of stresses with increasing drain bias [149]. Several

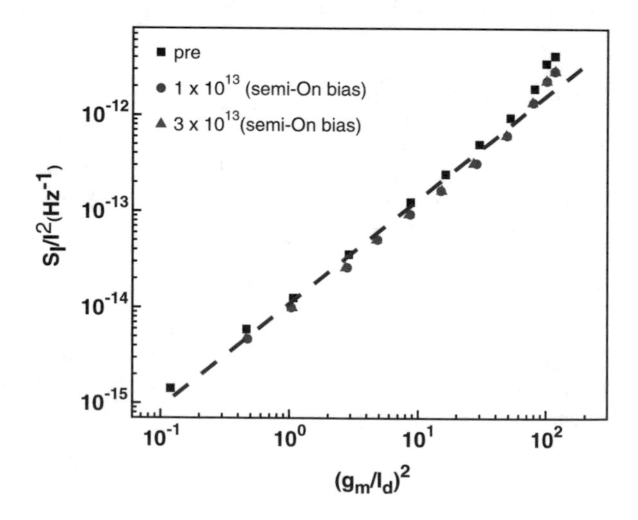

Fig. 13 Normalized drain-current noise-power spectral density S_I/I^2 at 10 Hz as a function of $(g_m/I_d)^2$ for the devices of Fig. 12, before and after proton irradiation in the semi-ON condition. Measurement uncertainties are comparable to the symbol sizes. The dashed line is an aid to the eye, with unity slope. (After Wang et al. [65], © 2017, IEEE)

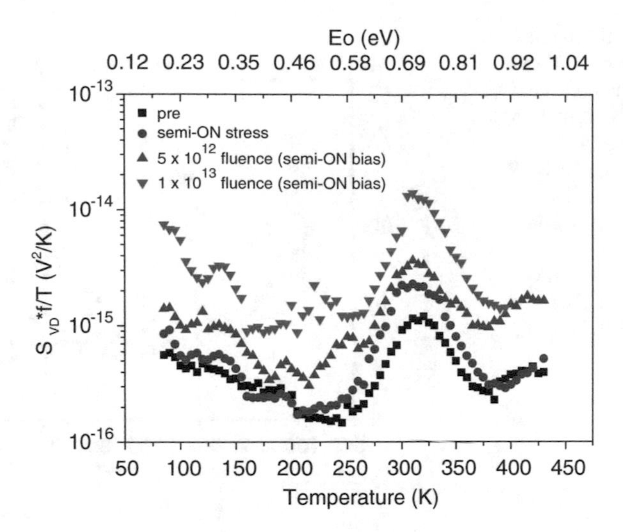

Fig. 14 Temperature-dependent noise measurements from 85 K to 445 K, for AlGaN/GaN HEMTs manufactured by Qorvo, Inc. Noise measurements were performed with $V_g - V_{th} = 0.4$ V and $V_d = 0.03$ V; magnitudes are shown at $f = 10$ Hz. The energy scale on the upper x-axis is derived from Eq. (13). Devices were stressed ($V_g = -2$ V, $V_d = 25$ V, 16 h) or irradiated under semi-ON bias conditions. Fluences are quoted in protons/cm^2; the flux was 1.25×10^{13} protons/h. (After Chen et al. [147], © 2015, IEEE)

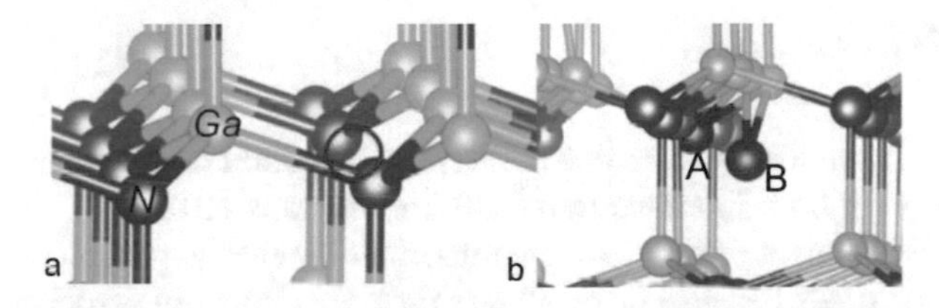

Fig. 15 Atomic structure of the defects related to the ~0.2 eV noise peak in Fig. 14. (**a**) The nitrogen vacancy position is highlighted by red circle. (**b**) An oxygen atom (shown in red) reconfigures from its interstitial position A to the DX center position B, with energy levels separated by ~ 0.2 eV. (After Chen et al. [147], © 2015, IEEE)

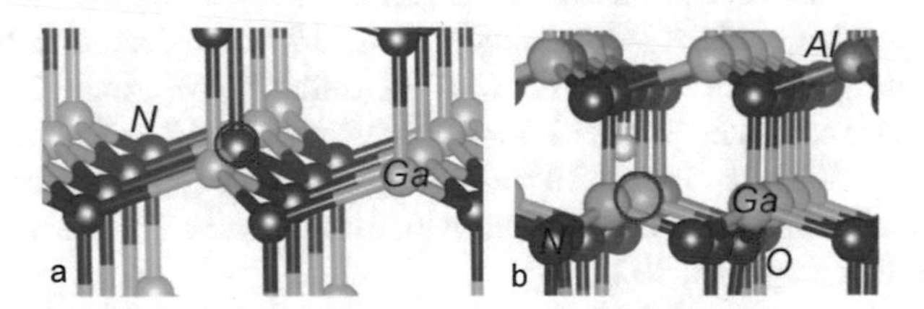

Fig. 16 Atomic structure of the defects potentially related to the ~ 0.7 eV peak in Fig. 14. (**a**) A nitrogen anti-site N_{Ga} is highlighted by the red circle. (**b**) A hydrogenated O_N complexed with a Ga vacancy H-O_N-V_{Ga} in GaN is shown, where the circle shows the position of the missing Ga atom, and O is shown in red. (After Chen et al. [147], © 2015, IEEE)

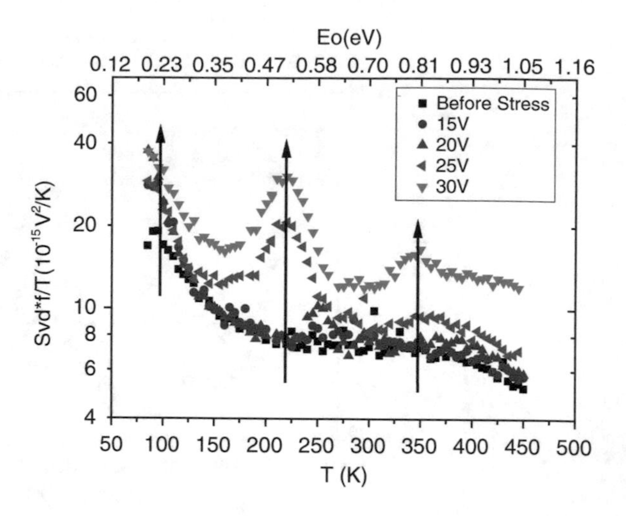

Fig. 17 Normalized noise magnitude at $f = 10$ Hz vs. temperature and high-field stress. The noise is measured in the linear region of response, with $V_{d_s} = 0.1$ V and V_{g_s}-$V_{th} = 2$ V. The normalization is consistent with the inferred defect-energy distribution $D(E_0)$ from Eq. (13). (After Chen et al. [149], © 2016, IEEE)

large defect peaks are observed, most notably a peak at ~0.5 eV that is most likely Fe-related [149]. More information on these defect and impurity centers is provided in [104, 147–153].

7 SiC MOS Devices

Wide-band-gap semiconductors like SiC typically exhibit low-frequency noise that is associated with both interface traps and border traps [102, 103, 154, 155]. The role of interface traps is relatively more important in wide-gap semiconductors than narrow-gap materials because much slower time constants are associated with deep interface traps in wide gap materials than for typical interface traps in Si or Ge [156, 157]. Figure 18 shows the excess input-referred gate-voltage noise power spectral densities $S_{V_g} = S_{V_d} (V_{g_s} - V_{th})^2/V_d^2$ and effective trap densities $D_t(E_f)$ at ~10 Hz as a function of temperature during measurements for SiC MOS devices with a 55 nm NO-nitrided oxide. The magnitude of the $1/f$ noise decreases by ~77% as the temperature increases from 85 K to 510 K. Using Eq. (8), the effective density of traps $D_t(E_f)$ is estimated to be ~ 2.3×10^{13} eV^{-1}cm^{-2} at $T = 85$ K, ~ 2.6×10^{12} eV^{-1}cm^{-2} at ~ 300 K, and ~1×10^{12} eV^{-1}cm^{-2} at ~ 510 K. The decrease in noise with increasing temperature is consistent with a decrease in the effective density of charged interface traps [48, 102, 103].

Figure 19 shows $\alpha = -\partial \ln S_V/\partial \ln f$ as a function of T for the data of Fig. 18. The overall shape of the measured $\alpha(T)$ curve is consistent with the Dutta–Horn model prediction. This enables the use of Eq. (12) to estimate the energy distribution of defects, as shown on the upper x-axis of Fig. 18. First principles calculations using DFT show that carbon vacancy clusters on the SiC side of the SiC/SiO_2

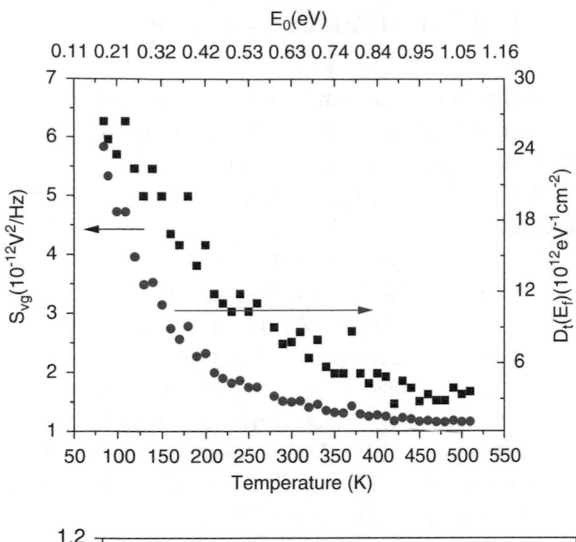

Fig. 18 Excess input-referred gate-voltage noise power spectral density S_{V_g} (left axis) $= S_{V_d}$ $(V_{g_s} - V_{th})^2/V_d^2$ and calculated effective density of traps $D_t(E_f)$ at ~ 10 Hz (right axis) vs. temperature from 85 K to 510 K, for SiC MOS devices with a 55 nm gate oxide that received a post-oxidation NO anneal at 1175 °C for 2 h. (After C. X. Zhang et al. [103], © IEEE, 2013)

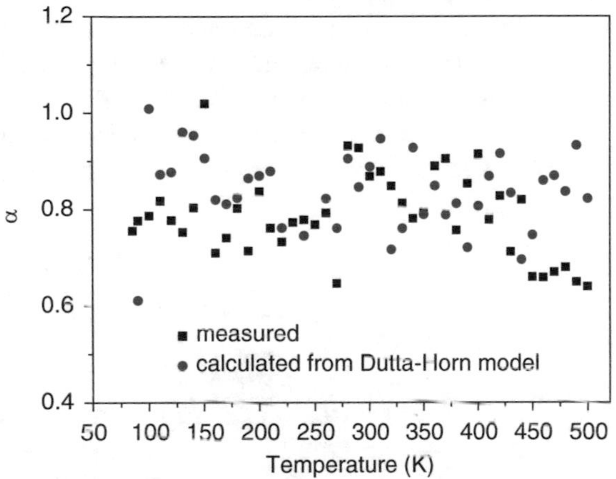

Fig. 19 Measured and predicted values of the frequency dependence of the noise, $\alpha = -\partial \ln S_V/\partial \ln f$, using the experimental results from Fig. 18 and Eq. (11) of the text. (After C. X. Zhang et al. [103], © 2013, IEEE)

interface have activation energy levels of ~0.1 to 0.2 eV [48, 103]. These defects appear to account for at least some of the increased noise at low temperature [102, 103]. Fluctuations in occupancy of N dopants [158–160] may also contribute to the increase in noise magnitude with decreasing temperature that is observed in Fig. 18. Hence, the defects that cause low-frequency noise in SiC MOS devices, especially at low temperatures, can be quite different both in location and in microstructure from those responsible for the noise in Si MOS devices. At higher temperatures, it is likely that O vacancy-related defects also contribute to the noise [103], similar to Si-based MOSFETs [48].

8 Two-Dimensional Materials

Low-frequency noise measurements are also useful for characterizing defects in MOS devices based on two-dimensional materials [47, 135, 161–168]. Here we discuss graphene [135] and black phosphorus (BP) [168] MOSFETs as illustrative examples of the responses of these devices, before and after exposure to ionizing radiation.

Figure 20 shows the excess low-frequency voltage-noise power-spectral density S_V versus irradiation dose and temperature for graphene devices with a ~4 nm Al_2O_3 passivation overlayer. The noise magnitude generally decreases with increasing noise magnitude before the device is irradiated. The right inset shows measured and predicted values of the frequency dependence of the noise, $\alpha = -\partial \ln S_V/\partial \ln f$, showing good agreement between measured values of the frequency dependence and predictions of the Dutta–Horn model. Broad peaks in noise magnitude at ~ 0.4 eV and ~ 0.7 eV are observed after 100 krad(SiO_2) irradiation. These peaks are not typically observed when Si-based MOS devices are irradiated [43, 48], suggesting that these peaks are characteristic of defects associated primarily with graphene and its interfaces, and not bulk SiO_2. DFT calculations for these structures show hydroxyl-related peaks ~ 0.3 eV and ~ 0.7 eV below the graphene Fermi level, showing the significance of hydrogen-related defects and impurities to the noise of graphene MOS devices [135]. It is likely that O vacancy-related defects also

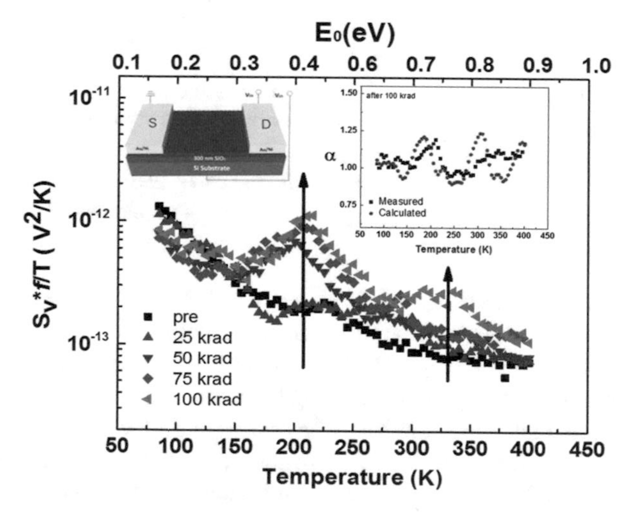

Fig. 20 Normalized low-frequency noise from 85 K to 400 K at $f = 10$ Hz before and after irradiation for graphene transistors built on 300 nm SiO_2 (see inset) with a ~4 nm Al_2O_3 overlayer irradiated under −5 V bias. Devices were biased at $V_g - V_{CNP} = -20$ V and $V_d = 0.02$ V during noise measurements. The insets show a schematic diagram of the device, as well as comparisons of the predicted and observed frequency responses, using Eq. (11) of the Dutta–Horn model. The effective defect distribution using Eq. (12) is shown on the upper x-axis. (After P. Wang et al. [135], © 2018, IEEE)

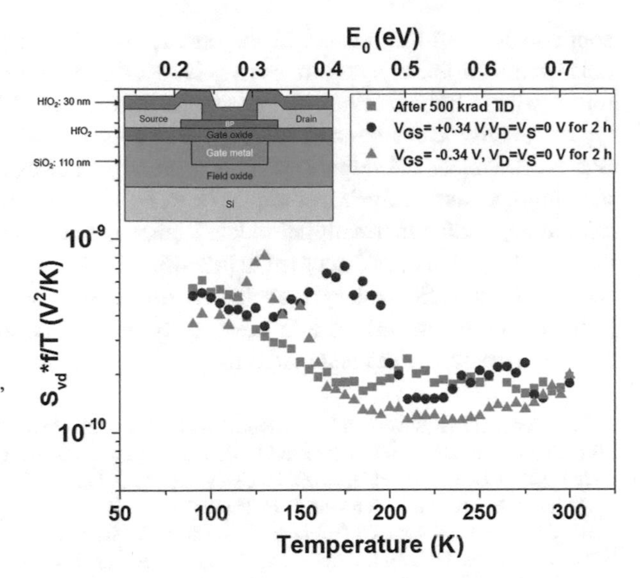

Fig. 21 Normalized low-frequency noise from 90 K to 300 K at $f = 10$ Hz for the irradiated BP transistors for devices (see inset for structure) irradiated to 500 krad(SiO_2) under 0.34 V bias, then annealed at the same bias, and finally annealed under the opposite bias. Irradiation and annealing was performed at room temperature. (After C. D. Liang et al. [168], © 2018, IEEE)

contribute to the noise in these devices [48], given that these devices are fabricated on a thick SiO_2 layer [135] (see left inset of Fig. 20).

Figure 21 shows noise measurements as a function of temperature for BP MOS devices [168] irradiated to 500 krad(SiO_2) under +0.34 V and annealed at room temperature in a vacuum cryostat $V_{gs} = \pm\ 0.34$ V for 2 h. After positive bias annealing, the normalized noise magnitude increases significantly in the range of 125 K to 200 K (~0.3 eV to 0.5 eV), relative to postirradiation levels. After negative bias annealing, the local maximum (peak) in the noise vs. temperature curve is observed now at an energy level of ~0.3 eV, in contrast to the peak in magnitude observed at ~0.4 eV after the positive-bias anneal. This behavior is attributed in [168] to the reversible motion and interactions of H^+ between the BP surface and defect sites in the HfO_2 gate dielectric. These results again emphasize the significance of hydrogen to the low-frequency noise of MOS devices using two-dimensional materials.

9 Summary and Conclusions

More than 60 years of investigation has demonstrated that the low-frequency noise of electronic materials and devices is caused primarily by defects and impurities. The number fluctuation model of McWhorter [14] provides useful first-order estimates of the effective densities of the defects responsible for the observed noise of many types of semiconductor devices. The Dutta–Horn model of low-frequency noise [24] enables estimates of the defect-energy distribution from measurements of the temperature and frequency dependence of the noise. These models have been applied successfully to assist in the characterization of the defects that cause the noise in a wide variety of systems [24, 29, 30, 48]. Other noise models are generally found to be of limited utility. To understand the origins of the low-frequency noise in typical electronic devices and materials, one must also understand the nature of their dominant defects and impurities. This typically requires the use of complementary measurement and analysis techniques (e.g., *I-V*, *C-V*, charge pumping, electron-spin resonance, and/or thermally stimulated current measurements, density functional theory calculations), as discussed in [43–46, 48, 129], for example. When combined with these complementary techniques, low-frequency noise measurements can provide significant insight into the quality, reliability, and radiation response of microelectronic devices and materials.

Acknowledgments Several of the discussions in this chapter are abridged from a recent review [48]. The interested reader is directed to this and other cited sources for additional details. The author thanks D. E. Beutler, J. Chen, C. Claeys, R. A. B. Devine, P. V. Dressendorfer, G. X. Duan, P. Dutta, S. A. Francis, T. Grasser, P. H. Handel, F. N. Hooge, P. M. Horn, N. Giordano, R. Jiang, S. Koester, Sh. M. Kogan, C. D. Liang, Z. Y. Lu, J. T. Masden, P. J. McWhorter, S. L. Miller, C. J. Nicklaw, S. T. Pantelides, J. Pelz, Y. Puzyrev, R. A. Reber, Jr., L. C. Riewe, T. Roy, J. H. Scofield, R. D. Schrimpf, M. R. Shaneyfelt, J. R. Schwank, E. Simoen, X. Shen, W. C. Skocpol, J. S. Speck, C. Surya, A. Tremblay, M. J. Uren, L. K. J. Vandamme, R. F. Voss, P. Wang, W. L. Warren, M. B. Weissman, P. S. Winokur, H. D. Xiong, C. X. Zhang, and E. X. Zhang for experimental assistance and/or stimulating discussions.

References

1. W.H. Press, Flicker noises in astronomy and elsewhere. Comments Astrophys. **7**(4), 103–119 (1978)
2. J. Timmer, M. Konig, On generating power-law noise. Astron Astrophys. **300**(3), 707–710 (1995)
3. J. Polygiannakis, P. Preka-Papadema, X. Moussas, On signal-noise decomposition of time-series using the continuous wavelet transform: Application to sunspot index. Mon. Not. R. Astron. Soc. **343**, 725–734 (2003)
4. M. Kobayashi, T. Musha, 1/*f* fluctuation of heartbeat period. IEEE Trans. Biomed. Eng. **29**(6), 456–457 (1982)
5. R.F. Voss, Evolution of long-range fractal correlations and 1/*f* noise in DNA-base sequences. Phys. Rev. Lett. **68**(25), 3805–3808 (1992)
6. K. Linkenkaer-Hansen, V.V. Nikouline, J.M. Palva, R.J. Ilmoniemi, Long-range temporal correlations and scaling behavior in human brain oscillations. J. Neurosci. **21**(4), 1370–1377 (2001)
7. M.D. Fox, A.Z. Snyder, J.L. Vincent, M.E. Raichle, Intrinsic fluctuations within cortical system account for inter-trial variability in human behavior. Neuron **56**(1), 171–184 (2007)
8. A.W. Lo, Long-term memory in stock market prices. Econometrica **59**(5), 1279–1313 (1991)
9. H. Niu, J. Wang, Quantifying complexity of financial short-term time series by composite multiscale entropy measure. Commun. Nonlinear Sci. Numer. Simul. **22**(1-3), 375–382 (2015)
10. I. Gvozdanovic, B. Podobnik, D. Wang, H.E. Stanley, 1/*f* behavior in cross-correlations between absolute returns in a US market. Physica A Stat. Mech. Appl. **391**(9), 2860–2866 (2012)

11. J.B. Johnson, The Schottky effect in low frequency circuits. Phys. Rev. **26**, 71–85 (1925)
12. J. Bernamont, Fluctuations de potentiel aux bornes d'un conducteur métallique de faible volume parcouru par un courant. Ann. de Physique **7**, 71–140 (1937)
13. A. van der Ziel, On the noise spectra of semiconductor noise and of flicker effect. Physica **XVI**, 359–372 (1950)
14. A.L. McWhorter, 1/*f* noise and germanium surface properties, in *Semiconductor Surface Physics*, (Univ. Pennsylvania Press, Philadelphia, 1957), pp. 207–228
15. C.T. Sah, F.H. Hielscher, Evidence of the surface origin of the 1/*f* noise. Phys. Rev. Lett. **17**, 956–958 (1966)
16. S.T. Hsu, D.J. Fitzgerald, A.S. Grove, Surface-state related 1/*f* noise in *p-n* junctions and MOS transistors. Appl. Phys. Lett. **12**, 287–289 (1968)
17. S. Christenson, I. Lundstrom, C. Svennson, Low-frequency noise in MOS transistors – Theory. Solid State Electron. **11**, 797–812 (1968)
18. F.N. Hooge, 1/*f* noise is no surface effect. Phys. Lett. **29A**(3), 139–140 (1969)
19. F.N. Hooge, A.M.H. Hoppenbrouwers, 1/*f* noise in continuous thin gold films. Physica **45**(3), 386–392 (1969)
20. R.F. Voss, J. Clarke, Flicker 1/*f* noise: Equilibrium temperature and resistance fluctuations. Phys. Rev. B **13**(2), 556–573 (1976)
21. F.N. Hooge, L.K.J. Vandamme, Lattice scattering causes 1/*f* noise. Phys. Lett. A **66**(4), 315–316 (1978)
22. A. van der Ziel, Flicker noise in electronic devices. Adv. Electron. Electron Phys. **49**, 225–296 (1979)
23. F.N. Hooge, T.G.M. Kleinpenning, L.K.J. Vandamme, Experimental studies on 1/*f* noise. Rep. Prog. Phys. **44**(5), 479–532 (1981)
24. P. Dutta, P.M. Horn, Low-frequency fluctuations in solids: 1/*f* noise. Rev. Mod. Phys. **53**, 497–516 (1981)
25. G. Reimbold, Modified 1/*f* noise trapping theory and experiments in MOS transistors biased from weak to strong inversion: Influence of interface states. IEEE Trans. Electron Devices **31**(9), 1190–1198 (1984)
26. K.S. Ralls, W.J. Skocpol, L.D. Jackel, R.E. Howard, L.A. Fetter, R.W. Epworth, D.M. Tennant, Discrete resistance switching in submicrometer Si inversion layers: Individual interface traps and low-frequency (1/*f*?) noise. Phys. Rev. Lett. **52**, 228–231 (1984)
27. S.M. Kogan, Low-frequency current noise with a 1/*f* spectrum in solids. Sov. Phys. Usp. **28**(2), 170–195. [*Usp. Fiz. Nauk*, vol. 145, pp. 285-328] (1985)
28. J. Pelz, J. Clarke, Quantitative local-interference model for 1/*f* noise in metal films. Phys. Rev. B **36**(8), 4479–4482 (1987)
29. M.B. Weissman, 1/*f* noise and other slow, nonexponential kinetics in condensed matter. Rev. Mod. Phys. **60**, 537–571 (1988)
30. N. Giordano, Defect motion and low-frequency noise in disordered metals. Rev. Solid State Science **3**(1), 27–69 (1989)
31. M.J. Kirton, M.J. Uren, Noise in solid-state microstructures: A new perspective on individual defects, interface states, and low-frequency (1/*f*) noise. Adv. Phys. **38**, 367–468 (1989)
32. A. Jayaraman, C.G. Sodini, A 1/*f* noise technique to extract the oxide trap density near the conduction band edge of Si. IEEE Trans. Electron Devices **36**(9), 1773–1782 (1989)
33. K.K. Hung, P.K. Ko, C. Hu, Y.C. Cheng, A unified model for the flicker noise in MOSFETs. IEEE Trans. Electron Devices **37**, 654–665 (1990)
34. G. Ghibaudo, O. Roux, C. Nguyen-Duc, F. Balestra, J. Brini, Improved analysis of low frequency noise in field-effect MOS transistors. Phys. Stat. Sol. (a) **124**, 571–581 (1991)
35. D.M. Fleetwood, P.S. Winokur, R.A. Reber Jr., T.L. Meisenheimer, J.R. Schwank, M.R. Shaneyfelt, L.C. Riewe, Effects of oxide, interface, and border traps on MOS devices. J. Appl. Phys. **73**, 5058–5074 (1993)
36. F.N. Hooge, 1/*f* noise sources. IEEE Trans. Electron Devices **41**(11), 1926–1935 (1994)
37. L.K.J. Vandamme, Noise as a diagnostic tool for quality and reliability of electronic devices. IEEE Trans. Electron Devices **41**(11), 2176–2187 (1994)

38. J.H. Scofield, N. Borland, D.M. Fleetwood, Reconciliation of different gate-voltage dependencies of 1/*f* noise in *n*MOS and *p*MOS transistors. IEEE Trans. Electron Devices **41**(11), 1946–1952 (1994)
39. D.M. Fleetwood, T.L. Meisenheimer, J.H. Scofield, 1/*f* noise and radiation effects in MOS devices. IEEE Trans. Electron Devices **41**, 1953–1964 (1994)
40. L.K.J. Vandamme, X.S. Li, D. Rigaud, 1/*f* noise in MOS devices: Mobility or number fluctuations? IEEE Trans. Electron Devices **41**(11), 1936–1945 (1994)
41. T.G.M. Kleinpenning, Low-frequency noise in modern bipolar transistors: Impact of intrinsic transistor and parasitic series resistances. IEEE Trans. Electron Devices **41**(11), 1981–1991 (1994)
42. E. Simoen, C. Claeys, On the flicker noise in submicron silicon MOSFETs. Solid State Electron. **43**, 865–882 (1999)
43. D.M. Fleetwood, H.D. Xiong, Z.Y. Lu, C.J. Nicklaw, J.A. Felix, R.D. Schrimpf, S.T. Pantelides, Unified model of hole trapping, 1/*f* noise, and thermally stimulated current in MOS devices. IEEE Trans. Nucl. Sci. **49**(6), 2674–2683 (2002)
44. T. Grasser, H. Reisinger, P.J. Wagner, B. Kaczer, Time-dependent defect spectroscopy for characterization of border traps in MOS transistors. Phys. Rev. B **82**(24), 245318 (2010)
45. T. Grasser, B. Kaczer, W. Goes, H. Reisinger, T. Aichinger, P. Hehenberger, P.J. Wagner, F. Schanovsky, J. Franco, M.T. Luque, M. Nehliebel, The paradigm shift in understanding the negative bias-temperature instability: From reaction-diffusion to switching oxide traps. IEEE Trans. Electron Devices **58**(11), 3652–3666 (2011)
46. T. Grasser, Stochastic charge trapping in oxides: From random telegraph noise to bias-temperature instabilities. Microelectron. Reliab. **52**(1), 39–70 (2012)
47. A.A. Balandin, Low-frequency 1/*f* noise in graphene devices. Nature Nanotechnol. **8**(8), 549–555 (2013)
48. D.M. Fleetwood, 1/*f* noise and defects in microelectronic materials and devices. IEEE Trans. Nucl. Sci. **62**(4), 1462–1486 (2015)
49. R.F. Voss, J.F. Clarke, 1/*f* noise in music and speech. Nature **258**(5533), 317–318 (1975)
50. R.F. Voss, J.F. Clarke, 1/*f* noise in music – Music from 1/*f* noise. J. Acoust. Soc. Amer. **63**(1), 258–263 (1978)
51. D.J. Levitin, P. Chordia, V. Menon, Musical rhythm spectra from Bach to Joplin obey a 1/*f* power law. Proc. Nat Acad. Sci. **109**(10), 3716–3720 (2002)
52. P. Bak, C. Tang, K. Wiesenfeld, Self-organized criticality: An explanation of 1/*f* noise. Phys. Rev. Lett. **59**(4), 381–384 (1987)
53. M. Paczuski, S. Maslov, P. Bak, Avalance dynamics in evolution, growth, and depinning models. Phys. Rev. A **53**(1), 414–443 (1996)
54. H.M. Jaeger, S.R. Nagel, Physics of the granular state. Science **255**(5051), 1523–1531 (1992)
55. M. Takayasu, H. Takayasu, 1/*f* noise in a traffic model. Fractals Compl. Geom. Patt. Scal. Nat. Soc. **1**(4), 860–866 (1993)
56. J. Beran, R. Sherman, M.S. Taqqu, W. Willinger, Long-range dependence in variable bit-rate video traffic. IEEE Trans. Commun. **43**(2-4), 1566–1579 (1995)
57. R.H. Riedi, M.S. Crouse, V.J. Ribeiro, R.G. Baraniuk, A multifractal wavelet model with application to network traffic. IEEE Trans. Inf. Theory **45**(3), 992–1018 (1999)
58. A. Sornette, D. Sornette, Self organized criticality and earthquakes. Europhys. Lett. **9**(3), 197–202 (1989)
59. K. Ito, M. Matsuzaki, Earthquakes as self-organized critical phenomena. J. Geophys. Res. Solid Earth Plan. **95**(B5), 6853–6860 (1990)
60. J.H. Scofield, N. Borland, D.M. Fleetwood, Temperature-independent switching rates for a random telegraph signal in a Si MOSFET at low temperatures. Appl. Phys. Lett. **76**, 3248–3250 (2000)
61. J.H. Scofield, D.M. Fleetwood, Physical basis for nondestructive tests of MOS radiation hardness. IEEE Trans. Nucl. Sci. **38**, 1567–1577 (1991)
62. C. Surya, T.Y. Hsiang, Theory and experiment on the $1/f^{\gamma}$ noise in *p*-channel MOSFETs at low drain bias. Phys. Rev. B **33**(7), 4898–4905 (1986)

63. E.G. Ioannidis, C.A. Dimitriadis, S. Haendler, R.A. Bianchi, J. Jomaah, G. Ghibaudo, Improved analysis and modeling of low-frequency noise in nanoscale MOSFETs. Solid State Electron. **76**, 54–59 (2012)
64. H. Wong, Y.C. Cheng, Study of the electronic trap distribution at the Si-SiO_2 interface utilizing the low-frequency noise measurement. IEEE Trans. Electron Devices **37**(7), 1743–1749 (1990)
65. P. Wang, R. Jiang, J. Chen, E.X. Zhang, M.W. McCurdy, R.D. Schrimpf, D.M. Fleetwood, 1/*f* noise in as-processed and proton-irradiated GaN/AlGaN HEMTs due to carrier-number fluctuations. IEEE Trans. Nucl. Sci. **64**(1), 181–189 (2017)
66. S.A. Francis, A. Dasgupta, D.M. Fleetwood, Effects of total dose irradiation on the gate-voltage dependence of the 1/*f* noise of *n*MOS and *p*MOS transistors. IEEE Trans. Electron Devices **57**(2), 503–510 (2010)
67. R.P. Jindal, A. van der Ziel, Phonon fluctuation model for flicker noise in elemental semiconductors. Appl. Phys. Lett. **52**(4), 2884–2888 (1981)
68. M. Mihaila, Phonon observations from 1/*f* noise measurements. Phys. Lett. **104A**(3), 157–158 (1984)
69. J. Xu, M.J. Deen, MOSFET 1/*f* noise model based on mobility fluctuation in linear region. Electron. Lett. **38**(9), 429–431 (2002)
70. M.E. Levinshtein, S.L. Rumantsev, R. Gaska, J.W. Yang, M.S. Shur, AlGaN/GaN HEMTs with low 1/*f* noise. Appl. Phys. Lett. **73**(8), 1089–1091 (1998)
71. F. Crupi, P. Magnone, S. Strangio, F. Iucolano, G. Meneghesso, Low-frequency noise and gate bias instability in normally off AlGaN/GaN HEMTs. IEEE Trans. Nucl. Sci. **63**(5), 2219–2222 (2016)
72. M.B. Weissman, Implications of mobility-fluctation descriptions of 1/*f* noise in semiconductors. Physica **100B**(2), 157–162 (1980)
73. M.B. Weissman, Survey of recent 1/*f* noise theories, in *Proc. Sixth Intl. Conf. Noise Phys. Syst*, ed. by M. D. Gaithersburg, P. H. E. Meijer, R. D. Mountain, R. J. Soulen Jr., (NBS Special Pub. No. 614, Washington, DC, 1981), pp. 133–142
74. D.M. Fleetwood, J.T. Masden, N. Giordano, 1/*f* noise in platinum films and ultrathin platinum wires: Evidence for a common, bulk origin. Phys. Rev. Lett. **50**(6), 450–453 (1983)
75. D.M. Fleetwood, N. Giordano, Direct link between 1/*f* noise and defects in metal films. Phys. Rev. B **31**(2), 1157–1159 (1985)
76. J.H. Scofield, J.V. Mantese, W.W. Webb, Temperature dependence of noise processes in metals. Phys. Rev. B **34**(2), 723–731 (1986)
77. N.M. Zimmerman, W.W. Webb, Microscopic scatterer displacements generate the 1/*f* resistance noise of H in Pd. Phys. Rev. Lett. **61**(7), 889–892 (1988)
78. J. Pelz, J. Clarke, W.E. King, Flicker (1/*f*) noise in copper films due to radiation-induced defects. Phys. Rev. B **38**(15), 10371–10386 (1988)
79. K.S. Ralls, R.A. Buhrman, Microscopic study of 1/*f* noise in metal nanobridges. Phys. Rev. B **44**, 5800–5817 (1991)
80. Y. Fisher, M. McGuire, R.F. Voss, M.F. Barnsley, R.L. Devaney, B.B. Mandelbrot, *The Science of Fractal Images* (Springer Science & Business Media, Berlin, 2012)
81. P.H. Handel, Quantum approach to 1/*f* noise. Phys. Rev. A **22**(2), 745–757 (1980)
82. A. van der Ziel, P.H. Handel, X. Zhu, K.H. Duh, A theory of the Hooge parameters of solid-state devices. IEEE Trans. Electron Devices **32**(3), 667–671 (1985)
83. P.H. Handel, Fundamental quantum l/*f* noise in semiconductor devices. IEEE Trans. Electron Devices **41**(11), 2023–2032 (1994)
84. G.S. Kousik, C.M. Van Vliet, G. Bosman, P.H. Handel, Quantum 1/*f* noise associated with ionized impurity scattering and electron-phonon scattering in condensed matter. Adv. Phys. **34**(6), 663–702 (2006)
85. L.B. Kiss, P. Heszler, An exact proof of the invalidity of Handel's quantum 1/*f* noise model. J. Phys. C **19**, L631–L633 (1986)
86. T.M. Nieuwenhuizen, D. Frenkel, N.G. van Kampen, Objections to Handel's quantum theory of 1/*f* noise. Phys. Rev. A **35**(6), 2750–2753 (1987)

87. G.S. Kousik, J. Gong, C.M. van Vliet, G. Bosman, W.H. Ellis, E.E. Carrol, P.H. Handel, Flicker noise fluctuations in alpha-radioactive decay. Canadian J. Phys. **65**(4), 365–375 (1987)
88. G. Concas, M. Lissia, Search for non-Poissonian behavior in nuclear β decay. Phys. Rev. E **55**(3), 2546–2550 (1997)
89. J.W. Eberhard, P.M. Horn, Excess (1/*f*) noise in metals. Phys. Rev. B **18**(12), 6681–6693 (1978)
90. P. Dutta, P. Dimon, P.M. Horn, Energy scales for noise processes in metals. Phys. Rev. Lett. **43**(9), 646–649 (1979)
91. C.D. Liang, P. Wang, S.M. Zhao, E.X. Zhang, M.L. Alles, D.M. Fleetwood, R.D. Schrimpf, R. Ma, Y. Su, S. Koester, Radiation-induced charge trapping in black phosphorus MOSFETs with HfO_2 gate dielectrics. IEEE Trans. Nucl. Sci. **65**(6), 1227–1238 (2018)
92. J.W. Martin, The electrical resistivity due to structural defects. Philos. Mag. **24**(189), 555–566 (1971)
93. S. Feng, P.A. Lee, A.D. Stone, Sensitivity of the conductance of a disordered metal to the motion of a single atom: Implications for 1/*f* noise. Phys. Rev. Lett. **56**, 1960–1963 (1986)
94. P.A. Lee, A.D. Stone, H. Fukuyama, Universal conductance fluctuations in metals: Effects of finite temperature, interactions, and magnetic field. Phys. Rev. B **35**(3), 1039–1070 (1987)
95. M.J. Kirton, M.J. Uren, Capture and emission kinetics of individual Si-SiO_2 interface states. Appl. Phys. Lett. **48**(19), 1270–1272 (1986)
96. B. Neri, P. Olivo, B. Ricco, Low-frequency noise in Si-gate MOS capacitors before oxide breakdown. Appl. Phys. Lett. **51**(25), 2167–2169 (1987)
97. P. Restle, Individual oxide traps as probes into sub-micron devices. Appl. Phys. Lett. **53**(19), 1862–1864 (1988)
98. M.J. Kirton, M.J. Uren, S. Collins, M. Schulz, A. Karmann, K. Scheffer, Individual defects at the Si-SiO_2 interface. Semicond. Sci. Technol. **4**(12), 1116–1126 (1989)
99. G. Ghibaudo, T. Boutchacha, Electrical noise and RTS fluctuations in advanced CMOS devices. Microelectron. Reliab. **42**, 573–582 (2002)
100. H.D. Xiong, D.M. Fleetwood, B.K. Choi, A.L. Sternberg, Temperature dependence and irradiation response of 1/*f* noise in MOSFETs. IEEE Trans. Nucl. Sci. **49**(6), 2718–2723 (2002)
101. S.A. Francis, C.X. Zhang, E.X. Zhang, D.M. Fleetwood, R.D. Schrimpf, K.F. Galloway, E. Simoen, J. Mitard, C. Claeys, Comparison of charge pumping and 1/*f* noise in irradiated Ge *p*MOSFETs. IEEE Trans. Nucl. Sci. **59**(6), 735–741 (2012)
102. C.X. Zhang, E.X. Zhang, D.M. Fleetwood, R.D. Schrimpf, S. Dhar, S.-H. Ryu, X. Shen, S.T. Pantelides, Origins of low-frequency noise and interface traps in 4H-SiC MOSFETs. IEEE Electron Device Lett. **34**(1), 117–119 (2013)
103. C.X. Zhang, X. Shen, E.X. Zhang, D.M. Fleetwood, R.D. Schrimpf, S.A. Francis, T. Roy, S. Dhar, S.H. Ryu, S.T. Pantelides, Temperature dependence and postirradiation annealing response of the 1/*f* noise of 4H-SiC MOSFETs. IEEE Trans. Electron Devices **60**(7), 2361–2367 (2013)
104. J. Chen, Y.S. Puzyrev, C.X. Zhang, E.X. Zhang, M.W. McCurdy, D.M. Fleetwood, R.D. Schrimpf, S.T. Pantelides, S.W. Kaun, E.C.H. Kyle, J.S. Speck, Proton-induced dehydrogenation of defects in AlGaN/GaN HEMTs. IEEE Trans. Nucl. Sci. **60**(6), 4080–4086 (2013)
105. Z. Celik, T.Y. Hsiang, Study of 1/*f* noise in *n*MOSFETs: Linear region. IEEE Trans. Electron Devices **32**(12), 2798–2802 (1985)
106. P.S. Winokur, M.M. Sokoloski, Comparison of interface-state buildup in MOS capacitors subjected to penetrating and non-penetrating radiation. Appl. Phys. Lett. **28**(10), 627–630 (1976)
107. S.K. Lai, Interface trap generation in SiO_2 when electrons are captured by trapped holes. J. Appl. Phys. **54**(5), 2540–2546 (1983)
108. M. Schulz, Interface states at the SiO_2-Si interface. Surf. Sci. **132**(1-3), 422–455 (1983)
109. Y. Nishioka, E.F. da Silva Jr., T.P. Ma, Radiation-induced interface traps in Mo/SiO_2/Si capacitors. IEEE Trans. Nucl. Sci. **34**(6), 1166–1171 (1987)

110. E.F. da Silva Jr., Y. Nishioka, T.P. Ma, Two distinct interface trap peaks in radiation-damaged metal/SiO_2/Si structures. Appl. Phys. Lett. **51**, 270–272 (1987)
111. T.L. Meisenheimer, D.M. Fleetwood, M.R. Shaneyfelt, L.C. Riewe, 1/*f* noise in *n*- and *p*-channel MOS devices through irradiation and annealing. IEEE Trans. Nucl. Sci. **38**, 1297–1303 (1991)
112. D.M. Fleetwood, M.J. Johnson, T.L. Meisenheimer, P.S. Winokur, W.L. Warren, S.C. Witczak, 1/*f* noise, hydrogen transport, and latent interface-trap buildup in irradiated MOS devices. IEEE Trans. Nucl. Sci. **44**, 1810–1817 (1997)
113. T. Grasser, K. Rott, H. Reisinger, M. Waltl, J. Franco, B. Kaczer, A unified perspective of RTN and BTI. Proc. IEEE Int. Reliab. Phys. Sympos., 4A.5.1–4A.5.7 (2014)
114. D.M. Fleetwood, J.H. Scofield, Evidence that similar point defects cause 1/*f* noise and radiation-induced-hole trapping in MOS devices. Phys. Rev. Lett. **64**, 579–582 (1990)
115. T.L. Meisenheimer, D.M. Fleetwood, Effect of radiation-induced charge on 1/*f* noise in MOS devices. IEEE Trans. Nucl. Sci. **37**, 1696–1702 (1990)
116. D.M. Fleetwood, M.R. Shaneyfelt, J.R. Schwank, Estimating oxide, interface, and border-trap densities in MOS transistors. Appl. Phys. Lett. **64**, 1965–1967 (1994)
117. D.M. Fleetwood, W.L. Warren, M.R. Shaneyfelt, R.A.B. Devine, J.H. Scofield, Enhanced MOS 1/*f* noise due to near-interfacial oxygen deficiency. J. Non Cryst. Solids **187**, 199–205 (1995)
118. F.J. Feigl, W.B. Fowler, K.L. Yip, Oxygen vacancy model for E'/E_1 defect in SiO_2. Solid State Commun. **14**(3), 225–229 (1974)
119. P.M. Lenahan, P.V. Dressendorfer, Hole traps and trivalent silicon centers in MOS devices. J. Appl. Phys. **55**(10), 3495–3499 (1984)
120. J.R. Schwank, D.M. Fleetwood, The effect of postoxidation anneal temperature on radiation-induced charge trapping in polycrystalline silicon gate metal-oxide-semiconductor devices. Appl. Phys. Lett. **53**(9), 770–772 (1988)
121. A.J. Lelis, T.R. Oldham, H.E. Boesch Jr., F.B. McLean, The nature of the trapped hole annealing process. IEEE Trans. Nucl. Sci. **36**, 1808–1815 (1989)
122. D.M. Fleetwood, S.L. Miller, R.A. Reber Jr., P.J. McWhorter, P.S. Winokur, M.R. Shaneyfelt, J.R. Schwank, New insights into radiation-induced oxide-trap charge through TSC measurement and analysis. IEEE Trans. Nucl. Sci. **39**, 2192–2203 (1992)
123. W.L. Warren, D.M. Fleetwood, M.R. Shaneyfelt, J.R. Schwank, P.S. Winokur, R.A.B. Devine, D. Mathiot, Links between oxide traps, interface traps, and border traps in high-temperature annealed Si/SiO_2 systems. Appl. Phys. Lett. **64**(25), 3452–3454 (1994)
124. R.A.B. Devine, W.L. Warren, J.B. Xu, I.H. Wilson, P. Paillet, J.L. Leray, Oxygen gettering and oxide degradation during annealing of Si/SiO_2/Si. J. Appl. Phys. **77**(1), 175–186 (1995)
125. P.M. Lenahan, J.F. Conley Jr., What can electron paramagnetic resonance tell us about the Si/SiO_2 system? J. Vac. Sci. Technol. B **16**(4), 2134–2153 (1998)
126. D.M. Fleetwood, Total ionizing dose effects in MOS and low-dose-rate sensitive linear-bipolar devices. IEEE Trans. Nucl. Sci. **60**(3), 1706–1730 (2013)
127. T. Grasser, K. Rott, H. Reisinger, M. Waltl, P. Wagner, F. Schanovsky, W. Goes, G. Pobegen, B. Kaczer, Hydrogen-related volatile defects as the possible cause for the recoverable component of NBTI. IEDM Tech. Dig., 409–412 (2013)
128. T. Grasser, W. Goes, Y. Wimmer, F. Schanovsky, G. Rzepa, M. Waltl, K. Rott, H. Reisinger, V.V. Afanase'ev, A. Stesmans, A.M. El-Sayed, A.L. Shluger, On the microscopic structure of hole traps in *p*MOSFETs. IEDM Tech. Dig., 530–533 (2014)
129. W. Goes, Y. Wimmer, A.-M. El-Sayed, G. Rzepa, M. Jech, A.L. Shluger, T. Grasser, Identification of oxide defects in semiconductor devices: A systematic approach linking DFT to rate equations and experimental evidence. Microelectron. Reliab. **87**(4), 286–320 (2018)
130. A. Kerber, E. Cartier, L. Pantisano, R. Degraeve, T. Kauerauf, Y. Kim, A. Hou, G. Groeseneken, H.E. Maes, U. Schwalke, Origin of the threshold voltage instability in SiO_2/HfO_2 dual layer gate dielectrics. IEEE Electron Device Lett. **24**(2), 87–89 (2003)
131. K. Xiong, J. Robertson, M.C. Gibson, S.J. Clark, Defect energy levels in HfO_2 high-dielectric-constant gate oxide. Appl. Phys. Lett. **87**(18), 183505 (2005)

132. J. Robertson, K. Xiong, S.J. Clark, Band gaps and defect levels in functional oxides. Thin Solid Films **496**, 1–7 (2006)
133. J.L. Gavartin, D.M. Ramo, A.L. Shluger, G. Bersuker, B.H. Lee, Negative oxygen vacancies in HfO_2 as charge traps in high-K gate stacks. Appl. Phys. Lett. **89**(8), 082908 (2006)
134. G.X. Duan, J.A. Hachtel, E.X. Zhang, C.X. Zhang, D.M. Fleetwood, R.D. Schrimpf, R.A. Reed, J. Mitard, D. Linten, L. Witters, N. Collaert, A. Mocuta, A.V.-Y. Thean, M.F. Chisholm, S.T. Pantelides, Effects of negative-bias-temperature instability on low-frequency noise in SiGe *p*MOSFETs. IEEE Trans. Dev. Mater. Reliab. **16**(4), 541–548 (2016)
135. P. Wang, C. Perini, A.O. Hara, B.R. Tuttle, E.X. Zhang, H. Gong, L. Dong, C. Liang, R. Jiang, W. Liao, D.M. Fleetwood, R.D. Schrimpf, E.M. Vogel, S.T. Pantelides, Radiation-induced charge trapping and low-frequency noise of graphene transistors. IEEE Trans. Nucl. Sci. **65**(1), 156–163 (2018)
136. D.M. Fleetwood, Border traps in MOS devices. IEEE Trans. Nucl. Sci. **39**(2), 269–271 (1992)
137. D.M. Fleetwood, M.R. Shaneyfelt, W.L. Warren, J.R. Schwank, T.L. Meisenheimer, P.S. Winokur, Border traps: Issues for MOS radiation response and long-term reliability. Microelectron. Reliab. **35**, 403–428 (1995)
138. D.M. Fleetwood, Fast and slow border traps in MOS devices. IEEE Trans. Nucl. Sci. **43**(6), 779–786 (1996)
139. D.M. Fleetwood, Border traps and bias-temperature instabilities in MOS devices. Microelectron. Reliab. **80**(1), 266–277 (2018)
140. T. Roy, E.X. Zhang, Y.S. Puzyrev, D.M. Fleetwood, R.D. Schrimpf, B.K. Choi, A.B. Hmelo, S.T. Pantelides, Process dependence of proton-induced degradation in GaN HEMTs. IEEE Trans. Nucl. Sci. **57**(6), 3060–3065 (2010)
141. A. Balandin, Gate-voltage dependence of low-frequency noise in AlGaN/GaN heterostructure field-effect transistors. Electron. Lett. **36**(10), 912–913 (2000)
142. J.A. Garrido, B.E. Foutz, J.A. Smart, J.R. Shealy, M.J. Murphy, W.J. Schaff, L.F. Eastman, E. Muñoz, Low-frequency noise and mobility fluctuations in AlGaN/GaN heterostructure FETs. Appl. Phys. Lett. **76**(23), 3442–3444 (2000)
143. S.L. Rumantsyev, N. Pala, M.S. Shur, R. Gaska, M.E. Levinshtein, P.A. Ivanov, M. Asif Khan, G. Simin, X. Hu, J. Yang, Concentration dependence of the 1/*f* noise in AlGaN/GaN HEMTs. Semicond. Sci. Technol. **17**, 476–479 (2002)
144. L.H. Huang, S.H. Yeh, C.T. Lee, High frequency and low frequency noise of AlGaN/GaN MOS HEMTs with gate insulator grown using photochemical oxidation method. Appl. Phys. Lett. **93**, 043511 (2008)
145. H. Rao, G. Bosman, Simultaneous low-frequency noise characterization of gate and drain currents in AlGaN/GaN HEMTs. J. Appl. Phys. **106**, 103712 (2009)
146. J. Peransin, P. Vignaud, D. Rigaud, L. Vandamme, 1/*f* noise in MODFETs at low drain biases. IEEE Trans. Electron Devices **37**(10), 2250–2253 (1990)
147. J. Chen, Y.S. Puzyrev, R. Jiang, E.X. Zhang, M.W. McCurdy, D.M. Fleetwood, R.D. Schrimpf, S.T. Pantelides, A. Arehart, S.A. Ringel, P. Saunier, C. Lee, Effects of applied bias and high field stress on the radiation response of GaN/AlGaN HEMTs. IEEE Trans. Nucl. Sci. **62**(6), 2423–2430 (2015)
148. T. Roy, E.X. Zhang, Y.S. Puzyrev, X. Shen, D.M. Fleetwood, R.D. Schrimpf, G. Koblmueller, R. Chu, C. Poblenz, N. Fichtenbaum, C.S. Suh, U.K. Mishra, J.S. Speck, S.T. Pantelides, Temperature dependence and microscopic origin of low frequency 1/*f* noise in GaN/AlGaN high electron mobility transistors. Appl. Phys. Lett. **99**, 203501 (2011)
149. J. Chen, Y.S. Puzyrev, E.X. Zhang, D.M. Fleetwood, R.D. Schrimpf, A.R. Arehart, S.A. Ringel, S.W. Kaun, E.C.H. Kyle, J.S. Speck, P. Saunier, C. Lee, S.T. Pantelides, High-field stress, low-frequency noise, and long-term reliability of AlGaN/GaN HEMTs. IEEE Trans. Device Mater. Reliab. **16**(3), 282–289 (2016)
150. C.G. de Walle, J. Neugebauer, First-principles calculations for defects and impurities: Applications to III-nitrides. J. Appl. Phys. **95**, 3851–3879 (2004)

151. Y. Puzyrev, T. Roy, E.X. Zhang, D.M. Fleetwood, R.D. Schrimpf, S.T. Pantelides, Radiation-induced defect evolution and electrical degradation of AlGaN/GaN high-electron-mobility transistors. IEEE Trans. Nucl. Sci. **58**(6), 2918–2924 (2011)
152. Y.S. Puzyrev, R.D. Schrimpf, D.M. Fleetwood, S.T. Pantelides, Role of Fe complexes in the degradation of GaN/AlGaN high-electron-mobility transistors. Appl. Phys. Lett. **106**, 053505 (2015)
153. X. Shen, Y.S. Puzyrev, D.M. Fleetwood, R.D. Schrimpf, S.T. Pantelides, Quantum mechanical modeling of radiation-induced defect dynamics in electronic devices. IEEE Trans. Nucl. Sci. **62**(5), 2169–2178 (2015)
154. M.E. Levinshtein, S.L. Rumyantsev, J.W. Palmour, D.B. Slater, Low frequency noise in 4H silicon carbide. J. Appl. Phys. **81**(4), 1758–1762 (1997)
155. M.E. Levinshtein, S.L. Rumyantsev, M.S. Shur, R. Gaska, M.A. Khan, Low frequency and 1/*f* noise in wide-gap semiconductors: SiC and GaN. IEE Proc Circ Devices Syst **149**(1), 32–39 (2002)
156. V.V. Afanas'ev, M. Bassler, G. Pensl, M. Schulz, Intrinsic SiC/SiO_2 interface states. Physica Status Solid A Appl. Mater. Sci. **162**(1), 321–337 (1997)
157. C. Raynaud, Silica films on SiC: A review of electrical properties and device applications. J. Non-Cryst. Solids **280**(1-3), 1–31 (2001)
158. G. Liu, A.C. Ahyi, Y. Xu, T. Isaacs-Smith, Y.K. Sharma, J.R. Williams, L.C. Feldman, S. Dhar, Enhanced inversion mobility on 4H-SiC (11$\underline{2}$0) using phosphorus and nitrogen interface passivation. IEEE Electron Device Lett. **34**(2), 181–183 (2013)
159. E. Arnold, Charge-sheet model for silicon carbide inversion layers. IEEE Trans. Electron Devices **46**(3), 497–503 (1999)
160. X. Shen, E.X. Zhang, C.X. Zhang, D.M. Fleetwood, R.D. Schrimpf, S. Dhar, S. Ryu, S.T. Pantelides, Atomic-scale origin of bias-temperature instabilities in SiC-SiO_2 structures. Appl. Phys. Lett. **98**(6), 063507 (2011)
161. S. Rumyantsev, G. Liu, W. Stillman, M. Shur, A.A. Balandin, Electrical and noise characteristics of graphene field-effect transistors: Ambient effects, noise sources and physical mechanisms. J. Phys. Condens. Matter **22**(39), 395302 (2010)
162. Y.M. Lin, P. Avouris, Strong suppression of electrical noise in bilayer graphene nanodevices. Nano Lett. **8**(8), 2119–2125 (2008)
163. Q. Shao, G. Liu, D. Teweldebrhan, A.A. Balandin, S. Rumyantsev, M.S. Shur, et al., Flicker noise in bilayer graphene transistors. IEEE Electron Device Lett. **30**(3), 288–290 (2009)
164. G. Liu, W. Stillman, S. Rumyantsev, Q. Shao, M.S. Shur, A.A. Balandin, Low-frequency electronic noise in the double-gate single-layer graphene transistors. Appl. Phys. Lett. **95**(3), 033103 (2009)
165. J. Renteria, R. Samnakay, S.L. Rumantsev, C. Jiang, P. Goli, M.S. Shur, A.A. Balandin, Low-frequency 1/*f* noise in MoS_2 transistors: Relative contributions of the channel and contacts. Appl. Phys. Lett. **104**, 153103 (2014)
166. X. Xie, D. Sarkar, W. Liu, J. Kang, O. Marinov, M.J. Deen, K. Banerjee, Low-frequency noise in bilayer MoS_2 transistor. ACS Nano **8**(6), 5633–5640 (2014)
167. C. Liang, Y. Su, E.X. Zhang, K. Ni, M.L. Alles, R.D. Schrimpf, D.M. Fleetwood, S.J. Koester, Total ionizing dose effects on HfO_2-passivated black phosphorus transistors. IEEE Trans. Nucl. Sci. **64**(1), 170–175 (2017)
168. C.D. Liang, R. Ma, Y. Su, A. O'Hara, E.X. Zhang, M.L. Alles, P. Wang, S.E. Zhao, S.T. Pantelides, S.J. Koester, R.D. Schrimpf, D.M. Fleetwood, Defects and low-frequency noise in irradiated black phosphorus MOSFETs with HfO_2 gate dielectrics. IEEE Trans. Nucl. Sci. **65**(6), 1227–1238 (2018)

Noise and Fluctuations in Fully Depleted Silicon-on-Insulator MOSFETs

Christoforos Theodorou and Gérard Ghibaudo

Abbreviations

CMF	Correlated mobility fluctuations
CNF	Carrier number fluctuations
C_{ox}	Gate oxide capacitance per area
C_{Si}	Silicon capacitance per area
FD	Fully depleted
f_e	Escape frequency
g_d	Drain voltage transconductance
g_m	Gate voltage transconductance
HMF	Hooge mobility fluctuations
n_1	Carrier density near trap when $E_T = E_F$
n_s	Carrier density near trap
PD	Partially depleted
PSD	Power spectral density
Q_i	Inversion charge density
Q_{i1}	Inversion charge density near trap when $E_T = E_F$
RCS	Remote Coulomb scattering
R_{sd}	Source/drain series resistance
S_{Id}	Drain current power spectral density
SOI	Semiconductor on insulator
S_{Vfb}	Flat-band voltage power spectral density
S_{Vg}	Input-referred gate voltage power spectral density

C. Theodorou (✉) · G. Ghibaudo
Institut de Microélectronique Electromagnétisme et Photonique et LAboratoire d'Hyperfréquences et de Caractérisation (IMEP-LAHC), Univ. Grenoble Alpes, Univ. Savoie Mont Blanc, CNRS, Grenoble INP, Grenoble, France
e-mail: christoforos.theodorou@grenoble-inp.fr

T. Grasser (ed.), *Noise in Nanoscale Semiconductor Devices*,
https://doi.org/10.1007/978-3-030-37500-3_2

t_{ox2}/t_{box}	Back-gate oxide
$V_{BG}/V_{G2}/V_B$	Back-gate voltage
V_t	Threshold voltage
x_t	Trap depth in oxide
α/α_{sc}	RCS coefficient
ν_{th}	Thermal velocity
σ	Trap cross section
σ_0	Trap cross section for interface trap
τ	Trap time constant
τ_c	Capture time
τ_e	Emission time
Ω	CMF factor

1 Introduction

The Semiconductor-on-Insulator (SOI) philosophy offers a great control over the transistor channel: the device body is isolated between two oxides: the gate oxide that all classical MOSFETs have and the so-called buried oxide (BOX) [1]. The first device parameter that was not at all controlled in classic Bulk MOSFETs was the channel thickness. Reducing the thickness of the silicon body provides a better subthreshold slope [2], a very efficient control of the short-channel effects (SCEs), as well as the drain-induced barrier level (DIBL). Furthermore, the SOI structure allows the application of back-bias to control the threshold voltage value. In addition, the threshold voltage variability can be dramatically reduced, because there is no need for strong doping in the channel. Actually, the film is so lightly doped that it is already partially depleted (PD) or fully depleted (FD) even before applying bias voltages. In this way, ultrathin body (UTB) SOI MOSFETs can be designed and produced, with a silicon film thickness from around 7 to 14 nm. However, as we will later discuss, the reduction of the film thickness can lead to serious electrostatic coupling effects between the front and back gate. Finally, by reducing the BOX thickness from the typical 145–300 nm value down to 8–25 nm, the device is called ultrathin body and buried-oxide (UTBB) FDSOI MOSFET [3] and offers an easier back-bias threshold voltage control.

FD-SOI technology is considered one of the best candidates for short-channel effect (SCE) control in future sub-28 nm CMOS generations, while it remains compatible with standard planar CMOS technology [3–7]. The use of a midgap/high-k metal gate stack with undoped SOI films allows great improvement of variability as compared to bulk technology [4, 8, 9]. The use of ultrathin body and buried oxide thickness (UTBB) also enables to enhance the technology scalability, providing an ideal subthreshold slope and better drain-induced barrier lowering (DIBL), as well as larger back-to-front gate coupling effect useful for threshold voltage V_{th} control.

Besides, low-frequency noise and RTS fluctuations, which scale with the reciprocal device area, become more important with technology scaling down. They are not only limiting the analog circuit operation, but they could also jeopardize the digital circuit functioning. They could even appear as an ultimate variability source due to the dynamic carrier trapping in undoped channel devices. For these reasons, the study of LFN and RTN in UTTB FDSOI is a key issue for technology evaluation and evolution.

Furthermore, it has been repeatedly shown that the LF noise in FDSOI devices can be influenced by the coupling effect between the back and front interfaces [10–14]. Due to this coupling, it is difficult to predict precisely the contribution of each interface to the measured noise. Moreover, application of a positive or negative bias voltage to the buried oxide can possibly lead to the appearance of either Lorentzian-type noise [10, 15, 16], or a significant increase of the flicker noise level [11, 17]. Thus, the analytical study of the noise sources and their dependence on the bias conditions are crucial for both device characterization and noise modeling.

2 Theoretical Background

2.1 Origin of Low-Frequency Noise in MOSFETs

Carrier Number Fluctuations and Correlated Mobility Fluctuations (CNF/CMF)

Within the classical carrier number fluctuation approach, the fluctuations in the drain current arise from the fluctuations of the inversion charge located at the gate dielectric interface, and stem from the variations of the interfacial charge due to trapping–detrapping of free carriers into slow dielectric traps. The interface charge variation δQ_{it} can be translated into a flat-band voltage variation as $\delta V_{\mathrm{fb}} = -\delta Q_{\mathrm{it}}/(\mathrm{WL}C_{\mathrm{ox}})$, C_{ox} being the gate dielectric capacitance and WL the device area [18]. A rigorous detailed analysis should also account for the correlated mobility fluctuations $\delta\mu_{\mathrm{eff}}$ due to the modulation of the Coulomb scattering rate by the interface charge fluctuations. The overall drain current fluctuations therefore become the linear operation region [19, 20].

$$\delta I_{\mathrm{d}} = -g_{\mathrm{m}}\delta V_{\mathrm{fb}} - \alpha I_d \mu_{\mathrm{eff}}\delta Q_{\mathrm{it}} \tag{1}$$

where g_{m} is the transconductance, μ_{eff} is the effective mobility, α is the Coulomb scattering coefficient ($\approx 10^4$ Vs/C for electrons and 10^5 Vs/C for holes [21, 22]).

This yields a normalized drain current noise given by [20]:

$$\frac{S_{\mathrm{Id}}}{I_{\mathrm{d}}^2} = S_{\mathrm{Vfb}}\left(\frac{g_{\mathrm{m}}}{I_d}\right)^2 (1 + \alpha\mu_{\mathrm{eff}}C_{\mathrm{ox}}I_{\mathrm{d}}/g_{\mathrm{m}})^2 \tag{2}$$

and to an input-referred gate voltage noise S_{Vg}, at strong inversion as

$$S_{\mathrm{Vg}} = \frac{S_{\mathrm{Id}}}{g_{\mathrm{m}}^2} = S_{\mathrm{Vfb1}}\left[1 + \alpha\mu_0 C_{\mathrm{ox}}\left(V_{\mathrm{g}} - V_{\mathrm{t}}\right)\right]^2 \tag{3}$$

where μ_0 is the low-field mobility, V_t the threshold voltage and

$$S_{\mathrm{Vfb}} = \frac{S_{\mathrm{Qit}}}{\mathrm{WL}C_{\mathrm{ox}}{}^2} \tag{4}$$

the flat-band voltage power spectral density (PSD), with S_{Qit} being the PSD of the dynamic interface charge per unit area. Note that S_{Vg} is different from S_{Vfb}, especially at strong inversion where CMF prevail.

The PSD of the interface charge is intimately related to the physical trapping mechanism into the gate dielectric. For a tunneling process (McWorther scheme [23]), the trapping probability decreases exponentially with dielectric depth, so that the PSD of the flat-band voltage reads for a uniform trap-depth profile [23, 24]:

$$S_{\mathrm{Vfb}} = \frac{q^2 kT\ \lambda\ N_{\mathrm{t}}}{\mathrm{WL}C_{\mathrm{ox}}{}^2 f} \tag{5}$$

where f is the frequency, λ is the tunnel attenuation distance ($\approx$0.1 nm in SiO_2), kT is the thermal energy, and N_t is the volumetric trap density (/eVcm^3). However, some experiments [25, 27] show that the time constants are either weakly or not at all correlated with depth. In that case, one has to consider that the trapping probability decreases exponentially with the activation energy of the defect cross section E_a, as in thermally assisted trapping processes (Dutta scheme [28]). This is also consistent with random telegraph noise data from [25–29]. Accounting for such a dependence of τ and E_a, the PSD of the flat-band voltage takes the form [30]:

$$S_{\mathrm{Vfb}} = \frac{q^2 k^2 T^2\ N_{\mathrm{st}}}{\mathrm{WL}C_{\mathrm{ox}}{}^2 f \varDelta E_{\mathrm{a}}} \tag{6}$$

where ΔE_a is the spread of the activation energy and N_{st} is the areal surface trap density (/eVcm^2).

The CNF–CMF derivation of Eq. (2) can be generalized within the gradual channel approximation to the nonlinear MOSFET operation region after integration of the local sheet conductivity σ fluctuations along the channel, yielding [31]:

$$\frac{S_{\mathrm{Id}}}{\mathrm{I}_{\mathrm{d}}^2} = S_{\mathrm{Vfb}} \frac{\int_0^{V_{\mathrm{d}}} \left(\frac{1}{\sigma}\frac{\partial\sigma}{\partial V_{\mathrm{fb}}} + \alpha\ \mu_{\mathrm{eff}} C_{\mathrm{ox}}\right)^2 \sigma\ \mathrm{d}U_{\mathrm{c}}}{\int_0^{V_{\mathrm{d}}} \sigma\ \mathrm{d}U_{\mathrm{c}}} \tag{7}$$

where $\sigma = q\mu_{\mathrm{eff}}.\ Q_i$ with Q_i being the inversion charge and U_c is the channel potential.

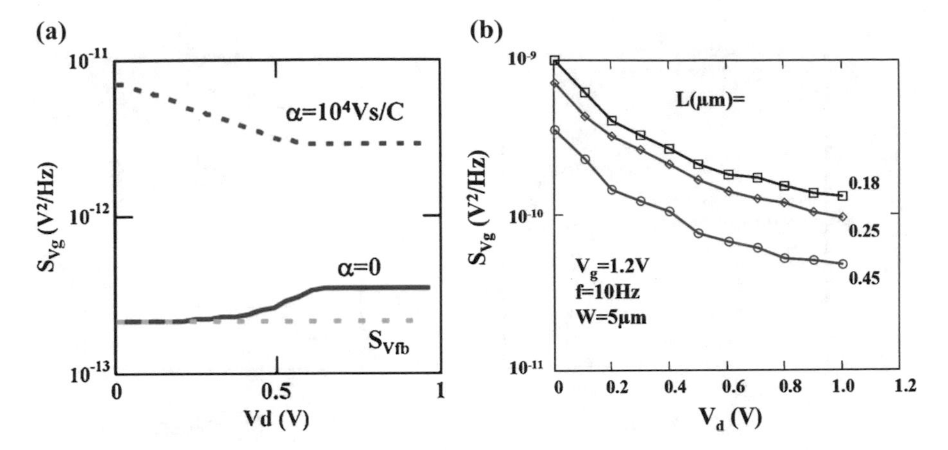

Fig. 1 (**a**) Theoretical $S_{\mathrm{Vg}}(V_{\mathrm{d}})$ characteristics as obtained from CNF–CMF model of Eq. (7) and (**b**) typical experimental $S_{\mathrm{Vg}}(V_{\mathrm{d}})$ characteristics (after [31])

As illustrated in Fig. 1a, the variation rate of the input gate voltage noise $S_{\mathrm{Vg}}(V_{\mathrm{d}})$ strongly depends on the CMF amplitude factor (α) and shows clear presence of CMF in the experimental results as in Fig. 1b.

Hooge Mobility Fluctuations (HMF)

According to the Hooge mobility model [32], the drain current noise comes from the fluctuations of the carrier mobility stemming from variations in the scattering probability due to the phonon number fluctuations [33, 34]. This results in a flicker noise with amplitude inversely proportional to the total number of carriers in the device. The normalized drain current noise in linear operation then reads [35]:

$$\frac{S_{\mathrm{Id}}}{I_{\mathrm{d}}^2} = \frac{q\alpha_{\mathrm{H}}}{\mathrm{WL}Q_i f} \tag{8}$$

where Q_i is the inversion charge and α_{H} is the Hooge parameter ($\approx 10^{-4}$–10^{-7}).

In the case of nonlinear operation region, it can be shown that, after integration along the channel, the normalized drain current noise becomes [35]:

$$\frac{S_{\mathrm{Id}}}{I_{\mathrm{d}}^2} = \frac{q\alpha_{\mathrm{H}}\langle\mu_{\mathrm{eff}}\rangle V_{\mathrm{d}}}{L^2 I_{\mathrm{d}} f} \tag{9}$$

where $<\mu_{\mathrm{eff}}>$ stands for the average mobility along the channel.

As the Hooge mobility fluctuations depend only on the phonon scattering rate [36], the Hooge mobility parameter should be modulated by its contribution among

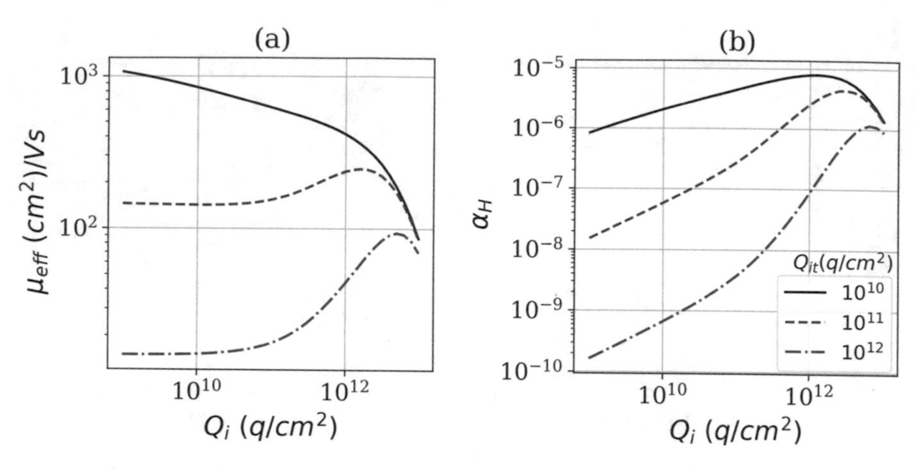

Fig. 2 Theoretical variations of μ_{eff} (**a**) and Hooge parameter (**b**) with MOSFET inversion charge Q_i for various interface charge Q_{it} levels modulating the Coulomb scattering rate ($\alpha_{\text{H0}} = 10^{-5}$)

other scattering mechanisms limiting the carrier mobility. Therefore, in the case of a MOSFET, the Hooge parameter should be written as:

$$\alpha_{\text{H}} = \alpha_{\text{H0}} \cdot \left(\frac{\frac{1}{\mu_{\text{ph}}}}{\frac{1}{\mu_{\text{ph}}} + \frac{1}{\mu_{\text{C}}} + \frac{1}{\mu_{\text{SR}}}} \right)^2 \tag{10}$$

where α_{H0} refers to the intrinsic Hooge parameter; μ_{ph}, μ_{C}, and μ_{SR} are, respectively, the phonon, Coulomb, and surface roughness scattering limited mobilities in the MOSFET inversion layer [37]. Accounting for the phonon, Coulomb, and surface roughness scattering universal mobility law [37, 38] against effective electric field, the dependence of the Hooge parameter given by Eq. (10) can be evaluated theoretically with the MOSFET inversion charge from weak to strong inversion as illustrated in Fig. 2. It clearly reveals that, in MOSFET, the Hooge parameter is far from being independent of inversion charge, i.e., gate voltage while varying from subthreshold to strong inversion region. The Hooge parameter is maximized when the phonon-scattering contribution prevails with respect to Coulomb and surface roughness scattering rates.

Figure 3 shows the impact of the Hooge parameter dependence with inversion charge (Eq. (10)) on the associated normalized drain current noise. In this situation, $S_{\text{Id}}/I_{\text{d}}^2$ is no longer simply inversely proportional to the inversion charge as it was the case for HMF model with constant mobility. Therefore, even if the Hooge model is empirical, the α_{H} parameter should not be considered constant with gate bias, due to its modulation by the interface charge Coulomb scattering.

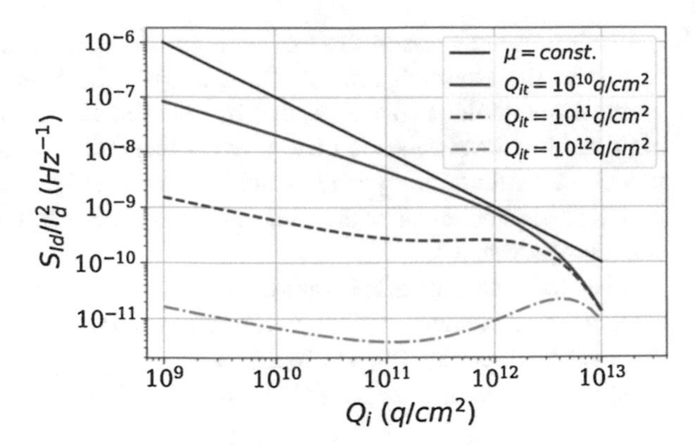

Fig. 3 Variation of normalized drain current noise S_{Id}/I_d^2 with inversion charge for HMF model with Hooge parameter of Fig. 2

Impact of Access Resistance

The impact of the access resistance R_{sd} on the LF noise can be obtained by adding to the channel drain current noise the LF noise component arising from the resistance of the source and drain access region. To this end, one should note that the sensitivity of the drain current to R_{sd} variation is simply given by

$$\frac{\frac{\partial I_d}{\partial R_{sd}}}{I_d} = g_d + \frac{g_m}{2} \tag{11}$$

where $g_d = dI_d/dV_d$ is the output conductance and assuming $R_{sd} = 2R_s = 2R_d$. Therefore, the extrinsic normalized drain current noise including the access region noise can be expressed as:

$$\frac{S_{Id}}{I_d^2} = \left(\frac{S_{Id}}{I_d^2}\right)_{channel} + \left(g_d + \frac{g_m}{2}\right)^2 S_{Rsd} \tag{12}$$

where S_{Rsd} is the PSD associated to the access resistance. In Eq. (12), the channel component is either calculated by the CNF/CMF Eqs. (2)–(7) or by the HMF Eq. (9) model.

Random Telegraph Noise (RTN)

The observation of RTN in small area MOS devices in the 1980s was attributed to elementary carrier trapping at the channel–gate dielectric interface [39, 40]. The drain current RTN was commonly interpreted as a conductance modulation due to the fluctuation in the carrier number with associated mobility fluctuations. The two-level RTN signal of a single trap is statistically characterized by three parameters: (1) the average amplitude of the drain current jump ΔI_d between the low and high levels, (2) the average time passed on the high level normally associated to the capture time τ_c, and (3) the average time passed on the low level normally related to the emission time τ_e. In this case, the histogram of the drain current amplitude exhibits two peaks at a distance equal to ΔI_d. The histogram of the high and low level times is generally exponentially distributed due to the Poisson process governing the dynamic trapping events [41]. In the case where more than one traps are active under the same gate bias, the result is a multilevel RTN signal (not discussed here).

The drain current RTN amplitude can be evaluated by considering that the trapping of an elementary charge q from the channel into an oxide defect modulates the local conductivity [20, 41]. In a first-order approximation, the relative drain RTN amplitude accounting for both CNF and CMF is given by [20, 42]:

$$\frac{\Delta I_d}{I_d} = \frac{g_m}{I_d}\frac{q}{\mathrm{WL}C_{ox}}\left(1-\frac{x_t}{t_{ox}}\right)\left(1+\frac{\alpha\mu_{eff}C_{ox}I_d}{g_m}\right) \quad (13)$$

where x_t is trap depth in the gate dielectric and t_{ox} is the gate oxide thickness. Therefore, the drain current RTN relative amplitude is maximized in the subthreshold region where the transistor gain g_m/I_d is larger.

Furthermore, in order to account for abnormally high RTN amplitudes in dramatically scaled devices, a parameter η should be added to Eq. (13) [43]:

$$\frac{\Delta I_{ds}}{I_{ds}} = \eta\frac{g_m}{I_{ds}}\frac{q}{\mathrm{WL}C_{ox}}\left(1-\frac{x_t}{t_{ox}}\right)\left(1+\Omega\frac{I_{ds}}{g_m}\right) \quad (14)$$

where η is unity for a homogeneous channel in the linear region and $\Omega = \alpha\mu_{eff}C_{ox}$ the CMF factor [44]. The large deviation of the parameter η from unity [45–48] could be explained by the huge variability of the threshold voltage associated with single charge trapping, which exhibits an exponential distribution due to the impact of trap location within the gate dielectric and facing the channel [48, 49].

Regarding the trap kinetics, in general, the RTN capture and emission times are governed by the Shockley–Read–Hall statistics [41] and reads:

$$\tau_c = \frac{1}{\sigma n_s v_{th}} \text{ (a)} \quad \text{and} \quad \tau_e = \frac{1}{\sigma n_1 v_{th}} \text{ (b)} \quad (15)$$

where v_{th} is the thermal velocity, σ is the trap cross section, n_s is the surface carrier concentration, and n_1 is the surface carrier concentration when the Fermi level E_f crosses the trap energy E_t. The trap cross section might depend on the trap depth into the oxide and on the temperature as $\sigma = \sigma_0\exp(-E_a/kT)\exp(-x_t/\lambda)$ [41].

However, when the trap is not located right at the oxide–channel interface, but at a depth x_t in the oxide, the apparent trap energy E_t depends on the band bending in the gate dielectric as [39, 41]:

$$E_t = E_{t0} - q\frac{x_t}{t_{ox}}\left(V_g - V_{fb} - V_s\right) \quad (16)$$

where V_s is the surface potential and V_g is the gate voltage. Another way to express this difference is through n_1 in Eq. (15) (b), if we replace it with:

$$n_1 = n_0 e^{\frac{\psi_{s1}}{kT}} = n_0 e^{\frac{\psi_{s1,0} - \Delta\psi_1}{kT}} \tag{17}$$

where ψ_{s1} is the surface potential for which E_t coincides with E_f, and $\Delta\psi_1 = x_t \cdot Q_i / \varepsilon_{ox}$ corresponds to the potential drop across the oxide, from the interface to the trap depth.

It should also be noted that the capture and emission times of Eq. (15) are evaluated within the classical statistics, i.e., using carrier volumetric concentration at the surface. They have to be updated when quantum mechanical effects become important in the MOSFET inversion layer, because n_s is canceled out at the surface. Indeed, the capture probability is proportional to the escape frequency of the electrons in the quantized sub-band and to the barrier tunneling transparency to reach the trap in the oxide. If in addition we take into consideration Eq. (17), the capture and emission time can be expressed in a way that accounts for the trap depth x_t and the single sub-band approximation:

$$\overline{\tau}_c = \frac{q}{\sigma\, f_e\, Q_i} \text{ (a)} \quad \text{and} \quad \overline{\tau}_e = \frac{q\, e^{\frac{x_t (Q_i + Q_d)}{kT \varepsilon_{ox}}}}{\sigma\, f_e\, Q_{i1}} \text{ (b)} \tag{18}$$

where f_e is the escape frequency ($\approx 2 \times 10^{13}$Hz), ε_{ox} is the oxide permittivity, Q_{i1} the inversion charge when the Fermi level E_f crosses the trap energy E_t, and Q_d is the depletion charge. In Fig. 4, typical variations of the capture and emission times, calculated with Eq. (18) as a function of gate voltage for a FDSOI structure, demonstrating the usual huge decrease of τ_c and the slight increase of τ_e with V_g in strong inversion, are illustrated. It should be emphasized that this formulation Eq. (18) of the capture and emission times will also be of great interest for compact modeling applied to circuit simulation (see Sect. 5).

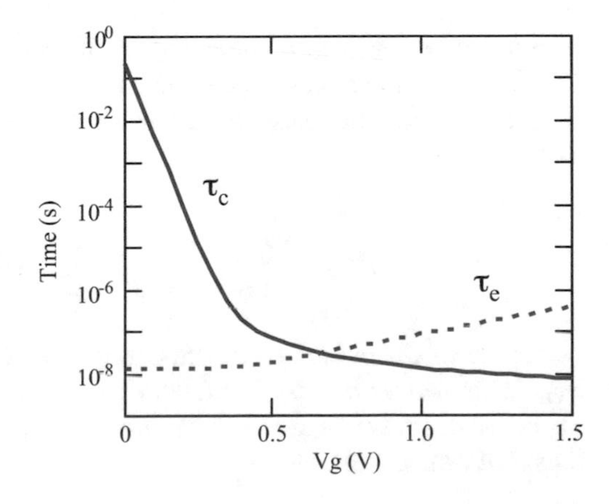

Fig. 4 Typical variations of capture and emission with gate voltage for a FDSOI structure ($Q_d = 0$) with parameters: $t_{ox} = 2$ nm, $x_t = 0.05.t_{ox}$, $\sigma_0 = 10^{-18}/\text{cm}^2$, $E_a = 0$)

Furthermore, in strong inversion, we can consider $Q_i = C_{ox}(V_g - V_t)$, so Eq. (18) becomes:

$$\overline{\tau}_c = \frac{q}{\sigma\, f_e\, C_{ox}\left(V_g - V_t\right)} \text{ (a) and } \overline{\tau}_e = \frac{q\, e^{\frac{x_t\left(V_g - V_t\right)}{kT\, t_{ox}}}}{\sigma\, f_e\, C_{ox}\left(V_{g,trap} - V_t\right)} \text{ (b)} \tag{19}$$

where $V_{g,trap}$ is the gate voltage bias for which $E_t = E_f$. Thus, it can be simply shown from Eq. (17)(b) that $d(\ln(\tau_e))/dV_g \approx qx_t/(t_{ox}kT)$ in strong inversion gives access to the trap depth. Similarly, one has $d(\ln(\tau_c/\tau_e))/dV_d \approx qy_t/(LkT)$ in weak inversion, providing a way to estimate the trap position from source along the channel [50].

It should also be mentioned that more sophisticated capture and emission time models have been proposed for a better description of the trapping–detrapping dynamics using multistate defects with multiphonon theory (see [51]).

Finally, it is worth recalling that the drain current power spectral density of a two-level RTN follows a Lorentzian spectrum given by [52]:

$$S_{Id} = 4\, A\, \Delta I_d^2 \frac{\tau}{1 + \omega^2\tau^2} \tag{20}$$

where $\tau = (1/\tau_c + 1/\tau_e)^{-1}$ is the effective time constant, $A = \tau/(\tau_c + \tau_e) = f_t(1 - f_t)$ is the space mark ratio, $\omega = 2\pi f$ is the angular frequency, and f_t is the trap occupancy factor, $f_t = 1/\{1 + \exp[(E_t - E_f)/kT]\}$.

The RTN phenomenon in MOSFETs and the various characterization methods are discussed in more detail in Chaps. 4 [53], 7 [54], 14 [55], and 18 [56] of this book.

2.2 *Noise Model Development and Challenges*

Multi-Interface CNF Approach

In multi-gate devices, trapping and de-trapping of carriers can occur at multiple channel/oxide interfaces. Therefore, if one applied the CNF model as described in Sect. 2.1 for this case, the normalized flicker drain current noise would be expressed as:

$$\left.\frac{S_{Id}}{I_d^2}\right|_{(multi-gate)} = \sum_{i=1}^{n} S_{Vfbi}\left(\frac{g_{mi}}{I_d}\right)^2 \tag{21}$$

where the subscript i refers to each interface, n is the total number of interfaces, and S_{Vfb} is defined by Eq. (5). In SOI devices, there are only two flicker noise sources: the interfaces between silicon/front-gate oxide (t_{ox}) and silicon/buried oxide (t_{BOX}), thus SOI can be expressed as:

$$\frac{S_{\mathrm{Id}}}{I_{\mathrm{d}}^2}_{\mathrm{(SOI)}} = S_{\mathrm{Vfb1}}\left(\frac{g_{\mathrm{m1}}}{I_{\mathrm{d}}}\right)^2 + S_{\mathrm{Vfb2}}\left(\frac{g_{\mathrm{m2}}}{I_{\mathrm{d}}}\right)^2 \tag{22}$$

where subscripts 1 and 2 refer to the top and back interface, respectively. It follows that $g_{\mathrm{m1}} = \mathrm{d}I_{\mathrm{d}}/\mathrm{dV_{G1}}$ and $g_{\mathrm{m2}} = \mathrm{d}I_{\mathrm{d}}/\mathrm{dV_{G2}}$ are the respective front and back-gate transconductance. Then, taking into account that only the oxide capacitance and trap densities differ between S_{Vfb1} and S_{Vfb2}, the front input referred noise $S_{\mathrm{Vg1}} = S_{\mathrm{Id}}/g_{\mathrm{m1}}{}^2$ can be expressed through:

$$S_{\mathrm{Vg1\,(SOI)}} = S_{\mathrm{Vfb1}}\left[1 + C_{21}^2\frac{N_{\mathrm{t2}}}{N_{\mathrm{t1}}}\left(\frac{C_{\mathrm{ox1}}}{C_{\mathrm{ox2}}}\right)^2\right] \tag{23}$$

where $C_{21} = g_{\mathrm{m2}}/g_{\mathrm{m1}}$. If the two oxides are made from the same dielectric material and with the same process, then the approximation $N_{\mathrm{t1}} = N_{\mathrm{t2}}$ can be considered. However, the high-k dielectric used in the front gate to reduce the leakage current degrades the interface quality compared to buried thermal oxide, resulting in N_{t1} values typically one decade higher than N_{t2} [17]. Nevertheless, from Eq. (23), it becomes clear that the defining factor for the coupling effect between the two noise sources is the product $(C_{21})^2 \times (N_{\mathrm{t2}}/N_{\mathrm{t1}}) \times (C_{\mathrm{ox1}}/C_{\mathrm{ox2}})^2$, revealing that the contribution of the back interface to the total drain current noise is not just depending on the back/front trap density ratio, but also on the back/front transconductance and oxide thickness ratios. From Eq. (23) we also conclude that in strong inversion, where $g_{\mathrm{m1}} \approx g_{\mathrm{m2}}$ [57], the front/back noise coupling only depends on the quality ratio of the two interfaces $(N_{\mathrm{t2}}/N_{\mathrm{t1}})$ and the ratio between the buried oxide and channel thicknesses.

To illustrate these effects, we run a series of TCAD simulations in FlexPDE (a fast and reliable partial differential equations solver), for FD-SOI MOSFETs with $t_{\mathrm{Si}} = 7$ nm, $t_{\mathrm{ox1}} = 1.5$ nm, and $t_{\mathrm{ox2}} = 25$ nm. Some results are shown in Fig. 5, where we varied the trap density ratio and in Fig. 6, where the simulations were repeated

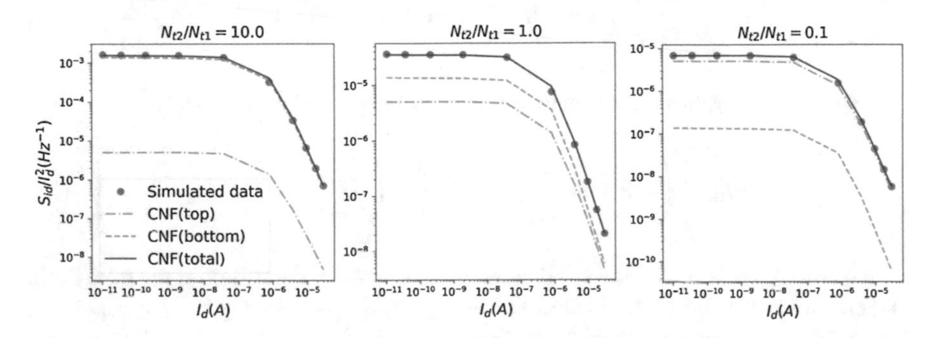

Fig. 5 TCAD simulations results of normalized noise for different trap density ratio between top and bottom interface. The dashed lines correspond to the two terms of the model expressed in Eq. (22) and the continuous line to their sum

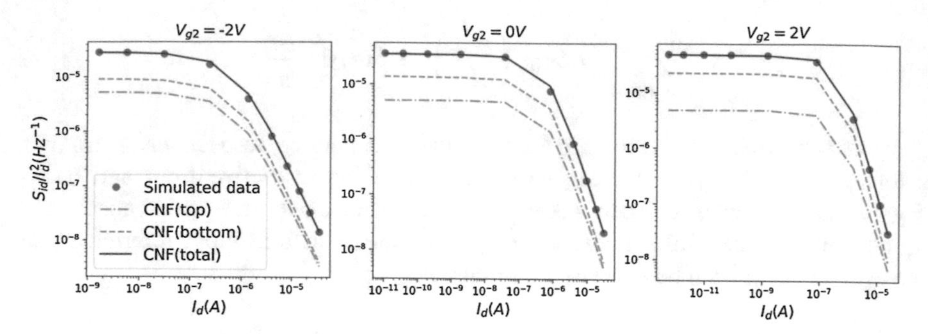

Fig. 6 TCAD simulations results of normalized drain current noise for different back-gate voltage values

for different back-gate voltage V_{G2}. As can be seen in Fig. 5, when the volumetric trap density values N_{t1} and N_{t2} are close to each other, both noise sources—front and bottom—contribute to the total noise level, while if one oxide has a much worse interface quality, its noise becomes dominant. This makes it very difficult to extract N_{t1} and N_{t2} from a single set of measurements in a reliable way. However, as shown in Fig. 6, the noise intensity of the bottom interface can be controlled by V_{G2}, resulting in a different total noise level for each back-bias value. Thus, one can extract the characteristic parameter values of each interface combining simultaneously noise measurement data at various bias conditions: front-gate mode, back-gate mode, back-accumulation mode, constant current, etc. [11, 58].

Two-Interface CNF/CMF Modeling

Now, if we also take into account the contribution from remote Coulomb scattering, we can add the CMF terms in Eq. (22) and obtain:

$$\frac{S_{\mathrm{Id}}}{I_{\mathrm{d}}^{2}}_{(\mathrm{SOI})} = S_{\mathrm{Vfb1}}\left(\frac{g_{\mathrm{m1}}}{I_{\mathrm{d}}}\right)^{2}\left(1+\Omega_{1}\frac{I_{\mathrm{d}}}{g_{\mathrm{m1}}}\right)^{2} + S_{\mathrm{Vfb2}}\left(\frac{g_{\mathrm{m2}}}{I_{\mathrm{d}}}\right)^{2}\left(1+\Omega_{2}\frac{I_{\mathrm{d}}}{g_{\mathrm{m2}}}\right)^{2} \quad (24)$$

or, expressed as input-referred gate voltage power spectral density [57]:

$$S_{\mathrm{Vg1}\,(\mathrm{SOI})} = S_{\mathrm{Vfb1}}\left[\left(1+\Omega_{1}\frac{I_{\mathrm{d}}}{g_{\mathrm{m1}}}\right)^{2} + C_{21}^{2}\frac{N_{\mathrm{t2}}}{N_{\mathrm{t1}}}\left(\frac{C_{\mathrm{ox1}}}{C_{\mathrm{ox2}}}\right)^{2}\left(1+\Omega_{2}\frac{I_{\mathrm{d}}}{g_{\mathrm{m2}}}\right)^{2}\right] \quad (25)$$

It becomes clear from Eq. (25) that the last parameters determining the total drain current noise are the CMF factors $\Omega_1 = \alpha_{\mathrm{sc1}}\mu_{\mathrm{eff1}}C_{\mathrm{ox1}}$ and $\Omega_2 = \alpha_{\mathrm{sc2}}\mu_{\mathrm{eff2}}C_{\mathrm{ox2}}$. Regarding these two factors, a relation between them can be found using the dependence of α_{sc} on the carriers' position. According to [59] (Chap. 3), α_{sc} depends on the distance x between the inversion charge centroid and the interface through:

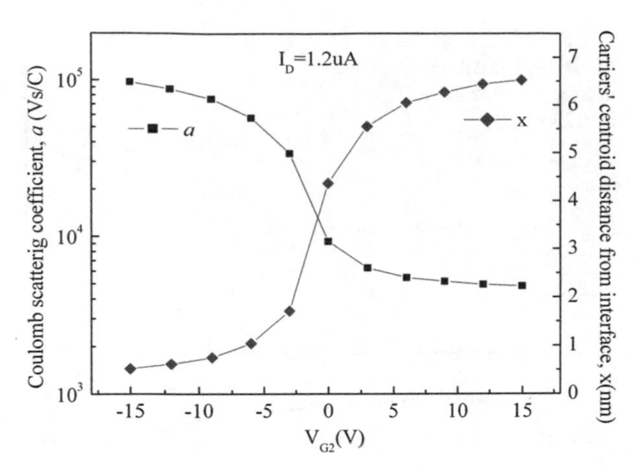

Fig. 7 Calculation results for carriers distribution centroid-front interface distance x and coulomb scattering coefficient α versus bias voltage V_{G2}, keeping the drain current constant [17]

$$\alpha_{sc} = \frac{\alpha_0}{\left(1 + \frac{x}{\lambda_c}\right)^2} \tag{26}$$

where $\alpha_0 = 10^5$ Vs/C approximately in ultrathin body SOI devices [17] and $\lambda_c = 1.2$ nm. As obtained by the numerical simulations shown in Fig. 7 (after [17]) for a constant drain current, the strong dependence of the charge centroid/interface distance on the back-bias voltage V_{G2} results in a significant increase of the RCS coefficient α_{sc} — through Eq. (26) — for negative values of V_{G2}. This occurs because as the bottom interface goes towards accumulation region, for the same total charge, the centroid moves much closer to the front interface.

Considering a distance x from the front interface and t_{Si}-x from the back interface, an expression for the ratio Ω_1/Ω_2 can be derived as [57]:

$$\frac{\Omega_1}{\Omega_2} = \frac{\frac{C_{ox1}}{\left(1+\frac{x}{\lambda_c}\right)^2}}{\frac{C_{ox2}}{\left(1+\frac{t_{Si}-x}{\lambda_c}\right)^2}} = \frac{\left(1 + \frac{t_{Si}-x}{\lambda_c}\right)^2}{\left(1 + \frac{x}{\lambda_c}\right)^2} \frac{t_{ox2}}{\varepsilon_{ox}} \frac{\varepsilon_{ox}}{t_{ox1}} = \left(\frac{\lambda_c + t_{Si} - x}{\lambda_c + x}\right)^2 \frac{t_{ox2}}{t_{ox1}} \tag{27}$$

Figure 8 (after [1]) shows a calculation example of Eq. (27) for devices with $t_{Si} = 8.7$ nm, $t_{ox1} = 1.2$ nm, and $t_{ox2} = 10$ nm. This figure gives a rough estimation of Ω_1 and Ω_2, demonstrating that when the charge is mainly concentrated near the front interface, then $\Omega_1 \gg \Omega_2$. Thus, Ω_2 can be considered negligible when the device is biased in front-gate (FG) mode without any back-bias.

Therefore, for the FG mode and $V_{G2} = 0$ V, Eq. (25) can be approximated as:

$$S_{Vg1\,(SOI)} = S_{Vfb1}\left[\left(1 + \Omega_1 \frac{I_d}{g_{m1}}\right)^2 + C_{21}^2 \frac{N_{t2}}{N_{t1}}\left(\frac{C_{ox1}}{C_{ox2}}\right)^2\right] \tag{28}$$

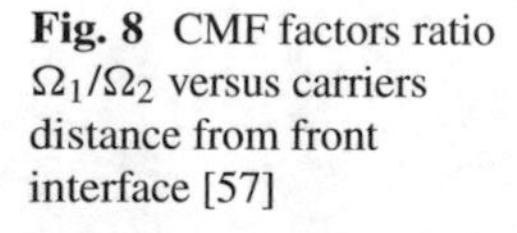
Fig. 8 CMF factors ratio Ω_1/Ω_2 versus carriers distance from front interface [57]

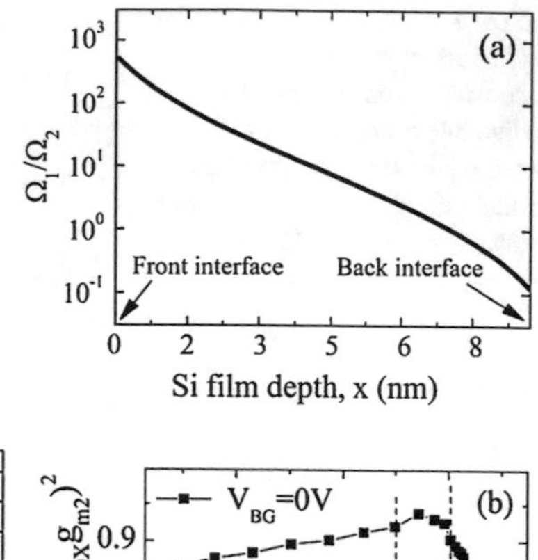

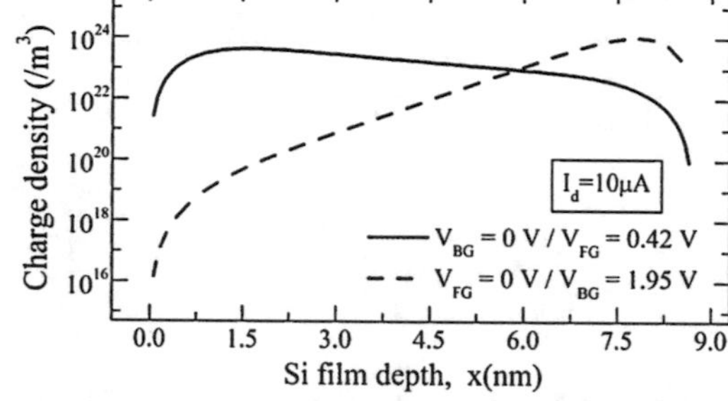

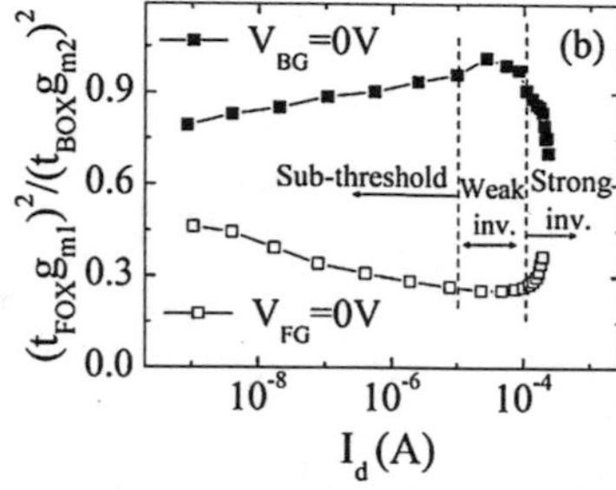

Fig. 9 Electron density versus silicon film depth at $V_{\mathrm{BG}} = 0$ V (solid line) and $V_{\mathrm{FG}} = 0$ V (dotted line) for the same total drain current $I_{\mathrm{d}} = 10$ μA ($W = 10$ μm, $L = 87$ nm) [57]

which, for fully depleted channel (ultrathin and lightly doped), reduces to the final CNF/CMF model approach for FD-SOI MOSFETs for the FG mode:

$$S_{V\mathrm{g1}\,(\mathrm{FD-SOI})} = S_{V\mathrm{fb1}}\left[\left(1+\Omega_1\frac{I_{\mathrm{d}}}{g_{\mathrm{m1}}}\right)^2+\frac{N_{\mathrm{t2}}}{N_{\mathrm{t1}}}\left(\frac{1}{1+C_{\mathrm{ox2}}/C_{\mathrm{Si}}}\right)^2\right] \tag{29}$$

Conversely, for back-gate (BG) mode and $V_{\mathrm{G1}} = 0$ V, we obtain:

$$S_{V\mathrm{g2}\,(\mathrm{FD-SOI})} = S_{V\mathrm{fb2}}\left[\left(1+\Omega_2\frac{I_{\mathrm{d}}}{g_{\mathrm{m2}}}\right)^2+\frac{N_{\mathrm{t1}}}{N_{\mathrm{t2}}}\left(\frac{1}{1+C_{\mathrm{ox1}}/C_{\mathrm{Si}}}\right)^2\right] \tag{30}$$

The above Eqs. (29) and (30) can be used to fit experimental data and extract the noise parameters N_{t1}, N_{t2}, and Ω_1 and Ω_2. The need of combining the results from both the FG and BG modes is demonstrated through TCAD simulations that illustrate the carrier concentration distribution in the silicon film channel [57]. As shown in Fig. 9a, while the total charge is exactly the same, one can observe a lack of symmetry in the distribution of the charge, depending on FG or BG operation mode. In FG mode, it is almost uniformly distributed across the channel, whereas for BG mode, most of the charge is concentrated near the bottom interface. This means that measuring in BG mode we have a better decoupling, leading to an easier extraction of N_{t2} and Ω_2, which then can be used in Eq. (29) to extract N_{t1} and Ω_1 using the

FG mode results. An experimental confirmation of this effect is shown in Fig. 9b [57], where a reduction of the coupling factor $(t_{ox1}g_{m1}/t_{ox2}g_{m2})^2$ was measured in BG mode, compared to FG mode

For practical reasons and simplicity of use in SPICE simulation, it is very helpful to transfer all the noise sources to the front gate, including the bottom interface component, as it was implemented in [60]. To this end, one could define the normalized flicker noise using a classic CNF/CMF model equation as in Eq. (2), but with front-gate input-referred effective bias-dependent values of $S_{Vfb,eff}$ and Ω_{eff}:

$$\frac{S_{Id}}{I_d^2} = S_{Vfb,eff}\left(\frac{g_{m1}}{I_d}\right)^2\left(1+\Omega_{eff}\frac{I_d}{g_{m1}}\right)^2 \tag{31}$$

$S_{Vfb,eff}$ can be obtained from weak inversion region limits where CNF dominates as [60]:

$$S_{Vfbeff} = \frac{q^2\lambda kT N_{t,eff}}{f\,WLC_{ox,1}^2} \tag{32}$$

where $N_{t,eff}$ is the effective slow oxide trap volumetric density, related the front and back interface trap densities via the coupling factor as $N_{t,eff} = N_{t1} + C_{21}{}^2N_{t2}\Omega_{eff}$ on the other hand can be obtained from strong inversion region limits where CMF prevails as [60]:

$$\Omega_{eff} \approx \sqrt{\left(\Omega_1{}^2 + \frac{S_{Vfb2}}{S_{Vfb1}}\Omega_2{}^2\right)/\left(1 + C_{21}{}^2\frac{S_{Vfb2}}{S_{Vfb1}}\right)} \tag{33}$$

where $S_{Vfb2}/S_{Vfb1} = (t^2{}_{ox2}\,N_{t2}/\,t^2{}_{ox1}\,N_{t1})$ revealing the effect of the front and back-gate oxide thickness and volumetric trap density ratio on the total noise level. These equations indicate that the effective parameters $S_{Vfb,eff}$ and Ω_{eff} are univocally related to both front and back interface parameters and coupling factor.

The model in Eq. (31), combined with Eqs. (32) and (33), can be easily used to fit drain current flicker noise data from various FD-SOI technology nodes, extract Ω_{eff} and $N_{t,eff}$, and use these parameters to perform accurate circuit noise simulations.

3 Noise Characterization in FD-SOI MOSFETs

3.1 *Flicker Noise in FD-SOI MOSFETs*

Front/Back Coupling Effects

As explained in Sect. 2, a very important parameter for the flicker noise analysis is the squared transistor gain $(g_{m1,2}/I_D)^2$. In front-gate mode, g_{m1} can be directly

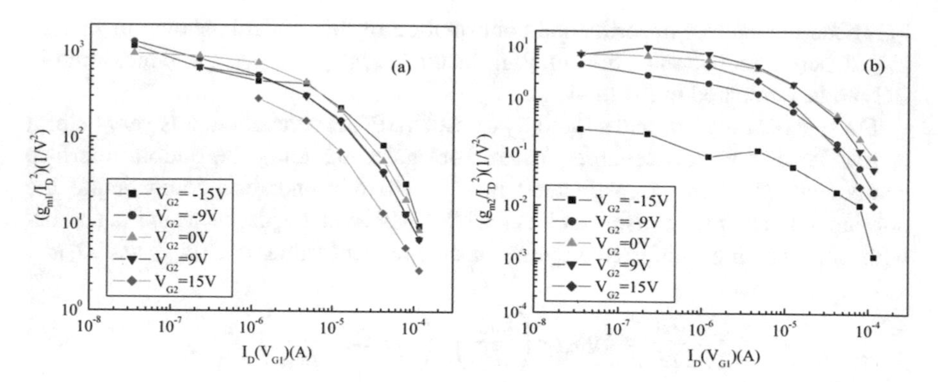

Fig. 10 $(g_{m1}/I_D)^2$ (**a**) and $(g_{m2}/I_D)^2$ (**b**) versus drain current for various V_{G2} values (after [17])

measured from the transfer characteristic at constant V_{G2} as $g_{m1} = dI_d/dV_{G1}$, whereas a supplementary static measurement under a nearby back-gate bias (i.e., $\Delta V_{G2} = 50$ mV for 10 nm BOX) is needed to extract $g_{m2} = dI_d/dV_{G2}$ at each value of V_{FG}. For the back-gate mode, the inverse procedure has to be followed. In Fig. 10, $(g_{m1,2}/I_D)^2$ are plotted for various back-bias values as a function of drain current in 28-nm FD-SOI MOSFETs [17]. As usual, $(g_m/I_D)^2$ exhibits a plateau in weak inversion, before dropping above threshold in strong inversion. The top gate gain factor seems to be affected by V_{G2} only for high positive values, where the position of the inversion channel is modified. However, $(g_{m2}/I_d)^2$ depends strongly on V_{G2}, becoming negligible for very negative back-gate bias, where accumulation is achieved.

The ratio (coupling factor) between the two gain factors is plotted in Fig. 11. As can be seen, it lies around $\approx$100 when the bottom interface is in depletion and weak inversion regimes ($V_{G2} > 0$ V), whereas it increases up to almost 10^4 for strong accumulation at the bottom ($V_{G2} \ll 0$ V). This feature indicates that the bottom interface LF noise contribution could be eliminated in the latter situation, allowing for an easier extraction of the front interface noise parameters. One can also notice that c_1 is almost independent of I_D in the depletion and weak inversion region. Nevertheless, when comparing the drain current PSD at different back-bias voltages (Fig. 12), we conclude that a high negative V_{G2} value can cause an increase of the flicker noise level, whereas a positive one the opposite.

The influence of the back interface coupling effect on the front-gate operation LF noise can be better analyzed by plotting the normalized drain current noise, measured at a constant drain current, as a function of back-gate voltage as shown in Fig. 13a. As can be seen, the normalized drain current noise is significantly higher when V_{G2} takes high negative value. It should also be noted, by comparing Figs. 11 and 13a, that the normalized drain current noise has a similar behavior with V_{G2} as that of the noise coupling factor c_1. This type of noise dependence on the back-bias voltage was also confirmed in [61] for the 22-nm FD-SOI node (Fig. 13b).

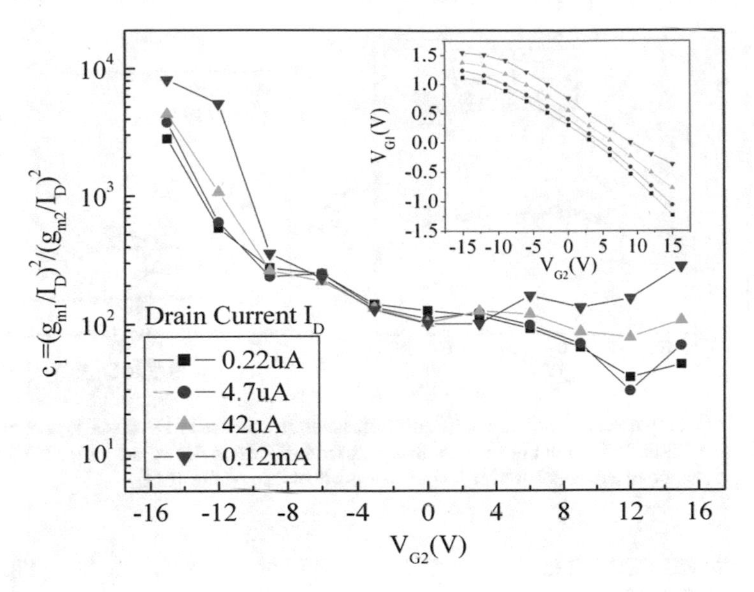

Fig. 11 Coupling factor c_1 and V_{G1} dependence on V_{G2} (inset) for various drain current values (after [17])

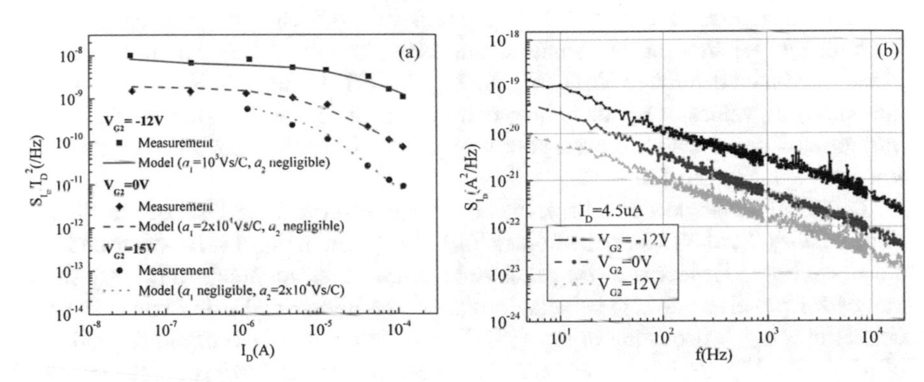

Fig. 12 (**a**) Normalized PSD at $f = 10$ Hz versus drain current for different V_{G2} values with CNF/CMF model and (**b**) Drain current PSD for $I_D = 4.5\ \mu$A measured at different V_{G2} (after [17])

If only the CNF model of Eq. (22) is taken into account without CMF, this behavior cannot be explained, because neither $(g_{m1}/I_D)^2$ nor $(g_{m2}/I_D)^2$ is increasing for negative V_{G2}. However, the CNF LF noise model applied in the positive V_{G2} range allows us to solve Eq. (22) considering negligible CMF contribution at low drain currents. By combining different bias conditions, one can extract constant values for S_{Vfb1} and S_{Vfb2}, from which the front and back-gate oxide trap densities can be determined using Eq. (5) ($N_{t1} = 9 \times 10^{17}$ cm^{-3} eV^{-1} and $N_{t2} = 2 \times 10^{17}$ cm^{-3} eV^{-1} for the 28-nm FDSOI in [17]). Note that the front interface shows higher oxide trap density as expected due to the high-k/metal gate

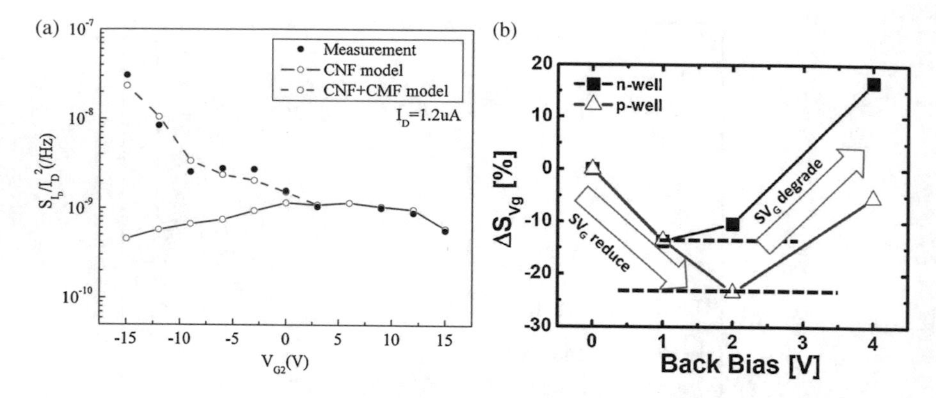

Fig. 13 (**a**) Comparison of experimental normalized drain current PSD versus V_{G2} results with the CNF and CNF/CMF model results, for samples from STMicroelectronics (after [17]) and (**b**) percentage change of S_{Vg} median over 30 dies versus back-bias (after [61])

stack, whereas the back interface trap density is lower due to the higher quality of a pure thermal oxide.

In order to interpret the abnormal LF noise behavior in the negative V_{G2} range, one has to account for the CMF dependence on the carriers position, through the influence of V_{G2} on the remote Coulomb scattering (RCS) parameter α_{sc}, as already explained through Eq. (26) and Fig. 7 [17]. Using this RCS formulation and choosing values of α_{sc} that correspond to the dependence shown in Fig. 7, the variation of noise with the back-gate voltage can be well approached (see fitting lines in Figs. 12a and 13a).

This CMF dependence on back-bias has been also confirmed for 14-nm FDSOI node MOSFETs of STMicroelectronics [60]. As shown in Fig. 14a, it was found that Ω_{eff} (see Eq. (31)) increases for negative back-gate bias and tends to be very small values for positive ones. This behavior cannot be interpreted using constant RCS coefficients $\alpha_{sc1,2}$ occurring in Eq. (33). Based on the carrier centroid dependence of $\alpha_{sc1,2}$ (see Fig. 7), it has been proposed to generalize the CMF coefficients under the symmetric forms [60]:

$$\Omega_1 = \alpha_{sc1}\,\mu_{eff}\,C_{ox1}\,X_{c1} + \alpha_{sc2}\,\mu_{eff}\,C_{ox2}\,(1 - X_{c1}) \tag{34}$$

and

$$\Omega_2 = \alpha_{sc1}\,\mu_{eff}\,C_{ox1}\,X_{c2} + \alpha_{sc2}\,\mu_{eff}\,C_{ox2}\,(1 - X_{c2}) \tag{35}$$

where $X_{c1,2}$ are the normalized carrier centroid positions for the front and back interface, respectively ($X_c = 1$ at the front interface and $X_c = 0$ at the back interface). Using these expressions in Eq. (33) and extracting the centroid position by solving Poisson's equation with TCAD simulations allows obtaining the simulated trend in Fig. 14a, justifying reasonably well the experimental data behavior.

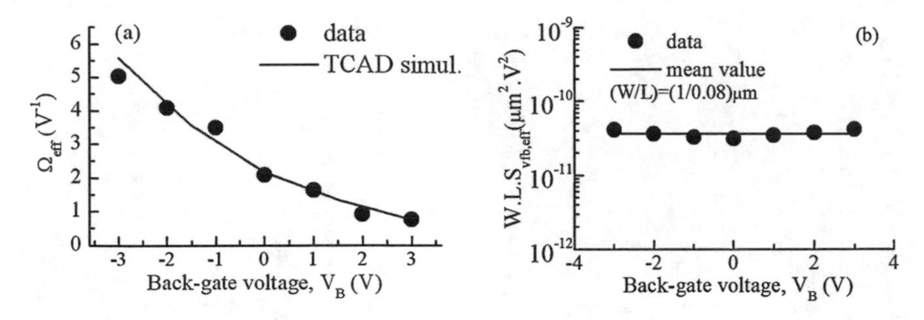

Fig. 14 Ω_{eff} (**a**) and W.L.$S_{Vfb,eff}$ (**b**) dependence for various back-gate biases for n-MOS 14-nm FDSOI node MOSFETs (after [60])

As already explained in Sect. 2.2.2, the combination of front- and back-gate modes of operation can be very useful for achieving different coupling conditions in order to modify the noise contribution of each interface and help the parameter extraction procedure. Indeed, as shown in Fig. 15a [58], in the subthreshold region $(g_{m1}/I_D)^2$ decreases when changing from front-gate ($V_{BG} = 0$ V) to back-gate ($V_{FG} = 0$ V) mode of operation, whereas $(g_{m2}/I_D)^2$ shows a relative increase as shown in Fig. 15b. The above behavior is typical for asymmetrical double-gate devices such as UTTB FDSOI MOSFETs [62], related to the different channel position in the silicon film under various bias conditions. A more direct way for an experimental representation of the carrier distribution at different bias conditions is the plot of the coupling factor $C = (g_{m1}/g_{m2})^2$ as a function of the drain current as described in the previous section. The dramatic reduction of C under the back-gate mode of operation (Fig. 15c) clearly shows that in that case the carriers have a higher concentration near the Si/BOX interface.

This can be taken advantage of and combine measurements from both front and back-gate modes of operation to extract a unique set of parameters that fits all data. An example is shown in Fig. 16, where the noise parameters S_{Vfb1}, S_{Vfb2}, Ω_1, and Ω_2 were obtained by fitting Eq. (24) simultaneously with all $1/f$ noise data for the front-gate ($V_{G2} = 0$ V and -5 V) and back-gate mode ($V_{G1} = 0$ V). Otherwise, fitting the noise data for a specific bias condition (e.g., $V_{G2} = 0$ V), may lead to a numerous number of model parameters. The difference between the "Measured ($f = 10$ Hz)" and "$1/f$ component" seen in Fig. 16 means that around 10 Hz there is a strong presence of Lorentzian-like noise, which needs to be removed in order to extract the flicker noise contribution (more on that in Sect. 3.2).

Furthermore, one can observe the dependence on the bias conditions of the Si/BOX interface contribution to the total noise level. In the front-gate mode, there is sufficient decoupling of the two interfaces in the subthreshold region, whereas in strong inversion, the back interface affects the noise level significantly. In back-gate mode, as expected, there is a strong influence on the noise of both back and front interfaces.

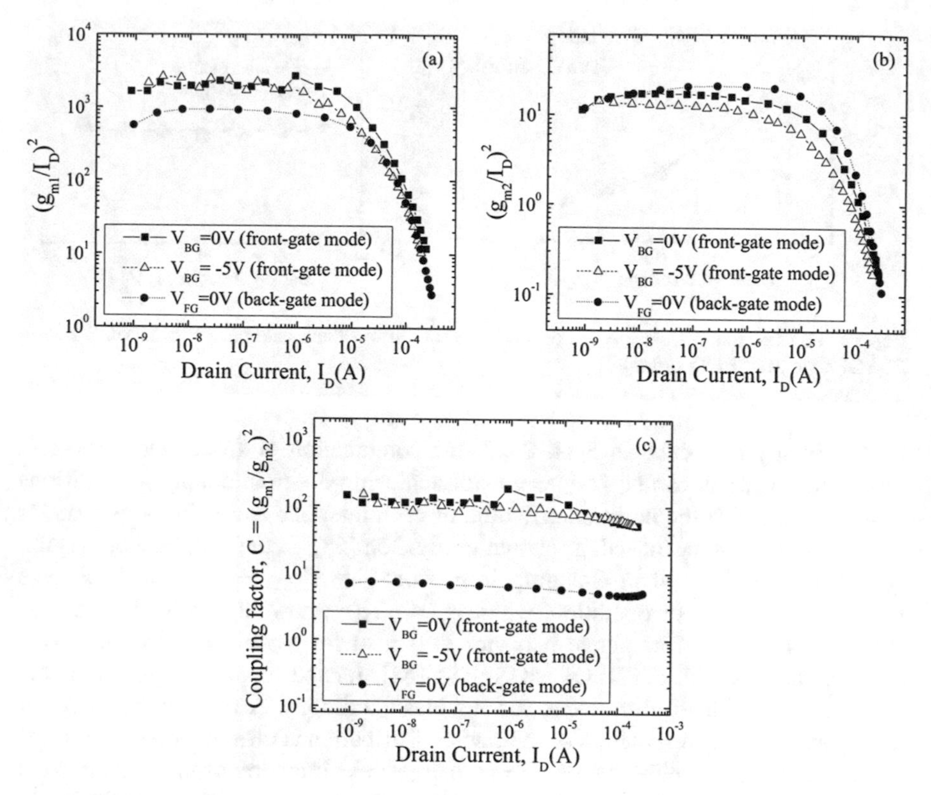

Fig. 15 CNF factors $(g_{m1}/I_D)^2$ (**a**), $(g_{m2}/I_D)^2$ (**b**), and coupling factor (**c**) versus drain current for n-MOS transistors issued from 20-nm FDSOI CMOS technology with $W = 10\ \mu\text{m}$ and $L = 87$ nm, $t_{ox} = 1.2$ nm, $t_{Si} = 8.7$ nm, and $t_{ox2} = 10$ nm (after [58])

Moreover, the involvement of the correlated mobility fluctuation is also bias dependent. In front-gate mode, the CMF takes place only at the front interface. Respectively, when the back-gate bias voltage V_{BG} is varying, the CMF appears only at the back interface. This behavior confirms the analysis regarding the Ω_1 and Ω_2 CMF factors, as already explained in Paragraph 11 through Eq. (27) and Fig. 8. The same technique was also used successfully in [57] (see Fig. 17), for both n- and p-channel FD-SOI MOSFETs.

Impact of Channel Geometry on RCS

In Fig. 18a–c [57], the plotted model lines have been derived using Eq. (29) as a fitting function. It is clear from Fig. 18a, b that for both 20-nm node n- and p-channel FDSOI devices of constant gate length and varying width, there is a strong dependence of Ω_1 on the gate width W. However, this does not seem to be the case

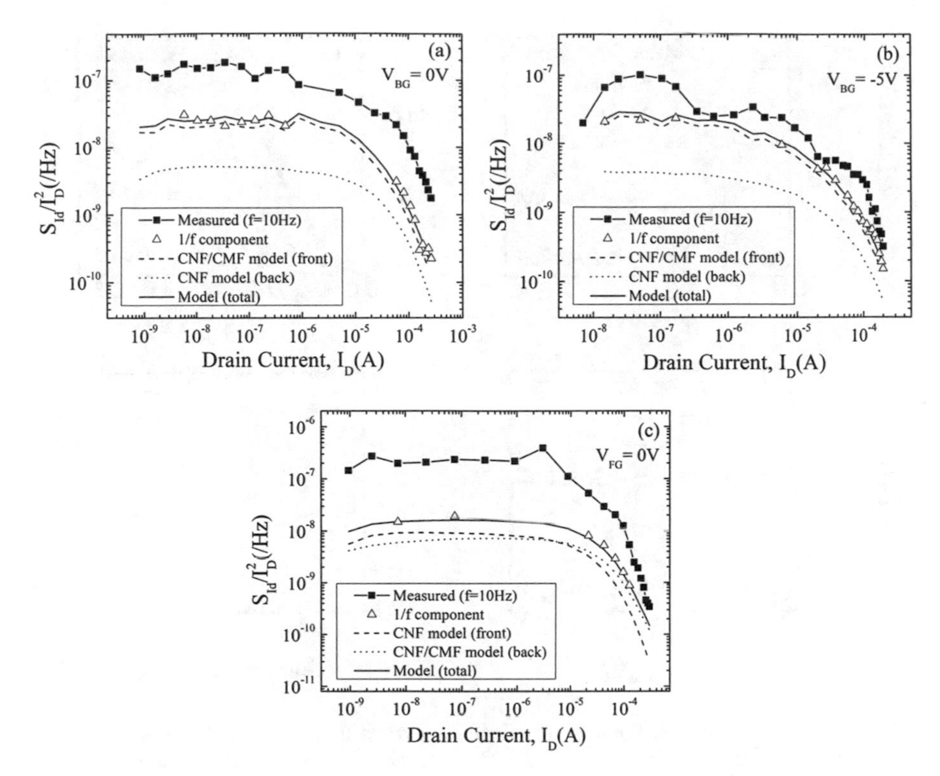

Fig. 16 Normalized power spectral densities, flicker noise component values, and CNF/CMF model fitting versus drain current for $V_{BG} = 0$ V (**a**), $V_{BG} = -5$ V (**b**) and $V_{FG} = 0$ V (**c**) for n-MOS transistors issued from 20-nm FDSOI CMOS technology with $W = 10$ μm and $L = 87$ nm, $t_{ox} = 1.2$ nm, $t_{Si} = 8.7$ nm, and $t_{ox2} = 10$ nm (after [58])

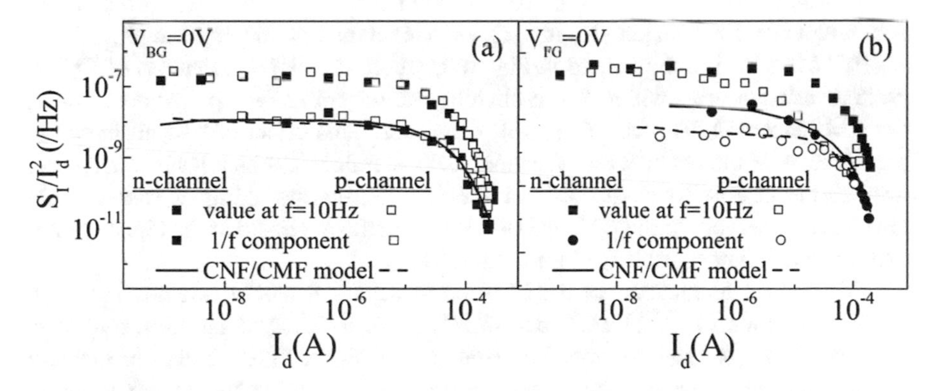

Fig. 17 Measured normalized drain current S_I/I_d^2 at $f = 10$ Hz, extracted $1/f$ component and CNF/CMF model versus drain current for n (solid symbols) and p (open symbols) channel devices at front-gate (**a**) and back-gate mode (**b**) with $W = 10$ μm and $L = 87$ nm (after [57])

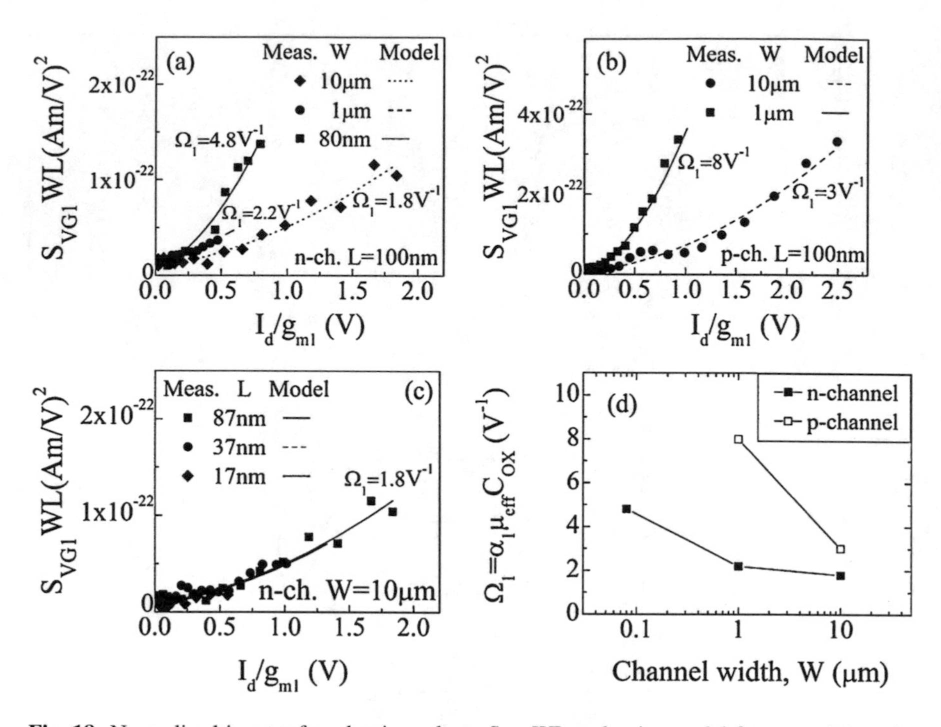

Fig. 18 Normalized input-referred noise voltage S_{VG1}WL and noise model fit versus I_d/g_{m1} for n-channel (**a**, **c**) and p-channel (**b**) devices, and extracted values of Ω_1 versus channel width W (**d**) for transistors issued from 20-nm FDSOI CMOS technology with $t_{ox} = 1.2$ nm, $t_{Si} = 8.7$ nm and $t_{ox2} = 10$ nm (after [57])

for devices of constant gate width. As shown in Fig. 18c, in devices with constant gate width and different gate lengths, Ω_1 is independent on the gate length.

This behavior is summarized in Fig. 18d, where the extracted values of Ω_1 are plotted versus the gate width W. It is obvious that for both n- and p-channel devices, the CMF factor Ω_1 is increasing as W decreases. This effect can be attributed to the decrease of the channel cross section, as W is reduced, which leads to a shorter average distance between carriers and interface. Furthermore, higher values of Ω_1 are clearly observed for the p-channel devices, which are related to the higher values of RCS coefficient α_{sc} observed for holes [22].

This lack of dependence of the CMF on the channel length was not, however, confirmed for the 14-nm FDSOI node MOSFETs of STMicroelectronics, studied in [60]. As shown in Fig. 19a, the Ω_{eff} parameter of Eq. (31) was found to increase in long channel devices. This trend is primarily related to the strong mobility dependence on channel length [63], which also follows a similar tendency, as shown in Fig. 19a. The small deviation can be attributed to the mobility degradation taking place in strong inversion, which was not taken into account. Nevertheless, contrary to the CMF dependence on channel geometry, no such impact was observed for the CNF factors S_{Vfb1} and S_{Vfb2}, meaning that the oxide trap density can be considered

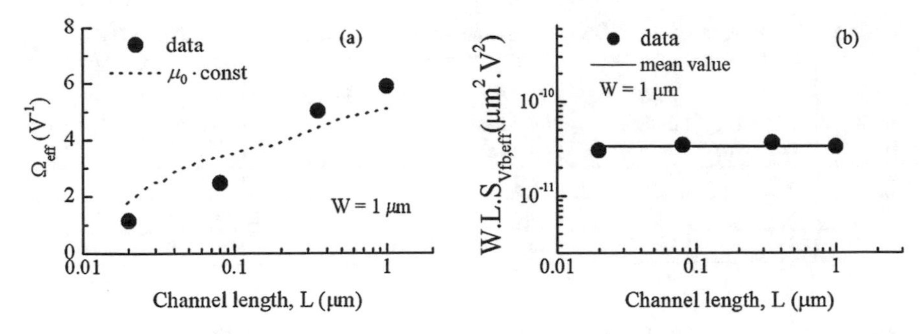

Fig. 19 Ω_{eff} (**a**) and W.L.$S_{Vfb,eff}$ (**b**) versus device length for n-MOS, respectively, for 14-nm FDSOI MOSFETs (after [60])

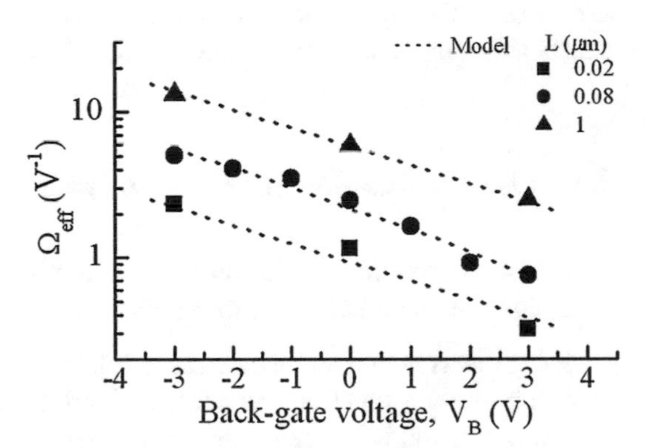

Fig. 20 Ω_{eff} dependence on back-gate bias for various channel lengths

constant with channel width and length, for both top and bottom interfaces. This is clear in Fig. 18a–c, where S_{VG1}WL is constant with W and L in weak inversion, but also in Figs. 14b and 19b, where the extracted $S_{Vfb,eff}$, normalized by the surface, is proven to stay constant with W and L, respectively.

For this effect of L on Ω_{eff}, as well as the impact of V_b (V_{G2}), a unified empirical equation that can fit to all data (as shown in Fig. 20) was proposed [60]:

$$\Omega_{eff}(V_b, L) = \frac{\Omega_0 e^{-V_b/V_0}}{(1 + L_0/L)} \tag{36}$$

where $\Omega_0 = 6.5\ V^{-1}$, $V_0 = 3.42\ V^{-1}$ and $L_0 = 0.12$ μm are some fitting parameters, with values that were extracted for the 14-nm FDSOI technology node devices in [60].

In another study [57], a comparative parameter extraction was done for three different FDSOI wafers. Table 1 presents the characteristic manufacturing details of the wafers, along with the extracted noise parameters. As shown, N_{t1} has almost the same value for all wafers, whereas N_{t2} depends slightly on t_{BOX}. In all wafers, the same front oxide high-k metal gate technology is used in terms of materials and thicknesses, explaining the constant N_{t1} values. However, the thickness and the

Table 1 FDSOI wafer characteristics and extracted flicker noise parameters (after [57])

Wafer number	Technology node (nm)	BOX thickness t_{BOX} (nm)	Si body thickness t_{Si} (nm)	Gate oxide trap density N_{t1}(cm^{-3} eV^{-1})	Buried oxide trap density N_{t2} (cm^{-3} eV^{-1})	CMF factor Ω_1 (V^{-1})
1	14	10	8.7	8.7×10^{17}	3.5×10^{17}	1.8
2	28	25	6.7	$9\cdot \times 10^{17}$	2.7×10^{17}	12.2
3	28	145	10	8×0^{17}	8.5×10^{16}	0.8

quality of the buried oxide change from wafer to wafer, thus affecting the value of N_{t2}. Moreover, the CMF factor $\Omega_1 = \alpha_{sc1}\mu_{eff}C_{ox}$ was found to decrease with the thickness of the silicon body, due to the increase in the average distance between interface–inversion charge centroid, as explained in Sect. 2.2.2.

3.2 Generation–Recombination Noise

The physical phenomenon that results in a Lorentzian type of spectrum is the carrier generation–recombination occurring when:

- A free electron and a free hole recombine.
- A pair of a free electron and a free hole are generated.
- A free electron is trapped at an empty trap.
- A free hole is trapped at a filled trap.

The random telegraph noise (explained in Sect. 2.1.4) is a case of g–r noise, for which there is only one discrete trap and not a group of defects.

In general, when the noise spectra are composed of 1/*f* and Lorentzian noise components due to several distinct and uncorrelated trap levels, the total current noise power spectral density S_{Id} can be written as:

$$S_{I_d} = \frac{K_f}{f} + \sum_{i=1}^{N} \frac{A_i}{1 + (f/f_{ci})^2}. \tag{37}$$

where K_f is the amplitude of the flicker noise, A_i is the Lorentzian plateau value, and f_{ci} is the corner frequency of each Lorentzian noise spectrum. Such an example can be seen in Fig. 21, where F corresponds to K_f, L_i to A_i and $T + B$ to the sum of thermal and background noise.

The origin of the g–r noise can be clarified by determining the dependence of the Lorentzian time constant ($\tau = 1/2\pi f_c$) on the gate bias. When the time constant remains unchanged with gate voltage, the g–r centers are uniformly distributed within the depletion region of the silicon body, whereas when it passes through a maximum, the g–r noise centers are defects located at the interface [64] or inside

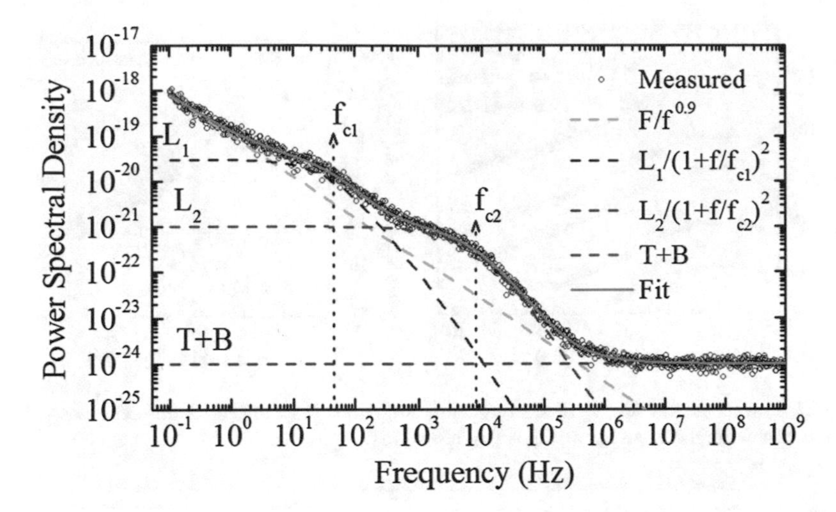

Fig. 21 Example of spectrum decomposition for a MOSFET drain current PSD measurement

the oxide region. A recent study has shown that in fully depleted devices, the g–r traps positioned in the Si film can have a misleading time constant behavior, due to its dependence on the carrier concentration above a gate voltage level [65]: the time constant maintains the same value as long as the Fermi level is below the trap energy level and τ starts decreasing with increasing gate voltage when the Fermi level exceeds the trap energy level.

In Fig. 22a [58], one can clearly notice that when $V_{\mathrm{BG}} = 0$ V, the total noise is composed of a flicker noise and a Lorentzian component. A direct way to ascertain if the g–r component is current dependent is to plot noise spectra, obtained at a constant drain current under different bias conditions, as shown in Fig. 22b. It is evident that the g–r component is mainly dependent on the carriers' position and not on the total channel charge, whereas the flicker noise component seems to change slightly. In addition, from Fig. 22b, one can safely conclude that the specific Lorentzian-type noise is not related to the front-gate interface.

After decomposing all spectra into their noise components, the Lorentzian noise parameters (time constants and plateau values) were extracted. In Fig. 23 (after [58]), these parameters are plotted as a function of the front- and back-gate voltages under back- and front-gate modes of operation, respectively, adjusted to correspond to the same drain current range. One can notice the different trends of the two sets of data. The time constant is not drain current related, but it is defined by the applied bias voltages. This finding leads to the conclusion that the time constant is related to the distribution of the carriers in the channel and not with the total charge.

As shown in Fig. 23a, for $V_{\mathrm{BG}} = 0$ V, there is no voltage dependence of the time constant and the g–r plateau reaches a constant value above threshold. However, in back-gate mode (Fig. 23b and in the voltage range above the maximum value of $S_{\text{g-r}}(0)$, the time constant decreases exponentially with increasing V_{BG}. The results of Fig. 23 can be explained considering that in FDSOI MOSFETs there is almost no

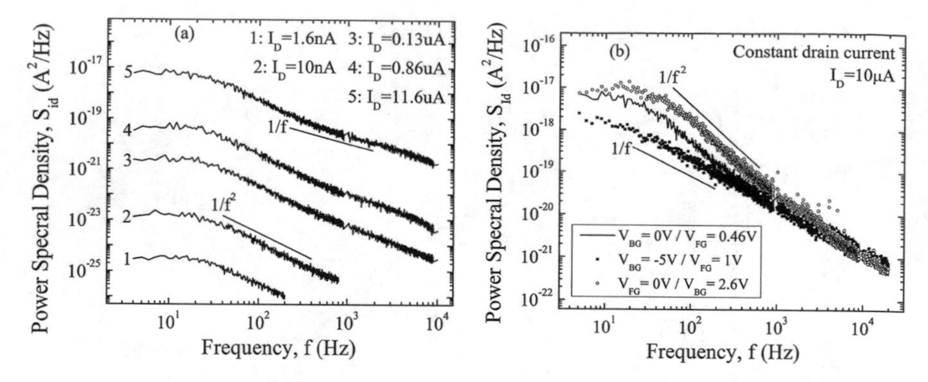

Fig. 22 Power spectral density versus frequency for $V_{\mathrm{BG}} = 0$ V (**a**) and for different bias voltages for a constant drain current $I_{\mathrm{D}} = 10\ \mu$A (**b**) (after [58])

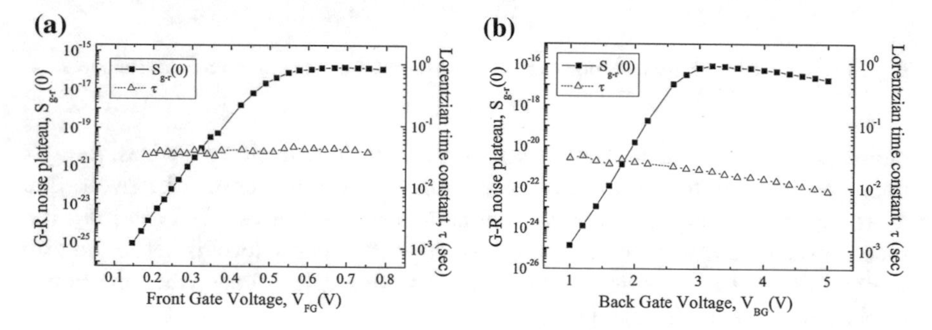

Fig. 23 Extracted values of the Lorentzian plateau (squares) and time constant (triangles) versus V_{FG} for $V_{\mathrm{BG}} = 0$ V (**a**) and versus V_{BG} for $V_{\mathrm{FG}} = 0$ V (**b**), plotted at the same drain current region (after [58])

depletion layer in the Si film when the inversion channel is created, as simulations have shown in [11, 65]. If the electron density n_1 (see Eq. (15)b at the position where the Fermi level is coinciding with the trap energy level in the Si film is higher than the density n of carriers all across the channel, the time constant of the g–r noise is dependent only on n_1 [65], thus it remains unchanged as shown in Fig. 23a. When n becomes comparable or larger than n_t, the time constant becomes inversely proportional to n, leading to the behavior of Fig. 23b. At low back-gate voltages τ is almost constant, but when $S_{\text{g-r}}(0)$ reaches a maximum value, the Fermi level approaches the trap energy level. After this point, the time constant decreases following the decrease of the g–r noise plateau. This observation leads to the conclusion that these defects are located inside the silicon channel, but near the bottom silicon/buried oxide interface. This explains why τ stays constant with the front-gate voltage.

3.3 Noise Variability

General Properties of LFN Variability

The reduction of the device area in advanced CMOS technologies, such as the FD-SOI structure, can lead to important issues regarding the device performance variability [66]. As far as the low-frequency noise (LFN) is concerned, one of its main properties is the scaling with the reciprocal of the device area [23] (Fig. 24), which has become a major concern for both analog and digital circuits operation, such as oscillators, RAM cells, inverters, and other mixed signal circuits [49]. Furthermore, the nanoscale-induced uncertainty in number of traps, the random telegraph noise (RTN) presence, common in small-area MOSFETs (<1 μm^2) [42], as well as generation–recombination (GR) noise centers in the Si film [67] directly lead to enormous levels of LFN variability (Fig. 24) that further limit the device performance and reliability and introduce a high inaccuracy in the prediction of circuit noise levels. Thus, a detailed study of the advanced FD-SOI LFN variability is crucial in order to better understand its origin and by turn its prediction and/or reduction.

An example of noise variability experimental results is shown in Fig. 25 (after [68]), where $S_{V\text{g}}$ is plotted versus frequency for all measured devices across a 300-mm wafer biased under the same gate voltage bias $V_\text{G} = 0.4$ V. Notice that few spectra are $1/f$-like, but there is also a significant number of Lorentzian-like spectra, commonly observed in such small area ($W \times L = 0.0012$ μm^2) devices, due to random telegraph noise (RTN), or due to generation–recombination (GR) centers located in the Si film [65, 69]. From Fig. 25 (after [68]), it is clear that the noise variability is strongly enhanced when Lorentzian noise is present.

Based on Eqs. (13) and (20), without taking into account the CMF component, this dispersion in the noise spectra can be expressed mathematically through [70, 71]:

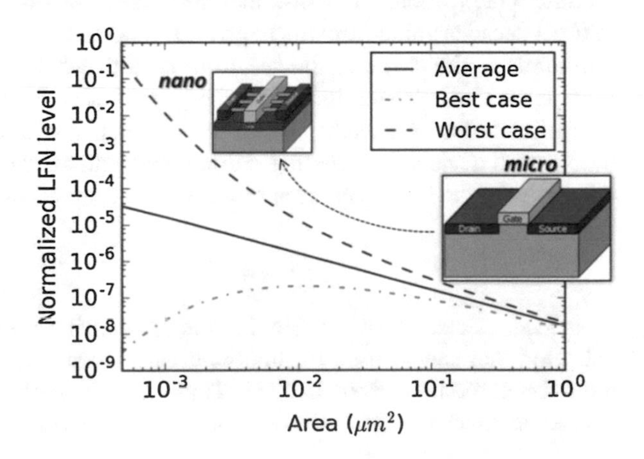

Fig. 24 Both LFN level and LFN variability scale reciprocally with the device area

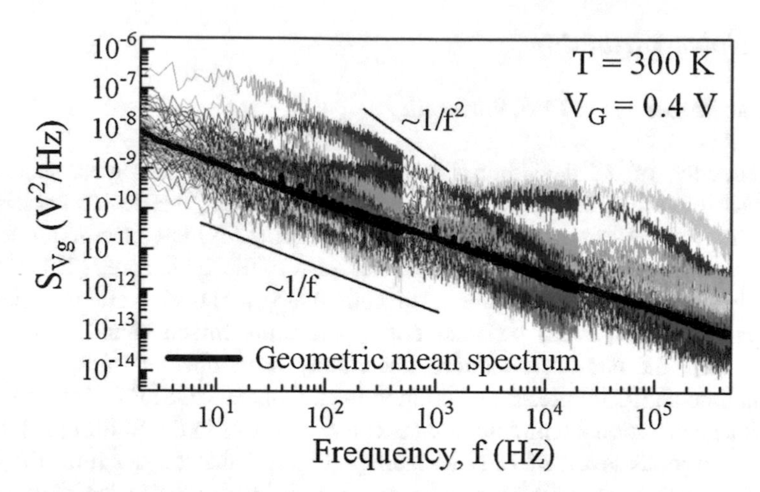

Fig. 25 Input-referred gate voltage noise spectra at $V_G = 0.4$ V for 85 dies (colored lines) and geometric mean spectrum (black line) for 14-nm FD-SOI MOSFETs with $L = 20$ nm and $W = 60$ nm (after [68])

$$S_{\mathrm{Vg},i} = \frac{q^2}{(\mathrm{WL}C_{\mathrm{ox}})^2}\sum_{k=1}^{N}\left(1 - \frac{x_{\mathrm{t},k}}{t_{\mathrm{ox}}}\right)4A_k\frac{\tau_k}{1+\omega^2\tau_k^2} \tag{38}$$

where τ refers to the effective trap time constant, $A = f_{\mathrm{t}}(1 - f_{\mathrm{t}})$ is the space mark ratio, x_{t} the trap depth, and N is the number of traps in the gate oxide within the energy range swept by the Fermi level ($N = \mathrm{W\,L}\,t_{\mathrm{ox}}\,N_{\mathrm{t}}\,\Delta E_{\mathrm{f}}$, N_{t} being the volumetric oxide trap density and ΔE_{f} the Fermi-level excursion). Therefore, since each device has a different number of traps N (i.e., RTN fluctuators) and each trap its own x_{t}, τ, and A, an induced variability increase is expected as N decreases dramatically, which is the case in nanoscale devices. This effect is further demonstrated in Fig. 26 (after [72]), where it is shown that the noise-level dispersion is enhanced by two to three decades going from the large (Fig. 26a) to the small (Fig. 26b) area device. Furthermore, the shape of the spectrum is noticeably changing from $1/f$ to more Lorentzian-like behavior, indicating the impact of RTN in LFN variability.

Furthermore, as already shown in [73, 74], the drain current noise data of ultra-scaled devices follow a log-normal distribution, thus the geometric mean (or, alternatively, the log-mean) spectrum can better represent the average noise:

$$\langle S\rangle_{\mathrm{geom.}} = \sqrt[n]{S_1 S_2 S_3 \cdots S_n} \tag{39}$$

Indeed, as can be seen in Fig. 25, the geometric mean spectrum has a clear $1/f$ behavior, that can be used for both interface characterization and noise modeling after the correction given in [73]. Thus, in this section, we always consider a geometric mean average in the noise variability analysis.

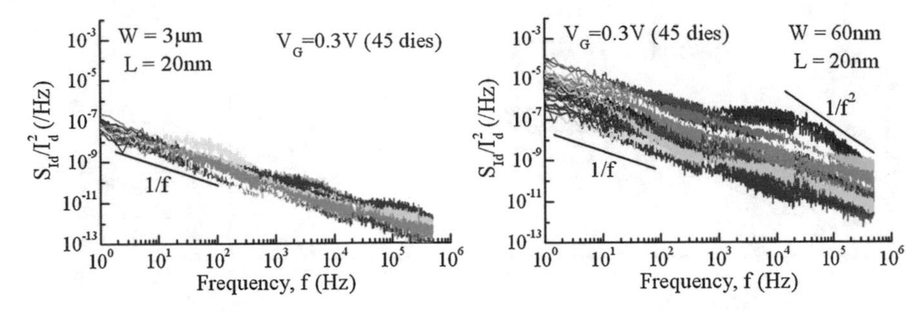

Fig. 26 Normalized drain current spectral density $S_{\mathrm{Id}}/I_{\mathrm{d}}^2$ versus frequency for large (left) and small (right) area n-channel 14-nm FDSOI MOSFETs (after [72])

Finally, in order to quantify the LFN variability effect and be able to make a variability parameter comparison between different technologies, one needs to calculate the standard deviation of the surface-normalized noise at a certain extraction frequency (usually at 10 Hz) [71]:

$$\text{Drain current LFN variability}: \ \sigma\left(\log\left(WL\frac{S_{\mathrm{Id}}}{I_{\mathrm{d}}^2}\right)\right) \tag{40}$$

or, even better, in order to eliminate the influence of the subthreshold swing or threshold voltage variability in the LFN variability estimation, one can use [68]:

$$\text{Input} - \text{referred gate voltage LFN variability}: \ \sigma\left(\log\left(\mathrm{WLS}_{V_{\mathrm{g}}}\right)\right) \tag{41}$$

where $S_{\mathrm{Vg}} = S_{\mathrm{Id}}/g_{\mathrm{m}}^2$ is the input-referred gate voltage noise.

Variability Comparison Between Different Technology Nodes

Due to the fact that the trap number variation is a local effect, if the normalized LFN standard deviation (Eq. (40) or (41)) is plotted versus $1/\sqrt{\mathrm{WL}}$, a Pelgrom's law [75] trend is observed, from which one can extract a noise-matching parameter A. An example is presented in Fig. 27 (after [71]), where the slope A seems to be changing, depending on the technology node.

Following this methodology, the noise variability in of the FDSOI technology [76] was found to be improved as compared to advanced bulk technology nodes, as shown in Fig. 27 (after [71, 72]). This was further verified in [72], where the 14-nm-node FDSOI technology [76] was found to be improved as compared to both 20-nm Bulk and 28-nm FDSOI nodes, as shown in Fig. 28 (after [72]). Interestingly, the noise-matching parameter A is following the same trend as the product of the extracted oxide trap density and the squared equivalent oxide thickness, $N_{\mathrm{t}} \times \mathrm{EOT}^2$, indicating that the use of a thinner oxide and better quality interface are the main reasons for this noise variability improvement [77].

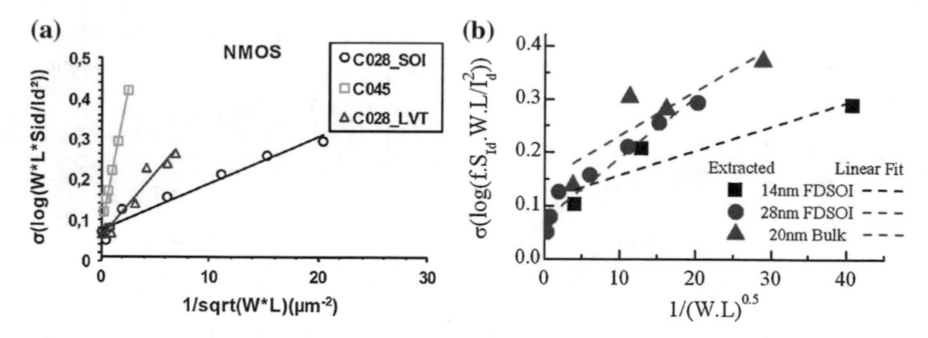

Fig. 27 Comparison of the drain current noise variability as a function of the inverse square root of the area between 28-nm SOI/Bulk and 45-nm Bulk (after [71]) (**a**) and between 28-nm/14-nm FDSOI and 20-nm bulk technology nodes (after [72]) (**b**)

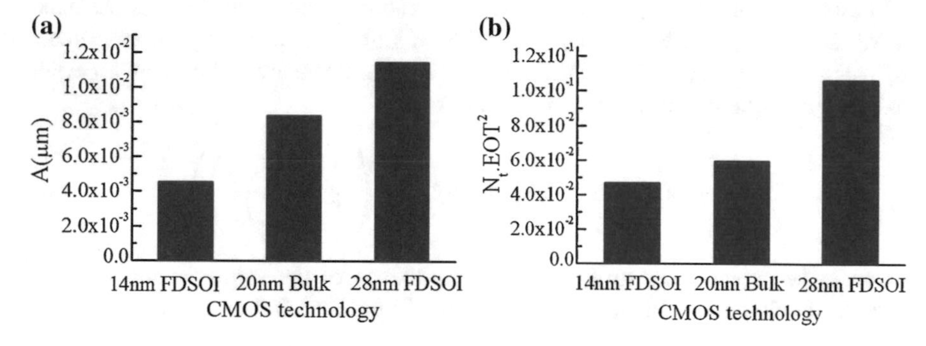

Fig. 28 (**a**) Comparison of the drain current noise matching parameter. (**b**) Comparison of the product N_t.EOT^2 for n-MOSFETs issued from 28-nm and 14-nm FDSOI and 20-nm bulk technology nodes (after [72])

Statistical LFN/RTN Characterization Methods

As shown in Fig. 26, when going from large to smaller area devices, not only we induce an increase in noise variability, but there is a change in the spectral shape, from 1/*f*-like to more Lorentzian-like, because the average number of active traps is decreased. Thus, the CNF/CMF flicker noise model, needed for the extraction of N_t, is very difficult to be applied in nanoscale MOSFETs, due to the lack of uniformity in the oxide trap distribution, which results in non-flicker noise spectra. However, if the statistical sample is large enough, the average noise spectrum will be 1/*f*-like (see Fig. 25) [29, 68], making it suitable for extracting the average trap density N_t [72]. Indeed, if we plot the extracted density values at 10 Hz from the spectra of Fig. 26 versus the drain current (see Fig. 29), the average LFN level follows the CNF model [72]. Therefore, with a sufficient number of measured dies, one can extract the average N_t for all gate oxide areas. An example of such an extraction is shown in Fig. 30a, where an almost constant state-of-the-art volumetric oxide trap density

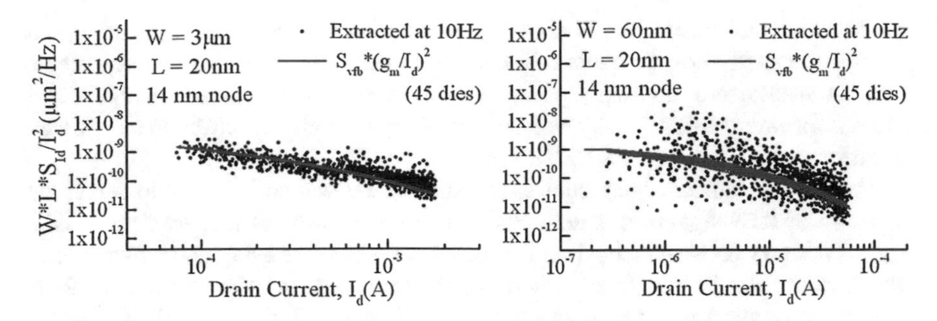

Fig. 29 Normalized drain current noise (symbols) and CNF model (lines) at 10 Hz for large (left) and small (right) area n-channel 14-nm FDSOI MOSFETs (after [72])

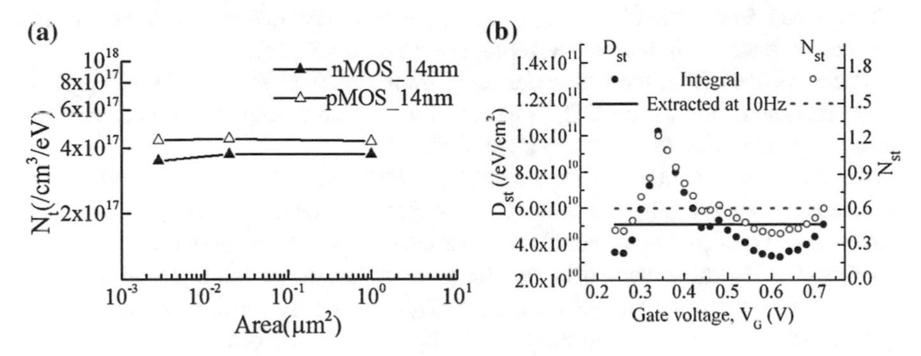

Fig. 30 (**a**) N_t versus device area for n- and p-MOS from 14-nm technology node and (**b**) full wafer mean trap areal density D_{st} and average absolute number per device N_{st} ($W = 60$ nm, $L = 20$ nm) (after [72])

is extracted for both n- and p-channel 14-nm FDSOI MOSFETs for all measured device areas.

Besides, Fig. 30b shows the full wafer mean trap areal density D_{st} and corresponding average absolute number per device, $N_{st} = D_{st}.(\mathrm{WL})$, values, as calculated using Eq. (42). The noise-induced threshold voltage standard deviation can be calculated using the spectra integrals (Eq. (43)). This way, a clear gate voltage dependence is revealed, due to specific trap-induced RTN.

$$D_{st} = \frac{\mathrm{W\,L}\,{C_{ox}}^2\,{\sigma_{Vt}}^2}{q^2kT} \tag{42}$$

where

$$\sigma_{Vt} = \sqrt{\int\limits_{f=f_{min}}^{f_{max}} S_{Vg}\,\mathrm{d}f} \tag{43}$$

Therefore, following this method, one can obtain a more representative and realistic illustration of the oxide trap density in the fully studied frequency range, i.e., the whole time constant space, as compared to the standard analysis that extracts the volume trap density N_t from a pure flicker noise spectrum in a restricted frequency range, e.g., around 10 Hz.

For certain studies, which requires to examine the trap activity and to detect the presence of RTN signals, it is not mandatory to extract the exact parameters (ΔI_d, τ_c, τ_e) of every RTN-inducing trap for the whole wafer and every gate voltage bias, which would also be very time-consuming. Alternatively, in order to probe the RTN impact on the spectrum at each gate voltage and measured die, a new RTN strength indicator is proposed [72]: the standard deviation of $\log(S_{Vg} f)$ vs. frequency, where $S_{Vg} = S_{Id}/{g_m}^2$. As one can see in Fig. 31, when a single RTN source is active, which occurs in a certain bias range, $\log(S_{Vg} f)$ strongly deviates from the flat line that corresponds to $1/f$-like noise, giving rise to $\sigma(\log(S_{Vg} f))$.

It can be seen from the full wafer cartography of Fig. 32 that $\sigma(\log(S_{Vg} f))$ is bias dependent, due to the RTN trap activity, and randomly distributed over the wafer map, revealing few RTN-dominated dies.

Considering the cumulative distribution function (CDF) for an arbitrarily selected gate voltage, Fig. 33a shows that the statistical distribution of $\sigma(\log(S_{Vg} f))$ can be easily modeled by a specific number (in this case 4) of randomly distributed additional RTN fluctuators over six frequency decades, superimposed to a $1/f$ background spectrum. Note also that $\sigma(\log(S_{Vg} f))$ is very sensitive to the number of RTN per decade, demonstrating the RTN statistical probing capability of this quantity. Figure 33b shows the full wafer cartography of the mean values of $\sigma(\log(S_{Vg} f))$ for all gate voltages. It is clear that they exhibit a smoother wafer distribution, meaning that $\sigma(\log(S_{Vg} f))$ follows well the convergence of many RTN-like spectra to a $1/f$ spectrum when averaged.

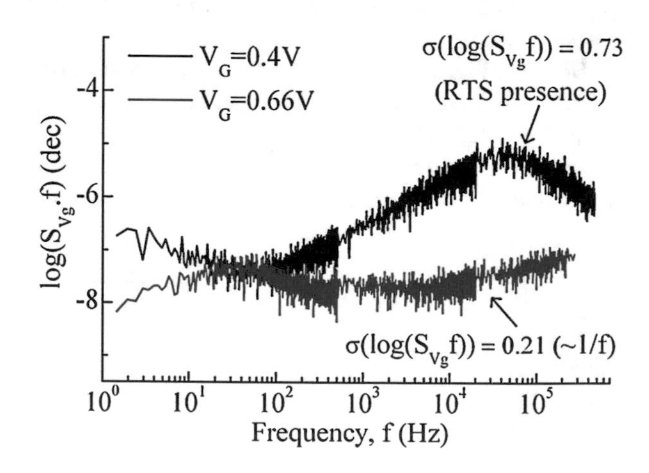

Fig. 31 $\log(S_{Vg} f)$ for $1/f$-like (red) and Lorentzian-like (black) spectra (n-MOS 14 nm FD-SOI, $W = 60$ nm, $L = 20$ nm) (after [72])

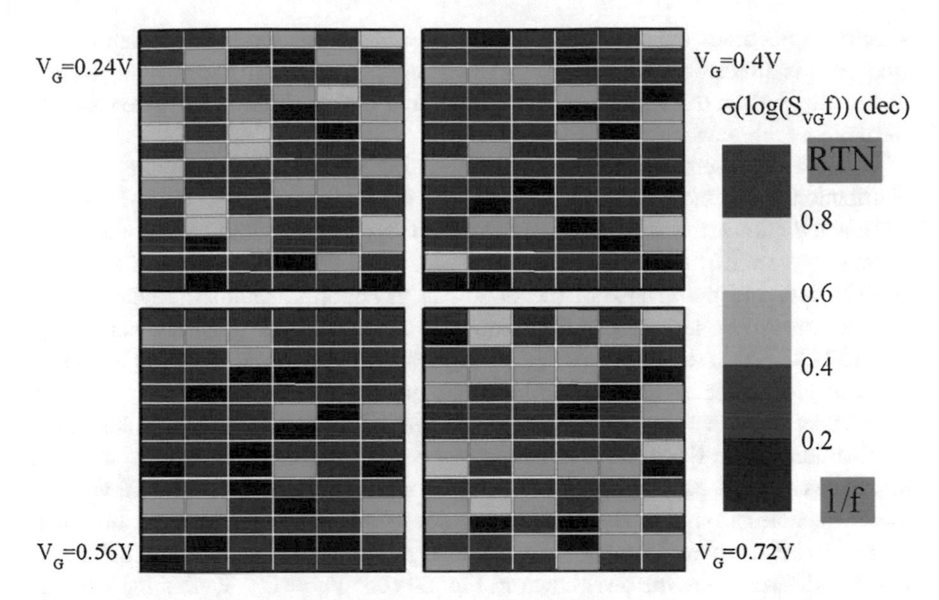

Fig. 32 Colored wafer maps based on the values of σ(log($S_{\mathrm{Vg}}\,f$)) (dec) for various gate voltage values (n-MOS 14-nm FD-SOI, $W = 60$ nm, $L = 20$ nm) (after [72])

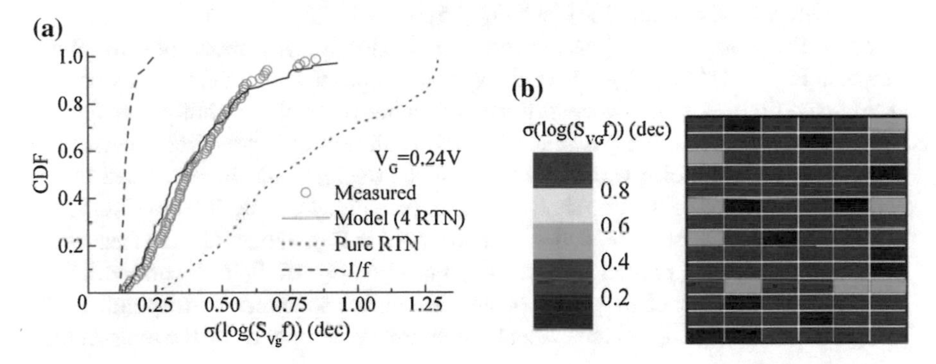

Fig. 33 (**a**) CDF of σ(log($S_{\mathrm{Vg}}\,f$)) (dec) (measurements and model) and (**b**) colored wafer maps based on the mean values of σ(log($S_{\mathrm{Vg}}\,f$)) (dec) (n-MOS 14-nm FD_SOI, $W = 60$ nm, $L = 20$ nm) (after [72])

Dependence on Frequency, Gate Bias, and Temperature

In Fig. 25, a dispersion of up to four decades of magnitude can be observed at high frequencies, while almost two decades around $f = 1$ Hz, revealing a possible frequency dependence of the standard deviation. Indeed, as shown in Fig. 34a (after [68]), where the measured spectra of 85 devices at $V_{\mathrm{G}} = 0.4$ V are plotted along with the noise standard deviation, a clear dependence of LFN variability with frequency is visible. In fact, in the frequency regions where $\sigma(\log(S_{\mathrm{Vg}}))$ is increased, there are Lorentzian-like spectra with significantly higher density levels than the remaining spectra. It becomes thus obvious that the dependence of the noise variability on frequency is directly related to the RTN or GR noise presence in many of the devices due to the random distribution of trap energy and trap position in the gate oxide or Si film for each device.

A more representative method for validating and analyzing the log-normal distribution is the use of the inverse error function (InvErf) of (2 CDF-1), where CDF is the cumulative distribution function. If the data distribution is indeed log-normal, the InvErf(2 CDF-1) versus the logarithm of the data should follow a straight line, and the inverse of the slope can provide the standard deviation. So, in order to further demonstrate the frequency dependence of the noise variability, InvErf(2 CDF-1) is shown versus S_{Vg} (in log scale) at $V_G = 0.26$ V for four different extraction frequencies in Fig. 34b. The log-normal distribution in all extraction frequencies is evident from the linear fit of InvErf(2 CDF-1) with S_{Vg} in log scale. Furthermore, from the difference observed between the linear fit slopes, it seems that at this specific gate voltage bias, the noise variability has its maximum value at low frequencies around 10 Hz, then it decreases with frequency reaching a minimum around 10 kHz, whereas it again increases at high frequencies around 250 kHz. This trend is different than the one shown in Fig. 34a for $V_G = 0.4$ V, thus the way the noise variability depends on the frequency changes with the device bias. Following the same data analysis methodology, a variability dependence on gate voltage and temperature was also found [68] (see Fig. 35).

To further probe the gate bias dependence of σ(log(S_{Vg})), a more intuitive plot is shown in Fig. 36 (after [68]), where σ(log(S_{Vg})) is plotted versus frequency and V_G at the same time. For all bias conditions, a spectral peak of variability at a specific frequency is observed, the position of which is shifted to higher frequencies with increasing V_G. The noise variability dependence on V_G could be explained by the voltage dependence of the trap time constants (see Fig. 4), as verified in Fig. 37a.

A similar time constant dependence is observed in Fig. 37b, where the frequency-normalized noise is plotted versus frequency at $V_G = 0.26$ V and different temperatures. The cutoff frequency of the Lorentzian spectrum corresponds to the maximum value of $f\,S_{Vg}$, which evidently increases with increasing the temperature.

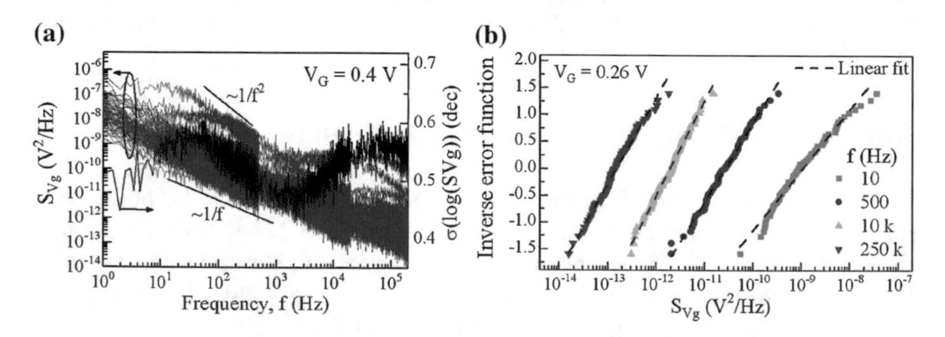

Fig. 34 (**a**) Input-referred gate voltage noise (left *Y*-axis) and standard deviation value (right Y-axis) at $V_G = 0.4$ V versus frequency and (**b**) inverse error function versus input-referred gate voltage noise at $V_G = 0.26$ V for four different frequencies (after [68])

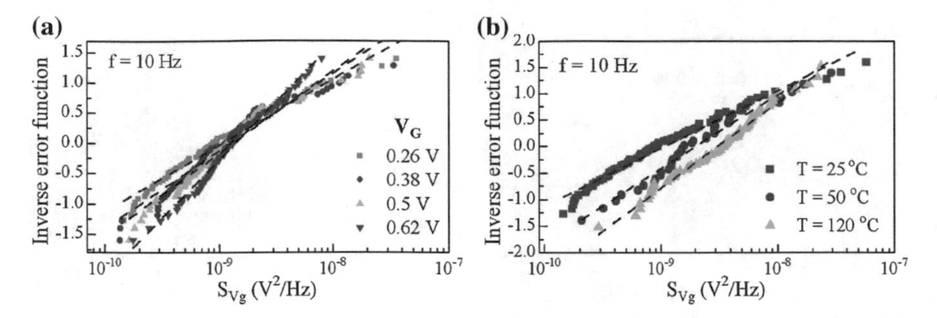

Fig. 35 Inverse error function versus input-referred gate voltage noise at $f = 10$ Hz for (**a**) four different gate voltage values under $T = 25$ °C and (**b**) for three different temperatures T under $V_G = 0.26$ V (after [68])

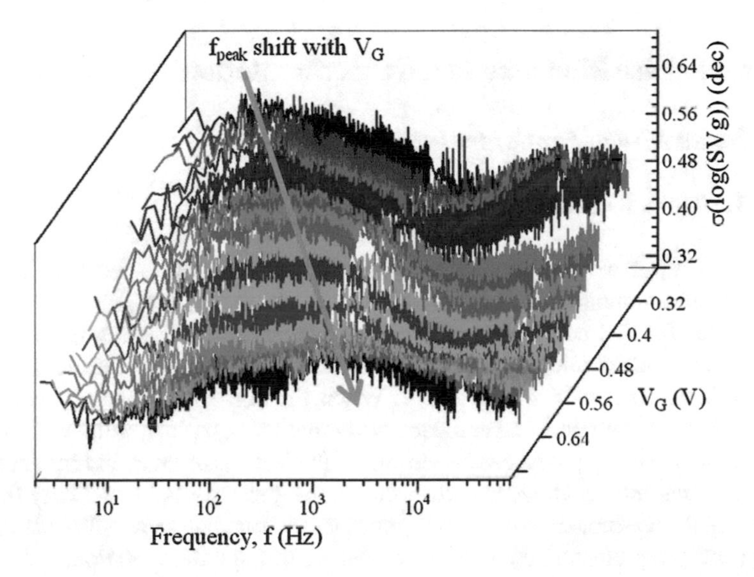

Fig. 36 Input-referred gate voltage noise standard deviation versus frequency for all measured gate voltage values (after [68])

This behavior is well known in RTN analysis, where the trapping process is thermally activated, resulting in a decrease of the trap time constant with temperature [78]. This translates into an augmentation of the Lorentzian cutoff frequency as the temperature rises, since $f_c = 1/2\pi\tau$. Regarding the impact of temperature on the noise variability, we can relate the behavior shown in Fig. 37b to the results of Fig. 35b. At $f = 10$ Hz, the noise is dominated by RTN at $T = 300$ K, whereas as the temperature increases, the noise becomes more and more $1/f$-like, resulting in a flat $f \cdot S_{Vg}$ around $f = 10$ Hz for $T = 400$ K. This can explain the reduction of $\sigma(\log(S_{Vg}))$ with increasing temperature at a given frequency for this particular wafer.

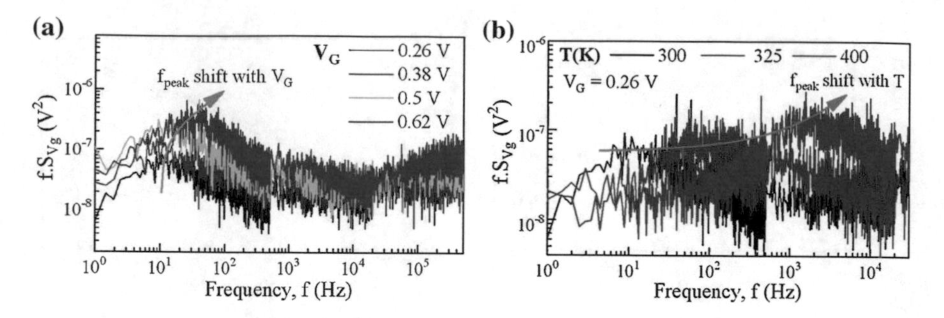

Fig. 37 Frequency-normalized input-referred gate voltage noise versus frequency (**a**) for four different gate voltage bias (one die example) at $T = 300$ K and (**b**) for three different temperatures T (one die example) under $V_G = 0.26$ V (after [68])

4 From Noise Modeling to Circuit Simulations

4.1 Noise Model Implementation

Using Verilog-A for Noise Modeling

Verilog-A [79] is a behavioral description language which is used by Cadence Spectre circuit simulator. It is a very powerful and useful tool, since it provides the potential for full description of a circuit netlist, as well as the behavior of a single device and its interface connections. In this section, two different ways to model a device behavior are presented. When the current–voltage characteristics are described by analytical and compact mathematical equations valid in all regions of operation, these equations can be written in the Verilog-A code and implemented as an analytical model. However, when the device behavior is known only through experimental measurements or TCAD simulations, one has to recall these data so that Spectre can predict all the in-between values through interpolation.

Provided that there is an analytical expression for the drain current as a function of both the gate and drain to source voltages, one can easily include a white or flicker noise source in the Verilog-A code. If the input-referred power spectral density S_{Vg} is constant, as in the case of the $1/f$ noise originating from carrier number fluctuations (CNF), then the device noise can be modeled by incorporating a voltage noise source at the transistor gate. This can be done by the following line of code, using the contributor symbol "<+" which adds a voltage signal on the declared node:

$$V(g) < +\text{flickernoise}\,(S_{\mathrm{vfb}}, 1, \text{“Svg”})\,;$$

where S_{Vfb} is the flat-band voltage power spectral density, the number 1 represents the value of the exponent γ and "Svg" is the name of the noise source. This voltage source will result to a drain current noise density equal to $S_{\mathrm{Vg}}{g_{\mathrm{m}}}^2$.

On the other hand, Verilog-A can also be used for time-dependent modules, since it gives access to the running time value of a transient simulation, as well as control over parameters such as permitted time step, etc. This is very useful for the simulation of RTN, but also LFN (through post-simulation FFT analysis) and BTI (time-dependent degradation) through defect-aware transient simulations, as will be shown in (c).

Implementing the Two-Interface CNF/CMF Model

If the values of the drain current or transconductance are needed for the noise expression, as in the case of correlated mobility fluctuations or thermal noise, the transconductance has to be calculated first. If the analytical drain current model provides a simple expression for the Id–Vg dependence, it is easy to obtain the derivative in the Verilog-A code as below:

$$\text{gm} = \text{ddx}\ (\text{Ids}, V(g))\,;$$

where gm is the variable name for the transconductance, ddx is the operator used for non-transient derivatives and Ids is the variable name for the drain current. Thus, the improved 1/*f* noise model that takes into account the correlated mobility fluctuations can be described as:

$$V(g) < +\text{flickernoise}\ (\text{Svfb} * \text{pow}\ ((1 + \text{alpha} * \text{ueff} * \text{Cox} * \text{Ids/gm})\,, 2)\,, 1, \text{“Svg”})\,;$$

where alpha, ueff, and Cox are the variable names of α, μ_{eff}, and C_{ox}, respectively. Using this method, a voltage noise source is being connected to the transistor gate, as illustrated in Fig. 38. As can be seen, both S_{V_g} source and the calculation of g_m regard the interior of the module.

The FDSOI noise model presented in paragraph Sect. 2.2.2 of this chapter takes into consideration the contribution of both front and back interfaces to the total noise

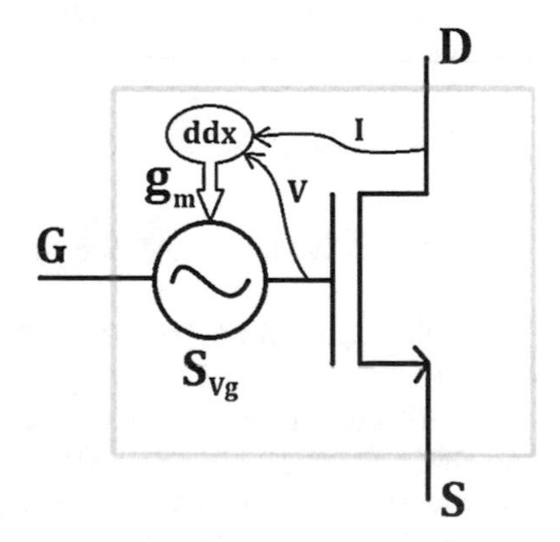

Fig. 38 Schematic representation of the Verilog-A CNF/CMF noise model implementation (after [80])

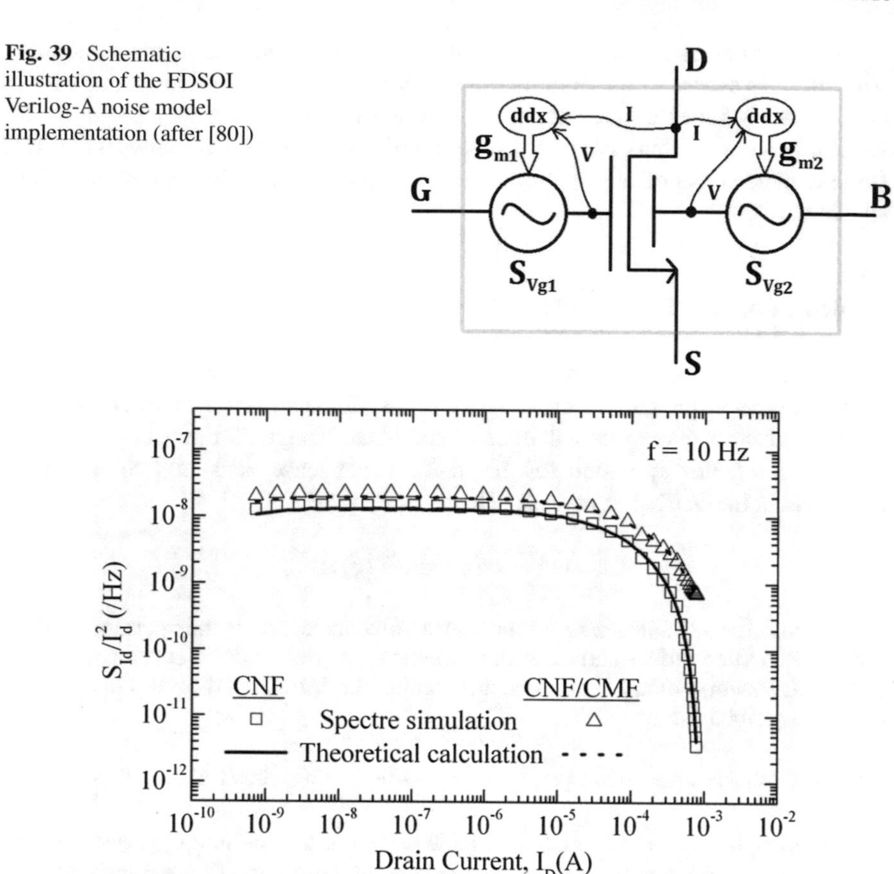

Fig. 39 Schematic illustration of the FDSOI Verilog-A noise model implementation (after [80])

Fig. 40 Normalized drain current noise S_{Id}/I_d^2 simulated (symbols) and theoretically calculated (lines) values versus drain current for CNF with $N_{t1} = 5 \times 10^{17}$ cm^{-3} and $N_{t2} = 10^{17}$ cm^{-3} and CNF/CMF with $\Omega_1 = 5$ V^{-1} and $\Omega_2 = 0.5$ V^{-1} (after [80])

level. Following the two-interface approach, the gate voltage noise sources can be declared as:

$$V(g) <+ \text{flickernoise}\,(\text{Svfb1} * \text{pow}\,((1 + \text{Omega1} * \text{Ids/gm1})\,, 2)\,, 1, \text{“Svg1”})\,;$$

$$V(b) <+ \text{flickernoise}\,(\text{Svfb2} * \text{pow}\,((1 + \text{Omega2} * \text{Ids/gm2})\,, 2)\,, 1, \text{“Svg2”})\,;$$

In this way, two noise sources are implemented in the module as illustrated in Fig. 39:

Figure 40 shows a comparison between Spectre simulation and theoretical calculation of the normalized drain current noise S_{Id}/I_d^2 values at 10 Hz versus drain current. The simulation data accurately follow the trend of the noise model, for both CNF and CNF/CMF cases.

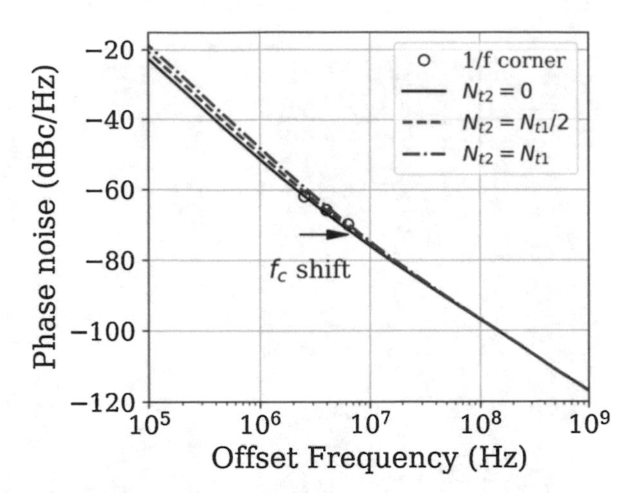

Fig. 41 Simulated three-stage oscillator ($W = 10$ μm, $L = 30$ nm) phase noise versus frequency for three different N_{t2}/N_{t1} ratios, using the model of Eq. (9). The flicker/thermal corner frequency points are noted with a circle

The above method is very efficient for circuit simulations in the frequency domain accounting for 1/*f* noise. A good example of such case is the phase noise, because in frequencies close to the oscillation frequency it is directly proportional to the LFN amplitude [81]. In order to demonstrate the importance of accurate LFN modeling in FDSOI circuits, we took the example of a three-stage ring oscillator circuit. Figure 41 shows three examples: one case where only the front interface noise is considered ($N_{t2} = 0$), one where $N_{t2} = N_{t1}/2$, and finally a case with equally defective front and back oxides ($N_{t2} = N_{t1}$). From the figure, it becomes clear that if the N_{t2} contribution is not taken into account, both the phase noise level and the 1/*f* corner frequency are underestimated by two to three times, which may lead to false design decisions.

Defect-Aware Time-Domain Module

Following the RTN modeling approach described in Sect. 2.1.4, one can build a self-contained defect-aware Verilog-A module for time-domain simulations [82, 83]. For the device static response, we can use the model presented in [84], because it is threshold voltage-based, allowing for a much more straightforward electrostatic trap impact declaration through an approximate relation:

$$\Delta \mathrm{V_t} = \frac{q\left(1 - \frac{x_\mathrm{t}}{t_\mathrm{ox}}\right)}{\mathrm{WL}C_\mathrm{ox}} \tag{44}$$

For simplification purposes, the CMF impact on the RTN amplitude was neglected, focusing on the feasibility of the module. The flowcharts for the initialization and the transient processes are shown in Fig. 42a and b, respectively, while the module parameters are shown in Table 2.

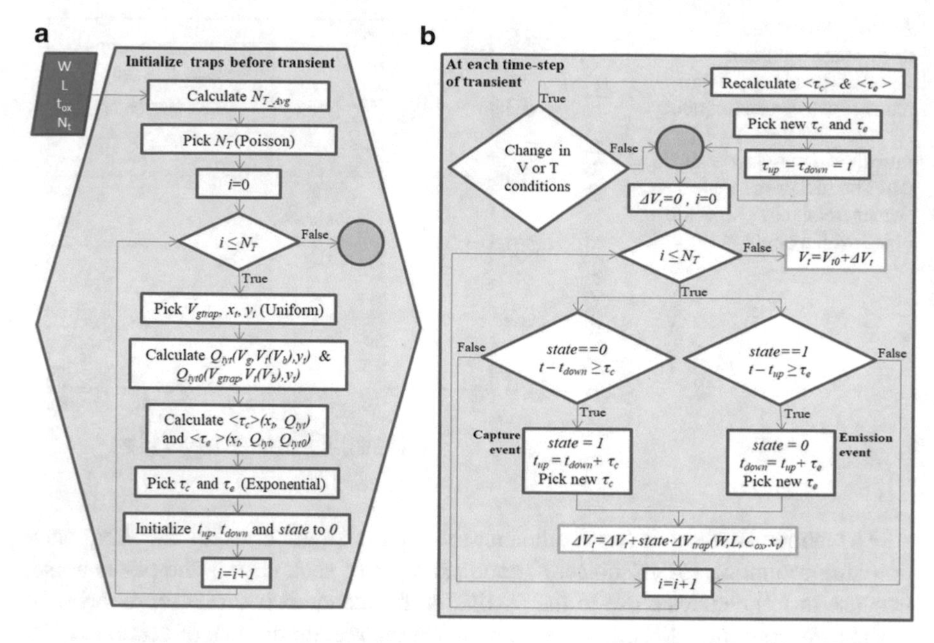

Fig. 42 Flowcharts of (**a**) trap initialization process and (**b**) trap activation process during transient simulation (after [82])

Table 2 Module parameters (after [82])

W	Gate width	V_{t0}	V_t for 0 active traps
L	Gate length	Q_{iyt}	Surface charge at trap position
t_{ox}	Gate oxide EOT	Q_{iyt0}	Value of Q_{iyt} when $E_T = E_F$
N_t	Trap density (/cm^3/eV)	τ_c	Capture time
N_{T_Avg}	Average # of traps	τ_e	Emission time
x_t	Trap depth in oxide	t_{up}	Time on capture event
y_t	Trap position across channel	t_{down}	Time on emission event
V_{gtrap}	V_{gt} for which $E_T = E_F$	State	0 when empty, 1 when occupied

As can be seen in Fig. 42a, right before the transient simulation starts, the number of active traps, is chosen from a Poisson distribution with an average value $N_T = WLt_{ox}N_t\Delta E_f$ based on the inputs (W, L, C_{ox}, and N_t). In [83], Kaczer et al. approached the $\overline{\tau}_c$ and $\overline{\tau}_e$ estimation picking values for $\overline{\tau}_{e,H}$ (high bias state of a digital circuit) uniformly distributed on the log scale between 10^{-9} and 10^9 and $\overline{\tau}_{e,L}$, $\overline{\tau}_{c,H}$, $\overline{\tau}_{c,L}$ calculated accordingly through different correlation levels. Regarding the intermediate voltages conditions, $\overline{\tau}_c$ and $\overline{\tau}_e$ were interpolated. In our work [82], a random vertical and lateral position is chosen for each trap assuming a uniform distribution in the oxide. Moreover, for the module to be functional under both stationary and nonstationary conditions, $\overline{\tau}_c$ and $\overline{\tau}_e$ are initialized and then recalculated at every change of voltage bias or temperature (see Fig. 42b), using Eq. (18) and the local inversion charge density Q_{iyt} calculated by the model [84].

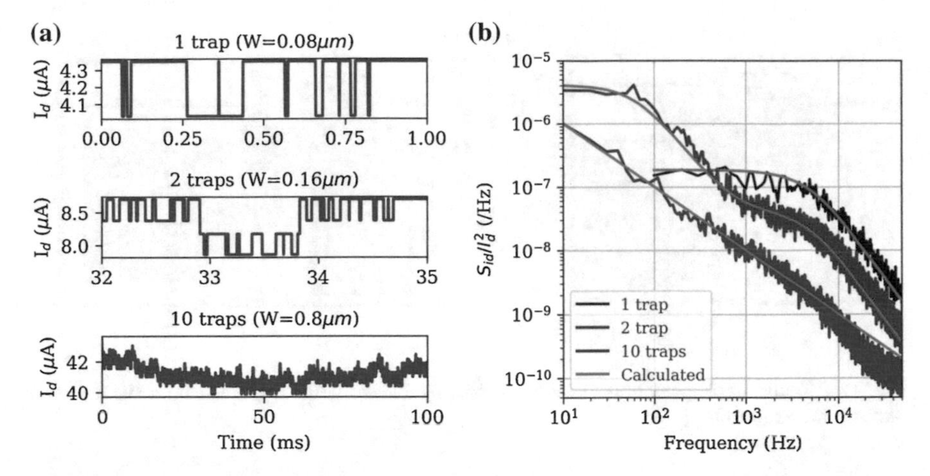

Fig. 43 (**a**) Simulated RTN signal examples for $N_{\mathrm{T}} = 1$, 2, and 10 and (**b**) corresponding normalized spectra through FFT (after [82])

This way, this module can be also used for simulating the recoverable part of the bias temperature instabilities (BTI) degradation (as shown in [82]).

A similar Verilog-A-based implementation approach is also presented in Chap. 4 [53] of this book.

In order to verify that our module can generate an RTN signal with a behavior analogous to the number of active traps N_{T}, we simulated three different cases (see Fig. 43). When only one trap is active, we obtain the typical two-level RTN pulse, whereas when $N_{\mathrm{T}} = 2$, it becomes four level and for $N_{\mathrm{T}} = 10$ the number of levels is 2^{10}, thus becoming indistinguishable. In the frequency domain, these cases translate into one Lorentzian-like LFN spectrum per trap or a $1/f$-like spectrum for $N_{\mathrm{T}} = 10$, exactly as expected and calculated using the traps input parameters.

In order to make use of our module completely generic and current model independent, we have to make it suitable to be used aside already existing PDK device instances. To do so, the simplest way would be to create a sub-circuit instance that contains the PDK transistor, together with a RTN voltage source in series with the transistor gate, as shown in Fig. 44a, in order that $V'_{\mathrm{g}} = V_{\mathrm{g}} + \Delta V_{\mathrm{g}}$, where $\Delta V_{\mathrm{g}} = -\Delta V_{\mathrm{t\_RTN}}$. However, since the industrial PDK device models usually do not provide access to the inversion charge values, we need to obtain the drain current, I_{d}, values during a transient so that we can calculate τ_{c} and τ_{e} for each trap. If we neglect, for simplicity reasons, the mobility degradation effects, one can express the inversion charge as a function of I_{d} as following [85]:

$$Q_i(t) = \frac{L}{W}\frac{I_{\mathrm{d}}(t)}{\mu_0 V_{\mathrm{d}}} \text{ (a)} \quad , \quad Q_{it} = \frac{L}{W}\frac{I_{\mathrm{dt}}}{\mu_0 V_{\mathrm{d}}} \text{ (b)} \tag{45}$$

There is a serious issue, however, with the above implementation approach: $I_{\mathrm{d}}(t)$ is the device current containing the defect activity through the RTN-induced ΔV_{t}

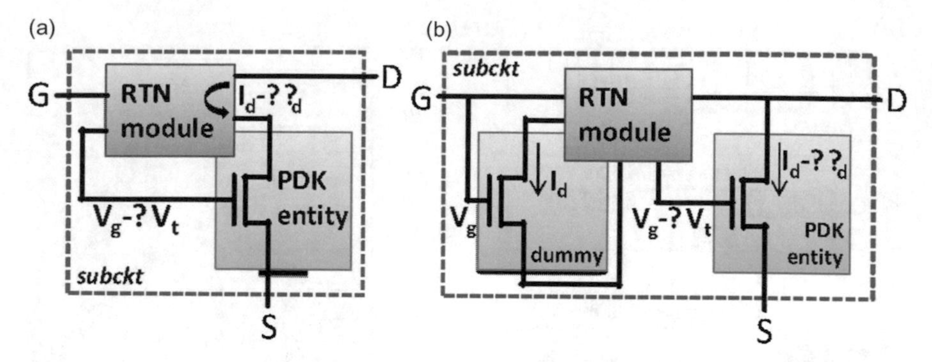

Fig. 44 Module implementation methods in existing PDKs: without (**a**) and with (**b**) noiseless dummy transistor (after [85])

that by turn causes a ΔI_d shift. Therefore, the capture time calculation during a transient simulation is affected by the trap occupancy itself, creating continuity errors. The solution we proposed in [85] is to utilize an ideal "dummy" transistor with no RTN inside the sub-circuit. Its role is to always provide the theoretical $I_d(t)$ values, when no traps are occupied, in all moments of the transient simulation. This method is illustrated schematically in Fig. 44b. Finally, regarding the energy levels, since we cannot express a charge–voltage relation as done in [82], they can be declared through a characteristic I_{dt}, for which E_t coincides with E_f and calculate Q_{it} through Eq. (45)(b), to use it in Eq. (18)(b) for the calculation of τ_e.

4.2 *Impact of LFN/RTN on Circuit Operation*

Phase Noise in FD-SOI Ring Oscillator Circuits

The frequency domain implementation of noise models, such as the one described in Sect. 4.1.2 of this chapter, can be very useful when it comes to performing circuit simulations in the frequency domain accounting for $1/f$ noise. A good example for the latter is the phase noise, because the $1/f$ noise is up-converted to phase noise around the oscillation frequency [81]. In order to demonstrate the importance of accounting for both interfaces contributions in FD-SOI circuits, we took the example of a three-stage ring oscillator circuit (schematic shown in Fig. 45). Here we present three cases: one where only the front interface noise is taken into account ($N_{t2} = 0$), one where $N_{t2} = N_{t1}/2$, and finally an example where the front and back oxide are equally defective ($N_{t2} = N_{t1}$). One can clearly observe in Fig. 45 that if the bottom interface contribution (N_{t2}) is not taken into account, both the phase noise level and the $1/f$ corner frequency are underestimated by two to three times [85]. This might cause wrong noise yield estimations or/and lead to false design decisions.

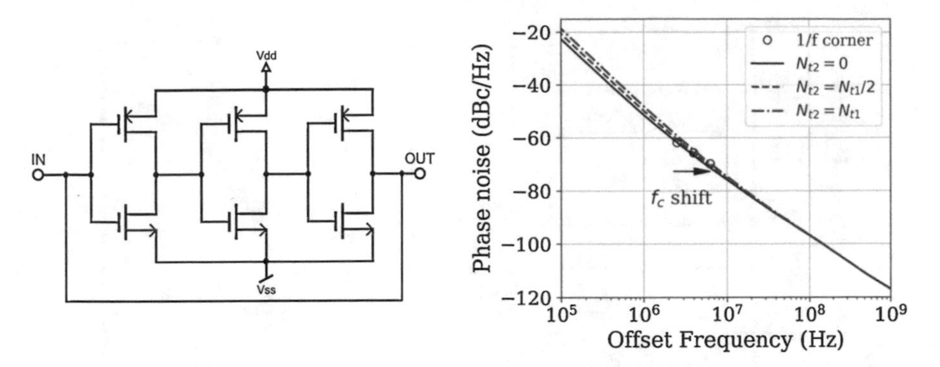

Fig. 45 Simulated three-stage oscillator ($W = 10$ μm, $L = 30$ nm) scematic (left) and phase noise versus frequency for three different N_{t2}/N_{t1} ratios (right) (after [85]), using the model of Eq. (9). The flicker/thermal corner frequency points are noted with a circle

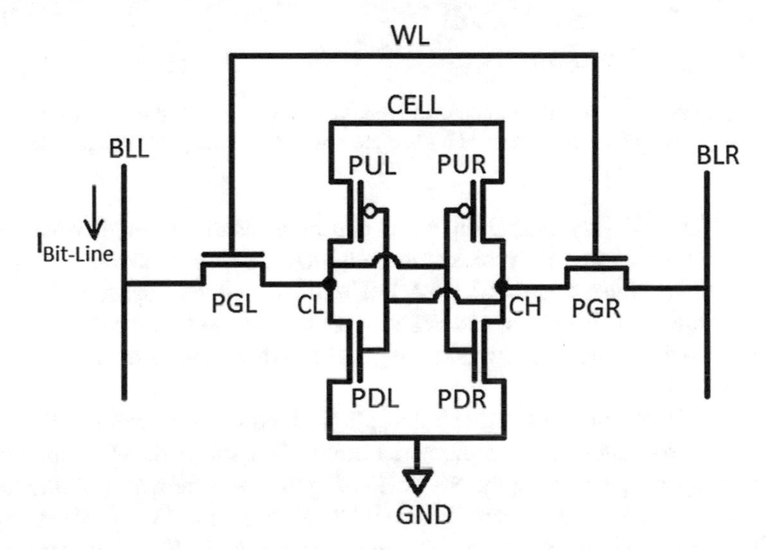

Fig. 46 Typical SRAM cell schematic: R stands for Right, L for left, PU for pull-up, PD for pull-down, PG for pass-gate, BL for bit-line, and WL for word-line

The SRAM Cell as a Circuit Reference

In order to examine the impact of LFN or RTN (or both) on a circuit operation, or verify any model approaches, we need a circuit that has a wide application, a well-known standard operation, and a noise-sensitive dynamic behavior. The typical 6T SRAM cell (Fig. 46) has all these characteristics and is very easy to design and simulate. The letters L and R refer to the left and right symmetrical sides, while BL stands for Bit-Line, PG for Pass-Gate, PD for Pull-Down, and PU for Pull-Up. The two bit versions (0 and 1) are written using the Write-Line (WL) and stored at the nodes CH and CL.

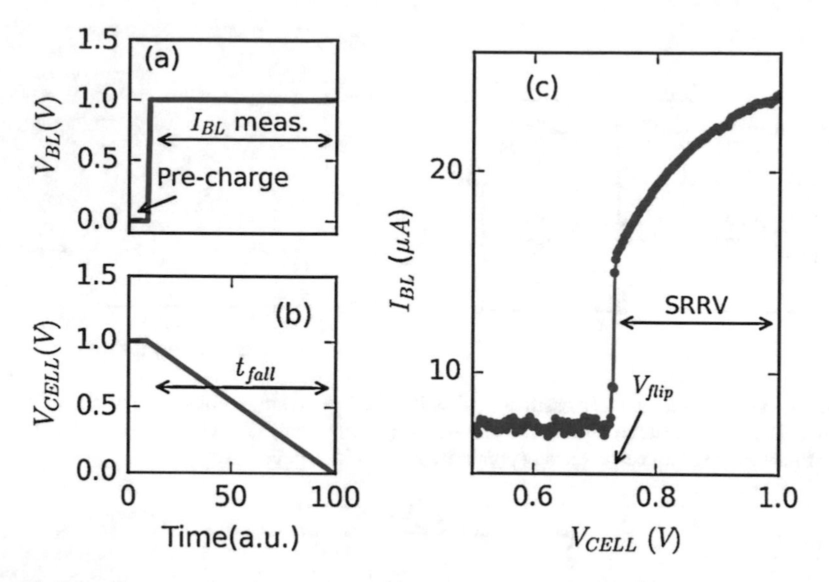

Fig. 47 Setup for measuring the supply read retention voltage: (**a**) Bit-line voltage and (**b**) cell supply voltage versus time. (**c**) Measured bit-line current versus cell supply voltage (after [87])

The SRAM cell's dynamic limitations can be measured through what we call "stability metrics". Some of the most common SRAM stability metrics are: the Read and Write Static Noise Margin (RSNM and WSNM) and the Supply and Wordline Read Retention Voltage (SRRV and WRRV) [86]. Here, we will focus on the SRRV and RSNM metrics that can provide a representative overview of the SRAM cell dynamic stability.

The SRRV is defined as the maximum allowed reduction of cell supply voltage for which the read stability is not affected. Figure 47a, b shows the time-domain bias setup for properly monitoring the SRRV. During the measurement (or simulation), the word-line (WL) and the right bit-line (BLR) are biased at $V_{\mathrm{WL}} = V_{\mathrm{BL(R)}} = 1$ V. An initial time is needed for the pre-charge of 0, for example, at the left bit node (CL), and then $\mathrm{V_{BL}}$ is constant at 1 V while V_{CELL} is ramped down in order to read the bit-line current $\mathrm{I_{BL}}$. As shown in Fig. 47c, at a certain point, the stored bit is flipped and $\mathrm{I_{BL}}$ drops suddenly, becoming significantly lower and equal to the Pass-Gate (PG) current. The V_{CELL} value at that point is called V_{flip} and SRRV is extracted as the difference between the power supply V_{DD} and V_{flip}.

The RSNM is an equally important stability metric for SRAM cells [86]. It represents the maximum tolerable DC noise voltage at each storage node before causing a read upset. Regarding its measurement, it requires a more difficult procedure, compared to the simple SRRV measurement setup, because it needs access to the internal nodes and it has to be performed in two steps. The RSNM simulation, nevertheless, is far more easy to perform than the SRRV and it can also be used to study the SRAM dynamic variability behavior.

Many RTN simulation studies have used these metrics to explore how the RTN amplitude and trap kinetics can affect the SRAM cell stability or even cause read/write errors, arguing this way on the importance of understanding and properly modeling the RTN behavior [49, 88–91]. Here, we will present some novel approaches, regarding the aspects of the periodic transient noise and the defect-aware time-domain circuit simulations.

"Periodic Transient Noise" Approach

Cadence Spectre incorporates a transient simulation option called "Transient Noise" (TN). When activated, this method generates time-domain current fluctuations that correspond to the noise power spectral density that has been declared in the model. Details of this simulation process can be found elsewhere [92, 93]. As a result, when a transient I_d–V_g simulation with a ramp gate voltage as an input is run repeatedly, it produces a set of curves as in Fig. 48a (after [94]), where we show an example of 100 multiple runs. The disadvantage of this simulation approach is that the noise bandwidth $\Delta f = f_{max} - f_{min}$ taken into account by the TN is limited by the rise time, i.e., the duration of each run, because $f_{min} = 1/t_r$, and by the timestep Δt through $f_{max} = 1/2.\Delta t$. This leads to a constant threshold voltage variation σ_{Vth} for all rise time values (see Fig. 48b), because the low-frequency part of the spectrum is not considered.

To resolve this issue, one can use the Periodic Transient Noise (PTN) approach [94], which means running a periodic transient simulation instead of multiple runs, with the TN module activated. This can enlarge the simulation duration and thus increase the noise bandwidth to lower frequencies as much as necessary. This way, the noise-induced threshold voltage variations are both speed- and duration-sensitive, as expressed in Sect. 3.3.3 through Eq. (43). Also, using the PTN method, the simulation stop time can approach the real duration of operation, obtaining more realistic results.

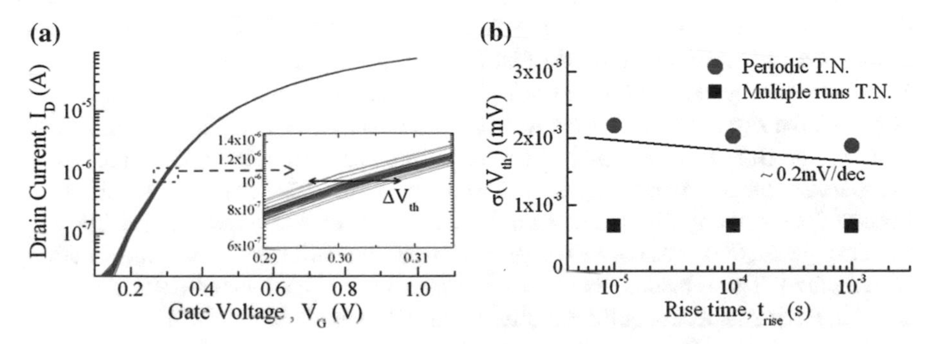

Fig. 48 (**a**) Transient noise simulation (100 multiple runs) example. (**b**) Dynamic standard deviation of V_{th} versus rise time for periodic and multiple runs transient noise simulation ($W = 80$ nm, $L = 30$ nm, $S_{Vg} = A/f$ with $A = 1.8 \times 10^{-8}$ V^2/Hz) (from [94])

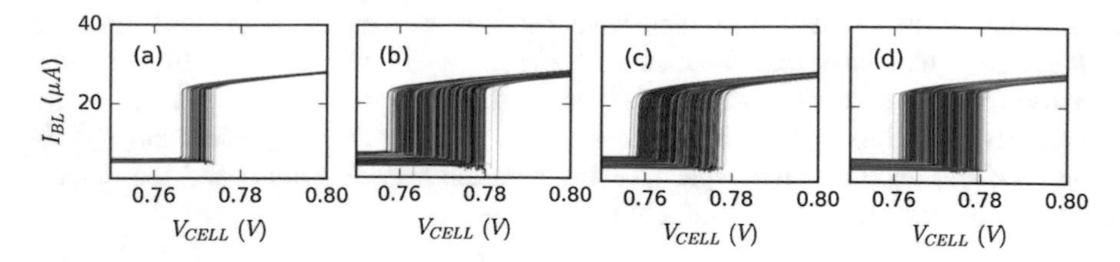

Fig. 49 Periodic transient noise simulation results of the bit-line current versus cell supply voltage for four different configurations: (**a**) Reference simulation, (**b**) S_{Vfb}·100, (**c**) t_{fall}/1000, and (**d**) t_{stop} ·100 (after [87])

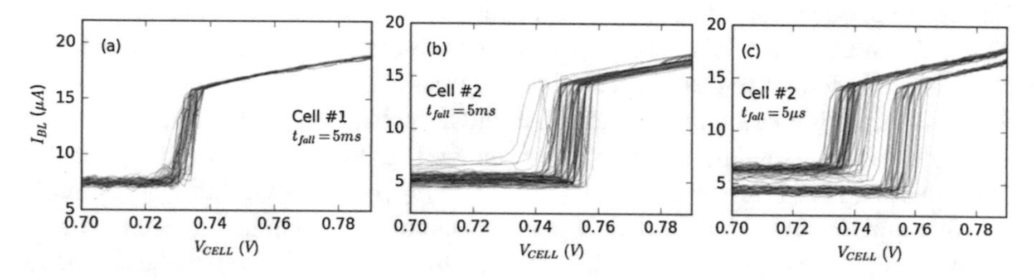

Fig. 50 Measured SRAM cell bit-line current versus cell supply voltage for three different cases: (**a**) Cell #1 with t_{fall} = 5 ms, (**b**) Cell #2 with t_{fall} = 5 ms, and (**c**) Cell #2 with t_{fall} = 5 μs (after [87])

Using the PTN method, we can easily simulate the noise-induced degradation of the SRRV metric [87] through a series of I_{BL}–V_{CELL} simulations (setup in Fig. 47), with the transistors LFN level being controlled by the flat-band voltage fluctuations power spectral density S_{Vfb} (Eq. (5)) declared in the flicker noise model (Sect. 4.1.2). A reference simulation with $S_{Vfb} = 10^{-8}$ V^2/Hz, stop time t_{stop} = 10 ms, and cell supply voltage fall time t_{fall} = 0.1 ms is shown in Fig. 49a (after [87]), where a small but not significant noise-induced variability of SRRV can be observed.

In the simulations of Fig. 49b–d, one of these three parameters was changed while keeping the rest constant, to examine the impact of each one separately. In Fig. 49b, the noise level was increased to $S_{Vfb} = 10^{-6}$ V^2/Hz, in accordance with the worst-case LFN level of nanoscale FDSOI MOSFETs [72]. In Fig. 49c, the V_{CELL} fall time was reduced down to t_{fall} = 1 μs to increase the speed of operation, and in Fig. 49d, the simulation stop time was increased to t_{stop} = 1 s to include the noise contribution from the lower part of the frequency bandwidth. In all the three cases, compared to the reference, a significant increase of the SRRV dynamic variability is observed, concluding that the noise-induced FDSOI SRAM dynamic variability is increasing with the transistor LFN level, the circuit operation speed, and the duration of operation. These results also verify the measured behavior of SRRV shown in Fig. 50, further underlining the validity of the PTN simulation method.

Regarding the PTN simulation of the LFN impact on the RSNM, some representative results are shown in Fig. 51 (after [87]). In order to capture both local and time-domain RSNM variations, Fig. 51a shows local variability results through

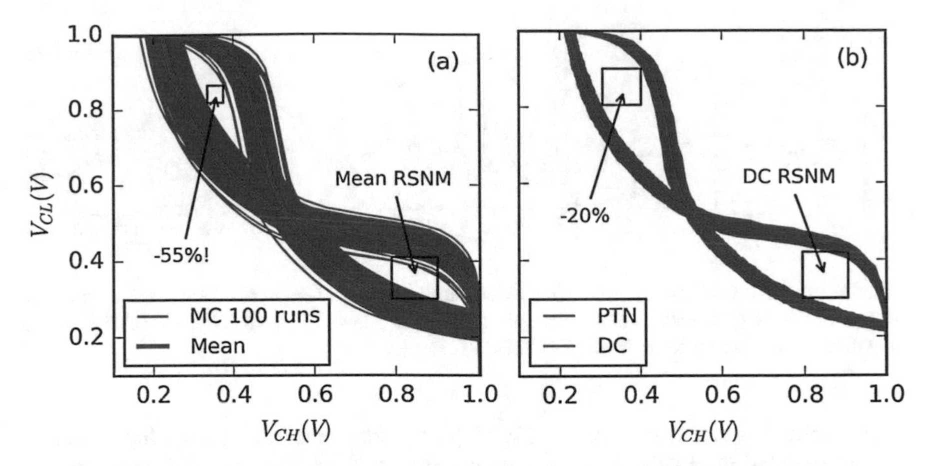

Fig. 51 (**a**) Impact of V_t mismatch on RSNM (MC: Monte Carlo) and (**b**) impact of LFN on RSNM (PTN: Periodic Transient Noise) (after [87])

mismatch-related Monte Carlo simulations. One can notice how the RSNM can be reduced by more than 50% from cell to cell, due to the threshold voltage static mismatch. The noise-induced dynamic variability of a single FDSOI SRAM cell, on the other hand, as shown in Fig. 51b, can lead to a RSNM reduction of up to 20%. For the latter simulation, the device noise level was boosted by two orders of magnitude ($S_{Vfb} = 10^{-6}$ V^2/Hz), in order to consider the worst-case scenario, which is to have either huge LFN variability or the presence of high-amplitude RTN. This alarming finding means that in FDSOI circuits consisting of nanoscale transistors, the LFN/RTN-induced variations can be comparable to the mismatch variations. Considering that these two sources of variations (topological and temporal) will in reality be added together and result to a variability level significantly higher than what is often taken into account in simulations (only process variations), we conclude that the accurate simulation of LFN and RTN becomes more and more important in circuit design.

Defect-Aware Time-Domain Simulations

Using the defect-aware module described in Sect. 4.1.3, the impact of RTN-TDV on the read static to noise margin (RSNM) of a SRAM cell was simulated. Figure 52a presents the 6T-cell that was used in our simulations, as well as the number of traps that was chosen by the simulator, considering a uniform trap distribution. The transistor width and length were fixed at $W = 80$ nm and $L = 30$ nm, respectively. In order to simulate the phenomenon, a periodic ramp voltage was applied first on the left node (L) and then on the right one (R), while monitoring the voltage at the opposite node.

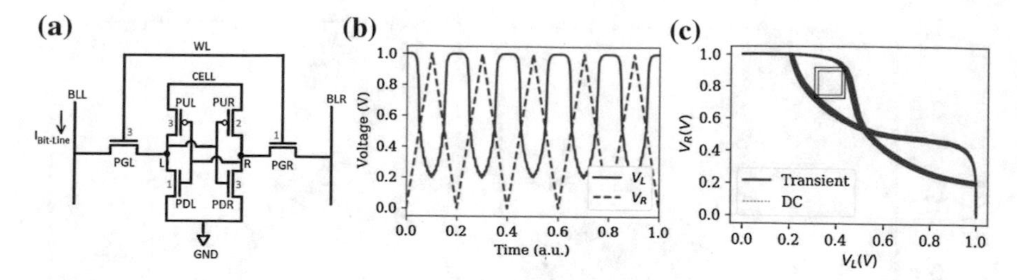

Fig. 52 (**a**) SRAM cell circuit used in the simulations. The numbers in red show the total number of traps per device, (**b**) voltage bias setup during transient simulation with RTN, and (**c**) right-node vs. left-node voltage for the extraction of RSNM (after [82])

A zoomed-in result is shown in Fig. 52b, where one can also note the presence of RTN-related fluctuations in the output voltage. A number of 1000 ramp cycles were chosen, with $t_{\text{rise}} = t_{\text{fall}} = 100\ \mu\text{s}$, in order to have a good statistical sample concerning the trap activity.

The final results are shown in Fig. 52c, where V_{R} is plotted versus V_{L} for both DC and transient simulations. The impact of the trap activity on the reduction of RSNM is visible: RSNM ≈ 0.13 V in the DC simulation, while it drops at around 0.1 V in the case of the defect-aware transient, revealing a more than 20% reduction of RSNM, only by trapping/de-trapping events in a very low number of traps, in line with the transient noise results presented in (c), where the same reduction was found by using a high level $1/f$ noise spectrum with the transient noise module in Spectre.

5 Conclusion

In this chapter, we presented a review of the most important aspects of our research work regarding the low-frequency noise behavior of Fully Depleted Silicon-on-Insulator MOSFETs.

First, we found it necessary to give a brief explanation of important terms and concepts regarding the fluctuations mechanisms in MOSFETs, providing some fresh points of view regarding RTN and the Hooge mobility fluctuations. Next, we presented the methodology we followed in order to derive the FDSOI LFN models, which were used on the one hand to extract the noise-related parameters for various FDSOI wafers and on the other hand to perform accurate simulations of noise-induced limitations of a circuit's performance. It was shown that neglecting the impact of the bottom interface on the total noise level or the latter's dependence on back-bias can lead to serious errors in the extraction of trap densities, etc.

Regarding the noise characterization of FDSOI MOSFETs, we analyzed the impact of front–back gate coupling on both carrier number and correlated mobility fluctuations, the impact of channel geometry (thickness, width, and length) on the strength of the remote Coulomb scattering, and the presence of generation–recombination noise centers in the semiconductor region of such ultrathin channels.

We demonstrated that reducing the channel cross section (width × thickness) or applying negative back-bias to increase the threshold voltage can lead to enhanced CMF levels.

Afterwards, we presented a thorough study of the LFN variability phenomenon and its properties. A comparison was shown between many different technology nodes, revealing a superiority of the FDSOI technology with respect to the LFN device-to-device variations. Moreover, some new methods of statistical LFN/RTN characterization were demonstrated, as well as the LFN variability dependences on gate bias, frequency, and temperature.

Finally, our methodology regarding the noise model implementation for circuit simulations was presented in detail. The "Periodic Transient Noise" simulation method was also explained thoroughly, demonstrating its ability to produce realistic LFN-aware time-domain simulations. Also, after showing in which way both frequency and time-domain noise model approaches can be implemented through Verilog-A, we showed how the LFN and RTN of a nanoscale device can affect the stability and thus the performance limitations of a SRAM cell.

We hope that the content of this chapter will contribute to the clarification of any misconceptions regarding the fluctuation mechanisms in FDSOI MOSFETs and help towards a better understanding of the LFN and RTN phenomena in general, as well as the adoption of the noise characterization, modeling, and simulation methods presented.

References

1. G.K. Celler, S. Cristoloveanu, Frontiers of silicon-on-insulator. J. Appl. Phys. (2003)
2. D.J. Wouters, J.P. Colinge, H.E. Maes, Subthreshold slope in thin-film SOI MOSFET's. IEEE Trans. Electron Devices (1990)
3. Q. Liu et al., Ultra-thin-body and BOX (UTBB) Fully Depleted (FD) device integration for 22nm node and beyond, in *Digest of Technical Papers—Symposium on VLSI Technology*, 2010
4. C. Fenouillet-Beranger et al., Fully-depleted SOI technology using high-K and single-metal gate for 32nm node LSTP applications featuring 0.179μm2 6T-SRAM bitcell, in *Tech. Dig.—Int. Electron Devices Meet, IEDM*, 2007, pp. 267–270
5. F. Andrieu et al., Low leakage and low variability Ultra-Thin Body and Buried Oxide (UT2B) SOI technology for 20nm low power CMOS and beyond, in *Digest of Technical Papers—Symposium on VLSI Technology*, 2010
6. C. Fenouillet-Beranger et al., FDSOI devices with thin BOX and ground plane integration for 32 nm node and below. Solid State Electron. (2009)
7. B. Doris et al., Extreme scaling with ultra-thin Si channel MOSFETs, in *IEDM*, 2002
8. O. Weber et al., High immunity to threshold voltage variability in undoped ultra-thin FDSOI MOSFETs and its physical understanding, in *Technical Digest—International Electron Devices Meeting, IEDM*, 2008
9. N. Sugii et al., Comprehensive study on vthvariability in silicon on thin BOX (SOTB) CMOS with small random-dopant fluctuation: Finding a way to further reduce variation, in *Technical Digest—International Electron Devices Meeting, IEDM*, 2008
10. E. Simoen, A. Mercha, C. Claeys, N. Lukyanchikova, N. Garbar, Critical discussion of the front-back gate coupling effect on the low-frequency noise in fully depleted SOI MOSFETs. IEEE Trans. Electron Devices **51**(6), 1008–1016 (2004)

11. L. Zafari, J. Jomaah, G. Ghibaudo, Low frequency noise in multi-gate SOI CMOS devices. Solid State Electron. **51**(2), 292–298 (2007)
12. W. Cheng, C. Tye, P. Gaubert, A. Teramoto, S. Sugawa, T. Ohmi, Suppression of 1/f noise in accumulation mode FD-SOI MOSFETs on Si(lOO) and (110) surfaces, in *AIP Conf. Proc.*, vol. 1129, no. May, 2009, pp. 337–340
13. J. El Husseini et al., New numerical low frequency noise model for front and buried oxide trap density characterization in FDSOI MOSFETs. Microelectron. Eng. **88**(7), 1286–1290 (2011)
14. L. Zafari, J. Jomaah, G. Ghibaudo, Modelling and simulation of coupling effect on low frequency noise in advanced SOI MOSFETs. Fluct. Noise Lett. (2008)
15. N. Lukyanchikova, N. Garbar, A. Smolanka, E. Sirnoen C. Claey, *Influence of an Accumulation Back-Gate Voltage on the Low-Frequency Noise Spectra of Fabricated on ELTRAN and UNIBOND Wafers*, pp. 357–360
16. N. Lukyanchikova, M. Petrichuk, N. Garbar, E. Simoen, C. Claeys, Back and front interface related generation-recombination noise in buried-channel SOI pMOSFET's. IEEE Trans. Electron Devices (1996)
17. C.G. Theodorou et al., Impact of front-back gate coupling on low frequency noise in 28 nm FDSOI MOSFETs. Eur. Solid-State Device Res. Conf. **2**(2), 334–337 (2012)
18. G. Ghibaudo, On the theory of carrier number fluctuations in MOS devices. Solid State Electron. **32**(7), 563–565 (1989)
19. K.K. Hung, P.-K. Ko, C. Hu, Y.C. Cheng, A unified model for the flicker noise in metal-oxide-semiconductor field-effect transistors. IEEE Trans. Electron Devices **37**(3), 654–665 (1990)
20. G. Ghibaudo, O. Roux, C. Nguyen-Duc, F. Balestra, J. Brini, Improved analysis of low frequency noise in field-effect MOS transistors. Phys. Status Solidi **124**(2), 571–581 (1991)
21. A. Emrani, F. Balestra, G. Ghibaudo, On the understanding of electron and hole mobility models from room to liquid helium temperatures. Solid State Electron. (1994)
22. K. Bennamane, G. Ghibaudo, C. Fenouillet-Beranger, F. Balestra, I. Ben Akkez, A. Cros, Mobility coupling effects due to remote coulomb scattering in thin-film FD-SOI CMOS devices. Electron. Lett. **49**(7), 490–492 (2013)
23. A.H. McWhorter, *McWorther Proc Physics of Semiconductor Surface*, 1957, p. 207
24. S. Christensson, I. Lundström, C. Svensson, Low frequency noise in MOS transistors-I theory. Solid State Electron. (1968)
25. T. Nagumo, K. Takeuchi, T. Hase Y. Hayashi, Statistical characterization of trap position, energy, amplitude and time constants by RTN measurement of multiple individual traps, in *Tech. Dig.—Int. Electron Devices Meet. IEDM*, 2010, pp. 628–631
26. J. Franco et al., RTN and PBTI-induced time-dependent variability of replacement metal-gate high-k InGaAs FinFETs, in *2014 IEEE International Electron Devices Meeting*, 2014, pp. 20.2.1–20.2.4
27. T. Grasser et al., On the microscopic structure of hole traps in pMOSFETs, in *Technical Digest—International Electron Devices Meeting, IEDM*, 2015
28. P. Dutta, P.M. Horn, Low-frequency fluctuations in solids: 1/f noise. Rev. Mod. Phys. (1981)
29. A.J. Scholten, L.F. Tiemeijer, R. Van Langevelde, R.J. Havens, A.T. Zegers-Van Duijnhoven, V.C. Venezia, Noise modeling for RF CMOS circuit simulation. IEEE Trans. Electron Devices (2003)
30. C. Surya, T.Y. Hsiang, A thermal activation model for 1/f^γ noise in Si-MOSFETs. Solid State Electron. (1988)
31. T. Boutchacha, G. Ghibaudo, Improved modeling of low-frequency noise in MOSFETs - focus on surface roughness effect and saturation region. IEEE Trans. Electron Devices **58**(9), 3156–3161 (2011)
32. F.N. Hooge, *l/f noise, Phys. B+C*, 1976
33. R.P. Jindal, A. van der Ziel, Model for mobility fluctuation 1/f noise.Pdf. Appl. Phys. Lett. (1981)
34. K.M. van Vliet, Noise in semiconductors and photoconductors. Proc. IRE **46**(6), 1004–1018 (1958)

35. I.M. Hafez, G. Ghibaudo, F. Balestra, A study of flicker noise in MOS transistors operated at room and liquid helium temperatures. Solid State Electron. (1990)
36. F.N. Hooge, 1/F noise sources. IEEE Trans. Electron Devices **41**(11), 1926–1935 (1994)
37. K. Takagi, T. Mizunami, T. Shiraishi M. Wada, Excess noise generation by carrier fluctuation in semiconductor devices, in *IEEE International Symposium on Electromagnetic Compatibility*, 1994
38. T.A. Karatsori, C.G. Theodorou, C.A. Dimitriadis, G. Ghibaudo, Influence of AC signal oscillator level on effective mobility measurement by split C-V technique in MOSFETs. Electron. Lett. **52**(17) (2016)
39. K.S. Ralls et al., Discrete resistance switching in submicrometer silicon inversion layers: Individual interface traps and low-frequency (1/f) noise. Phys. Rev. Lett. **52**(3), 228–231 (1984)
40. M.J. Uren, D.J. Day, M.J. Kirton, 1/F and random telegraph noise in silicon metal-oxide-semiconductor field-effect transistors. Appl. Phys. Lett. **47**(11), 1195–1197 (1985)
41. M.J. Kirton, M.J. Uren, S. Collins, M. Schulz, Individual defects at the Si:SiO_2 interface. Semicond. Sci. Technol. **4**, 1116–1126 (1989)
42. O.R. dit Buisson, G. Ghibaudo, J. Brini, Model for drain current {RTS} amplitude in small-area {MOS} transistors. Solid State Electron. **35**(9), 1273–1276 (1992)
43. E. Simoen, B. Dierickx, B. De Canne, F. Thoma, C. Claeys, On the gate- and drain-voltage dependence of the RTS amplitude in submicron MOSTs. Appl. Phys. A Solids Surfaces (1994)
44. E.G. Ioannidis et al., Analytical low-frequency noise model in the linear region of lightly doped nanoscale double-gate metal-oxide-semiconductor field-effect transistors. J. Appl. Phys. **108**(6), 064512 (2010)
45. T.A.T.A. Karatsori et al., Study of hot-carrier-induced traps in Nanoscale UTBB FD-SOI MOSFETs by low-frequency noise measurements. IEEE Trans. Electron Devices **63**(8), 1–7 (2016)
46. J.P. Campbell et al., Large random telegraph noise in sub-threshold operation of nano-scale nMOSFETs, in *2009 IEEE International Conference on Integrated Circuit Design and Technology, ICICDT 2009*, 2009
47. L.K.J. Vandamme, D. Sodini, Z. Gingl, On the anomalous behavior of the relative amplitude of RTS noise. Solid State Electron. **42**(6), 901–905 (1998)
48. A. Subirats, X. Garros, J. El Husseini, C. Le Royer, G. Reimbold, G. Ghibaudo, Impact of single charge trapping on the variability of ultrascaled planar and trigate FDSOI MOSFETs: Experiment versus simulation. IEEE Trans. Electron Devices **60**(8), 2604–2610 (2013)
49. K. Takeuchi, T. Nagumo, S. Yokogawa, K. Imai Y. Hayashi, Single-charge-based modeling of transistor characteristics fluctuations based on statistical measurement of RTN amplitude, in *2009 Symp. VLSI Technol*, 2009, pp. 54–55
50. G. Ghibaudo, T. Boutchacha, Electrical noise and RTS fluctuations in advanced CMOS devices. Microelectron. Reliab. **42**(4–5), 573–582 (2002)
51. T. Grasser, Stochastic charge trapping in oxides: From random telegraph noise to bias temperature instabilities. Microelectron. Reliab. **52**(1), 39–70 (2012)
52. S. Machlup, Noise in semiconductors: Spectrum of a two-parameter random signal. J. Appl. Phys. (1954)
53. C. Marquez, O. Huerta, A.I. Tec-Chim, F. Guarin, E.A. Gutierrez-D F. Gamiz, in *Systematic Characterization of Random Telegraph Noise and its Dependence with Magnetic Fields in MOSFET Devices*, ed. by T. Grasser (2019)
54. B. Stampfer, A. Grill M. Waltl, in *Advanced Electrical Characterization of Single Oxide Defects utilizing Noise Signals*, ed. by T. Grasser (2019)
55. J. Martin-Martinez, R. Rodriguez M. Nafria, in *Advanced Characterization and Analysis of Random Telegraph Noise in CMOS Devices*, ed. by T. Grasser (2019)
56. A.S.M.S. Rouf, Z. Çelik-Butler, in *Oxide Trap-Induced RTS in MOSFETs*, ed. by T. Grasser (2019)
57. C.G. Theodorou et al., Low-frequency noise sources in advanced UTBB FD-SOI MOSFETs. IEEE Trans. Electron Devices **61**(4), 1161–1167 (2014)

58. C.G. Theodorou et al., Low-frequency noise behavior of n-channel UTBB FD-SOI MOSFETs, in *2013 22nd International Conference on Noise and Fluctuations, ICNF 2013*, 2013
59. A. Nazarov et al., *Semiconductor-on-Insulator Materials for Nanoelectronics Applications*, 2011
60. E.G. Ioannidis, C.G. Theodorou, T.A. Karatsori, S. Haendler, C.A. Dimitriadis, G. Ghibaudo, Drain-current flicker noise modeling in nMOSFETs from a 14-nm FDSOI technology. IEEE Trans. Electron Devices **62**(5), 1574–1579 (2015)
61. L. Pirro et al., RTN and LFN Noise Performance in Advanced FDSOI Technology, in *2018 48th European Solid-State Device Research Conference (ESSDERC)*, 2018, pp. 254–257
62. T. Skotnicki, Competitive SOC with UTBB SOI, in *SOI Conference (SOI), 2011 IEEE International*, 2011, pp. 1–61
63. M. Shin et al., Magnetoresistance mobility characterization in advanced FD-SOI n-MOSFETs, in *Solid State Electron*, (2015)
64. N.B. Lukyanchikova, M.V. Petrichuk, N.P. Garbar, E. Simoen, C. Claeys, Non-trivial GR and 1/f noise generated in the p-Si layer of SOI and SOS MOSFETs near the inverted front or buried p-Si/SiO2 interface. Semicond. Sci. Technol. **14**(9), 775–783 (1999)
65. A. Luque Rodríguez et al., Dependence of generation-recombination noise with gate voltage in FD SOI MOSFETs. IEEE Trans. Electron Devices **59**(10), 2780–2786 (2012)
66. D.J. Frank, Y. Taur, Design considerations for CMOS near the limits of scaling. Solid State Electron. (2002)
67. E. Simoen et al., Towards single-trap spectroscopy: Generation-recombination noise in UTBOX SOI nMOSFETs. Phys. Status Solid Curr. Top. Solid State Phys. **12**(3), 292–298 (2015)
68. C.G. Theodorou, E.G. Ioannidis, S. Haendler, E. Josse, C.A. Dimitriadis, G. Ghibaudo, Low frequency noise variability in ultra scaled FD-SOI n-MOSFETs: Dependence on gate bias, frequency and temperature. Solid State Electron. **117**, 88–93 (2016)
69. S.D. Dos Santos et al., Low-frequency noise assessment in advanced UTBOX SOI nMOSFETs with different gate dielectrics. Solid State Electron. **97**, 14–22 (2014)
70. G. Ghibaudo, O. Roux-dit-Buisson, J. Brini, Impact of scaling down on low frequency noise in silicon MOS transistors. Phys. Status Solidi **132**(2), 501–507 (1992)
71. E.G. Ioannidis et al., Low frequency noise variability in high-k/metal gate stack 28nm bulk and FD-SOI CMOS transistors, in *Technical Digest—International Electron Devices Meeting, IEDM*, 2011, pp. 449–452
72. E.G. Ioannidis et al., Low frequency noise statistical characterization of 14nm FDSOI technology node, in *EUROSOI-ULIS 2015–2015 Joint International EUROSOI Workshop and International Conference on Ultimate Integration on Silicon*, 2015
73. E.G. Ioannidis, C.G. Theodorou, S. Haendler, C.A. Dimitriadis, G. Ghibaudo, Impact of low-frequency noise variability on statistical parameter extraction in ultra-scaled CMOS devices. Electron. Lett. **50**(19) (2014)
74. T.H. Morshed, M.V. Dunga, J. Zhang, D.D. Lu, A.M. Niknejad C. Hu, Compact modeling of flicker noise variability in small size MOSFETs, in *Technical Digest—International Electron Devices Meeting, IEDM*, 2009
75. M.J.M. Pelgrom, A.C.J. Duinmaijer, A.P.G. Welbers, Matching properties of MOS transistors. IEEE J. Solid State Circuits **24**(5), 1433–1439 (1989)
76. O. Weber et al., 14nm FDSOI technology for high speed and energy efficient applications, in *Digest of Technical Papers—Symposium on VLSI Technology*, 2014
77. E.G. Ioannidis, S. Haendler, C.G. Theodorou, S. Lasserre, C.A. Dimitriadis, G. Ghibaudo, Evolution of low frequency noise and noise variability through CMOS bulk technology nodes from 0.5 nm down to 20 nm. Solid State Electron. **95**, 28–31 (2014)
78. M.J. Kirton, M.J. Uren, Capture and emission kinetics of individual Si:SiO_2 interface states. Appl. Phys. Lett. (1986)
79. G. Depeyrot, Verilog-A compact model coding whitepaper. Memory **2**(June), 821–824 (2010)

80. C.G. Theodorou, *Low-Frequency Noise in Advanced CMOS/SOI Nanoscale Multi-Gate Devices and Noise Model Development for Circuit Applications* (Aristotle University of Thessaloniki, 2013)
81. A. Hajimiri, T.H.T.H.T.H. Lee, A general theory of phase noise in electrical oscillators. IEEE J. Solid State Circuits **33**(2), 179–194 (1998)
82. C.G. Theodorou, G. Ghibaudo, A self-contained defect-aware module for realistic simulations of LFN, RTN and time-dependent variability in FD-SOI devices and circuits, in *IEEE S3S Conference*, 2018
83. B. Kaczer et al., Atomistic approach to variability of bias-temperature instability in circuit simulations, in *IEEE International Reliability Physics Symposium Proceedings*, 2011
84. T.A. Karatsori et al., Analytical compact model for lightly doped nanoscale ultrathin-body and box SOI MOSFETs with back-gate control. IEEE Trans. Electron Devices **62**(10), 3117–3124 (2015)
85. C. Theodorou and G. Ghibaudo, Low-frequency noise and random telegraph noise in nanoscale devices: modeling and impact on circuit operation, in *International Conference on Noise and Fluctuations*, 2019
86. Z. Guo, A. Carlson, L.T. Pang, K.T. Duong, T.J.K. Liu, B. Nikolić, Large-scale SRAM variability characterization in 45 nm CMOS. IEEE J. Solid State Circuits (2009)
87. C.G. Theodorou, M. Fadlallah, X. Garros, C. Dimitriadis G. Ghibaudo, Noise-induced dynamic variability in nano-scale CMOS SRAM cells, in *European Solid-State Device Research Conference*, vol. 2016-Oct, 2016, pp. 256–259
88. L. Brusamarello, G.I. Wirth, R. da Silva, Statistical RTS model for digital circuits. Microelectron. Reliab. **49**(9–11), 1064–1069 (2009)
89. L. Gerrer et al., Modelling RTN and BTI in nanoscale MOSFETs from device to circuit: A review. Microelectron. Reliab. **54**(4), 682–697 (2014)
90. V.V.A. Camargo, B. Kaczer, T. Grasser, G. Wirth, Circuit simulation of workload-dependent RTN and BTI based on trap kinetics. Microelectron. Reliab. **54**(11), 2364–2370 (2014)
91. K. Takeuchi et al., Direct observation of RTN-induced SRAM failure by accelerated testing and its application to product reliability assessment. Dig. Tech. Pap. Symp. VLSI Technol., 189–190 (2010)
92. P. Bolcato and R. Poujois, A new approach for noise simulation in transient analysis, in *[Proceedings] 1992 IEEE International Symposium on Circuits and Systems*, vol. 2, pp. 887–890
93. J.-J. Sung, G.-S. Kang, S. Kim, A transient noise model for frequency-dependent noise sources. IEEE Trans. Comput. Des. Integr. Circuits Syst. **22**(8), 1097–1104 (Aug. 2003)
94. C.G. Theodorou, E.G. Ioannidis, S. Haendler, C.A. Dimitriadis, G. Ghibaudo, Dynamic variability in 14nm FD-SOI MOSFETs and transient simulation methodology. Solid State Electron. **111**(September), 100–103 (2015)

Noise in Resistive Random Access Memory Devices

F. M. Puglisi

1 Introduction

Noise has always been a remarkably important feature of electronic devices, and, for several decades, it has been looked at from many different perspectives. Noise itself, in its various forms and paradigms, has always been simultaneously a fascinating fundamental phenomenon [1–3], a versatile in-line tool to evaluate the defectiveness of materials and the quality of device production processes [4–6], and a serious concern from the standpoint of circuit reliability [1, 7–10]. As one could expect, the emerging resistive random access memory (RRAM) technology makes no exceptions. Ever since the first demonstration of a working RRAM device in 2008 [11], noise in RRAM has been the subject of several investigations [12–27]. Particularly, the main noise type that is of interest in such devices is the low-frequency noise, mainly in the form of "Flicker" ($1/f$) and random telegraph noise (RTN) [12–27], seen as a detrimental phenomenon limiting reliability. Pioneering investigations about low-frequency noise in RRAM devices, although delivering only preliminary results, clearly outlined the crucial role played by defects [12–14]. For instance, an optimal level of intrinsic defectiveness of the dielectric layer was found to be vital for a correct and reliable resistive switching in RRAM devices based on transition metal oxides (TMOs) [28]. From this standpoint, low-frequency noise analysis constitutes an invaluable tool to estimate the switching layer defectiveness. Soon after the first demonstration of a working RRAM, which came along with a first physical interpretation of the switching mechanism, a variety of RRAM devices based on different materials (selected from a wide landscape) appeared [12–28]. From then, the electronics community assisted to the full bloom

F. M. Puglisi (✉)
Dipartimento di Ingegneria "Enzo Ferrari", Università di Modena e Reggio Emilia, Modena, Italy
e-mail: francescomaria.puglisi@unimore.it

T. Grasser (ed.), *Noise in Nanoscale Semiconductor Devices*,
https://doi.org/10.1007/978-3-030-37500-3_3

of research in this field (as testified by the explosion in the rate of published papers about RRAM devices). Simultaneously, the need to develop a more refined physical understanding of the mechanisms underlying resistive switching emerged. Even in this case, noise analysis has been a valuable tool to provide precious insights about the physics of RRAM devices [15–21]. Nevertheless, the study of low-frequency noise, and particularly of RTN, in RRAM devices always constituted a mere support tool until recently. It is only in approximately the last 5 years that few complete and thorough analyses of RTN were published in the literature, acknowledging the importance of RTN and providing important details about the nature of the defects involved in resistive switching and in the noise phenomenon itself [15–21, 27]. However, the number of such contributions has been limited, because the complexity of the phenomenon and of the physics governing RRAM devices made the analysis of RTN an intricate matter. These contributions are precious for a variety of reasons. Firstly, understanding the atomistic nature of the defects involved in RTN is necessary to attempt at mitigating its effects. Secondly, the insights and conclusions derived in such contributions can frequently be extended outside the RRAM domain, as many other electron devices are based on the same (or similar) materials. Thirdly, in recent times, RRAM devices are considered the most attractive option not only for nonvolatile memory (NVM) applications [29, 30], but also for neuromorphic circuits (playing the role of synapses) [31, 32], logic-in-memory (LiM) solutions [33, 34], physical unclonable functions (PUFs) for security and authentication [35, 36], and random number generator (RNG) circuits [37, 38]. Investigating the effect of RTN in circuits devoted to such applications is hence mandatory. Finally, a refined comprehension of RTN may lead to the exploitation of a number of its peculiar and interesting properties, potentially culminating in the creation of new circuits and concepts that exploit RTN as a resource, as RTN-based RNG circuits and temperature sensors. In the light of such recent ideas, the bad reputation of RTN can be somewhat redeemed. Indeed, while tickling the minds of device physicists and engineers, RTN has always been seen as a tremendous limitation by circuit designers, who were forced to deal with this unpleasant phenomenon by shrinking the already tiny design margins, as they could not benefit from compact models of RRAM devices that included the effects of RTN. Mentioning a review paper on the topic (although not focused on RRAM devices) [39], RTN is "a device physicist's dream" and "a circuit designer's nightmare". As we will see throughout this chapter, this last sentence though not being an overstatement, can be slightly revised to make RTN not only a nightmare but also an opportunity for the circuit designers of tomorrow.

This chapter will be devoted to RTN in RRAM devices, as it constitutes the most important form of noise in such devices. To make the journey self-consistent and enjoyable, we will cover all the most relevant aspects related to RTN, starting from its mathematical and statistical properties. Appropriate methodologies to measure and analyze RTN will be presented and discussed, including a variety of practices and procedures frequently used for these purposes (that, incidentally, apply to a much broader range of devices, not only RRAM). These tools will then be extensively used to analyze and characterize RTN in RRAM devices. Because RRAM devices can be realized with many materials (that could show remarkable differences), we will focus on perhaps the most popular and widely studied of such devices: the HfO_2-based RRAM. Dedicated RTN analyses together with refined simulations will allow shedding light on the mechanisms behind RTN and on the atomistic properties of the defects involved. Finally, a compact model of RTN in RRAM devices that can be easily used by circuit designers will be presented, and few examples of the opportunities that it can unleash will be given, with the hope of making RTN in RRAM less of a nightmare and more of an opportunity.

2 Random Telegraph Noise: Measurement and Analysis Tools

RTN is frequently observed in a wide variety of devices [40–45], notwithstanding the differences in materials and manufacturing. Its origin is commonly ascribed to charge trapping and de-trapping into/from defects in the device [12–27, 40–45]. The analysis of RTN characteristics has been devised as a fundamental tool to gather important information about the properties of the defects and of the device itself, especially when facing novel structures that prevent using classical characterization techniques (e.g., C–V in scaled structures, charge pumping in devices that exhibit no bulk contact [45]). Particularly, RRAMs seem to be an interesting platform to perform RTN investigations, as their operation principles are based on the ability to rearrange defects in the device structure using electrical pulses [12–27]. As such, significant efforts are currently addressed toward the development of automated RTN measurement systems for RRAMs [46], procedures to separate RTN from other concurrent noise contributions [47], and methodologies to estimate the statistical parameters associated with RTN signals [48–50]. In this context, enabling RTN analysis as a characterization tool for material- and device-specific properties requires discussing the choice of measurement conditions that guarantee reliable results, although this is rarely addressed.

In the following, a classification of the RTN typologies will be presented, and the relevant characteristics of RTN signals that are useful to their characterization will be discussed. The discussion will then be focused on the different methodologies currently used to analyze RTN signals. Finally, the importance of considering the limitations imposed by the measurement instrumentation will be stressed, as such limitations can have a detrimental impact on the reliability of the results obtained with the aforementioned analysis methodologies. Concurrently, a rule of thumb to obtain reliable results will be devised and discussed.

2.1 RTN Statistics

The study of RTN signals presented in this chapter starts with the analysis of their statistical characteristics. In its simplest form, an RTN signal is associated with the random switching of an observable quantity, e.g., current, between two discrete states [51, 52], labeled here as "state 1" and "state 0," as shown in Fig. 1.

The probability per unit time for a transition to happen from state 0 to state 1 is given by $1/\tau_0$, with $1/\tau_1$ being the same quantity for the transition from state 1 to state 0. This implies that the time spent in either states follows the exponential distribution. Indeed, the probability for the system, initially assumed in a well-defined state (either 0 or 1) at time $t = 0$, to still be in the same state (while no transitions happened) at time t can be expressed as in Eq. (1).

$$p_0(t) = \frac{1}{\tau_0} e^{\left(-\frac{t}{\tau_0}\right)} \qquad p_1(t) = \frac{1}{\tau_1} e^{\left(-\frac{t}{\tau_1}\right)} \tag{1}$$

The simple two-level RTN signal is fully characterized by three parameters: the average time spent in either states (τ_0 and τ_1), and the amplitude of the RTN fluctuation (ΔI, i.e., the difference between the discrete levels), which can be estimated from experimental data. The corresponding power spectral density $S(f)$ can be derived by evaluating the autocorrelation function and exploiting the Wiener–Khintchine theorem [51], yielding Eq. (2).

$$S(f) = \frac{4(\Delta I)^2}{(\tau_0 + \tau_1)\left[\left(\frac{1}{\tau_0} + \frac{1}{\tau_1}\right)^2 + (2\pi f)^2\right]} \tag{2}$$

Frequently called "Lorentzian" or "$1/f^2$" spectrum, it is characterized by a plateau at relatively low frequencies ($2\pi f << 1/\tau_0 + 1/\tau_1$), a knee at a corner frequency, $2\pi f_C = 1/\tau_0 + 1/\tau_1$, and then a negative slope of $1/f^2$, see Fig. 2.

The RTN parameters generally depend on the operation conditions and on the device physics. These dependences have led many authors to analyze RTN in order to establish a link between the signal properties and the physical properties of the

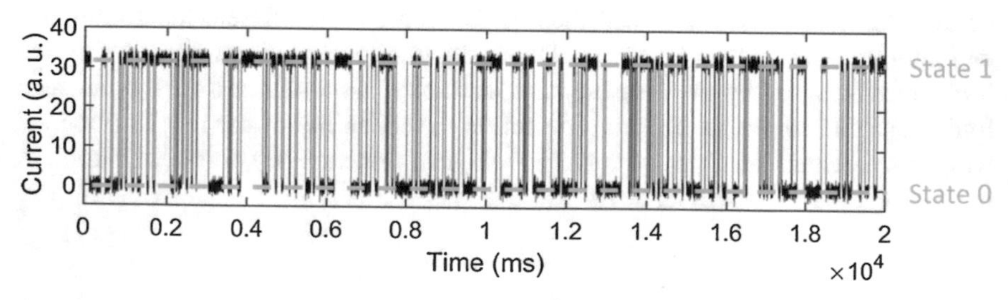

Fig. 1 Example of a simple two-level RTN signal. The two discrete states are evidenced

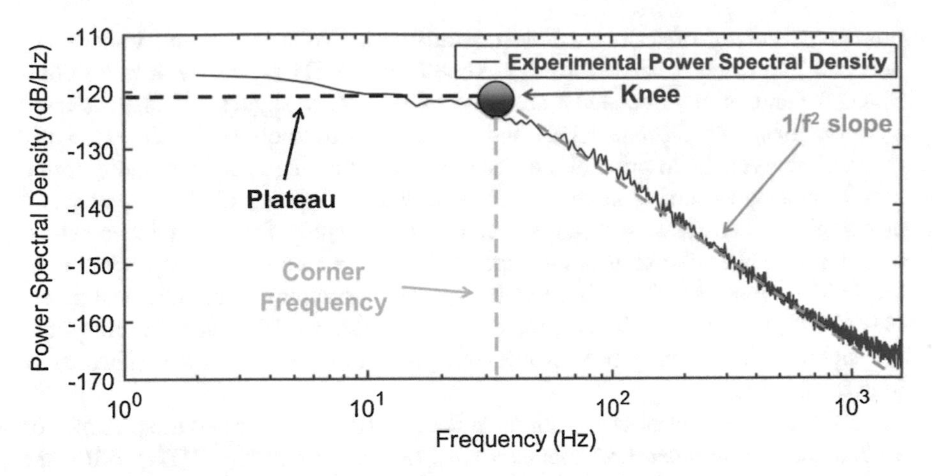

Fig. 2 Example of a Lorentzian spectrum. Its characteristics are evident: the plateau at low frequencies (black line), the knee point (red circle) at the corner frequency (green line), and the $1/f^2$ slope (orange line)

defects associated with RTN [12–27]. As such, RTN characterization is currently exploited to study the structural defects, revealing information on their energy level, capture cross section and physical location. In this scenario, since the transitions between the discrete levels are associated with charge capture/emission from/by a defect, τ_1 and τ_0 are usually relabeled as capture, τ_c, and emission, τ_e, time. Typically, a two-level RTN signal is associated with charge trapping and de-trapping into/from an individual defect. Multilevel RTN signals, instead, are associated with charge trapping and de-trapping events occurring at more than one defect, or with the possible existence of multistate/amphoteric defects. In the frequency domain, the summation of independent RTN processes is given by the superposition of the corresponding Lorentzian spectra. It has been demonstrated that under mild conditions on the number of superimposed Lorentzian spectra (i.e., if the number of Lorentzian spectra is not too low and if their corner frequency values are not too close to each other), the overall spectrum tends to the form of Eq. (3) [51]:

$$S(f) = \frac{A}{f} \tag{3}$$

where A is a parameter representing the power spectral density at the unit frequency.

2.2 *RTN Classifications*

Although conceptually simple, RTN signals can be catalogued according to many classifications. For instance, one criterion could simply be the number of discrete levels that the signal exhibits. In its simplest form, an RTN signal appears as the random switching of an observable quantity (e.g., voltage, current, impedance) between two discrete states [50–52]. Nonetheless, RTN signals with more complicated features are frequently observed. These RTN signals are characterized by more than two discrete levels and are labeled as multilevel RTN, typically (but not necessarily, in principle) associated with trapping and de-trapping events simultaneously occurring at more than one defect. As long as the trapping and de-trapping processes at such defects can be considered as mutually independent, multilevel RTN can be seen as the superposition of a number of independent two-level RTN signals, each associated with trapping/de-trapping at an individual defect [41, 47, 48, 50, 53]. In this case, the observed multilevel RTN signal results from the superposition of many two-level RTN signals (*components*), each related to an individual defect.

However, a number of observations in the last 10 years revealed the presence of RTN signals with *anomalous* characteristics (anomalous RTN—aRTN) in RRAM devices [54, 55]. Both the aRTN and the multilevel RTN signals are characterized by more than two discrete levels, but the aRTN cannot be modeled as a superposition of independent two-level RTN fluctuations. Specifically, the two-level RTN components that sum up to the observed aRTN signal clearly show some degree of correlation, i.e., the characteristics of one *component* at a given instant of time are determined also by the state in which the other *components* are found, see Fig. 3. Clearly, this complicates the analysis and requires refined methodologies.

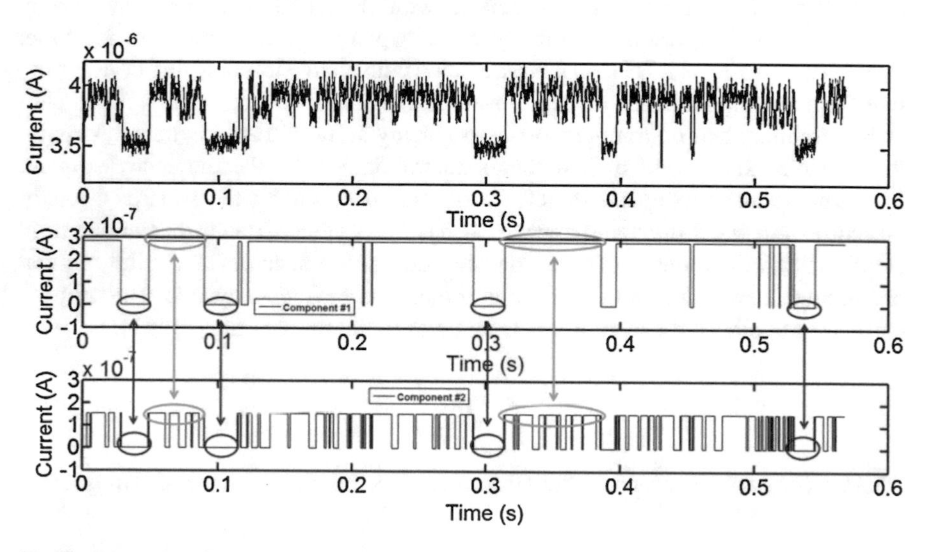

Fig. 3 (Upper panel) Example of an aRTN signal, which exhibits three discrete current levels (resulting from the superposition of two components). The components are separately shown in the middle and lower panels. A clear correlation between the two components is observed. Particularly, the switching rates of components #2 (lower panel) are influenced by the state in which component #1 is found

Besides two-level, multilevel, and anomalous RTN fluctuations, experimental observations [1, 55] showed the occurrence of temporary phenomena. Coarsely, we can classify these phenomena as follows:

1. Temporary RTN (tRTN)—a two-level RTN fluctuation randomly appearing and disappearing during RTN measurements. It differs from aRTN because it shows no correlation with other two-level components, Fig. 4.
2. Mutant RTN (mRTN)—a two-level RTN fluctuation that randomly and temporarily changes its statistical characteristics, sometimes in an iterative fashion.

For the sake of clarity, it is worth saying that some of these features were already detected many years ago (and more recently as well) in RTN signals recorded on different devices than RRAM [56–58]. As such, the terminology used above may differ from that adopted by other research groups. For instance, Grasser et al., studying the characteristics of MOSFET devices after electrical stress, identified the presence of RTN signals that had the property of being visible for a relatively short amount of time and then disappeared, and they were consistently termed "temporary RTN" signals [57]. The same group identified in MOSFET devices RTN signals that exhibited features resembling those of what is termed here "mutant RTN," and called the phenomenon "defect volatility" [58]. Nevertheless, we will keep the above-mentioned terminology throughout the chapter.

Such temporary phenomena on the one hand are difficult to analyze, but on the other hand offer unique opportunities to investigate the details of the physics involved, which can be essential for a more refined understanding of the microscopic nature of the defects and of the physical phenomena in which they are involved. Interestingly, defects are also responsible for many reliability issues in electronic devices (such as Bias Temperature Instability, Dielectric Breakdown, Stress-Induced Leakage Current), which are currently seen as major concerns, as elucidated in [1–3, 7–10] and references within. From this perspective, RTN analysis offers a matchless chance to refine the comprehension of such issues.

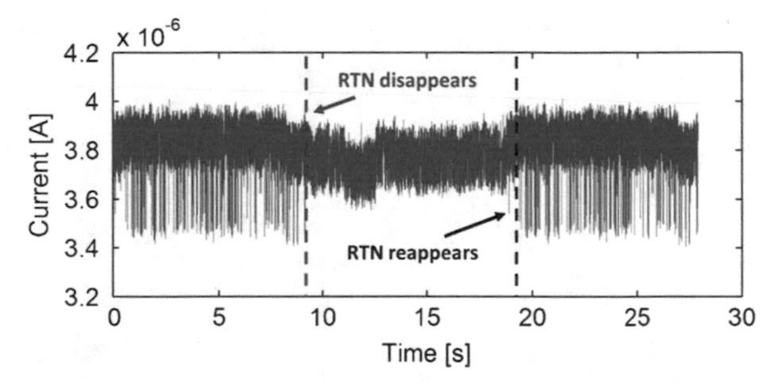

Fig. 4 Example of temporary RTN. The RTN fluctuation randomly disappears at $t \approx 9$ s, to reappear at $t \approx 19$ s with the same statistical characteristics

2.3 RTN Analysis Tools

This section of the chapter is devoted to the analysis tools that are typically employed to study RTN signals and to retrieve their statistical properties. Indeed, since in recent times RTN has gained much attention as a reliability issue, some research groups focused on RTN data analysis techniques, i.e., the algorithms allowing the extraction of the RTN parameters from the measured data. Both qualitative and quantitative methods have been developed. A review of the most commonly used methods is presented here, underlining the advantages and drawbacks of each methodology. Finally, some considerations about open issues and unresolved matters are presented.

Histogram and Time-Lag Plots

The characterization of RTN data aims at extracting the statistical features of the signal starting from unprocessed experimental data. In the simple case of a two-level RTN, as the one reported in Fig. 5a, a representative approach frequently used in the literature relies on the fitting of the histogram of the measured RTN signal, as shown in Fig. 5c. By fitting the resulting histogram with a bi-Gaussian distribution model, it is possible to estimate the two discrete levels (i.e., the average values

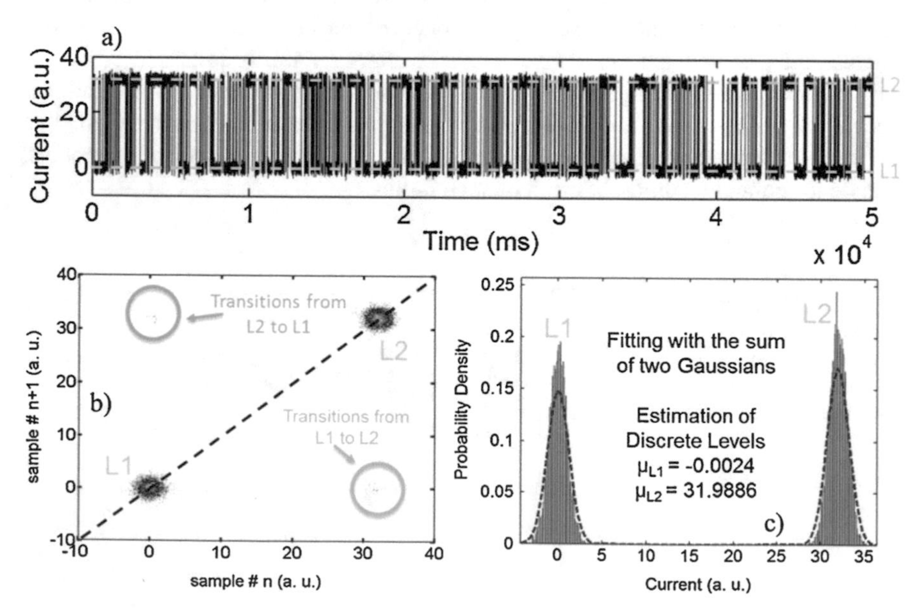

Fig. 5 (**a**) A two-level RTN generated by one defect. (**b**) Its time-lag plot representation, where the discrete levels and the transitions are highlighted. (**c**) Its histogram representation with a double-Gaussian fitting, revealing the two discrete levels

of the two Gaussian distributions). The main assumption of this procedure is that the background noise due to additional noise sources and to noise introduced by the measurement instrumentation can be assumed Gaussian. Another tool that has been recently employed for RTN characterization (together with or in place of the histogram) is the time-lag plot (TLP) [40, 41], shown in Fig. 5b.

TLP is frequently used to assess the randomness of a time series, and has been refined for RTN characterization in tools like the "color-coded time-lag plots" [41], and the "weighted time-lag plots" [59]. These topics are discussed also in Chaps. 4, 7, 11, and 14 of this book [60–63]. TLPs can be used to estimate the number of discrete levels (and their values) by counting the number of spots (and their positions) on the plot's main diagonal (transitions are represented by off-diagonal clouds). Then, each signal's point can be associated with one of the two discrete levels, and the different realizations of the capture and emission times can be estimated. Their average values are easily extracted as the average of the corresponding exponential distributions. To the extent of the identification of the discrete levels, TLPs and histograms can be helpful also when facing multilevel RTN, Fig. 6a.

However, in this case, it is challenging to retrieve the statistical properties (ΔI, τ_c, and τ_e) of *each individual RTN component (defect)* of the measured multilevel RTN. For instance, the four levels of the RTN signal in Fig. 6a are correctly identified using the TLP, Fig. 6b, and/or the histogram fitting, Fig. 6c. However, it is tricky (though technically possible) to translate the four discrete levels into the two amplitudes (ΔI) of the superimposed RTN *components*. Likewise, although it

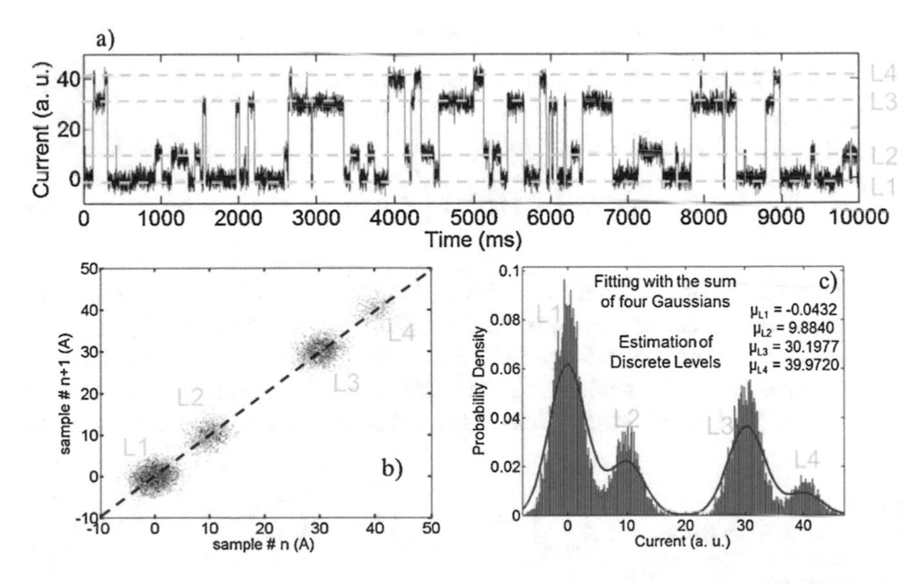

Fig. 6 (**a**) A multilevel RTN generated by two defects, resulting in four discrete levels. (**b**) Its time-lag plot representation, where the discrete levels and the transitions are highlighted. (**c**) Its histogram representation with a quadruple-Gaussian fitting, revealing the four discrete levels

is easy to determine the average time spent in each state, it is hard to convert this information into the capture and emission times of the two RTN *components* [41, 50, 52].

Hidden Markov Model

More advanced methodologies exploit the Markovian nature of the RTN signals to the aim of evaluating their statistical parameters. The most popular and versatile of such approaches is the Hidden Markov Model (HMM), a powerful tool commonly used in pattern recognition and statistical signal analysis. The same topic is further discussed in Chap. 7 of this book [61]. In HMM, the system is assumed a Markov (i.e., memoryless) process with hidden states, i.e., the discrete RTN levels. Each state is associated with a set of transition probabilities defining how likely it is for the system, being in a given state at a given instant of time, to switch to another possible state (including itself) at the following instant of time.

In HMM, a sequence of observations $\{Y_t\}$ $t = 1 \ldots T$ is modeled by specifying a probabilistic relation between the observations and a set of hidden (unknown) states $\{S_t\}$ through a Markov transition structure linking the states. In this framework, the state is represented by a random variable assuming one out of N values at each instant of time. The HMM approach relies on two conditional independence assumptions:

1. S_t only depends on S_{t-1} (known as the first-order Markov or "memoryless" property).
2. Y_t is independent from all other observations $Y_1,\ldots,Y_{t-1}$, $Y_{t+1},\ldots,Y_T$ given S_t.

The joint probability for the state sequence and observations can be formalized as in Eq. (4).

$$P\left(S_t|Y_t\right) = P\left(S_1\right)\ P\left(Y_1|S_1\right) \prod_{t=2}^{T} P\left(S_t|S_{t-1}\right)\ P\left(Y_t|S_t\right) \tag{4}$$

A schematic representation of the HMM is given in Fig. 7, where the Markov property is evidenced. According to the formalism used by Rabiner [52], an HMM is completely defined as a 5-tuple (N, M, A, B, π) where N is the number of hidden states, S, in the model (i.e., the number of discrete current levels to be found in RTN); because observations assume discrete values, M is defined as the

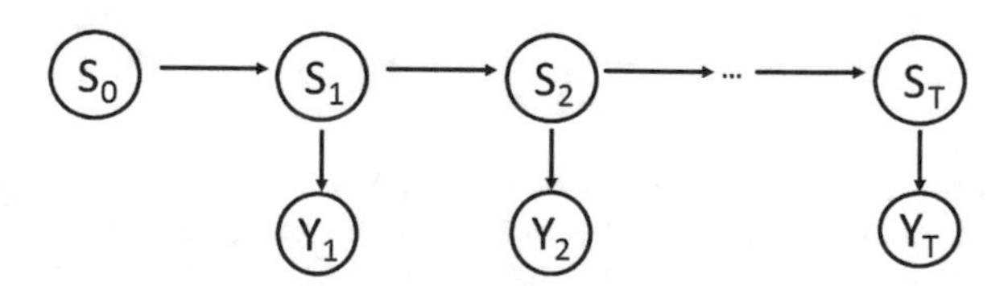

Fig. 7 Graphical representation of an HMM. At each instant of time t, each output Y_t is related only to the current state S_t of the Markov chain defining the model

number of distinct observable symbols (i.e., the possible current values assumed by RTN); A is an N-by-N matrix defining the transition probabilities among states and B is a N-by-M matrix defining the observation probability of each observable symbol in each hidden state; π is a vector defining the initial state probability distribution. The inference problem in this model consists in finding the most likely set of probability of hidden states, given the observations. This is achieved through a maximum likelihood estimate of the HMM parameters, given the observations using the forward–backward algorithm [52]. Then the most likely sequence of hidden states representing the dynamics of the observations can be achieved via the "Viterbi" algorithm, a dynamic programming paradigm [52]. As a result, HMM analysis can efficiently estimate the discrete current levels and the best sequence of states representing RTN data, as shown in Fig. 8.

The capability of the HMM approach to solve for the statistical properties of RTN has been widely assessed [15, 19, 21, 41, 43, 50, 53, 64–66] and is shown as follows. Firstly, a two-level RTN signal with Additive White Gaussian Noise (AWGN) is generated. The parameters used to generate the signal are summarized in Table 1. Then, the HMM algorithm is executed, with the initial assumption on the number of hidden levels, which is $N = 2$. The results of the analysis are summarized in Table 1 and Fig. 8. The HMM routine correctly identifies the hidden levels of the generated RTN signal, together with their sequence, which describes the time evolution of the Markov model behind the observations. The amplitude of the fluctuation is obtained as the difference between the discrete levels. It is also possible to estimate the duration of the time slots that the signal spends in each of the states, as well as their average values (average of the respective exponential distributions). The comparison

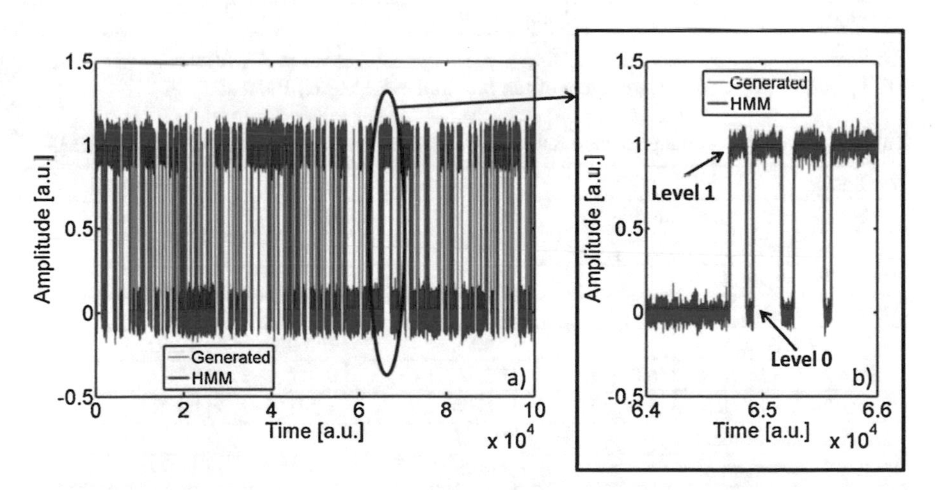

Fig. 8 Example of a two-level RTN (blue) and HMM fitting (red). The hidden levels and the most likely state sequence are correctly retrieved. The distinctive features of the two-level RTN (amplitude of the fluctuation and average capture/emission times) can be easily extracted. The zoomed-in section highlights the high accuracy of the HMM

between the generated and retrieved signal properties evidenced in Table 1 shows the suitability of the HMM approach when dealing with RTN signals.

Notwithstanding its potential, the HMM approach suffers from similar limitations as those affecting TLPs and histograms when facing the challenge of characterizing multilevel RTN. For instance, a four-level RTN signal with AWGN can be synthesized and characterized by means of the HMM, assuming $N = 4$. The results of the analysis are summarized in Table 2 and Fig. 9. The HMM routine correctly identifies the hidden levels of the generated RTN signal, together with their sequence. Again, it is possible to estimate the duration of the time slots that the signal spends in each of the states, as well as their average values. However, contrary to the previous case of a two-level RTN, it is impossible to solve for the average dwell times and the fluctuation amplitude *of the two independent Markov chains* (*components*) that make up the signal.

Indeed, the average time spent by the system in each level (*that is correctly retrieved*) is not useful to the estimation of the dwell times of the two Markov chains that are causing the observed signal (*components*). Thus, even though being effective in capturing the Markov dynamics of RTN, HMM is not best suited to characterize multilevel RTN. Although the HMM analysis is correctly identifying

Table 1 Parameters of the generated two-level RTN signal and extracted values with the HMM

Parameter	Generated	HMM	Precision %
Level 0 (see Fig. 8b)	0	0.025	97.5
Level 1 (see Fig. 8b)	1	0.991	99.1
Amplitude (level 1–level 0)	1	0.966	96.6
Average τ_0	500	476.2	95.24[a]
Average τ_1	500	480.6	96.12[a]

[a]The precision depends on the number of samples in the signal, i.e., 100 k samples

Table 2 Parameters of the generated four-level RTN signal and extracted values with the HMM

Parameter	Generated	HMM	Precision %
Defect 1—level 0 (level 0 in Fig. 7)	0[a]	0.009	99.10
Defect 1—level 1 (level 1 in Fig. 7)	0.452[b]	0.449	99.34
Defect 1—amplitude	0.452	–	–
Defect 1—average τ_0	500	407.95	81.59
Defect 1—average τ_1	500	408.02	81.60
Defect 2—level 0 (level 2 in Fig. 7)	0.855[c]	0.862	99.18
Defect 2—level 1 (level 3 in Fig. 7)	1.314[d]	1.301	99.01
Defect 2—amplitude 2	0.855	–	–
Defect 2—average τ_0	3000	371.14	**12.37**
Defect 2—average τ_1	3000	407.21	**13.57**

[a]Assumed to be 0 for defect 1
[b]Assumed to be the amplitude of the defect for defect 1
[c]Assumed to be the amplitude of the defect for defect 2
[d]Assumed to be sum of the amplitudes of the defects 1 and 2

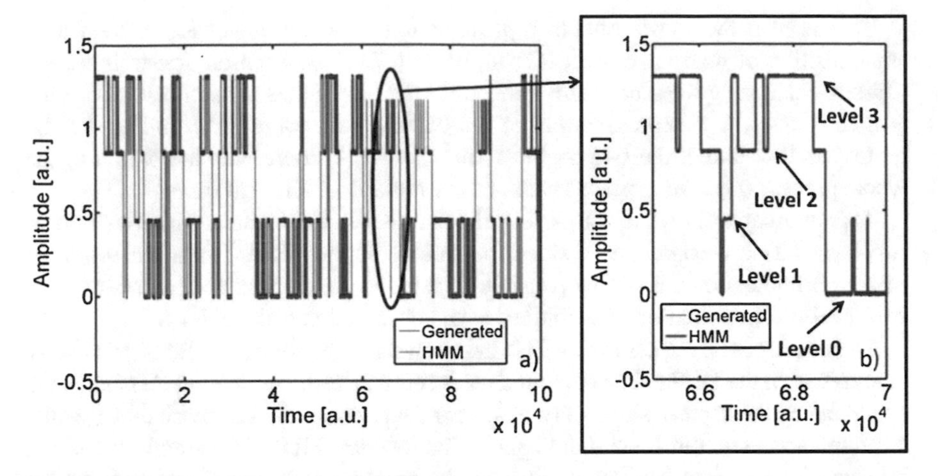

Fig. 9 Example of a multilevel RTN (blue) and HMM fitting (red). Though successful in characterizing the hidden states and their durations, HMM output is insufficient to achieve a comprehensive characterization of all the defects leading to the observed noise

the hidden states of the multilevel RTN and their most likely sequence, it is generally impossible to separately define the amplitudes of fluctuations and dwell times for each single defect contributing to the RTN. This implies that even though the characterization of the RTN signal is achieved, it is impossible to retrieve the distinctive features of each defect contributing to the observed signal. Subsequently, we will see how this limitation can be overcome by using a more refined HMM-based concept, namely the Factorial Hidden Markov Model (FHMM).

Factorial Hidden Markov Model

The FHMM [50, 65, 67] extends the HMM by considering the hidden state as a collection of K state variables, instead of a single random variable, each potentially assuming one out of N values at each instant of time (i.e., K different and parallel Markov chains) [50, 65, 67]. This results in a space state having a dimension of N^K. If no constraints are applied to the model, it can potentially take into account all the possible interdependencies among the K Markov chains, resulting in a high computational burden. However, a natural approach consists in assuming that each of the K Markov chains evolves independently from the other chains, resulting in a significant reduction of the problem complexity. This is formalized as in Eq. (5).

$$P\left(S_t|S_{t-1}\right) = \prod_{k=1}^{K} P\left({S_t}^k|{S_{t-1}}^k\right) \tag{5}$$

This is also the most suitable representation of a multilevel RTN, seen as a superposition of many two-level RTNs [50, 65]. This assumption constraints each Markov chain state to assume only one out of the two values at each instant of time (which is $N = 2$). A representation of the FHMM concept is given in Fig. 10: S_t^k represents the state of the k-th chain at time t, while Y_t represents the output of the whole process (i.e., the expected value of the multilevel RTN) at time t.

This approach allows decomposing the multilevel RTN into a superposition of two-level RTNs (*components*): since the output of the FHMM is a collection of two-level fluctuations, it is now possible to retrieve the distinctive features of each two-level RTN fluctuation contributing to the observed multilevel RTN.

To this purpose, we perform an FHMM analysis of the four-level RTN previously analyzed with the HMM. The K parameter is set to 2, as the multilevel RTN signal is generated by the superposition of two independent two-level Markov chains. Results are summarized in Table 3 and in Fig. 11. The four-level RTN is correctly separated into two independent Markov chains, the parameters of which (amplitude of the fluctuation and average dwell times) are estimated with high accuracy.

The implementation of either HMM or FHMM suffers from a trade-off between the computational burden and the fitting accuracy [50, 64, 67]. Indeed, a more complex model (higher number of hidden states in HMM or higher number of Markov chains in FHMM) results in higher time-to-solution. Regrettably, as in HMM, the number of hidden states is an input parameter for the model, so the number of parallel Markov chains (i.e., the number of defects contributing to the observed RTN) should be somehow estimated ahead of time in the case of FHMM (for instance, exploiting time-lag plots or histograms). However, this issue can be solved by feeding the algorithm a reasonably large number of expected defects (though resulting in a more time-consuming task): the chains related to defects

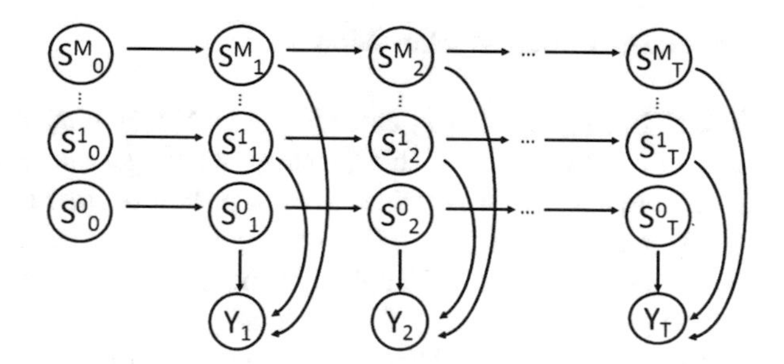

Fig. 10 Graphical representation of an FHMM. At each instant of time t, each output Y_t is related to the superposition of the states $S^0{}_t, \ldots, S^M{}_t$ of M independent and parallel Markov chains

Table 3 Parameters of the generated four-level RTN signal and extracted values with the FHMM

Parameter	Generated	FHMM	Precision %
Defect 1—amplitude	0.452	0.452	100.00
Defect 1—average τ_0	500	469.12	93.82
Defect 1—average τ_1	500	461.04	92.21
Defect 2—amplitude 2	0.855	0.855	100.00
Defect 2—average τ_0	3000	2855.9	95.20
Defect 2—average τ_1	3000	2675.5	89.18

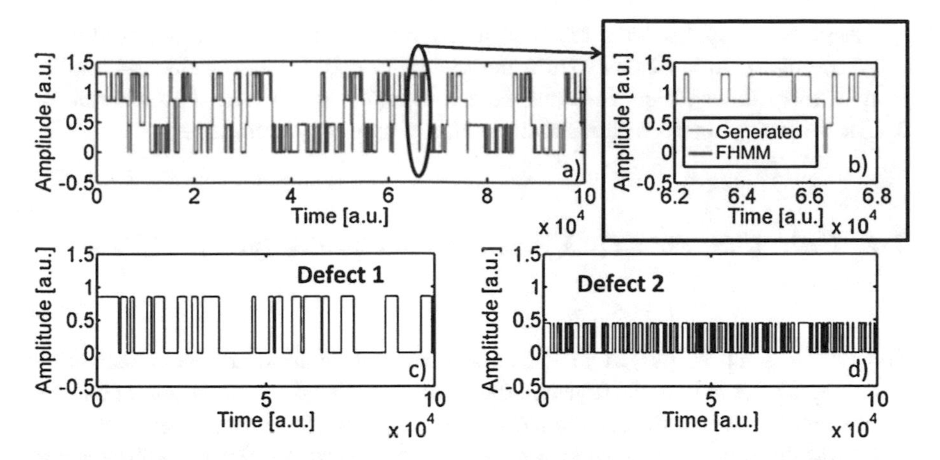

Fig. 11 (**a**) A four-level RTN generated by two defects with superimposed Gaussian noise (blue) and FHMM fitting (red). (**b**) Zoomed-in view of a portion of the RTN signal in (**a**). (**c–d**) The RTN signal is correctly decomposed in two distinct two-level fluctuations. Amplitudes of fluctuations and state sequences of all defects can be easily inferred

that are unnecessary to match the input RTN will be characterized by negligible amplitude and can easily be discarded after the analysis. The advantage of the FHMM over HMM is evident even in this respect: a too large estimation of the number of hidden states in HMM causes the algorithm to be forced to identify more hidden levels than the actual number (over segmentation). Instead, in the FHMM approach, using a large number of parallel chains is not affecting the goodness of fitting, making the FHMM a self-consistent and accurate tool.

Other Approaches and Open Challenges

Besides HMM and FHMM, other approaches have been used in the literature to perform the same task. Particularly, an effective one uses a Markov Chain Monte Carlo (MCMC) approach and employs the Gibbs sampling algorithm [48]. Akin to FHMM, the algorithm can be used to decompose the measured RTN sequence to retrieve the time constants and the amplitude of each *component*. Currently, the main issues all these methods (HMM, FHMM, MCMC) are facing are:

1. Limitation of accuracy when the superimposed noise is strong. Such algorithms show a tendency to over-segment (i.e., to associate the existence of a transition between two discrete levels with regions of the signal in which the superimposed noise is strong enough).
2. Possible non-Gaussianity of superimposed noise. Typically, all these algorithms assume a Gaussian distribution of the superimposed noise, which is not always true. Particularly, the superposition of a number of RTN processes whose typical switching frequencies are at the edge of the useful bandwidth of the

measurement system may result in a relatively strong $1/f$ superimposed noise whose distribution deviates from the Gaussian one. Similarly, under sampling may give similar effects and contribute additional random non-Gaussian noise.
3. The difficulty to deal with instabilities and temporary phenomena.

2.4 Guidelines for RTN Measurement and Analysis in RRAM Devices

The typical setup employed in RTN measurements consists of tools that allow acquiring signals in both the time and the frequency domain, such as semiconductor parameter analyzers, scopes, dynamic signal analyzers. However, at least in the case of RRAM devices, time-domain characterization is more useful than the frequency-domain counterpart is, as testified by the number of works in the literature that follow this paradigm [15–27]. This is ascribed to two main reasons. The first is related to the small feature size of RRAM devices, which limits the number of defects in the device. As such, the analysis of the corresponding time-domain signals can deliver, in principle, insights into the physics of individual RTN processes. Particularly, the time-domain paradigm offers the advantage of time locality, which is extremely useful in the case of RRAMs as their operating principles rely on the controlled motion of defects. Indeed, if during a frequency-domain RTN measurement some structural modifications (e.g., motion of a defect associated with charge transport) occur in the device at given time (i.e., t^*), the entire spectrum will be affected, while a time-domain measurement will only be partially affected (from time t^* on). The second reason is that the frequency-domain approach suffers from a missing degree of freedom. Although both the time series and the full spectrum of a two-level RTN are a function of three statistical parameters (ΔI, τ_c, and τ_e), the magnitude spectrum can be entirely defined by only two parameters, namely A and f_C, such that $S(f) = A/(1 + (f/f_C)^2)$. As such, different time-domain signals could exhibit the same magnitude spectrum: for instance, two different RTN signals with $\Delta I_1 = \Delta I_2$, $\tau_{c1} = \tau_{e2}$, and $\tau_{e1} = \tau_{c2}$ will show the same magnitude spectrum. The resulting indetermination in the fitting problem requires assuming an unknown relation between two out of three parameters. As such, the time-domain paradigm is currently preferred to provide a link between the electrical response of the device (RTN) and the physical properties of the defects in RRAM devices. However, possible inadequate measurement conditions may introduce errors in the estimation of RTN parameters that propagate to the estimation of defects properties. As such, a careful choice of measurement conditions is mandatory to obtain reliable and accurate results. Indeed, the measured RTN (i.e., the actual result of the measurement procedure) may be somewhat different from the signal one intends to measure, because measuring is the result of a sampling process that needs to be properly taken into account. Particularly, sampling implies the discretization of the signal over time (which limits the bandwidth) and limits the duration of the measured signal to a finite amount (a finite number of samples can be acquired, hence a finite number of transitions). As such, the RTN parameter estimation is intrinsically affected by inaccuracies, some of which are determined by the choice of measurement conditions. The accuracy in the estimation of ΔI is limited by the amount of superimposed noise, in part related to the additional noise produced by the measurement instrumentation [68–71]. Contrarily, the accuracy in the estimation of the average capture and emission times is constrained by the values of the sampling time, τ_S, and by the number of acquired samples, N. Naturally, large N results, on average, in a higher accuracy as it is more likely to detect more transitions between the discrete states. This is confirmed in Fig. 12, where the same synthesized RTN signal is sampled and analyzed considering three different N values. Clearly, the $\tau_{c,e}$ estimation error increases with decreasing N. This effect was already pointed out in the early days of RTN analysis by Kirton and Uren [2], who introduced a minimum value ($N_{min} = 20$ k) to ensure a $\tau_{c,e}$ average estimation error <10%.

The crucial role of the choice of τ_S has been nicely analyzed in [71], where the authors show that a reliable extraction of the RTN $\tau_{c,e}$ is constrained by the choice of τ_S. The latter must be much smaller than the RTN $\tau_{c,e}$ to prevent errors in the $\tau_{c,e}$ estimates that could be as large as more than one order of magnitude. In the opposite case (i.e., τ_S and $\tau_{c,e}$ are comparable), some portions of the signal could experience under-sampling, resulting in a distorted waveform [68–71]. The

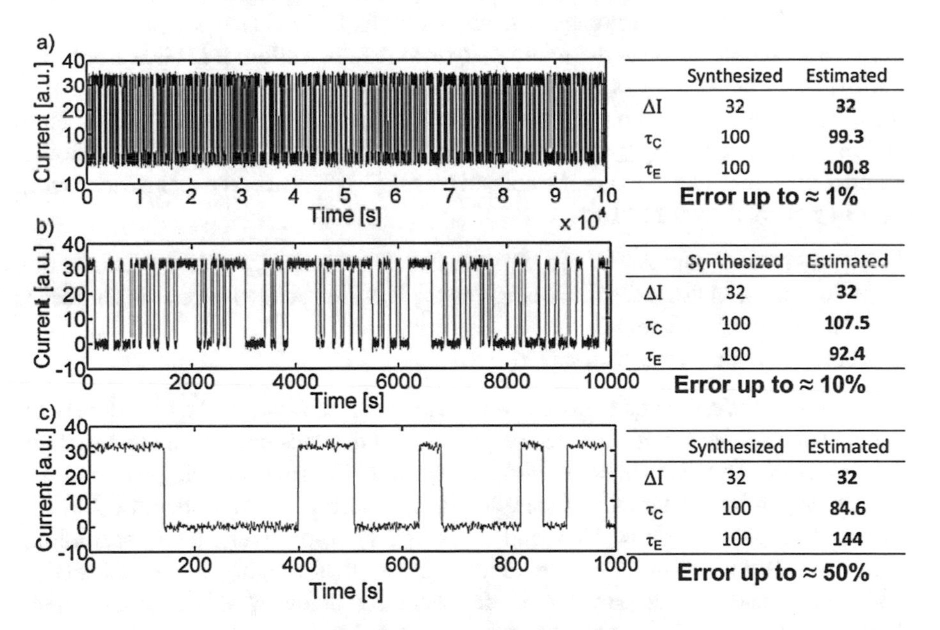

Fig. 12 A generated two-level RTN signal with AWGN considering $\tau_S = 1$ s and (**a**) $N = 100$ k, (**b**) $N = 10$ k, (**c**) $N = 1$ k. The statistical properties of the synthesized signals are reported in the tables next to each signal, together with the estimated values. Clearly, the accuracy in the estimation of $\tau_{c,e}$ increases for increasing N values. The accuracy in ΔI estimation, conversely, shows no dependence on N

authors proposed to perform different measurements of the same RTN signal at different τ_s to double-check the reliability of the extracted $\tau_{c,e}$ values. While being theoretically appropriate, a different methodology would be better suited for RTN measurements in RRAMs: as the electric field in RRAM devices tend to be relatively high (approximately some MV/cm), the stress delivered to the device while measuring RTN may alter the RTN signal itself. For instance, the defect involved in RTN may diffuse, move, or recombine, under the action of the applied electric field so that consecutive measurements (at different τ_s) may not correspond to different observations of the same RTN signal. Moreover, this approach significantly increases the number of RTN traces to be analyzed.

However, it is possible to devise a strategy to evaluate the correctness of an RTN measurement based on an a posteriori evaluation of the RTN parameters estimation. Indeed, an under-sampled RTN signal shows peculiar properties that are useful to identify the signal as such. To better understand the relations among the choice of the sampling time, the actual $\tau_{c,e}$ values, and the related estimation errors, we synthesized different symmetric ($\tau_c = \tau_e$) two-level RTN signals with AWGN. All synthesized signals are composed of 100 k samples, and the time interval between two consecutive samples is arbitrarily fixed at 1 s (with no loss of generality). The generated signals have different $\tau_c = \tau_e$ spanning several orders of magnitude (from 1 s to 10^5 s). To mimic the measurement operation, all signals were resampled with $\tau_S = 100$ s that represents the sampling time. The resulting signals are composed of only $N = 100$ k/$\tau_S = 1$ k samples. The results of the FHMM analysis (the extracted $\tau_{c,e}$) were used to calculate the relative error in the estimation. It is clear from Fig. 13 that the relative error is fairly constant only in a "safe zone", i.e., when $\tau_{c,e} \ll\gg \tau_S$, in agreement with previous reports. However, this analysis allows roughly defining the relation $\tau_{c,e} \geq H\ \tau_S$, where H is the "threshold" and depends on the number of samples acquired in the measurement (N). The average estimation error, E, is in general determined by:

1. The number of samples of the RTN signal (N), when $\tau_{c,e} \geq H\ \tau_S$ (i.e., in the "safe zone"). In this case, the average error, E, is independent from the choice of τ_S and only depends on N.
2. The choice of τ_S, when $\tau_{c,e} < H\ \tau_S$.

As such, if the estimated $\tau_{c,e}$ are in the "safe zone," then we can be confident with the obtained results within an error margin, E, that depends merely on the length of the time series (N). This is confirmed by repeating the analysis using synthesized signals with a larger number of samples (i.e., 1 M and 10 M instead of 100 k), which corresponds to $N = 10$ k and $N = 100$ k, respectively, after resampling with $\tau_S = 100$ s. In all cases, the relative error stays roughly constant inside a "safe zone," and its average value, E, decreases for higher N values, as expected. Also, the threshold, H, is found to increase with N. Therefore, given N, both H and E can be determined, providing a quick metric to evaluate the correctness of the RTN measurement and the average expected error in the evaluation of its statistical parameters. For a detailed examination of issues related to the measurement of RTN signals, the reader is encouraged to refer to [70].

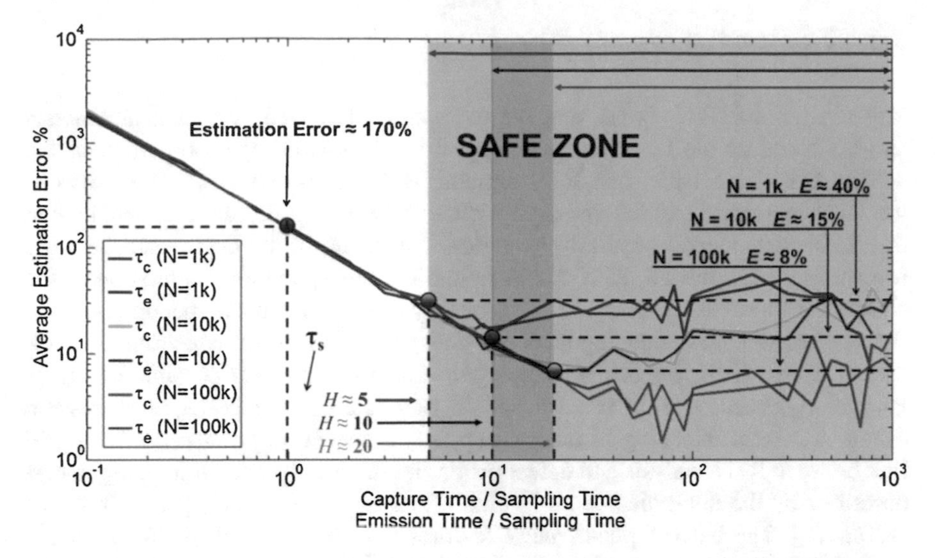

Fig. 13 The relative percentage average error in the estimation of capture (red, green, and purple) and emission (blue, black, and brown) times of two-level RTN signals vs. the ratio between their actual $\tau_{c,e}$ and the sampling time, τ_s. RTN signals with $\tau_{c,e}/\tau_s$ spanning from 1 to 10^3 and with different lengths ($N = 1$ k, 10 k, 100 k) are sampled. When under-sampling occurs ($\tau_{c,e} \approx \tau_s$) the relative error in the estimation of RTN parameters may be significant. A "safe zone" ($\tau_{c,e} > H\ \tau_s$) can be identified, in which the average relative error E is constant. Both H and E only depend on the length of the signal (N)

3 Statistical Investigation of the RTN Physical Mechanisms

As outlined previously in this chapter, the RTN disturb represents one of the most significant limitations for the full exploitation of RRAM technology. At the same time, it shows interesting features that could leverage the birth of new circuits. As such, understanding the physical mechanisms behind the RTN phenomenon in RRAM is mandatory. However, this is far from being a trivial task. RTN is indeed an alteration of the charge transport mechanism, probably related to trapping and de-trapping at defects. The physical description of RTN must encompass also a deep comprehension and modeling of charge transport, which differs depending on the resistive state. Subsequently, starting from the analysis of the role of defects for charge transport and RTN, we will explore the possible mechanisms behind RTN through an extensive noise characterization of RRAM devices. By retrieving the statistical properties of traps contributing to RTN and comparing the results with the outcomes of physics-based multi-scale simulations, we will shed light on the physics of the phenomenon.

3.1 The Role of Atomic Defects

The analysis of RTN signals and the extraction of related statistical parameters can be useful as the latter can be mapped to the microscopic properties of the defects associated with such RTN signals. However, providing a link between the RTN properties and the underlying defects requires a thorough understanding of all the fundamental physical phenomena occurring in the device and possibly involved in RTN. Indeed, RTN can be naturally seen as a random alteration of the charge transport mechanism [15, 19], probably due to some phenomenon involving defects: as such, understanding RTN physics requires a full self-consistent description of charge transport, charge trapping/de-trapping, and their possible interplay. Particularly, because RTN is supposed to be related to charge capture/emission events, a careful modeling of charge trapping is needed to provide a meaningful link between RTN analysis and defect properties. Currently, charge trapping is best described by the non-radiative multi-phonon trap-assisted tunneling (TAT) theory [1, 72, 73]. The latter explains many features that the classical Shockley–Read–Hall (SRH) approach cannot describe by including the exchange of phonons with the lattice in the description of charge capture and emission, accompanied by the atomic lattice relaxation, i.e., the displacement of the atoms surrounding the defect to accommodate the trapped charge. The relaxation process heavily depends on the microscopic properties of the defect and is mainly described by the relaxation energy, E_{REL} [72, 73]. The latter, together with the defect thermal ionization energy, E_T, is by far the most important microscopic parameter that dictates the behavior of the defect from the charge-trapping standpoint. In this framework, experimental RTN data can be exploited to understand not only the nature of the defect involved in the phenomenon, but also the interaction between the trapped charge and the charge transport mechanism that, together with charge trapping and de-trapping, determines RTN.

3.2 Charge Transport in RRAM Devices

Because RTN results in an alteration of the charge transport mechanism in RRAM, it is mandatory to precisely understand the microscopic features of charge transport in RRAM devices, which is, in general, a nontrivial task. An exceedingly large number of different RRAM devices (i.e., based on different materials) have been demonstrated in the literature, each displaying unique features as well as common traits with other RRAM devices. As such, dozens of more or less specific models have been presented [74–84]. The reader is encouraged to consult [85] for a broad discussion of the similarities and differences among a variety of RRAM structures.

In the context of this chapter, we will focus the discussion on one of the most popular RRAM devices: the TiN/Ti/HfO_2/TiN RRAM. This device exhibits filamentary switching [75–79, 80, 81, 83], i.e., the resistive switching between a high-resistive (HRS) and a low-resistive (LRS) state is ascribed to the partial rupture and reconstruction of a conductive filament (CF) within the dielectric (HfO_2) layer. Although many physical explanations for the observed electrical behavior have been proposed, at the present stage, there is a consensus on some fundamental aspects related to the physics of the CF formation and of its structural changes during switching, as well as those related to charge transport in the two resistive states. Particularly, the formation of the CF is associated with field- and temperature-driven Hf–O bond breaking that results in the generation of a defect pair (i.e., an oxygen vacancy, Vo^+, and an oxygen atom/ion, O). Although the oxygen atom easily diffuses away from the generation spot, the Vo^+ constitutes a charge-trapping center at which electrons can tunnel in and out. This mechanism causes localized power dissipation at the Vo^+ site, which enhances the local temperature resulting in a higher probability of additional defect generation close to the existing Vo^+. Eventually, a thermal runaway process is triggered, which causes the rapid formation of a CF (made of tightly packed Vo^+ defects). This process is only limited by the imposed current compliance, I_C, that determines the lateral size of the CF. The details of the CF process formation are described in [80, 81]. In the LRS, the device is therefore characterized by the presence of a full CF and it is accepted that charge transport is due to the delocalized electrons drift in the CF [80, 81]. Driving the device into the HRS (i.e., performing the reset operation) promotes the drift of O ions toward the Vo^+ making up the CF. The recombination between Vo^+ and O leads to the partial reoxidation of the CF, creating a dielectric barrier [80, 81]. The dielectric barrier thickness (t_b) is determined by the reset conditions, i.e., the maximum absolute value of the applied voltage, V_{RESET}, and the temperature, T. Due to the stochasticity of the Vo^+/O recombination process, the dielectric barrier will contain some Vo^+ and O defects. The HRS current is dominated by the electron TAT at Vo^+ in the reoxidized portion of the CF (dielectric barrier) [73, 80, 81].

3.3 *Physics of RTN in HRS*

In this section, we explore the microscopic mechanisms responsible for RTN fluctuations in RRAM. The investigation of the RTN physical mechanisms is performed by a thorough analysis of a large RTN data set, and is supported by self-consistent physics-based Kinetic Monte Carlo (KMC) charge-transport simulations [86, 87].

The investigation of RTN properties is performed by measuring RTN traces in TiN/5 nm Ti/3.4 nm HfO_x/TiN RRAM devices in high-resistive state (HRS). Devices under test were initially electroformed imposing a current compliance $I_C = 100\ \mu A$, the same adopted also during the set operations in the subsequent switching cycles. HRS is reached through the reset operation, performed by applying a negative voltage sweep to the top electrode. The maximum absolute value of the voltage reached during the reset operation, V_{RESET}, determines the thickness of the dielectric barrier at the end of the reset operation [83], and is used as a parameter to achieve different barrier thickness (i.e., associated with different HRS resistance). The device is switched between HRS and LRS 100 times, and RTN traces are recorded over time with the device in HRS, by applying a small $V_{READ} = 100$ mV. Then, the experiment is repeated at different temperatures and V_{RESET} (i.e., different barrier thickness).

In order to understand the physical mechanisms responsible for the RTN, we propose a microscopic explanation that allows describing its statistical features, i.e., the ΔI and $\tau_{c,e}$ distributions. The HRS current is due to the electron TAT at Vo^+ defects in the reoxidized tip of the CF (dielectric barrier), with each individual Vo^+ defect carrying a share of the overall current. We speculated that the RTN current fluctuations are due to the random activation and deactivation of individual Vo^+ defects assisting charge transport [15, 18–21]. To test the hypothesis, we simulated the distribution of the current fraction driven by each individual Vo^+ defect randomly distributed within the barrier, Fig. 14. Charge transport simulations include the TAT description of charge transport and the microscopic parameters for the Vo^+ defects are taken from the results of ab initio studies [88–90]. The thickness and the cross section of the dielectric barrier used in simulations are extracted from the experimental I–V curves of the device using a compact model [75, 77, 83]. The uniform distribution of Vo^+ defects ($N_T = 2 \times 10^{21}$ cm^{-3}, in agreement with ab initio calculations [91]) considered in the simulations allows accurately reproducing the ΔI distributions measured on RRAM devices consecutively cycled 100 times in different reset conditions ($V_{RESET} = 1.1$–1.5 V, corresponding to different barrier thickness values), validating the hypothesis.

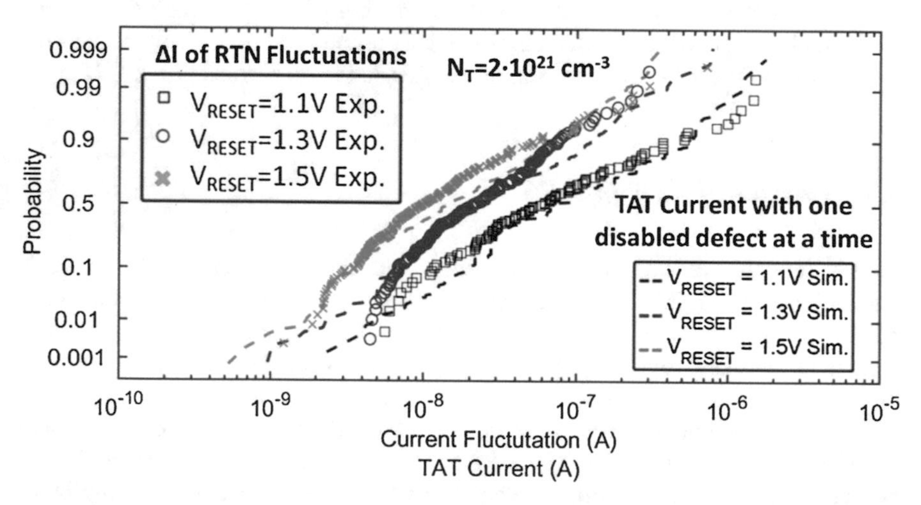

Fig. 14 Experimental (symbols) and simulated (lines) distributions of the RTN current fluctuations, ΔI, detected in an RRAM device in HRS, reset by applying different voltages. ΔI is simulated as the TAT current due to individual oxygen vacancy defects randomly distributed within the reoxidized portion of the CF (dielectric barrier)

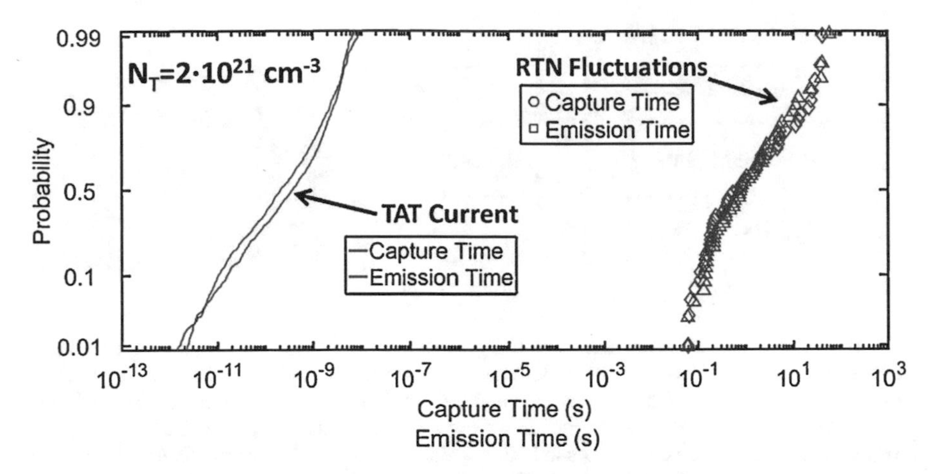

Fig. 15 Experimental (symbols) RTN capture and emission times measured in RRAM device in HRS reset at $V_{RESET} = 1.3$ V. The capture and emission times associated with the TAT at oxygen vacancy defects are also shown for comparison (solid lines)

The simulations of the RTN ΔI statistics, which confirms that TAT is the dominant charge transport mechanism for the HRS current, allow calculating the distributions of electron capture and emission times at Vo^+ defects, which are in the sub-nanoseconds regime, Fig. 15. This is expected, as fast charge trapping and de-trapping are required to support a relatively large defect-assisted current as the one observed in these devices. Simulated τ_c and τ_e are orders of magnitude smaller than those experimentally extracted from RTN current fluctuations, Fig. 15, indicating that the same defect, i.e., Vo^+, cannot be responsible for both charge transport and RTN. Hence, a different mechanism must be responsible for the activation/deactivation of the Vo^+ defects. In this scenario, we identified two possible candidates:

1. The presence of a defect of different atomic nature (i.e., a "slow" defect with higher E_{REL} compared to a Vo^+ leading to much higher capture/emission times), located in proximity to a Vo^+ defect assisting the TAT current, which interferes with the charge transport due to Coulomb interaction.
2. The possibility for Vo^+ defects to have both a fast/metastable (assisting charge transport) and a slow/stable (unable to assist charge transport) defect configuration, corresponding to different arrangements of the electron cloud around the atoms surrounding the defect location [1].

The first hypothesis is based on the strong dependence of the current driven by a TAT-supporting defect on the local electric field/potential. The perturbation of the Coulomb potential generated by the charge trapped at a "slow" defect located in the surroundings of the "fast" Vo^+ assisting the HRS charge transport modifies significantly the alignment between the electrons and the ground state of the "fast" oxygen vacancy defect, Fig. 16. In order to quantify the effect of this Coulomb

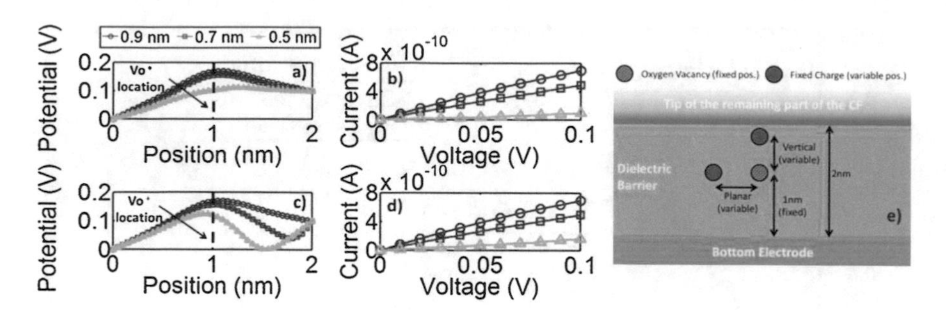

Fig. 16 Simulations of the vertical potential profile (**a**, **c**) and the TAT current (**b**, **d**) flowing through a single Vo^+ located in the middle (at 1 nm, see the dashed line) of a 2 nm-thick HfO_x dielectric barrier, representing the CF portion oxidized during the reset operation. The effect of the electric field due to a charge trapped at a slow defect is included by a fixed negative charge, whose distance from the Vo^+ is varied in both the planar (**a**, **b**) and the vertical (**c**, **d**) directions by considering three values (see the legend). The potential profile is severely altered by the presence of the negative charge, which causes a significant reduction of the TAT current assisted by the Vo^+. The positions of the fixed charge (variable) and of the Vo^+ (fixed) are also highlighted in a schematic representation (**e**) in which the planar and vertical directions are indicated for clarity

interaction, we simulated the effect of a fixed negative charge (i.e., electron trapped at a "slow" defect) on the current driven by a single Vo^+ defect as a function of their mutual distance, Fig. 16. Simulations show that the current driven by the Vo^+ defect is strongly reduced when the distance from the fixed negative charge is below 1 nm, as the Coulomb interaction becomes significant. As a result, the charge capture/emission into/from "slow" defects, together with the Coulomb interaction, inhibits the electrons TAT through the Vo^+ causing the observed RTN current fluctuations. This is confirmed by the KMC simulations, Figs. 16 and 17: the TAT current assisted by an oxygen vacancy defect is affected by the charge trapping at "slow" defects in their proximity. Noticeably, KMC simulations also reproduce the typical white (shot) noise resulting from the fast capture and emission events involved in the TAT charge transport, Fig. 17.

To estimate the microscopic properties of the "slow" defect possibly responsible for RTN fluctuations, RTN traces were recorded on a single device at different voltages, paying attention not to change the state of the device (i.e., by applying low voltages without cycling the device). The statistical properties of each RTN trace were then estimated (ΔI and average τ_c and τ_e). The ΔI can be regarded, in the approximation of a strong Coulomb interaction, as the current driven by a Vo^+ being activated and deactivated by the electron trapping in a neighboring "slow" defect. In this framework, the Coulomb interaction of the trapped charge at the "slow" defect on the Vo^+ causes the (almost) complete suppression of the TAT current driven by the Vo^+ defect. The ΔI can hence be considered equal to the current driven by the Vo^+ defect when active, i.e., when the "slow" defect is empty and no Coulomb interaction occurs. Performing simulations in which a single Vo^+ defect is present in the dielectric barrier allows identifying the position and energy of the Vo^+ defect, x_T and E_T respectively, by reproducing the ΔI dependence on the applied voltage.

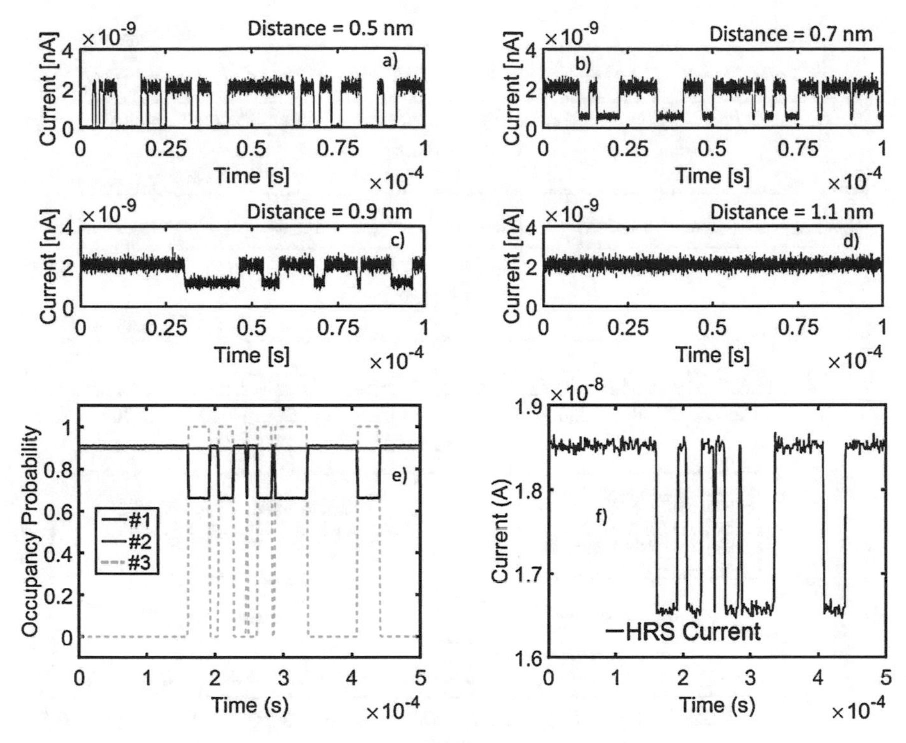

Fig. 17 KMC simulations of RTN via an individual Vo^+ placed at different distances, (**a**) 0.5 nm, (**b**) 0.7 nm, (**c**) 0.9 nm, (**d**) 1.1 nm, from the "slow" defect. The superimposed white noise is due to the average effect of the Vo^+ capture and emission processes. KMC simulations of (**e**) the occupation probability and (**f**) the TAT current driven by two Vo^+ (#1 and #2) and a "slow" defect (#3), close to the Vo^+ #1. The occupancy of the Vo^+ #2 is not perturbed provided that #3 is sufficiently far from #2 (the distance must be $>\approx 1$ nm). The occupancy of the Vo^+ #1 changes over time due to charge trapping at #3 and to Coulomb interaction. The resulting TAT current exhibits RTN

Fig. 18 shows an example of x_T and E_T estimation as performed on two RRAM devices reset in different conditions (i.e., characterized by different barrier thickness values) and at different temperatures. The relaxation energy considered for the Vo^+ defect is $E_{REL} = 0.7$ eV, consistent with ab initio studies [88–90] and the Marcus theory [92]. In both cases, the position of the Vo^+ defect is found to be close to the middle of the barrier, in agreement with the predictions of the TAT theory [73].

To extract relevant information about the microscopic properties of the "slow" defect, we perform additional simulations that consider the presence of the "slow" defect. For simplicity, we assume that the "slow" defect lies at the same vertical position as the Vo^+ defect. The voltage dependence of τ_c and τ_e extracted from RTN fluctuations is correctly reproduced by simulations that include a "slow" defect having a relaxation energy $E_{REL} = 2.65$ eV, which agrees with the value predicted by ab initio calculations for the neutral interstitial oxygen defect [88–90], being

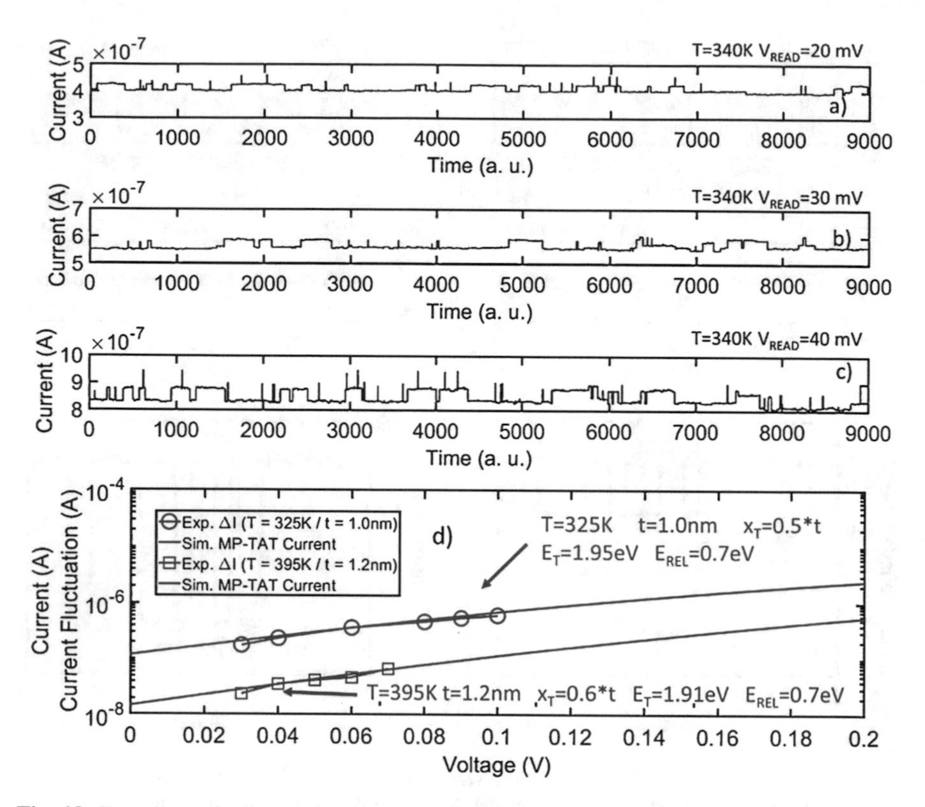

Fig. 18 Experimental RTN traces recorded at different voltages (**a**–**c**) on a device in HRS, without altering its state, are used to retrieve the statistical properties of a single two-level RTN fluctuation (through the FHMM) as a function of the applied voltage. (**d**) Experimental (symbols) and simulated (solid lines) TAT current fluctuations due to an individual Vo^+ defect, ΔI, extracted from the analysis of RTN traces recorded at different temperatures and voltages. The simulation of the ΔI–V curves allows extracting the position of the individual Vo^+ defect within the barrier x_T, and its thermal ionization energy, E_T, t is the barrier thickness, extracted using the compact RRAM model in [75, 77, 83]

far from that of any other oxygen vacancy configuration. Nicely, this agreement is verified in more than one device, and in different reset conditions, as shown in Fig. 19. The results of the analysis indicate that the Coulomb interaction between a Vo^+ defect (assisting charge transport) and the charge trapped at a "slow" defect (e.g., neutral interstitial oxygen) in the proximity of the Vo^+ could be responsible for the RTN current fluctuations.

The second physical mechanism that can possibly explain RTN fluctuations is based on the possibility for the Vo^+ defect to exhibit both a stable and a metastable defect configuration [1]. Metastable states for oxygen vacancy defects have been reported for SiO_2 [1] but are still unidentified in the case of HfO_2. Nevertheless, the mechanism can be explored based on the 3-state oxygen vacancy model shown in Fig. 20. The Vo^+ defect is supposed to have two different positively charged

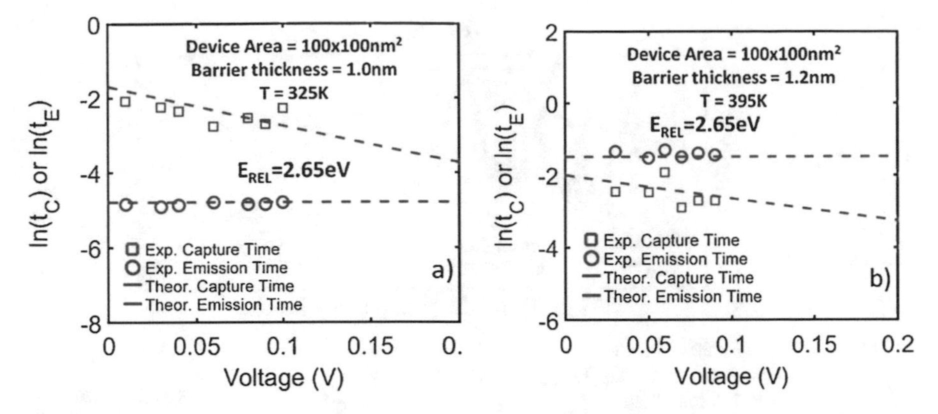

Fig. 19 Simulated (dashed lines), and experimental (symbols) RTN capture (red) and emission (blue) times extracted at different voltages for the same two devices considered in Fig. 18. Simulations consider the charge trapping at a "slow" defect in the proximity of the Vo^+ defect, which is responsible for the TAT current in Fig. 18. The relaxation energy considered for the "slow" defects is 2.65 eV. The thickness of the HfO_x dielectric barrier is extracted from the *I–V* curves using the compact RRAM model in [75, 83] for both devices (**a**) and (**b**)

states (labeled as stable and meta-stable). These two states are separated by a relatively large energy barrier, which makes the transition between the two states rather slow. Notably, these transitions are thermally activated and are not associated with charge trapping or de-trapping. Conversely, the transition from the positive meta-stable to the neutral state (electron trapping) and vice versa (electron emission) is significantly faster due to the lower energy barrier, providing the mechanism that could assist the TAT current. The results of KMC simulations accounting for the existence of metastable defect configurations confirm that the transition of the defect into the positive stable state effectively suppresses the TAT current, resulting in the observed RTN fluctuations.

Results suggest that a ~0.8-eV barrier between the stable and metastable state of the Vo^+ defect is required to obtain RTN fluctuations exhibiting capture and emission times that are compatible with observations. These results may be useful to try to identify the corresponding possible atomic configuration of the Vo^+ defect.

3.4 *Physics of RTN in LRS*

The RTN phenomenon has been frequently detected not only in HRS, but also in LRS. Nevertheless, it has been shown [43, 93] that the relative fluctuation amplitude ($\Delta I/I$) tends to be much smaller than in HRS, making the RTN issue less threatening in LRS than in HRS. Since the HRS and the LRS are characterized by different charge transport modes, RTN in LRS is expected to show at least a few different features as compared to the HRS case. However, the physics of the RTN in the two

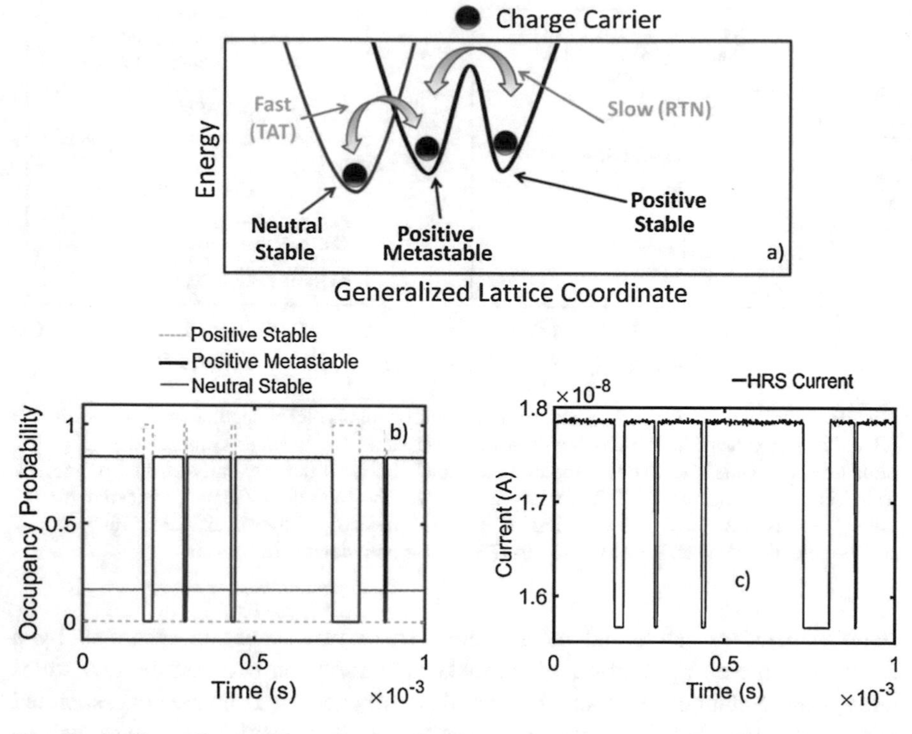

Fig. 20 (**a**) Generalized configuration coordination diagram showing the 3-state oxygen vacancy model explaining the RTN current fluctuation. The TAT current is given by the transitions occurring upon charge trapping between the neutral and the positive metastable states of the oxygen vacancy. The RTN current fluctuations are related to the relaxation of the positive metastable state into a stable configuration: this is a "slow" process, dominated by a relatively high thermal barrier, temporarily inhibiting the TAT charge transport and causing RTN. (**b–c**) KMC simulations of the evolution over time of (**b**) the occupancy probability and (**c**) the corresponding TAT current via an oxygen vacancy defect modeled as a 3-state defect. In (**c**), the absence of white noise on the lower current level is due to the fact that only one defect is considered in this simulation

resistive states could also show important similarities. Indeed, a similar mechanism to the one responsible for the RTN in HRS could also be responsible for the RTN fluctuations in LRS, as schematically illustrated in Fig. 21, despite the different charge transport mode in LRS (i.e., electrons drift through a conductive sub-band resulting from the very high defect density in the CF).

Because of the resistive switching physics, the CF is expected to be surrounded with additional defects (potentially both Vo^+ and oxygen atoms/ions). Electron trapping and emission to and from a defect located in close proximity of the CF may affect the local CF conductivity due to the effect of the Coulomb potential of the trapped charge, thus affecting the charge transport (i.e., the current in the CF) [23, 43, 93]. In this case, the screening effect of the trapped charge on the charge transport along the CF would be much smaller than that observed in HRS because

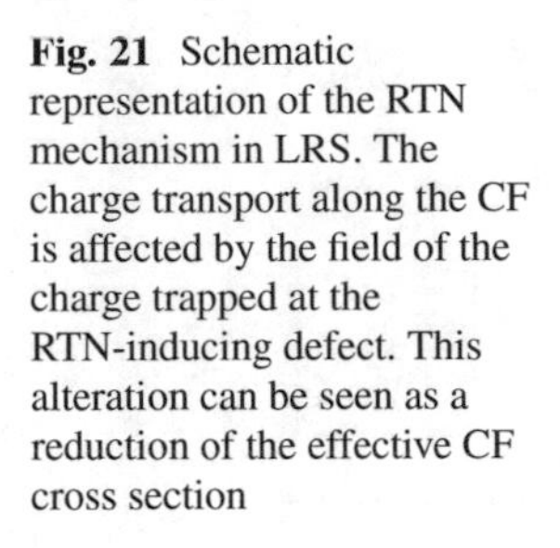

Fig. 21 Schematic representation of the RTN mechanism in LRS. The charge transport along the CF is affected by the field of the charge trapped at the RTN-inducing defect. This alteration can be seen as a reduction of the effective CF cross section

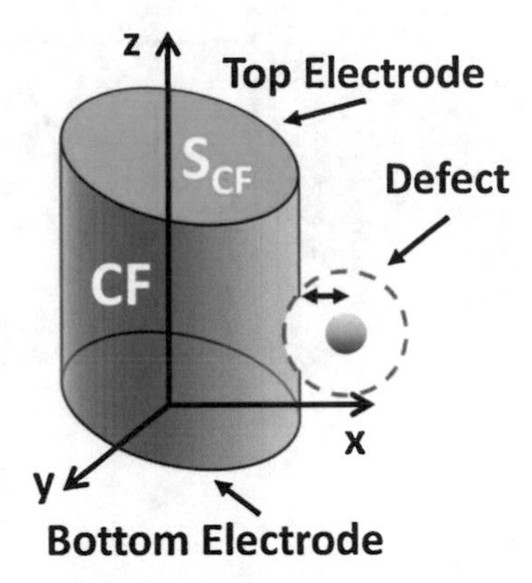

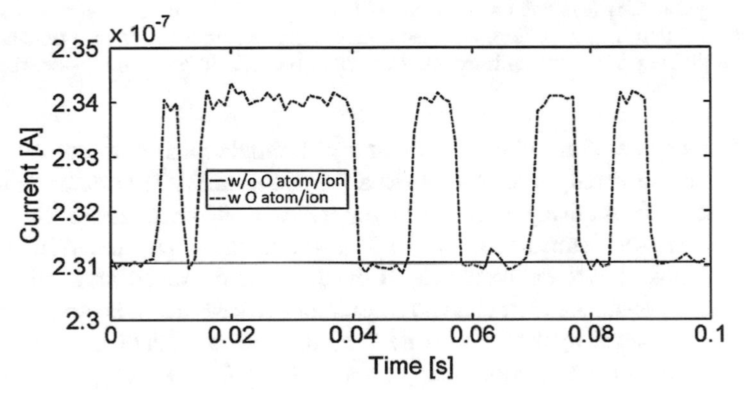

Fig. 22 KMC simulation of a 5-nm thick CF with a 10 nm^2 cross section. The density of Vo^+ defects within the CF is set to $N_D = 2 \times 10^{22}$ cm^{-3}. The current evolution over time is simulated with and without an RTN-inducing defect (O atom/ion) in the proximity of the CF, indicated with a dashed and a solid line, respectively. Trapping and de-trapping at the defect in close proximity to the CF—green spot in Fig. 21a—are simulated using a TAT model that accounts for the 3D local electric field and temperature profile

of the quasi-metallic character of the CF that increases the immunity to the potential perturbation induced by the trapped charge.

In order to verify this idea, we performed full 3D RTN simulations of a RRAM device in LRS using a KMC approach. Simulations include the effect of the trapped charge potential perturbation and of the localized power dissipation in the CF (together with the associated temperature gradient) due to the charge transport, calculated in the framework of the Landauer formalism [94]. Figure 22 shows the results of KMC simulations of a 5-nm thick CF with a narrow cross-section $S = 10$ nm^2. The CF is composed of Vo^+ defects with a very high density $N_D = 2 \times 10^{22}$ cm^{-3} [91]. An extra defect near the CF is also included to verify whether charge trapping at its site can efficiently affect the charge transport in the CF. The electron trapping at this defect results in the apparent RTN current fluctuation, confirming a picture similar to that proposed for the HRS.

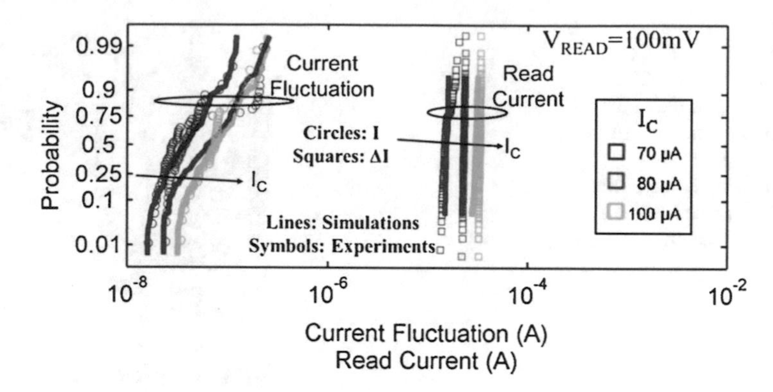

Fig. 23 Experimental (symbols) and simulated (lines) probability plots of current fluctuation amplitude, ΔI (left, empty circles), and read current extracted from I to V data (middle, empty squares) at different current compliances (I_C from 70 to 100 μA, different colors—see the legend)

To further verify the model, we employed simulations to reproduce the ΔI distributions measured on RRAM devices formed at different current compliance levels, each consecutively cycled 100 times. At each switching cycle, the CF cross section was estimated from the R_{LRS} resistance extracted from the I–V curve, and was used to define the CF in the simulation. As shown in Fig. 23, the agreement between measurements and simulations is accurate, indicating that the ΔI distribution is caused by the defects (inducing RTN current fluctuations) distributed around the CF with a relaxation energy $E_{REL} = 1.2 \pm 0.2$ eV (compatible with that of Vo^+ defects). The same distribution allows reproducing also the distributions of the average capture, τ_c, and emission, τ_e, times detected in the experiments, as shown in Fig. 23.

Both τ_c and τ_e do not depend, as expected, on the current compliance [23, 93], which in turn control the filament cross section, see Fig. 24. Charge trapping at defect sites in the surroundings of the CF is simulated by considering all possible charge transitions (including trap-to-trap transitions), also including tunneling from/into the defects constituting the CF itself. However, it has to be noticed that the above simulations can reproduce the capture and emission times distributions of RTN fluctuations measured using a 10-μs sampling time and a 0.1-s total measurement time. Nevertheless, RTN fluctuations in LRS were also detected in different time ranges, from several milliseconds to several tens of seconds [18, 23, 43, 93]. These slower RTN fluctuations can hardly be associated with Vo^+ defects as they exhibit a relatively small relaxation energy ($E_{REL} = 1.2$ eV). Such slower RTN fluctuations may be determined by the obvious presence of oxygen atoms/ions around the CF, due to their larger relaxation energy ($E_{REL} = 2.65$ eV) [15, 19, 88–90].

Also in the case of RTN in LRS, the possible presence of metastable states for Vo^+ defects could deliver an alternative explanation for RTN. Considering the defects constituting the CF, a thermally activated transition from the metastable state

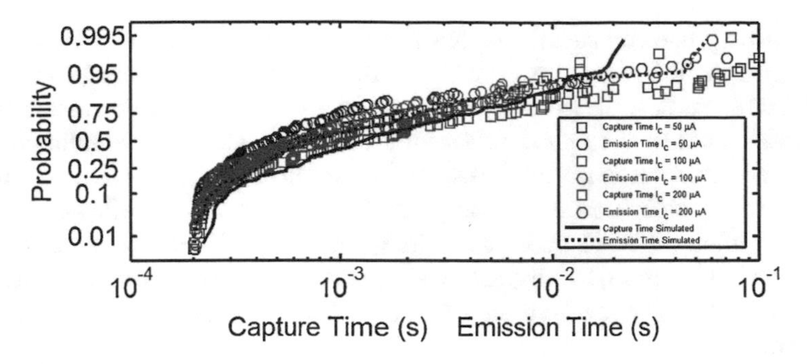

Fig. 24 Experimental (symbols) and simulated (lines) probability plots of capture, τ_c, and emission, τ_e, times measured on devices formed at different current compliances (only three compliances are reported for clarity). Both capture and emission time distributions are not sensitive to the current compliance

to the stable state could effectively suppress the current contribution driven by a defect, as shown earlier for the HRS case. Nevertheless, since the CF is composed of very densely packed defects, their representation as isolated entities loses its validity, hence it becomes challenging to predict the effects of such transitions on the charge transport in LRS. Another possibility is that the Vo^+ defects in the surroundings of the CF can experience a transition to a neutral stable state, which corresponds to holding a trapped electron for a relatively long time causing a screening effect on the CF identical to the one discussed above.

4 Compact Model of RTN

As RRAM technology is approaching the industrial stage, RTN mechanisms must not only be well understood but also formalized in a compact fashion to be included in RRAM device compact models [74–84]. This is vital to evaluate the technological possibilities and limitations for definite applications such as nonvolatile memory (NVM) [29, 30], neuromorphic computing [31, 32], LiM [33, 34], PUF [35, 36], and RNG [37, 38]. For instance, RTN may reduce the margin between neighboring bit distributions (which is detrimental especially for multibit NVM elements). RTN can cause unpredictable variations of the synaptic weight associated with the device when used in neuromorphic applications. RNGs and PUFs can also be affected by RTN, which contributes to worsen their figures of merit. On the other hand, the intrinsic randomness of RTN attracted the interest of some people in the community who suggested to realize RTN-based PUFs and RNGs [35–38], which could exploit RTN in a favorable manner as an entropy source. The strong temperature dependence of RTN in RRAM has also been envisioned as a way to implement RTN-based temperature sensors [95]. From the circuit designer perspective, it is therefore imperative to include RTN effects into RRAM models.

However, only one compact model has been introduced in the literature up to date [96, 97]. In the following, we present and review this compact model for RTN in RRAM, which is derived according to the physics of RTN delineated in this chapter. The model provides a compact representation of the fluctuation amplitude, of the capture and emission times, and considers the intrinsic variability in the number of defects and on their location in energy and space. As such, it allows simulating complex multilevel RTN traces over time, enabling RTN-aware simulations for advanced circuit design. Finally, practical uses of the proposed model are suggested by exploring the RTN-aware design of: (1) RRAM NVMs and (2) an RTN-based RNG circuit.

4.1 Statistical Model for the Fluctuation Amplitude

Developing a complete RTN compact model requires capturing the dependence of the RTN statistical properties not only on the microscopic properties of the defects but also on the operating conditions and on the physical state of the device (i.e., HRS or LRS). Deriving a compact formulation of such dependences requires formalizing the problem differently for the two resistive states, as they are characterized by different dominant charge transport mechanisms.

The Fluctuation Amplitude in HRS

The physical mechanism responsible for RTN in HRS is likely due to the temporary (de-)activation of Vo^+ defects, distributed in the barrier, assisting charge transport. The (de-)activation mechanism is itself due to charge (de-)trapping in additional “slow” defects not participating to charge transport (allegedly oxygen atoms/ions). A significant insight that helps building a compact formulation of the RTN ΔI statistics is given by physics-based simulations of charge transport across an HfO_2 dielectric barrier, Fig. 25. Simulations that account for the presence of defects and of the trapped charge are performed considering different applied voltage, barrier thickness, and temperature values. Defects are considered randomly distributed in space and energy [80, 86, 87], and we estimate $\Delta I/I$ by deactivating an individual, randomly selected, defect at a time. ΔI coincides with the current contribution associated with an individual Vo^+ active defect [15, 19], and the overall current, I, is the sum of all the contributions from all the defects involved in charge transport.

From the results in Fig. 25, it is apparent that only a very small fraction (magenta spheres) of all the Vo^+ defects in the barrier (represented by black squares) is responsible for almost the entire current (i.e., >95%), regardless of the barrier thickness and of the operating conditions (i.e., voltage and temperature). This result is expected because of the physics of TAT transport, according to which the defects in the middle of the barrier and energetically aligned with the emitting electrode (the bottom one for positive applied voltage) drive exponentially larger current shares.

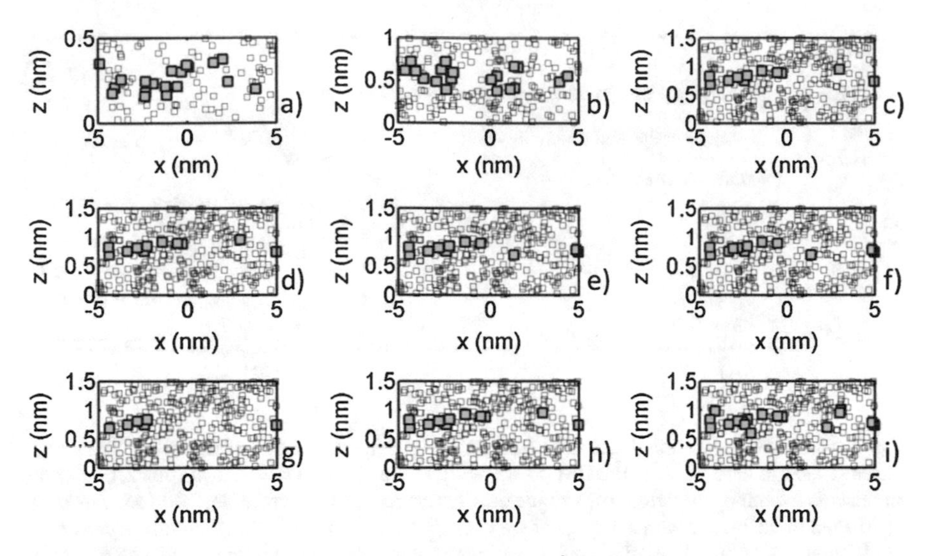

Fig. 25 Simulation of charge transport in a 10 × 10 nm^2 barrier. Projection of the 3D defects distribution on a two-dimensional frontal cross section. Vo^+ defects (squares) are randomly distributed in the barrier with a density of $N_T = 2{\cdot}10^{21}$ cm^{-3} [88–91]. Few Vo^+ defects (light blue squares) in the middle of the barrier (±0.2 nm) assist >95% of the total current, regardless of the operating conditions. **(a–c)** $t_b = 0.5$ nm–1.0 nm–1.5 nm at $V_{READ} = 100$ mV and $T = 25$ °C. **(d–f)** $t_b = 1.5$ nm at $V_{READ} = 100$ mV and $T = 25$ °C–75 °C–125 °C. **(g–i)** $t_b = 1.5$ nm at $V_{READ} = 50$ mV–100 mV–150 mV and $T = 25$ °C

Because the scenario depicted in Fig. 25 is invariant with operating conditions, the average $\Delta I/I$ (or equivalently $\Delta R/R$) is expected to show the same invariance (i.e., it is expected to show no dependence on the operating condition). This is confirmed by the results reported in Fig. 26, where the lognormal probability distribution of the $\Delta I/I$ extracted from the simulations is found to be invariant with operating conditions.

According to the results in Figs. 25 and 26, notwithstanding the significant intrinsic variability and the complexity of the RTN mechanism, it is possible to derive a compact formulation for the $\Delta I/I$ (or equivalently $\Delta R/R$) lognormal statistics, completely described by its median (M) and its standard deviation, (σ), as in Eq. (6).

$$M\left(\frac{\Delta R}{R}_{\text{HRS}}\right) \cong \frac{1}{2} \qquad \sigma\left(\frac{\Delta R}{R}_{\text{HRS}}\right) \cong 0.6 \tag{6}$$

The Fluctuation Amplitude in LRS

In contrast to the HRS, the relative resistance change ($\Delta R/R = \Delta I/I$) depends on the CF geometry, i.e., its radius (r_{CF}) (assuming a cylindrical CF), and its thickness (t_{ox}). For a large CF (i.e., $r_{CF} >> \lambda$, with λ being the electron Debye length), the

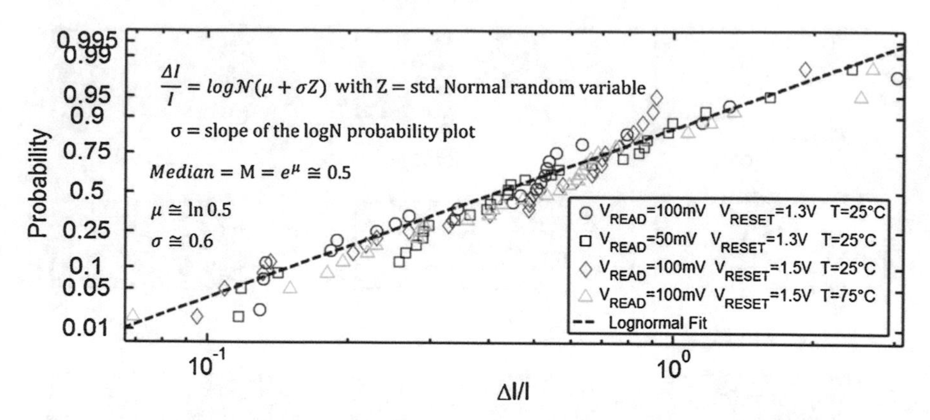

Fig. 26 Distribution of simulated $\Delta I/I$ in HRS obtained using KMC simulations. The I-time traces related to a 10×10 nm^2 MIM device with TiN electrodes with a 1-nm-thick barrier are simulated, including the effect of Vo$^+$ defects activation and deactivation. Vo$^+$ are randomly distributed in the barrier with a density of $N_T = 2{\cdot}10^{21}$ cm^{-3} [88–91]. Simulations are performed in different conditions (V_{READ}, t_b, and T) and are repeated on 10 different device realizations by randomizing defects positions and energies (40 simulations). $\Delta I/I$ distribution is insensitive to operating conditions in agreement with experimental reports [15, 19, 21]. Its statistical properties (μ and σ) are reported in the figure

CF area being screened is much smaller than the whole CF area, resulting in a relatively small resistance change [23, 42, 93]. For a smaller CF area, the ratio of the screened portion of the CF to the whole CF area increases, resulting in a larger relative average resistance change, as also reported in [23]. While the average effect is CF size-dependent, the deviations from the average are CF-size independent, as they are controlled by the distance between the defect and the CF border. So, the lognormal $\Delta I/I$ distribution is expected to have a σ value that is invariant with the CF size [23], as opposite to the median value that changes with the CF size, as in Eq. (7).

$$M\left(\frac{\Delta R}{R}_{\text{LRS}}\right) = f\left(S_{\text{CF}}\right) \qquad \sigma\left(\frac{\Delta R}{R}_{\text{HRS}}\right) \cong 0.3 \tag{7}$$

Henceforth, the discussion will be focused on the description of the relation between the median of the distribution and S_{CF}. Deriving such relation is very complex without introducing a simplified geometric model (based on reasonable assumptions). The latter accounts for the effect of a trapped charge at a defect close to the CF on the transport in LRS, in terms of relative resistance fluctuation, i.e., $\Delta R/R$. A linear charge transport in the CF is assumed, in line with the linear I–V relation observed in LRS. In addition, a constant CF resistivity (ρ_{CF}) in space and a geometrically uniform CF are assumed. To simplify the problem, here we assume an individual defect at the border of the CF (i.e., we assume zero distance between the defect and the CF border), as shown in Fig. 27. This defect induces a complete screening on a portion of the CF when filled with an electron. Moreover, we consider

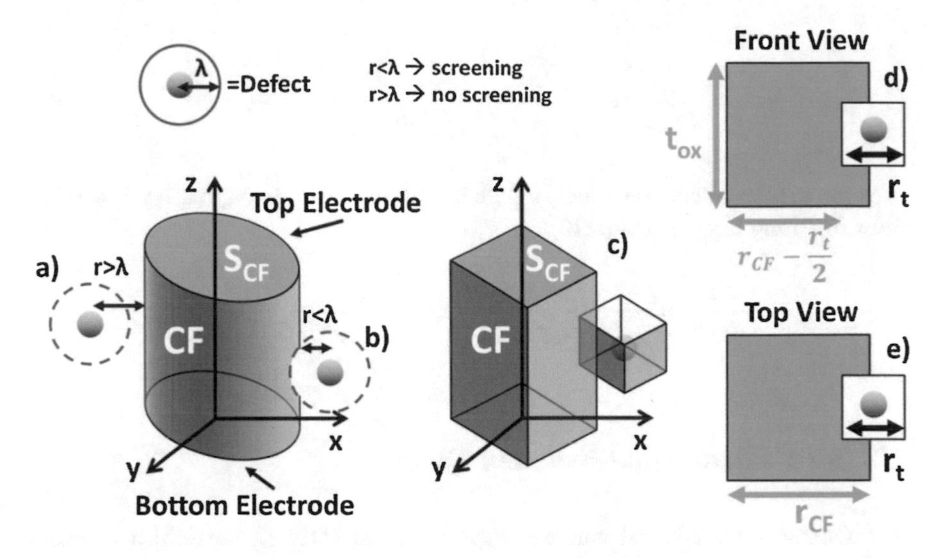

Fig. 27 Schematic of RTN mechanism in LRS. Defects close to the CF ($r < \lambda$) cause a screening effect on a portion of the CF, inducing a resistance change. (**a**) No screening ($r > \lambda$). (**b**) Screening ($r < \lambda$). (**c**) Simplified geometrical framework with (**d**) its front view and (**e**) its top view

that the trapped charge produces uniform and complete screening (the screened portion of the CF is completely unavailable to charge transport) in a region around itself, and no screening beyond this region. Such region is supposed to be spherical, but we simplify the spherical symmetry of the electric field of the trapped charge to a cube with side r_t, a parameter accounting for the effective screening length of the trapped charge in this simplified framework, related to λ [23, 43, 96, 97]. In this context, r_t can be seen as a free parameter, compensating for the assumption of zero distance between the CF and the trapped charge. The full CF resistance with no screening effect (R_{LRS}), and the relative resistance change induced in the CF by the trapped charge ($\Delta R/R_{LRS}$), can be written as in Eq. (8).

$$R_{LRS} = \frac{\rho_{CF}\, t_{ox}}{S_{CF}} \qquad \frac{\Delta R}{R_{LRS}} = \frac{R_{LRS}{}^{*} - R_{LRS}}{R_{LRS}} \tag{8}$$

where $R_{LRS}{}^{*}$ is the resistance of the CF when the screening effect occurs. In the case of a cylindrical filament, calculating this term is complex.

To simplify calculations, Fig. 27, we consider the CF to be a prism with a square base and cross-section S_{CF}, i.e., side r_{CF}. This allows rewriting Eq. (8) as $R_{LRS} = \rho_{CF}\, t_{ox}/r_{CF}{}^2$. Under these conditions, the $\Delta R/R_{LRS}$ in Eq. (8) can be written in terms of r_t and of the CF properties (t_{ox}, r_{CF}, ρ_{CF}) as in Eq. (9).

$$\frac{\Delta R}{R_{\mathrm{LRS}}} = \frac{\left[\frac{\rho_{\mathrm{CF}}\,(t_{\mathrm{ox}}-r_t)}{r_{\mathrm{CF}}{}^2} + \frac{\rho_{\mathrm{CF}}\,r_t}{r_{\mathrm{CF}}{}^2 - \frac{r_t{}^2}{2}}\right] - \frac{\rho_{\mathrm{CF}}\,t_{\mathrm{ox}}}{r_{\mathrm{CF}}{}^2}}{\frac{\rho_{\mathrm{CF}}\,t_{\mathrm{ox}}}{r_{\mathrm{CF}}{}^2}} \tag{9}$$

Some mathematical operations, together with the assumption that $r_{\mathrm{CF}}{}^2 \gg r_t{}^2/2$, allow rewriting Eq. (9) as Eq. (10).

$$\frac{\Delta R}{R_{\mathrm{LRS}}} = \frac{r_t{}^3}{2t_{\mathrm{ox}}\left(r_{\mathrm{CF}}{}^2 - \frac{r_t{}^2}{2}\right)} \cong \frac{r_t{}^3}{2t_{\mathrm{ox}}\,r_{\mathrm{CF}}{}^2} = \frac{r_t{}^3}{2t_{\mathrm{ox}}\,S_{\mathrm{CF}}} \tag{10}$$

4.2 RTN Capture and Emission Times

To complete the statistical compact model for the RTN fluctuation, a compact formulation of the τ_{c} and τ_{e} of each defect must be included. This is essential to reproduce the RTN signals over time and to perform transient simulations including RTN and its dynamics. Regardless of the resistive state, the physical mechanism causing RTN is associated with charge (de-)trapping and is best represented in the framework of the TAT formalism. As such, τ_{c} and τ_{e} can be calculated with the simplified TAT equations derived in [98], as in Eq. (11).

$$\tau_{\mathrm{C}} \propto \mathrm{e}^{\left(\frac{d}{\lambda_{\mathrm{c}}}\right)}\mathrm{e}^{\left(\frac{E_{\mathrm{C}}(V)}{kT}\right)};\ \tau_{\mathrm{e}} \propto \mathrm{e}^{\left(\frac{d}{\lambda_{\mathrm{e}}}\right)}\mathrm{e}^{\left(\frac{E_{\mathrm{e}}}{kT}\right)} \tag{11}$$

These formulae encompass the temperature and voltage dependence of the capture and emission times, as well as their dependence on the physical location of the defect (the distance from the electrodes, d) and on the microscopic properties of the atomic species considered. In particular, the thermal ionization energy and the relaxation energy of the defects must be specified. In these formulae, they are effectively lumped in the typical capture (λ_{c}) and typical emission (λ_{e}) lengths, and the capture (E_{c}) and emission (E_{e}) activation energies [73, 80, 98]. These parameters are calculated exploiting ab initio calculations (i.e., density functional theory or molecular dynamics) [88–91]. Since charge carriers can be captured from (and emitted to) either the top or the bottom electrodes (notice that in HRS the role of one electrode is played by the interface between the barrier and the remaining portion of the CF), τ_{c} and τ_{e} are calculated for both cases. Two possible capture times and two possible emission times are so considered, and the minimum value is chosen for both τ_{c} and τ_{e}, as it is representative of the most likely event between the two.

4.3 Compact Model Validation

The effectiveness of the compact RTN model is validated against an extensive experimental dataset (Fig. 28). Five RRAM devices (exhibiting a 5-nm HfO_2 layer) formed at different I_C values (40–150 μA) were used. To demonstrate the ability of the model to catch the intrinsic variability of the RTN parameters, all devices have been cycled through HRS and LRS for 100 times, collecting RTN data in both resistive states. This allows also accounting for the cycle-to-cycle and device-to-device variability simultaneously. In addition, to confirm the ability of the model to reproduce the dependence of RTN on the operating conditions, the experiment was repeated for all devices at different V_{RESET} and T. The experimental $\Delta I/I$ vs. I scatter plot shown in Fig. 28 comprises all the data from all the devices in all operating conditions. Notably, the compact model predictions are in excellent agreement with experimental data. It is noteworthy how the model not only captures the average behavior, but also demonstrates its validity on a statistical basis to high precision. Indeed, almost the entire experimental data set falls within the $\pm 3\sigma$ predictions reported as gray dashed lines.

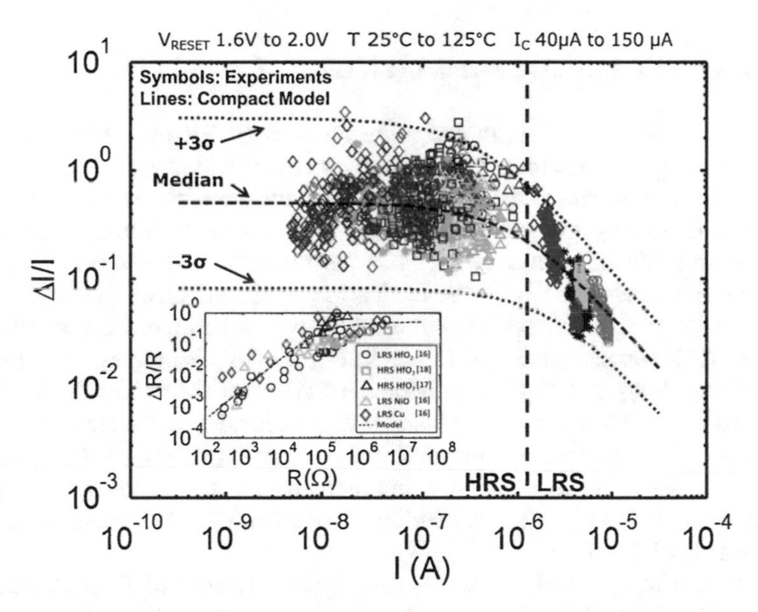

Fig. 28 Experimental $\Delta I/I$ vs. I (symbols) in HRS (left) and LRS (right) at different operating conditions (I_C, V_{RESET}, T) for five devices (different symbols) with a 5-nm HfO_2 layer. Model predictions (lines) are in excellent agreement with experimental data. The extracted σ values of the normal distributions associated with the lognormal $\Delta I/I$ distributions in HRS and LRS agree with the values extracted from simulations and used in the compact model (the median prediction is the dashed line, the $\pm 3\sigma$ predictions are the dotted lines). Inset—The compact model median prediction plotted as $\Delta R/R$ vs. R (dashed line) agrees with RTN data in HRS and LRS taken from the literature (symbols) [16–18]. Data from devices based on different materials (i.e., NiO and Cu) are included, showing a good agreement with the model predictions

The compact model predictions are also in excellent agreement with data found in the literature for HfO_2 RRAM devices, as well as for devices employing different materials for the switching layer (e.g., Cu and NiO) [16–18], which is consistent with the compact model not depending on the CF resistivity. This finding suggests that the same (or a similar) picture for the RTN mechanism may hold also for such devices. The compact model proposed here has been readily integrated into existing RRAM device compact models [75, 83, 84] as it can be implemented in Verilog-A. The reader is encouraged to consult [96, 97] for the details of the model implementation.

4.4 Compact Model Applications and Advanced Circuit Design

The proposed model allows considering the presence of RTN in circuit simulations, which is advantageous in the design of circuits for many applications. Here we illustrate two examples: (1) the design of RRAM memory circuits considering RTN and (2) the design of an RTN-based RNG.

Design of RRAM Circuits Considering RTN

One of the major constraints in designing an array of RRAM devices for memory application is to guarantee a sufficient memory window, i.e., to ensure the distinguishability of stored bits during the read operation at any time and in any condition. Obviously, the presence of RTN may reduce such margin, especially when dealing with multi-bit cells and when fast readout is required [97]. Therefore, including the effect of RTN in the evaluation of the effective memory window is mandatory. In this context, the effect of RTN can be quantified by using the proposed RTN compact model, as shown in Fig. 29. To this purpose, we simulated 100 full switching cycles of an individual RRAM cell for two different V_{RESET} values, namely 1.0 V and 1.6 V, with and without considering the presence of RTN fluctuations. The resistance of the device is evaluated at each cycle by applying a small read voltage $V_{READ} = 100$ mV and using an averaging time of 100 μs. The corresponding HRS resistance distributions (that account for the cycling variability) are reported in Fig. 29.

Regardless of the reset condition, tails appear because of RTN fluctuations, whose effect is detrimental for multi-bit memory applications. The resulting shrinking of the margin between adjacent bit distributions can indeed lead to read failures (i.e., the impossibility to distinguish unambiguously the stored bits). In the example of Fig. 29, simulations that do not account for the RTN fluctuations predict a small but nonzero margin between the two HRS resistance distributions obtained in the two reset conditions $V_{RESET} = 1.0$ V and $V_{RESET} = 1.6$ V. However, including the effect of RTN clearly shows that the two distributions largely overlap due to the presence of RTN-related tails, so that the two reset conditions ($V_{RESET} = 1.0$ V and

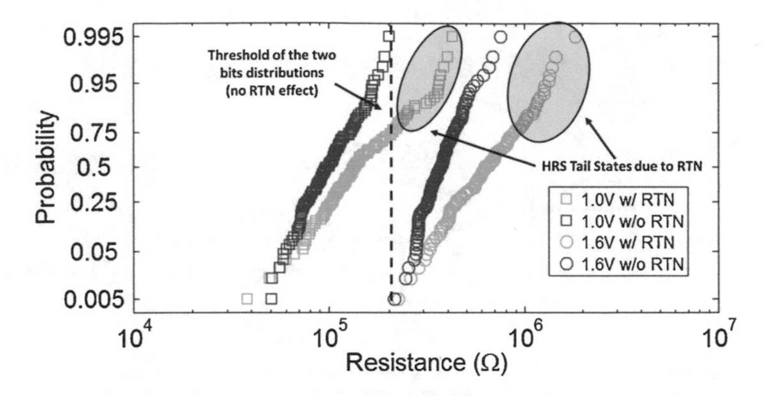

Fig. 29 HRS resistance (read at $V_{READ} = 100$ mV) distributions obtained by cycling an individual cell 100 times with (light blue symbols) and without (violet symbols) RTN, for $V_{RESET} = 1.0$ V (squares) and 1.6 V (circles). The presence of RTN broadens the resistance distribution and results in tail states

$V_{RESET} = 1.6$ V) cannot be used to drive the device in two distinct logic states that can be unambiguously resolved. In general, RTN reduces the number of bits that can be stored in a single cell.

Design of an RTN-Based Random Number Generator Circuit

Recently, the idea of exploiting RTN as a resource has drawn significant attention, as it seemingly represents a source of true physical randomness (entropy) that could be exploited in the realization of true RNG circuits [37, 38] for security, encryption, and authentication. Of course, the design of such circuits is constrained to the availability of appropriate compact models. Hereafter, we demonstrate how the proposed RTN compact model can be used to design an RNG circuit. The circuit topology we explore, Fig. 30, comprises an RRAM device in HRS (with $t_b = 1$ nm) and a transistor connected in series, together with a buffer, a simple first-order high-pass filter, and a Schmitt trigger. Simulations have been run in Cadence Virtuoso™, using a compact RRAM model [75, 83, 84] extended with the proposed compact RTN model. The results of a transient simulation are shown in Fig. 30. The application of a small read voltage, V_{READ}, to the top electrode of the RRAM device causes RTN fluctuations in the current, and, consequently, on the voltage at the drain of the transistor. The RTN pattern is buffered to avoid any load effect on the RRAM device. The high-pass filter eliminates the undesired DC and low-frequency components. Finally, the AC RTN signal is fed to the Schmitt trigger. The amount of hysteresis that the trigger can withstand is regulated by the reference voltage V_{REF}. The randomness in the RTN signal is successfully exploited to generate a random bit stream at the output of the trigger characterized by a noteworthy randomness figure of merit of 50.8% (with 50% being perfect randomness) [34]. This example is not intended, of course, as a correct and robust design process of an RNG based on

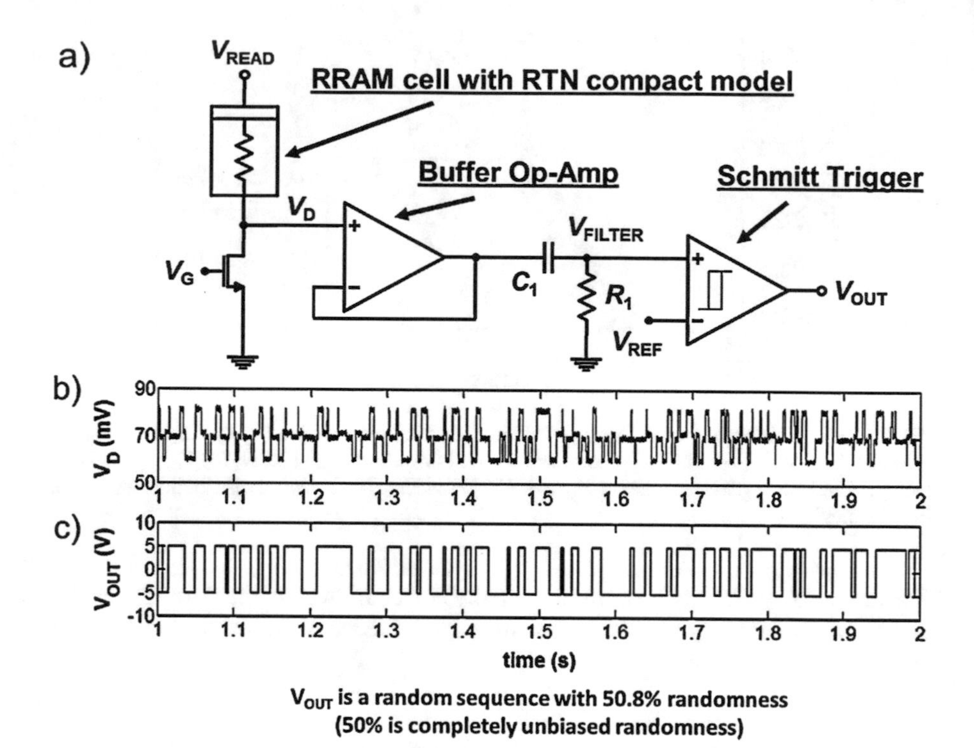

Fig. 30 (**a**) Schematic of the RTN-based RNG circuit. An RRAM device in HRS with $t_b = 1$ nm is considered. The compact device model used here is found in [75, 83, 84] and has been extended with the proposed RTN compact model. (**b**) RTN voltage fluctuations simulated at the transistor drain node. (**c**) The random RTN pattern is reproduced at the circuit output, giving a random stream with an excellent randomness (50.8%)

RTN, but clearly shows that the compact RTN model allows the design of advanced RTN-based circuits, paving the way to the design of a new class of circuits for a variety of applications.

5 Conclusions

Noise issues in RRAM devices (primarily in the form of RTN) are still a relevant challenge for the development of the RRAM technology for several applications. In recent times, significant progress has been made by several research groups toward the assessment of the physical mechanisms ruling over the phenomenon. At the same time, additional efforts are required to refine the understanding of, for instance, temporary and anomalous phenomena. From the device physicist's viewpoint, RRAM has been and still is an excellent playground for RTN analysis, being a device in which many interesting aspects emerge clearly, though at the cost of facing additional complexity. Nevertheless, such difficulties helped in finding new approaches and more refined "cross-platform" analysis techniques (i.e., applicable to a large variety of devices), eventually leading to the development of a deeper understanding of the RTN phenomenon itself, as well as of the microscopic properties of a variety of materials and associated defects. Recent advancements in the field have also attracted the attention of the community working on the characterization of the materials at the nanoscale. Indeed, RTN investigations are currently performed on blanket films employing tools like the conductive-atomic force microscopy and the scanning tunneling microscopy [99–102], besides conventional device-level characterization. In addition, prototypical RRAM devices based on emerging 2D materials (e.g., graphene, boron nitride, and molybdenum disulfide) are being actively investigated [102–105]. On the one hand, the advanced tools for RTN analysis developed so far will surely help in understanding the (sometimes) surprising properties of such emerging materials; on the other hand, their characterization will in turn assist further developments of the RTN analysis tools.

From the standpoint of the circuit designer, it is finally the time to redeem RTN by observing it from a different perspective, thanks to the availability of new tools (e.g., RTN compact model for RRAM devices). As RRAM technology is entering the industrial phase, the availability of new resources translates to a tremendous opportunity for circuit designers, who can more easily cope with the RTN "nightmare" and, concurrently, start exploiting RTN properties to devise a new class of circuits for a variety of applications.

Acknowledgments The author would like to express gratitude to Prof. Paolo Pavan and Prof. Luca Larcher (Università di Modena e Reggio Emilia, Italy) for uncountable useful discussions and for providing excellent insights; Dr. Andrea Padovani and Dr. Luca Vandelli (MDLSoft Inc., USA) for continuous support and for the help with advanced simulation tools; Dr. Nagarajan Raghavan (Singapore University of Technology and Design, Singapore) for constant support and fruitful collaboration; Dr. Gennadi Bersuker and Dr. Dmitry Veksler (The Aerospace Corporation, USA) for providing samples and for insightful discussion; Mr. Nick Hill for useful technical advices and Ms. Valentina Reda for valuable support.

References

1. T. Grasser, Stochastic charge trapping in oxides: From random telegraph noise to bias temperature instabilities. Microelectron. Reliab. **52**, 39–70 (2009). https://doi.org/10.1016/j.microrel.2011.09.002
2. M.J. Kirton, M.J. Uren, Noise in solid-state microstructures: A new perspective on individual defects interface states and low-frequency (1/ f) noise. Adv. Phys. **38**(4), 367–468 (1989)
3. E. Simoen, C. Claeys, Random telegraph signal: A local probe for single point defect studies in solid-state devices. Mater. Sci. Eng. B **91–92**, 136–143 (2002)
4. C. Claeys, E. Simoen, Noise as diagnostic tool for semiconductor material and device characterization. J. Electrochem. Soc. **145**(6), 2058–2067 (1998). https://doi.org/10.1149/1.1838597

5. L.K.J. Vandamme, Opportunities and limitations to use low-frequency noise as a diagnostic tool for device quality, in *Proceedings of the 17th International Conference Noise and Fluctuations ICNF 2003*, Prague 2003, pp. 735–748
6. Z. Chobola, Noise as a tool for non-destructive testing of single-crystal silicon solar-cells. Microelectr. Reliab. **41**, 1947–1952 (2001)
7. Y. Mori, K. Ohyu, K. Okonogi, R.-I. Yamada, The origin of variable retention time in DRAM, in *Proc. IEEE Int. Electron Devices Meeting*, Washington, DC, USA, 2005, pp. 1034–1037
8. F.M. Puglisi, F. Costantini, B. Kaczer, L. Larcher, P. Pavan, Probing defects generation during stress in high-κ/metal gate FinFETs by random telegraph noise haracterization, in *2016 46th European Solid-State Device Research Conference (ESSDERC)*, Lausanne, 2016, pp. 252–255. doi: https://doi.org/10.1109/ESSDERC.2016.7599633
9. H. Miki et al., Quantitative analysis of random telegraph signals as fluctuations of threshold voltages in scaled flash memory ells, in *Proc. 45th Int. Rel. Phys. Symp.*, Phoenix, AZ, USA, 2007, pp. 29–35
10. F.M. Puglisi, F. Costantini, B. Kaczer, L. Larcher, P. Pavan, Monitoring stress-induced defects in HK/MG FinFETs using random telegraph noise. IEEE Electron Device Letters **37**(9), 1211–1214 (2016). https://doi.org/10.1109/LED.2016.2590883
11. D.B. Strukov, G.S. Snider, D.R. Stewart, R.S. Williams, The missing memristor found. Nature **453**, 80–83 (2008). https://doi.org/10.1038/nature06932
12. J.K. Lee et al., Conduction and low-frequency noise analysis in Al/TiO_X/Al bipolar switching resistance random access memory devices. IEEE Electron Device Letters **31**(6), 603–605 (2010). https://doi.org/10.1109/LED.2010.2046010
13. M. Terai, Y. Sakotsubo, Y. Saito, S. Kotsuji, H. Hada, Memory-state dependence of random telegraph noise of Ta_2O_5/TiO_2 stack ReRAM. IEEE Electron Device Letters **31**(11), 1302–1304 (2010). https://doi.org/10.1109/LED.2010.2068033
14. S. Yu, R. Jeyasingh, Yi Wu, H.S. Philip Wong, Understanding the conduction and switching mechanism of metal oxide RRAM through low frequency noise and AC conductance measurement and analysis, in *2011 International Electron Devices Meeting*, Washington, DC, 2011, pp. 12.1.1–12.1.4. doi: https://doi.org/10.1109/IEDM.2011.6131537
15. F.M. Puglisi, L. Larcher, A. Padovani, P. Pavan, A complete statistical investigation of RTN in HfO2-based RRAM in high resistive state. IEEE Trans. Electron Devices **62**(8), 2606–2613 (2015). https://doi.org/10.1109/TED.2015.2439812
16. S. Ambrogio, S. Balatti, A. Cubeta, A. Calderoni, N. Ramaswamy, D. Ielmini, Understanding switching variability and random telegraph noise in resistive RAM, in *IEEE Proc. of IEEE Proc. of International Electron Device Meeting*, 2013, pp. 31.5.1–4, 9–11. doi: https://doi.org/10.1109/IEDM.2013.6724732
17. N. Raghavan, R. Degraeve, A. Fantini, L. Goux, S. Strangio, B. Govoreanu, D.J. Wouters, G. Groeseneken, M. Jurczak, Microscopic origin of random telegraph noise fluctuations in aggressively scaled RRAM and its impact on read disturb variability, in *Proc. of IEEE International Reliability Physics Symposium (IRPS)*, Anaheim, CA, 2013, pp. 5E.3.1–5E.3.7. doi: https://doi.org/10.1109/IRPS.2013.6532042
18. D. Veksler, G. Bersuker, L. Vandelli, A. Padovani, L. Larcher, A. Muraviev, B. Chakrabarti, E. Vogel, D.C. Gilmer, P.D. Kirsch, Random telegraph noise (RTN) in scaled RRAM devices, in *Proceedings of the IEEE International Reliability Physics Symposium (IRPS)*, 2013, MY.10.1–4. doi: https://doi.org/10.1109/IRPS.2013.6532101
19. F.M. Puglisi, P. Pavan, L. Vandelli, A. Padovani, M. Bertocchi, L. Larcher, A microscopic physical description of RTN current fluctuations in HfOx RRAM, in *Proc. of IEEE International Reliability Physics Symposium (IRPS),* Monterey, CA, 2015, pp. 5B.5.1-5B.5.6. doi: https://doi.org/10.1109/IRPS.2015.7112746
20. D. Veksler, G. Bersuker, B. Butcher, D. Gilmer, K. Matthews, S. Deora, Evaluation of variability and RTN in scaled RRAM, in *2014 IEEE International Integrated Reliability Workshop Final Report (IIRW)*, South Lake Tahoe, CA, 2014, pp. 52–52. doi: https://doi.org/10.1109/IIRW.2014.7049509

21. F.M. Puglisi, P. Pavan, L. Larcher, Random telegraph noise in HfOx Resistive Random Access Memory: From physics to compact modeling, in *2016 IEEE International Reliability Physics Symposium (IRPS)*, Pasadena, CA, 2016, pp. MY-8-1-MY-8-5. doi: https://doi.org/10.1109/IRPS.2016.7574624
22. S. Choi, Y. Yang, W. Lu, Random telegraph noise and resistance switching analysis of oxide based resistive memory. Nanoscale **6**(1), 400–404 (2014). https://doi.org/10.1039/c3nr05016e
23. S. Ambrogio, S. Balatti, A. Cubeta, A. Calderoni, N. Ramaswamy, D. Ielmini, Statistical fluctuations in HfOx resistive-switching memory: Part II—Random telegraph noise. IEEE Trans. Electron Devices **61**(8), 2920–2927 (2014)
24. Z. Chai et al., RTN-based defect tracking technique: Experimentally probing the spatial and energy profile of the critical filament region and its correlation with HfO_2RRAM switching operation and failure mechanism, in *2016 IEEE Symposium on VLSI Technology*, Honolulu, HI, 2016, pp. 1–2. doi: https://doi.org/10.1109/VLSIT.2016.7573402
25. D. Veksler et al., Methodology for the statistical evaluation of the effect of random telegraph noise (RTN) on RRAM characteristics, in *2012 International Electron Devices Meeting*, San Francisco, CA, 2012, pp. 9.6.1–9.6.4. doi: https://doi.org/10.1109/IEDM.2012.6479013RTN in RRAM
26. P. Huang et al., RTN based oxygen vacancy probing method for Ox-RRAM reliability characterization and its application in tail bits, in *2017 IEEE International Electron Devices Meeting (IEDM)*, San Francisco, CA, 2017, pp. 21.4.1–21.4.4. doi: https://doi.org/10.1109/IEDM.2017.8268435
27. J. Ma et al., Investigation of preexisting and generated defects in nonfilamentary a-Si/TiO2 RRAM and their impacts on RTN amplitude distribution. IEEE Trans. Electron Devices **65**(3), 970–977 (2018). https://doi.org/10.1109/TED.2018.2792221
28. M. Lanza, G. Bersuker, M. Porti, E. Miranda, M. Nafría, X. Aymerich, Resistive switching in hafnium dioxide layers: Local phenomenon at grain boundaries. Appl. Phys. Lett. **101**, 193502 (2012). https://doi.org/10.1063/1.4765342
29. M.-C. Hsieh, Y.-C. Liao,Y.-W. Chin, C.-H. Lien, T.-S. Chang, Y.-D. Chih, S. Natarajan, M.-J. Tsai, Y.-C. King, C.J. Lin, Ultra high density 3D via RRAM in pure 28nm CMOS process, in *IEEE International Electron Devices Meeting (IEDM)* (Tech. Digest, 2013), pp. 10.3.1–10.3.4. doi: https://doi.org/10.1109/IEDM.2013.6724600
30. X.P. Wang, Z. Fang, X. Li, B. Chen, B. Gao, J.F. Kang, Z.X. Chen, A. Kamath, N.S. Shen, N. Singh, G.Q. Lo, D.L. Kwong, Highly compact 1T-1R architecture (4F2 footprint) involving fully CMOS compatible vertical GAA nano-pillar transistors and oxide-based RRAM cells exhibiting excellent NVM properties and ultra-low power operation, in *IEEE International Electron Devices Meeting (IEDM)* (Tech. Digest, San Francisco, CA, 2012), pp. 20.6.1–20.6.4. doi: https://doi.org/10.1109/IEDM.2012.6479082
31. S. Park, H. Kim, M. Choo, J. Noh, A. Sheri, S. Jung, K. Seo, J. Park, S. Kim, W. Lee, J. Shin, D. Lee, G. Choi, J. Woo, E. Cha, J. Jang, C. Park, M. Jeon, B. Lee, B.H. Lee, H. Hwang, RRAM-based synapse for neuromorphic system with pattern recognition function, in *IEEE Proc. of International Electron Device Meeting*, 2012, pp.10.2.1–4, 10–13. doi: https://doi.org/10.1109/IEDM.2012.6479016
32. Z. Chen, B. Gao, Z. Zhou, P. Huang, H. Li, W. Ma, D. Zhu, L. Liu, X. Liu, Optimized learning scheme for grayscale image recognition in a RRAM based analog neuromorphic system, in *IEEE International Electron Devices Meeting (IEDM)* (Tech. Digest, Washington, DC, USA, 2015), pp. 17.7.1–17.7.4. doi: https://doi.org/10.1109/IEDM.2015.7409722
33. R. Liu, H. Wu, Y. Pang, H. Qian, S. Yu, Experimental characterization of physical Unclonable function based on 1 kb resistive random access memory arrays. IEEE Electron Device Letters **36**(12), 1380–1383 (2015). https://doi.org/10.1109/LED.2015.2496257
34. A. Chen, Comprehensive assessment of RRAM-based PUF for hardware security applications, in *Proc. of IEEE International Electron Devices Meeting (IEDM)*, Washington D.C., USA, 2015, pp. 10.7.1–10.7.4. doi: https://doi.org/10.1109/IEDM.2015.7409672

35. S. Kvatinsky, G. Satat, N. Wald, E.G. Friedman, A. Kolodny, U.C. Weiser, Memristor-based material implication (IMPLY) logic: Design principles and methodologies. IEEE Trans. Very Large Scale Integr. Syst. **22**(10), 2054–2066 (2014). https://doi.org/10.1109/TVLSI.2013.2282132
36. F.M. Puglisi, L. Pacchioni, N. Zagni, P. Pavan, Energy-efficient logic-in-memory 1-bit full adder enabled by a Physics-based RRAM compact model, in European Solid State Device Research Conference (ESSDERC) (2018)
37. J. Yang, J. Xu, B. Wang, X. Xue, R. Huang, Q. Zhou, J. Wu, Y. Lin, A low cost and high reliability true random number generator based on resistive random access memory, in *2015 IEEE 11th International Conference on ASIC (ASICON)*, Chengdu, 2015, pp. 1–4. doi: https://doi.org/10.1109/ASICON.2015.7516996
38. T. Figliolia, P. Julian, G. Tognetti and A.G. Andreou, A true Random Number Generator using RTN noise and a sigma delta converter, in *2016 IEEE International Symposium on Circuits and Systems (ISCAS)*, Montreal, QC, 2016, pp. 17–20. doi: https://doi.org/10.1109/ISCAS.2016.7527159
39. E. Simoen et al., Random telegraph noise: From a device Physicist's dream to a Designer's nightmare. ECS Trans. **39**(1), 3–15 (2011)
40. S. Realov, K.L. Shepard, Random telegraph noise in 45 nm CMOS: Analysis using an on-chip test and measurement system. IEDM Tech. Dig., 624–627 (2010)
41. F.M. Puglisi, P. Pavan, A. Padovani, L. Larcher, G. Bersuker, RTS noise characterization of HfOx RRAM in high resistive state. Solid State Electron. **84**, 160–166 (2013)
42. M. Luo et al., Impacts of random telegraph noise (RTN) on digital circuits. IEEE Trans. Electron Devices **62**(6), 1725–1732 (2015)
43. F.M. Puglisi, P. Pavan, L. Larcher, A. Padovani, Analysis of RTN and cycling variability in HfO_2 RRAM devices in LRS, in *Proceedings of the 44th European Solid State Device Research Conference (ESSDERC)*, 2014, pp. 246–249
44. B.H. Hong et al., Temperature dependent study of random telegraph noise in gate-all-around PMOS silicon nanowire field-effect transistors. IEEE Trans. Nanotechnol. **9**(6), 754–758 (2010)
45. N. Conrad et al., Low-frequency noise and RTN on near-ballistic III–V GAA nanowire MOSFETs, in *IEEE International Electron Devices Meeting*, 2014, pp. 20.1.1–20.1.4, 15–17
46. K.K. Hung, P.K. Ko, C. Hu, Y.C. Cheng, An automated system for measurement of random telegraph noise in metal-oxide-semiconductor field-effect transistors. IEEE Trans. Electron Devices **36**(6) (1989)
47. G. Giusi, F. Crupi, C. Pace, An algorithm for separating multi-level random telegraph signal from 1/f noise. Rev. Sci. Instrum. **79**(2), 024701 (2008)
48. H. Awano et al., Multi-Trap RTN Parameter Extraction Based on Bayesian Inference, in *14th International Symposium on Quality Electronic Design (ISQED)*, 2013, pp. 597–602, 4–6
49. H. Miki et al., Statistical measurement of random telegraph noise and its impact in scaled-down high-κ/metal-gate MOSFETs, in *2012 IEEE International Electron Devices Meeting (IEDM)*, 2012, pp.19.1.1–19.1.4, 10–13
50. F.M. Puglisi, P. Pavan, RTN analysis with FHMM as a tool for multi-trap characterization in HfOx RRAM, in *IEEE International Conference of Electron Devices and Solid-State Circuits (EDSSC)*, 2013, pp. 1–2
51. G. Zanella, Superposition of similar single-sided RTN processes and 1/f noise, [Online]. arXiv:physics/0504164
52. L. Rabiner, A tutorial on hidden Markov models and selected applications in speech recognition. Proc. IEEE **77**(2), 257–286 (1989)
53. F.M. Puglisi, L. Larcher, P. Pavan, A. Padovani, G. Bersuker, Instability of HfO2 RRAM devices: Comparing RTN and cycling variability, in *Proceedings of the IEEE International Reliability Physics Symposium (IRPS)*, 2014, pp. MY.5.1–5. doi: https://doi.org/10.1109/IRPS.2014.6861160

54. F.M. Puglisi, L. Larcher, A. Padovani, P. Pavan, Characterization of anomalous random telegraph noise in resistive random access memory, in *2015 45th European Solid State Device Research Conference (ESSDERC)*, Graz, 2015, pp. 270–273. doi: https://doi.org/10.1109/ESSDERC.2015.7324766
55. F.M. Puglisi, L. Larcher, A. Padovani, P. Pavan, Anomalous random telegraph noise and temporary phenomena in resistive random access memory. Solid State Electron. **125**, 204–213 (2016)
56. M.J. Uren, M.J. Kirton, S. Collins, Anomalous telegraph noise in small-area silicon metal-oxide-semiconductor field-effect transistors. Phys. Rev. **B 37**, 8346 (1988). https://doi.org/10.1103/PhysRevB.37.8346
57. T. Grasser, H. Reisinger, P. Wagner, F. Schanovsky, W. Goes, B. Kaczer, The time dependent defect spectroscopy (TDDS) for the characterization of the bias temperature instability, in *2010 IEEE International Reliability Physics Symposium*, Anaheim, CA, 2010, pp. 16–25. doi: https://doi.org/10.1109/IRPS.2010.5488859
58. T. Grasser et al., On the volatility of oxide defects: Activation, deactivation, and transformation, in *2015 IEEE International Reliability Physics Symposium*, Monterey, CA, 2015, pp. 5A.3.1-5A.3.8. doi: https://doi.org/10.1109/IRPS.2015.7112739
59. J. Martin-Martinez, J. Diaz, R. Rodriguez, M. Nafria, X. Aymerich, New weighted time lag method for the analysis of random telegraph signals. IEEE Electron Device Letters **35**(4), 479–481 (2014). https://doi.org/10.1109/LED.2014.2304673
60. C. Marquez, O. Huerta, A.I. Tec-Chim, F. Guarin, E.A. Gutierrez-D, F. Gamiz, *Systematic Characterization of Random Telegraph Noise and Its Dependence with Magnetic Fields in MOSFET Devices* (this volume)
61. B. Stampfer, A. Grill, M. Waltl, Advanced Electrical Characterization of Single Oxide Defects Utilizing Noise Signals (this volume)
62. T. Tsuchiya, Detection and Characterization of Single Defects in MOSFETs (this volume)
63. J. Martin-Martinez, R. Rodriguez, M. Nafria, *Advanced Characterization and Analysis of Random Telegraph Noise in CMOS Devices* (this volume)
64. F.M. Puglisi, P. Pavan, A. Padovani, L. Larcher, G. Bersuker, Random telegraph signal noise properties of HfOx RRAM in high resistive state, in *2012 Proceedings of the European Solid-State Device Research Conference (ESSDERC)*, Bordeaux, 2012, pp. 274–277. doi: https://doi.org/10.1109/ESSDERC.2012.6343386
65. F.M. Puglisi, P. Pavan, Factorial hidden Markov model analysis of random telegraph noise in resistive random access memories. ECTI Trans. Electr. Eng. Electr. Commun. **12**(1), 24–29 (2014)
66. F.M. Puglisi, Measuring and analyzing random telegraph noise in nanoscale devices: the case of resistive random access memories, in *2017 17th Non-Volatile Memory Technology Symposium (NVMTS)*, Aachen, 2017, pp. 1–5. doi: https://doi.org/10.1109/NVMTS.2017.8171308
67. Z. Ghahramani, Jordan, Factorial Hidden Markov Models. M.I. Machine Learn. **29**, 245 (1997). https://doi.org/10.1023/A:1007425814087
68. F.M. Puglisi, Random telegraph noise analysis as a tool to link physical device features to electrical reliability in nanoscale devices, in *2016 IEEE International Integrated Reliability Workshop (IIRW)*, South Lake Tahoe, CA, 2016, pp. 13–17. doi: https://doi.org/10.1109/IIRW.2016.7904891
69. F.M. Puglisi, A. Padovani, L. Larcher, P. Pavan, Random telegraph noise: Measurement, data analysis, and interpretation, in *2017 IEEE 24th International Symposium on the Physical and Failure Analysis of Integrated Circuits (IPFA)*, Chengdu, 2017, pp. 1–9. doi: https://doi.org/10.1109/IPFA.2017.8060057
70. F.M. Puglisi, P. Pavan, Guidelines for a reliable analysis of random telegraph noise in electronic devices. EEE Trans. Instrument. Measure. **65**(6), 1435–1442 (2016). https://doi.org/10.1109/TIM.2016.2518880
71. G. Kapila, V. Reddy, Impact of sampling rate on RTN time constant extraction and its implications on Bias dependence and trap spectroscopy. IEEE Trans. Device Mater. Reliab. **14**(2) (2014)

72. K. Huang, A. Rhys, Theory of light absorption and non-radiative transitions in F-centres. Proc. R. Soc. Lond. A **204**, 406–423 (1950). https://doi.org/10.1098/rspa.1950.0184
73. L. Vandelli et al., A physical model of the temperature dependence of the current through SiO2/HfO2 stacks. IEEE Trans. Electron Devices **58**(9), 2878–2887 (2011)
74. X. Lian et al., Quantum point contact model of filamentary conduction in resistive switching memories, in *2012 13th International Conference on Ultimate Integration on Silicon (ULIS)*, Grenoble, 2012, pp. 101–104. doi: https://doi.org/10.1109/ULIS.2012.6193367
75. F.M. Puglisi, L. Larcher, G. Bersuker, A. Padovani, P. Pavan, An empirical model for RRAM resistance in low-and high-resistance states. IEEE Electron Device Letters **34**(3), 387 (2013). https://doi.org/10.1109/LED.2013.2238883
76. Z. Jiang et al., A compact model for metal–oxide resistive random access memory with experiment verification. IEEE Trans. Electron Devices **63**(5), 1884–1892 (2016). https://doi.org/10.1109/TED.2016.2545412
77. F.M. Puglisi, P. Pavan, A. Padovani, L. Larcher, A compact model of hafnium-oxide-based resistive random access memory, in *Proceedings of 2013 International Conference on IC Design & Technology (ICICDT)*, Pavia, 2013, pp. 85–88. doi: https://doi.org/10.1109/ICICDT.2013.6563309
78. P. Huang et al., A physics-based compact model of metal-oxide-based RRAM DC and AC operations. IEEE Trans. Electron Devices (12), 4090–4060, 4097 (2013). https://doi.org/10.1109/TED.2013.2287755
79. U. Russo, D. Ielmini, C. Cagli, A.L. Lacaita, Self-accelerated thermal dissolution model for reset programming in unipolar resistive-switching memory (RRAM) devices. IEEE Trans. Electron Devices **56**(2), 193–200 (2009). https://doi.org/10.1109/TED.2008.2010584
80. A. Padovani, L. Larcher, O. Pirrotta, L. Vandelli, G. Bersuker, Microscopic modeling of HfOx RRAM operations: From forming to switching. IEEE Trans. Electron Devices **62**(6), 1998–2006 (2015)
81. G. Bersuker, D.C. Gilmer, D. Veksler, P.D. Kirsch, L. Vandelli, A. Padovani, L. Larcher, K. McKenna, A. Shluger, V. Iglesias, M. Porti, M. Nafria, Metal oxide resistive memory switching mechanism based on conductive filament microscopic properties. J. Appl. Phys. **110**, 124518 (2011). https://doi.org/10.1063/1.3671565
82. J.P. Strachan et al., State dynamics and modeling of tantalum oxide Memristors. IEEE Trans. Electron Devices **60**(7), 2194–2202 (2013). https://doi.org/10.1109/TED.2013.2264476
83. L. Larcher, F.M. Puglisi, P. Pavan, A. Padovani, L. Vandelli, G. Bersuker, A compact model of program window in HfOx RRAM devices for conductive filament characteristics analysis. IEEE Trans. Electron Devices **61**(8), 2668–2673 (2014). https://doi.org/10.1109/TED.2014.2329020
84. F.M. Puglisi, L. Larcher, A. Padovani, P. Pavan, Bipolar resistive RAM based on HfO_2: Physics, compact modeling, and variability control. IEEE J. Emerg. Select. Topics Circuits Syst. **6**(2), 171–184 (2016). https://doi.org/10.1109/JETCAS.2016.2547703
85. H.-S.P. Wong et al., Metal–oxide RRAM. Proc. IEEE **100**(6), 1951–1970 (2012)
86. L. Larcher, F.M. Puglisi, A. Padovani, L. Vandelli, P. Pavan, Multiscale modeling of electron-ion interactions for engineering novel electronic devices and materials, in *2016 26th International Workshop on Power and Timing Modeling, Optimization and Simulation (PATMOS)*, Bremen, 2016, pp. 128–132. doi: https://doi.org/10.1109/PATMOS.2016.7833676
87. A. Padovani, L. Larcher, F.M. Puglisi, P. Pavan, Multiscale modeling of defect-related phenomena in high-k based logic and memory devices, in *2017 IEEE 24th International Symposium on the Physical and Failure Analysis of Integrated Circuits (IPFA)*, Chengdu, 2017, pp. 1–6. doi: https://doi.org/10.1109/IPFA.2017.8060063
88. D. Muñoz Ramo, J.L. Gavartin, A.L. Shluger, G. Bersuker, Spectroscopic properties of oxygen vacancies in monoclinic HfO2 calculated with periodic and embedded cluster density functional theory. Phys. Rev. B **75**, 205336 (2007)
89. J. Robertson, R. Gillen, Defect densities inside the conductive filament of RRAMs. Microelectron. Eng. **109**, 208–210 (2013)

90. A.S. Foster, A.L. Shluger, R.M. Nieminen, Mechanism of interstitial oxygen diffusion in Hafnia. Phys. Rev. Lett. **89**(22), 225901 (2012)
91. A.S. Foster, F. Lopez Gejo, A.L. Shluger, R.M. Nieminen, Vacancy and interstitial defects in hafnia. Phys. Rev. B **65**, 174117 (2002)
92. R.A. Marcus, Electron transfer reactions in chemistry. Theory and experiment. Rev. Mod. Phys. **65**, 599–610 (1993)
93. F.M. Puglisi, P. Pavan, L. Larcher, A. Padovani, Statistical analysis of random telegraph noise in HfO2-based RRAM devices in LRS. Solid State Electron. **113**, 132–137 (2015). https://doi.org/10.1016/j.sse.2015.05.027
94. R. Landauer, Spatial variation of currents and fields due to localized Scatterers in metallic conduction. IBM J. Res. Dev. **1**(3), 223–231 (1957)
95. L.S. Chang, C.Y. Huang, Y.H. Tseng, Y.C. King, C.J. Lin, Temperature sensing scheme through random telegraph noise in contact RRAM. IEEE Electron Device Letters **34**(1), 12–14 (2013). https://doi.org/10.1109/LED.2012.2226137
96. F.M. Puglisi, N. Zagni, L. Larcher, P. Pavan, A new verilog-A compact model of random telegraph noise in oxide-based RRAM for advanced circuit design, in *2017 47th European Solid-State Device Research Conference (ESSDERC)*, Leuven, 2017, pp. 204–207. doi: https://doi.org/10.1109/ESSDERC.2017.8066627
97. F.M. Puglisi, N. Zagni, L. Larcher, P. Pavan, Random telegraph noise in resistive random access memories: Compact modeling and advanced circuit design. IEEE Trans. Electron Devices. https://doi.org/10.1109/TED.2018.2833208
98. L. Larcher, A. Padovani, P. Pavan, Leakage current in HfO_2 stacks: From physical to compact modeling. Nanotechnology **2**, 809–814 (2012)
99. R. Thamankar et al., Single vacancy defect spectroscopy on HfO_2 using random telegraph noise signals from scanning tunneling microscopy. J. Appl. Phys. **119**, 084304 (2016). https://doi.org/10.1063/1.4941697
100. R. Thamankar et al., Localized characterization of charge transport and random telegraph noise at the nanoscale in HfO2 films combining scanning tunneling microscopy and multi-scale simulations. J. Appl. Phys. **122**, 024301 (2017). https://doi.org/10.1063/1.4991002
101. F.M. Puglisi, U. Celano, A. Padovani, W. Vandervorst, L. Larcher, P. Pavan, Scaling perspective and reliability of conductive filament formation in ultra-scaled HfO2 resistive random access memory, in *2017 IEEE International Reliability Physics Symposium (IRPS)*, Monterey, CA, 2017, pp. PM-8.1–PM-8.5. doi: https://doi.org/10.1109/IRPS.2017.7936390
102. A. Ranjan et al., Random telegraph noise in 2D hexagonal boron nitride dielectric films. Appl. Phys. Lett. **112**, 133505 (2018). https://doi.org/10.1063/1.5022040
103. F.M. Puglisi et al., 2D h-BN based RRAM devices, in *2016 IEEE International Electron Devices Meeting (IEDM)*, San Francisco, CA, 2016, pp. 34.8.1–34.8.4. doi: https://doi.org/10.1109/IEDM.2016.7838544
104. C.B. Pan et al., Coexistence of grain-boundaries-assisted bipolar and threshold resistive switching in multilayer hexagonal boron nitride. Adv. Funct. Mater. **27**, 1604811 (2017)
105. Y. Shi et al., Coexistence of volatile and non-volatile resistive switching in 2D h-BN based electronic synapses, in *2017 IEEE International Electron Devices Meeting (IEDM)*, San Francisco, CA, 2017, pp. 5.4.1–5.4.4. doi: https://doi.org/10.1109/IEDM.2017.8268333

Systematic Characterization of Random Telegraph Noise and Its Dependence with Magnetic Fields in MOSFET Devices

Carlos Marquez, Oscar Huerta, Adrian I. Tec-Chim, Fernando Guarin, Edmundo A. Gutierrez-D, and Francisco Gamiz

1 Introduction

In order to satisfy the permanent demand of incremental integration density according to Moore's law [1], the semiconductor industry has faced substantial challenges applying innovative solutions [2–4]. Albeit reducing the gate oxide thickness and using alternative gate oxide elements (such as high-k materials) have been some technological improvements used to boost the transistor performance, they have reduced the channel-to-gate interface quality. Furthermore, the scaling of the device area has resulted in the decrease of the current signals down to levels comparable with their fluctuations. In this regard, due to poorer oxide interfaces and lower current levels, random telegraph noise (RTN) has arisen as one of the most relevant reliability issues in highly scaled semiconductor devices.

Figure 1 reinforces the previous statements: coupled with the progressive down-scaling of field-effect transistor (FET) dimensions and the associated higher integration density, an increasing relevance of the reliability issues (concretely random telegraph noise) is noticed. To complicate things even more, new alternative channel materials (III–V compounds, graphene, transition metal dichalcogenides, and other counterparts) or advanced technological implementations such as silicon-on-insulator (SOI), multi-gate (MG), or fin field-effect transistors would trigger

C. Marquez · F. Gamiz (✉)
CITIC-UGR and Department of Electronics, University of Granada, Granada, Spain
e-mail: carlosmg@ugr.es; fgamiz@ugr.es

O. Huerta · A. I. Tec-Chim · E. A. Gutierrez-D
Department of Electronics, National Institute for Astrophysics, Optics and Electronics (INAOE), San Andres Cholula, Mexico

F. Guarin
Reliability Group, GLOBALFOUNDRIES, Albany, NY, USA

T. Grasser (ed.), *Noise in Nanoscale Semiconductor Devices*,
https://doi.org/10.1007/978-3-030-37500-3_4

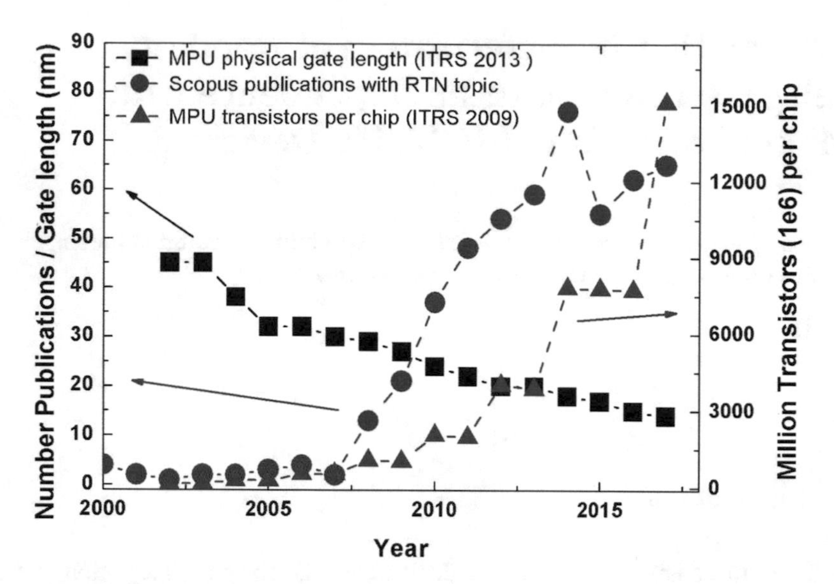

Fig. 1 Scientific contributions including random telegraph noise as subject (source: www.scopus.com) compared to transistor gate length dimensions and integration density according to ITRS (International Technology Roadmap for Semiconductors) reports [3, 4]

the instability phenomena and their subsequent implication in the transistor reliability [5, 6]. Throughout this chapter, especial attention will be given to silicon-on-insulator technology and how random telegraph noise instability affects the transistor performance under different back-bias conditions. The introduction of SOI technology has opened the path for controlling the threshold voltage of the transistors by using an implanted ground plane or even its dynamic adjustment through the substrate or back(-gate) bias [7, 8]. However, the impact of additional insulator interfaces, partially and fully depleted channel, and back voltage dependence on the reliability of the transistors will still have to be completely studied and characterized.

Besides novel structures and materials employed in transistor fabrication, other alternatives have been proposed to improve the performance of the devices, some of them involving the presence of magnetic fields [9, 10]. Regrettably, despite the fact that some works have already made important advances in the understanding of the magnetic field effect on MOSFETs [11–15], the influence of these magnetic fields on the reliability of the transistor operation is not entirely comprehended [13]. This lack of information is even more noticeable regarding the dependence of the magnetic field on random telegraph signals. Additionally, the advances in devices such as magnetoresistive random-access memories, spin-field effect transistors, or even regular transistors working under unintentional external magnetic fields require more extensive characterization of the instability phenomena [16–18].

Despite the relevance of this instability effect, there is a significant lack of experimental approaches in terms of systematic characterization methods or protocols.

This is probably due to the ostensible random characteristic of this disturbing phenomenon which may indistinctly affect devices in different time scales and voltage bias conditions. In this sense, the following pages will provide some advances in the electrical characterization of random telegraph noise and novel results dealing with silicon-on-insulator technology and low static magnetic fields. In Sect. 2, the theoretical framework and the time scale experimental characterization of random telegraph noise are depicted. In Sect. 3, low-frequency noise characterization is introduced as an approach to determine the noise sources. Subsequently, in Sect. 4, a systematic characterization protocol is implemented by using both time and low-frequency characterizations to facilitate the determination of the bias regime and number of traps. In addition, the synthesized experimental method is employed to determine the characteristic and distribution of traps in silicon-on-insulator on-wafer transistors. Finally, in Sect. 5, trap phenomena are examined under the presence of weak static magnetic fields at room temperature.

2 RTN Time Signature Characterization

Random telegraph noise can be explained as a stochastic trapping/detrapping phenomenon of the channel carriers into the switching oxide traps (or border traps) of the gate dielectric [19–21]. A trap in the oxide can occasionally capture a charge carrier from the channel, and the captured carrier can be emitted back to the channel after a period of time represented as a current fluctuation (Fig. 2a). The characteristic time of the captured and emitted states are denoted as τ_e (mean time at low current level) and τ_c (mean time at high current level), respectively. Likewise, threshold voltage (V_{TH}) shows a binary fluctuation caused by a single trap (Fig. 2c) up to ten times the simple electrostatic expectation calculated by a charge sheet approximation as $-q/C_{ox}$, where q is the electron charge and C_{ox} the oxide capacitance [22]. In fact, some works [6, 23] suggest that the fluctuations in the threshold voltage due to a single trap in the oxide can be expressed as $\Delta V_{TH} = \frac{q}{A_{eff} C_{ox}}$, where A_{eff} is the effective channel area. Since RTN causes threshold voltage variation and increases rapidly if V_G is not properly scaled, this instability becomes a major challenge which has to be overcome, especially in semiconductor memories where this effect negatively impacts on the noise margin in SRAM blocks [6, 24] or induces fluctuating readout states in floating-gate non-volatile memories [25]. Further information regarding the theoretical framework of RTN can be found in the initial chapters of this book.

From an experimental point of view, the signature of random telegraph noise is extracted by monitoring the drain current (I_D) on time slots and sampling rates suitable to capture the trap–carrier interactions. In case of relatively slow traps, as the ones presented in this work, the characterization can be carried out by using a semiconductor analyzer equipped with high-resolution source-measurement units (SMU) allowing maximum sampling rates around 2ms. Dealing with faster traps, below millisecond scale, requires *quicker* characterization tools which usually

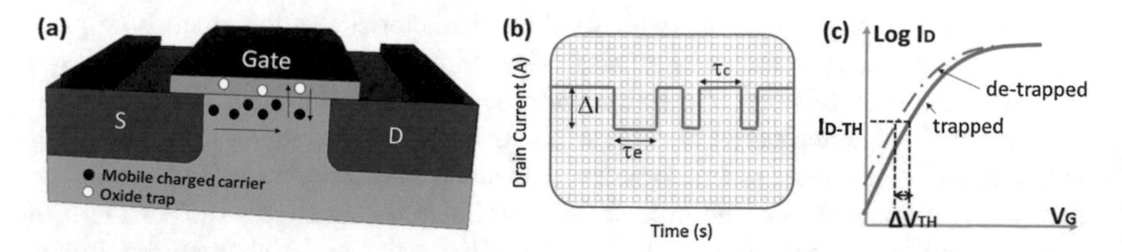

Fig. 2 (**a**) Schematic illustration of a mobile charged carrier trapped/detrapped in the gate oxide. (**b**) Trapping/detrapping events result in a two-level current characteristic where the capture and emission times are determined from the average high and low current levels, respectively. (**c**) Threshold voltage shift observed between trapped (solid line) and detrapped (dashed line) carrier levels

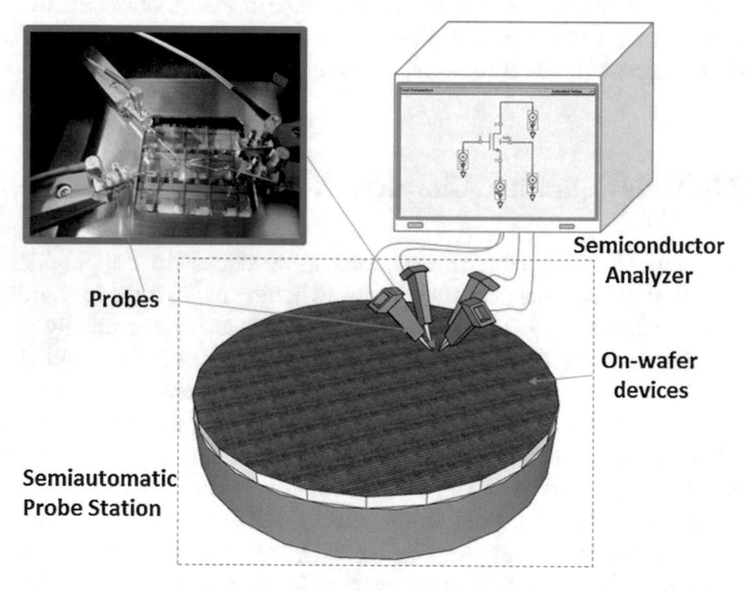

Fig. 3 Schematic illustration of the setup employed for an experimental on-wafer characterization

present lower current and voltage resolutions, such as the pulsed I–V ultra-fast measure unit (PMU, Keithley) or the waveform generator and fast measurement unit (WGFMU, Keysight). These analyzers, regardless of the sampling rate and the manufacturer, allow biasing the transistor terminals while simultaneously monitoring repeatedly the drain current at a properly selected sampling rate (scheme shown in Fig. 3). It is worth mentioning that depending on the selected sampling rate and slot time only a determined range of traps (with characteristic time lower than the sampling rate) can be characterized [26]. Figure 4 depicts the experimental characterization of a transistor affected by random telegraph noise. The selected device was a fully depleted (FD)-SOI MOS-transistor fabricated in a 22 nm process at CEA-LETI [27, 28]. In Fig. 4a, the characteristic time τ_e and τ_c are detectable for the three different gate voltages. Note that for different bias conditions these

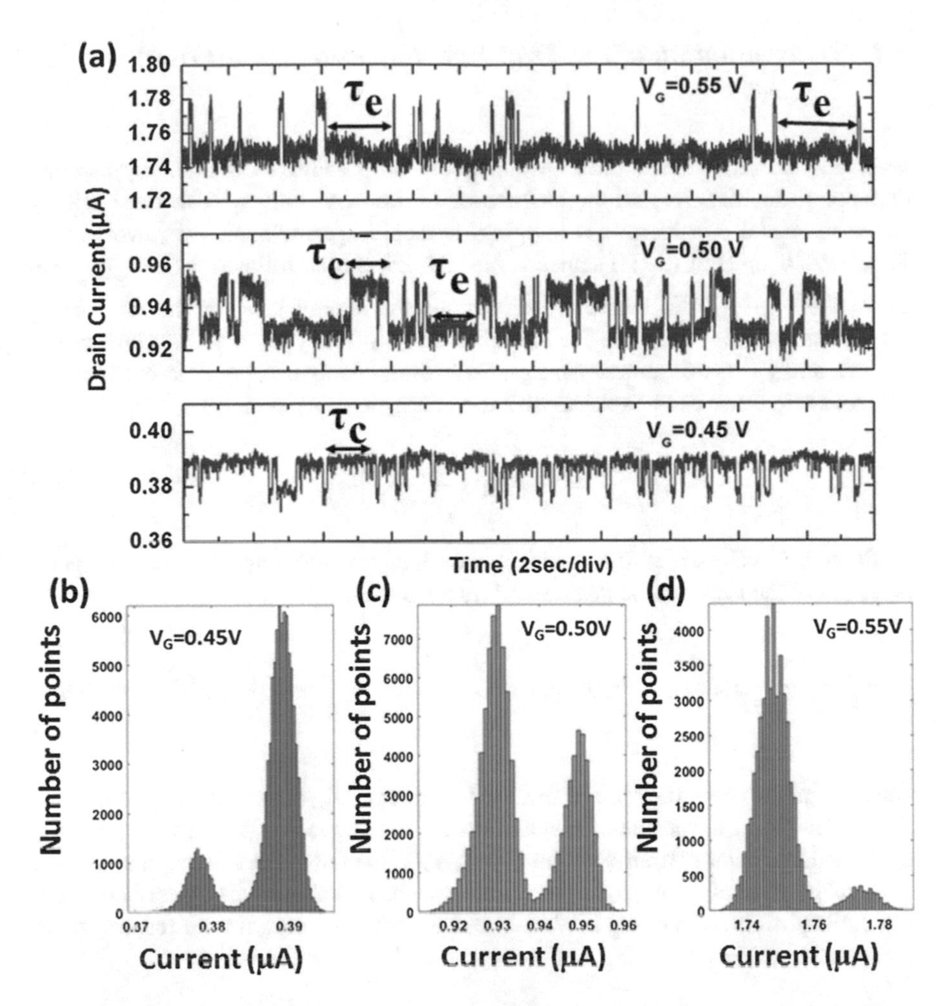

Fig. 4 (**a**) Time signature of a 100 nm length and 80 nm width silicon-on-insulator n-MOS transistor affected by one oxide trap. Current histogram for gate voltages of (**b**) 0.45 V, (**c**) 0.50 V, and (**d**) 0.55 V. Drain and source voltages are set to 100 mV and 0 V, respectively. The sampling rate was 2 ms

characteristic times change in time showing a different signature. In this specific case, τ_c is reduced as gate voltage increases and inversely in the case of τ_e. This fact is more noticeable by representing the current histogram for the different gate voltages (Fig. 4b–d) where the peaks reveal the portion of time that the current remains in each of the two levels.

2.1 Determination of the Trap Location and Back-Biased Devices

Over the years, different approaches have been proposed to model the trapping and detrapping fluctuations and the characteristic times, τ_c and τ_e. The most widely extended model considers that trapping/detrapping phenomena are governed by Shockley–Read–Hall (SRH) statistics [19–21, 29, 30] as follows: $\tau_c = \frac{1}{(nv\sigma)}$ and $\tau_\mathrm{e} = \frac{\exp[(E_\mathrm{F}-E_\mathrm{T})/k_\mathrm{BT}]}{gnv\sigma}$, where E_T represents the energy level of the trap; k_B is the Boltzmann constant; v is the carrier average velocity; σ is the capture cross-section; and g is the degeneracy factor. Following this theory, the ratio between the characteristic times can be related with the energy level of the trap:

$$\frac{\tau_c}{\tau_e} = g \times \exp\left(\frac{E_T - E_F}{k_B T}\right) \quad (1)$$

From the analysis of the device band diagram, the physical and energetic positions of the traps can be determined through [31]:

$$\ln\frac{\tau_c}{\tau_e} = -\frac{1}{k_B T}[(E_{Cox} - E_T) - (q\phi_S - \chi_{ox} - q\psi_S) + \frac{x_T}{t_{EOT}}(|V_{FB}| + V_G - q\psi_S)] \quad (2)$$

where ϕ_s is the work function of the semiconductor; χ_{ox}, the electronic affinity of the oxide; E_{Cox}, the insulator conduction band energy; ψ_s, the surface potential of the semiconductor (front-gate interface); V_{FB}, the flat-band voltage; and x_T, the physical position of the trap measured from the channel/silicon–film interface. From Eq. 2, the position of the trap is worked out by differentiating it with respect to the front-gate bias V_G:

$$x_T = \frac{t_{EOT}\left(\frac{k_B T}{q}\frac{d\ln\frac{\tau_c}{\tau_e}}{dV_G} + \frac{d\psi_s}{dV_G}\right)}{\frac{d\psi_s}{dV_G} - 1} \quad (3)$$

Once the physical position of the trap is known from Eq. 3, its energetic position is evaluated from Eq. 2. Note that Eq. 3 is valid as long as the trap is located inside the SiO_2 cap layer. However, this is the most common case: it is not easy to observe RTN signals caused by traps in the high-k dielectric, because there are few traps close to the silicon Fermi level due to the high conduction band offset between the hafnium-based oxide and the SiO_2. Even in the case of traps located close to the conduction band edge of the hafnium-based oxide, the electrons would be prone to escape due to fast thermionic emission to the metal contact of the gate [32].

Figure 5a shows typical τ_c and τ_e trends acquired from a transistor with the same physical dimensions as the one shown in Fig. 4. The capture time shows a

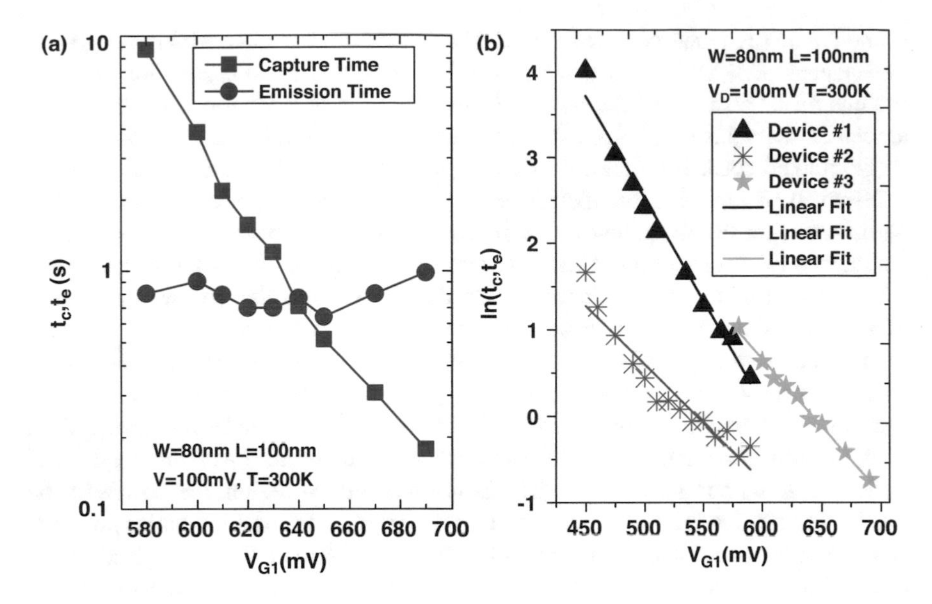

Fig. 5 (**a**) Dependence of τ_c and τ_e with the gate voltage. (**b**) Logarithm of the ratio between capture and emission times as a function of the gate voltage (symbols) and linear fitting (lines) according to Eq. 2 for three devices belonging to the same SOI set

dependence with the gate bias, the larger the carrier concentration of the inversion channel, the larger the probability for a carrier to be captured. In contrast, the emission time is independent of the gate bias. This absence of electric field dependence points at the thermionic emission as the primary mechanism for the electrons to escape from the trap. Experimental results of the logarithmic ratio between τ_c and τ_e with respect to the gate voltage for three different devices are shown in Fig. 5b. The physical position of the traps can be determined following this initial approach through Eq. 3. The extracted results are $x_T = 0.79$ nm for device #1, $x_T = 0.45$ nm for device #2, and $x_T = 0.52$ nm for device #3. Note that for simplicity in Eq. 3, the surface potential variations contribution ($\frac{d\psi_s}{dV_G}$) for gate bias conditions higher than threshold voltage has been neglected.

Although SRH approach is an extended theory to determine the location of traps, recent works have demonstrated that this approach is insufficient to reproduce the complex behavior of trapping and detrapping events in ultra-scaled transistors, specifically in those cases with several traps and large dependence of the capture time on gate voltage and on temperature conditions [33–40]. In Zanolla et al. [33], the standard SRH model was modified in order to fit the extracted larger dependence of the capture time on V_{GS} through: (a) an empirical V_{GS} dependence of the capture cross-section, accounting for the bias dependence of tunneling probability [41] (This issue is discussed in more detail in Chapter 18 of this book [42]), or (b) a modification of the trap energy due to Coulomb blockade [43]. It is worth noting that in this work [33], two traps which feature quite close values for x_T, E_T, and σ, one of them requires a modification to the SRH model in order to cope with the large bias dependence of τ_c while the other not, proving that there is no correlation between the energy or position of the trap and the adequacy of the basic SRH model. In agreement with this fact, Fig. 6a shows the capture and emission times for one of our single-trap silicon-on-insulator devices featuring a t_{EOT} of 1.3 nm and silicon and BOX layers of 7 and 145 nm thicknesses, respectively. τ_c follows a gate dependence while emission events follow a thermionic process according to the constant τ_e characteristic as a function of the gate voltage. Figure 6b shows the capture time as a function of the drain current. According to SRH model, linear dependence of τ_c on the drain current is expected; however, a current power-law dependence with b factor close to 2 is extracted in this case. In accordance with the model initially proposed in [34] and modified in [33], the larger dependence of capture time with bias conditions, observed in Fig. 6b, can be explained by the quantization of the conduction band into discrete energy levels in strong inversion conditions. Following this premise, in MOSFET biased in inversion region, the profile of the conduction band in the vicinity of the Si–SiO_2 interface forms a triangular potential well that induces substantial quantization effects (subbands). This effect is modeled through introducing a factor ΔE_0 in the inversion charge equation to take into account the energy distance between the subbands and the conduction band energy edge. However, in our particular devices, the large gate voltage dependence is observed when devices operate in a weak inversion regime (other characterized devices, not shown, present the same behavior), instead of at inversion charge condition. This fact suggests that, in first instance, this approximation does not properly describe the phenomenon in our back-biased SOI nMOS transistors. In this regard, Grasser et al. have developed over the years a non-radiative multiphonon model which would explain the very large bias dependence mentioned [39, 40]. This model includes an internal barriers concept associated with the trap atomic reconfiguration or structural relaxation during the trapping/detrapping kinetics. This approach suggests that the capture and emission processes are controlled by tunneling between the carrier reservoir and the trap itself, as well as internal thermal barriers, with both components depending on gate bias V_G, the trap energy level E_T and the depth x_T, and the thermal barriers (determined by two defined parameters: S and R). Specifically, τ_c and τ_e are determined by (1) the tunneling barrier between the carrier reservoir (inversion layer in MOSFET) and the trap and (2) the defined barriers between the unoccupied and occupied states. These barriers are responsible for the strong temperature dependence of τ_c and τ_e being also V_G dependent.

Another approach, proposed in [38, 43], considers capture time also follows a current power law with factors $b > 1$ when the device is operated in the linear regime from weak to strong inversion. In these cases, the Coulomb blockade effect has been proposed to explain these deviations in standard MOSFET devices. This phenomenon suggests that when an electron is trapped at an oxide or interface center, in order to maintain the charge neutrality, fractional image charges are created in the semiconductor substrate and the gate, overall in weak inversion [38, 44]. Although this approach could agree with the presented result in silicon-on-insulator technology, the significant higher current power law factor $b = 3.7$,

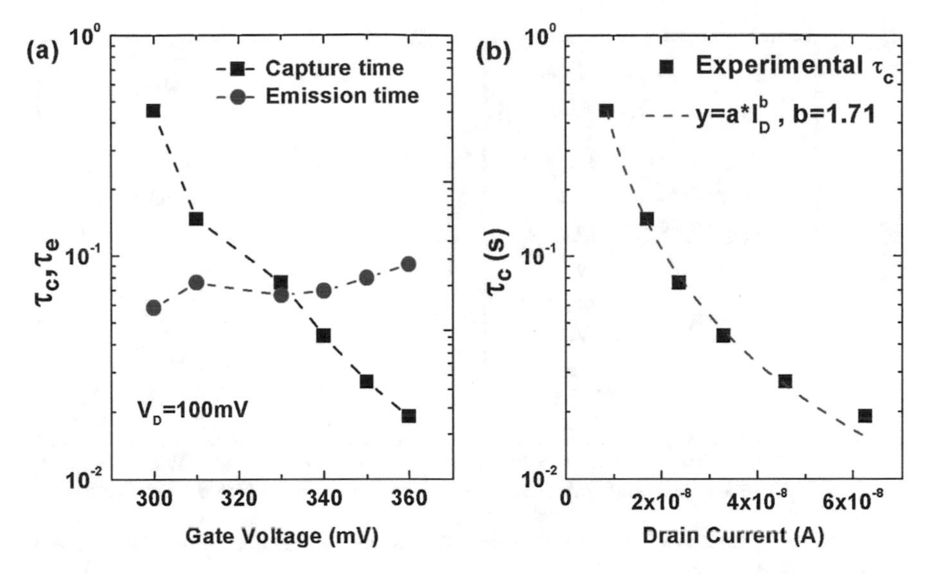

Fig. 6 (**a**) τ_c and τ_e ratio as a function of the gate voltage. (**b**) τ_c as a function of the drain current. A device from the silicon-on-insulator set featuring $t_{ox} = 1.3$ nm, $t_{si} = 7$ nm, $t_{BOX} = 145$ nm

reported in these works, generates discrepancies with the result shown in Fig. 6b. Also the combination of both theories have been reported in [43] where a complete model has been developed to take into account these effects on the capture time, regrettably with fitting factors still too higher respected to the ones observed in our silicon-on-insulator transistors.

An extra possible explanation is developed in [36]. This work proposes a model for capture kinetics following a quadratic inverse dependence on gate voltage (with MOSFET in linear operation) instead of the inverse carrier density dependence predicted by the standard Shockley–Read–Hall theory. This model, labeled as generation-recombination-tunneling (GRT), is based on the assumption that the transition phenomenon consists of three interactions: (a) a trapping transition between the carrier at the conduction band and an interface trap with energy within a few kT of the surface Fermi level E_F, (b) a tunneling transition between the interface trap and the oxide trap as second step, and (c) a quantum transition between the oxide trap and the conduction band. Following this hypothesis, increasing the concentration of charge carriers causes a higher interface trap occupation and τ_c varies inversely with the square of current when the quasi-Fermi level is below the trap level E_t. This is in agreement with the results depicted in Fig. 6b. In order to corroborate this latter model, different substrate voltages have been applied. In silicon-on-insulator technology, the inversion charge in the front interface as well as the threshold voltage varies depending on the back-gate applied voltage [45]. Figure 7 shows the values of τ_c and τ_e as a function of the overdrive voltage V_{ov} (same inversion charge conditions) for two devices. From the analysis of this plot we can note that for a given overdrive voltage, the capture time (τ_c) decreases

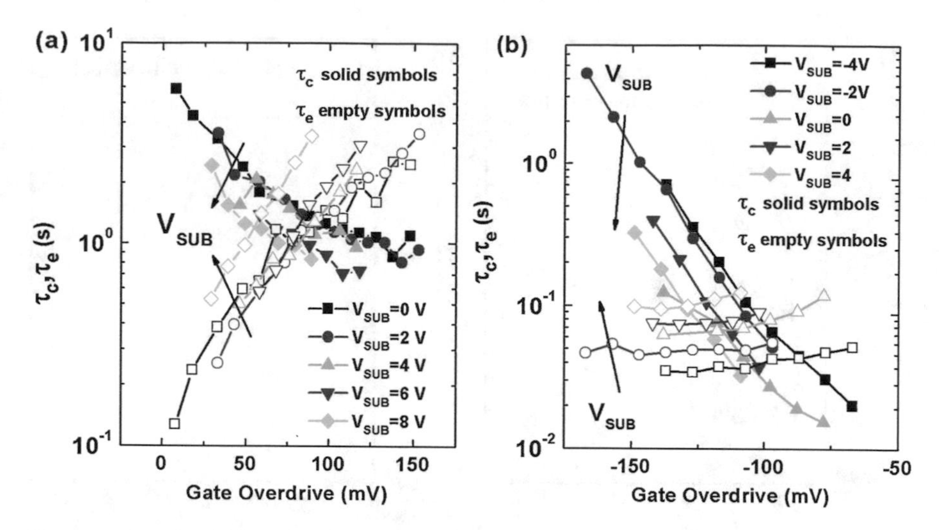

Fig. 7 τ_c and τ_e ratios as a function of the gate overdrive voltage for different substrates biases in the same devices as in Fig. 6

whereas the emission time (τ_e) increases as V_{SUB} increases, in other words, the carriers remain trapped for longer time. This effect is contrary to the one expected from the application of classical RTN models which rely on the inversion charge and the decrease of the electric field in the Si-film while increasing the substrate bias [45]. Consequently, these results are compatible with the fact that the carrier is emitted to the metal gate contact since, for this mechanism, the lower the gate oxide electric field intensity, the larger the probability of the trap to be filled [46]. These explanations could be partially in agreement with the two-state model previously described, where the tunneling effect is introduced, but in this case from the oxide trap to the gate metal. Otherwise, this result could also be explained by configuration changes occurring at the defect site following the trapping or emission event (in this particular case depending on the substrate voltage) and modeled through multiphonon charge transfer mechanisms [39]. However, extensive investigations are required to corroborate this hypothesis and to develop a model.

Additionally, the behavior observed in Fig. 8 cannot be explained only by the shift of the threshold voltage induced by the back-bias according to the Lim–Fossum model [45]:

$$V_{TH} = V_{TH}^{I} - \frac{C_{Si}C_{BOX}}{C_{ox}(C_{Si} + C_{BOX})}\left(V_{SUB} - V_{SUB}^{I}\right) \tag{4}$$

where C_{Si}, C_{BOX}, C_{ox} are the silicon, buried oxide, and front-gate oxide capacitance, respectively. V_{SUB} is the substrate voltage and V_{TH}^{I} is the threshold voltage of the front-channel when the back interface is inverted and V_{SUB}^{I} is the substrate voltage to achieve inversion at the back interface [45].

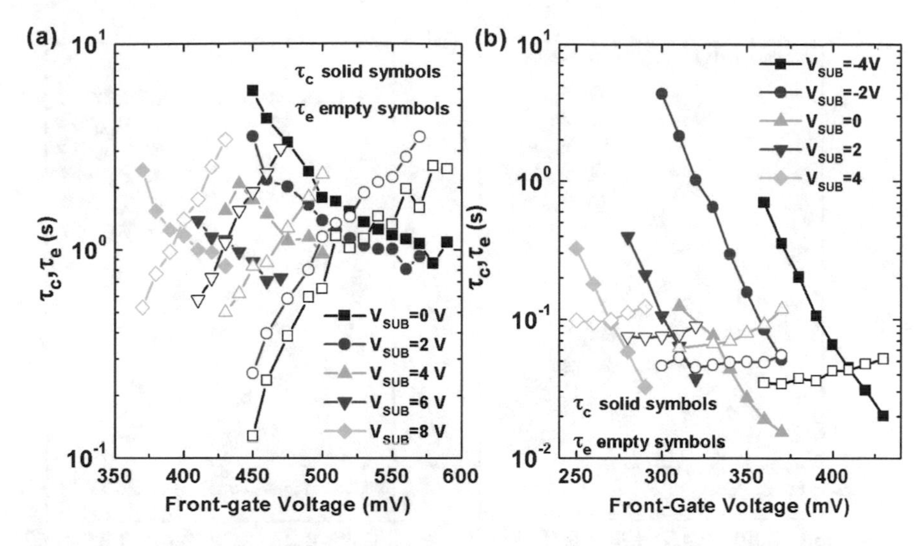

Fig. 8 Capture, τ_c, and emission, τ_e, evolution as a function of the gate voltage for different substrate biases (V_{SUB}) for two different devices

The V_{TH} shift predicted by Eq. 4 is not enough to correlate the curves in Fig. 8: this (direct) correction would correspond only to a lateral shift of the curves (displacement in the V_G axis) but, for example, it cannot explain the decrease observed in τ_e. This fact is even clearer if the τ_c/τ_e ratios, used to extract the physical parameters of the traps, are plotted as a function of the gate overdrive, illustrated in Fig. 9. Although the curves are represented as a function of the overdrive voltage, the τ_c/τ_e ratio depends on the particular value of V_{SUB} (even though the inversion charge is the same in all cases). Nevertheless, Eq. 4 can be included in Eq. 2 to capture the influence of the substrate bias in the physics of the process.

The accuracy of this model has been tested by reproducing the experimental value of $\ln\frac{\tau_c}{\tau_e}$ for two different transistors (Fig. 9) under different substrate bias conditions. The model provides a reasonable prediction of the electrical characteristics as long as Eq. 4 remains valid, i.e., depletion regime of the back interface.

Once the vertical electric field (V_G) dependence on the kinetic of the trapping/detrapping events is studied, the location of the trap along the channel and its dependence on the lateral electric field (V_D) have to be evaluated. The characteristic time might change depending on the position and the effective length of the channel, modulated by the lateral electric field. This dependence can be calculated by extracting τ_c and τ_e as a function of V_D (for a given V_G) and performing measurements under different bias configurations (forward and reserve) of the device terminals. In Illarionov et al. [47], a robust method for the extraction of the lateral position of traps in nanoscale MOSFETs is proposed, demonstrating that the extraction uncertainty decreases for devices with smaller channel length. Figure 10a shows one device presenting a strong asymmetry in the values τ_c and τ_e when the roles of source and drain are exchanged. This asymmetry is the consequence

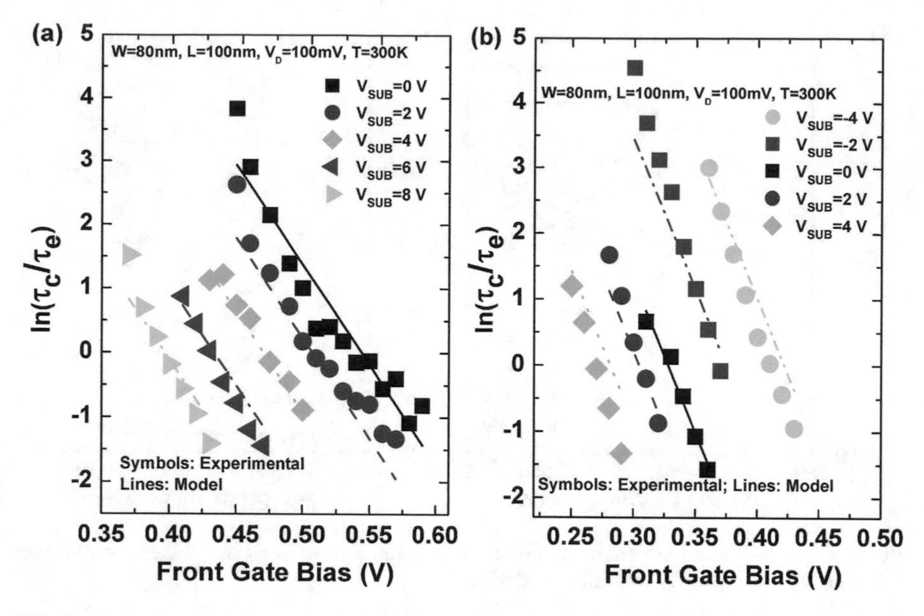

Fig. 9 Comparison of the experimental values of $ln(\tau_c/\tau_e)$ (symbols) and the proposed model (lines) for two devices (**a** and **b**)

of the non-centered position of the trap along the length of the channel [41, 42]. The trap is located close to the source since a strong dependence of τ_c on the applied voltage is observed in the source configuration. However, in Fig. 10b, the symmetrical response reveals that the trap is near the center of the channel. In the case of traps away from the drain terminal, their characteristic times are independent of the drain bias. Note that especial attention to this issue should be paid if the dependence on other effects, such as substrate bias influence, are studied.

3 Low-Frequency Noise Characterization

Trapping and detrapping phenomena can be characterized not only through the analysis in the time domain ($I_D(t)$, $V_D(t)$, $V_G(t)$, etc.), but also through the power spectral density measurements in the frequency domain. The fundamental sources of noise observed in MOSFET devices are thermal noise, generation–recombination noise, flicker noise (or 1/f noise), and shot noise [48, 49], besides other less common contributions such as the series or access resistance, when high contact contributions are demonstrated [50]. This issue is discussed in more detail in Chapter 1 [51], Chapter 2 [52], Chapter 7 [53], and Chapter 15 [54] of this book. Figure 11 depicts the power spectral density of the current experimentally characterized for multiple transistors affected by various sources of noise. The inset illustrates the spectral decomposition of different noise sources which generally affect the behavior of the transistor.

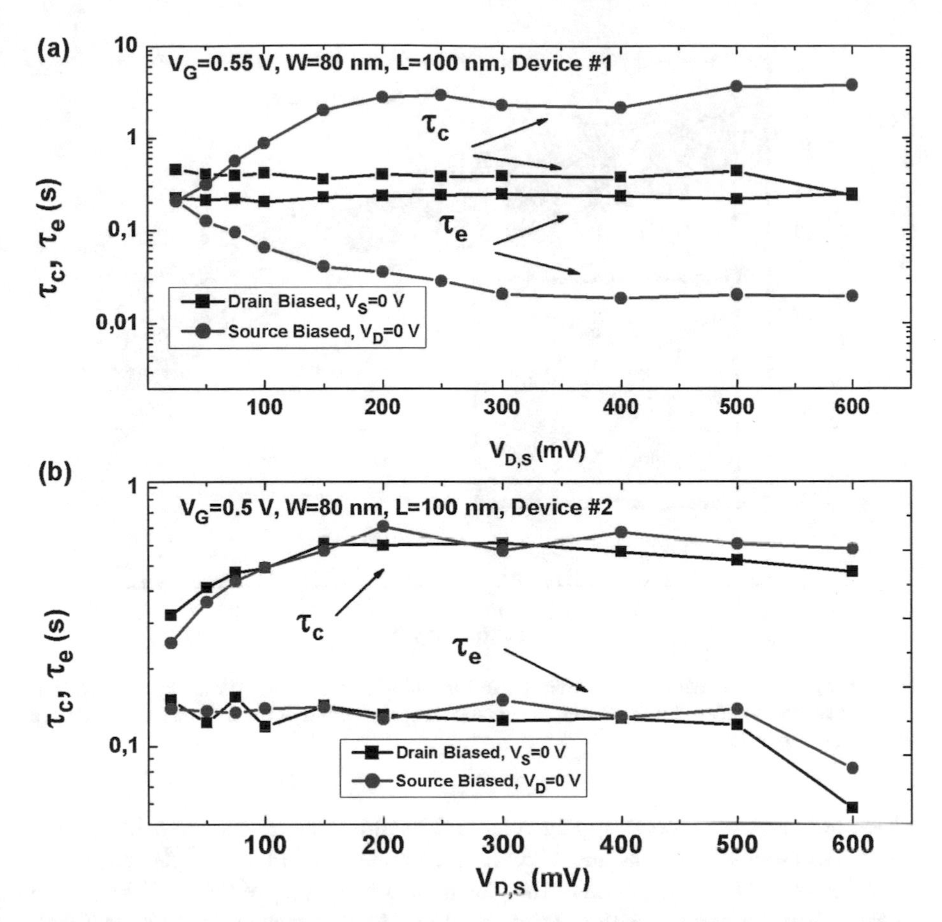

Fig. 10 τ_c and τ_e dependence with the drain/source bias applied **(a)** to a device where an asymmetry appears between drain and source configuration (oxide trap away from the channel center) and **(b)** to a device where drain and source configuration present similar results (oxide trap close to the channel center)

In this work, we focus our attention on the generation–recombination noise (G–R) as it is related to trap interactions. This noise is caused by the fluctuation of the number of carriers due to random generation and recombination of electron-hole pairs in defects [55, 56]. The normalized noise level corresponding to the fluctuations of a single trap depends on the square of the current as follows:

$$\frac{S_{ID}}{I_D^2} = \frac{\overline{\Delta N^2}}{N^2} \times \frac{4\tau_r}{1 + 4\pi^2\tau_r^2 f^2} \tag{5}$$

where $\tau_r = (1/\tau_c + 1/\tau_e)^{-1}$ is the carrier lifetime or effective time constant. The variance $\overline{\Delta N^2}$ of the carrier number N is proportional to the concentration of

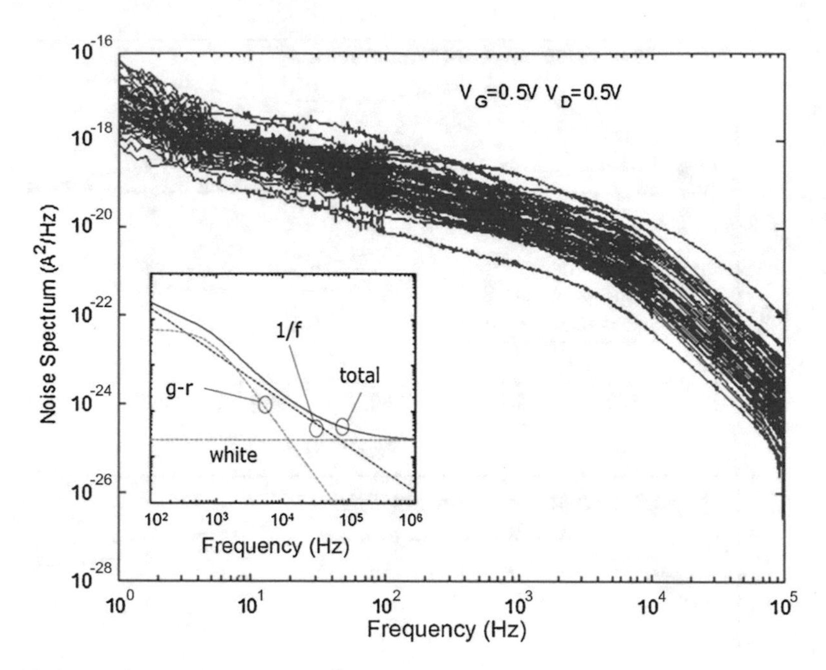

Fig. 11 Spectral characteristic of current noise for multiple transistors affected by the superposition of several sources of noise. Inset: Schematic frequency trend of the different noise sources affecting the transistor

generation–recombination-trapping centers. The noise corresponds to a *Lorentzian spectrum* composed of a plateau at low frequency followed by a $1/f^2$ decrease. The cutoff or corner frequency (end of the plateau and beginning of the $1/f^2$ trend, f_c) is related to the carrier lifetime as $\tau_r = 1/f_c$. In this regard, random telegraph noise (RTN), which is attributed to individual carrier trapping and detrapping phenomenon at the interface or oxide traps, can be modeled as a particular case of G–R noise with a single oxide trap. The power spectral density of this noise, induced by a single type of trap, also depicts a plateau at low frequency followed by a $1/f^2$ decrease. The mathematical model can be described as follows [56]:

$$S_{ID} = 4\Delta I_D^2 \times \frac{\tau_r}{\tau_e + \tau_c} \times \frac{\tau_r}{1 + (2\pi\tau_r f)^2} \tag{6}$$

However, when many traps with different time constants coexist, the superposition of their individual RTN signals results in a $1/f$ noise, typically known as Flicker noise. Flicker or low-frequency noise is associated with imperfections of the fabrication process and material defects and appears mostly at low frequencies. This kind of noise is observed in all semiconductor devices. The power density spectrum is characterized by a 1decade/1decade slope in the frequency domain. In transistors, the trend was described by McWhorther [57], who attributed the source of 1/f noise to random trapping and detrapping events in charged traps near to the gate oxide, ruled by the carrier tunneling between the inversion channel and slow oxide traps N_{ot} located at the Si–SiO_2 interface following the expression [55, 57, 58]:

$$S_{V_{fb}} = \frac{q^2 N_{ot}}{LWC_{OX}^2 f} \tag{7}$$

being $N_{ot} = kTN_t\lambda_{OX}$ where N_t is the density of oxide traps per unit of volume and energy, λ is the McWhorther tunneling parameter (average tunneling length), f the frequency, and the other symbols have their usual meaning. The concentration of slow traps, N_t, and interface traps, D_{it}, may be correlated with a first-order approximation $D_{it} \approx \lambda_{OX} N_t$ [8]. However, it has been demonstrated in [35] that there is no correlation between τ and x_T, indicating that the elastic tunneling model is inadequate for explaining RTN in FETs with thin gate dielectric [59]. For a thermally activated trapping process, the trapping probability decreases exponentially with the cross-section activation energy E_a and the flat band voltage noise follows [50]:

$$S_{V_{fb}} = \frac{q^2 k^2 T^2 N_{it}}{LWC_{OX}^2 f \Delta E_a} \tag{8}$$

where ΔEa is the amplitude of the activation energy dispersion and N_{it} is the oxide trap surface state density. If the mobility fluctuations due to trapping/detrapping events are taken into account, the input gate voltage noise is modeled as follows [50, 60]:

$$S_{V_G} = (1 + \alpha \mu_{eff} C_{ox} (V_G - V_{TH}))^2 (Sv_{fb}) \tag{9}$$

where μ is the effective mobility and α the Coulomb scattering coefficient. The equivalent drain current noise density S_{ID} can be expressed through the transconductance following the expression:

$$S_{I_D} = S_{V_G} \times g_m^2 \tag{10}$$

Finally, the normalized drain current noise is modeled as [50, 60]:

$$\frac{S_{I_D}}{I_D^2} = (1 + \alpha \mu_{eff} C_{ox} I_D / g_m)^2 \left(\frac{g_m}{I_D} \right)^2 Sv_{fb} \tag{11}$$

To experimentally carry out this characterization, the spectral density of the current noise, S_{ID}, should be measured as a function of gate voltage, V_G. For this purpose, the equipment employed in the electrical characterization consists of a low-noise current (or voltage) amplifier connected with a high-resolution A/D converter and a spectrum analyzer. A schematic example of our particular setup is shown in Fig. 12. This experimental setup employs an automatic low-frequency noise measuring

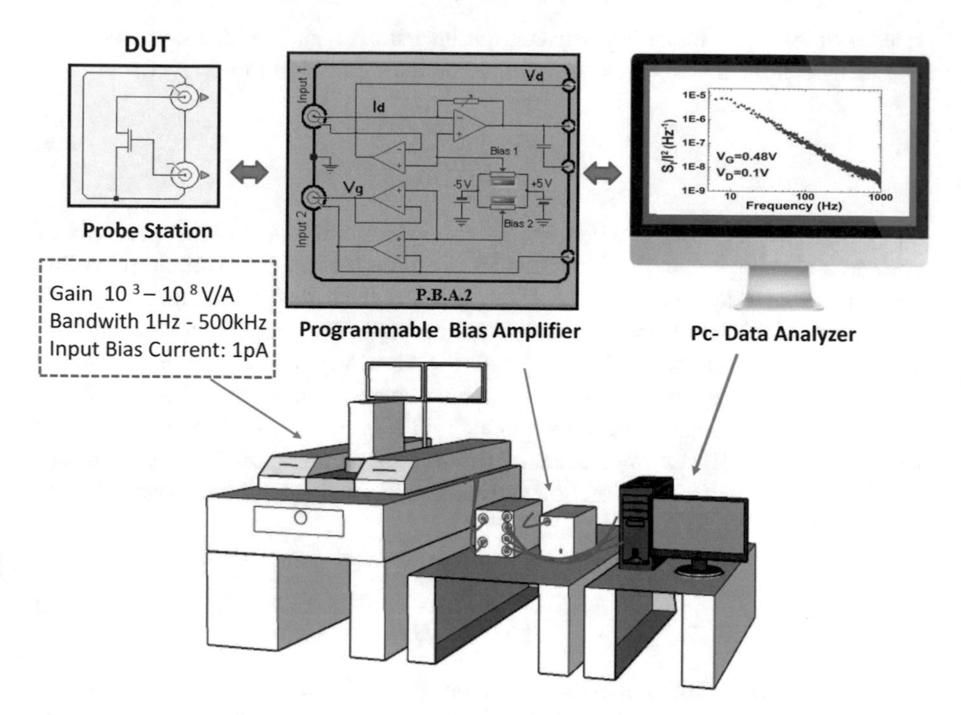

Fig. 12 Experimental setup used to carry out the on-wafer low-frequency noise characterization

system which consists of a low-noise programmable bias amplifier (PBA) controlled by a computer with a dedicated software, digital I/O, and data acquisition cards. This system biases the device under test (if on wafer through a probe station), measures the currents and their fluctuations with accuracy, carries out the Fourier analysis and, finally, records and displays the data.

Additionally, the log–log scale representation of the normalized drain current noise versus drain current characteristic corroborates the noise sources. If the noise level varies with the drain current as the transconductance to drain current ratio squared, the main contributor to the noise is the carrier number fluctuation [60]. As well, this representation could determine the impact of the access series resistance, which induces an increase at high drain current in the normalized drain current ratio, due to the increase contribution of the series resistance (contacts) against the channel resistance, and modeled as [50, 61]:

$$\frac{S_{ID}}{I_D^2} = \frac{S_{ID}}{I_D^2}_{\text{(channel)}} + \frac{I_D}{V_D}^2 S_{Rsd} \tag{12}$$

Figure 13 shows two experimentally characterized cases where transistors are affected by different noise sources. In Fig. 13a, the normalized spectral density of the noise follows the transconductance to drain current ratio squared, making

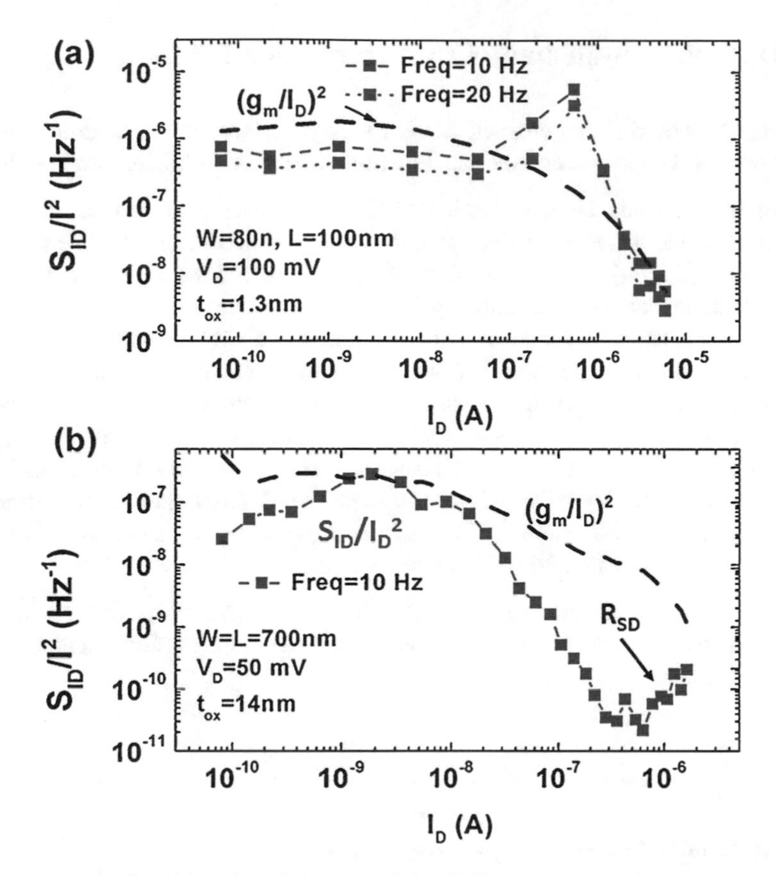

Fig. 13 Normalized power spectral density of the drain current as a function of the drain current for two different devices, (**a**) silicon-on-insulator transistor showing a clear carrier number fluctuation noise source and (**b**) a thicker standard bulk MOSFET transistor showing a high series resistance contribution, R_{SD} (noise level increasing with the drain current in the high inversion regime)

carrier number fluctuation as the dominating source according to Eq. 11. A peak close to the inversion regime current is observed showing a noise increment due to trapping/detrapping events. This case corresponds to a SOI transistor fabricated in a 22 nm process featuring a 1.3 nm thick oxide and 7 nm thick silicon layer. In Fig. 13b, an n-MOSFET device fabricated in a 250 nm process and featuring 14 nm-thick oxide shows a power noise spectral density where not only carrier number fluctuation must be taken into account, but also a series resistance contribution responsible for the noise increase at high current levels (Eq. 12). These results reveal that depending on the device technology different instability sources may affect the transistor operation.

4 Optimized Systematic Characterization Protocol

To date, the experimental characterization used to determine the random telegraph noise temporal signature and the trap location entails the following limitations:

- The time domain characterization of the electrical noise requires large time windows due to the extensive quantity of events needed to obtain meaningful statistics [26, 62]. In addition, depending on the selected sampling rate in the temporal characterization, only a portion of interactions can be captured. Events outside this temporal range, slower or faster, would not be detected.
- Even if an appropriate sampling rate is selected, some trapping/detrapping phenomena could be not visible because of the bias conditions in the characterization are outside the voltage range where the characteristic times are discernible. Figure 14 shows the experimental case of three transistors fabricated following the same process but being affected by traps in a different gate bias condition.
- Identifying the real number of traps in a trapping phenomenon is tedious and requires extensive analysis of long data streams.

The Fourier frequency transform (FFT) of the current signals can act as a measurement tool to solve some of these limitations in the random telegraph noise characterization.

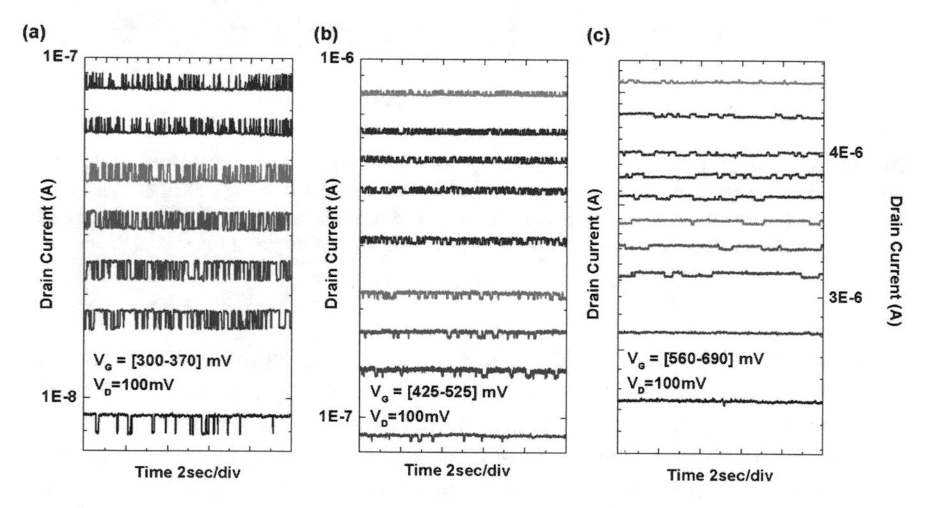

Fig. 14 Drain current traces as a function of the time from three different transistors affected by random telegraph noise at different gate voltages. Device affected for gate voltage ranges from (**a**) 300 to 370 mV, (**b**) 425 to 525 mV, (**c**) 560 to 690 mV. Silicon-on-insulator transistors fabricated in a 22 nm process featuring 1.3 nm thick oxide, 7 nm thick silicon channel, and with length and width of 100 and 80 nm, respectively

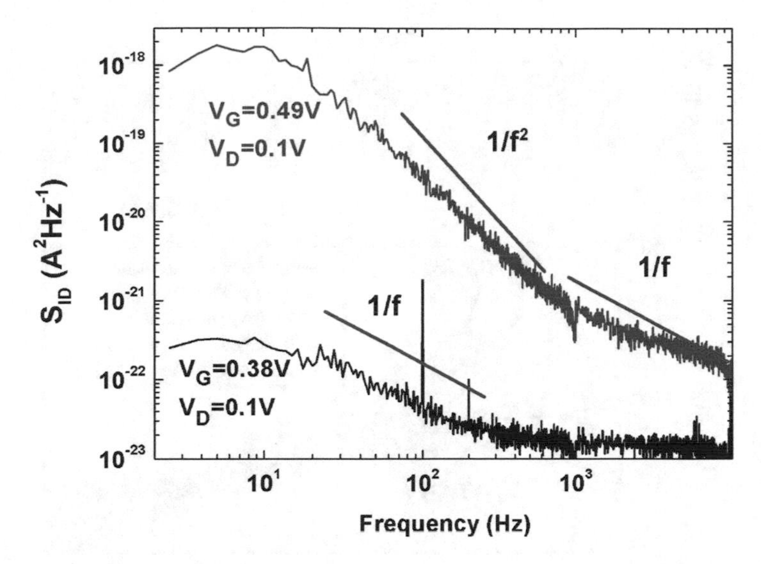

Fig. 15 Power spectral noise density of the drain current (S_{ID}) for two different gate voltages (V_G). Same device as in Fig. 14b

4.1 Determination of the Optimum Bias Condition

The low-frequency noise characteristic of a device affected by random telegraph noise (a single trap) was given in Eq. 6. According to this, both the plateau power spectral magnitude and the cutoff frequency are τ_r dependent, i.e., proportional to the number of trapping/detrapping events. Higher number of interactions implies shorter characteristic times and higher noise spectral densities in sub-cutoff frequencies. In fact, this trend can be used to determine the bias conditions where the device is affected by the trapping/detrapping phenomenon whose power spectral density would present higher magnitude than in case where the interactions rarely appear.

This characterization method, spectral scanning by gate bias (SSGB) [63] consists of:

1. Initially, the spectral noise density of the drain current (S_{ID}) is characterized. As an experimental example, Fig. 15 shows the case of the device shown in Fig. 14b at two different bias conditions. While for the highest gate voltage, a plateau followed by a square frequency dependence reveals the existence of generation–recombination noise generated by traps, the lowest gate voltage only shows a 1/f trend according to regular flicker noise.
2. The corner (cutoff) frequency of the noise spectrum can be determined by the sum of the inverse of the characteristic times ($f_c = 1/\tau_e + 1/\tau_c$) [20] as it has been shown in Sect. 3.

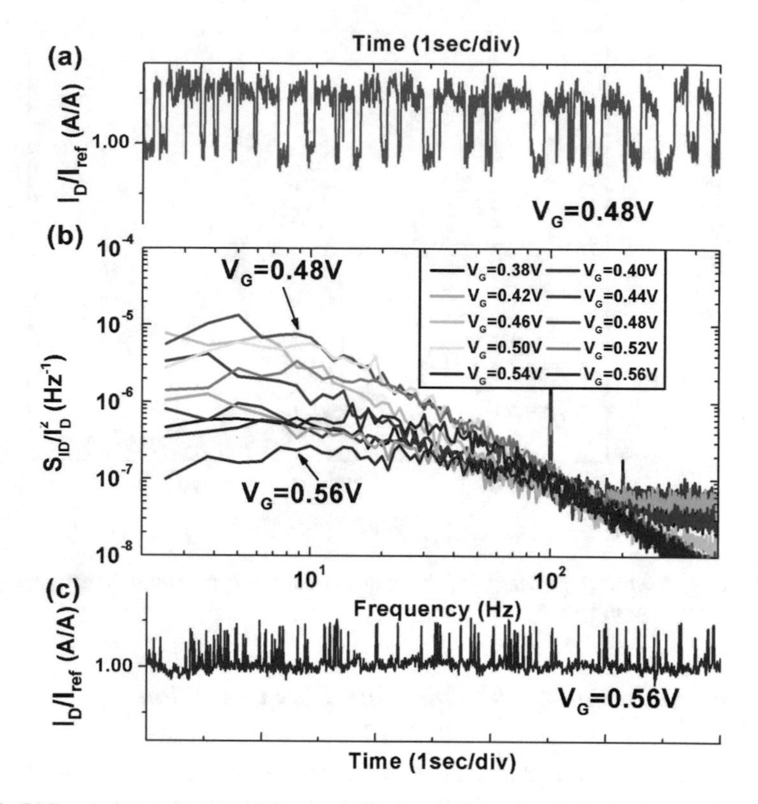

Fig. 16 (**b**)Spectral dependence of the normalized power density of the drain current (S_{ID}/I_D^2) for different gate voltages (V_G). $I_D - t$ traces for the highest and the lowest power spectral density observed correspond to (**a**) $V_G = 0.48$ V and (**c**) $V_G = 0.56$ V, respectively. Devices from the same set as in Fig. 14

3. Once the corner frequency is located, the normalized spectral density of the current noise is estimated (S_{ID}/I_D^2). Figure 16 shows the power spectral density of the noise around the corner frequency when the device is affected by random telegraph noise.
4. The next step is to depict the normalized spectral density of the current (S_{ID}/I_D^2) as a function of the gate voltage (V_G) at the corner frequency, leading to the $S_{ID}/I_D^2 - V_G$ curve shown in Fig. 17.

In this case, different frequencies around the corner frequency are shown to demonstrate that accurate calculations in step 4 are not required. The bell shaped characteristic identifies the bias range where the RTN will be easily observable (V_G ∈[0.35 V, 0.60 V] for this particular case). This result is consistent with the drain current signals shown in Fig. 14b, where the fluctuations are visible in the same voltage range given by the SSGB plot in Fig. 17. Additionally, the SSGB approach allows the quickly determination of the bias condition where $\tau_c = \tau_e$ (used to determine the energy position of the trap in the oxide in Eq. 2) as the bias condition which maximizes the noise in the bell shaped curve.

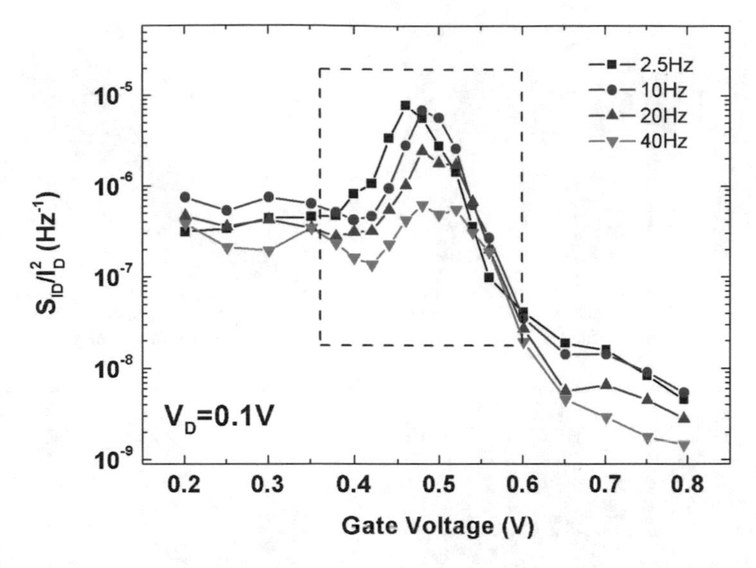

Fig. 17 Normalized current noise power (S_{ID}/I_D^2) as a function of the gate voltages (V_G) for different frequencies around the corner frequency. The high normalized noise power levels on the left of the bell shape are due to the limited resolution of the characterization equipment for these low drain current levels (at the weak inversion charge below threshold voltage)

4.2 *Identifying the Number of Traps in the RTN Signals*

SSGB protocol allows the determination of the gate bias range where the device is affected by trap interactions through the increment in the noise level. This noise excess is attributed to the current fluctuations which increase according to the trapping/detrapping events. Identifying the RTN signals corresponding to a single trap is one of the most challenging tasks. Different approaches have been implemented to overcome this challenge. Hidden Markov model (HMM) is widely employed to solve the statistical properties of multi-level RTN caused by multiple traps. The essence of a Markov process resides in its memory-less characteristic which in this context means that the next transition (due to trapping/detrapping) depends solely on the current state, regardless of the defect state. In these cases, discrete current level and number of traps can be mathematically extracted [39, 64, 65]. Despite HHM is a robust approach to determine the number of traps involved, in this work, time log plot (TLP) approach [66, 67] has been employed. This method presents less resources demanding characteristic and a trivial implementation, although as counterpart, demonstrates poorer resolutions on noisy data. This issue is discussed in more detail in Chapter 14 of this book [68]. In the weighted-TLP (w-TLP) approach, each point of the TLP space (events) is given by the sample of the current at a specific time, and the subsequent sample $((I_D(i), I_D(i+1)))$. Then, the TLP space event is color-weighted by an *appearance* function which accounts for the number of events inside a definite area of the TLP space (mathematical framework

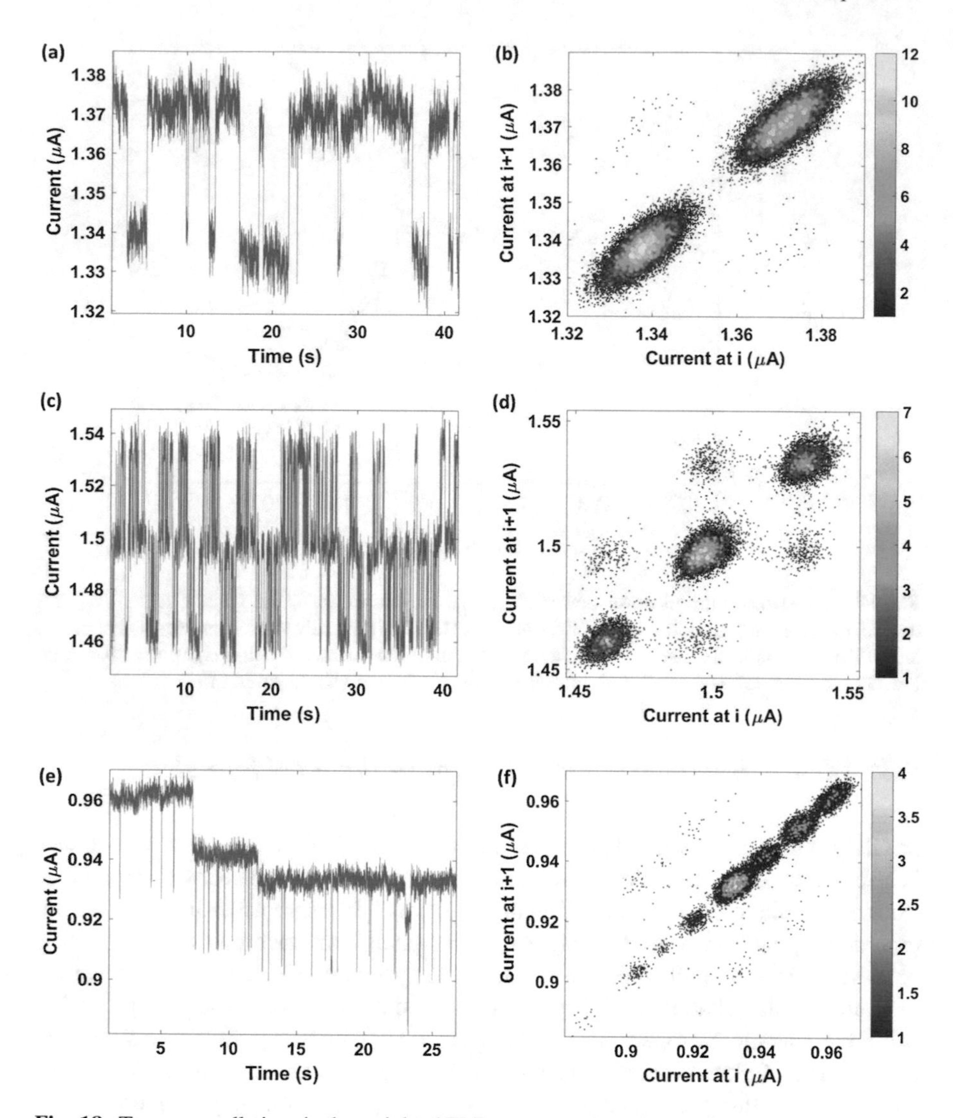

Fig. 18 Traps constellations in the weighted TLP space together with the drain current signals: (**a**) and (**b**) transistor with two lobes (states) in the TLP constellation result from a single active trap. (**c**) and (**d**) Transistor with a three-lobe constellation identifying the characteristic signature of two traps. (**e**) and (**f**) Spread cloud corresponding to a transistor with multi-trap events

can be found in [63]). This color-weighted approach of the TLP method allows to identify the RTN levels as populated regions in the diagonal of the TLP space, while populated regions outside the diagonal are related to the transitions between states. Figure 18 shows the results of the application of the w-TLP method. When the device is affected by a single trap (Fig. 18a), the appearance of a constellation with two lobes indicates the presence of a single trap (Fig. 18b); a three-lobe constellation (Fig. 18d) reveals the existence of three predominant current levels, two traps (Fig. 18c); when multiple current levels appear (Fig. 18e), a spread lobe indicates a multi-trap situation (Fig. 18f). It is worth noting that complex random telegraph signals as observed in [39, 40, 69] with anomalous RTN or meta-stable states can disturb the appropriate identification method. Additionally, appropriate sample-rating and slot times should be selected to capture the main active traps.

4.3 Systematic Experimental Protocol

Figure 19 depicts a scheme summarizing the systematic protocol proposed. The low-frequency noise (LFN) of the device for different gate biases is analyzed through the spectral scanning by gate bias approach. The output of this first stage provides the optimum bias condition where the device is affected by RTN. Then, the transient monitoring of the drain current and the analysis through the weighted time lag plot method determine the number of active traps in the device. This selection of the appropriate bias condition and the identification of the number of traps involved in the device can be an advantage to implement automatic testers and the extraction of statistical result.

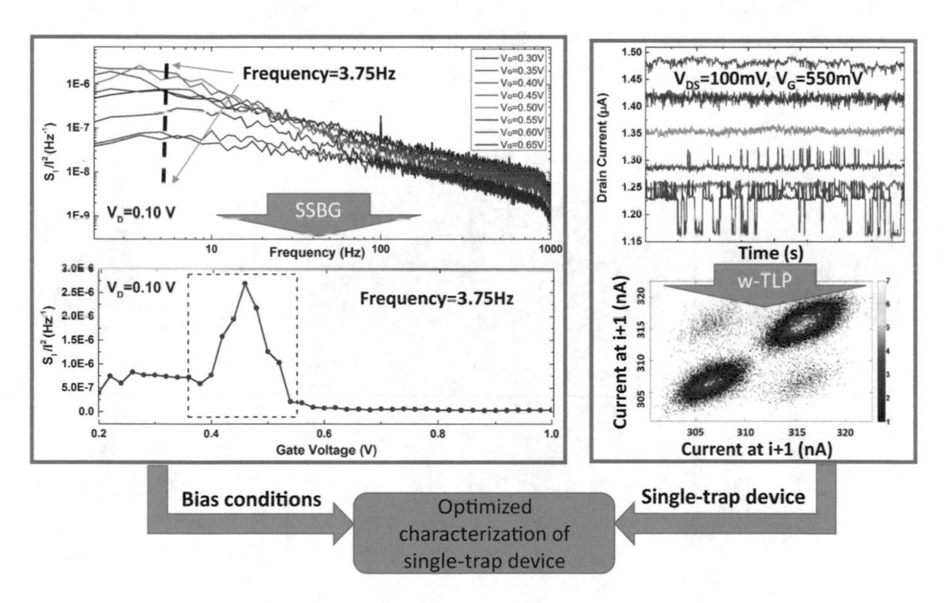

Fig. 19 The experimental method proposed for the single-trap random telegraph noise characterization in transistors. The method permits the optimum RTN characterization of single-trap devices combining the spectral scanning by gate bias technique and the time lag plot method

4.4 On-Wafer Trap Distribution

The effectiveness of the method is fully exploited when it is combined with an automatic probe station. The distribution of traps over the transistors on the wafer can be determined by the automatic monitoring of the drain current of specific transistors and the simultaneous application of the w-TLP method previously described. Figure 20 shows the map for traps in 100 nm length and 80 nm width transistors fabricated in each die of the wafer (336 transistors in total) when the gate voltage is close to the threshold voltage ($V_D = 0.1$ V, $V_G = 0.5$ V). The map represents the number of traps detected in each transistor: failed transistors, for which the automatic contact was missed or simply because they were faulty (black); transistors with only one drain current level, i.e., transistor without trap (green); transistors with two current levels, i.e., single-trap devices (blue); transistors with three current levels, i.e., two-trap devices (red); and transistors with more than two active traps (pink). Examples of the current level for each type of transistor are shown in the annexed time domain representations in Fig. 20. Figure 21 presents the histogram summarizing the trap count over the wafer. For this specific dimensions studied (one transistor per die with dimensions of $L = 80$ nm and $W = 100$ nm), one out of three of the transistors is affected by one trap, and virtually the same number of transistors is not affected by traps leading to RTN. The case of devices with three current levels, i.e., two traps, represents the 7.5% of the devices characterized on the wafer. Finally, the case of multi-trap devices (transistors with more of three current levels) involves 12.5% of the devices. The previous result

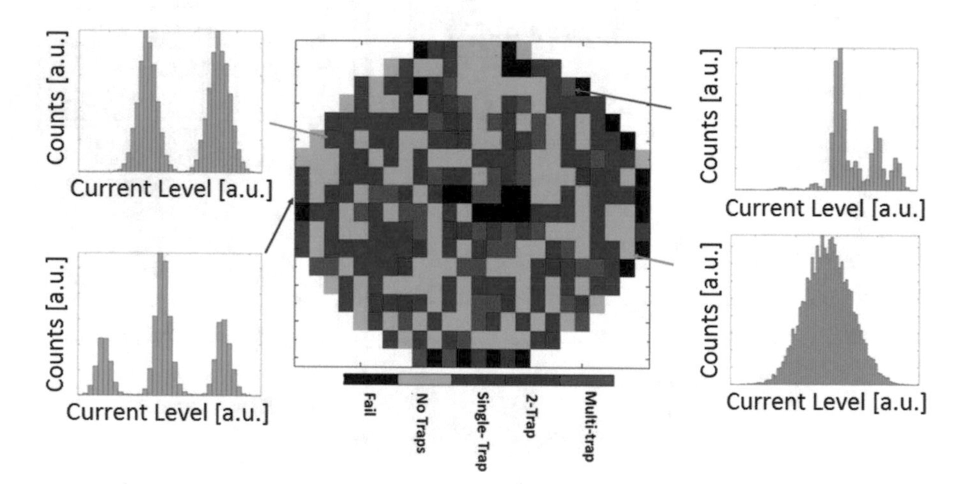

Fig. 20 Distribution of the number of current levels detected in the transistors over the wafer at $V_D = 0.1$ V and $V_G = 0.5$ V. (Left) Examples of transistors with two and three current levels (top and down, respectively). (Right) Examples of multi-current level (>3) and RTN-free transistors (top and down, respectively). Silicon-on-insulator device with $t_{ox} = 1.3$ nm. Drain current histograms have also been included

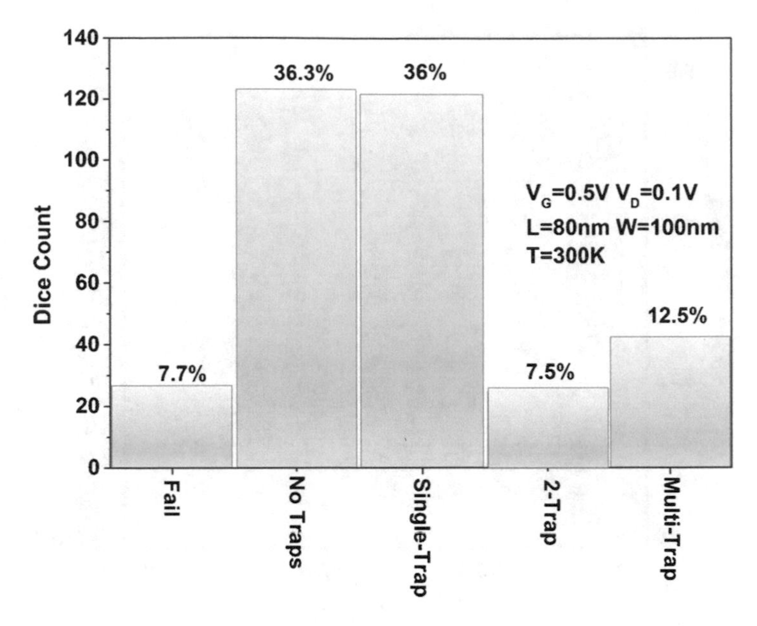

Fig. 21 Histogram of the number of dice presenting a determined number of traps in the device under study detected from the RTN characterization ($L = 100$ nm, $W = 80$ nm, one transistor per die). Same bias than in Fig. 20

demonstrates the large percentage of single-trap devices within this wafer run. This information constitutes a useful feedback for process engineers in the path of the optimization of the technology.

5 Magnetic Field Effect

The systematic measuring protocol defined in Sect. 4 has been employed to characterize single-trap devices from two sets of SiO_2 n-MOSFETs fabricated at Global Foundries facilities in a 250 nm process, both in the presence and absence of low static magnetic fields at room temperature. The first series of devices, *Set A*, features a $t_{ox} = 14$ nm, with gate length and width of 700 nm. The second one, *Set B*, presents a $t_{ox} = 7$ nm, with gate length and width of 400 nm. To carry out the experimental measurements, the devices have been diced from the wafer to be directly characterized in a Janis ST-500 probe station where perpendicular-to-the-channel magnetic fields were induced through interchangeable magnetic rings with definite values in a range from -200 to $+200$ mT. The setup will be described below in Sect. 5.1.

The determination of the number of traps involved and the optimum bias range where the device is affected by random telegraph noise can be carried out by the color-weighted approach (w-TLP) and the spectral scanning by gate bias (SSGB)

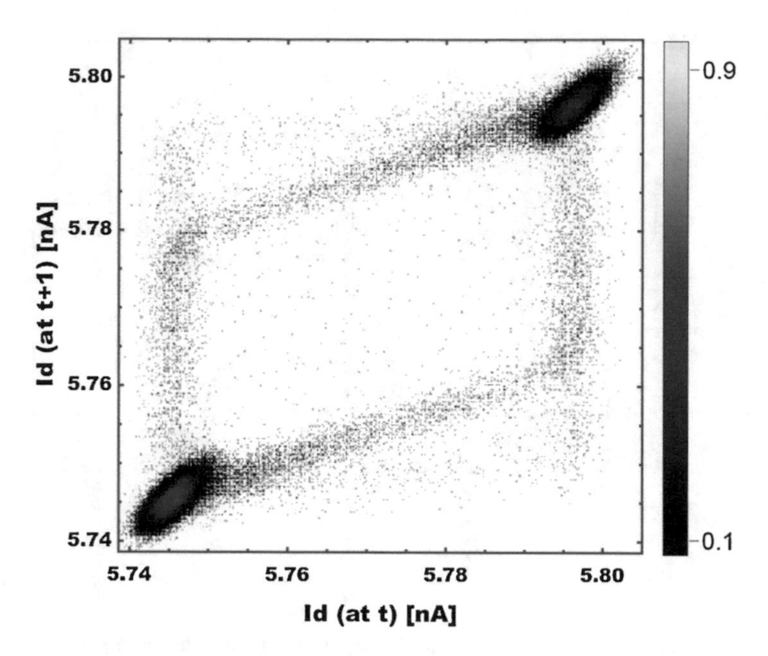

Fig. 22 Constellation in the w-TLP space at $V_G = 360\,\text{mV}$ for a device from *Set A* ($t_{ox} = 14\,\text{nm}$). The two defined states indicate one transition, i.e., a single oxide trap is activated (the color scale indicates the weighted number of events)

approximations, respectively. The results for a device belonging to Set A are shown in Figs. 22 and 23. Figure 22 demonstrates, with the appearance of two lobes in the diagonal, that the device is affected by single-trap events or random telegraph noise. Figure 23a corroborates this fact showing the power spectral density of the drain current for different gate voltages. A well-defined *Lorentzian spectrum* composed of a plateau at low frequency followed by a $1/f^2$ decrease is observed in accordance with the expected noise generated by random telegraph signals. Following the SSGB protocol, Fig. 23b determines the range of gate voltages where the device is affected by the trapping/detrapping events. A semi-bell like curve is seen in the range between 250 and 500 mV. This particular shape is attributed to the limitations in the accuracy imposed by the experimental setup for gate biases below 200 mV (very low drain currents for these devices with $V_{TH} \approx 0.5\,\text{V}$).

Once a single-trap device is localized, an extended characterization of the device can be carried out through time and spectral measurements. Time signature permits the extraction of the characteristic times, τ_c and τ_e (Fig. 24). The trap location inside the oxide can be extracted through the logarithm of the characteristic time ratio (Fig. 24b) following Eq. 3, and the energy of the trap through Eq. 2 at $\tau_c = \tau_e$. Note that the bias condition needed for $\tau_c = \tau_e$ also implies the highest power spectral density point in the peak of the semi-bell shape in Fig. 23 ($V_G \approx 350\,\text{mV}$).

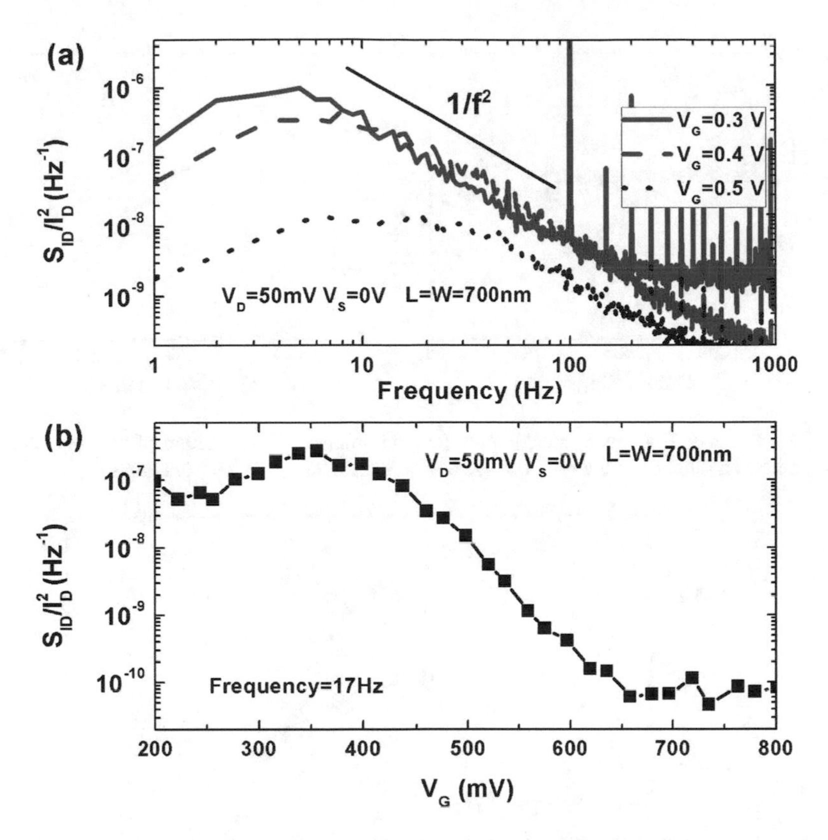

Fig. 23 (**a**) Normalized power spectral density of the drain current for three different gate voltages. (**b**) A semi-bell shaped $S_{ID} - V_G$ curve obtained at the corner frequency, f_c, shows the bias range where the device is affected by random telegraph noise. Results concerning the same device from *Set A* ($t_{ox} = 14$ nm) as in Fig. 22

On the other hand, the low-frequency noise characterization might reveal complementary details on the trapping events. Figure 25 depicts the normalized noise spectral density as a function of the drain current for the same single-trap device. According to the discussion previously reported in Sect. 3, the sources of noise can be clearly discerned between carrier number fluctuations and series resistance contribution through the trend shown in the $S_{ID}/I_D^2 - I_D$ curve. In addition, the squared transconductance to current ratio has been plotted as Eq. 11 suggests. Even neglecting the increment in the noise observed at the current levels where the trapping/detrapping events are observable (nanoAmps in this case), the normalized noise does not fit with the carrier fluctuations trend ($(g_m/I_D)^2$) and an increase in the normalized noise is observed at high current levels according to series access resistance contribution.

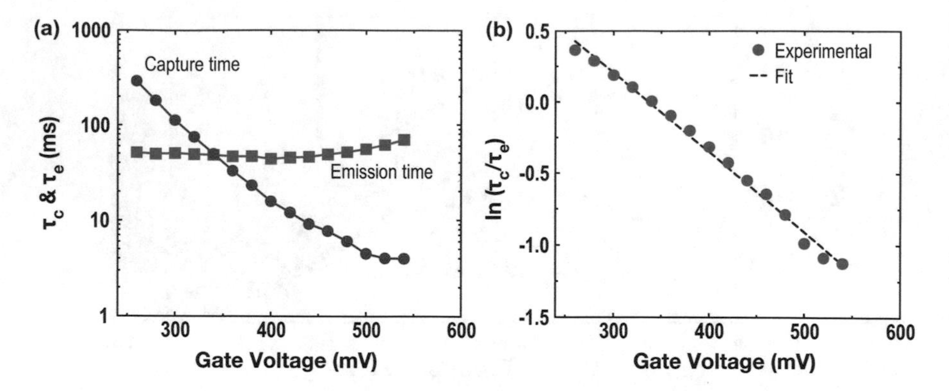

Fig. 24 (**a**) Dependence of τ_c and τ_e with the gate voltage. (**b**) Logarithm of the characteristic times ratio as a function of the gate voltage. $V_D = 50$ mV. Same device from *Set A*

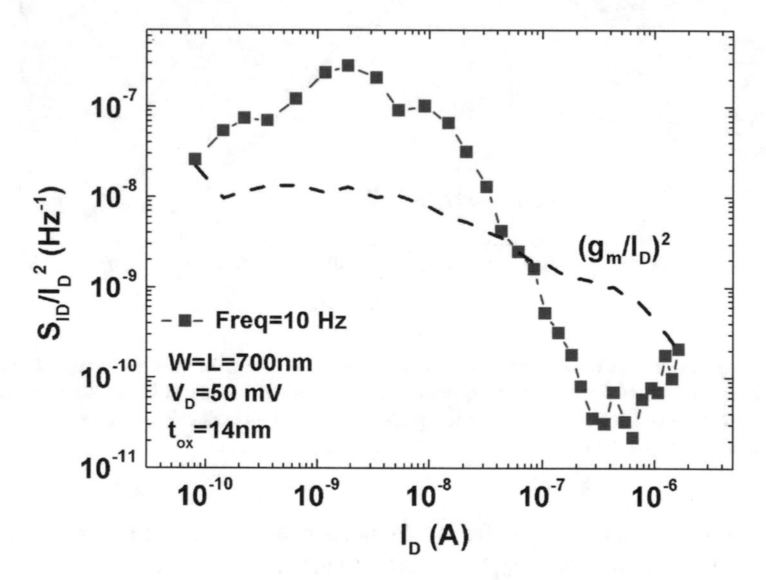

Fig. 25 Normalized noise spectral density (S_{ID}/I_D^2) as a function of the drain current (I_D) for the same single-trap device from Set A

5.1 Magnetic Field Effect on Trapping Dynamics

In order to determine the behavior of the trapping phenomena under the influence of magnetic fields, different low magnitude static and permanent magnetic fields have been set in the two possible orientations (referred to as positive and negative). The magnetic flux is perpendicular to the wafer surface and it is applied, thanks to interchangeable neodymium rings magnetized through thickness. A picture and schematic illustration are shown in Fig. 26. At first, the w-TLP method has been applied to determine if the number of traps involved in the phenomenon changes

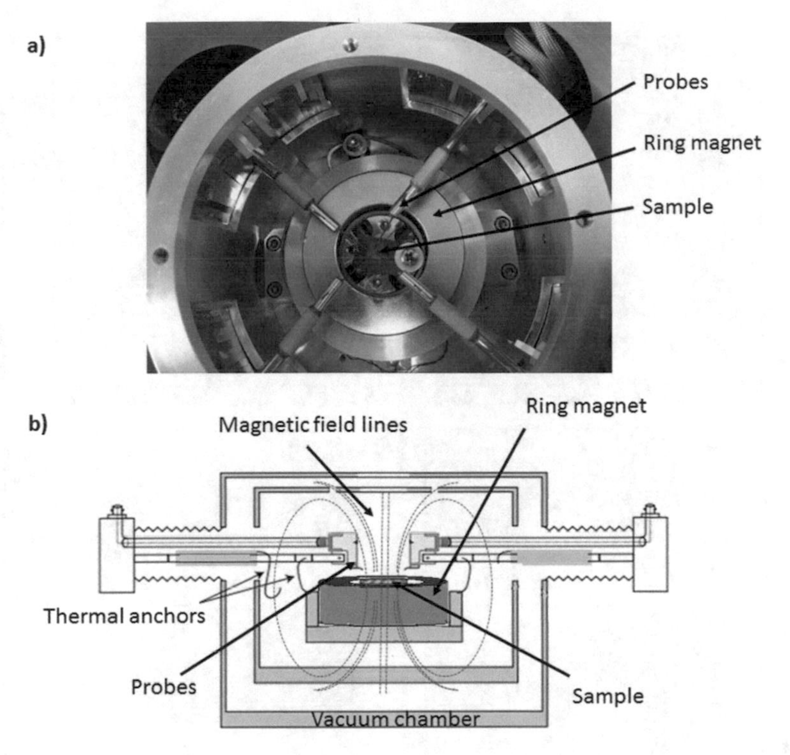

Fig. 26 Image of the probe station used for the electrical characterization under the influence of magnetic fields: (**a**) picture of the sample stage area, (**b**) cross-sectional diagram

as a function of the magnetic field (Fig. 27). The results clearly show two lobes in the diagonal of the space demonstrating single-trap events under the influence of the four magnetic fields, in this voltage range. It is worth mentioning that inferring more conclusions apart from the number of traps involved from this w-TLP depiction is difficult.

From the frequency domain point of view, the results for the SSBG approximation are shown in Fig. 28 focusing on the gate voltage dependence. Note that the magnetic field does not induce a significant impact on the bell shaped curve where the device is affected by random telegraph noise, between 0.25 and 0.50 V as it was previously corroborated through the TLP analysis in Fig. 27. Curves are similar in power density terms at this voltage range and no significant shift is detected, demonstrating that the signature of the traps is in principle unalterable by the weak magnetic fields applied here. Note that for the case of applying a magnetic field of $B = -110$ mT, green line in the graph, a slight shift is observed, inducing a small deviation in the voltages where the device is affected by random telegraph noise. This behavior has been only observed for this value of the magnetic field. However, an unexpected phenomenon is observed at high gate bias for the particular case of $B = -200$ mT: a noise power peak appears between 0.6 and 0.75 V.

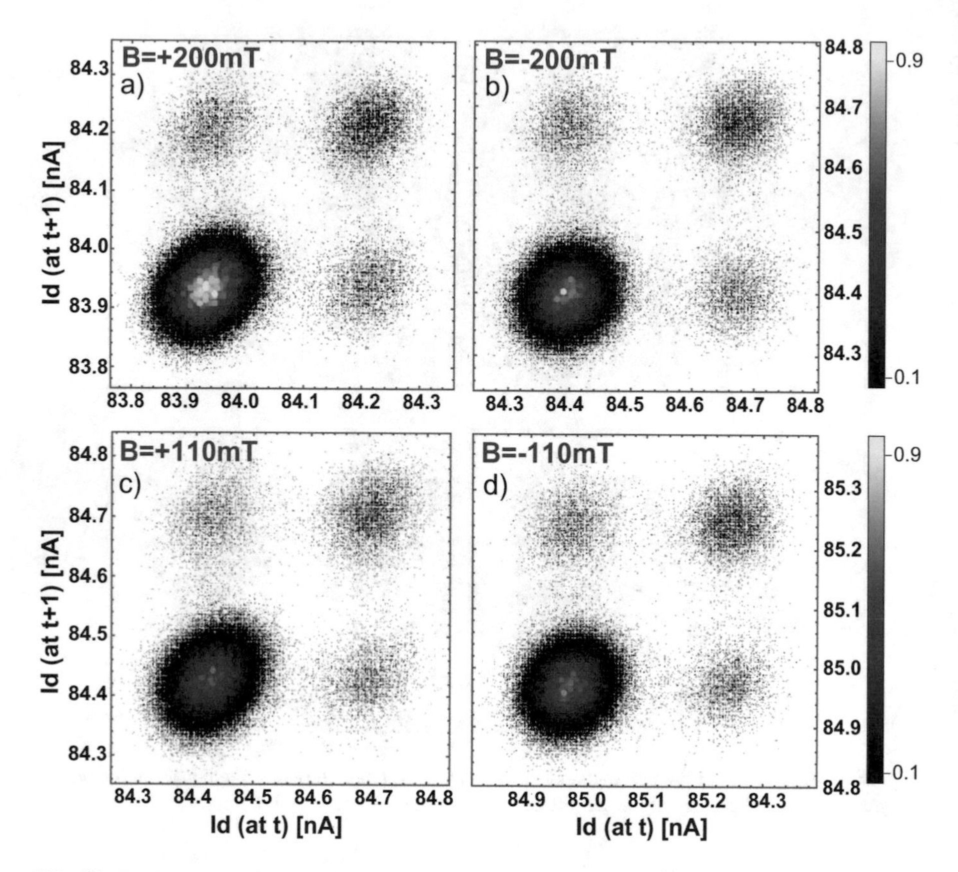

Fig. 27 RTN response is shown in the w-TLP space at (**a**) $B = +200\,\text{mT}$, (**b**) $B = -200\,\text{mT}$, (**c**) $B = +110\,\text{mT}$, and (d) $B = -110\,\text{mT}$, applying a constant gatè voltage of 500 mV on the selected device from *Set A*. Transition between two states is clear for the four cases. The color scale indicates the weighted number of events

To clearly understand this effect, Fig. 29 shows the $S_{ID} - f$ characteristic for this device at the voltage which maximizes the noise inside this unexpected peak ($V_G = 0.7\,\text{V}$). The significant variation among spectra is evidenced. The spectrum corresponding to $-200\,\text{mT}$ shows a noise excess following an almost quadratic inverse dependence with the frequency. This result suggests either several trapping/detrapping events caused by traps with close characteristic times or a single-trap activation with especially high current fluctuation contribution, as the possible sources of the noise excess. However, the rest of the spectra depict practically a linear inverse dependence with the frequency pointing out the flicker noise (contribution of traps with different characteristic times) as the fundamental source of the noise in this voltage range.

In-depth analysis has to be carried out due to the novelty of the results regarding the fact that magnetic field could activate new oxide traps. In addition, this

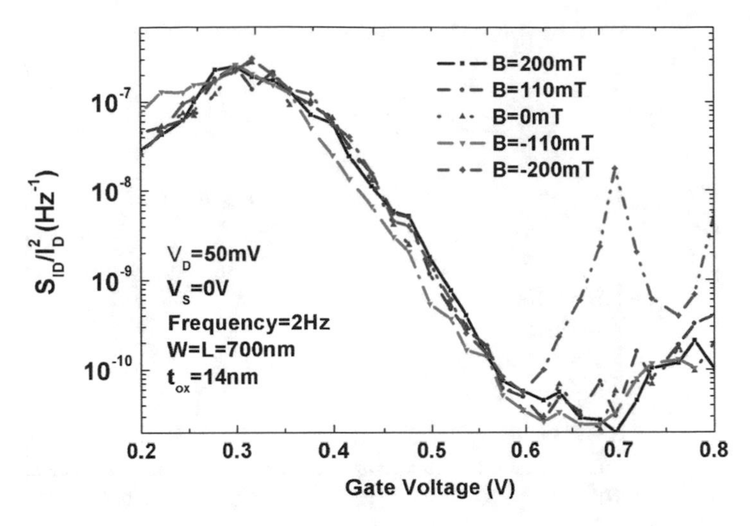

Fig. 28 Normalized spectral density of the noise (S_{ID}/I_D^2) as a function of the gate voltage (V_G) for the selected single-trap device from Set A in the absence of magnetic field (blue) and under the influence of four perpendicular-to-channel magnetic fields

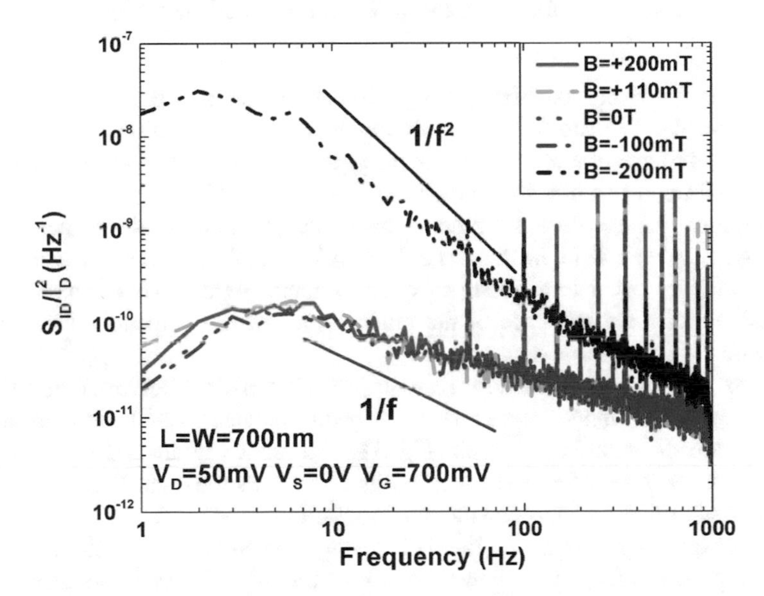

Fig. 29 Normalized spectral density of the noise (S_{ID}/I_D^2) as a function of the frequency (f) for the selected single-trap device in the absence of a magnetic field (blue) and under the influence of four magnetic fields (perpendicular to channel) at the gate voltage of $V_G = 0.7$ V

phenomenon also might be attributed to the increment of the series resistance appreciable at high gate voltages (strong-inversion condition). To test that, Fig. 30 shows the S_{ID}/I_D^2 dependence with I_D for the same bias conditions as in Fig. 28.

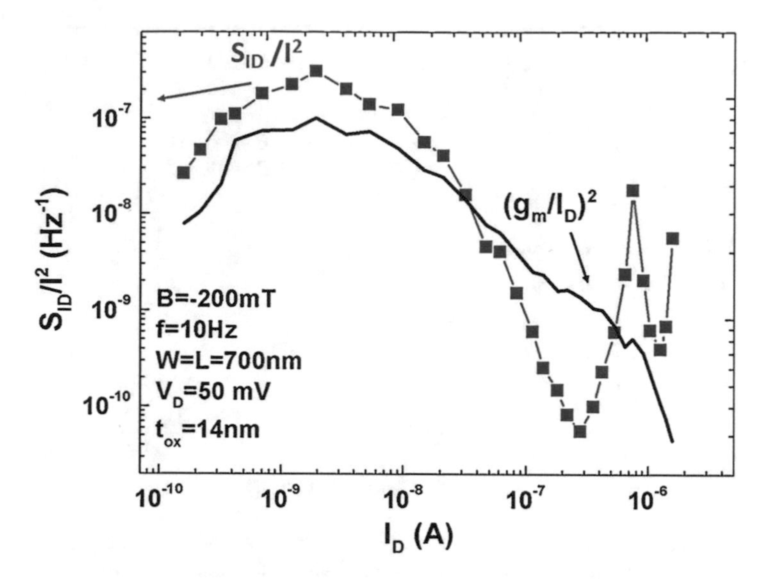

Fig. 30 Normalized noise spectral density (S_{ID}/I_D^2) as a function of the drain current (I_D) for the selected single-trap device under a 200 mT perpendicular magnetic field in negative orientation

In this case, it is worth mentioning that there is a reduction of the noise level before the increment related to the series resistance contribution, discarding the series resistance as source of noise at this bias range as it should always show a monotonic incremental trend as a function of the drain current (Eq. 12).

Therefore, the conclusion extracted from the previous results is that a negative high magnetic field in this specific single-trap device induces a new trapping/detrapping event in a strong inversion charge regime. This behavior is not detectable under the influence of the other values of the perpendicular magnetic field tested in the same device.

To analyze if this phenomenon is also appreciable in other devices, a device from *Set B* showing a single-trap behavior has been characterized. Time analysis and signature are shown in Fig. 31: while Fig. 31a shows the capture and emission times (an $I_D(t)$ trace is included in the inset), Fig. 31b shows the logarithmic ratio of the characteristic times to determine the trap location into the oxide.

Further investigations are carried out through the SSGB approach on the same single-trap device. Figure 32 shows the result of low-frequency noise characterization. Contrary to the previous device, only small differences among the magnetic fields are observed in this case, inferring a lack of magnetic field dependence. Note that the trend at low frequencies almost fits the quadratic dependence with frequency. The bell shaped curve is shown in Fig. 33 where, for this specific device, the left part of the curve is not visible due to the trapping/detrapping events taking place at such small gate voltages that the low current levels prevent showing reliable results. However, this behavior agrees with Fig. 31: the bias condition where $\tau_c = \tau_e$, i.e., the maximum of the noise in the bell shaped curve, apparently occurs

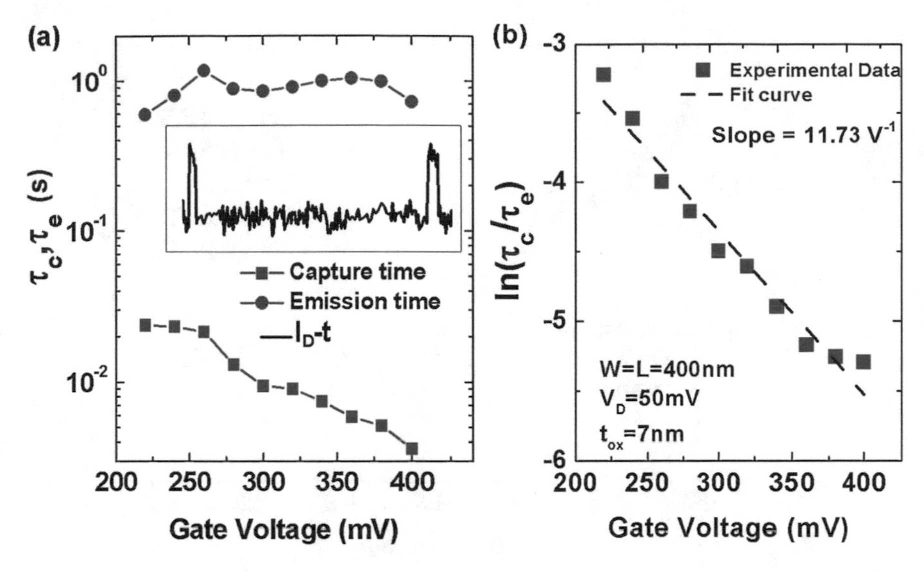

Fig. 31 (**a**) Dependence of τ_c and τ_e on the gate voltage. Inset: trace of 0.5 s of time depicting an $I_D(t)$ characteristic at $V_G = 300$ mV. (**b**) Logarithm of the ratio between capture and emission times with respect to gate voltages (symbols) and linear fitting (lines) according to Eq. 2. Single-trap device of Set B

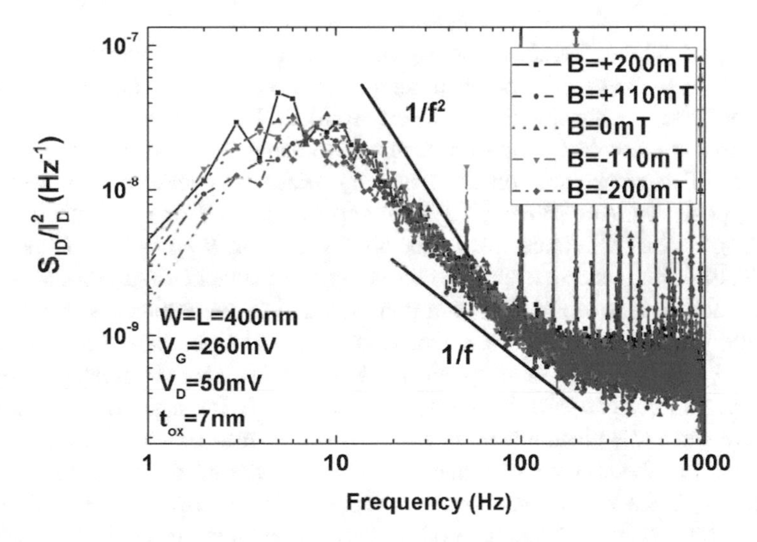

Fig. 32 Normalized noise spectral density (S_{ID}/I_D^2) as a function of the drain current (I_D) for the selected single-trap device from Set B in the absence of a magnetic field

at extremely low gate voltage. In addition, Fig. 33 suggests neither a dependence on the gate voltage with the magnetic field nor the presence of a series resistance noise contribution in the strong inversion regime.

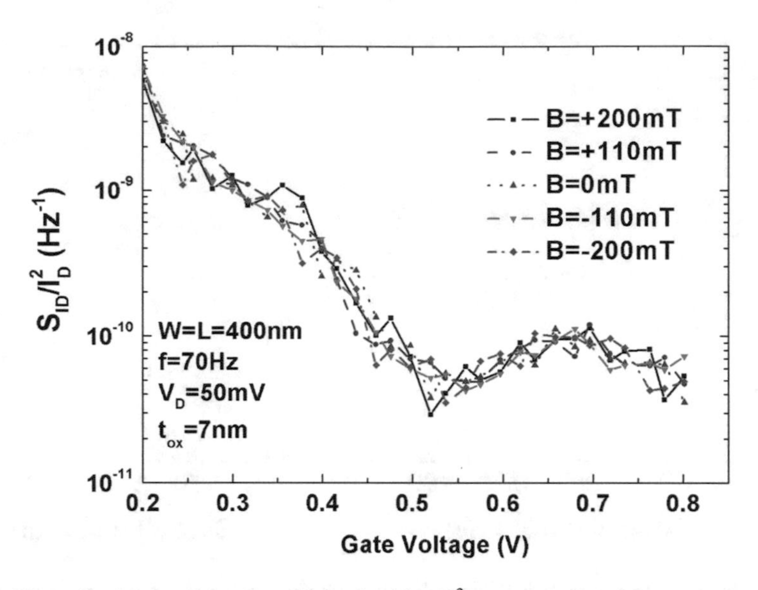

Fig. 33 Normalized spectral density of the noise (S_{ID}/I_D^2) as a function of the gate voltage (V_G) for the selected single-trap device of Set B in the absence of a magnetic field (blue) and under the influence of four magnetic fields (perpendicular to channel)

As a part of the characterization protocol, Fig. 34 shows the w-TLP and the time domain signature of these phenomena when different perpendicular magnetic fields (positive, negative, and null B) are applied. The analysis of the RTN trace reveals that a secondary transition occurs associated with the presence of a new trap state. This remarkable observation only arises at specific magnitudes of the B field, positive ($B = 200$ mT, Fig. 34a) and negative ($B = -110$ mT, Fig. 34d), exhibiting a non-monotonic trend (not observable for $B = -200$ mT and $B = 110$ mT, Fig. 34b,c, respectively). Additionally, the magnetic field introduces a shift in the drain current (up to 3%) as a consequence of the variations in any of the operating parameters of the transistor (mobility variations have been previously reported in the presence of magnetic fields [70, 71]), also observable in Fig. 27. However, this current shift, by itself, does not justify the appearance of the new trap state since this phenomenon is not observed in measurements, even at higher gate biases, in the absence of a magnetic field (regarding the same device). At a first glance, these results may seem contrary to those shown in Figs. 33 and 32 from the frequency analysis, but both cases can be simultaneously in agreement. The frequency domain graphs do not show any differences among magnetic fields in power spectral density terms; however, this does not mean that the traps are the same, but that the correspondent spectral density of the noise generated by the trap events is approximately the same, independently of the number of traps. Finally, note in the case of the 7 nm-thick oxide device, the time signature through the TLP method demonstrates new oxide trap events while the frequency domain does not identify this change. In the case of the 14 nm-thick oxide device this

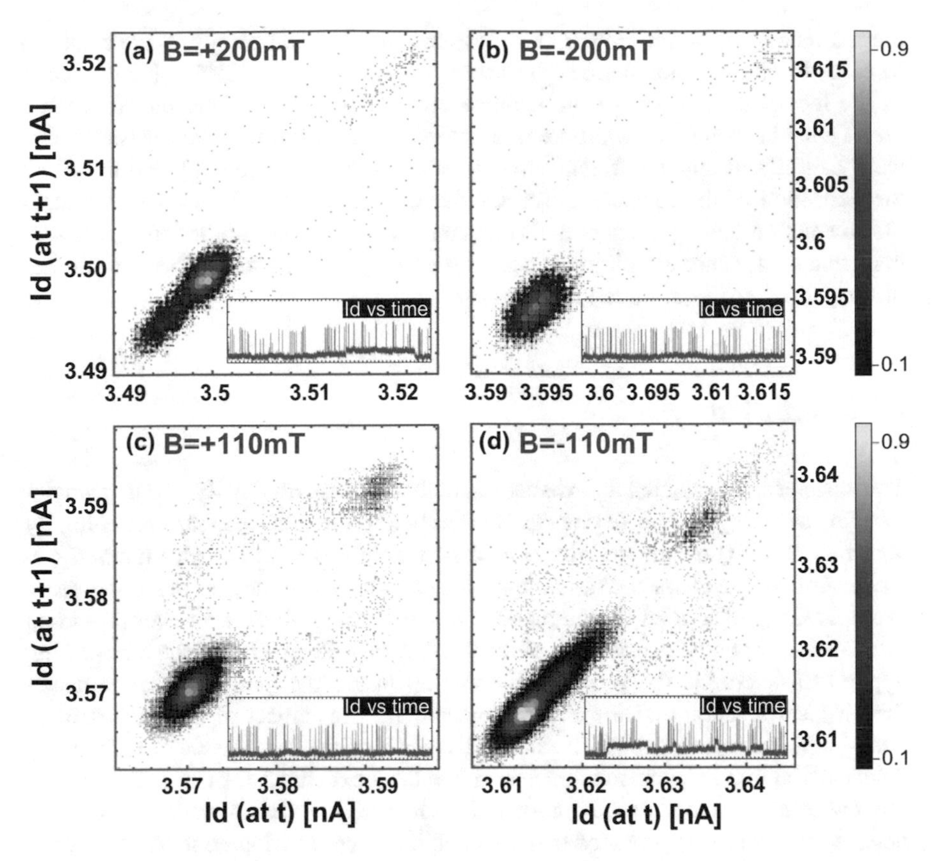

Fig. 34 RTN response is shown in the w-TLP space at (**a**) $B = +200\,\text{mT}$, (**b**) $B = -200\,\text{mT}$, (**c**) $B = +110\,\text{mT}$, and (**d**) $B = -110\,\text{mT}$, applying a constant gate voltage of 260 mV on a device from *Set B*. Only one transition between two states is clear when $B = -200\,\text{mT}$ (**a**) and $B = +110\,\text{mT}$ (**c**). An additional transition is induced, indicating a second trap joins in the trapping phenomena. This new state arises only at certain magnitudes of the B field, as in (**a**) and (**d**). The color scale indicates the weighted number of events. Inset of plots (**b**) and (**c**) shows the two-level drain current in the time domain when only one trap is present, and the inset of plots (**a**) and (**d**) indicates the appearing of a new trap as a three-level drain current in the time domain. Magnetic field in Tesla

phenomenon occurs contrary. This remarkable behavior only reinforces the evidence that the characterization under magnetic field influence is highly challenging and the characterization protocol can fail if only frequency or time domains are taken into account.

To date, some works regarding the magnetic field effect on transistor performance suggest that the effect can be related to the Zeeman effect [11, 12, 14], in which a single state splits up into two states, each with different spin orientation. However, the energy difference between the two new levels would be $\sim 20\,\mu\text{eV}$, corresponding to the highest field of $B = 200\,\text{mT}$. The small energy gap, compared to the thermal energy at room temperature, implies that low temperatures are required in order to observe modulations in parameters affecting the RTN signature, such as the recombination rate or the capture cross-section. Therefore, the noticeable effect must be studied by a different approach, considering that the measurements were carried out under different circumstances. One possible explanation is that the deflection of the carriers caused by the Lorentz force [70, 71] may facilitate the interaction among carriers and new oxide traps (depending on the particular magnetic field) hence affecting the trapping/detrapping dynamics. However, further investigations are required to justify this statement.

6 Conclusion

This chapter has evaluated an exhaustive method to experimentally identify single-trap random telegraph noise devices by combining the noise spectral scanning by gate bias (SSGB) technique with a modified weighted time lag plot method (w-TLP). At first, the characterization protocol has been implemented in an automatic probe station and applied for the massive test of devices all over a wafer providing information about the distribution of traps. Shockley–Read–Hall models have been applied to determine the position of the trap inside the gate oxide in terms of the physical location and the energetic position. The impact of the substrate (or back-) bias is studied on the RTN characteristics for silicon-on-insulator technology devices. The trap mechanisms involved have been examined for these devices and models have been updated to predict the characteristic times under positive or negative substrate bias provided that the back interface is in depletion. Additionally, the influence of low static and perpendicular-to-channel magnetic fields has been analyzed in 250 nm processed MOSFETs at room temperature. The results suggest that depending on the specific magnetic field applied, new oxide traps can be activated which then induce a new current level in the temporal and spectral signatures. The experimental findings could be partially explained by the magnetic field deflecting the carriers in the surface, consequently changing the dynamic interactions with the oxide traps, but due to the novelty of these results further exhaustive investigations are required to model the magnetic field dependence on trapping phenomena.

Acknowledgements The authors would like to thank Dr. O. Faynot and Dr. F. Andrieu from CEA-LETI for supplying samples. This work has been partially funded by Spanish Government through project TEC-2014-59730, by Andalusian Regional Government through P12-TIC-1996, by European Commission under Grant REMINDER 687931, and by the Mexican Government under Grant CONACyT 291212. The author Carlos Marquez has received funding by Spanish "Ministerio de Educacion, Cultura y Deporte" through "Estancias de Movilidad Jose Castillejo CAS18/00460."

References

1. G. Moore, Electronics **38**(8), 114 (1965). https://doi.org/10.1109/JPROC.1998.658762
2. H. Iwai, Microelectron. Eng. **86**(7–9), 1520 (2009). https://doi.org/10.1016/j.mee.2009.03.129
3. ITRS, International Technology Roadmap for Semiconductors: 2013 Edition Executive Summary. Technical Report (2013)
4. ITRS, International Technology Roadmap for Semiconductors 2.0: Executive Report. Technical Report (2015)
5. E. Simoen, C. Claeys, Solid State Electron. **39**(7), 949 (1996). https://doi.org/10.1016/0038-1101(95)00427-0
6. N. Tega, H. Miki, M. Yamaoka, in *2008 IEEE International Reliability Physics Symposium* (IEEE, Piscataway, 2008), pp. 541–546. https://doi.org/10.1109/RELPHY.2008.4558943
7. J.P. Colinge, *Silicon-on-Insulator Technology: Materials to VLSI* (Springer, Boston, 2004). https://doi.org/10.1007/978-1-4419-9106-5
8. S. Cristoloveanu, S.S. Li, *Electrical Characterization of Silicon-on-Insulator Materials and Devices* (Springer, Boston, 1995). https://doi.org/10.1007/978-1-4615-2245-4
9. L.A. Larsson, M. Larsson, E.S. Moskalenko, P.O. Holtz, Nanoscale Res. Lett. **5**(7), 1150 (2010). https://doi.org/10.1007/s11671-010-9618-x
10. S.P. Giblin, M. Kataoka, J.D. Fletcher, P. See, T. Janssen, J.P. Griffiths, G.A.C. Jones, I. Farrer, D.A. Ritchie, Nat. Commun. **3**, 930 (2012). https://doi.org/10.1038/ncomms1935
11. M. Xiao, I. Martin, H.W. Jiang, Phys. Rev. Lett. **91**(7), 078301 (2003). https://doi.org/10.1103/PhysRevLett.91.078301
12. F.A. Baron, Y. Zhang, M. Bao, R. Li, J. Li, K.L. Wang, Appl. Phys. Lett. **83**(4), 710 (2003). https://doi.org/10.1063/1.1596381
13. E. Prati, in *AIP Conference Proceedings*, vol. 780 (AIP, College Park, 2005), pp. 221–224. https://doi.org/10.1063/1.2036736
14. I. Martin, D. Mozyrsky, H.W. Jiang, Phys. Rev. Lett. **90**(1), 018301 (2003). https://doi.org/10.1103/PhysRevLett.90.018301
15. C. Navarro, S.J. Chang, M. Bawedin, F. Andrieu, B. Sagnes, S. Cristoloveanu, ECS Trans. **66**(5), 251 (2015). https://doi.org/10.1149/06605.0251ecst
16. E.A. Gutierrez-D, E.P. de los A, V.H. Vega-G, F. Guarin, in *2012 8th International Caribbean Conference on Devices, Circuits and Systems (ICCDCS)* (IEEE, Piscataway, 2012), pp. 1–4. https://doi.org/10.1109/ICCDCS.2012.6188885
17. J. Daughton, in *Magnetoelectronics* (Academic Press, Cambridge, 2004), pp. 205–229. https://doi.org/10.1016/B978-012088487-2/50005-4
18. R.N. Gurzhi, A.N. Kalinenko, A.I. Kopeliovich, A.V. Yanovsky, E.N. Bogachek, U. Landman, Appl. Phys. Lett. **83**(22), 4577 (2003). https://doi.org/10.1063/1.1630839
19. M.J. Kirton, M.J. Uren, Appl. Phys. Lett. **48**(19), 1270 (1986). https://doi.org/10.1063/1.97000
20. Z. Shi, J.P. Mieville, M. Dutoit, IEEE Trans. Electron Devices **41**(7), 1161 (1994)
21. K. Ralls, W. Skocpol, L. Jackel, R. Howard, L. Fetter, R. Epworth, D. Tennant, Phys. Rev. Lett. **52**(3), 228 (1984). https://doi.org/10.1103/PhysRevLett.52.228
22. J. Franco, B. Kaczer, M. Toledano-Luque, M.F. Bukhori, P.J. Roussel, T. Grasser, A. Asenov, G. Groeseneken, IEEE Electron Device Lett. **33**(6), 779 (2012). https://doi.org/10.1109/LED.2012.2192410
23. M. Kirton, M. Uren, Adv. Phys. **38**(4), 367 (1989). https://doi.org/10.1080/00018738900101122
24. M.L. Fan, V.P.H. Hu, Y.N. Chen, P. Su, C.T. Chuang, in *Proceedings of IEEE International Reliability Physics Symposium* (IEEE, Piscataway, 2012), pp. 1–6. https://doi.org/10.1109/IRPS.2012.6241886
25. S.M. Joe, J.H. Yi, S.K. Park, H. Shin, B.G. Park, Y.J. Park, J.H. Lee, IEEE Trans. Electron Devices **58**(1), 67 (2011). https://doi.org/10.1109/TED.2010.2088126

26. M. Waltl, P.J. Wagner, H. Reisinger, K. Rott, T. Grasser, in *2012 IEEE International Integrated Reliability Workshop Final Report* (IEEE, Piscataway, 2012), pp. 74–79. https://doi.org/10.1109/IIRW.2012.6468924
27. F. Andrieu, O. Weber, J. Mazurier, O. Thomas, J.P. Noel, C. Fenouillet-Beranger, J.P. Mazellier, P. Perreau, T. Poiroux, Y. Morand, T. Morel, S. Allegret, V. Loup, S. Barnola, F. Martin, J.F. Damlencourt, I. Servin, M. Casse, X. Garros, O. Rozeau, M.A. Jaud, G. Cibrario, J. Cluzel, A. Toffoli, F. Allain, R. Kies, D. Lafond, V. Delaye, C. Tabone, L. Tosti, L. Brevard, P. Gaud, V. Paruchur, K. Bourdelle, W. Schwarzenbach, O. Bonnin, B.Y. Nguyen, B. Doris, F. Boeuf, T. Skotnicki, O. Faynot, in *2010 Symposium on VLSI Technology*, San Francisco (2010), pp. 57–58. https://doi.org/10.1109/VLSIT.2010.5556122
28. C. Fenouillet-Beranger, P. Perreau, L. Pham-Nguyen, S. Denorme, F. Andrieu, L. Tosti, L. Brevard, O. Weber, S. Barnola, T. Salvetat, X. Garros, M. Casse, C. Leroux, J. Noel, O. Thomas, B. Le-Gratiet, F. Baron, M. Gatefait, Y. Campidelli, F. Abbate, C. Perrot, C. De-Buttet, R. Beneyton, L. Pinzelli, F. Leverd, P. Gouraud, M. Gros-Jean, A. Bajolet, C. Mezzomo, C. Leyris, S. Haendler, D. Noblet, R. Pantel, A. Margain, C. Borowiak, E. Josse, N. Planes, D. Delprat, F. Boedt, K. Bourdelle, B. Nguyen, F. Boeuf, O. Faynot, T. Skotnicki, in *2009 IEEE International Electron Devices Meeting (IEDM)* (IEEE, Piscataway, 2009), pp.1–4. https://doi.org/10.1109/IEDM.2009.5424251
29. C.M. Chang, S.S. Chung, Y.S. Hsieh, L.W. Cheng, C.T. Tsai, G.H. Ma, S.C. Chien, S.W. Sun, in *Technical Digest – International Electron Devices Meeting, IEDM* (IEEE, Piscataway, 2008), pp. 8–11. https://doi.org/10.1109/IEDM.2008.4796815
30. J.W. Lee, B.H. Lee, H. Shin, J.H. Lee, IEEE Trans. Electron Devices **57**(4), 913 (2010). https://doi.org/10.1109/TED.2010.2041871
31. C. Marquez, N. Rodriguez, F. Gamiz, R. Ruiz, A. Ohata, Solid State Electron. **117**, 60 (2016). https://doi.org/10.1016/j.sse.2015.11.022
32. S. Lee, H.-J. Cho, Y. Son, D.S. Lee, H. Shin, in *2009 IEEE International Electron Devices Meeting (IEDM)* (IEEE, Baltimore, 2009), pp. 1–4. https://doi.org/10.1109/IEDM.2009.5424227
33. N. Zanolla, D. Siprak, P. Baumgartner, E. Sangiorgi, C. Fiegna, in *2008 9th International Conference on Ultimate Integration of Silicon* (IEEE, Piscataway, 2008), pp. 137–140. https://doi.org/10.1109/ULIS.2008.4527158
34. R. Siergiej, M. White, N. Saks, Solid State Electron. **35**(6), 843 (1992). https://doi.org/10.1016/0038-1101(92)90287-M
35. T. Nagumo, K. Takeuchi, T. Hase, Y. Hayashi, *2010 International Electron Devices Meeting* (IEEE, Piscataway, 2010), pp. 28.3.1–28.3.4. https://doi.org/10.1109/IEDM.2010.5703437
36. J. Sikula, J. Pavelka, M. Tacano, M. Toita, in *2009 International Conference on Microelectronics – ICM* (IEEE, Piscataway, 2009), pp. 296–299. https://doi.org/10.1109/ICM.2009.5418625
37. N.V. Amarasinghe, Z. Çelik-Butler, Solid-State Electron. **44**, 1013 (2000). https://doi.org/10.1016/S0038-1101(99)00324-X
38. E. Simoen, C. Claeys, Mater. Sci. Eng. B **91–92**, 136 (2002). https://doi.org/10.1016/S0921-5107(01)00963-1
39. T. Grasser, Microelectron. Reliab. **52**(1), 39 (2012). https://doi.org/10.1016/j.microrel.2011.09.002
40. B. Kaczer, J. Franco, P. Weckx, P.J. Roussel, V. Putcha, E. Bury, M. Simicic, A. Chasin, D. Linten, B. Parvais, F. Catthoor, G. Rzepa, M. Waltl, T. Grasser, Microelectron. Reliab. **81**(January), 186 (2018). https://doi.org/10.1016/j.microrel.2017.11.022
41. Z. Celik-Butler, P. Vasina, N. Vibhavie Amarasinghe, IEEE Trans. Electron Devices **47**(3), 646 (2000). https://doi.org/10.1109/16.824742
42. A.S.M.S. Rouf, Z. Çelik-Butler, Oxide trap-induced RTS in MOSFETs, in *Noise in Nanoscale Semiconductor Devices*, ed. by T. Grasser (Springer, Cham, 2019). https://doi.org/10.1007/978-3-030-37500-3_17
43. N.B. Lukyanchikova, M.V. Petrichuk, N.P. Garbar, E. Simoen, C. Claeys, Appl. Phys. Mater. Sci. Process. **70**(3), 345 (2000). https://doi.org/10.1007/s003390050058

44. H. Mueller, M. Schulz, IEEE Trans. Electron Devices **44**(9), 1539 (1997). https://doi.org/10.1109/16.622612
45. H.K. Lim, J. Fossum, IEEE Trans. Electron Devices **30**(10), 1244 (1983)
46. X. Ji, Y. Liao, C. Zhu, J. Chang, F. Yan, Y. Shi, Q. Guo, in *2013 IEEE International Reliability Physics Symposium (IRPS)* (IEEE, Piscataway, 2013), pp. XT.7.1–XT.7.5. https://doi.org/10.1109/IRPS.2013.6532122
47. Y. Illarionov, M. Bina, S. Tyaginov, K. Rott, B. Kaczer, H. Reisinger, T. Grasser, IEEE Trans. Electron Devices **62**(9), 2730 (2015). https://doi.org/10.1109/TED.2015.2454433
48. H. Nyquist, Phys. Rev. **32**(1), 110 (1928)
49. J.B. Johnson, Phys. Rev. **32**(1), 97 (1928)
50. J. Jomaah, F. Balestra, G. Ghibaudo, Res. J. Telecommun. Inf. Technol. **1**(1), 24 (2005)
51. D.M. Fleetwood, Origins of 1/f noise in electronic materials and devices: a historical perspective, in *Noise in Nanoscale Semiconductor Devices*, ed. by T. Grasser (Springer, Cham, 2019). https://doi.org/10.1007/978-3-030-37500-3_1
52. C. Theodorou, G. Ghibaudo, Noise and fluctuations in fully depleted silicon-on-insulator mosfets, in *Noise in Nanoscale Semiconductor Devices*, ed. by T. Grasser (Springer, Cham, 2019). https://doi.org/10.1007/978-3-030-37500-3_2
53. B. Stampfer, A. Grill, M. Waltl, Advanced electrical characterization of single oxide defects utilizing noise signals, in *Noise in Nanoscale Semiconductor Devices*, ed. by T. Grasser (Springer, Cham, 2019). https://doi.org/10.1007/978-3-030-37500-3_7
54. T.H. Both, M. Banaszeski, G.I. Wirth, H.P. Tuinhout, A.Z. van Duijnhove, J. Croon, A. Scholten, An overview on statistical modeling of random telegraph noise in the frequency domain, in *Noise in Nanoscale Semiconductor Devices*, ed. by T. Grasser (Springer, Cham, 2019). https://doi.org/10.1007/978-3-030-37500-3_15
55. H. Haddara (ed.), *Characterization Methods for Submicron MOSFETs*, *The Kluwer International Series in Engineering and Computer Science*, vol. 352 (Springer, Boston, 1996). https://doi.org/10.1007/978-1-4613-1355-7
56. S. Machlup, J. Appl. Phys. **25**(3), 341 (1954). https://doi.org/10.1063/1.1721637
57. A.L. McWhorter, *1/f Noise and Germanium Surface Properties*. Semiconductor Surface Physics (1957)
58. F. Balestra, G. Ghibaudo, J. Jomaah, Int. J. Numer. Modell. Electron. Networks Devices Fields **28**(6), 613 (2015). https://doi.org/10.1002/jnm.2052
59. J. Campbell, J. Qin, K. Cheung, L. Yu, J. Suehle, A. Oates, K. Sheng, in *2009 IEEE International Reliability Physics Symposium* (IEEE, Montreal, 2009), pp. 382–388. https://doi.org/10.1109/IRPS.2009.5173283
60. G. Ghibaudo, T. Boutchacha, Microelectron. Reliab. **42**(4–5), 573 (2002). https://doi.org/10.1016/S0026-2714(02)00025-2
61. S.L. Rumyantsev, Y. Deng, S. Shur, M.E. Levinshtein, M. Asif Khan, G. Simin, J. Yang, X. Hu, R. Gaska, Semicond. Sci. Technol. **18**(6), 589 (2003). https://doi.org/10.1088/0268-1242/18/6/333
62. M.J. Uren, D.J. Day, M.J. Kirton, Appl. Phys. Lett. **47**(11), 1195 (1985). https://doi.org/10.1063/1.96325
63. C. Marquez, N. Rodriguez, F. Gamiz, A. Ohata, Solid State Electron. **128**, 115 (2017). https://doi.org/10.1016/j.sse.2016.10.031
64. F.M. Puglisi, P. Pavan, Trans. Electr. Eng. Electron. Commun. **12**(1), 24 (2014)
65. S. Realov, K.L. Shepard, *Technical Digest – International Electron Devices Meeting*, IEDM (212), 28.2.1 (2010). https://doi.org/10.1109/IEDM.2010.5703436
66. T. Nagumo, K. Takeuchi, S. Yokogawa, K. Imai, Y. Hayashi, *2009 IEEE International Electron Devices Meeting (IEDM)* (2009). https://doi.org/10.1109/IEDM.2009.5424230
67. J. Martin-Martinez, J. Diaz, R. Rodriguez, M. Nafria, X. Aymerich, IEEE Electron Device Lett. **35**(4), 479 (2014). https://doi.org/10.1109/LED.2014.2304673

68. J. Martin-Martinez, R. Rodriguez, M. Nafria, Advanced characterization and analysis of random telegraph noise in CMOS devices, in *Noise in Nanoscale Semiconductor Devices*, ed. by T. Grasser (Springer, Cham, 2019). https://doi.org/10.1007/978-3-030-37500-3_14
69. M.J. Uren, M.J. Kirton, Appl. Surf. Sci. **39**(1–4), 479 (1989). https://doi.org/10.1016/0169-4332(89)90464-9
70. A. Beer, Solid State Electron. **9**(5), 339 (1966). https://doi.org/10.1016/0038-1101(66)90148-1
71. A. Vandooren, S. Cristoloveanu, D. Flandre, J. Colinge, Solid State Electron. **45**(10), 1793 (2001). https://doi.org/10.1016/S0038-1101(01)00207-6

Principles and Applications of I_g-RTN in Nano-scaled MOSFET

Steve S. Chung and E. R. Hsieh

1 Introduction

In the history of the development of high-k dielectrics, it is known that either the intrinsic or stress-induced traps are important to the understanding of CMOS reliability. Random telegraph noise (RTN), in the form of digital waveform with two or several levels, is actually caused by the generated traps in the gate dielectric [1–4]. I_g-RTN is the current fluctuation observed from the gate terminal [5]. While I_d-RTN is the current fluctuation observed from the drain terminal [6]. By measuring the capture time and emission time of I_g-RTN, we can extract the oxide trap location (depth) and energy levels. RTN traps can significantly impact the degradation of device reliabilities when CMOS devices have suffered from the bias–temperature instability [7–11] or breakdown [12]. In this chapter, we will thoroughly discuss each domain of I_g-RTN. Firstly, we will derive the equations of depth–position extraction from the RTN traps in gate dielectrics. Then, we will elucidate the characteristics of RTN trap before and after soft breakdown of gate dielectrics. Based on these experimental observations, an approach to profiling the breakdown path has been developed with I_g-RTN measurement. Through this methodology, we investigate different types of breakdown paths; moreover, the evolution and growth of breakdown paths have been depicted in more detail, and a new mechanism of dielectric breakdown has been discovered, named dielectric-fuse breakdown (dFuse). Finally, we will demonstrate the design of one-time programming (OTP) memory cell based on the dielectric-fuse breakdown as an application to I_g-RTN.

Therefore, this chapter will be organized as follows: in the next section, the preparation of devices under tests will be described. In Sect. 3, we will explain the

S. S. Chung (✉) · E. R. Hsieh
Department of Electronics Engineering, National Chiao Tung University, Hsinchu City, Taiwan
e-mail: schung@cc.nctu.edu.tw

T. Grasser (ed.), *Noise in Nanoscale Semiconductor Devices*,
https://doi.org/10.1007/978-3-030-37500-3_5

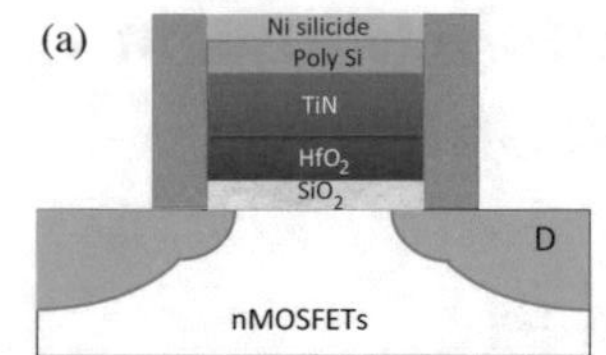

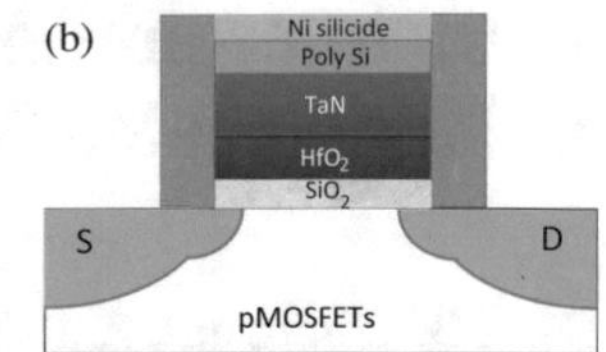

Fig. 1 Device structure: (**a**) 28-nm high-k nMOSFET with TiN as metal gate and (**b**) pMOSFET with TaN as metal gate

theory on how to identify the positions of RTN traps in the gate dielectric. Section 4 is to evaluate the characteristics of I_g-RTN during the development of dielectric soft breakdown. Then, a methodology has been established to profile the breakdown path of gate dielectrics by using I_g-RTN measurement. Furthermore, the evolution and types of breakdown paths have been analyzed as well. Finally, in Sect. 5, an OTP cell based on the new breakdown mechanism will be proposed. In the end, the conclusions will be given in Sect. 6.

2 Device Preparations

Figure 1 shows the schematic cross section of HKMG devices with two splits of E.O.T. equivalent to 1.2 nm (2-nm HfO_2/0.8-nm SiO_2) and 3 nm (5-nm HfO_2/2-nm SiO_2), respectively. For the process of gate stacks, a high-quality SiO_2 thermal oxide was deposited on the wafers, followed by atomic layer deposition (ALD) HfO_2. Then, ALD TiN capping layer was grown on HfO_2 with different concentrations of nitrogen ions. After that, the dummy poly gate electrode and source/drain activation have been processed. Following removal of the dummy electrode, the sputtered work-function gate metal was formed on the capping layer, and finally the post metal anneal at 500 °C for 5 min has been applied to passivate the bulk traps in HK dielectrics. The conventional polysilicon gate/SiO_2 CMOS devices with two splits of 3-nm and 1.2-nm E.O.T have been prepared for comparisons. Several device areas have been prepared for the measurement of the threshold voltage variations in this work.

3 Methodology of Extracting Trap Depth and Energy Level

3.1 I_g-RTN

When there is sufficient number of carriers in the channel, there is a high possibility for those carriers to be trapped by the slow trap in the dielectric. If an electron is trapped, the coulomb barrier caused by filled trap will hence suppress the gate current, as shown in Fig. 2a. It looks like a big stone lying in the flow of river so the flow rate is apparently rolling off. Otherwise, if the trap in the dielectric is empty,

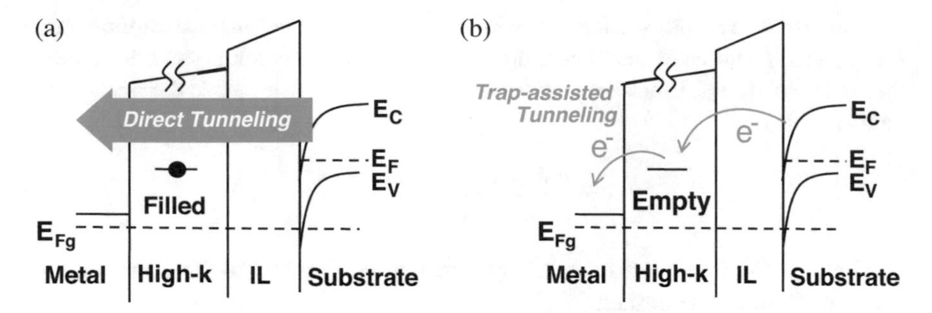

Fig. 2 The schematic shows the concept of capturing and emission of RTN trap, which determines the magnitude of the gate current through the high-k/IL dielectric (**a**) electrons are tunneled directly through the gate when the trap is filled by electron and (**b**) electrons transport through the gate via trap-assisted tunneling when trap is empty

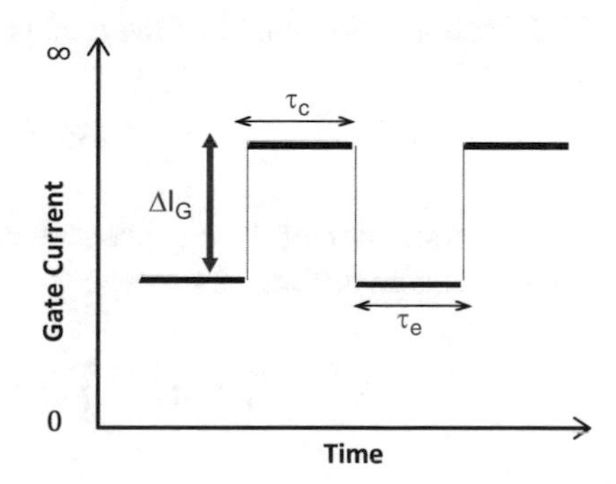

Fig. 3 The two-level RTN signal observed from gate terminal. τ_c represents time to capture; τ_e represents time to emit

the electrons can transport through the *trap-assistance tunneling mechanism*, which boosts the level of gate current, as shown in Fig. 2b. The definitions of τ_c and τ_e in the measured gate current (I_g) are described in Fig. 3. Based on the Shockley–Read–Hall (SRH) statistics, the carrier capture rate $1/\tau_c$ can be written in terms of the carrier density (per unit volume), n, the average velocity of the carrier, v, and the average capture cross section σ in Eq. (1), where the capture cross section is given by Eq. (2), [13–16], i.e.,

$$\tau_c = \frac{1}{nv\sigma} \tag{1}$$

and

$$\sigma = \sigma_0 \exp\left(-\frac{\Delta E_B}{KT}\right). \tag{2}$$

Here, σ_0 is the cross-section prefactor, and ΔE_B is the thermal activation energy for capture. T and v are usually the equilibrium lattice temperature and the average thermal velocity v_{th}. Emission time τ_e is given by Eq. 3, where g is the degeneracy factor:

$$\tau_e = \frac{\exp\left[(E_F - E_T)/K_B T\right]}{gnv\sigma}. \tag{3}$$

The $(E_F–E_T)$ term represents the trap energy with respect to the Fermi energy. K_B is the Boltzmann constant.

3.2 Single-Layer Oxide

The fractional occupancy of the oxide trap is governed by

$$\frac{\tau_c}{\tau_e} = \exp\left[(E_T - E_F)/K_B T\right]. \tag{4}$$

The expression of the capture time and emission time in terms of the position of the trap can be derived as follows:

$$k_B T \ln\left(\frac{\tau_c}{\tau_e}\right) = \Phi - [(E_C - E_T) + E_x] \tag{5}$$

and

$$E_x = \left|q \frac{X_T}{T_{ox}} V_{ox}\right| \tag{6}$$

Here, Φ is the difference between substrate Fermi level and the standard energy level (dash line), E_c is the conduction band edge of the SiO_2, q is the elementary charge, T_{ox} is the oxide thickness, X_T is the position of the trap in the oxide from the substrate, and V_{ox} is the oxide voltage drop, which is the same as the applied bias by neglecting the surface band bending. By differentiating Eq. (5) with respect to the applied bias, X_T can be derived:

$$X_T = \frac{k_B T}{q} \frac{\partial}{\partial V}\left[\ln \frac{\tau_c}{\tau_e}\right] T_{ox} \tag{7}$$

as shown in Fig. 4.

3.3 Bilayer Oxide

For high-k dielectric, the presence of interfacial layer between high-k layer and substrate is inevitable. Thus, a bilayer system must be considered. Based on the preview report [17], the surface band bending and bilayer system were included in the derivation. However, the location of the trap needs to be corrected as a result of the changes of the bilayer permittivity. Let us start with a simple case; as shown in Fig. 4, assume a trap is located in the interfacial layer and exchanges electrons with substrate Fermi level. Then, the X_t now represents the distance from trap location to substrate surface. E_c–E_T is the difference of trap energy level and HfO_2 conduction band. V_g represents the applied positive voltage. Moreover, in order to unify the electric field within the entire oxide, we replace the oxide with effective SiO_2 thickness. The reason to take SiO_2 as an effective oxide is according to the trap location, and in this case the trap is located in the interfacial layer as shown in Fig. 5.

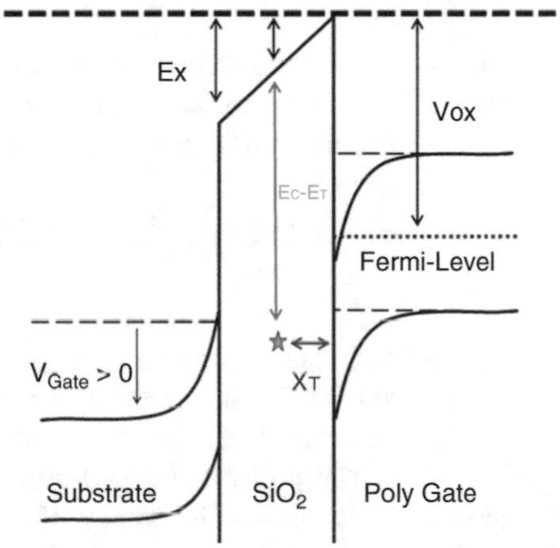

Fig. 4 The band diagram of a single layer oxide with applying positive gate voltage. X_T represents the depth from substrate to trap; E_C–E_T represents the energy difference from conduction band to trap energy level

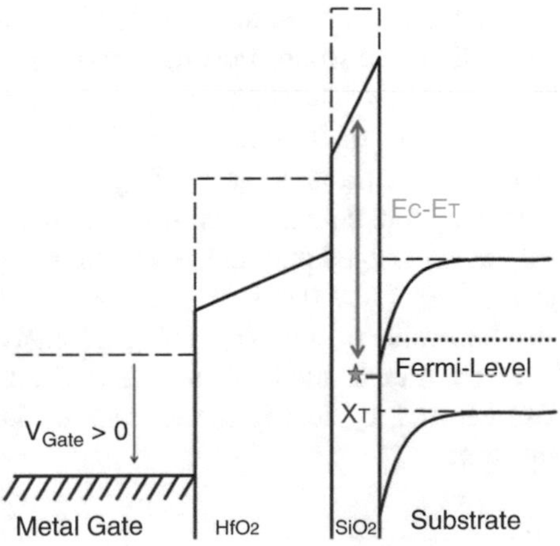

Fig. 5 The band diagram of a bilayer oxide when applying positive gate voltage

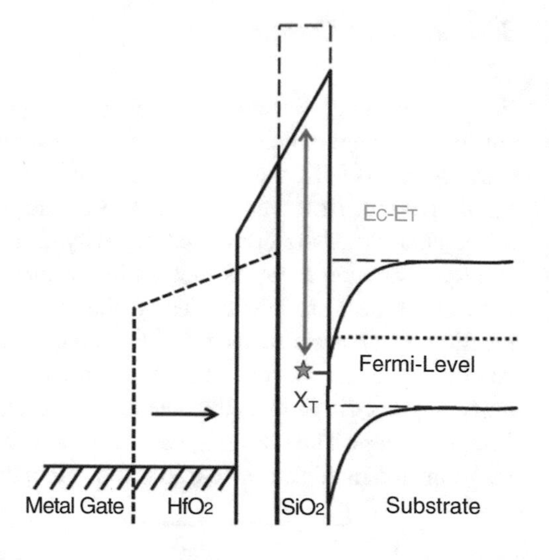

Fig. 6 Because the trap is located in the interfacial layer, we need to transform hafnium dioxide thickness into effective silicon oxide thickness

Finally, we can easily pick up a similar triangle with the modified band diagram, as shown in Fig. 6, because the electrical field is unified. Then, the equation of trap depth X_T can be derived by applying the law of similar triangle as given below:

$$X_T = \left(T_{SiO_2} + \frac{\varepsilon_{ox1}}{\varepsilon_{ox2}} T_{HfO_2}\right)\left(\frac{KT}{q}\frac{d\ln(\tau_c/\tau_e)}{dV_{gs}} + \frac{d\phi_S}{dV_{gs}}\right) / \left(\frac{d\phi_S}{dV_{gs}} - 1\right). \quad (8)$$

Since RTN is measured when the trap is in equilibrium state with the Fermi level of gate electrode or silicon substrate, carriers can be captured only if the trap energy is close to the Fermi level. There are two trends in the slope of $\ln(\tau_c/\tau_e)$ versus gate voltages. The positive slope corresponds to trap interact with gate electrode. Because as V_g increases, the energy of the trap will be raised up and away from the gate Fermi level; the time needed to capture is longer and the time to emit electron is shorter. On the other hand, negative slope is the case of traps interacting with silicon substrate since the trap energy will be pulled down close to the Fermi level of silicon. The time needed to capture electrons is shorter and time to emit is longer. Either the trap is located in the interfacial layer or hafnium dioxide layer, and either the trap interacts with gate metal or interacts with silicon substrate, there are totally four equations as shown as Table 1. Based on Table 1, we can further plot a diamond-shaped diagram as shown in Fig. 7. For those traps that interact with the gate terminal, the depth of them can be identified by the value of $\ln(\tau_c/\tau_e)$. If the absolute value of $\ln(\tau_c/\tau_e)$ is higher than the value of the cross point of the upper lines, the trap should be located in the interfacial layer. Similarly, the same way can also be used to identify the location of those traps which interact with the silicon substrate.

Table 1 Four equations corresponding to different trap location and different Fermi level which interact with RTN traps

1. *Interaction of electrons between traps and Si—substrate*
In SiO_2 : $X_T = \left(T_{SiO_2} + \frac{\varepsilon_{SiO_2}}{\varepsilon_{HfO_2}} T_{HfO_2}\right)\left(\frac{kT}{q}\frac{d\ \ln(\tau_c/\tau_e)}{dV_{gs}} + \frac{d\phi_s}{dV_{gs}}\right) / \left(\frac{d\phi_s}{dV_{gs}} - 1\right)$
In HfO_2 : $X_T = \left(T_{HfO_2} + \frac{\varepsilon_{HfO_2}}{\varepsilon_{SiO_2}} T_{SiO_2}\right)\left(\frac{kT}{q}\frac{d\ \ln(\tau_c/\tau_e)}{dV_{gs}} + \frac{d\phi_s}{dV_{gs}}\right) / \left(\frac{d\phi_s}{dV_{gs}} - 1\right) + \left(1 - \frac{\varepsilon_{HfO_2}}{\varepsilon_{SiO_2}}\right) T_{SiO_2}$
$\frac{d\ \ln(\tau_c/\tau_e)}{dV_{gs}} > 0$
2. *Interaction of electrons between traps and metal—gate*
In SiO_2 : $X_{T1} = \left(T_{SiO_2} + \frac{\varepsilon_{SiO_2}}{\varepsilon_{HfO_2}} T_{HfO_2}\right)\left(1 - \frac{kT}{q}\frac{d\ \ln(\tau_c/\tau_e)}{dV_{gs}}\right) / \left(1 - \frac{d\phi_s}{dV_{gs}}\right)$
In HfO_2 : $X_{T2} = \left(T_{SiO_2} + \frac{\varepsilon_{HfO_2}}{\varepsilon_{SiO_2}} T_{HfO_2}\right)\left(1 - \frac{kT}{q}\frac{d\ \ln(\tau_c/\tau_e)}{dV_{gs}}\right) / \left(1 - \frac{d\phi_s}{dV_{gs}}\right) + \left(1 - \frac{\varepsilon_{HfO_2}}{\varepsilon_{SiO_2}}\right) T_{SiO_2}$
$\frac{d\ \ln(\tau_c/\tau_e)}{dV_{gs}} < 0$

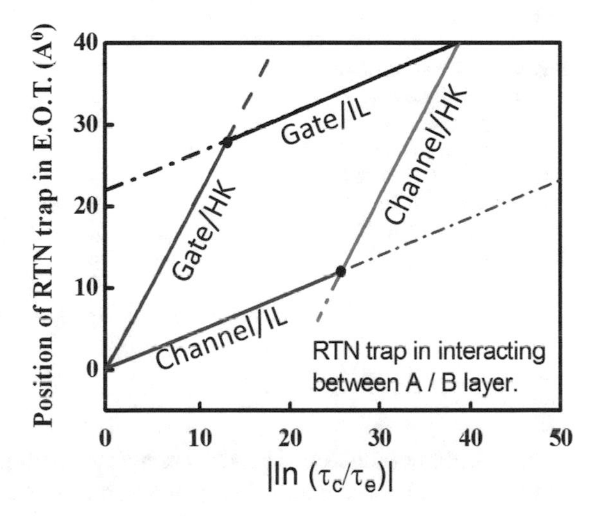

Fig. 7 The diamond-shaped diagram. The black/red lines represent traps exchanging with gate metal. The blue/green lines represents traps exchanging with silicon substrate from the point of cross section, we can determine the extracted RTN trap in the interfacial layer or high k layer

4 Characteristics of I_g-RTN During Development of Dielectric Soft Breakdown

4.1 Signals of I_g-RTN Waveforms

Figure 8 shows the gate current versus time, in which RTN signals due to traps in dielectrics before (left) and after soft breakdown (SB) (right) were observed. For the measurement in Fig. 8a before the soft breakdown, gate leakage is hundreds of pA and exhibits a fluctuation with an order of tens of pA. The capture and/or emission is obviously dependent on the gate bias. Here, τ_c is the capture time and τ_e is the emission time. After soft breakdown, Fig. 8b, we can see a huge increase of the leakage current (with a third order of magnitude increase) and a longer time

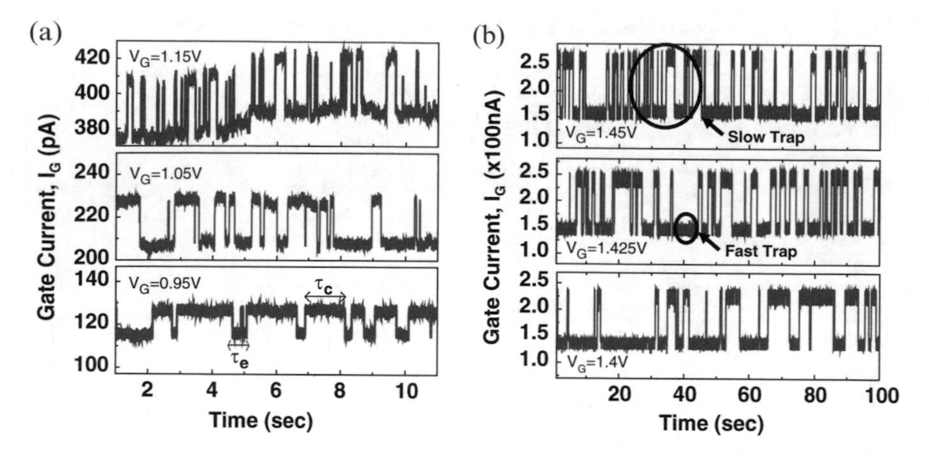

Fig. 8 I_g-RTN measurement for the device (**a**) before SRD and (**b**) after soft breakdown. Capture and emission times are defined in the plot. Note that the gate leakage shows a third-order difference in magnitude

Fig. 9 Calculated capture and emission time as a function gate bias for the devices before and after the soft breakdown. Note that emission time is field dependent after the soft breakdown

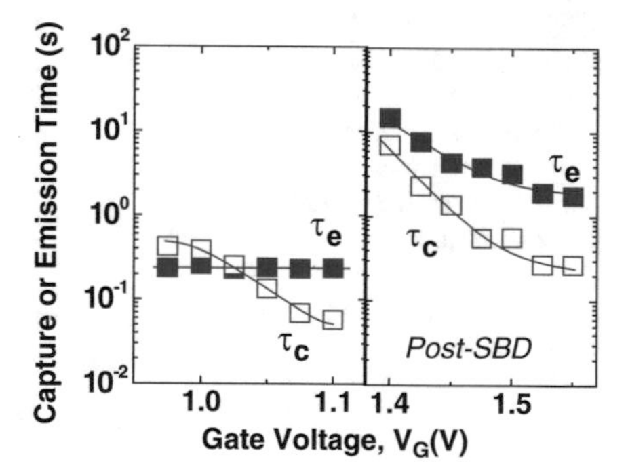

for capturing or emission. The soft breakdown path performs as a channel to open or close one or more conductive paths across the dielectrics. As a consequence, electron trapping/detrapping in defects near (or inside) the SB weak spot will induce some local changes in the barrier height as well as the tunneling probability between two neighboring traps [18]. A single electron trapped in a site inside or very close to the conductive path could in principle block the majority of the current flowing through the path as a result of coulomb repulsion. In addition, a fraction of the conductive paths is not always available for conduction but can be switched ON and OFF, giving rise to the large current fluctuations.

Figure 9 shows the average emission (τ_e) and capture time (τ_c) as a function of V_{gs}. The average capture and emission time are calculated by analyzing numerous switching events for each RTS trace. For pre-soft breakdown, one can see that τ_c for the dielectric traps is an exponential function of gate bias, given by [19],

$$\tau_c = 1/(nv_{th}\sigma) \tag{9}$$

where n is the surface carrier concentration, v_{th} is the thermal velocity, and σ is the capture cross section of the trap involved. The surface potential ψ_s is a linear function of V_{gs} and $n \sim \exp(q\psi_s/kT)$. Higher carrier concentration or larger V_{gs} implies more electrons participation and reduces the time to capture. On the other hand, time constant of post-soft breakdown is relatively longer than the case of pre-soft breakdown, i.e., τ_c is much longer after device post-soft breakdown. According to Eq. (9), v_{th} is V_{gs} independent and n increases with gate bias. As a result, the possibility for the increase of capture time in post-soft breakdown is attributed to a smaller σ; i.e., after the formation of conduction path via the device soft breakdown, the effective cross section of traps reduces since some of the traps have been used to build up the soft breakdown conduction path. As a consequence, a smaller cross section for post-soft breakdown gives rise to a longer capture time comparing to the pre-soft breakdown ones.

4.2 I_g-RTN After Soft Breakdown of MOSFET

The capture time is influenced by not only capture cross section but also the electric field. Figure 10 shows the temperature dependence of τ_c, which is obviously shorter at higher temperature. Channel carriers are heated at high temperature and more easily surpass the barrier of the capture [20]. The activation energy of capture, E_a, i.e., the slope of $\ln(\tau_c)$ versus $1/kT$ plot, is calculated as shown in the plots. The cross section, σ, can be further related to E_a as [21]

$$\sigma = \sigma_0 \exp\left(-E_a/kT\right) \tag{10}$$

where σ_0 is the pre-factor of capture cross section. The activation energy is an index to reflect the capture cross section variation as shown in Eq. (10). Before soft breakdown, negligible change of E_a at different V_{gs} can be seen, i.e., σ is assumed a constant reasonably. After the soft breakdown, Fig. 10b, we can see a strong dependence of E_a on the gate bias. The main cause is that the capture probability has been enhanced via the soft breakdown path. At high V_{gs}, E_a decreases and is believed to effectively increase the cross section, according to Eq. (2). This amounts to a decrease of the capture time. Hence, the activation energy shows a gate bias dependence feature after the soft breakdown.

One major difference can be seen from the comparison of τ_e in Fig. 9, in which τ_e shows completely different behavior for pre-soft -breakdown and post-soft breakdown. For pre-soft breakdown, τ_e *for the dielectric trap is completely independent of* V_{gs}. This accounts for the electron escaping through thermionic emission instead of the gate bias; i.e., it shows no field dependence and hence the possibility for direct tunneling back to the substrate or gate is ruled out. In comparison, emission of the electrons trapped within the conventional oxide is electric field dependent [22]. The main difference is that high-k dielectric has larger physical thickness where electrons captured cannot easily escape through direct

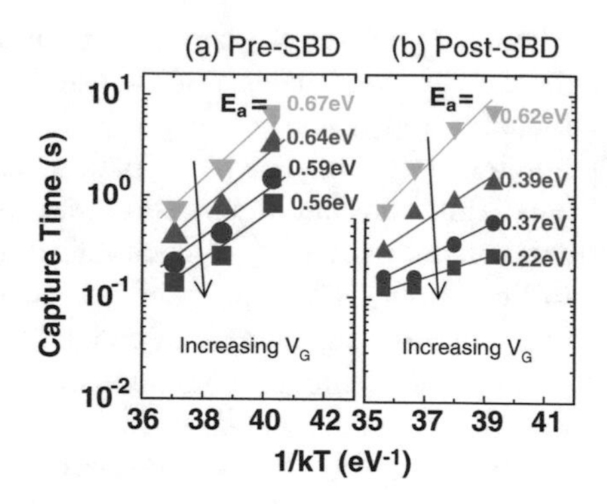

Fig. 10 The dependence of capture time with temperature: (**a**) pre-soft breakdown; (**b**) post-soft breakdown. The activation energy E_a shows a huge change for devices after the soft breakdown

tunneling. In other words, time to emission through thermionic emission is much shorter than the time through direct tunneling over the wide barrier.

The emission time is given by [23]

$$\tau_e = \frac{\exp\left[\left(E_F - E_T\right)/k_B T\right]}{g\sigma v_{th} n}. \tag{11}$$

Here, E_F–E_T is the distance from the Fermi energy level to the trap energy level and g is the degeneracy factor, which is usually considered as unity for electrons. Also, from the Arrhenius plot [24, 25], the emission time can be represented by

$$\ln \tau_e T^2 = -\ln \gamma\sigma + \left(E_{Cd} - E_T\right)/kT. \tag{12}$$

Here, E_{Cd}–E_T is the distance from dielectric conduction band to trap energy level. $\gamma = (v_{th}/T^{1/2})(N_c/T^{3/2}) = 3.25 \times 10^{21}(m_n/m_0)\ \text{cm}^{-2}\text{s}^{-1}\text{K}^{-2}$, where m_n is the electron density-of-state effective mass [26]. For the post-soft breakdown, the emission time is no longer a constant but varies with gate bias, as shown in Fig. 9b. It was found that the root cause is the field-dependent property of the capture cross section. It can be justified from the plot in Fig. 11 and Eq. (12). Prior to soft breakdown, the lines in $\tau_e T^2$–$1/kT$ plots have the same slopes and interception with Y-axis, which means E_{Cd}–E_T is fixed and capture cross section is independent of gate bias. After the formation of soft breakdown path, E_{Cd}–E_T is roughly identical at distinct gate bias. Nevertheless, capture cross section varies with gate bias. It shows capture cross section decreases with increasing gate bias, from Eq. (12). More importantly, τ_e is largely increased for post-soft breakdown, comparing to pre-soft breakdown, as a result of this decreased cross section, σ.

In this section, I_g-RTN has been utilized to examine the high-k gate dielectric traps in MOSFET. The capture and emission mechanisms have been evaluated, especially for a study on the soft breakdown. Several conclusions can be drawn from this study: (1) The gate leakage has been drastically increased as a result of the soft breakdown with a formed conductive path; (2) τ_e and τ_c take longer time since the capture cross section becomes smaller; (3) the major difference between

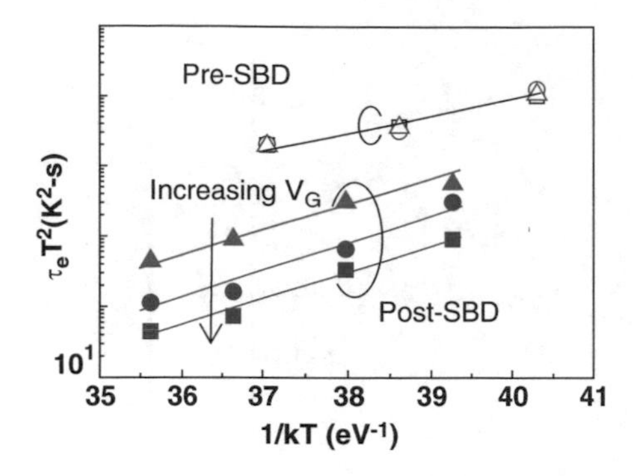

Fig. 11 Arrhenius plot for pre- and post-soft breakdown. It shows no changes of electron with gate bias, at a specific temperature in pre-soft breakdown case. However, it shows gate bias dependence for post-soft breakdown case. This is attributed to a decrease in the cross section as a result of the formation of a soft breakdown–induced conductive path

pre-soft breakdown and post-soft breakdown lies in the fact that cross section σ is field dependent for the latter and field independent for the former, shown in Fig. 11, while in most of the studies for SiO_2, σ is considered to be a constant.

5 BTI-Induced RTN Trap Depth

5.1 Observation on RTN Trap Distribution in an nMOSFET During BTI Stress

Figure 12a, b shows the distribution of traps during short-time BTI stresses for nMOSFET and pMOSFET respectively, under the stress condition of $|V_{gs}| = 2$ V, 85 °C, and 50 s. The trap positions were extracted for the condition of measurement at different values of V_{gs}. The results show that the traps are located around the Fermi level, which is because the probability that electrons population in Si substrate or gate electrode is the highest around the Fermi level.

Also, we can find wider regions of trap distribution from the Si substrate to the gate electrode, which allows the tunneling electrons to "jump" on each trap by trap-assisted tunneling mechanism. Thus, at the very beginning for the evolution of gate leakage paths, several traps are generated in the dielectric along the Fermi level, and the conduction of leakage current is migrated from pure direct or FN-tunneling to trap-assisted tunneling. Figure 12c shows the results of normalized amplitude of variation of the gate current ($\Delta I_g/I_g$) for trap A, trap B, and trap C, labeled in Fig. 12a. Although these traps are very close to each other, their characteristics of $\Delta I_g/I_g$ fluctuate in a very wide range, which cannot be explained by the conventional scattering concept applied to the $\Delta I_d/I_d$ [27, 28]. Thus, we suggest another scenario to explain this experimental result, as shown in Fig. 12d. In this schematic, the gate

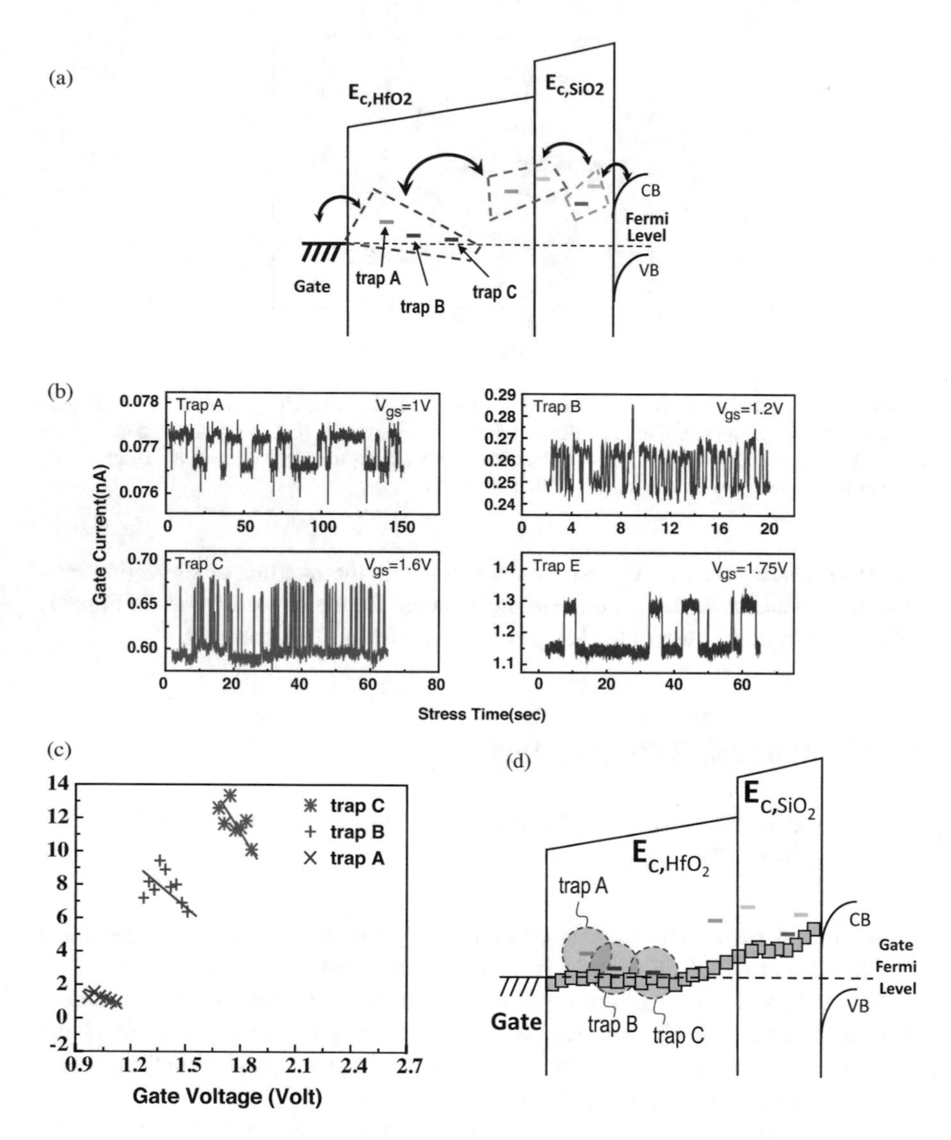

Fig. 12 (**a**) The experimental results of the trap location in the energy-band diagram for an n-channel MOSFET after PBTI stress at $V_{gs} = 2.4$ V, 25 °C during 50 s. (**b**) The experimental results of the trap location in the energy-band diagram for a p-channel MOSFET after NBTI stress at $V_{gs} = -2.4$ V, 25 °C during 50 s. (**c**) The normalized amplitudes of gate current induced by trap A, trap B, and trap C, labeled in (**a**). (**d**) The schematic to address how the potential of RTN traps affects the variation of gate current. It is assumed that the gate current path is formed in proximity to the traps because of the trap-assisted tunneling process

leakage current path, shown in the yellow-squared, and the RTN traps are generated around this path. Since the gate current conducts via trap-assisted tunneling of electrons, the path of gate current will be reasonably assumed to be near the traps.

However, according to the different distances from the coverage of the trap potential to the gate current path, the fluctuation of gate current affected by those traps corresponds to the distance of traps to gate current path. If the trap is very close to the path, the fluctuation will be very large, or it will be smaller. Hence, it is believed that the trap A is closer to the path than trap C. More importantly, the gate current path can thereby be traced by the traps. As a result, as the stress time progresses, the gate leakage current path will grow near the traps in the dielectric, and the gate current increases with larger amplitudes of fluctuations.

5.2 Profile of the RTN-Trap Paths in MOSFET During BTI Stress

Based on the understandings as mentioned previously, an experiment can be designed to depict the gate current path during BTI stress, as shown in Fig. 13a, where the constant voltage stress is applied on devices to accelerate the formation of traps in a periodic duration; after each duration, the stress will be paused, and the devices will be tested and tried with a lower level of gate voltage to measure the signals of I_g-RTN. This process will be repeated until the dielectrics reach breakdown.

Figure 13b, c shows the experimental results for nMOSFET after PBTI stress of $V_{gs} = 2.4$ V at 25 °C and $V_{gs} = 2.45$ V at 125 °C respectively. The results of both figures reveal completely different trends, which is originated from the environment of temperature but not for slight difference of stressing voltages. Furthermore, in Fig. 13b, for a device under a weaker stress at lower temperature, 25 °C, the first trap was initially generated near the interface of HfO_2 and SiO_2 and entangled at this interface to form an internal path; then this path was gradually extended to the interface of SiO_2 and Si substrate. Finally, the path penetrated into the bulk of dielectrics, followed by the formation of breakdown. It was found that, for PBTI at 25 °C, the integrity of gate dielectrics can sustain a weak intensity of stressing electric field; hence, the traps tend to be generated at the interface between the bulks of dielectrics. Furthermore, as the integrity of very thin interfacial layer, SiO_2, is weaker than that of HfO_2, the traps are generated gradually in SiO_2 layer during the preceding of stress. And, owing to thicker thickness and good integrity of bulk HfO_2 layer, the traps are measured in this region finally. Until all of the gate dielectrics cannot endure the applying electric field, the breakdown occurs.

On the other hand, Fig. 13c shows a different path, in which the trap was generated at the interface of SiO_2 and silicon substrate, and the following traps immediately created a critical path from the Si substrate to the gate electrode, resulting in the formation of breakdown. Figure 13b, c shows much different forms of the gate leakage path. The former reveals that the gate current is constructed by the accumulated traps, centered around the interface of dielectrics, which is more

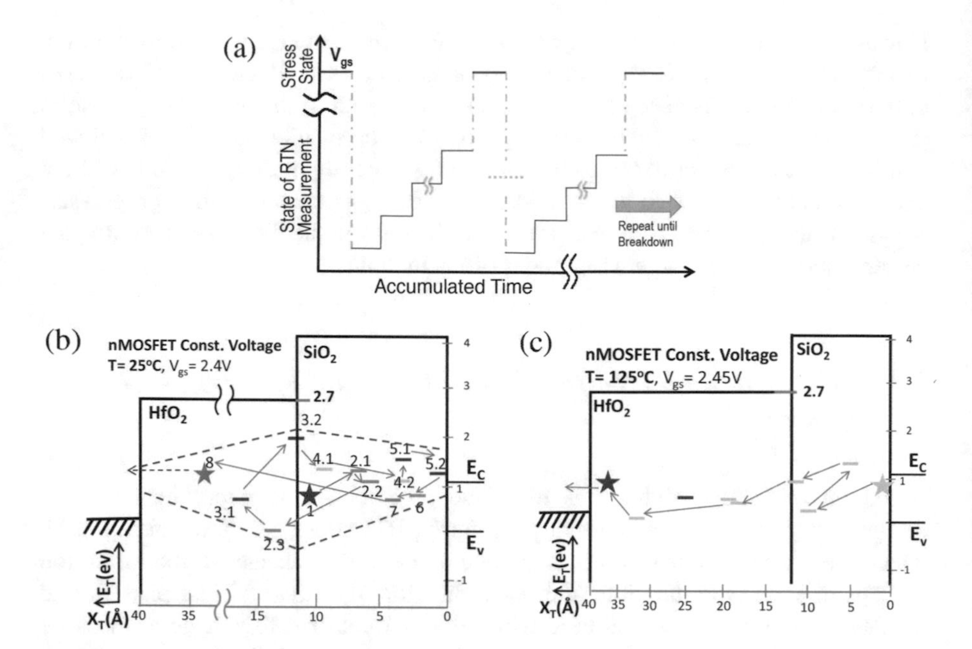

Fig. 13 (**a**) The setup of experiments to measure the positions of BTI-induced traps before dielectric breakdown happens. The constant gate voltage stress is applied periodically and is paused to wait for the measurement of I_g-RTN signals. (**b**) The generation of traps for n-channel MOSFET after PBTI stress at $V_{gs} = 2.4$ V and 25 °C until the dielectric reaches breakdown. (**c**) The generation of traps for n-channel MOSFET after PBTI stress at $V_{gs} = 2.4$ V and 125 °C until the dielectric reaches breakdown

like a shape of "spindle"; the latter exhibits a snaking walking behavior, named as "snake" path. However, the common correlation of both paths is formed by traps.

Figure 14a shows two curves, the black and the red, of transient gate current for the same stress conditions of Fig. 13b, c, respectively. It can be clearly observed that, for the black curve in Fig. 13a, there are many levels of gate current fluctuation, contributed by corresponding RTN traps generated in the dielectrics before breakdown happens. Note that after each trap is formed, the level of gate current shows a staircase, which is because the trap-assisted tunneling is enhanced by new generated trap. Since the integrity of the dielectrics can sustain such intensity of stress field, the time to breakdown is reasonably longer. On the other hand, the red curve of Fig. 14a shows few levels of gate current variation, which fits into the result shown in Fig. 13c, and its time to breakdown is shorter accordingly. Figure 13c summarizes our experiments from the lifetime prediction of dielectric time-to-breakdown. Two significant parameters affect the formations of gate current paths: temperature and the intensity of stressing electric field. During a BTI stress with temperature higher than 125 °C, it was found that only "snake" shape of current paths is identified, and during the stress at temperature lower than 25 °C, only "spindle" shape is observed. However, between these boundaries, the intensity of

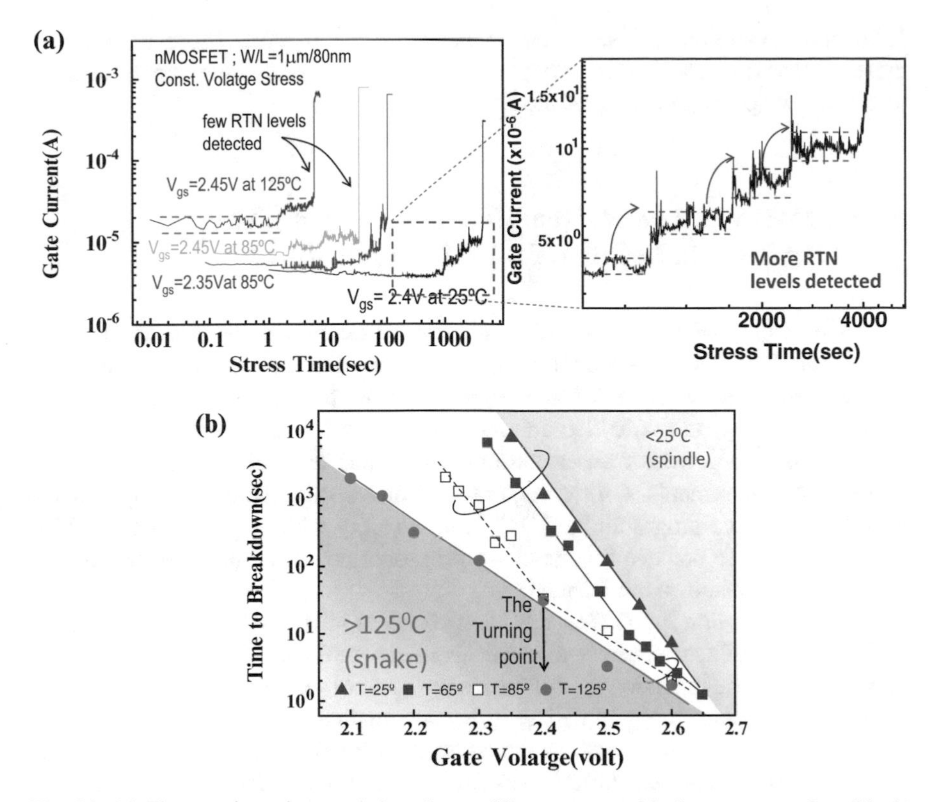

Fig. 14 (**a**) The transient characteristics of two different types of leakage current. One (black) shows many levels of gate current with longer time-to-breakdown. The other (red) exhibits fewer levels of gate current with shorter time-to-breakdown. (**b**) The experimental results in the prediction of time-to-breakdown lifetime. Two-slope curve has been identified. Temperature and intensity of stressing electric field dominate the magnitude of the induced gate currents

electric field dominates. There is a turning point defining this two-slope curve. If the intensity of field is lower than the turning point, the spindle paths of the gate current are measured, leading to a longer prediction of lifetime, otherwise, the snake paths happen along with much shorter lifetime if the intensity of field is over this point.

In summarizing this section, I_g-RTN has been employed to analyze the evolution of gate current paths in HK MOSFET. By observing the generated RTN traps under the BTI stress, a concept has been developed to explain the correlation between gate current leakage path and the traps. It was found that the potential of traps in dielectrics will cause the fluctuation of measured gate current, which also provides the evidence that the gate current path can be traced by the traps. Further, two different shapes of gate current paths have been found experimentally. The spindle shape of current paths is induced by a weaker condition of BTI stress, and the snake one is by a stronger condition. As a consequence, two different kinds of breakdown, the hard and soft breakdowns actually can be gauged by the same mechanisms about the interactions between traps and gate current paths. More importantly, these mechanisms can precisely explain the interesting two-slope curves of time-to-breakdown prediction as published in reported articles.

6 Discovery of a New Breakdown: Dielectric-Fuse Breakdown in MOSFET and its Applications

With the scaling of CMOS devices and the operating voltage, the information applications have been extended from the traditional PC to the wearable and portable devices, that allows each device to work together and form a seamless computation environment, e.g., the so-called Internet of Things (IoT), in which the variant applications, low-power wireless communication, and firm privacy security are the three basic requirements. On the requirement of security, there is strong demand to design one-time programming (OTP) functionality with very low cost but high reliability [29]. To achieve this goal, we propose a new OTP memory based on a newly found breakdown mechanism.

The dielectric-*fuse breakdown* has been observed in HKMG and poly-Si CMOS devices, called dFuse. dFuse is different from conventional hard breakdown (e.g., dielectric anti-fuse breakdown). For the comparison of their differences, in Fig. 15, after dFuse happens, the gate dielectric became open and the gate current will be pulled down to a very low level; on the other hand, it will be raised to a higher level as anti-fuse occurred, i.e., short. It was found that, different from the conventional

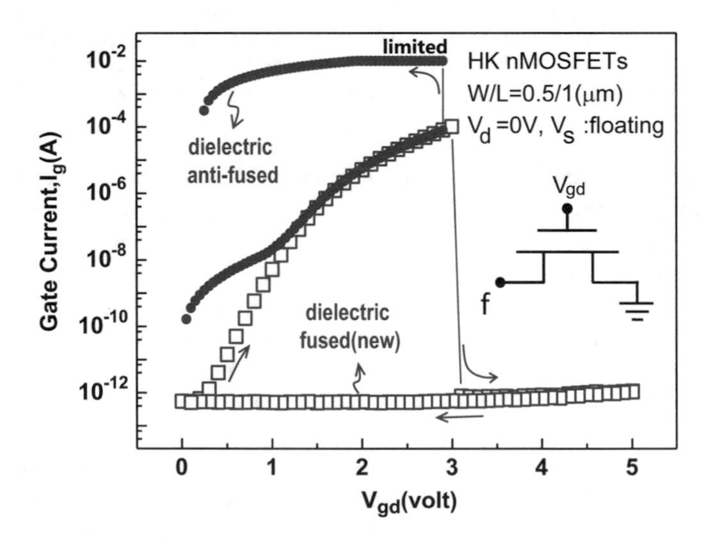

Fig. 15 The characteristics of gate current for the conventional gate dielectric anti-fuse and the newly found dielectric fuse mechanism. After anti-fusing, the gate current increases rapidly (P1); after fusing, it is blocked (P2)

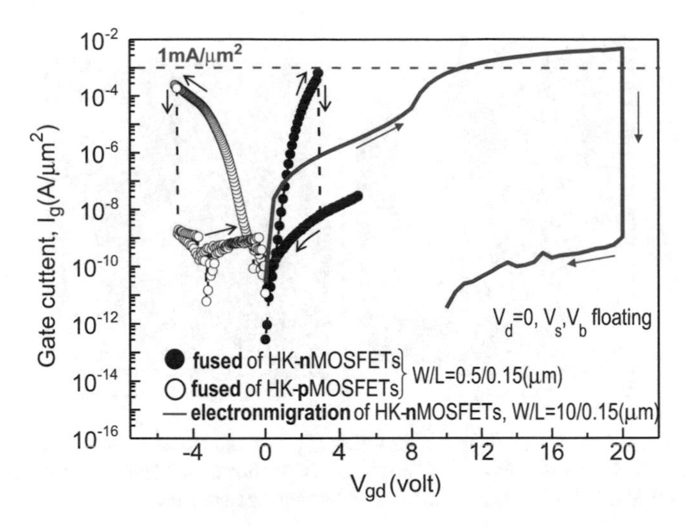

Fig. 16 The fuse is intrinsically different from the electromigration. The former happens in the dielectric and blocks current paths in condition of lower voltage and smaller current (<1 mA/μm^2); the latter happens in the poly or metal liner through high density of current

anti-fuse dielectric breakdown, such as the hard and soft breakdowns, this new fuse-breakdown behavior exhibits a typical property of an open gate and can be operated in much lower programming current (<50 μA), fast speed (~20 μs), and excellent data retention, in comparison to the other fuse mechanisms. dFuse phenomenon opens up an opportunity for us to design a feasible OPT memory. It is completely different from the electromigration, which uses the breakdown of dielectric film by ultrahigh current density, Fig. 16.

6.1 *Investigation of Dielectric-Fuse Breakdown*

In order to visibly observe the differences between dielectric-fuse and anti-fuse breakdowns, OBIRCH (optical beam-induced resistance change) technique has been used to inspect hot spots induced by the gate current, Fig. 17. It was found that only the anti-fused one shows hot spots, but the fused one does not, which indicates that the current is cut off after the device is fused. To further understand the path development during the breakdown, I_g-RTN can be utilized to measure the physical location of traps. Thus, we can monitor the generation of traps via the measurement of RTN signals.

From the measurement of the breakdown path profiling by using I_g-RTN [30], Fig. 18a shows the profiling results after dFuse breakdown. The result shows that traps are almost generated at the interface of HK/interfacial layer (interfacial layer). But if the breakdown path was developed under high temperature, Fig. 18b, the

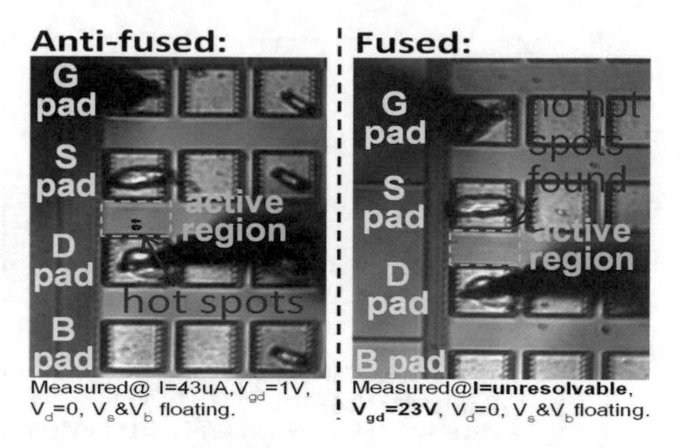

Fig. 17 The location of hot spots is identified by optical beam induced resistance change (OBIRCH) technique after anti-fusing, but there is no hot spot identified after fusing, which confirms the fuse process will not result in a conducting leakage path

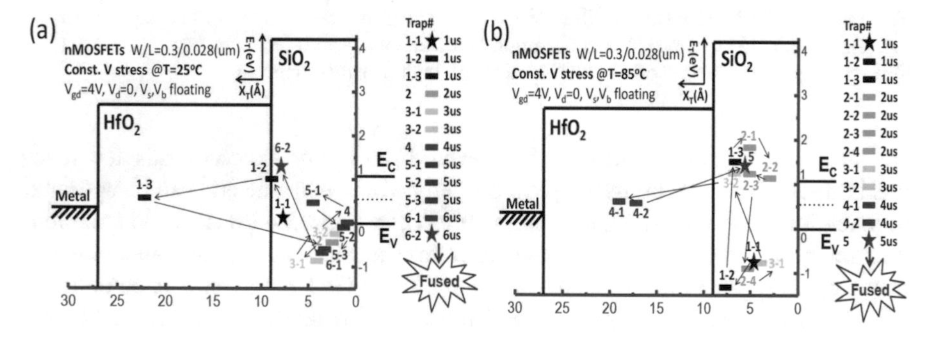

Fig. 18 Dielectric-fuse breakdown: (**a**) During the fuse breakdown, RTN traps are generated in the interfacial layer and highly concentrated near the interface between the interfacial layer and channel. It is believed that, by applying very high-intensity electric field, the oxygen ions in SiO_2 layer will be pulled out, and the oxygen vacancies are left as traps. With the growing of trap numbers, a gap will be created in the SiO_2 interfacial layer and open the current path, resulting in the fuse breakdown. (**b**) The fuse breakdown at 85 °C. Note that the traps are located in the middle region of interfacial layer to form a porous structure in SiO_2 to block the current path

trap migrates from the HK/interfacial layer interface to the middle of the interfacial layer. As a result, it is believed that the accumulated traps in interfacial layer develop a similar structure of porous structure in SiO_2 or create a burnout of SiO_2, in which the SiO2 layer is considered to be open under such circumstances. In other words, this interfacial layer, SiO_2, was burned out and the current measured between gate-to-source (drain) or gate-to-substrate was blocked, resulting in a sudden reduction of gate current.

Figure 19 shows the multilevel I_g-RTN during the dFuse breakdown process, which is an indication of more traps generated in the interfacial layer and forms an internal closed loop, i.e., traps are generated only inside the interfacial layer

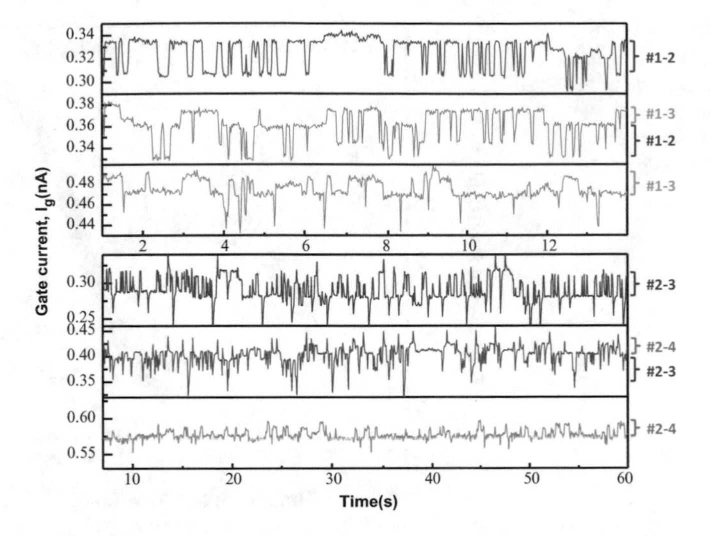

Fig. 19 RTN signals, measured at $V_{gd} = 4$ V, temperature = 85 °C, during the fuse process. It was found that RTN traps interact with each other, exhibiting multilevels, i.e., there is an internal closed loop in the SiO_2 layer, but this loop has never reached to the gate before fuse (dFuse) breakdown

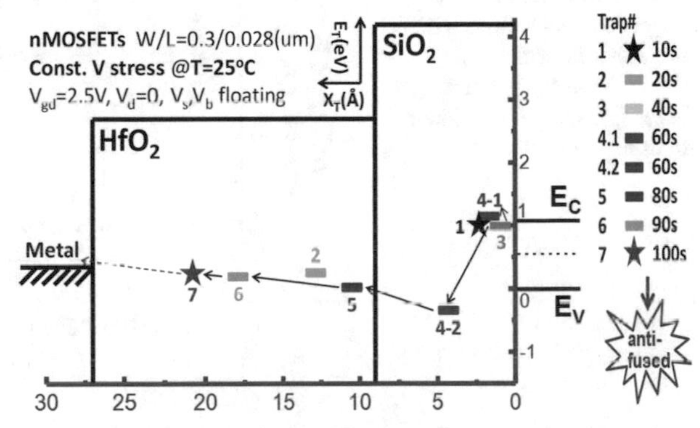

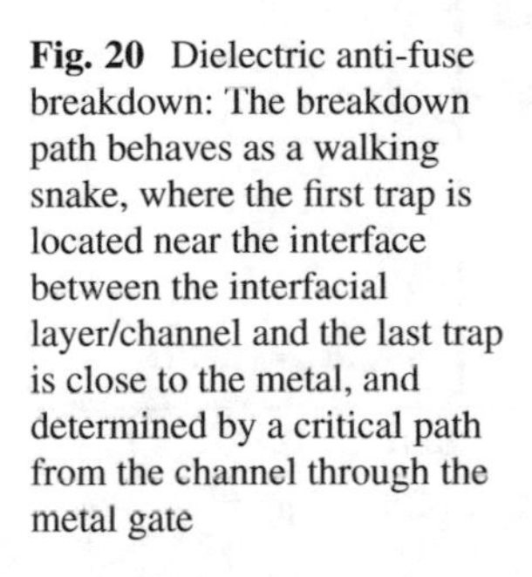
Fig. 20 Dielectric anti-fuse breakdown: The breakdown path behaves as a walking snake, where the first trap is located near the interface between the interfacial layer/channel and the last trap is close to the metal, and determined by a critical path from the channel through the metal gate

since its permittivity is much lower than that of high-k layer. However, this closed loop did not grow to connect the gate metal and channel before the dFuse occurred. In comparison, in Fig.20, the path of anti-fuse breakdown is also plotted. The hard breakdown path is determined by a critical path from the channel through the metal gate and looks like a *walking snake*. More importantly, it is conducting paths different from the dielectric-fuse breakdown path. Figure 21 shows the time-to-dielectric-breakdown curves of the fuse and anti-fuse. It was found that the time to fuse breakdown is much faster than anti-fuse breakdown.

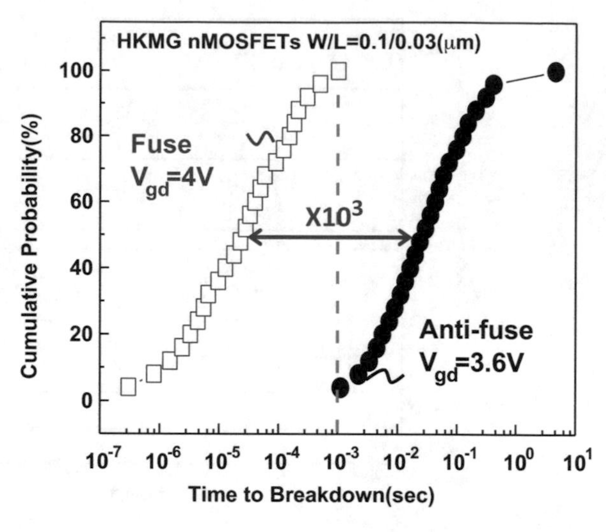

Fig. 21 Under higher intensity of stress field, the time to breakdown of the fuse is much faster, up to 10^3 times, than that of the anti-fuse breakdown

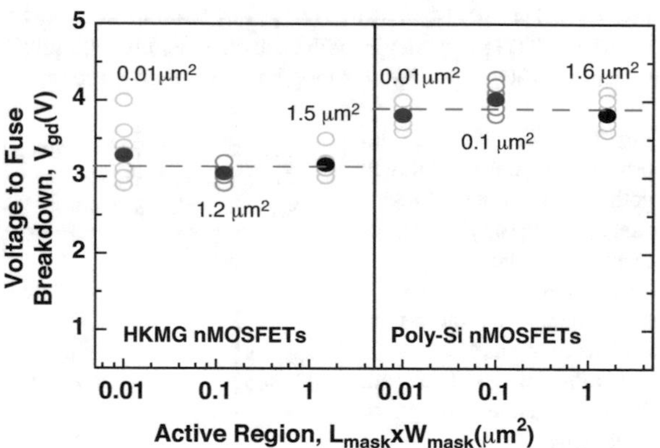

Fig. 22 The comparisons of the voltage to be fused for poly-Si and HKMG nMOSFET. The HK one requires lower voltages than the poly one, since the thickness of SiO_2 interfacial layer for HK one is thinner than that of poly-Si

6.2 *Dielectric Fuse Breakdown of HKMG and Poly-Si nMOSFET*

Figure 22 compares the dFuse voltage of high-k and poly nMOSFETs, and the dFuse voltage is independent of the device active regions. Furthermore, the dFuse voltage of HK devices is smaller than the poly-Si ones, which is because SiO_2 thickness of HK devices is thinner than that of poly-Si devices. Figure 23 compares the anti-fuse and fuse voltage as a function of temperature. It was found that the anti-fuse voltage is dependent on the temperature, but the dFuse voltage is independent of the temperature, indicating that the anti-fuse *obeys thermionic emission* but the dFuse does not. Finally, Fig. 24 shows the I_d and I_g–V_{gs} curves of HK and Poly-Si nMOSFETs before/after fuse breakdown, in which I_d cannot be controlled by gate anymore after fuse breakdown and keeps at a high level. This is because the electrostatic of the gate to channel is blocked by the fused gap existing in the interfacial layer, i.e., an open gate.

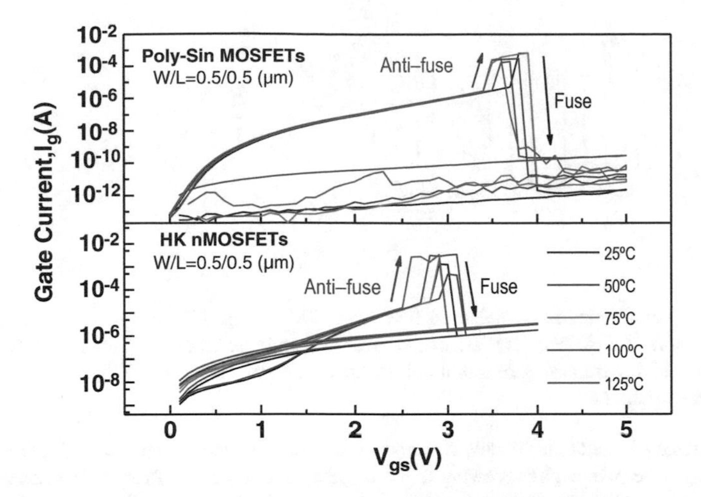

Fig. 23 Extraction of the voltage to fuse or hard breakdown. Note that the anti-fuse one is dependent on the temperature, while the fuse one is independent of the temperature, indicating different mechanisms

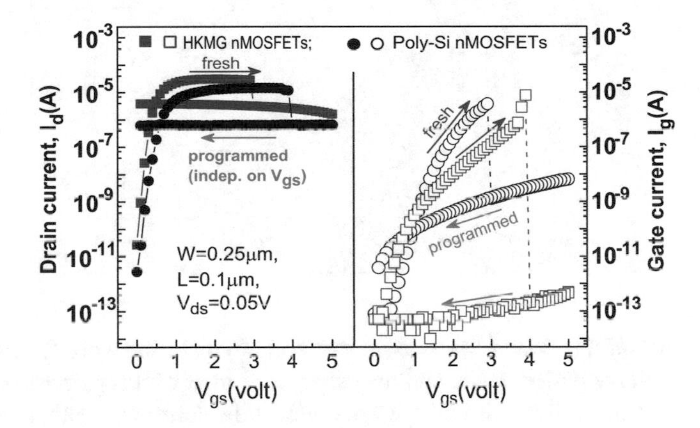

Fig. 24 The $I_d V_{gs}$ and $I_g V_{gs}$ curves before/after the fuse breakdown. After fused, I_g is shut down, and I_d shows a large source-to-drain current, independent of the gate electrostatic

6.3 Dielectric-Fuse Breakdown OTP Cells

To achieve high density and lower cost, nMOSFETs are series connected in a column as a NAND type, where three transistors are considered as a pair, Fig. 25. The middle one is the control transistor, while the two side ones are considered as the storage. So, each pair has two bits. The select transistors are a complementary transmission gate with thick dielectric layer to provide high-voltage driving ability. Furthermore, a new paired read scheme is proposed, Table 2.

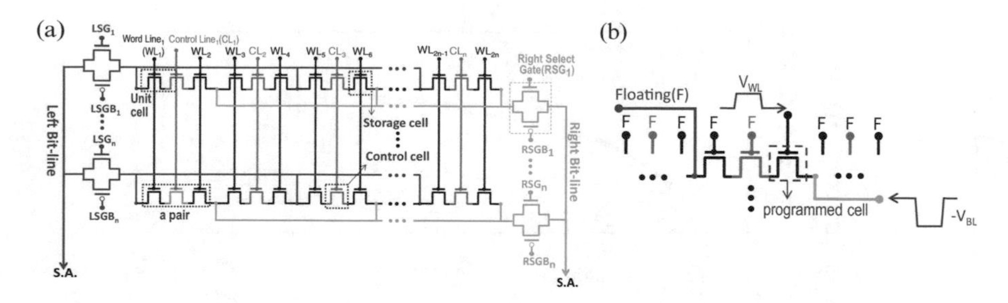

Fig. 25 (**a**) The schematic of dielectric fuse OTP 128-bit array is composed of a unit cell with three transistors. Two OTP cells share one control cell and one bit line can dramatically reduce the layout area. (**b**) Programming of a unit cell by the edge-FN tunneling on the right transistor of a three-transistor unit cell

Table 2 A new paired read scheme is introduced to read the test cell by a set of one control cell connecting to the two neighbors owing to the large source-to-drain current of the fused cell since the gate-to-channel electrostatic is disabled after the cell is fused

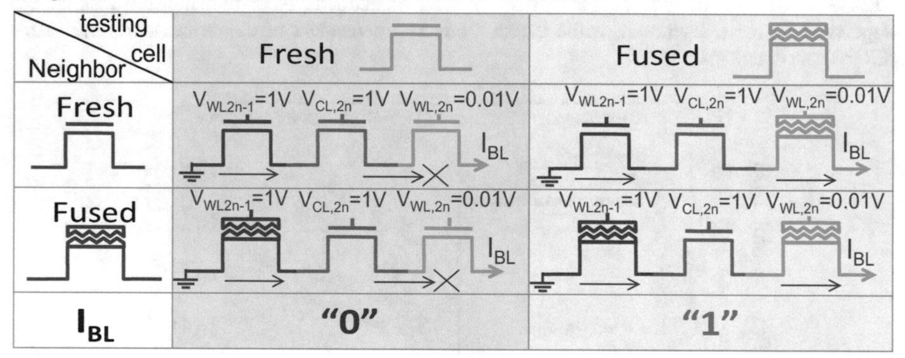

Neighbor \ testing cell	Fresh	Fused
Fresh	V_{WL2n-1}=1V $V_{CL,2n}$=1V $V_{WL,2n}$=0.01V I_{BL}	V_{WL2n-1}=1V $V_{CL,2n}$=1V $V_{WL,2n}$=0.01V I_{BL}
Fused	V_{WL2n-1}=1V $V_{CL,2n}$=1V $V_{WL,2n}$=0.01V I_{BL}	V_{WL2n-1}=1V $V_{CL,2n}$=1V $V_{WL,2n}$=0.01V I_{BL}
I_{BL}	"0"	"1"

For instance, if one would like to read one cell of a pair, the word lines (WL) of the selected cell are pulled down, and the other word lines of this pair are pulled up. The unselected one in this pair will always conduct the current no matter whether it is fused or not. Then the selected cell with pull-down WL will be read as "1" if it is fused, or "0" if it is fresh. Figure 26 shows the programming transient results of the fuse OTP array. It can reach 20 μs at V_{gs} = 4 V. Figure 27 shows the read on–off ratio for the fuse and anti-fuse OTP array. The on–off statistical distribution of the fuse is much narrower, and its window is much broader than that of the anti-fuse. Finally, Fig. 28 is the retention test for over one month at 125 °C while still kept very Stable 1 and 0 states.

In short, in this section, by the profiling technique of breakdown paths based on the measurement of I_g-RTN, a new dielectric-fuse breakdown has been investigated, which is intrinsically distinct from the well-known fuse mechanisms, such as electromigration in dielectric film, and quite different from the traditional dielectric anti-fuse breakdown. The dFuse breakdown is caused by the porous structure of SiO_2 layer under very high electric field. The time-to-breakdown of the dFuse

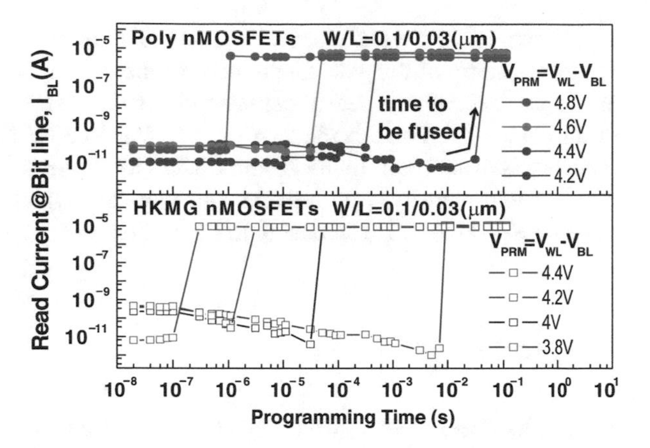

Fig. 26 Time to fuse breakdown for poly and HK splits. It was found that the dielectric can be fused just during 50 ms at $V_{gs} = 4$ V

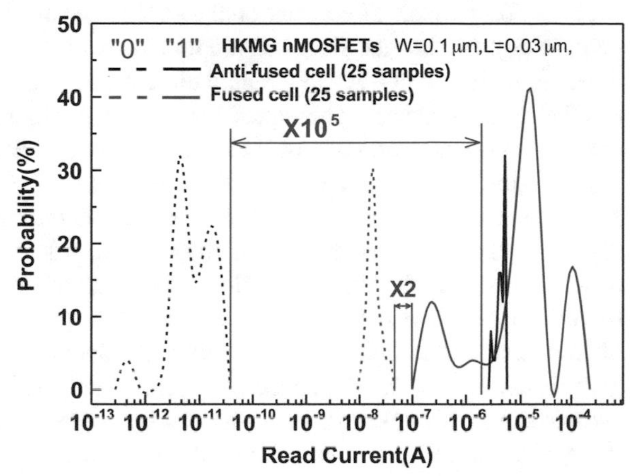

Fig. 27 Read on/off ratio characteristics for anti-fuse and fuse breakdown. The window of the fused one is much larger than that of the anti-fused one, up to 10^5 times

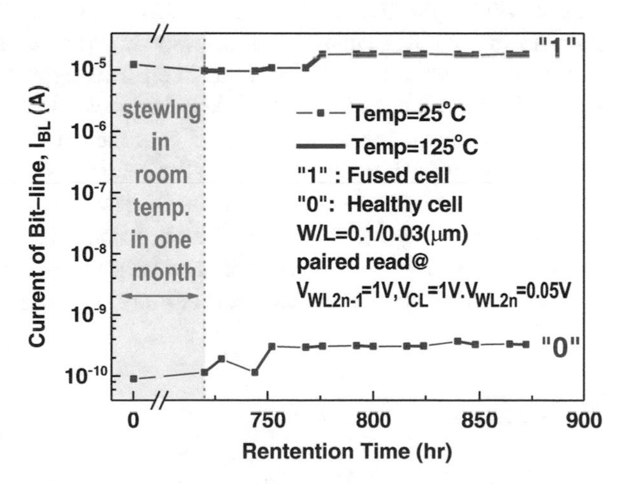

Fig. 28 The retention of the fused cell. First, the fused cell was stored at the room environment for over a month and then baked 8 h at 125 °C, followed by storing for more than 16 h at 85 °C repeatedly for a week

breakdown is very weakly dependent on the temperature. After dFuse breakdown, the electrostatic of the gate cannot control the conduction of the channel anymore, as we can see from a nearly constant channel resistance between the source and drain. Thus, based on this mechanism, a new paired NAND OTP memory has been demonstrated successfully by 28-nm CMOS technology. This memory array has the merit of high density, very reliable, and low cost, which meets the requirement of the embedded CMOS OTP design in the IoT era.

7 Conclusions

I_g-RTN has been a very successful RTN measurement since its inception in 2008 [26]. In comparison to conventional I_d-RTN, the RTN signals recorded from the gate leakage on the gate electrode are more sensitive and accurate in response to the dynamic behavior of traps in the gate dielectrics. In this chapter, we widely discuss I_g-RTN in all respects. The equations of trap positions for RTN traps have been derived in consideration of single dielectric layer and double dielectric layers. Then, we have observed the phenomena of RTN traps before and after soft breakdown of gate dielectrics. The amplitude of RTN signals dramatically increases, and thus the capture cross section reduces after the soft breakdown happened. Based on these observations, we construct a profiling technique of the dielectric breakdown paths by using a *transient* I_g-RTN measurement technique. It was found that the potential of traps inside the dielectrics will cause the fluctuation of measured gate current. Two different shapes of gate current paths have been found experimentally. The spindle shape of current paths is induced by a weaker condition of BTI stress, and the snake one is by a stronger condition. As a consequence, two different kinds of breakdown, the hard and soft breakdown can actually be gauged by the same mechanisms about the interactions between traps and gate current paths. Ultimately, a new dielectric-fuse breakdown has been investigated. The dFuse breakdown is caused by the porous structure of SiO_2 layer under very high electric field. After dFuse breakdown, the electrostatic of gate cannot control the conduction of the channel anymore. Further, a 4-kb OTP memory based on the dFuse scheme has also been demonstrated on a 28-nm CMOS technology platform successfully [31]. This memory array has the merit of high density, very reliable, and low cost. All these results should be very valuable and helpful for the researchers and engineers to discover and to create new concepts or more applications of I_g-RTN in the future.

References

1. M. Toledano-Luque, B. Kaczer, E. Simon, R. Degraeve, J. Franco, et al., Correlation of single trapping and detrapping effects in drain and gate currents of nanoscaled nFETs and pFETs, in *IEEE Int. Rel. Phys. Sym.(IRPS)*, Anaheim, USA 2012, pp. XT.5.1-XT.5.6

2. P. Ren, P. Hao, C. Liu, R. Wang, X. Jiang, et al., New observations on complex RTN in scaled high-κ/metal-gate MOSFETs—The role of defect coupling under DC/AC condition, in *IEEE Int. Elec. Dev. Meet.(IEDM)*, San Francisco, USA, 2013, pp. 778–781
3. F. M. Puglisi, A. Padovani, L. Larcher, P. Pavan, Random telegraph noise: Measurement, data analysis, and interpretation, in *IEEE Sym. on Phy. & Fail. Analysis of IC (IPFA)*, Chengdu, China, (2017), pp. 5–14
4. S. Guo, R. Wang, D. Mao, Y. Wang, R. Huang, Anomalous random telegraph noise in nanoscale transistors as direct evidence of two metastable states of oxide traps. Nature Sci. Rep. **7**, 6239 (2017)
5. S. S. Chung, C.M. Chang, Y.S. Hsieh, et al., The investigation of capture/emission mechanism in high-gate dielectric soft breakdown by gate current random telegraph noise approach. API Appl. Phys. Lett. **93**, 213502 (2008)
6. E. R. Hsieh, Y. L. Tsai, S. S. Chung, et al., The understanding of multi-level RTN in trigate MOSFETs through the 2D profiling of traps and its impact on SRAM performance: A new failure mechanism found, in *IEEE Int. Elec. Dev. Meet.(IEDM)*, San Francisco, USA (2012), pp. 19.2.1–19.2.4
7. T. Grasser, H. Reisinger, W. Goes, Th. Aichinger, Ph. Hehenberger, et al., Switching oxide traps as the missing link between negative bias temperature instability and random telegraph noise, in *IEEE Int. Ele. Dev. Meet. (IEDM)*, Baltimore, USA (2009), pp. 729–732
8. C. Yu, Chen, H.-J. Cho, A. Kerver, et al., Correlation of Id- and Ig-random telegraph noise to positive bias temperature instability in scaled high-κ/metal gate n-type MOSFETs, in *IEEE Int. Rel. Phys. Sym.(IRPS)*, Monterey, USA (2011), pp. 190–195
9. T. Grasser, K. Rott, H. Reisinger, M. Waltl, J. Franco, et al., A unified perspective of RTN and BTI, in *IEEE Int. Rel. Phys. Sym. (IRPS)*, Waikoloa, USA (2014), pp. 4A.5.1–4A.5.7
10. G. Rzepa, M. Waltl, W. Goes, B. Kaczer, T. Grasser, Microscopic oxide defects causing BTI, RTN, and SILC on high-k FinFETs, in *Int. Con. on Sim. of Semi. Proc. & Dev.(SISPAD)*, Washington DC, USA (2015), pp. 144–147
11. F.M. Puglisi, F. Costantini, B. Kaczer, L. Larcher, P. Pavan, Monitoring stress-induced defects in HK/MG FinFETs using random telegraph noise. IEEE Elec. Dev. Let.(EDL) **37**(9), 1211–1214 (2016)
12. W. Liu, A. Padovani, L. Larcher, N. Raghavan, K.L. Pey, et al., Analysis of correlated gate and drain random telegraph noise in post-soft breakdown TiN/HfLaO/SiOx nMOSFETs. IEEE Elec. Dev. Let.(EDL) **35**(2), 157–160 (2014)
13. J. Franco, B. Kaczer, N. Waldron, Ph. J. Roussel, A. Alian, M. A. Pourghaderi, Z. Ji, T. Grasser, T. Kauerauf, S. Sioncke, N. Collaert, A. Thean, G. Groeseneken, RTN and PBTI-induced time-dependent variability of replacement metal-gate high-k InGaAs FinFETs, in *IEEE Int. Ele. Dev. Meet. (IEDM)*, San Francisco, USA (2014), pp. 506–509
14. W. Goes, M. Waltl, Y. Wimmer, G. Rzepa, T. Grasser, Advanced modeling of charge trapping: RTN, 1/f noise, SILC, and BTI, in *IEEE SISPAD*, Yokohama, Japan (2014), pp. 77–80
15. E. Bury, R. Degraeve, M. J. Cho, B. Kacer, W. Goes, T. Grasser, N. Horiguchi, G. Groeseneken, Study of (correlated) trap sites in SILC, BTI and RTN in SiON and HKMG devices, in *IEEE IPFA*, Marina Bay Sands, Singapore (2014), pp. 250–253
16. L. Larcher, Statistical simulation of leakage currents in MOS and flash memory devices with a new multiphonon trap-assisted tunneling model. IEEE Trans. on Ele. Dev. (TED) **50**, 1246–1253 (2003)
17. S. Lee, H.-J. Cho, Y. Son, D. S. Lee, H. Shin, Characterization of oxide traps leading to RTN in high-k and metal ate MOSFETs, in *International Electron Devices Meeting* (2009), p. 763
18. F. Crupi, R. Degraeve, G. Groeseneken, T. Nigam, H.E. Maes, On the properties of the gate and substrate current after soft breakdown in ultrathin oxide layers. IEEE Trans. Electron Devices **45**, 2329 (1998)
19. M.J. Kirton, M.J. Uren, Capture and emission kinetics of individual Si:SiO_2 interface states. Appl. Phys. Lett. **48**, 1270 (1986)
20. S.H. Ho, T.C. Chang, C.W. Wu, W.H. Lom, C.E. Chen, et al., Investigation of an anomalous hump in gate current after negative-bias temperature-instability in HfO_2/metal gate p-channel

metal-oxide-semiconductor field-effect transistors, AIP. J. Appl. Phys. **102**, 012103 (2013)
21. H. J. Lim, Y. Kim, I. S. Jeon, J. Yeo, et. al., Impact of the crystallization of the high-k dielectric gate oxide on the positive bias temperature instability of the n-channel metal-oxide-semiconductor field emission transistor, AIP, J. Appl. Phys. 102, 232909 (2013)
22. M.H. Tsai, H. Muto, T.P. Ma, Random telegraph signals arising from fast interface states in metal-SiO2-Si transistors. Appl. Phys. Lett. **61**, 1691 (1992)
23. M.J. Kirton, M.J. Uren, Noise in solid-state microstructures: A new perspective on individual defects, interface states and low-frequency (1/f) noise. Adv. Phys. **38**, 367 (1989)
24. D.K. Schroder, *Semiconductor Material and Device Characterization*, 3rd edn. (Wiley, New York, 2006), p. 259
25. G.M. Martin, A. Mitonneau, A. Mircea, Electron traps in bulk and epitaxial GaAs crystals. Electron. Lett. **13**, 191 (1977)
26. D.V. Lang, H.G. Grimmeiss, E. Meijier, M. Jaros, Complex nature of gold-related deep levels in silicon. Phys. Rev. B **22.3917**(22), 3917 (1980)
27. A. Avellan, W. Krautschneider, S. Schwantes, A strong analogy between the dielectric breakdown of high-K gate stacks and the progressive breakdown of ultrathin oxides. AIP, Appl. Phys. Lett. **78**, 2790 (2001)
28. S. Kobayashi, M. Saitoh, K. Uchida, Id Fluctuations by Stochastic Single-Hole Trappings in High-κ Dielectric P-MOSFET, IEEE, Symp. VLSI Tech. Dig. (2008), p. 78
29. S. Babar, A. Stango, N. Prasad, J. Sen, R. Prasad, Proposed embedded security framework for internet of things (IoT), in *Wireless-VITAE, IEEE Digest of* (2011), p. 90
30. S.S. Chung, C.M. Chang, The investigation of capture/emission mechanism in high-k gate dielectric soft breakdown by gate current random telegraph noise approach. Appl. Phys. Lett. **93**, 213502 (2008)
31. E. R. Hsieh, C. W. Chang, C. C. Chuang, H. W. Chen, S. Chung, The demonstration of gate dielectric-fuse 4kb OTP memory feasible for embedded applications in high-k metal-gate CMOS generations and beyond, in *Symp. on VLSI Technology*, Kyoto (2019), pp. C208–209

Random Telegraph Noise in Flash Memories

Alessandro S. Spinelli, Christian Monzio Compagnoni, and Andrea L. Lacaita

1 Introduction

The first observation of random telegraph noise (RTN) in MOS transistors dates back to the 1980s [1], when low-frequency noise investigations were carried out on sub-μm devices. The abrupt two-level fluctuation in drain current I_D there observed has received considerable attention since then, prompting early investigators [1–9] to recognize the role played by capture/emission processes of single electrons by silicon/oxide interface traps and derive an expression for the RTN amplitude [2]:

$$\frac{\Delta I_D}{I_D} = \frac{q}{C_{ox} W L (V_G - V_T)} + \frac{\Delta \mu_n}{\mu_n} \frac{\Delta A}{W L}, \tag{1}$$

where q is the electron charge, C_{ox} the oxide capacitance per unit area, and W and L the channel width and length, respectively. In (1), μ_n is the electron mobility in the inversion layer, which is modified by a quantity $\Delta \mu_n$ because of the additional scattering due to the trapped charge, taking place over an area ΔA.

Subsequent investigations of RTN in MOS devices unveiled a more complex picture, in which both the amplitude and the time behavior turned out not to be consistent with the above theory [2, 5, 10–12]. After a few tentative works [5, 10–12], a convincing explanation was proposed in [13], where non-uniformity in the conduction channel was singled out as the reason for the large RTN amplitude observed: the band energy profile of the channel in an MOS transistor is now seen as being locally modulated by trapped oxide and interface charges, resulting in current percolation and filamentary current flow. This makes it easier for an RTN trap to shut

A. S. Spinelli (✉) · C. M. Compagnoni · A. L. Lacaita
Dipartimento di Elettronica, Informazione e Bioingegneria, Politecnico di Milano, Milano, Italy
e-mail: alessandro.spinelli@polimi.it; christian.monzio@polimi.it; andrea.lacaita@polimi.it

T. Grasser (ed.), *Noise in Nanoscale Semiconductor Devices*,
https://doi.org/10.1007/978-3-030-37500-3_6

off one of these paths, resulting in a large drain current fluctuation. First statistical investigations [14] showed indeed fluctuations larger than 10%.

Starting from the 90 nm technology node, RTN began to be investigated in Flash memories [15–21] and in SRAMs [22–24], taking advantage of the data collection capabilities made possible by the array organization. Flash memories in particular proved to be an ideal test vehicle for RTN characterization, because of the analog nature of their operation and the larger value of the oxide thickness when compared to logic devices, enhancing ΔI_D (see (1)). The Flash array arrangement allows to monitor a very large number of devices, making possible a true statistical characterization of the phenomenon that is unattainable with measurements on discrete devices. On the other hand, read operations on Flash memories can be slower than on single transistors, because of the array organization and the related circuitry, preventing the investigation of fast phenomena.

The first measurements of RTN in Flash memories showed current fluctuations up to 60% [20] and threshold voltage (V_T) shifts up to 700 mV [16], reviving the interest in the phenomenon and its underlying physics. By that time, however, a new source of non-uniformities in the channel conduction had emerged: random dopant fluctuations (RDF) [25–28], which had been shown to have a significant impact on RTN [29–32].

In the following, we will review in detail the experimental data on RTN in Flash memories and the physical models developed to quantitatively account for them. We will then address the impact of RTN on the array operation and finally discuss RTN in novel 3D Flash memories. Please note that here we will not review the operating principle of Flash memories nor the other reliability concerns that affect the technology development, referring the interested reader to the many publications in the field [33–38].

2 Experimental Data

The first published results of an experimental characterization of the RTN amplitude in 90 nm NAND Flash memory arrays [15] are shown in Fig. 1. The authors employed an experimental technique [15, 16, 18, 21] relying on measuring the minimum and maximum V_T of each cell upon the application of repetitive ascending and descending gate voltage ramps, thereby sampling the maximum RTN amplitude. Note that the majority of cells show only negligible V_T fluctuations, while an increasingly small number of them exhibit increasingly large fluctuations. Also, the number of cells affected by RTN is shown to increase with program/erase (P/E) cycling.

The abovementioned characterization is useful to assess the maximum RTN fluctuation that, however, may not be the true parameter of interest from a device operation viewpoint, where what really matters is the V_T fluctuation between two successive read operations, not necessarily equal to its maximum value. To obtain such a result, the scheme highlighted in Fig. 2 was first employed in [17]: all

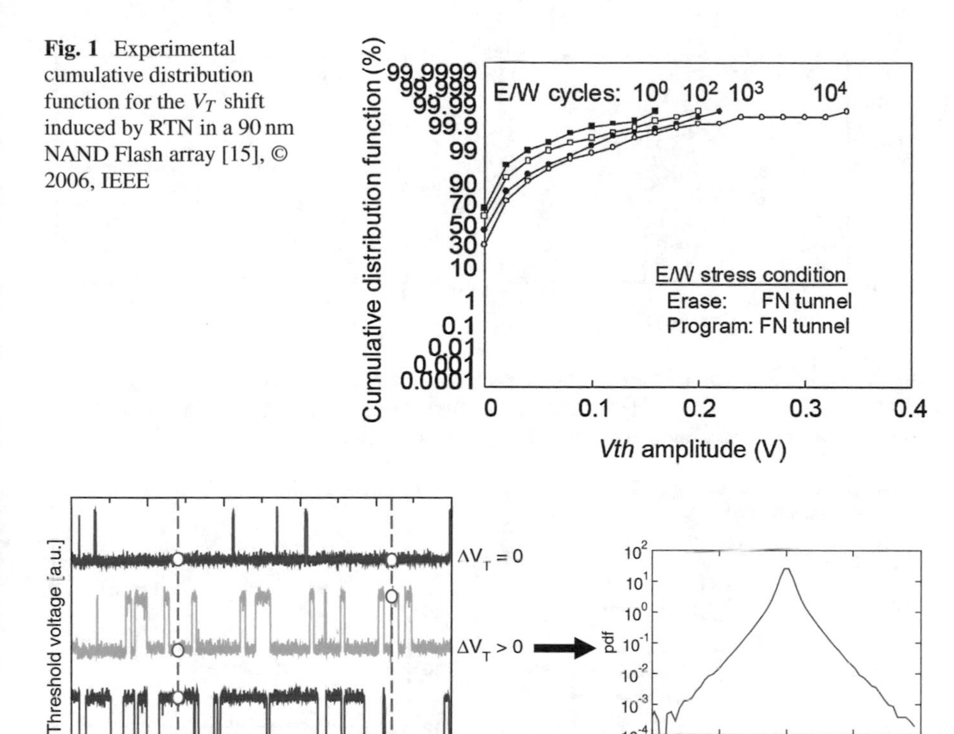

Fig. 1 Experimental cumulative distribution function for the V_T shift induced by RTN in a 90 nm NAND Flash array [15], © 2006, IEEE

Fig. 2 Experimental scheme for RTN measurement. Array cells V_T are sampled at different times and the difference $\Delta V_T = V_T(t_2) - V_T(t_1)$ is computed, collecting the resulting statistical distribution. Note that both positive and negative ΔV_T values are possible

the cells in the array are read at two different times and the RTN amplitude for each cell is now defined as the difference between the two measured V_T values, $\Delta V_T = V_T(t_2) - V_T(t_1)$. The RTN amplitude probability density function (pdf) can then be constructed, showing both positive and negative values, depending on the initial state of the measured cells.

Typical results obtained with this technique are shown in Fig. 3 for 65 nm NOR (left) and 60 nm NAND (right) Flash arrays [17, 39, 40]. For better readability, figures show the cumulative density function (cdf) and its complement to one, demonstrating that the RTN amplitude distribution follows a clear exponential behavior down to the lowest measurable probability values for both positive and negative ΔV_T values. Note that the region around $\Delta V_T = 0$ is affected by Gaussian noise related to the read operation, masking the cells with very low RTN amplitude. Differences between NOR and NAND results, such as the slope of the exponential tail and the number of cells affected by RTN, are a result of the different cell design and operating conditions, as discussed further on. We will also defer to the next

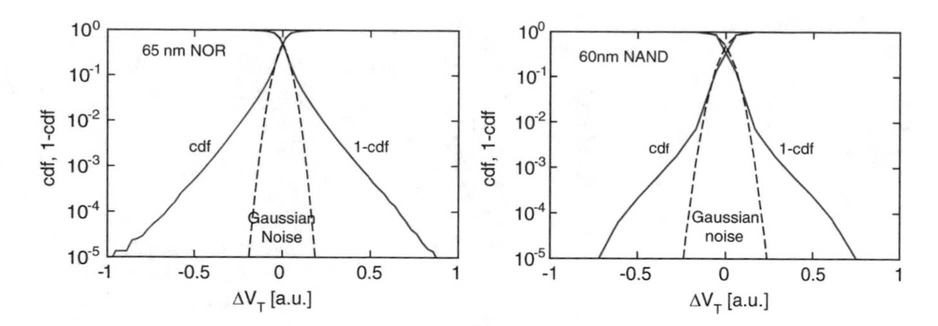

Fig. 3 Experimental results for the ΔV_T statistical distribution measured on 512 kb of a 65 nm NOR (left) and 60 nm NAND (right) Flash arrays. Both the cumulative distribution function (cdf) and its complement are displayed to highlight the exponential tails

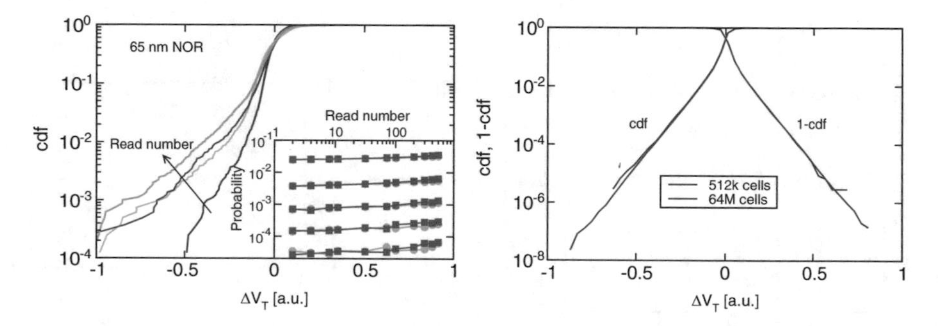

Fig. 4 Left: RTN cdf distribution in a 65 nm NOR array for different read operations [41]. The inset shows the probability to achieve a set value of $|\Delta V_T|$ as a function of the number of read operations [39]. Right: RTN cdf for two arrays, with size of 512 kb and 64 Mb [42]

section the discussion on the physical origin of such exponential tails and proceed here with collecting the experimental evidences.

Two more features of the RTN, namely the effect of time and of the sample size, are shown in Fig. 4: The figure on the left shows the effect on the RTN amplitude of elapsed time between the read operations in Fig. 2, here expressed as the number of reads [17, 39, 41]. To this aim, successive read operations are performed on the array and the RTN is defined as ΔV_T measured between the nth and the first read operation. Note the slow increase in the tail amplitude (cdf only is shown for clarity), which is reported in the inset as the probability to achieve a fixed value of $|\Delta V_T|$, featuring a logarithmic dependence on time.

Figure 4 (right) shows instead the effect of the sample size, by comparing data collected on arrays of 512 kb and 64 Mb [42]. Note that a neat exponential behavior is maintained, showing no sign of saturation in the RTN amplitude. If one considers that we are showing the amplitude distribution among the cell arrays, this result is tantamount to saying that there is an exponentially decreasing probability of finding a cell with a given RTN amplitude.

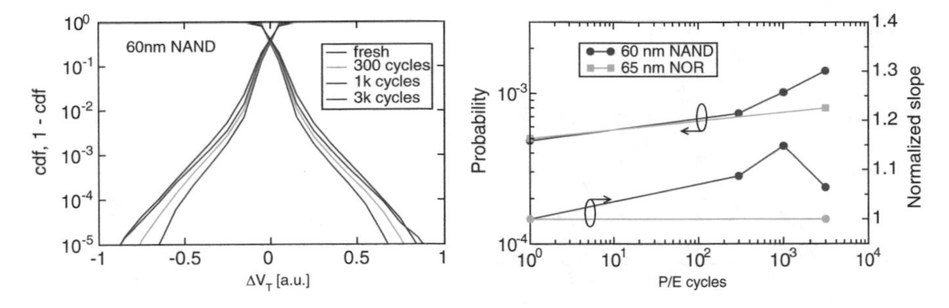

Fig. 5 Left: Experimental data for the RTN distribution in NAND arrays for different values of the number N_C of P/E cycles, © 2008 IEEE. Right: RTN tail height at a fixed $|\Delta V_T|$ value and slope as a function of N_C for a 60 nm NAND and a 65 nm NOR array [40]

Since RTN is known to be related to the presence of border traps, it is expected to get worse with P/E cycling, as indeed reported by the initial investigation (see Fig. 1) and by later studies [19, 40, 43–45]. Figure 5 (left) shows experimental data for RTN in a 60 nm NAND array subjected to different cycling conditions [40], resulting in an increase of the exponential tails. This result is also reported in Fig. 5 (right), where the probability at a fixed $|\Delta V_T|$ value and the slope λ of the RTN tails (measured in a semilog scale, with units of mV/decade) are shown as a function of the number of cycles N_C for NOR and NAND technologies. The figure makes clear that cycling results in a negligible dependence of λ on N_C, i.e., in a parallel shift of the RTN tails. The increase in tail height follows a power-law dependence with exponent 0.29 for NAND and 0.05 for NOR arrays [40]. This difference is most likely a consequence of the different cell design and cycling conditions, and prompted several works on the optimization of the cell process/architecture in order to reduce the extent of the RTN [42, 46–56]. Due to the more complex arrangement of cells in the NAND array, moreover, additional dependences of the RTN amplitude have been reported, such as on the cell position and state along the string [57, 58] (because of the different transconductances, depending on the source and drain series resistances) and on the state of cells on adjacent bitlines [59, 60] (because of the modification in the electron density profile induced by electrostatic interference).

The temperature dependence of RTN was investigated in [45], with reference to both cycling and read conditions. Figure 6 (left) shows the RTN distributions for different values of N_c when cycling is performed via two schemes involving different times and temperatures, collecting then the RTN distribution at room temperature. The figure clearly demonstrates that no significant difference exists in the RTN cdfs, meaning that the high-temperature phase applied during the 500 h scheme does not anneal any defect responsible for the RTN. The right-hand side of Fig. 6 shows instead the results for one cycling condition and different read temperatures. Even in this case, no dependence appears when the temperature is changed from room temperature to 55 °C. This result appears surprising at first, as capture and emission times are temperature dependent [58, 61, 62], and will be discussed in the next section.

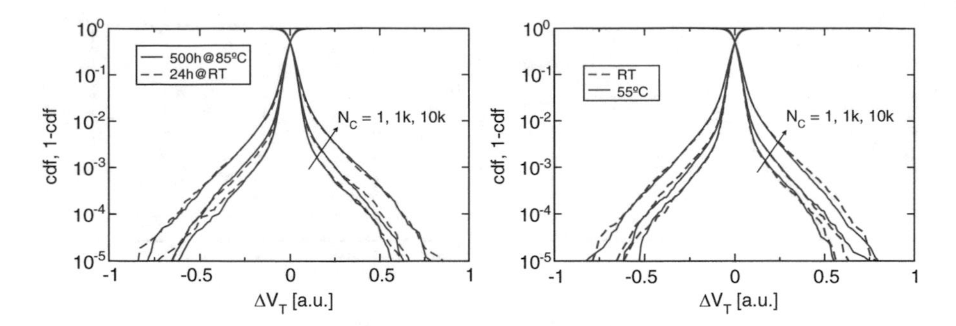

Fig. 6 Left: RTN cdfs for two cycling schemes and room temperature read in a 1X NAND technology, © 2015, IEEE. Right: RTN cdfs for the same technology after one cycling scheme and different read temperatures, © 2015, IEEE. No significant differences appear in all cases

3 Models

A thorough model of RTN would entail a description of the ΔV_T amplitude distribution as well as its dependences on time, while the account of the operating conditions (read, cycling) of the cells can be achieved via parameter tailoring. To tackle these issues, we recall that RDF had been identified as a potential source of significant RTN amplitude since 2003 [31, 32]. However, those pioneering works were mainly concerned with the variance associated with the different fluctuation sources, and not with the reliable extraction of the RTN cdf. This task was undertaken in [63, 64], where 3D Monte Carlo simulations of electron transport in a Flash memory cell were carried out, accounting for random dopant and trap positions. To focus on the electrostatic contribution of RDF, which is the dominant term (as discussed in [65–68]), drift-diffusion simulations with constant mobility were employed, while charge quantization at the Coulomb wells was added via a density-gradient approach [69].

Figure 7 shows simulation results for the current density profile at the silicon-oxide interface of a decananometer Flash memory biased at threshold [64]. The figure on the left shows the case of an empty trap: note that current transport is not uniform, but has a percolative nature because of the localized energy barriers induced by the individual dopants and the effect of field intensification at the edges of the active area. If a "strategic" RTN trap (black square in the right figure) is located exactly above one main percolation path, close to one edge in this case, its charging will shut off such path, leading to a large variation in current and V_T. On the other hand, if the trap is located above a region where the current density is low, its charging and discharging will barely affect the overall current flow. This explains why very large ΔV_T can be achieved, if a particular combination of dopant and trap position occurs in a cell.

Monte Carlo results for the statistical distribution of the V_T shift due to a single RTN trap are shown in Fig. 8 (left): note that an exponential distribution

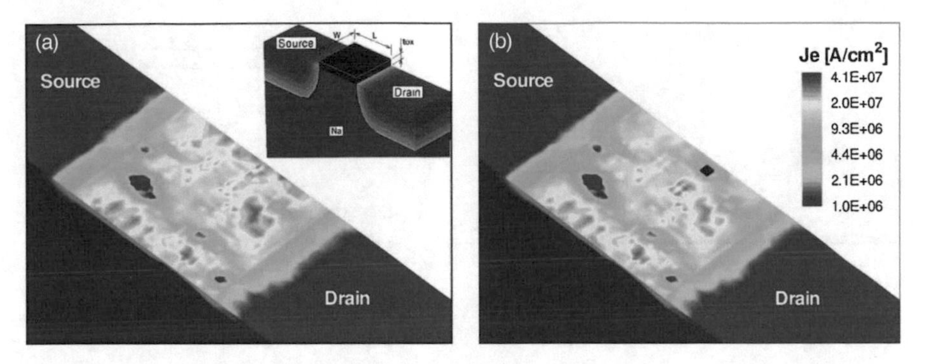

Fig. 7 (**a**) Current density profiles in the channel of a decananometer Flash memory cell (shown in the inset) without any charged trap, © 2009 IEEE. (**b**) Same but with a charged RTN trap, whose position is marked by the black square, © 2009 IEEE

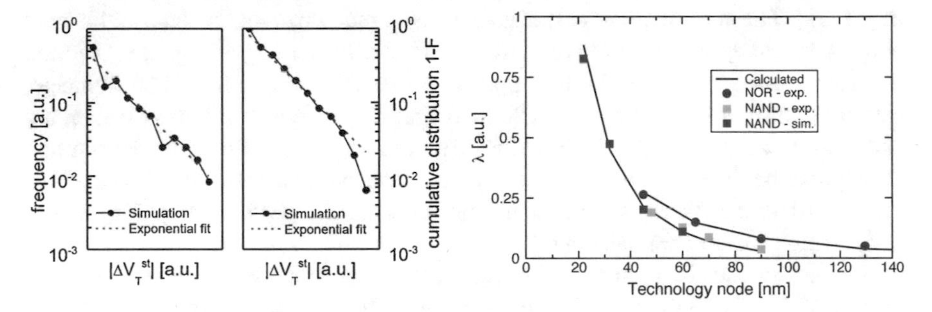

Fig. 8 Left: Probability density function and complementary of the cumulative distribution for the V_T shift determined by the charging of a single interfacial trap in a decananometer NAND Flash cell [63], © 2008, IEEE. Right = Comparison between (2) and experimental data over different technologies, © 2009, IEEE

was obtained, which can be easily characterized by its slope λ, usually expressed in mV/decade. A thorough parametric analysis of λ was carried out in [64, 70], resulting in the following dependence:

$$\lambda = \frac{K}{\alpha_G} \frac{t_{ox}^{\alpha} \sqrt{N_A}}{W\sqrt{L}}, \tag{2}$$

where t_{ox} is the tunnel oxide thickness, N_A the average substrate doping, α_G the control-gate coupling coefficient, and K is a constant dependent on device design, doping profile, and other device parameters. This relation allows to explain experimental evidence such as the increase in the RTN tails with doping [71] and has been very successful in matching the experimental results for the RTN slope over several technologies, as demonstrated by the comparison in Fig. 8 (right) [64, 70].

When these results are to be compared against experimental data, where array cells V_T are measured at two different times and then subtracted to obtain the RTN

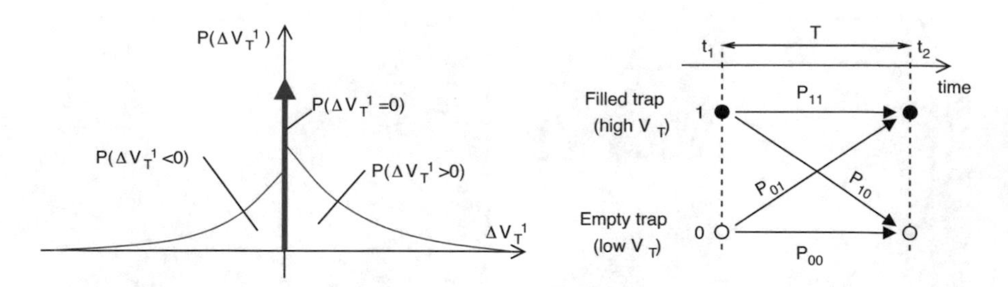

Fig. 9 Left = pdf of the single-trap threshold voltage shift, $P(\Delta V_T^1)$. Right = Markov process view of the single-trap RTN waveform between times t_1 and $t_2 = t_1 + T$

amplitude, such a static, single-trap analysis must be complemented by two more factors: the effect of multiple traps and the effect of time, that were first discussed in [17, 39]. The starting point of the analysis is to recognize that when a cell is read at times t_1 and t_2, we can observe a positive V_T shift if an empty trap at t_1 is found filled at t_2 and a negative shift if the opposite is true. If the trap state does not change, a zero V_T shift holds (see Fig. 2). The final time-dependent distribution is then the one shown in Fig. 9 (left), where the positive and negative parts of the distribution are exponential in shape, with a slope obtained from simulation results. The areas of such distributions represent the probabilities to observe a positive, negative, or zero ΔV_T^1, which will be now calculated.

To this aim, the single-trap RTN process can be described by a two-state Markov process [72, 73], as depicted in Fig. 9 (right). The two states correspond to the empty (low-V_T, state 0) and filled (high-V_T, state 1) trap case. Considering constant capture and emission average times τ_c and τ_e, the transition probabilities P_{ij} between the initial state "i" and the final state "j" over a time interval $T = t_2 - t_1$ become:

$$P_{11}(\tau_c, \tau_e) = \frac{\tau_c}{\tau_c + \tau_e}\left(\mathrm{e}^{-T/\tau_{\mathrm{eq}}} + \frac{\tau_e}{\tau_c}\right) \tag{3}$$

$$P_{10}(\tau_c, \tau_e) = 1 - P_{11} = \frac{\tau_c}{\tau_c + \tau_e}\left(1 - e^{-T/\tau_{\mathrm{eq}}}\right) \tag{4}$$

$$P_{00}(\tau_c, \tau_e) = \frac{\tau_e}{\tau_c + \tau_e}\left(e^{-T/\tau_{\mathrm{eq}}} + \frac{\tau_c}{\tau_e}\right) \tag{5}$$

$$P_{01}(\tau_c, \tau_e) = 1 - P_{00} = \frac{\tau_e}{\tau_c + \tau_e}\left(1 - e^{-T/\tau_{\mathrm{eq}}}\right), \tag{6}$$

where $\tau_{eq}^{-1} = \tau_c^{-1} + \tau_e^{-1}$. Note that these probabilities depend on the ratio between the elapsed time T and the time constants: only when T becomes comparable with τ_{eq} we begin to observe significant RTN transitions. Besides, it is worth pointing out that these expressions describe the probabilities to find the trap in a certain state after a time interval T and a given initial condition, irrespective of the actual number of transitions taking place during T.

From (3) to (6), the probability for a single trap to provide a V_T shift, ΔV_T^1, can be easily calculated:

$$P\left(\Delta V_T^1 > 0|\tau_c, \tau_e\right) = P_0(t_1)P_{01}(\tau_c, \tau_e) \quad (7)$$

$$P\left(\Delta V_T^1 < 0|\tau_c, \tau_e\right) = P_1(t_1)P_{10}(\tau_c, \tau_e) \quad (8)$$

$$P\left(\Delta V_T^1 = 0|\tau_c, \tau_e\right) = P_0(t_1)P_{00}(\tau_c, \tau_e) + P_1(t_1)P_{11}(\tau_c, \tau_e), \quad (9)$$

where $P_i(t_1)$ is the probability for the trap to be in state $i = 0, 1$ at time t_1.

These results describe the RTN dynamics with time for given values of τ_c, τ_e. If we want a statistical distribution of the RTN, we need first to assess the statistical distribution of τ_c and τ_e associated with traps in the memory cell. Such a task requires, in principle, a detailed model for tunneling in/out of the traps, which should account for structural relaxation effects and the microscopic nature of defects [62, 74–77]. However, such an approach is not strictly necessary for the development of our model: a much simpler solution is to leave this part to investigations into the fundamental nature of RTN defects [78–87] and start right from the time-constant distributions, without explicitly linking them to any particular tunneling model nor to any defect position or energy in the oxide, in view of the fact that many factors such as device variability [88–90] and structural relaxation of defects [62, 91] have questioned such a link, also suggesting that only a very small correlation exists between the two time constants [92]. Such an approach was followed in [45], where a constant pdf on a log-time scale was assumed for both τ_c and τ_e, in analogy with the time constant distribution that was used to account for charge detrapping after cycling [93–96] and neglecting any correlation between the two time constants.

Once the probability for a trap to possess a given set of time constants, $p(\tau_c, \tau_e)$, is known, the final expression for the single-trap V_T shift pdf, $P(\Delta V_T^1)$, becomes

$$P\left(\Delta V_T^1\right) = \int \int P\left(\Delta V_T^1|\tau_c, \tau_e\right) p(\tau_c, \tau_e) d\tau_c d\tau_e, \quad (10)$$

which only leaves us with the final task of accounting for multiple traps. If we assume statistical independence of their V_T shifts [64], the total ΔV_T is obtained by simply adding the individual contributions, i.e., by performing a convolution operation on the respective pdfs. For N_t independent traps, each with the same pdf $P(\Delta V_T^1)$, we then have

$$P\left(\Delta V_T^{N_t}\right) = \bigotimes_1^{N_t} P(\Delta V_T^1). \quad (11)$$

Finally, since the number of traps per cell can be assumed to be statistically distributed according to Poisson statistics $p(N_t)$, we get the RTN pdf as:

$$P(\Delta V_T) = \sum_{N_t=1}^{\infty} P\left(\Delta V_T^{N_t}\right) p(N_t). \tag{12}$$

To allow a thorough comparison against experimental data, this distribution is convoluted with the Gaussian noise contribution apparent in Fig. 3.

Results of a comparison between model predictions and experimental data are reported in Fig. 10 for NAND [45] and NOR [39] arrays. The left-hand side shows that a good agreement can be achieved for the entire RTN distribution in different technologies and cycling conditions, using as only unknown parameter the average number of traps N_t, as the exponential distribution of the single-trap RTN amplitude is obtained from simulation results. In particular, because of the exponential character of $P(\Delta V_T^1)$, it is worth noting that its slope λ reproduces the slope obtained from array data, as the convolution operations in (11)–(12) do not usually affect such a shape. The right-hand side is instead related to data reported in the inset of Fig. 4 (left), where RTN distributions for increasing read operations (i.e., time) are reported, and shows the increase in the RTN tails with time, i.e., the increase in the probabilities to find a fixed $|\Delta V_T|$ values as a function of time (i.e., the number of read operations). Again, the increase in the RTN tails is well reproduced by the model and can now be explained with reference to (3)–(6): As time T between the read operations increases, more and more traps will start to show RTN fluctuations, as their τ_{eq} become comparable with T. This increases the randomness of V_T and the magnitude of the RTN tails.

Finally, the temperature-independent RTN behavior observed in [45] can be explained within the frame of the adopted distribution of time constants: if we consider for simplicity a unique activation energy of time constants, it is easy to

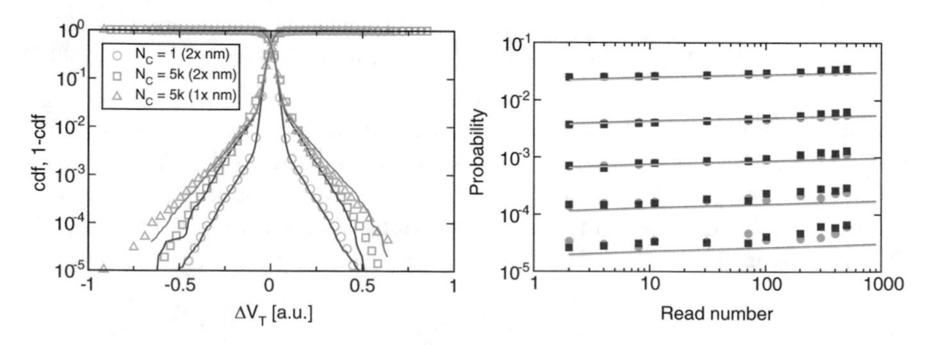

Fig. 10 Left: Comparison between experimental data and model for the RTN distribution in 1x and 2x NAND arrays with different N_C (symbols = experimental data, lines = model) [45]. Right: Similar comparison for the tail probability at a fixed $|\Delta V_T|$ as a function of the number of reads for a 65 nm NOR Flash array (same data as in the inset of Fig. 4, left) [39]

see that $p(\tau_c, \tau_e)$ is not affected by temperature, as the number of traps active in a certain stretch of time T remains the same: some of them are accelerated and will become too fast to be detected, but others that were too slow to be detected will now contribute to RTN. Note that this conclusion remains true even if the activation energy is not constant, provided it is not correlated with the values of the time constants.

4 Effect on Programmed V_T Distribution

The trend toward ever-increasing storage capacity has long pushed Flash memory to store more than one bit per cell. Today, all major NAND manufacturers place three bits per cell [97–99] while four-bit-per-cell technology is being pushed to market [100]. Under these conditions, the separation between adjacent V_T levels becomes increasingly small and any fluctuation in V_T may compromise the read operation.

To investigate the role played by RTN in this context, we must first recall that Flash memories are programmed using the *incremental step-pulse programming* (ISPP) algorithm [101, 102], in which fast pulses whose amplitude is increased every time by a fixed quantity V_{step} are used to inject a controlled amount of charge into the storage layer, resulting in a V_T shift per pulse almost equal to the incremental amplitude V_{step}. After each pulse, a *program verify* (PV) phase follows, in which V_T is monitored and the program operation on a cell is stopped as soon as it exceeds the verify V_T level, V_{PV}.

Figure 11 (left) shows the programmed V_T cdf after the PV operation and after the first read operation performed on the array. Note that no cell exists below the PV level right after program, as a consequence of the ISPP algorithm. However, cells V_T fluctuate because of RTN, meaning that their value can increase or decrease according to the exponential distribution already discussed. Any negative V_T shift

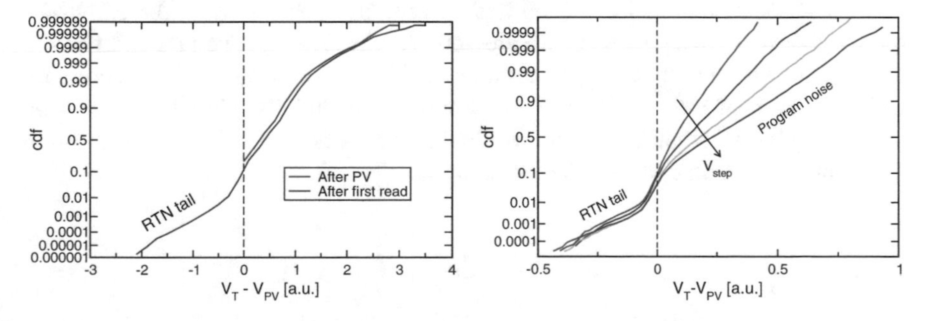

Fig. 11 Left: V_T cdf in a Flash array at the end of the ISPP operation and after a successive read operation. Right: Dependence of the cdf after read on the step voltage used in the program operation [103, 106], © 2015, IEEE

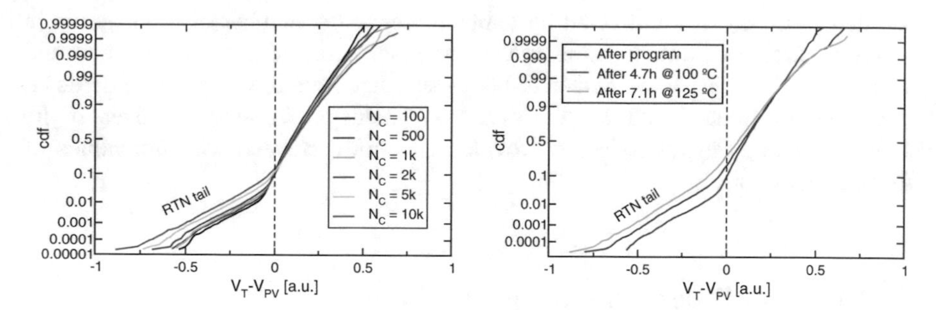

Fig. 12 Left: V_T cdf in a 1x-nm NAND array after a read operation following program. Results are shown for different number of cycles, © 2015, IEEE. Right: same but for different bake times after cycling [106]. © 2015, IEEE

results in the emergence of a tail in the distribution, due to cells moving below the verify level [103]. Note also that the high-V_T side of the distribution does not appear to be affected by RTN: this is because other effects—namely, program noise [104, 105], that will not be discussed here—come into play in widening the distribution toward positive values, thus masking the appearance of the positive RTN tail.

The effect of different values of V_{step} is reported in Fig. 11 (right): increasing its value does not affect the extent of the RTN tail, while the program noise increases with the square root of V_{step} [106]. In proportion, then, RTN affects more the width of the programmed V_T distribution when small values of V_{step} are used, and constitutes an ultimate limitation to the width of the programmed distribution when more complex programming algorithms [107] that reduce the program noise are used.

Figure 12 shows the impact of RTN on the programmed V_T distribution after cycling and bake. The figure on the left shows the V_T distribution following the first read operation after program when the array is subjected to an increasing number of cycles. Note that more and more cells fall below the PV level as cycling increases, as a consequence of the additional defectivity generated by cycling that results in a larger RTN, as reported in Figs. 5 and 6. Moreover, the slope of the RTN tails remains constant, in agreement with previous results. The figure on the right shows the effect of a bake time on the programmed distribution that was read after program and after each of the two bake phases applied sequentially. Note once again an increase in the RTN tail amplitude with bake, that is however due to the retention time, being not related to the bake temperature.

5 3D Cells

Silicon nanowire structures have long been recognized as the "ultimate" silicon transistor, because of the better electrostatic control exerted by the gate with respect to conventional planar devices. In the Flash memory field, such structures provide

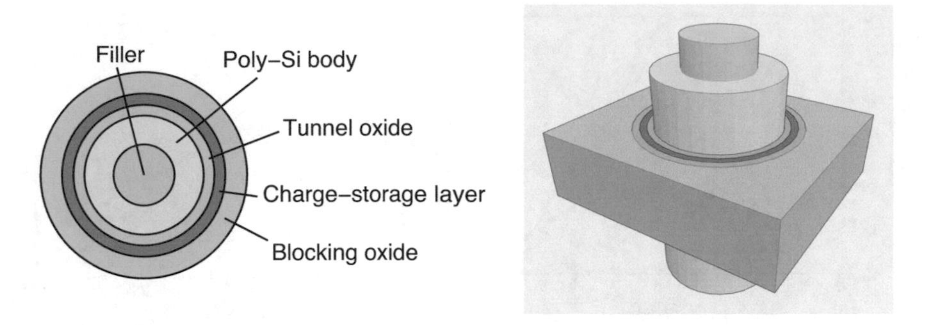

Fig. 13 Cross-section (left) and pictorial view (right) of the most common memory cell used in 3D NAND arrays

an important additional advantage, i.e., the possibility to unfold the NAND strings along the vertical direction, leading to true 3D arrays. This possibility, along with its promise of a huge increase in storage capacity, has pushed the manufacturers from early proposals [108–111] to the first cost-effective solution [112], leading to commercial products [113, 114] culminating in today's leading-edge technology exploiting 96-layer chips [99] with four-bit-per-cell storage [100].

One of the main cell geometries employed in 3D NAND Flash memories consists of a vertical-channel transistor with polysilicon body, usually shaped as a hollow cylinder filled with dielectric material [115], whose cross-section and view are depicted in Fig. 13. The advantages of such a structure with respect to the conventional nanowire lie in the thinning of the silicon body and in the removal of the central silicon region, which is affected by a large defectivity that degrades the subthreshold slope and the array performance [115]. The gate stack can be either a floating gate [97, 116–120], similar to planar NAND devices, or a charge-trap stack similar to an oxide/nitride/oxide (ONO) layer [100, 114, 121–123]. Moreover, it is worth pointing out that 3D memory cells have a larger footprint than 2D ones, allowing to relieve many reliability issues without compromising the storage density.

5.1 Experimental

Single-trap characterization [124–126] as well as statistical data for RTN [127–129] has been presented in the literature and is reported in Fig. 14 (left) for different temperatures from −10 to 125 °C [130, 131]. Note that the RTN distribution retains an exponential behavior, whose slope is nevertheless greatly reduced with respect to planar technologies [130, 132, 133]. This is due not only to the larger cell size of 3D arrays, but also to the different doping of the channel and nature of the electron percolative conduction in the subthreshold region, shifting from a 2D conduction in planar devices to a bulk one in 3D cells [134].

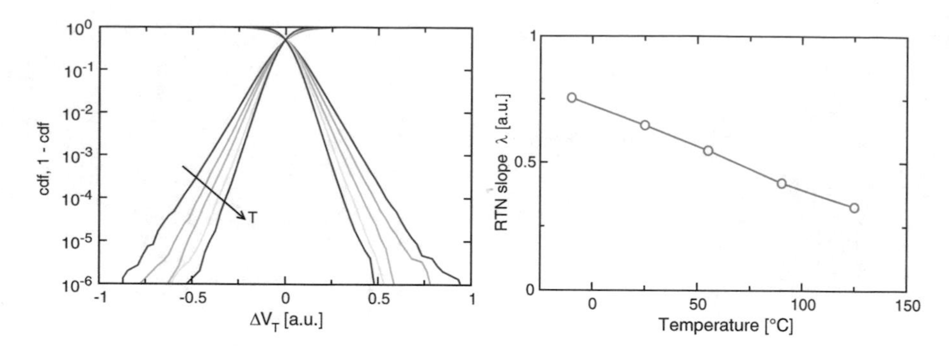

Fig. 14 Left: RTN cdf for a 3D NAND for different temperatures [130, 131], © 2018, IEEE. Right: RTN slope λ as a function of temperature

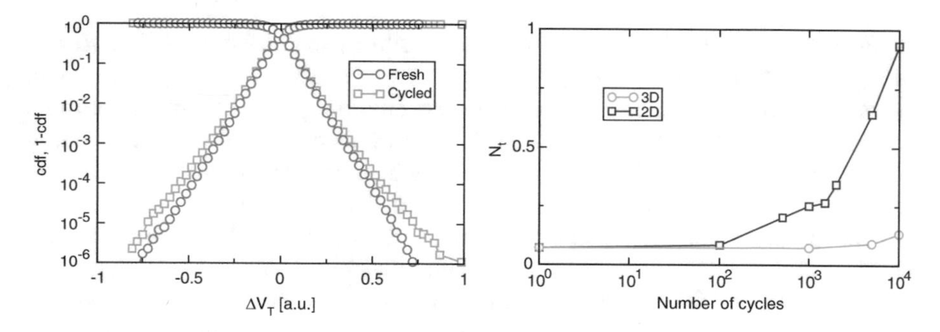

Fig. 15 Left: RTN cdf for a 3D NAND before and after cycling to 10k cycles [131], © 2018, IEEE. Right: Average number of RTN traps as a function of cycling, extracted by fitting the distributions with the model presented in Sect. 3 [131]

Figure 14 also shows another difference with respect to planar devices, i.e., the temperature dependence: while λ is basically independent of temperature in 2D cells (see Fig. 6), 3D cells show an improved RTN figure as temperature is increased, as reported in Fig. 14, right. Similar dependences were also reported in [135, 136]. This finding is in stark contrast with the common wisdom suggesting that high-temperature operation is the worst from a reliability standpoint, and must be given careful consideration. It is also worth mentioning that RTN in 3D NAND does not feature an exponential tail (due to cells actually showing RTN) departing from a central Gaussian distribution (due to cells not showing RTN in the observation time) as in 2D arrays, but rather exhibits a unique exponential distribution. This suggests that the large majority of cells in the 3D array are affected by RTN.

The effect of cycling on RTN in 3D arrays has been investigated in [131] and is reported in Fig. 15. The RTN ΔV_T cdfs shown on the left for a fresh and a heavy-cycled ($N_C = 10^4$) array show only a minimal increase in the magnitude of the phenomenon, different once again from what reported in planar devices (see Figs. 5 and 6). This is confirmed by the figure on the right, depicting the average number of traps $\langle N_t \rangle$ extracted from fitting the RTN distributions with the model reported in Sect. 3. Note that a constant (native) density of RTN traps is extracted in 3D Flash cells, up to a very high cycling dose, meaning that cycling-induced defects do not significantly affect practical array operation. This conclusion is also supported by Fig. 16 (left), where the RTN slope shown in Fig. 14 (right) is also reported after cycling, featuring no difference over the entire range investigated. The impact of RTN on the V_T distribution width is shown on the right of Fig. 16 [133], making clear that a significant improvement is achieved with 3D cells.

From the viewpoint of the physical interpretation, discussed in detail in the next section, it is worth mentioning that the polycrystalline channel of 3D NAND cells is composed of many single-crystal regions called grains, having different orientations and being separated by grain boundaries. Such regions are highly defective [137] and behave as very effective generation–recombination centers, controlling low-frequency noise [138, 139] in resistors or the charging/discharging of bitline current in NANDs [140, 141]. The dependence of RTN on the grain size is reported in Fig. 17 [135, 142], where an improvement is observed for larger cells (which can be explained by simple electrostatic considerations) and smaller average grain size. Other works also reported poor reliability and higher variability in 3D NAND for

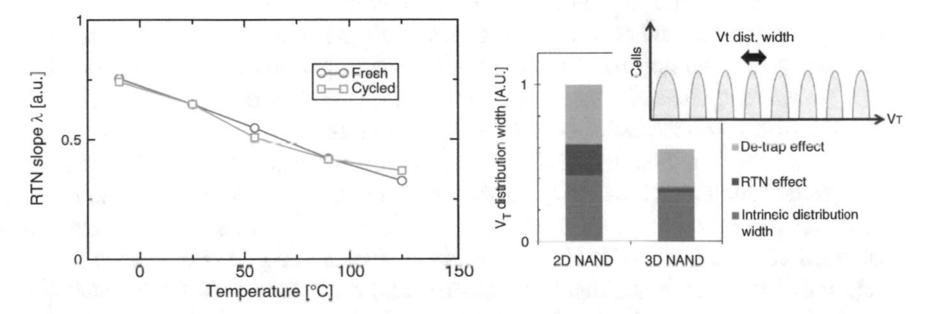

Fig. 16 Left: RTN slope as a function of temperature for fresh and cycled arrays [131]. Right: Contributions to the V_T distribution width in planar and 3D NAND arrays [133], © 2015, IEEE

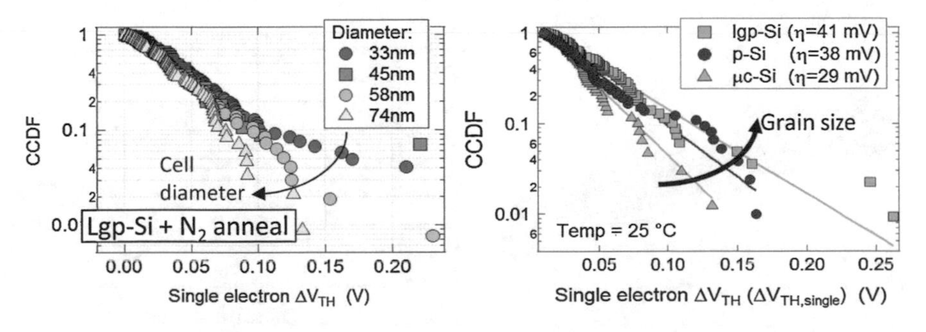

Fig. 17 RTN cdf for vertical poly-Si channel transistors having different channel diameter (left) and poly-Si grain size (right) [142], © 2012, IEEE

large poly-Si grain size [124, 143]. This is ascribed to the larger number of grains associated with a smaller size, resulting in a self-averaging effect, as discussed in [144] for the case of a polysilicon gate in a planar transistor.

5.2 Models

A careful analysis of RTN in 3D NAND devices should not overlook their detailed conduction mechanism, for 2D devices have neatly demonstrated a strong connection between microscopic non-uniform transport along the channel and RTN amplitude distribution. In particular, because of the current fabrication process, the vertical string that make up the channel of 3D NAND Flash cells is built of polycrystalline silicon. This material has long been studied and several models for current conduction have been proposed, based on either a drift-diffusion [145–147] or a thermionic [137, 148, 149] approach. Numerical models of conduction have been used to investigate the effect of grain boundaries on variability in nanowires [150–155] and 3D NAND devices [156, 157]. However, several important features of such models still have to be assessed, such as the grain size [158, 159], the density and energy distribution of grain boundary traps [160, 161], and the mobility degradation and conduction process at the grain boundaries [162, 163].

Recently, detailed numerical models for the investigation of electron conduction in 3D polysilicon channels have been developed, based on either a drift-diffusion approach with a reduced mobility at the grain boundaries [155] or a physically based transport model including drift-diffusion within the grains and thermionic emission at the grain boundaries [164–167]. In the latter model, a random configuration of grains was generated for each structure (see Fig. 18, left) and double-exponential profiles of acceptor and donor states were introduced at the grain boundaries [168–171], with values calibrated against experimental data. Figure 18 (right) shows a typical conduction-band profile along the channel of a 3D NAND string for increasing values of the control-gate bias. Note that the profile is not smooth, featuring

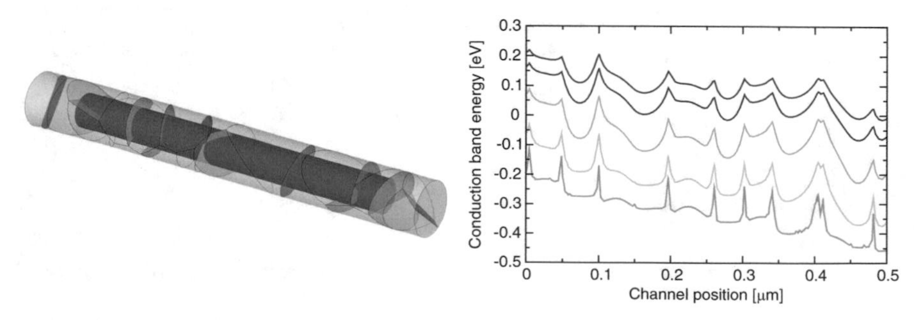

Fig. 18 Left: Pictorial view of the simulated granular structure of the polysilicon channel in a 3D NAND string (polysilicon and filler oxide are only shown). Right: Conduction-band profile along the channel of a 3D NAND string for different values of the control-gate bias [166]

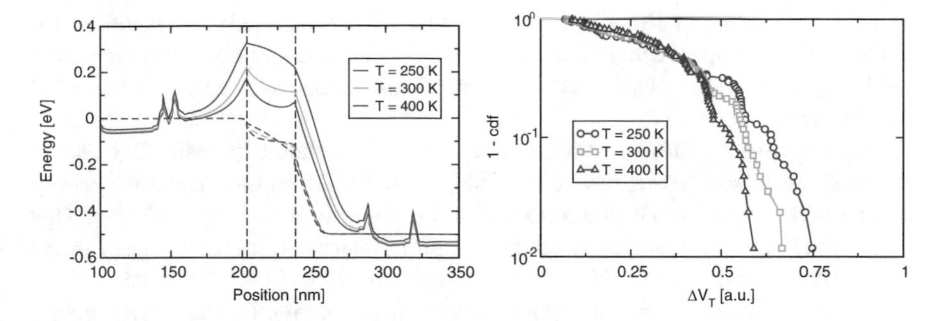

Fig. 19 Left: Conduction-band (solid lines) and quasi-Fermi energy (dashes) along the position of a NAND cell for different temperatures [167], © 2018, IEEE. Right: Simulation results for RTN in 3D polysilicon cells for different temperatures [167], © 2018, IEEE

peaks due to the charging of the highly defective interfaces in correspondence of the grain boundaries. As bias is increased, the band bending lowers the conduction-band profile, increasing the localized trap occupation and sharpening the peaks, which become the true bottlenecks of conduction [166].

The polysilicon behavior can also be investigated as a function of temperature [167]: Fig. 19 (left) shows typical conduction-band and quasi-Fermi energy profiles along a cell in a NAND string, at the same conduction current but different temperatures. The positions of two randomly generated grain boundaries are also highlighted by dashed vertical lines. Note that when temperature is reduced, the wordline bias must be increased to retain the same conduction current, resulting in an increased grain boundary trap occupation and in the emergence or sharpening of the respective peaks in the conduction band. The role played by the grain boundaries in limiting the current conduction is also demonstrated by the quasi-Fermi energy profiles (dashed curves in Fig. 19, left): as T is decreased, the gradient of the quasi-Fermi levels becomes more and more localized at the grain boundary positions.

Given the above results, it is not surprising that the critical trap positions from the viewpoint of RTN are those placed exactly at the grain boundaries, rather than at one of the silicon/oxide interfaces [172]. The simulation tool just described was then employed to model the RTN V_T shift resulting from such charged traps. Typical results as a function of temperature are shown in Fig. 19 (right): the model can predict the correct temperature behavior, supporting the correctness of the physical picture.

6 Conclusions

RTN represents one of the major issues in limiting cell size scaling in planar Flash technologies, while being a powerful scientific tool for probing the microscopic properties of silicon/oxide traps. For these reasons, it has been widely studied and understood, leading Flash memory manufacturers to develop effective solutions for dealing with it. Experimental data as well as physical interpretation and numerical modeling of RTN that paved the way for such improvements have been reviewed in this chapter.

Notwithstanding those advancements, however, it should be said that one of the most successful moves toward a curbing of the RTN impact on Flash memory operation has been the recent adoption of 3D cells, with their larger cell size. This new technology has nonetheless brought to prominence novel issues related to the polycrystalline nature of the NAND string, which have been also reviewed.

The capability of the scientific and technological communities to work together and put efforts spanning from the understanding of the key physical properties of RTN to the development of successful product solutions remains the utmost factor for the continuous development of the NAND Flash memory technology.

Acknowledgements The authors would like to acknowledge A. Visconti, P. Tessariol, E. Camerlenghi, and A. Goda from Micron Technology Inc. for the fruitful and long-lasting collaboration. Support from our former Ph.D. students C. Miccoli, G. Paolucci, D. Resnati and G. Nicosia is also gratefully acknowledged.

References

1. K.S. Ralls, W.J. Skocpol, L.D. Jackel, R.E. Howard, L.A. Fetter, R.W. Epworth, D.M. Tennant, Phys. Rev. Lett. **52**, 228 (1984). https://doi.org/10.1103/PhysRevLett.52.228
2. M.J. Kirton, M.J. Uren, Adv. Phys. **38**, 367 (1989). https://doi.org/10.1080/00018738900101122
3. K.K. Hung, P.K. Ko, C. Hu, Y.C. Cheng, in *Technical Digest, International Electron Devices Meeting* (1988), pp. 34–37. https://doi.org/10.1109/IEDM.1988.32743
4. K.K. Hung, P.K. Ko, C. Hu, Y.C. Cheng, IEEE Electron Device Lett. **11**, 90 (1990). https://doi.org/10.1109/55.46938
5. A. Ohata, A. Toriumi, M. Iwase, K. Natori, J. Appl. Phys. **68**, 200 (1990). https://doi.org/10.1063/1.347116
6. Z.M. Shi, J.P. Mieville, J. Barrier, M. Dutoit, in *International Electron Devices Meeting 1991 [Technical Digest]* (1991), pp. 363–366. https://doi.org/10.1109/IEDM.1991.235378
7. P. Fang, K.K. Hung, P.K. Ko, C. Hu, IEEE Electron Device Lett. **12**, 273 (1991). https://doi.org/10.1109/55.82058
8. E. Simoen, B. Diericks, C.L. Claeys, G.J. Declerck, IEEE Trans. Electron Devices **39**, 422 (1992). https://doi.org/10.1109/16.121702
9. M.H. Tsai, T.P. Ma, T.B. Hook, IEEE Electron Device Lett. **15**, 504 (1994)
10. M.J. Uren, M.J. Kirton, S. Collins, Phys. Rev. B **37**, 8346 (1988). https://doi.org/10.1103/PhysRevB.37.8346
11. M.J. Kirton, M.J. Uren, S. Collins, M. Schultz, A. Karmann, K. Scheffer, Semicond. Sci. Technol. **4**, 1116 (1989)
12. H. Nakamura, N. Yasuda, K. Taniguchi, C. Hamaguchi, A. Toriumi, Jpn. J. Appl. Phys. **28**, L2057 (1989)
13. H.H. Mueller, M. Schulz, J. Appl. Phys. **79**, 4178 (1996). https://doi.org/10.1063/1.361785
14. H.H. Mueller, M. Schulz, in *Proceedings of 14th International Conference Noise in Physical Systems and 1/F Fluctuations* (1997), pp. 195–200

15. H. Kurata, K. Otsuga, A. Kotabe, S. Kajiyama, T. Osabe, Y. Sasago, S. Narumi, K. Tokami, S. Kamohara, O. Tsuchiya, in *Symposium on VLSI Circuits Digest of Technical Papers* (2006), pp. 112–113. https://doi.org/10.1109/VLSIC.2006.1705335
16. N. Tega, H. Miki, T. Osabe, A. Kotabe, K. Otsuga, H. Kurata, S. Kamohara, K. Tokami, Y. Ikeda, R. Yamada, in *International Electron Devices Meeting Technical Digest* (2006), pp. 491–494. https://doi.org/10.1109/IEDM.2006.346821
17. R. Gusmeroli, C. Monzio Compagnoni, A. Riva, A.S. Spinelli, A.L. Lacaita, M. Bonanomi, A. Visconti, in *International Electron Devices Meeting Technical Digest* (2006), pp. 483–486. https://doi.org/10.1109/IEDM.2006.346819
18. H. Kurata, K. Otsuga, A. Kotabe, S. Kajiyama, T. Osabe, Y. Sasago, S. Narumi, K. Tokami, S. Kamohara, O. Tsuchiya, IEEE J. Solid State Circuits **42**, 1362 (2007). https://doi.org/10.1109/JSSC.2007.897158
19. K. Fukuda, Y. Shimizu, K. Amemiya, M. Kamoshida, C. Hu, in *International Electron Devices Meeting Technical Digest* (2007), pp. 169–172. https://doi.org/10.1109/IEDM.2007.4418893
20. P. Fantini, A. Ghetti, A. Marinoni, G. Ghidini, A. Visconti, A. Marmiroli, IEEE Electron Device Lett. **28**, 1114 (2007). https://doi.org/10.1109/LED.2007.909835
21. H. Miki, T. Osabe, N. Tega, A. Kotabe, H. Kurata, K. Tokami, Y. Ikeda, S. Kamohara, R. Yamada, in *International Reliability Physics Symposium Proceedings* (2007), pp. 29–35. https://doi.org/10.1109/RELPHY.2007.369864
22. N. Tega, H. Miki, M. Yamaoka, H. Kume, T. Mine, T. Ishida, Y. Mori, R. Yamada, K. Torii, in *Proceedings of International Reliability Physics Symposium* (2008), pp. 541–546. https://doi.org/10.1109/RELPHY.2008.4558943
23. S.O. Toh, Y. Tsukamoto, Z. Guo, L. Jones, T.J. King Liu, B. Nikolic, in *International Electron Devices Meeting Technical Digest* (2009), pp. 767–770. https://doi.org/10.1109/TED.2009.5424228
24. M. Tanizawa, S. Ohbayashi, T. Okagaki, K. Sonoda, K. Eikyu, Y. Hirano, K. Ishikawa, O. Tsuchiya, Y. Inoue, in *Symposium on VLSI Circuits Digest of Technical Papers* (2010), pp. 95–96. https://doi.org/10.1109/VLSIT.2010.5556184
25. H.S. Wong, Y. Taur, in *International Electron Devices Meeting Technical Digest* (1993), pp. 705–708. https://doi.org/10.1109/IEDM.1993.347215
26. A. Asenov, IEEE Trans. Electron Devices **45**, 2505 (1998). 10.1109/16.735728
27. A. Asenov, A.R. Brown, J.H. Davies, S. Saini, IEEE Trans. Comput. Aided Des. **18**, 1558 (1999). https://doi.org/10.1109/43.806802
28. A. Asenov, S. Saini, IEEE Trans. Electron Devices **46**, 1718 (1999). https://doi.org/10.1109/16.777162
29. L.K.J. Vandamme, D. Sodini, Z. Gingl, Solid State Electron. **42**, 901 (1998). https://doi.org/10.1016/S0038-1101(98)00105-1
30. A. Asenov, R. Balasubramaniam, A.R. Brown, J.H. Davies, S. Saini, in *International Electron Devices Meeting 2000. Technical Digest* (2000), pp. 279–282. https://doi.org/10.1109/IEDM.2000.904311
31. A. Asenov, R. Balasubramaniam, A.R. Brown, J.H. Davies, IEEE Trans. Electron Devices **50**, 839 (2003). https://doi.org/10.1109/TED.2003.811418
32. A. Asenov, A.R. Brown, J.H. Davies, S. Kaya, G. Slavcheva, IEEE Trans. Electron Devices **50**, 1837 (2003). https://doi.org/10.1109/TED.2003.815862
33. C. Zambelli, P. Olivo, in *Inside Solid-State Drives (SSD)s*, ed. by R. Micheloni, A. Marelli, K. Eshghi (Springer, Berlin, 2013), chap. 8, pp. 203–231. https://doi.org/10.1007/978-94-007-5146-0_8
34. R. Shirota, in *Advances in Non-volatile Memory and Storage Technology*, ed. by Y. Nishi (Elsevier, Amsterdam 2014), chap. 2, pp. 27–74. https://doi.org/10.1533/9780857098092.1.27
35. S. Aritome, *NAND Flash Memory Technologies* (John Wiley & Sons, Hoboken, 2015). https://doi.org/10.1002/9781119132639

36. R. Micheloni (ed.), *3D Flash memories* (Springer, Berlin, 2016). https://doi.org/10.1007/978-94-017-7512-0
37. C. Monzio Compagnoni, A. Goda, A.S. Spinelli, P. Feeley, A.L. Lacaita, A. Visconti, Proc. IEEE **105**, 1609 (2017). https://doi.org/10.1109/JPROC.2017.2665781
38. A.S. Spinelli, C. Monzio Compagnoni, A.L. Lacaita, Computers **6**, 16 (2017). https://doi.org/10.3390/computers6020016
39. C. Monzio Compagnoni, R. Gusmeroli, A.S. Spinelli, A.L. Lacaita, M. Bonanomi, A. Visconti, IEEE Trans. Electron Devices **55**, 388 (2008). https://doi.org/10.1109/TED.2007.910605
40. C. Monzio Compagnoni, A.S. Spinelli, S. Beltrami, M. Bonanomi, A. Visconti, IEEE Electron Device Lett. **29**, 941 (2008). https://doi.org/10.1109/LED.2008.2000964
41. C. Monzio Compagnoni, R. Gusmeroli, A.S. Spinelli, A.L. Lacaita, M. Bonanomi, A. Visconti, in *Proceedings of International Reliability Physics Symposium* (2007), pp. 161–166. https://doi.org/10.1109/IRPS.2007.RELPHY.2007.369886
42. A.S. Spinelli, C. Monzio Compagnoni, R. Gusmeroli, M. Ghidotti, A. Visconti, Jpn. J. Appl. Phys. **47**, 2598 (2008). https://doi.org/10.1143/JJAP.47.2598
43. K. Seidel, R. Hoffmann, D.A. Löhr, T. Melde, M. Czernohorsky, J. Paul, M.F. Beug, V. Beyer, in *Proceedings of Non-Volatile Memory Technology Symposium (NVMTS)* (2009), pp. 67–71. https://doi.org/10.1109/NVMT.2009.5429788
44. B.-S. Jo, H.J. Kang, S.-M-Joe, M.K. Jeong, K.R. Han, S.K. Park, B.G. Park, J.H. Lee, Jpn. J. Appl. Phys. **52**, 04CA07 (2013). https://doi.org/10.7567/JJAP/52.04CA07
45. C. Miccoli, G.M. Paolucci, C. Monzio Compagnoni, A.S. Spinelli, A. Goda, in *Proceedings of International Reliability Physics Symposium* (2015), pp. MY.9.1–MY.9.6. https://doi.org/10.1109/IRPS.2015.7112812
46. W. Kwon, Y.H. Song, Y. Cai, W. Ryu, Y. Jang, S. Shin, J. Jun, S.A. Lee, C.K. Park, W.S. Lee, in *Proceedings of Non-Volatile Semiconductor Memory Workshop* (2008), pp. 20–21. https://doi.org/10.1109/NVSMW.2008.11
47. R.V. Wang, Y.H. Lee, Y.L.R. Lu, W. McMahon, S. Hu, A. Ghetti, IEEE Trans. Electron Devices **56**, 2107 (2009). https://doi.org/10.1109/TED.2009.2026116
48. H. An, K. Kim, S. Jung, H. Yang, Y. Song, Jpn. J. Appl. Phys. **49**, 114302 (2010). https://doi.org/10.1143/JJAP.49.114302
49. C.H. Lee, I.C. Yang, C. Lee, C.H. Cheng, L.H. Chong, K.F. Chen, J.S. Huang, S.H. Ku, N.K. Zous, I.J. Huang, T.T. Han, M.S. Chen, W.P. Lu, K.C. Chen, T. Wang, C.Y. Lu, in *Proceedings of International Reliability Physics Symposium* (2011), pp. 6B.3.1–6B.3.5. https://doi.org/10.1109/IRPS.2011.5784549
50. T. Kim, D. He, K. Morinville, K. Sarpatwari, B. Millemon, A. Goda, J. Kessenich, IEEE Electron Dev. Lett. **32**, 999 (2011). https://doi.org/10.1109/LED.2011.2152362
51. T. Kim, N. Franklin, C. Srinivasan, P. Kalavade, A. Goda, IEEE Electron Dev. Lett. **32**, 1185 (2011). https://doi.org/10.1109/LED.2011.2159573
52. A. Ghetti, S.M. Amoroso, A. Mauri, C. Monzio Compagnoni, IEEE Trans. Electron Devices **59**, 309 (2012). https://doi.org/10.1109/TED.2011.2175399
53. S.M. Amoroso, A. Ghetti, A.R. Brown, A. Mauri, C. Monzio Compagnoni, A. Asenov, IEEE Trans. Electron Devices **59**, 2774 (2012). https://doi.org/10.1109/TED.2012.2208224
54. F.H. Li, R. Shirota, Jpn. J. Appl. Phys. **52**, 074201 (2013). https://doi.org/10.7567/JJAP.52.074201
55. S. Kim, M. Lee, G.B. Choi, J. Lee, Y. Lee, M. Cho, K.O. Ahn, J. Kim, in *Proceedings of International Reliability Physics Symposium* (2015), pp. MY.8.1–MY.8.4
56. T. Tomita, K. Miyaji, Jpn. J. Appl. Phys. **54**, 04DD02 (2015). https://doi.org/10.7567/JJAP.54.04DD02
57. S.M. Joe, J.H. Yi, S.K. Park, H.I. Kwon, J.H. Lee, IEEE Electron Device Lett. **31**, 635 (2010). https://doi.org/10.1109/LE.2010.2047235
58. S.M. Joe, J.H. Yi, S.K. Park, H. Shin, B.G. Park, Y.J. Park, J.H. Lee, IEEE Trans. Electron Devices **58**, 67 (2011). https://doi.org/10.1109/TED.2010.2088126

59. S.M. Joe, M.K. Jeong, B.S. Jo, K.R. Han, S.K. Park, J.H. Lee, IEEE Trans. Electron Devices **59**, 3568 (2012). https://doi.org/10.1109/TED.2012.2219866
60. S.M. Joe, J.H. Bae, C.H. Park, J.H. Lee, Semicond. Sci. Technol. **29**, 125013 (2014). https://doi.org/10.1088/0268-1242/29/12/125013
61. H. Miki, N. Tega, M. Yamaoka, D.J. Frank, A. Bansal, M. Kobayashi, K. Cheng, C.P. D'Emic, Z. Ren, S. Wu, J.B. Yau, Y. Zhu, M.A. Guillorn, D.G. Park, W. Haensch, E. Leobandung, K. Torii, in *International Electron Devices Meeting Technical Digest* (2012), pp. 450–453. https://doi.org/10.1109/IEDM.2012.6479071
62. T. Grasser, Microelectron. Reliab. **52**, 39 (2012). https://doi.org/10.1016/j.microrel.2011.09.002
63. A. Ghetti, M. Bonanomi, C. Monzio Compagnoni, A.S. Spinelli, A.L. Lacaita, A. Visconti, in *Proceedings of International Reliability Physics Symposium* (2008), pp. 610–615. https://doi.org/10.1109/RELPHY.2008.4558954
64. A. Ghetti, C. Monzio Compagnoni, A.S. Spinelli, A. Visconti, IEEE Trans. Electron Devices **56**, 1746 (2009). https://doi.org/10.1109/TED.2009.2024031
65. C.L. Alexander, A.R. Brown, J.R. Watling, A. Asenov, IEEE Trans. Nanotechnol. **4**, 339 (2005). https://doi.org/10.1109/TNANO.2005.846929
66. C. Alexander, A.R. Brown, J.R. Watling, A. Asenov, Solid State Electron. **49**, 733 (2005). https://doi.org/10.1016/j.sse.2004.10.012
67. U. Kovac, C. Alexander, G. Roy, C. Riddet, B. Cheng, A. Asenov, IEEE Trans. Electron Devices **57**, 2418 (2010). https://doi.org/10.1109/TED.2010.2062517
68. U. Kovac, C. Alexander, A. Asenov, in *Proceedings of International Workshop on Computational Electronics* (2010), pp. 235–238. https://doi.org/10.1109/IWCE.2010.5677971
69. A. Asenov, G. Slavcheva, A.R. Brown, J.H. Davies, S. Saini, IEEE Trans. Electron Devices **48**, 722 (2001). https://doi.org/10.1109/16.915703
70. A. Ghetti, C. Monzio Compagnoni, F. Biancardi, A.L. Lacaita, S. Beltrami, L. Chiavarone, A.S. Spinelli, A. Visconti, in *International Electron Devices Meeting Technical Digest* (2008), pp. 835–838. https://doi.org/10.1109/IEDM.2008.4796827
71. K. Abe, A. Teramoto, S. Watabe, T. Fujisawa, S. Sugawa, Y. Kamata, K. Shibusawa, T. Ohmi, Jpn. J. Appl. Phys. **49**, 04DC07 (2010). https://doi.org/10.1143/JJAP.49.04DC07
72. S. Machlup, J. Appl. Phys. **25**, 341 (1954). https://doi.org/10.1063/1.1721637
73. Y. Yuzhelevski, M. Yuzhelevski, G. Jung, Rev. Sci. Instrum. **71**, 1681 (2000). https://doi.org/10.1063/1.1150519
74. W. Goes, M. Karner, V. Sverdlov, T. Grasser, IEEE Trans. Dev. Mat. Reliab. **8**, 491 (2008). https://doi.org/10.1109/TDMR.2008.2005247
75. D. Veksler, G. Bersuker, H. Park, C. Young, K.Y. Lim, W. Taylor, in *Proceedings of International Integrated Reliability Workshop* (2009), pp. 102–105. https://doi.org/10.1109/IRWS.2009.5383021
76. J. G, D. Popescu, P. Lugli, M.J. Häufel, W. Weinreich, A. Kersch, Phys. Rev. B **85**, 045303–1–045303–8 (2012). https://doi.org/10.1103/PhysRevB.85.045303
77. K.P. McKenna, J. Blumberger, Microelectron. Eng. **147**, 235 (2015). https://doi.org/10.1016/j.mee.2015.04.009
78. J.P. Campbell, J. Qin, K.P. Cheung, L.C. Yu, J.S. Suehle, A. Oates, K. Sheng, in *Proceedings of International Reliability Physics Symposium* (2009), pp. 382–388
79. J.P. Campbell, L.C. Yu, K.P. Cheung, J. Qin, J.S. Suehle, A. Oates, K. Sheng, in *Proceedings of International Conference on IC Design & Technology* (2009), pp. 17–20
80. K. Abe, A. Teramoto, S. Sugawa, T. Ohmi, in *Proceedings of International Reliability Physics Symposium* (2011), pp. 381–386. https://doi.org/10.1109/IRPS.2011.5784503
81. A. Yonezawa, A. Teramoto, T. Obara, R. Kuroda, S. Sugawa, T. Ohmi, in *Proceedings of International Reliability Physics Symposium* (2013), pp. XT.11.1–XT.11.6. https://doi.org/10.1109/IRPS.2013.6532126
82. S. Guo, P. Ren, R. Wang, Z. Yu, M. Luo, X. Zhang, R. Huang, in *Proceedings of International Reliability Physics Symposium* (2014), pp. XT.14.1–XT.14.4. https://doi.org/10.1109/IRPS.2014.6861191

83. T. Grasser, K. Rott, H. Reisinger, M. Waltl, J. Franco, B. Kaczer, in *Proceedings of International Reliability Physics Symposium* (2014), pp. 4A.5.1–4A.5.7. https://doi.org/10.1109/IRPS.2014.6860643
84. J. Chen, Y. Higashi, K. Kato, Y. Mitani, in *Symposium on VLSI Technology (VLSI-Technology): Digest of Technical Papers* (2014), pp. 164–165. https://doi.org/10.1109/VLSIT.2014.6894418
85. M.J. Chen, K.C. Tu, H.H. Wang, C.L. Chen, S.Y. Lai, Y.S. Liu, IEEE Trans. Electron Devices **61**, 2495 (2014). https://doi.org/10.1109/TED.2014.2323259
86. M.J. Chen, K.C. Tu, L.Y. Chuang, H.H. Wang, IEEE Electron Dev. Lett. **36**, 217 (2015). https://doi.org/10.1109/LED.2015.2388787
87. W. Goes, Y. Wimmer, A.M. El-Sayed, G. Rezpa, M. Jech, A.L. Schluger, T. Grasser, Microelectron. Reliab. **87**, 286 (2018). https://doi.org/10.1016/j.microrel.2017.12.021
88. N. Castellani, C. Monzio Compagnoni, A. Mauri, A.S. Spinelli, A.L. Lacaita, IEEE Trans. Electron Devices **59**, 2488 (2012). https://doi.org/10.1109/TED.2012.2202910
89. C. Monzio Compagnoni, N. Castellani, A. Mauri, A.S. Spinelli, A.L. Lacaita, IEEE Trans. Electron Devices **59**, 2495 (2012). https://doi.org/10.1109/TED.2009.2203412
90. F. Adamu-Lema, C. Monzio Compagnoni, S.M. Amoroso, N. Castellani, L. Gerrer, S. Markov, A.S. Spinelli, A.L. Lacaita, A. Asenov, IEEE Trans. Electron Devices **60**, 833 (2013). https://doi.org/10.1109/TED.2012.2230004
91. T. Grasser, B. Kaczer, W. Goes, H. Reisinger, T. Aichinger, P. Hehenberger, P.J. Wagner, F. Schanovsky, J. Franco, M.T. Luque, M. Nelhiebel, IEEE Trans. Electron Devices **58**, 3652 (2011). https://doi.org/10.1109/TED.2011.2164543
92. T. Grasser, W. Goes, Y. Wimmer, F. Schanovsky, G. Rzepa, M. Waltl, K. Rott, H. Reisinger, in *International Electron Devices Meeting Technical Digest* (2014), pp. 530–533. https://doi.org/10.1109/IEDM.2014.7047093
93. G.M. Paolucci, C. Monzio Compagnoni, C. Miccoli, A.S. Spinelli, A.L. Lacaita, A. Visconti, IEEE Trans. Electron Devices **61**, 2802 (2014). https://doi.org/10.1109/TED.2014.2327661
94. G.M. Paolucci, C. Monzio Compagnoni, C. Miccoli, A.S. Spinelli, A.L. Lacaita, A. Visconti, IEEE Trans. Electron Devices **61**, 2811 (2014). https://doi.org/10.1109/TED.2014.2327149
95. D. Resnati, G. Nicosia, G.M. Paolucci, A. Visconti, C. Monzio Compagnoni, IEEE Trans. Electron Devices **63**, 4753 (2016). https://doi.org/10.1109/TED.2016.2617888
96. D. Resnati, G. Nicosia, G.M. Paolucci, A. Visconti, C. Monzio Compagnoni, IEEE Trans. Electron Devices **63**, 4761 (2016). https://doi.org/10.1109/TED.2016.2617890
97. T. Tanaka, M. Helm, T. Vali, R. Ghodsi, K. Kawai, J.K. Park, S. Yamada, F. Pan, Y. Einaga, A. Ghalam, T. Tanzawa, J. Guo, T. Ichikawa, E. Yu, S. Tamada, T. Manabe, J. Kishimoto, Y. Oikawa, Y. Takashima, H. Kuge, M. Morooka, A. Mohammadzadeh, J. Kang, J. Tsai, E. Sirizotti, E. Lee, L. Vu, Y. Liu, H. Choi, K. Cheon, D. Song, D. Shin, J.H. Yun, M. Piccardi, K.F. Chan, Y. Luthra, D. Srinivasan, S. Deshmukh, K. Kavalipurapu, D. Nguyen, G. Gallo, S. Ramprasad, M. Luo, Q. Tang, M. Incarnati, A. Macerola, L. Pilolli, L. De Santis, M. Rossini, V. Moschiano, G. Santin, B. Tronca, H. Lee, V. Patel, T. Pekny, A. Yip, N. Prabhu, P. Sule, T. Bemalkhedkar, K. Upadhyayula, C. Jaramillo, in *International Solid-State Circuits Conference Digest Technical Papers* (2016), pp. 142–143. https://doi.org/10.1109/ISSCC.2016.7417947
98. C. Kim, J.H. Cho, W. Jeong, I.H. Park, H.W. Park, D.H. Kim, D. Kang, S. Lee, J.S. Lee, W. Kim, J. Park, Y.L. Ahn, J. Lee, J.H. Lee, S. Kim, H.J. Yoon, J. Yu, N. Choi, Y. Kwon, N. Kim, H. Jang, J. Park, S. Song, Y. Park, J. Bang, S. Hong, B. Jeong, H.J. Kim, C. Lee, Y.S. Min, I. Lee, I.M. Kim, S.H. Kim, D. Yoon, K.S. Kim, Y. Choi, M. Kim, H. Kim, P. Kwak, J.D. Ihm, D.S. Byeon, J.Y. Lee, K.T. Park, K.H. Kyung, in *International Solid-State Circuits Conference Digest Technical Papers* (2017), pp. 202–204. https://doi.org/10.1109/ISSCC.2017.7870331
99. H. Maejima, K. Kanda, S. Fujimura, T. Takagiwa, S. Ozawa, J. Sato, Y. Shindo, M. Sato, N. Kanagawa, J. Musha, S. Inoue, K. Sakurai, N. Morozumi, R. Fukuda, Y. Shimizu, T. Hashimoto, X. Li, Y. Shimizu, K. Abe, T. Yasufuku, T. Minamoto, H. Yoshihara, T. Yamashita, K. Satou, T. Sugimoto, F. Kono, M. Abe, T. Hashiguchi, M. Kojima, Y. Sue-

matsu, T. Shimizu, A. Imamoto, N. Kobayashi, M. Miakashi, K. Yamaguchi, S. Bushnaq, H. Haibi, M. Ogawa, Y. Ochi, K. Kubota, T. Wakui, D. He, W. Wang, H. Minagawa, T. Nishiuchi, H. Nguyen, K.H. Kim, K. Cheah, Y. Koh, F. Lu, V. Ramachandra, S. Rajendra, S. Choi, K. Payak, N. Raghunathan, S. Georgakis, H. Sugawara, S. Lee, T. Futatsuyama, K. Hosono, N. Shibata, T. Hisada, T. Kaneko, H. Nakamura, in *International Solid-State Circuits Conference Digest Technical Papers* (2018), pp. 336–338. https://doi.org/10.1109/ISSCC.2018.8310321

100. S. Lee, C. Kim, M. Kim, S. m. Joe, J. Jang, S. Kim, K. Lee, J. Kim, J. Park, H.J. Lee, M. Kim, S. Lee, S. Lee, J. Bang, D. Shin, H. Jang, D. Lee, N. Kim, J. Jo, J. Park, S. Park, Y. Rho, Y. Park, H. j. Kim, C.A. Lee, C. Yu, Y. Min, M. Kim, K. Kim, S. Moon, H. Kim, Y. Choi, Y. Ryu, J. Choi, M. Lee, J. Kim, G.S. Choo, J.D. Lim, D.S. Byeon, K. Song, K.T. Park, K. h. Kyung, in *International Solid-State Circuits Conference Digest Technical Papers* (2018), pp. 340–342. https://doi.org/10.1109/ISSCC.2018.8310323
101. G.J. Hemink, T. Tanaka, T. Endoh, S. Aritome, R. Shirota, in *Symposium on VLSI Technology. Digest of Technical Papers* (1995), pp. 129–130. https://doi.org/10.1109/VLSIT.1995.520891
102. R. Micheloni, L. Crippa, in *Advances in Non-volatile Memory and Storage Technology*, ed. by Y. Nishi (Woodhead Publishing, Sawston, 2014), chap. 3, pp. 75–119. https://doi.org/10.1533/9780857098092.1.75
103. C. Monzio Compagnoni, M. Ghidotti, A.L. Lacaita, A.S. Spinelli, A. Visconti, IEEE Electron Dev. Lett. **30**, 984 (2009). https://doi.org/10.1109/LED.2009.2026658
104. C. Monzio Compagnoni, A.S. Spinelli, R. Gusmeroli, S. Beltrami, A. Ghetti, A. Visconti, IEEE Trans. Electron Devices **55**, 2695 (2008). https://doi.org/10.1109/TED.2008.2003230
105. C. Monzio Compagnoni, R. Gusmeroli, A.S. Spinelli, A. Visconti, IEEE Trans. Electron Devices **55**, 3192 (2008). https://doi.org/10.1109/TED.2008.2003332
106. G.M. Paolucci, C. Monzio Compagnoni, A.S. Spinelli, A.L. Lacaita, A. Goda, IEEE Trans. Electron Devices **62**, 1491 (2015). https://doi.org/10.1109/TED.2015.2414711
107. S.H. Shin, D.K. Shim, J.Y. Jeong, O.S. Kwon, S.Y. Yoon, M.H. Choi, T.Y. Kim, H.W. Park, H.J. Yoon, Y.S. Song, Y.H. Choi, S.W. Shim, Y.L. Ahn, K.T. Park, J.M. Han, K.H. Kyung, Y.H. Jun, in *Symposium on VLSI Circuits Technical Digest Papers* (2012), pp. 132–133. https://doi.org/10.1109/VLSIC.2012.6243825
108. T. Endoh, K. Kinoshita, T. Tanigami, Y. Wada, K. Sato, K. Yamada, T. Yokoyama, N. Takeuchi, K. Tanaka, N. Awaya, K. Sakiyama, F. Masuoka, in *International Electron Devices Meeting Technical Digest* (2001), pp. 33–36. https://doi.org/10.1109/IEDM.2001.979396
109. S.M. Jung, J. Jang, W. Cho, H. Cho, J. Jeong, Y. Chang, J. Kim, Y. Rah, Y. Son, J. Park, M.S. Song, K.H. Kim, J.S. Lim, K. Kim, in *International Electron Devices Meeting Technical Digest* (2006), pp. 37–40. https://doi.org/10.1109/IEDM.2006.346902
110. E.K. Lai, H.T. Lue, Y.H. Hsiao, J.Y. Hsieh, C.P. Lu, S.Y. Wang, L.W. Yang, T. Yang, K.C. Chen, J. Gong, K.Y. Hsieh, R. Liu, C.Y. Lu, in *International Electron Devices Meeting Technical Digest* (2006), pp. 41–44. https://doi.org/10.1109/IEDM.2006.346903
111. K.T. Park, D. Kim, S. Hwang, M. Kang, H. Cho, Y. Jeong, Y.I. Seo, J. Jang, H.S. Kim, S.M. Jung, Y.T. Lee, C. Kim, W.S. Lee, in *International Solid-State Circuits Conference – Digest of Technical Papers* (2008), pp. 510–511. https://doi.org/10.1109/ISSCC.2008.4523281
112. H. Tanaka, M. Kido, K. Yahashi, M. Oomura, R. Katsumata, M. Kito, Y. Fukuzumi, M. Sato, Y. Nagata, Y. Matsuoka, Y. Iwata, H. Aochi, A. Nitayama, in *Symposium on VLSI Technology Technical Digest* (2007), pp. 14–15. https://doi.org/10.1109/VLSIT.2007.4339708
113. K.T. Park, J.M. Han, D. Kim, S. Nam, K. Choi, M.-S. Kim, P. Kwak, D. Lee, Y.H. Choi, K.M. Kang, M.H. Choi, D.H. Kwak, H.W. Park, S.W. Shim, H.J. Yoon, D. Kim, S.W. Park, K. Lee, K. Ko, D.K. Shim, Y.L. Ahn, J. Park, J. Ryu, D. Kim, K. Yun, J. Kwon, S. Shin, D. Youn, W. tae Kim, T. Kim, S.J. Kim, S. Seo, H.G. Kim, D.S. Byeon, H.J. Yang, M. Kim, M.S. Kim, J. Yeon, J. Jang, H.S. Kim, W. Lee, D. Song, S. Lee, K.H. Kyung, J.H. Choi, in *International Solid-State Circuits Conference Digest of Technical Papers* (2014), pp. 334–335. https://doi.org/10.1109/ISSCC.2014.6757458

114. K.T. Park, S. Nam, D. Kim, P. Kwak, D. Lee, Y.H. Choi, M.H. Choi, D.H. Kwak, D.H. Kim, M.S. Kim, H.W. Park, S.W. Shim, K.M. Kang, S.W. Park, K. Lee, H.J. Yoon, K. Ko, D.K. Shim, Y.L. Ahn, J. Ryu, D. Kim, K. Yun, J. Kwon, S. Shin, D.S. Byeon, K. Choi, J.M. Han, K.H. Kyung, J.H. Choi, K. Kim, IEEE J. Solid State Circuits **50**, 204 (2015). https://doi.org/10.1109/JSSC.2015.2352293
115. Y. Fukuzumi, R. Katsumata, M. Kito, M. Kido, M. Sato, H. Tanaka, Y. Nagata, Y. Matsuoka, Y. Iwata, H. Aochi, A. Nitayama, in *International Electron Devices Meeting Technical Digest* (2007), pp. 449–452. https://doi.org/10.1109/IEDM.2007.4418970
116. S. Whang, K. Lee, D. Shin, B. Kim, M. Kim, J. Bin, J. Han, S. Kim, B. Lee, Y. Jung, S. Cho, C. Shin, H. Yoo, S. Choi, K. Hong, S. Aritome, S. Park, S. Hong, in *International Electron Devices Meeting Technical Digest* (2010), pp. 668–671. https://doi.org/10.1109/IEDM.2010.5703447
117. Y. Noh, Y. Ahn, H. Yoo, B. Han, S. Chung, K. Shim, K. Lee, S. Kwak, S. Shin, I. Choi, S. Nam, G. Cho, D. Sheen, S. Pyi, J. Choi, S. Park, J. Kim, S. Lee, S. Aritome, S. Hong, S. Park, in *Symposium on VLSI Technology Technical Digest* (2012), pp. 19–20. https://doi.org/10.1109/VLSIT.2012.6242440
118. S. Aritome, S. Whang, K. Lee, D. Shin, B. Kim, M. Kim, J. Bin, J. Han, S. Kim, B. Lee, Y. Jung, S. Cho, C. Shin, H. Yoo, S. Choi, K. Hong, S. Park, S. Hong, Solid State Electron. **79**, 166 (2013). https://doi.org/10.1016/j.sse.2012.07.005
119. S. Aritome, Y. Noh, H. Yoo, E.S. Choi, H.S. Joo, Y. Ahn, B. Han, S. Chung, K. Shim, K. Lee, S. Kwak, S. Shin, I. Choi, S. Nam, G. Cho, D. Sheen, S. Pyi, J. Choi, S. Park, J. Kim, S. Lee, S. Hong, S. Park, T. Kikkawa, IEEE Trans. Electron Devices **60**, 1327 (2013). https://doi.org/10.1109/TED.2013.2247606
120. K. Parat, C. Dennison, in *International Electron Devices Meeting Technical Digest* (2015), pp. 48–51. https://doi.org/10.1109/IEDM.2015.7409618
121. M. Sako, Y. Watanabe, T. Nakajima, J. Sato, K. Muraoka, M. Fujiu, F. Kouno, M. Nakagawa, M. Masuda, K. Kato, Y. Terada, Y. Shimizu, M. Honma, A. Imamoto, T. Araya, H. Konno, T. Okanaga, T. Fujimura, X. Wang, M. Muramoto, M. Kamoshida, M. Kohno, Y. Suzuki, T. Hashiguchi, T. Kobayashi, M. Yamaoka, R. Yamashita, in *International Solid-State Circuits Conference – (ISSCC) Digest of Technical Papers* (2015), pp. 128–129. https://doi.org/10.1109/ISSCC.2015.7062959
122. D. Kang, W. Jeong, C. Kim, D.H. Kim, Y.S. Cho, K.T. Kang, J. Ryu, K.M. Kang, S. Lee, W. Kim, H. Lee, J. Yu, N. Choi, D.S. Jang, J.D. Ihm, D. Kim, Y.S. Min, M.S. Kim, A.S. Park, J.I. Son, I.M. Kim, P. Kwak, B.K. Jung, D.S. Lee, H. Kim, H.J. Yang, D.S. Byeon, K.T. Park, K.H. Kyung, J.H. Choi, in *International Solid-State Circuits Conference – (ISSCC) Digest of Technical Papers* (2016), pp. 130–131. https://doi.org/10.1109/ISSCC.2016.7417941
123. S. Lee, J.Y. Lee, I.H. Park, J. Park, S.W. Yun, M.S. Kim, J.H. Lee, M. Kim, K. Lee, T. Kim, B. Cho, D. Cho, S. Yun, J.N. Im, H. Yim, K.H. Kang, S. Jeon, S. Jo, Y.L. Ahn, S.M. Joe, S. Kim, D.K. Woo, J. Park, H.W. Park, Y. Kim, J. Park, Y. Choi, M. Hirano, J.D. Ihm, B. Jeong, S.K. Lee, M. Kim, H. Lee, S. Seo, H. Jeon, C.H. Kim, H. Kim, J. Kim, Y. Yim, H. Kim, D.S. Byeon, H.J. Yang, K.T. Park, K.H. Kyung, J.H. Choi, in *International Solid-State Circuits Conference (ISSCC) Digest Technical Papers* (2016), pp. 138–139. https://doi.org/10.1109/ISSCC.2016.7417945
124. M.K. Jeong, S.M. Joe, C.S. Seo, K.R. Han, E. Choi, S.K. Park, J.H. Lee, in *Symposium on VLSI Technology (VLSIT) Technical Digest* (2012), pp. 55–56. https://doi.org/10.1109/VLSIT.2012.6242458
125. M.K. Jeong, S.M. Joe, B.S. Jo, H.J. Kang, J.H. Bae, K.R. Han, E. Choi, G. Cho, S.K. Park, B.G. Park, J.H. Lee, in *International Electron Devices Meeting Technical Digest* (2012), pp. 207–210. https://doi.org/10.1109/IEDM.2012.6479010
126. D. Kang, C. Lee, S. Hur, D. Song, J.H. Choi, in *International Electron Devices Meeting Technical Digest* (2014), pp. 367–370. https://doi.org/10.1109/IEDM.2014.7047052
127. E. Nowak, J.H. Kim, H. Kwon, Y.G. Kim, J.S. Sim, S.H. Lim, D.S. Kim, K.H. Lee, Y.K. Park, J.H. Choi, C. Chung, in *Symposium on VLSI Technology (VLSIT) Technical Digest* (2012), pp. 21–22. https://doi.org/10.1109/VLSIT.2012.6242441

128. Y.L. Chou, T. Wang, M. Lin, Y.W. Chang, L. Liu, S.W. Huang, W.J. Tsai, T.C. Lu, K.C. Chen, C.Y. Lu, IEEE Electron Dev. Lett. **37**, 998 (2016). 10.1109/LED.2016.2585860
129. C.C. Hsieh, H.T. Lue, T.H. Hsu, P.Y. Du, K.H. Chiang, C.Y. Lu, in *Symposium on VLSI Technology Technical Digest* (2016), pp. 63–64. https://doi.org/10.1109/VLSIT.2016.7573386
130. D. Resnati, A. Goda, G. Nicosia, C. Miccoli, A.S. Spinelli, C. Monzio Compagnoni, IEEE Electron Dev. Lett. **38**, 461 (2017). 10.1109/LED.2017/2675160
131. G. Nicosia, A. Goda, A.S. Spinelli, C.Monzio Compagnoni, IEEE Electron Dev. Lett. **39**, 1175 (2018). https://doi.org/10.1109/LED.2018.2847341
132. P. Cappelletti, in *International Electron Devices Meeting (IEDM) Technical Digest* (2015), pp. 241–244. https://doi.org/10.1109/IEDM.2015.7409666
133. A. Goda, C. Miccoli, C. Monzio Compagnoni, in *International Electron Devices Meeting (IEDM) Technical Digest* (2015), pp. 374–377. https://doi.org/10.1109/IEDM.2015.7409699
134. G.M. Paolucci, A.S. Spinelli, C. Monzio Compagnoni, P. Tessariol, IEEE Trans. Electron Devices **63**, 1871 (2016). https://doi.org/10.1109/TED.2016.2543605
135. M. Toledano-Luque, R. Degraeve, P.J. Roussel, V. Luong, B. Tang, J.G. Lisoni, C.L. Tan, A. Arreghini, G. Van den bosch, G. Groeseneken, J. Van Houdt, in *International Electron Devices Meeting Technical Digest* (2013), pp. 562–565. https://doi.org/10.1109/IEDM.2013.6724676
136. C.M. Lee, B.Y. Tsui, in *Proceedings of International Symposium on VLSI Technology, Systems and Application (VLSI-TSA)* (2013), pp. 47–48. https://doi.org/10.1109/VLSI-TSA.2013.6545598
137. J.Y.W. Seto, J. Appl. Phys. **46**, 5247 (1975). https://doi.org/10.1063/1.321593
138. T. Hashimoto, M. Aoki, T. Yamanaka, Y. Kamigaki, T. Nishida, in *Proceedings of 1994 VLSI Technology Symposium Technical Digest* (1994), pp. 87–88. 10.1109/VLSIT.1994.324442
139. R. Brederlow, W. Weber, C. Dahl, D. Schmitt-Landsiedel, R. Thewes, in *International Electron Devices Meeting 1998. Technical Digest* (1998), pp. 89–92. https://doi.org/10.1109/IEDM.1998.746286
140. H.J. Kang, M.K. Jeong, S.M. Joe, J.H. Seo, S.K. Park, S.H. Jin, B.G. Park, J.H. Lee, in *Symposium on VLSI Technology (VLSI-Technology): Digest of Technical Papers* (2014), pp. 24–25. https://doi.org/10.1109/VLSIT.2014.6894348
141. W.J. Tsai, W.L. Lin, C.C. Cheng, S.H. Ku, Y.L. Chou, L. Liu, S.W. Hwang, T.C. Lu, K.C. Chen, T. Wang, C.Y. Lu, in *International Electron Devices Meeting. Technical Digest* (2016), pp. 11.3.1–11.3.4. https://doi.org/10.1109/IEDM.2016.7838395
142. M. Toledano-Luque, R. Degraeve, B. Kaczer, B. Tang, P.J. Roussel, P. Weckx, J. Franco, A. Arreghini, A. Suhane, G.S. Kar, G. Van den bosch, G. Groeseneken, J.V. Houdt, in *International Electron Devices Meeting Technical Digest* (2012), pp. 203–206. https://doi.org/10.1109/IEDM.2012.6479009
143. S.Y. Kim, J.K. Park, W.S. Hwang, S.J. Lee, K.H. Lee, S.H. Pyi, B.J. Cho, J. Nanosci. Nanotech. **16**, 5044 (2016). https://doi.org/10.1166/JNN.2016.12251
144. A.R. Brown, G. Roy, A. Asenov, IEEE Trans. Electron Devices **54**, 3056 (2007). https://doi.org/10.1109/TE.2007.907802
145. M. Peisl, A.W. Wieder, IEEE trans. Electron Devices **30**, 1792 (1983). https://doi.org/10.1109/T-ED.1983.21447
146. D.M. Kim, A. Khondker, S.S. Ahmed, R.R. Shah, IEEE Trans. Electron Devices **31**, 480 (1984). https://doi.org/10.1109/T-ED.1984.21554
147. A.N. Khondker, D.M. Kim, S.S. Ahmed, R.R. Shah, IEEE Trans. Electron Devices **31**, 493 (1984). https://doi.org/10.1109/T-ED.1984.21555
148. G. Baccarani, B. Riccò, G. Spadini, J. Appl. Phys. **49**, 5565 (1978). https://doi.org/10.1063/1.324477
149. N.C.C. Lu, L. Gerzberg, C.Y. Lu, J.D. Meindl, IEEE Trans. Electron Devices **28**, 818 (1981). https://doi.org/10.1109/T-ED.1981.20437

150. Y.H. Hsiao, H.T. Lue, W.C. Chen, C.P. Chen, K.P. Chang, Y.H. Shih, B.Y. Tsui, C.Y. Lu, in *International Electron Devices Meeting Technical Digest* (2012), pp. 609–612. https://doi.org/10.1109/IEDM.2012.6479111
151. Y.H. Hsiao, H.T. Lue, W.C. Chen, K.P. Chang, Y.H. Shih, B.Y. Tsui, K.Y. Hsieh, C.Y. Lu, IEEE Trans. Electron Devices **61**, 2064 (2014). https://doi.org/10.1109/TED.2014.2318716
152. C.W. Yang, P. Su, IEEE Trans. Electron Devices **61**, 1211 (2014). https://doi.org/10.1109/TED.2014.2308951
153. J. Kim, J. Lee, H. Oh, T. Rim, C.K. Baek, M. Meyyappan, J.S. Lee, in *Proceedings of International Nanoelectronics Conference (INEC)* (2014), pp. 66–68. https://doi.org/10.1109/INEC.2014.7460420
154. P.Y. Wang, B.Y. Tsui, IEEE Trans. Electron Devices **62**, 2488 (2015). https://doi.org/10.1109/TED.2015.2438001
155. R. Degraeve, S. Clima, V. Putcha, B. Kaczer, P. Roussel, D. Linten, G. Groeseneken, A. Arreghini, M. Karner, C. Kernstock, Z. Stanojevic, G. Van den bosch, J. Van Houdt, A. Furnemont, A. Thean, in *International Electron Devices Meeting (IEDM) Technical Digest* (2015), pp. 121–124. https://doi.org/10.1109/IEDM.2015.7409636
156. Z. Lun, L. Shen, Y. Cong, G. Du, X. Liu, Y. Wang, in *Proceedings of Silicon Nanoelectronics Workshop (SNW)* (2015), pp. 35–37
157. H. Oh, J. Kim, J. Lee, T. Rim, C.K. Baek, J.S. Lee, Microelectron. Eng. **149**, 113 (2016). https://doi.org/10.1016/J.MEE.2015.09.018
158. R. Degraeve, M. Toledano-Luque, A. Arreghini, B. Tang, E. Capogreco, J. Lisoni, P. Roussel, B. Kaczer, G. Van den bosch, G. Groeseneken, J. Van Houdt, in *International Electron Devices Meeting Technical Digest* (2013), pp. 558–561. https://doi.org/10.1109/IEDM.2013.6724675
159. W.S. Yoo, T. Ishigaki, T. Ueda, K. Kang, N.Y. Kwak, D.S. Sheen, S.S. Kim, M.S. Ko, W.S. Shin, B.S. Lee, S.J. Yeom, S.K. Park, in *Proceedings of Non-Volatile Memory Technology Symposium (NVMTS)* (2014), pp. 44–47. https://doi.org/10.1109/NVMTS.2014.7060843
160. I. Amit, D. Englander, D. Horvitz, Y. Sasson, Y. Rosenwaks, Nano Lett. **14**, 6190 (2014). https://doi.org/10.1021/NL5024468
161. A. Shamir, I. Amit, D. Englander, D. Horvitz, Y. Rosenwaks, Nanotechnology **26**, 355201 (2015). https://doi.org/10.1088/0957-4484/26/35/355201
162. D. He, N. Okada, C.M. Fortmann, I. Shimizu, J. Appl. Phys. **76**, 4728 (1994). https://doi.org/10.1063/1.357240
163. A.J. Walker, S.B. Herner, T. Kumar, E.H. Chen, IEEE Trans. Electron Devices **51**, 1856 (2004). https://doi.org/10.1109/TED.2004.837388
164. D. Resnati, A. Mannara, G. Nicosia, G.M. Paolucci, P. Tessariol, A.L. Lacaita, A.S. Spinelli, C. Monzio Compagnoni, in *International Electron Devices Meeting (IEDM) Technical Digest* (2017), pp. 103–106. https://doi.org/10.1109/IEDM.2017/8268329
165. G. Nicosia, A. Mannara, D. Resnati, G.M. Paolucci, P. Tessariol, A.L. Lacaita, A.S. Spinelli, A. Goda, C. Monzio Compagnoni, in *International Electron Devices Meeting (IEDM) Technical Digest* (2017), pp. 521–524. https://doi.org/10.1109/IEDM.2017.8268434
166. D. Resnati, A. Mannara, G. Nicosia, G.M. Paolucci, P. Tessariol, A.S. Spinelli, A.L. Lacaita, C. Monzio Compagnoni, IEEE Trans. Electron Devices **65**, 3199 (2018). https://doi.org/10.1109/TED.2018.2838524
167. G. Nicosia, A. Mannara, D. Resnati, G.M. Paolucci, P. Tessariol, A.S. Spinelli, A.L. Lacaita, A. Goda, C. Monzio Compagnoni, IEEE Trans. Electron Devices **65**, 3207 (2018). https://doi.org/10.1109/TED.2018.2839904
168. G. Fortunato, P. Migliorato, Appl. Phys. Lett. **49**, 1025–1027 (1986). https://doi.org/10.1063/1.97460
169. M. Hack, J.G. Shaw, P.G. LeComber, M. Willums, Jpn. J. Appl. Phys. **29**, 2360 (1990). https://doi.org/10.1143/JJAP.29.L2360
170. M.D. Jacunski, M.S. Shur, M. Hack, IEEE Trans. Electron Devices **43**, 1433 (1996). https://doi.org/10.1109/16.535329

171. M. Valdinoci, L. Colalongo, G. Baccarani, A. Pecora, I. Policicchio, G. Fortunato, F. Plais, P. Legagneux, C. Reita, D. Priba, Solid-State Electron. **41**, 1363 (1997). https://doi.org/10.1016/S0038-1101(97)00130-5
172. R. Degraeve, M. Toledano-Luque, A. Suhane, G. Van den bosch, A. Arreghini, B. Tang, B. Kaczer, P. Roussel, G.S. Kar, J. Van Houdt, G. Groeseneken, in *International Electron Devices Meeting Technical Digest* (2011), pp. 287–290. https://doi.org/10.1109/IEDM.2011.6131540

Advanced Electrical Characterization of Single Oxide Defects Utilizing Noise Signals

Bernhard Stampfer, Alexander Grill, and Michael Waltl

1 Introduction

A variety of electrical measurement techniques for investigating the functionality and performance of single semiconductor transistors has evolved over the recent decades. Among these, the simplest method is to measure an I_D-V_G characteristic of a pristine or stressed device from which important measures for the device quality like (1) the sub-threshold slope, (2) the leakage current, and (3) the on-state resistance can be extracted, as shown in Fig. 1. The change of parameters may not seem particularly critical for a single transistor, but it is of fundamental importance in the context of complex integrated circuits as any transistor operating outside of specification might lead to failure of the circuit.

Although the manufacturing processes for semiconductor structures used in modern circuits are continuously being optimized, their electrical properties still remain determined by a large number of electrically active defects. Such defects can either be located in the insulator, at the insulator semiconductor interface, or in the semiconductor bulk material. In order to characterize the prevalent defect densities and their trap levels, measurement techniques such as capacitance voltage (CV) measurements [1], charge pumping (CP) [2, 3], deep level transient spectroscopy (DLTS) [4], direct-current current voltage (DCIV) measurements [5, 6], extended stress and recovery measurements (eMSM) [7], and noise measurements [8] are widely used. While most of these measurements require switching or sweeping

B. Stampfer (✉) · M. Waltl
Christian Doppler Laboratory for Single-Defect Spectroscopy in Semiconductor Devices at the Institute for Microelectronics, Vienna, Austria
e-mail: stampfer@iue.tuwien.ac.at; waltl@iue.tuwien.ac.at

A. Grill
Institute for Microelectronics, Vienna, Austria
e-mail: alexander.grill@imec.be

T. Grasser (ed.), *Noise in Nanoscale Semiconductor Devices*,
https://doi.org/10.1007/978-3-030-37500-3_7

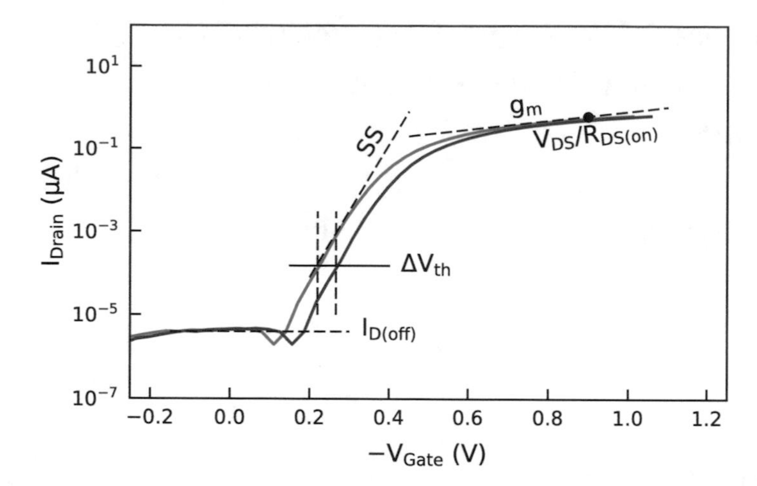

Fig. 1 Schematic of an I_D-V_G characteristics for a fresh (blue) and a stressed (red) device. Shown are the sub-threshold slope, the off-state leakage current, the on-state resistance, and threshold voltage shift

between different bias levels, noise measurements are considered an equilibrium measurement technique as the device current is recorded at constant biases.

To analyze the measurement data gathered from noise, either the time- or frequency-domain can be used. In the most general sense, the noise spectra of conventional MOS transistors typically follow a 1/f behavior, which is often referred to as flicker or pink noise. It was first proposed by McWorther that the nowadays well-known 1/f noise characteristics arise from defects in the transistors which dynamically change their charge state during operation [8]. However, at that time the feature size of transistors was too large to observe single charge transition events in the device current. With the continued decrease of the gate area of MOS devices it was observed that the 1/f noise power is roughly inversely proportional to the gate area [9]. Ralls et al. later showed that the origin for 1/f noise can be linked to individual defects modulating the resistance of the inversion channel depending on their charge state [10]. They observed that the changes in the charge state can be so large that they can be measured as discrete steps in the device current. This outstanding feature opens the avenue for device reliability characterization at the single defect level and can be directly observed on all small enough devices where only a few tens of single defects are active.

By examining the experimental data of the noise measurements on small MOS structures, one can observe many remarkable features of single defects which are not measurable on large MOS structures. The diversity of the behavior of the single defects poses a particular challenge for the development of physical models for device simulators. However, exact physical models can only be ensured if the various characteristics of individual processes can be exactly reproduced.

1.1 Random Telegraph Noise

For the most common case, an RTN signal may look as depicted in Fig. 2a. At a constant gate- and drain-source bias, the amplitude of the current through the conducting channel exhibits a step-like decrease as soon as the active defect captures a charge, and increases by the same amount as soon as the same defect emits its charge. The time period from the last discrete current decrease to the next increase is termed *charge capture time*, and the reverse is called the *charge emission time*. It has to be noted that both, the charge capture and the charge emission processes, are stochastic events. The time periods between the charge capture and successive charge emissions events follow an exponential distribution.

$$p(t) = \exp(-t/\tau_e), \tag{1}$$

with the average emission time τ_e, usually referred to simply as *emission time*. Likewise, the average time between charge emission events and successive charge capture events is called the *capture time* (τ_c), and follows an exponential distribution too. An interesting feature of the charge transition times is that the values of τ_c and τ_e depend on the applied gate bias, as shown in Fig. 2a. The average step height—which is the third parameter which can be extracted from an RTN signal—

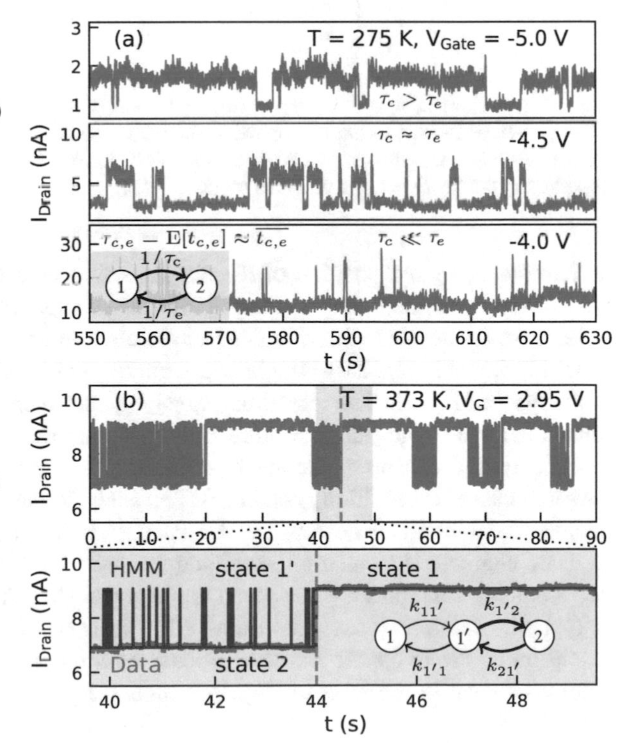

Fig. 2 Exemplary RTN traces for a simple defect (**a**) and a more complex defect (**b**). The currents shown in (**a**) decrease once the defect captures a charge and increases again once it emits. The average dwelling times of the RTN signals change with gate voltage. The more complex defect affecting the current in (**b**) shows two different capture times in its neutral state, effect termed anomalous RTN (aRTN). Markov models as shown in the insets can be used to model such defects. Modified from [11]

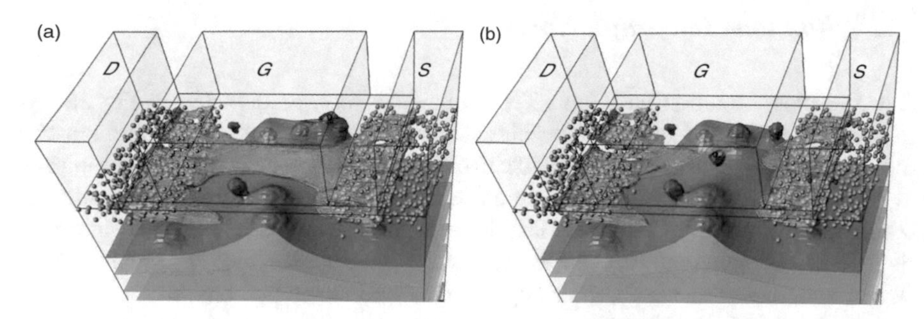

Fig. 3 TCAD simulation showing the current percolation path in a MOS transistor for two defect configurations (**a,b**). In a small-area device, a single defect changing its charge state may heavily influence the percolation path (orange) and thus the conductivity of the device. Cyan balls: source and drain donor ions, red dots: charged defects. From [12]

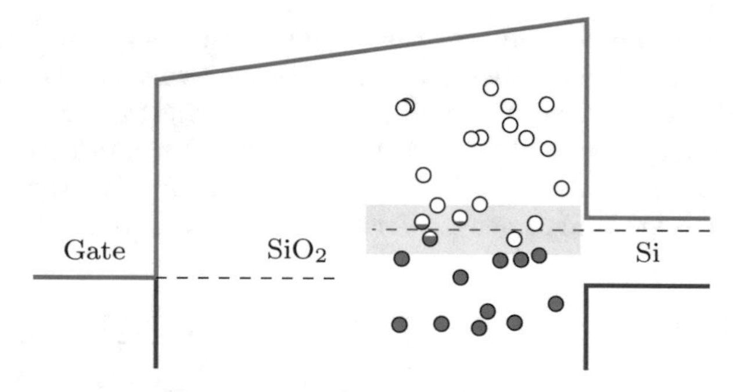

Fig. 4 Band diagram illustrating the energetical window of defects producing RTN signals at a given gate bias. Defects located far above or far below the Fermi level remain neutral or charged during the entire measurement, while defects located closer to the Fermi level have an occupancy approaching 0.5 (at $E_T = E_F$), and thus produce RTN

is determined by the position of the single defect with respect to the percolation path of the inversion channel (see Fig. 3). On average, the step height is observed to increase for devices with a smaller active channel area [13]. To model the defect behavior shown in Fig. 2a, a simple two-state Markov process can be used, as shown in the inset of Fig. 2a. The transition rates ($k_{c,e}$) between the two states are directly related to the average charge capture and emission time by $k_{c,e} = 1/\tau_{c,e}$.

Defects that can be explained by two-state Markov processes are among the simplest cases. In addition, detailed RTN analysis has shown that the temporal behavior of individual defects can be very complex, as shown in Fig. 2b. Such intricate characteristics can be attributed to defects which can exist in more than one atomistic configuration at a given charge state [14]. A signal as given in Fig. 2b is typically referred to as *anomalous RTN* (aRTN). To model this behavior, an extension of the two-state Markov process to a three- or even four-state Markov chain as shown in the inset of Fig. 2b is necessary and will be discussed in more detail in Sect. 4.

The effect of RTN for a given defect is most pronounced if the defect captures and emits its charge at a similar rate (i.e., $\tau_c \approx \tau_e$), which is the case if the energetic trap level of the defect is close to the Fermi level of the channel or the gate at the given gate bias, as indicated by the shaded area in Fig. 4.

In the case of $E_T \approx E_F$, the probability of the defect being charged (referred to as *occupancy*) becomes $\approx$0.5 and thus, its average capture time equals its average emission time, which at the same time maximizes the noise power of the signal for the given defect. Changing the gate bias from this point shifts the relative position of the defect with respect to the Fermi level, which in turn impacts the defect's capture and emission rates and thus its occupancy. By varying the gate bias, the bias dependence of the charge transition times, i.e., $\tau_c(V_G)$ and $\tau_e(V_G)$, can be determined. This voltage dependence provides important information on the location and the atomistic nature of single defects.

1.2 Link Between RTN and 1/f Noise

In Fig. 5, typical noise measurements recorded from a large-area and a small-area device are compared. As stated earlier, RTN can be seen as the manifestation of 1/f noise in small devices—or likewise—1/f noise as the sum of many single RTN signals. To show this relation, the power spectral density (PSD) of a two-state RTN signal is calculated from its auto-correlation function using the Wiener–Khintchine theorem, which yields [15, 16]:

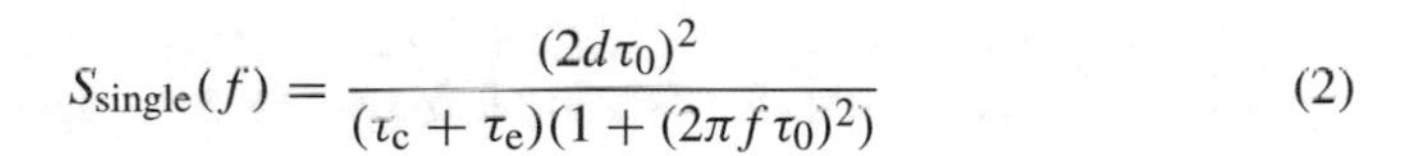

$$S_{\text{single}}(f) = \frac{(2d\tau_0)^2}{(\tau_c + \tau_e)(1 + (2\pi f \tau_0)^2)} \tag{2}$$

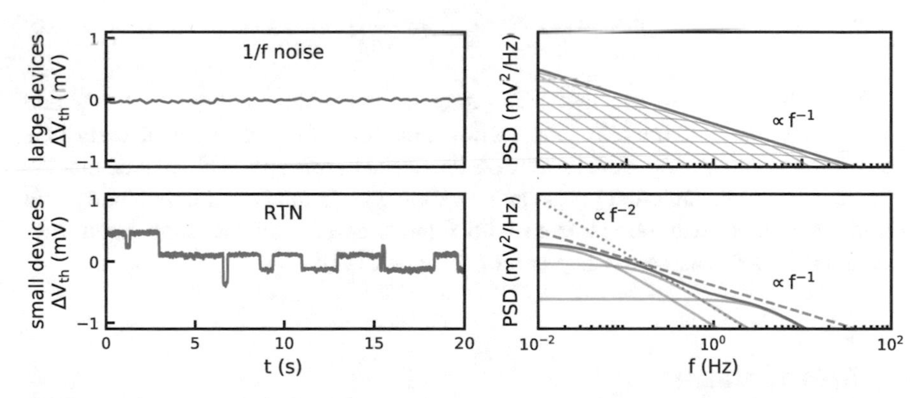

Fig. 5 Exemplary noise measurements shown for a large-area (top) and a small-area (bottom) device at static operating conditions. Illustrations of the frequency behavior are given to show the link between small and large area transistors. In large devices, many Lorentzian power spectral densities (PSDs) produced by individual defects add up to 1/f noise

with the step height d and the mean frequency

$$1/\tau_0 = 1/\tau_c + 1/\tau_e. \tag{3}$$

The resulting Lorentzian PSD is depicted for three defects in the bottom right plot in Fig. 5. Assuming the defects as causing only a small perturbation to device electrostatics, the noise spectral density for a number of defects is given by the superposition

$$S(f) = \sum_i \frac{(2d_i \tau_{0,i})^2}{(\tau_{c,i} + \tau_{e,i})(1 + (2\pi f \tau_{0,i})^2)}. \tag{4}$$

For a large number of defects with a uniform distribution of $\log(\tau)$, as is apparently often the case for large-area transistors, the distribution $S(f)$ converges to a 1/f shape, as indicated in Fig. 5. This can be shown by simplifying Equation 4 by replacing all defects within ranges of $\Delta \log(\tau_0)$ by equivalent defects with $\tau_c = \tau_e = \tau$ and the summed step heights of $d = d_e$. This simplification leads to:

$$S(f) \approx \frac{d_e^2}{2} \sum_i \frac{\tau_i}{1 + (\pi f \tau_i)^2} \tag{5}$$

Now, at any given frequency f, the PSD of one defect will be dominating the sum. Thus, we can simplify the equation further by considering only these dominant defects. Maximizing the inner function with τ_i as the parameter gives $\tau_m = 1/(\pi f)$, i.e., the dominant defect is the defect with its corner frequency at πf. This approximation then reveals the overall 1/f frequency dependence:

$$S(f) \approx \frac{d_e^2 \tau_m}{4} = \frac{d_e^2}{4\pi f}. \tag{6}$$

From this, one can see that RTN may be characterized both in time- and frequency-domain. In the following sections, the RTN and single defect analysis is performed using the time-domain data. Advantages of this approach include that the charge capture and emission times can be extracted separately and individually, and multi-state defects can be extracted with relative ease. Analysis using frequency-domain data is discussed in Chapter 4 of this book [17].

2 Measurements

To characterize RTN, either general purpose instruments or specialized measurement configurations as shown in Fig. 6 are commonly used. In the first configuration depicted in Fig. 6a, the biases applied at the gate and drain terminals of the transistor

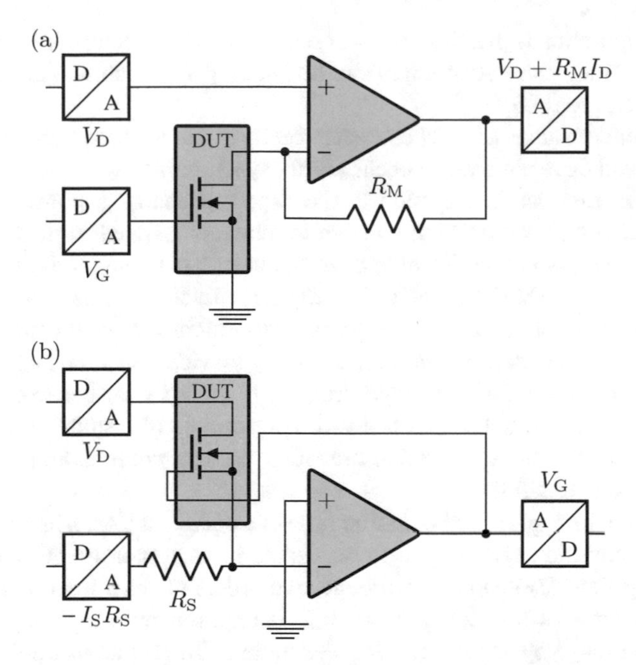

Fig. 6 Simplified measurement configurations used for RTN (and TDDS) measurements: (**a**) constant gate voltage or (**b**) constant source current. Practical designs will typically exhibit several feedback resistors in parallel to switch between drain current ranges, include additional passive components to adjust the bandwidth of the op-amp stage to achieve a maximum signal-to-noise ratio (SNR), and may further include secondary stage amplifiers. Notice that in the case of the constant current configuration (**b**) both the DUT and the operational amplifier feed back into each other, forming a closed loop. This may lead to instabilities if the circuit is not carefully designed

are held at a constant voltage during the experiment. If one considers an ideal operational amplifier (op-amp), with zero input currents on the positive and negative terminals ($I_+ = I_- = 0\,\mathrm{A}$), infinite amplification, and zero offset voltage between the terminals ($V_{+-} = 0\,\mathrm{V}$), the current through the feedback loop equals the drain-source current, and the output voltage can be calculated as $V_{\mathrm{out}} = V_{\mathrm{D}} + R_{\mathrm{M}} I_{\mathrm{D}}$. Given a measurement circuit where these conditions can be matched closely, this voltage can be recorded and the drain current can be calculated. As the degradation of the device is typically considered in terms of an equivalent shift of the threshold voltage, the recorded drain current is subsequently mapped to a ΔV_{th} using a pre-recorded I_{D}-V_{G} characteristic [18, 19].

The second widely used configuration is shown in Fig. 6b, where the MOS transistor is operated at a constant source current. This can be achieved by placing the DUT in the feedback loop of the operational amplifier (op-amp) with the gate voltage being directly controlled by the op-amp [20]. The constant current configuration has the big advantage that the threshold voltage shift can be measured directly, which may lead to slightly different values for ΔV_{th} as compared to those measured using the first configuration [21]. A serious disadvantage of the constant current configuration is that the transistor being part of the feedback loop can lead to oscillations. Thus, more attention has to be put on part selection and design before the scheme can be used.

One major advantage of RTN characterization is that the recording can be started at any time and does not have to occur highly synchronized with any bias switches. The sampling rate can be adjusted to the expected charge capture and emission times which allows to optimize the system for maximum signal-to-noise ratio (SNR) and highest step resolution. To maximize the usable bias range, the sampling unit recording the trace should be able to sample relatively fast, as well as to record a large number of points, which is necessary to monitor defects with significantly different charge transition times, i.e., $\tau_c \ll \tau_e$ or vice versa. A large data buffer of the sampling stage allows to measure at gate voltages farther from $E_T = E_F$, where higher ratios of τ_c/τ_e are observed. An example of a suitable sampling unit are high speed oscilloscopes which are often used in combination with custom-designed circuits [22, 23].

To estimate the capture and emission time of a defect, at least a handful of charge capture and emission events have to be visible in each measurement trace, while the sampling time (T_S) should be around two orders of magnitude [24] below the lowest time constant for reliable results. This requirement effectively reduces the measurement range by about three decades in time. To give an example, consider a measurement system with a sampling buffer size (n) of one million samples and a defect with a charge capture and emission time at the intersection point ($\tau_c = \tau_e$) of around a second. In order to obtain a large characterization range (in bias), one could choose a sampling time of around $t_S = 1$ ms, leading to a total measurement time of $nT_S = 1$ ks. This effectively allows to characterize RTN signals with transition times in the range of $\approx$100 ms to $\approx$100 s.

2.1 Number of Observations

The error made by observing a limited number of events can be predicted using a chi-square distribution, which gives the one-sigma confidence limit for the estimated time constant [25]. For 4, 10, and 100 observations (N), the measured time constant will lie approximately within a 50, 30, and 10% error margin, respectively, 68% of the time. Alternatively, one may use the width of the Gaussian distribution to obtain the relative error as $1/\sqrt{N}$, which gives values close to the errors calculated using a chi-square distribution for all but very small numbers of N. To reflect this in the design of experiment, a minimum cumulative measurement time of $t_m = N(\tau_c + \tau_e)$ should be used.

3 Defect Parameter Extraction

A number of methods have been developed to extract the properties of RTN signals [26, 27]. The method of choice depends on (1) the parameters which should be extracted, for instance, step height and/or (ratio of) transition times, (2) the number of defects visible in the measurements, and (3) their signal-to-noise ratio. In the following, the classical methods such as the histogram and time lag plot methods are discussed briefly, before more advanced methods based on edge detection and hidden Markov models (HMMs) are presented.

3.1 *Histogram and Time Lag Plots*

The most straightforward method for RTN parameter extraction is the histogram method [28]. For this, a histogram of the samples recorded from the drain current, or equivalent the ΔV_{th} values, is drawn in a first step, as shown in Fig. 7a. Given a large enough SNR, Gaussian peaks, which correspond to the individual charge

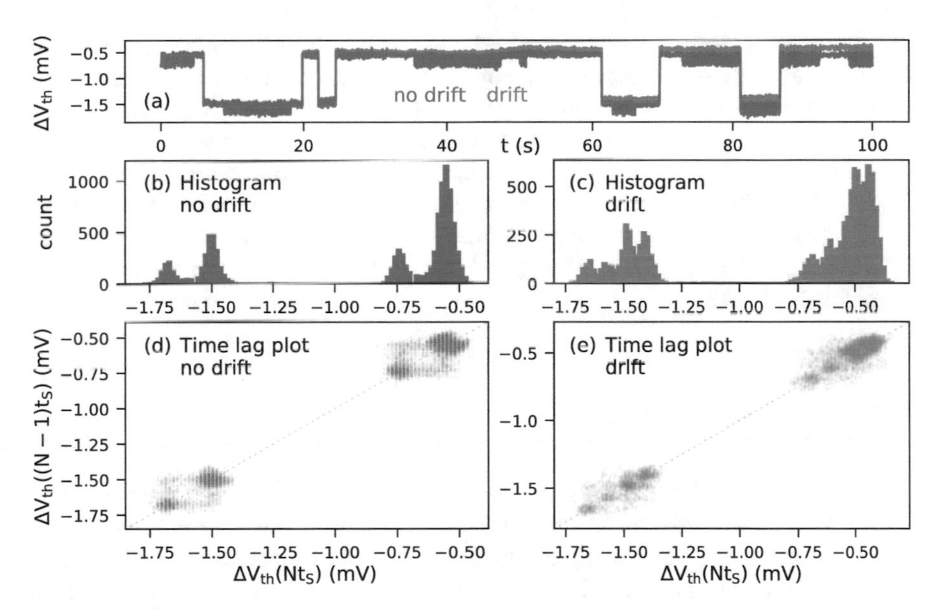

Fig. 7 Histogram (**b**) and time lag plot (**d**) of a measured RTN signal (**a**) (blue figures). The histogram exhibits two defects, one with a smaller and one with a larger step height. A clearer separation between the distinct levels can be observed for the time lag plot. Additionally, the time lag plot shows the transitions in the off-diagonals. To show the influence of drift on these methods, a linear drift of 150 μV was added to the signal (orange plots). It can be seen that both the histogram (**c**) and time lag plot (**e**) can only distinguish the larger defect with the drift present. Note that the smaller of the defects shows abnormal RTN behavior, which these methods do not reflect

states of the single defects, should become visible in the histogram. The step height of the defect can then be extracted directly from the distance between two peaks. To determine the charge transition times, threshold values can be defined to assign regions of drain current to the charge states of the defect(s). The charge transition times can then be extracted by counting and averaging the time spent in each charge state. The ratio of the charge transition times can be obtained directly from the areas of the peaks in the histogram.

An improvement of the histogram method is the time lag plot method [26], where instead of a simple histogram, a scatterplot is created from the samples, with the x- and y-values of the points defined by each two consecutive drain current steps (see Fig. 7b). This leads to a better separation of the (now two-dimensional) peaks, and thus enables a more reliable extraction of the defect parameters, especially in the case of multiple defects. In the time lag plot representation, transitions between the states show as off-diagonal clusters.

The largest drawback of the histogram and time lag plot methods is that a relatively large SNR is required for reliable assignment of drain current values to defect states, as few erroneous assignments may lead to large errors of the extracted charge transition times. This can be mitigated to some extent by the removal of outliers or circumvented by just obtaining the ratio of the transition times which will be much less affected by a small number of wrong assignments. Another disadvantage of these methods is their reliance on the absolute drain current values for any combination of charge states. Any measurement drift, low frequency noise, or slow defects which are not subject of the analysis will widen the extracted step height distributions and make the extraction unreliable. This is a serious challenge especially when long-term RTN measurements in the kilo second range or even larger have to be analyzed. A recent improvement called the weighted time lag method has been proposed by Martin-Martinez et al., it allows the extraction of RTN at higher noise levels [29]. The weighted time lag method is presented in Chapter 14 of this book [30].

3.2 Edge Detection

Especially when analyzing measurement data which is superimposed by a slow drift, it is favorable to consider the first derivative of the dataset. The most straightforward approach seems to be to directly calculate the first derivative; however, this direct approach is disadvantageous for two reasons. First, a step occurring during a sampling period might be distributed over two samples, causing a lower derivative, and second any measurement noise will fully contribute to the result. A more reliable approach is to utilize an edge detection filter such as the Canny filter, which is widely used in digital image processing [31]. This filter allows to find the positions of the discrete switching events in the trace even for signals with low SNR. Once the positions of the discrete steps are identified, the step heights can be obtained by subtracting the I_D values before and after the steps from the original data, or

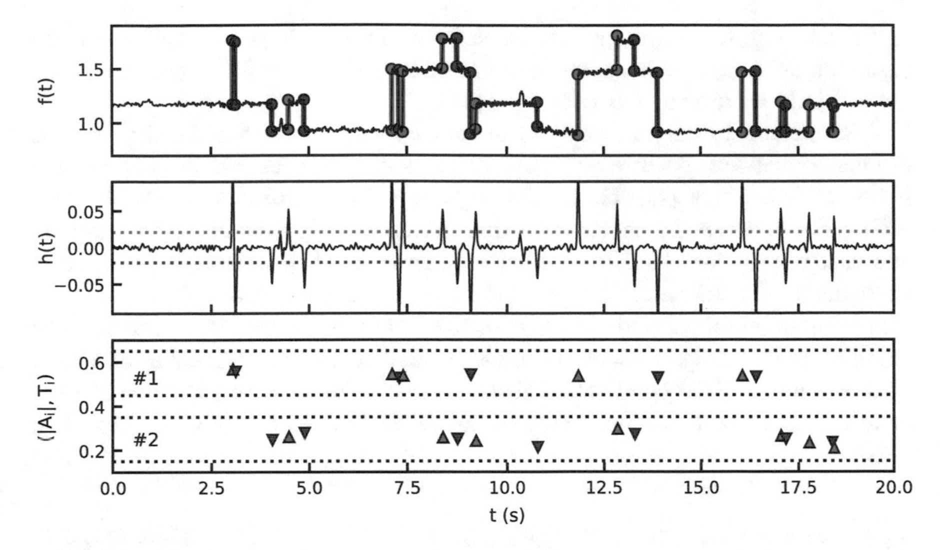

Fig. 8 Working principle of the Canny edge detection algorithm used for RTN parameter extraction, shown on an exemplary measurement trace. The original signal $f(t)$ is convoluted with the first derivative of a Gaussian to obtain the filtered signal $h(t)$. The positions of the local extrema in $h(t)$ above a certain threshold give the positions of the steps. This method is insensitive to most drifts in the measurement trace

alternatively by integrating the derivatives. Afterwards, steps with similar heights can be considered for further evaluation, and the time intervals between the step-up and step-down transitions can be analyzed. An illustration showing a trace analyzed with the Canny edge detection algorithm can be seen in Fig. 8.

To calculate the positions of the edges in the signal $f(t) = \Delta V_{\mathrm{th}}(t)$, we first calculate the derivative $\partial \Delta \mathrm{V}_{\mathrm{th}}(t)/\partial t$ and convolute it with a Gaussian kernel $g(t, \sigma)$, or equivalently, convolute the original signal with the derivate of the Gaussian kernel

$$h(t) = f(t) * \frac{\partial g(t, \sigma)}{\partial t}. \tag{7}$$

The positions of the relative maxima in the signal above a noise threshold m then give the positions of the steps:

$$\mathbf{T} = \mathbf{argrelmax}(|h(t)| > m) \tag{8}$$

With (**argrelmax**(*x*,*y*)) defined as a function yielding the x positions of the local maxima in y (c.f. *findpeaks* in matlab). The step heights can be obtained by taking the difference of the measurement values before and after the step:

$$\mathbf{A} = f(\mathbf{T} + \delta) - f(\mathbf{T} - \delta) \tag{9}$$

or alternatively by integrating over the segment in $h(t)$. Improved variants of the algorithm exist, i.e., methods which use multiple values of σ [32] or estimation of the threshold value using Otsu's method [33].

After detection, the steps can be binned by step height and finally the time differences between positive and negative steps can be measured. If only a single two-state defect is responsible for the steps in a bin, the mean values of the time differences between the charge transition events directly give the charge capture and charge emission time. If a defect with more than two states is observed, a histogram of the differences between the charge transition events may be used to fit the transition times. Similarly, if two defects fall within a single bin, a state machine might be used to assign individual steps to the defects given additional information, as shown in Fig. 9 for correlated defects. At this point, however, it might already be favorable to use the method based on hidden Markov models which is discussed next.

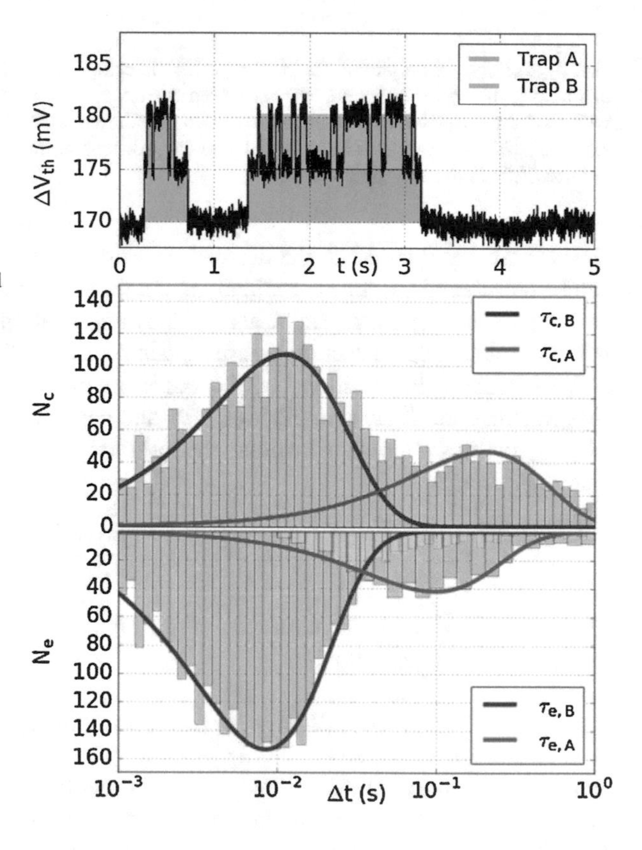

Fig. 9 Two defects with similar step heights, or the charge transition times of a single defect with more than two states, may be distinguished by fitting the exponential distributions to the differences in step-up and step-down times. The example shows steps obtained from a measurement trace with two correlated defects where defect B was only active when defect A captured charge. This feature allowed to clearly separate the extracted steps for each defect. For regular defects with similar step height, extraction using HMMs might be necessary to obtain their parameters. (Picture taken from [34])

3.3 Hidden Markov Models

The last and most sophisticated method which is presented here is based on maximum likelihood estimation using hidden Markov models (HMMs). A Markov model as described in Sect. 4 is constructed from the suspected defects visible in the measured data and its parameters are typically trained to the measured data using the Baum–Welch algorithm [35].

Given the observed data $O(t) = I_{\mathrm{D}}(t)$ (or $\Delta V_{\mathrm{th}}(t)$) and the structure of the Markov chain, we want to find the parameters of the model θ, whereby θ consists of:

- the transition matrix $\boldsymbol{K} = \{k_{ij}\}$ with the transition rates between all defect states i and j,
- the observed mean drain current $\boldsymbol{M} = \{\mu_i\}$ for each state i, and
- the standard deviation σ of the drain current noise.

The Baum–Welch algorithm is a variant of an expectation–maximization algorithm and each iteration consists of two steps:

- In the expectation step, the probabilities of being in each state for each point in time, as well as the average number of transitions between the states are calculated given the observations and the model parameters from the previous iteration.
- In the maximization step, new model parameters are calculated from these statistics and the observations to maximize the likelihood of the expectations.

Once the algorithm has converged, the parameters and observations may also be used to calculate a most likely sequence of states using the Viterbi algorithm [36].

The main advantage of the HMM method is that it is suitable for signals exhibiting a lower SNR than required for the previously presented methods. Thus, this method can be used to extract parameters from defects which are barely visible in the measurement [37]. Another advantage of this method is that it is possible to distinguish defects with similar step heights and extract defects with more than two states. The method, however, is negatively affected by any slow drift present in the measurement data, just like the histogram and time lag plot method. To mitigate this, a method for baseline correction can be implemented within the iterative Baum–Welch solver. Suitable solutions comprise spline smoothing, local regression [38], and asymmetric least squares [39]. For a large set of measurement data, this method is less suitable than, for example, the Canny edge detection method as the defects have to be defined prior to extraction, requiring manual interaction. Furthermore, the method has to be used carefully, as it regularly converges to wrong results with good scores (*likelihood*). For instance, overfitting may occur if too large a number of defects is chosen, often with better likelihood than a regular fit. It can also happen that a single real defect is fitted using multiple defects, or measurement noise is fitted with defects. The method is also computationally expensive as the computing time increases exponentially with the number of states in the Markov model. Thus, an efficient implementation is required if traces with multiple defects should be analyzed. The basic algorithm can be found in [40], and various implementations are available, e.g., for python [41, 42]. To train a HMM to multiple defects, additional pre- and post-processing is necessary to assemble and disassemble the individual Markov chains before and after each iteration as described by Frank et al. in [43]. A refined approach to fitting multiple defects was proposed by Puglisi et al. in [44] where they used a factorial hidden Markov model (FHMM) [45] to fit the individual defects which decreases the computational complexity. This approach, instead of compiling a single Markov chain representing all defects, uses separate Markov chains for each defect and optimizes their state variables in a self-consistent manner.

4 Oxide Defect Modeling

Before discussing the identification of physical defect properties from the obtained signal parameters, we should take a closer look at the defects themselves. The main culprit giving rise to RTN (and BTI/TDDS) is thought to be defects located within the oxide of the devices. Each defect can capture or emit a charge from the channel or the gate carrier reservoirs. Such a charge transition event changes the Coulomb potential in the vicinity of the defect, and as a consequence affects the surface potential along the semiconductor/insulator interface, and further perturbs the current flux through the transistor [46]. Due to the amorphous nature of the oxide material, a number of atomistic arrangements serving as possible defect candidates exist [47]. The most likely defect candidates for SiO_2 are the hydrogen bridge and the hydroxyl E′ center, which both exhibit a number of different, but stable, atomistic configurations, as shown in Fig. 10. The neutral, puckered, and charged

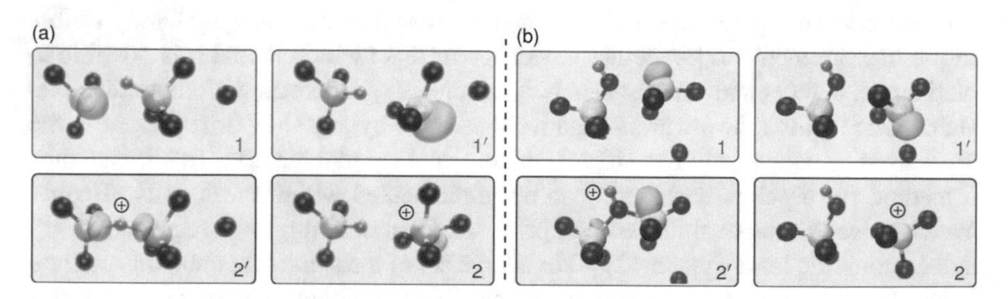

Fig. 10 The hydrogen bridge (**a**) and the hydroxyl E′ center (**b**), two possible defect candidates in SiO_2. Consistent with experimental data, these defects are able to change their atomistic configuration while maintaining the same net charge, leading to multiple states. In this figure, the yellow balls represent silicon atoms, red balls oxygen, gray balls hydrogen, and the blue bubbles represent a negative charge distribution. The numbers 1 and 2 describe the neutral and charged configurations, respectively, and a prime (′) denotes a metastable state. It can be seen that the transition between the stable and metastable states commonly involves the silicon atom transitioning through the plane spanned by three of its oxygen neighbors to enter a so-called puckered configuration [48, 49]. Adapted from [50]

configurations can be determined using computational expensive density functional theory (DFT) calculations. Recent calculations suggest that such defects might be (in)activated by the interaction with the various hydrogenous species always present in any device [50, 51].

The transition between the different states of the defects can be considered memoryless, meaning that the probability of the defect changing state depends only on the current defect state and potential distribution, and not on the previous state of the defect or the device. This is crucial for modeling as so the defects can be modeled as Markov chains where each Markov state represents one of the possible atomistic configurations. In this context, the simplest model for an oxide defect or interface state is given by a two-state Markov chain, representing the defect in its charged and neutral state. To some extent, the two-state description is sufficient to describe the charge trapping behavior observed in many RTN measurements. However, these measurements also reveal defects hibernating in one of the two charge states between phases of RTN, a behavior which is termed anomalous RTN as shown in Fig. 2b. To describe this trapping feature, a Markov chain with at least three states is required. An additional remarkable observation is that defects show a kind of volatility in TDDS measurements, meaning that they can disappear for a number of measurements, and later reappear [52]. This means that the defect becomes stuck in a more stable atomic configuration for a certain amount of time. Furthermore, defects were found to either show strongly bias dependent charge capture and emission times, the so-called switching traps, or bias dependent charge capture times and bias independent emission times, the so-called fixed traps. To explain all these effects, two additional states have been introduced in the model, which leads to the so-called four-state model as shown in Fig. 11 [14, 53].

The probability $P_i(t)$ of the defect being in any state i can be described by the master equation, a set of first-order differential equations

$$\frac{\partial P_i}{\partial t} = \sum_{j \neq i} \left(P_j k_{ji} - P_i k_{ij} \right) \tag{10}$$

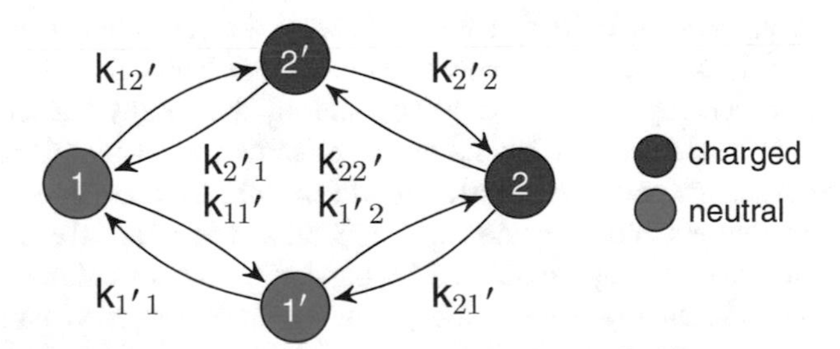

Fig. 11 Markov chain representing an oxide defect modeled using the 4-state model. States 1 and 1′ represent the defect in a neutral state, states 2 and 2′ in a charged state. States marked with ′(prime) represent metastable states

with the sum of all probabilities being 1 ($\sum_i P_i = 1$) and $k = 1/\tau$ denoting the transition rates.

In the four-state model, each of the two charge states is represented by both a "stable" and a "metastable" state. This allows for two types of transitions or reactions to occur: those without carrier exchange, and those with. Transitions without charge transfer can be modeled as thermally activated reactions, where overcoming a thermal barrier allows the defect to switch its configuration. The transition rates are typically modeled in the form of Arrhenius equations:

$$k = k_0 \mathrm{e}^{-\beta \mathcal{E}} \tag{11}$$

with $\beta = (k_\mathrm{B} T)^{-1}$, a pre-factor k_0 and an energetic barrier between the states $\mathcal{E}$.

To describe transitions between different charge states (charge transfer reactions), the tunneling probability for the carrier involved, as well as the availability of carriers or target states for capture and emission, respectively, has to be considered. This process can be described using non-radiative multi-phonon (NMP) theory [54–56] and yields equations in the form of [57–59]:

$$k_{\mathrm{c},12'} = k_0 \int D_\mathrm{p}(E) f_\mathrm{p}(E) \lambda(d, E) \mathrm{e}^{-\beta \mathcal{E}_{12'}(E_\mathrm{T}, E)} \mathrm{d}E \tag{12}$$

$$k_{\mathrm{e},2'1} = k_0 \int D_\mathrm{p}(E) f_\mathrm{n}(E) \lambda(d, E) \mathrm{e}^{-\beta \mathcal{E}_{2'1}(E_\mathrm{T}, E)} \mathrm{d}E, \tag{13}$$

e.g., for the exchange rates of holes between a defect and the valence band for the $1 \leftrightarrow 2'$ transition. Using the assumption that most of the exchange happens close to the band edge, the equations can be simplified to:

$$k_{\mathrm{c},12'} = k_0 p \lambda \mathrm{e}^{-\beta \mathcal{E}_{12'}(E_\mathrm{T}, E_\mathrm{v})} \tag{14}$$

$$k_{\mathrm{e},2'1} = k_0 N_\mathrm{v} \lambda \mathrm{e}^{-\beta \mathcal{E}_{2'1}(E_\mathrm{T}, E_\mathrm{v})}. \tag{15}$$

Here, k_0 denotes a pre-factor, λ a tunneling coefficient (commonly calculated using the WKB approximation), D the density of states, f the Fermi factor, p the density of holes, N_v the effective density of states in the valence band, E_t and E_v the defect and valence band energy level, respectively, and $\mathcal{E}_{ij}$ the energy barrier heights of the reaction. The barrier heights depend on the effective defect energy and the band energy of the exchange point, which strongly determines the bias dependence of the rates. In the simplest and most common approximation, the potential energy surfaces of the individual states are approximated by parabolic functions along the reaction coordinates, and the energy barriers are given by the intersection point between them. A more elaborate choice of the potential for the investigated defect might improve the physical accuracy and one example for such a potential is the Morse potential [60].

From the experimental data, we can only distinguish between charged or discharged states, resulting in effective capture and emission times being measured,

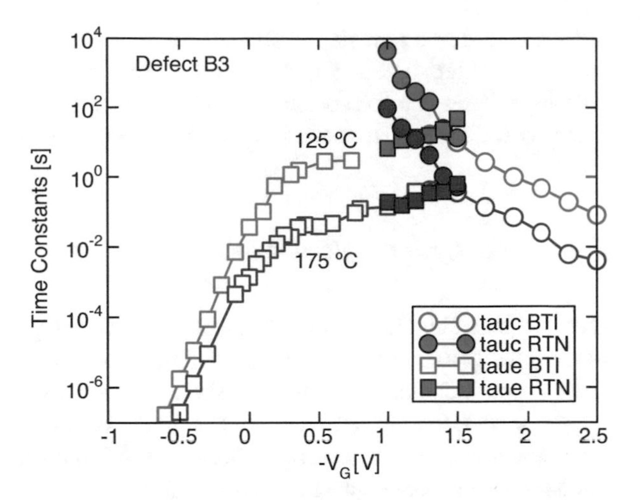

Fig. 12 Gate voltage dependence of effective charge capture and charge emission time characteristics for a defect characterized using TDDS (open symbols) and RTN (filled symbols) measurements, at two temperatures. The voltage ranges for both measurement methods overlap and the obtained data show good agreement. From [16]

as shown in Fig. 12. Depending on the gate voltage, usually one of the two possible transitions with charge transfer will be dominant. The capture and emission times obtained from the experiments will then be either the inverse rates between a stable state of one charge and metastable states of the other charge, or the first passage times between the two stable states, depending on the defect and bias. The expectation value of the first passage time from, e.g., state 2 to state 1 over state 2′ can be calculated as:

$$\tau_{e,2\to 2'\to 1} = \frac{\tau_{22'} + \tau_{2'1} + \tau_{2'2}}{\tau_{22'}\tau_{2'1}} \tag{16}$$

The atomistic aspects of oxide defects can be studied with DFT calculations. A more detailed discussion of this aspect of reliability engineering can be found in Chapter 19 of this book [61].

5 Defect Characterization

With the step height and the charge transition times at a number of bias points, conclusions about a defect parameters such as its position and energy can be drawn using either first-order calculations or TCAD simulations.

5.1 TCAD Simulation

To perform TCAD simulations, an abstracted version of the DUT reflecting the essential geometry and doping details of the real device is designed in the simulation software, and in a first step calibrated using I_D-V_G and/or C-V measurement data. In a second step, a defect as described in Sect. 4 is placed in the oxide. Its charge transition times can then be simulated at the bias points of the measurement and in that way the model parameters calibrated to reproduce the experimental data.

5.2 First-Order Calculations

Due to the fact that RTN measurements give both charge capture and emission times of the defect close to the intersection point, RTN data allows us to draw conclusions from simple calculations, as we will show below. This proves useful if, for example, large sets of defects have to be analyzed, if TCAD simulation is not yet available for the technology, or as a starting point for TCAD calibration. Starting from the rate equations for charge capture and emission

$$k_{\mathrm{c,e}} = k_{0\mathrm{c,e}} \exp\left(-\frac{\mathcal{E}_{\mathrm{c,e}}}{k_{\mathrm{B}}T}\right), \tag{17}$$

the change in defect energy with gate voltage can be obtained from: [27]

$$\frac{\partial \ln k_{\mathrm{e}}}{\partial V_{\mathrm{G}}} - \frac{\partial \ln k_{\mathrm{c}}}{\partial V_{\mathrm{G}}} = \frac{1}{k_{\mathrm{B}}T}\frac{\partial(\mathcal{E}_{\mathrm{c}} - \mathcal{E}_{\mathrm{e}})}{\partial V_{\mathrm{G}}} = \frac{1}{k_{\mathrm{B}}T}\frac{\partial E_{\mathrm{T}}}{\partial V_{\mathrm{G}}} \tag{18}$$

assuming a constant carrier density in the channel. Further assuming a homogeneous electric field in the oxide gives a link to the position of the defect

$$\partial E_{\mathrm{T}} = -\frac{d}{t_{\mathrm{ox}}}\partial V_{\mathrm{G}}. \tag{19}$$

The vertical position of the defect can then be estimated as

$$d = -t_{\mathrm{ox}}\frac{\partial E_{\mathrm{T}}}{\partial V_{\mathrm{G}}} = -t_{\mathrm{ox}}k_{\mathrm{B}}T\left(\frac{\partial \ln \tau_{\mathrm{c}}/\tau_{\mathrm{e}}}{\partial V_{\mathrm{G}}}\right). \tag{20}$$

Integrating Equation 19 gives the energetic position of the defect with a still unknown integration constant C

$$E_{\mathrm{T}}(V_{\mathrm{G}}) = k_{\mathrm{B}}T\left(\frac{\partial \ln \tau_{\mathrm{c}}/\tau_{\mathrm{e}}}{\partial V_{\mathrm{G}}}\right)V_{\mathrm{G}} + C. \tag{21}$$

At $V_{\mathrm{G}} = V_{\mathrm{G},i}$, the defect's capture and emission time are equal, which puts its energy at the Fermi level, i.e.:

$$E_{\mathrm{T}}(V_{\mathrm{G},i}) = E_{\mathrm{F}} \tag{22}$$

From this, the integration constant can be obtained to yield:

$$E_{\mathrm{T}}(V_{\mathrm{G}}) = k_{\mathrm{B}}T\left(\frac{\partial \ln \tau_{\mathrm{c}}/\tau_{\mathrm{e}}}{\partial V_{\mathrm{G}}}\right)(V_{\mathrm{G}} - V_{\mathrm{G},i}) + E_{\mathrm{F}} \tag{23}$$

The equations can be easily adapted for gate stacks consisting of more than one material as shown in [34].

6 TDDS

In the previous sections, we discussed RTN, which can be observed if a small-area device is operated in equilibrium. Another way of obtaining the capture and emission times of a defect is to measure not at static bias conditions as discussed above but to record the response of the device to transient bias conditions. For this, the gate bias is switched from one value to another, usually chosen in a manner to push the defect far above or far below the Fermi level. This forces defects in the device to charge or discharge starting with the bias change, with a defect capturing or emitting on average after τ_{c} or τ_{e} at the new voltage. The affected region in the oxide is called active energy region, as illustrated in Fig. 13.

This type of measurement is the basis for both the TDDS characterization method in small devices and stress-recovery measurements in large devices. An illustration is given in Fig. 14, where an overview of traces recorded at transient and static bias conditions on large and small devices is shown. The link between the static behavior of small-area and large-area devices was already discussed in the introduction. For the transient case, the link between the responses of large- and small-area devices can easily be seen as the result of a superposition of numerous charge transition events of single defects, thereby smoothing the drain currents.

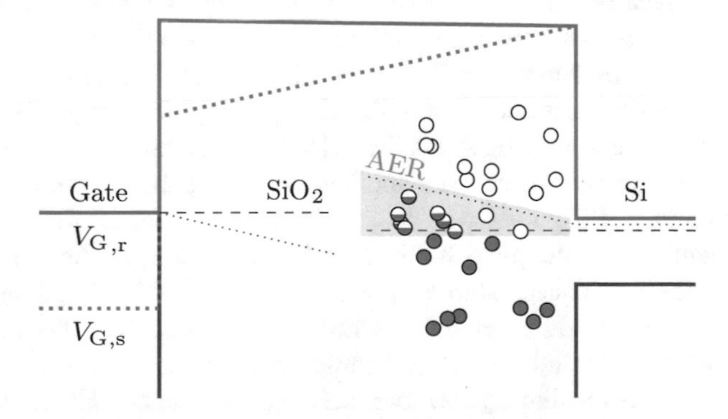

Fig. 13 Band diagram illustrating the scanning window of TDDS—the active energy region—for a specific set of stress and recovery biases. Defects located, e.g., below the Fermi level during stress conditions and located above the Fermi level during recovery conditions change their occupation after switching the bias, enabling their characterization

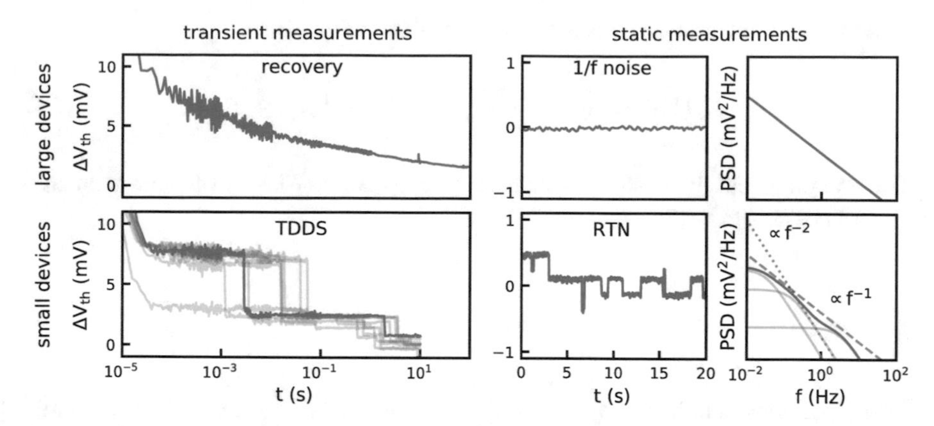

Fig. 14 Exemplary transient measurements recorded after the devices have been stressed (left) and noise measurements (static, right) shown for large-area (top) and small-area (bottom) devices. It can be consistently observed that for large-area devices continuous trends are observed, while a discrete behavior is obtained for the small-area counterparts. Both techniques can be used to characterize individual defects in small-area devices. For the static measurements, illustrations of the frequency behavior are given to show the link between small- and large-area devices. In large-area devices, many Lorentzian power spectral densities (PSDs) produced by individual defects add up to 1/f noise

6.1 Measurement

For TDDS measurements, the same setup as shown in Fig. 6 for RTN measurements can be used, but now the measurement sequence consists of two separate phases. The first phase is the stress or charging phase where the device is usually operated in strong inversion for a defined stress (or charging) time. In this phase, no sampling is typically done, as no drain current is available due to $V_{\mathrm{D,str}} = 0\,\mathrm{V}$. After the stress phase, the gate bias is switched to a voltage in the sub-threshold regime, and a small drain bias of typically $V_{\mathrm{D,rec}} = 50\,\mathrm{mV}$ is applied and the data sampling started. The emission time is then considered the time interval between the bias switch and the occurrence of the charge transition event. In contrast, the capture time cannot be measured directly but is deducted indirectly by doing measurements at the same bias conditions but with varying stress times [14]. Compared to RTN measurements where linear sampling is required, the application of a logarithmic sampling scheme with, for instance, 200 points per decade is preferred for TDDS. This lowers the requirement on the sampling buffer size but now an exact timing of the bias switches and the subsequent start of sampling is inevitable to achieve accurate charge emission times. As previously mentioned, the change in the occupancy of a defect between stress and relaxation conditions determines the probability of an emission event to occur during the measurement sequence. This entails that the stress time should be chosen such that the occupancy of the defect is as close as possible to unity after stress. This can also be achieved by the modification of the stress bias; however, the bias is limited by the oxide breakdown voltage. Regarding the error due to a limited number (N) of observations, the same considerations as for RTN discussed in Sect. 2 apply. To achieve a given amount of observations, a total number of measurements $N_\mathrm{m} = \frac{N}{o_\mathrm{str}}$, with the occupancy after stress, o_str, have to be recorded.

6.2 Measurement Limitations

RTN and TDDS measurements differ in the gate voltage range where defects can be characterized. An overview of the limits for both RTN and TDDS measurements shown for an exemplary defect and device is given in Fig. 15. In the figure, it can be seen that TDDS allows to measure the charge capture time (red area) in a voltage range limited by the breakdown voltage for high fields and limited by low defect occupancy for low fields. The minimum and maximum time constants which can be measured depend on the minimum time a stress bias can be applied and the maximum time one is willing to spend. The range where the charge emission time (blue area) may be measured is limited by the defect occupancy and measurement resolution of larger drain currents, i.e., the defect visibility in the current data for high fields, and measurement sensitivity and noise in the depletion region. The time constants which can be measured depend on the minimum sampling time and the time required for the bias switch, and again the maximum measurement time one is willing to use determines the upper limit.

RTN (orange area) is limited to the voltage range where both capture and emission times are in the range between the sampling time and the total recording time of the measurement traces. Note that by modifying the device temperature, the defect's charge transition times move to shorter or longer times. Thus, the temperature can be used to cleverly adjust the measurement window.

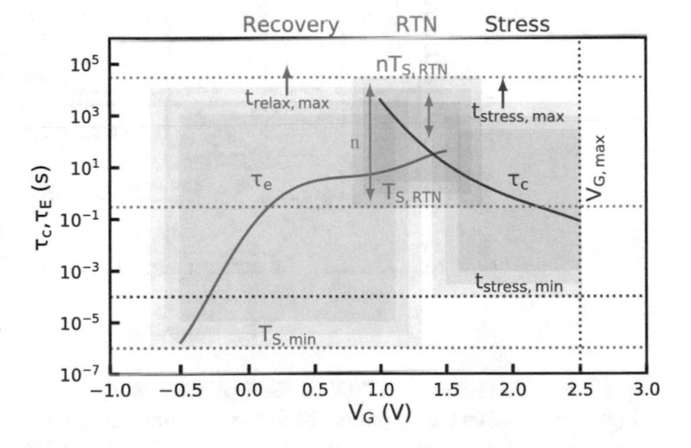

Fig. 15 Limits for RTN and TDDS measurements shown for an exemplary defect and device. $T_{S,\mathrm{min}}$, $t_{\mathrm{stress,min}}$, and n are limitations of the measurement setup, while the upper limit of stress bias $V_{G,\mathrm{max}}$ is usually given by the device. Notice that the position of the RTN window can be shifted by the choice of $T_{S,\mathrm{RTN}}$

6.3 Defect Parameter Extraction

For TDDS measurements, the recovery traces are analyzed using a step-detection algorithm such as the Canny algorithm discussed in Sect. 3. The extracted charge emission times and step heights are then plotted in the charge emission time versus step height plane, which is called spectral map, as shown in Fig. 16. As can be seen, the single charge emission events tend to form a cluster in the spectral map for each identified defect. The emission time and step height of a defect can then be calculated by taking the average values of the points belonging to one cluster. The capture events can usually not be observed directly as there is no current during stress. To indirectly obtain the capture time, the measurement has to be repeated at varying stress times. One then plots the probability of a charge emission event to occur (i.e., the number of points in each cluster divided by the number of measurements) over the stress times, which can be described by an exponential CDF which then finally gives the charge capture time.

This technique can also be used for "mixed" measurement traces recorded at a recovery bias where the defects also produce RTN. In this case, only the first charge emission event obtained for each defect should be used for the construction of the spectral map. The remaining points may then be used for RTN extraction. Comparing parameter extraction from RTN and TDDS data, TDDS has the advantage that

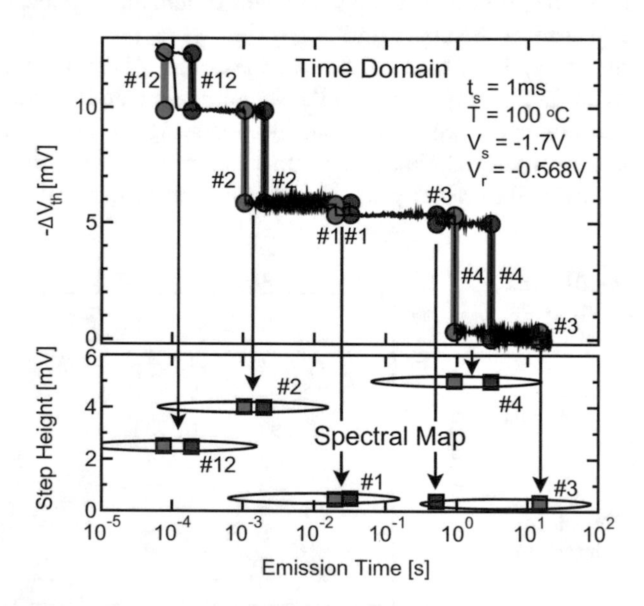

Fig. 16 Working principle of TDDS. Steps in ΔV_{th} (top) are marked in a scatterplot showing step height over emission time (bottom). Multiple measurements produce clusters in the scatterplot with exponentially distributed emission times and step heights normally distributed around a mean, which itself is exponentially distributed. Each cluster corresponds to an individual defect. The emission time of the defect (at $V_{\mathrm{G,rec}}$) can be obtained directly from the measurement. To obtain the capture time (at $V_{\mathrm{G,str}}$) multiple such measurements at varying stress times have to be conducted. From [14]

due to the bias switch, which enforces the occupational change of the defects, both step height and emission time can be directly obtained from the measurement. This allows to easily distinguish between defects even with similar step height as they will form separate clusters, provided their emission times are sufficiently different.

7 Link Between RTN and TDDS

While we presented both RTN and TDDS as methods to characterize oxide defects, the question whether both methods characterize the same sets of (oxide) defects has been left open. Grasser et al. reported in [16] a number of defects initially identified using TDDS measurements, all of which could also be observed using RTN measurements (see Fig. 17). Based on observations on around a hundred nanoscale devices, they concluded that both RTN and the recoverable component of BTI are due to the same defects.

Considering this from a theoretical perspective, oxide defects may be separated in three groups. The first group comprises defects whose charge capture time decreases relative to the charge emission time towards stress conditions. Their equilibrium occupancy is higher at stress bias than at recovery bias. This is the kind of defect which is most commonly found, and due to its decrease in occupancy readily emits during the recovery phase in TDDS measurements. The second group consists of defects whose charge capture times increase relative to their charge emission times towards stress bias. This can be defects close to the gate, positioned below the Fermi level during stress and above during recovery (for NBTI, and vice versa for PBTI). After switching to recovery conditions, they capture a charge instead of emitting it and are thus visible as inverse steps in TDDS measurements. Both

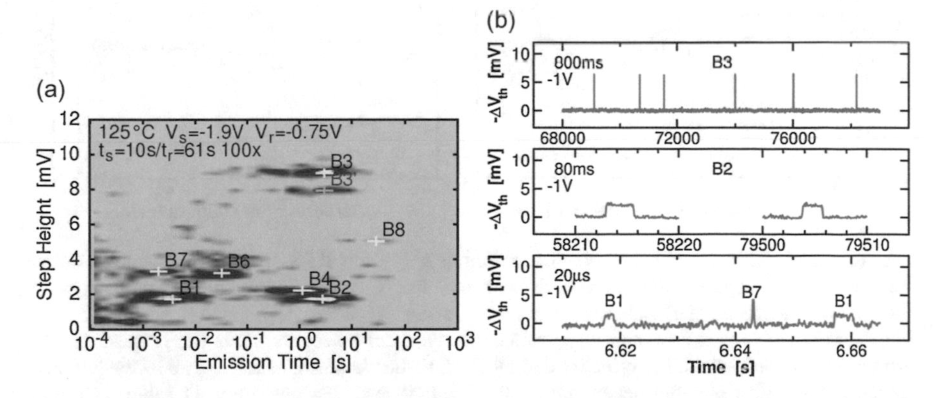

Fig. 17 Spectral map from TDDS measurements (**a**) showing a number of defects in a device and RTN traces (**b**) recorded on the same device, showing some of the same defects. All defects found in (**a**) could also be measured using RTN measurements and vice versa, indicating that RTN and TDDS identify the same type of defects. From [16]

groups of defects can also be found in RTN measurements, if the experiments are performed at suitable conditions in between stress and recovery conditions, as their thermodynamic trap level has to cross the Fermi level between stress and recovery conditions. The last group of defects are the ones whose charge capture to charge emission time ratio does not change significantly with gate bias. Their occupancies thus stay relatively constant, usually close to zero or close to one both during stress and recovery. This might be defects located energetically outside the TDDS measurement range or very close to the interface. They do not show up in TDDS measurements but are still able to sporadically emit or capture a charge for a short period of time. While these defects still produce RTN, most will not be visible in regular RTN measurements due to the large factor between their charge transition times and the limited time span of the RTN measurement window. Simulated band diagrams illustrating the suspected defect bands in SiO_2 together with measurement windows for TDDS and RTN are given in Fig. 18 for *P*- and NMOS devices (or NBTI- and PBTI-like measurements). Note that while parts of the defect bands

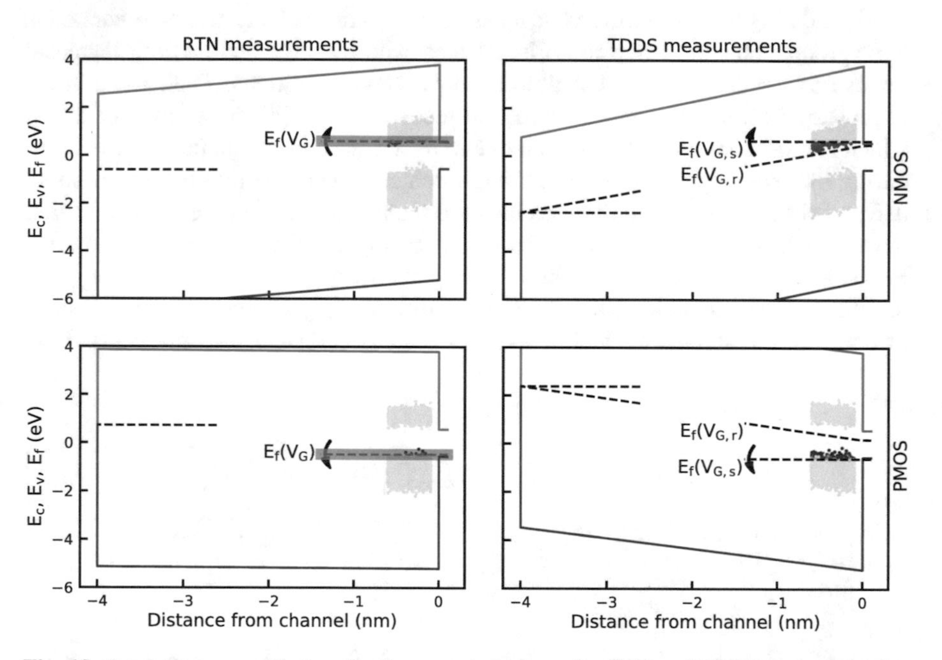

Fig. 18 Band diagrams with the effective scanning ranges for RTN and TDDS characterization for positive and negative bias conditions with exemplary defect distributions plotted close to the channel interface. TDDS measurements are sensitive over a trapezoidal area spanned by the interface Fermi levels at $V_{G,str}$ and $V_{G,rec}$ called the active energy region, while RTN measurements are sensitive to defects with an effective energy close to the channel and gate Fermi levels (marked as red bars), with the width depending on the limitations of the measurement setup. For both methods, modification of the applied bias changes the scanned area, and thus allows to pinpoint the located defects. As TDDS measurements are usually performed at bias conditions far beyond operating specifications, defects located outside the scanning ranges are unlikely to affect device performance at operating conditions. Values for the locations of the defect bands are taken from [62]

are inaccessible to the electrical measurements presented here, those are also very unlikely to affect device performance during operating conditions.

8 Summary

Non-ideal device behavior during operation is mainly determined by defects located in the oxide or at the oxide/semiconductor interface. In order to identify these defects, RTN or TDDS measurements can be performed. From these measurements, the charge transition times, i.e., charge capture time and charge emission time, and the contribution of the defect to the overall ΔV_{th}, i.e., the defect's step height, can be extracted. Afterwards the charge transition times can be analyzed using advanced computer simulations and a possible link to defect candidates using DFT simulations can be established.

For highest accuracy of the extracted data, elaborate experiments have to be conducted. This can be achieved by either using general purpose instruments or by using custom-design hardware based on the circuits shown initially. The advantage of custom solutions is that they can be optimized for highest SNR and performance. For the case of custom solutions data acquisition is typically performed using high speed digital storage oscilloscopes. To maximize the measurement ranges of RTN and TDDS measurements, the sampling tool should on the one hand be able to allow for fast and precise sampling and on the other hand be able to record a large number of samples for a long time period. To keep the number of samples in a feasible range, logarithmic time intervals for data sampling are typically used for TDDS. The next step is to analyze the measurement data. For this, several methods like the histogram and time lag plot methods, parameter extraction using the Canny edge detection algorithm, and maximum likelihood estimation using hidden Markov models have been presented. The method of choice depends among others on the amount of measurement sets, the signal-to-noise ratio, and the complexity of the recorded data. Finally, the extracted charge transition time characteristics can be used to calibrate physics-based models like the four-state NMP model, which is implemented in modern microelectronic device simulators.

In summary, in this chapter we have briefly discussed the characterization as well as the evaluation of the measurement data and the modeling of individual defects. The main focus is put on defect characterization by means of noise measurements, i.e., RTN measurements, and TDDS measurements. The careful execution of the experiments, whether with custom-build circuit or general purpose instruments, allows a deep insight into the physical mechanisms of charge trapping in MOS transistors, which is responsible for device altering even in extremely scaled technologies.

Acknowledgements This work was supported in part by the Austrian Science Fund (FWF) under Project P 26382-N30, Project P 23958-N24, and Project I2606-N30, in part by the European Union FP7 Project ATHENIS 3-D under Grant 619246, and in part by the Austrian Research Promotion Agency (FFG, Take-Off Program) under Project 861022 and Project 867414.

References

1. L. Terman, An investigation of surface states at a silicon/silicon oxide interface employing metal-oxide-silicon diodes. Solid State Electron. **5**(5), 285–299 (1962)
2. J.S. Brugler, P.G.A. Jespers, Charge pumping in MOS devices. IEEE Trans. Electron Devices **16**, 297–302 (1969)
3. G. Groeseneken, H.E. Maes, N. Beltran, R.F. De Keersmaecker, A reliable approach to charge-pumping measurements in MOS transistors. IEEE Trans. Electron Devices **31**, 42–53 (1984)
4. D.V. Lang, Deep-level transient spectroscopy: a new method to characterize traps in semiconductors. J. Appl. Phys. **45**(7), 3023–3032 (1974)
5. A. Neugroschel, C.-T. Sah, K.M. Han, M.S. Carroll, T. Nishida, J.T. Kavalieros, Y. Lu, Direct-current measurements of oxide and interface traps on oxidized silicon. IEEE Trans. Electron Devices **42**(9), 1657–1662 (1995)
6. C.-T. Sah, A. Neugroschel, K. M. Han, and J. T. Kavalieros, Profiling interface traps in MOS transistors by the DC current-voltage method. IEEE Electron Device Lett. **17**(2), 72–74 (1996)
7. B. Kaczer, T. Grasser, J. Roussel, J. Martin-Martinez, R. O'Connor, B. O'sullivan, G. Groeseneken, Ubiquitous relaxation in BTI stressing-new evaluation and insights, in *2008 IEEE International Reliability Physics Symposium*, pp. 20–27 (IEEE, Piscataway, 2008)
8. A.L. McWhorter et al., *1/f Noise and Related Surface Effects in Germanium* (MIT Lincoln Laboratory, Cambridge, 1955)
9. H. Mikoshiba, 1/f noise in n-channel silicon-gate MOS transistors. IEEE Trans. Electron Devices **29**, 965–970 (1982)
10. K.S. Ralls, W.J. Skocpol, L.D. Jackel, R.E. Howard, L.A. Fetter, R.W. Epworth, D.M. Tennant, Discrete resistance switching in submicrometer silicon inversion layers: individual interface traps and low-frequency ($\frac{1}{f}$) noise. Phys. Rev. Lett. **52**, 228–231 (1984)
11. B. Stampfer, F. Zhang, Y.Y. Illarionov, T. Knobloch, P. Wu, M. Waltl, A. Grill, J. Appenzeller, T. Grasser, Characterization of single defects in ultrascaled MoS2 field-effect transistors. ACS Nano **12**(6), 5368–5375 (2018). PMID: 29878746
12. M. Bina, O. Triebl, B. Schwarz, M. Karner, B. Kaczer, T. Grasser, Simulation of reliability on nanoscale devices, in *Proceedings of the 16th International Conference on Simulation of Semiconductor Processes and Devices (SISPAD)*, pp. 109–112 (2012)
13. A. Ghetti, C. Compagnoni, A. Spinelli, A. Visconti, Comprehensive analysis of random telegraph noise instability and its scaling in deca–nanometer flash memories. IEEE Trans. Electron Devices **56**(8), 1746–1752 (2009)
14. T. Grasser, H. Reisinger, P.-J. Wagner, W. Goes, F. Schanovsky, B. Kaczer, The time dependent defect spectroscopy (TDDS) for the characterization of the bias temperature instability, in *IEEE International Reliability Physics Symposium*, pp. 16–25 (2010)
15. S. Machlup, Noise in semiconductors: spectrum of a two-parameter random signal. J. Appl. Phys. **25**(3), 341–343 (1954)
16. T. Grasser, K. Rott, H. Reisinger, M. Waltl, J. Franco, B. Kaczer, A unified perspective of RTN and BTI, in *2014 IEEE International Reliability Physics Symposium*, pp. 4A.5.1–4A.5.7 (2014)
17. C. Marquez, O. Huerta, A.I. Tec-Chim, F. Guarin, E.A. Gutierrez-D, F. Gamiz, Systematic characterization of random telegraph noise and its dependence with magnetic fields in MOSFET devices, in *Noise in Nanoscale Semiconductor Devices*, ed. by T. Grasser (Springer, Cham, 2019). https://doi.org/10.1007/978-3-030-37500-3_4
18. B. Kaczer, V. Arkhipov, R. Degraeve, N. Collaert, G. Groeseneken, M. Goodwin, Disorder-controlled-kinetics model for negative bias temperature instability and its experimental verification, in *IEEE International Reliability Physics Symposium, 2005. Proceedings.*, pp. 381–387 (2005)

19. A. Kerber, K. Maitra, A. Majumdar, M. Hargrove, R. J. Carter, E. A. Cartier, Characterization of fast relaxation during BTI stress in conventional and advanced CMOS devices with HfO_2/TiN gate stacks, IEEE Trans. Electron Devices **55**(11), 3175–3183 (2008)
20. H. Reisinger, O. Blank, W. Heinrigs, A. Mühlhoff, W. Gustin, C. Schlünder, Analysis of NBTI degradation- and recovery-behavior based on ultra fast V_{th}-measurements, in *IEEE International Reliability Physics Symposium Proceedings*, pp. 448–453 (2006)
21. B. Ullmann, K. Puschkarsky, M. Waltl, H. Reisinger, T. Grasser, Evaluation of advanced MOSFET threshold voltage drift measurement techniques. IEEE Trans. Device Mater. Reliab. **19**(2), 358–362 (2019)
22. G.A. Du, D.S. Ang, Z.Q. Teo, Y.Z. Hu, Ultrafast measurement on NBTI. IEEE Electron Device Lett. **30**, 275–277 (2009)
23. M. Maestro, J. Diaz, A. Crespo-Yepes, M. Gonzalez, J. Martin-Martinez, R. Rodriguez, M. Nafria, F. Campabadal, X. Aymerich, New high resolution random telegraph noise (RTN) characterization method for resistive RAM. Solid State Electron. **115**, 140–145 (2016). Selected papers from the EUROSOI-ULIS conference
24. G. Kapila, V. Reddy, Impact of sampling rate on RTN time constant extraction and its implications on bias dependence and trap spectroscopy. IEEE Trans. Device Mater. Reliab. **14**(2), 616–622 (2014)
25. H. Reisinger, The time-dependent defect spectroscopy, in *Bias Temperature Instability for Devices and Circuits* (Springer, Berlin, 2014), pp. 75–109
26. T. Nagumo, K. Takeuchi, S. Yokogawa, K. Imai, Y. Hayashi, New analysis methods for comprehensive understanding of random telegraph noise, in *2009 IEEE International Electron Devices Meeting (IEDM)*, pp. 1–4 (2009)
27. T. Nagumo, K. Takeuchi, T. Hase, Y. Hayashi, Statistical characterization of trap position, energy, amplitude and time constants by RTN measurement of multiple individual traps, in *2010 IEEE International Electron Devices Meeting (IEDM)*, pp. 28–3 (IEEE, Piscataway, 2010)
28. Y. Yuzhelevski, M. Yuzhelevski, G. Jung, Random telegraph noise analysis in time domain. Rev. Sci. Instrum. **71**(4), 1681–1688 (2000)
29. J. Martin-Martinez, J. Diaz, R. Rodriguez, M. Nafria, X. Aymerich, New weighted time lag method for the analysis of random telegraph signals. IEEE Electron Device Lett. **35**, 479–481 (2014)
30. J. Martin-Martinez, R. Rodriguez, M. Nafria, Advanced characterization and analysis of random telegraph noise in CMOS devices, in *Noise in Nanoscale Semiconductor Devices*, ed. by T. Grasser (Springer, Cham, 2019). https://doi.org/10.1007/978-3-030-37500-3_14
31. J. Canny, A computational approach to edge detection, in *Readings in Computer Vision* (Elsevier, Amsterdam, 1987), pp. 184–203
32. P. Bao, L. Zhang, X. Wu, Canny edge detection enhancement by scale multiplication, IEEE Trans. Pattern Anal. Mach. Intell. **27**, 1485–1490 (2005)
33. N. Otsu, A threshold selection method from gray-level histograms. IEEE Trans. Syst. Man Cybern. **9**, 62–66 (1979)
34. A. Grill, B. Stampfer, M. Waltl, K.-S. Im, J.-H. Lee, C. Ostermaier, H. Ceric, T. Grasser, Characterization and modeling of single defects in GaN/AlGaN fin-MIS-HEMTs, in *IEEE International Reliability Physics Symposium Proceedings* (2017), pp. 3B–5
35. L.E. Baum, T. Petrie, G. Soules, N. Weiss, A maximization technique occurring in the statistical analysis of probabilistic functions of Markov chains. Ann. Math. Stat. **41**(1), 164–171 (1970)
36. A. Viterbi, Error bounds for convolutional codes and an asymptotically optimum decoding algorithm, IEEE Trans. Inf. Theory **13**(2), 260–269 (1967)

37. H. Miki, N. Tega, M. Yamaoka, D.J. Frank, A. Bansal, M. Kobayashi, K. Cheng, C.P. D'Emic, Z. Ren, S. Wu, J. Yau, Y. Zhu, M.A. Guillorn, D.-G. Park, W. Haensch, E. Leobandung, K. Torii, Statistical measurement of random telegraph noise and its impact in scaled-down high-k/metal-gate MOSFETs, in *2012 International Electron Devices Meeting*, pp. 19.1.1–19.1.4 (2012)
38. W.S. Cleveland, Robust locally weighted regression and smoothing scatterplots. J. Am. Stat. Assoc. **74**(368), 829–836 (1979)
39. P.H. Eilers, A perfect smoother. Anal. Chem. **75**(14), 3631–3636 (2003)
40. L.R. Rabiner, A tutorial on hidden Markov models and selected applications in speech recognition. Proc. IEEE **77**(2), 257–286 (1989)
41. R. Weiss, S. Du, J. Grobler, S. Lebedev, and G. Varoquaux, hmmlearn 0.2.2 (2017)
42. J. Schreiber, Pomegranate: fast and flexible probabilistic modeling in python. J. Mach. Learn. Res. **18**(1), 5992–5997 (2017)
43. D.J. Frank, H. Miki, Analysis of oxide traps in nanoscale MOSFETs using random telegraph noise, in *Bias Temperature Instability for Devices and Circuits* (Springer, Berlin, 2014), pp. 111–134
44. F.M. Puglisi, P. Pavan, Factorial hidden Markov model analysis of random telegraph noise in resistive random access memories. ECTI Trans. Electr. Eng. Electron. Commun. **12**(1), 24–29 (2014)
45. Z. Ghahramani, M.I. Jordan, Factorial hidden Markov models, in *Advances in Neural Information Processing Systems* (1996), pp. 472–478
46. E. Simoen, B. Dierickx, C.L. Claeys, G.J. Declerck, Explaining the amplitude of RTS noise in submicrometer MOSFETs. IEEE Trans. Electron Devices **39**, 422–429 (1992)
47. A.-M. El-Sayed, Y. Wimmer, W. Goes, T. Grasser, V.V. Afanas' ev, A.L. Shluger, Theoretical models of hydrogen-induced defects in amorphous silicon dioxide. Phys. Rev. B **92**(1), 014107 (2015)
48. A.L. Shluger, K.P. McKenna, Models of oxygen vacancy defects involved in degradation of gate dielectrics, in *2013 IEEE International Reliability Physics Symposium (IRPS)* (2013), pp. 5A.1.1–5A.1.9
49. S. Pantelides, Z.-Y. Lu, C. Nicklaw, T. Bakos, S. Rashkeev, D. Fleetwood, R. Schrimpf, The E' center and oxygen vacancies in SiO_2. J. Non-Cryst. Solids **354**(2–9), 217–223 (2008)
50. Y. Wimmer, A.-M. El-Sayed, W. Goes, T. Grasser, A.L. Shluger, Role of hydrogen in volatile behaviour of defects in SiO_2-based electronic devices, in *Proceedings of the Royal Society A*, vol. 472 (The Royal Society, London, 2016), p. 20160009
51. T. Grasser, M. Waltl, Y. Wimmer, W. Goes, R. Kosik, G. Rzepa, H. Reisinger, G. Pobegen, A. El-Sayed, A. Shluger, et al., Gate-sided hydrogen release as the origin of "permanent" NBTI degradation: from single defects to lifetimes, in *IEEE International Electron Devices Meeting (IEDM)* (IEEE, Piscataway, 2015), pp. 20–21
52. T. Grasser, M. Waltl, W. Goes, Y. Wimmer, A.-M. El-Sayed, A. Shluger, B. Kaczer, On the volatility of oxide defects: activation, deactivation and transformation, in *IEEE International Reliability Physics Symposium Proceedings* (2015), pp. 5A.3.1–5A.3.8
53. M.J. Uren, M.J. Kirton, S. Collins, Anomalous telegraph noise in small-area silicon metal-oxide-semiconductor field-effect transistors. Phys. Rev. B **37**, 8346–8350 (1988)
54. K. Huang, A. Rhys, Theory of light absorption and non-radiative transitions in f-centres, in *Proceedings of the Royal Society of London A: Mathematical, Physical and Engineering Sciences*, vol. 204 (1950), pp. 406–423
55. D. Lang, C. Henry, Nonradiative recombination at deep levels in GaAs and GaP by lattice-relaxation multiphonon emission. Phys. Rev. Lett. **35**(22), 1525 (1975)
56. M. Kirton, M. Uren, Noise in solid-state microstructures: a new perspective on individual defects, interface states and low-frequency (1/f) noise. Adv. Phys. **38**(4), 367–468 (1989)
57. T. Grasser, Stochastic charge trapping in oxides: From random telegraph noise to bias temperature instabilities. Microelectron. Reliab. **52**(1), 39–70 (2012)

58. W. Goes, F. Schanovsky, T. Grasser, Advanced modeling of oxide defects, in *Bias Temperature Instability for Devices and Circuits* (Springer, Berlin, 2014), pp. 409–446
59. W. Goes, Y. Wimmer, A.-M. El-Sayed, G. Rzepa, M. Jech, A. Shluger, T. Grasser, Identification of oxide defects in semiconductor devices: a systematic approach linking DFT to rate equations and experimental evidence. Microelectron. Reliab. **87**, 286–320 (2018)
60. Y. Wimmer, Hydrogen Related Defects in Amorphous SiO_2 and the Negative Bias Temperature Instability. Dissertation, Technische Universität Wien (2017)
61. D. Waldhoer, A.-M. El-Sayed, Y. Wimmer, M. Waltl, T. Grasser, Atomistic modeling of oxide defects, in *Noise in Nanoscale Semiconductor Devices*, ed. by T. Grasser (Springer, Cham, 2019). https://doi.org/10.1007/978-3-030-37500-3_18
62. G. Rzepa, J. Franco, B. O'Sullivan, A. Subirats, M. Simicic, G. Hellings, P. Weckx, M. Jech, T. Knobloch, M. Waltl, P. Roussel, D. Linten, B. Kaczer, T. Grasser, Comphy – a compact-physics framework for unified modeling of BTI. Microelectron. Reliab. **85**, 49–65 (2018)

Measurement and Simulation Methods for Assessing SRAM Reliability Against Random Telegraph Noise

Kiyoshi Takeuchi

1 Introduction

If we measure the drain-to-source current of a metal–insulator–semiconductor (MIS) field-effect transistor (FET) applying dc bias, excess low-frequency noise, called 1/*f* noise, is usually observed, whose power spectral density is approximately inversely proportional to frequency. A widely accepted explanation of 1/*f* noise today is that it is caused by superposition of multiple random telegraph noise (RTN) signals [1, 2]. If a trap site in the gate insulator of a MISFET captures or emits an electron, the conductance of the MISFET will change, depending on the charge state of the trap. An example of single-trap RTN waveform is shown in Fig. 1a. The signal can be characterized by three parameters: amplitude A, and time constants τ_0 and τ_1, where τ_0 and τ_1 are defined here as the mean time of stay in states 0 and 1, respectively. The power spectral density of the signal is given by [3]

$$S(f) = \frac{4A^2}{(\tau_0 + \tau_1)\left\{1/\tau^2 + (2\pi f)^2\right\}}, \quad \frac{1}{\tau} \equiv \frac{1}{\tau_0} + \frac{1}{\tau_1}. \tag{1}$$

By adding many such signals, whose τ value is distributed uniformly per $\log\tau$ (i.e., the expected number of traps in the range of 1–10 Hz is the same as 10–100 Hz) for many orders, a 1/*f* power spectrum is generated. Therefore, when the channel length *L* and width *W* of a MISFET are large, in which many traps are expected to exist, 1/*f* noise will be observed. However, as MISFETs are scaled down, it becomes more and more likely that a limited number of traps exist in a FET. In such situations, RTN signals become apparent (Fig. 1).

K. Takeuchi (✉)
Institute of Industrial Science, The University of Tokyo, Tokyo, Japan
e-mail: takeuchi@nano.iis.u-tokyo.ac.jp

T. Grasser (ed.), *Noise in Nanoscale Semiconductor Devices*,
https://doi.org/10.1007/978-3-030-37500-3_8

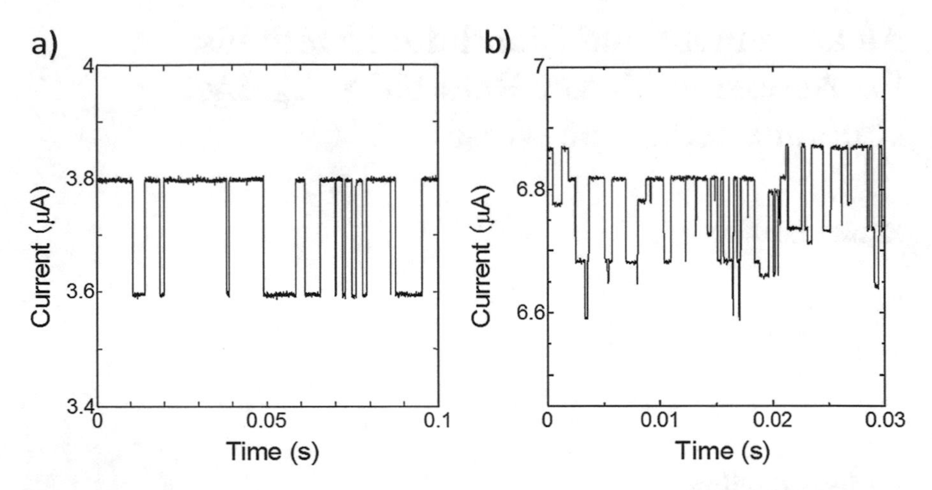

Fig. 1 Random telegraph noise waveforms for single trap (**a**) and multiple (three) traps (**b**) cases

While the expected number of traps per transistor will decrease by shrinking the channel area LW, on the contrary, the impact of RTN on transistor characteristics becomes more serious [4–6]. This is because the sensitivity of MISFET characteristics on a single charge carrier is increased in proportion to q/C_{OX}, where C_{OX} is the MISFET gate capacitance. As a result, recently, low-frequency noise has become a concern, not only for analog circuits but also for digital circuits using miniaturized transistors. RTN was first reported as a practical reliability problem for Flash memories [7–9], in which C_{OX} of the memory cell transistors is smaller than logic FETs, owing to the thicker gate dielectric and smaller LW used. Today, in state-of-the-art complementary metal–oxide–semiconductor (CMOS) integrated circuit technologies, L and W on the order of 10 nm are used, and hence C_{OX} has become extremely small, even for logic transistors. To realize reliable and error-free digital circuits using such advanced technologies, RTN must be taken into account [10–15].

To deal with RTN, a statistical approach is indispensable, since there is large inter-device "variability of noise." Depending on the number of traps, as well as the characteristics of each of the traps, one transistor exhibits statistically different noise waveforms from others. In such situations, effects of noise on MISFET circuits cannot be judged only from the averaged $1/f$ noise characteristics, which are obtained by measuring large transistors. Therefore, the author and coworkers proposed a set of methods for assessing "RTN reliability" of static random access memories (SRAMs) [16, 17], where "accelerated" SRAM measurements and fast Monte Carlo simulations are combined. SRAM cells are considered to be most vulnerable to RTN in logic integrated circuits, since the cell transistors are smaller than other logic transistors. In this chapter, these methods will be reviewed and given more thorough descriptions. Works in [6] are also briefly reviewed in the next section, which will serve as an additional introduction and motivation part of this chapter.

2 Individual FET Characterization

A simple and straightforward way of statistically characterizing RTN would be to measure noise waveforms for many individual transistors [5, 18–21]. For example, in [19], hundreds of n-channel and p-channel MISFETs were measured, applying dc bias on all the terminals. Then, the FETs exhibiting single-trap RTN signals were selected, and the parameters A, τ_0, and τ_1 for each RTN signal were extracted from the measured current vs. time waveforms, varying the dc bias conditions. By focusing on only such single-trap FETs, the determination of the trap parameters was simplified. This was possible, since single-trap RTN signals were found in a sufficiently large portion (20–30%) of all the FETs, thanks to the miniaturization. From such measurements, it was possible to discuss in detail the statistical distributions of the trap time constants, energy levels, and even vertical locations in the gate oxide. Correlation between the parameters was also easily examined. In [20, 21], transient measurements were also used, which is effective for covering a wider trap energy range than dc biasing. An advantage of this approach, i.e., simply measuring many individual transistors, is that it can be combined with almost any measurement and analysis methods of any sophistication. Detailed information on each trap can be obtained, e.g., by manipulating bias conditions or changing temperature. However, there is a disadvantage that single transistor measurements are usually time-consuming, and therefore it is difficult to measure a sufficient number of devices necessary for revealing detailed statistical distributions of the trap parameters in the low quantile range.

To alleviate this problem, addressable transistor arrays were used in [6]. Today, since variability in scaled down transistors is significant, characterization of variability in threshold voltage, drain current, etc. is indispensable, and therefore addressable transistor arrays are commonly used for this purpose [22–25]. By using addressable arrays, a large number of devices to be tested can be accommodated in a small area, by sharing area consuming pads by many devices. If appropriately designed, such arrays can be reused for characterizing RTN for a large number of FETs. The arrays used in [6] integrate 1024 identically designed MISFETs, where the gate and drain terminals of one selected FET out of the 1024 FETs, specified by an address value, are connected to external pads via FET switches (Fig. 2). Kelvin connections are used for the drain terminals (not shown) to reduce any voltage drop by parasitic resistance. Common source and body pads are also provided. Using this configuration, dc characterization including low-frequency noise measurements of each FET, one-by-one, using standard parametric testers is possible. However, if these 1024 FETs are measured sequentially, the measurement time per device will be essentially the same with single FET measurement cases, though the number of

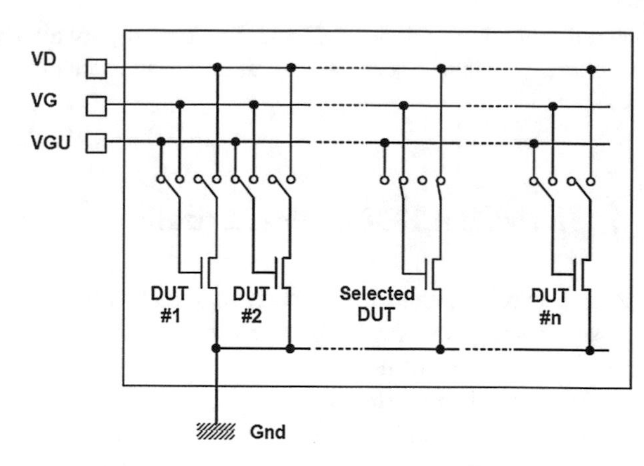

Fig. 2 Addressable transistor array configuration

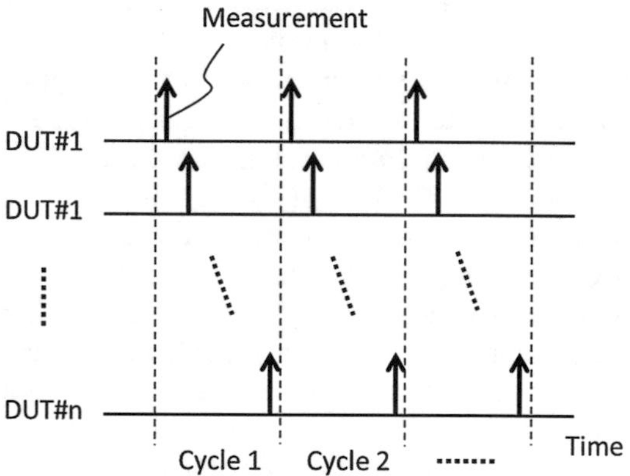

Fig. 3 Quasi-parallel measurement scheme

available FETs is increased. Therefore, to measure more FETs, while avoiding the unacceptable increase of measurement time, a quasi-parallel measurement scheme, as shown in Fig. 3, was adopted. First, only DUT #1 (DUT: device under test) is connected to the pads VD and VG, and its drain current is measured. Then, the connection is switched to DUT #2, and then to DUT #3, and so on, and finally to DUT #1024. After all the FETs are measured, the address goes back to #1, and the procedure is repeated for a desired number of times. Constant voltage is applied to all the pads VD, VG, and VGU throughout the measurements. By applying the same voltage on VG and VGU, transient trapping/de-trapping by DUT switching was avoided.

One significant advantage of parallel measurement is its ability to perform long time measurements for a large number of FETs, and hence information of slow traps can be obtained. Figure 4 shows examples of slow RTN signals found by the quasi-parallel scheme. It can be confirmed that traps with time constants on the order of hour actually exist. As will be discussed later, such slow traps are problematic in that their existence is not easily detected by short-time screening tests applicable to production. Drawback of the quasi-parallel scheme, compared with true parallel

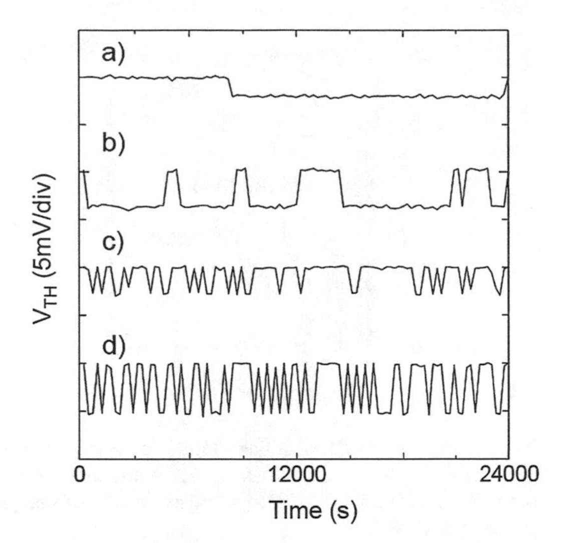

Fig. 4 Single-trap waveform examples obtained by quasi-parallel measurements. Waveforms (**a**) to (**d**) were taken from four different FETs in the same array. © 2009 the Japan Society of Applied Physics (JSAP). Reprinted, with permission, from [6]

measurements, is that the sampling interval becomes very long (at least 1024 times longer in this particular example). Because of the sparse sampling, obtained current vs. time data do not fully track true current vs. time RTN waveforms, unless the RTN time constants are much larger than the long sampling interval. Even if the measured current values for two consecutive samples are equal, it does not guarantee that the current was constant between the sampling events; it is possible that the current has travelled to a different value, and came back to the original one. In spite of this, if the measurements are repeated for many cycles, fast RTN signals whose time constants are much shorter than the interval can be detected, on a condition that the signal satisfies the following conditions: (1) Both f_0 and f_1 are sufficiently large compared to $1/N$, where N is the total number of current sampling per FET,

$$f_0 \equiv \frac{\tau_0}{\tau_0 + \tau_1}, \quad f_1 \equiv 1 - f_0 = \frac{\tau_1}{\tau_0 + \tau_1}, \tag{2}$$

and (2) both τ_0 and τ_1 are long enough compared to the time resolution of the source measure units (SMUs). From the data thus obtained, statistical distributions of trap number and amplitude could be efficiently determined, without spending the cost of true parallel measurements using 1024 sets of SMUs.

The measured source-to-drain current (I_{DS}) was first translated into an effective threshold voltage (V_{TH}) defined as

$$V_{TH} \equiv V_{GS} - I_{DS}/g_m, \tag{3}$$

where V_{GS} is the constant gate-to-source voltage, and g_m is the transconductance. The g_m value used for each FET was determined by individually measuring current–voltage characteristics of the same FET, to reduce the effects of g_m variability. Then, from the N sampled V_{TH} values per FET, the number of traps in the FET

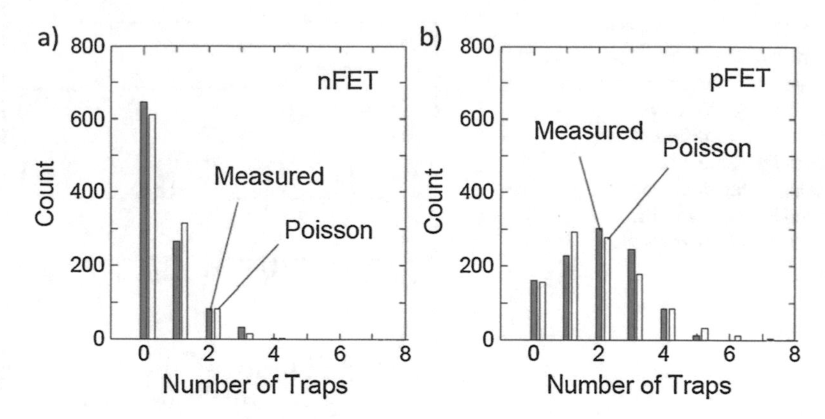

Fig. 5 Trap number distributions obtained by using addressable transistor array for nFETs (**a**) and pFETs (**b**). Poisson distributions whose mean values are equal to the measured values (0.52 for the nFETs, 1.90 for the pFETs) are also shown for comparison. © 2009 JSAP. Reprinted, with permission, from [6]

was determined using a simple algorithm: (1) Sort the N V_{TH} values for each FET, (2) find discontinuity of sorted V_{TH} values which is larger than a certain value (e.g., 0.5 mV), and (3) determine the number of traps n as

$$n = \text{ceil}\left(\log_2(m+1)\right), \tag{4}$$

where m is the number of discontinuities. Figure 5 shows the number distributions of detected traps for both n-channel and p-channel FETs, which were SRAM cell transistors fabricated by a conventional poly-Si/SiON gate stack technology. More traps were found in pFETs than nFETs in this example. Poisson distributions, whose expected values are set equal to the measured sample means, are also shown for comparison. Figure 5 shows that it is reasonable to assume Poisson distributions for the trap number in a FET. The slight disagreement could be attributed to the limited number of FETs measured and incompleteness of the simple number determination algorithm.

Next, RTN signal amplitudes were determined. To do this, similar to [19], only those FETs containing one trap were selected and used, to unambiguously determine the amplitude associated with a single trap. Figure 6 shows cumulative distributions of the amplitude of effective threshold voltage shift ΔV_{TH} (normalized by channel area LW) thus determined. It was found that the distributions can be approximated by exponential distributions, shown by the straight lines, which were also found in Flash memory cells [9]. For plotting the data in Fig. 6, care was taken to account for the fact that traps with too small amplitudes cannot be detected. Note that, in Fig. 6, there is a very small gap between $\Delta V_{TH} = 0$ and the lowest ΔV_{TH} plot. Since the number of the traps below this detection limit is unknown, that number was used as a fitting parameter. Denoting the number of detected traps n_1, and the number of undetected ones n_2, accounting for n_2 results in shifting all the plots

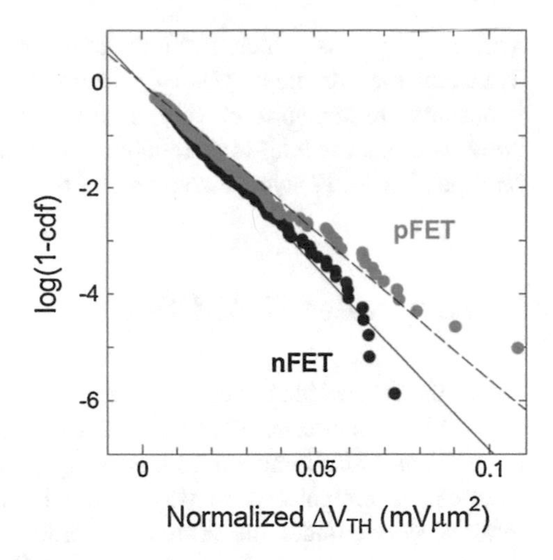

Fig. 6 Cumulative distributions of RTN amplitude obtained by using an addressable transistor array for n-channel FETs (black marks) and p-channel FETs (gray marks). Amplitude is normalized by transistor channel area. "Log" here is the natural logarithm. © 2009 JSAP. Reprinted, with permission, from [6]

(quantile values) in Fig. 6 downward by a constant ratio $n_1/(n_1 + n_2)$. The fitting parameter n_2 was determined, such that each straight regression line crosses the vertical axis $\Delta V_{TH} = 0$ at cdf $= 0$, where cdf is cumulative distribution function of ΔV_{TH}. In this way, two fitting parameters were used for determining the straight lines in Fig. 6, though exponential distribution itself contains only one parameter. Similar fit could be obtained by using naturally two parameter distributions, such as lognormal distributions [5]. However, assuming nonzero n_2 seems to be more adequate, since it can decouple the effect of the measurement limitation.

So far, methods for collecting statistical information of RTN by measuring individual transistors have been discussed. It was shown that the measurement efficiency can be improved by using addressable transistor arrays. However, a question arises. Is it practically possible to fully understand or describe RTN phenomena by straightforwardly accumulating such statistical information (number, amplitude, time constants, as well as their voltage and temperature dependence) of the traps? To obtain such information, the problem of limited measurement window must be considered. If we perform current sampling measurements with an interval T and sampling count N, only those traps whose time constants are sufficiently larger than T, and sufficiently smaller than NT, can be detected. On the other hand, to fully predict product reliability, it would be necessary to know about those traps, whose time constants are between around the operation clock cycle (e.g., 1 ns) and product lifetime (e.g., 10 years). To straightforwardly achieve this, measurements of many devices (e.g., 1000) must be continued for more than 10 years, which is practically impossible. Even if all such information is available, there is another problem. How can we predict product failure probability? It would be possible to simulate product failures using Monte Carlo circuit simulations [26–28], using the perfect RTN parameter sets. However, it is not realistic to repeat transient simulations of 10-year duration for a large number of devices. Therefore, some additional methods must be provided that can link individual trap characterization and product reliability.

Various proposals, which will be useful for achieving this goal, are already made. Transient measurements [29, 30] originally used in the field of bias temperature instability are known to be very effective for enlarging the measurement windows. New methods for long-term reliability simulations are also reported [31, 32]. The methods in [16, 17] were also intended for solving these problems.

3 Accelerated SRAM Test

As for the first problem mentioned above, the same also applies to other phenomena concerning long-term reliability, such as time-dependent dielectric breakdown (TDDB) and bias temperature instability (BTI). To determine TDDB lifetime, for example, a straightforward way would be to measure several devices for more than 10 years, under the normal operation conditions. However, practically this is not possible. Therefore, a common practice is to predict lifetime by combining accelerated tests and lifetime extrapolation [33, 34]. For example, time to dielectric breakdown (TBD) is measured by applying higher than normal voltage, to make TBD short enough to measure. By obtaining TBD values for several different such accelerated voltage conditions, TBD under normal operation voltage is estimated by extrapolation. A similar method would also be required for RTN. As for the second problem, it is considered that directly measuring circuit failures caused by RTN would be a good solution. As already mentioned, a circuit that will be most easily affected by RTN (or any random variability) in logic integrated circuits is static random access memory (SRAM). The reasons are that an SRAM cell uses smaller transistors than logic circuits, and that it is essentially an analog circuit relying on a subtle balance between the transistors constituting the memory cells. Based on these considerations, it was proposed to directly measure SRAM failures caused by RTN, by applying accelerated bias conditions [16, 17]. The work will be reviewed in the following.

Figure 7 shows a typical SRAM cell, consisting of six transistors. Transistors p_1 and d_1 form a first CMOS inverter, and p_2 and d_2 form a second. The two CMOS inverters are cross-coupled, and constitute a bistable latch. Transistors a_1 and a_2 serve as pass gates to connect the internal nodes n_1 and n_2 to the bit lines BL and BL$'$. Usually, an SRAM cell is disconnected from the bit lines, by turning off the pass gates. In this situation (retention state), the cell is very stable, owing to the near-ideal CMOS inverter transfer curves (Fig. 8a). The high node (either n_1 or n_2) voltage (V_1 or V_2) is close to the power supply voltage V_{CC}, and the low (the other) node voltage is close to zero. However, when the cell content needs to be read out, the pass gates are turned on by raising the word line (WL) voltage, while setting the voltage of the bit lines equal to V_{CC}. This deforms the inverter characteristics as shown in Fig. 8b. The low node voltage is pulled up by the bit line voltage. In this situation (read disturb state), the cell becomes less stable. SRAM cell failure due to RTN will occur, if any, almost certainly during this read disturb state. A static noise margin (SNM) [35, 36] is defined as the edge length of a square that nests

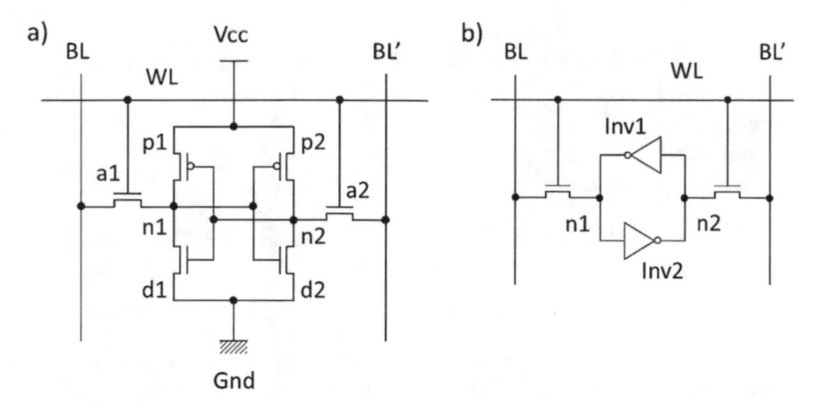

Fig. 7 Six-transistor CMOS SRAM cell (**a**) and its equivalent circuit (**b**)

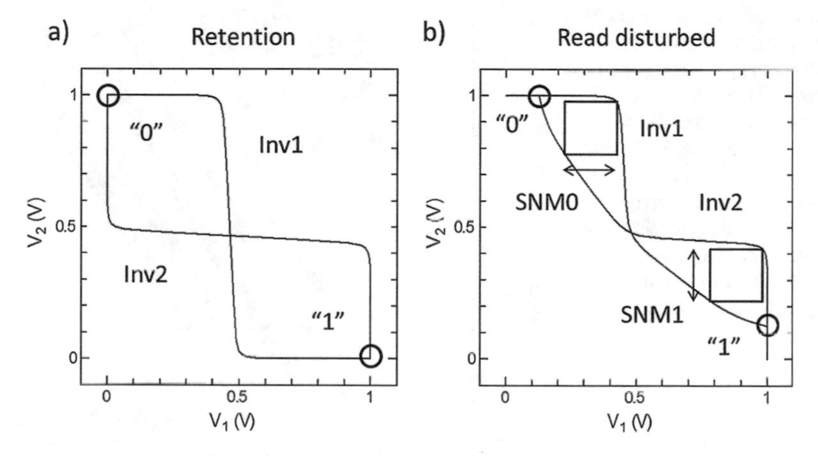

Fig. 8 Butterfly curves for retention (**a**) and read disturbed (**b**) states. V_1 and V_2 are the voltage at nodes n_1 and n_2, respectively. Circles show stable crossing points of two inverter transfer curves, which correspond to memory states "0" and "1," respectively. If SNM1 < 0, crossing point "1" disappears, and memory "1" is lost upon reading

in the "butterfly curves" in this read disturb state (see Fig. 8b). This definition can be modified to allow negative SNM, by defining SNM as the maximum distance between the two transfer curves (divided by the square root of two for compatibility with the original definition). A failure occurs if SNM becomes negative.

Figure 9 shows the concept of SRAM-accelerated test. In recent scaled SRAMs, there is large transistor variability. Usually, the threshold voltage of SRAM cell transistors is normally distributed, with a standard deviation of around a few tens of millivolts. As a result, SNM is also nearly normally distributed, because the nonlinearity between SNM and transistor threshold voltage is weak. To avoid yield loss due to the variability, SRAM cells are designed such that the mean SNM is larger than at least around six times the standard deviation σ (six sigma) of SNM. In this situation, SRAM cells whose SNM is small enough to be diminished by RTN is

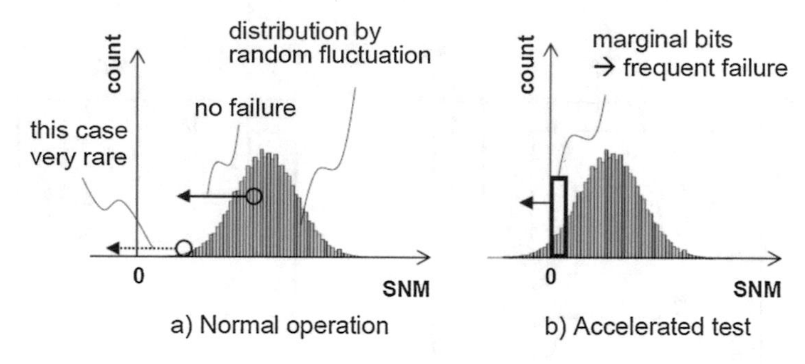

Fig. 9 Concept of accelerated SRAM test. Histogram of SNM in normal operation (**a**) is shifted to (**b**) by adjusting bias conditions. © 2010 IEEE. Reprinted, with permission, from [16]

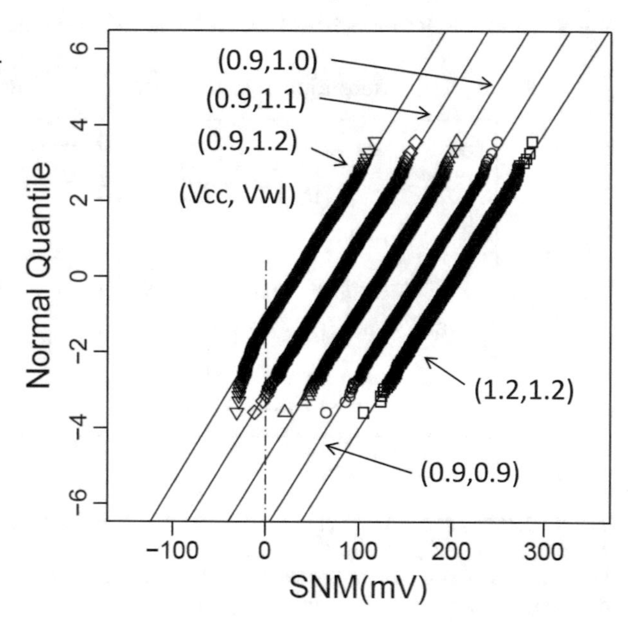

Fig. 10 Quantile–quantile plots of SNM distributions for various combinations of cell power supply voltage (V_{CC}) and word line voltage (V_{WL}), obtained by Monte Carlo circuit simulations. Sixty-five nanometer technology transistor models were used. A straight line corresponds to a normal distribution. SNM distributions can be shifted keeping the same shape

extremely rare, since the amplitude of SNM change due to RTN (σ is on the order of mV) is small compared to the SNM variability range (σ is a few tens of mV). Hence, in normal operation of properly designed SRAMs, RTN failure can hardly be detected (Fig. 9a). Therefore, to observe many SRAM failure events by RTN experimentally, it was proposed to intentionally reduce SNM by applying a special bias to the cell (Fig. 9b). In this situation, there will be a large number of cells whose SNM is negative due to variability, and always fail. In addition, there are cells whose SNM is so small that RTN can easily cause their failure. As a result, SRAM cell failure events due to RTN can be frequently observed. Fortunately, almost ideal parallel shift of the SNM distribution as schematically shown in Fig. 9 is possible, by simply applying a voltage higher than the nominal value to the word lines, and/or setting V_{CC} to a lower than nominal value, as demonstrated in Fig. 10.

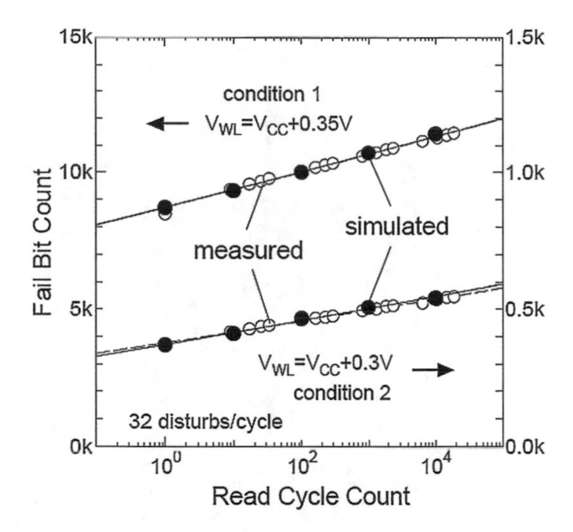

Fig. 11 Measured (open circles) and simulated (closed circles) fail bit count vs. read cycle count for two different bias (i.e., acceleration) conditions. Here, V_{WL} is constant. © 2011 JSAP. Reprinted, with permission, from [17]

For the accelerated SRAM measurements, a 512-k bit (32-k words × 16 bits) test SRAM cell array was used, which was capable of setting the high-state word line voltage (V_{WL}) and V_{CC} independently. A 40-nm bulk CMOS technology was used for the fabrication. First, zero was written to all the 512-k bits. Then, the SRAM array was repeatedly read out, one word (16 bits) at a time, by scanning the address starting from 0 to 32,767 (= 32 × 1024 – 1). Note the similarity to the quasi-parallel measurements discussed in the previous section. If a cell operates normally, zero will be read out every time. However, if a failure occurs in a cell during a read disturb period, the bit state is flipped from 0 to 1. Once a failure occurs, the bit state will never return to 0, since state 1 should be more stable for that specific bit, and 1 is read out for all the readings after the failure. Figure 11 shows results of such an accelerated SRAM read test, where the number of 1 bits (i.e., fail bit count, FBC) vs. read cycle count is plotted. It can be seen that, even at the very first reading, many cells fail. This is because, as a result of the intentional margin reduction, there are a large number of cells whose SNM is negative owing to the variability. This initial FBC should stay constant, no matter how many times the reading process is applied, if there is no noise. However, the fact is that the FBC monotonically increased, as the number of read cycles increased. This shows that some noise source that causes bit failures certainly exists. It is suspected that the noise may be caused by some external source, such as that generated by the memory tester. Therefore, the same measurement for the same chip was repeated three times, to check the reproducibility. Figure 12 shows maps of those bits that failed at the first reading (gray marks), and those that did not fail at the first reading, but failed at the second or later readings (solid marks). Similarity of the locations of the solid marks between the three trials is apparent. The time-dependent failures tend to occur repeatedly at the same bits. Those failing bits seem to be randomly distributed over the chip area. Absence of any positional correlation suggests that the failures are not caused by any deficiency of the array circuitry. These results strongly suggest that

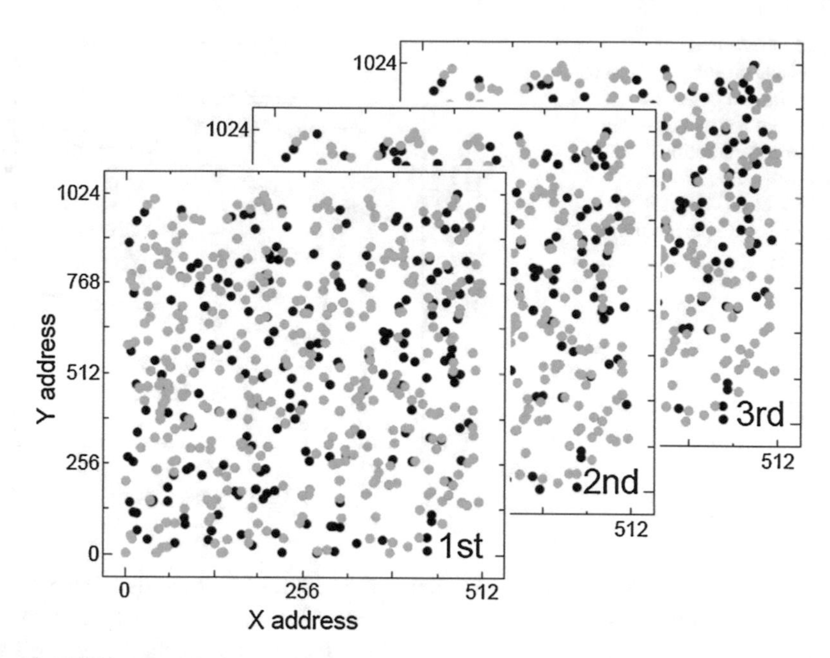

Fig. 12 Fail bit maps obtained by repeatedly measuring the same chip three times. Gray circles show always failing bits, and black ones show initially passed but finally failed bits. © 2010 IEEE. Reprinted, with permission, from [16]

the noise source exists in each SRAM cell, which should be RTN. Table 1 shows the statistics of the results. Among the bits that exhibited a time-dependent failure at least once, 60% consistently showed first-pass, last-fail behavior in all the three measurements. However, the repeatability is not perfect. This is natural considering that the failures caused by RTN will be statistical.

4 RTN Monte Carlo Simulation

By using SRAM arrays, direct observations of RTN failures seem to be possible. However, still, the link between the measurements and product reliability is missing. To solve this problem, realizing numerical simulations of the SRAM failures was considered, to understand the failure mechanism. Let us first discuss what is happening in a cell during the accelerated SRAM measurements. As already pointed out, a cell stays in the retention state for almost all the time, in which the cell is very stable, and no failure is expected to occur. However, trapping and de-trapping do occur in the cell transistors, and their threshold voltage (V_{TH}) will change over time. Since the applied bias does not change in the retention state, each FET is in a similar situation as during dc bias RTN measurements. This situation is occasionally interrupted by the read operations, which will cause some transient

Table 1 Statistics of bit failures for three repeated accelerated SRAM tests

Case	Bit count	Ratio (%)
Bias condition 1		
0–1	28	12.33
0–2	15	6.61
0–3	141	62.11
1–3	23	10.13
2–3	20	8.81
Subtotal	227	100.00
3–3	344	
Total	571	
Bias condition 2		
0–1	328	8.93
0–2	248	6.76
0–3	2252	61.35
1–3	461	12.56
2–3	381	10.38
1–2	1	0.03
Subtotal	3671	100.00
3–3	8086	
Total	11757	

Case *x*–*y* means that the bit failed *x* out of three times at the first reading, and *y* times at the last reading. 3–3 means the bit always failed, and 0–3 means the bit always passed at the first reading, but always failed at the last reading.

effects. However, since the duration of a read disturb state is very short (typically, on the order of 0.1–10 ns, for product SRAMs), it is assumed that the slow traps of interest for long-term reliability will be frozen during the reading operation, and the trap states of the cell established during the retention state will be sampled. Since SNM is a function of $V_{\mathrm{TH},i}$ ($i = 1,2,\ldots,6$), where $V_{\mathrm{TH},i}$ is the threshold voltage of the ith cell transistor, SNM changes over time depending on the trap states of all the cell transistors. The sampling result will be either 0 (SNM ≥ 0, pass) or 1 (SNM < 0, fail).

Based on these considerations, a simple and fast Monte Carlo (MC) simulation method was proposed [16], which will be described below. The basic idea is that, given the parameters (amplitude and time constants) of all the traps in a cell, and assuming that the amplitudes are additive, it would be possible to estimate the worst-case amplitude that is likely to occur in N sampling events, by simple analytical considerations, even if N is a very large number (e.g., $N = 3 \times 10^{14}$ for 1-M samples/s $\times$ 10 years). First, let us introduce a linear approximation

$$\Delta\mathrm{SNM} = \sum_{i=1}^{6} a_i \Delta V_{\mathrm{TH},i}, \tag{5}$$

where ΔSNM is SNM deviation, and $\Delta V_{\mathrm{TH},i}$ is threshold voltage deviation of the ith transistor ($i = 1,2, \ldots, 6$) from the respective reference values. The set of coefficients a_i can be determined by iteratively searching for the most probable failure point (MPFP) using circuit simulations, applying SRAM cell variability design methods [37, 38]. We also assume that trap amplitudes are additive. That is,

$$\Delta V_{\mathrm{TH},i} = \sum_{j=1}^{n_i} \Delta V_{\mathrm{TH},i,j}, \tag{6}$$

where n_i is the number of traps in the ith transistor, $\Delta V_{\mathrm{TH},i,j}$ is the threshold voltage deviation caused by the jth trap in the ith transistor. By combining Eqs. (5) and (6), the SNM shift is now expressed as a simple linear combination of the V_{TH} shifts caused by all the traps. Here, $\Delta V_{\mathrm{TH},i,j}$ denotes the V_{TH} change from some reference value, and is not necessarily equal to the amplitude ($\equiv A_{i,j}$) of the trap (i, j), but can be either 0 (no trap state change) or $\pm A_{i,j}$ (low V_{TH} to high V_{TH} transition, and vice versa). For simulating the time-dependent FBC increase as in Fig. 11, it would be natural to select the initially sampled (read disturbed) state as the reference. If $\mathrm{SNM} + \Delta\mathrm{SNM} < 0$ at the moment of any read disturb (including the first), the cell will fail. Note that SNM is also randomly distributed due to variability, and differs from cell to cell.

Noting that an SRAM read failure is determined only by the trap states at the moment of a read disturb, a simple time domain MC simulation would be, in principle, possible. The probability that a trap is in state 0 at time zero, and is found in state 0 (denoted P_{00}) or 1 (denoted P_{01}) after a time t is given by [3]

$$P_{00} = f_0 + f_1 \exp\left(-\frac{t}{\tau}\right), \; P_{01} = f_1 \left(1 - \exp\left(-\frac{t}{\tau}\right)\right). \tag{7}$$

Using this formula, it is possible to track the state of a trap over time by MC simulations. That is, starting from an initial state, the state of a trap at the next reading is probabilistically determined using Eq. (7). Then, collecting all the states of the traps in a cell, Eqs. (5) and (6) are calculated to judge if SNM of the cell is negative or not. This can be repeated for a desired number of times to simulate the behavior as in Fig. 11. This discretized time domain simulation, based on a Markov chain model, is much more efficient than industry standard general purpose transient circuit simulations. However, even using this method, simulations of a sufficiently large number of cells (e.g., 1-M cells) for a sufficiently large number of readings (e.g., $N \sim 10^6$ for 1 s, not to mention $N \sim 3 \times 10^{14}$ for 10 years) is still computationally too demanding. It should also be noted that the number of traps in a cell to be simulated is much larger than the measured numbers shown in Fig. 5, since traps with small probability of transition, which did not fall within the measurement window, must be taken into account. Therefore, in [16, 17], an even simpler method of MC simulation was adopted. That is, the largest amplitude of a cell that is likely to occur in N reading events with a constant interval T was directly calculated from the trap parameters. The procedure is as shown below. In the following, the state of a trap, which corresponds to the better SNM (good state), will be denoted state 0, and the other (bad state) will be denoted state 1.

(S1) Assign an SNM value to cell, according to random variability of transistors.
(S2) Assign number of traps n to each transistor.
(S3) Assign amplitude A and two time constants τ_0 and τ_1 to each trap.
(S4) Randomly select initial state of all the traps, according to the ratio of τ_0 and τ_1.
(S5) Find a combination of traps which maximizes $-\Delta$SNM, while the probability of finding all the traps simultaneously in state 1 after N read disturb events is high enough.
(S6) If SNM + ΔSNM<0, judge that the cell will fail.

By repeating this for many cells, and for several N values, FBC vs. N relationship can be simulated. Note that the simulation time using this method does not depend on N, and is much more simplified than [27, 28]. Therefore, simulations of 10 years operation can be easily performed.

It can be noticed that, by following this procedure, a cell failure is deterministic. That is, a cell is always assumed to fail, if N exceeds a certain value. However, in reality, a failure of the same cell may occur much earlier or later. This simplification (i.e., use of an expected lifetime for a given set of trap parameters) could be justified by the following reasons. Firstly, the simulation will be performed for a large number of cells. Because of the simplification, simulating 1-M bits is easily accomplished. As a result, the stochastic difference between real and simulated time-to-failure will be averaged, and its effect on FBC will be reduced. This is supported by the fact that almost the same FBC vs. cycle count results are obtained by measuring the same array for three times (Fig. 11). Second, consideration of stochastically different waveforms for only a single bit introduces additional dimension of variability. At an early stage of study, removal of such complication would be desirable.

For the MC simulations, normal distributions for the random variability of SNM, Poisson distributions for trap number, and exponential distributions for trap amplitude are assumed, taking into account the results shown in Sect. 2. As for the time constants, it is assumed that the occupancy ratio follows Fermi–Dirac type relationship [1, 2].

$$\frac{\tau_1}{\tau_0 + \tau_1} = \frac{1}{1 + \exp\left(\frac{E - E_{\mathrm{F}}}{k_{\mathrm{B}} T}\right)}, \tag{8}$$

where the trap energy E is uniformly and symmetrically distributed around E_{F}. It is also assumed that the effective time constant τ defined as

$$\frac{1}{\tau} \equiv \frac{1}{\tau_0} + \frac{1}{\tau_1} \tag{9}$$

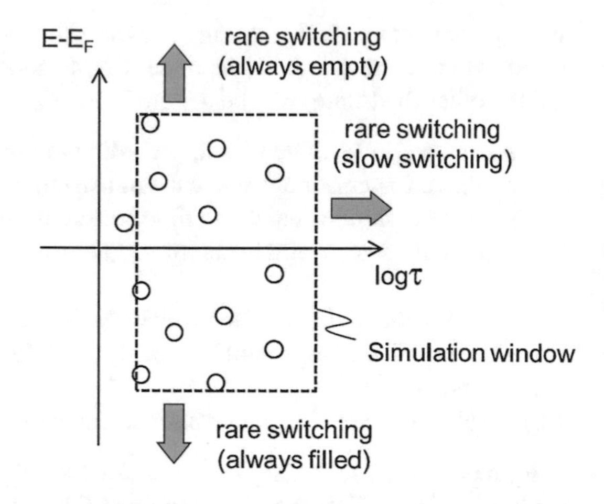

Fig. 13 Energy vs. effective time constant plane. Traps are assumed to be uniformly distributed in this plane. Traps in a sufficiently large rectangle are considered for simulation

is uniformly distributed per logτ, as already mentioned in connection with 1/*f* noise. Then, τ_0 and τ_1 are given as

$$\tau_0 = \tau\left\{1 + \exp\left(\frac{E - E_{\rm F}}{k_{\rm B}T}\right)\right\},\ \tau_1 = \tau\left\{1 + \exp\left(\frac{E_{\rm F} - E}{k_{\rm B}T}\right)\right\}. \tag{10}$$

The uniform distributions were chosen here as crude initial assumption. The lower bound of τ was set equal to the duration of read disturb, whereas, the upper bound of τ was selected to be large enough compared with the time range to be simulated. The upper and lower bounds of E–$E_{\rm F}$ were selected to be sufficiently away from zero, so that the probability of switching outside these bounds is negligible. In other words, it was assumed that traps are uniformly distributed in a rectangle placed in the E vs. logτ plane (Fig. 13); the rectangle is selected to be large enough so that switching outside its top, bottom, and right edges is so rare and can be ignored. The mean number of traps λ and their mean amplitude Λ were selected by fitting to SRAM measurement results, using the values obtained by the addressable transistor array measurements as initial guess. Note that the number of traps to be fed to the simulator should not be equal to the actually measured number. Since the area of the rectangle in the E vs. logτ plane used for the simulation is much larger than that covered by measurement windows, the measured number should be multiplied by the ratio of the areas (simulated over measured). It was assumed that Λ, E, and τ, are independent, and that SNM variability and RTN are also independent, according to our measurement results, part of which are reported in [19].

FBC vs. cycle count simulated using the above described method is overlaid in Fig. 11. The MC simulation could reproduce the accelerated SRAM measurement results quite well, using reasonable trap parameters consistent with individual transistor measurements. For the comparison, the fact that a cell suffers from 32 read disturbs in a cycle was taken into account. This is because a word line is shared by 512 bits, while only 16 bits are read out at a time. Therefore, N is set equal to 32 times the cycle count. Because of this shared word line architecture of the array, which is customary for SRAMs, the first reading does not necessarily correspond to the first disturb, but may be any of the 1st to 32nd disturb. This will cause some overestimation of SNM variability, since any failure at the first reading is regarded as caused by SNM variability, not RTN. This inaccuracy is ignored here, expecting that its impact on long-term results will be small. Assuming that SNM is normally distributed, the mean SNM (μ) normalized by its standard deviation σ can be estimated from the initial fail bit ratio (FBR = FBC/512 k), using a relationship

$$1 - \Phi\left(\frac{\mu}{\sigma}\right) = \text{FBR}, \; \Phi(x) \equiv \frac{1}{\sqrt{2\pi}} \int_{-\infty}^{x} \mathrm{e}^{-\frac{x^2}{2}} \mathrm{d}x. \tag{11}$$

Since SNM of the memory cells in the array was not directly measurable, it was decided to assume that mean SNM estimated using Eq. (11) is the true mean SNM. This results in automatic alignment of the measured and simulated first FBC. The increase of FBC with the number of cycles depends on the trap parameters. It was found that good agreement between the measurement and simulation can be obtained without much effort of parameter fitting. The results for the two different bias conditions were reproduced by the same set of trap parameters without bias dependence, in spite of the 50-mV V_{CC} difference. It was also found that different sets of parameters can yield almost the same simulation results. This suggests that, for the modeling of RTN failures, a simplified noise model with reduced number of parameters, e.g., by assigning traps to only one or two transistors, could be used. In the following, to save simulation time, only the three most relevant transistors for read stability were taken into account, while keeping the parameters in the range reasonably consistent with individual FET measurements.

5 Reliability Extrapolation

With the aid of MC simulations, it is now possible to discuss, in more detail, what is happening during the accelerated SRAM measurements. Figure 14 shows the simulated probability density function (pdf) of worst RTN amplitude (i.e., $-\Delta$SNM determined in step (S5) of the simulation, or largest amplitude expected to be found in at least one of N read disturbs). As the number of disturbs N increases, it becomes more likely that a cell encounters a larger SNM amplitude, since the probability of simultaneous switching of more traps towards the bad direction increases. This is because slow traps with large τ or rarely switching traps with large $|E\text{–}E_{\text{T}}|$ contributes to the amplitude. As a result, the mode (peak position) of the amplitude pdf increases with N, while its variance also increases. It was found that these simulated pdfs can be approximated by gamma distributions. This is expected, since exponential distributions are assumed for the single-trap amplitude, and that there is a close relationship between exponential and gamma distributions.

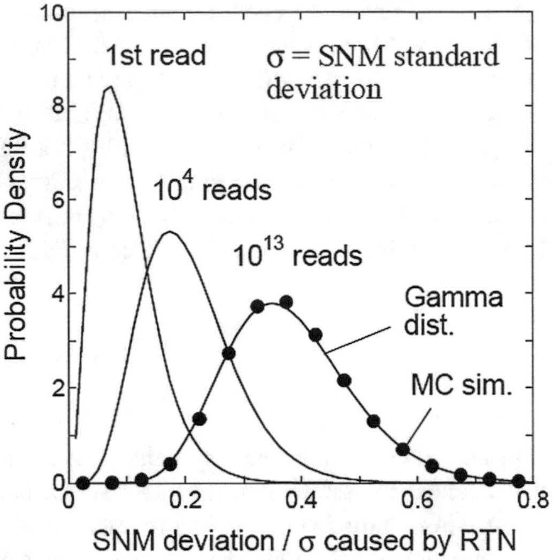

Fig. 14 Simulated worst RTN amplitude distributions, normalized by SNM standard deviation. © 2011 JSAP. Reprinted, with permission, from [17]

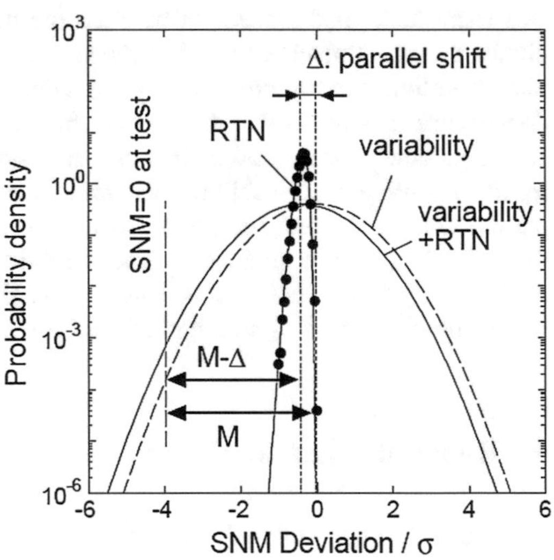

Fig. 15 Explanation of probability distribution shift of worst-case SNM, caused by coexistence of variability (variability without time dependence) and RTN (time-dependent variability). Original normal distribution of variability is shifted horizontally. © 2011 JSAP. Reprinted, with permission, from [17]

It should be noted that there is large SNM variability, which is not time-dependent. RTN amplitude calculated in Fig. 14 is still much smaller than the SNM variability. Let us now consider what happens when variability and RTN coexist. Assuming that SNM variability and RTN are independent, the pdf of the sum of the two can be calculated by convolution of the respective pdfs. Figure 15 shows an exemplary result of such convolution obtained numerically. It can be seen that the resulting summed amplitude distribution is almost equal to the original SNM normal distribution with a shifted mean. Although the RTN amplitude pdf has some width, since the distribution is much narrower than SNM variability, the original normal distribution is scarcely broadened.

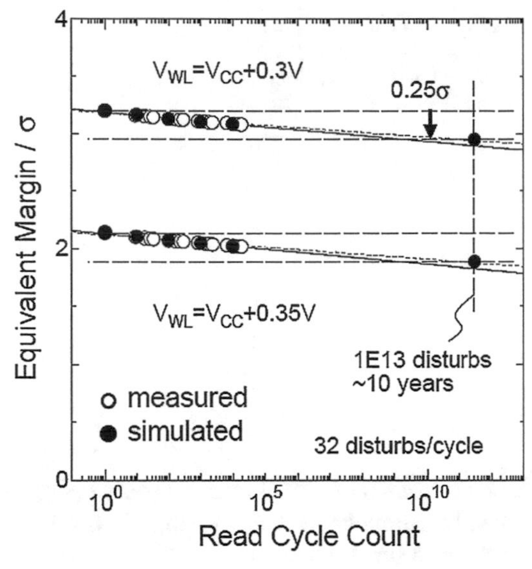

Fig. 16 Measured (open circles) and simulated (closed circles) effective margin vs. read cycle count. © 2011 JSAP. Reprinted, with permission, from [17]

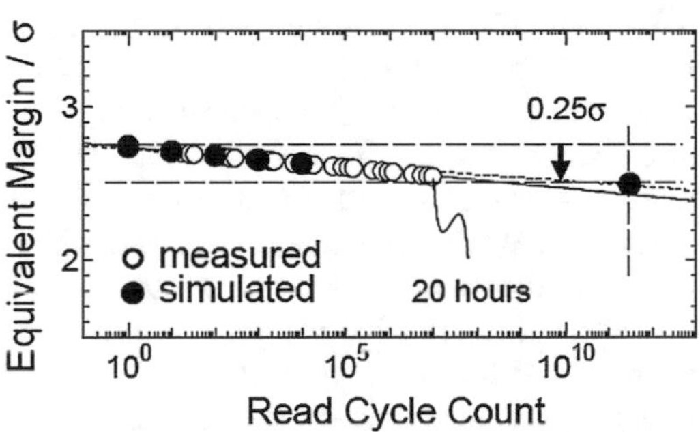

Fig. 17 Measured (open circles) and simulated (closed circles) effective margin vs. read cycle count for an extended number of cycles. © 2011 JSAP. Reprinted, with permission, from [17]

Considering that RTN effectively shifts the SNM distribution as the number of read disturbs N increases, we can define an effective SRAM margin M as a function of N. $M(N)$ normalized by the SNM standard deviation σ is calculated from the measured fail bit ratio (FBR) at the Nth read disturb using Eq. (11), where $M(N)$ is obtained as μ, which satisfies Eq. (11). By using this translation of FBR to $M(N)$, we can plot effective SRAM margin vs. read disturb counts, as shown in Fig. 16. It can be seen that the effective margin linearly decreases with $\log(N)$. The fact that the slope of the decrease does not depend on the initial FBC (i.e., the degree of acceleration) supports the assumption that the effective SNM distribution is shifted in parallel. It has to be pointed out here that the linear relationship is a result of assuming a uniform trap number distribution in the energy vs. $\log\tau$ plane. If the real distribution deviates from this assumption, a nonlinear dependence should be observed. To confirm the linearity for a larger N range, an accelerated measurement for extended number of cycles was performed. A linear relationship up to 10-M cycles (320-M disturbs) was confirmed (Fig. 17). These results suggest that effective

Fig. 18 Explanation of guard banding (GB). © 2011 JSAP. Reprinted, with permission, from [17]

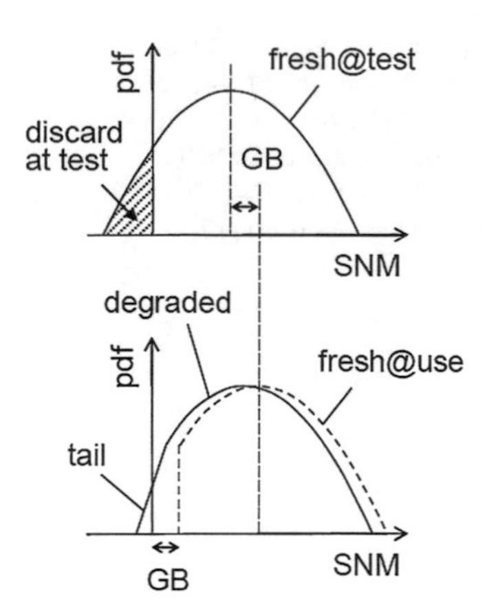

margin loss due to RTN after 10 years operation can be empirically estimated by linearly extrapolating $M(N)$ vs. $\log(N)$. If a 1-M byte (= 8-M bits) SRAM in which each word line is shared by 512 bits, and all the word lines are evenly activated by a 500-MHz clock, a cell will be disturbed 30-k times per second (=500 MHz/8-M bits $\times$ 512 bits). Then, the margin loss ΔM after 10 years at $N = 1 \times 10^{13}$ (~30-k disturbs/s $\times$ 10 years) is estimated to be around 0.25σ, both by extrapolation and direct simulation. Though this value is much smaller than the usually assumed worst-case variability of 6σ, it is not negligible, since this degradation is time-dependent. While time-independent variability only degrades the product yield at shipment, the margin loss is relevant to reliability in the field.

Though the effective margin loss can now be estimated by the extrapolation technique, it has to be pointed out that the tolerance necessary for guaranteeing reliability would be somewhat larger than the ΔM value thus obtained, as discussed below. Consider a situation as shown in Fig. 18, where SRAMs, whose cells suffer from SNM variability, are tested before shipment, and a certain part of the products are discarded because of a failure caused by negative SNM. Since RTN should effectively shift the SNM variability distribution to the left during use, this testing condition should be stricter than the real use conditions. Then, the SNM distribution of an SRAM in use that passed the test would look like the dashed line in Fig. 18b. It can be assured that, if RTN is ignored, the worst SNM in the SRAM is greater than zero by at least a certain amount. This tolerance will be called a guard band (GB). If RTN simply shifts this truncated SNM distribution in parallel, similarly to Fig. 15, a GB width slightly larger than the extrapolated ΔM would suffice. However, this is too optimistic. Figure 19 shows a result of convoluting a truncated normal distribution (corresponding to variability) with a gamma distribution (corresponding to RTN). In addition to a parallel shift as in Fig. 15, a tail emerges at the left side, extending further than the parallel shift, owing to the nonzero width of the gamma distribution. The GB width should be determined such that the probability of failure during use does not exceed a certain acceptable limit (may be 1 ppm or 0.1%, depending on the applications). In doing so, existence of the tail portion

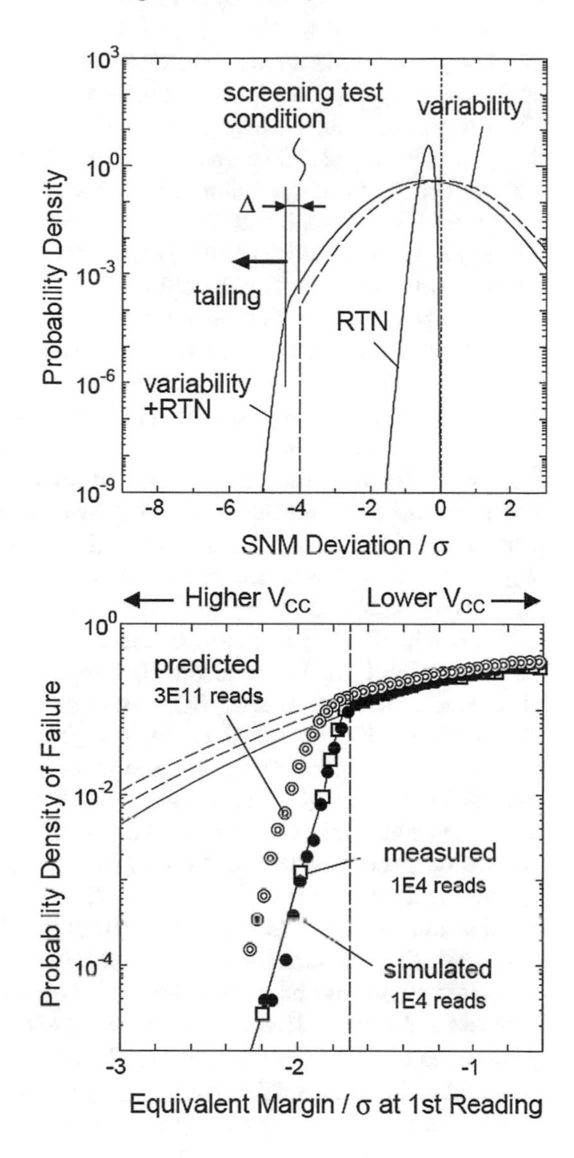

Fig. 19 Explanation of probability distribution change of worst-case SNM, caused by coexistence of variability and RTN, where the initial variability distribution is truncated by a screening test. In addition to a horizontal parallel shift, additional tailing caused by RTN amplitude variability emerges. The screening condition -4σ is too strict practically, and is chosen here for illustrative purpose. © 2011 JSAP. Reprinted, with permission, from [17]

Fig. 20 Measured (open squares) and simulated (closed circles) distributions of worst-case SNM, corresponding to Fig. 19. Double circles show simulation results corresponding to 10-years operation ($=3 \times 10^{11}$ reads $= 1 \times 10^{13}$ disturbs). © 2011 JSAP. Reprinted, with permission, from [17]

should be taken into account. In the following, it will be shown that the shape of the distribution tail in Fig. 19 can be estimated, again using both the accelerated SRAM measurements and MC simulations, as shown in Fig. 20.

The measured plots in Fig. 20 were obtained as follows. The accelerated SRAM measurements of 10-k cycles were repeated for the same array, by changing the cell power supply voltage V_{CC} with a small interval from 0.88 V to 0.98 V, while keeping the word line voltage V_{WL} constant at a higher than nominal value (1.3 V). At these bias conditions, the mean SNM normalized by σ, or margin M (Fig. 15), is moved to around 1–2, depending on the V_{CC} value. Lower V_{CC} corresponds to lower M. For each condition, the address values of all the fail bits at the first and the last readings were recorded. Let us denote fail bit counts for the first and last cycles FBC_1 and FBC_2, respectively ($FBC_1 < FBC_2$). By translating FBC_1 into a SNM deviation using Eq. (11), M for a certain V_{CC} value can be estimated. That is, the μ/σ that satisfies Eq. (11) is an estimate of M for each V_{CC}. Since FBC_1 was measured for several V_{CC} values, it was also possible to approximately determine the pdf of FBC_1 (pdf_1) by calculating (ΔFBC1/512 k)/ΔM, where Δx means the difference of x between two adjacent V_{CC} conditions and 512 k is the total number of bits. By definition, pdf_1 vs. $-M$ plots should fall on a standard normal distribution $N(0,1)$. Similarly, pdf_2 can be obtained from FBC_2, which will fall on a shifted normal distribution $N(-\Delta M,1)$. Note that the negative sign accompanying M comes from the fact that in the evaluation procedure here, the shape of the pdf is obtained by changing M and counting bits whose margin deviation is smaller than $-M$ (i.e., SNM < 0). By increasing M, the pdf information at a more negative deviation $-M$ is accessed. To obtain the tail distribution, the measured data were further manipulated. First, one of the V_{CC} condition ($V_{CC} = 0.92$ V, $M = 1.7\sigma$ denoted M_{REF}) was selected as a reference and a special fail bit count FBC_2' was defined. FBC_2' is the number of those bits that failed after 10-k cycles at the respective V_{CC} (or M) condition, but passed at the reference condition at the first reading. Since the entire fail bit maps were recorded, determination of FBC_2' is straightforward. Then, similar to other cases, FBC_2' was converted to pdf_2'. The open marks in Fig. 20 show pdf_2' vs. $-M$ thus obtained. The vertical line shows the reference condition. This mimics a shipment test discussed earlier, though the condition is unrealistically strict such that many failure bits can be measured (i.e., this is an accelerated condition). On the left-hand side of the line, an exponentially decaying tail was obtained, which should correspond to the tail in Fig. 19. These tail bits passed a stricter test ($M = M_{REF} = 1.7\sigma$) at the first reading, but failed under a looser condition ($M > 1.7\sigma$) after 10-k cycles (= 320-k disturbs).

In Fig. 20, MC simulation results are also shown (solid circles). It is first mentioned that the procedures for obtaining the measured and simulated plots in Fig. 20 would look quite different. This is because while the measured pdf is determined by scanning it at SNM = 0 by shifting the mean of the pdf by sweeping V_{CC}, the simulated pdf is determined directly. The simulations were performed by modifying the procedure in Sect. 4, by replacing steps (S5) and (S6) with the following.

(S5′) Find ΔSNM for $N = N_1$ ($\equiv\Delta SNM_1$), using the procedure as in (S5).
(S6′) Find ΔSNM for $N = N_2$ ($\equiv\Delta SNM_2$), using the procedure as in (S5).
(S7′) If $(SNM + \Delta SNM_1 - \mu)/\sigma < -M_{REF}$, mark the cell as rejected at test.
(S8′) Record the values of $SNM + \Delta SNM_1$ and $SNM + \Delta SNM_2$.

The simulated solid marks in Fig. 20 were obtained by creating a histogram of $SNM + \Delta SNM_2$ for a large number of bits (i.e., MC simulation runs), and normalizing it both horizontally and vertically. The screening operation at $M = M_{REF}$ was taken into account by assigning an out-of-range value to $SNM + \Delta SNM_2$ of the rejected bits. For the horizontal normalization, an SNM + ΔSNM_2 value must be translated into a normalized SNM deviation. To do this, a seemingly natural choice is to use the transformation

$$f(x) = \frac{x - \mu}{\sigma}, \tag{12}$$

where μ and σ are the mean and standard deviation of SNM determined in step (S1) of the MC simulations. However, in real measurements, it is not possible to perfectly separate the effects of variability and RTN. Some randomness induced by RTN is always mixed up with variability. This is also the case for the measured plots in Fig. 20, where the normalized deviation was estimated from measured FBC_1 data, which is certainly affected by RTN. Therefore, for the horizontal normalization, μ and σ in Eq. (12) were replaced by those of SNM + ΔSNM_1 to make the results consistent with the measurements. In addition, N_1 in step (S5′) was set to 16 (i.e., average number of disturbs before the first reading in the SRAM measurements). By properly taking into account the effects of RTN on the first reading in this way, the good agreement between the measurements and simulations was obtained, using the same set of parameters that reproduce FBC vs. cycle count measurements. Without such measures, the simulated extension of the tail portion will be overestimated. It is also pointed out that, if we perform multiple tests, or increase N_1, the extension of the tail can be reduced, because more vulnerable bits can be screened. Once the MC simulation is calibrated in this way, the tail distribution after 10 years operation can be estimated using the simulation, as shown by the dotted circle plots in Fig. 20.

6 Conclusion and Remarks

A systematic set of methods for assuring SRAM reliability against RTN proposed earlier [16, 17] has been reviewed, supplementing additional details. It was pointed out that there is similarity between RTN reliability and other long-term reliability issues (e.g., TDDB, BTI), owing to the existence of extremely rarely switching traps, and that something similar to "lifetime extrapolation" is necessary. Another important point is that there is large inter-device variability of RTN waveforms and statistical evaluation of a large number of devices is mandatory. It was argued that these requirements could be met by using accelerated SRAM array measurements, combined with an extremely simplified and fast Monte Carlo simulation. A procedure for evaluating SRAM read failure probability in a product lifetime (e.g., 10 years) was described.

It is considered that, regarding the methods presented, there are still problems that remain to be addressed. Following are some of them that seem to be important. (1) The methods should be extended to cover SRAM write stability, which is equally important as read stability. (2) Validity of the MC simulation algorithm should be proved mathematically. The agreement with the measurements is still empirical. (3) The trap parameter distributions assumed, which strongly affect extrapolation results, seems to be too simplified. More trustworthy parameter settings based on experimental and/or theoretical studies are desired.

Acknowledgments The author would like to express sincere gratitude to Toshiharu Nagumo, Koichi Takeda, Shinobu Asayama, Shinji Yokogawa, Kiyotaka Imai, Yoshihiro Hayashi, and Takashi Hase. The work could not be accomplished without their collaboration and/or support.

References

1. K.S. Ralls, W.J. Skocpol, L.D. Jackel, R.E. Howard, L.A. Fetter, R.W. Epworth, D.M. Tennant, Discrete resistance switching in submicrometer silicon inversion layers: Individual interface traps and low-frequency (1/f?) noise. Phys. Rev. Lett. **52**(3), 228–231 (1984)
2. M.J. Kirton, M.J. Uren, Noise in solid-state microstructures: A new perspective on individual defects, interface states and low-frequency (1/f) noise. Adv. Phys. **38**(4), 367–468 (1989)
3. S. Machlup, Noise in semiconductors: Spectrum of a two-parameter random signal. J. Appl. Phys. **25**(3), 341–343 (1954)
4. M. Tsai, T. Ma, The impact of device scaling on the current fluctuations in MOSFET's. IEEE Trans. Electron Devices **41**(11), 2061–2068 (1994)
5. N. Tega, H. Miki, F. Pagette, D.J. Frank, A. Ray, M.J. Rooks, W. Haensch, K. Torii, Increasing threshold voltage variation due to random telegraph noise in FETs as gate lengths scale to 20 nm, in *Symposium on VLSI Technology (VLSIT)*, 2009, pp. 50–51
6. K. Takeuchi, T. Nagumo, S. Yokogawa, K. Imai, Y. Hayashi, Single-charge-based modeling of transistor characteristics fluctuations based on statistical measurement of RTN amplitude, in *Symposium on VLSI Technology (VLSIT)*, 2009, pp. 54–55
7. K. Fukuda, Y. Shimizu, K. Amemiya, M. Kamoshida, C. Hu, Random telegraph noise in flash memories—model and technology scaling, in *IEEE International Electron Devices Meeting (IEDM)*, 2007, pp. 169–172
8. H. Kurata, K. Otsuga, A. Kotabe, S. Kajiyama, T. Osabe, Y. Sasago, S. Narumi, K. Tokami, S. Kamohara, O. Tsuchiya, Random telegraph signal in flash memory: Its impact on scaling of multilevel flash memory beyond the 90-nm node. IEEE J. Solid State Circuits **42**(6), 1362–1369 (2007)
9. C.M. Compagnoni, R. Gusmeroli, A.S. Spinelli, A.L. Lacaita, M. Bonanomi, A. Visconti, Statistical model for random telegraph noise in flash memories. IEEE Trans. Electron Devices **55**(1), 388–395 (2008)
10. N. Tega, H. Miki, M. Yamaoka, H. Kume, T. Mine, T. Ishida, Y. Mori, R. Yamada, K. Torii, Impact of threshold voltage fluctuation due to random telegraph noise on scaled-down SRAM, in *IEEE International Reliability Physics Symposium (IRPS)*, 2008, pp. 541–546
11. S.O. Toh, Y. Tsukamoto, Z. Guo, L. Jones, T.K. Liu, B. Nikolić, Impact of random telegraph signals on Vmin in 45nm SRAM, in *IEEE International Electron Devices Meeting (IEDM)*, 2009, pp. 767–770
12. S.O. Toh, T.K. Liu, B. Nikolić, Impact of random telegraph signaling noise on SRAM stability, in *Symposium on VLSI Technology (VLSIT)*, 2011, pp. 204–205
13. M. Yamaoka, H. Miki, A. Bansal, S. Wu, D.J. Frank, E. Leobandung, K. Torii, Evaluation methodology for random telegraph noise effects in SRAM arrays, in *IEEE International Electron Devices Meeting (IEDM)*, 2011, pp. 745–758

14. M. Fan, V.P. Hu, Y. Chen, P. Su, C. Chuang, Analysis of single-trap-induced random telegraph noise on FinFET devices, 6T SRAM cell, and logic circuits. IEEE Trans. Electron Devices **59**(8), 2227–2234 (2012)
15. T. Matsumoto, K. Kobayashi, H. Onodera, Impact of random telegraph noise on CMOS logic circuit reliability, in *IEEE Custom Integrated Circuits Conference (CICC)*, 2014, pp. 1–8
16. K. Takeuchi, T. Nagumo, K. Takeda, S. Asayama, S. Yokogawa, K. Imai, Y. Hayashi, Direct observation of RTN-induced SRAM failure by accelerated testing and its application to product reliability assessment, in *Symposium on VLSI Technology (VLSIT)*, 2010, pp. 189–190
17. K. Takeuchi, T. Nagumo, T. Hase, Comprehensive SRAM design methodology for RTN reliability, in *Symposium on VLSI Circuits (VLSIC)*, 2011, pp. 130–131
18. T. Nagumo, K. Takeuchi, S. Yokogawa, K. Imai, Y. Hayashi, New analysis methods for comprehensive understanding of random telegraph noise, in *IEEE International Electron Devices Meeting (IEDM)*, 2009, pp. 759–762
19. T. Nagumo, K. Takeuchi, T. Hase, Y. Hayashi, Statistical characterization of trap position, energy, amplitude and time constants by RTN measurement of multiple individual traps, in *IEEE International Electron Devices Meeting (IEDM)*, 2010, pp. 628–631
20. H. Miki, N. Tega, Z. Ren, C.P. D'Emic, Y. Zhu, D.J. Frank, M.A. Guillorn, D. Park, W. Haensch, K. Torii, Hysteretic drain-current behavior due to random telegraph noise in scaled-down FETs with high-κ/metal-gate stacks, in *IEEE Electron Devices Meeting (IEDM)*, 2010, pp. 620–623
21. H. Miki, N. Tega, M. Yamaoka, D.J. Frank, A. Bansal, M. Kobayashi, K. Cheng, C.P. D'Emic, Z. Ren, S. Wu, J.-B. Yau, Y. Zhu, M.A. Guillorn, D.-G. Park, W. Haensch, E. Leobandung, K. Torii, Statistical measurement of random telegraph noise and its impact in scaled-down high-κ/metal-gate MOSFETs, in *IEEE International Electron Devices Meeting (IEDM)*, 2012, pp. 450–453
22. N. Izumi, H. Ozaki, Y. Nakagawa, N. Kasai, T. Arikado, Evaluation of transistor property variations within chips on 300-mm wafers using a new MOSFET array test structure. IEEE Trans. Semicond. Manuf. **17**(3), 248–254 (2004)
23. K. Agarwal, F. Liu, C. McDowell, S. Nassif, K. Nowka, M. Palmer, D. Acharyya, J. Plusquellic, A test structure for characterizing local device mismatches, in *Symposium on VLSI Circuits (VLSIC)*, 2006, pp. 67–68
24. K.Y. Doong, T.J. Bordelon, L. Hung, C. Liao, S. Lin, S.P. Ho, S. Hsieh, K.L. Young, Field-configurable test structure array (FC-TSA): Enabling design for monitor, model, and manufacturability. IEEE Trans. Semicond. Manuf. **21**(2), 169–179 (2008)
25. T. Tsunomura, A. Nishida, T. Hiramoto, Verification of threshold voltage variation of scaled transistors with ultralarge-scale device matrix array test element group. Jpn. J. Appl. Phys. **48**(12R), 124505 (2009)
26. M. Tanizawa, S. Ohbayashi, T. Okagaki, K. Sonoda, K. Eikyu, Y. Hirano, K. Ishikawa, O. Tsuchiya, Y. Inoue, Application of a statistical compact model for random telegraph noise to scaled-SRAM Vmin analysis, in *Symposium on VLSI Technology (VLSIT)*, 2010, pp. 95–96
27. K.V. Aadithya, A. Demir, S. Venugopalan, J. Roychowdhury, SAMURAI: An accurate method for modelling and simulating non-stationary random telegraph noise in SRAMs, in *Design, Automation & Test in Europe (DATE)*, 2011, pp. 1–6
28. K. Aadithya, S. Venogopalan, A. Demir, J. Roychowdhury, MUSTARD: A coupled, stochastic/deterministic, discrete/continuous technique for predicting the impact of random telegraph noise on SRAMs and DRAMs, in *ACM Design Automation Conference (DAC)*, 2011, pp. 292–297
29. H. Reisinger, T. Grasser, W. Gustin, C. Schlünder, The statistical analysis of individual defects constituting NBTI and its implications for modeling DC- and AC-stress, in *IEEE International Reliability Physics Symposium (IRPS)*, 2010, pp. 7–15
30. T. Grasser, K. Rott, H. Reisinger, P.J. Wagner, W. Goes, F. Schanovsky, M. Waltl, M. Toledano-Luque, B. Kaczer, Advanced characterization of oxide traps: The dynamic time-dependent defect spectroscopy, in *IEEE International Reliability Physics Symposium (IRPS)*, 2013, pp. 2D. 2.1–2D. 2.7

31. P. Weckx, B. Kaczer, M. Toledano-Luque, T. Grasser, P.J. Roussel, H. Kukner, P. Raghavan, F. Catthoor, G. Groeseneken, Defect-based methodology for workload-dependent circuit lifetime projections-application to SRAM, in *IEEE International Reliability Physics Symposium (IRPS)*, 2013, pp. 3A. 4.1–3A. 4.7
32. K. Giering, C. Sohrmann, G. Rzepa, L. Heiß, T. Grasser, R. Jancke, NBTI modeling in analog circuits and its application to long-term aging simulations, in *IEEE International Integrated Reliability Workshop (IIRW)*, 2014, pp. 29–34
33. E.Y. Wu, J. Suñé, Power-law voltage acceleration: A key element for ultra-thin gate oxide reliability. Microelectron. Reliab. **45**(12), 1809–1834 (2005)
34. Z. Ji, L. Lin, J.F. Zhang, B. Kaczer, G. Groeseneken, NBTI lifetime prediction and kinetics at operation bias based on ultrafast pulse measurement. IEEE Trans. Electron Devices **57**(1), 228–237 (2010)
35. E. Seevinck, F.J. List, J. Lohstroh, Static-noise margin analysis of MOS SRAM cells. IEEE J. Solid State Circuits **22**(5), 748–754 (1987)
36. A.J. Bhavnagarwala, X. Tang, J.D. Meindl, The impact of intrinsic device fluctuations on CMOS SRAM cell stability. IEEE J. Solid State Circuits **36**(4), 658–665 (2001)
37. Y. Tsukamoto, K. Nii, S. Imaoka, Y. Oda, S. Ohbayashi, T. Yoshizawa, H. Makino, K. Ishibashi, H. Shinohara, Worst-case analysis to obtain stable read/write DC margin of high density 6T-SRAM-array with local Vth variability, in *IEEE/ACM International Conference on Computer-Aided Design (ICCAD)*, 2005, pp. 398–405
38. D. Khalil, M. Khellah, N. Kim, Y. Ismail, T. Karnik, V.K. De, Accurate estimation of SRAM dynamic stability. IEEE Trans. Very Large Scale Integr. Syst. **16**(12), 1639–1647 (2008)

Random Telegraph Noise Under Switching Operation

Kazutoshi Kobayashi, Mahfuzul Islam, Takashi Matsumoto, and Ryo Kishida

1 Introduction

The physical feature size of a transistor has been reduced continually over time. Leading edge products have a feature size of 7 nm in 2018. Due to the device miniaturization, the number of transistors in one processor becomes as much as 6.9 billion in 2018 [1], which is almost same as the population on earth. Designing reliable systems has become a big challenge in recent years [2–5]. One of the dominant issues is transistor performance variation. It can be classified into static variation and dynamic variation. Static variation originates from manufacturing process variation [6], while dynamic variation is caused by low-frequency noise ($1/f$) and random telegraph noise (RTN). RTN and low-frequency noise are the results of trap (capture) and de-trap (emission) of carriers into the Si–SiO_2 interface

K. Kobayashi (✉)
Department of Electronics, Graduate School of Science and Technology, Kyoto Institute of Technology, Kyoto, Japan
e-mail: kazutoshi.kobayashi@kit.ac.jp

M. Islam
Department of Electrical Engineering, Graduate School of Engineering, Kyoto University, Kyoto, Japan
e-mail: islam.akmmahfuzul.3w@kyoto-u.ac.jp

T. Matsumoto
VLSI Design and Education Center (VDEC), The University of Tokyo, Tokyo, Japan
e-mail: takashi.matsumoto@cad.t.u-tokyo.ac.jp

R. Kishida
Department of Electrical Engineering, Faculty of Science and Technology, Tokyo University of Science, Noda, Japan
e-mail: kishida@rs.tus.ac.jp

T. Grasser (ed.), *Noise in Nanoscale Semiconductor Devices*,
https://doi.org/10.1007/978-3-030-37500-3_9

[7]. Capture and emission of carriers causes dynamic variation that causes jitter in oscillators and timing violation in logic circuits [8, 9].

RTN has been reported to cause large delay fluctuations especially when the transistors operate in the weak inversion region [10]. RTN has attracted much attention in these years due to the continuous technology scaling. It has been reported that the impact of RTN-induced fluctuation may exceed manufacturing process variation in 22 nm technology [11]. Fifty percent of delay fluctuation has been estimated for a 14 nm high-k metal gate extremely-thin silicon-on-insulator process [12]. On the other hand, [13] reported that RTN is not a significant limitation for circuit design in a 14 nm FinFET technology. One of the key reasons of large RTN amplitudes is the surface potential fluctuation in the channel [10, 14]. In order to assess the impact of RTN amplitude on circuit reliability, the following three phenomena need to be understood and modeled accordingly. Firstly, an appropriate distribution function of threshold voltages caused by RTN must be estimated. Empirically, exponential distributions are widely used to model threshold voltage fluctuation induced by a single trap, ΔV_{th} [15]. On the other hand, there are reports mentioning that lognormal distributions better represent ΔV_{th} distributions [12]. Secondly, we need to extract how the distributions change according to gate bias and temperature. Thirdly, the predictability of RTN for a single delay path across various gate bias and temperature conditions. As RTN is a dynamic phenomenon unlike the static process variation, low correlation of delay fluctuation across the operating conditions means that post-silicon timing correction cannot be applied by methods such as delay tuning.

This chapter introduces four approaches to estimate and model RTN. Section 2 discusses the impact of RTN on CMOS logic circuits. In order to measure the impact of RTN on circuit operations, ring oscillators (ROs) are utilized. RTN causes a temporal fluctuation in oscillation frequencies of ROs. Continuous measurements can capture trapping and de-trapping events as frequency fluctuations. From measurement results, low-voltage operations increase delay fluctuations becoming over 16.8% at 4σ. Section 3 investigates the effect of backgate bias and ambient temperature on RTN. Measurement results show that delay fluctuation distribution gets smaller with the increase of temperature, which means that RTN has a large impact at low temperature region.

In Sect. 4, a topology-reconfigurable RO is proposed that incorporates the concept of inhomogeneity to estimate the delay of a particular stage. Section 5 discusses how RTN is modeled in circuit-level simulations. A Verilog-AMS implementation is proposed to replicate RTN in transient simulations. Finally, Sect. 6 summarizes this chapter.

2 Impact of Random Telegraph Noise on CMOS Logic Circuit Reliability

RTN has attracted much attention in these years due to the continuous technology scaling [16]. RTN appears as a temporal transistor performance fluctuation. It has been reported that the impact of RTN-induced fluctuation may exceed manufacturing process variation in 22 nm technology [11]. Fifty percent of delay fluctuation has been estimated for a 14 nm high-k metal gate process [12]. This section summarizes our results from 40 nm test chips [9, 17, 18].

Sections 2.1–2.4 describe the state-of-the-art understanding of the mechanism of RTN. Sections 2.5–2.9 describe the impact of RTN on CMOS logic circuit reliability.

2.1 *Low-Frequency Noise and RTN*

Low-frequency noise or $1/f$ noise is observed in various systems [19]. As for Si-based semiconductor devices, $1/f$ noise can be observed as a drain current fluctuation in a transistor with moderately large gate area. On the other hand, RTN is observed in a small transistor with a few oxide traps. RTN is characterized by a $1/f^2$ spectrum that is called the Lorentzian power spectrum. One possible interpretation is that the superposition of various RTN (with a broad distribution of activation energies in RTN process) generates $1/f$ noise [20]. Mobile charged carriers in a transistor channel can be randomly trapped into or de-trapped from oxide traps (Fig. 1). The capture and emission of one carrier induces a two-state RTN. Figure 2 is the typical example of a measured drain current fluctuation in a commercial 40 nm CMOS transistor in our test chip that has a large two-state RTN. τ_c and τ_e are defined as the period when the drain current stays at high-current state (H-state) and low-current state (L-state), respectively. The fluctuation amplitude is defined as ΔI_{ds}. RTN is an intrinsically random phenomenon. The parameters such as τ_c, τ_e, and ΔI_{ds} differ by transistor. Thus, statistical characterization is required for the correct RTN modeling [15, 21, 22].

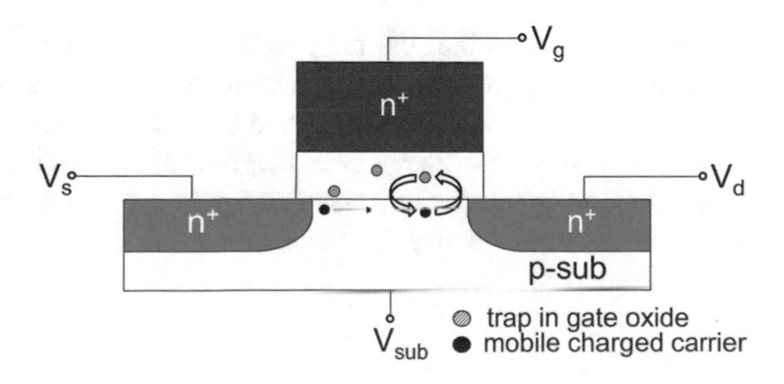

Fig. 1 Capture and emission of carriers by gate oxide traps (nMOS)

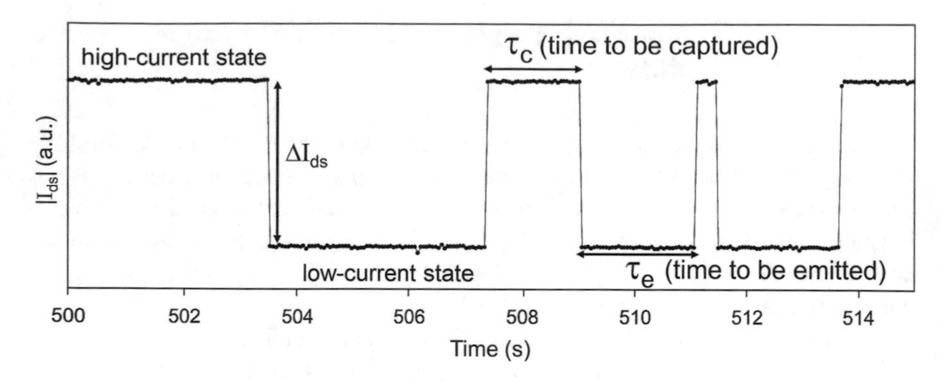

Fig. 2 RTN-induced drain current fluctuation in a pMOS

2.2 *RTN Time Constant*

When a two-state RTN is measured for some period, distributions of τ_c and τ_e are obtained. The average of τ_c and τ_e is denoted as $\langle\tau_c\rangle$ and $\langle\tau_e\rangle$, respectively. The probability of a transition from H-state to L-state per unit time is $1/\langle\tau_c\rangle$ and from L-state to H-state per unit time is given by $1/\langle\tau_e\rangle$. $A(t)$ is defined as the probability that a transition from H-state to L-state does not happen after time t. Then,

$$A(t + \mathrm{d}t) = A(t)\left(1 - \frac{\mathrm{d}t}{\langle\tau_c\rangle}\right) \tag{1}$$

is obtained. Integrating Eq. (1) with $A(0) = 1$,

$$A(t) = \exp(-t/\langle\tau_c\rangle). \tag{2}$$

As a result, the probability, $P_H(t)$, that the transition from H-state to L-state does not happen for time t, and then happens between time t and $t + \mathrm{d}t$ is given by

$$P_H(t) = \frac{1}{\langle\tau_c\rangle}\exp(-t/\langle\tau_c\rangle). \tag{3}$$

Equation (3) shows that the time constants, τ_c and τ_e, follow exponential distributions [7]. It is experimentally shown later that time constants actually follow exponential distribution for a two-state logic delay fluctuation (Sect. 2.7).

Figure 3 shows the energy band diagram of an nMOS transistor. The Fermi level is denoted by E_F and the trap energy level is denoted by E_T. Traps below E_F (filled circle) are filled and above E_F (open circle) are empty. Several traps close to E_F can

Fig. 3 Energy band diagram of an nMOS transistor

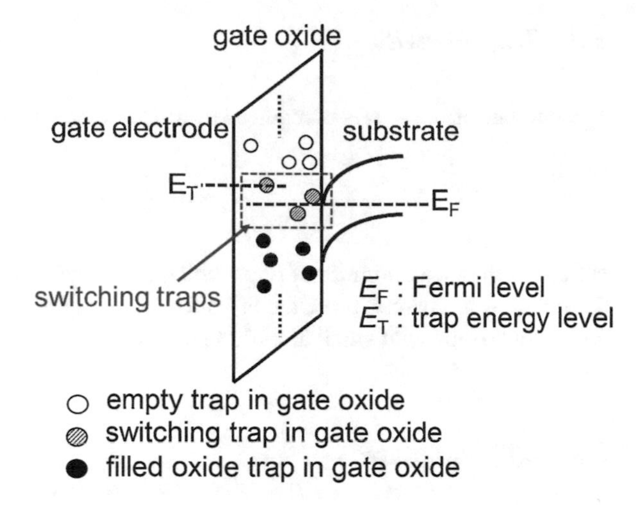

act as switching traps. When only a single trap exists, 2-state discrete drain current fluctuation is observed. If there are n switching traps, 2^n-state discrete fluctuation can be observed.

Historically, the tunneling mechanism has been thought to be the determinant factor in the time to capture, τ_c. However, recent observations show that the tunneling mechanism-based model is inadequate to explain the time constants. Oxide relaxation is now considered to be the determinant factor for a carrier to be trapped [23, 24]. The average time to capture, $\langle\tau_c\rangle$, is modeled as

$$\langle\tau_c\rangle = \frac{1}{n_s\,\sigma\,\langle v_T\rangle} = \frac{1}{n_s\,\sigma_0\,\langle v_T\rangle}\exp\left(\frac{E_B}{k_B\,T}\right). \tag{4}$$

Here, n_s is the channel carrier density, $\langle v_T\rangle$ is mean thermal voltage of carriers, and σ is capture cross-section. σ_0 is a scaling parameter. k_B is the Boltzmann constant and T is the absolute temperature. E_B is the energy barrier for a carrier to be trapped. With the increase of temperature, an exponential decrease of τ_c is expected according to the model. Time to emission, τ_e, is modeled as

$$\langle\tau_e\rangle = \langle\tau_c\rangle\,\exp\left(-\frac{E_T - E_F}{k_B\,T}\right). \tag{5}$$

Here, E_T is trap energy level, E_F is Fermi level. It is also reported that the above model fails to predict the measured τ_e in some cases. Trapping and de-trapping models incorporating intermediate states are proposed to describe the different mechanisms [25, 26].

2.3 Trap Density

The number of traps, N_T, is reported to follow a Poisson distribution [15].

$$a_{N_T} = \frac{e^{-\lambda}\lambda^{N_T}}{N_T!}. \tag{6}$$

Here, λ is the mean number of traps which scales with channel area. However, with the increase of channel area, the effect of traps on the amplitude becomes smaller. As a result, traps with small amplitude may not be characterized [11].

2.4 RTN Amplitude

Trapping of a charge results in fluctuation of the surface potential, which in turn modulates the channel carrier density. In an ideal transistor where the silicon surface potential is flat, the change of threshold voltage, ΔV_{th}, due to the trapping of a carrier into the oxide can be expressed by Eq. (7).

$$\Delta V_{th} = \frac{q}{C_{ox}WL}. \tag{7}$$

Here, q is the elementary charge, C_{ox} is the oxide capacitance per area, W is channel width, and L is channel length. The capture and emission of a carrier also induces the threshold voltage fluctuation, ΔV_{th}. Equation (7) shows the impact of one charged carrier on ΔV_{th} becomes larger as the gate area shrinks. However, the simple expression of Eq. (7) cannot explain the large amplitudes reported in the literature. RTN-induced ΔV_{th} distribution has a long tail. [10] has shown by simulation that the reason for large amplitude originates from the surface potential, ϕ_s unevenness. Sonoda et al. [14] has given a simple model that models the relationship between RTN amplitude variability and surface potential, ϕ_s, variability by Eq. (8).

$$\Delta V_{th} = \frac{q}{C_{ox}WL}\exp\left(\frac{q\left(V_{th} - V_{thj}\right)}{\eta kT}\right). \tag{8}$$

Here, V_{thj} is analogous to local threshold voltage when the channel is virtually divided into many smaller channels. Because of the surface potential fluctuation, V_{thj} varies randomly in the channel. V_{thj} variation is related to ϕ_s by Eq. (9) [27].

$$\sigma_{V_{thj}} = \eta\sigma_{\phi_s}. \tag{9}$$

σ_{ϕ_s} is related to random dopant fluctuation induced threshold voltage variability, $\sigma_{V_{th}}$. If ϕ_s distribution can be considered to be Gaussian, Eq. (8) implies that single trap induced ΔV_{th} distribution would follow a lognormal distribution. However, in real devices, other factors such as drain bias effect and trap locations may play a significant role. Thus, it is essential to measure in-situ RTN effects for accurate modeling.

After knowing the trap density distribution and single trap induced RTN amplitude distribution, the probability distribution function (PDF) of the overall amplitude with varying number of traps can be statistically obtained from Eqs. (6) and (10)–(13).

$$P_1(x) = \frac{1}{x\sigma\sqrt{2\pi}} \exp\left(-\frac{(\ln x - \mu)^2}{2\sigma^2}\right), \tag{10}$$

$$P_N(x) = \int_{-\infty}^{\infty} P_{N-1}(x-t)P_1(t)dt, \tag{11}$$

$$P(x) = a_0\delta(x) + \sum_{i=1}^{N} a_i P_i(x), \tag{12}$$

$$\mu = \log\left(\frac{\mathrm{q}}{C_{\mathrm{ox}}WL}\right), \sigma = \frac{\mathrm{q}}{\mathrm{k_B}T} \times \sigma_{\phi_\mathrm{s}}. \tag{13}$$

Here, $P(x)$ is the PDF of overall RTN amplitude distribution. μ and σ are the parameters for single trap induced RTN amplitude lognormal distribution. It would be useful for the designers to approximate $P(x)$ with an equivalent closed form function that does not need statistical simulation.

When the operating voltage of a circuit decreases, the impact of ΔV_{th} also becomes larger. Recent studies show that ΔV_{th} caused by RTN grows more rapidly than the threshold variation caused by random dopant fluctuation.

2.5 Impact on Logic Circuit

RTN in transistors is a critical issue for digital circuits. RTN already has a serious impact on CMOS image sensors [28], flash memories [29], and SRAMs [30–32]. These circuits use small area devices and the integration density is extremely high. A logic path consists of multiple logic gates with transistors of various gate widths. Some delay paths have a higher activity rate whereas some may have a very low activity rate. As multiple transistors are included in a delay path, the probability of RTNs with large amplitudes gets higher.

Fully depleted SOI (FDSOI) MOSFETs are one of the attractive devices for the present and the future planar CMOS technology [33]. As one of the FDSOI devices, the silicon on thin buried (SOTB) oxide is being developed because of the superior device characteristics for ultra-low-voltage operations and the suppression of device variability caused by dopant fluctuation [34]. RTN amplitude is also considered to be suppressed by FDSOI MOSFETs compared to bulk MOSFETs because large channel potential fluctuation in the bulk device is suppressed in the FDSOI device [35]. Multi-gate transistors such as tri-gate device are also attractive and have already been applied to the advanced SoC in 22 nm technology [36]. RTN in multi-gate device and its impact on circuit will further be investigated in the future [37, 38].

We have reported that RTN induces performance fluctuation to logic circuits [8, 39]. The impact of RTN can be a serious problem even for logic circuits when they are operated under low supply voltage [9]. Circuit designers can change various parameters such as operating voltage, transistor size, number of logic stages, logic gate type, and substrate bias. However, the impact of such parameters on RTN is not well understood at the circuit level [40]. This impact is clarified based on our measurement results in Sect. 2.8 [17]. Finally, comparison of RTN and process variation is described in Sect. 2.9 [18].

Furthermore, the effective trap density increases under wide temperature range as some traps may be active at low power and some traps may be active at high temperature. This phenomenon is confirmed in our measurement results as will be discussed in Sect. 3.4. The total delay variation is the result of complex RTN occurring either in a transistor or single RTNs over multiple transistors. In order to estimate the delay variation, the scaling effect on trap density and amplitude need to be modeled.

In the following Sects. 2.6–2.9, the impact of RTN on CMOS logic circuit reliability is described based on our measurement results from 40 nm test chips [9, 17, 18].

2.6 *Test Structure for RTN Impact Evaluation*

In this section, a test structure for the statistical characterization of RTN-induced logic delay fluctuation is described. A logic path exists between two registers in a typical synchronous circuit structure (Fig. 4). Figure 5 shows the simplest test structure that can emulate the synchronous circuit operation of Fig. 4. Combinational

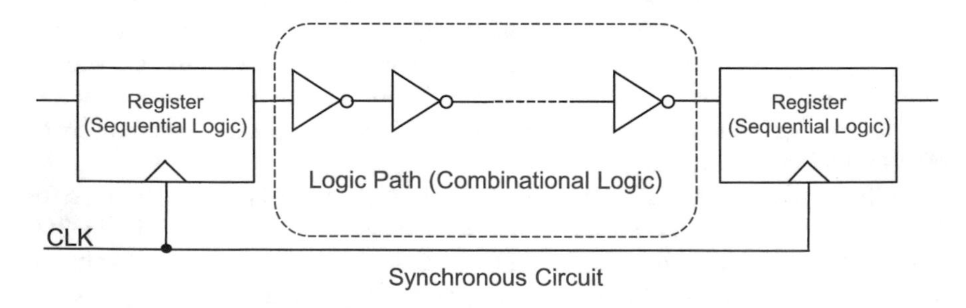

Fig. 4 Typical synchronous circuit structure

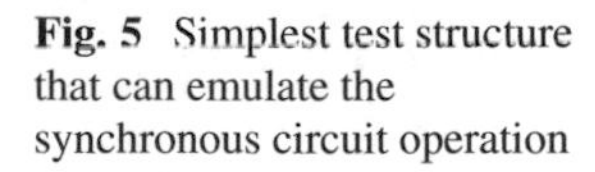

Fig. 5 Simplest test structure that can emulate the synchronous circuit operation

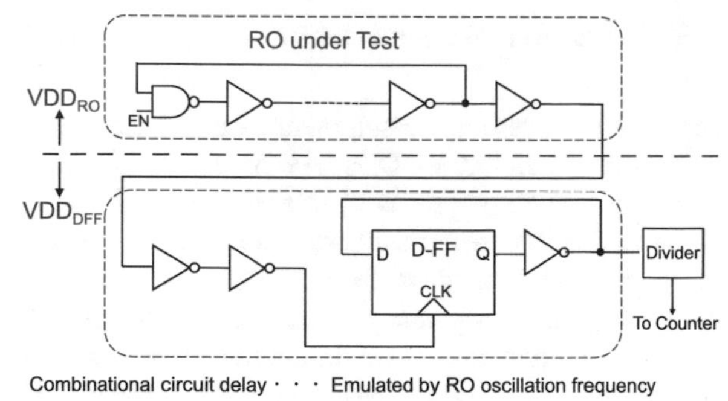

RO-Array Chip
28 sections
30 sections
Section Control Logic
Section
RO1
RO2
RO47
RO48
RO49
RO96
control circuit
RO array (840 ROs / 2mm²)
·40nm CMOS Technology
·Various ROs / section
·Statistical nature of RTN can be evaluated.
·RO power supply can be separately controlled.
·Substrate bias can be separately controlled.
RTN-induced RO frequency fluctuation is evaluated.

Fig. 6 Whole test structure for RTN measurement. One test structure contains 840 ROs

circuit delay is emulated by ring oscillator (RO) oscillation frequency. All logic gates except NAND2 with EN input are homogeneous in this section. Sequential circuit operation is emulated by D flip-flop (DFF) toggled by the RO output. The power supply for RO (VDD_{RO}) and DFF (VDD_{DFF}) can be independently supplied. We can also control the substrate bias for pMOS and nMOS. Figure 6 shows the whole test structure for the RTN measurement. RTN-induced delay fluctuation is measured by the RO frequency fluctuation. Various types of ROs are included in one circuit unit, which is called a section as depicted in Fig. 6. There are 840 sections with the same structure on a 2 mm^2 area. The statistical nature of RTN can be evaluated by the RO array. This chip is fabricated in a commercial 40 nm CMOS technology. All measurements are done at room temperature.

2.7 Measurement Results of Logic Delay Fluctuation

Figure 7a shows the measurement result of the oscillation frequency of a 7-stage RO for about 80 s at $VDD_{RO} = 0.65$ V. The size of the inverter (INV) is smallest in this technology. The body bias for pMOS ($V_{bs\text{-}pMOS}$) and nMOS ($V_{bs\text{-}nMOS}$) is set to 0 V. Measurement results show the large step-like frequency fluctuation. Here, F_{max} is defined as the maximum oscillation frequency and ΔF is defined as the maximum frequency fluctuation as shown in Fig. 7a. When the number of trapped charges becomes minimum during the measurement period, RO oscillation frequency is F_{max}. When the number of trapped charges becomes maximum during the measurement period, RO oscillation frequency fluctuation is considered to be ΔF. We use $\Delta F/F_{max}$ as a measure for the impact of RTN-induced frequency fluctuation for logic delay. It is 10.4% for one RO (Fig. 7a). However, significant fluctuation is not observed for another RO (Fig. 7b). Figure 7a, b is plotted with the same frequency range. Although large fluctuation such as Fig. 7a is a rare event, it has a large impact on circuit performance. Figure 8 shows typical measurement data of a 7-stage RO at VDD_{RO}=0.65 V where a large 2-state fluctuation is observed. $VDD_{RO} = 0.65$ V is close to a minimum operating voltage in this technology. Time constants τ_c and τ_e represent the time when the RO stays at high-frequency state and low-frequency state, respectively. The PSD of Fig. 8 is obtained by quantizing the measurement data of Fig. 8 into the 2-state waveform. Figure 9 shows Lorentzian power spectrum obtained from Fig. 8. Figure 10 shows time constant (τ_c, τ_e) distributions of Fig. 8. It is found that both distributions for τ_c and τ_e follow an exponential distribution ($e^{-t/\tau}$). A Lorentzian PSD ($1/f^2$) and $e^{-t/\tau}$ distribution are

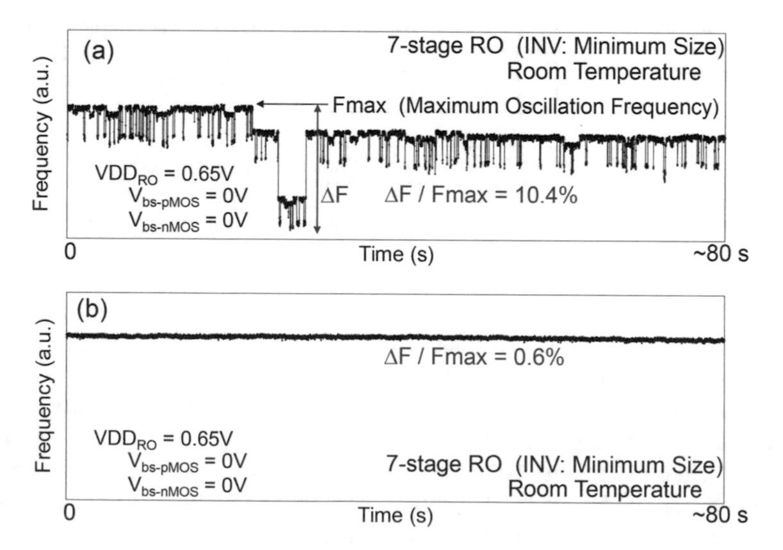

Fig. 7 Measurement result of RTN-induced RO frequency fluctuation. (**a**) $\Delta F/F_{max} = 10.4\%$. (**b**) $\Delta F/F_{max} = 0.6\%$

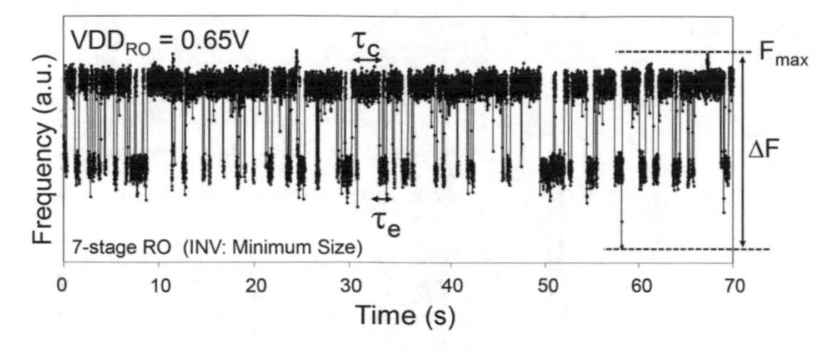

Fig. 8 Measurement results that show 2-state fluctuation

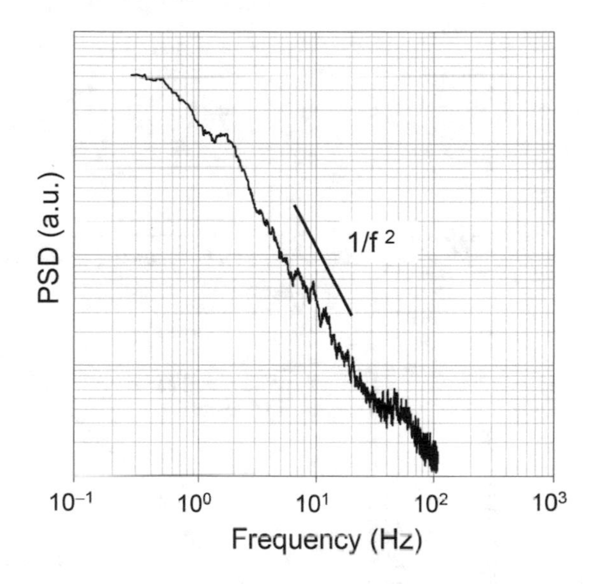

Fig. 9 Power spectral density of Fig. 8

observed for the case of a transistor where a single defect causes RTN fluctuation. It indicates that the RTN fluctuation of Fig. 8 is caused by a single defect in a specific transistor in the 7-stage RO.

2.8 Impact of RTN on Logic Circuit Reliability

In this section, the impact of RTN on logic circuit reliability is described. The distribution of RO frequency ($F_{\max}$) variation for the 7-stage ROs follows a normal distribution when data are collected from the whole test structure of Fig. 6 over 15 chips (12,600 ROs) at 0.65 V operation. The distribution of $\Delta F/F_{\max}$ for the same ensemble follows a log-normal distribution above the 50% level of the cumulative probability (Fig. 11). The maximum value of $\Delta F/F_{\max}$ becomes 16.8%. It is found that a small number of samples have a large RTN-induced fluctuation. If

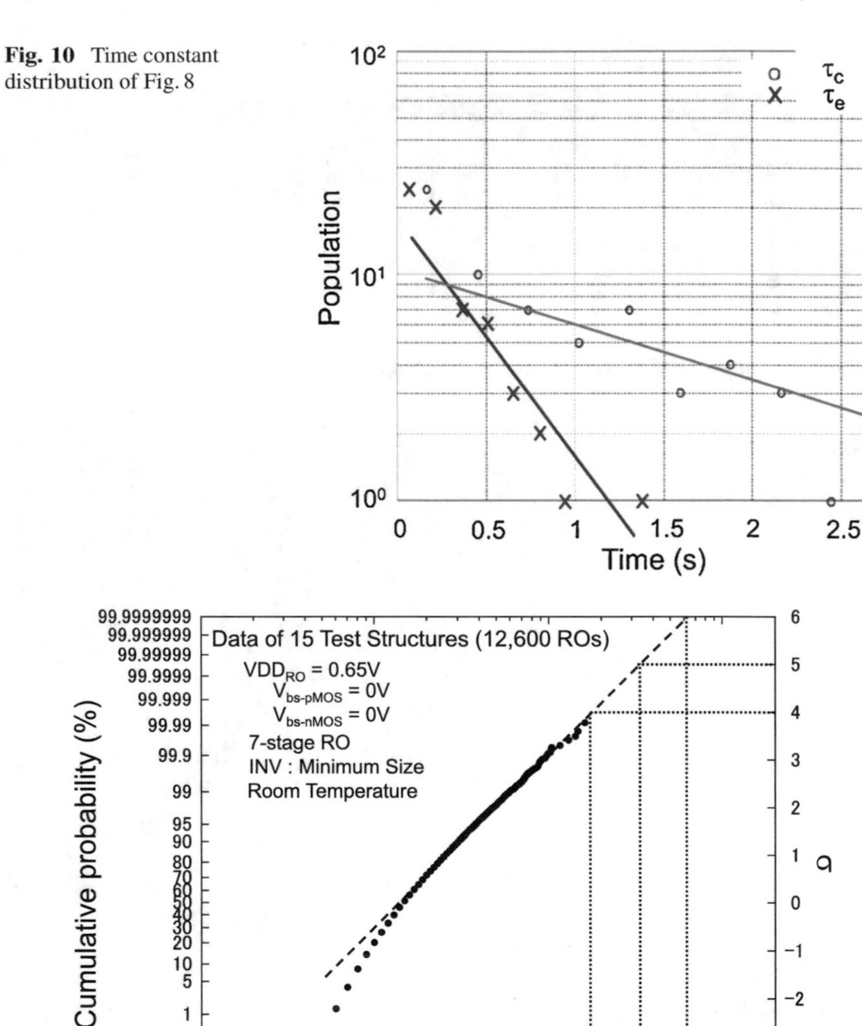

Fig. 10 Time constant distribution of Fig. 8

Fig. 11 Cumulative probability of RO frequency fluctuation ($\Delta F/F_{\max}$) which follows a log-normal distribution

$\Delta F/F_{\max}$ follows a log-normal distribution up to the 6σ level, $\Delta F/F_{\max}$ becomes as much as 60%. Our results suggest the impact of RTN-induced fluctuation compared with the frequency variation caused by manufacturing process increases. When the supply voltage decreases, the impact of ΔV_{th} caused by RTN becomes larger (Fig. 12). Log-normal plot of $\Delta F/F_{\max}$ for $\mathrm{VDD_{RO}} = 1.0\,\mathrm{V}$, 0.75 V, and 0.65 V over the same 840 ROs indicates the rapid increase of $\Delta F/F_{\max}$ towards

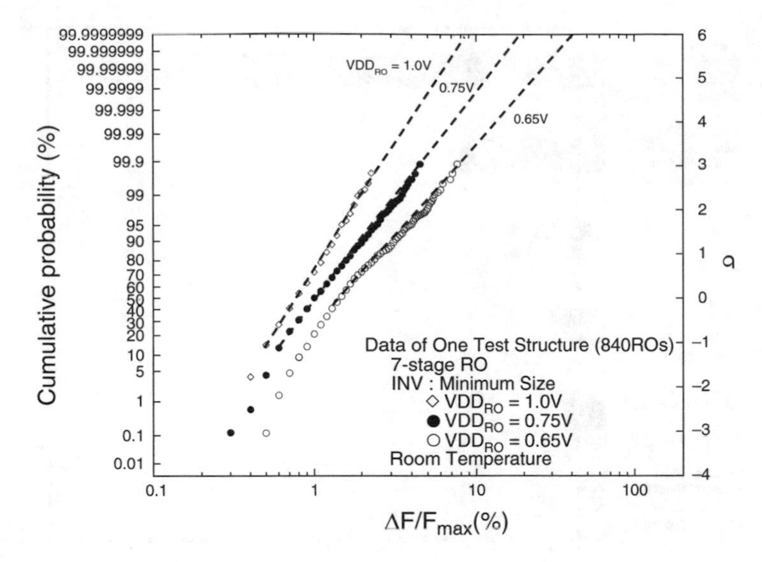

Fig. 12 Cumulative probability of $\Delta F/F_{\mathrm{max}}$ for various $\mathrm{VDD_{RO}}$ which follows a log-normal distribution

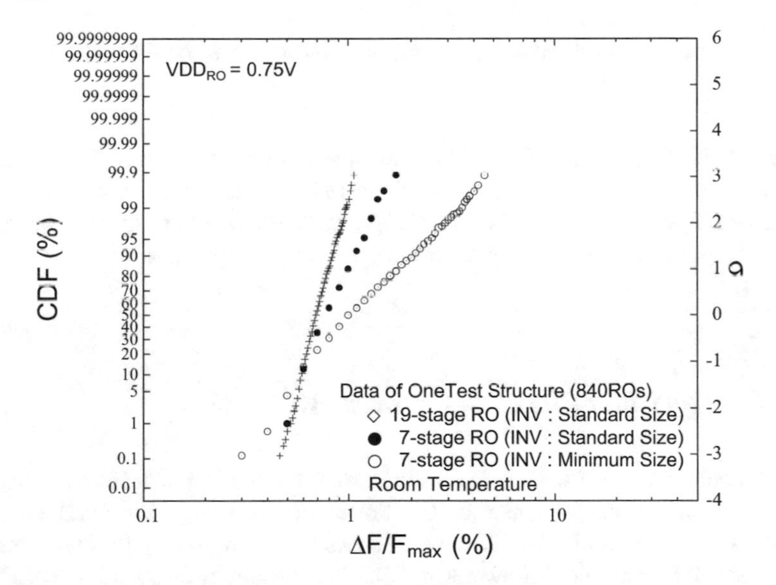

Fig. 13 Impact of gate area and number of stages on $\Delta F/F_{\mathrm{max}}$

lower $\mathrm{VDD_{RO}}$. The impact of RTN also becomes larger as the gate area shrinks and the number of stages decreases. Figure 13 indicates more than 50% reduction of $\Delta F/F_{\mathrm{max}}$ (at 95% level in cumulative probability) can be achieved by increasing the INV size for 7-stage RO under 0.75 V operation. Here, the ratios of pMOS and nMOS gate areas ($W \times L$) of the minimum size INV to the standard size INV are

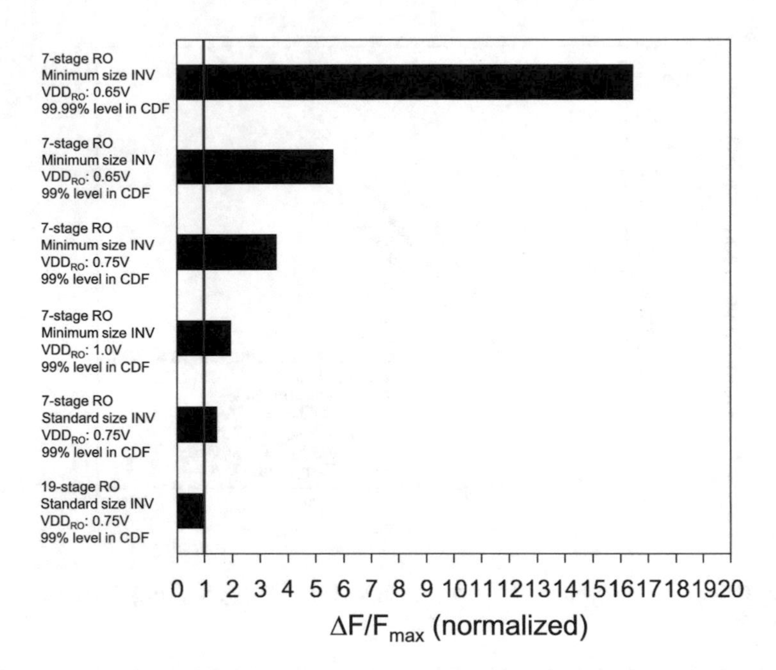

Fig. 14 The impact of supply voltage, gate area, and number of stages on $\Delta F/F_{\mathrm{max}}$ (normalized)

0.21 and 0.30, respectively. Figure 13 also shows that the impact of RTN becomes larger as the number of stages decreases from 19 to 7. The impact of supply voltage, gate area, and number of stages on $\Delta F/F_{\mathrm{max}}$ is summarized in Fig. 14. The impact of RTN is drastically reduced by increasing supply voltage, gate area, and number of stages.

2.9 Comparison of RTN and Process Variation

The horizontal axis in Fig. 15 is the drain current fluctuation on a linear scale. The vertical axis is the normal quantile. The dotted line is a large MOSFET case and the solid line is a small MOSFET case. Process variation usually follows a normal distribution. RTN does not follow normal distribution and its distribution has a long-tail part. The impact of RTN dominates process variation at the cross point depicted as red circle in Fig. 15.

In a circuit with a large number of small transistors, the impact of RTN-induced temporal fluctuation is considered to increase when it is compared with the static frequency variation caused by manufacturing process. Figure 16 shows the distribution of RO frequency (F_{max}) variation for the 7-stage ROs. It follows a normal distribution when data are collected from the whole test structure of Fig. 6

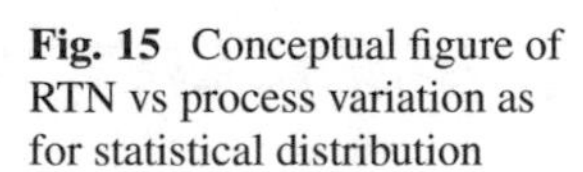

Fig. 15 Conceptual figure of RTN vs process variation as for statistical distribution

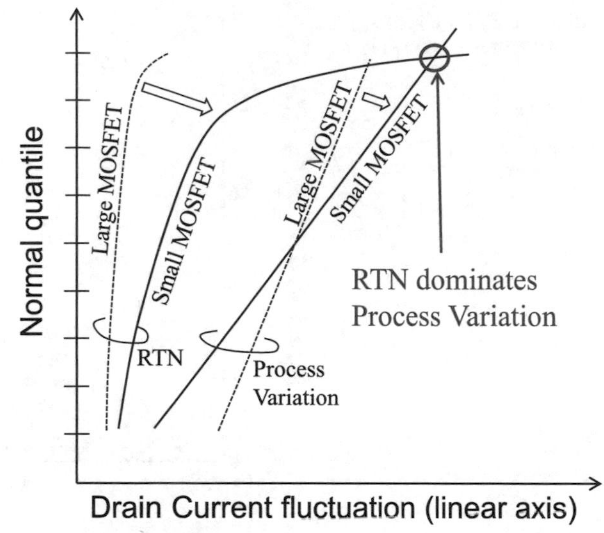

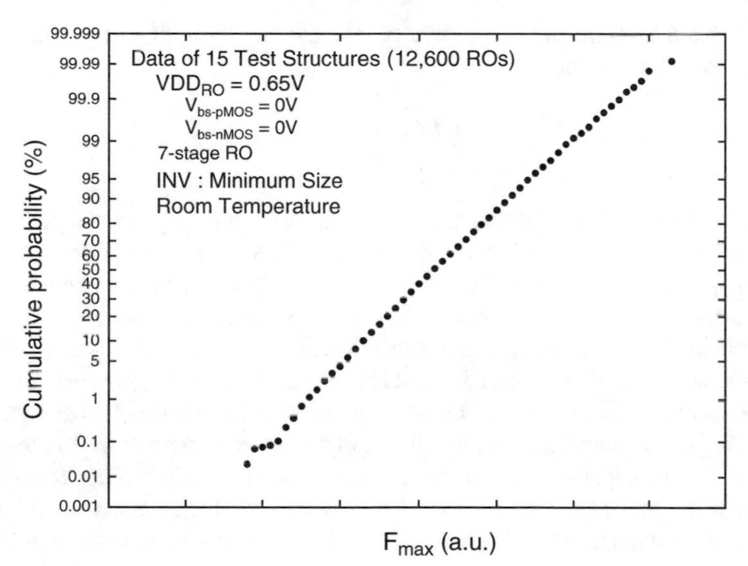

Fig. 16 Normal distribution plot of RO frequency ($F_{\max}$) variation caused by process variation

over 15 chips (12,600 ROs) at 0.65 V operation. Figure 17 shows a $\Delta F/F_{\max}$ versus $F_{\max}$ plot over 12,600 ROs. It represents how the distribution of the impact of RTN correlates with process variation distribution. The triangle shape distribution suggests that there is no or weak correlation between RTN and process variations. Figure 18 can be obtained by plotting the vertical axis of Fig. 17 with log scale. The circle shape distribution suggests that there is no or weak correlation between the variation of the impact of RTN and process variation. Figure 19 shows the impact

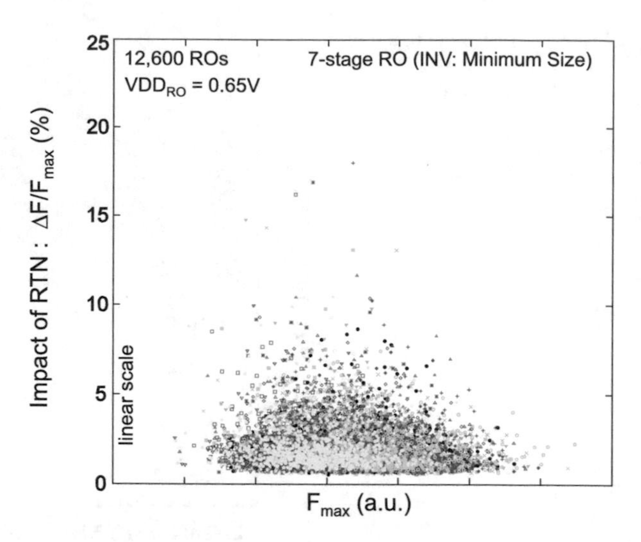

Fig. 17 $\Delta F/F_{\max}$ versus $F_{\max}$ plot over 12,600 ROs

of RTN when it is compared with that of process variation. The impact of RTN on process variation is defined as

$$\frac{(\Delta F/F_{\max})_{n\sigma}}{(n\sigma/\mu)}. \tag{14}$$

To clarify the meaning of above definition, how to obtain Fig. 19 is explained. The plot for the minimum size 7-stage RO at 0.65 V ($\times$) can be obtained as follows. $\Delta F/F_{\max}$ follows a lognormal distribution. $(\Delta F/F_{\max})_{\mathrm{n}\sigma}$ can be obtained for each σ using the lognormal distribution. $F_{\max}$ follows normal distribution. $(\mathrm{n}\sigma/\mu)$ can be obtained for each σ using the normal distribution. The dotted line is estimated from measured distributions of both RTN and process variation when lognormal distribution for RTN is assumed up to 7σ value. For the minimum size 7-stage RO at 0.65 V ($\times$), the impact grows exponentially when σ is increased. It is found that RTN becomes comparable to process variation around 7σ value. When the operating voltage is slightly increased to 0.75 V ($\triangle$, $\circ$), the RTN impact decreases rapidly. Finally, when the transistor size is increased from the minimum to the standard size at 0.75 V, RTN has small (and almost constant) impact on process variation ($\circ$).

2.10 Summary

This section describes the impact of RTN on CMOS logic circuits based on the measurement results from 40 nm test chips. Recent researches on RTN and its impact on circuits are briefly summarized. Even for a combinational circuit, its operation under low supply voltage is found to be affected seriously by RTN. Statistical nature of RTN-induced delay fluctuation is described by measuring 12,600 ROs fabricated in

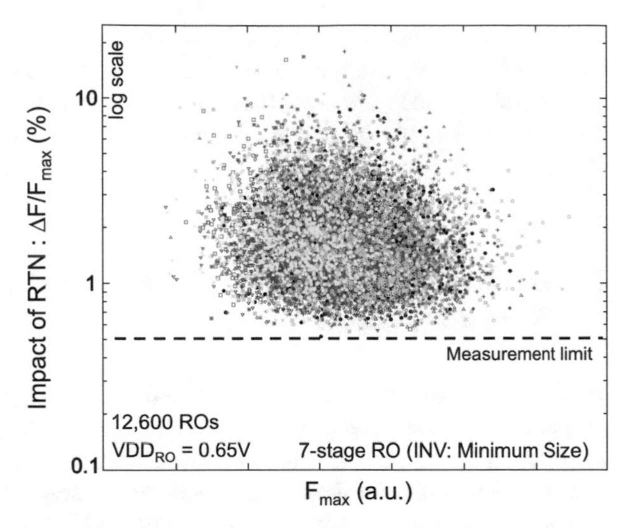

Fig. 18 $\Delta F/F_{\max}$ versus $F_{\max}$ plot over 12,600 ROs (vertical axis: log-scale)

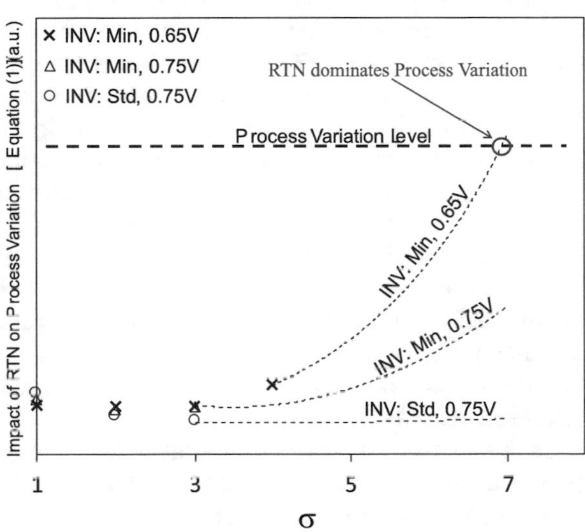

Fig. 19 The impact of RTN on process variation

a commercial 40 nm CMOS technology. RTN-induced delay fluctuation ($\Delta F/F_{\max}$) becomes as much as 16.8% of nominal oscillation frequency under 0.65 V operation. By increasing the transistor size from the minimum to the standard size, more than 50% reduction of $\Delta F/F_{\max}$ can be achieved at 95% level in cumulative probability under 0.75 V operation. RTN-induced delay fluctuation also decreases rapidly with increasing supply voltage. From the measurement results, $\Delta F/F_{\max}$ increases log-normally when the number of logic circuits increases.

The impact of RTN can be a serious problem even for digital circuits when they are densely integrated and operated under a low supply voltage. The impact of RTN on process variation with respect to combinational circuit delay is estimated to grow exponentially by the experimental data. It suggests that there is no or weak correlation between RTN variation and process variation. It is found that the impact of RTN fluctuation can be drastically reduced by slightly increasing the supply voltage and gate size. It is possible to decrease the impact of RTN on a combinational circuit by tuning design parameters such as supply voltage and gate size.

3 Substrate Bias and Temperature Effect on Random Telegraph Noise

In the case of static random dopant fluctuation (RDF) based V_{th} variability, detailed measurements have been performed showing that V_{th} distribution follows Gaussian distribution up to 5σ [41]. An adaptive substrate bias control has been widely used to compensate for die-to-die parameter variations [42] and to improve aging degradation such as bias temperature instability [43, 44]. However, the impact of the substrate bias on RTN at the circuit level has not been well understood. Furthermore, it has been investigated that temperature has very small effect on static random ΔV_{th} variability [45]. In a conventional DC I–V based RTN characterization, the drain bias is kept small so that the channel carrier density remains nearly flat throughout the channel. However, during a switching operation of transistors on a digital circuit, transistor gate voltage switches from low to high and vice versa. Furthermore, strong drain bias applies making delay prediction difficult. It is reported that RTN amplitude varies depending on the previous state of gate bias [12]. For the case of RTN, there are few results on the temperature effect on RTN amplitudes. We find that depending on the temperature, not only the time constants of a trap change but also appearing and disappearing of traps occur in our measurement time scale [12, 46]. As a result, low correlations between delay fluctuations across temperatures have been observed.

In this section, we describe substrate bias effect on RTN in Sect. 3.1. Then we present detailed measurement results of temperature effect on random telegraph noise (RTN) induced delay fluctuation. Skewed ring oscillators (ROs) are used to evaluate pMOFSET and nMOSFET specific RTN effects. Furthermore, threshold voltage distributions have been extracted such that the simulated delay distribution matches with the measured delay distributions. We describe our test structure and design methodology of ROs in Sect. 3.3. Section 3.4 gives detailed measurement results and some discussions on the possible reasons behind the observations.

3.1 Substrate Bias Effect on RTN

In this section, the impact of the substrate bias on RTN at the circuit level is described based on our results from 40 nm test chips [9, 17, 40]. The test structure shown in Fig. 6 of Sect. 2 is used. RTN time constant can be affected by the substrate

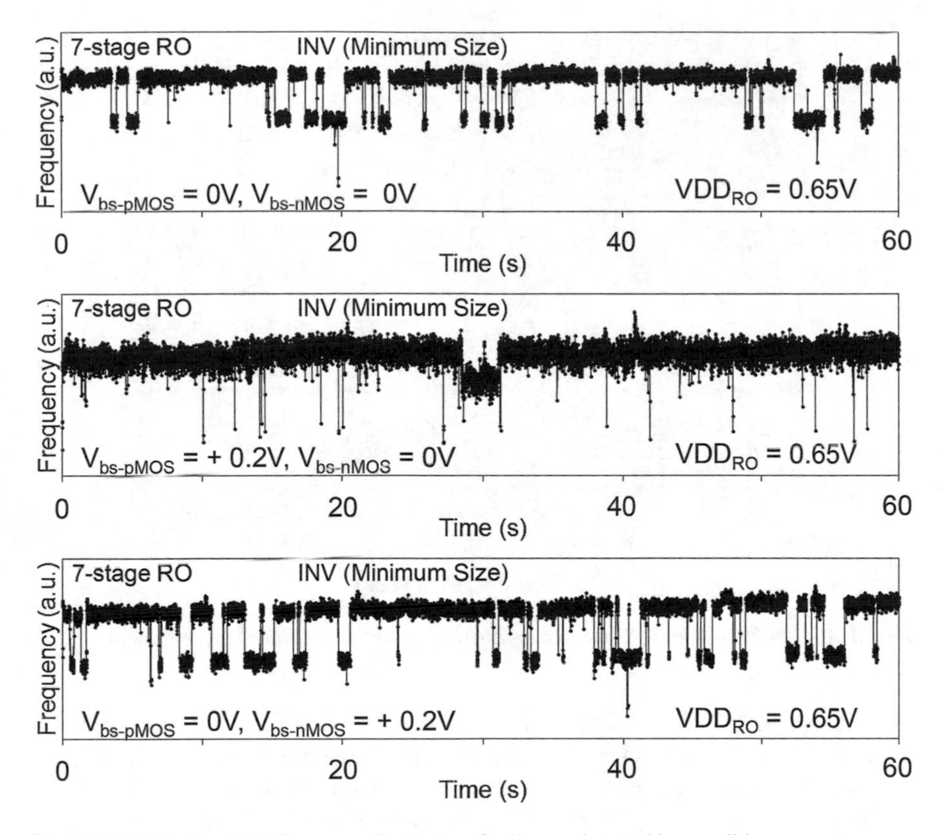

Fig. 20 RTN-induced RO frequency fluctuation for three substrate bias conditions

bias. Figure 20 shows the measurement results of frequency fluctuation of one RO for 60 s under three substrate bias conditions. For this sample, the time constant is modulated considerably only when the pMOS substrate bias is changed from 0 V to +0.2 V (middlemost figure). Then PSDs for the same sample of Fig. 20 for five substrate bias conditions are calculated (Fig. 21). When the pMOS substrate bias is changed from 0 V to +0.2 V (c, e), the large two-state fluctuation ($\tau_c \gg \tau_e$) rarely happens. We observe the effect of one trap at the pMOS transistor in the RO that induces large noise at the circuit level (a, b, d). Figure 22 shows the frequency fluctuation of RO location 1 (RO1 in Fig. 6) for three substrate bias conditions. Four-state fluctuation due to two traps is clearly observed for the zero substrate bias ($V_{bs\text{-}pMOS} = 0$ V, $V_{bs\text{-}nMOS} = 0$ V) case. The effect of one of two traps disappears only when nMOS transistor is forward biased by 0.2 V (middlemost figure). The seemingly disappeared two-state fluctuation is caused by a single trap in a specific nMOS transistor in the RO.

Substrate bias conditions are categorized as the reverse bias case ($V_{bs\text{-}pMOS} = -0.2$ V, $V_{bs\text{-}nMOS} = 0$ V), zero bias case ($V_{bs\text{-}pMOS} = 0$ V, $V_{bs\text{-}nMOS} = 0$ V), and forward bias case ($V_{bs\text{-}pMOS} = +0.2$ V, $V_{bs\text{-}nMOS} = +0.2$ V). To evaluate the

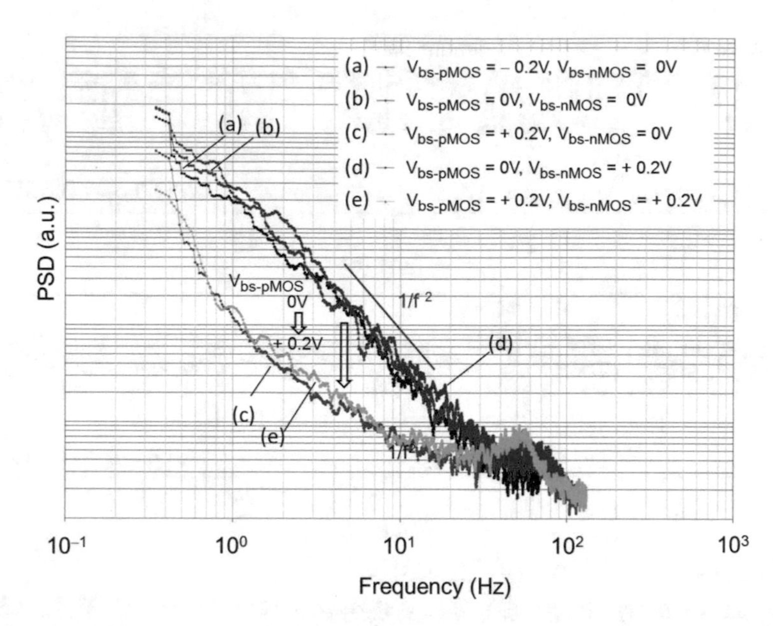

Fig. 21 PSD of RTN-induced RO frequency fluctuation for five substrate bias conditions

forward body-bias effect on large $\Delta F/F_{\max}$ samples, ROs that have more than 4% fluctuation at the reverse bias case (28 ROs) are shown in Fig. 23. When substrate bias is changed from the reverse bias case to the forward bias case, $\Delta F/F_{\max}$ tends to decrease monotonically due to $F_{\max}$ increase. However, it does not decrease monotonically in the case of the RO location "68," "160," and "219" when substrate bias is changed from the reverse bias case to the forward bias case. It is because the impact of substrate bias on RTN appears individually by ROs. It must be considered when forward substrate bias is applied. Figure 24 shows average and sigma values of $\Delta F/F_{\max}$ calculated from Fig. 23. Next, $\Delta F/F_{\max}$ for one test structure under three substrate bias conditions is plotted in log-normal way (Fig. 25). The impact of RTN-induced delay fluctuation can be statistically reduced by the forward substrate bias control.

3.2 *Temperature Effect on RTN*

Temperature effects on low-frequency noise in scaled silicon-on-insulator MOSFETs are reported in [47]. According to the report, the noise variability decreases with the increase of temperature. The noise variability also depends on the operating frequency. Especially for a logic gate, the RTN amplitude may differ depending on the idle time of the operation [12]. Thus, hysteric effect may occur. As a delay path consists of multiple transistors, in-situ measurements with ROs are helpful in understanding the RTN effects.

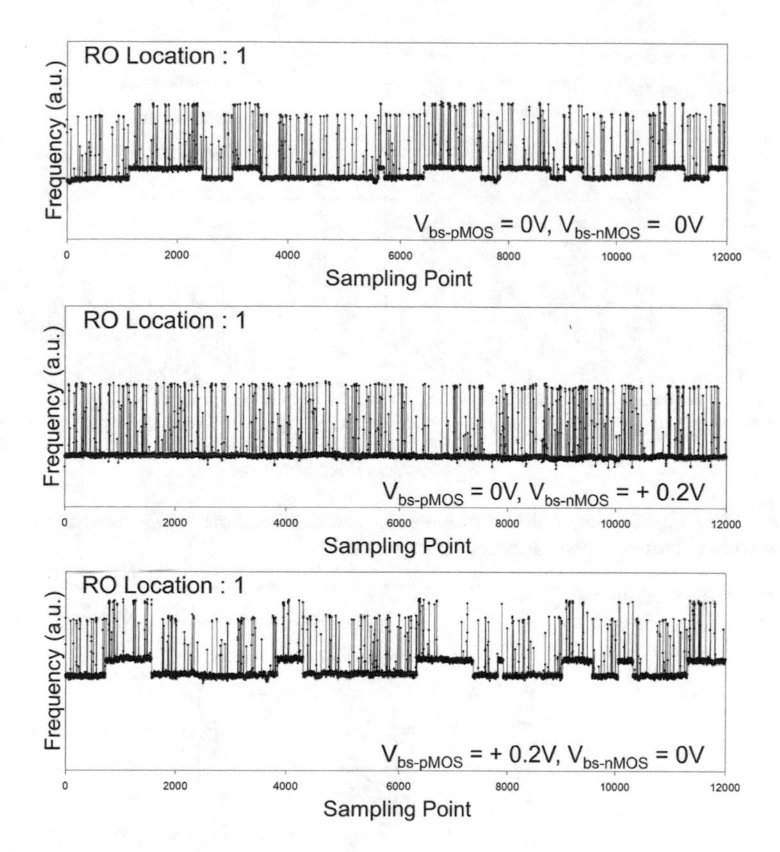

Fig. 22 RTN-induced RO frequency fluctuation for three substrate bias conditions (RO Location 1)

3.3 Design of Test Structure

Test Chip

We use ROs of different gate sizes to evaluate RTN-induced delay variations. Figure 26 shows the test structure implemented in a 65 nm thin buried-oxide (BOX) fully depleted silicon-on-insulator (FDSOI) process [48]. We set the supply voltage to be 0.6 V which ensures the transistors operate at strong inversion region. The test chip consists of a 16 × 13 array of RO blocks. Several ROs with different gate widths are implemented in each block. Therefore, we obtain 208 samples of frequency measurements for each RO type from a chip. All ROs are 7-staged where the first stage is a NAND gate. Frequency of each RO sample is measured over a time period of 10 s with an integration time of 1 ms.

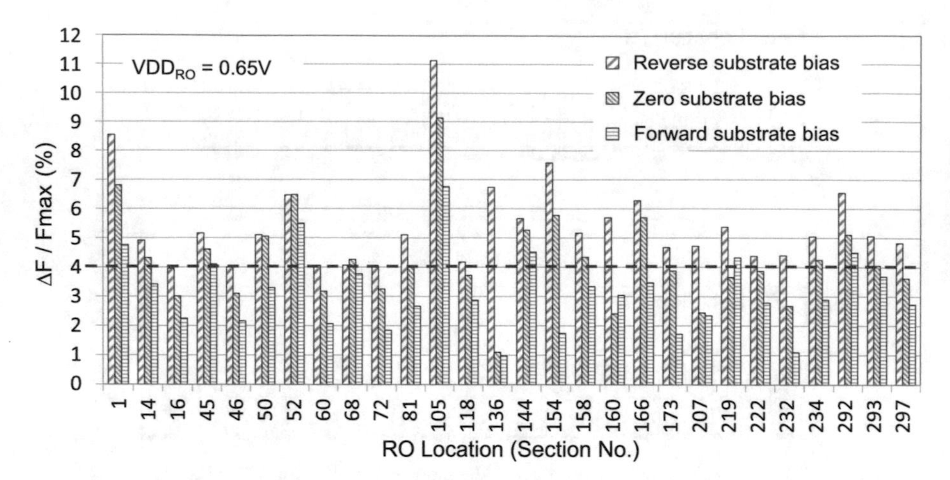

Fig. 23 $\Delta F/F_{max}$ of different ROs for three substrate bias conditions. ROs that have more than 4% fluctuation at reverse substrate bias case are shown

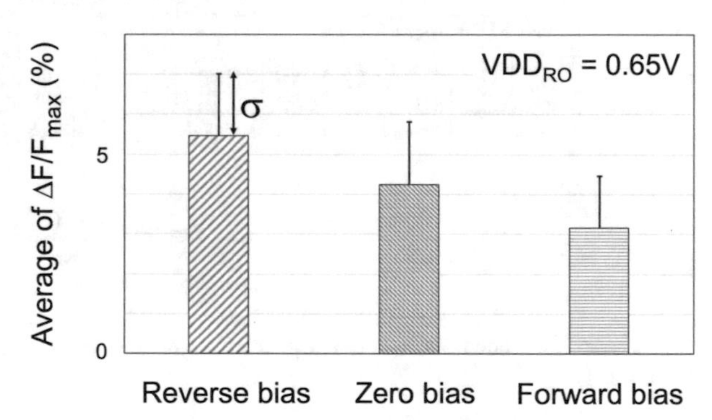

Fig. 24 Average and sigma value of $\Delta F/F_{max}$ calculated from Fig. 23

RO Design

To overcome the limitations of separating pMOSFET- and nMOSFET-specific characteristics, we have designed the following three ROs as shown in Fig. 26 where the gate widths of pMOSFET and nMOSFET differ.

1. RO #1: nMOSFET gate width is 75 times larger than pMOSFET. Minimum gate width is used for pMOSFET.
2. RO #2: pMOSFET gate width is 48 times larger than nMOSFET. Minimum gate width is used for nMOSFET.
3. RO #3: nMOSFET gate width is the same as pMOSFET. Minimum gate width is used here.

In the case of RO #1, it is expected that the delay distribution will be dominated by the much smaller pMOSFETs. We name this RO as p-dominant RO. Similarly, the delay distribution will be dominated by the much smaller nMOSFETs in the case of RO #2. We name this RO as n-dominant RO. In the case of RO #3, both of the

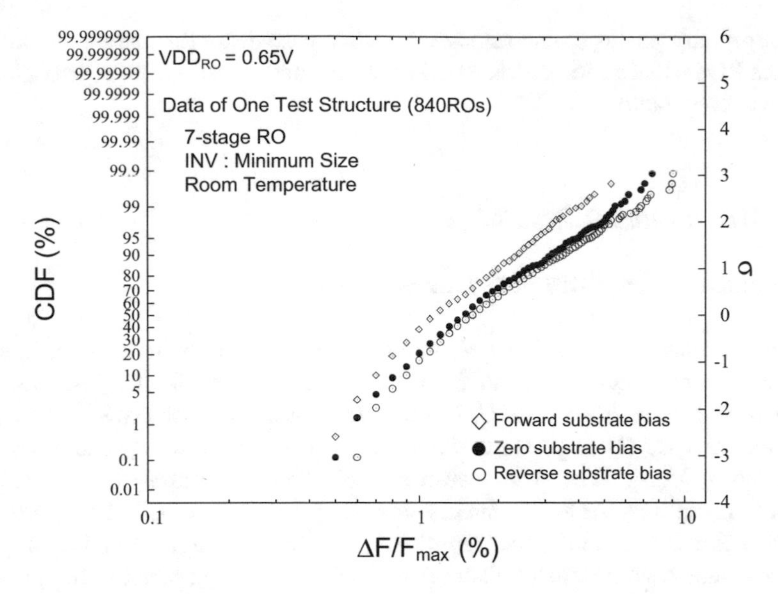

Fig. 25 Log-normal distribution plot of $\Delta F/F_{\max}$ for one test structure under three substrate bias conditions

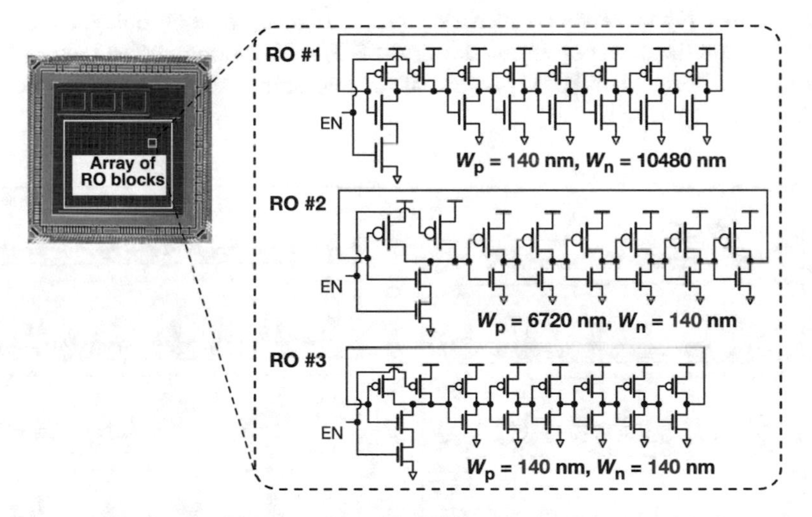

Fig. 26 Test structure containing an array of identical RO blocks. Each block contains ROs with different gate widths

pMOSFETs and nMOSFETs will contribute to the delay distribution. RO #3 here therefore represents a typical delay path of a digital circuit. ROs #1 and #2 here represent the two extreme cases where either the nMOSFETs or the pMOSFETs contribute to the delay variation. The gate widths of MOSFETs in the NAND gate are the same as those used in the inverter gates.

Our primary goal is to characterize the worst-case delay fluctuation due to RTN for each RO instance over a certain period of time and investigate the correlation for different temperatures.

3.4 Measurement Results

Delay Fluctuations at Different Temperatures

Figure 27a shows measured $\Delta d/d$ against time for a p-dominant RO sample for 5 different temperatures of 0 °C, 20 °C, 40 °C, 60 °C, and 80 °C. Discrete delay fluctuations are observed at 0 °C. Overall delay fluctuation is 0.6% in this case. With the increase of temperature, the average time to capture, τ_c, decreases as is suggested by Eq. 4. Thus, delay waveform will be more flicker-like than RTN at higher temperatures. At 80 °C, the discrete delay levels become indistinguishable, and the delay fluctuation is reduced to 0.2%. When the trapping and de-trapping time constants become smaller than the switching time, the impact of trapping and de-trapping gets averaged out. On the other hand, traps with longer time constants, which may not be detectable at low temperatures, may become detectable at higher temperatures. Figure 28 shows such a sample. In Fig. 28, we show delay fluctuations over time and the power spectrum density (PSD) for a p-dominant RO where large amplitude is observed at 80 °C because of the appearance of new traps. Figure 28a,

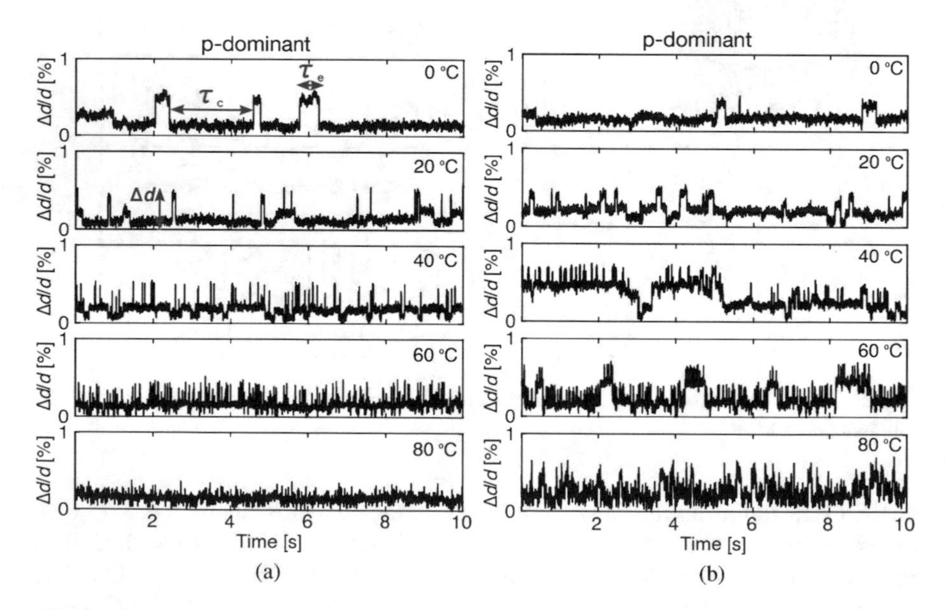

Fig. 27 RTN-induced delay fluctuation for p-dominant ROs at different temperatures. (**a**) Sample 1. (**b**) Sample 2

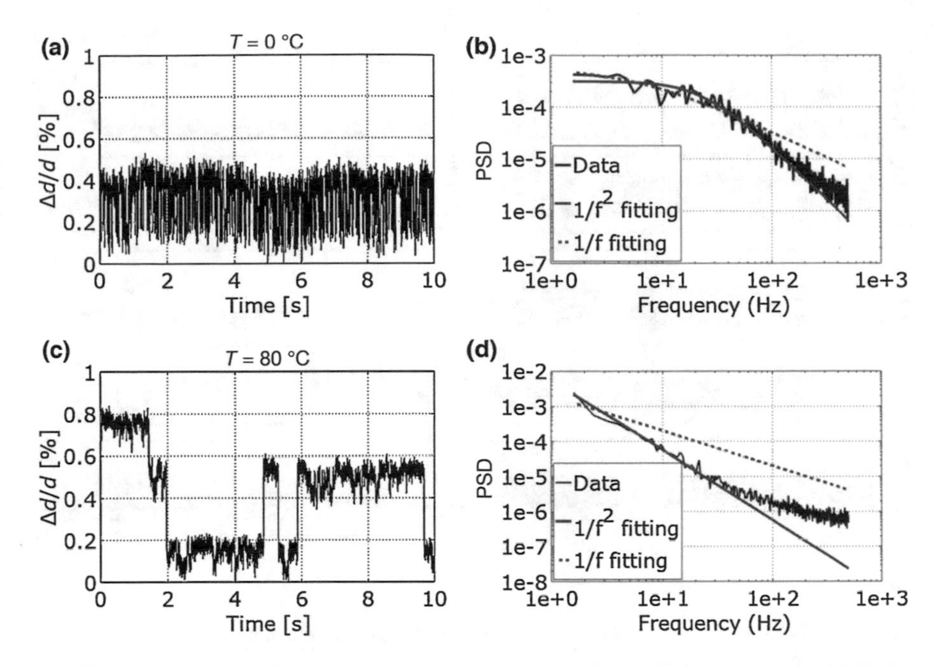

Fig. 28 A p-dominant RO sample with a new trap appearing at high temperature. (**a**) Delay waveform at 0 °C, (**b**) Power spectral density (PSD) for the waveform at 0 °C, (**c**) delay waveform at 80 °C, and (**d**) PSD for the waveform at 80 °C

b represents delay fluctuations and its PSD for 0 °C. Figure 28c, d represents delay fluctuations and its PSD for 80 °C. From the PSD results, we observe that both the delay waveforms at 0 °C and 80 °C follow $1/f^2$ at frequency range below 100 kHz.

Figure 27b shows RTN profiles at 5 different temperatures for another sample of a p-dominant RO. In the figure, new traps have appeared at temperatures of 20 °C, 40 °C, and 60 °C. The number of traps involved increases compared with that at 0 °C. In this case, large delay fluctuations are observed at the higher temperatures compared with that at 0 °C. Thus, because of the difference in trap profiles and the number of active traps at different temperatures, the delay fluctuations of a particular delay path are expected to show less correlation across the temperatures.

Figure 29a, b shows RTN profiles for two samples of n-dominant ROs. Similar to the case of pMOSFET, τ_c gets smaller with the increase of temperature. In Fig. 29b, new traps are being observed at the higher temperatures of 60 °C and 80 °C.

Correlation Across Temperatures

From the above discussions, it is clear that not only the trap characteristics change over temperature, but also appearing or disappearing of traps occur. As a result, low correlation between delay fluctuation amplitude is expected for a particular delay across temperatures. Figure 30 shows the correlation between delay fluctuations at two different temperatures of 0 °C and 80 °C for three different ROs. No significant

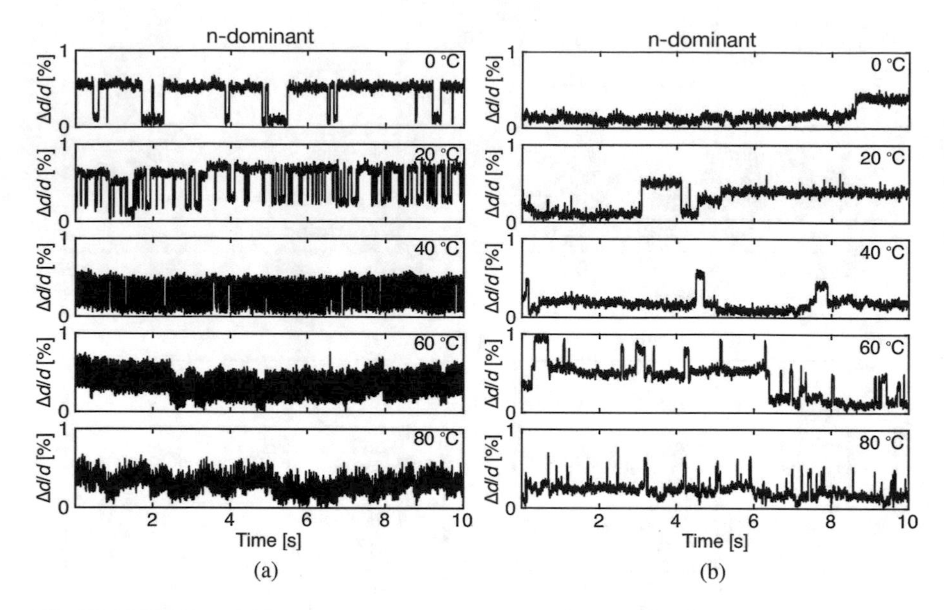

Fig. 29 RTN-induced delay fluctuation for n-dominant ROs at different temperatures. (**a**) Sample 1. (**b**) Sample 2

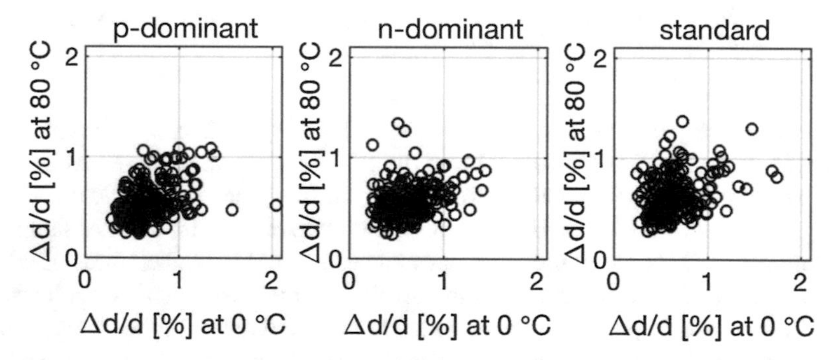

Fig. 30 Correlation between RTN-induced delay fluctuations between 0 °C and 80 °C for three different ROs. Correlation coefficient varies from 0.3 to 0.4

correlation is found for the p-dominant RO. The delay fluctuation is much larger at 0 °C compared with that at 80 °C. The n-dominant ROs, and ROs where both the pMOSFETs and nMOSFETs have small gate widths show a similar trend. No significant correlation across temperatures has been observed.

Distribution of Delay Fluctuation

As explained in Sect. 3.2, one key concern is the distribution of overall RTN amplitudes. Figure 31 shows the quantile–quantile plot of delay variations for temperature range from 0 °C to 80 °C. The delay distribution gets slightly smaller

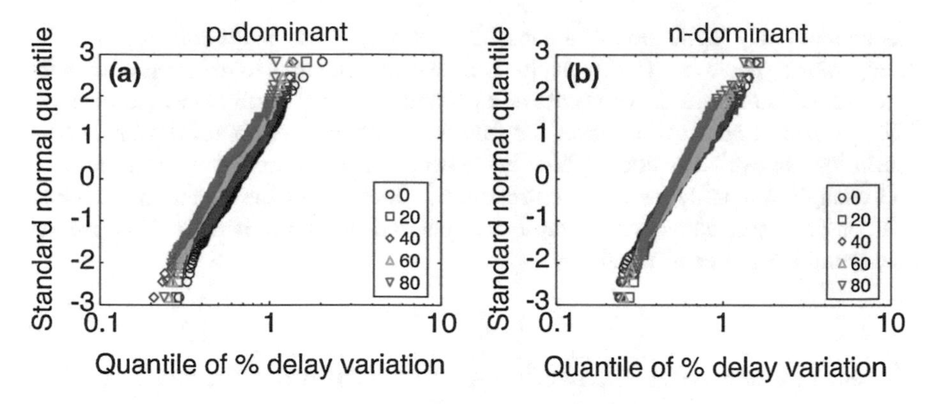

Fig. 31 Measured distributions of $\Delta d/d$ at different temperatures, (**a**) p-dominant RO, (**b**) n-dominant RO

with the increase of temperature. There are two reasons for the decrease of delay fluctuation. One is the lowering of threshold voltage with the increase of temperature causing less delay fluctuation for the same amount of ΔV_{th}. The other is the lowering of the RTN amplitude ΔV_{th} itself. One key observation here is that delay distributions follows lognormal distributions at all the temperatures.

Impact on Circuit Reliability

The measurement results show that increase in temperature decreases RTN-induced delay fluctuation to a small extent. The worst-case RTN effect occurs at the low temperature region. The dynamic nature of RTN results in very low correlation of delay fluctuation of a delay path at different temperatures. The low correlation has severe effect on the use of spare cells or critical-path replicas for performance compensation.

3.5 Summary

Measurement results from a 40 nm process technology reveal substrate bias effect on RTN. The impact of RTN-induced logic delay fluctuation can be statistically reduced by the forward substrate bias control. Then measurement results from a 65 nm thin-BOX FDSOI process reveal that RTN with significant discrete delay levels occurs quite frequently over a wide temperate range. A maximum delay fluctuation of 2% has been observed at 0 °C. In general, τ_c and τ_e decrease rapidly with the increase of temperature. However, for some samples, new slow traps appear at high temperatures. Also, the number of active traps differs from temperature to temperature, but in general more traps are active at a higher temperature. All these mean that delay fluctuation of a particular delay path has small correlation across temperature. The overall delay fluctuation distribution tends to get smaller with the increase of temperature. Worst-case delay fluctuation may occur at low temperature. The measured observation poses the following impacts on circuit design. Firstly, statistical analysis considering low temperature distribution is required and may be enough. Secondly, as delay correlation is low across the temperatures, post-silicon performance compensation techniques will not be reliable if RTN become the dominant source of variability.

4 RTN Parameter Extraction Under Switching Operation

Measurement results illustrated in Sects. 2 and 3 help us to assess the impact of RTN on CMOS circuit's reliability under different operating conditions. Incorporating RTN behavior into circuit simulation will allow us to estimate the impact of RTN on different circuits. To incorporate RTN in the simulation, RTN parameters of amplitude, trap density, and time constant are required. Although characterizations on these parameters are discussed in detail for discrete MOSFETs under DC bias, we need to extract these parameters for MOSFETs under switching operation. To facilitate gate-level delay evaluation, we develop a topology-reconfigurable RO structure which incorporates the concept of inhomogeneity to estimate the delay of a particular inverter stage. Then, we discuss an extraction methodology of amplitude and trap density from the measured data.

4.1 Gate Delay Evaluation

Gate-by-gate delay evaluation can be performed by utilizing a reconfigurable delay path. Figure 32 shows a reconfigurable structure where each inverter stage can be configured to follow either of the path-1 or path-2. Path-1 has a much larger delay

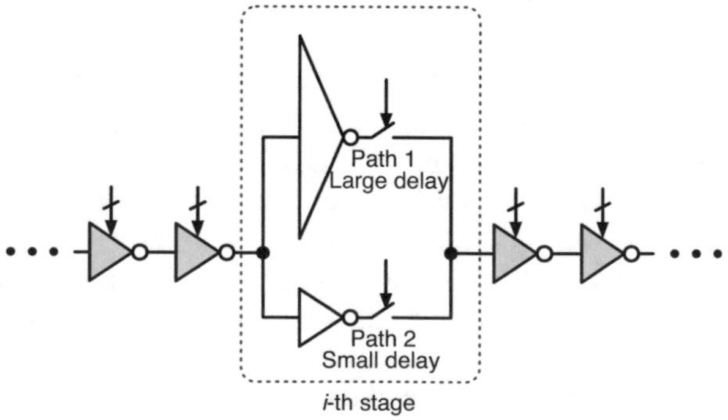

Fig. 32 Reconfigurable RO structure for gate-by-gate delay evaluation. Each inverter gate can be configured to have either of the large or small delay paths. The delay difference between the configurations gives us an approximate delay of the path-1 given that path-1 delay is much larger than path-2 delay

than that of path-2. As a result, changing the ith stage to the two configurations and measuring the corresponding delays, that delay difference gives us an approximation of the path-1 delay of the ith stage. Neglecting the path-2 delay, the overall delay, D^i, can be expressed as follows:

$$D^i = D_0^i + d^i. \tag{15}$$

Here, d^i is the delay for the ith stage. D_0^i is the delay contribution from all the inverter delays except the ith stage fall delay. Thus, the delay of the target ith stage can be obtained as follows if D_0^i can be estimated.

$$d^i = D^i - D_0^i. \tag{16}$$

Although we can only measure D^i with our circuit, D_0^i can be approximated with D_0 which is the delay when all the inverter stages are configured to follow path-2. If D_0 does not show any discrete fluctuation, then the discrete delay fluctuations observed in D^i can be attributed to the delay of path-1 of the ith stage. As a result, RTN-induced delay fluctuation of path-1 of the ith stage is evaluated by measuring ΔD^i over time.

$$\Delta d^i \approx \Delta D^i. \tag{17}$$

In summary, $(D^i - D_0)$ gives the static process variation for the ith stage, and ΔD^i over time provides the RTN-induced fluctuation for the ith stage. Furthermore, making the inverter delay sensitive to only pMOSFET or nMOSFET provides independent pMOSFET and nMOSFET characterizations.

4.2 *Test Structure*

Figure 33a shows a test chip which comprises four modules of reconfigurable ROs. Each module contains three different reconfigurable ROs as shown in Fig. 33b. The three ROs differ in MOSFET gate width. Figure 33c shows an inverter topology that can be configured to have two paths with large delay difference. When “C0” is “LOW” and “C1” is “HIGH,” the topology of Fig. 33d is realized whose delay characteristics are the same in a standard inverter topology. Here, standard topology refers to a topology with equal rise and fall delay and the delays are small. When “C0” is “HIGH” and “C1” is “LOW,” the topology of Fig. 33e is realized whose fall delay characteristic is the same as that in Fig. 33d topology, but the rise delay characteristic is different. Here, inserting a pMOSFET pass-gate before the pull-up pMOSFET lowers the gate-overdrive of the pull-up pMOSFET. As a result, the current through the pull-up pMOSFET decreases exponentially, and the corresponding delay becomes much larger.

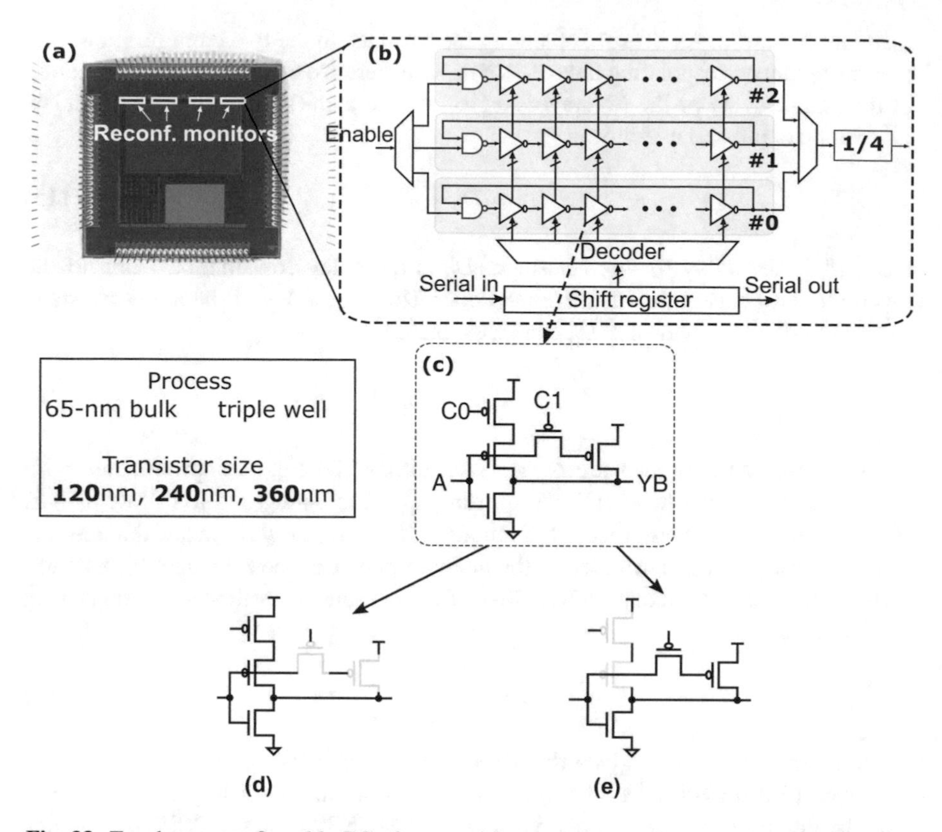

Fig. 33 Topology-reconfigurable ROs for gate delay variation measurement implemented in a 65-nm test chip. ROs with three different gate widths of 120, 240, and 360 nm are implemented. (**a**) Chip photograph, (**b**) reconfigurable RO test structure, (**c**) topology-reconfigurable inverter cell, (**d**) typical inverter configuration, and (**e**) p-dominant configuration

ROs of three different gate widths of 120, 240, and 360 nm are implemented for demonstration. For detailed understanding and modeling, gate widths with a wider range are preferable. Minimum gate length of the process is used for all the transistors. Each RO is 127 staged. The first stage is a NAND gate to turn ON and OFF the oscillation, and the last stage is used as a buffer. Thus, from a single RO, 125 samples for pMOSFET dominated delay are measured by reconfiguring the topology of each stage. Finally, as the four modules contain the same ROs, we obtain 500 samples of pMOSFET-dominated delay from a chip. Thus, we obtain 500 samples for each of the gate widths. Two control signals are used to configure the topology to achieve the following delay characteristics:

1. equal rise/fall delay as shown in Fig. 33d, and
2. larger rise delay sensitive to pMOSFET as shown in Fig. 33e.

The supply voltage V_{dd} is set to 0.8 V. Because pMOSFET pass-gate can only pull down the pull-up pMOSFET's gate down to its threshold voltage, the gate-source voltage of the pull-up pMOSFET is reduced resulting in weak inversion operation. As a result, the delay of the inhomogeneous gate when configured as Fig. 33e is multiple times larger than delays of other gates.

4.3 RTN Parameter Extraction

To characterize RTN parameters of trap number, time constant, and amplitude, discrete levels in the measured values of current or delay need to be detected. RTN is considered to be discrete fluctuations of drain current due to trapping and de-trapping of charges at the gate oxide interface. The power spectrum of RTN is a Lorentzian or $1/f^2$ spectrum, whereas flicker noise or $1/f$ noise is considered to be a superposition of multiple RTN events [7]. Detecting discrete fluctuations is difficult in the presence of background noise. The background noise consists of flicker noise, thermal noise, supply noise, and environmental noise.

Detecting discrete fluctuations from measured samples over time is challenging in the presence of background noise. Although several detection mechanisms are proposed in the literature, there is a limitation on the detection resolution and accuracy for every mechanism. We illustrate these phenomena using two examples that are shown in Fig. 34. Background noise is assumed to follow a normal distribution with a standard deviation of 13 ps. In Fig. 34a, large delay fluctuation due to one trap is considered. As the measured delay distribution is a convolution between the background noise PDF and RTN PDF, the resulted distribution becomes a bimodal PDF. Here, the dotted line shows the convoluted distribution. Open circles refer to the delay distribution from an RO that is obtained using our test chip. The two distributions overlap each other. A simple peak detection mechanism using

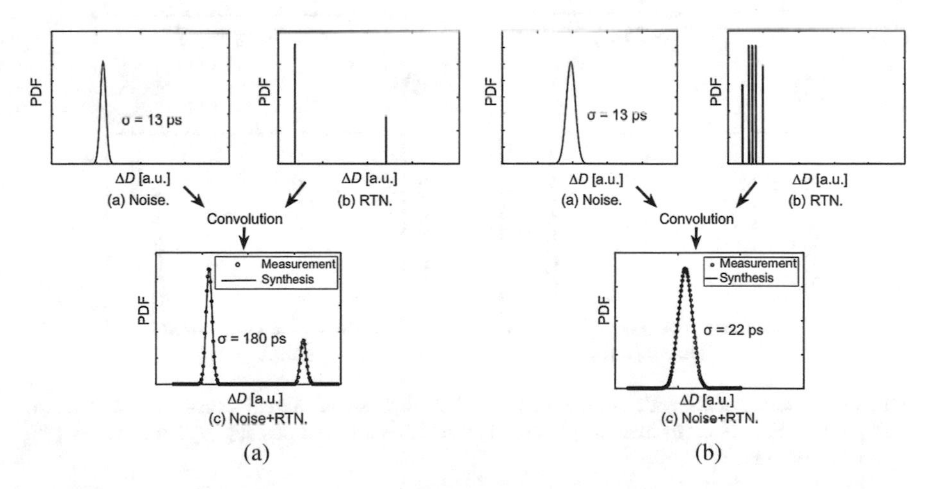

Fig. 34 Delay fluctuation distributions with different RTN levels. Convolutions of noise and RTN are compared with the delay distributions. (**a**) Distinguishable RTN level. (**b**) Non-distinguishable RTN level

a kernel density estimation will suffice here for RTN detection [49]. However, when multiple RTN states are very close to each other, as shown in Fig. 34b, the resulting PDF follows a normal distribution where no second peak exists. However, even though there is only one peak, the variance of the delay distribution increases indicating undetectable RTN events.

The Gaussian mixture model (GMM) based on expectation–maximization (EM) algorithm is used to find the best PDF [50]. A large number of iterations are performed in this case. Markov chain-based technique to filter out the noises has also been proposed in [51]. Weighted techniques to distinguish RTN states are proposed in [52]. Reference [13] uses a kernel density-based approach to detect the peaks. Here, we discuss the kernel density-based method to extract the parameters of amplitude and trap density from measurement samples. In the delay measurement, depending on the measurement time and the time constants of the traps, intermediate delay values between two discrete states are also measured. Figure 35a shows one such example. Here, many intermediate delay values are observed which can happen when the time constant is smaller than the measurement time. Using a simple 1-dimensional kernel density-based peak detection thus can give many pseudo-RTN states as shown in Fig. 35c. If we utilize a time lag plot as shown in Fig. 35b, and

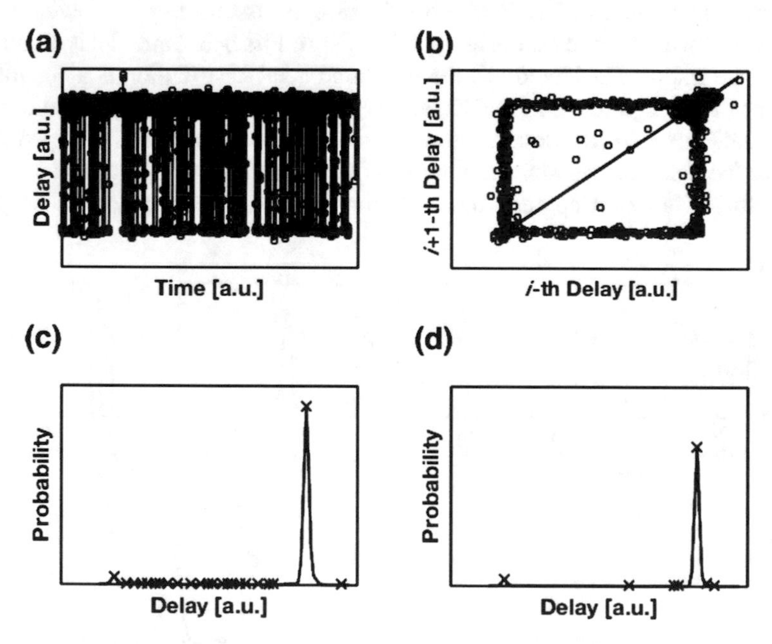

Fig. 35 Detecting RTN with 1-dimensional and 2-dimensional kernel density estimation. (**a**) Delay versus time plot, (**b**) time lag plot, (**c**) 1-dimensional kernel density, (**d**) kernel density on $y = x$ line of time lag plot

concentrate on the $y = x$ line of the 2-dimensional kernel density, the pseudo-RTN states are mostly eliminated which is shown in Fig. 35d. Therefore, a 2-dimensional kernel density estimation is effective in eliminating the pseudo-RTN states so that true RTN events are detected. Note that pseudo-RTN states may nonetheless remain whereas some true RTN states may be missed. The above discussed detection method reduces the probability of detecting the pseudo-RTN states significantly.

4.4 RTN Examples

Discrete gate delay fluctuations are then measured for each gate width to evaluate RTN caused by pMOSFET and nMOSFET separately. Discrete states are detected using the method presented in Sect. 4.3. Several measurement samples for pMOSFET RTN are shown in Fig. 36, which shows measurement samples of oscillation period observed over time for three different inhomogeneous stages. Figure 36a shows a sample with probability density having a single peak. We term this kind of samples as non-detectable samples rather than treating them as zero. Samples that show only one peak in the distribution imply that their discrete fluctuations are not detected. In our evaluation, the percentages of samples showing at least two peaks are 74, 76, and 85% for pMOSFET gate widths of 120, 240, and 360 nm, respectively. Figure 36b shows a sample with probability density having two distinguishable peaks. In Fig. 36c, samples with multiple peaks are shown. With the proposed circuit and characterization system, we observe RTN of various amplitudes, time constants, and different numbers of traps.

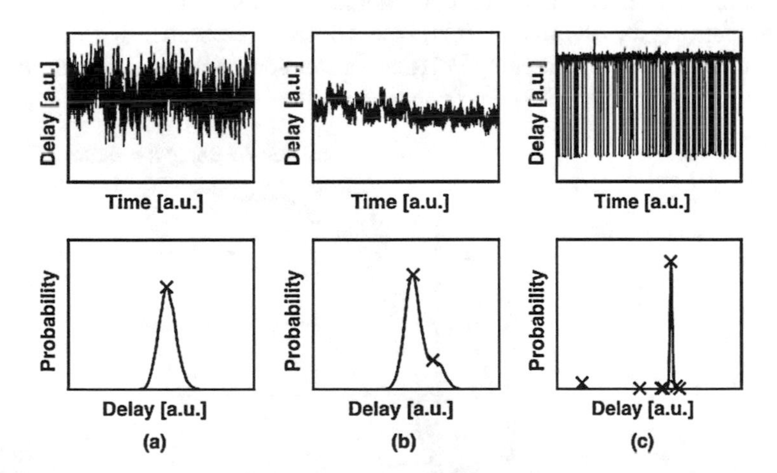

Fig. 36 Samples with different RTN profiles. (**a**) Delay waveform sample with probability density having a single peak, (**b**) delay waveform sample with probability density having two distinguishable peaks, and (**c**) delay waveform sample corresponding probability density having multiple peaks

4.5 Amplitude Distribution

We show the distributions of delay fluctuations for all the gate widths of pMOSFETs and nMOSFET. Figure 37 shows $\Delta d/d$ distributions due to RTN in the pMOSFETs. Distributions of three different gate widths are shown in the figure. As expected, long tails are observed in all the three distributions. We observe an RTN-induced delay fluctuation of 40% for a gate width of 120 nm.

4.6 Trap Density

We show the distributions of the number of detected traps for all the gate widths of pMOSFETs and nMOSFETs. In theory, the number of traps is believed to follow a Poisson distribution. Sum of two Poisson distributions is also a Poisson distribution and the parameter λ of the resulting Poisson distribution is the sum of λs of the underlying distributions. The number of traps N is calculated from the detected number of states, n, as follows:

$$N = \text{ceil}(\log_2 \text{n}). \tag{18}$$

λ of trap number distribution for a discrete MOSFET can be approximated by dividing the λ by 2 as two MOSFETs are involved here.

Figure 38 shows the distributions of trap number for three different gate widths of pMOSFET. The distributions of trap numbers are then fitted to a Poisson distribution. The fitted values of the parameter λ are 1.40, 1.22, and 1.56 for gate widths of 120, 240, and 360 nm, respectively. We observe that the extracted distributions can be expressed by Poisson distributions. However, the mean value,

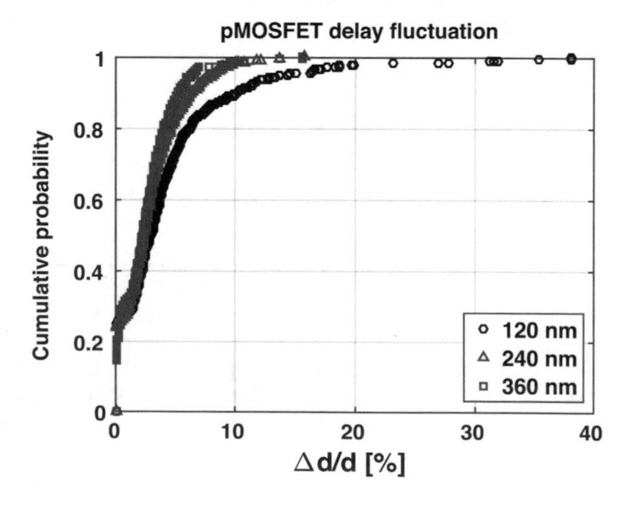

Fig. 37 Cumulative probability of detected RTN-induced $\Delta d/d$ of pMOSFET dominant gates

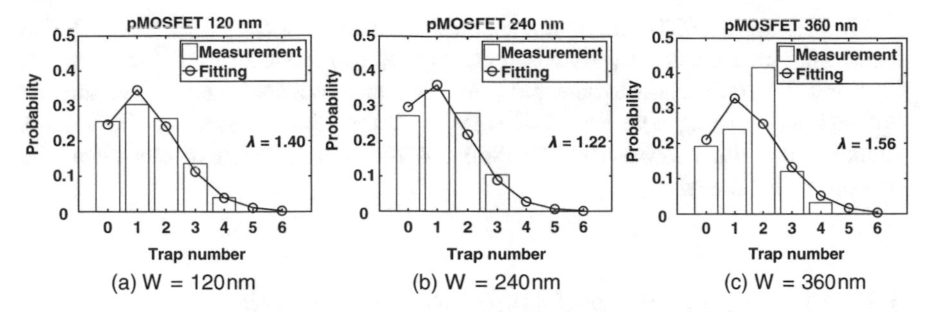

Fig. 38 Number of traps for pMOSFETs of three gate widths of, (**a**) 120 nm, (**b**) 240 nm, and (**c**) 360 nm

λ, of the distributions does not follow the theory, that is, larger gate-width MOSFET exhibits higher λ value. As explained in Sect. 4.3, the extraction accuracy of discrete states is affected by the background noise of flicker and thermal noises. MOSFETs with larger gate width generate more background noise than the smaller gate-width counterpart. As a result, the detection accuracy decreases with the increase of gate width. With a much larger gate width, we end up with all the samples with only one peak. Nonetheless, the extracted trap number is an important parameter to develop models for RTN simulation.

4.7 Summary

Using a topology-reconfigurable RO architecture, RTN parameters of amplitude and trap density can be extracted. Characterization using an RO-based test structure enables us to extract the RTN parameters under switching operation. The characterization results can then be incorporated into transistor-level simulation to evaluate the reliability. We present an RTN simulation technique in Sect. 5.

5 Replication of Random Telegraph Noise by Using a Physical-Based Verilog-AMS Model

It is easy to get RTN-induced fluctuation over much longer periods than seconds by measuring fabricated chips. But it is hard to obtain fluctuation in short period for a few nano-second because RTN is averaged out during measurement. On the other hand, it is possible to replicate RTN-induced frequency fluctuations with the same tendency as the measurement results by transient simulations in a short period less than the interval of the actual measurement.

We propose an RTN simulation method by using Verilog-AMS [53, 54]. We explain the mechanism of RTN based on physics and models of RTN in Sect. 5.1. Section 5.2 describes how to control V_{th} that manipulate RTN-induced drain current fluctuations by using a variable DC voltage source. The behavior of the voltage source is described in Sect. 5.3. Section 5.4 shows simulation results using the proposed RTN model.

5.1 RTN Mechanism and Model Based on Physics

RTN is caused when defects in gate dielectrics trap or emit carriers in the channel as shown in Fig. 1. It can be represented by the threshold voltage shift ΔV_{th} in a transistor model.

Figure 2 shows a V_{th} fluctuation caused by RTN from a single defect. Time constants $\langle\tau_c\rangle$ and $\langle\tau_e\rangle$ are defined as the average time to capture and emit carriers, respectively. They depend on gate voltage (V_G). As V_G increases, $\langle\tau_c\rangle$ and $\langle\tau_e\rangle$ become short and long, respectively [55]. As shown in Fig. 2, V_{th} has two states. When a defect captures or emits a carrier, V_{th} becomes the high or low state, respectively. A threshold voltage shift ΔV_{th_d} is constant in each defect [56]. If multiple defects exist in the gate oxide, V_{th} fluctuates among multiple states. This phenomenon is explained by the charge trapping model (CTM) [57]. In this work, we propose an RTN simulation method based on CTM.

In recent deca-nanometer processes, high-k (HK) materials such as HfSiO and HfO_2 are used in gate dielectrics to decrease leakage current (Fig. 39). The interface layer (IL, e.g., SiO_2) is fabricated between Si and HK because HK on Si causes threshold voltage pinning and phonon scattering. It is found that the defect distribution characteristics are different between HK and IL dielectrics [58]. For this reason, we must consider the bimodal CTM which is adequate to utilize two different defects in HK and IL. When the gate dielectric consists of the interface layer without HK, we apply the unimodal model instead of the bimodal model.

CTM has parameters n, ΔV_{th_d}, $\langle\tau_c\rangle$, and $\langle\tau_e\rangle$, where n is the number of defects in the gate oxide. n is different for each transistor and follows the Poisson distribution [59].

The probability density function (PDF) of the Poisson distribution $P(n)$ is expressed as in Eq. (19).

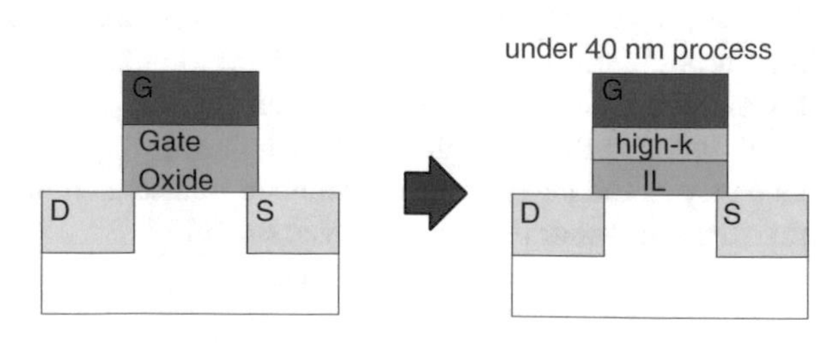

Fig. 39 The high-k (HK) materials are used in gate dielectrics to decrease leakage current

$$P(n) = \frac{N^n \exp(-\lambda)}{n!} . \tag{19}$$

Here, λ is the expected value of n and explained as in Eq. (20).

$$N = D\,LW , \tag{20}$$

where D is the number of defects per gate area. We assume $D_{\mathrm{HK}} = 4.0 \times 10^{-17}\,\mathrm{cm}^{-2}$ and $D_{\mathrm{IL}} = 2.0 \times 10^{-18}\,\mathrm{cm}^{-2}$ [59, 60].

$\Delta V_{\mathrm{th\_d}}$ and τ follow an exponential distribution [57, 61]. The PDF of the distribution of $\Delta V_{\mathrm{th\_d}}$ is described by Eq. (21).

$$P(\Delta V_{\mathrm{th\_d}}, \eta) = \frac{1}{\eta} \exp\left(\frac{\Delta V_{\mathrm{th\_d}}}{\eta}\right) , \tag{21}$$

where η is the expected value of $\Delta V_{\mathrm{th\_d}}$ explained as in Eq. (22).

$$\eta = \frac{s}{LW} , \tag{22}$$

where s is a coefficient of η. We assume $s_{\mathrm{HK}} = 9\,\mathrm{V{\cdot}nm^2}$ and $s_{\mathrm{IL}} = 90\,\mathrm{V{\cdot}nm^2}$ [15]. τ follows the logarithmic uniform distribution from 10^{-9} to 10^{9} s [61]. It depends on V_{GS} and changes exponentially as in Eq. (23) [39, 62].

$$\tau = \tau_0 \exp\left(B\, V_{\mathrm{GS}}\right) , \tag{23}$$

where τ_0 is the time constant at $V_{\mathrm{GS}} = 0$ and B is the sensitivity to V_{GS}. As mentioned above, τ_0 varies from 10^{-9} to 10^{9} s for each defect. The sensitivity B is distributed from 1 to 10 [62, 63]. In the proposed RTN model, B is defined as 5.

5.2 *Charge Trapping Model to MOSFET*

During transient circuit simulations, we must temporally fluctuate V_{th} which is a voltage source to replicate drain current fluctuation by RTN. Moreover, as mentioned above, τ depends on V_{GS}. It is impossible to use a set of voltage waveforms prepared prior to transient simulations.

V_{th} is shifted by changing device parameters. Standard transistor models are generally used such as BSIM (Berkeley short channel IGFET model). However, we cannot change device parameters during transient simulations in those models. As shown in Fig. 40a, we replicate the RTN-induced threshold voltage fluctuation to connect a variable DC voltage source implemented by using Verilog-AMS attached to the gate terminal. We call the voltage source an RTN module. $V_{\mathrm{th}}(t)$ is changed by the gate overdrive voltage (V_{OV}). $\Delta V_{\mathrm{th}}(t)$ corresponds to the V_{th} shift value at time t during transient analysis.

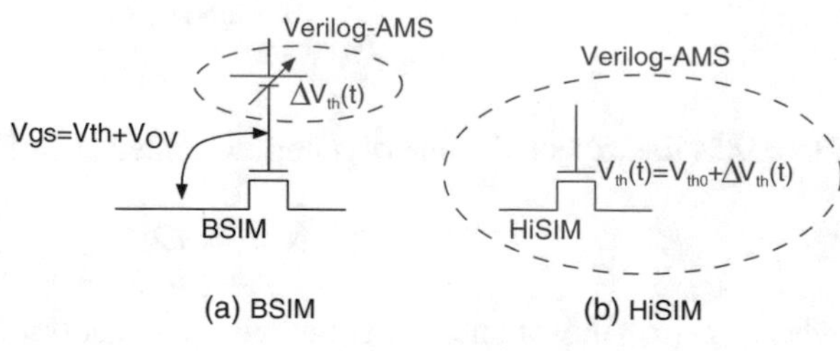

Fig. 40 $V_{\rm th}$ shift method in a circuit-level transient simulation **(a)** BSIM. **(b)** HiSIM

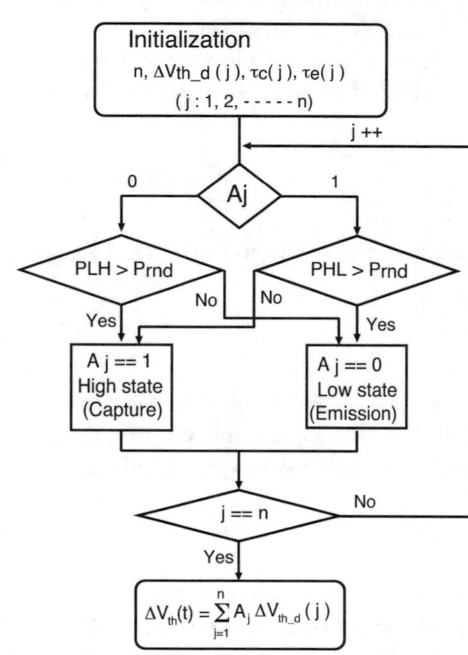

Fig. 41 Flowchart of RTN module

In the case of HiSIM (Hiroshima-University STARC IGFET MODEL) [64], we can directly change $V_{\rm th}$ because it is usually described in Verilog-AMS as shown in Fig. 40b. $V_{\rm th0}$ means the threshold voltage at time 0.

The $I_{\rm DS}$-$V_{\rm DS}$ characteristics up to $\Delta V_{\rm th}$ within $\pm 3\eta$ at the body bias from -1.0 to 0.5 V are exactly the same as those obtained from our voltage source and $\Delta V_{\rm th}$ at 0 s in the transistor models of a commercial 40 nm CMOS technology.

5.3 RTN Circuit Simulation Method Using CTM

In this section, we explain the detail of the RTN module. Figure 41 shows the flowchart to compute RTN-induced $\Delta V_{\rm th}$. Parameters to calculate RTN are shown in Table 1.

Table 1 Parameters to calculate RTN

Name	Explanation
L	Gate length
W	Gate width
n	Number of defects in the gate oxide
N	Expected value of n
D	Defects of gate oxide per unit area
$\Delta V_{\mathrm{th\_d}}$	Threshold voltage fluctuation per defect
η	Expected value of $\Delta V_{\mathrm{th\_trap}}$
s	Coefficient of η
$\langle \tau_{\mathrm{c}} \rangle$	Time to capture a carrier
$\langle \tau_{\mathrm{e}} \rangle$	Time to emission a carrier
T_{step}	Time step of transient analysis
P_{LH}	Probability to capture a carrier
P_{HL}	Probability to emit a carrier

First, n, $\Delta V_{\mathrm{th\_d}}$, and τ are initialized. Then, a carrier is emitted or captured according to the Markov process without hysteresis. Finally, ΔV_{th} is increased by $\Delta V_{\mathrm{th\_d}}$ if the carrier is captured. This process is repeated for all defects.

A_j in Fig. 41 stands for the defect-capture state. "High state" and "Low state" mean that a carrier is captured and emitted, respectively. If the state is "High," A_j is equal to 1 while the state is "Low" when A_j is equal to 0. P_{LH} is the transition probability from "Low" to "High." P_{HL} is the reverse transition of P_{LH}. P_{LH} and P_{HL} are expressed as in Eqs. (24) and (25), respectively.

$$P_{\mathrm{LH}} = 1 - \exp\left(\frac{T_{\mathrm{step}}}{\langle \tau_{\mathrm{c}} \rangle}\right), \tag{24}$$

$$P_{\mathrm{HL}} = 1 - \exp\left(\frac{T_{\mathrm{step}}}{\langle \tau_{\mathrm{e}} \rangle}\right), \tag{25}$$

where T_{step} is the time step on transient analysis. The defect state is determined by comparing P_{LH} or P_{HL} with P_{rnd}, which is a random number from 0 to 1. After the states of all defects are fixed, ΔV_{th} is calculated by Eq. (26).

$$\Delta V_{\mathrm{th}}(t) = \sum_{j=1}^{n} A_j \Delta V_{\mathrm{th\_d}}(j), \tag{26}$$

where $\Delta V_{\mathrm{th\_d}}(j)$ is the threshold voltage shift by the jth defect.

In the bimodal case, $\Delta V_{\mathrm{th\_HK}}$ and $\Delta V_{\mathrm{th\_IL}}$ are calculated in HK and IL dielectrics separately.

5.4 *Simulation Results of RTN-Induced Drain Current and Frequency Fluctuation*

In section "RTN-Induced Drain Current Fluctuation of NMOSFETs", we analyze RTN-induced drain current fluctuations of NMOSFETs. Section 5.4 describes the distribution of RTN-induced frequency fluctuations in ring oscillators. Model parameters of a 40 nm CMOS are used in BSIM. In the case of HiSIM, we use a commercial 65 nm FDSOI technology because no HiSIM model is available for the 40 nm CMOS technology.

RTN-Induced Drain Current Fluctuation of NMOSFETs

We perform transient analysis to replicate RTN-induced current fluctuation. The simulation conditions are shown in Table 2. Figure 42 shows simulation results of drain currents in a single NMOSFET by BSIM. The upper figure shows the drain current fluctuations and the lower one is the number of defects capturing carriers. The drain current fluctuates according to the number of captured defects. Figure 43 shows the drain current fluctuation of 100 NMOSFETs from the RTN module attached to the gate terminal. The drain current fluctuates down to 5% at maximum in 1 μs. Figure 44 shows RTN-induced drain current fluctuations in two NMOSFETs. Figure 44a, b are the simulation results on BSIM, HiSIM, respectively. The amplitudes of RTN and timing of fluctuation are different for each MOSFET. Therefore, we confirm that the proposed RTN models can successfully replicate the temporal drain current fluctuation.

The Distribution of RTN-induced Frequency Fluctuation in Ring Oscillators (ROs)

RTN affects the oscillation frequency of ROs. Here, F_{max} is the maximum oscillation frequency and ΔF is the maximum frequency fluctuation as in Sect. 2.7. The

Table 2 Simulation conditions of NMOSFET

Explanation	Parameters	Value
Gate length	L	44 nm (BSIM)
		60 nm (HiSIM)
Gate width	W	88 nm (BSIM)
		260 nm (HiSIM)
Gate-source voltage	V_{GS}	1.0 V
Drain-source voltage	V_{DS}	1.0 V
Source voltage	V_S	0 V
Backgate voltage	V_B	0 V
Simulation time		1 μs

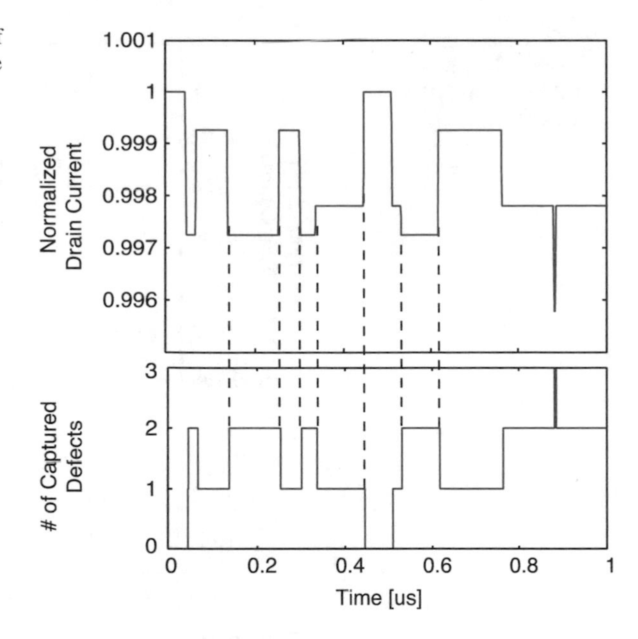

Fig. 42 Simulation results of drain currents (upper) and the number of captured defects (lower) in a single NMOSFET

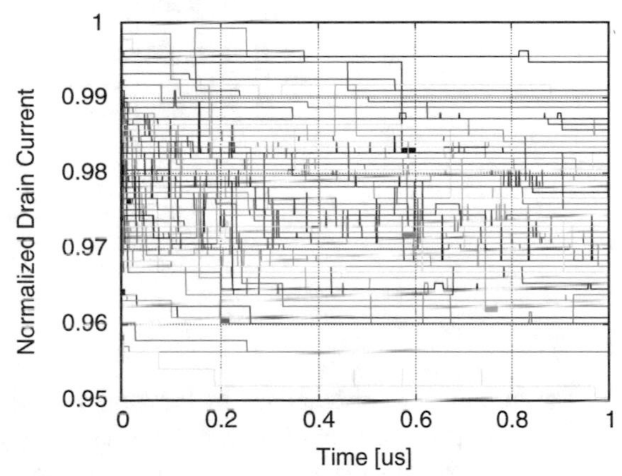

Fig. 43 Simulation results of drain currents of 100 NMOSFETs

distribution of $\Delta F/F_{\max}$ of 840 ROs fabricated in a 65 nm FDSOI process is shown in Fig. 45a. It has a long-tail distribution.

Figure 45b is the simulation results of 840 ROs by the proposed model. The simulated circuit and conditions are shown in Fig. 46. These conditions are different from the measurement results. ROs with 3 stages are used to amplify RTN-induced $\Delta F/F_{\max}$ [56]. The RTN module is connected to the gate terminals of each MOSFET. Note that every RTN module is independent. $\Delta V_{\text{th}}(t)$ in all transistors are computed from different random values as already shown in Fig. 43.

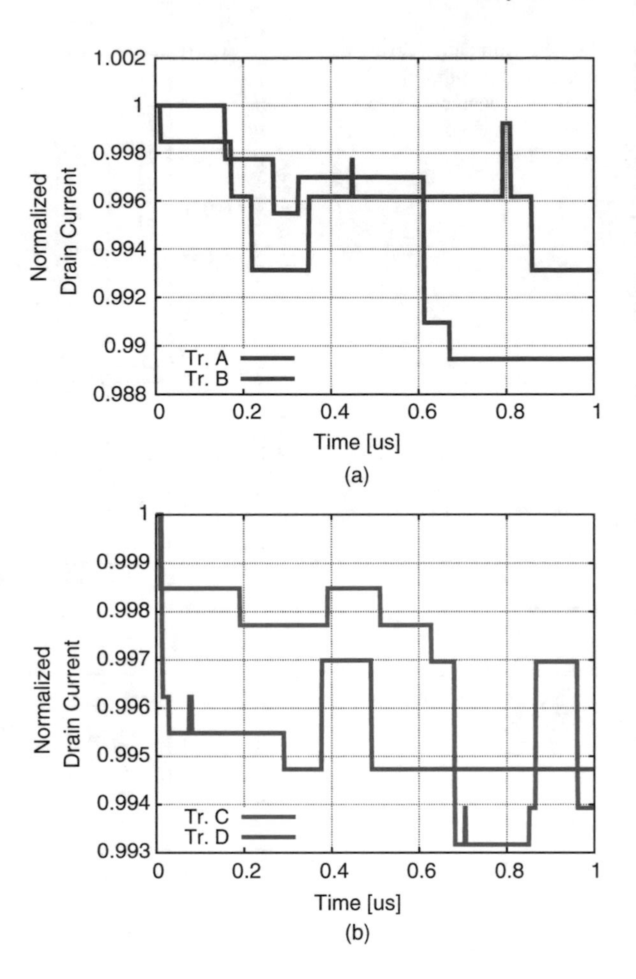

Fig. 44 Drain current fluctuations in two NMOSFETs. (**a**) BSIM. (**b**) HiSIM

The "unimodal" distribution in Fig. 45b is one of CTM which includes only IL characteristics. It cannot replicate the frequency distribution of the measured one in Fig. 45a. On the other hand, the measured distribution can be replicated by the "bimodal" model. We found that the proposed model well replicates the measurement results of a HK process.

The proposed RTN module implemented by Verilog-AMS can replicate RTN-induced $\Delta F/F_{\max}$ distribution in the unimodal and bimodal models as in [65].

Simulation times by using the proposed RTN module are 2.0x and 3.1x longer in the unimodal and bimodal models than those without the RTN module, respectively. Although it takes longer simulation time due to Verilog-AMS module, RTN-induced fluctuation is replicated by using proposed RTN model.

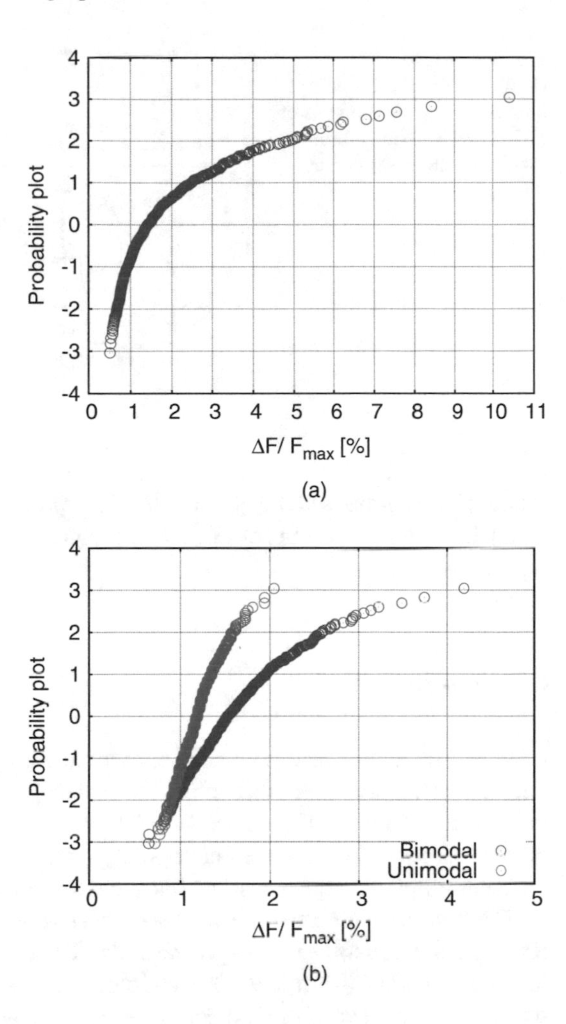

Fig. 45 The distribution of $\Delta F/F_{\max}$. (**a**) Measurement results of 840 ROs [56]. (**b**) Simulation results of 840 ROs by BSIM

5.5 *Summary*

We propose an RTN simulation method to implement a variable DC voltage source which fluctuates threshold voltages (V_{th}) temporally by using Verilog-AMS for BSIM and HiSIM models. We construct the bimodal charge trapping model (CTM) which considers the defect distribution characteristics of both HK and IL. In the case of BSIM, we attach an independent voltage source implemented by using Verilog-AMS to the gate terminal because device parameters cannot be changed dynamically. V_{th} is shifted directly in HiSIM implemented by Verilog-AMS. The proposed models can be applied to both of the unimodal and bimodal CTM. We obtain the RTN-induced drain current fluctuation in MOSFETs. The distribution of $\Delta F/F_{\max}$ in ROs obtained from measurement results is replicated by the bimodal

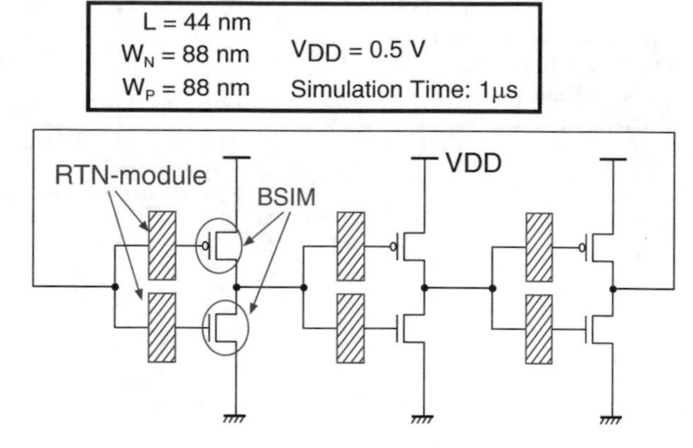

Fig. 46 3-stage ring oscillator. The RTN module is connected to gate terminals of all MOSFETs. Simulation conditions are described in this figure

model in the proposed RTN module. The proposed method can be applied to estimate the temporal impact of RTN for digital analog circuits including multiple transistors.

6 Conclusion

In this chapter, we have presented measurement results of RTN under switching operation using RO-based test chips. The test chips were fabricated in three different processes of 65 nm bulk, 65 nm FDSOI, and 40 nm bulk. The measurements are performed for ROs of different topology, gate width, and stage number under different supply voltage, substrate bias, and temperature.

The measurement results give us several valuable insights into the impact of RTN on the reliability of logic circuits. By measuring 12,600 ROs fabricated in a commercial 40 nm CMOS technology, a maximum of 16.8% of frequency fluctuation has been observed for a 7-stage RO at a 0.65 V supply voltage. The impact of RTN on process variation with respect to combinational circuit delay is estimated by the experimental data. It suggests that there is no or weak correlation between RTN variation and process variation. It is found that the impact of RTN fluctuation can be drastically reduced by slightly increasing the supply voltage and gate size. In a 65 nm process, a maximum of 40% of delay fluctuation for an inverter operating in weak inversion has been observed. The results confirm that digital circuits can be severely affected by RTN at low-voltage operation, and thus circuit design for low-voltage operation need to consider RTN. In the case of a 65 nm bulk process, 74% of inverters show RTN events which suggest that RTN occurs frequently. Measurement results under different substrate bias and temperature reveal that appearing and disappearing of switching traps occur under different operating conditions. When temperature changes, the delay distribution remains the same for different temperatures which are good news from the modeling perspective. However, the correlation under different temperatures is found to be low. The low correlation implies that some delay paths may encounter large RTN events at one temperature, while at some different temperature, some other paths may encounter large RTN. Low correlation poses a challenge in post-silicon tuning and testing if RTN is the dominant source of variability. To optimize a circuit under RTN-induced variability, designers can tune the design parameters of gate width, stage number, and supply voltage. We have shown the variability change according to gate width, stage number, and supply voltage. This quantitative evaluation helps us choosing the optimum parameters. Using the results, models for gate size and supply voltage dependency of RTN-induced delay fluctuation can be developed that will help the designers to optimize their circuits.

We then have presented a test structure design methodology to extract the RTN parameters so that models for simulation can be developed. The presented test structure enables evaluating gate-level delay fluctuation utilizing inhomogeneity and reconfigurability in a delay path. We have presented a simulation framework using a physical-based Verilog-AMS model for BSIM and HiSIM. Frequency fluctuations of an RO are then simulated using the proposed framework. The distribution of the simulated frequency fluctuations follows a lognormal distribution. Incorporating the RTN parameters of the number of traps, time constants, and amplitude will allow us to evaluate critical paths of a design to assess the worst-case delay.

The observations and methods discussed in the chapter will provide insights into the designers in assessing the impact of RTN on digital circuits. The ROs used in the test structures here can be used as a guideline to design test structures to characterize and model RTN effects for a target process.

References

1. N. Summers, *Apple's A12 Bionic is the first 7-nanometer smartphone chip* (2018). https://www.engadget.com/2018/09/12/apple-a12-bionic-7-nanometer-chip/. Accessed 10 Dec 2018
2. S. Borkar, Designing reliable systems from unreliable components: the challenges of transistor variability and degradation. IEEE Micro **25**, 10 (2005)
3. M. Alam, Reliability- and process-variation aware design of integrated circuits. Microelectron. Reliab. **48**, 1114 (2008)
4. H. Onodera, Variability modeling and impact on design, in *IEEE International Electron Devices Meeting (IEDM)* (2008), p. 701
5. G.F. Taylor, Where are we going? Product scaling in the system on chip era, in *IEEE International Electron Devices Meeting (IEDM)* (2013), pp. 441–443
6. D. Boning, S. Nassif, Models of process variations in device and interconnect, in *Chapter 6 of Design of High-Performance Microprocessor Circuits*, ed. by A. Chandrakasan, W. Bowhill, F. Fox (IEEE, Piscataway, 2001), pp. 98–115
7. M.J. Kirton, M.J. Uren, Noise in solid-state microstructures: a new perspective on individual defects, interface states and low-frequency ($1/f$) noise. Adv. Phys. **38**(4), 367–468 (1989)
8. K. Ito, T. Matsumoto, S. Nishizawa, H. Sunagawa, K. Kobayashi, H. Onodera, The impact of RTN on performance fluctuation in CMOS logic circuits, in *IEEE International Reliability Physics Symposium (IRPS)* (2011), pp. 710–713

9. T. Matsumoto, K. Kobayashi, H. Onodera, Impact of random telegraph noise on CMOS logic delay uncertainty under low voltage operation, in *IEEE International Electron Devices Meeting (IEDM)* (2012), p. 581
10. A. Asenov, R. Balasubramaniam, A.R. Brown, J.H. Davies, RTS amplitudes in decananometer MOSFETs: 3-D simulation study. IEEE Trans. Electron Devices **50**(3), 839–845 (2003)
11. N. Tega, H. Miki, Z. Ren, C.P. D'Emic, Y. Zhu, D.J. Frank, M.A. Guillorn, D.-G. Park, W. Haensch, K. Torii, Impact of HK/MG stacks and future device scaling on RTN, in *IEEE International Reliability Physics Symposium (IRPS)* (2011), pp. 630–635
12. H. Miki, M. Yamaoka, D.J. Frank, K. Cheng, D.G. Park, E. Leobandung, K. Torii, Voltage and temperature dependence of random telegraph noise in highly scaled HKMG ETSOI nFETs and its impact on logic delay uncertainty, in *Symposium on VLSI Technology*, vol. 12 (2012), pp. 137–138
13. S. Dongaonkar, M.D. Giles, A. Kornfeld, B. Grossnickle, J. Yoon, Random telegraph noise (RTN) in 14 nm logic technology : high volume data extraction and analysis, in *Symposium on VLSI Technology* (2016), pp. 176–177
14. K. Sonoda, K. Ishikawa, T. Eimori, O. Tsuchiya, Discrete dopant effects on statistical variation of random telegraph signal magnitude. IEEE Trans. Electron Devices **54**(8), 1918–1925 (2007)
15. K. Takeuchi, T. Nagumo, S. Yokogawa, K. Imai, Y. Hayashi, Single-charge-based modeling of transistor characteristics fluctuations based on statistical measurement of RTN amplitude, in *Symposium on VLSI Technology* (2009), pp. 54–55
16. S. Kiamehr, M.B. Tahoori, L. Anghel, Manufacuturing threats, in *Part I of Dependable Multicore Architectures at Nanoscale* ed. by M. Ottavi, D. Gizopoulos, S. Pontarelli (Springer, Berlin, 2018), pp. 3–36
17. T. Matsumoto, K. Kobayashi, H. Onodera, Impact of random telegraph noise on CMOS logic circuit reliability, in *IEEE Custom Integrated Circuits Conference (CICC)* (2014), pp. 14.4.1–14.4.8
18. T. Matsumoto, K. Kobayashi, H. Onodera, Impact of RTN-induced temporal performance fluctuation against static performance variation, in *IEEE Electron Devices Technology and Manufacturing Conference (EDTM)* (2017), pp. 31–32
19. P. Dutta, P.M. Horn, Low-frequency fluctuations in solids: $1/f$ noise. Rev. Mod. Phys. **53**(3), pp. 497–516 (1981)
20. K.S. Ralls, W.J. Skocpol, L.D. Jackel, R.E. Howard, L.A. Fetter, R.W. Epworth, D.M. Tennant, Discrete resistance switching in submicrometer silicon inversion layers: individual interface traps and low-frequency ($1/f$?) noise. Phys. Rev. Lett. **52**(3), pp. 228–231 (1984)
21. A. Teramoto, T. Fujisawa, K. Abe, S. Sugawa, T. Ohmi, Statistical evaluation for trap energy level of RTS characteristics, in *Symposium on VLSI Technology* (2010), pp. 99–100
22. H. Miki, N. Tega, M. Yamaoka, D.J. Frank, A. Bansal, M. Kobayashi, K. Cheng, C.P. D'Emic, Z. Ren, S. Wu, J. Yau, Y. Zhu, M.A. Guillorn, D. Park, W. Haensch, E. Leobandung, K. Torii, Statistical measurement of random telegraph noise and its impact in scaled-down high-κ/metal-gate MOSFETs, in *IEEE International Electron Devices Meeting (IEDM)* (2012), p. 19
23. D. Veksler, G. Bersuker, H. Park, C. Young, K.Y. Lim, S. Lee, H. Shin, The critical role of defect structural relaxation in interpreting noise measurements in MOSFETs introduction and motivation, in *IEEE International Integral Reliability Working Final Report* (2009), pp. 102–105
24. T. Nagumo, K. Takeuchi, T. Hase, Y. Hayashi, Statistical characterization of trap position, energy, amplitude and time constants by RTN measurement of multiple individual traps, in *IEEE International Electron Devices Meeting (IEDM)* (2010), pp. 28.3.1–28.3.4
25. T. Grasser, H. Reisinger, P. Wagner, F. Schanovsky, W. Goes, B. Kaczer, The time dependent defect spectroscopy (TDDS) for the characterization of the bias temperature instability, in *IEEE International Reliability Physics Symposium (IRPS)* (2010), pp. 16–25
26. Y. Son, T. Kang, S. Park, H. Shin, A simple model for capture and emission time constants of random telegraph signal noise. IEEE Trans. Nanotechnol. **10**(6), 1352–1356 (2011)

27. G. Slavcheva, J.H. Davies, A.R. Brown, A. Asenov, Potential fluctuations in metal-oxide-semiconductor field-effect transistors generated by random impurities in the depletion layer. J. Appl. Phys. **91**(7), 4326–4334 (2002)
28. X. Wang, P. Rao, A. Mierop, A. Theuwissen, Random telegraph signal in CMOS image sensor pixels, in *IEEE International Electron Devices Meeting (IEDM)* (2006), pp. 115–118
29. H. Kurata, K. Otsuga, A. Kotabe, S. Kajiyama, T. Osabe, Y. Sasago, S. Narumi, K. Tokami, S. Kamohara, O. Tsuchiya, Random telegraph signal in flash memory: its impact on scaling of multilevel flash memory beyond the 90-nm node. IEEE J. Solid-State Circ. **42**(6), 1362–1369 (2007)
30. M. Yamaoka, H. Miki, A. Bansal, S. Wu, D. Frank, E. Leobandung, K. Torii, Evaluation methodology for random telegraph noise effects in SRAM arrays, in *IEEE International Electron Devices Meeting (IEDM)* (2011), pp. 745–748
31. K. Takeuchi, T. Nagumo, K. Takeda, A. Asayama, S. Yokogawa, K. Imai, Y. Hayashi, Direct observation of RTN-induced SRAM failure by accelerated testing and its application to product reliability assessment, in *Symposium on VLSI Technology* (2010), pp. 189–190
32. K. Takeuchi, T. Nagumo, T. Hase, Comprehensive SRAM design methodology for RTN reliability, in *Symposium on VLSI Technology* (2011), pp. 130–131
33. F. Arnaud, N. Planes, O.Weber, V. Barral, S. Haendler, P. Flatresse, F. Nyer, Switching energy efficiency optimization for advanced CPU thanks to UTBB technology, in *IEEE International Electron Devices Meeting (IEDM)* (2012), pp. 48–51
34. H. Makiyama, Y. Yamamoto, H. Shinohara, T. Iwamatsu, H. Oda, N. Sugii, K. Ishibashi, T. Mizutani, T. Hiramoto, Y. Yamaguchi, Suppression of die-to-die delay variability of silicon on thin buried oxide (SOTB) CMOS circuits by balanced P/N drivability control with back-bias for ultralow-voltage (0.4 V) operation, in *IEEE International Electron Devices Meeting (IEDM)* (2013), pp. 822–825
35. T. Hiramoto, A. Kumar, T. Mizutani, J. Nishimura, T. Saraya, Statistical advantages of intrinsic channel fully depleted SOI MOSFETs over bulk MOSFETs, in *IEEE Custom Integrated Circuit Conference (CICC)* (2011), p. 5.2
36. C.-H. Jan, U. Bhattacharya, R. Brain, S.-J. Choi, G. Curello, G. Gupta, W. Hafez, M. Jang, M. Kang, K. Komeyli, T. Leo, N. Nidhi, L. Pan, J. Park, K. Phoa, A. Rahman, C. Staus, H. Hashiro, C. Tsai, P. Vandervoorn, L. Yang, J.-Y. Yeh, P. Bai, A 22 nm SoC platform technology featuring 3-D tri-gate and high-k/metal gate, optimized for ultra low power, high performance and high density SoC applications, in *IEEE International Electron Devices Meeting (IEDM)* (2012), pp. 44–47
37. Y.F. Lim, Y.Z. Xiong, N. Singh, R. Yang, Y. Jiang, D.S.H. Chan, W.Y. Loh, L.K. Bera, G.Q. Lo, N. Balasubramanian, D.-L. Kwong, Random telegraph signal noise in Gate-all-around Si-FinFET with ultranarrow body. IEEE Electron Device Lett. **27**(9), 765–768 (2006)
38. M.-L. Fan, V.P.-H. Hu, Y.-N. Chen, P. Su, C.-T. Chuang, Analysis of single-trap-induced random telegraph noise on FinFET devices, 6T SRAM cell, and logic circuits. IEEE Trans. Electron Devices **59**(8), 2227–2234 (2012)
39. K. Ito, T. Matsumoto, S. Nishizawa, H. Sunagawa, K. Kobayashi, H. Onodera, Modeling of random telegraph noise under circuit operation -simulation and measurement of RTN-induced delay fluctuation, in *IEEE International Symposium on Quality Electronic Design (ISQED)* (2011), pp. 22–27
40. T. Matsumoto, K. Kobayashi, H. Onodera, Impact of body-biasing technique on random telegraph noise induced delay fluctuation. Jpn. J. App. Phys. **52**, 04CE05 (2013)
41. T. Tsunomura, A. Nishida, F. Yano, A.T. Putra, K. Takeuchi, S. Inaba, S. Kamohara, K. Terada, T. Hiramoto, T. Mogami, Analyses of 5σ 5th fluctuation in 65 nm-MOSFETs using Takeuchi plot, in *Symposium on VLSI Technology* (2008), pp. 156–157
42. J.W. Tschanz, J.T. Kao, S.G. Narendra, R. Nair, D.A. Antoniadis, A.P. Chandrakasan, V. De, Adaptive body bias for reducing impacts of Die-to-die and within-die parameter variations on microprocessor frequency and leakage. IEEE J. Solid-State Circ. **37**(11), 1396–1402 (2002)

43. R. Kishida, K. Kobayashi, Degradation caused by negative bias temperature instability depending on Body Bias on NMOS or PMOS in 65 nm bulk and thin-BOX FDSOI processes, in *IEEE Electron Devices Technology and Manufacturing Conference (EDTM)* (2017), pp. 122–123
44. J. Franco, B. Kaczer, G. Eneman, P.J. Roussel, T. Grasser, J. Mitard, L.-A. Ragnarsson, M. Cho, L. Witters, T. Chiarella, M. Togo, W.-E. Wang, A. Hikavyy, R. Loo, N. Horiguchi, G. Groeseneken, Superior NBTI reliability of SiGe channel pMOSFETs: replacement gate, FinFETs, and impact of Body Bias, in *International Electron Devices Meeting* (2011), pp. 18.5.1–18.5.4
45. T. Tsunomura, A. Nishida, T. Hiramoto, Investigation of threshold voltage variability at high temperature using takeuchi plot. Jpn. J. Appl. Phys. **49**(5), 054101 (2010)
46. M. Toledano-Luque, B. Kaczer, P. Roussel, M. Cho, T. Grasser, G. Groeseneken, Temperature dependence of the emission and capture times of SiON individual traps after positive bias temperature stress. J. Vac. Sci. Technol. B Microelectron. Nanometer Struct. **29**, 01 (2011)
47. C.G. Theodorou, E.G. Ioannidis, S. Haendler, E. Josse, C.A. Dimitriadis, G. Ghibaudo, Low frequency noise variability in ultra scaled FD-SOI n-MOSFETs: dependence on gate bias, frequency and temperature. Solid. State. Electron. **117**, 88–93 (2016)
48. Y. Morita, R. Tsuchiya, T. Ishigaki, N. Sugii, T. Iwamatsu, T. Ipposhi, H. Oda, Y. Inoue, K. Torii, S. Kimura, Smallest Vth variability achieved by intrinsic silicon on thin BOX (SOTB) CMOS with single metal gate, in *Symposium on VLSI Technology* (2008), pp. 166–167
49. B. Silverman, Using kernel density estimates to investigate multimodality. J. R. Stat. Soc. Ser. B Methodol. **43**(1), 97–99 (2018)
50. Z. Zhang, S. Guo, X. Jiang, R. Wang, R. Huang, Investigation on the amplitude distribution of random telegraph noise (RTN) in nanoscale MOS devices, in *IEEE International Nanoelectronics Conference (INEC)* (2016), pp. 5–6
51. H. Awano, H. Tsutsui, H. Ochi, T. Sato, Bayesian estimation of multi-trap RTN parameters using markov chain Monte Carlo method, in *IEICE Transaction on Fundamentals of Electronics, Communications and Computer Sciences*, vol. E95.A(12) (2012), pp. 2272–2283
52. J. Martin-Martinez, J. Diaz, R. Rodriguez, M. Nafria, X. Aymerich, New weighted time lag method for the analysis of random telegraph signals. IEEE Electron Device Lett. **35**(4), 479–481 (2014)
53. T. Komawaki, M. Yabuuchi, R. Kishida, J. Furuta, T. Matsumoto, K. Kobayashi, Replication of random telegraph noise by using a physical-based Verilog-AMS model. IEICE Trans. Fundam. Electron. Commun. Comput. Sci. **E100.A**(12), 2758–2763 (2017)
54. T. Komawaki, M. Yabuuchi, R. Kishida, J. Furuta, T. Matsumoto, K. Kobayashi, Circuit-level simulation methodology for random telegraph noise by using Verilog-AMS, in *IEEE International Conference on IC Design and Technology (ICICDT)* (2017), pp. 1–4
55. M. Tanizawa, S. Ohbayashi, T. Okagaki, K. Sonoda, K. Eikyu, Y. Hirano, K. Ishikawa, O. Tsuchiya, Y. Inoue, Application of a statistical compact model for Random Telegraph Noise to scaled-SRAM Vmin analysis, in *Symposium on VLSI Technology* (2010), pp. 95–96
56. T. Matsumoto, K. Kobayashi, H. Onodera, Impact of random telegraph noise on CMOS logic circuit reliability, in *IEEE Custom Integrated Circuits Conference (CICC)* (2014), pp. 1–8
57. B. Kaczer, T. Grasser, P. Roussel, J. Franco, R. Degraeve, L.-A. Ragnarsson, E. Simoen, G. Groeseneken, H. Reisinger, Origin of NBTI variability in deeply scaled pFETs, in *IEEE International Reliability Physics Symposium (IRPS)* (2010), pp. 26–32
58. A. Oshima, P. Weckx, B. Kaczer, K. Kobayashi, T. Matsumoto, Impact of random telegraph noise on ring oscillators evaluated by circuit-level simulations, in *IEEE International Conference on IC Design and Technology (ICICDT)* (2015), pp. 1–4
59. M. Toledano-Luque, B. Kaczer, J. Franco, P. Roussel, T. Grasser, T. Hoffmann, G. Groeseneken, From mean values to distributions of BTI lifetime of deeply scaled FETs through atomistic understanding of the degradation, in *Symposium on VLSI Technology* (2011), pp. 152–153

60. T. Grasser, B. Kaczer, W. Goes, H. Reisinger, T. Aichinger, P. Hehenberger, P.-J. Wagner, F. Schanovsky, J. Franco, P. Roussel, M. Nelhiebel, Recent advances in understanding the bias temperature instability, in *IEEE International Electron Devices Meeting (IEDM)* (2010), pp. 4.4.1–4.4.4
61. H. Reisinger, T. Grasser, W. Gustin, C. Schlunder, The statistical analysis of individual defects constituting NBTI and its implications for modeling DC- and AC-stress, in *IEEE International Reliability Physics Symposium (IRPS)* (2010), pp. 7–15
62. M. Nour, M. Mahmud, Z. Celik-Butler, D. Basu, S. Tang, F.-C. Hou, R. Wise, Variability of random telegraph noise in analog MOS transistors, in *IEEE International Conference on Noise and Fluctuations (ICNF)* (2013), pp. 1–4
63. K. Ito, T. Matsumoto, S. Nishizawa, H. Sunagawa, K. Kobayashi, H. Onodera, Modeling of Random Telegraph Noise under circuit operation—Simulation and measurement of RTN-induced delay fluctuation, in *IEEE International Symposium on Quality Electronic Design (ISQED)* (2011), pp. 1–6
64. M. Miura-Mattausch, H. Ueno, M. Tanaka, H. Mattausch, S. Kumashiro, T. Yamaguchi, K. Yamashita, N. Nakayama, HiSIM: a MOSFET model for circuit simulation connecting circuit performance with technology, in *IEEE International Electron Devices Meeting (IEDM)* (2002), pp. 109–112
65. A. Oshima, T. Komawaki, K. Kobayashi, R. Kishida, P. Weckx, B. Kaczer, T. Matsumoto, H. Onodera, Physical-based RTN modeling of ring oscillators in 40-nm SiON and 28-nm HKMG by bimodal defect-centric behaviors, in *IEEE International Conference on Simulation of Semiconductor Processes and Devices (SISPAD)* (2016), pp. 327–330

Low-Frequency Noise in III–V, Ge, and 2D Transistors

Mengwei Si, Xuefei Li, Wangran Wu, Sami Alghamdi, and Peide Ye

1 Introduction

The performance of CMOS-integrated circuits has been enhanced by the scaling of MOS transistors for the past few decades. As of today (2019), the gate length of a single transistor in advanced IC product is close to 7 nm. The downscaling of MOS transistors is approaching the physical limit together with many challenges, including short channel effects, power constraint, fabrication process, etc. New semiconductor materials such as III–V, Ge, and 2D start to emerge and develop as potential alternative channel materials to silicon. III–V and Ge, for example, are well-known because of their high electron and hole mobilities for high-speed and low-power digital integrated circuits. 2D semiconductors such as molybdenum disulfide (MoS_2) and black phosphorous (BP) offer ideal electrostatic control for stronger immunity to short channel effects because of the atomically thin-layered structure. Accurate and reliable measurement of interface, dielectrics, channel, contacts on such novel devices and novel materials is required for device characterization, process optimization, and the understanding of carrier transport mechanism. However, conventional characterization methods, such as C–V method and charge pumping method, cannot be used on these novel devices because of the ultra-scaled dimensions and novel device architectures, while noise measurement is not limited by the small gate capacitance. It is also important to evaluate the noise performance and identify noise sources for transistors made of new material systems and to understand the carrier transport mechanism. Therefore, low-frequency noise measurement is an important approach in the characterization of devices with novel materials and architectures.

M. Si · X. Li · W. Wu · S. Alghamdi · P. Ye (✉)
School of Electrical and Computer Engineering, Purdue University, West Lafayette, IN, USA
e-mail: yep@purdue.edu

T. Grasser (ed.), *Noise in Nanoscale Semiconductor Devices*,
https://doi.org/10.1007/978-3-030-37500-3_10

In this chapter, we summarize recent progress in low-frequency noise measurement on transistors with novel channel materials. The first section discusses the low-frequency noise measurement on III–V gate-all-around (GAA) nanowire MOSFETs down to 20-nm channel length. The second section discusses the low-frequency noise measurement on Ge nanowire MOSFETs down to 40-nm channel length. These two sections also study the impact of channel length scaling on the low-frequency noise performance on transistors with high mobility channel materials and with advanced 3D structures. Finally, low-frequency noise on 2D field-effect transistors (FETs) including MoS_2 and BP are examined.

2 Low-Frequency Noise in Near-Ballistic III–V MOSFETs

InGaAs has been considered as one of the promising channel materials for future CMOS logic circuit because of its large electron injection velocity [1]. In the last decade, tremendous efforts have been spent on the development of high-performance InGaAs transistors with both competitive on- and off-state performance to replace silicon in low-power and high-speed applications [2–10]. In particular, InGaAs gate-all-around MOSFETs have been demonstrated to offer large drive current and excellent immunity to short channel effects down to deep sub-100-nm channel length (L_{ch}) [9]. However, one of the bottlenecks that prevents InGaAs MOSFETs from being applied in mainstream industry is the defective interface and gate stack [11–13]. Accurate measurement of these defects is very important but very challenging due to the ultra-scaled dimensions and the gate-all-around structures. Low-frequency noise can be utilized to investigate these defects, taking the advantage of noise measurements in small structures [14–23]. It has been generally admitted that the low-frequency noise in MOSFETs can be well described by carrier number fluctuation model or mobility fluctuation model [24]. Random telegraph noise (RTN) is attributed to the trapping and de-trapping event in a single defect. $1/f$ noise is the superposition of a number of individual RTNs in the carrier number fluctuation theory. On the other hand, classical theories suggest that $1/f$ noise increases inversely with decreasing channel length [25–32]. If true, this may negate some of the performance gain of short channel transistors. In this section, we examine the origin of low-frequency noise on highly scaled InGaAs GAA MOSFETs and systematically study the properties of low-frequency noise and RTN characteristics in near-ballistic InGaAs GAA nanowire MOSFETs. Mobility fluctuation is identified to be the source of $1/f$ noise. The $1/f$ noise was found to decrease as the channel length scaled down from 80 to 20 nm comparing with classical theory, indicating the near-ballistic transport in highly scaled InGaAs GAA MOSFET [14, 15].

Figure 1 shows the schematic diagram and cross-sectional view of an InGaAs GAA MOSFET. The details of the top-down fabrication process can be found in [9]. Samples A and B have a 0.5-nm Al_2O_3/4-nm $LaAlO_3$ stack (EOT = 1.2 nm), where Al_2O_3 was grown before $LaAlO_3$ for sample A and vice versa for sample

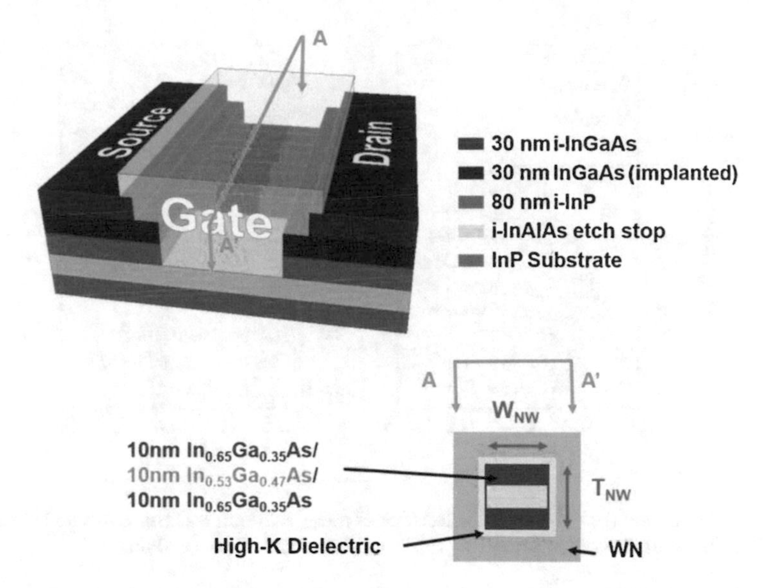

Fig. 1 Schematic and cross section of the tested InGaAs gate-all-around MOSFETs

B. Sample C has 3.5-nm Al_2O_3 as gate dielectric (EOT = 1.7 nm). The InGaAs channel layer consists of a single 10-nm $In_{0.53}Ga_{0.47}As$ layer sandwiched by two 10 nm $In_{0.65}Ga_{0.35}As$ layers. Devices with L_{ch} varying from 20 to 80 nm, W_{NW} varying from 20 to 35 nm, nanowire thickness (T_{NW}) of 30 nm and nanowire length (L_{NW}) of 200 nm are measured. L_{NW} is the physical length of the nanowire, while L_{ch} is the channel length defined by implantation.

Source current power spectral density (S_{Is}) was measured in the linear region of operation ($V_{ds} = 50$ mV). The gate voltage (V_{gs}) is supplied by a digital controllable voltage source. A Stanford SR570 battery-powered current amplifier is used as source voltage supply, monitor, and amplifier for the source current (I_s). The SR570 current amplifier output is directly connected to a Tektronix TDS5032B oscilloscope to record the RTN signal and an Agilent 35670A dynamic signal analyzer to obtain the power spectrum density (PSD) of I_s noise at the same time. All noise measurements were performed at $V_{ds} = 50$ mV and at V_{gs} from −0.2 to 0.4 V and at room temperature unless otherwise specified.

Figure 2a, b shows a typical output and transfer characteristics of an InGaAs GAA MOSFET measured with $L_{ch} = W_{NW} = 20$ nm. Figure 3 shows (a) a I_s histogram and (b) an RTN signal in the time domain on a $L_{ch} = 20$ nm, $W_{NW} = 25$ nm device of Sample B, (c) a time segment inside (b), showing the superposition of two switching level signal and a Gaussian-like current distribution. In the device without an RTN signal, the I_s histogram shows only a Gaussian-like current distribution and with a $1/f$ noise spectrum. This phenomenon suggests mobility fluctuation (rather than number fluctuation) is the origin of the $1/f$ noise on devices without RTN signal. In classical noise theory, $1/f$ noise in MOSFETs can

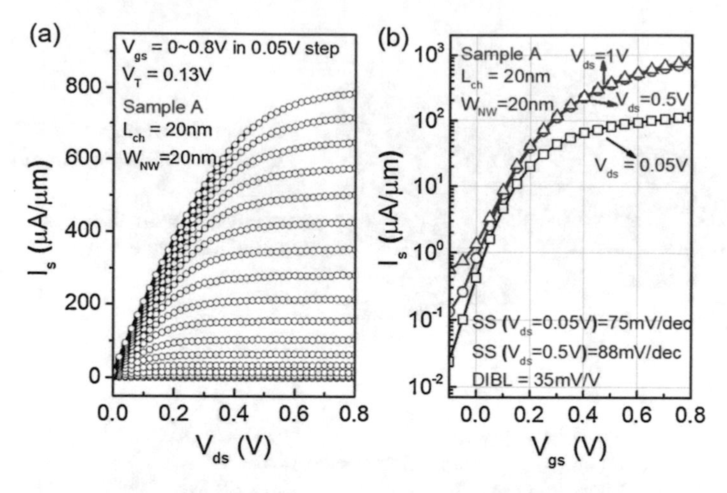

Fig. 2 (**a**) Output and (**b**) transfer characteristics of a $L_{ch} = 20$ nm InGaAs GAA MOSFET with $Al_2O_3/LaAlO_3$ gate dielectric (Sample A, EOT = 1.2 nm) and $W_{NW} = 20$ nm

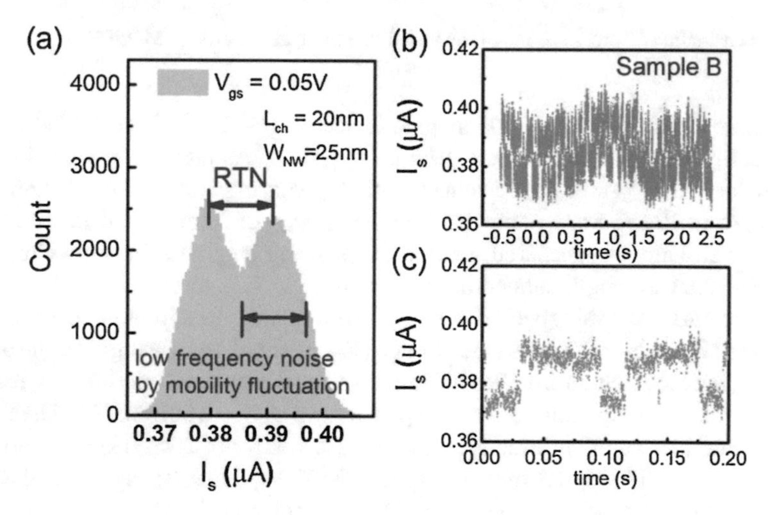

Fig. 3 (**a**) Histogram of an RTN signal of sample B with $L_{ch} = 20$ nm and $W_{NW} = 25$ nm. (**b**, **c**) RTN signals in the time domain of the same signal as (**a**) and (**c**) is a time segment inside (**b**)

be well described by the carrier number fluctuation or mobility fluctuation models. For carrier number fluctuation theory, current fluctuation in a MOSFET is attributed to the trapping and de-trapping events, which induce a V_T shift. Each individual trapping and de-trapping event has a Lorentzian spectrum. The superposition of Lorentzian spectrum of a few traps is not likely to be enough to support a $1/f$ spectrum. So, the observation of RTN is one of the evidences that the $1/f$ spectrum is not from the number fluctuation. By estimating the electron trapping events that are responsible for the hysteresis, we can estimate the number of defects in the oxide. The estimated number of active defects is calculated as $AC_{\mathrm{ox}}\Delta V_{\mathrm{T}}/q$, where A is the gate area for the devices, C_{ox} is the gate capacitance calculated from EOT, ΔV_{T} is the hysteresis V_{T} shift, q is the elementary charge. ΔV_{T} is on the order of several mV with a maximum of $V_{\mathrm{gs}} = 0.8$ V [14], corresponding to several active traps. Furthermore, not all traps will affect the current fluctuation because of the distribution of trap energy levels [33], in other words, not all oxide traps are active at a certain gate voltage. Moreover, the inversion charges in this work are farther from the interface than planar MOSFETs due to the volume inversion nature of GAA MOFSETs, as suggested by simulation results [34]. It could potentially reduce the interaction between oxide traps and inversion charges, as suggested in low-frequency noise study in silicon nanowire MOSFETs [30].

Therefore, for those devices with 1/*f* spectrum in which RTN signal cannot be observed, mobility fluctuation is the source of low-frequency noise. Figure 4 shows the comparison of the noise spectrum between a device with an RTN signal and a device without RTN. The two devices share the same dimensions with $L_{\mathrm{ch}} = 20$ nm, $W_{\mathrm{NW}} = 25$ nm and 3.5 nm Al_2O_3 as gate dielectric. It is clear that noise spectrum of the device without RTN shows 1/*f* characteristic, while the noise spectrum of the device with RTN is the superposition of 1/*f* noise spectrum and Lorentzian spectrum. This is not the only evidence for the identification of noise source as mobility fluctuation in this work. This fact will be further discussed in the following part.

Figure 5a shows the source current power spectrum density normalized by I_{s}^2 ($S_{I\mathrm{s}}/I_{\mathrm{s}}^2$) versus I_{s} at $f = 10$ Hz, $W_{\mathrm{NW}} = 20$ nm, and various channel lengths for Sample B. $S_{I\mathrm{s}}/I_{\mathrm{s}}^2$ versus I_{s} shows a weak dependence on L_{ch}, which is opposite to the classical noise L_{ch} scaling characteristics ($S_{I\mathrm{s}}/I_{\mathrm{s}}^2 \sim 1/L_{\mathrm{ch}}$ at a given current). Figure 5b shows the input gate noise ($S_{V\mathrm{g}}$) normalized by channel area ($WLS_{V\mathrm{g}}$) vs. I_{s}. It can be seen clearly that the normalized input noise is reduced by scaling down the channel length, while in classical theory, $WLS_{V\mathrm{g}}$ should be independent of channel

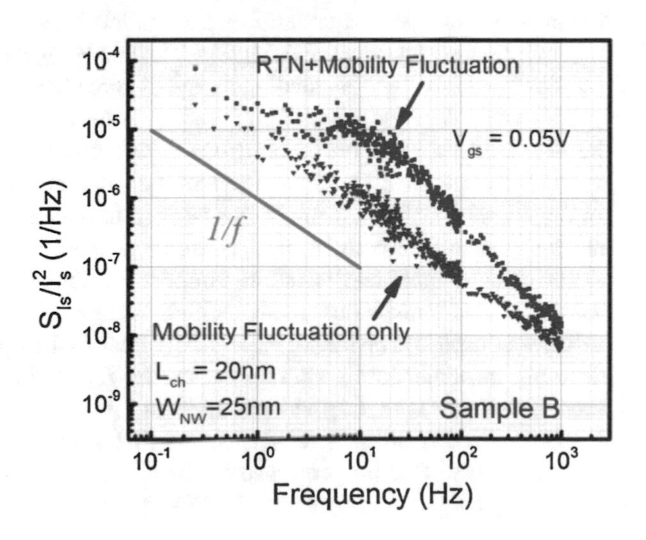

Fig. 4 Normalized I_{s} noise of Sample B devices with RTN signal and without RTN signal. Noise spectrum of device without RTN is attributed to mobility fluctuation

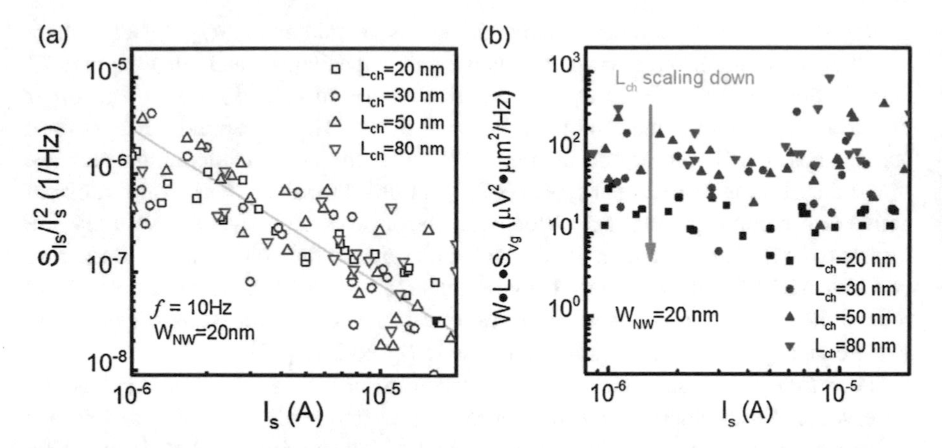

Fig. 5 (**a**) PSD of I_s normalized by I_s^2 at $f = 10$ Hz for various L_{ch} and $W_{NW} = 20$ nm at $V_{ds} = 0.05$ V on Sample B. (**b**) Input gate voltage noise normalized by channel width times channel length of the same data set as (**a**)

area [24]. In both classical carrier number fluctuation model and mobility fluctuation model, S_{Vg} and S_{Is}/I_s^2 are inversely proportional to L_{ch}. The experimental data show the I_s noise measured in InGaAs GAA MOSFETs is reduced at short channel devices compared with classical theory. This phenomenon cannot be explained by carrier number fluctuation theory because the V_T shift caused by a single trapping and de-trapping event can be estimated simply by $\Delta V_T = q/(AC_{ox})$, where A is proportional to L_{ch}. We will have larger ΔV_T at smaller channel length so that amplitude of the single trapping and de-trapping event is increased. As current fluctuation is the combination of a number of single trapping and de-trapping events, noise will be increased while channel length scaling down if carrier number fluctuation noise dominates the S_{Is}/I_s^2. Thus, the anomalous scaling trend in Fig. 5 also indicates that carrier number fluctuation is not the source of 1/f noise in the highly scaled InGaAs GAA MOSFETs. The Hooge's parameter is estimated as $3 \times 10^{-4} - 1.2 \times 10^{-3}$ depending on the channel length [35].

This weak dependence of S_{Is}/I_s^2 versus I_s on L_{ch} leads to the conclusion that the near-ballistic transport of electrons in the channel is achieved through noise study. Although mobility fluctuation model also suggests $1/L_{ch}$ scaling metrics for S_{Is}/I_s^2, this conclusion might not be true in devices with short channel length and long mean free path in high mobility channel materials. As electrons from source cannot equilibrate to lattice temperature immediately at drain contact, the classical mobility fluctuation model, which assumes uniform carrier distribution and diffusive transport, is no longer valid. In our near-ballistic InGaAs GAA MOSFETs, electrons encounter less scattering at smaller L_{ch} during transport from source to drain. Therefore, scattering-induced mobility fluctuation decreases at small L_{ch} so that normalized I_s noise is reduced at small L_{ch} compared with the prediction of classical theory. This property also confirms that mobility fluctuation is the origin of low frequency for highly scaled InGaAs MOSFETs.

3 Low-Frequency Noise in Near-Ballistic Ge MOSFETs

Ge channel is one of the promising candidates for future ultimate CMOS applications because Ge has high and balanced electron and hole mobility, good compatibility with Si large-scale integration technologies and great potential for voltage scaling [36]. Ge CMOS has been intensively studied in the past decade and highly scaled Ge NW CMOS has been demonstrated to offer excellent performance [37–39]. However, advanced high-k/Ge gate stacks with scaled EOT and superior MOS interfaces are still needed to develop Ge CMOS manufacturing technology with high reliability [40]. Low-frequency noise and RTN can be used on the characterization of Ge nanowire transistors similarity to previous section on InGaAs transistors [21, 41–48]. In this section, we examine the origin of the $1/f$ low-frequency noise and systematically study the properties of low-frequency noise on Ge NW nMOSFETs with various NW geometries (NW width, W_{NW}, NW height, H_{NW}) and channel lengths (L_{ch}) and process conditions.

Figure 6 illustrates the schematic image of the Ge NW nMOSFETs. The cross-sectional view shows the accumulation mode (AM), inversion mode (IM) nMOSFETs, and the key geometry parameters of devices. Ge NW nMOSFETs with

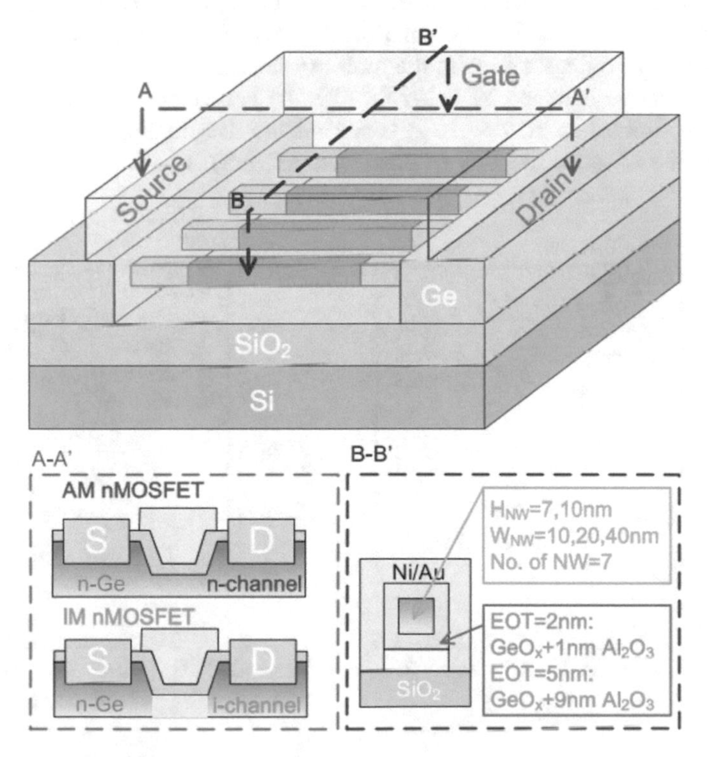

Fig. 6 Schematic diagram of the Ge NW nMOSFETs. The accumulation mode (AM) and inversion mode (IM) nMOSFETs with various key geometry parameters were prepared

W_{NW} from 10 to 40 nm, H_{NW} of 7 and 10 nm, L_{ch} from 40 to 80 nm and EOT of 2 and 5 nm are used for RTN and low-frequency noise characteristics. The channel width is calculated from $W_{ch} = (2H_{NW} + W_{NW}) \times$ (number of wires) and each device has seven nanowires. AM NW nMOSFETs are measured unless otherwise specified. Keysight B1500A with B1530A Waveform Generator/Fast Measurement Unit is applied for characterization. All the measurements are performed at room temperature at a drain voltage (V_{ds}) of 50 mV.

Figure 7 shows the (a) transfer and (b) output characteristics of a NW nMOSFET with 40-nm L_{ch} and 2-nm EOT. Subthreshold slope of 91 mV/dec is achieved at V_{ds} of 0.5 V and the drain-induced barrier lowering (DIBL) is 106 mV/V, indicating that the device was well fabricated. Because the electrons in the inversion channel populate very close to the interface, the electron transportation is significantly affected by the scattering centers (traps, dopants, surface roughness, etc.) near the interface. The low-frequency noise may originate from the carrier number fluctuation or the carrier mobility fluctuation in the conventional theory.

The normalized S_{Id} (S_{Id}/I_d^2) between two devices, sharing the same device dimension, with and without RTN signal is compared in Fig. 8a. The noise spectrum of the device without RTN follows 1/*f* characteristics. Figures 8b, 9 and 10 are obtained in devices without RTN at a frequency of 10 Hz and each group of the data contains experimental results from several devices sharing the same device dimension. Figure 8b shows the normalized S_{Id} (S_{Id}/I_d^2) as a function of I_d. The clear $1/I_d$ dependence agrees well with the mobility fluctuation model [49]. The Hooge parameter can be obtained by linear fit with the knowledge of effective mobility and it is estimated as 2.59×10^{-3} based on the data in Fig. 8b [47]. We have measured the effective mobility of planar Ge MOSFETs with the same gate stack

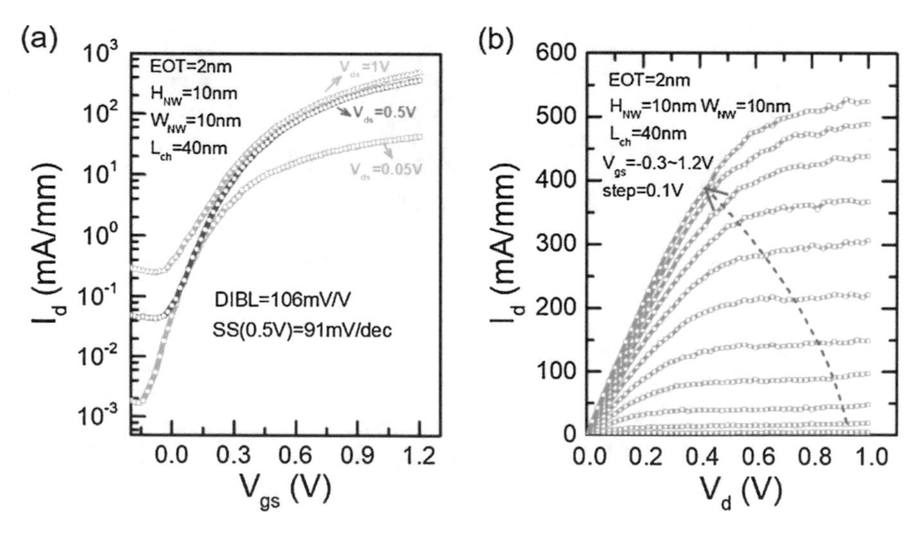

Fig. 7 (**a**) Transfer and (**b**) output characteristics of a 40-nm L_{ch} NW nMOSFET with an EOT of 2 nm

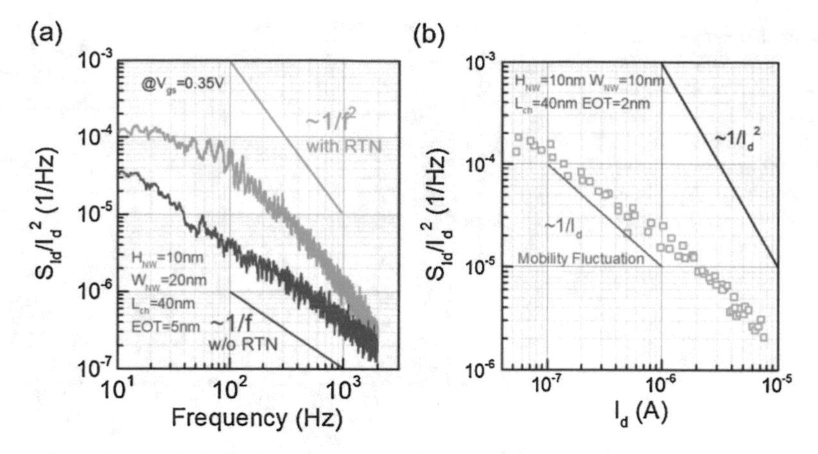

Fig. 8 (**a**) Normalized S_{Id} of two devices, having the same device dimension, with and w/o RTN. S_{Id} in device w/o RTN shows $1/f$ characteristics. (**b**) Normalized S_{Id} versus I_d of devices w/o RTN. $1/I_d$ dependence indicates mobility fluctuation-induced noise

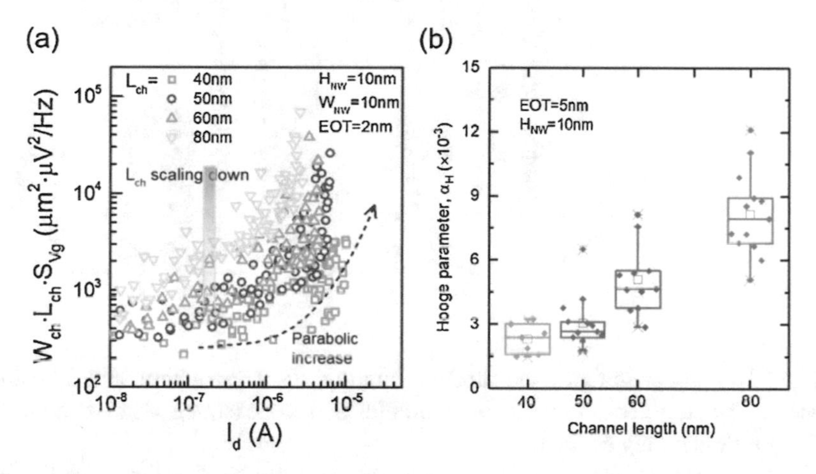

Fig. 9 (**a**) S_{Vg} normalized by $W_{ch} \times L_{ch}$ of devices (w/o RTN) with different values of L_{ch}. Normalized S_{Vg} decreases with scaling down of L_{ch}. (**b**) The box plot of Hooge parameters of devices with channel length from 40 to 80 nm. The Hooge parameter decreases with the scaling down of L_{ch}

but larger gate area [50]. Also, according to the transconductance of devices with various EOTs and device dimensions [39], the variation of effective mobility is very small compared with that of the slope of the $S_{Id}/I_d^2 \sim 1/I_d$ curves. A constant electron mobility of 200 cm^{-2}/V s is assumed to extract the Hooge parameters in this study, since it is hard to precisely extract the effective carrier mobility in the ultra-scaled Ge NW MOSFETs. The Hooge parameter of Ge NW MOSFETs used in this study is higher compared with the state-of-art Si device and similar to the value of high-k HfO_2/Si devices. The typical value of Hooge parameter in Si device ranging from

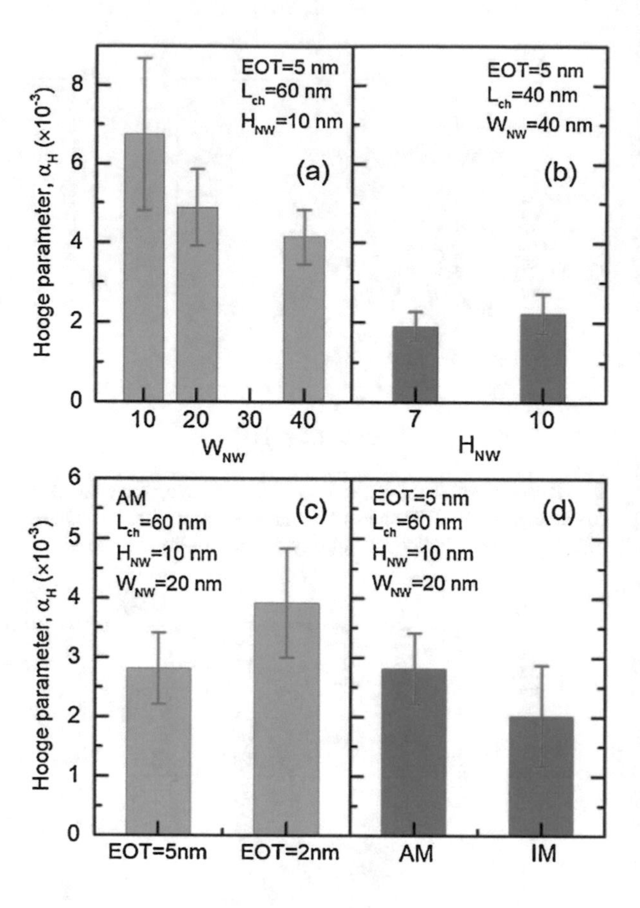

Fig. 10 Hooge parameters in Ge NW nMOSFETs with (**a**) W_{NW} of 10, 20, 40 nm and (**b**) H_{NW} of 7, 10 nm. Hooge parameters in Ge NW nMOSFETs with (**c**) EOT of 5, 2 nm and (**d**) AM Ge NW nMOSFETs and IM Ge NW nMOSFETs

10^{-3} to 10^{-6}, depending on the substrate quality, device structure, and fabrication process. The suppression of Hooge parameter and low-frequency noise in Ge NW nMOSFETs is highly desired.

In the frame of the conventional carrier mobility fluctuation model, S_{Id}/I_d^2 increases with the scaling down of channel length and S_{Vg} normalized by $W_{ch} \times L_{ch}$ should be independent of channel length [14, 15]. However, the normalized $W_{ch}L_{ch}S_{Vg}$ in the Ge NW nMOSFETs decreases when L_{ch} scales down from 80 to 40 nm (Fig. 9a). The Hooge parameters are extracted by a linear fit in order to give a deep insight on the anomalous decline of low-frequency noise in ultra-scaled Ge NW nMOSFETs. Figure 9b shows the box plot of Hooge parameters as a function of the channel length. It is clearly shown that the Hooge parameter decreases with the scaling down of channel length. The Hooge parameter reflects the mobility fluctuations induced by the electron-scattering processes during the transportation in the channel. The reduced Hooge parameter indicates the attenuation in scattering when the electrons travel through the short channel. The Hooge parameter of devices with 40 nm L_{ch} is only one-third of those with 80 nm L_{ch}. This is attributed to the near-ballistic transport of electrons because of the long mean free path of electrons in Ge. The decreasing low-frequency noise in shorter channel devices is also observed in highly scaled InGaAs MOSFETs as in the previous section. The Hooge parameter results provide a direct evidence for the near-ballistic transport in ultra-scaled Ge nMOSFETs. It also shows that the low-frequency noise declines when L_{ch} decreases in highly scaled Ge nMOSFETs, which is a unique advantage for the application in scaled low-noise integrated circuits.

Since the Hooge parameter could represent the carrier-scattering process, it is sensitive to the substrate quality, gate stack properties, and substrate doping concentration. The dry etching process used in the formation of Ge NW causes damage in the NW side walls. Thus, electrons beneath the side wall are subject to a lot of scatterings. Figure 10 illustrates the dependence of Hooge parameter on the device geometries (W_{NW}, H_{NW}, EOT, doping). The Hooge parameter decreases when W_{NW} increases from 10 to 40 nm because of the smaller proportion of NW side wall contributed to the whole channel. The Hooge parameters are smaller in devices with H_{NW} of 7 nm because of the shorter dry etching time and approaching volume inversion condition for Ge. Furthermore, assuming the electrons are evenly distributed in the inversion layer under the side and top wall, the Hooge parameter of electrons beneath the side wall is about two times larger than that of the top wall according to the experimental results in Fig. 10a. The severe scattering of electrons caused by the poor side wall quality not only elevates the low-frequency noise but also suppresses the electron mobility. Thus, the dry etching process and side wall quality have to be carefully optimized.

Figure 10c shows that the Hooge parameters in devices with smaller EOT are higher. The severer scattering in devices with 2-nm EOT might arise from the degradation in GeO_x interfacial layer because the Al_2O_3 capping layer is too thin. Since Coulomb scattering is also an important source of mobility scattering, the Hooge parameters are smaller in IM NW nMOSFETs because the relatively lower channel doping induces less Coulomb scattering (Fig. 10d). The low-frequency noise results provide a direct insight into the electron-scattering process in ultra-scaled Ge NW devices.

In summary, low-frequency noise in highly scaled Ge NW nMOSFETs with various NW geometries, channel length, EOT, and channel doping are comprehensively studied. The low-frequency noise arises from the carrier mobility fluctuation and it decreases in devices with shorter channels. This anomalous dependence is ascribed to the near-ballistic transport of electrons. The electron-scattering process is examined by evaluating the Hooge parameters. NW side wall and gate oxide optimization are required for the suppression of scattering and mobility enhancement. From the perspective of device performance, the ultra-scaled Ge NW nMOSFETs simultaneously promises the enhancement of on-state performance and the suppression of low-frequency noise.

4 Low-Frequency Noise in 2D Transistors

4.1 Low-Frequency Noise in MoS_2 transistors

MoS_2, a two-dimensional layered transition metal dichalcogenide material of Mo atoms sandwiched between two layers of S atoms, has generated considerable interest in recent years for its unusual electronic sand optical properties [51–56]. Unlike graphene, atomically thin MoS_2 is a semiconductor with a bandgap from 1.2 eV of bulk MoS_2 to 1.8 eV of monolayer MoS_2 [57]. Moreover, large area MoS_2 growth techniques by CVD have been developed [58], making it a suitable candidate for practical electronic device applications, such as thin film logic circuits and amplifiers with high gain. Fundamental studies on low frequencies noise mechanisms [58–64] are important for evaluating the intrinsic electrical properties of MoS_2. Large uncertainties exist in the presence of significant extrinsic factors, such as contacts and gating, and more importantly, the lack of systematic studies on a series of transistors with long-term stability through thermal cycles. These factors combined together lead to significant deviations from the intrinsic properties of MoS_2 and inaccurate projection of its performance potential. In order to address these problems, in this section, we perform systematic studies on its precision limit determined by the low-frequency noise, which is an important factor in future nanoelectronic applications.

MoS_2 devices studied here are based on a 90-nm SiO_2 back-gated field-effect transistor (FET) configuration, as depicted by the schematic view in Fig. 11a. The MoS_2 flake was identified and measured by atomic force microscopy (AFM) as shown in Fig. 11b, the thickness of which is around 6 nm, corresponding to about nine layers. Details about the device fabrication process can be found in [51]. Arrays of devices with four different channel lengths from 1 μm down to 100 nm were fabricated. All the measurements were conducted in vacuum ($\leq 10^{-5}$ Torr) to minimize the impact from the environment. Figure 12 shows the direct current (DC) output and transfer characteristics of a representative 100-nm channel length

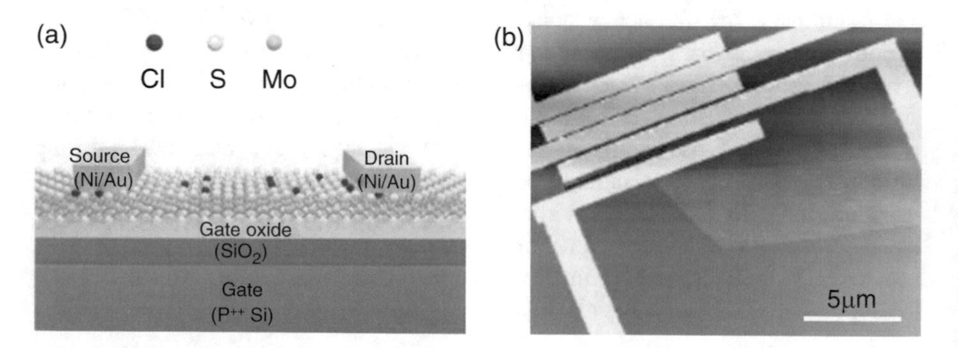

Fig. 11 (**a**) Schematic of the MoS_2 back-gate FET. The gate dielectric is 90-nm SiO_2 and the thickness of MoS_2 is 6 nm. (**b**) AFM image of the few-layer MoS_2 FETs on a flake

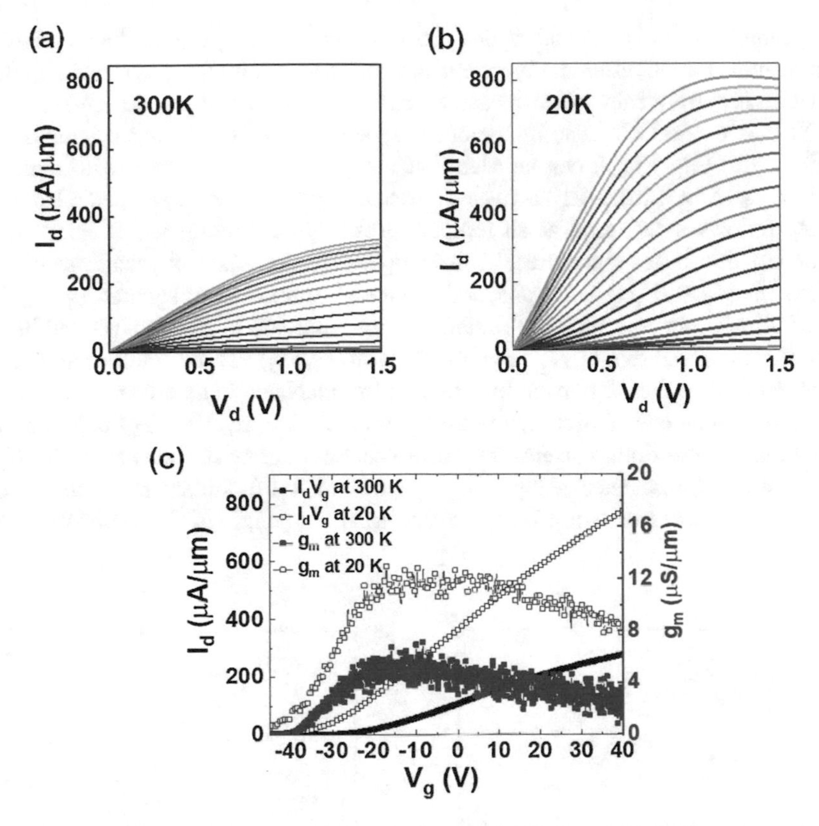

Fig. 12 Output characteristics of a 100-nm device at (**a**) 300 K and (**b**) 20 K and a back-gate voltage from −45 to 40 V. (**c**) Transfer characteristics of the same device at $V_d = 1$ V

MoS_2 FET at 300 and 20 K, with a drain voltage from 0 to 1.5 V and back-gate voltage from −45 to 40 V. At room temperature, the maximum drain current is 334 μA/μm for this device, and the highest current among all seventeen 100-nm devices measured in this work reaches 550 μA/μm at room temperature. When the temperature decreases to 20 K, the drain current of this device further increases by 240% to 800 μA/μm, suggesting that an ohmic contact is achieved instead of a Schottky contact due to Cl doping.

Low-frequency noise generated in the electronic devices determines how small signals can be detected and resolved without error in circuits. Maintaining a high signal to noise ratio becomes challenging for deeply scaled devices, which have a stringent operation voltage window requirement and severe short channel effects. In modern communications, the low-frequency noise could up-convert to phase noise at radio frequencies and hampers their usability [24]. In the view of scientific significance and practical applications, a thorough understanding of the low-frequency noise mechanisms in MoS_2 FET is of great importance. However, previous results always yield scattered, rather poor noise level partially because of the nonoptimized contact and device process. Furthermore, the mechanisms of the dominant factors in different output current and temperature regions remain unclear.

Here, low-frequency noise measurement of a 200-nm MoS_2 device from 300 to 20 K is carried out. The temperature dependence of the S_{Id}/I_d^2 of this device is plotted in Fig. 13a. It can be seen that the noise level decreases with temperature, suggesting a thermally activated process. The current noise spectral density nicely follows a 1/*f* trend at all temperatures without emergence of generation–recombination bulge signatures. To investigate the dominant physical mechanism of the 1/*f* noise, S_{Id}/I_d^2 and transconductance to drain current squared $(g_m/I_d)^2$ as functions of drain current I_d of the same device from 300 to 20 K are plotted in Fig. 13b. It is observed that S_{Id}/I_d^2 is proportional to $(g_m/I_d)^2$ at the small drain current level, indicating the carrier number fluctuation mechanism as the main source of 1/*f* noise, which is consistent with other reports on few-layer MoS_2 FETs [62, 63]. However, at high drain current, the noise spectral density deviates from $(g_m/I_d)^2$. The noise of homogeneous layers irrespective of the dominant noise model can be typically expressed using Hooge's empirical formula $S_{Id}/I_d^2 = \alpha_H/(fN)$, where

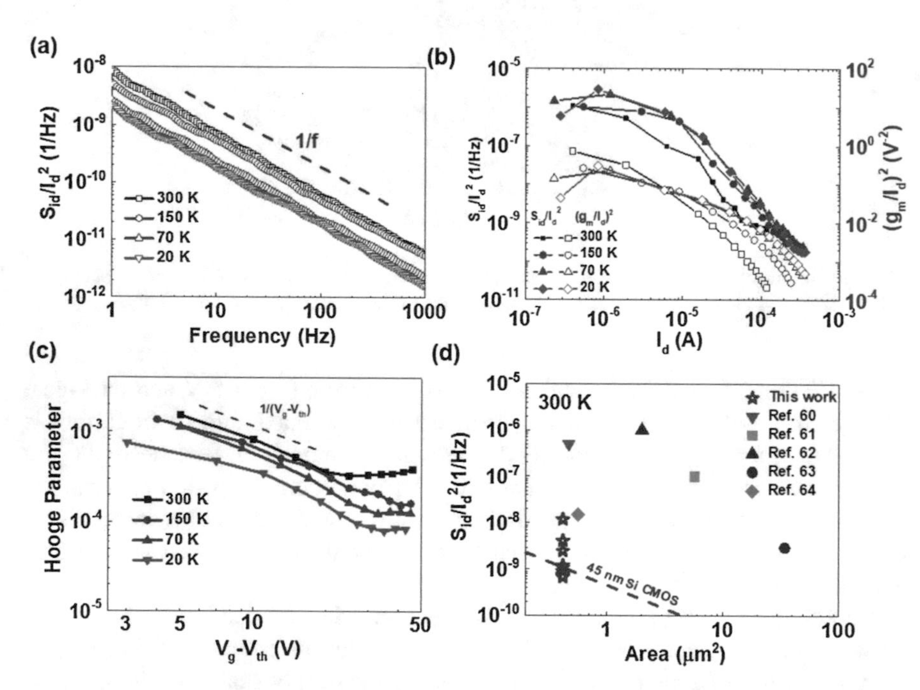

Fig. 13 (**a**) Current noise spectral density (S_{Id}/I_d^2) as a function of frequency with $V_g = 40$ V at various temperatures for the 200-nm MoS_2 FET. (**b**) Normalized drain current noise spectral density (S_{Id}/I_d^2) and transconductance to drain current ratio squared $[(g_m/I_d)^2]$ versus drain current for the MoS_2 device from 300 to 20 K. (**c**) Hooge parameter versus V_g–V_{th} at various temperatures from 300 to 20 K. (**d**) Benchmark of current noise spectral density S_{Id}/I_d^2 versus channel area for different MoS_2 FETs. The 1/*f* noise requirement of 45-nm node precision analog/RF driver of mixed-signal Si CMOS technology is added as a reference

α_H is the Hooge parameter, N is the total number of carriers. In linear region ($V_d = 0.2$ V) under gate overdrive conditions $V_g – V_{th} > 0$, N can be approximated by the gate capacitance. Therefore, considering both carrier number fluctuation and mobility fluctuation, a correlated mobility fluctuation model can also be applied here. This issue is discussed in more detail in chapter "Noise and Fluctuations in Fully Depleted Silicon-On-Insulator MOSFETs" of this book [64]. The extracted Hooge parameter α_H of the same device at various temperatures is shown in Fig. 13c, exhibiting a clear gate voltage and temperature dependence. The minimum α_H is 3.3×10^{-4} at room temperature, which is about one order of magnitude smaller than previous results without Cl doping [60, 61, 63, 65]. At all temperatures, the Hooge parameter follows the same trend, indicating that the dominant noise mechanism does not change with temperature. α_H exhibits $1/(V_g – V_{th})$ dependence when $V_g – V_{th}$ is smaller than 25 V at all temperatures, which is in agreement with carrier number mechanism. For $V_g – V_{th}$ larger than 25 V at all temperatures, α_H remains constant with minimal gate voltage dependence, consistent with the mobility fluctuation model. This result shows that noise mechanism of few-layer MoS_2 FET is dominated by mobility fluctuation at high drain current level in strong inversion, while number carrier fluctuation noise takes over at low drain current. To benchmark the performance of few-layer MoS_2 devices, we compare the drain current noise level S_{Id}/I_d^2 against the device channel area $W \times L$ with other MoS_2 FETs in literature, as shown in Fig. 13d. The MoS_2 device in this work has shown significant improvement in the noise level, which is down to 2.8×10^{-10} μm^2/Hz at 10 Hz as a result of much improved contact and on-state current. This result is comparable to the $1/f$ noise requirement of 45-nm node precision analog/RF driver mixed signal in Si CMOS technology listed by the International Technology Roadmap for Semiconductors [66].

In conclusion, we have systematically investigated the low-frequency $1/f$ noise in high-performance scaled MoS_2 devices at various temperatures down to 20 K using a robust Cl-doping technique. These devices exhibit a remarkable low noise level, reaching a low current noise spectral density of 2.8×10^{-10} μm^2/Hz at 10 Hz. These results demonstrate the great potential of MoS_2 for future nanoelectronic applications.

4.2 *Low-Frequency Noise in BP Transistors*

Recently, black phosphorus (BP) has been found to be a p-type layered material with a tunable bandgap ranging from ~0.3 to ~2.0 eV [67–71], which is key to realize complementary metal-oxide semiconductor (CMOS) and functional circuits. Field-effect transistors (FETs) based on few-layer BP show encouraging results with high hole mobility up to 1000 cm^2/V/s and even higher with hexagonal boron nitride passivation [67, 68, 72–74]. In addition, the potential to operate in the ambipolar region could greatly reduce the process complexity and cost [75–77]. On the other hand, BP transistors, similar to other 2D semiconductors, suffer from high contact resistance, which plays an important role not only in output current as studied by many, but also in its noise floor, which determines the signal-to-noise ratio and remains an important target for every technology node on the CMOS roadmap. As a result, it becomes critical to understand the metal–BP contact and its impact on the ambipolar electronic transport and associated noise behavior. Thus, a systematic temperature-dependent study of the noise behavior and mechanisms of the BP nFET, and its comparison with that of the pFET, as we address in this section, serves as a necessary part for better understanding BP ambipolar transistors [78].

Here, we study hole and electron transport as well as the low-frequency noise in multilayer BP nFETs and pFETs from 300 to 20 K. The contact resistance of nFETs is much larger than that of pFETs, which limits its performance, including drain current and noise level. We also observe that the dominant 1/*f* noise mechanism of multilayer BP nFETs and pFETs are mobility fluctuation and carrier number fluctuations with correlated mobility fluctuations, respectively.

The BP transistor structure is depicted in Fig. 14a, with the channel lengths ranging from 0.1 to 2 μm. An Al_2O_3-capping layer deposited by atomic layer deposition is used to protect the BP FETs. The thickness of BP used in this work is 8.6 nm, measured by AFM. Figure 14b shows transfer characteristics of

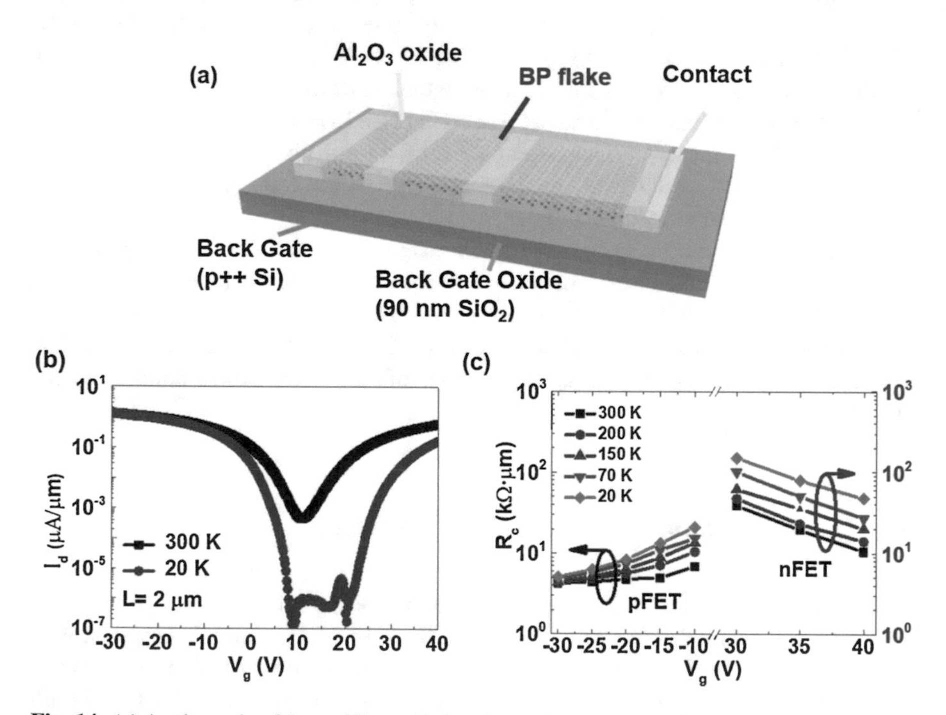

Fig. 14 (**a**) A schematic of the multilayer (8.6 nm by AFM) BP device. (**b**) Transfer characteristics of the $L_{ch} = 2$-μm BP ambipolar transistor with $V_d = -0.05$ V at 300 and 20 K. (**c**) Contact resistance as a function of back-gate voltage extracted from I_d–V_d curves of the BP nFET and pFET at different temperatures

the BP device with a channel length $L = 2\ \mu m$ at 300 and 20 K. The device exhibits clear ambipolar behavior. I_{on}/I_{off} of about 3.3×10^3 and 1.3×10^3 at $V_d = -0.05$ V are obtained for the BP pFET and nFET at 300 K, respectively. When the temperature decreases to 20 K, the off-current decreases significantly and the I_{on}/I_{off} is over 10^6 for both branches. The contact resistance for both BP nFET and pFET are extracted by transmission line measurements (TLM). A clear dependence on the back-gate bias is observed for the contact resistance at all temperatures and the contact resistance increases at lower temperatures, indicating the dominant thermionic injection of the Schottky barrier formed at metal–BP contacts. The R_c of BP FETs at the p-side is smaller than that of the n-side, which is due to the Ni metal Fermi level being closer to the valence band.

We performed low-frequency noise measurements on the BP nFET and pFET in the linear region at $V_d = 0.2$ V from 300 to 20 K on the 2-μm device. The noise spectra follow the typical $1/f$ dependence. To investigate the noise mechanism, S_{Id}/I_d^2 at $f = 100$ Hz and the corresponding $(g_m/I_d)^2$ at 300 K as a function of drain current for the BP pFET and nFET are plotted in Fig. 15a, b, respectively. Figure 15a shows the S_{Id}/I_d^2 and the corresponding $(g_m/I_d)^2$ of the BP pFET at $f = 100$ Hz as a

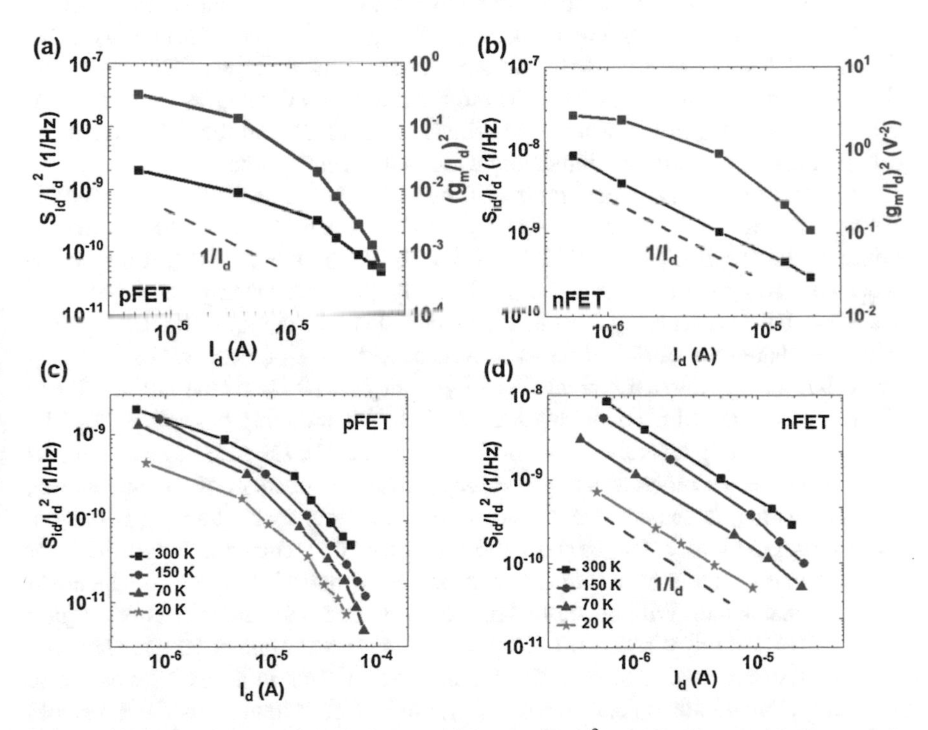

Fig. 15 Normalized drain current noise spectral density (S_{Id}/I_d^2) and the transconductance to drain current ratio squared [$(g_m/I_d)^2$] at $V_d = 0.2$ V and $f = 100$ Hz versus drain current at 300 K for the BP (**a**) pFET and (**b**) nFET. Normalized drain current noise spectral density (S_{Id}/I_d^2) at $V_d = 0.2$ V and $f = 100$ Hz as a function of drain current from 300 to 20 K for the BP (**c**) pFET and (**d**) nFET

function of drain current at 300 K. We can see that the S_{Id}/I_d^2 and the corresponding $(g_m/I_d)^2$ follow the same trend over a wide drain current range. This indicates that the carrier number fluctuation model is responsible for the BP pFET device. The deviation of the noise level from the $(g_m/I_d)^2$ observed at strong inversion can be attributed to correlated mobility fluctuations. This is consistent with the previous reports on few-layer BP pFETs [79]. Meanwhile, for the nFET as shown in Fig. 15b, it is clear that there is substantial deviation between the S_{Id}/I_d^2 and the $(g_m/I_d)^2$ over a broad drain current range. And the observed S_{Id}/I_d^2 follows the $1/I_d$ trend, indicating that a mobility fluctuation model governs [29, 80]. To further investigate the mechanisms, Fig. 15c shows that S_{Id}/I_d^2 curves of pFETs at low temperatures also exhibit the same trend as in 300 K. In the pFET, low-frequency noise originates from the trapping and releasing of the carrier near the BP/SiO_2 interface, which dominates the noise mechanism. On the other hand, as shown in Fig. 15d, S_{Id}/I_d^2 at $f = 100$ Hz as a function of drain current from 300 to 20 K for the BP nFET, all follow the $1/I_d$ trend in the whole range, indicating the temperature-independent noise mechanism. In the nFET, the main noise model is mobility fluctuation induced by lattice and impurity scattering. It has been previously reported that the Al_2O_3 passivation layer deposited at low temperature contains large amounts of positive fixed charge, inducing an ambipolar behavior and decreasing the hole mobility of BP pFETs [76]. Therefore, it is likely that the 1/f noise source in the nFET is from the electron impurity scattering centers in the interface with the Al_2O_3 capping layer. In addition, the magnitude of noise amplitude of the nFET and the pFET decreases with decreasing temperature, indicating a thermal activated process.

The extracted Hooge parameter α_H of the 2-μm device at various temperatures is shown in Fig. 16a, exhibiting clear temperature dependence. At 300 K, the α_H values of the BP nFET and pFET are 1.4×10^{-2} and 9.6×10^{-3}, respectively. When decreasing the temperature down to 20 K, the values change to 3.3×10^{-3} and 6.0×10^{-4} accordingly. This confirms that the performance of the BP nFET is inferior to that of the pFET, which is consistent with the electrical results. To study the scaling of $1/f$ noise, a typical $W \times S_{Id}/I_d^2$ at $f = 10$ Hz of the BP nFET and pFET as a function of channel length at 300 K when sweeping the gate voltage V_g from weak to strong inversion is shown in Fig. 16b, c. For the nFET in Fig. 16b, $1/f$ noise decreases with increasing gate voltage and is independent of channel length, which is mainly attributed to the contact-dominated transport at strong inversion. The lowest noise level at $f = 10$ Hz at 300 K is 7.0×10^{-8} $\mu m^2/Hz$ for the nFET. For the pFET in Fig. 16c, at threshold gate voltage, the $1/f$ noise is large and independent of the channel length. With increase in the gate voltage to the strong inversion region ($V_g = -30$ V for the pFET), the noise level decreases over one order of magnitude and shows strong dependence on the channel length. This indicates that the noise from the source/drain contact makes a significant contribution at the threshold regime, while the channel noise dominates in the strong inversion region for the pFET [81].

In conclusion, we investigated the ambipolar output current and noise properties of the multilayer BP transistors from 300 to 20 K. The low-frequency noise mechanisms of the multilayer BP nFET and pFET are found to be mobility fluctuation and

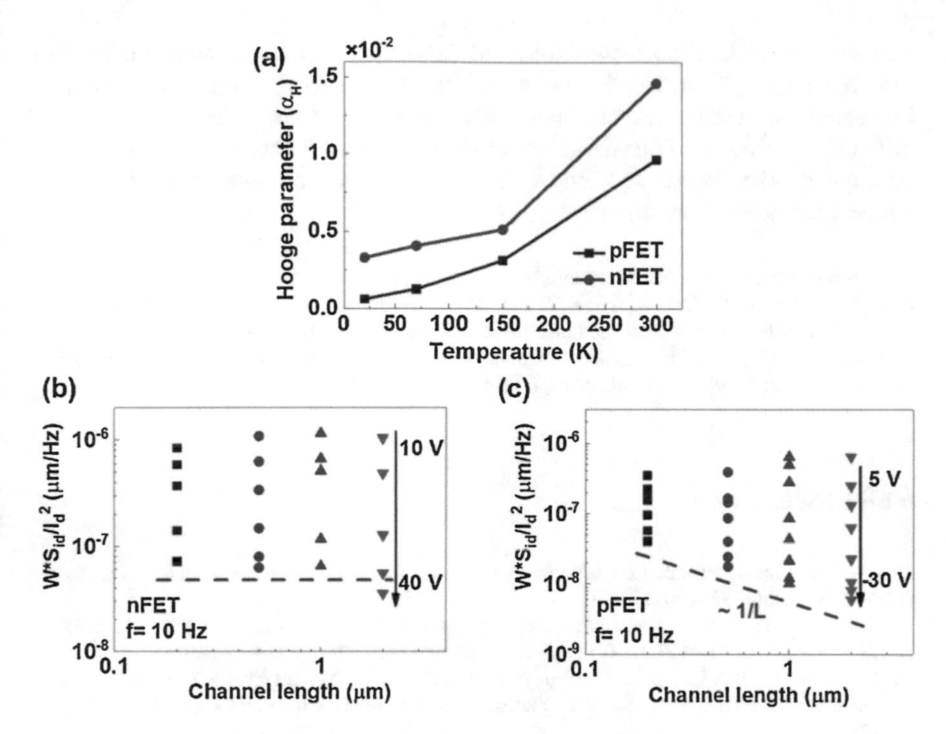

Fig. 16 (**a**) Hooge parameter α_H from 300 to 20 K for the BP nFET and pFET. Normalized current spectral density $W \times S_{Id}/I_d^2$ versus channel length for different BP transistors at 300 K in this work for (**b**) nFETs and (**c**) pFETs

carrier number fluctuation with correlated mobility fluctuation, respectively. These results not only demonstrated high-performance ambipolar operation based on multilayer BP transistors with low noise, but also provide fundamental insights into the electron and hole transport and related $1/f$ mechanisms at various temperatures.

5 Conclusion

In summary, low-frequency noise measurements on MOSFETs with novel channel materials such as III–V, Ge, and 2D materials are presented and analyzed. The impact of device scaling on the low-frequency noise is discussed. Low-frequency noise behaves differently on different material systems and different process conditions such as NW geometries, channel length, EOT, and channel doping, etc. In particular, low-frequency noise in ultra-scaled high-mobility MOSFETs shows a unique scaling trend, in which mobility fluctuation can be suppressed by the ballistic transport of carriers. This is confirmed by the systematical measurement on deep sub-100 nm 3D InGaAs and Ge nanowire transistors. Meanwhile, low-frequency noise 2D FETs using MoS_2 and BP as channel materials are studied. The low-frequency noise mechanisms of the 2D transistors are discussed. Both mobility fluctuation and carrier number fluctuation contribute to the noise performance at different working conditions. So, a correlated mobility fluctuation model can also be applied. The impact of Schottky barrier is investigated and studied by low-temperature noise measurement.

Acknowledgments The authors gratefully acknowledge the contributions and the discussions from N. J. Conrad, S. Shin, J. Gu, J. Zhang, M. A. Alam, H. Wu, L. Yang, Y. Du at Purdue University in USA, in particular the noise measurements on BP transistors carried out at Wuhan National High Magnetic Field Center in China with Y. Wu. The work at Purdue University was supported in part by NSF, AFOSR, ARO, DTRA, SRC and DARPA.

References

1. J.A. Del Alamo, Nanometre-scale electronics with III-V compound semiconductors. Nature **479**(7373), 317–323 (2011)
2. J. Lin, X. Cai, Y. Wu, D.A. Antoniadis, J.A. Alamo, Record maximum transconductance of 3.45 mS/μm for III–V FETs. IEEE Electron Device Lett. **37**(4), 381–384 (2016)
3. J. Zhang, M. Si, X.B. Lou, W. Wu, R.G. Gordon, P.D. Ye, InGaAs 3D MOSFETs with drastically different shapes formed by anisotropic wet etching, in *IEDM Tech. Dig.*, 2015, pp. 12–15
4. N. Waldron et al., InGaAs gate-all-around nanowire devices on 300mm Si substrates. IEEE Electron Device Lett. **35**(11), 1097–1099 (2014)
5. C.B. Zota, L.E. Wernersson, E. Lind, $In_{0.53}Ga_{0.47}As$ multiple-gate field-effect transistors with selectively regrown channels. IEEE Electron Device Lett. **35**(3), 342–344 (2014)
6. S. Lee et al., Record extrinsic transconductance (2.45 mS/μm at $V_{DS} = 0.5$ V) $InAs/In_{0.53}Ga_{0.47}As$ channel MOSFETs using MOCVD source-drain regrowth, in *VLSI Tech. Dig.*, 2013, pp. 246–247
7. Y. Sun et al., Self-aligned III-V MOSFETs: towards a CMOS compatible and manufacturable technology solution, in *IEDM Tech. Dig.*, 2013, pp. 48–51
8. R.T.P. Lee et al., VLSI processed InGaAs on Si MOSFETs with thermally stable, self-aligned Ni-InGaAs contacts achieving: enhanced drive current and pathway towards a unified contact module, in *IEDM Tech. Dig.*, 2013, pp. 44–47
9. J.J. Gu et al., 20-80nm channel length InGaAs gate-all-around nanowire MOSFETs with EOT=1.2nm and lowest SS=63mV/dec, in *IEDM Tech. Dig.*, 2012, pp. 633–636
10. M. Radosavljevic et al., Electrostatics improvement in 3-D tri-gate over ultra-thin body planar InGaAs quantum well field effect transistors with high-K gate dielectric and scaled gate-to-drain/gate-to-source separation, in *IEDM Tech. Dig.*, 2011, pp. 765–768
11. S. Deora et al., Positive bias instability and recovery in InGaAs channel nMOSFETs. IEEE Trans. Device Mater. Reliab. **13**(4), 507–514 (2013)
12. J. Franco et al., Suitability of high-k gate oxides for III-V devices: a PBTI study in $In_{0.53}Ga_{0.47}As$ devices with Al_2O_3, in *IEEE International Reliability Physics Symposium Proceedings*, 2014, p. 6A.2
13. S. Shin et al., Origin and implications of hot carrier degradation of gate-all-around nanowire III-V MOSFETs, in *IEEE International Reliability Physics Symposium Proceedings*, 2014, p. 4A.3
14. M. Si et al., Low-frequency noise and random telegraph noise on near-ballistic III-V MOSFETs. IEEE Trans. Electron Devices **62**(11), 3508–3515 (2015)

15. N. Conrad et al., Low-frequency noise and RTN on near-ballistic III-V GAA nanowire MOSFETs, in *IEDM Tech. Dig.*, 2014, pp. 502–505
16. P. Ren et al. New observations on complex RTN in scaled high-κ/metal-gate MOSFETs—the role of defect coupling under DC/AC condition, in *IEDM Tech. Dig.*, 2013, pp. 778–781
17. T. Grasser et al., The paradigm shift in understanding the bias temperature instability: from reaction–diffusion to switching oxide traps. IEEE Trans. Electron Devices **58**(11), 3652–3666 (2011)
18. N. Tega et al., Increasing threshold voltage variation due to random telegraph noise in FETs as gate lengths scale to 20 nm, in *VLSI Tech. Dig.*, 2009, pp. 50–51
19. J.P. Campbell et al., The origins of random telegraph noise in highly scaled SiON nMOSFETs, in *IEEE International Reliability Physics Symposium Proceedings*, 2008, pp. 105–109
20. J. Chen, Y. Higashi, I. Hirano, Y. Mitani, Experimental study of channel doping concentration impacts on random telegraph signal noise and successful noise suppression by strain induced mobility enhancement, in *VLSI Tech. Dig.*, 2013, pp. 184–185
21. T. Nagumo, K. Takeuchi, T. Hase, Y. Hayashi, Statistical characterization of trap position, energy, amplitude and time constants by RTN measurement of multiple individual traps, in *IEDM Tech. Dig.*, 2010, pp. 628–631
22. H. Miki et al., Statistical measurement of random telegraph noise and its impact in scaled-down high-κ/metal-gate MOSFETs, in *IEDM Tech. Dig.*, 2012, pp. 450–453
23. W. Feng et al., Fundamental origin of excellent low-noise property in 3D Si-MOSFETs ~ Impact of charge-centroid in the channel due to quantum effect on 1/f noise ~, in *IEDM Tech. Dig.*, 2011, pp. 630–633
24. M. Haartman, M. Östling, *Low-Frequency Noise in Advanced MOS Devices* (Springer Science & Business Media, Dordrecht, 2007)
25. B. Min et al., Low-frequency noise in submicrometer MOSFETs with HfO_2, HfO_2/Al_2O_3 and $HfAlO_x$ gate stacks. IEEE Trans. Electron Devices **51**(8), 1315–1322 (2004)
26. D. Lopez, S. Haendler, C. Leyris, G. Bidal, G. Ghibaudo, Low-frequency noise investigation and noise variability analysis in high-k/metal gate 32-nm CMOS transistors. IEEE Trans. Electron Devices **58**(8), 2310–2316 (2011)
27. S.C. Tsai et al., Low-frequency noise characteristics for various ZrO_2-added HfO_2-based 28-nm high-k/metal-gate nMOSFETs. IEEE Electron Device Lett. **34**(7), 834–836 (2013)
28. M. Valenza, A. Hoffmann, D. Sodini, A. Laigle, F. Martinez, D. Rigaud, Overview of the impact of downscaling technology on 1/f noise in p-MOSFETs to 90 nm, in *IEE Proceedings—Circuits, Devices and Systems*, vol. 151(2), 2004, pp. 102–110
29. L.K.J. Vandamme, F.N. Hooge, What do we certainly know about 1/f noise in MOSTs? IEEE Trans. Electron Devices **55**(11), 3070–3085 (2008)
30. C. Wei et al., Investigation of low-frequency noise in silicon nanowire MOSFETs in the subthreshold region. IEEE Electron Device Lett. **30**(6), 668–671 (2009)
31. C.J. Delker, Y. Zi, C. Yang, D.B. Janes, Current and noise properties of InAs nanowire transistors with asymmetric contacts induced by gate overlap. IEEE Trans. Electron Devices **61**(3), 884–889 (2014)
32. K.W. Chew, K.S. Yeo, S.F. Chu, Effect of technology scaling on the 1/f noise of deep submicron PMOS transistors. Solid State Electron. **48**(7), 1101–1109 (2004)
33. T. Grasser et al., Gate-sided hydrogen release as the origin of "permanent" NBTI degradation: from single defects to lifetimes, in *IEDM Tech. Dig.*, 2015, pp. 535–538
34. J.J. Gu, H. Wu, Y. Liu, A.T. Neal, R.G. Gordon, P.D. Ye, Size-dependent-transport study of $In_{0.53}Ga_{0.47}As$ gate-all-around nanowire MOSFETs: impact of quantum confinement and volume inversion. IEEE Electron Device Lett. **33**(7), 967–969 (2012)
35. F.N. Hooge, 1/f noise is no surface effect. Phys. Lett. A **29**(3), 139–140 (1969)
36. A. Toriumi et al., Material potential and scalability challenges of germanium CMOS, in *IEDM Tech. Dig.*, 2011, pp. 24–28
37. C.-T. Chung, C.-W. Chen, J.-C. Lin, C.-C. Wu, C.-H. Chien, G.-L. Luo, First experimental Ge CMOS FinFETs directly on SOI substrate, in *IEDM Tech. Dig.*, 2012, pp. 14–16

38. R. Pillarisetty, Academic and industry research progress in germanium nanodevices. Nature **479**(7373), 324 (2011)
39. H. Wu, W. Wu, M. Si, P. Ye, First demonstration of Ge nanowire CMOS circuits: lowest SS of 64 mV/dec, highest gmax of 1057 μS/μm in Ge nFETs and highest maximum voltage gain of 54 V/V in Ge CMOS inverters, in *IEDM Tech. Dig.*, 2015, pp. 16–19
40. R. Zhang, P.-C. Huang, J.-C. Lin, N. Taoka, M. Takenaka, S. Takagi, High-mobility Ge p-and n-MOSFETs with 0.7-nm EOT using $HfO_2/Al_2O_3/GeO_x$/Ge gate stacks fabricated by plasma postoxidation. IEEE Trans. Electron Devices **60**(3), 927–934 (2013)
41. T. Grasser et al., Switching oxide traps as the missing link between negative bias temperature instability and random telegraph noise, in *IEDM Tech. Dig.*, 2009, pp. 681–684
42. E. R. Hsieh et al., The experimental demonstration of the BTI-induced breakdown path in 28nm high-k metal gate technology CMOS devices, in *VLSI Tech. Dig.*, 2014
43. M.-L. Fan, V.P.-H. Hu, Y.-N. Chen, P. Su, C.-T. Chuang, Analysis of single-trap-induced random telegraph noise on FinFET devices, 6T SRAM cell, and logic circuits. IEEE Trans. Electron Devices **59**(8), 2227–2234 (2012)
44. J. Zou et al., Deep understanding of AC RTN in MuGFETs through new characterization method and impacts on logic circuits, in *VLSI Tech. Dig.*, 2013, pp. T186–T187
45. E. Simoen et al., Low-frequency noise characterization of strained germanium pMOSFETs. IEEE Trans. Electron Devices **58**(9), 3132–3139 (2011)
46. W. Fang et al., Low-frequency noise characterization of GeO_x passivated germanium MOSFETs. IEEE Trans. Electron Devices **62**(7), 2078–2083 (2015)
47. W. Wu et al., Mobility fluctuation-induced low-frequency noise in ultrascaled Ge nanowire nMOSFETs with near-ballistic transport. IEEE Trans. Electron Devices **65**, 2573–2577 (2018)
48. W. Wu, H. Wu, M. Si, N. Conrad, Y. Zhao, P. Ye, RTN and low frequency noise on ultra-scaled near-ballistic Ge nanowire nMOSFETs, in *VLSI Tech. Dig.*, 2016
49. G. Ghibaudo, T. Boutchacha, Electrical noise and RTS fluctuations in advanced CMOS devices. Microelectron. Reliab. **42**(4–5), 573–582 (2002)
50. W. Wu, H. Wu, J. Zhang, M. Si, Y. Zhao, P.D. Ye, Carrier mobility enhancement by applying back-gate bias in Ge-on-insulator MOSFETs. IEEE Electron Device Lett. **39**(2), 176–179 (2018)
51. L. Yang et al., Chloride molecular doping technique on 2D materials: WS_2 and MoS_2. Nano Lett. **14**(11), 6275–6280 (2014)
52. M. Si et al., Steep-slope hysteresis-free negative capacitance MoS_2 transistors. Nat. Nanotechnol. **13**, 24–29 (2018)
53. H. Wang et al., Integrated circuits based on bilayer MoS_2 transistors. Nano Lett. **12**(9), 4674–4680 (2012)
54. H. Liu, A.T. Neal, P.D. Ye, Channel length scaling of MoS_2 MOSFETs. ACS Nano **6**(10), 8563–8569 (2012)
55. C.D. English, G. Shine, V.E. Dorgan, K.C. Saraswat, E. Pop, Improved contacts to MoS_2 transistors by ultra-high vacuum metal deposition. Nano Lett. **16**(6), 3824–3830 (2016)
56. B. Radisavljevic, A. Radenovic, J. Brivio, V. Giacometti, A. Kis, Single-layer MoS_2 transistors. Nat. Nanotechnol. **6**(3), 147–150 (2011)
57. K.F. Mak, C. Lee, J. Hone, J. Shan, T.F. Heinz, Atomically thin MoS_2: a new direct-gap semiconductor. Phys. Rev. Lett. **105**(13), 136805 (2010)
58. Y. Zhan, Z. Liu, S. Najmaei, P.M. Ajayan, J. Lou, Large-area vapor-phase growth and characterization of MoS_2 atomic layers on a SiO_2 substrate. Small **8**(7), 966–971 (2012)
59. X. Li et al., Performance potential and limit of MoS_2 transistors. Adv. Mater. **27**(9), 1547–1552 (2015)
60. X. Xie et al., Low-frequency noise in bilayer MoS_2 transistor. ACS Nano **8**(6), 5633–5640 (2014)
61. V.K. Sangwan, H.N. Arnold, D. Jariwala, T.J. Marks, L.J. Lauhon, M.C. Hersam, Low-frequency electronic noise in single-layer MoS_2 transistors. Nano Lett. **13**(9), 4351–4355 (2013)

62. J. Renteria et al., Low-frequency 1/f noise in MoS_2 transistors: relative contributions of the channel and contacts. Appl. Phys. Lett. **104**(15), 153104 (2014)
63. H.-J. Kwon, H. Kang, J. Jang, S. Kim, C.P. Grigoropoulos, Analysis of flicker noise in two-dimensional multilayer MoS_2 transistors. Appl. Phys. Lett. **104**(8), 83110 (2014)
64. C. Theodorou, G. Ghibaudo, Noise and fluctuations in fully depleted silicon-on-insulator MOSFETs, this book
65. J. Na et al., Low-frequency noise in multilayer MoS_2 field-effect transistors: the effect of high-k passivation. Nanoscale **6**(1), 433–441 (2014)
66. International Technology Roadmap for Semiconductors (ITRS), 2007 Edition
67. L. Li et al., Black phosphorus field-effect transistors. Nat. Nanotechnol. **9**(5), 372 (2014)
68. H. Liu et al., Phosphorene: an unexplored 2D semiconductor with a high hole mobility. ACS Nano **8**(4), 4033–4041 (2014)
69. F. Xia, H. Wang, Y. Jia, Rediscovering black phosphorus as an anisotropic layered material for optoelectronics and electronics. Nat. Commun. **5**, 4458 (2014)
70. X. Ling, H. Wang, S. Huang, F. Xia, M.S. Dresselhaus, The renaissance of black phosphorus. Proc. Natl. Acad. Sci. **112**(15), 4523–4530 (2015)
71. L. Kou, C. Chen, S.C. Smith, Phosphorene: fabrication, properties, and applications. J. Phys. Chem. Lett. **6**(14), 2794–2805 (2015)
72. L. Li et al., Quantum oscillations in a two-dimensional electron gas in black phosphorus thin films. Nat. Nanotechnol. **10**, 608–613 (2015)
73. X. Chen et al., High-quality sandwiched black phosphorus heterostructure and its quantum oscillations. Nat. Commun. **6**, 7315 (2015)
74. N. Gillgren et al., Gate tunable quantum oscillations in air-stable and high mobility few-layer phosphorene heterostructures. 2D Mater **2**(1), 11001 (2014)
75. J.D. Wood et al., Effective passivation of exfoliated black phosphorus transistors against ambient degradation. Nano Lett. **14**(12), 6964–6970 (2014)
76. H. Liu, A.T. Neal, M. Si, Y. Du, P.D. Ye, The effect of dielectric capping on few-layer phosphorene transistors: tuning the Schottky barrier heights. IEEE Electron Device Lett. **35**(7), 795–797 (2014)
77. W. Zhu et al., Flexible black phosphorus ambipolar transistors, circuits and AM demodulator. Nano Lett. **15**(3), 1883–1890 (2015)
78. X. Li et al., Mechanisms of current fluctuation in ambipolar black phosphorus field-effect transistors. Nanoscale **8**, 3572–3578 (2016)
79. J. Na et al., Few-layer black phosphorus field-effect transistors with reduced current fluctuation. ACS Nano **8**(11), 11753–11762 (2014)
80. L.K.J. Vandamme, X. Li, D. Rigaud, 1/f noise in MOS devices, mobility or number fluctuations? IEEE Trans. Electron Devices **41**(11), 1936–1945 (1994)
81. Y. Lai, H. Li, D.K. Kim, B.T. Diroll, C.B. Murray, C.R. Kagan, Low-frequency (1/f) noise in nanocrystal field-effect transistors. ACS Nano **8**(9), 9664–9672 (2014)

Detection and Characterization of Single Defects in MOSFETs

Toshiaki Tsuchiya

1 Analysis of Single Interface Defects by the Charge Pumping Method

1.1 Introduction

As the dimensions of metal–oxide–semiconductor field-effect transistors (MOSFETs) are scaled into the nanometer regime, fluctuations in device characteristics due to variations in gate length, discrete dopant fluctuations, and line-edge roughness have become serious problems [1–3]. Moreover, it is expected that interface defects will cause various reliability and variability problems in future nano- and atomic-scale devices. Since the number of interface defects contained in such nanoscale devices (so-called More Moore devices) becomes countable, the randomness of the number and even the position of interface defects induce fluctuations in their electrical characteristics such as the threshold voltage [4]. In addition to the randomness of the number and position of defects, the randomness of the energy level of defects is also important, since a defect may or may not be charged depending on the energy level, which will affect device characteristics, as shown in Fig. 1. Also, the defects may cause additional current noise in analog/RF devices (so-called More than Moore devices), and the capture/emission of electrons in defects may interfere with normal operation in future atomistic devices (so-called Beyond CMOS devices) using one or a few electrons.

Thus, understanding the physics of interface defects in nanoscale and future MOS-based devices will become increasingly important from the viewpoints of reliability, variability, and stability of circuit operations. Therefore, studies on the

T. Tsuchiya (✉)
Research Institute of Electronics, Shizuoka University, Hamamatsu, Japan
e-mail: tsuchiya.toshiaki@shizuoka.ac.jp

T. Grasser (ed.), *Noise in Nanoscale Semiconductor Devices*,
https://doi.org/10.1007/978-3-030-37500-3_11

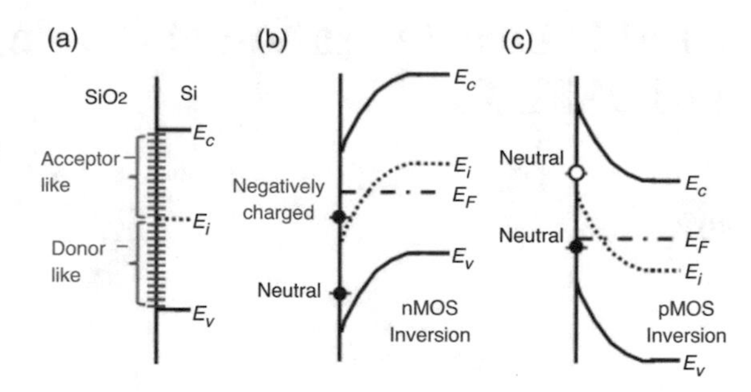

Fig. 1 (**a**) Electrical properties of interface defects, i.e., interface defects are acceptor-like in the upper part of the Si bandgap and donor-like in the lower part. An example of the charging conditions of defects in nanoscale (**b**) nMOSFETs and (**c**) pMOSFETs, which involve only one acceptor-like state and one donor-like state. In such cases, there are possible cases, for instance, that one of the two states is negatively charged in nMOS, but there are no charged states in pMOS, depending on their energy levels

detection and characterization of individual defects will become important. In particular, it is essential to gain information of the defects on an atomistic way, e.g., to count their actual number, to determine their location on an atomic scale, and to evaluate electrical properties such as energy levels and capture cross sections of individual defects.

The charge pumping (CP) technique [5, 6] is known to be a highly precise method for evaluating interface defects even in extremely small-area MOSFETs, and many studies on the in-depth exploration of defects near the MOS interface have been reported using the CP method [7–23]. Some studies on the emission properties of interface defects have been reported [14, 18], including some using the CP method [6, 7]. However, the development of good methods is essential for identifying individual interface defects, i.e., the actual number of defects involved, the carrier capture/emission properties of individual defects, and their fluctuations.

In this section, a systematic characterization procedure using the CP technique with a high resolution (on the order of sub fA) is introduced, and the detection and fundamental characterization of individual single interface defects in MOSFETs are successively carried out. Based on the results, the essential nature of *genuine* single interface defects is discovered [24], the density of states of single interface defects is clarified, and fundamental refinement of CP theory is described [25].

In this work, small-size poly-Si-gate nMOSFETs were mainly used. Their gate lengths (L) and widths (W) were 100–130 nm and 110–300 nm, respectively, and the thickness of the gate oxide grown by rapid thermal oxidation was 4 nm.

The CP measurements were performed mainly using the variable pulse-base-level method [6] while keeping the pulse height constant. In a typical measurement condition, the pulse period t_P was 5 μs (the pulse frequency f is 200 kHz), the pulse width t_W was 2.5 μs, both the pulse risetime t_r and fall time t_f were 120 ns, the pulse height V_p was 2 V, and the reverse-bias to the source and drain junctions V_j was 0.05 V. Besides the typical measurement, various measurements were systematically performed in order to investigate details, controlling gate pulse parameters such as on-time, off-time, t_r, t_f, and f. In addition to the variable pulse-base-level method, the variable pulse-height method [6] while keeping the base level fixed in accumulation was also carried out complementarily. A Keysight B1500A Semiconductor Device Analyzer with a very high resolution of 0.1 fA was used to measure the CP current accurately.

1.2 Conventional CP Theory and Some Useful Applications of the CP Method

The CP technique is recognized as a very high-precision method for evaluating interface defects between the gate oxide and the semiconductor surface in MOSFETs. The CP measurement configuration is shown in Fig. 2 for nMOSFETs. A pulsed voltage, as shown in Fig. 3, is applied to the gate to alternately form inversion and accumulation layers. Electrons are captured in interface defects during inversion, and the trapped electrons recombine with holes coming from the substrate during accumulation. The CP current arises from the capture and recombination of these electrons and is therefore proportional to the density of interface defects. The CP current can be measured on the source/drain junction as well as the bulk contact. Both current values should be equal but opposite in sign.

Typical CP characteristics for a large gate area are shown in Fig. 4, which were obtained by the variable pulse-base-level method while keeping the pulse height constant. As shown in the figure, there is a plateau region of the maximum CP current, where both strong inversion during the high level and strong accumulation during the base level of the gate pulse are satisfied.

The CP current (I_{CP}) in a MOSFET containing N interface defects contributing to the CP current has been described by

$$I_{\mathrm{CP}} = fq \sum_{i=1}^{N} \left\{ 1 - \exp\left(-\frac{t_{\mathrm{high}}}{\tau_{\mathrm{cni}}}\right) \right\} \left\{ 1 - \exp\left(1 - \frac{t_{\mathrm{low}}}{\tau_{\mathrm{cpi}}}\right) \right\} \tag{1}$$

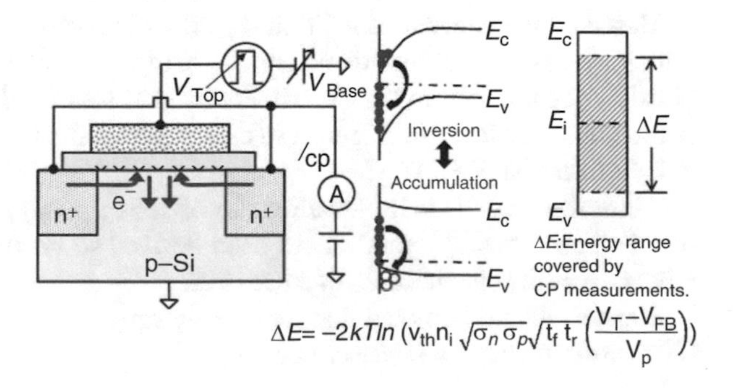

Fig. 2 Configuration of a basic CP measurement for nMOSFETs

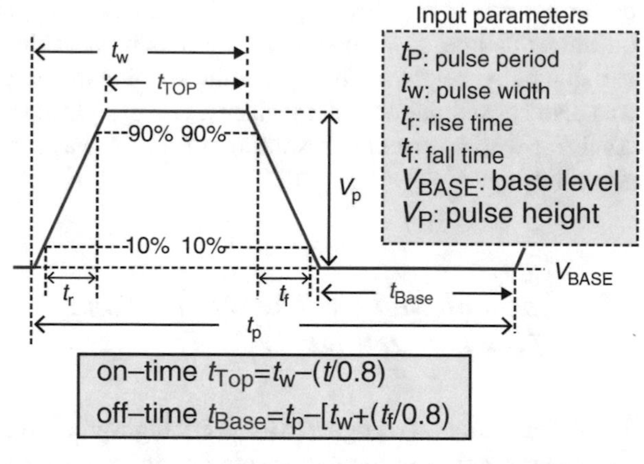

Fig. 3 Waveform of a pulsed voltage applied to the gate in CP measurements

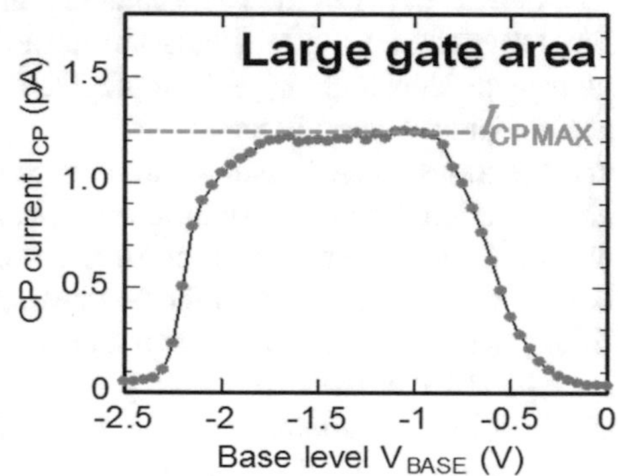

Fig. 4 Typical CP characteristics for an nMOSFET with a large gate area

directly based on the Shockley–Read–Hall (SRH) theory [26, 27], where f is the gate pulse frequency, q is the electron charge, t_{high} and t_{low} are the times when the pulse voltage is at its high level and the time when the pulse voltage is at its low level, respectively. τ_{cni} is the capture time constant for electrons of defect i when the pulse voltage is at its high level, τ_{cpi} is the capture time constant for holes of defect i when the pulse voltage is at its low level. When $t_{high} \gg \tau_{cni}$ and $t_{low} \gg \tau_{cpi}$, Eq. (1) leads to the maximum CP current $I_{CPMAX} = fqN$. Therefore, if only one single defect contributes to the CP current, I_{CPMAX} should be fixed at exactly fq.

Based on the conventional CP theory, the observation of single interface defects with I_{CPMAX} of fq has been demonstrated using submicron MOSFETs [13, 14]. Gate pulse frequency used in the experiment in [14] was 1 MHz; therefore, I_{CPMAX} per defect should be 0.16 pA. Figure 1 in the reference shows that there are two interface defects in the MOSFET.

Moreover, based on the relationship of $I_{CPMAX} = fqN = fqA_GN_{it}$, the density of interface defects N_{it} (cm^{-2}) has been derived so far from $N_{it} = I_{CPMAX}/(fqA_G)$, where A_G is the channel area of MOSFET.

Here, it should be noted that the energy range in the bandgap, ΔE, covered by CP measurements can be described by [13]

$$\Delta E = -2kT \ln \left\{ v_{\text{th}} n_{\text{i}} \sqrt{\sigma_{\text{n}}\sigma_{\text{p}}} \sqrt{t_{\text{f}} t_{\text{r}}} \left(\frac{V_{\text{T}} - V_{\text{FB}}}{V_{\text{P}}} \right) \right\} \tag{2}$$

where v_{th} is the thermal velocity of the carriers, n_{i} is the intrinsic carrier concentration, σ_{n} and σ_{p} are the capture cross sections of electrons and holes, respectively, V_{T} is the threshold voltage, V_{FB} is the flat-band voltage, and V_{P} is the pulse height. Therefore, the average density of states of interface defects D_{it} ($\text{eV}^{-1}\ \text{cm}^{-2}$) can be derived from $D_{\text{it}} = N_{\text{it}}/\Delta E$. ΔE depends on fall time t_{f} and risetime t_{r} of the gate pulse and decreases with increasing fall time or risetime, which will be discussed in more detail. ΔE at room temperature is approximately ± 0.3 eV around the midgap.

Dependences of CP current upon risetime and fall time of the gate pulse are very important not only for deeper understanding of the CP processes but also to obtain the density of states profile of interface defects. Nonsteady-state hole emission from defects to the valence band in the rising portion of the gate pulse and nonsteady-state electron emission from defects to the conduction band in the falling portion are especially important [6].

The maximum CP current decreases with increasing risetime. As shown in Fig. 5, holes are captured in interface defects during accumulation (i.e., electrons trapped in interface defects recombine with holes), and then electrons are captured in interface defects during inversion. But, it should be noted that during risetime, some of the trapped holes can be emitted to the valence band (i.e., the nonsteady-state hole emission to the valence band), and the number of such holes increases for slower risetimes. That is why the number of electrons trapped during inversion decreases for slower risetimes, which leads to a decrease of the maximum CP current with increasing risetime. The change in the maximum CP current due to the change in risetime is caused by interface defects in the energy range of ΔE_{r} in the lower part of the bandgap as shown in the figure. Moreover, ΔE_{r} increases with increasing temperature because the emission probability increases. Therefore, using the risetime and temperature dependences, the density of states profile of interface defects in the lower part of the bandgap can be obtained.

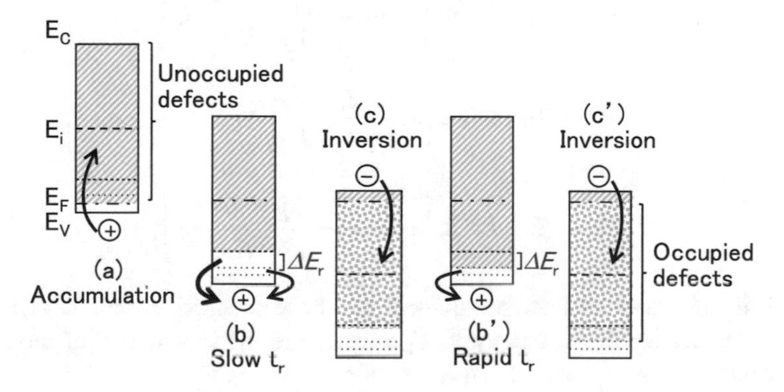

Fig. 5 Changes in the energy band diagram explaining risetime-dependent CP characteristics. (**a**)–(**b**)–(**c**) shows CP process for slow t_{r}, and (**a**)–(**b′**)–(**c′**) for rapid t_{r}

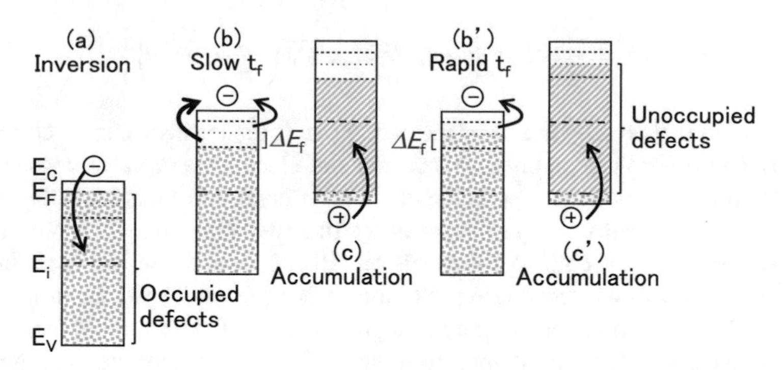

Fig. 6 Changes in the energy band diagram explaining fall-time-dependent CP characteristics. (**a**)–(**b**)–(**c**) shows CP process for slow t_f, and (**a**)–(**b′**)–(**c′**) for rapid t_f

Similarly, during fall time, some of the trapped electrons can be emitted to the conduction band (i.e., the nonsteady-state electron emission to the conduction band), as shown in Fig. 6. The number of such emitted electrons increases for slower fall times. That is why the number of holes captured during accumulation decreases for smaller fall times, and the maximum CP current decreases with increasing fall time. The change in the maximum CP current due to the change in fall time is caused by interface defects in the energy range of ΔE_f in the upper part of the bandgap, as shown in the figure. Therefore, by measuring fall time dependences for various temperatures, the density of states profile of interface defects in the upper part of the bandgap can be obtained.

Thus, by measuring risetime and fall time dependences for various temperatures, and using the following equations [6],

$$D_{it}(E_1) = -\frac{t_r}{fqA_G kT}\frac{dI_{CP}}{dt_r} \tag{3}$$

$$E_1 = E_i + kT \ln\left[v_{th}\sigma_p n_i \frac{|V_{FB} - V_T|}{|\Delta V_G|} t_r\right] \tag{4}$$

$$D_{it}(E_2) = -\frac{t_f}{fqA_G kT}\frac{dI_{CP}}{dt_f} \tag{5}$$

$$E_2 = E_i - kT \ln\left[v_{th}\sigma_n n_i \frac{|V_{FB} - V_T|}{|\Delta V_G|} t_f\right] \tag{6}$$

the density of states of interface defects can be calculated, where $D_{it}(E)$ is the interface defect density at energy E, E_1 and E_2 are the boundaries of the energy range which is scanned, and $|\Delta V_G|$ is the gate pulse height.

There are some other useful applications of the CP method: The single-pulse CP method [28–30] yields the CP current during accumulation, inversion, and the transition between those phases in the time domain. In the time domain current, a current spike due to the minority-carrier capture in interface defects from the inversion layer can be observed during the rising portion of the gate pulse. During the pulse fall time, recombination of captured carriers with majority carriers from the substrate can be observed separately. Therefore, the density of interface defects can be obtained by integration of the current spike caused by either minority- or majority-carrier capture processes. A suitable application of the single-pulse CP method is the observation of the first recovery of interface defects. It has been reported that the recovery of stress-induced interface defects can be observed at the microsecond scale [28].

The three-voltage-level CP method is useful to examine the capture cross sections for electrons and holes, and the energy levels of particular interface defects [8, 17].

Low-frequency CP measurements allow for the observation of near-interface oxide defects, i.e., slow defects in the oxide near the interface. Such oxide defect contributions to the CP current become significant for lower gate pulse frequency [12]. It has been reported that gate pulse frequency smaller than approximately 175 kHz can be applicable to detect the slow defects [31]. By the low-frequency CP method, the depth concentration profiles of slow oxide defects within approximately 1 nm tunneling distance from the interface have been obtained by a simple model using an average capture cross section exponentially decreasing with the oxide depth [32].

As mentioned above, the CP method is extremely useful to evaluate interface defects (and near-interface oxide defects). However, the conventional CP theory will be fundamentally refined based on the remarkable findings in this work, by the detection and characterization of *genuine* single interface defects described in the following section.

1.3 Detection and Characterization of Single Interface Defects

Fundamental Defect Counting: Separation of CP Current into Components from each Individual Defect

It is typically assumed that nanoscale MOSFETs involve only a few interface defects. Suppose that the gate length and width of a MOSFET are 50 and 60 nm, respectively, and the density of interface defects is 1×10^{11} cm^{-2}, then the average number of defects involved in a MOSFET is approximately 3. However, in order to count the exact number of defects contributing to the CP current, and to judge if the CP current is due to a single interface defect, we need a procedure to separate the CP current into components from each individual defect.

An easy and effective procedure is the observation of the differences in the threshold voltage, ΔV_T, or the flatband voltage, ΔV_{FB}, in the local area where each defect is located. A schematic total impurity charge profile along the channel surface in nMOSFETs used in this work is shown in Fig. 7, which leads to differences in

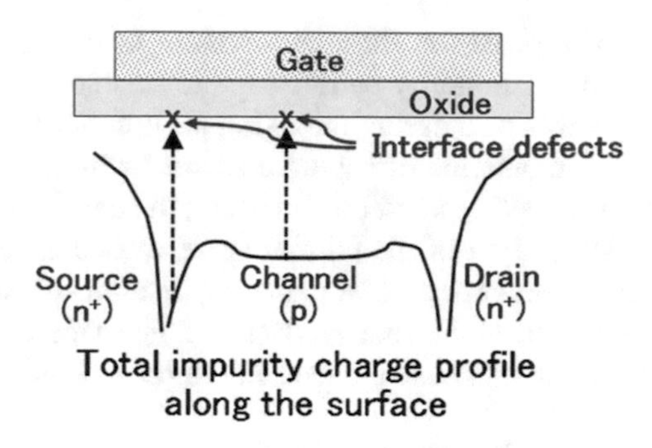

Fig. 7 Total impurity charge profile along the channel surface in nMOSFETs used in the experiments with the halo and source/drain extension regions, which leads to differences in the total concentration of the impurity charges under the local area (i.e., the difference in threshold voltage) where each defect is located

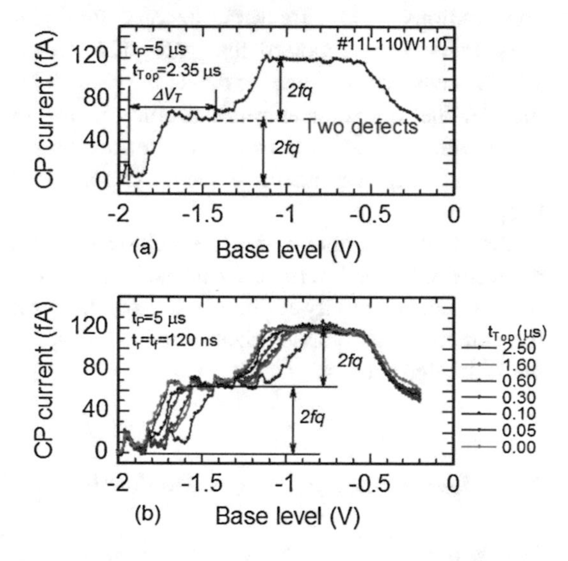

Fig. 8 (**a**) CP characteristics separated into components from individual defects (two defects) due to the difference in threshold voltage ΔV_T in the local area where each defect is located, and (**b**) the dependences of the CP characteristics upon on-time t_{Top} of the gate pulse

the total impurity charge concentration under the local area (i.e., ΔV_T and ΔV_{FB}) where each defect is located. A good example of CP characteristics suitable for this procedure is shown in Fig. 8a. We can judge from the figure that the CP characteristics are clearly separated into two portions by ΔV_T, and this sample involves two interface defects.

The dependences of the CP characteristics upon on-time t_{Top} and off-time t_{Base} of the gate pulse are other effective procedures for separating the CP current into components from each individual defect, because the carrier capture/emission processes in individual defects can be observed from the dependences [24].

A transient phenomenon in the rising portion of the CP characteristics has been reported [22], which depends on the pulse width (t_W), or the on-time of the gate pulse, as shown in Fig. 9. In the rising portion, the CP current at a given base level

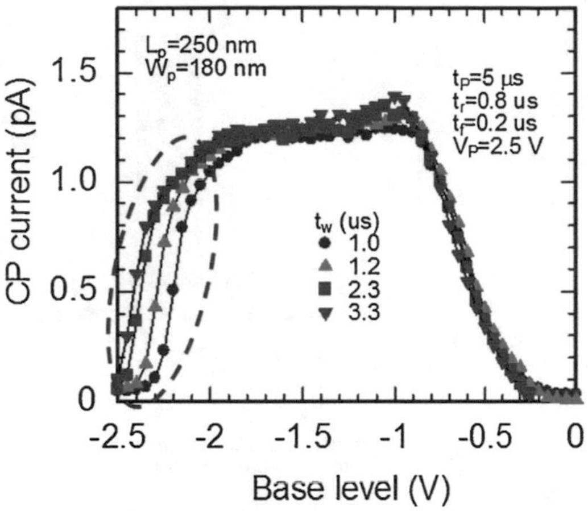

Fig. 9 Transient behavior appearing in the rising portion of the CP characteristics, which depends on the width of the gate pulse, i.e., the on-time of the gate pulse

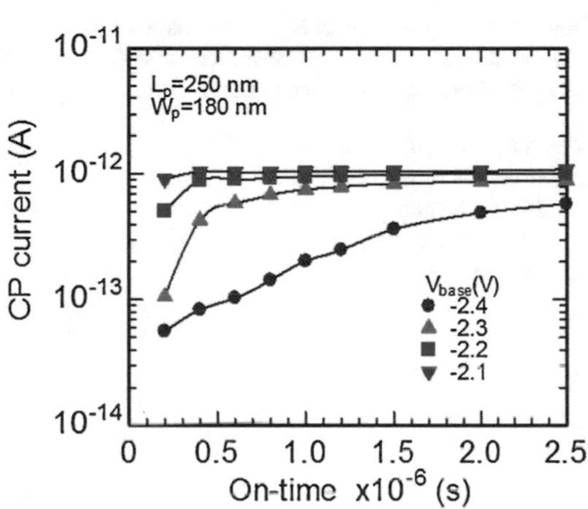

Fig. 10 Saturation behavior of CP current at a constant base level (V_{BASE}) as a function of on-time in the rising portion of the CP characteristics

V_{BASE} of the gate pulse increases with increasing on-time. The CP current at a constant V_{BASE} as a function of the on-time is shown in Fig. 10 with V_{BASE} as a parameter, which indicates a saturation behavior. The behavior can be considered to be related to the electron capture process by interface defects as follows: Changes in the energy band diagram during this process are shown in Fig. 11. Electrons trapped in the interface defects recombine with holes from the accumulation layer formed while at the base level (i.e., during accumulation) as shown in Fig. 11a, and some of the electrons in the inversion layer are captured by interface defects during the on-time as shown in Fig. 11b, b′. However, in this case, inversion is less strong because of the rising portion in the CP characteristics, and only a small fraction of the interface defects can capture electrons if the on-time is insufficient, as shown in Fig. 11b.

The capture rate for electrons c_n can be described as $c_n = \sigma_n v_{\mathrm{TH}} n$, and the time constant τ_{T} related to the capture process can be described as $1/c_n$, where n is the density of electrons in the less strong inversion layer. The relatively low

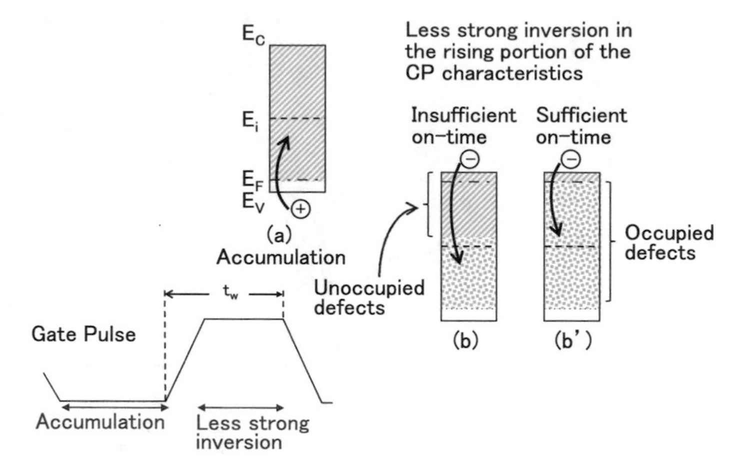

Fig. 11 Changes in the energy band diagram during the rising portion of the CP characteristics. (**a**)–(**b**) shows CP process for insufficient on-time, and (**a**)–(**b**′) for sufficient on-time. Only a small fraction of interface defects can capture electrons if the on-time is insufficient in the rising portion

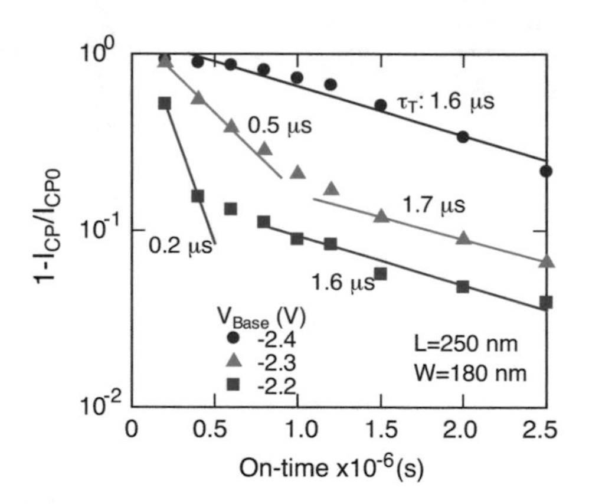

Fig. 12 $1 - I_{CP}/I_{CP0}$ on a log scale as a function of the on-time of the gate pulse with V_{BASE} as a parameter, where I_{CP0} is a saturated CP current

value of n (i.e., lower c_n and longer τ_T) in the rising portion is the reason why the direct observation of the capture process shown in Figs. 9 and 10 can be realized. Therefore, the saturation behavior in the capture process can be described as $I_{CP} = I_{CP0}[1 - \exp(-t/\tau_T)]$ using τ_T, where I_{CP0} is a saturated CP current; thus $1 - I_{CP}/I_{CP0} = \exp(-t/\tau_T)$. $1 - I_{CP}/I_{CP0}$ on a log scale as a function of the on-time is shown in Fig. 12 with V_{BASE} as a parameter. Some straight lines based on the least-squares method are also drawn in Fig. 12. From the figure, it is found that the value of τ_T $(=1/(\sigma_n v_{TH} n))$ depends on V_{BASE}, which is mainly because n is a function of V_{BASE} (the gate voltage during the inversion is $V_{BASE} + V_P$ (pulse height)). Moreover, it is found from the figure that the value of τ_T changes with the on-time even at a constant V_{BASE} (i.e., under a fixed n), which is considered to be because the values of σ_n for individual interface defects contained are distributed.

Unfortunately, since τ_T is sensitive to n, it is not easy to obtain the accurate value of σ_n at present.

However, at least, it is considered that each interface defect has an individual value of σ_n and shows individual electron capture process in the rising portion of CP characteristics.

The dependence of the CP characteristics upon on-time t_{TOP} for the sample used in Fig. 8a is shown in Fig. 8b. The electron capture processes for each defect can clearly and separately be seen by the two separate changes in CP current. From Fig. 8b, it is more clearly found that the sample involves two defects.

Figure 13a shows an ordinarily seen CP characteristics measured by the typical (ordinary) pulse condition described in Sect. 1.1, from which it is impossible to judge how many defects are involved. However, as shown in Fig. 13b, the t_{TOP} dependences of the rising portion in the CP characteristics clearly reveal that two defects are involved, and the two defect property becomes more apparent with decreasing t_{TOP}, i.e., one defect with $I_{CPMAX} = 2fq$, and the other with $I_{CPMAX} = fq$. Another example is shown in Fig. 14a, b. From Fig. 14a measured by the typical pulse condition, it is impossible to know the number of defects involved. However, the t_{TOP} dependences reveal that this MOSFET involves five interface defects, as shown in Fig. 14b.

Moreover, the dependence of CP characteristics upon off-time t_{Base} is also an effective procedure. An example of such dependences is shown in Fig. 15. In the procedure, the hole capture (or electron emission) processes can be seen in the falling portion of the CP characteristics, where accumulation is less strong, and

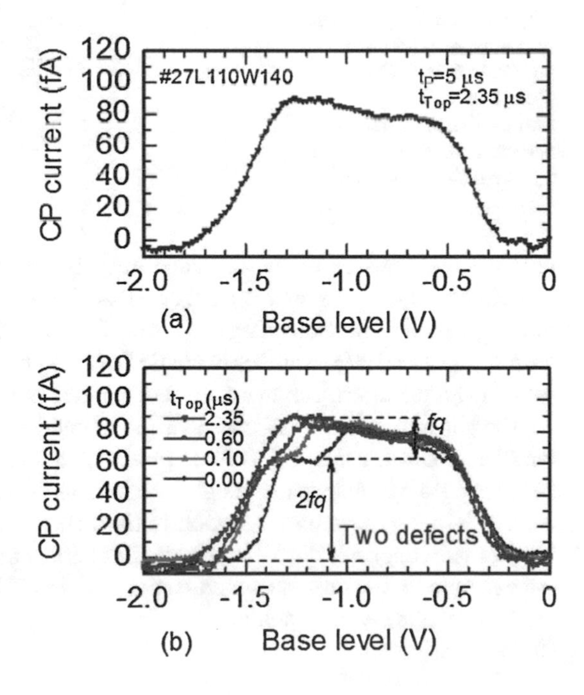

Fig. 13 (**a**) CP characteristics measured by the typical (ordinary) CP pulse condition described in Sect. 1.1, from which it is impossible to judge how many defects are involved, (**b**) the dependences of the CP characteristics upon on-time t_{Top}, which emphasizes the electron capture processes in the rising portion of the CP characteristics in individual defects, revealing that there are two defects involved

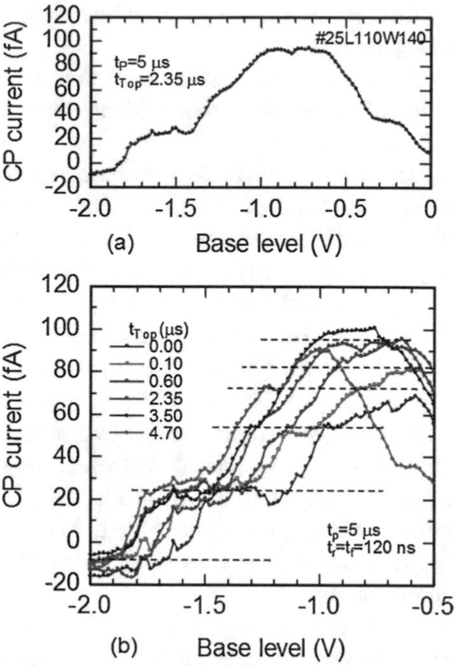

Fig. 14 (**a**) Example of the CP characteristics for a sample involving more than three defects, which was obtained using the typical (ordinary) CP measurement conditions described in Sect. 1.1, (**b**) dependences of the CP characteristics upon on-time t_{Top} for the same sample, which reveals that this sample involves five interface defects

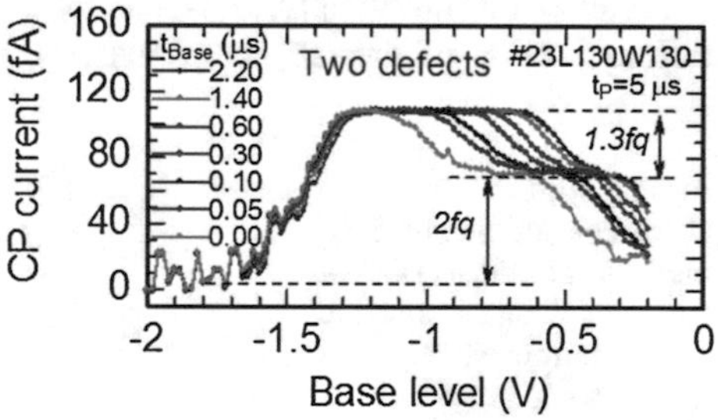

Fig. 15 Dependences of the CP characteristics upon off-time t_{Base}, which emphasizes the hole capture (or electron emission) processes in the falling portion of the CP characteristics in individual defects, revealing that there are two defects involved

only a small fraction of the interface defects can capture holes if the off-time is insufficient. It can be seen from Fig. 15 that there are two defects and this becomes more apparent with decreasing t_{Base}, one defect with $I_{\mathrm{CPMAX}} = 2fq$, and the other with $I_{\mathrm{CPMAX}} = 1.3fq$. Another example for this procedure is shown in Fig. 16. It is found from the figure that there are four defects in the sample.

The abovementioned on- and off-time dependences are based on the CP characteristics obtained using the variable pulse-base-level method. Similar dependences based on the variable pulse-height method are also effective to separate the CP current into components from each individual defect, because the variable pulse-height method reveals the difference in local threshold voltage of individual defects. An example for the on-time dependences is given in Fig. 17, which shows that there are two defects involved and the individual electron capture process for each defect can be clearly seen.

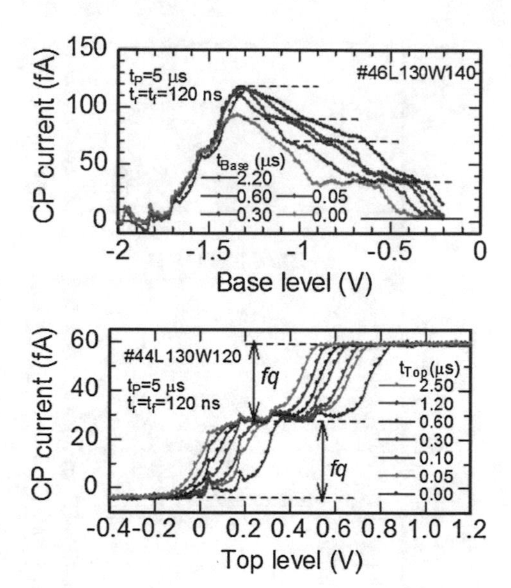

Fig. 16 Another example for the dependences of the CP characteristics upon off-time t_{Base}, which reveals that there are four defects in the sample

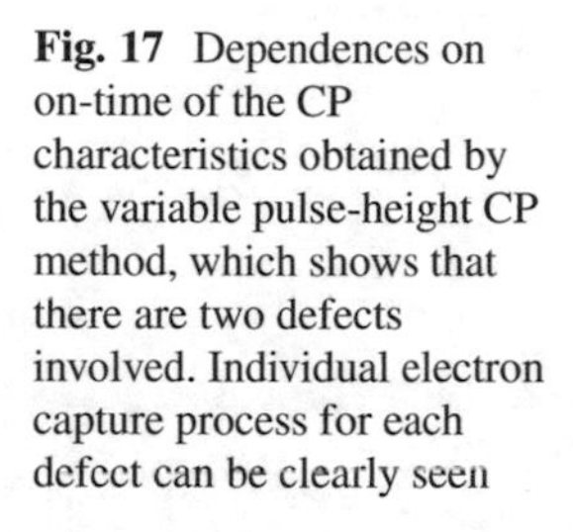

Fig. 17 Dependences on on-time of the CP characteristics obtained by the variable pulse-height CP method, which shows that there are two defects involved. Individual electron capture process for each defect can be clearly seen

In addition, measuring the dependences of the CP characteristics upon risetime t_{r} and fall time t_{f} is also sometimes helpful for counting the number of interface defects involved, although these dependences are more important to evaluate the energy levels of individual interface defects, as described later.

The abovementioned effective and systematic measurement procedure to count the number of defects (and to evaluate the energy levels of defects, described in detail later) are summarized in Fig. 18. Here, (e) frequency dependences are effective to distinguish between interface defects and oxide defects (slow defects). Since interface defects are fast states, the CP current from interface defects is proportional to the pulse frequency, and the maximum CP current normalized by frequency is independent of frequency, as shown in Fig. 18e. However, oxide defects are slow states, as described in Sect. 1.2, because they are considered to be accompanied by tunneling phenomena in their electron capture/emission processes. Therefore, the charge per CP cycle I_{CP}/f of oxide defects is not independent of frequency, but decreases with increasing frequency. An example of CP characteristics containing the contribution of an oxide defect is shown in Fig. 19. It can be seen clearly from the figure that there is a frequency-dependent component, which shows that the sample involves an oxide defect as well as interface defects.

As described above, the CP current can be separated into components from each individual defect involved, and therefore it can also be judged by utilizing the systematic procedure whether CP current is due to only a single defect.

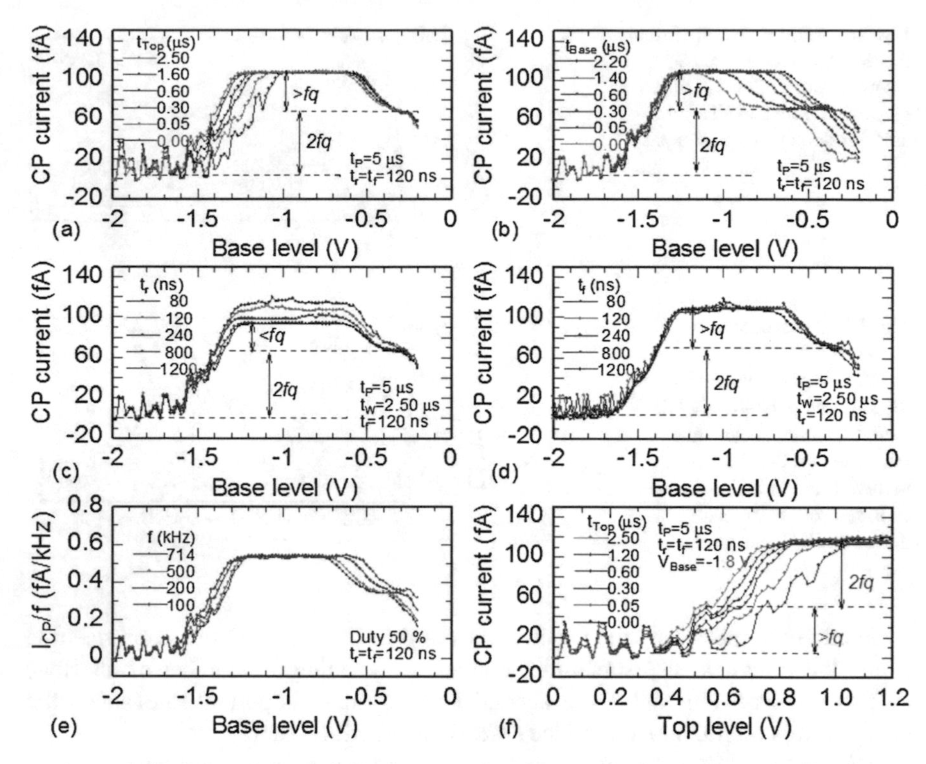

Fig. 18 Summary of systematic characterization procedure of interface defects. (**a**) On-time and (**b**) off-time dependences to count the number of defects, (**c**) risetime and (**d**) fall time dependences, helpful to count the number of interface defects and more important to evaluate the energy levels of individual defects, (**e**) frequency dependences to distinguish between interface defects and oxide defects (slow defects), and (**f**) on-time dependences in the variable pulse height CP measurements, also helpful to count the number of defects

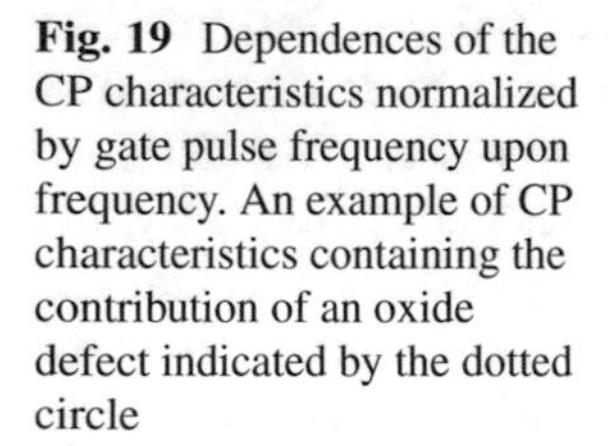

Fig. 19 Dependences of the CP characteristics normalized by gate pulse frequency upon frequency. An example of CP characteristics containing the contribution of an oxide defect indicated by the dotted circle

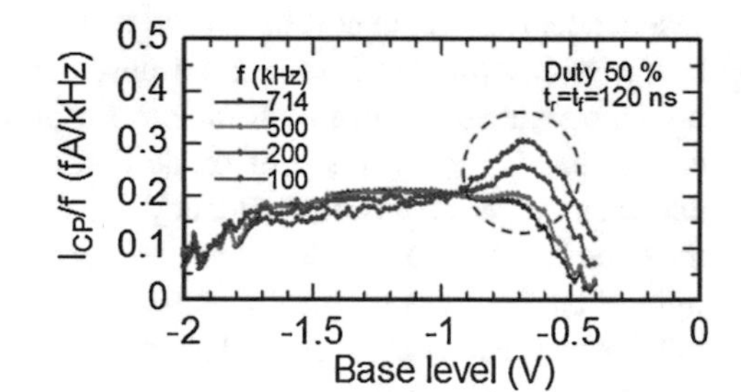

CP Current from a *Genuine* Single Interface Defect

Based on the systematic procedure for counting the number of defects contributing to the CP current, various CP characteristics due to only a *genuine* single interface defect were obtained for the first time. The dependences of the CP characteristics upon t_{Top} and t_{Base} for four typical samples are shown in Figs. 20–23. All of these

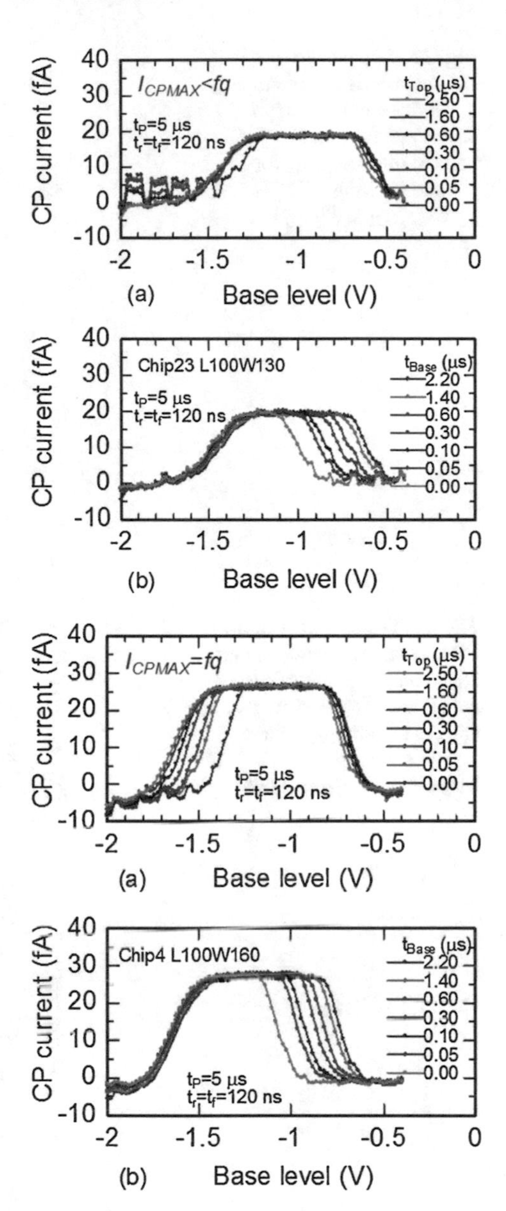

Fig. 20 Dependences of the CP characteristics due to only a *genuine* single interface defect upon (**a**) on-time t_{Top} and (**b**) off-time t_{Base} for a sample showing $I_{\mathrm{CPMAX}} < fq$ (=32 fA)

Fig. 21 Dependences of the CP characteristics due to only a *genuine* single interface defect upon (**a**) on-time t_{Top} and (**b**) off-time t_{Base} for a sample showing $I_{\mathrm{CPMAX}} = fq$ (=32 fA)

CP characteristics cannot be separated into any components by the dependences. Therefore, it can be concluded that each sample in the figures involves only a *genuine* single interface defect. Since the gate pulse frequency f is 200 kHz in the measurements, fq is 32 fA. The sample in Fig. 20 shows $I_{\mathrm{CPMAX}} \leq fq$, and $I_{\mathrm{CPMAX}} = fq$ in Fig. 21, $fq < I_{\mathrm{CPMAX}} \leq 2fq$ in Fig. 22, and $I_{\mathrm{CPMAX}} = 2fq$ in Fig. 23. It is impressed to find the cases of $I_{\mathrm{CPMAX}} = 2fq$ and $fq < I_{\mathrm{CPMAX}} \leq 2fq$. Based on the first-order CP theory, i.e., $I_{\mathrm{CPMAX}} = fqN$ as described in Sect. 1.2, I_{CPMAX}

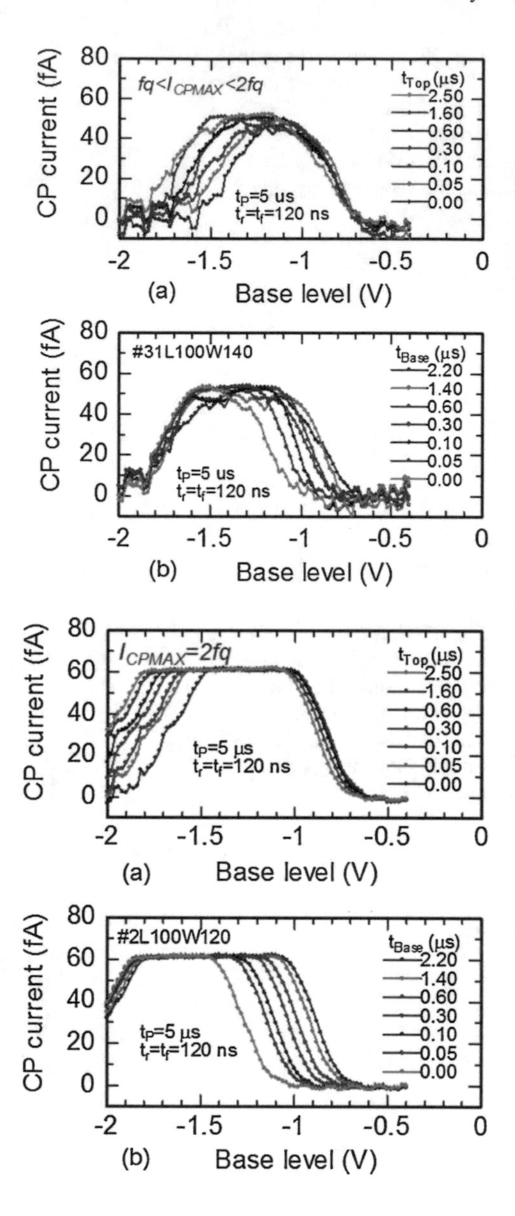

Fig. 22 Dependences of the CP characteristics due to only a *genuine* single interface defect upon (**a**) on-time t_{Top} and (**b**) off-time t_{Base} for a sample showing $fq < I_{\mathrm{CPMAX}} < 2fq$ ($fq = 32$ fA)

Fig. 23 Dependences of the CP characteristics due to only a *genuine* single interface defect upon (**a**) on-time t_{Top} and (**b**) off-time t_{Base} for a sample showing $I_{\mathrm{CPMAX}} = 2fq$ ($fq = 32$ fA)

for a single defect ($N = 1$) should be fixed at exactly fq. Based on the second-order CP theory, previous reports give a case of $I_{\mathrm{CPMAX}} \leq fq$ for a single defect [13–15, 18] depending upon its energy level, risetime, and fall time. I_{CPMAX} can be reduced because of statical reduction of the pumped charge, which represents the recombination between an electron and a hole, due to so-called nonsteady-state hole emission to the valence band during the risetime, or the nonsteady-state electron emission to the conduction band during the fall time. However, $I_{\mathrm{CPMAX}} = 2fq$ and $fq < I_{\mathrm{CPMAX}} \leq 2fq$ were observed for the first time. In this work, 70 samples were

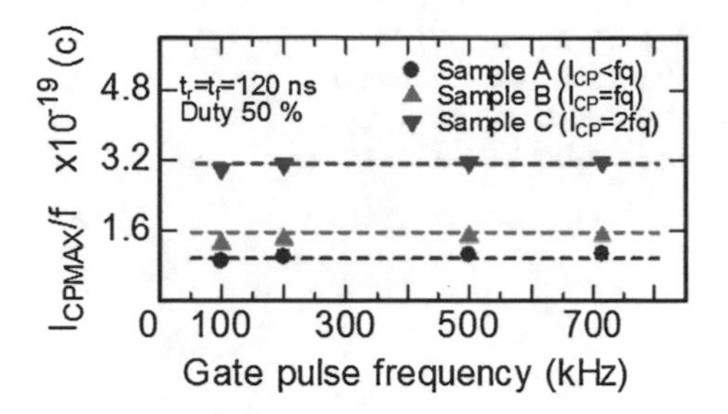

Fig. 24 I_{CPMAX} normalized by the gate pulse frequency f as a function of f for the typical samples involving only a single defect, showing $I_{CPMAX} < fq$, $I_{CPMAX} = fq$, and $I_{CPMAX} = 2fq$, respectively

measured, and among them 27 samples showed single defect properties, and it was found that the CP current from a single defect is in the range of 0–2fq, but not a fixed value of fq. The reason for this will be explained later in detail.

In Fig. 24, I_{CPMAX}/f is plotted as a function of frequency f up to about 700 kHz for the typical samples involving only a single defect. It can be seen that I_{CPMAX}/f is independent of f, which verifies that the defects are interface defects, but not oxide defects. It is interesting that the interface defects indicating $I_{CPMAX} = fq$ or $2fq$ show $I_{CPMAX}/f = q$ or $2q$ constant, respectively, even in the high-frequency region. This may possibly be used as a quantum standard for electrical current, which has recently become a topic of interest [33].

Amphoteric Nature of Interface Defects

The origin of interface defects is considered to be P_b centers, i.e., silicon dangling bonds (or trivalent silicon) at the Si/SiO_2 interface [34–38]. While the (111) Si/SiO_2 interface exhibits one defect, called a P_b center, the (100) Si/SiO_2 interface has two P_b variants called P_{b0} and P_{b1} [39, 40]. Approximately, 80% are considered to be P_{b0} centers. A schematic illustration of the P_{b0} and P_{b1} densities of states and P_{b0} occupancy is shown in Fig. 25 [40, 41]. P_{b0} centers in the lower part of the silicon bandgap are donor-like, and acceptor-like in the upper part. When a positively charged donor-like P_{b0} center accepts an electron, it becomes neutral, and if this traps a further electron, it becomes negatively charged, giving it an acceptor-like nature.

Therefore, if only a single interface defect is involved in a MOSFET, it can be supposed that two energy levels are formed in the device, one from the donor-like, and the other from the acceptor-like P_{b0} center.

Such supposed typical pairs of energy levels of P_{b0} centers and the corresponding maximum CP current I_{CPMAX} expected from each pair for a single interface defect are shown in Fig. 26. ΔE shown in the figure is the energy range covered by the CP measurements, and is described by Eq. (2). ΔE at room temperature (RT) is approximately ± 0.3 eV around the midgap E_i. It should be noted that ΔE decreases with increasing risetime t_r or fall time t_f of the gate pulse according to Eq. (2), and the upper boundary level of ΔE moves slightly toward E_i with increasing t_f, and the lower boundary moves slightly toward E_i with increasing t_r, as shown in Fig. 26.

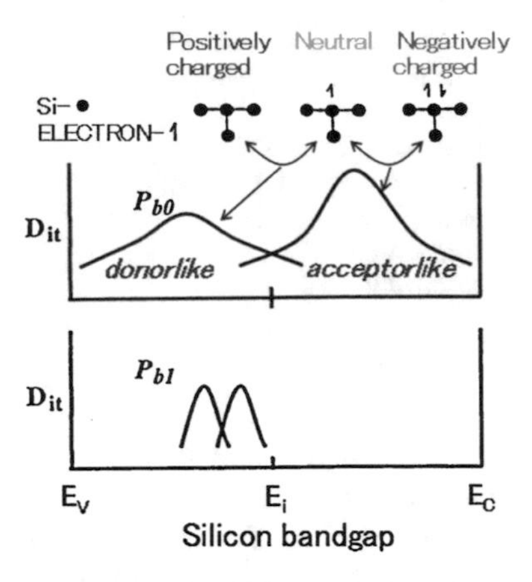

Fig. 25 Schematic illustration of the P_{b0} and P_{b1} densities of states and P_{b0} occupancy [40, 41]

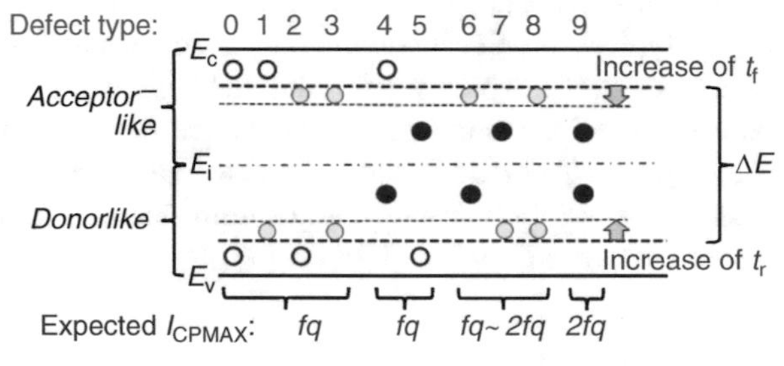

Fig. 26 Typical pairs of energy levels supposed from the nature of the P_{b0} centers and the corresponding expected maximum CP current I_{CPMAX}

Therefore, it can be expected that I_{CPMAX} from a single interface defect may depend on the pairs of energy levels of the P_{b0} center involved as follows: when both energy levels are located outside ΔE (defect type 0 as shown in Fig. 26), I_{CPMAX} should be 0. When one of the two energy levels is located inside ΔE but near the boundary level, its contribution to I_{CPMAX} may be smaller than fq due to the statistical nonsteady-state emission. Therefore, for defect types 1–2, it may be that $I_{CPMAX} \leq fq$, and it is expected that I_{CPMAX} decreases with increasing t_r or t_f. I_{CPMAX} for type 1 would decrease with increasing t_r, but would be independent of t_f. For type 2, I_{CPMAX} would decrease with increasing t_f, but would be independent of t_r. Even for defect type 3, where each energy level is located near each boundary level, I_{CPMAX} can be smaller than fq. In this case, I_{CPMAX} would decrease with increasing t_r and/or t_f. For defect types 4 and 5, one of the two energy levels is located fully inside ΔE, and the other is outside ΔE; therefore, I_{CPMAX} should have a fixed value of fq, and it should be independent of both t_r and t_f. Similarly, it is considered that I_{CPMAX} for defect types 6–7 would be in the range $fq < I_{CPMAX} \leq 2fq$. For type 6, I_{CPMAX} would decrease with increasing t_f, but would be independent of t_r. For type 7, I_{CPMAX} would decrease with increasing t_r, but would be independent of t_f. Even for defect type 8, I_{CPMAX} can be in the range $fq < I_{CPMAX} \leq 2fq$, and it would decrease with increasing t_r and/or t_f. Finally, defect type 9, for which both energy levels are located fully inside ΔE, should give a fixed value of $I_{CPMAX} = 2fq$, and it should be independent of both t_r and t_f.

Thus, it is expected from the nature of the P_{b0} centers that the maximum CP current from single Si/SiO_2 interface defects would be in the range of $0 \leq I_{CPMAX} \leq 2fq$, which corresponds well to the experimental results described in section "CP Current from a *Genuine* Single Interface Defect."

Two Different Energy Levels in a Single Interface Defect

As described in Sect. 1.2, the density of states of interface defects can be calculated using the dependences of CP characteristics upon fall time and risetime.

Suppose that only one single interface defect is involved in a sample, and an electron is captured in the defect during inversion and the trapped electron recombines with a hole during accumulation. Note that the events are happening during fall time. If the energy level of the defect is located near the boundary level ΔE in the upper part of the bandgap, the trapped electron can be statistically emitted to the conduction band, as shown in Fig. 27. The energy range ΔE_f in the figure is the difference in ΔE caused by the difference in fall time t_f. Therefore, if the CP current from a single interface defect would decrease due to the increase of t_f, we can find the existence of an energy level in the energy range ΔE_f. Thus, from the fall time dependences, energy levels of single interface defects in the range ΔE_f in the upper part of the bandgap can be evaluated.

Similarly, if the CP current from a single interface defect would decrease due to the increase of risetime t_r, we can find the existence of an energy level in the energy range ΔE_r, as shown in Fig. 28. Thus, from risetime dependences, energy levels of single interface defects in the range ΔE_r in the lower part of the bandgap can be evaluated.

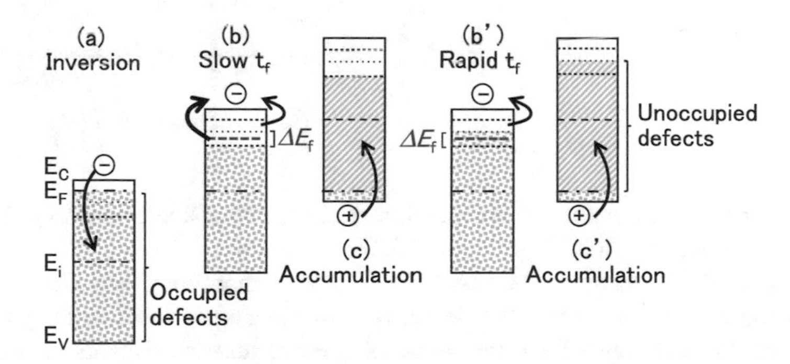

Fig. 27 Changes in the energy band diagram during fall time t_f. (**a**)–(**b**)–(**c**) shows CP process for slow t_f, and (**a**)–(**b′**)–(**c′**) for rapid t_f. The energy range ΔE_f is the difference in ΔE (the energy range in the bandgap covered by CP measurements) caused by the difference in t_f. If CP current from a single interface defect would decrease due to the increase of t_f, it can be found that there is an energy level of the single defect in the range of ΔE_f

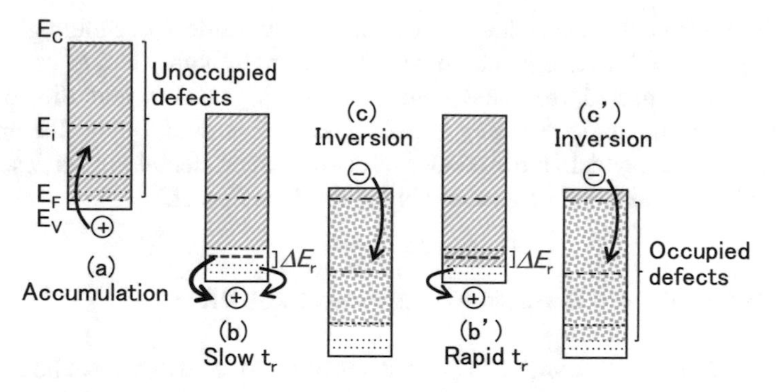

Fig. 28 Changes in the energy band diagram during risetime t_r. (**a**)–(**b**)–(**c**) shows CP process for slow t_r, and (**a**)–(**b′**)–(**c′**) for rapid t_r. The energy range ΔE_r is the difference in ΔE caused by the difference in t_r. If CP current from a single interface defect would decrease with increasing t_f, it can be found that there is an energy level of the single defect in the range of ΔE_r

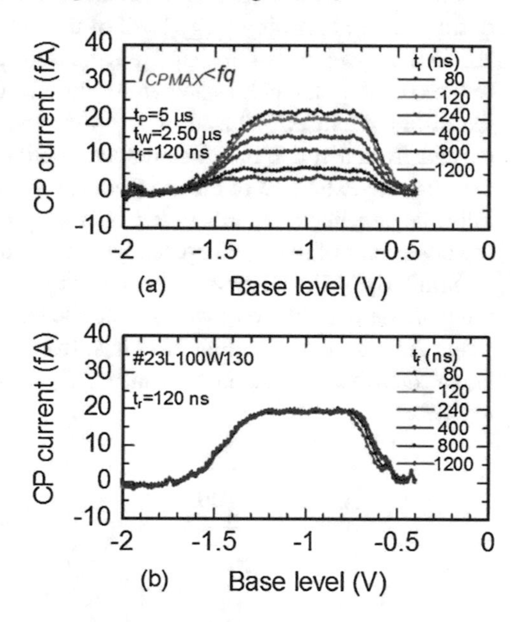

Fig. 29 Dependences of the CP characteristics upon (**a**) risetime t_r and (**b**) fall time t_f for a single interface defect. $I_{CPMAX} < fq$ (=32 fA), and I_{CPMAX} decreases with increasing t_r, but is independent of t_f. These features correspond to defect type 1 in Fig. 26

The dependences of the CP characteristics for various single interface defects upon t_r and t_f are shown in Figs. 29, 30, 31, 32, 33, 34, and 35.

In this study, 70 samples were measured. Among them there were 27 samples with a single interface defect. The defect type of the single defect in each sample can be easily determined from the value of I_{CPMAX} and the dependences of the CP current on t_r and/or t_f, as described above. The details on the number of samples with each single defect type are summarized in Table 1. These results indicate that I_{CPMAX} from *genuine* single interface defects is indeed in the range of $0 \leq I_{CPMAX} \leq 2fq$.

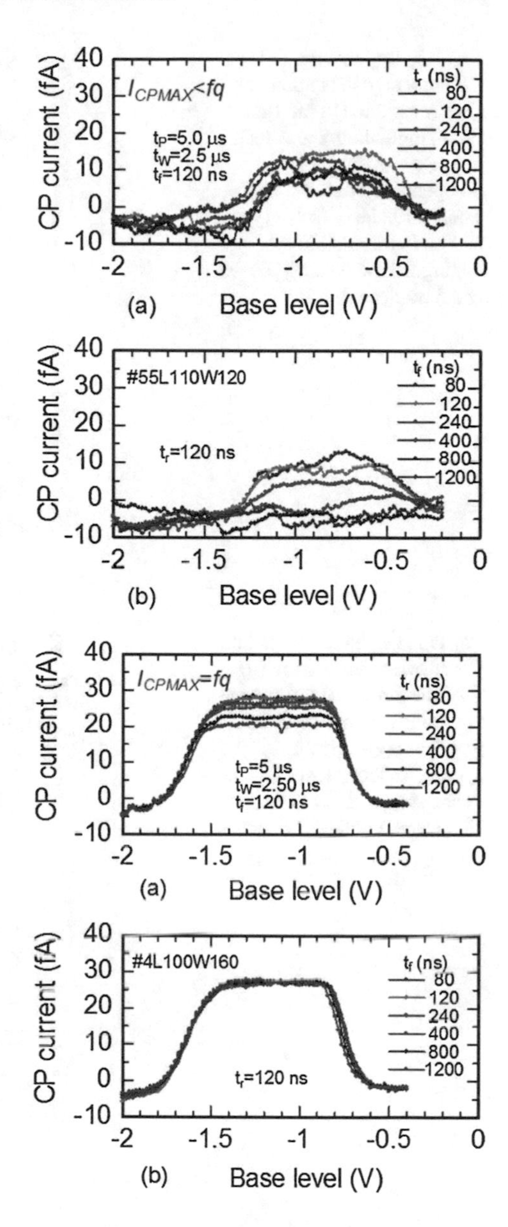

Fig. 30 Dependences of the CP characteristics upon (**a**) risetime t_r and (**b**) fall time t_f for a single interface defect. $I_{CPMAX} < fq$. Some noise and instability in the current are involved probably due to the voltage noise from the pulse generator. But, I_{CPMAX} is unquestionably independent of t_r, and decreases with increasing t_f. These features correspond to defect type 2

Fig. 31 Dependences of the CP characteristics upon (**a**) risetime t_r and (**b**) fall time t_f for a single interface defect. $I_{CPMAX} = fq$, and I_{CPMAX} decreases with increasing t_r, but is independent of t_f. Therefore, the defect in the sample is determined to be defect type 1

Therefore, it can be concluded that two different energy levels in a single interface defect could participate in the CP process, and hence I_{CPMAX} from a single defect is not fixed at fq, but can range from 0 to $2fq$ depending upon the energy levels of the two states. Although it has been believed that I_{CPMAX} is given simply by fqN, where N is the total number of defects contributing to the CP current, the above results unveiled that this belief is valid only in some particular conditions. A fundamental refinement of CP theory will be discussed in Sect. 1.5.

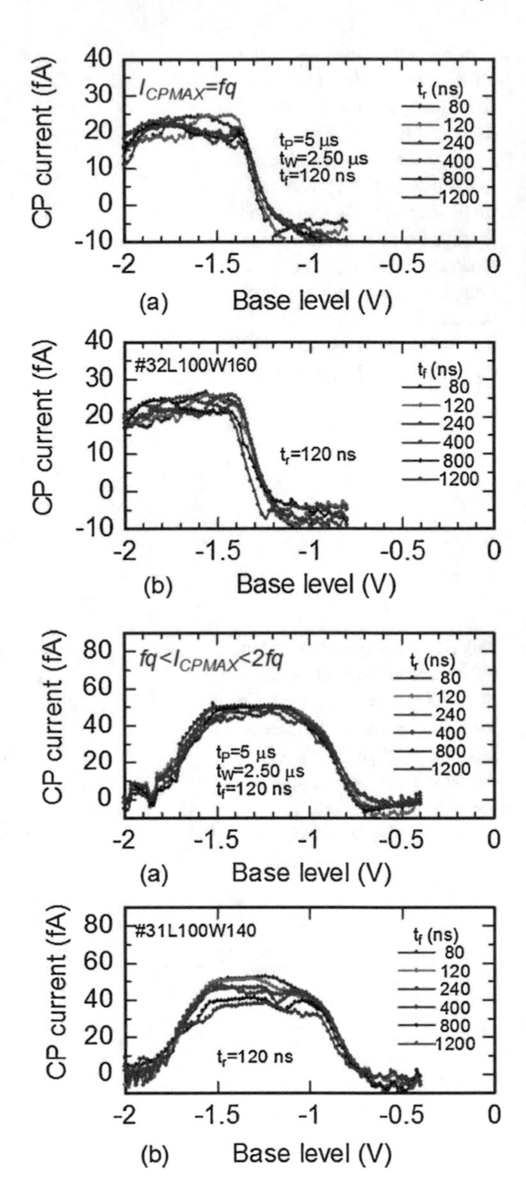

Fig. 32 Dependences of the CP characteristics upon (**a**) risetime t_r and (**b**) fall time t_f for a single interface defect. $I_{CPMAX} = fq$ and I_{CPMAX} independent of both t_r and t_f. Therefore, the defect is type 4 or 5. It is impossible to distinguish between types 4 and 5 at present

Fig. 33 Dependences of the CP characteristics upon (**a**) risetime t_r and (**b**) fall time t_f for a single interface defect. $fq < I_{CPMAX} < 2fq$ and I_{CPMAX} independent of t_r, but decreasing with increasing t_f. Therefore, the defect is type 6

Density of States of Single Interface Defects

The distribution of the energy levels of single interface defects (the density of states, DOS) can be obtained by measuring individually a large number of single defects. The results of Table 1 give us the number of energy levels in the bandgap for each defect type, as shown in Fig. 36. From the results, the density of the two energy levels of each individual interface defect can be roughly estimated, as shown in Fig. 37. Some people might accept that the DOS of the interface defects is a symmetrical

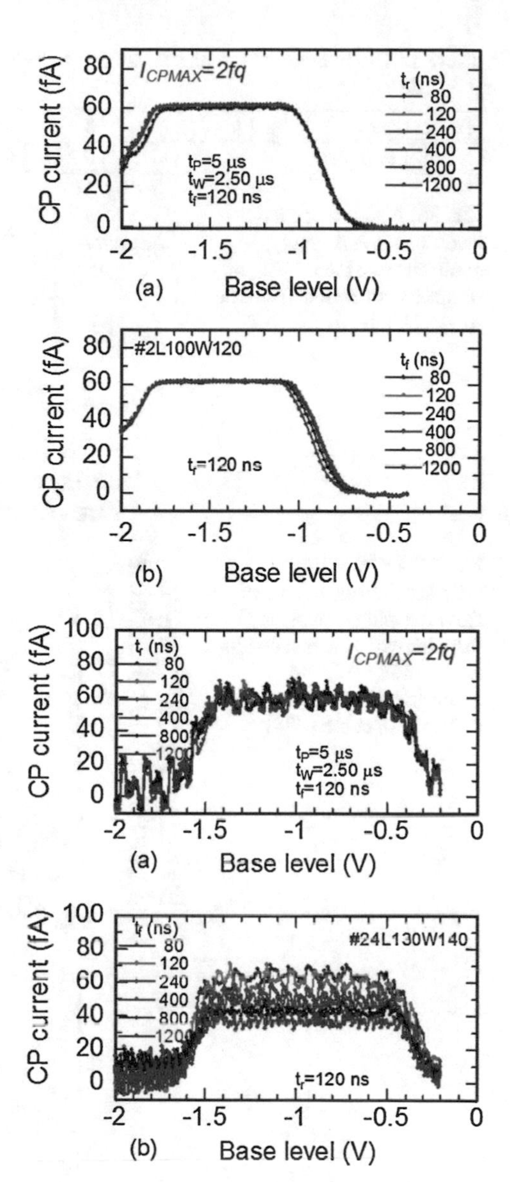

Fig. 34 Dependences of the CP characteristics upon (**a**) risetime t_r and (**b**) fall time t_f for a single interface defect. $I_{CPMAX} = 2fq$ and I_{CPMAX} independent of both t_r and t_f. Therefore, the defect is type 9

Fig. 35 Dependences of the CP characteristics upon (**a**) risetime t_r and (**b**) fall time t_f for a single interface defect. $I_{CPMAX} = 2fq$. The figure also involves some noise in the current, but, I_{CPMAX} is independent of t_r, and clearly decreases with increasing t_f. Therefore, the defect is determined to be type 6

"U-shape." However, it is clear that the DOS of individual interface defects detected by the CP method has two peaks located at approximately ±0.3 eV around the midgap. For comparison, a schematic illustration of the DOS of P_{b0} centers [40] is also shown in Fig. 37. They are reasonably similar, i.e., two peaks are located at approximately ±0.3 eV around the midgap, and the peak in the upper part of the bandgap is higher than that in the lower part (the area of each peak is the same), which strongly suggests that the origin of the interface defects measured by the CP

Table 1 The number of samples with each defect type for 27 samples showing single defect properties

Defect type	1	2	3	4 or 5	6	7	8	9
Number of defects	6	8	0	6	3	1	1	2

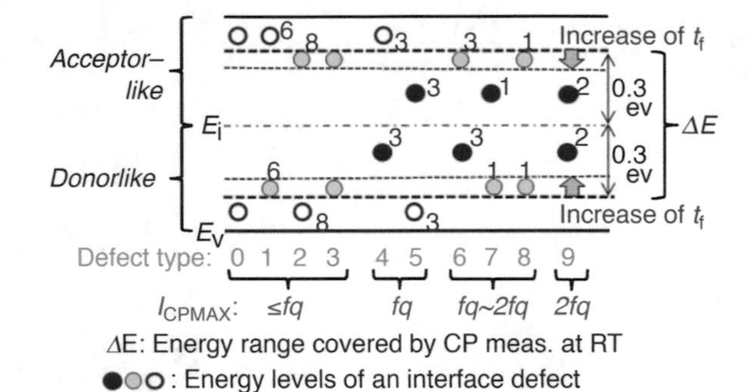

Fig. 36 Number of energy levels for each defect type, obtained from the number of samples with each defect type shown in Table 1

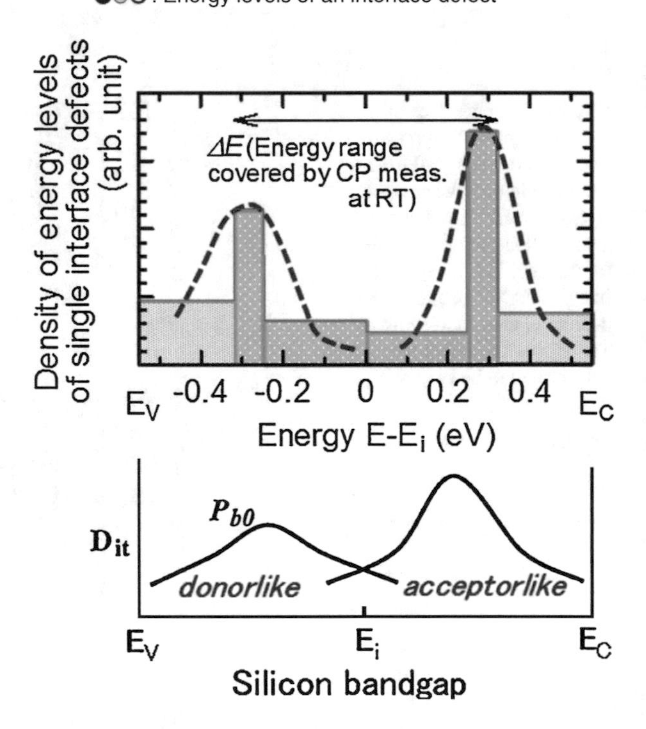

Fig. 37 DOS of single interface defects vs. energy from the midgap, estimated from the experimental results shown in Fig. 36, and schematic illustration of the DOS of P_{b0} centers [38]

method is a P_{b0} center with two energy levels, one donor-like and the other acceptor-like per center.

From this viewpoint, the DOS of interface defects reported so far in the literature [37, 42–52] were rechecked. Most studies show the changes in DOS after e-beam or γ-irradiation, or FN or BT stress for MOS structures on (111)– or (100)–Si substrates, and in those studies, the DOS is usually evaluated using the high- and

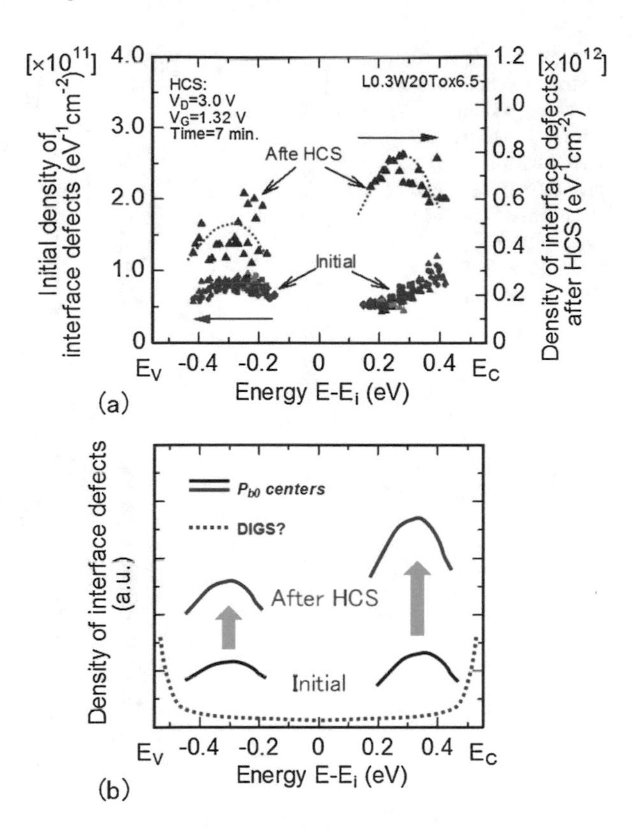

Fig. 38 (**a**) Interface defect distributions before and after hot-carrier stress (HCS) application, measured by the CP method. HCS: $V_D = 3.0$ V, $V_G = 1.32$ V, stress time = 7 min at RT. (**b**) Schematic illustrations of each component from P_{b0} centers expected from the results in (**a**), and the "U-shaped" DOS (DIGS?)

low-frequency capacitance–voltage (CV) method. Various DOS profiles have been reported including those showing clear two peaks [37]. Note that CV measurements are considered valid between $E_V + 0.2$ eV and $E_V + 0.9$ eV (E_V is the valence band edge) because of the free-carrier capacitance [50].

On the other hand, the DOS of interface defects before and after hot-carrier stress (HCS) was evaluated in this work using the CP method on nMOSFETs ($L = 0.3$ μm, $W = 20$ μm, $T_{OX} = 6.5$ nm) on a (100)–Si substrate. Measurement temperatures were 200–393 K. Hot carrier stress was applied at $V_D = 3.0$ V and $V_G = 1.32$ V for 7 min at RT. The results are shown in Fig. 38a, which were obtained on the basis of the conventional CP theory. It can be clearly seen from the figure that the two peaks at approximately $E_i \pm 0.3$ eV arose after stress application, which is consistent with the results shown in Fig. 37 based on *single* interface defects. Therefore, it is considered that even before stress application, the P_{b0} center component is more dominant than the U-shaped states (e.g., disorder-induced gap state (DIGS) model [53]), and this tendency becomes more significant owing to the generation of P_{b0} centers after stress application, as schematically illustrated in Fig. 38b. However, note that even when the P_{b0} center component is dominant, the DOS looks U-shape-like when the CP measurements are performed only at room temperature, i.e., in the range of ±0.3 eV around the midgap.

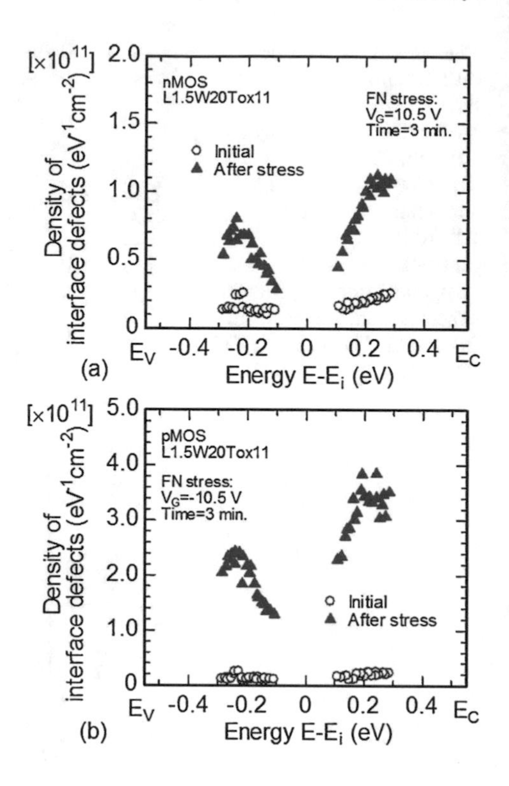

Fig. 39 (**a**) Interface defect distribution before and after Fowler–Nordheim (FN) stress application for (**a**) an nMOSFET, and (**b**) a pMOSFET, measured by the CP method. FN stress: $|V_G| = 10.5$ V, $V_D = V_S = 0$ V, stress time = 3 min at RT

Moreover, the DOS of interface defects before and after Fowler–Nordheim (FN) stress were also evaluated in this work by the CP method for nMOS and pMOSFETs with $L = 1.5$ μm, $W = 20$ μm, $T_{OX} = 11$ nm on a (100)–Si substrate. FN stress for nMOS and pMOS was applied at $|V_G| = 10.5$ V and $V_D = V_S = 0$ V for 3 min at RT. The DOS of interface defects before and after FN stress is shown in Fig. 39a for an nMOS and (b) for a pMOSFET, respectively. It can clearly be seen from the figures that the two peaks at approximately $E_i \pm 0.25$ eV arose after the FN stress, which is again consistent with Fig. 37 (the DOS of *single* interface defects) and Fig. 38 (the DOS of interface defects for an nMOSFET before and after hot-carrier stress).

In addition to the investigation of interface defects, it might be necessary to investigate the contribution of oxide defects in nMOSFETs caused by HCS. It is generally believed that hot-carrier injection into the gate oxide leads to carrier trapping in the bulk of the oxide layer, generation of interface defects (P_b centers), and in some cases generation of oxide defects (i.e., slow defects, which generate additional random telegraph noises). The origin of the oxide defects is not clear, but E' centers (i.e., oxygen-deficient silicon defects in the oxide layer) may be a possible candidate. Recently, though, it has been shown by electron spin resonance (ESR) measurements that (1) hot-electron injection generates P_b centers at the (111)–Si/SiO_2 interface, but it does not lead to any observable generation of E' centers [54], whereas (2) channel hot-hole injection in nMOSFETs generates E' centers in addition to P_{b0} centers at the (100)–Si/SiO_2 interface. Therefore, the generation of E' centers is considered to be associated with holes [55–57].

For the hot-carrier experiments in this work, hot electrons were injected, and hot-hole injection was considered to be negligible, as will be explained below. The change in the ratio of the substrate current I_{SUB} to the source current I_S during the hot-carrier stress application is shown in Fig. 40. The current ratio strongly depends on the electric field near the drain. The figure shows that the ratio increases with stress time, which indicates the generation of negative charges in the gate oxide layer near the drain, i.e., the generation of interface defects and/or trapping of the injected electrons [58–60]. Therefore, the carriers injected into the oxide layer near the drain in this experiment were almost exclusively electrons, and the number of generated E' centers associated with holes was considered to be negligible. Indeed, as shown in Fig. 41, the ratio of the maximum CP current I_{CPMAX} to the gate pulse frequency f for the stressed nMOSFET is almost independent of f in the range of 10^2–10^6 Hz, which demonstrates that the contribution of interface defects to the measured CP current is certainly more dominant than that of slow oxide defects [12, 61].

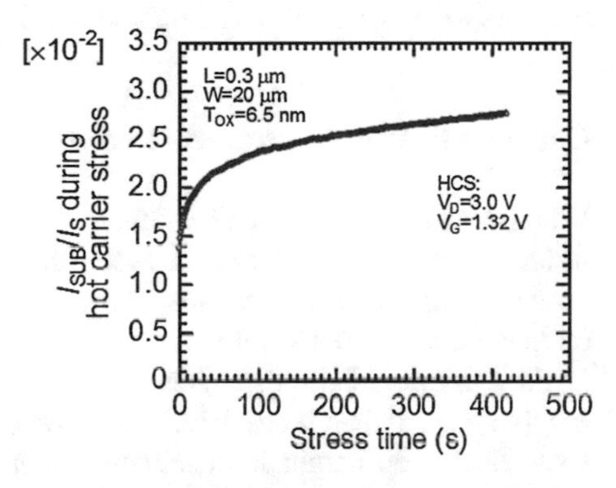

Fig. 40 Change in the ratio of the substrate current I_{SUB} to the source current I_S during the hot-carrier stress in the experiment of Fig. 38

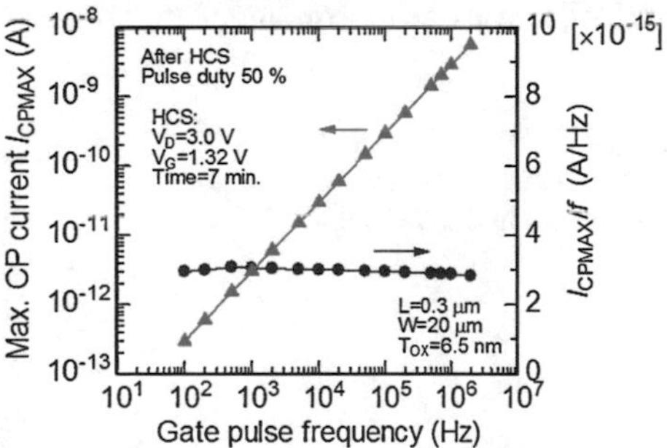

Fig. 41 The maximum CP current I_{CPMAX}, and the ratio of I_{CPMAX} to the gate pulse frequency f as a function of f for the nMOSFET after hot-carrier stress in the experiment of Fig. 38

1.4 Actual Number of Interface Defects Involved in MOSFETs

Comparison Between the Actual Number of Defects and the Values Determined by the Conventional CP Theory

By applying the systematic measurement procedure to count the number of defects, the *actual* number of defects involved in nanoscale MOSFETs was evaluated. Some results have been shown in Figs. 14b and 16. From the clear steps due to individual defects in these figures, it is found that the *actual* numbers are 5 and 4, respectively.

In the same way, the *actual* number of defects, N_T, in 40 samples was evaluated. The variation in the *actual* number is shown in Fig. 42 as a function of gate width at a constant gate length of 130 nm, which is the first example giving the variability in the *actual* number of defects. It is found from the figure that the variability is quite large. A quantitative comparison between N_T and the values determined on the basis of the conventional CP theory (i.e., the values based on I_{CPMAX}/fq) is shown in Fig. 43. They are quite different, and the values determined by the conventional CP theory are usually smaller (in some samples, markedly smaller) than the values of N_T for $N_T > 5$, which is considered to be mainly due to Coulomb interactions between defects, as will be explained next.

Coulomb Interaction Between Defects

When more than three interface defects are involved in a sample, it may be necessary to take Coulomb interactions between the defects in the electron capture/emission processes into account, because I_{CPMAX} per defect may be reduced due to the interaction for any defect type.

An example of the dependence of t_{Top} on the CP characteristics for a sample involving more than three defects is shown in Fig. 44a, where five steps are clearly seen. Each step height is considered to indicate the I_{CPMAX} from an individual defect, and it can be concluded that the correct number of interface defects involved

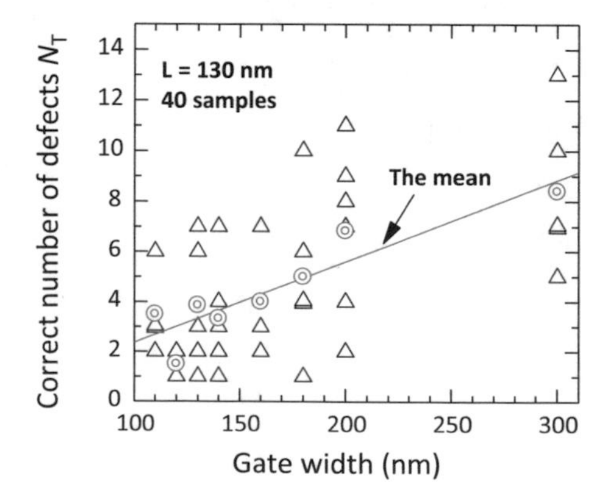

Fig. 42 Variation in the *actual* number of interface defects N_T in the MOSFETs

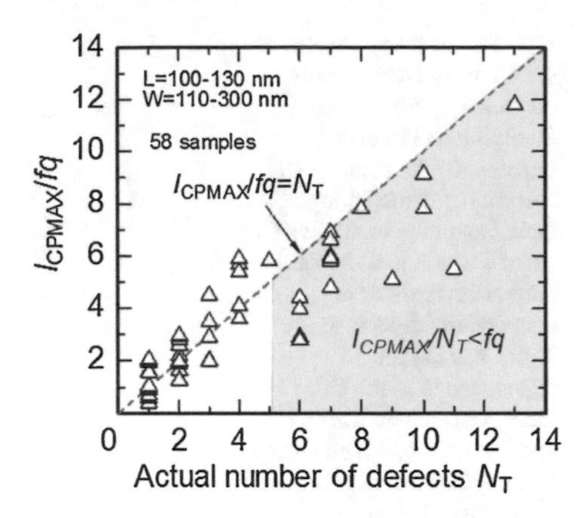

Fig. 43 Quantitative comparison between the *actual* number of defects N_T and the values determined on the basis of the conventional CP theory (i.e., I_{CPMAX}/fq)

in this sample is 5. The capture time constant of a defect for electrons τ_{cn} is given by $1/(\sigma_n v_{th} n)$, based on the Shockley–Read–Hall (SRH) model, where n is the density of electrons in the inversion layer during the top level of the gate pulse. As the base level of the gate pulse increases, the value of n increases, and thus τ_{cn} decreases. As a result, the CP current of each defect reaches its saturated value with increasing base level, because t_{Top} becomes sufficiently longer than τ_{cn}, even when t_{Top} is set at a small value. Therefore, each step height is regarded as the I_{CPMAX} from each corresponding defect.

The average CP current per defect $I_{CPMAXAV}$ and the cumulative CP current obtained from Fig. 44a are shown in Fig. 44b, as a function of the increasing number of contributing defects defined in Fig. 44a. The I_{CPMAX} of defect No. 1, which appears first, is equal to fq (i.e., no interactions), and $I_{CPMAXAV}$ monotonically decreases with the increasing number of contributing defects, which is speculated to be due to the interactions between defects. Incidentally, the dependences of the CP characteristics upon t_r and t_f are shown in Fig. 44c. It can be seen from the figure that the CP current from defect Nos. 1–4 is independent of both t_r and t_f, but only that of defect No. 5 depends upon t_f, and decreases with increasing t_f. This indicates that defect No. 5 is defect type 2 or 3, i.e., an acceptor-like defect level is located near the boundary level of ΔE. Therefore, the reason that I_{CPMAX} is less than fq for defect Nos. 2–4 is demonstrated to be due to the effect of the interactions between defects, and that for defect No. 5 is partly due to the effect of the defect location near the boundaries of ΔE, and partly to the effect of the interactions.

The relationship of $I_{CPMAX}/fq < N_T$ mentioned in the previous section means $I_{CPMAX}/N_T < fq$, i.e., the average *actual* CP current from individual defects is smaller than fq. Regarding the reason for this, besides the effect of Coulomb interactions between defects, it is necessary to consider the effect of defects located near the boundaries of ΔE. As mentioned above, such an effect can be checked by measuring the dependences of the CP characteristics on t_r and t_f, because the CP current component from a defect located near the boundaries of ΔE decreases with increasing t_r and/or t_f.

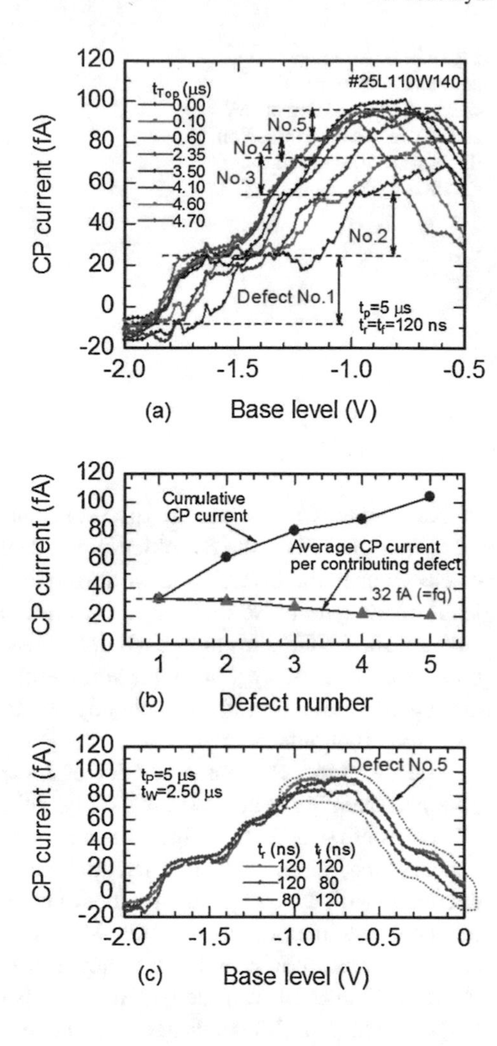

Fig. 44 (**a**) Dependences of the CP characteristics upon on-time t_{Top} for a sample involving five interface defects, (**b**) the average CP current per contributing defect and the cumulative CP current as a function of the increasing number of contributing defects defined in (**a**), and (**c**) the dependences of the CP characteristics on risetime t_{r} and fall time t_{f}, which shows that the CP current from defect Nos. 1–4 is independent of both t_{r} and t_{f}, but only the CP current from defect No. 5 decreases with increasing t_{f}

Thus, we can distinguish between the effect of the Coulomb interactions and that of the defect location near the boundaries of ΔE. The average distance between defects D_{AV} estimated by a simple assumption of $2\sqrt{(L_{\mathrm{eff}}W_{\mathrm{eff}}/\pi N_{\mathrm{T}})}$ for the samples with $N_{\mathrm{T}} \geq 2$ in Fig. 43 is shown in Fig. 45 as a function of N_{T}. From the figure, it is found that I_{CPMAX}/fq is smaller than N_{T}, and also D_{AV} is smaller than approximately 50 nm for the samples with $N_{\mathrm{T}} > 5$. This means that the CP current per defect is smaller than fq, when the distance between the defects is shorter than approximately 50 nm. It was confirmed that from the abovementioned CP current dependences on t_{r} and t_{f}, the effect of the defect location is clearly negligible at least for the samples denoted by $\bullet^{\#}$ in Fig. 45. Therefore, it is considered that Coulomb interactions between defects in the capture/emission processes are significant for D_{AV} smaller than approximately 50 nm.

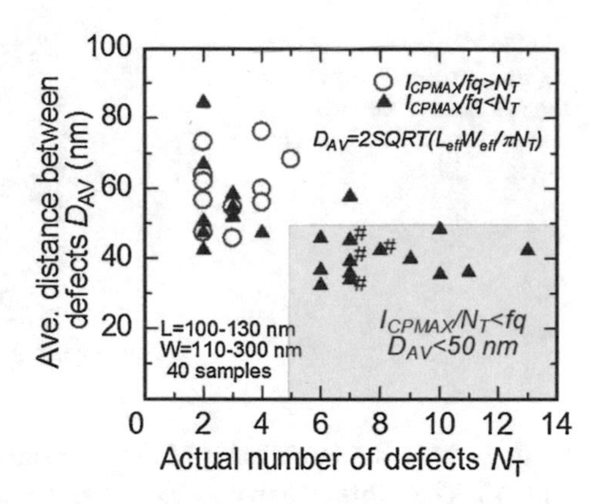

Fig. 45 Estimated average distance between defects D_{AV} as a function of the actual number of defects N_T

1.5 Fundamental Refinement of CP Theory

In the conventional CP theory based on the Shockley–Read–Hall (SRH) model [26, 27], the CP current I_{CP} for a single defect is given by

$$I_{CP} = fq\left[1 - \exp\left(-\frac{t_{Top}}{\tau_{cn}}\right)\right]\left[1 - \exp\left(-\frac{t_{Base}}{\tau_{cp}}\right)\right] \tag{7}$$

which is based on the electron capture/emission process of a defect with a *single* energy level, where τ_{cn} and τ_{cp} are the capture time constants of the electrons and holes, respectively, and t_{Top} and t_{Base} are the time duration of high-level and low-level gate pulse, respectively. Therefore, when $t_{Top} \gg \tau_{cn}$ and $t_{Base} \gg \tau_{cp}$, both the electron capture rate and hole capture (i.e., electron emission) rate become unity, then the maximum CP current I_{CPMAX} is given by

$$I_{CPMAX} = fq \text{ (constant)} \tag{8}$$

This is fundamentally incorrect. Based on the experimental results and discussions in Sect. 1.3, the CP current from a *genuine single* defect should be described by the sum of components from *two* energy levels, one in the donor-like state and the other in the acceptor-like state, as expressed by

$$\begin{aligned} I_{CP} &= \alpha_D fq\left[1 - \exp\left(-\tfrac{t_{Top}}{\tau_{cnD}}\right)\right]\left[1 - \exp\left(-\tfrac{t_{Base}}{\tau_{cpD}}\right)\right] \\ &+ \alpha_A fq\left[1 - \exp\left(-\tfrac{t_{Top}}{\tau_{cnA}}\right)\right]\left[1 - \exp\left(-\tfrac{t_{Base}}{\tau_{cpA}}\right)\right] \end{aligned} \tag{9}$$

where the subscript D and A denote donor-like and acceptor-like levels, respectively. Moreover, a factor α ($\leq$1), which depends on the positions of the two energy levels, has to be introduced, because even when $t_{Top} \gg \tau_{cn}$ and $t_{Base} \gg \tau_{cp}$, the states outsides of ΔE do not contribute to the CP current (i.e., $\alpha = 0$), and the contributions

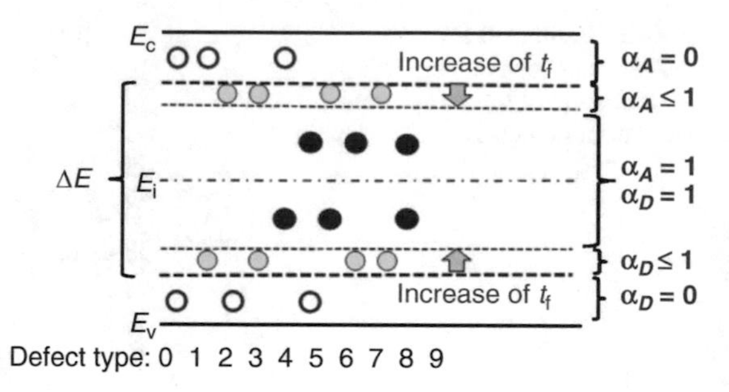

Fig. 46 Values of the factors α_A and α_D, which depend on the positions of the two energy levels of each single interface defect

of those near the boundaries of ΔE are smaller than fq (i.e., $\alpha \leq 1$), as shown in Fig. 46. Therefore, when $t_{Top} \gg \tau_{cn}$ and $t_{Base} \gg \tau_{cp}$, the maximum CP current is given by

$$I_{CPMAX} = \alpha_D fq + \alpha_A fq = 0 \sim 2fq \tag{10}$$

which means that the CP current per defect varies in the range of 0–2*fq*, depending on the positions of the two energy levels.

For multiple defects involved in MOSFETs, another factor β ($\leq$1), which depends on the Coulomb interactions between defects, has to be introduced for each state, and the CP current of multiple defects is given by

$$\begin{aligned} I_{CP} = &\sum_{i=1}^{N} \left\{ \alpha_{Di}\beta_{Di} fq \left[1 - \exp\left(-\frac{t_{Top}}{\tau_{cnDi}} \right) \right] \left[1 - \exp\left(-\frac{t_{Base}}{\tau_{cpDi}} \right) \right] \right\} \\ &+ \sum_{i=1}^{N} \left\{ \alpha_{Ai}\beta_{Ai} fq \left[1 - \exp\left(-\frac{t_{Top}}{\tau_{cnAi}} \right) \right] \left[1 - \exp\left(-\frac{t_{Base}}{\tau_{cpAi}} \right) \right] \right\} \end{aligned} \tag{11}$$

Detailed descriptions of the factors α and β or determination of their values is difficult to date; however, it is considered that Eqs. (9)–(11) fundamentally express the essential nature of the interface defects and help us understand the CP current.

2 Characterization of Individual Oxide Defects Using Charging History Effects in RTN

2.1 *Introduction*

Current noise in semiconductor devices has been the subject of investigation for a long time now [62]; however, it becomes a serious problem for recent scale-down MOSFETs [63–65]. Although noise has usually been recognized as a problem in analog circuits, it has recently become a problem in digital circuits as noise signals have become comparable to device signals [66]. Important sources of noise in semiconductor devices are thermal noise, shot noise, generation–recombination noise, and flicker (1/*f*) noise. The 1/*f* noise is the dominant type of noise in silicon MOSFETs [67], and is considered to be determined by the electronic properties near the MOS interface. Discrete drain current fluctuations, called random telegraph noise (RTN), are observed in MOSFETs under constant bias conditions and are considered to be caused by charge transport fluctuations due to the capture/emission of carriers at single oxide defects. The 1/*f* noise can be viewed as a superposition of all the electrically active RTN [68, 69].

Recently, anomalously large fluctuations in threshold voltage due to RTN in floating-gate memories and static random access memories have been reported, and it has been pointed out that RTN will become one of the greatest reliability issues in scaled-down digital devices [63, 66, 70–73]. Therefore, proper characterization of the oxide defects is important in order to understand the RTN phenomenon and reduce its influence on device reliability.

In this section, a novel method is proposed to characterize the oxide defects that contribute to RTN by using the effects of the charging history on the defects [74]. In this method, the variation in the frequency of the high/low drain current derived from RTN with the charging history is monitored instead of the timescale parameters, such as the capture/emission times, that are usually used. Moreover, another method is also proposed to determine the number and charging conditions of the defects. These methods are particularly effective for multi-defect RTN, which is quite difficult to characterize using the conventional method.

2.2 *Evaluation of Individual Oxide Defects Using Drain Current Histograms*

Validity of the Histogram Method

Polycrystalline–silicon–gate nMOSFETs were used in this study. Gate pattern lengths L_p and gate pattern widths W_p of mainly used devices were 110–130 and 140–180 nm, respectively. The actual gate lengths L_G measured by scanning electron microscopy (SEM) were approximately $L_G = L_P$–55 nm, and the electrically measured effective channel widths W_{eff} were approximately $W_{eff} = W_p$–40 nm. The thickness of the gate oxide T_{OX}, grown by rapid thermal oxidation, was 4 nm.

RTN characteristics were measured at a drain voltage $V_D = 1.0$ V with various gate voltages V_G. All the measurements were carried out at room temperature.

Firstly, the validity of a method to evaluate RTN using drain current I_D histograms derived from single-defect RTN waveforms is demonstrated. Typical single-defect RTN waveforms are shown in Fig. 47a–c for three values of V_G. Figure 47d–f is drain current histograms obtained by dividing the x-axis of the corresponding RTN waveforms by $\Delta t = 1$ ms. The ratio of the average capture to emission time, τ_c/τ_e, obtained from Fig. 47a–c, is shown in Fig. 48 as a function

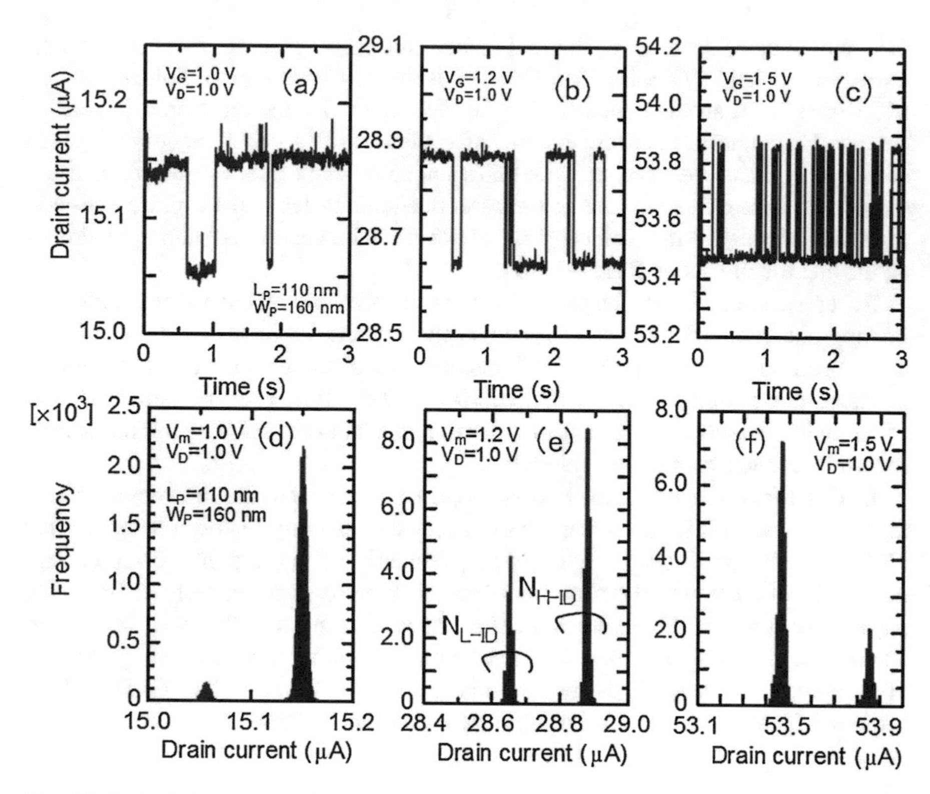

Fig. 47 Typical single-defect RTN characteristics. (**a–c**) Waveforms, (**d–f**) drain current histograms obtained from the waveforms

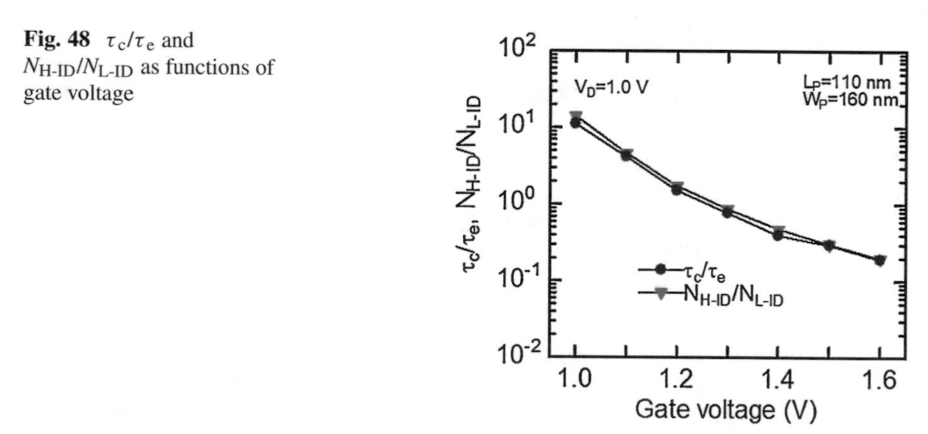

Fig. 48 τ_c/τ_e and $N_{H\text{-}ID}/N_{L\text{-}ID}$ as functions of gate voltage

of V_G. The ratio of the frequencies of high to low current, $N_{H\text{-}ID}/N_{L\text{-}ID}$, obtained from Fig. 47d–f, is also shown in Fig. 48. As can be seen, they agree precisely.

The defect temporal properties are determined by tunneling between the carrier reservoir and the defect itself, as well as internal thermal barriers of the defect, related to atomic reconfiguration based on the switching oxide defect model [75–78]. However, from the dependence of τ_c/τ_e upon V_G, the defect position X_T (the distance from the Si/SiO$_2$ interface), and the defect energy level $E_{T0} - E_F$ (E_F: the Fermi energy) can be roughly extracted using the following equations [65], although these are not sufficient for identifying the nature of the defect.

$$X_T/T_{OX} = -(kT/q)\,\Delta \ln(\tau_c/\tau_e)\,/\Delta V_G \quad (12)$$

$$E_{T0} - E_F = q\,(V_{G0} + V_0)\,X_T/T_{OX} \quad (13)$$

where V_{G0} is defined as the value of V_G at $\tau_c = \tau_e$, and $V_0 = -V_{FB} - \phi_S$ (V_{FB} is the flat band voltage, and ϕ_S is the surface potential at inversion). Therefore, the values of X_T and $E_{T0} - E_F$ can be obtained using $N_{\text{H-ID}}/N_{\text{L-ID}}$ instead of τ_c/τ_e from these equations.

Application to Two-Defect RTN: Simplicity and Usefulness of the Histogram Method

The method using histograms was applied to two-defect RTN waveforms, and its validity was confirmed as follows: two-defect waveforms for three values of V_G are shown in Fig. 49a–c, with corresponding histograms being shown in Fig. 49d–f. Each current level is labeled a–d in the figures. Considering the magnitudes of the current levels and their V_G dependences, the relationship between the current levels and the occupied/unoccupied states (charging conditions) of the two defects (named defect A and defect B, where $E_{TA} > E_{TB}$) can be determined as shown in Table 2. Here, it should be noted that in Fig. 49a–c, the transition between levels a and b (i.e., repeated electron capture and emission in defect A under an *unoccupied* defect B) is dominant, while the transition between levels c and d (i.e., repeated electron capture and emission in defect A under an *occupied* defect B) clearly appears to be held separate on the time axis from the transition between the levels a and b. This separation is based on the fact that τ_c and τ_e for defect B are very much larger than those for defect A.

Therefore, it is possible to independently evaluate the electron capture/emission processes in defect A under an *unoccupied* defect B (process A1) from N_b/N_a, and those under an *occupied* defect B (process A2) from N_d/N_c. The V_G dependence of the frequency ratio for each process is shown in Fig. 50. It is clear from the figure that the processes in defect A are affected by the electrical condition of defect B, which indicates the existence of an interaction between the oxide defects. X_T and $E_{T0} - E_F$ of defect A for each of the processes, A1 and A2, can be derived from Eqs. (12) and (13), using the frequency ratios instead of τ_c/τ_e, and are as follows (assuming $V_{FB} = -1$ V and $\phi_S = 0.9$ V, therefore $V_0 = -V_{FB} - \phi_S = 0.1$ V):

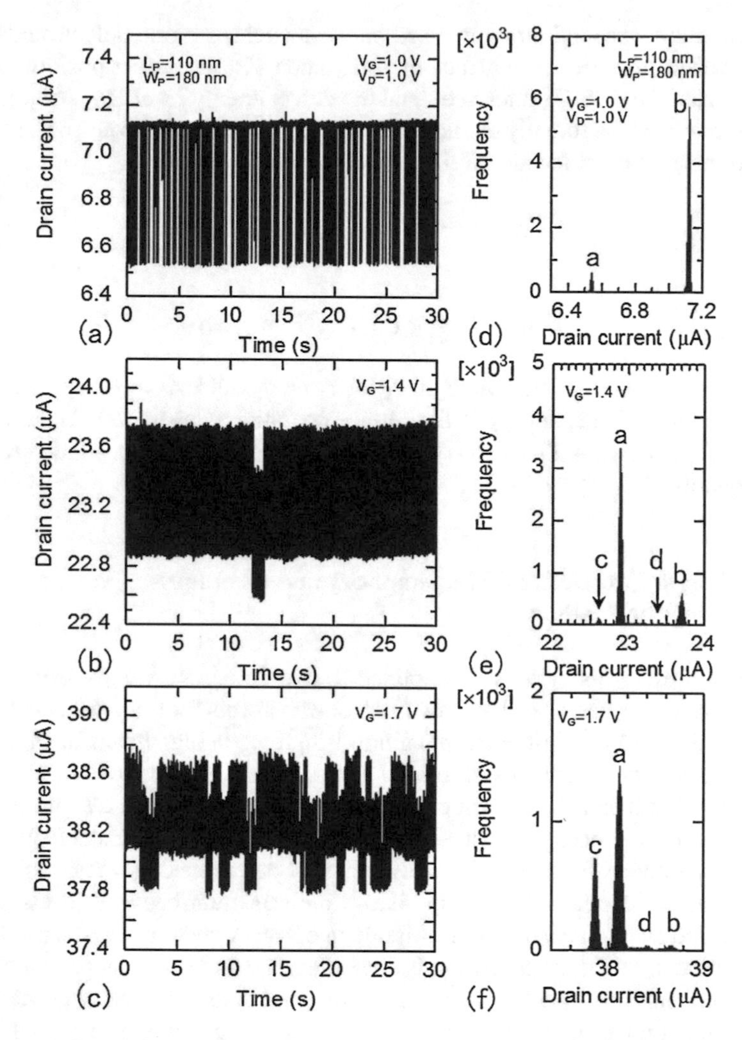

Fig. 49 (**a–c**) Two-defect RTN waveforms for three values of V_G, (**d–f**) corresponding drain current histograms

Table 2 Correspondence between each current level and charging conditions of the two defects A and B

Current level	Defect A	Defect B
a	●	○
b	○	○
c	●	●
d	○	●
	$E_{TA} > E_{TB}$	

•: occupied by an electron, ○: empty

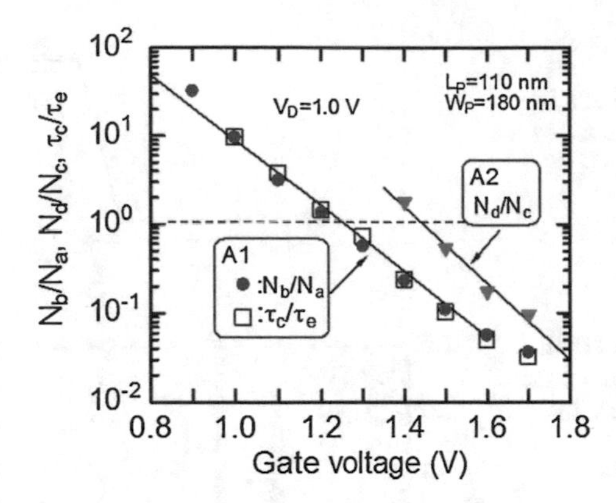

Fig. 50 N_b/N_a and N_d/N_c as functions of V_G for processes A1 and A2. τ_c/τ_e is also shown for process A1 only, since it was impossible to obtain the values for process A2 from the RTN waveforms, because the duration of process A2 was too short. However, N_d/N_c can be easily obtained even for process A2. A1: electron capture/emission process in defect A under an *unoccupied* defect B. A2: electron capture/emission process in defect A under an *occupied* defect B

$$\text{A1}: X_T = 0.97\ \text{nm},\quad E_{T0} - E_F = 0.32\ \text{eV},\quad \text{A2}: X_T = 1.03\ \text{nm},$$
$$E_{T0} - E_F = 0.40\ \text{eV}.$$

When defect B is occupied by an electron, the increases in $E_{T0} - E_F$ and X_T of defect A are 0.08 eV and 0.06 nm, respectively. These values may be inaccurate, but they indeed readily verify the existence of an interaction between the oxide defects at room temperature. Such Coulomb repulsive interaction between oxide defects has already been reported [79–81], but the histogram method can easily verify the existence of this interaction, which demonstrates the validity of the method.

2.3 *Charging History RTN Method to Characterize Individual Oxide Defects*

The sequence of gate voltages V_G used in a novel method (named the Charging History (CH) RTN method) is shown in Fig. 51. After accumulation at V_a for a period t_a (s), V_i is applied to the gate for a period t_i (s), representing an RTN period that will be evaluated in the following period at V_m. The history of the charging conditions of the oxide defects at the end of the period at V_i (i.e., the charging history) is evaluated from the drain current histogram at the beginning of the period at V_m. The charging conditions at the end of the period at V_i should be determined by the probabilities based on τ_c and τ_e. Therefore, the drain current histograms at the beginning of the period at V_m, obtained from repetitive measurements of

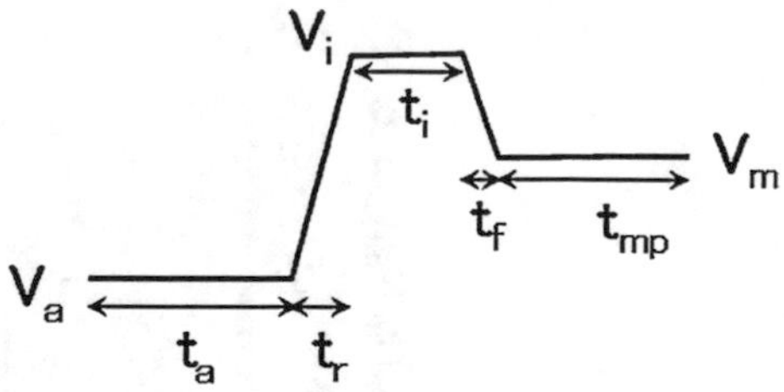

Fig. 51 Sequence of gate voltages in the novel characterization method using the charging history of oxide defects contributing to RTN

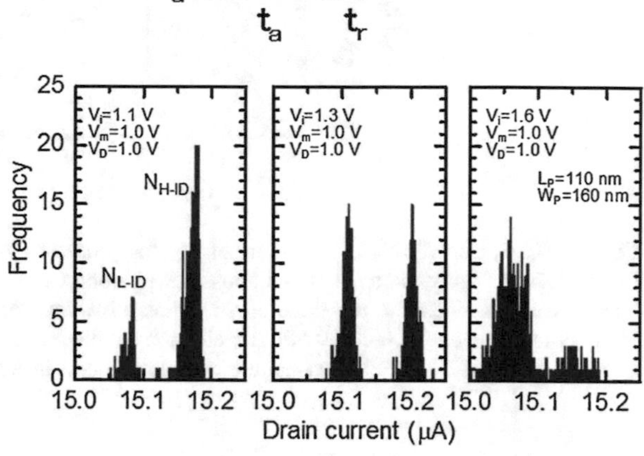

Fig. 52 Histograms of drain current at the beginning of the period at V_m for three values of V_i. $V_m = 1.0$ V, 512 measurements for each V_i

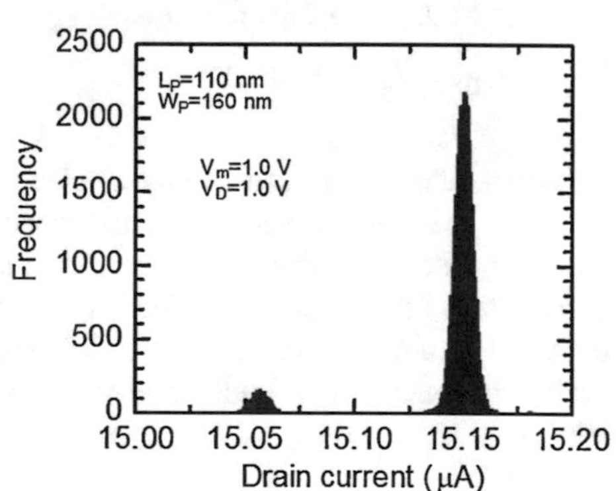

Fig. 53 Drain current histograms obtained from an RTN waveform ordinarily measured at $V_G = 1.0$ V (which corresponds to V_m in Fig. 52)

the V_G sequence, are considered to be equivalent to the probabilities based on τ_c and τ_e. Examples of such histograms for three values of V_i are shown in Fig. 52 at $V_m = 1.0$ V, which were obtained from 512 measurements for each V_i. These histograms are quite different from those obtained from RTN waveforms that are ordinarily measured at $V_G = 1.0$ V (which corresponds to V_m in Fig. 52), as shown in Fig. 53. The differences are due to electron trapping during the period at V_i.

The ratio of the frequency of high to low current, $N_{H\text{-}ID}/N_{L\text{-}ID}$, obtained from Fig. 52 is shown in Fig. 54 as a function of V_i, with the dependence of τ_c/τ_e upon V_G obtained from RTN waveforms measured in the conventional way. They agree quite well, which demonstrates the validity of the CH-RTN method. Therefore, we can easily estimate the defect position and the energy level using the dependence of $N_{H\text{-}ID}/N_{L\text{-}ID}$ upon V_i instead of the dependence of τ_c/τ_e upon V_G. Moreover, the

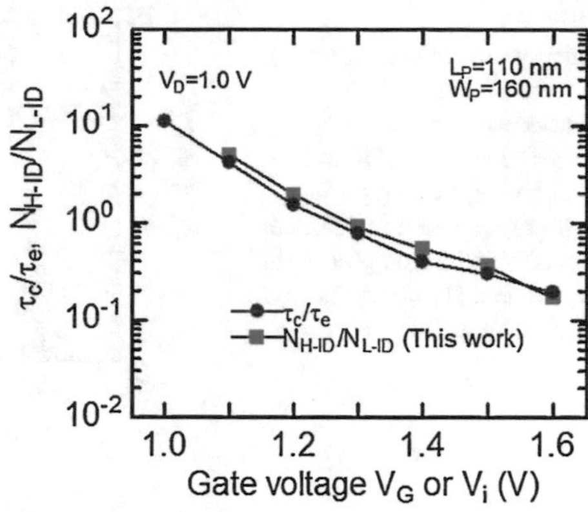

Fig. 54 The dependence of $N_{\text{H-ID}}/N_{\text{L-ID}}$ at the beginning of the period at V_m upon V_i, and the dependence of τ_c/τ_e upon V_G obtained from conventional RTN waveforms

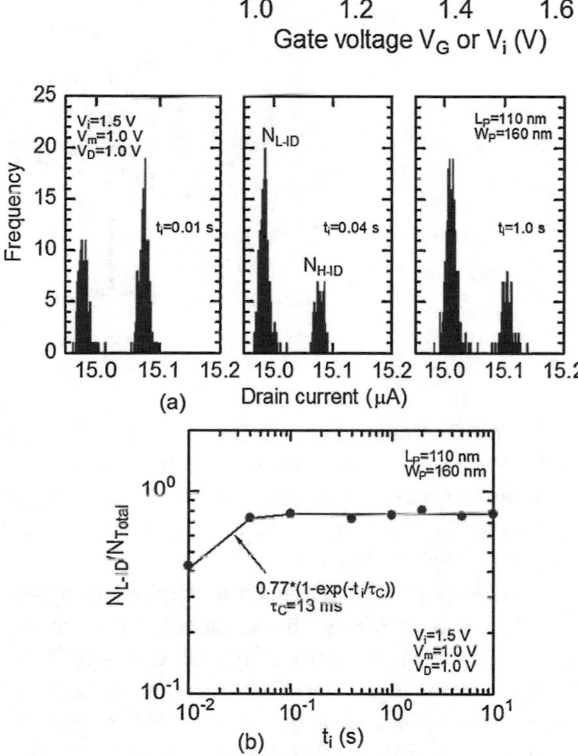

Fig. 55 (**a**) Drain current histograms at the beginning of the period at V_m for various t_i, and (**b**) the dependence of $N_{\text{L-ID}}/N_{\text{Total}}$ upon t_i, which indicates an electron capture process in the defect ($V_i = 1.5$ V, $V_m = 1.0$ V)

electron capture and emission processes in the defect can also be evaluated from the dependences of $N_{\text{L-ID}}/N_{\text{Total}}$ upon t_i and upon t_{mp}, as shown in Figs. 55b and 56b, respectively. More details on the electron capture process will be described later.

As shown above, repetitive measurements are necessary for the CH-RTN method to obtain the drain current histograms. So, it was investigated how many times the repetitive measurements are required in order to obtain appropriate histograms. The changes in the histograms at the beginning of the period at V_m with an increasing number of measurements from 128 to 4096 are shown in Fig. 57a for $V_i = 0.7$ V, $V_m = 0.3$ V. There are three main histograms (low, middle, and high, as shown in

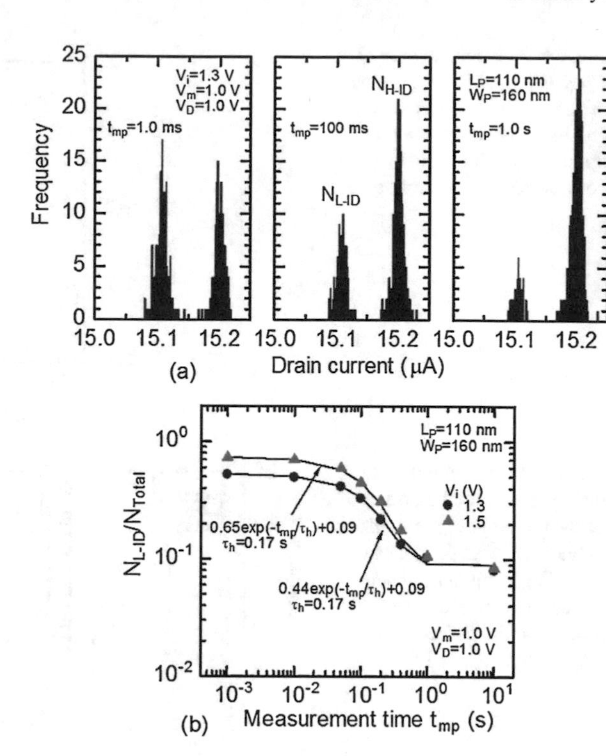

Fig. 56 (**a**) Drain current histograms at the beginning of the period at V_m for various t_{mp} ($V_i = 1.3$ V, $V_m = 1.0$ V) and (**b**) the dependence of $N_{L\text{-}ID}/N_{Total}$ upon t_{mp}, which indicates an electron emission process in the defect ($V_i = 1.3$ and 1.5 V, $V_m = 1.0$ V)

the figure). The ratio of each to the total number is shown in Fig. 57b as a function of the number of measurements. From the figure, it is found that 512 measurements are sufficient, and more quantitatively we need repetitive measurements that give at least a count of 10 even for the drain current level indicating the lowest frequency (i.e., $N_{High} > 10$ in Fig. 57).

Next, the CH-RTN method was applied to a two-defect (named defect α and β) RTN, as a relatively simple case. It will be explained how to evaluate the contributing oxide defects through the example. The variation in the drain current histogram at the beginning of the period at V_m with respect to V_i is shown in Fig. 58. There are two states, c and d at $V_i = 1.0$ V (i.e., two-level RTN), and two more states a and b appear at $V_i = 1.8$ V (i.e., four-level RTN). Schematic time lag plots [81] for the sample at $V_i = 1.8$ V are shown in Fig. 59. Judging from the transitions between the current levels, the charging conditions of the two defects α and β for each state are determined as shown in Table 3. Since the histogram of state b, N_b, is negligibly small at $V_i < 1.8$ V, the transition probability between states a and b, and also that between states b and d are considered to be negligible. Therefore, it is assumed that N_d/N_c and N_c/N_a correspond to τ_c/τ_e for defects α and β, respectively. The dependences of N_d/N_c and N_c/N_a upon V_i are shown in Fig. 60. From the figure and

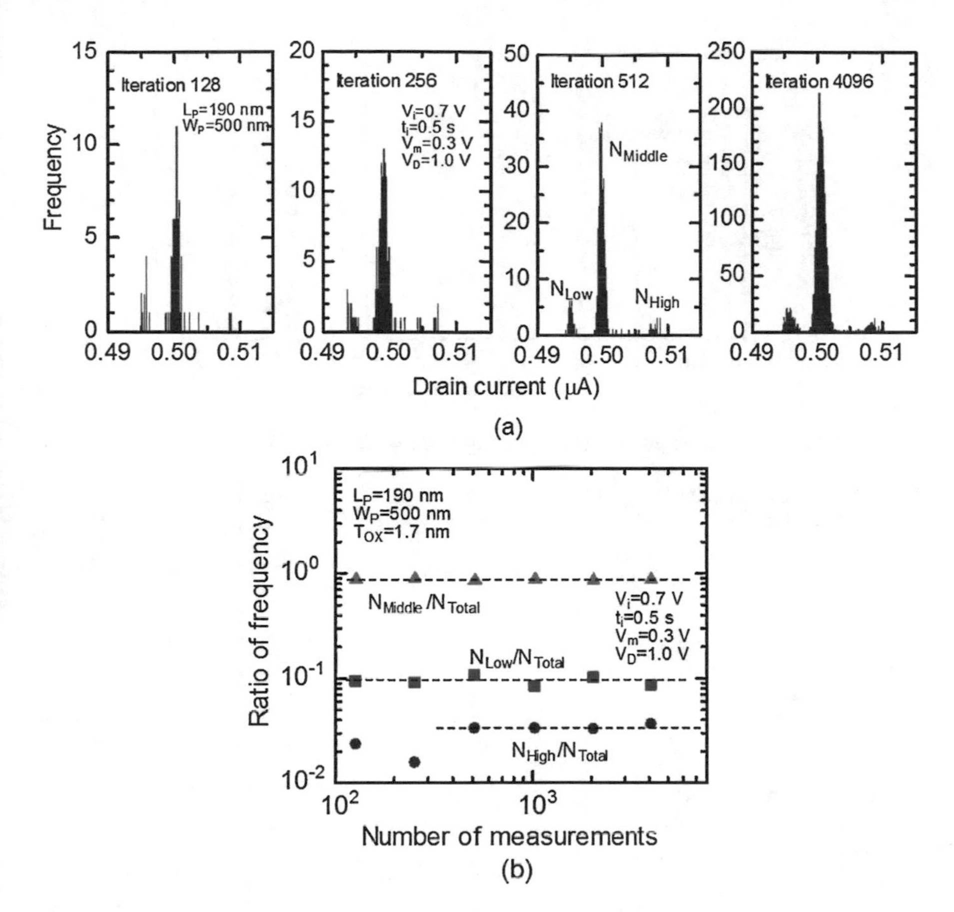

Fig. 57 (**a**) Variation in the drain current histograms at the beginning of the period at V_m due to the change in the number of measurements and (**b**) the frequency ratio for each current level as a function of the number of measurements ($V_i = 0.7$ V, $V_m = 0.3$ V)

Eqs. (12) and (13), the values of X_T and $E_{T0} - E_F$ for each defect are calculated to be as follows: defect α: $X_T = 0.39$ nm, $E_{T0} - E_F = 0.11$ eV, defect β: $X_T = 1.19$ nm, $E_{T0} - E_F = 0.58$ eV.

Thus, the defect position and energy level can be estimated from the histograms at the beginning of the period at V_m using the CH-RTN method.

Moreover, electron capture processes in the contributing defects can be evaluated by the CH-RTN method. It is found from Table 3 that the electron capture rates for defects α and β can be evaluated from $(N_a + N_c)/N_{Total}$ and $(N_a + N_b)N_{Total}$, respectively. Drain current histograms at the beginning of the period at V_m for various values of t_i are shown in Fig. 61. From the figures, the dependences of $(N_a + N_c)/N_{Total}$ and $(N_a + N_b)N_{Total}$ upon t_i (i.e., electron capture processes in

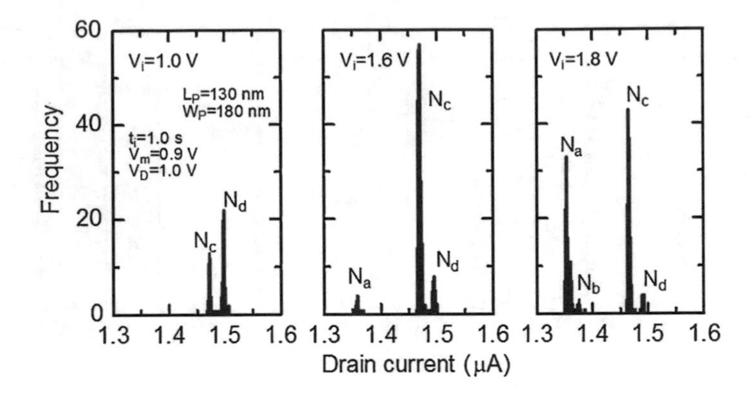

Fig. 58 Drain current histograms at the beginning of the period at V_{m} for various V_{i}

Fig. 59 Schematic time lag plots with $V_{\mathrm{i}} = 1.8$ V

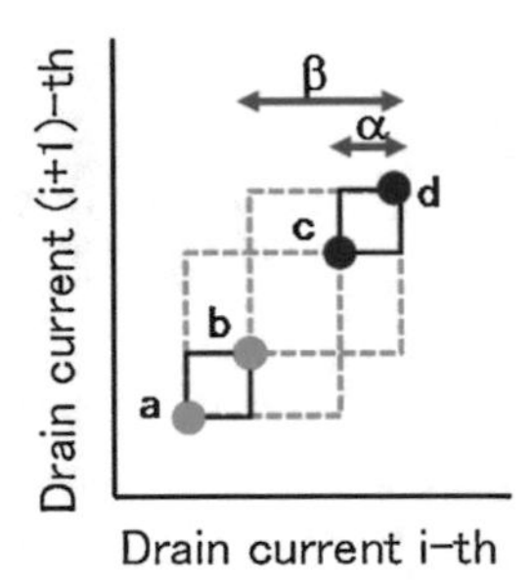

Table 3 Charging conditions of the two defects α and β for each state

State	Defect α	Defect β
d	○	○
c	●	○
b	○	●
a	●	●

Defect α : $\tau_c/\tau_e = N_d/N_c$
Defect β : $\tau_c/\tau_e = N_c/N_a$

•: occupied by an electron, ○: empty

defects α and β) are obtained as shown in Fig. 62, and the electron capture times for defects α and β can be derived as 40 μs and 0.2 s, respectively. For this sample, $\tau_{\mathrm{c}} = 38$ μs was obtained from the RTN waveform at $V_{\mathrm{G}} = 1.8$ V for defect α only using the ordinary method, which agrees well with the value obtained from the CH-RTN method.

Thus, the CH-RTN method is a promising way to characterize individual oxide defects contributing to RTN.

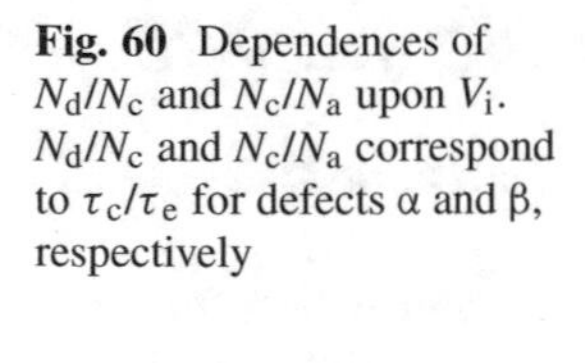

Fig. 60 Dependences of N_d/N_c and N_c/N_a upon V_i. N_d/N_c and N_c/N_a correspond to τ_c/τ_e for defects α and β, respectively

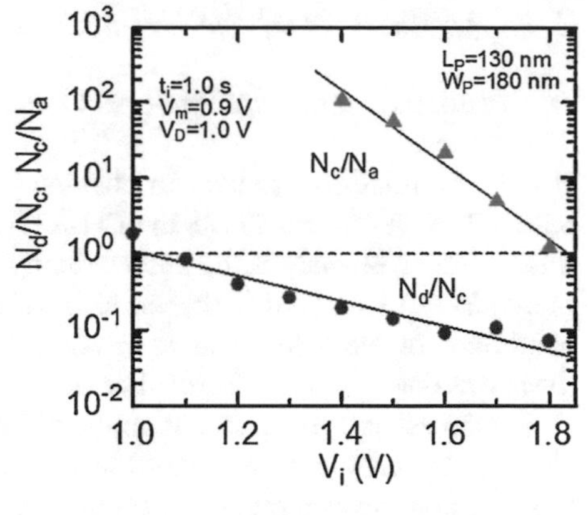

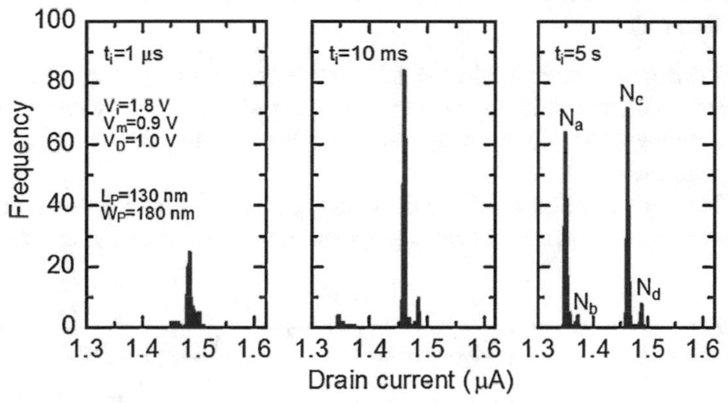

Fig. 61 Drain current histograms at the beginning of the period at V_m for various t_i

Fig. 62 Dependences of $(N_a + N_c)/N_{Total}$ and $(N_a + N_b)N_{Total}$ upon t_i, which indicate the electron capture processes in defects α and β, respectively

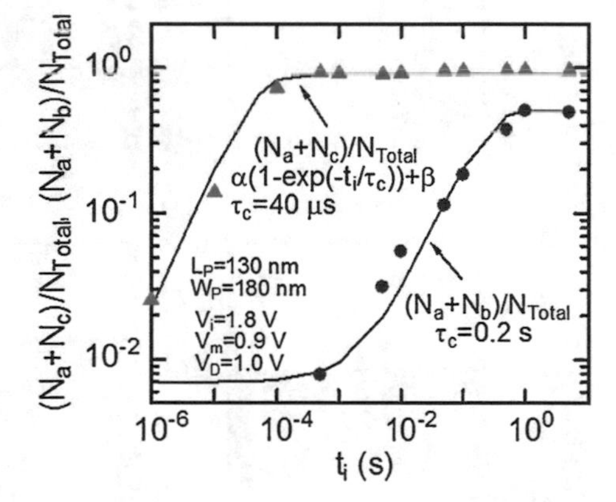

2.4 Application of the CH-RTN Method to Multi-Defect RTN

Determination of the Number and Charging Conditions of Oxide Defects

It is very difficult in general to characterize oxide defects contributing to multi-defect RTN. As shown above, the CH-RTN method has great potential in achieving this. In order to analyze the defects using the CH-RTN method, it is a necessary prerequisite to determine the number of participating defects and the charging conditions of the defects for each current level in RTN. A method to determine these was developed as described next.

Examples of V_G-dependent multi-defect RTN waveforms are shown in Fig. 63. It is not easy even to count the number of current levels. Using the sample, the procedure to determine the number and charging conditions of the defects is described as follows.

Step 1: Measure the variation in the RTN waveform with V_G (Fig. 63), and draw detailed drain current histograms and time lag plots for each value of V_G (Fig. 64). Semilog plots for the histograms are better than linear ones in order to recognize each peak.

Step 2: From the detailed V_G-dependent histograms and time lag plots, determine the observed current levels (states a–g) and transitions between any two states,

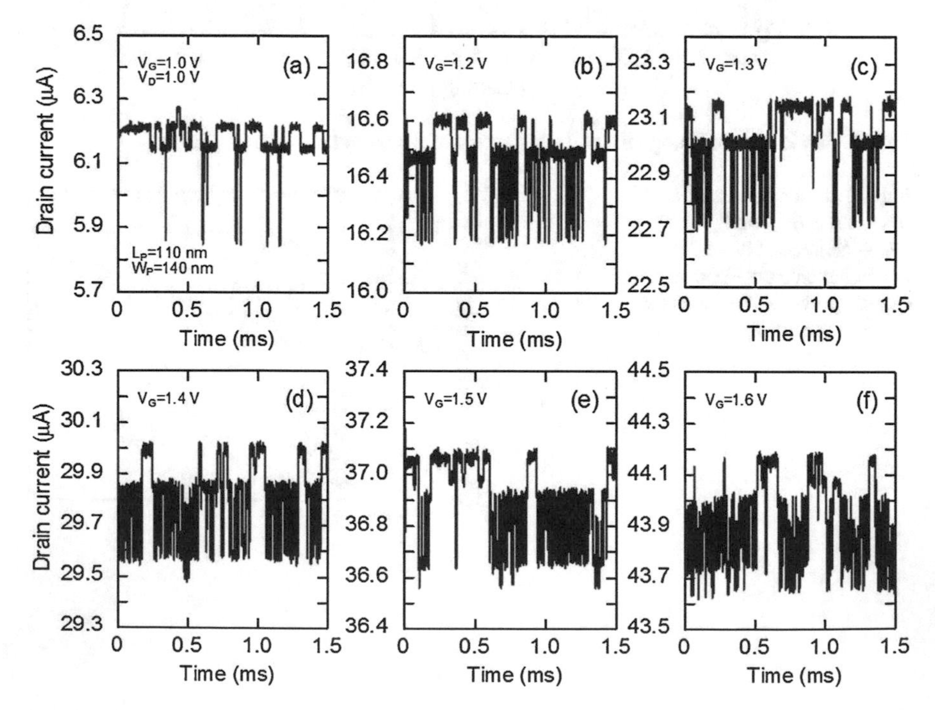

Fig. 63 Multi-defect RTN waveforms. V_D = 1.0 V, and V_G = (**a**) 1.0 V, (**b**) 1.2 V, (**c**) 1.3 V, (**d**) 1.4 V, (**e**) 1.5 V, (**f**) 1.6 V

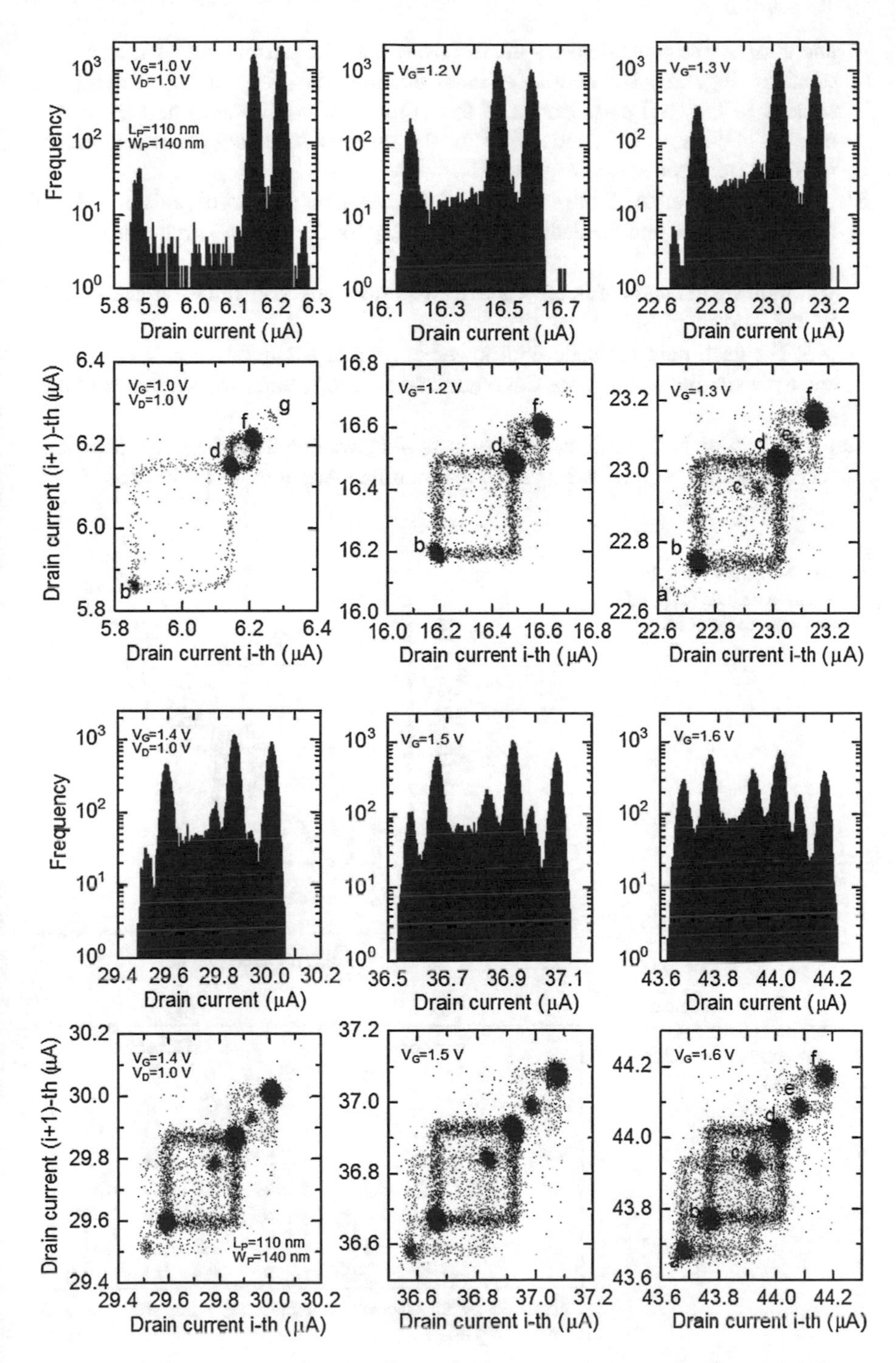

Fig. 64 Drain current histograms and time lag plots for various values of V_G. $V_G = 1.0$–1.6 V

and classify the transitions by the differences in the current between each pair of states (Fig. 65). The number of classifications (i.e., the number of transition squares in Fig. 65) corresponds to the number of contributing defects m (the number of states M is 7, and $m = 3$ for the sample). The value of m can also be obtained from $2m - 1 < M \leq 2m$ [71].

Step 3: Make up a table (Table 4a) with the states a–g on the vertical axis in order of the current level, and the defect names (α, β, γ for the sample) on the horizontal axis.

Step 4: Indicate the pairs of states where transitions were observed using the symbol] (Table 4a).

Step 5: For each pair, the state with lower current is occupied by an electron, so enter the symbol •, and since the other one is empty, enter the symbol ∘ (Table 4a).

Step 6: Other defects in each pair of states should have identical charging conditions for each pair; therefore, enter the corresponding symbol ∘ or • (Table 4b).

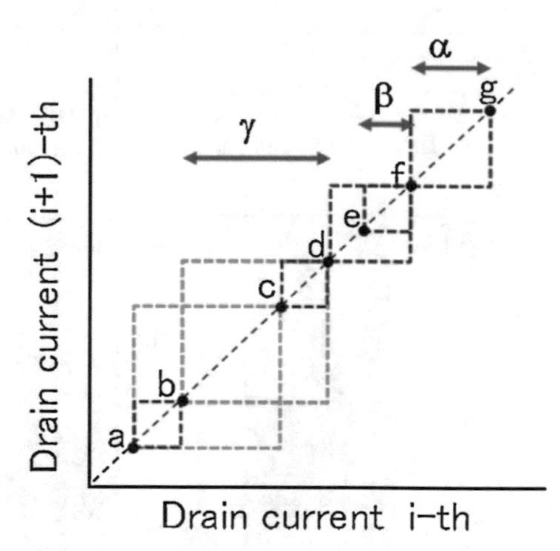

Fig. 65 Schematic time lag plots. Three defects (α, β, and γ) contribute to the RTN

Table 4 Correspondence between each state and charging conditions of the three defects

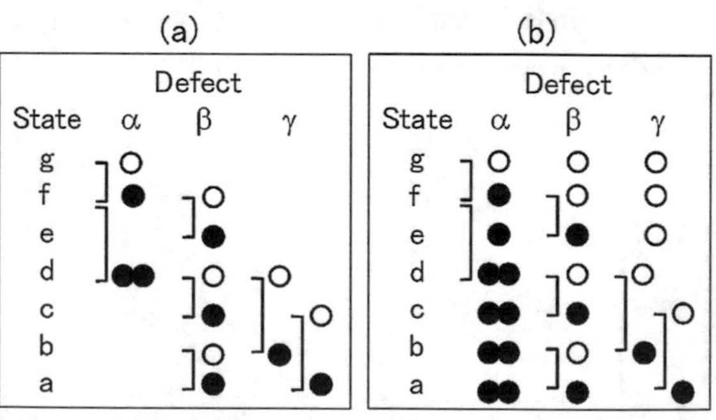

(a)

State	α	β	γ
g	○		
f	●	○	
e		●	
d	●●	○	○
c		●	○
b		○	●
a		●	●

(b)

State	α	β	γ
g	○	○	○
f	●	○	○
e	●	●	○
d	●●	○	○
c	●●	●	○
b	●●	○	●
a	●●	●	●

•: occupied by an electron, ∘: empty. (a) After step nos. 3–5. (b) After step no. 6

As mentioned above, the number of participating defects and the charging conditions of the defects for each current level in multi-defect RTN can be systematically and completely determined.

It will be necessary to briefly explain the reason why defect α is judged as being a double-charged defect (or a metastable defect) [68, 80, 82], as shown in Table 4. A feature of defect α is that the difference in current level between states g and f is identical to that between states f and d. There are two other possible models to explain such behavior by introducing one more defect δ. One model is that the two defects α and δ cause an identical change in drain current [83]. The other model is that the two defects α and δ are adjacent, and the trapping of incoming electrons by a defect is prevented as a result of the Coulomb repulsive force due to another occupied defect [84]. However, it should be noted that state g disappears with increasing V_G, although states f and d are clearly seen, as shown in Fig. 64. State g can be explained by either model if both α and δ are empty, as shown in Tables 5 and 6. For the former model, state f corresponds to two conditions: one is that α is occupied and δ is empty, and the other is that α is empty and δ is occupied, as shown in Table 5. For the latter model, on the other hand, states f and d correspond to α or δ being occupied and δ or α being empty, respectively, as shown in Table 6. Therefore, even when V_G increases, there is a sufficient possibility that the defects α and δ are empty, which contradicts the fact that state g, which corresponds to both α and δ being empty, disappears. Therefore, the two models are not applicable for the sample.

Thus, a detailed consideration of the change in states with V_G is important in order to determine the charging conditions of the defects.

To verify this method, it was applied to another sample. Further examples of multi-defect RTN waveforms are shown in Fig. 66, and the variations in the drain current histogram and the time lag plot with V_G are shown in Fig. 67. Schematic

Table 5 Correspondence between each state (g, f, d) and charging conditions of the defects α and δ, assuming that the two defects α and δ cause an identical change in drain current

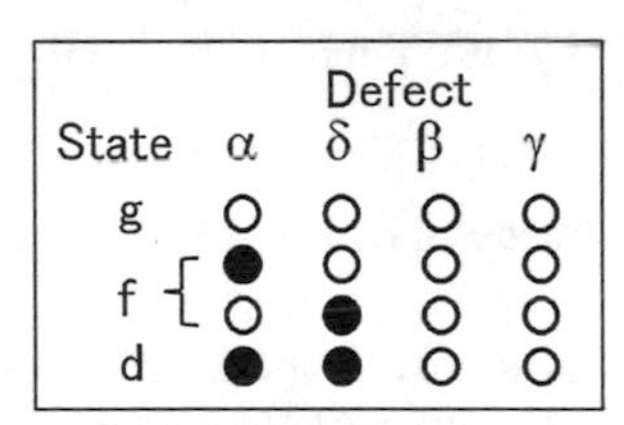

State	Defect α	δ	β	γ
g	○	○	○	○
f	●	○	○	○
	○	●	○	○
d	●	●	○	○

•: occupied by an electron,
o: empty

Table 6 Correspondence between each state (g, f, d) and charging conditions of the defects α and δ, assuming a Coulomb repulsive force between the two defects α and δ

State	Defect α	δ	β	γ
g	○	○	○	○
f	●	○	○	○
d	○	●	○	○

•: occupied by an electron,
o: empty

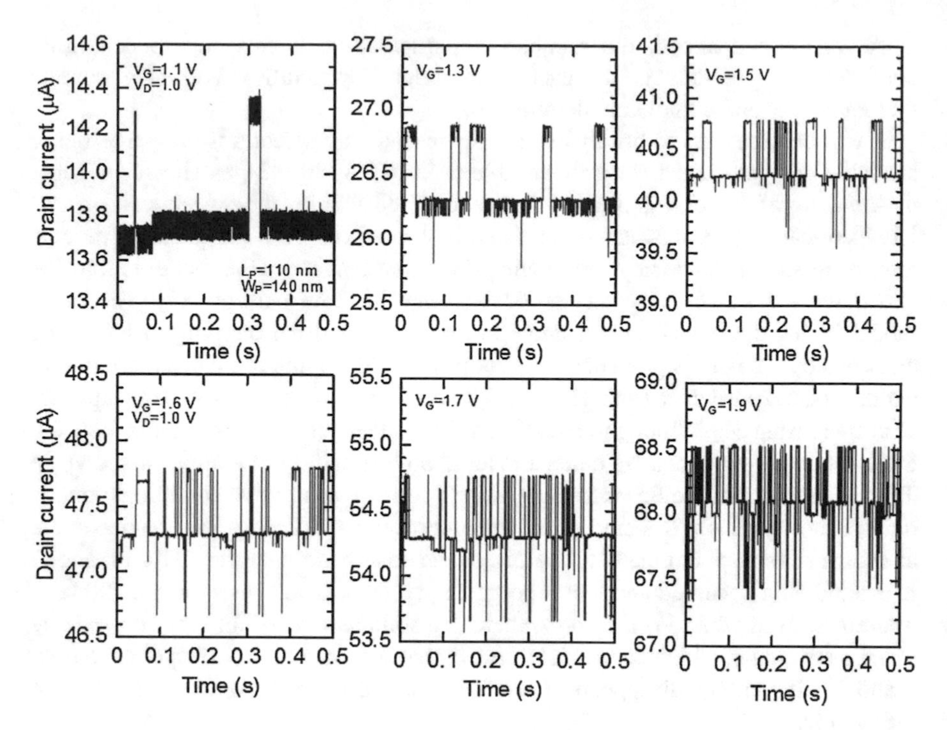

Fig. 66 Further examples of multi-defect RTN waveforms. $V_G = 1.1$–1.9 V, $V_D = 1.0$ V

time lag plots and the transition squares obtained from Fig. 67 are shown in Fig. 68 for three different values of V_G. It is found that there are three defects participating in the RTN. Following steps 3–5 of the procedure, the correspondence between each state and the charging conditions of the three defects are obtained as shown in Table 7. For the sample, the determination procedure has been completed without step 6.

Characterization of Each Individual Oxide Defects in Multi-Defect RTN

It is almost impossible to characterize individual oxide defects participating in multi-defect RTN by conventional timescale analysis. In this section, the usefulness and effectiveness of the CH-RTN method for analyzing individual defects will be shown using data obtained from the same sample shown in Figs. 63–65, which is a three-defect RTN. In the sample, the number of participating defects and the charging conditions of the defects for each current level have already been clarified.

Drain current histograms obtained from an RTN waveform, which was ordinarily measured at $V_G = 1.0$ V, are shown in Fig. 69, where states b, d, and f are mainly observed. The variation in the drain current histogram at the beginning of the period at V_m (=1.0 V) with t_i, obtained using the CH-RTN method with $V_i = 1.6$ V, is shown in Fig. 70a–e. Although all the histograms in Figs. 69 and 70 are obtained

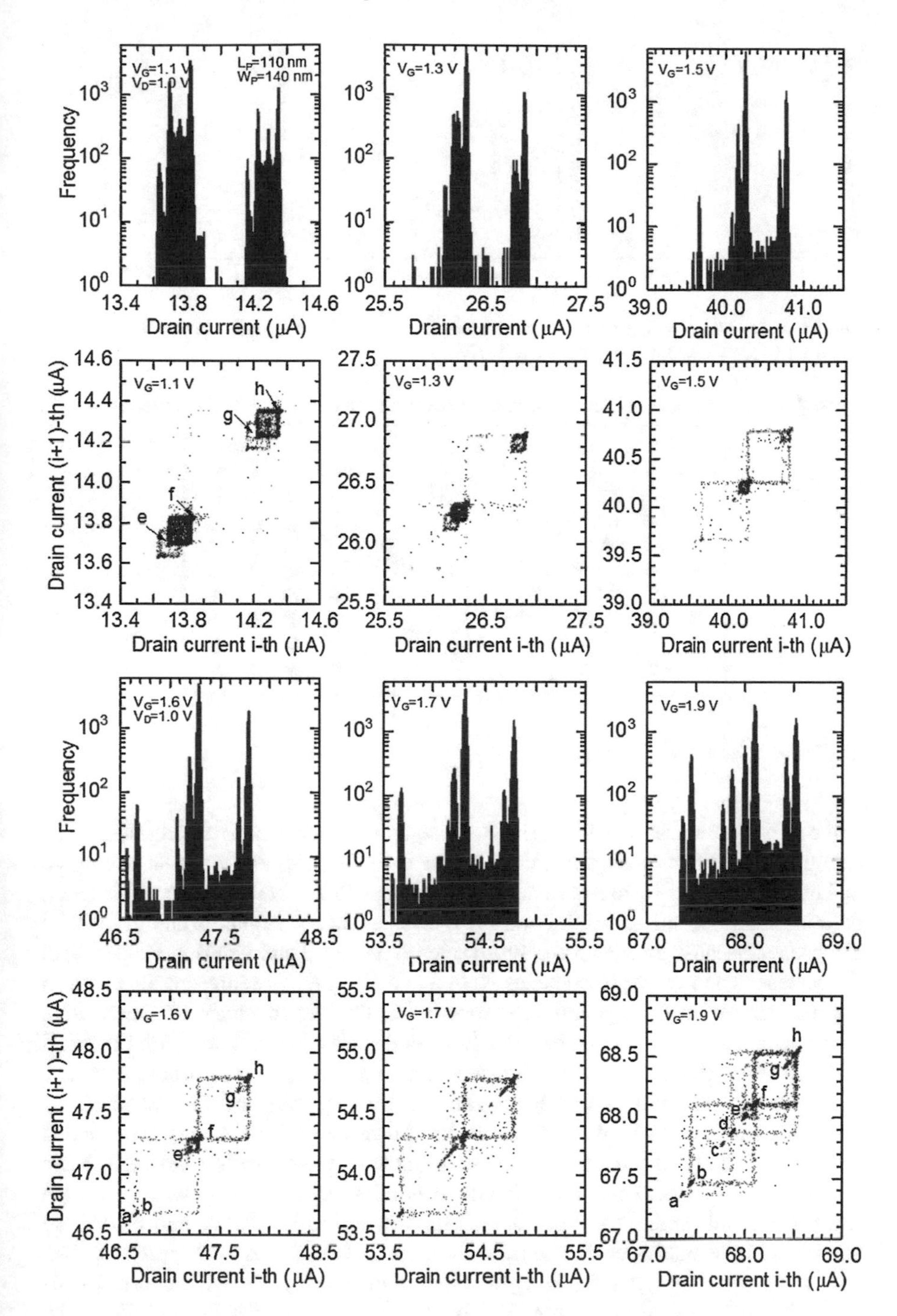

Fig. 67 Drain current histograms and time lag plots for various values of V_G. $V_G = 1.1$–1.9 V

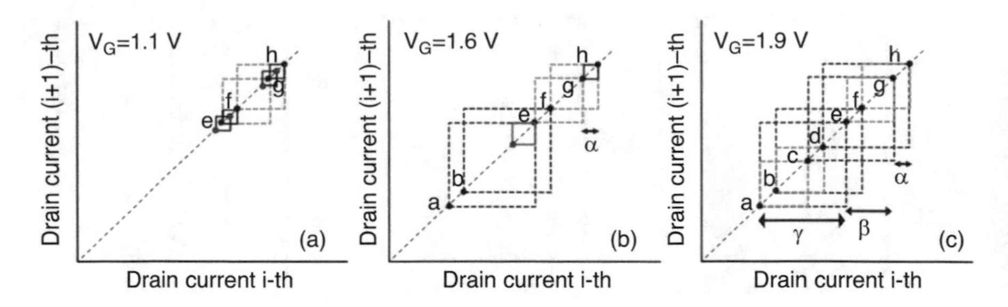

Fig. 68 Schematic time lag plots. Three defects (α, β, and γ) are contributing to the RTN. (**a**) $V_G = 1.1$ V, (**b**) $V_G = 1.6$ V, and (**c**) $V_G = 1.9$ V

Table 7 Correspondence between each state and charging conditions of the three defects

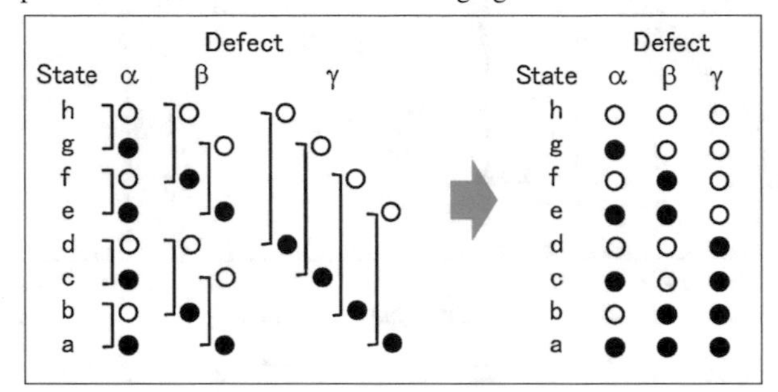

State	Defect α	Defect β	Defect γ
h	○	○	○
g	●	○	○
f	○	●	○
e	●	●	○
d	○	○	●
c	●	○	●
b	○	●	●
a	●	●	●

•: occupied by an electron, o: empty

at the same gate voltage of 1.0 V, there are significant differences among them. The charging conditions of the three defects for states b, c, d, and f are shown again in Fig. 70f. N_d and N_f are particularly prominent in Fig. 69. On the other hand, with increasing t_i in Fig. 70, N_b and N_c increase, and N_f decreases and eventually disappears, because the electron capture rates of defects β and γ during the period at V_i increase. The relative frequencies N_c/N_{Total} and N_b/N_{Total} represent the electron capture rates of defects β and γ, respectively. The dependences of the relative frequencies upon t_i are shown in Fig. 71. The electron capture time τ_c for defects β and γ can be determined from the figure and is found to be 10 μs for both defects.

Moreover, the drain current histograms at the beginning of the period at V_m (=1.0 V) for various V_i were obtained as shown in Fig. 72. States b, c, and d are predominantly observed. As shown in Table 4, the transition between states c and d, and that between states b and d is allowed, but that between states b and c is not allowed. Therefore, from Fig. 70f, the ratios N_d/N_c and N_d/N_b are considered to be equivalent to the ratio τ_c/τ_e for defects β and γ, respectively. The dependences of N_d/N_c and N_d/N_b upon V_i are shown in Fig. 73, which corresponds to the dependences of τ_c/τ_e upon V_G for defects β and γ, respectively. From the figure, the defect positions and energy levels for the defects are estimated to be as follows: defect β: $X_T = 0.24$ nm, $E_{T0} - E_F = 0.07$ eV, defect γ: $X_T = 0.69$ nm, $E_{T0} - E_F = 0.28$ eV.

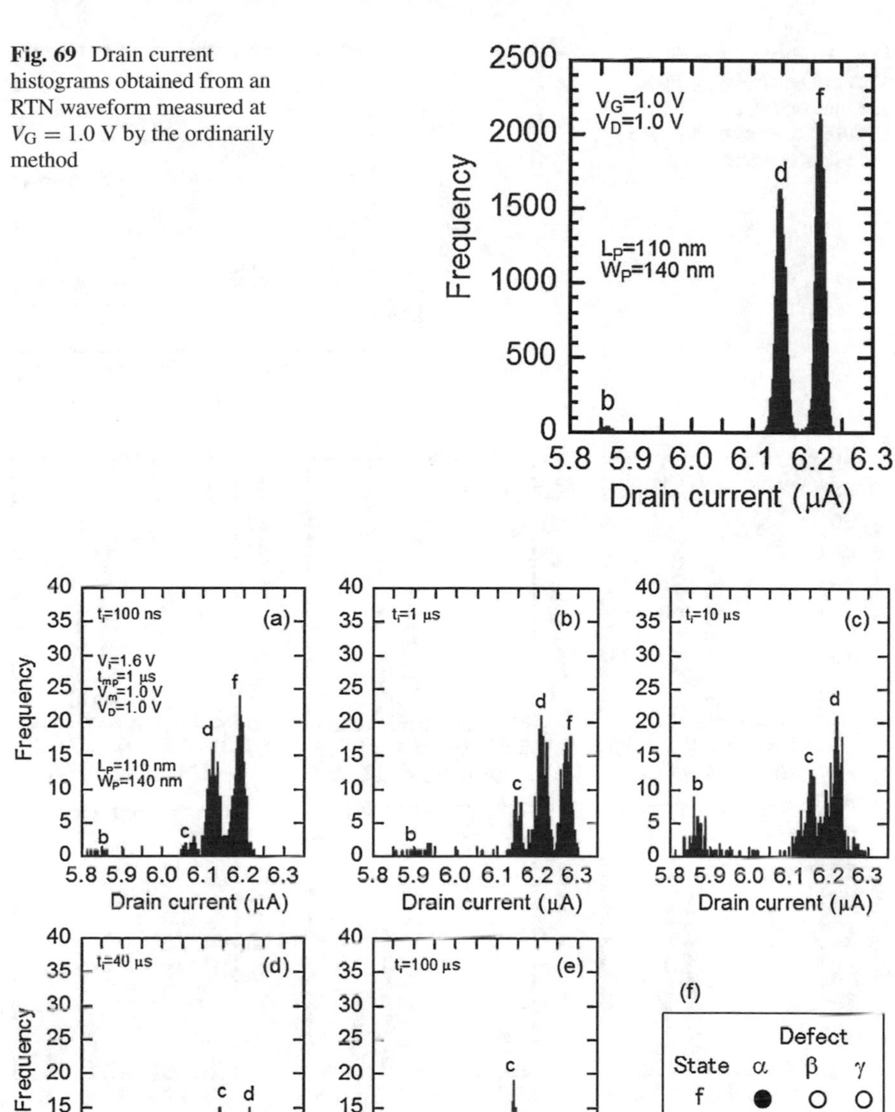

Fig. 69 Drain current histograms obtained from an RTN waveform measured at $V_G = 1.0$ V by the ordinarily method

Fig. 70 (**a–e**) Drain current histograms at the beginning of the period at V_m for various t_i and (**f**) charging conditions of the three defects for states b, c, d, and f. •: occupied by an electron, ○: empty

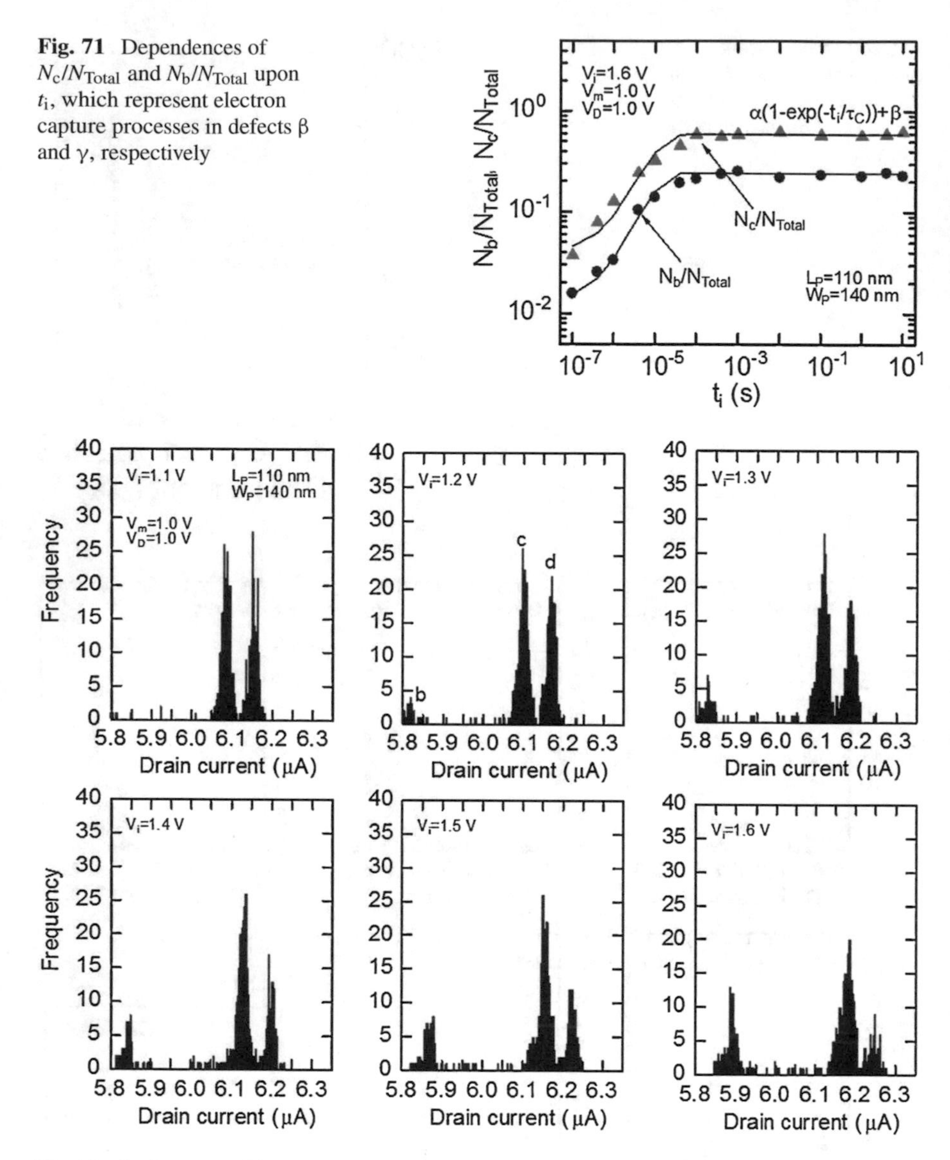

Fig. 71 Dependences of N_c/N_{Total} and N_b/N_{Total} upon t_i, which represent electron capture processes in defects β and γ, respectively

Fig. 72 Drain current histograms at the beginning of the period at V_m for various values of V_i

In addition, τ_e for defects β and γ can be estimated using the respective values of τ_c/τ_e (i.e., the values of N_d/N_c and N_d/N_b obtained with $V_i = 1.6$ V in Fig. 70, respectively) and τ_c (=10 μs for both defects) with $V_i = 1.6$ V: $\tau_e = 37$ μs (defect β) and 17 μs (defect γ).

From the experimental results described above, the CH-RTN method is a simple and effective method for characterizing individual defects especially in multi-defect RTN. Unfortunately, the method is not flawless. In fact, it is hard to characterize

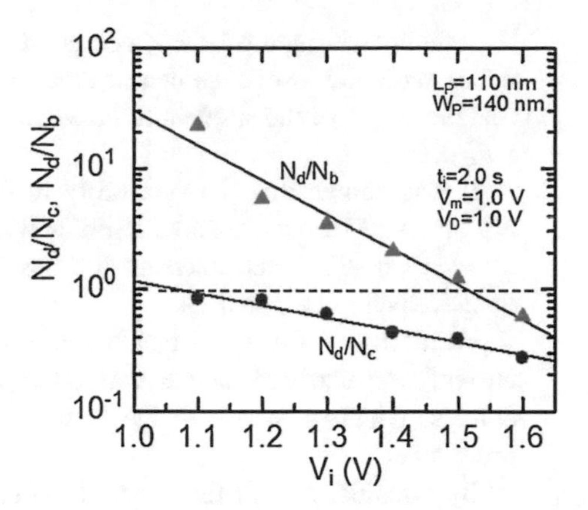

Fig. 73 Dependences of N_d/N_c and N_d/N_b upon V_i, which corresponds to the dependence of τ_c/τ_e upon V_G for defects β and γ, respectively

defect α in the above sample because of its deep energy level. Also, when the drain histograms overlap (i.e., the differences in drain current between the states are very small), the CH-RTN method is not applicable. Nevertheless, the simplicity, practicality, and usefulness of the method for multi-defect RTN are considered to be very appealing qualities.

3 Summary

1. A high-resolution (on the order of sub fA) systematic technique for detecting and characterizing *genuine single* MOS interface defects has been established using the CP method, in which the CP current is systematically evaluated based on understanding the physics of its dependences on the various gate pulse parameters such as the pulse on-time, off-time, risetime, fall time, and frequency.

   Using the systematic technique, it was found for the first time that two energy levels participate in electron capture/emission processes in a Si/SiO_2 single interface defect, and the maximum CP current (I_{CPMAX}) from a single defect is not fixed, i.e., fq, but is in the range of $0 \leq I_{CPMAX} \leq 2fq$ depending on the positions of the two energy levels. Although it is widely believed that I_{CPMAX} is given by fqN, where N is the total number of defects contributing to the CP current, the findings established that this belief is fundamentally incorrect.

   The density of states (DOS) of the *genuine single* interface defects was successfully estimated for the first time by measuring a large number of single defects individually. It was found that two DOS peaks are located at approximately ±0.3 eV around the midgap, and the peak in the upper part of the bandgap is higher than that in the lower part (however, the area under each peak is the same).

   These discoveries on the *genuine single* interface defects, i.e., two energy levels per defect and the asymmetric peak height of the DOS, are consistent with the amphoteric nature of P_{b0} centers at the (100)–Si/SiO_2 interface, observed using magnetic resonance measurements in literatures. Therefore, it can be said that the origin of the single interface defects detected by the CP measurements is a P_{b0} center.

It was shown that the variability in the *actual* number of interface defects N_T involved in nanoscale MOSFETs is quite large, and the values of N_T and $N = I_{CPMAX}/fq$ determined on the basis of the conventional CP theory were quantitatively compared.

Furthermore, the effect of Coulomb interactions between defects on the CP current was clarified, and it was attempted to estimate the distance between defects, which give rise to the Coulomb interactions in the capture/emission processes.

By considering the fundamental nature of the amphoteric defects (i.e., the presence of two energy levels per defect) and the Coulomb interaction between defects, a widely held misconception was corrected, and a fundamental refinement of the CP theory was introduced.

2. A novel CH-RTN method was proposed to characterize each individual oxide defect contributing to RTN. This method utilizes the charging history effects in defects, and the variations in the drain current histogram derived from the RTN with respect to the charging history are monitored instead of the timescale parameters that are usually used.

   Moreover, a method to determine the number and charging conditions of the defects for each current level in RTN was also proposed, which is a necessary prerequisite to put the CH-RTN method into practice.

   Actually, these methods were applied to three-defect RTN, and successfully derived the electron capture time, position (distance from the Si/SiO_2 interface), and energy level of each individual defect. These methods are simple and particularly effective for characterizing individual defects in such *multi-defect* RTN. Characterization of such *multi-defect* RTN by conventional timescale analysis is almost impossible, which typically requires the use of sophisticated hidden Markov models [83].

   These novel methods are considered to be a valuable asset for determining the nature of the defects, and to improve device fabrication processes in order to reduce the defects and current noise.

Acknowledgments The author would like to thank Yukinori Ono (Shizuoka Univ.) and Patrick M. Lenahan (The Pennsylvania State Univ.) for the invaluable discussions, and Tohru Mogami and Yuzuru Ohji (former Selete, Semiconductor Leading Edge Technologies, Inc.) for their cooperation in device fabrication. Furthermore, useful discussions with Naoyoshi Tamura and Visiting Researchers of former STARC (the Semiconductor Technology Academic Research Center) are gratefully acknowledged. This work was partially supported by the Grants-in-Aid for Scientific Research from the JSPS, and also by the former STARC.

References

1. T. Mizuno, J. Okamura, A. Toriumi, IEEE Trans. Electron Devices **41**, 2216 (1994)
2. K. Takeuchi, T. Fukai, T. Tsunomura, A.T. Putra, A. Nishida, S. Kamohara, T. Hiramoto, in *IEDM Tech. Dig.*, Washington, DC, 2007, p. 467
3. A. Asenov, A. Cathibnol, B. Cheng, K.P. McKenna, A.R. Brown, A.L. Shluger, D. Chanemougame, K. Rochereau, G. Ghibaudo, IEEE Electron Device Lett. **29**, 913 (2008)
4. Y.-Y. Chiu, Y. Li, H.-W. Cheng, in *Ext. Abstr. Solid State Devices and Materials*, 2011, p. 126
5. J.S. Brugler, P.G.A. Jespers, IEEE Trans. Electron Devices **16**, 297 (1969)
6. G. Groeseneken, H.E. Maes, N. Beltran, R.F. DeKeersmaecker, IEEE Trans. Electron Devices **31**, 42 (1984)
7. P. Heremans, J. Witters, G. Groesenekn, H.E. Maes, IEEE Trans. Electron Devices **36**, 1318 (1989)
8. N.S. Saks, M.G. Ancona, IEEE Electron Device Lett. **11**, 339 (1990)
9. G. Van den Bosch, G.V. Groeseneken, P. Heremans, H.E. Maes, IEEE Trans. Electron Devices **38**, 1820 (1991)
10. G. Van den bosch, G. Groeseneken, H.E. Maes, IEEE Electron Device Lett. **14**, 107 (1993)
11. M. Tsuchiaki, H. Hara, T. Morimoto, H. Iwai, IEEE Trans. Electron Devices **40**, 1768 (1993)
12. R.E. Paulsen, M.H. White, IEEE Trans. Electron Devices **41**, 1213 (1994)
13. G.V. Groeseneken, I. De Wilf, R. Bellens, H.E. Maes, IEEE Trans. Electron Devices **43**, 940 (1996)
14. N.S. Saks, G. Groeseneken, I. DeWolf, Appl. Phys. Lett. **68**, 1383 (1996)
15. N.S. Saks, Appl. Phys. Lett. **70**, 3380 (1997)
16. D. Bauza, Y. Maneglia, IEEE Trans. Electron Devices **44**, 2262 (1997)
17. L. Militaru, P. Masson, G. Guegan, IEEE Electron Device Lett. **23**, 94 (2002)
18. L. Militaru, A. Souifi, Appl. Phys. Lett. **83**, 2456 (2003)
19. M. Masuduzzaman, A.E. Islam, M.A. Alam, IEEE Trans. Electron Devices **55**, 3421 (2008)
20. D. Bauza, IEEE Trans. Electron Devices **56**, 70 (2009)
21. D. Bauza, IEEE Trans. Electron Devices **56**, 78 (2009)
22. T. Tsuchiya, Y. Mori, Y. Morimura, T. Mogami, Y. Ohji, Jpn. J. Appl. Phys. **49**, 064001 (2010)
23. T. Tsuchiya, Y. Morimura, Y. Mori, ECS Trans. **35**(4), 3 (2011)
24. T. Tsuchiya, Y. Ono, Jpn. J. Appl. Phys. **54**, 04DC01 (2015)
25. T. Tsuchiya, P.M. Lenahan, Jpn. J. Appl. Phys. **56**, 031301 (2017)
26. W. Shockley, W.T. Read Jr., Phys. Rev. **87**, 835 (1952)
27. R.N. Hall, Phys. Rev. **87**, 387 (1952)
28. L. Lin, Z. Ji, J.F. Zhang, W.D. Zhang, B. Kaczer, S.D. Gendt, G. Groeseneken, IEEE Trans. Electron Devices **58**, 1490 (2011)
29. M. Hori, T. Watanabe, T. Tsuchiya, Y. Ono, Appl. Phys. Lett. **105**, 261602 (2014)
30. M. Hori, T. Watanabe, T. Tsuchiya, Y. Ono, Appl. Phys. Lett. **106**, 041603 (2015)
31. N.L. Cohen, R.E. Paulsen, M.H. White, IEEE Trans. Electron Devices **42**, 2004 (1995)
32. Y. Maneglia, D. Bauza, J. Appl. Phys. **79**, 4187 (1996)
33. L. Fricke, M. Wulf, B. Kaestner, F. Hohls, P. Mirovsky, B. Mackrodt, R. Dolata, T. Weimann, K. Pierz, U. Siebner, H.W. Schumacher, Phys. Rev. Lett. **112**, 226803 (2014)
34. Y. Nishi, Jpn. J. Appl. Phys. **10**, 52 (1971)
35. P.M. Lenahan, P.V. Dressendorfer, Appl. Phys. Lett. **41**, 542 (1982)
36. P.M. Lenahan, P.V. Dressendorfer, Appl. Phys. Lett. **44**, 96 (1984)
37. E.H. Poindexter, G.J. Gerardi, M.-E. Rueckel, P.J. Caplan, N.M. Johnson, D.K. Biegelsen, J. Appl. Phys. **56**, 2844 (1984)
38. P.M. Lenahan, P.V. Dressendorfer, J. Appl. Phys. **55**, 3495 (1984)
39. P.M. Lenahan, J.F. Conley Jr., J. Vac. Sci. Technol. B **16**, 2134 (1998)
40. P.M. Lenahan, T.D. Mishima, J. Jumper, T.N. Fogarty, R.T. Wilkins, IEEE Trans. Nucl. Sci. **48**, 2131 (2001)
41. J.P. Campbell, P.M. Lenahan, Appl. Phys. Lett. **80**, 1945 (2002)

42. P.V. Gray, D.M. Brown, Appl. Phys. Lett. **8**, 31 (1966)
43. E.H. Nicollian, A. Goetzberger, Bell Syst. Tech. J. **46**, 1055 (1967)
44. M.R. Boudry, Appl. Phys. Lett. **22**, 530 (1973)
45. T.P. Ma, G. Scoggan, R. Leone, Appl. Phys. Lett. **27**, 61 (1975)
46. S.K. Lai, Appl. Phys. Lett. **39**, 58 (1981)
47. N.M. Johnson, D.K. Biegelsen, M.D. Moyer, S.T. Chang, E.H. Poindexter, P.J. Caplan, Appl. Phys. Lett. **43**, 563 (1983)
48. F.-C. Hsu, K.-Y. Chiu, in *IEDM Tech. Dig.*, 1984, p. 96
49. M.V. Fischetti, Z.A. Weinberg, J.A. Calise, J. Appl. Phys. **57**, 418 (1985)
50. G.J. Gerardi, E.H. Poindexter, P.J. Caplan, Appl. Phys. Lett. **49**, 348 (1986)
51. Y. Hiruta, H. Iwai, F. Matsuoka, K. Hama, K. Maeguchi, K. Kanzaki, IEEE Trans. Electron Devices **36**, 1732 (1989)
52. Z. Liu, H.-J. Wann, P.K. Ko, C. Hu, Y.C. Cheng, IEEE Electron Device Lett. **13**, 402 (1992)
53. H. Hasegawa, H. Ohno, J. Vac. Sci. Technol. B **4**, 1130 (1986)
54. R.E. Mikawa, P.M. Lenahan, J. Appl. Phys. **59**, 2054 (1986)
55. J.T. Krick, P.M. Lenahan, Appl. Phys. Lett. **59**, 3437 (1991)
56. J.W. Gabrys, P.M. Lenahan, W. Weber, Microelectron. Eng. **22**, 273 (1993)
57. C.A. Billman, P.M. Lenahan, W. Weber, Microelectron. Eng. **36**, 271 (1997)
58. T. Tsuchiya, J. Frey, IEEE Electron Device Lett. **6**, 8 (1985)
59. T. Tsuchiya, T. Kobayashi, S. Nakajima, IEEE Trans. Electron Devices **34**, 386 (1987)
60. T. Tsuchiya, IEEE Trans. Electron Devices **34**, 2291 (1987)
61. R.E. Paulsen, R.R. Siergiej, M.L. French, M.H. White, IEEE Electron Device Lett. **13**, 627 (1992)
62. F.N. Hooge, IEEE Trans. Electron Devices **41**, 1926 (1994)
63. N. Tega, H. Miki, F. Pagette, D.J. Frank, A. Ray, M.J. Rooks, W. Haensch, K. Torii, in *Tech. Dig. Symp. VLSI Technology*, Kyoto, 2009, p. 50
64. H. Miki, N. Tega, Z. Ren, C.P. D'Emic, Y. Zhu, D.J. Frank, M.A. Guillorn, D.-G. Park, W. Haensch, K. Torii, in *IEDM Tech. Dig.*, 2010, p. 620
65. T. Nagumo, K. Takeuchi, T. Hase, Y. Hayashi, in *IEDM Tech. Dig.*, 2010, p. 628
66. N. Tega, H. Miki, T. Osabe, A. Kotabe, K. Otsuga, H. Kurata, S. Kamohara, K. Tokami, Y. Ikeda, R. Yamada, in *IEDM Tech. Dig.*, San Francisco, CA, 2006, p. 491
67. J. Chang, A.A. Abidi, C.R. Viswanathan, IEEE Trans. Electron Devices **41**, 1965 (1994)
68. M.J. Kirton, M.J. Uren, Adv. Phys. **38**, 367 (1989)
69. M.H. Tsai, T.P. Ma, IEEE Trans. Electron Devices **41**, 2061 (1994)
70. K. Takeuchi, T. Nagumo, S. Yokogawa, K. Imai, Y. Hayashi, in *Tech. Dig. Symp. VLSI Technology*, Kyoto, 2009, p. 54
71. T. Nagumo, K. Takeuchi, S. Yokogawa, K. Imai, Y. Hayashi, in *IEDM Tech. Dig.*, Baltimore, MD, 2009, p. 759
72. S. Lee, H.J. Cho, Y. Son, D.S. Lee, H. Shin, in *IEDM Tech. Dig.*, Baltimore, MD, 2009, p. 763
73. N. Tega, H. Miki, Z. Ren, C.P. D'Emic, Y. Zhu, D.J. Frank, J. Cai, M.A. Guillorn, D.G. Park, W. Haensch, K. Torii, in *IEDM Tech. Dig.*, Baltimore, MD, 2009, p. 771
74. T. Tsuchiya, N. Tamura, A. Sakakidani, K. Sonoda, M. Kamei, S. Yamakawa, S. Kuwabara, ECS Trans. **58**(9), 265 (2013)
75. T. Grasser, H. Reisinger, W. Goes, Th. Aichinger, Ph. Hehenberger, P.-J. Wagner, M. Nelhiebel, J. Franco, B. Kaczer, in *IEDM Tech. Dig.*, 2009, p. 729
76. T. Grasser, B. Kaczer, W. Goes, H. Reisinger, T. Aichinger, P. Hehenberger, P.-J. Wagner, F. Schanovsky, J. Franco, M.T. Luque, M. Nelhiebel, IEEE Trans. Electron Devices **58**, 3652 (2011)
77. T. Grasser, Microelectron. Reliab. **52**, 39 (2012)
78. B. Kaczer, J. Franco, P. Weckx, P.J. Roussel, V. Putcha, E. Bury, M. Simicic, A. Chasin, D. Linten, B. Parvais, F. Catthoor, G. Rzepa, M. Waltl, T. Grasser, Microelectron. Reliab. **81**, 186 (2018)
79. A. Ohata, A. Toriumi, Jpn. J. Appl. Phys. **38**, 2473 (1999)
80. M.J. Uren, M.J. Kirton, S. Collins, Phys. Rev. B **37**(14), 8346 (1988)

81. J. Chan et al., Phys. Rev. B **80**, 033402 (2009)
82. H. Nakamura, N. Yasuda, K. Taniguchi, C. Hamaguchi, A. Toriumi, Jpn. J. Appl. Phys. **28**, L2057 (1989)
83. H. Miki, N. Tega, M. Yamaoka, D.J. Frank, A. Bansal, M. Kobayashi, K. Cheng, C.P. D'Emic, Z. Ren, S. Wu, J-B. Yau, Y. Zhu, M.A. Guillorn, D.-G. Park, W. Haensch, E. Leobandung, K. Torii, in *IEDM Tech. Dig.*, 2012, p. 450
84. E.R. Hsieh, Y.L. Tsai, S.S. Chung, C.H. Tsai, R.M. Huang, C.T. Tsai, in *IEDM Tech. Dig.*, 2012, p. 454

Random Telegraph Noise Nano-spectroscopy in High-κ Dielectrics Using Scanning Probe Microscopy Techniques

Alok Ranjan, Nagarajan Raghavan, Kalya Shubhakar, Sean Joseph O'Shea, and Kin Leong Pey

1 Introduction

At the heart of semiconductor devices lies the ability to have precise control on the manipulation and engineering of electronic defects, which allows for the realization of diverse electronic properties across a wide range of materials used in the microelectronics industry. With CMOS scaling trending towards the sub-10 nm technology nodes, the role of individual defects is becoming more important as they give rise to performance variability and other potential reliability issues [1, 2]. High-κ gate dielectric, used in transistors, is one such example where stringent control of the interfacial and bulk defects (also called traps) is important for the reliable operation of the device [3, 4]. These pre-existing interfacial and bulk defects (also called process induced traps (PITs)), if not precisely controlled, can lead to potential reliability issues including the presence of a dominant low-frequency noise (LFN) [5–7] and bias temperature instability (BTI) [8–10] during device operations as discussed in chapter "Defect-Based Compact Modeling of Combined

A. Ranjan
Engineering Product Development Pillar, Singapore University of Technology and Design, Singapore, Singapore

Institute of Materials Research and Engineering, Agency for Science Technology and Research (A∗STAR), Singapore, Singapore
e-mail: alok_ranjan@sutd.edu.sg

N. Raghavan · K. Shubhakar · K. L. Pey (✉)
Engineering Product Development Pillar, Singapore University of Technology and Design, Singapore, Singapore
e-mail: peykinleong@sutd.edu.sg

S. J. O'Shea
Institute of Materials Research and Engineering, Agency for Science Technology and Research (A∗STAR), Singapore, Singapore

T. Grasser (ed.), *Noise in Nanoscale Semiconductor Devices*,
https://doi.org/10.1007/978-3-030-37500-3_12

Random Telegraph Noise and Bias Temperature Instability." In addition, during the operation of a transistor, additional defects are stochastically generated in the bulk and interfacial regions of gate oxide layers due to electrical stress (called stress induced traps (SITs)), which can further increase the gate leakage current and eventually trigger early dielectric breakdown [11–15]. Therefore, characterization of these defects (including both PITs and SITs) in the dielectric is crucial.

In the last few decades, several studies were aimed at developing transient electrical tests to characterize these defects in the dielectrics at device level (by using a transistor or a capacitor test structure) as discussed in previous chapters. These experimental methods include charge pumping measurements [16–19], stress-induced leakage current (SILC) spectroscopy [20–24], and measurement of dielectric relaxation current [25–28]. Each of these methods/techniques has their own limitations due to the nature and dynamics of defects that they effectively probe. For example, charge pumping measurements have been demonstrated to effectively measure the interfacial defects between the semiconductor and the high-κ dielectric layer, but the technique works well only for thicker dielectric layers where leakage currents are relatively low [16–19]. Recently, the charge pumping technique has also been demonstrated to measure the bulk defects in high-κ thin films. However, such an analysis is limited to the cases where the contribution from the interfacial oxide layers is minimal [29]. SILC spectroscopy has shown the ability to probe the stress-induced defects in dielectrics when the contribution of trap-assisted tunneling (TAT) current components (due to stress induced defects) exceeds the intrinsic prestress leakage current (due to process-induced bulk and interfacial defects) [20–24]. The measurement of relaxation currents only provides the transient dynamics of the defects [25–28].

Random telegraph noise (RTN)-based spectroscopy has been successful in probing both bulk and interface defects in dielectrics [30–43]. However, most of the RTN measurements reported so far have been carried out at the device level where the areal dimensions vary from a few micrometers to ~10 nm at the best, as discussed in the previous chapters. This approach, in the majority of instances, measures the responses of multiple defects present within the volume of the device being tested, thereby limiting the capability of this technique to directly probe and measure the responses of individual defects. Given the in-depth physical insights on the defect that can be obtained from a simple RTN trace measurement at different voltage and temperature, it is desirable to observe RTN at the same atomic/nanometer resolution in which they occur. A detailed analysis of RTN can provide defect properties including trap energy level (E_T), relaxation energy (E_R) capture cross section (A), its physical location (x–y–z coordinates), average trap density (N_T), and activation (E_a) energy.

In this chapter, we summarize the recent developments of scanning probe microscopy (SPM) techniques to isolate individual defects and characterize them using RTN, taking the HfO_2 high-κ dielectric sample as the specimen.

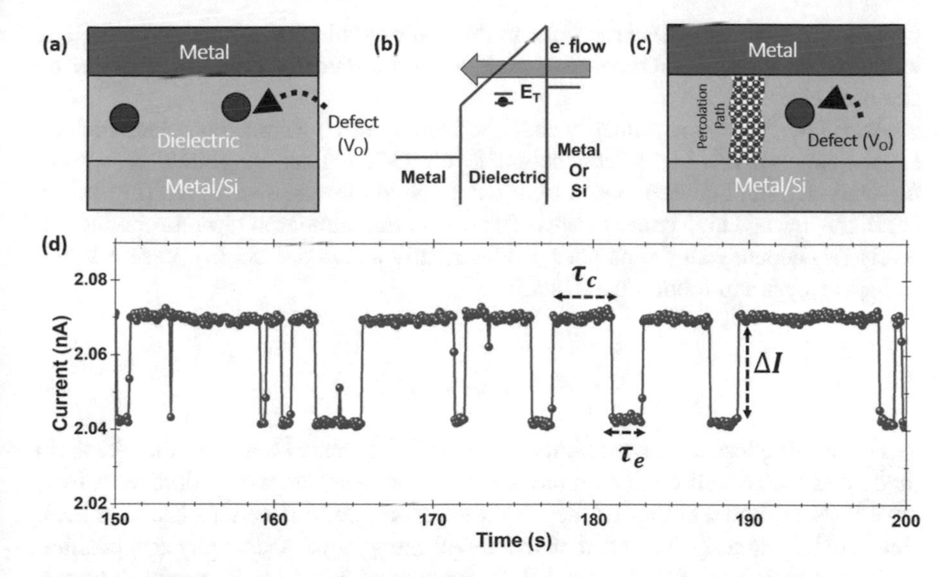

Fig. 1 (**a**) Schematic showing a standard MOS capacitor with intrinsic bulk defects and (**b**) corresponding simplified energy band diagram with trap energy levels in the dielectric. (**c**) Schematic showing a typical scenario after soft dielectric breakdown in which a defect closer to a vacancy-rich percolation path can trigger RTN. (**d**) Plot of a two-level RTN trace observed in ultrathin SiO_2 ($t_{SiO_2} = 2.2$ nm) capacitor showing the time constants ($\tau_{c,e}$) and amplitude of current fluctuations (ΔI)

2 Random Telegraph Noise in Dielectrics and Its Characterization

In the context of dielectrics, a random telegraph noise, in short called "RTN", refers to the stochastic jumps in the tunneling/leakage currents between discrete conduction levels when the dielectric is subjected to a small voltage bias. RTN has been clearly observed among a wide range of dielectric materials including SiO_2 and HfO_2 and is one of the primary causes for intrinsic performance variations (e.g., threshold voltage (V_{th}) and on current (I_{on}) variations), in ultra-scaled logic and memory devices [30–43]. The origin of RTN in dielectrics is attributed to the stochastic capture and emission of electrons occurring at the individual defect sites during electron tunneling [30, 31, 34, 36, 37, 39]. In the case of SiO_2 and HfO_2 (see Fig. 1), oxygen vacancy (V_O) defects, present in the bulk as well as in the interfacial regions, have been identified to be responsible for random trapping/de-trapping of tunneling electrons during low-voltage bias. Recently, *ab-initio* calculations have provided evidence that hydrogen-related defects could also potentially give rise to similar LFN in these dielectrics [44], as discussed in chapter "The Langevin-Boltzmann Equation for Noise Calculation."

When a defect captures an electron, it perturbs the electrostatic environment locally by raising the potential in the vicinity of the defect site, thereby reducing overall tunneling leakage current due to the coulomb blockade effect [41, 43]. The tunneling leakage current recovers back to its initial state when the defect emits the captured electron.

An RTN spectrum is primarily characterized by three parameters which include (a) the number of discrete conduction levels (N), (b) the amplitude of current fluctuations (ΔI) between these discrete conduction levels, and (c) the dwell time (τ) in the low and high current states. The maximum number of discrete conduction levels (N) depends on the number of electrically active defects (n) present in the dielectric layer and follows the relation:

$$N = 2^n \quad (n = 1, 2, 3, \ldots) \tag{1}$$

For the simplest case, where only one electrically active defect in the dielectric modulates the overall carrier capture/emission process, we would observe a two-level RTN as shown in Fig. 1. The amplitude of current fluctuations (ΔI) has been shown to be primarily dependent on the defect energy level and its physical position in the dielectric layer [35, 36, 42, 43]. In the case of transistors, it has been shown that the discrete random dopants present in the channel region could also contribute to the overall low-frequency noise fluctuations [45, 46].

For an RTN spectrum, the time spent in the low conduction state is referred to as the time to emission (τ_e) or down-time (τ_{down}) and time spent in the high conduction state is denoted as time to capture (τ_c) or up-time (τ_{up}). While the time constants $\tau_{c,e}$ are randomly distributed, ΔI is almost invariant for a given defect. When the time constants ($\tau_{c,e}$) are extracted for reasonably long periods, they follow an exponential probability distribution function (P_1) given by:

$$P_1(t) = \frac{1}{\tau_{C,E}} \exp\left(-\frac{\tau_{c,e}}{\tau_{C,E}}\right) \tag{2}$$

The experimental values of the time constants ($\tau_{c,e}$) can be fitted to Eq. (2) to extract their average values ($\tau_{C,E}$). Average time constants ($\tau_{C,E}$) are sensitive to the electrical bias and bias-dependent RTN data sets can be used to extract the physical position of the defect in the dielectric [47].

The time traces of the RTN spectra can be analyzed in the frequency domain as well by applying a Fourier transform. The frequency domain analysis helps to measure the response of the defect in different frequency regimes and extract various frequency-domain parameters of the defect. The Fourier transform of a standard two-level RTN shows a Lorentzian trend and its power spectral density (S_I) is described by [39, 48]:

$$S_I = \frac{2(\Delta I)^2 \tau_o}{\left[4 + (\omega \tau_o)^2\right]} \tag{3}$$

where ΔI is the RTN amplitude, τ_o is the characteristic time constant, and ω ($=2\pi f$) is the radial frequency. Equation (3) implies that the noise power spectral density (S_I) has a cut-off frequency $f_c = \frac{1}{\tau_o}$, above which S_I rolls off with $\frac{1}{f^2}$. However, in practice, it is found that S_I rolls off as $\frac{1}{f^\alpha}$ above f_c with $1 \leq \alpha \leq 2$.

For small area devices, wherein a handful of countable defects participate in the capture/emission of electrons, the noise power spectral density shows a $\frac{1}{f^\alpha}$ ($1 < \alpha \leq 2$) dependency above f_c. As the number of defects increases (typically >3–4 defects corresponding to 8–16 discrete conduction levels) for large area devices, which host large number of defects, the noise spectra result in a $\frac{1}{f}$ ($\alpha \approx 1$) dependency arising from the superposition of many Lorentzian spectra with different cut-off frequencies associated with simultaneous capture/emission events at each defect site [49].

3 Challenges of Device-Level RTN Spectroscopy

Traditionally, most RTN studies reported so far have been performed using either a transistor or a capacitor test structure using a standard electrical probe station setup as discussed in chapters "Systematic Characterization of Random Telegraph Noise and Its Dependence with Magnetic Fields in MOSFET Devices, Principles and Applications of I_g-RTN in Nano-scaled MOSFET, Random Telegraph Noise in Flash Memories, Advanced Electrical Characterization of Single Oxide Defects Utilizing Noise Signals, Measurement and Simulation Methods for Assessing SRAM Reliability Against Random Telegraph Noise and Random Telegraph Noise Under Switching Operation" [30–44, 49]. The RTN signature is sensed by applying a low constant voltage (so as not to induce new defects) to one of the metal/semiconductor electrodes and electrically grounding the other electrode. Typically, for sensing the RTN signals, an experimentalist must try multiple sensing voltages (trial and error), to find the optimum sense voltages at which RTN spectra are clearly observed within a reasonable measurement time frame. Generally, such a measurement setup is handy for device reliability analysis and technology qualifications. However, these are not suitable for single vacancy defect spectroscopy studies in many instances due to the following reasons:

(a) *Spatial resolution*: The RTN signals measured using a probe station setup are spatially averaged over the entire volume of the stressed device and thus provide only a cumulative response of multiple traps (distributed in space and energy) present within the device. For example, in case of large area devices where multiple defects (typically more than three) are undergoing the random electron capture and emission events, the noise spectra simply show a *1/f* dependence and decoding the contribution of individual defects in these cases may not be reliable and feasible.
(b) *Interfacial layers*: The presence of a top metal electrode in transistor/capacitor test structures, used for RTN analysis, creates an additional interfacial layer

between the dielectric and the metal layer, which usually has high defect densities compared to the bulk dielectric layer [50]. Therefore, the RTN spectra measured in such cases could be nonstationary and complex due to the convoluted effect of the interface and bulk defects, which cannot be avoided, thereby making the defect spectroscopy analysis challenging.

(c) *Microstructural analysis*: It has been experimentally demonstrated that annealing of HfO_2 leads to a change in its microstructure from amorphous to polycrystalline form, implying the existence of grain and grain boundaries (GB) [51–54]. Furthermore, it has also been shown using *ab-initio* calculations [55, 56] and experiments [57, 58] that GBs act as a thermodynamic sink for oxygen vacancy defects and have a higher defect density compared to grain sites. The device-level measurements mask the effect of microstructure and only provide spatially averaged information, which makes it challenging for one to study the differences (if any) in the trapping kinetics of grain and GB defects.

(d) *Signal-to-noise ratio*: The tunneling leakage current through a dielectric consists of two components: (a) intrinsic leakage current component (tunneling through the ideal defect-free bulk dielectric) and (b) a dynamic RTN current component. The intrinsic tunneling leakage current scales with the area of the device; however, the RTN current component depends on the defect, i.e., trap location and its energy level, and is usually independent of the area of the device. Therefore, in typical large area devices, when only a handful of traps are present, a large intrinsic leakage current component can easily mask the RTN current component in many scenarios and hinder RTN analysis.

(e) *Mechanical stress*: During device-level electrical measurements, usually a hard-mechanical contact is made between the probe and the device bond pad. Usually, it is assumed that the contact forces are low enough so that they do not transfer extra mechanical stress leading to plastic deformation of the dielectric layer. It has been shown experimentally that mechanical stress can create additional electric defects in the dielectric layer [59, 60]. Thus, the probe station setup may allow defect generation due to mechanical stress because in a standard probe station there is no metrology setup to measure the force exerted by the probe tip on the bond pad when the user tries to establish a mechanical contact. An alternate approach to control the mechanical contact forces between the probe and the electrode during electrical testing is to use wire-bonded devices. This approach requires an additional packaging step and is typically used for the noise analysis at circuit-level for technology qualifications.

Given the atomic scale interactions between a defect and the tunneling electrons during RTN, it is desirable to map this phenomenon with an atomic spatial resolution in order to isolate and probe the kinetics, energetics, and dynamics of single bulk defects in high-κ dielectrics.

4 Sample and Instrumentation Requirements for Defect Nano-Spectroscopy Using SPM

Scanning probe microscopy (also referred to as SPM), with their atomic/nanometer resolution, has shown the potential to address some of the above challenges in the measurement of RTN. Table 1 compares the basic attributes of the SPM and device-level approaches. Recently, we have demonstrated the successful measurement of RTN with nanoscale precision in ultrathin HfO_2 dielectrics using both scanning tunneling microscopy (STM) and conductive atomic force microscopy (CAFM) techniques [61–64]. These experimental findings demonstrate the feasibility of successful isolation and mapping of individual defects using RTN techniques in existing 3D (bulk) as well as emerging 2D dielectrics [65, 66].

Table 1 A summary comparing device-level and SPM-based RTN spectroscopy

	Device	CAFM	STM
Spatial resolution	Typically, 100 nm–few μm (limited by electrode dimensions)	Limited by the probe dimensions (typically CAFM tip radius <50 nm) and contact forces	Limited by the probe dimensions (typically, STM tip radius <10 nm) and vacuum tunnel gap
Interfacial defects	Two defective interfacial layers are present	Only bottom interfacial layer is present	Only bottom interfacial layer is present
Top electrode	Required	Not required	Not required
Probing defects at microstructures	Not possible	Possible to probe nanometer defects	Possible to probe atomic defects
Contact forces	Not possible to measure and control in commercially available electrical probe station setups	Possible to precisely control and measure contact forces (from pN to μN) using suitable cantilevers	Possible to indent the tip into the surface but the forces are not known
Timescales of measurement *(at room Temperature)*	Suitable for extended timescale RTN measurements ($t > 10^3$ s)	Preferable for short–moderate timescale RTN measurements (limited by thermal drift between sample and tip) ($t < 10^3$ s)	Suitable for very short RTN measurements (limited by thermal drift between sample and tip) ($t < 10$ s)
Measurement environment	Generally done in ambient conditions or at elevated temperatures	UHV preferable to avoid surface contamination and humidity	UHV preferable to avoid tunneling current instabilities, surface contamination, and humidity
Isolation of defect/vacancy rich percolation path	Not possible	Possible	Possible

In a simplified description, SPM consists of a suite of techniques in which interaction (contact, near field, or far field) between the tip and the sample is used to characterize a wide range of material properties (including mechanical, chemical, electrical, magnetic, optical). Two of these SPM techniques, namely STM and CAFM, have been instrumental in the investigation of surface and sub-surface/buried electrical defects (point, line, edge) in metallic, semiconducting, and insulating thin films [67–72]. In particular, both techniques have been used by the dielectric reliability community for the characterization of electrical defects, measurement of intrinsic charge transport, time-dependent dielectric breakdown (TDDB) measurements, and investigation of resistive switching phenomenon in a wide range of dielectric thin films [61–66, 73–80]. In this section, we summarize the sample requirements and experimental setup for a successful RTN spectroscopy for both STM and CAFM.

The sample usually consists of a thin dielectric film (typically <10 nm) deposited on a conductive substrate (usually highly doped silicon or metal film deposited on silicon). A high surface uniformity and minimum interfacial oxide thickness are desirable for effective probing of the bulk defects during RTN spectroscopy using CAFM and STM. Figure 2 shows a schematic of the experimental setup and electrical connections for the detection of RTN signals using an STM [61–63]. Once the bias has been applied between the tip and the sample and a desired set point of tunneling current (usually a few pA–nA) has been selected, the tip is brought to a tunneling distance (<1 nm) from the sample using the feedback-controlled piezoelectric scanners. For the measurement of RTN signals, the z-feedback loop is switched off and the current–time (*I*–*t*) traces are acquired at different voltages. The z-feedback loop is re-enabled after the acquisition of the RTN spectra. Practically, it is strongly recommended to measure RTN spectra on timescales over which thermal drift of the STM setup does not affect the measurements [62]. The STM tips used here are generally prepared either by mechanical cutting or electrochemical etching of refractory metals/alloys (e.g., W, Pt/Ir) [61–63, 81–83]. In our STM setup operating at room temperature, electrical bias is applied to the sample substrate, while the tip is electrically grounded. Measurements are preferably carried out in ultrahigh vacuum (UHV) environments to avoid surface contamination of the sample from the ambient, which can lead to tunneling current instabilities. It is to be noted that the voltage drop across the dielectric is only a fraction of the

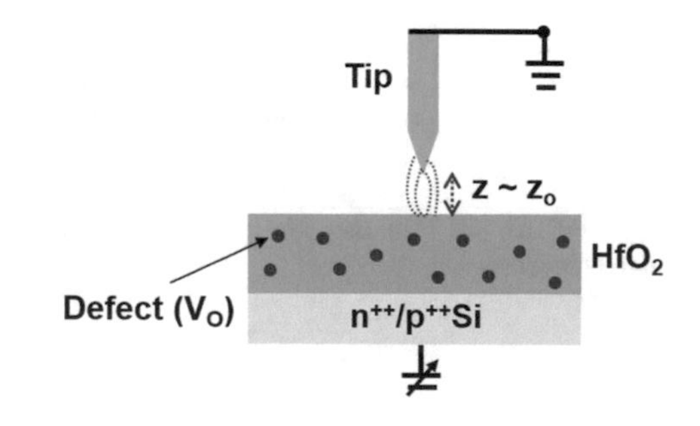

Fig. 2 Schematic diagram of the STM setup for RTN spectroscopy. The tunneling current between the STM tip and the sample is established only when the distance between the tip and sample approaches a threshold tunneling distance, z_o (typically $z_o < 1$ nm)

voltage applied to the STM tip as some part of the voltage drops across the tunnel vacuum gap (z_o) [62]. For cases where the feedback loop is kept turned off during acquisition of the RTN spectrum for relatively long periods of time, the vacuum tunnel gap could change considerably due to thermal drift of the instrumentation. This is a significant limitation on the longer timescale RTN measurement using room temperature STM. However, the contribution of thermal drift can be reduced significantly by using a cryogenic STM.

The basic configuration for the detection of RTN using CAFM is shown in Fig. 3 [64]. Akin to standard CAFM measurements, the tip is brought into physical contact (unlike STM where the tip is maintained at a tunneling distance) with the sample using piezoelectric scanners. The force applied by the CAFM tip can be precisely controlled, thereby allowing decoupling of the mechanical stress on defect generation. Once mechanical contact has been established between tip and sample, appropriate voltage biases are applied to measure the RTN signals. For cases where we want to measure RTN from pre-existing process-induced traps, caution must be taken in the selection of voltage bias, as higher voltage can generate additional SITs and can even trigger an early dielectric breakdown [64]. A wide range of conducting AFM tips (thin film coatings of noble metals, doped diamond, diamond-like carbon, metal silicides, and all metal wire cantilevers) and cantilever (soft and stiff) combinations are commercially available. Notably, sharp (R_{tip} < 20 nm) all-metal wire levers with low lateral stiffness have shown very promising results for

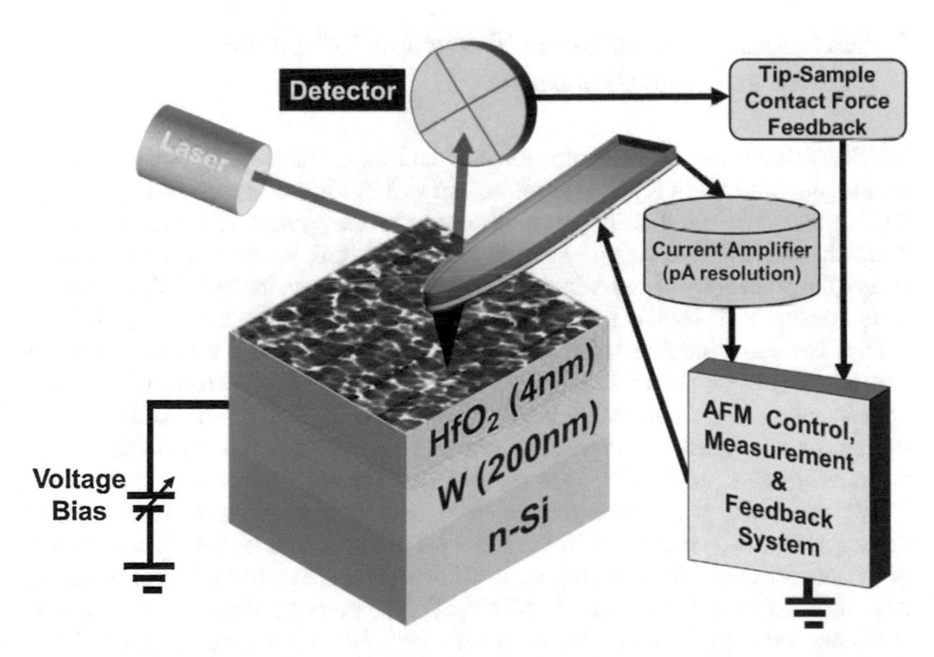

Fig. 3 Schematic diagram showing the CAFM setup and electrical connections for RTN measurements. (Reproduced with permission from [59], IEEE Publishers 2016)

long timescale RTN measurements as they are very resistant to electrical wear out [64–66] compared to the coated metal tips.

The timescales for the localized RTN measurement using room temperature CAFM setup is limited by thermal drift of the instrumentation [84]. Experimentally, we have found that for a room temperature STM and CAFM working in UHV, CAFM is preferable to STM for longer timescale RTN measurements. The nanoscale contact area, torsional force of the cantilever, and shank length of the AFM tip control the conditions under which the tip will begin to slide over the surface from a resting position. Hence, these same parameters control the time taken for the tip to drift (slide) away from a location during an RTN measurement in CAFM and can be engineered to give long timescales and ensure the stability of nanoscale contacts in RTN measurement [84].

The sample used for our STM trap spectroscopy consists of ~4 nm HfO_2 deposited on a n^{++}Si (100) substrate by atomic layer deposition (ALD) using H_2O and Tetrakis(dimethylamino)hafnium (TDMAH) as precursors at 250 °C [62–64]. The sample is further annealed in an N_2 ambient at 400 °C for ~40 min to passivate defects and to induce polycrystallinity in HfO_2 films. For CAFM-based RTN measurements, we use ~4 nm HfO_2 deposited using ALD techniques on a ~200 nm Tungsten (W) bottom electrode on n-Si substrate [64]. Annealing of the samples is carried out under similar conditions as for the STM sample above.

5 Measurement of Random Telegraph Noise from Process- and Stress-Induced Defects

We first demonstrate STM application for the measurement of RTN signals from pre-existing defects in a polycrystalline HfO_2 thin film as shown in Fig. 4 [62]. During the STM-based RTN measurements, the set point current, measurement duration, and effective voltage applied across the sample are kept low, so as to reduce the probability of stress-induced defect generation at the tested site, allowing us to measure RTN signals from process-induced defects present in HfO_2. A well-defined two-level and four-level telegraph noise signal can be observed across the two tested sites as shown in Fig. 4a, b. As the number of defects (n) and the corresponding discrete current levels (N) in the RTN spectra are related by $N = 2^n$, the data clearly shows the presence of one ($n = 1, N = 2$) and two ($n = 2, N = 4$) electrically active pre-existing traps at these probed site.

We now show a set of experiments to controllably generate stress-induced defects and subsequently probe the RTN signature that includes their presence. We preferably use CAFM for this purpose as it allows longer timescale RTN sensing due to lower thermal drift compared to STM [84]. The detailed sequence of sense and stress methodology to control the dielectric degradation at various stages of SILC and soft breakdown (SBD) is shown in Fig. 5 [64]. A ramp voltage stress is used to stress the dielectric to different current compliance levels and arrest the degradation

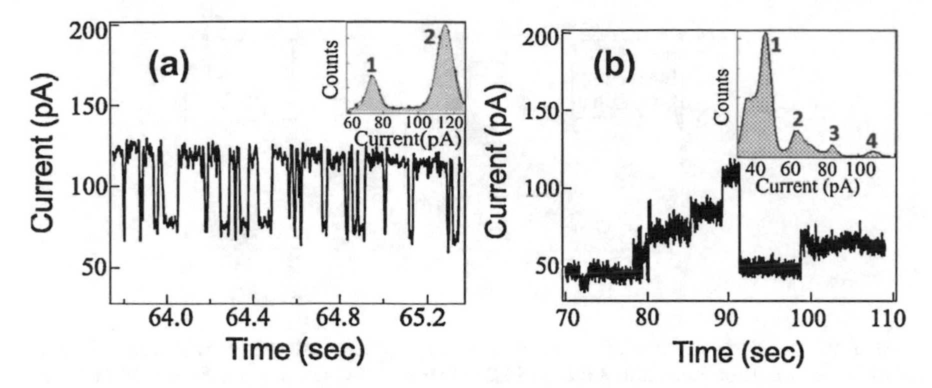

Fig. 4 RTN signals measured from fast traps in HfO_2 using a STM setup in UHV operating at room temperature. Noise signals in (**a**) and (**b**) correspond to two different locations on the sample where a two-level and four-level RTN signature is observed. The tunneling conditions for the measurement are 3.5 V/50 pA. (Reproduced with permission from [62], AIP Publishers, 2016)

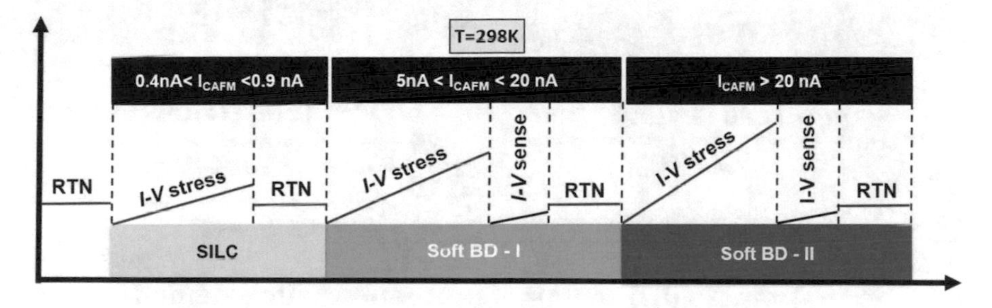

Fig. 5 The sequential sense and stress technique for measuring RTN during various stages of dielectric degradation, including SILC and soft breakdown (SBD). (Reproduced with permission from [64], IEEE Publishers 2016)

at the desired level. A subsequent sense *I–V* with low-voltage conditions is used to assess the extent of dielectric degradation and determine the optimum voltage required for RTN measurements.

Figure 6 shows the successive *I–V* patterns for the SILC and SBD stages for two such randomly probed locations. For effectively probing SILC defects during *I–V* stressing, the degradation is arrested at a low current compliance (I_{comp} < 1 nA). Similarly, SBD is classified in two regimes depending on the current compliance used: (a) SBD-I (5 nA < I_{comp} < 20 nA) and (b) SBD-II (I_{comp} > 20 nA). In other words, breakdown is harsher for SBD-II, while still not in the hard breakdown (Ohmic conduction) regime yet.

The noise trends for the fresh, SILC and SBD (I and II) stages for the same location are shown in Fig. 7. Prior to any electrical stress, the probed location shows white noise due to the absence of electrically active defects at the probed site. We observed a consistent two-level RTN in both SILC and SBD-I stages, although with an increase in the base level of the leakage current. This is expected as additional

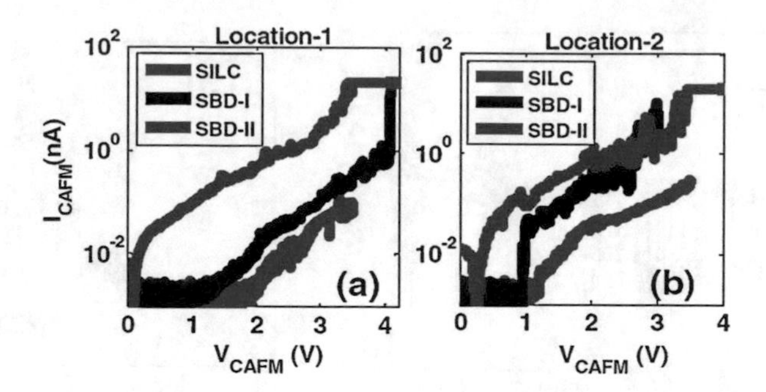

Fig. 6 (**a**, **b**) *I*–*V* trends at two different locations on HfO_2 tested using the CAFM setup. The plots show dielectric degradation arrested at various stages of SILC (I_{comp} < 1 nA), soft breakdown-I (5 nA < I_{comp} < 20 nA), and soft breakdown-II (I_{comp} > 20 nA). (Reproduced with permission from [64], IEEE Publishers 2016)

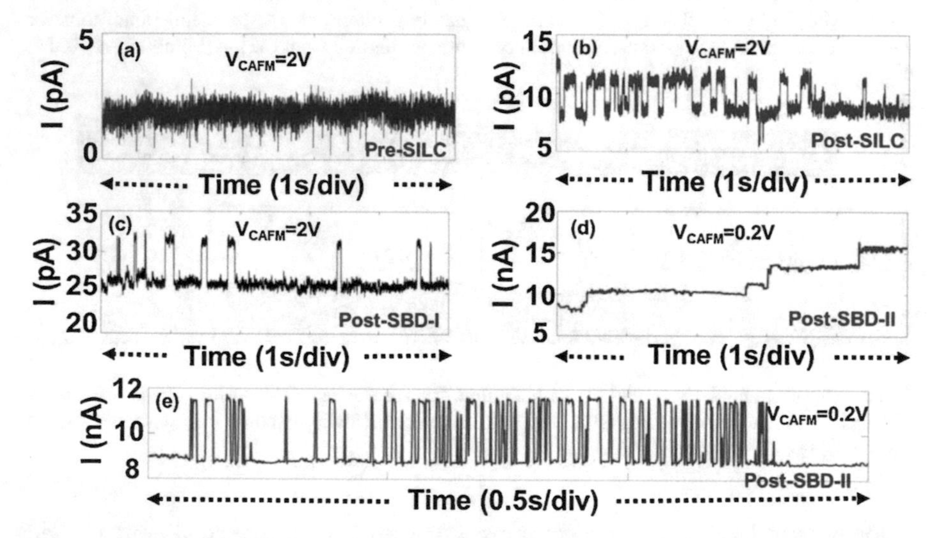

Fig. 7 RTN measured in different stages of dielectric degradation using CAFM on HfO_2. (**a**) RTN measured at +2 V prior to any electrical stress shows white noise. RTN measured after (**b**) SILC and (**c**) SBD-I at +2 V shows a clear trend of two-level RTN implying a single defect/trap in the capture/emission event at the probed site. (**d, e**) The *I*–*t* traces measured post-SBD-II show two orders of magnitude increase in the leakage current and signatures of nonstationary RTN patterns when measured at a low bias of +0.2 V. (Reproduced with permission from [64], IEEE Publishers 2016)

defects generated at the SBD-I stage can increase the overall trap-assisted tunneling (TAT) leakage current at the tested site.

DC leakage current changes by almost ~2 orders of magnitude after SBD-II, even for a low sensing voltage of ~0.2 V, as shown in Fig. 7e. This change in the DC current levels suggests the formation of a localized percolation path [40, 85, 86]. LFN spectra measured after SBD-II show anomalous RTN patterns, where a two-level RTN signal can be measured at the start of the spectra, but the remaining portions show a stepwise increase in current. Similar conductivity trends have been observed at the device-level measurements [87]. In that study, the breakdown spot has been modeled using a quantum point contact (QPC) model and it has been demonstrated that the conductivity of the percolation path is strongly modulated by the number of oxygen vacancy (V_O) defects at the constriction region of the percolation path.

6 Bias-Dependent RTN Spectroscopy

The theoretical formulations of the RTN phenomenon are described using quantum physics and they have been applied to extract the defect parameters in a range of dielectric materials [30–43]. It has been found that average capture and emission times in an RTN spectrum depends on the electrical bias at which they are measured, and thus by measuring the voltage dependency of RTN spectra, it is possible to estimate the trap energy level and its physical position within the dielectric layer [88–90].

To obtain the trap parameters, RTN spectra at the test site are measured for at least three different voltage bias (V_G) and the time constants for both capture and emission phenomenon are extracted. A simplified relation between the trap position and average time constants ($\tau_{C,E}$) is given by [41, 91]:

$$\frac{\partial \ln\left(\frac{\tau_C}{\tau_E}\right)}{\partial V_G} = x_T\left(\lambda + \mu x_T V_G\right) \tag{4}$$

where λ and μ depend on the thermal ionization and relaxation energies as well as conduction band offset and effective oxide thickness of the dielectric, and x_T is the distance of the trap from the bottom interface of the dielectric. It is also advisable to measure RTN signals as a function of temperature to obtain the activation energy of the traps. For the detailed derivation of equations of bias dependency of RTN, readers are advised to refer to the following articles [34, 36, 92].

Figure 8a–c shows sections of RTN measured at three different voltages using CAFM at the same location on a HfO_2/W/n-Si stack. The average time constants are extracted from the RTN spectra and plotted in Fig. 8d, which shows that τ_C increases with applied voltage while τ_E remains insensitive to the change in applied bias. Figure 8e shows the plot of $\ln(\tau_C/\tau_E)$ versus V_{CAFM} for this data set, which shows a positive slope. This suggests that the defect causing RTN for this case is located closer to the HfO_2–W interface.

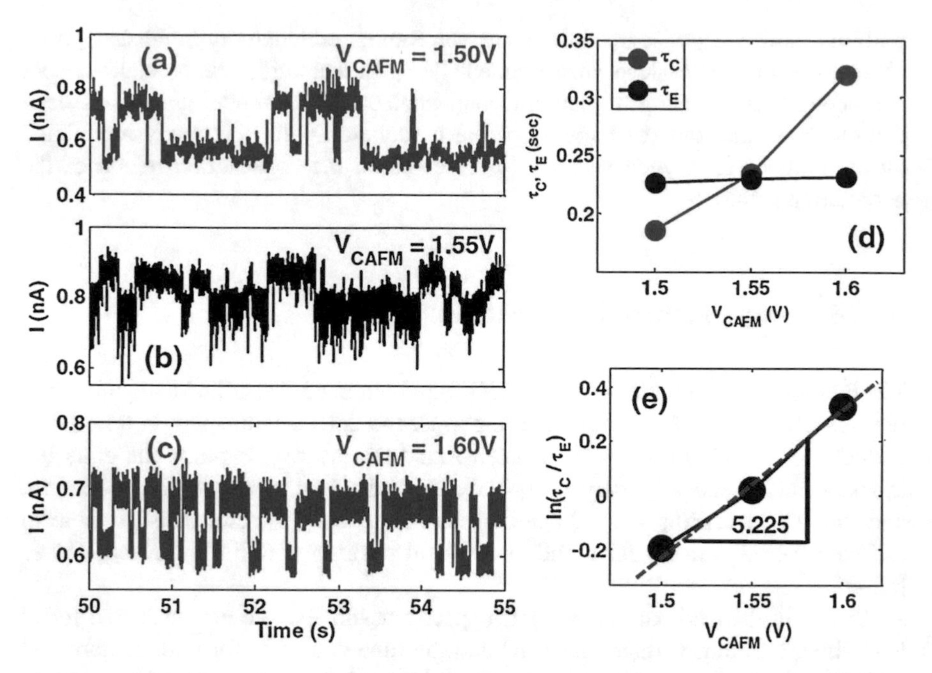

Fig. 8 (**a–c**) RTN traces measured on HfO_2/W/Si stack at three different voltage bias (1.50, 1.55, 1.60 V) after SBD-I. (**d**) Extracted average values of the capture and emission time constants ($\tau_{C,\,E}$) as a function of applied bias. (**e**) Plot of $\ln(\tau_C/\tau_E)$ as a function of applied bias (V_{CAFM}) where the slope can be used to estimate the physical location of the trap. (Reproduced with permission from [64], IEEE Publishers 2016)

7 Metastability of Oxygen Vacancy (V_O) Defects

By metastability, we refer to the tendency of the oxygen vacancy to temporarily transit to an intermediate charge state or structural configuration [93–100]. The metastability of a defect is often characterized by various permissible "state configurations," allowing occasional switching between these different defect "state configurations," with an associated time spent in each "state configuration." While *ab-initio* calculations have pointed to the metastability of oxygen vacancy defects in SiO_2 and HfO_2 as discussed in chapter "The Langevin–Boltzmann Equation for Noise Calculation," experimentally it has been challenging to probe these various possible state configurations (structural/electrical) of the defect [101–106]. Recently, researchers have used RTN spectroscopy to examine the metastability of defects in SiO_2 and HfO_2 in greater detail [94–100, 102–106]. Given the sensitivity of RTN signals to the local atomic potential surrounding the defect, an RTN spectrum provides a real time, direct means to observe metastability of the defect. When a defect changes its "state configuration," it does so by either changing its structure (possibly due to lattice relaxation or phonon events) or charge state

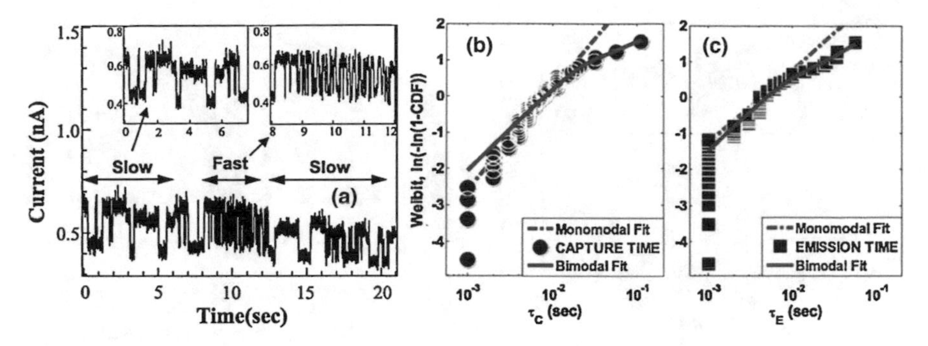

Fig. 9 (**a**) Temporal plot of a metastable RTN signal measured with a STM setup on HfO_2 dielectric. The RTN signal initially shows a slow switching rate, changes to a faster rate, before reverting back to its previous slow rate. (**b, c**) The plot of capture and emission time constants ($\tau_{c,\,e}$) of the RTN signal in (**a**) shows bimodal distribution on the exponential probability scale, indicating the metastable nature of the defect. (Reproduced with permission from [62], AIP Publishers 2016)

[95–107]. This, in turn, changes the local atomic potential for the tunneling electrons, which gives a unique signature to the measured RTN spectra.

Figure 9 plots a noise spectrum measured on HfO_2 films using an STM in which an anomalous RTN trend is seen. While the high and low current levels remain unvaried throughout the noise spectrum, the frequency of transition between the two current levels changed noticeably for certain duration [62]. The time series data show a slow switching rate at the initial and final sections, and a faster switching rate in the middle part. The time constants ($\tau_{c,\,e}$) of the data are extracted and plotted on a Weibull scale, as shown in Fig. 9b, c. Analysis suggests a bimodal distribution for time constants ($\tau_{c,\,e}$). This bimodality in the distribution of $\tau_{c,\,e}$ cannot be explained by a static single defect model. The bimodality is attributed to the existence of the metastable state of V_O defect arising from a temporary change in its structure or charge state [95–102].

Another example using a CAFM where metastability of the defect is observed in the RTN spectra is shown in Fig. 10a, b [64]. Within the same RTN signal, in different portions, the fraction of time spent by the defect in the low and the high current state changes, although the magnitude of the low and high current levels remains the same. For the initial portion as shown in Fig. 10a, the defect preferably spends more time in the high current state, while for the latter portion of the signal, the defect mostly stays in the low current level (Fig. 10b). Analysis of the time constants ($\tau_{c,\,e}$) again shows a bimodal distribution as shown in Fig. 10c, d, suggesting the possible existence of a metastable defect.

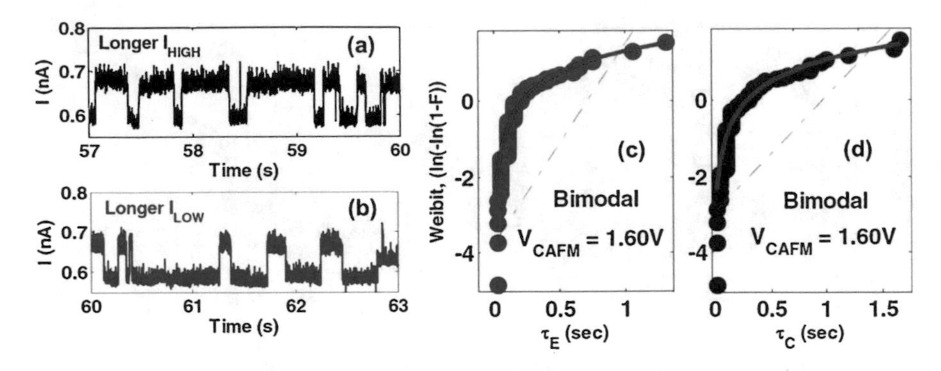

Fig. 10 (**a**, **b**) Two portions of an RTN signal measured using CAFM on HfO_2 at $V_{CAFM} = 1.6$ V showing an inversion in the electron occupation probability in the two portions of the spectra. While for (**a**), the defect spends more time in the higher current state, for (**b**) the defect preferably resides for longer time in the low current state. (**c**, **d**) Exponential probability plot of time constants ($\tau_{c,\,e}$) for the RTN signals measured at $V_{CAFM} = 1.6$ V. Data is fitted using the expectation–maximization (EM) algorithm, which shows a bimodal distribution as a best fit, implying the existence of a metastable defect. (Reproduced with permission from [64], IEEE Publishers 2016)

8 RTN Analysis at Microstructural Sites

In this section, we highlight the capability of SPM for simultaneous high-resolution imaging and spectroscopy of structural/electronic defects, effectively probing defects with spatially resolved RTN nano-spectroscopy. This approach provides a unique capability of isolating and probing defects at different microstructural sites in real time.

We present a case study on mapping grains and grain boundaries (GB) in polycrystalline HfO_2 thin films using UHV–STM and measurement of RTN at the grain sites. Amorphous HfO_2 films used in silicon-based logic and memory devices are prone to poly-crystallization when subjected to annealing or high thermal processes used in CMOS technology [51–54]. The poly-crystallization creates distinct microstructural defects in HfO_2 films (grains and GB). The GBs act as accumulation sites for V_O defects and thus these sites have higher defect concentrations compared to grains in HfO_2 [55–58]. With a highly localized concentration of vacancy defects, the GB acts as precursors for early breakdown of dielectric films used in logic devices [58]. While a polycrystalline dielectric film is not desirable for logic applications, there are emerging fields where they can have potential applications. For instance, GBs in polycrystalline HfO_2 can be used as a preferential path for conductive filament formation in resistive random access memory devices due to its low energy barrier, which in turn can reduce variability in the switching voltage statistics and also reduce the overall energy requirement for the forming process [108]. Therefore, it is of technological importance to characterize the defects at the grain and grain boundary sites both for failure analysis and to engineer them for suitable applications.

To map grain and grain boundary contours in the polycrystalline HfO_2 film deposited on n-Si, we simultaneously acquire the topography and corresponding differential *dI/dV* maps[109]. To acquire the *dI/dV* map, a lock-in amplifier is configured with the STM setup. As *dI/dV* is directly proportional to the surface local density of states, the simultaneous acquisition of topography and *dI/dV* map helps to correlate the microstructure to its electronic properties. Figure 11a, b are the topography and *dI/dV* map of one such area on a HfO_2 film, showing the contours of grains and GB [63]. The bright areas in the topography map correspond to the dark areas (lower conductivity) in the *dI/dV* map. Figure 11c shows the line scan across both topography and *dI/dV* map, where a correlation can be clearly observed. While the grain areas have lower conductivity, the GB shows enhanced conductivity due to the accumulation of vacancy defects. It is also found that while *I–V* breakdown spectra (not shown here for brevity) at GB show enhanced leakage current at low voltages and gradual BD trends, the grain sites show low leakage current and abrupt

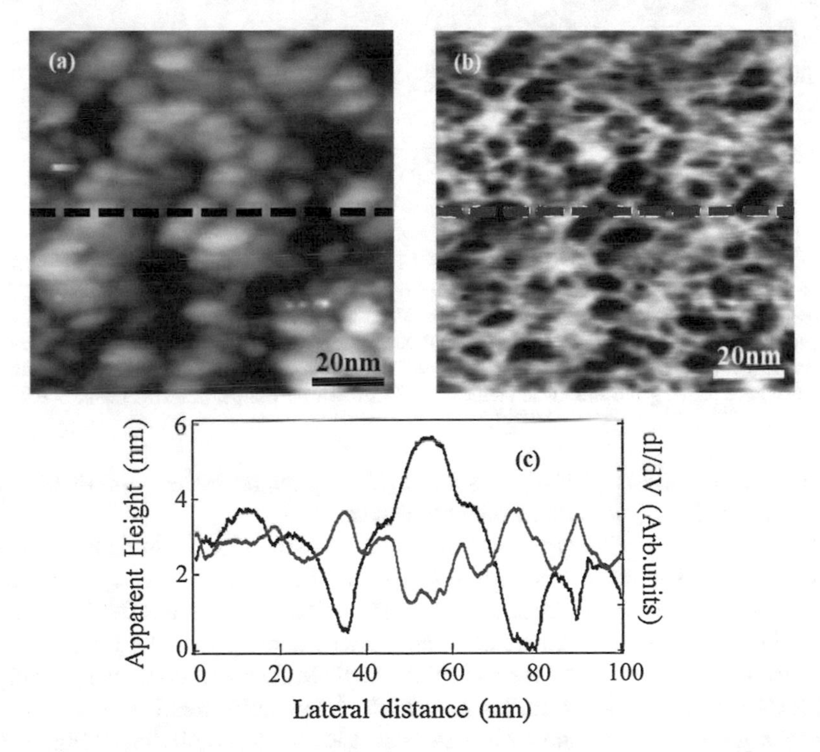

Fig. 11 (**a**) STM topography map and its corresponding (**b**) *dI/dV* map of a polycrystalline HfO_2 thin film. The bright regions in the topography map correspond to the dark regions (lower conductivity) in the *dI/dV* map. Scanning parameters are 3.5 V/50 pA and lock-in parameters are 2.7 kHz/10 mV_{ac}. (**c**) A line scan across both topography and *dI/dV* map showing the correlation of the physical and electronic properties. (Reproduced with permission from [63], AIP Publishers 2017)

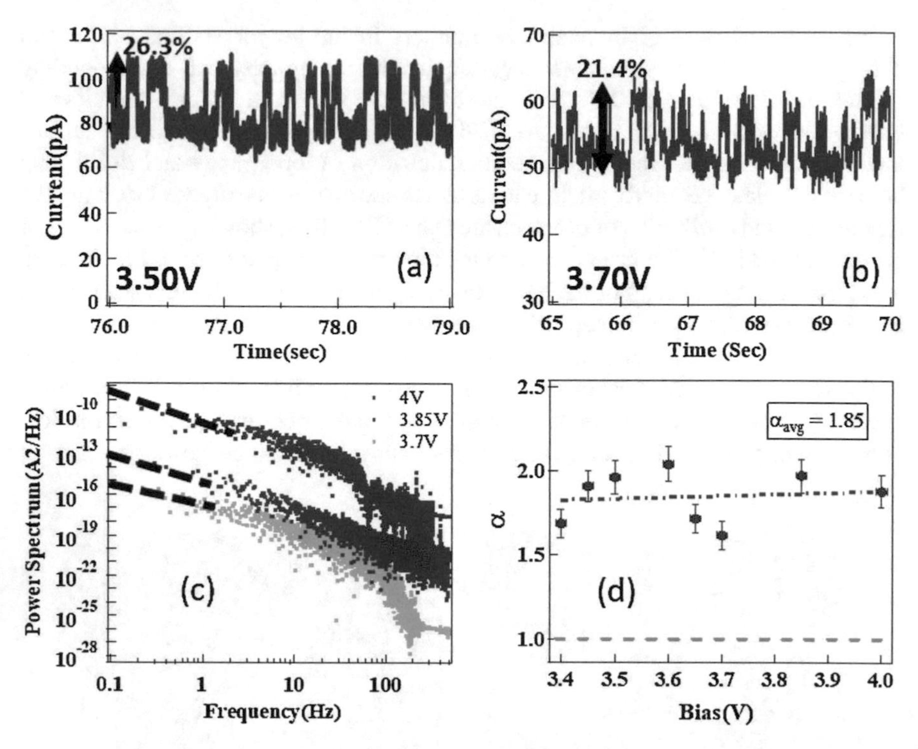

Fig. 12 (**a**, **b**) Noise patterns measured with STM across two isolated grains of HfO_2 at different voltages clearly showing a two-level RTN. (**c**) Plot of power spectral density (PSD) of the RTN spectra measured across multiple grain sites. The slope (α) of the PSD for each of the spectra in (**c**) is extracted and plotted with respect to the applied bias in (**d**). For all the RTN spectra measured across the grain sites, the average value of $\alpha = 1.85$, suggesting that a discrete defect is being probed at each tested grain site. (Reproduced with permission from [63], AIP Publishers 2017)

breakdown characteristics [63]. This finding further supports the claim that GBs in HfO_2 have higher defect density than grain sites.

To probe RTN in polycrystalline thin films, we preferably use grain sites as they have a lower defect density compared to GB and hence provide an opportunity to isolate and individually probe the intrinsic defects. To analyze the RTN signals from the grain, we first acquire topography and *dI/dV* maps to find the grain and GB contours. Once the grain sites are identified, the STM tip is moved to these locations and RTN spectra are acquired. Figure 12a, b shows noise trends measured from two such grain locations and a clear two-level RTN can be observed. Figure 12c shows the power spectral density (PSD) plots for the RTN signals measured across different grain sites. The slope of the PSD is extracted and is plotted in Fig. 12d, which shows an average value of $\alpha = 1.85$, suggestive of Lorentzian behavior from one predominant defect only (refer to Eq. (3)).

9 Summary and Conclusions

In this chapter, we report recent developments in scanning probe microscopy techniques, namely conduction atomic force microscopy (CAFM) and scanning tunneling microscopy (STM), for the measurement of random telegraph noise (RTN) signals in ultrathin high-κ dielectric films. The chapter provides an overview on the challenges associated with the existing probe station-based device-level RTN spectroscopy and the potential of CAFM and STM techniques to effectively address some of these challenges. We have demonstrated the effectiveness of using CAFM- and STM-based RTN nanospectroscopy to study the kinetics of individual/clustered defects in thin high-κ dielectric films. This approach has been used to isolate and spatially identify the individual percolation paths at different stages of dielectric degradation (by controlling the current compliance) and measure the noise patterns in each stage of dielectric degradation. Furthermore, the bias dependency of RTN has been used to infer the physical position of the defects. We have also presented experimental evidence at the nanoscale on the metastability of the oxygen vacancy (V_O) defects in HfO_2. Finally, as a proof of concept, we have demonstrated the application of STM for the measurement of spatially resolved RTN spectra from grain and grain boundaries in HfO_2, which provides an opportunity to directly correlate the electrical defects/traps at the microstructural sites and their effective noise contributions.

References

1. S. Borkar, Designing reliable systems from unreliable components: the challenges of transistor variability and degradation. IEEE Micro **25**(6), 10–16 (2005)
2. S. Nassif et al., High performance CMOS variability in the 65nm regime and beyond, in *IEEE International Electron Devices Meeting*, 2007, pp. 569–571
3. G. Ribes et al., Review on high-κ dielectrics reliability issues. IEEE Trans. Device Mater. Reliab. **5**(1), 5–19 (2005)
4. A. Kerber et al., Reliability challenges for CMOS technology qualifications with hafnium oxide/titanium nitride gate stacks. IEEE Trans. Device Mater. Reliab. **9**(2), 147–162 (2009)
5. H.D. Xiong et al., Characterization of electrically active defects in high-κ gate dielectrics by using low frequency noise and charge pumping measurements. Microelectron. Eng. **84**(9–10), 2230–2234 (2007)
6. B.C. Wang et al., Comparison of the trap behavior between ZrO_2 and HfO_2 gate stack nMOSFETs by 1/f noise and random telegraph noise. IEEE Electron Device Lett. **34**(2), 151–153 (2013)
7. E. Simoen et al., On the oxide trap density and profiles of 1-nm EOT metal-gate last CMOS transistors assessed by low-frequency noise. IEEE Trans. Electron Devices **60**(11), 3849–3855 (2013)
8. K. Onishi et al., Bias-temperature instabilities of polysilicon gate HfO_2 MOSFETs. IEEE Trans. Electron Devices **50**(6), 1517–1524 (2003)
9. S. Kalpat et al., BTI characteristics and mechanisms of metal gated HfO_2 films with enhanced interface/bulk process treatments. IEEE Trans. Device Mater. Reliab. **5**(1), 26–35 (2005)

10. S. Pae et al., BTI reliability of 45 nm high-κ + metal-gate process technology, in *IEEE International Reliability Physics Symposium*, 2008, pp. 352–357
11. Z. Xu et al., Constant voltage stress induced degradation in HfO_2/SiO_2 gate dielectric stacks. J. Appl. Phys. **91**(12), 10127 (2002)
12. L. Vandelli et al., Microscopic modeling of electrical stress-induced breakdown in polycrystalline hafnium oxide dielectrics. IEEE Trans. Electron Devices **60**(5), 1754–1762 (2013)
13. C. Mannequin et al., Stress-induced leakage current and trap generation in HfO_2 thin films. J. Appl. Phys. **112**(7), 074103 (2012)
14. R. Degraeve et al., Degradation and breakdown of 0.9 nm EOT SiO_2 ALD HfO_2 metal gate stacks under positive constant voltage stress, in *IEEE International Electron Devices Meeting*, 2005, pp. 408–411
15. V.L. Lo et al., Multiple digital breakdowns and its consequence on ultrathin gate dielectrics reliability prediction, in *IEEE International Electron Devices Meeting*, 2007, pp. 497–500
16. J.S. Brugler et al., Charge pumping in MOS devices. IEEE Trans. Electron Devices **16**(3), 297–302 (1969)
17. G. Van den Bosch et al., Spectroscopic charge pumping: a new procedure for measuring interface trap distributions on MOS transistors. IEEE Trans. Electron Devices **38**(8), 1820–1831 (1991)
18. M. Cho et al., Interface trap characterization of a 5.8 Å EOT p-MOSFET using high-frequency on-chip ring oscillator charge pumping technique. IEEE Trans. Electron Devices **58**(10), 3342–3349 (2011)
19. J.T. Ryan et al., Frequency-modulated charge pumping with extremely high gate leakage. IEEE Trans. Electron Devices **62**(3), 769–775 (2015)
20. K. Sakakibara et al., Identification of stress-induced leakage current components and the corresponding trap models in SiO_2 films [MOS transistors]. IEEE Trans. Electron Devices **44**(12), 2267–2273 (1997)
21. W.L. Chang et al., Role of interface layer in stress-induced leakage current in high-κ/metal-gate dielectric stacks, in *IEEE International Reliability Physics Symposium*, 2010, pp. 787–791
22. R. O'Connor et al., SILC defect generation spectroscopy in HfSiON using constant voltage stress and substrate hot electron injection, in *IEEE International Reliability Physics Symposium*, 2008, pp. 324–329
23. N. Raghavan et al., Spectroscopy of SILC trap locations and spatial correlation study of percolation path in the high-κ and interfacial layer, in *IEEE International Reliability Physics Symposium*, 2015, pp. 5A.2.1–5A.2.7
24. K. Joshi et al., A fast reliability screening technique for identification of trap generation, in *IEEE International Reliability Physics Symposium*, 2016, pp. DI-3-1–DI-3-5
25. C. Mannequin et al., Dielectric relaxation in hafnium dioxide: a study of transient currents and admittance spectroscopy in HfO_2 metal-insulator-metal devices. J. Appl. Phys. **110**(10), 104108 (2011)
26. C.H. Yang et al., Charge detrapping and dielectric breakdown of nanocrystalline zinc oxide embedded zirconium-doped hafnium oxide high-κ dielectrics for nonvolatile memories. Appl. Phys. Lett. **96**(19), 192106 (2010)
27. B.H. Lee et al., Transient charging and relaxation in high-κ gate dielectrics and their implications. Jpn. J. Appl. Phys. **44**(Part 1, 4S), 2415–2419 (2005)
28. M.S. Rahman et al., Dielectric relaxation and charge trapping characteristics study in germanium based MOS devices with HfO_2/Dy_2O_3 gate stacks. IEEE Trans. Electron Devices **58**(10), 3549–3558 (2011)
29. D. Veksler et al., Analysis of charge pumping data for identification of dielectric defects. IEEE Trans. Electron Devices **60**(05), 1514–1522 (2013)
30. M.J. Uren et al., 1/f and random telegraph noise in silicon metal-oxide-semiconductor field-effect transistors. Appl. Phys. Lett. **47**(11), 1195 (1985)
31. K.K. Hung et al., Random telegraph noise of deep-submicrometer MOSFETs. IEEE Electron Device Lett. **11**(2), 90–92 (1990)

32. E. Simoen et al., Explaining the amplitude of RTS noise in submicrometer MOSFETs. IEEE Trans. Electron Devices **39**(2), 422–429 (1992)
33. A. Avellán et al., Modeling random telegraph signals in the gate current of metal–oxide–semiconductor field effect transistors after oxide breakdown. J. Appl. Phys. **94**(1), 703 (2003)
34. C.M. Chang et al., The observation of trapping and detrapping effects in high-κ gate dielectric MOSFETs by a new gate current random telegraph noise (I_G-RTN) approach, in *IEEE International Electron Devices Meeting*, 2008, pp. 1–4
35. C. Monzio Compagnoni et al., Statistical model for random telegraph noise in flash memories. IEEE Trans. Electron Devices **55**(1), 388–395 (2008)
36. S. Lee et al., Characterization of oxide traps leading to RTN in high-κ and metal gate MOSFETs, in *IEEE International Electron Devices Meeting*, 2009, pp. 1–4
37. X. Li et al., The physical origin of random telegraph noise after dielectric breakdown. Appl. Phys. Lett. **94**(13), 132904 (2009)
38. T. Nagumo et al., Statistical characterization of trap position, energy, amplitude and time constants by RTN measurement of multiple individual traps, in *International Electron Devices Meeting*, 2010, pp. 28.3.1–28.3.4
39. E. Simoen et al., Random telegraph noise: from a device physicist's dream to a designer's nightmare. ECS Trans. **39**(1), 3–15 (2011)
40. N. Raghavan et al., Microscopic origin of random telegraph noise fluctuations in aggressively scaled RRAM and its impact on read disturb variability, in *IEEE International Reliability Physics Symposium*, 2013, pp. 5E.3.1–5E.3.7
41. W. Liu et al., Analysis of correlated gate and drain random telegraph noise in post-soft breakdown TiN/HfLaO/SiO_x nMOSFETs. IEEE Electron Device Lett. **35**(2), 157–159 (2014)
42. F.M. Puglisi et al., A complete statistical investigation of RTN in HfO_2-based RRAM in high resistive state. IEEE Trans. Electron Devices **62**(8), 2606–2613 (2015)
43. Z. Chai et al., Probing the critical region of conductive filament in nanoscale HfO_2 resistive-switching device by random telegraph signals. IEEE Trans. Electron Devices **64**(10), 4099–4105 (2017)
44. T. Grasser et al., On the microscopic structure of hole traps in pMOSFETs, in *IEEE International Electron Devices Meeting*, San Francisco, CA, 2014, pp. 21.1.1–21.1.4
45. A. Asenov, Random dopant induced threshold voltage lowering and fluctuations in sub-0.1μm MOSFET's: a 3-D "atomistic" simulation study. IEEE Trans. Electron Devices **45**(12), 2505–2513 (1998)
46. A. Asenov et al., Simulation of intrinsic parameter fluctuations in decananometer and nanometer-scale MOSFETs. IEEE Trans. Electron Devices **50**(9), 1837–1852 (2003)
47. T. Nagumo et al., New analysis methods for comprehensive understanding of random telegraph noise, in *IEEE International Electron Devices Meeting*, 2009, pp. 1–4
48. M.J. Kirton et al., Individual defects at the Si:SiO_2 interface. Semicond. Sci. Technol. **4**(12), 1116 (1989)
49. J.P. Campbell et al., Random telegraph noise in highly scaled nMOSFETs, in *IEEE International Reliability Physics Symposium*, 2009, pp. 382–388
50. J. Robertson et al., High dielectric constant oxides. Eur. Phys. J. Appl. Phys. **28**(3), 265–291 (2004)
51. J. Aarik et al., Texture development in nanocrystalline hafnium dioxide thin films grown by atomic layer deposition. J. Cryst. Growth **220**(1–2), 105–113 (2000)
52. H. Kim et al., Effects of crystallization on the electrical properties of ultrathin HfO_2 dielectrics grown by atomic layer deposition. Appl. Phys. Lett. **82**(1), 106 (2003)
53. G.D. Wilk et al., Correlation of annealing effects on local electronic structure and macroscopic electrical properties for HfO_2 deposited by atomic layer deposition. Appl. Phys. Lett. **83**(19), 3984 (2003)
54. D. Triyoso et al., Impact of deposition and annealing temperature on material and electrical characteristics of ALD HfO_2. J. Electrochem. Soc. **151**(10), F220–F227 (2004)

55. K. McKenna et al., Electronic properties of defects in polycrystalline dielectric materials. Microelectron. Eng. **86**(7–9), 1751–1755 (2009)
56. K. McKenna et al., The interaction of oxygen vacancies with grain boundaries in monoclinic HfO_2. Appl. Phys. Lett. **95**(22), 222111 (2009)
57. G. Bersuker et al., Grain boundary-driven leakage path formation in HfO_2 dielectrics, in *Proceedings of the European Solid State Device Research Conference*, 2010, pp. 333–336
58. K. Shubhakar et al., Study of preferential localized degradation and breakdown of HfO_2/SiO_x dielectric stacks at grain boundary sites of polycrystalline HfO_2 dielectrics. Microelectron. Eng. **109**, 364–369 (2013)
59. Y. Kim et al., Mechanical control of electroresistive switching. Nano Lett. **12**(9), 4068–4074 (2013)
60. A. Ranjan et al., Analysis of quantum conductance, read disturb and switching statistics in HfO_2 RRAM using conductive AFM. Microelectron. Reliab. **64**, 172–178 (2016)
61. A. Ranjan et al., Localized random telegraphic noise study in HfO_2 dielectric stacks using scanning tunneling microscopy—analysis of process and stress-induced traps, in *IEEE International Symposium on the Physical and Failure Analysis of Integrated Circuits*, 2015, pp. 458–462
62. R. Thamankar et al., Single vacancy defect spectroscopy on HfO_2 using random telegraph noise signals from scanning tunneling microscopy. J. Appl. Phys. **119**(8), 084304 (2016)
63. R. Thamankar et al., Localized characterization of charge transport and random telegraph noise at the nanoscale in HfO_2 films combining scanning tunneling microscopy and multi-scale simulations. J. Appl. Phys. **122**(2), 024301 (2017)
64. A. Ranjan et al., CAFM based spectroscopy of stress-induced defects in HfO_2 with experimental evidence of the clustering model and metastable vacancy defect state, in *IEEE International Reliability Physics Symposium*, 2016, pp. 7A-4-1–7A-4-7
65. A. Ranjan et al., Random telegraph noise in 2D hexagonal nitride dielectric films. Appl. Phys. Lett. **112**(13), 133505 (2018)
66. A. Ranjan et al., Conductive atomic force microscope study of bipolar and threshold resistive switching in 2D hexagonal boron nitride films. Sci. Rep. **8**(1), 2854 (2018)
67. G. Binning et al., Surface studies by scanning tunneling microscopy. Phys. Rev. Lett. **49**(1), 57–61 (1982)
68. J.V. Barth et al., Scanning tunneling microscopy observations on the reconstructed Au (111) surface: atomic structure, long-range superstructure, rotational domains, and surface defects. Phys. Rev. B **42**(15), 9307, 57–61 (1990)
69. U. Diebold et al., Evidence for the tunneling site on transition-metal oxides: TiO_2 (110). Phys. Rev. Lett. **77**(7), 1322–1325 (1996)
70. S. Marchini et al., Scanning tunneling microscopy of graphene on Ru (0001). Phys. Rev. B **76**(7), 075429 (2007)
71. A. Hofer et al., Analysis of crystal defects on GaN-based semiconductors with advanced scanning probe microscope techniques. Thin Solid Films **544**, 139–143 (2013)
72. C. Couso et al., Conductance of threading dislocations in InGaAs/Si stacks by temperature-CAFM measurements. IEEE Electron Device Lett. **37**(5), 640–643 (2016)
73. M.E. Welland et al., Spatial location of electron trapping defects on silicon by scanning tunneling microscopy. Appl. Phys. Lett. **48**(11), 724–726 (1986)
74. M.P. Murrell et al., Spatially resolved electrical measurements of SiO_2 gate oxides using atomic force microscopy. Appl. Phys. Lett. **62**(7), 786–788 (1992)
75. S.J. O'Shea et al., Conducting atomic force microscopy study of silicon dioxide breakdown. J. Vac. Sci. Technol. B **13**(5), 1945–1952 (1995)
76. Y.C. Ong et al., Bilayer gate dielectric study by scanning tunneling microscopy. Appl. Phys. Lett. **91**(10), 102905 (2007)
77. K.L. Pey et al., Physical analysis of breakdown in high-κ/metal gate stacks using TEM/EELS and STM for reliability enhancement. Microelectron. Eng. **88**(7), 1365–1372 (2011)
78. K. Shubhakar et al., Grain boundary assisted degradation and breakdown study in cerium oxide gate dielectric using scanning tunneling microscopy. Appl. Phys. Lett. **98**(7), 072902 (2011)

79. K.S. Yew et al., Bimodal Weibull distribution of metal/high-κ gate stack TDDB—insights by scanning tunneling microscopy. IEEE Electron Device Lett. **33**(2), 146–148 (2012)
80. B.J. Choi et al., Resistive switching mechanism of TiO_2 thin films grown by atomic-layer deposition. J. Appl. Phys. **98**(3), 033715 (2015)
81. J.P. Ibe et al., On the electrochemical etching of tips for scanning tunneling microscopy. J. Vac. Sci. Technol. A **8**(4), 3570–3375 (1990)
82. J.P. Song et al., A development in the preparation of sharp scanning tunneling microscopy tips. Rev. Sci. Instrum. **64**(4), 99–903 (1993)
83. S. Kerfriden et al., The electrochemical etching of tungsten STM tips. Electrochim. Acta **43**(12–13), 1939–1944 (1998)
84. A. Ranjan et al., The interplay between drift and electrical measurement in conduction atomic force microscopy. Rev. Sci. Instrum. **90**(7), 073701 (2019)
85. E. Miranda et al., The quantum point-contact memristor. IEEE Electron Device Lett. **33**(10), 1474–1476 (2012)
86. N. Raghavan et al., Modeling the impact of reset depth on vacancy-induced filament perturbations in HfO_2 RRAM. IEEE Electron Device Lett. **34**(5), 614–616 (2013)
87. R. Degraeve et al., Hourglass concept of RRAM: a dynamic and statistical device model, in *21st IEEE International Symposium on the Physical and Failure Analysis of Integrated Circuits*, 2014
88. Z. Celik-Butler et al., A method for locating the position of oxide traps responsible for random telegraph signals in submicron MOSFETs. IEEE Trans. Electron Devices **47**(3), 646–648 (2000)
89. H. Cho et al., Observation of slow oxide traps at MOSFETs having metal/high-κ gate dielectric stack in accumulation mode. IEEE Trans. Electron Devices **57**(10), 2697–2703 (2010)
90. Y. Li et al., Random-telegraph-signal noise in AlGaN/GaN MIS-HEMT on silicon. Electron. Lett. **49**(2), 156–157 (2013)
91. L. Larcher et al., Leakage current in HfO_2 stacks: from physical to compact modeling, in *Proceedings of the International Workshop on Compact Modeling*, 2012, pp. 809–814
92. H.-J. Cho et al., Extraction of trap energy and location from random telegraph noise in gate leakage current (I_g RTN) of metal-oxide semiconductor field effect transistor (MOSFET). Solid State Electron. **54**, 362–367 (2010)
93. F.M. Puglisi et al., RTS noise characterization of HfO_x RRAM in high resistive state. Solid State Electron. **84**, 160–166 (2013)
94. M.J. Uren et al., Anomalous telegraph noise in small-area silicon metal-oxide-semiconductor field-effect transistors. Phys. Rev. B Condens. Matter **37**(14), 8346–8350 (1988)
95. T. Grasser et al., Switching oxide traps as the missing link between negative bias temperature instability and random telegraph noise, in *IEEE International Electron Devices Meeting*, 2009, pp. 1–4
96. T. Grasser et al., The time dependent defect spectroscopy (TDDS) for the characterization of the bias temperature instability, in *IEEE International Reliability Physics Symposium*, 2010, pp. 16–25
97. T. Grasser, Stochastic charge trapping in oxides: from random telegraph noise to bias temperature instabilities. Microelectron. Reliab. **52**(1), 39–70 (2012)
98. F.M. Puglisi et al., Characterization of anomalous random telegraph noise in resistive random access memory, in *Proceedings of the 45th European Solid-State Device Research Conference*, 2015, pp. 270–273
99. F.M. Puglisi et al., Anomalous random telegraph noise and temporary phenomena in resistive random access memory. Solid State Electron. **125**, 204–213 (2016)
100. S. Guo et al., Anomalous random telegraph noise in nanoscale transistors as direct evidence of two metastable states of oxide traps. Sci. Rep. **7**(1), 6239 (2017)
101. C.J. Nicklaw et al., The structure, properties, and dynamics of oxygen vacancies in amorphous SiO_2. IEEE Trans. Nucl. Sci.Trans. **49**(6), 2667–2673 (2002)

102. J. Ji et al., New framework for the random charging/discharging of oxide traps in HfO_2 gate dielectric: *ab-initio* simulation and experimental evidence, in *IEEE International Electron Devices Meeting*, 2014, pp. 21.4.1–21.4.4
103. Y. Qiu et al., Deep understanding of oxide defects for stochastic charging in nanoscale MOSFETs, in *Silicon Nanoelectronics Workshop*, 2014, pp. 1–2
104. K.P. McKenna, D.M. Ramo, Electronic and magnetic properties of the cation vacancy defect in m-HfO_2. Phys. Rev. B **92**(20), 205124 (2015)
105. D.M. Fleetwood, 1/f noise and defects in microelectronic materials and devices. IEEE Trans. Nucl. Sci.Trans. **62**(4), 1462–1486 (2015)
106. T. Grasser et al., Advanced characterization of oxide traps: the dynamic time-dependent defect spectroscopy, in *IEEE International Reliability Physics Symposium*, 2013, pp. 2D-2-1–2D-2-7
107. A. Padovani et al., A microscopic mechanism of dielectric breakdown in SiO_2 films: an insight from multi-scale modeling. J. Appl. Phys. **121**(15), 155101 (2017)
108. M. Lanza et al., Grain boundaries as preferential sites for resistive switching in the HfO_2 resistive random access memory structures. Appl. Phys. Lett. **100**(12), 123508 (2012)
109. C.J. Chen, *Introduction to Scanning Tunneling Microscopy* (Oxford University Press, New York, 1993)

RTN and Its Intrinsic Interaction with Statistical Variability Sources in Advanced Nano-Scale Devices: A Simulation Study

F. Adamu-Lema, C. Monzio Compagnoni, O. Badami, V. Georgiev, and A. Asenov

1 Introduction

The physical dimension of the conventional MOSFETs and other novel devices currently under mass production has reached decananometer scales. Major semiconductor manufacturers have already started shipping bulk FinFETs of 14 and 22 nm technology process with FIN widths of 7–10 nm [1, 2]. In such small devices, the interface imperfections (defect states) and trapped charges in the gate oxide make the reliability concerns worse and their impact became undesirable on some important device parameters. For example, trapping of a single charge carrier in defect states near the Si/SiO_2 interface and the related local modulation in carrier density and/or mobility [3–5] has an impact on an area comparable with the characteristic device dimensions, with a consequent profound impact on the drain

We thank S.M. Amoroso for his insightful comments and constructive discussions, which greatly improved the manuscript.

F. Adamu-Lema (✉) · O. Badami · V. Georgiev · A. Asenov
University of Glasgow, Glasgow, UK
e-mail: fikru.adamu-lema@glasgow.ac.uk; Oves.Badami@glasgow.ac.uk; Vihar.Georgiev@glasgow.ac.uk; Asen.Asenov@glasgow.ac.uk

C. Monzio Compagnoni
Dipartimento di Elettronica, Informazione e Bioingegneria, Politecnico di Milano, Milan, Italy
e-mail: christian.monzio@polimi.it

T. Grasser (ed.), *Noise in Nanoscale Semiconductor Devices*,
https://doi.org/10.1007/978-3-030-37500-3_13

and gate current [6]. Corresponding RTN fluctuations with amplitudes larger than 50% the on-current have already been reported at room temperature in nano-scale devices [7].

Current fluctuations on such a scale are a serious issue, not only as a source of excessive low-frequency noise in analogue and mixed-mode circuits [8, 9] but also in dynamic memories and potentially in digital applications. Depending on the device geometry a single [10] or few discrete charges [11] (see Fig. 1) trapped in hot carrier or radiation created defect states will be sufficient to cause a pronounced degradation in deeply scaled semiconductor devices. As shown in Fig. 1 the impact of RTN is detrimental to the current flow of the transistor. The figure is a 2D potential profile showing how a single trapped charge which is located in the centre of the device (Si/SiO_2 interface) is splitting the current path and blocking the current flow by creating two percolative paths. On the other hand, the bottom picture shows the current density when no trap is introduced in a continuously doped device simulation.

Many experimental studies have been reported that characterize RTN [8, 12–14]. However, simulation can help to forecast the RTN amplitudes that should be expected in decananometer devices. In this work, we exploit the capability of 3D statistical variability aware modelling and simulation tools to study the wide range

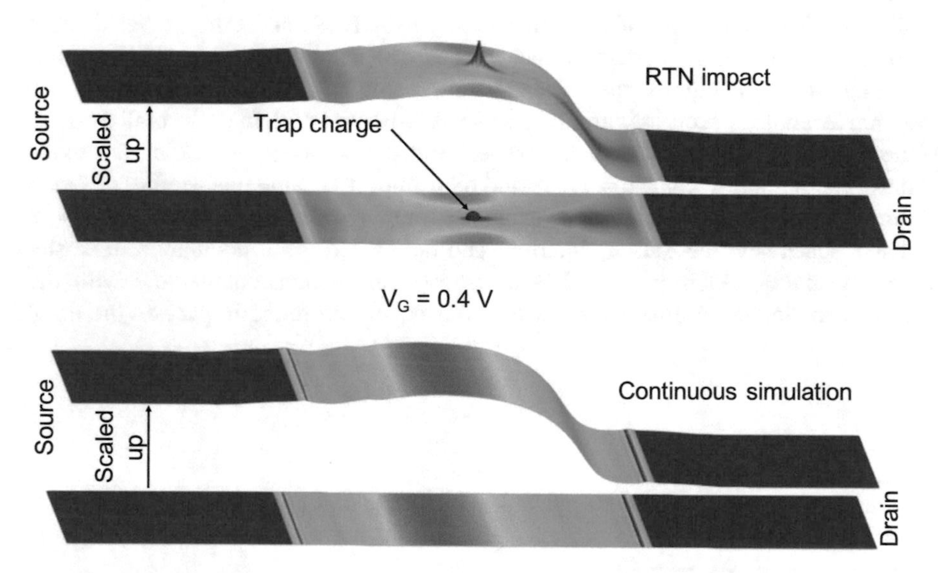

Fig. 1 A 2D potential profile showing the impact of a single trapped charge located in the centre of a MOSFET (Si/SiO_2 interface). The top current density profile is when a single charge is blocking the current flow by creating a percolative path. Due to the percolative nature of the source-to-drain conduction, the trapped charge will induce different values of RTN related amplitudes depending on its position over the channel and on the underlying RDF in the substrate. This fluctuation will have a negative impact on device and circuit operation, compromising the achievement of a reliability-robust design. On the other hand, the bottom picture shows a current density when no trap is introduced

of RTN amplitudes observed in otherwise identical devices [15] and particularly their statistical distribution and interaction with the underlining discreteness of dopant and trapped charges.

There are suggestions that due to surface potential fluctuations and channel non-uniformity strategically located traps influence the magnitude and the spreading of RTN amplitudes [8, 15, 16]. However, such potential fluctuations have been associated mainly with oxide non-uniformity [17] and fixed and trapped interface charges [18]. Additionally, the impact of the random discrete dopants, which are one of the major sources of fluctuations in decananometer conventional bulk MOSFETS [19], has already been shown in [20]. In this chapter, we extend the investigation of RTN effects in FinFETs and SOI and to the latest conventional MOSFETs, by considering other major sources of statistical variability.

The interaction (intrinsic interplay) [21] between RTN traps and the statistical variability sources in advanced MOS devices impacts not only the amplitude of the resulting RTN fluctuation in terms of threshold voltage and drain current, but also the RTN fluctuation time constants and the possibility to extract from them the trap position z_T in the gate oxide. The variability typically resulting in the capture and emission time constants of RTN traps, in particular, results in inaccuracies in the spectroscopic analysis of the traps and has also unwanted outcomes on some experimental setups used to measure and characterize RTN.

The estimation of z_T cannot, in fact, make use of the formula given in [22, 23]:

$$\ln\left(\frac{\tau_c}{\tau_e}\right) = K - \frac{q}{kT}\left[\left(1 - \frac{z_T}{T_{ox}}\right)\Psi_s + \frac{z_T}{t_{ox}} V_G\right], \tag{1}$$

where q is the electronic charge, kT is the thermal energy, t_{ox} is the oxide thickness.

The previous equation, in fact, is valid under the assumption of a continuously doped (uniform) device electrostatics. However, in nano-scale devices, many variability sources introduce randomness not only in the value of the surface potential (Ψ_s) and the quasi-Fermi level for electrons ($E_{F,n}$) at each point of a nano scale device, but also in the ratio between the τ_c and τ_e of an RTN trap randomly located over device channel. As a consequence, relying on the change with (V_G) of the ratio τ_c / τ_e to extract the position of an RTN trap in an advanced MOS device results in relevant errors coming from the approximations involved in (1), which cannot be improved due to the intrinsic variability involved in device physics. In Sect. 4, we will discuss further the implication of formula (1) in estimating the trap position, z_T.

The remaining section of this chapter is organized as follows. In Sect. 2 we navigate through the methodology employed to simulate RTN and its interaction with statistical variability sources. In Sect. 3 we provide a brief background discussion of RTN and statistical variability sources with simulation results on ΔV_T and further results and analysis on capture and emission time constants will be discussed in Sect. 4 and concluding remarks will be drawn in the final section.

2 Computational Scheme and Device Specifications

This section describes the overall methodology of the simulation framework employed to produce results and the devices we used in this chapter. The core of the framework is the 3D quantum corrected drift-diffusion (DD) simulator GARAND https://www.synopsys.com/silicon/tcad.html featuring physics-based modelling of the major sources of variability, reliability module capable of describing reliability issues of semiconductor devices was added to the DD. Description of device structure and the simulation methodology will be presented in the next sections.

2.1 Template Devices

Three device architectures (conventional bulk transistor, FDSOI, and FinFET) which are representatives of the current CMOS technology process have been used in the simulations. An ensemble of atomistic device simulations, which are microscopically different, were performed using the 3D simulator GARAND https://www.synopsys.com/silicon/tcad.html. A 3D simulation image of devices is shown in Fig. 2 including all variability sources and a single trap in the centre of the Si/SiO_2 interface with a 3D potential profile and an equi-concentration contour of interface region. The first of the three devices is a well-scaled 25 nm gate length template conventional MOSFET, featuring a thin silicon dioxide with a thickness of $t_{ox} = 0.85$ nm and a metal gate.

A source-and-drain doping figure is given in Table 1. Lightly doped regions extending 5 nm under the gate were assumed. A retrograde doping profile was considered in the substrate, aiming at reducing the number of dopants close to the channel surface, and in turn, the impact of random discrete dopant (RDD) on device variability. The second device is the FDSOI MOSFET also shown in Fig. 2. Important physical and doping parameters are listed in Table 1. The third test-bed transistor used in this study is representative of the 10 nm technology generation bulk n-channel FinFETs and is described in greater detail elsewhere [24]. Two cases are considered for each type of device simulations. The first case is a simulation of atomistic devices with all major sources of variability considered. The second case

Table 1 Selected physical and doping parameters of all three devices

Bulk	FDSOI	FinFET
$t_{ox} = 0.85$ nm	$t_{ox1} = 1.2$ nm	$t_{ox} = 2$ nm
$L_G = 25$ nm	$t_{ox2} = 0.8$ nm	$h_{Fin} = 42$ nm
$N_A = 1 \times 10^{20}$ cm^{-3}	$L_G = 24$ nm	$W_{Fin} = 8$ nm; $L_G = 18$ nm
$N_{CH} = 5 \times 10^{18}$ cm^{-3}	$t_{Si} = 0.7$ nm	$N_D = 1 \times 10^{20}$ cm^{-3}
		$N_{CH} = 1 \times 10^{17}$ cm^{-3}

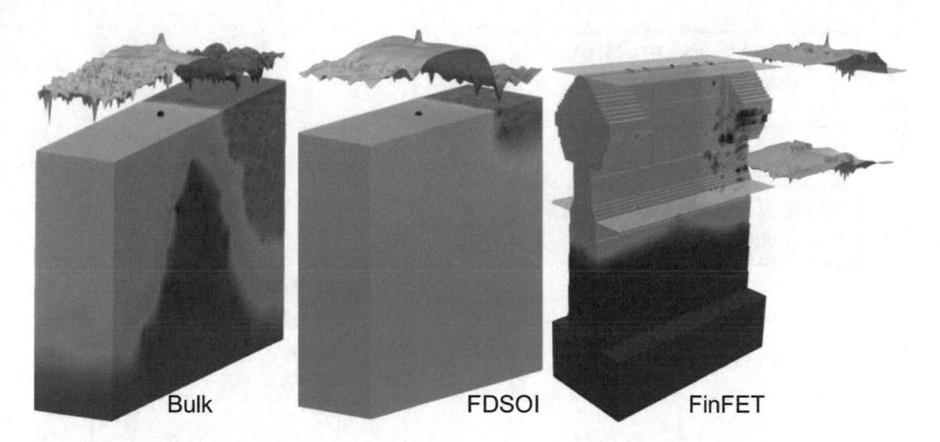

Fig. 2 A 3D-potential distribution in the conventional bulk MOSFET, fully depleted silicon on insulator 3rd generation FinFET technology [1]. In the case of the bulk and FDSOI, the top picture is a 2D potential profile that is extracted at the interface of Si/SiO_2, while in the FinFET case the two profiles are extracted at the top and bottom of the fin

is a combination of RTN and statistical variability at high and low drain biases. This results in a total of 12,000 device simulation runs. This sample size will make sure that the distributions and calculated figure of merit characteristic properties reflect the experimental findings.

2.2 The 3D Atomistic Simulator and RTN Trap Introduction

The simulation methodology emanates from the 3D drift-diffusion simulator GARAND, featuring density-gradient corrections and physics-based modelling of the major sources of statistical variability [25], to which the modelling of oxide trap dynamics was added, as described in [26]. An accept/reject algorithm based on the continuous doping profile of the device was used to implement random discrete dopant (RDD) [27], in both the substrate and source/drain regions. On the other hand, the metal grain granularity (MGG) effects were included by generating random grain patterns and changing the metal gate work function [28, 29] between $\Phi_m^{Bulk} = 3.8523$–4.0523 eV for the conventional transistor, $\Phi_m^{Fin} = 4.3681$–4.5681 eV for FinFET, and $\Phi_m^{SOI} = 4.58$–4.78 eV for FDSOI, with the occurrence probability of 60% and 40%, respectively, based on the experimental data reported in [30]. Finally, line edge roughness (LER) was implemented based on a 1-D Fourier synthesis technique [31, 32], by generating gate patterns at the source and drain edges of the device from power spectrum of a Gaussian auto-correlation function with root mean square of 1.33 nm and correlation length of 30 nm [33, 34].

The computational scheme of evaluating the statistical variability in τ_c, τ_e, and ΔV_T was done, following the algorithm shown in Fig. 3. In order to map the

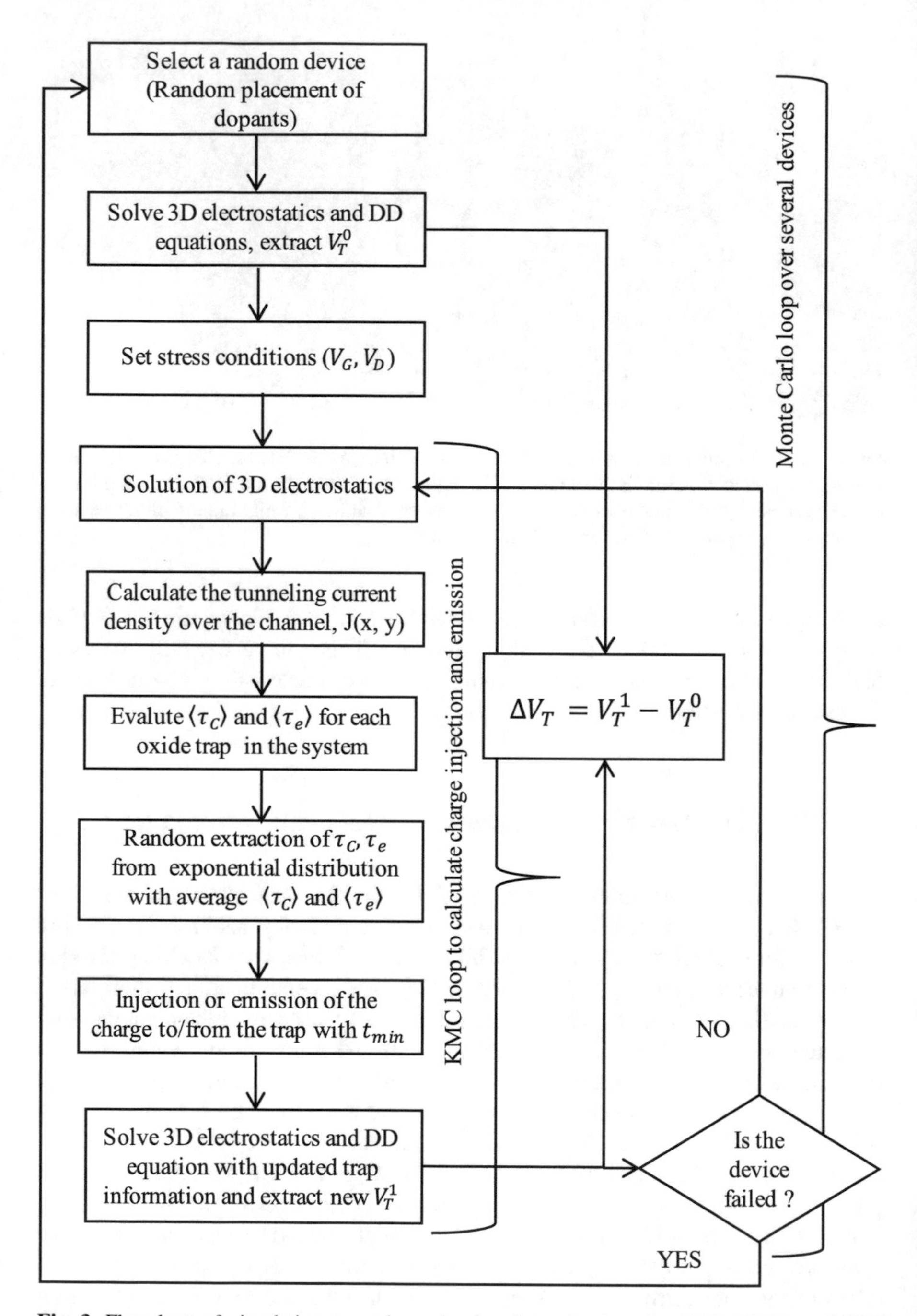

Fig. 3 Flowchart of simulation procedure, developed to simulate the RTN, NBTI, and TAT phenomena showing the incorporation of kinetic Monte Carlo (KMC) module to account for trap dynamics, [40]

features associated with an individual trap and its interplay with the statistical variability sources, we define an auxiliary lateral grid to capture the locations of traps at position (z_T) above the Si/SiO$_2$ interface. The following steps of the computational loop shown in Fig. 3 are executed for each trap location (x_T, y_T, z_T) of the auxiliary grid, after placing a trap in it. First, 3D device electrostatic and drift-diffusion equations are solved self consistently without a charge in the trap, to obtain the initial threshold voltage V_T. Then, the direct tunnelling current density J(x,y) reaching the trap from the channel is computed using the WKB approximation. The corresponding average capture time of the trap is calculated by integrating J(x,y) over the cross-section (σ_s) of the trap, and applying a multi-phonon activation energy term to take into account the experimentally observed temperature dependence [8, 26]

$$\langle \tau_c \rangle = \exp\left(\frac{E_a}{kT}\right) \frac{q}{\int J(x, y, z)\, dxdy} \tag{2}$$

The exponential factor in the trap cross-section takes into account the activation energy (E_a) of the trapping process, the latter confirmed by many experimental works to be an inelastic multi-phonon assisted charge transfer from channel to oxide trap [35]. Please note that, in this work, the dependence of the activation energy from the electric field [36, 37] and corrective factors due to Coulomb blockade effects [38] are neglected and the corresponding average emission time constant $\langle \tau_e \rangle$ is obtained from the SRH statistics, [39]:

$$\langle \tau_e \rangle = \tau_c \exp\left(\frac{E_T - E_{F,n}}{kT}\right) \tag{3}$$

In (3), E_T is the trap energy level, and $E_{F,n}$ is the electron quasi-Fermi level. Note that the trap energy level with respect to the oxide conduction band $E_{T,0} = 3.12\,\text{eV}$, the trap cross-section $\sigma_s = 10^{-14}\,\text{cm}^2$, and the activation energy $E_a = 0.5\,\text{eV}$ are kept constant in all the simulations reported in section four. Finally, a charge is placed in the trap, and the 3D device electrostatic and drift-diffusion equations are solved self-consistently again, to obtain the corresponding single-trap-induced threshold voltage shift ΔV_T. Equations (2) and (3) are solved assuming capture (emission) from (to) the substrate, and from (to) the gate. Knowing the time constants for both cases allows us to model the TAT current through an individual trap, which is out of the scope of this chapter.

3 Random Telegraph Noise and Statistical Variability

In this section we discuss briefly the background of RTN and its interaction with statistical variability. Results related to the threshold shift and the amplitude of the drain current variation will be presented.

3.1 The RTN Phenomena in Devices

RTN occurs as a consequence of the cyclic trapping and de-trapping of electrons or holes (charging and discharging) [8] in single defect at particular gate bias conditions. Due to the percolative nature of the source-to-drain conduction, the trapped charge will induce different values of RTN ΔV_T amplitude depending on its position over the channel and on the underlying random dopant fluctuation in the channel [41]. This fluctuation will have a negative impact on device and circuit operations, compromising the achievement of a reliability-robust design. Figure 4 shows an energy band diagram of an MOS structure with arbitrary trap position in the oxide and also illustrates the possible trapping mechanisms and a sequence of charge transfer from the channel to gate or vice versa via the trap.

A typical trace of RTN as a function of time is shown in Figs. 5, 6, 7, and 8b for all three devices. Two cases, single trap RTN and multiple traps are presented in the document. In the first case we select a randomly located single trap that gives the maximum ΔV_T (Figs. 5, 7a and b). The other three figures show (Figs. 6, 8a and b) when multiple traps are considered in the simulation (increasing the fluctuation amplitude). Note that the maximum ΔV_T is the sum of the threshold voltage shifts by the corresponding traps involved. As we expect the cumulative effect of the V_T shift increases with the number of traps involved in the trapping and de-trapping mechanisms.

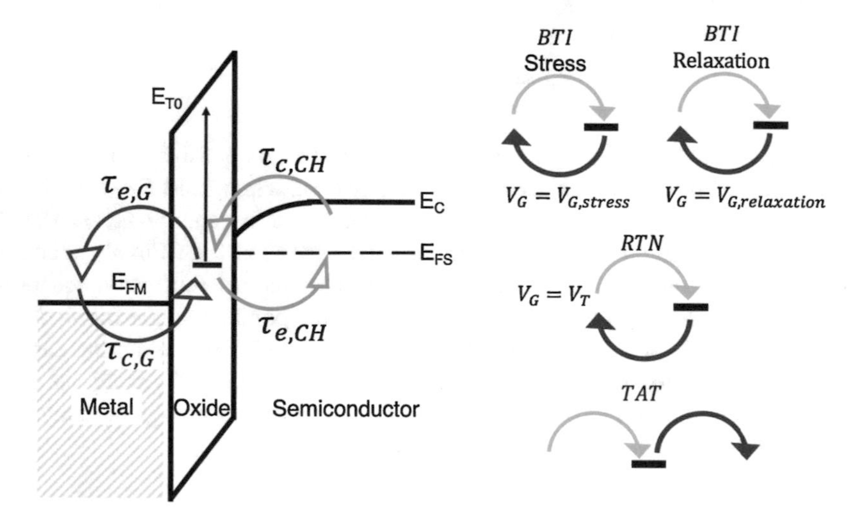

Fig. 4 The energy band diagram of an arbitrary trap position in the oxide (left) illustrating the trapping ad de-trapping mechanisms and a sequence of charge transfer from the channel to the gate via the trap

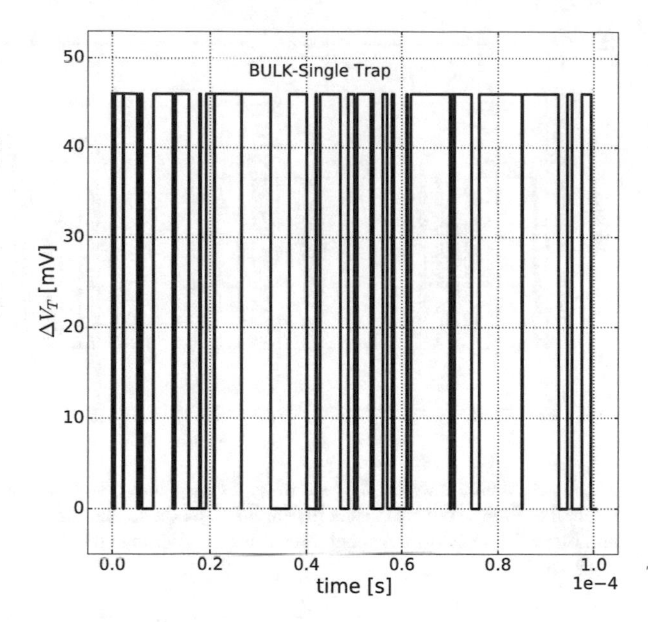

Fig. 5 Simulated track of RTN showing maximum threshold shift ($\Delta V_{T\max}$) for the case of the conventional bulk MOSFET. This result is a trace of single trap over a specified time interval

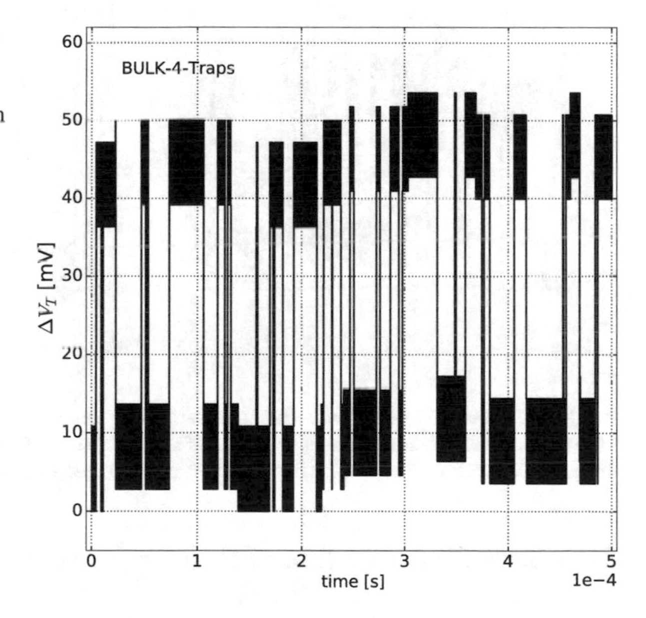

Fig. 6 Simulated RTN trace of a trapping/de-trapping process when multiple traps are involved in the simulation of the conventional bulk MOSFET. Note that the maximum amplitude is the sum of all four ΔV_Ts from the four traps which are traced over the time interval

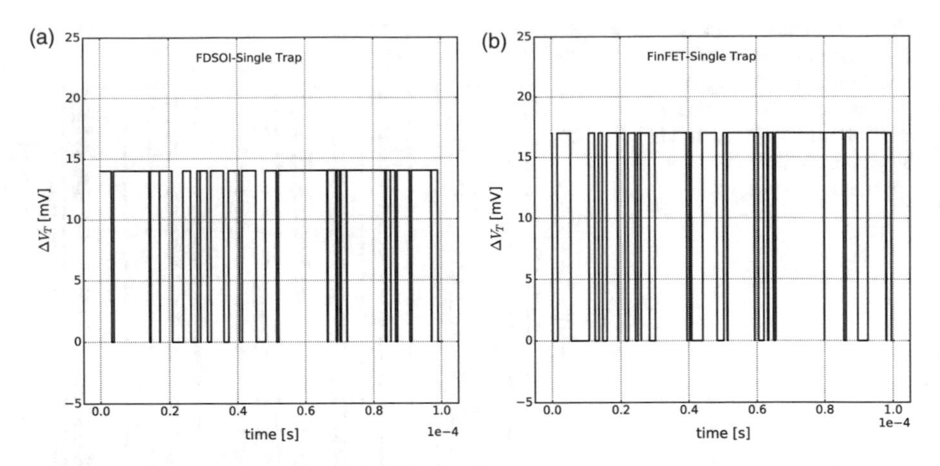

Fig. 7 Simulated track of RTN showing $\Delta V_{T\max}$ as a result of a single trap over a specified time interval in FDSOI device (**a**) and in FinFET (**b**). Notice that the maximum V_T shifts in both FDSOI and FinFET transistors are about three times smaller than the conventional bulk transistor as shown in Fig. 5

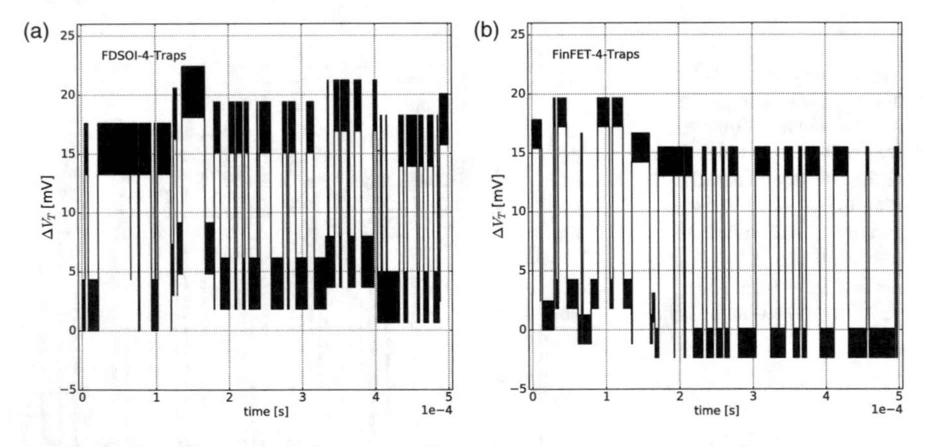

Fig. 8 Simulated RTN trace of a trapping/de-trapping process when more than one traps are involved in the simulation of the FDSOI (**a**) and FinFET (**b**). Note that the maximum amplitude is the sum of all four ΔV_Ts from the four traps which are traced over the time interval

3.2 Intrinsic Interaction of RTN and Statistical Variability

In order to investigate the interplay between RTN and statistical variability, we employed 3D 'atomistic' device simulation with all sources of variability in the template devices described in Sect. 2.1. Figure 9 shows simulated I_dV_g curves, at the low drain bias of 50 mV for 1000 devices in each device architecture. The data include all sources of variability and randomly placed RTN single trap in all three devices. The threshold voltage shift and drain current amplitude information is

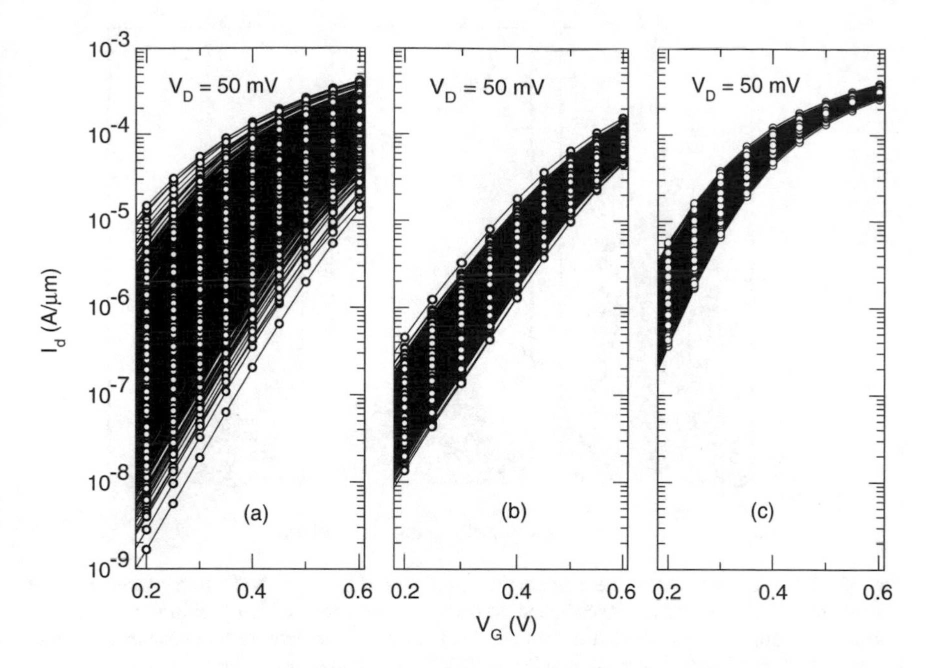

Fig. 9 Simulated $I_d V_g$ curves at the low drain bias of 50 mV for 1000 devices of each architecture. The data include all sources of variability and randomly located RTN single trap in all three devices. (**a**) Bulk MOSFET. (**b**) FD SOI. (**c**) FinFET

extracted at both high and low drain biases. The obtained results show the simulated characteristics of the impact of spatially resolved statistical variability sources on device electrostatics and threshold voltage while underestimating I_D variability in the ON-state regime due to the lack of transport-related variations [42].

As expected the variability in the bulk transistor is more severe when compared to the other two template devices. The main reason is that both SOI and FinFET have channel doping, as listed in Table 1, 'near' to the level of the intrinsic silicon concentration.

The impact of a single negative charging in the RTN trap on the device V_T is shown in Fig. 10 as a function of trap position in the source-to-drain direction. The data is obtained from an ensemble of 3000 (only for the high drain bias case) devices subject to all sources of variability and a single trap with a random (x, y, z) location. A large statistical dispersion of ΔV_T appears for traps placed over the channel as a result of percolative conduction [21] for the bulk MOSFETs, while in the case of conventional and FDSOI devices, much lower values and less variable ΔV_T are obtained for traps over the source-and-drain overlap with the gate. The fact that FDSOI and FinFET devices did not show comparable ΔV_T fluctuation with the bulk MOSFETs is due to much less doping in the channel region where the trap charge interacts with underlining random discrete dopants in the channel (see Eq. 4). The direct correlation of the standard deviation of V_T and doping concentration N_A

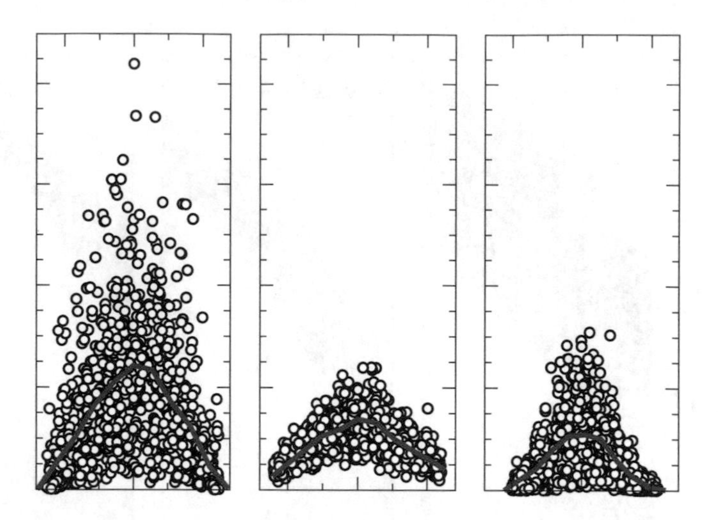

Fig. 10 ΔV_T resulting from the negative charging of the single RTN trap placed in each microscopic realization of the investigated devices as a function of trap position in the source-to-drain direction. Results refer to the case where all sources of variability are considered. The red line represents a calculated average of ΔV_T. (**a**) Bulk. (**b**) FDSOI. (**c**) FinFET

is given in (4), where t_{ox} is the oxide thickness, ϵ_{ox} is the permittivity of the gate oxide, and L_{eff} and W_{eff} are the effective channel length and width, respectively [20, 43, 44].

$$\sigma V_T = \frac{q t_{ox}}{\epsilon_{ox}} \sqrt{\frac{N_A W_d}{4 L_{eff} W_{eff}}} \tag{4}$$

3.3 *Dispersion of* ΔV_T *and Drain Current*

Figure 10 shows the dispersion of ΔV_T induced by a single electron trapped at a varying location within the extent of the gate, for the transistor identified in Fig. 2. The 'envelope' shape is due to the 2D non-uniform potential distribution between the source and drain, and reflects the well-known fact that a trap near the peak of the potential barrier has the largest influence on V_T. Besides, also the dispersion of V_T is the largest in the middle of the channel as shown in Fig. 10a. This is entirely due to RDD in the substrate of the transistors. ΔV_T as a function Fin-height is presented in Fig. 11. As we can observe from the data, there is no correlation of ΔV_T with the trap location along the Fin-height.

A large statistical dispersion of ΔV_T appears for traps placed over the channel as a result of percolative conduction [35] in all three devices. However, much lower and

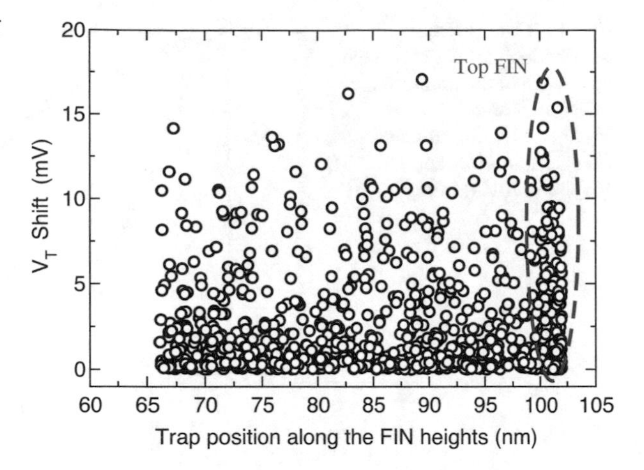

Fig. 11 ΔV_T as a function of Fin-height. Unlike the trap location along the carrier transport direction, it is clear that there is no interdependence (correlation) between ΔV_T and the trap position along the Fin-height

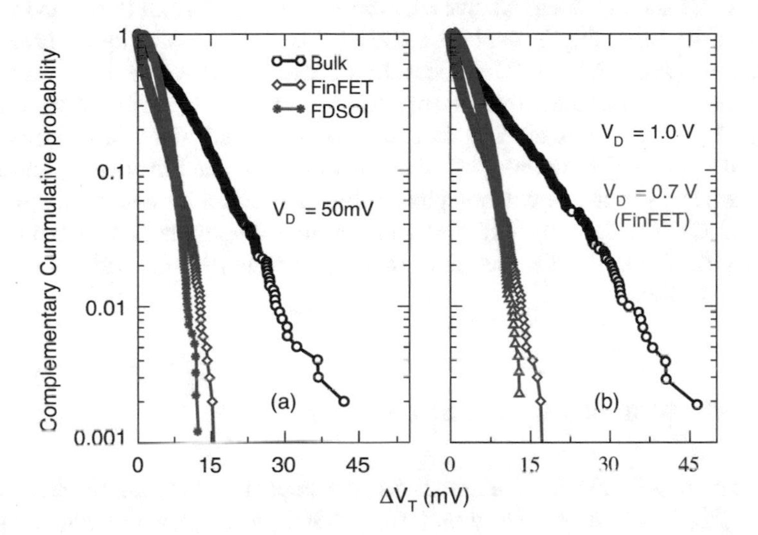

Fig. 12 Complementary cumulative distribution plot of ΔV_T, at low drain voltage (**a**) and high drain voltage (**b**), for all three devices using a sample of 1000 devices for each type

less variable ΔV_T values are obtained for traps over the source-and-drain overlap with the gate. Note that the dispersion of ΔV_T is much lower for FinFET and FDSOI when compared to bulk MOSFET, which is consistent with data in Fig. 4, where the level of channel doping plays a significant role. The envelop shape of the curve is more uniform in the case of a continuously doped channel as reported in [23], which is due to the 2D non-uniform potential distribution between the source and drain. A similar envelope shape can be perceived in Fig. 10 as well, however, the dispersion in the case of Fig. 10a is much broader, while in (b) and (c) relatively less dispersion is visible.

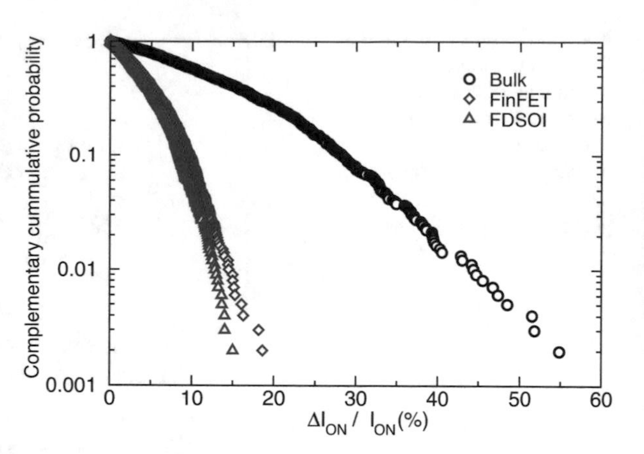

Fig. 13 Cumulative distribution plot of percentage difference of the on-current at the drain base of $V_D D = 1.0V$ for bulk and FDSOI transistors and $V_{DD} = 0.70V$ for the FinFET

Figure 12 shows the cumulative distribution of ΔV_T for all three devices. It is clear that the variability in the bulk MOSFET is much higher (up to three times) than in the FDSOI and FinFET devices. In all cases, RDD variability is responsible for the extreme cases and governs, therefor, the tail of ΔV_T distribution as shown in Fig. 12. These results confirm that the typical exponential tail, experimentally observed in ΔV_T distributions [45], is mainly due to dopant fluctuation and in turn to the variability of the percolation paths ruling the source-to-drain carrier transport (conduction). The impact of RTN on the current amplitude is shown in Fig. 13 depicting the tail of the distribution of relative percentage ΔI_{on} with respect to I_{on} is more than 50%.

4 Fluctuation of τ_c, τ_e, and z_T

By means of 3-D numerical simulations, the impact of random discrete dopants (RDDs) [20, 27, 46], metal gate granularity (MGG) [28, 29], and line edge roughness (LER) [34, 47] on RTN time constants and z_T estimation is first studied for different gate overdrives. Results reveal that RDD and MGG significantly contribute to broaden the statistical distribution of the RTN time constants coming from 3-D electrostatics, resulting, in turn, in a reduction of the accuracy of z_T. The accuracy may increase with the gate overdrive used in (1) with, however, the possible drawback of enhancing, as a result is, the electron emission rate toward the gate. The simultaneous trap interaction with the channel and the gate represents a further degree of complexity for the spectroscopic investigation of RTN traps in thin-oxide devices, leading to the possibility of significant errors in z_T estimation.

In this section we investigate the possibility of inaccuracies surrounding the analyses of RTN traps and their characteristic parameters when using Eq. (5). In the introduction of this chapter we mention that Eq. (1) will provide inaccurate spectroscopic analysis of RTN. To highlight this problem and demonstrate the accuracy and issues of the spectroscopic analysis of RTN, we present simulation data obtained from the conventional bulk MOSFET which is described in Sect. 2.

4.1 Randomness of Trap Location

In order to investigate trap position from the RTN results, we used Eqs. (5) and (6) which are deduced from (1). As reported in [23, 48] using these equations introduces significant errors when estimating trap depth in Flash memory cells and decananometer MOSFETs, respectively. Moreover, the error displays a strong dependence on trap position over the channel area, due to the fluctuations introduced by 3D electrostatics and atomistic doping in the surface potential, the quasi-Fermi level for electrons, and the τ_c/τ_e ratio.

$$\frac{z_{\mathrm{T}}^{calc,ch}}{t_{\mathrm{ox}}} = \frac{-\frac{kT}{q}\frac{\partial \ln(\tau_{\mathrm{c}}/\tau_{\mathrm{e}})}{\partial V_{\mathrm{G}}} + \frac{\partial V}{\partial V_{\mathrm{G}}} - \frac{\partial \Psi_{\mathrm{s}}}{\partial V_{\mathrm{G}}}}{1 - \frac{\partial \Psi_{\mathrm{s}}}{\partial V_{\mathrm{G}}}} \quad (5)$$

The calculated trap position was indicated as z_T^{calc} to distinguish it from the real trap distance from the channel surface $z_T = 0.3$ nm. Although Eq. (5) already involves a first degree of approximation in that it comes from simplified 1-D calculations, the lack of knowledge on Ψ_s and $E_{F,n}$ at the trap position can be overcome only with further simplifying assumptions, namely, considering a strong inversion condition and a low applied drain bias (V_D). In this way, the terms $\partial V/\partial V_G$ and $\partial \Psi_s/\partial V_G$ in (5) are usually neglected. Further simplification of Eq. (5) as shown in [48] will give us Eq. (6) to estimate trap depth in the oxide.

$$\frac{z_T^{calc}}{t_{ox}} = -\frac{kT}{q}\frac{\partial \ln(\tau_c/\tau_e)}{\partial V_G} \quad (6)$$

Problems surrounding the accuracy of Eq. (6) have been reported in [48, 49] mainly in Flash Memory cells investigation. In this work, we extend the analysis on the accuracy of Eq. (5) and look in to the case of deeply scaled conventional MOSFETs by including a detailed discussion on the impact of the main variability sources affecting the performance of these devices at the nano-scale and highlighting some additional issues specific of thin-oxide devices related to RTN phenomena for which the physical parameters are listed in Table 1.

The statistical distribution of z_T^{calc} obtained when applying (5) in a 200-mV range around the gate overdrive voltage giving $I_D = 1.0$ and 1.5×10^{-4} A/μm is shown in Fig. 14. From this figure, not only a quite large disagreement between the average z_T^{calc} and z_T but also a quite large statistical dispersion of z^{calc} appears at $I_D = 1.0$ and 1×10^{-4} A/μm, with z_T^{calc} ranging from 0.45 to 0.625 nm. A reduction of both the average value and the spread of z_T^{calc} is observed when moving from $I_D = 1.0$ to 1.5×10^{-4} A/μm, confirming that the accuracy of trap spectroscopy

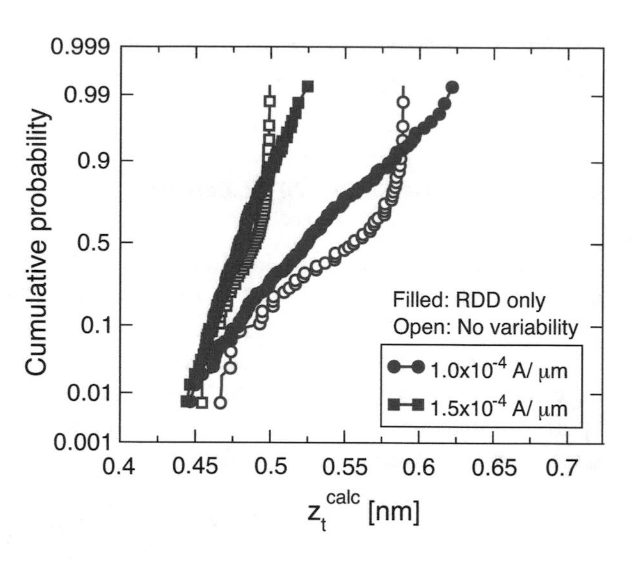

Fig. 14 Cumulative distribution of z_T^{calc} of a single RTN trap placed in each microscopic realization of the investigated MOSFET, in the case when RDD is the only variability sources is accounted for. Results for two gate overdrive conditions giving $I_D = 1.0$ and 1.5×10^{-4} A/μm are shown. Note that the real trap distance from the channel surface is $z_T = 0.3$ nm

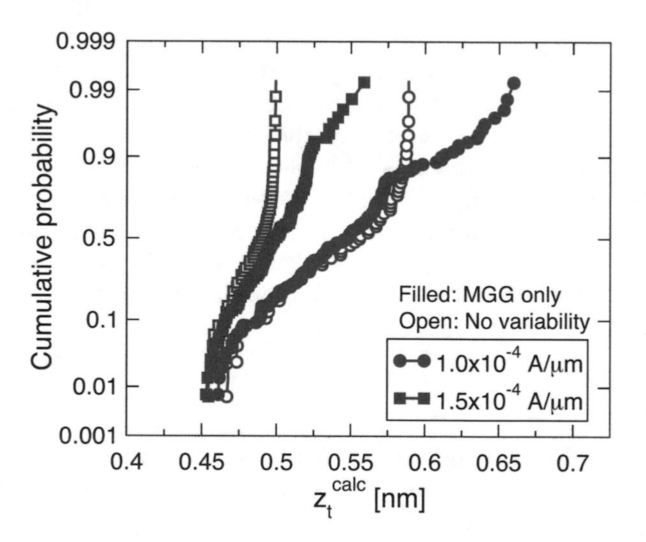

Fig. 15 Cumulative distribution of z_T^{calc} of a single RTN trap placed in each microscopic realization of the investigated MOSFET, in the case when only MGG source of variability is accounted for. Results for two gate overdrive voltages giving $I_D = 1.0$ and 1.5×10^{-4} A/μm are shown

increases at higher gate overdrives. Fig. 15 shows the impact of MGG on RTN time constants in nanoscale MOSFETs. MGG enlarges the τ_c distribution with respect to the distribution when no variability is included. This is due to the change in gate work function over the active area of the device whose effect on τ_c is similar to the change of V_G at the trap position. Finally, note that the relevance of RDD on the accuracy of trap spectroscopy appears in Fig. 14 when comparing these results with those obtained neglecting all the fundamental MOSFET variability sources, in agreement with the discussions about the distribution of τ_c of a single RTN trap located in the channel region of the transistor as shown in Fig. 16.

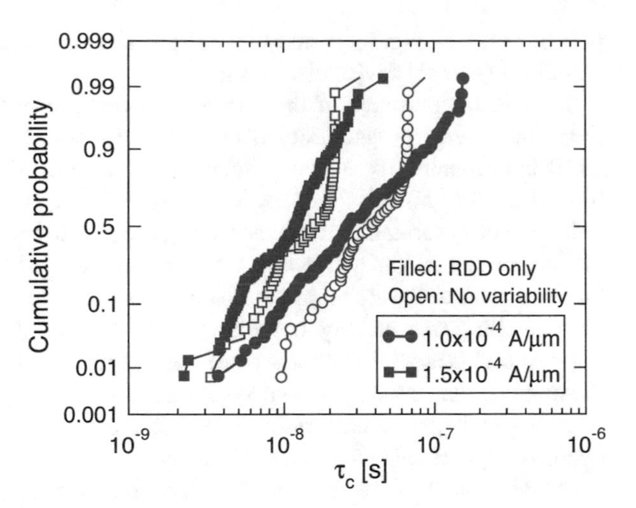

Fig. 16 Cumulative distribution of the τ_C of a single RTN trap placed in each microscopic realization of the investigated MOSFET, in the case when RDD is the only source of variability accounted for and in the case no other variability case is activated. Results for two gate overdrive voltages giving $I_D = 1.0$ and 1.5×10^{-4} A/μm are shown

4.2 Randomness of Capture and Emission Time Constants

The effect of variability in device electrostatics and of random trap placement over the active area on the RTN time constants is addressed in Fig. 16, where the cumulative distribution of τ_c reported at different gate overdrives giving $I_D = 1.0$ and 1.5×10^{-4} A/μm and corresponding to $V_G - V_T \simeq 250$ and 450 mV, respectively. Results obtained when including RDD as the only variability source (filled symbols) are compared in this figure with those obtained when no MOSFET variability source is accounted for (open symbols), i.e., doping is treated as continuous, and MGG and LER are not activated. In this latter case, a statistical dispersion of the results comes only from the random trap placement over the active area in the presence of 3-D electrostatics. Note, first of all, that extremely small average values were obtained in all cases, due to the choice to neglect a complete physical description of the electron capture dynamics, which may require lattice relaxation [36, 50], Coulomb blockade [38], or different capture cross-sections [51] to be accounted for in the simulations. Notwithstanding this point, results of Fig. 16 correctly highlight the statistical dispersion of τ_c coming from RDD and 3-D device electrostatics.

In the case when no MOSFET variability source is activated (doping is considered as continuous and shown in Fig. 16 as open symbols), results give a ratio between the maximum and minimum τ_c's nearly equal to 1 decade for both the investigated overdrive voltage values. The statistical dispersion comes as solely from the random location of traps over the active region of the transistor in the presence of 3D electrostatics. Note, first of all, that extremely short average values were obtained in all cases, due to the choice to neglect a complete physical description of the electron capture dynamics, which may require advanced models to be accounted for in the simulations as mentioned in Sect. 4.3. Notwithstanding this point, results of Fig. 16 correctly highlight the statistical dispersion of τ_c coming from RDD and 3D device electrostatics.

Despite the presence of the retrograde doping profile of the bulk transistor, a large increase of τ_c variability with respect to these results clearly appears when RDD is accounted for in the simulations, with a maximum-to-minimum τ_c ratio nearly equal to 50 and 30 at $I_D = 1.0$ and 1.5×10^{-4} A/μm, respectively. The reduction of τ_c variability when increasing the gate overdrive reveals a more uniform electrostatics over the active area when V_G is increased and, in particular, a more uniform electron inversion at the channel surface. Note, in fact, that variability in the tunnelling transparency determining the electron flux from the channel to the trap position is negligible (not shown) and fluctuations in the electron concentration feeding this flux is the main source spread, as shown in [22].

A larger degree of uniformity in device electrostatics when increasing the gate overdrive appears also from Fig. 17a showing the τ_c/τ_e ratio when RDD is the only variability source accounted for. A reduction of the τ_c/τ_e dispersion appears when V_G is increased, with the curves merging together in a narrower beam. This involves a different slope of τ_c/τ_e with V_G for the different Monte Carlo cases and, in turn, a variability in the spectroscopic results coming from (6).

Figure 18 shows that MGG represents an additional important variability source for the RTN time constants in nano-scale MOSFETs, enlarging the τ_c distribution with respect to the case of no variability included. This is due to the change in the metal gate work function over the active area, whose effect on τ_c is that of an equivalent change of V_G at the trap position. This clearly appears from Fig. 17b, where the τ_c/τ_e curves are grouped into two distinct families (the MGG has bi-modal statistical distribution) that are horizontally shifted by 200 mV, corresponding to the difference in the Φ_m values of the metal gate work function. Similar to Figs. 16, 18 confirms that variability coming from MGG impacts the accuracy of

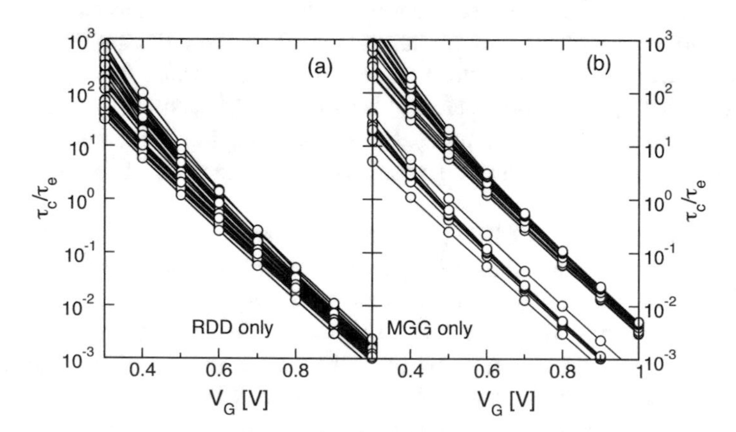

Fig. 17 The dependence of the ratio τ_c/τ_e when including only (**a**) RDD or (**b**) MGG as a function of V_G. In the case of MGG, it is clear from the data in (**b**) that the work function difference due to the two grain boundaries creates a bi-modal distribution of τ_c/τ_e ratio

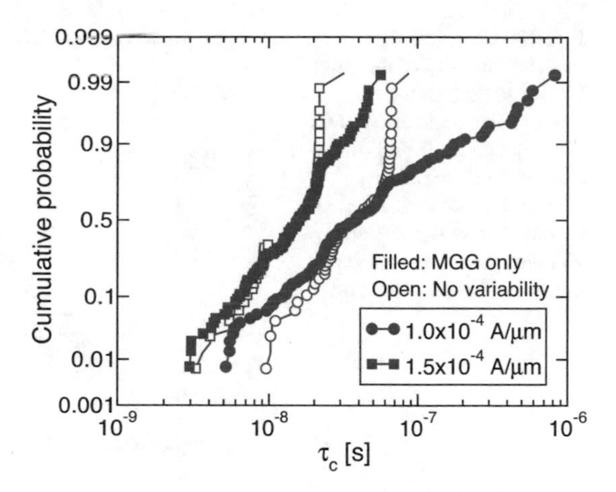

Fig. 18 Only MGG variability is activated

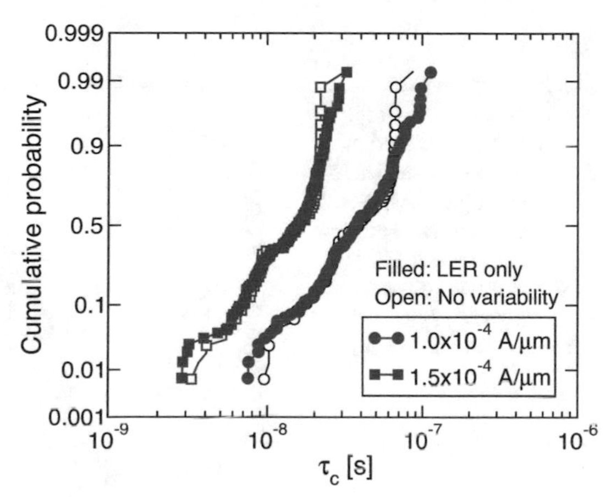

Fig. 19 Only LER variability is activated

trap spectroscopy by a magnitude that depends on the selected gate overdrive, with, moreover, the possibility for quite high z_T^{calc} (>0.6 nm) to occur. High z_T^{calc}, i.e., large errors in z_T estimation, result when the high value of the metal gate work function is present at the trap position, leading to a lower electron concentration at this position.

Differently from RDD and MGG, LER appears as a negligible variability source for both RTN time constants and trap spectroscopy: Fig. 19 shows, in fact, that the τ_c distributions are not significantly displaced from the results obtained when no variability source is activated. This can be understood considering that LER changes device electrostatics at the gate edges over the source and drain junctions, but traps placed over these regions may only give rise to very small ΔV_T, which are statistically negligible (see Fig. 10) and are excluded from our analysis.

Finally, Figs. 20 and 21 show the τ_c and z_T^{calc} statistical distributions, respectively, obtained when all the variability sources are simultaneously included in the simulations. A quite large spread of more than 2 and 1 decade appears for τ_c

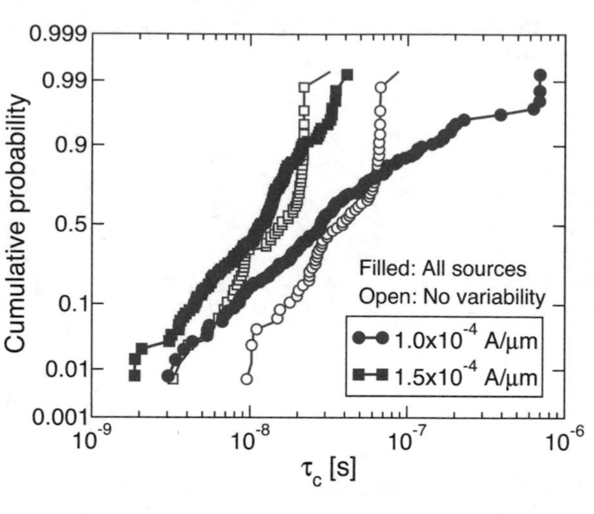

Fig. 20 Cumulative distribution of τ_c of the single RTN trap placed in each microscopic realization of the investigated MOSFET, in the case when all variability sources are accounted for. Results for two gate overdrive voltages giving $I_D = 1.0$ and 1.5×10^{-4} A/μm are shown

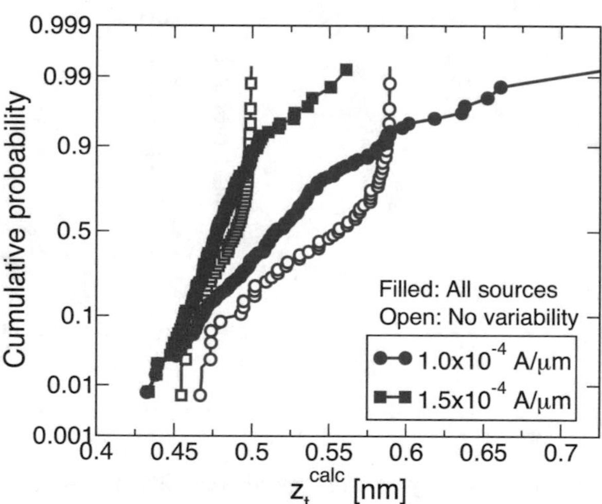

Fig. 21 Cumulative distribution of z_T^{calc} of the single RTN trap placed in each microscopic realization of the investigated MOSFET, in the case when all sources of variability are considered. Results for two gate overdrive voltages giving $I_D = 1.0$ and 1.5×10^{-4} A/μm are shown

at $I_D = 1.0$ and 1.5×10^{-4} A/μm, respectively. More importantly, a significant displacement of the average z_T^{calc} from z_T appears in Fig. 21 with a large variability in the results, spanning in a region of width equal to the 25% and the 12% of the total oxide thickness at low and high gate overdrives, respectively.

4.3 Trap Interaction with Channel and Gate

Results shown in the previous section revealed that the accuracy of (6) is too low to allow a meaningful investigation of trap depth in thin-oxide devices, also due to the

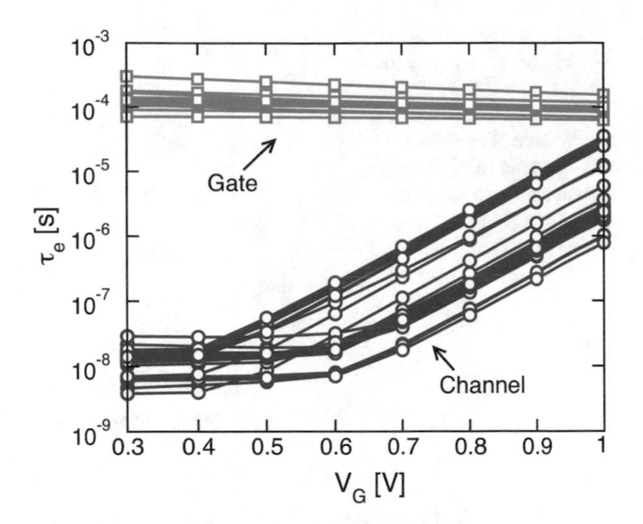

Fig. 22 Simulated τ_e^{ch} and τ_e^{g} for some of the Monte Carlo cases including all sources of variability sources investigated in this work as a function of V_G

large variability of z_T^{calc} obtained for traps having the same z_T. Some benefits for the spectroscopic accuracy appear from Fig. 21 when the gate overdrive is increased, owing to a more uniform and V_G independent Ψ_S. Figure 22 shows, however, that τ_e for electron emission toward the channel (τ_e^{ch}; open circle) gets longer when V_G increases, with the possibility for this time to become comparable to the electron emission time toward the gate (τ_e^{g}; square). τ_e^{g} was calculated from the electron capture time from the gate (τ_c^{g}) according to (7):

$$\tau_e^{g} = \tau_c^{g}\, e^{-(E_T - E_{F,g})/kT}, \tag{7}$$

where $E_{F,g}$ is the Fermi level in the gate and τ_c^{g} was calculated from the integral of the tunnelling electron flux reaching the trap from the gate. When τ_e^{ch} becomes longer than τ_e^{g}, the RTN trap starts to interact simultaneously with the channel and the gate, capturing electrons from the channel and emitting them to the gate, compromising the use of (6). To explore this situation in more detail, we considered RTN traps with $z_T = 0.6$ nm and different positions over the channel area of the same microscopic MOSFET.

Figure 23 shows the calculated time constants at $I_D = 1.0 \times 10^{-4}$ A/μm, revealing that, for the selected z_T, the conditions $\tau_c^{ch} < \tau_c^{g}$ and $\tau_e^{ch} > \tau_e^{g}$ are met at lower gate overdrives than in the case of $z_T = 0.3$ nm. Figure 24b shows that, as a consequence of this trap interaction with both the channel and the gate, the ratio between the total capture time $\tau_c^{tot} = (1/\tau_c^{ch} + 1/\tau_c^{g})^{-1}$ and the total emission time $\tau_e^{tot} = (1/\tau_e^{ch} + 1/\tau_e^{g})^{-1}$ displays a weak dependence on the gate voltage, much weaker than that resulting for τ_c^{g}/τ_e^{g} (Fig. 24a) and τ_c^{ch}/τ_e^{ch} (Fig. 24c).

The use of (6) to estimate z_T from the $\tau_c^{tot}/\tau_e^{tot}$ results of Fig. 24b leads, therefore, to the wrong conclusion that the RTN traps are placed very close to the channel surface, as shown in Fig. 25. Note that this conclusion cannot be excluded

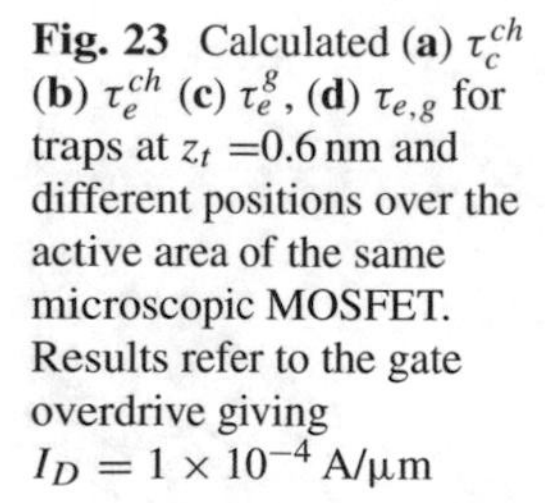

Fig. 23 Calculated (**a**) τ_c^{ch} (**b**) τ_e^{ch} (**c**) τ_e^{g}, (**d**) $\tau_{e,g}$ for traps at z_t =0.6 nm and different positions over the active area of the same microscopic MOSFET. Results refer to the gate overdrive giving $I_D = 1 \times 10^{-4}$ A/μm

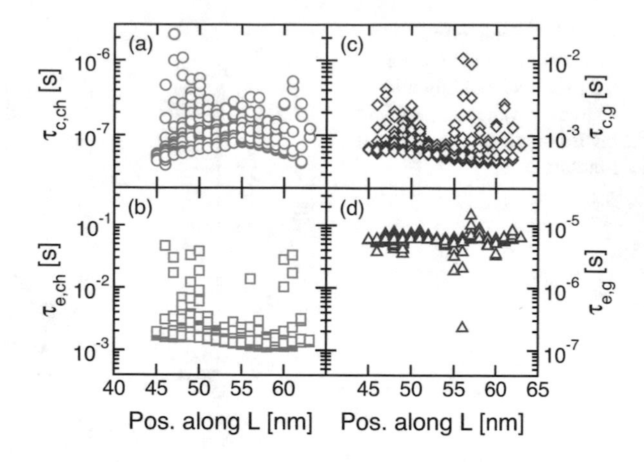

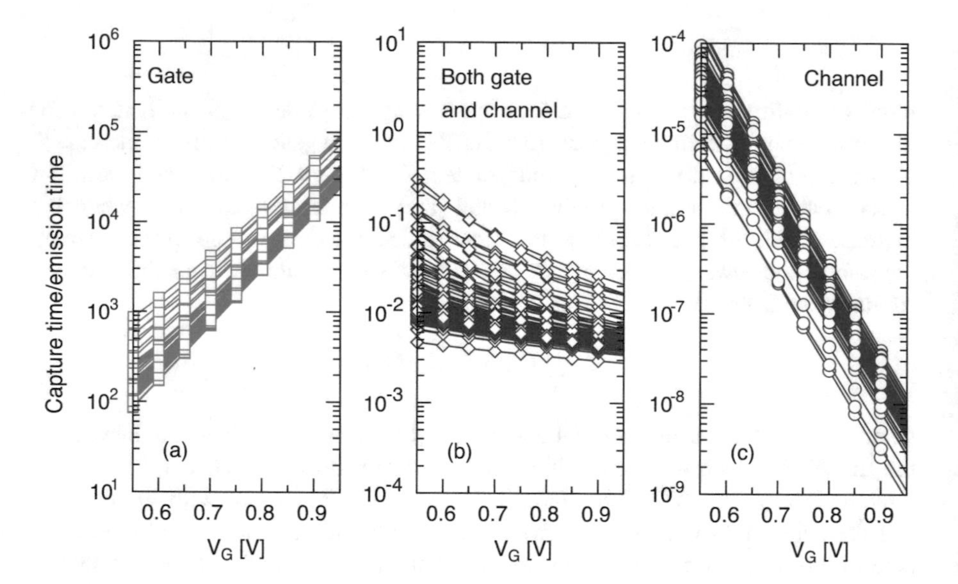

Fig. 24 V_G dependence of the ratio between the capture and emission time constants for traps at $Z_T = 0.6$ nm assuming interaction (**a**) only with the gate, (**b**) with both the gate and channel, and (**c**) only with the channel

by the direct observation of the τ_c values, which should, in principle, be quite short for traps very close to the interface. Not only the fundamental MOSFET variability sources but also the statistical dispersion of physical trap parameters such as trap capture cross-section, lattice relaxation energy [36, 50], and Coulomb blockade energy [38], may, in fact, result in traps having very long capture and emission times, as reported in [52] and [53] where the latter shows a large τ_c of up $10^4 s$. As a consequence, large errors in trap spectroscopy may arise for traps interacting with

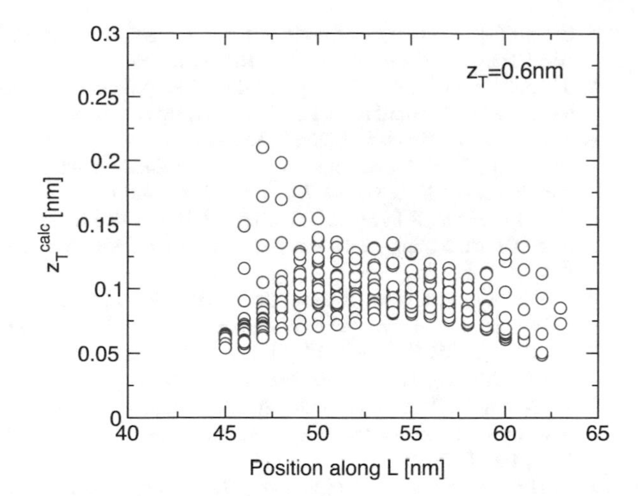

Fig. 25 z_T^{calc} as a function of the trap position along the channel direction for traps with $z_T = 0.6$ nm and a different position over the active area of the same microscopic MOSFET

both the channel and the gate electrode, and this depends not only on z_T of the traps but also on their energy depth in the oxide.

5 Summary

RTN and its intrinsic interaction with the sources of statistical variability in conventional bulk MOSFE, FDSOI, and FinFET transistors is presented. The dispersion of ΔV_T is more severe in the conventional bulk ($\Delta V_T^{bulk} \approx 50$ mV), while both FDSOI and FinFET show $\Delta V_T^{\max} \approx 14$ mV. In addition to ΔV_T variation, τ_c, τ_e, and estimated z_T^{calc} fluctuate significantly. This fluctuation highlights the problem with the inaccuracy of RTN trap spectroscopy. Moreover, the combined effects of 3D electrostatics, RDD, and MGG show severe limitation to the accuracy of z_T extraction.

References

1. S. Natarajan et al., A 14 nm logic technology featuring 2nd-generation FinFET air-gapped interconnects self-aligned double patterning and a 0.0588 μm^2 SRAM cell size, in *International Electron Devices Meeting Technical Digests IEDM*, San Francisco (2014), pp. 71–73
2. C. Auth, et al., A 10nm high performance and low-power CMOS technology featuring 3rd generation FinFET transistors, self-aligned quad patterning, contact over active gate and cobalt local interconnects, in *International Electron Devices Meeting Technical Digests IEDM*, San Francisco (2017), pp. 674–676
3. K.K. Hung, P.K. Ko, C. Hu, Y.C. Cheng, Random telegraph noise of deep-submicrometer MOSFET's. IEEE Electron Device Lett. **11**, 90–92 (1990)

4. Z. Shi, J.-P. Mieville, M. Dutoit, Random telegraph signals in deep submicron n-MOSFET's. IEEE Trans. Electron Devices **41**, 1161–1168 (1994)
5. S.T. Martin, G.P. Li, E. Worley, J. White, The gate bias and geometry dependence of random telegraph signal amplitudes. IEEE Electron Device Lett. **18**, 444–446 (1997)
6. A. Avellan, W. Krautschneider, S. Schwantes, Observation and modeling of random telegraph signals in the gate and drain currents of tunnelling metal-oxide-semiconductor field-effect transistors, Appl. Phys. Lett. **78**, 2790–2792 (2001)
7. H.M. Bu, Y. Shi, X.L. Yuan, Y.D. Zheng, S.H. Gu, H. Majima, H. Ishicuro, T. Hiramoto, Impact of the device scaling on the low frequency noise in n-MOSFET's, Appl. Phys. A **71**, 133–136 (2000)
8. M.J. Kirton, M.J. Uren, Noise in solid state microstructures A new perspective on individual defects, interface states and low frequency ($1/f$) noise. Adv. Phys. **38**, 367–468 (1989)
9. P.J. Restle, J.W. Park, B.F. Lloyd, DRAM variable retention time, in *International Electron Devices Meeting Technical Digests IEDM* (1992), pp. 807–810
10. A. Asenov, R. Balasubramaniam, A.R. Brown, J.H. Davies, Effect of single-electron interface trapping in decanano MOSFETs: A. 3D atomistic simulation study. Superlatt. Microstruct. **27**, 411–416 (2000)
11. K. Hess, A. Haggag, W. McMahon, B. Fischer, K. Cheng, J. Lee, J. Lyding, Simulation of Si-SiO defects generation in CMOS chips: from atomistic structure to chip failure rates, in *International Electron Devices Meeting Technical Digests IEDM* (2000), pp. 93–96
12. A. Avellan, D. Schroeder, W. Krautschneider, Modeling random telegraph signals in the gate current of metal-oxide-semiconductor field effect transistors after oxide breakdown. J. Appl. Phys. **94**, 703 (2003)
13. K. Kandiah, M.O. Deighton, F.B. Whiting, A physical model for random telegraph signal currents in semiconductor devices. J. Appl. Phys. **66**, 937 (1989)
14. Z. Celik-Butler, P. Vasina, N.V. Amarasinghe, A method for locating the position of oxide traps responsible for random telegraph signals in submicron MOSFET's. IEEE Trans. Electron Devices, **47**(3) (2000)
15. M.-H. Tsai, T.-P. Ma, The impact of device scaling on the current fluctuations in MOSFET's. IEEE Trans. Electron Devices **41**, 2061–2068 (1994)
16. K.S. Ralls, W.J. Skocpol, L.D. Jackel, R.E. Howard, L.A. Fetter, R.W. Epworth, D.M. Tennant, Discrete resistance switching in submicron silicon inversion layers: individual interface traps and low frequency ($1/f$) noise. Phys. Rev. Lett. **52**, 228–231 (1984)
17. E. Simoen, B. Dierick, C.L. Claeys, G.J. Declerck, Explaining the amplitude of RTS noise in submicrometer MOSFET's. IEEE Trans. Electron Devices **39**, 422–429 (1992)
18. H.H. Mueller, M. Schulz, Random telegraph signal: an atomic probe of the local current in field-effect transistors. J. Appl. Phys. **83**, 1734–1741 (1998)
19. A. Asenov, R. Balasubramaniam, A.R. Brown, J.H. Davies, S. Saini, Random telegraph signal amplitudes in sub 100 nm (Decanano) MOSFETs: A 3D 'Atomistic' simulation study, in *International Electron Devices Meeting 2000. Technical Digests IEDM* (2000), pp. 279–282
20. A. Asenov, Random dopant induced threshold voltage lowering and fluctuations in sub 0.1 micron MOSFETs: a 3-D "atomistic" simulation study. Trans. Electron Devices **45**, 2505–2513 (1998)
21. S.M. Amoroso, L.Gerrer, F.Adamu-Lema, S. Markov, A. Asenov, Statistical study of bias temperature instability by means of 3D 'Atomistic' simulation, in ed. by T. Grasser, *Bias Temperature Instability for Device and Circuits* (Springer, New York, 2014)
22. A. Mauri, N. Castellani, C. Monzio Compagnoni, A. Ghetti, P. Cappelletti, A.S. Spinelli, A.L. Lacaita, Impact of atomistic doping and 3D electrostatics on the variability of RTN time constants in Flash memories, in *Proceedings of IEEE 2011 International Electron Devices Meeting IEDM* (2011), pp. 405–408
23. F. Adamu-LemaAccuracy, C.M. Compagnoni, S.M. Amoroso, N. Castell, L. Gerrer, S. Markov, A.S. Spinelli, A.L. Lacaita, A. Asenov, Issues of the spectroscopic analysis of RTN traps in nanoscale MOSFETs. IEEE Trans. Electron Devices **60**(2), 833–839 (2012)

24. F. Adamu-Lema, X. Wang, S.M. Amoroso, C. Riddet, B. Cheng, L. Shifren, R. Aitken, S. Sinha, G. Yeric, A. Asenov, Performance and variability of doped multi-threshold FinFETs for 10nm CMOS. IEEE Trans. Electron Devices **61**(10), 3372–3378 (2014)
25. G. Roy, A. Brown, F. Adamu-Lema, S. Roy, A. Asenov, Simulation study of individual and combined sources of intrinsic parameter fluctuations in conventional nano-MOSFETs. IEEE Trans. Electron Devices **53**, 3063 (2006)
26. S. Amoroso, A. Maconi, A. Mauri, C.M. Compagnoni, A.S. Spinelli, A.L. Lacaita, Three-dimensional simulation of charge-trap memory programming-part I: average behavior. IEEE Trans. Electron Devices **58**, 1864 (2011)
27. A. Asenov, A.R. Brown, J.H. Davies, S. Kaya, G. Slavcheva, Simulation of intrinsic parameter fluctuations in decananometer and nanometer-scale MOSFETs. IEEE Trans. Electron Devices **50**(9), 1837–1852 (2003)
28. A. Brown, J. Watling, A. Asenov, Intrinsic parameter fluctuations due to random grain orientations in high-k gate stacks. J. Comput. Electron. **5**(4), 333–336 (2006)
29. A.R. Brown, N.M. Idris, J.R. Watling, A. Asenov, Impact of metal gate granularity on threshold voltage variability: a full-scale three-dimensional statistical simulation study. IEEE Electron Device Lett. **31**(11), 1199–1201 (2010)
30. H. Dadgour, K. Endo, V. De, K. Banerjee, Modeling and analysis of grain-orientation effects in emerging metal-gate devices and implications for SRAM reliability, in *Proceedings of 2008 IEEE International Electron Devices Meeting IEDM* (2008), pp. 705–708
31. A. Lannes, E. Anterrieu, K. Bouyoucef, Fourier interpolation and reconstruction via Shannon-type techniques. J. Modern Opt. **41**(8), 1537–1574 (1994)
32. A. Lannes, E. Anterrieu, K. Bouyoucef, Fourier interpolation and reconstruction via Shannon-type techniques II. Technical developments and applications. J. Modern Opt. **43**(1), 105–138 (1996)
33. S. Kaya, A.R. Brown, A. Asenov, D. Magot, T. Linton, Analysis of statistical fluctuation due to line edge roughness in sub-0.1μm MOSFETs, in *Simulation of Semiconductor Processes and Devices 2001: SISPAD 01* (2001), pp 78–81
34. D. Reid, C. Millar, S. Roy, A. Asenov, Understanding LER-induced MOSFET VT variability: part I: three-dimensional simulation of large statistical samples. IEEE Trans. Electron Devices **57**(11), 2801–2807 (2010)
35. A. Ghetti, C.M. Compagnoni, A. Spinelli, A. Visconti, Comprehensive analysis of random telegraph noise instability and its scaling in deca-nanometer flash. IEEE Trans. Electron Devices **56**, 1746 (2009)
36. T. Grasser, H. Reisinger, W. Goes, T. Aichinger, P. Hehenberger, P.-J. Wagner, M. Nelhiebel, J. Franco, B. Kaczer, Switching oxide traps as the missing link between negative bias temperature instability and random telegraph noise, in *2009 IEEE International Electron Devices Meeting Technical Digests IEDM* (2009), pp. 729–732
37. T. Grasser, H. Reisinger, P.-J. Wagner, F. Schanovsky, W. Goes, B. Kaczer, The time dependent defect spectroscopy (TDDS) for the characterization of the bias temperature instability, in *2010 IEEE International Reliability Physics Symposium* (2010), pp. 16–25
38. A. Palma, A. Godoy, J.A. Jimenez-Tejada, J.E. Carceller, J.A. Lopez-Villanueva, Quantum two-dimensional calculation of time constants of random telegraph signals in metal-oxide-semiconductor structures. Phys. Rev. B **56**, 9565–9574 (1997)
39. W., Shockley, W. Read, Statistics of recombination of holes and electrons. Phys. Rev. **87**, 835 (1952)
40. S.M. Amoroso, L. Gerrer, A. Asenov, 3D TCAD statistical analysis of transient charging in BTI degradation of nanoscale MOSFETs, in *2013 International Conference on Simulation of Semiconductor Processes and Devices (SISPAD)* (2013), pp. 5–6
41. K. Takeuchi, T. Tatsumi, A. Furukawa, channel engineering for the reduction of random-dopant-placement-induced threshold voltage fluctuation, in *International Electron Devices Meeting Technical Digests IEDM* (1997), pp. 841–844

42. U. Kovac, C. Alexander, G. Roy, C. Riddet, B. Cheng, A. Asenov, Hierarchical simulation of statistical variability: from 3-D MC with 'ab initio' ionized impurity scattering to statistical compact models. IEEE Trans. Electron Devices **57**(10), 2418–2426 (2010)
43. K. Nishiohara, N. Shiguo, T. Wada, Effects of mesoscopic fluctuations in dopant distributions on MOSFET threshold voltage. IEEE Trans. Electron Devices **39**, 634–639 (1992)
44. P.A. Stolk, D.B.M. Klaasen, The effect of statistical dopant fluctuations on MOS device performance, in *1996 IEEE International Electron Devices Meeting. Technical Digest IEDM* (1996)
45. T. Fischer, E. Amirante, K. Hofmann, M. Ostermayr, P. Huber, D. Schmitt-Landsiedel, A 65 nm test structure for the analysis of NBTI induced statistical variation in SRAM transistors, in *Proceedings of ESSDERC 2008-38th European Solid-State Device Research Conference*, Edinburgh (2008), pp. 51–5
46. H.-S. Wong, Y. Taur, Three-dimensional 'atomistic' simulation of discrete random dopant distribution effects in sub-0.1 μm MOSFET's, in *Proceedings of IEEE International Electron Devices Meeting Technical Digests IEDM* (1993), pp. 705–708
47. A. Asenov, S. Kaya, A.R. Brown, Intrinsic parameter fluctuations in decananometer MOSFETs introduced by gate line edge roughness. IEEE Trans. Electron Devices **50**(5), 1254–1260 (2003)
48. C. Monzio Compagnoni, N. Castellani, A. Mauri, A.S. Spinelli, A.L. Lacaita, 3-Dimensional electrostatics- and atomistic doping induced variability of RTN time constants in nanoscale MOS devices-Part II: Spectroscopic implications. IEEE Trans. Electron Devices **59**(9), 2495–2500 (2012)
49. N. Castellani, C. Monzio Compagnoni, A. Mauri, A.S. Spinelli, A.L. Lacaita, 3-Dimensional electrostatics- and atomistic doping induced variability of RTN time constants in nanoscale MOS devices-part I: physical investigation. IEEE Trans. Electron Devices **59**(9), 2488–2494 (2012)
50. G. Bersuker, RTN analysis for defect identification in advanced gate stacks, in *Proceedings of 2013 IEEE International Reliability Physics Symposium IRPS*, Anaheim (2011), IRPS Tutorial No. 113
51. N. Zanolla, D. Siprak, P. Baumgartner, E. Sangiorgi, C. Fiegna, Measurement and simulation of gate voltage dependence of RTS emission and capture time constants in MOSFETs, in *Proceedings of 2008 9th International Conference on Ultimate Integration of Silicon ULIS*, 2008, pp. 137–140.
52. J. P. Campbell, J. Qin, K.P. Cheung, L.C. Yu, J.S. Suehle, A. Oates, K. Sheng, Random telegraph noise in highly scaled nMOSFETs, in *Proceedings of 2009 IEEE International Reliability Physics Symposium IRPS* (2009), pp. 382–388
53. T. Grasser, K.Rott, H. Reisinger, M. Waltl, J. Franco, B. Kaczer, A unified perspective of RTN and BTI, in *Proceedings of 2014 IEEE International Reliability Physics Symposium (IRPS)*, Waikoloa (2014)

Advanced Characterization and Analysis of Random Telegraph Noise in CMOS Devices

J. Martin-Martinez, R. Rodriguez, and M. Nafria

1 Introduction

Random telegraph noise (RTN) is a well-known phenomenon in electron devices, which has been extensively studied for more than 30 years [1–3]. Since its first observation in 1984 by Ralls et al. [4], hundreds of publications have dealt with RTN, with an explosion in the last years due to the relevance that it has gained in the most recent nodes, in both bulk and SOI technologies [5–8].

RTN is a consequence of trapping/detrapping of charges in/from defects (or, equivalently, traps) in the gate dielectric and/or its interfaces, which lead to sudden and random changes in the device current (see an illustrative example in Fig. 1). When considering the drain current in a MOSFET, these changes are associated with shifts in the threshold voltage (V_{th}) of the device induced by the trapping/detrapping of charges. These random and transient changes in the device V_{th} can be actually a source of device time-dependent variability (TDV). Actually, for extremely scaled technologies, the RTN contribution to the variability of CMOS logic devices and circuits may overtake that of random dopant fluctuations (RDFs) as its dominant source [9].

The transient variability introduced by RTN in the threshold voltage of the devices will be translated into transient shifts of the circuit electrical parameters, negatively impacting the circuit performance and reliability. In logic circuits, RTN can seriously influence the energy-delay trade-off in digital applications [10, 11]. In memories, RTN causes variable retention times in DRAM [12, 13], bit cell failure in SRAM [14], and in NAND flash memories, since its effect becomes more pronounced as the number of bits per cell increases [15]. SRAM cells

J. Martin-Martinez · R. Rodriguez (✉) · M. Nafria
Departament d'Enginyeria Electrònica, Escola d'Enginyeria, Edifici Q, Universitat Autònoma de Barcelona (UAB), Barcelona, Spain
e-mail: Rosana.Rodriguez@uab.es

T. Grasser (ed.), *Noise in Nanoscale Semiconductor Devices*,
https://doi.org/10.1007/978-3-030-37500-3_14

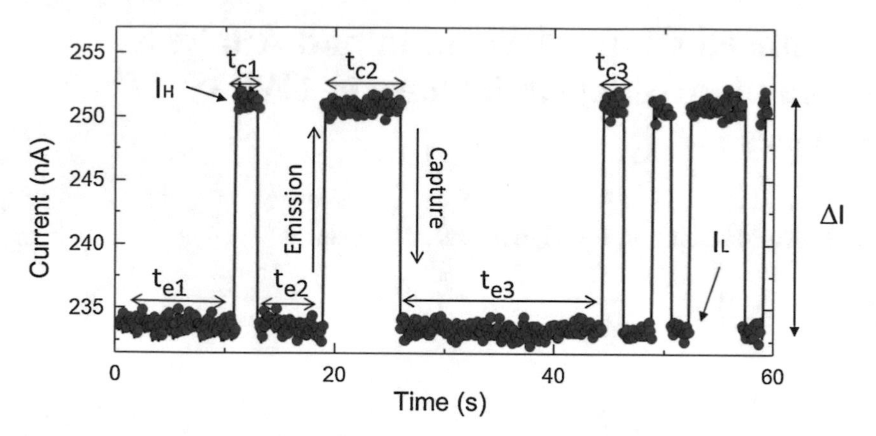

Fig. 1 Typical two-level RTN trace (obtained through simulation) caused by the charge trapping/detrapping in/from a single defect located in the transistor gate oxide. Background Gaussian noise has been added to the RTN signal, to mimic the real situation

fabricated in SOI technologies are also sensitive to RTN [16, 17], which can cause intermittent errors contributing to bitcell failure. Note that the ultrathin body and buried oxide fully depleted silicon-on-insulator (UTBB FD-SOI) technology offers new advantageous design opportunities thanks to its very efficient body-biasing capabilities [18], introducing the body of the device as a second gate. However, from an RTN perspective, both front and back gate (FG and BG) of the device could act as RTN sources. Still in the memory field, emerging memory architectures, such as ReRAM [19, 20], can also be affected by RTN. In these memories, a gate voltage applied to a metal–insulator–metal (MIM) or metal–insulator–semiconductor (MIS) structure may reversibly change the resistance of the dielectric between a high-resistance state (HRS) and a low-resistance state (LRS). RTN can be observed in the current through the device, especially in its high-resistance state (HRS) [21], where it can cause very wide current fluctuations, potentially inducing read failures [22]. Though the defects and physics behind RTN in this kind of memories probably differ from those based in MOSFETs, RTN coming from all types of devices are qualitatively equivalent.

From the above, it can be concluded that, with current CMOS technology downscaling, the time-dependent variability introduced by RTN must be taken into account in the design of memories, digital and analog VLSI-integrated circuits (ICs), to implement reliability-aware circuits and systems. To fill the gap between the device and circuit levels, the TDV introduced by RTN must be introduced into the device compact models. Then, the development of physics-based compact models that accurately describe the observed shifts of the electrical parameters of the devices is a must. Moreover, suitable and accurate model parameters extraction procedures must be developed in order to account for the specific characteristics of the technology under study. In this regard, because of the particular features of RTN, which will be described in the next section, the characterization and data analysis needed for RTN parameter extraction in advanced devices are challenging.

1.1 RTN Phenomenology

As described previously, RTN is related to the trapping/detrapping of charges in/from traps in the gate oxide and/or its interfaces. A detailed explanation about the physics related to the RTN phenomenon can be found in chapter "Oxide Trap-Induced RTS in MOSFETs" of this book [23]. In MOSFETs, these trapping events lead to changes in the threshold voltage of the device [24]. This is shown in Fig. 2, where TCAD simulations of MOSFETs with trapped charges at a location of the channel interface are shown. Clearly, the charges introduce changes in the potential profile of the device (Fig. 2a), which are translated into a shift of its I_D–V_G transfer characteristics (Fig. 2b). Less current is driven by the device for a fixed V_G, which can be interpreted as a shift in V_{th}. If the charges are detrapped, the initial characteristics would be recovered, i.e., the initial V_{th}. As a consequence, when a constant voltage is applied, RTN is manifested as discrete jumps in the drain current between two levels, as defined by the trap state (neutral or charged). As an example, Fig. 1 schematically shows the RTN signal associated to the simplest case, which corresponds to the presence of only one defect in the device. Experimentally, RTN signals will be always accompanied by the background experimental noise, so that the current/threshold voltage levels could not always be precisely defined (see, e.g., Fig. 4a). The drain current switches between two current levels, I_H (when charges are detrapped) and I_L (when charges have been trapped), separated by a distance ΔI. The times t_e and t_c are the times that a charge needs to be emitted or trapped, respectively, which have been observed to be statistically distributed, following an exponential distribution with averages τ_e and τ_c [25]. Three parameters are required to describe a two-level RTN signal like the one shown in Fig. 1; the mean emission (τ_e) and capture (τ_c) times and the magnitude of the current shift (ΔI). The last parameter, ΔI, can be obtained directly from the RTN trace. These parameters

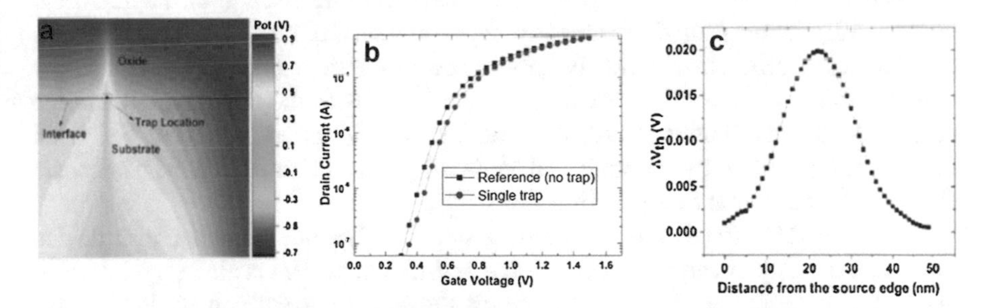

Fig. 2 2D TCAD simulations of a MOSFET, when a single charged trap is considered at its interface: (**a**) Potential profile. (**b**) I_D–V_G of the device when the trap is neutral (black squares) or charged (red circles). Charge trapping leads to a change of the potential in the channel and a shift of the transfer curve towards the right. This shift means smaller currents for the same bias, which can be interpreted as an increase of the threshold voltage. (**c**) The threshold voltage shift depends on the location of the trap along the device channel. In this 50-nm long channel device, the impact is maximum when it is located close to the center of the channel

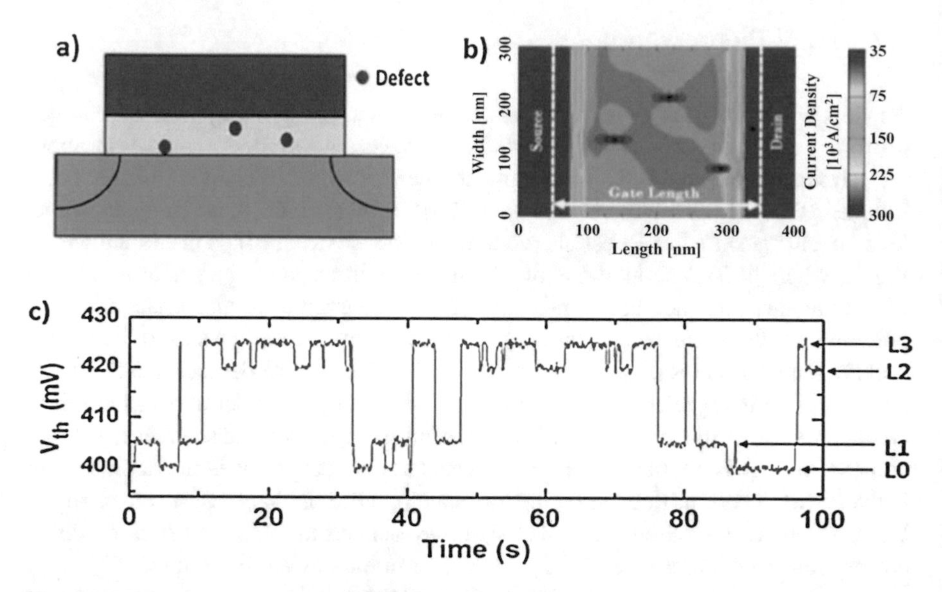

Fig. 3 (**a**) Schematic front view of a MOSFET with three defects located at different depths in the gate oxide and location along the channel. (**b**) TCAD simulation of the current density in a device with three defects. Charged defects locally affect the conduction across the channel. (**c**) Example of a typical RTN trace (obtained through simulation) in a device with two defects. The charging and discharging of the defects lead to multiple levels in the RTN (L0–3). Background noise could mask these shifts

are strongly related to the physical characteristics of the defect (i.e., energy level, location in the oxide and along the channel…). As an example, Fig. 2c shows TCAD simulation of the threshold voltage shift when charged traps are located at different locations along the channel. In this short channel device, charges close to the center of device have a larger impact, whereas those close to the source/drain have a smaller effect [26]. Consequently, depending on the position of the trap along the channel, a different ΔI will be introduced in the device drain current. Other authors have also shown the trap impact dependence with the trap position and the random dopant distribution (RDD) [27] and it is also discussed in chapter "RTN and Its Intrinsic Interaction with Statistical Variability Sources in Advanced Nanoscale Devices: A Simulation Study" of this book [28].

Two-level RTN traces are related to a single defect but, actually, in a device more than one defect can be active, located at different positions along the interface and/or oxide depth (see Fig. 3a). Figure 3b shows the current density in a device where three defects are present and charged. It is evident that each of these defects will lead to a different ΔV_{th} (or, equivalently ΔI), since they are located at different positions at the interface, and could also be characterized by different time constants (τ_e, τ_c, and ΔI) [29]. Then, trapping/detrapping events in the defects will lead to multilevel RTN traces. In the ideal situation (absence of noise, of interaction among traps, …), the number of levels is determined by the number of defects in the device: for N defects, 2^N levels could be distinguished. As an example, Fig. 3c shows an RTN trace with four levels, which results from the presence of two defects in the device. However, the number of defects in a device inversely depends on the device area, so that in advanced nodes this number will be typically small [30].

From the discussion above, we can conclude that RTN is an intrinsically stochastic process that introduces random and transient V_{th} changes in the device. Moreover, it will change from device to device because:

- The number of defects per device in a set of devices is statistically distributed. It is commonly assumed that this number follows a Poisson distribution [29].
- The drain current change (ΔI) or, equivalently, the V_{th} shift (ΔV_{th}) associated with each defect is exponentially distributed [31].
- τ_e and τ_c are widely distributed, expanding over several time decades, ranging from μs to seconds. Note, however, that this estimation has been done taking into account typical experimental windows, so that smaller and larger time constants cannot be ruled out.

Taking into account the characteristics of the phenomena, it becomes clear that (1) a statistical characterization of the phenomenon becomes mandatory (many devices, long-lasting traces) and (2) advanced automatic analysis methodologies are needed, in order to process the huge amount of data that will be generated from such characterization. These two issues will be the main focus in this work.

Before going into RTN characterization and analysis, it must be noted that a similar physical framework has been proposed to describe the phenomenon of bias temperature instability (BTI), one of the most relevant aging mechanisms in MOSFETs that limits the circuit reliability in current technologies. As RTN, BTI is observed in the device electrical characteristics as a change in the device threshold voltage. However, in the case of BTI, two separate components can be distinguished in the threshold voltage shift, i.e., a permanent component (related to the charging of defects that are not discharged in the experimental window) and a recoverable component (related to the discharging of a defect when the gate bias is decreased or removed). The observation of RTN and/or BTI is strongly dependent on the operation conditions (voltage and temperature) of the device [32]. RTN and BTI effects can be observed simultaneously, the first one as fast increase/decrease events in the device current, the second one as slower or permanent changes in the current. Actually, nowadays it is believed that RTN is an important factor in the understanding of BTI reliability issues in advanced CMOS technologies [33]. In this regard, it has been proposed that RTN and the recoverable component of BTI are very likely related to the same kind of defects [34]. Then, (1) all defects that contribute to the BTI recovery can also become spontaneously charged, leading to an RTN signal, (2) most RTN defects would also contribute to BTI recovery, (3) the distributions of RTN and BTI characteristic parameters are similar for the two phenomena, and (4) both RTN and BTI defects are volatile, meaning that they can disappear and reappear [34]. A very important consequence of the similarities between RTN and BTI is that they must be analyzed and guardbanded against together. In this direction, models that are able to describe both phenomena simultaneously have been proposed [35]. Though RTN and BTI are closely related, in this work, RTN will be the main focus.

2 The Weighted Time Lag Method

In this section, an advanced methodology suitable for the extraction of the RTN current/threshold voltage levels from RTN traces will be described in detail. The method has a wide applicability, provided that the signals to be analyzed show the typical RTN signature. In addition, it overcomes the detection limitation introduced by background noise, which can impede the precise detection of levels (i.e., the defect parameters), since their contribution can be hidden. The method here presented for the RTN analysis extends the time lag plot (TLP) procedure [36], being easily implementable and robust even when the background noise is large. The method will be first explained and illustrated using as example a numerically generated RTN caused by only one defect. After that, the presented method is applied to identify the contribution of defects associated to experimental multilevel RTN.

To illustrate the method, numerically generated RTN signals have been considered. Then, Monte Carlo simulations of two-level RTN waveforms were performed by considering two-state Markov processes [37]. The black line in Fig. 4a shows an example of a simulated current RTN without noise with levels $I_L = 3$ nA and $I_H = 5$ nA. Assuming that the responsible defect of this RTN is occupied at I_L and empty at I_H, its mean capture and emission times were chosen to be $\langle\tau_c\rangle = 10$ s and $\langle\tau_e\rangle = 6.6$ s, respectively. Dots in Fig. 4a show the waveform obtained when a Gaussian noise with standard deviation $\sigma = 0.9$ nA (to represent the experimental background noise) is added to the RTN. Due to the large unfavorable ratio between the noise and the RTN amplitude, an accurate direct I_L and I_H identification is unfeasible from Fig. 4a. The RTN histogram (dots in Fig. 4b) cannot help, because the two peaks associated with I_L and I_H are hidden by the background noise [38]. Another solution is to draw the TLP (Fig. 4c), which is constructed by plotting the ith point of the RTN on the x-axis and the $(i + 1)$th point in the y-axis for the full RTN trace [36]. If the background noise is low compared to the RTN amplitude, using the TLP, the RTN levels can be identified as populated regions in the diagonal, while populated regions outside the diagonal are related to the transitions between states [36] (these regions are indicated with crosses in Fig. 4c). However, in the TLP constructed from the RTN trace in Fig. 4a, these regions again overlap because of the background noise.

The weighted time lag (w-TLP) method [39] tries to extend the TLP by minimizing the effect of the noise in the RTN, so that a more accurate defect parameters extraction is possible. The starting point of the w-TLP method is a point of the TLP with coordinates (I_i, I_{i+1}). For this point, we define the φ_i function as

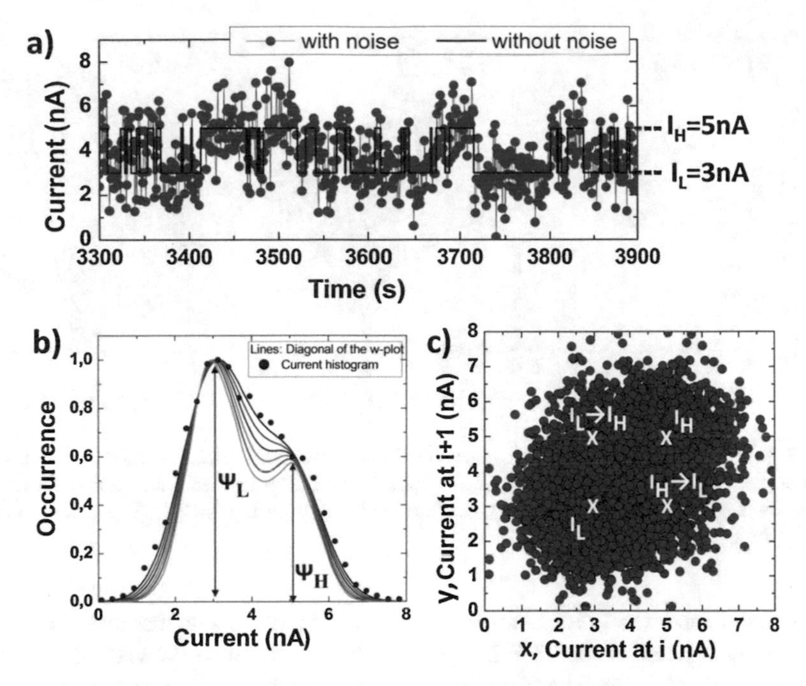

Fig. 4 (**a**) Monte Carlo–generated RTN signal with and without noise; the sample rate considered was 1 s and the number of points in the RTN is 10,000. (**b**) Dots indicate the current probability distribution obtained directly from the RTN signal in (**a**); lines correspond to the diagonal of the w-TLP (Fig. 5c) obtained for different values of α, which is the standard deviation of φ_i (Eq. 1). (**c**) TLP of the RTN; the background noise hides the current levels and transition regions (marked with crosses)

$$\varphi_i\,(x,y) = \frac{1}{2\pi\alpha^2}\exp\left(\frac{-\left[(I_i-x)^2+(I_{i+1}-y)^2\right]}{2\alpha^2}\right) \tag{1}$$

where x and y are the coordinates of the space where the TLP is considered. Note that φ_i is a normal bivariate distribution with standard deviation α and correlation coefficient 0. Then, $\varphi_i(x, y)$ represents the probability that the point with coordinates (I_i, I_{i+1}) corresponds to a level or to a transition in the a location (x, y) of the TLP space. Figure 5a shows the 2D plot of $\varphi_i(x, y)$ in log scale for a single point of the TLP. After the φ_i definition, we define the weighted time lag function Ψ as:

$$\Psi\,(x,y) = K\sum_{i=1}^{N-1}\varphi_i \tag{2}$$

being K a normalization constant chosen so that the maximum value of Ψ is 1 and N the number of points in the RTN. If Ψ is plotted for few points (Fig. 5b), two local maximums are roughly defined whose values are closer to I_L and I_H.

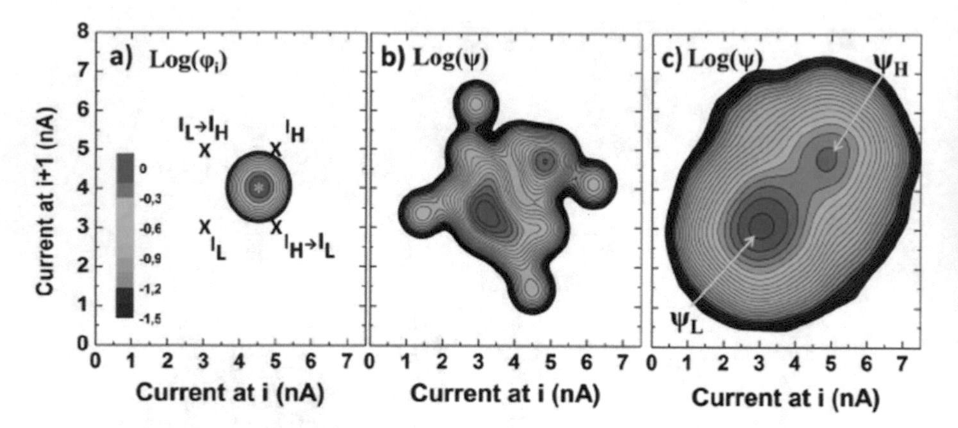

Fig. 5 (**a**) Representation of Log(φ_i), normalized to its maximum value, obtained for one point of the noisy RTN in Fig. 4a. The position of this point in the TLP is marked with * and the levels and transitions regions with crosses. (**b**) Representation of Log(Ψ) using the first 30 points of the RTN and (**c**) the full RTN trace

To understand why these peaks are revealed, we have to note that the contribution to Ψ of each point of the TLP is weighted by the distance between the position of this point and (x, y). Therefore, the function Ψ takes higher values in the most populated regions of the TLP. This is more evident in Fig. 5c where the function Ψ is plotted for all the points of the TLP. In the following, we will refer to the plots of the type of Fig. 5c as weighted TLP (w-TLP). In the diagonal of the w-TLP of Fig. 5c two well-defined local maximums can be detected (Ψ_L and Ψ_H) in the positions 3.05 and 4.89 nA. These values correspond to the two levels of the RTN, which are very close to the nominal $I_L = 3$ nA and $I_H = 5$ nA values considered to generate this signal (Fig. 4a). Then, the construction of the function Ψ is valid to detect levels of the RTN that cannot be determined from conventional methods, such as the RTN histogram (Fig. 4b) or the TLP (Fig. 4c), when the background noise is relevant. The ratio τ_e/τ_c, which in most of the cases is enough to obtain the relevant physical information of the defect [40], can be easily calculated from the w-TLP by the evaluation of the ratio of the two local maximums found in the w-TLP (Ψ_H/Ψ_L). In this case, we obtain $\Psi_H/\Psi_L = 0.63$, again close to the nominal value imposed during the generation of the RTN signal ($\tau_e/\tau_c = 0.66$).

In the weighted time lag method, the chosen value of α (the standard deviation of φ_i), which is the only fitting parameter, is a key point to correctly identify the levels of the RTN. Lines in Fig. 4b show the cross section through the diagonal of the w-TLP obtained when Ψ is constructed using different α values. For high α values, only one peak (as in the case of the histogram, dots in Fig. 4b) appears, and the two peaks are only revealed when α is decreased. However, if α is very low, this procedure fails because each point only contributes to Ψ in a region very close to it, leading in the limit, to a conventional histogram with a small bin size. Then, the proper selection of α is crucial to correctly determine and identify the defect in the RTN. This point will be further discussed later.

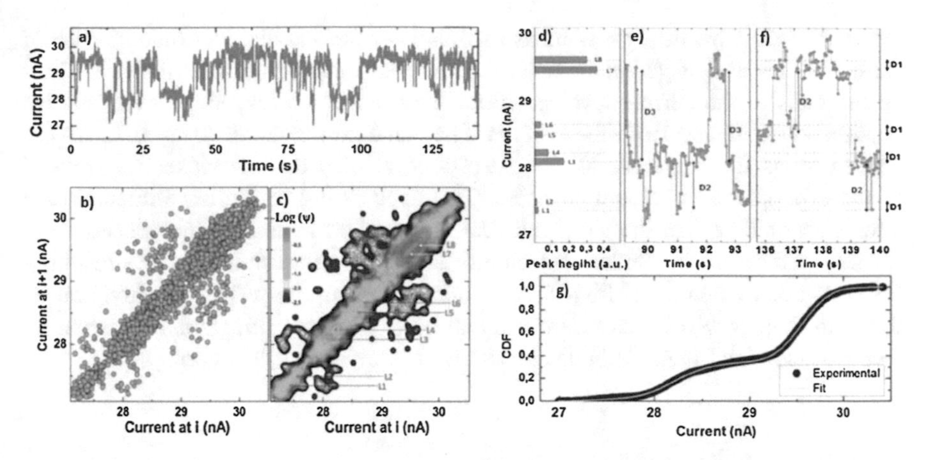

Fig. 6 (**a**) Experimental multilevel RTN in the drain current of a pMOS transistor. Conventional TLP (**b**) and w-TLP (**c**) obtained from the RTN of (**a**). (**d**) Position and height of the peaks detected in the diagonal of the w-TLP in (**c**). (**e, f**) Zooms of the experimental RTN trace in (**a**). Switchings of the three defects are indicated. D2 and D3 are evident but those of D1 cannot be clearly appreciated because of the background noise. (**g**) Experimental cumulative distribution function of the RTN in (**a**) and the fitting obtained from the data in (**d**)

The w-TLP method can be applied to multilevel signals without any change. To show this point, we considered an experimental RTN signal with a sample rate of 34.2 ms in the drain current of a pMOS transistor when it is biased with a gate voltage of −0.6 V and drain voltage of −200 mV during 140 s. The obtained current trace (Fig. 6a) shows a clear RTN where at least two discrete steps can be detected, which indicates that at least one defect is active in this device. However, a large background noise is present, making the accurate identification of the resulting current levels difficult. Moreover, other defects that could provoke smaller current steps could be hidden by the noise.

Figure 6b shows the TLP of the RTN plotted in Fig. 6a. Note that, when using the TLP, the current levels in the signal are difficult to distinguish. However, with the w-TLP (Fig. 6c), eight local maximums (L1–L8) can be detected. This suggests the existence of eight current levels in the RTN, and therefore at least three active defects in the device. The position and height of the eight current levels observed in the w-TLP are plotted in Fig. 6d. Considering the position of the current levels, three defects with different switching amplitudes have been identified (D1, D2, and D3). The current shifts between the levels associated with D2 and D3 are large enough to detect the transition regions in the TLP and w-TLP (regions outside the diagonal in Fig. 6b, c), and also can be detected in a zoom of the RTN (Fig. 6e, f). However, as in the case of the generated signal in Fig. 4, the small current shift associated with D1 compared to the amplitude of the background noise impedes the correct identification of D1 directly from the RTN.

Finally, the continuous line in Fig. 6g shows the cumulative distribution function (CDF) of the experimental RTN (symbols), which can be well reproduced using the data of Fig. 6d. This fitting is done by associating a normal distribution with each of the current levels detected in the w-TLP. The mean value of the normal distributions corresponds to the current level position; their height corresponds to the relative peak height obtained from the w-TLP. The standard deviation ($\sigma = 0.22$ nA) is chosen to be the same for all the current level distributions since it is related to the experimental background noise. The good fitting of the experimental data allows concluding that the weighted time lag method is an efficient procedure to identify defects in the device and obtain an accurate description of the resulting RTN. Another example of RTN signals analysis using the w-TLP method can be found in chapter "Systematic Characterization of Random Telegraph Noise and Its Dependence with Magnetic Fields in MOSFET Devices" of this book [41].

3 RTN Experimental Characterization Challenges

Precise RTN characterization in modern CMOS technologies is a key point towards the design and fabrication of reliable CMOS integrated circuits (ICs). However, the particular features of RTN traces pose important requirements to the experimental setup needed for such characterization. As can be concluded from the previous section, low-noise systems are required in order to discern the presence of defects whose contribution to the device current could be hidden by the unavoidable background noise. But other considerations are also to be accounted for. On the one hand, transients that are characterized by time constants in the microsecond range may appear. Then, systems with a large time resolution must be used to capture those fast defects. On the other hand, due to the stochastic nature of these phenomena, a statistical characterization is mandatory, so that a large number of devices must be measured to obtain statistically relevant results. In the next subsections, we address some of the approaches that can help to fulfill these requirements.

3.1 Ultrafast Measurement Techniques

Usually, standard characterization equipment has been commonly used to characterize RTN, such as semiconductor parameter analyzers (SPAs), with typical measurement time resolution in the ms range. To improve the time resolution, high-cost equipment can be used, such as the ultrafast modules that can be added to a standard SPA (as, e.g., Keithley 4225-PMU ultrafast I–V module [42] or Keysight B1530A waveform generator/fast measurement unit [43]). Here, we propose a low-cost, custom-made experimental setup to measure RTN, which provides a higher time resolution than that available in standard characterization equipment and comparable to the one provided by more expensive systems (in the microsecond range). As will be shown, when the results with this setup are analyzed with the w-TLP method, additional and relevant information about the RTN phenomenon can be obtained. As an application example, the setup has been used to characterize an

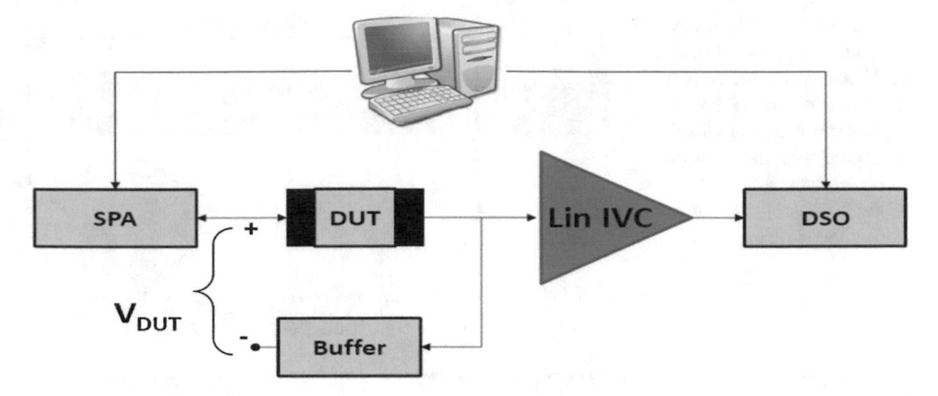

Fig. 7 Schematics of the experimental setup developed to measure RTN signals with large time resolution. V_{DUT} is the voltage drop across the DUT

RTN signal coming from RRAM devices. Though RTN signals can be observed in both resistance states (HRS [44] and LRS [45]), this work is focused on the analysis at HRS, because RTN becomes more relevant in the case of larger dielectric resistance values [46], i.e., lower currents through the dielectric.

$5 \times 5\ \mu m^2$ Ni/HfO_2 (20 nm)/Si RRAM devices were characterized. In order to get a stable conduction, the devices were first subjected to 30 resistive switching (RS) cycles (where the first cycle corresponds to the forming process [47]), changing successively the dielectric conductivity between a high- and a low-resistance state, fixing a current limit of 10 μA when switching to the LRS [47]). After this initial cycling, when the device was at HRS, several constant voltages were applied until a clear RTN signal was found. To allow comparisons, the RTN signal was measured simultaneously with SPA and the new setup. Figure 7 shows a schematic of the experimental setup developed to measure RTN with large temporal resolution. The voltage at which RTN appears was applied to one of the terminals of the DUT by the SPA (Agilent 4156C in our case), which also allows measuring the current through this terminal. The other terminal of the DUT was connected to a current-to-voltage converter (Lin-IVC), which is connected to a digital storage oscilloscope (DSO) in order to register data at different time scales. This second DUT terminal is also connected to a high impedance buffer, so that a precise value of the voltage drop across the DUT can be measured. By changing the timescale of the DSO, the current through the DUT can be easily measured with different time resolutions, so that fast events can be captured.

Figure 8 shows the flowchart of the RTN measurement process. Once a suitable value of voltage has been selected, the HRS current is measured as a function of time. While the SPA is continuously measured during a predetermined time, several oscilloscope captures are registered using different time scales. When the SPA measurement finishes, the sequence defined in Fig. 8 starts again. In this work, about 325 SPA sequential measurements were done and 21 oscilloscope captures were obtained for each SPA measurement.

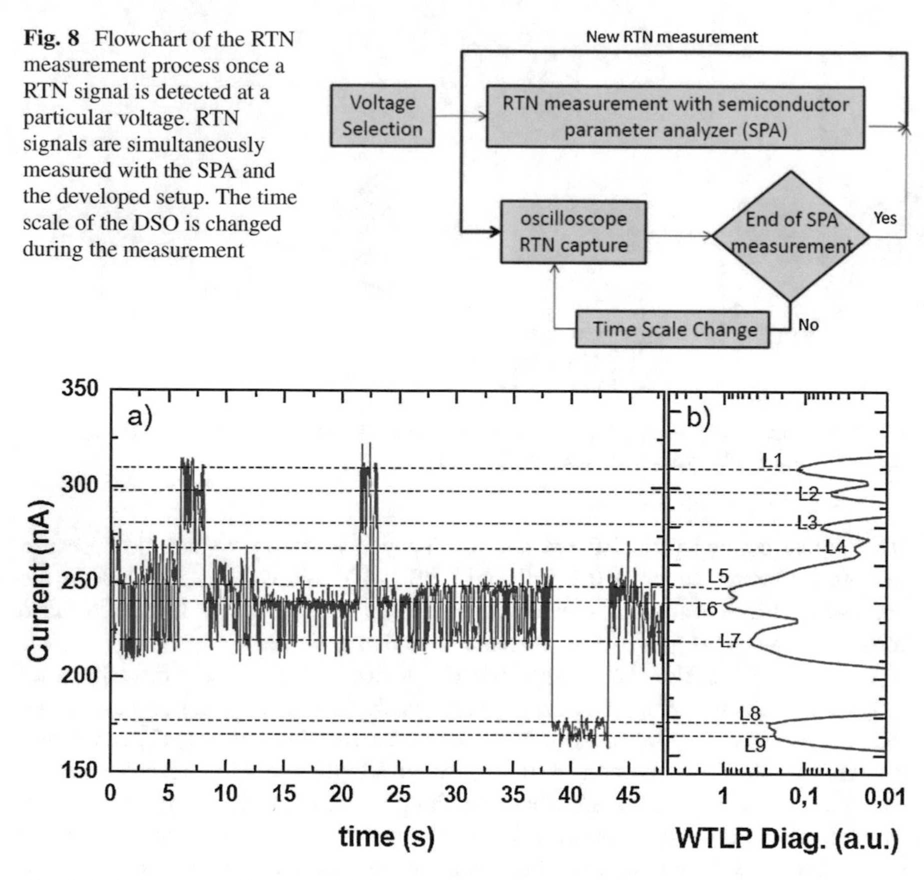

Fig. 8 Flowchart of the RTN measurement process once a RTN signal is detected at a particular voltage. RTN signals are simultaneously measured with the SPA and the developed setup. The time scale of the DSO is changed during the measurement

Fig. 9 (**a**) Typical multilevel RTN signal measured with a semiconductor parameter analyzer. Applied voltage = 1.25 V, step time ~6 ms, and number of measured points 8000. (**b**) Current levels obtained by using the w-TLP method

In Fig. 9a, a typical multilevel RTN signal measured with the SPA in a RRAM, with a time resolution of 6 ms, is shown. At naked eye, five different current levels are evidenced. However, by applying the w-TLP method to analyze this RTN signal, more levels are encountered, concretely nine different levels (L1–L9) (Fig. 9b). Figure 10 shows three oscilloscope captures of the same signal using several time scales, or equivalently time resolutions of (a) 40 μs, (b) 20 μs, and (c) 1 μs, recorded simultaneously at different time slots. In the three cases, some charge trapping and detrapping events are detected with switching times lower than 6 ms. These fast current fluctuations are not detected with the SPA (Fig. 9). Figure 10 shows that some current levels coincide with those in Fig. 9b (L2, L3, L5, L6, and L7). However, in Fig. 10b and Fig. 10c, there is a new level (L10) that does not appear in Fig. 9a, because of the fast transitions between states.

To show these points clearly, the w-TLP method was used. In Fig. 11, the w-TLPs corresponding to the RTN signals measured with the SPA and the oscilloscope

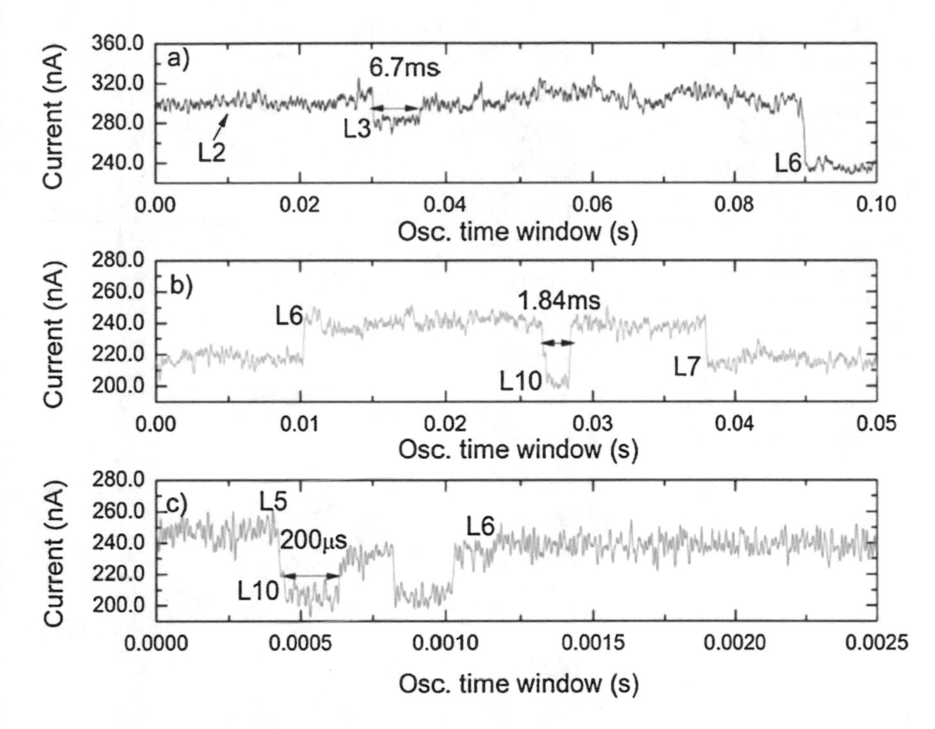

Fig. 10 Traces captured with the oscilloscope in the interval of: (**a**) 7.26–9.68 s of the SPA measurement shown in Fig. 9a with 40-μs time resolution. (**b**) 9.68–12.15 s, with a time resolution of 20 μs. (**c**) 19.5–21.93 s with a time resolution of 1 μs

are obtained. Figure 11a shows the w-TLP of the complete RTN signal measured with the SPA presented in Fig. 9a. A profile of the diagonal is plotted in Fig. 9b where all the detected levels are sequentially numbered. From the analysis of the w-TLP diagonal, nine peaks and therefore nine current levels (labeled as L1–L9) are obtained. Note that the number of current levels as determined by the w-TLP method is larger than that encountered at first sight. The guidelines plotted in Fig. 9 help to observe a good match between the peaks in the diagonal of the w-TLP and the current levels in Fig. 9a.

Figure 11b–d shows the w-TLPs of the three RTN signals captured with the DSO, and, as in the case of Fig. 11a, the existing traps levels are identified. Comparing these figures with Fig. 11a, we can recognize levels 2, 3, 5, 6, and 7. But from Fig. 11c, d, a new trap level at ~200 nA, labeled L10, appears, as mentioned before. This latter state does not appear in the SPA measurement because of the equipment resolution. However, the larger time resolution of the developed setup highlights its presence and it can be clearly visualized in the corresponding w-TLP.

In summary, combining high time resolution equipment and accurate parameters extraction, a precise characterization of the RTN current fluctuations observed

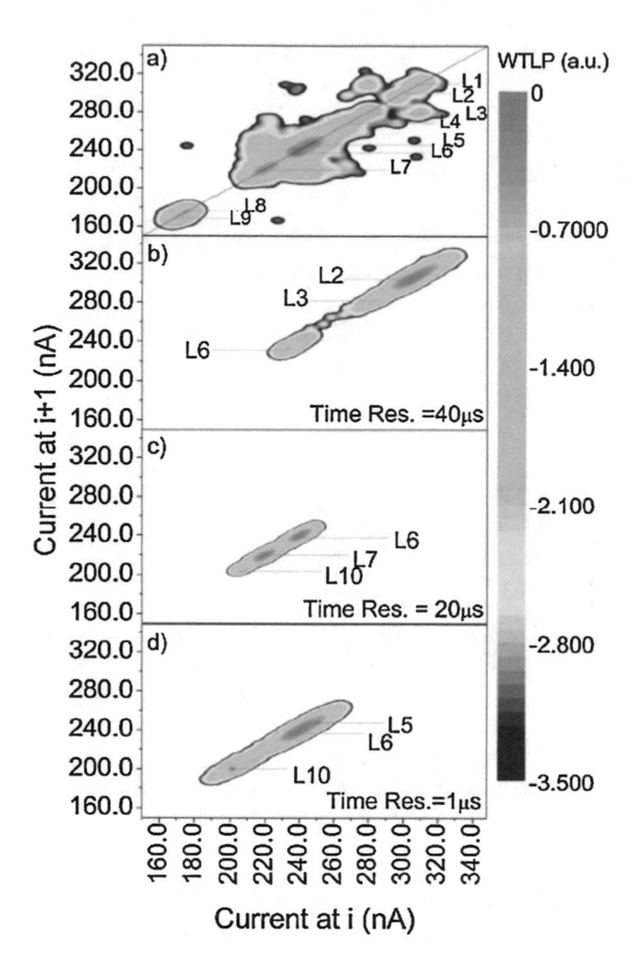

Fig. 11 w-TLP method applied to (**a**) the RTN measured by the SPA (Fig. 9a). The line shows the diagonal of the plot, whose maxima correspond to the RTN current levels. Nine trap levels are detected. The RTN captured with the oscilloscope with time resolutions of (**b**) 40 μs (Fig. 10a), three trap levels appear; (**c**) 20 μs, with three trap levels (Fig. 10b); (**d**) the larger time resolution (1 μs), showing three trap levels (Fig. 10c)

in devices can be carried out using the w-TLP method, the RTN levels can be accurately identified.

3.2 Statistical RTN Characterization: Array-Based Solutions

The stochastic nature of the RTN phenomenon requires the investigation of a very large number of samples to evaluate the statistical parameters required for a reliable estimation of its impact on the circuit functionality. The customary used approach of direct testing of devices on a wafer (i.e., direct connection of the measurement instrumentation to the contact pads of the device in the wafer on a probe station) would then become unfeasible. On the one hand, direct measurement of RTN traces with typical durations of 100–1000 s in a large number of samples will require impractical measurement times. The automation of the full characterization process would partially overcome this problem, but still the testing time would be very large. One solution is the parallelization of the measurement procedure (i.e., several devices at a time), but this requires a carefully designed wafer layout (e.g., devices could not share contact pads) and/or multiple measurement equipment, increasing the cost of the setup. Then, the reduction of the RTN testing time is a challenge. On the other hand, individual, isolated devices in a wafer would be very expensive in terms of silicon area, since most of the area of the wafer would be occupied by the contact pads of the devices. To reduce the silicon area, a feasible solution is the fabrication of ICs that contain arrays of devices, where device contact pads are shared and a specific circuitry for device selection is included. Such kind of device arrangement has been proven successful for the evaluation of process variability and aging of devices [48, 49]. This access capability brings the advantage that the total silicon area used will be considerably reduced, as compared to the probe station approach. Moreover, with this arrangement, automatic characterization of thousands of DUTs can be carried out, reducing the required testing time. It must be mentioned, however, that though the simultaneous biasing of several devices is possible (to reduce the stress time when performing reliability tests), usually only currents through one device can be measured at a time.

However, though conceptually simple, designing a transistor array for this type of measurement is not a simple task. A set of requirements must be fulfilled to carry out a proper characterization of the phenomena. For example, when current is flowing through a DUT, a voltage difference appears between the DUT terminals and the externally applied voltage on the corresponding IC pad. This voltage drop is caused by the series resistance of the chip metal paths, the access circuitry, and the chip pads. Therefore, for accurate measurements, force-and-sense techniques are necessary, meaning that independent force-and-sense paths are required to access the DUT terminals where current is flowing. With this access structure, the voltage at the DUT terminal is sensed through the sense path, while the external instrumentation adjusts the force voltage, until the desired voltage value is set at the DUT terminal. Also, calibration techniques are required to compensate the leakage currents coming from the access circuitry that are added to the current of the measured transistor.

Recently, several examples of this device arrangement have been presented, with different advantages and/or disadvantages, mainly developed for the statistical characterization of process variability and device aging, but only few of them have been used for RTN characterization [50, 51]. For example, the chip presented in [50] contains 96 $\times$ 18 cells, each including 48 DUTs and an A/D converter that serially digitizes the current of each DUT. The chip, which contains devices with various channel doping and lengths (with $W = 80$ nm and $L = 30$ or 86 nm), was fabricated in a 28-nm HKMG bulk process. Due to the selected architecture, no force-and-sense scheme is necessary, although a calibration is performed for leakage compensation.

The recently developed ENDURANCE transistor array chip [52], fabricated on a 65-nm technology, contains 3136 MOS transistors (nMOS and pMOS) of different geometries, to allow statistically significant results for electrical characterization,

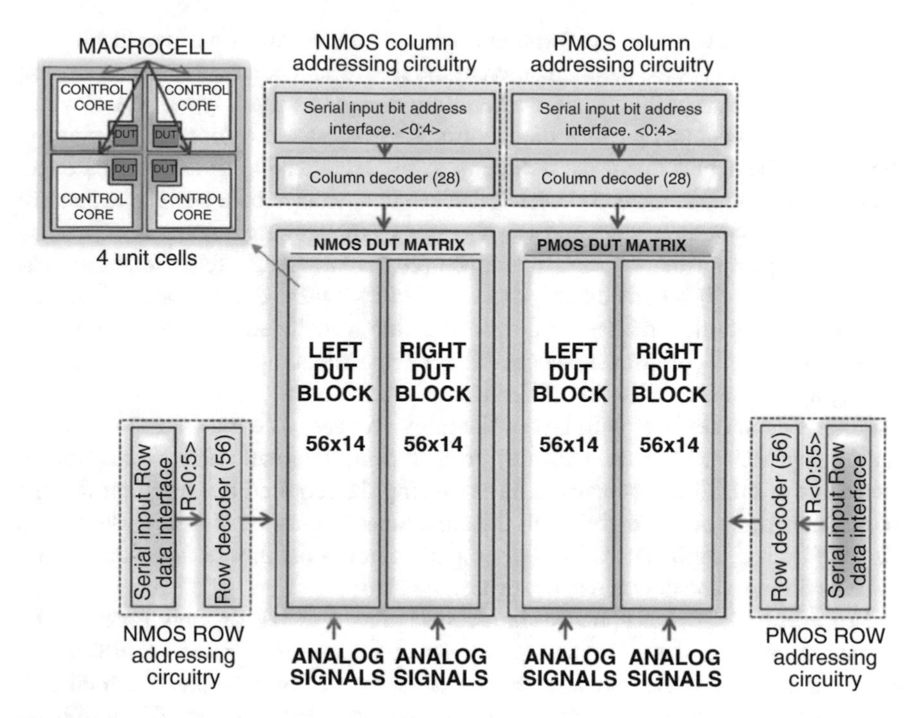

Fig. 12 Architecture of the ENDURANCE chip

not only of RTN, but also to evaluate process variability, BTI and HCI aging using a dedicated IC (Fig. 12). The chip has demonstrated a great versatility, providing parallelized accelerated aging tests with accurate timing, and a force and sense technique that compensates IR drops. Using this array chip, a full RTN characterization can be automatically done. As an example, Fig. 13 shows a set of several current traces displaying RTN effects. This figure demonstrates that the system is capable of capturing current fluctuations associated with charge trapping/detrapping in/from defects in the analyzed devices, with different emission and capture times. Multilevel signals, corresponding to more than one defect in the transistor, can also be observed. The control logic included in the chip allows the automation of the tests, partially reducing the experimental time needed for the characterization of a large number of devices [53].

4 Other Applications of the w-TLP Method

In Sects. 2 and 3, we have shown the potentiality of the w-TLP method for the evaluation of the current levels in the RTN signals, which could come from either RRAM or MOSFETs. In this section, we will describe two other application

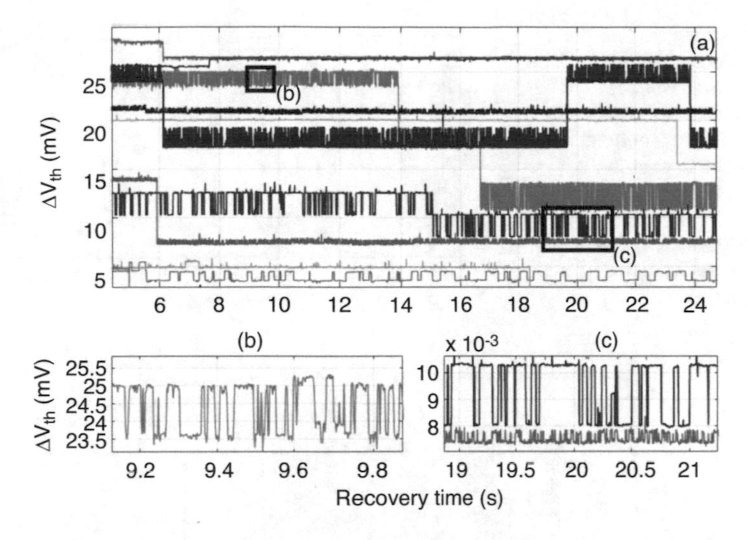

Fig. 13 (**a**) Examples of RTN tests on seven DUTs using the ENDURANCE chip [52]. (**b, c**) Zooms of RTN traces

examples of the w-TLP method. First, the method will be used to evaluate the intervals of variation of the drain current introduced by RTN in the presence of noise, which can be helpful from a circuit design point of view [54]. As a second example, the method will be used to separate the RTN and BTI components in relaxation traces in CMOS devices after their stress, to facilitate the extraction of the required information for BTI compact modeling [55].

4.1 Evaluating the Intervals of Variations of the Drain Current in MOSFETs

As demonstrated previously, from a physics point of view, the w-TLP method is useful to obtain information on the properties of the defects involved in the RTN. In this section, we will show that it is also able to provide useful information from a circuit point of view such as the margins of variation of the drain current in a MOSFET, introduced by RTN. The knowledge of these variations during the operation of the device can help to establish precise safety margins for the circuit operation. However, as demonstrated before, the identification of the RTN current levels, which will affect the current limits, can be difficult because of the presence of a background noise, due to other active physical processes in the device or to experimental factors. This background noise, which is unknown in experimental RTN traces, can mask the current levels, especially if the background noise is large and the levels are very close [38]. Hereafter, we will show that the w-TLP can be used to obtain the rms value, σ, of the additional noise, improving the

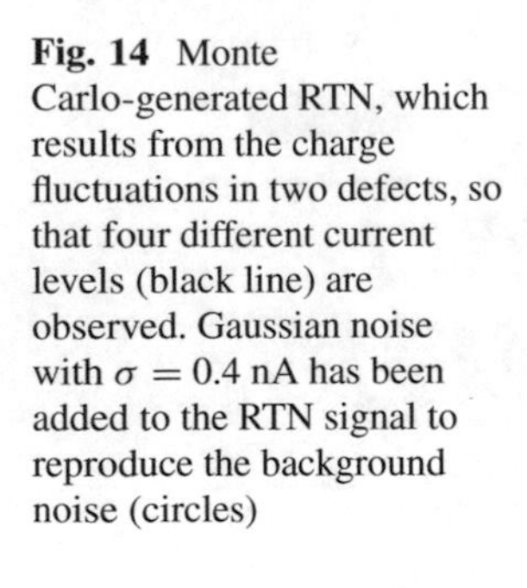

Fig. 14 Monte Carlo-generated RTN, which results from the charge fluctuations in two defects, so that four different current levels (black line) are observed. Gaussian noise with $\sigma = 0.4$ nA has been added to the RTN signal to reproduce the background noise (circles)

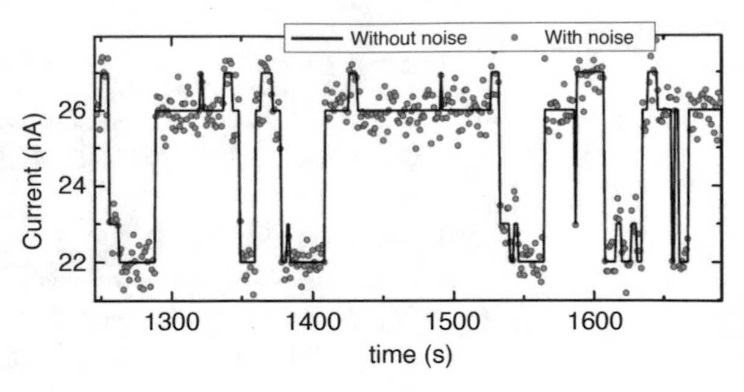

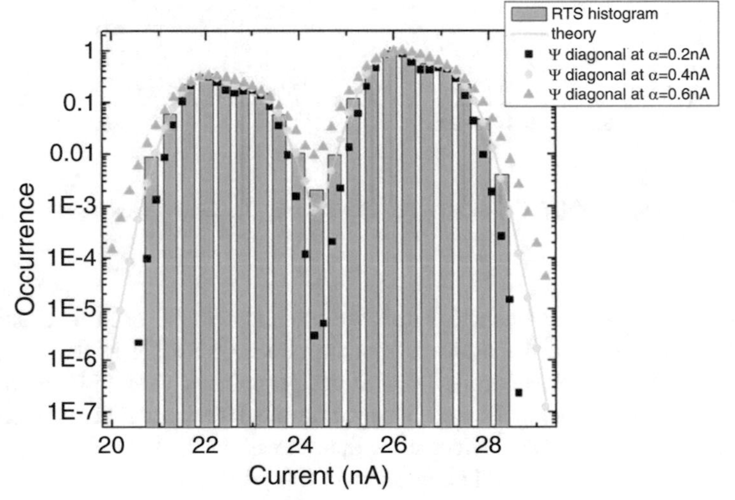

Fig. 15 Histogram of the currents in the RTN signal in Fig. 14 (bars). The yellow line corresponds to the theoretical *F*(*I*) function (Eq. 3). The symbols show the w-TLP Ψ diagonal function with $\alpha = 0.2$ nA (black), 0.4 nA (yellow), and 0.6 nA (green)

evaluation of the amplitude of the RTN drain current fluctuations. As in Sect. 2, first, a numerically generated RTN signal will be considered as an example and, afterwards, RTN coming from experimental measurements in pMOS transistors will be studied.

Figure 14 illustrates a typical RTN signal, in this case numerically built. The black solid line in Fig. 14 shows a Monte Carlo-generated RTN signal caused by two defects, which leads to four current levels, centered at 22, 23, 26, and 27 nA. To get a more realistic RTN, a background Gaussian noise with standard deviation $\sigma = 0.4$ nA has been added (circles in Fig. 14).

The histogram of the RTN currents in Fig. 14 is plotted in Fig. 15 (bars). Since we are dealing with a numerically generated RTN, it is straightforward to deduce the analytical equation that describes this histogram. For each of the current levels of the generated RTN, a Gaussian distribution is built, with average $I_{L,i}$ (the value of each current level). The standard deviation is the same for all the distributions and equal to the standard deviation of the background noise, i.e., $\sigma = 0.4$ nA. The height of each Gaussian distribution is proportional to the probability of occurrence of the associated current level. These probabilities can be easily calculated from the (known, in this case) input parameters used to generate the RTN. Therefore, the analytical probability density function obeys Eq. (3):

$$F(I) = \frac{K}{2\pi\sigma^2} \sum_{i=1}^{L} A_i \exp\left(\frac{-\left(I - I_{\mathrm{L},i}\right)^2}{2\sigma^2}\right) \tag{3}$$

being L the number of levels in the RTN, i an index that denotes the level, I the current at each point of the RTN, σ the standard deviation of the background noise, A_i the peak height, and $I_{\mathrm{L},i}$ the current of each RTN level. "K" is a normalization constant, whose value has been chosen so that the maximum of $F(I)$ is 1. The red line in Fig. 15 corresponds to the analytical probability distribution obtained from Eq. (3). But in a realistic case, the parameters of the Gaussian distributions are unknown, so $F(I)$ cannot be built. The w-TLP method, however, can be used to fit the current histogram. To show this point, symbols in Fig. 15 show the Ψ functions, calculated according to Eq. (2) for the case $x = y$, i.e., the diagonal of the function, for different values of α. From now on, we will refer to this function as "the w-TLP Ψ diagonal function." Note that the best match between the w-TLP Ψ diagonal function and the RTN histogram is obtained for $\alpha = 0.4$ nA, that is, when α is equal to the standard deviation of the background noise, σ.

Since σ is unknown in experimental RTN traces, these results indicate that the noise σ can be calculated using the w-TLP method, as the α value of the w-TLP Ψ diagonal function that better fits the RTN histogram. Additionally, the w-TLP Ψ diagonal function with $\alpha = \sigma$ (yellow circles) also matches the analytical probability density function (yellow line), even outside the histogram limits. This leads to a second conclusion: the w-TLP Ψ diagonal function that best fits the RTN histogram can be used to evaluate the intervals of variation of the drain current in the device, even outside the experimental window, that is, for longer operation times.

In the following, this methodology will be applied to determine the intervals of variation of the drain current from RTN traces experimentally obtained, where all the parameters that define the RTN parameters are unknown. As examples, Fig. 16 shows the RTN traces obtained in high-k/metal gate pMOS transistor biased at a gate voltage $V_{\mathrm{G}} = -0.6$ V and drain voltages ranging from -50 mV (bottom) to -300 mV (top). The discrete sudden changes in the current reveal the presence of defects in the device, leading to RTN in the drain current.

As explained before, for a correct evaluation of the limits of the drain current fluctuations, the α value, which is indicative of the standard deviation of the background noise (σ) and unknown in experimental measurements, must be carefully chosen. To evaluate the suitable α, we use the property demonstrated in Fig. 15, that is, the best fit of the RTN histogram with the w-TLP Ψ diagonal function is obtained when $\alpha = \sigma$. So, the w-TLP Ψ diagonal function has been calculated for a wide

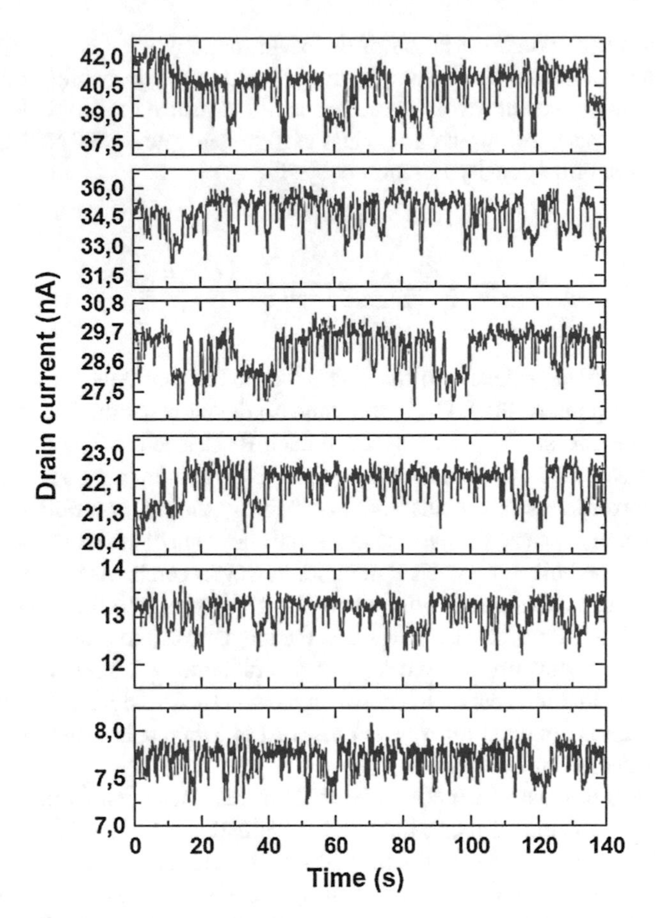

Fig. 16 RTN traces measured in the drain current of a high-k/metal gate pMOS transistor. Different drain voltages were applied, ranging from −50 mV (bottom) to −300 mV (top) with −50 mV intervals, with a gate voltage of −0.6 V

range of α values, and the difference between this function and the corresponding histograms (the error) has been calculated. As an example, Fig. 17 (top) shows the envelope of the histogram of the RTN obtained at $V_D = -250$ mV (dots). Figure 17 (bottom) shows the fitting error, as a function of the α values. The minimum of the error function in Fig. 17 (bottom) corresponds to $\alpha = \sigma$ (approximately 150 pA in this case). The corresponding w-TLP Ψ diagonal function has also been plotted in Fig. 17 (top), showing a good fit to the experimental data.

Figure 18 shows the envelopes of the current histograms of all the RTN traces in Fig. 16 and the corresponding best-fit curves. Extrapolating the w-TLP Ψ diagonal function to low occurrence probabilities provides the information about the expected lower and upper limits of the drain current variations caused by RTN for large operation times of the transistors [54]. Note that the function is very steep, leading to well-defined limits for the drain current.

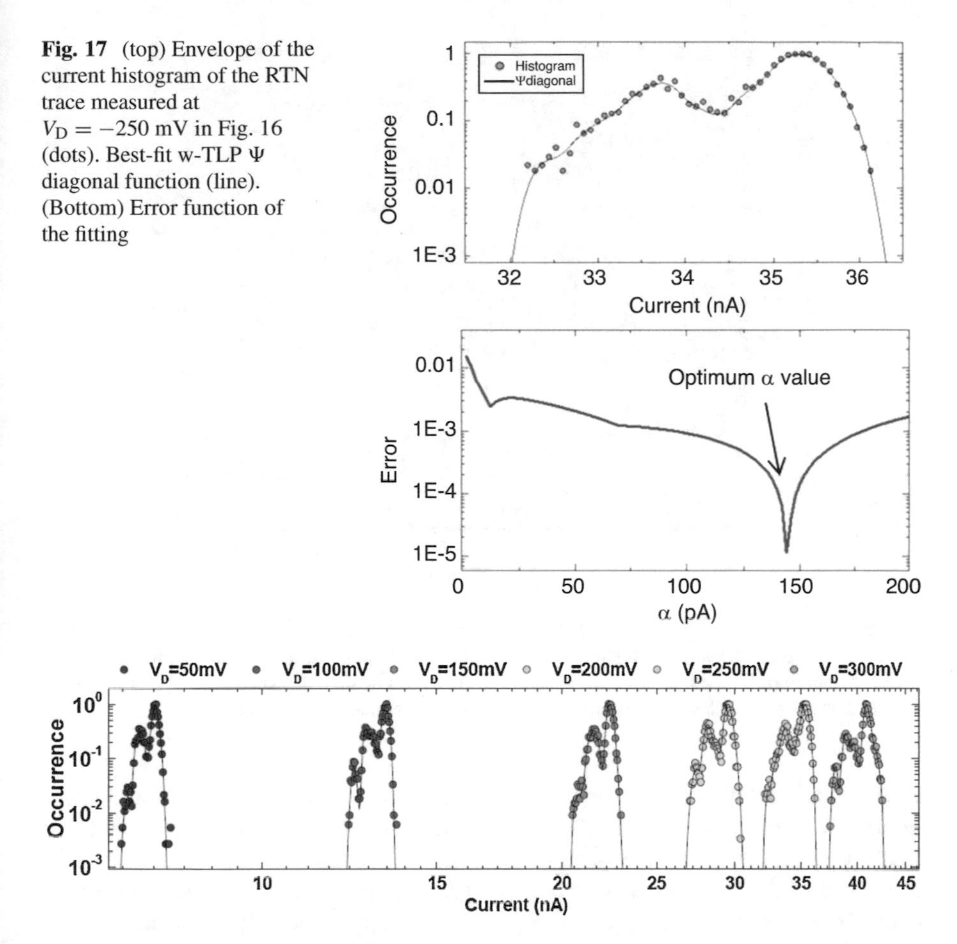

Fig. 17 (top) Envelope of the current histogram of the RTN trace measured at $V_D = -250$ mV in Fig. 16 (dots). Best-fit w-TLP Ψ diagonal function (line). (Bottom) Error function of the fitting

Fig. 18 Current histograms of the RTN traces in Fig. 16 (symbols) and their best-fit w-TLP Ψ diagonal function (lines). The extrapolation to low-probability occurrence allows determining the upper and lower bounds of the drain current

4.2 RTN and BTI Phenomena Identification in CMOS Devices

As described in Sect. 1, nowadays it is accepted that RTN and BTI are two manifestations of the same phenomenon (i.e., the trapping/detrapping of charges in/from defects in the device), being the main difference the time constants of the defects involved, faster in the case of RTN. BTI is expected at larger voltages and consists of a permanent and a recovery component. For the characterization of the aging, the recovery component is usually analyzed, from which the defect parameters can be evaluated. Actually, aging can be modeled by analyzing the emission times (τ_e) and related ΔV_{th} (η) of defects during the recovery phases, after the application of overvoltage stress [35]. Figure 19a shows a typical experimental recovery trace after a BTI test, after the conversion of the measured current into the

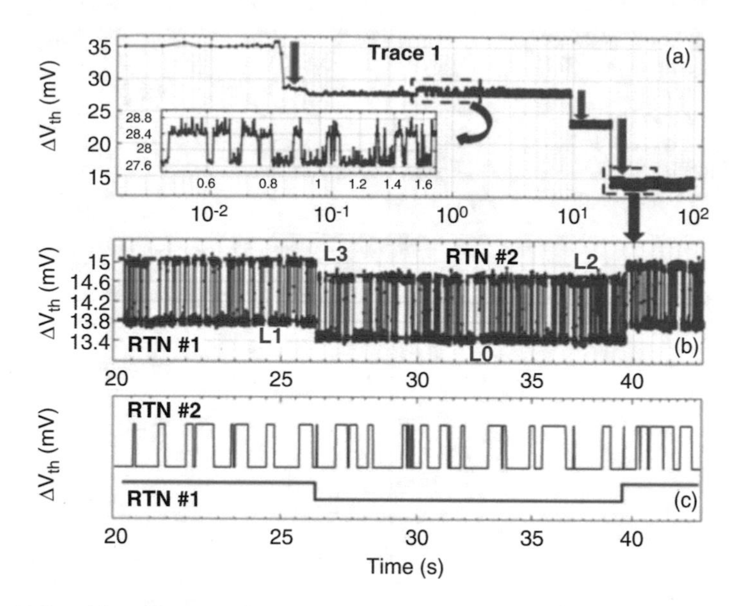

Fig. 19 (**a**) Resulting ΔV_{th} trace from the measured I_{DS} recovery data showing three clear slow defect discharges. The inset in figure (**a**) shows RTN and background noise on the top of a ΔV_{th} level. (**b**) Zoom of 23 s where four different ΔV_{th} levels can be distinguished corresponding to two fast defects charge/discharges. (**c**) Illustration of the RTN signals in (**b**) uncoupled

equivalent threshold voltage change [56]. Charge detrapping from oxide defects can be clearly seen during the recovery period as abrupt current jumps. The zoom in Fig. 19b shows the simultaneous presence of RTN in the traces. Depending on the amplitude of the RTN, the net current changes related to the BTI aging component could be masked, so, for the correct BTI modeling, it is critical to distinguish between current changes that are linked to BTI or to RTN. With this purpose, the w-TLP method will be used. Note that BTI is also a stochastic process, so that many traces as that shown in Fig. 19a will have to be analyzed to get statistically meaningful information.

As an example, the trace in Fig. 19a is considered. In this figure, three slow emissions (denoted with green arrows) without any capture counterpart are observed, linked to BTI and whose associated τ_e and η must be evaluated. However, a detailed inspection reveals that RTN is superposed to some of the levels. The inset in Fig. 19a exposes fast transitions between two separated ΔV_{th} levels on top of a slow emission. Figure 19b reveals fast capture/emission transitions mixed with background noise, switching between four different ΔV_{th} levels (i.e., L0–L3). A visual inspection of Fig. 19b reveals the join contribution of two individual RTN signals, which have been separated in Fig. 19c in order to visually distinguish them.

The w-TLP method allows the identification of the fast and slow defects, by identifying the number, i.e., L0 to $\#V_{th\text{-LEVELS}}$-1 and value of the ΔV_{th} shifts present in the trace. Figure 20a shows the w-TLP resulting from the trace in Fig. 19a. By

Fig. 20 (**a**) w-TLP resulting from the analysis of Trace #1 in Fig. 19a. (**b**) Zoom of the region marked with a red square in the w-TLP in (**a**) showing, for the two RTN signals in Fig. 19b, the ΔV_{th} levels in the diagonal and the corresponding transitions between ΔV_{th} levels outside the diagonal

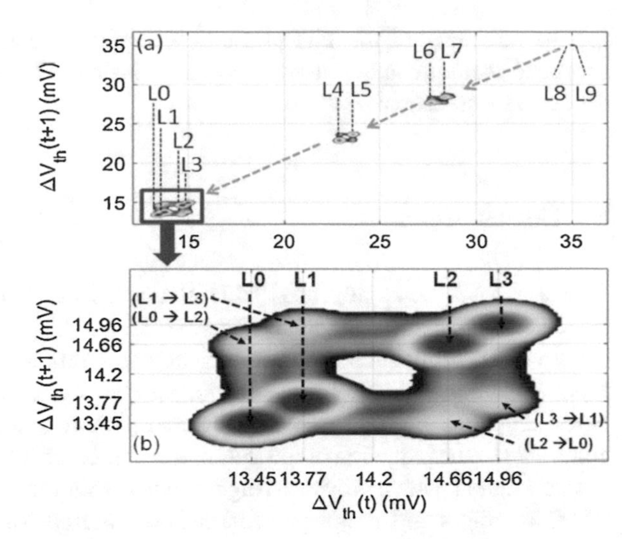

analyzing the counter diagonal of the w-TLP, four groups of populated data regions, separated in the figure by green dashed arrows, can be distinguished. Each of the data groups corresponds to a different ΔV_{th} level present in the recovery trace. Transitions from one data group to the next one are considered as slow emissions at a specific τ_{e} and the difference between the two levels corresponds to the η value of the associated defect.

The fast capture/emissions on top of each ΔV_{th} level are clearly distinguished as red populated regions in the counter diagonal (i.e., ΔV_{th} levels L0–L9) with other less-populated regions outside the counter diagonal (i.e., ΔV_{th}-level transitions). Figure 20b shows a zoom in of the w-TLP, between $\Delta V_{\text{th}}(t) = 13$ mV and $\Delta V_{\text{th}}(t+1) = 15$ mV. The zoom clearly shows four different ΔV_{th} levels resulting from the combination of the charge/discharges of two fast defects, shown in the zoom of "Trace #1" in Fig. 19c as individual RTN signals:

- RTN #1: As shown in Fig. 19b, the two associated main ΔV_{th} levels are located at L0 and L1. Transitions between these two ΔV_{th} levels occur from level L0 to L1 and vice versa.
- RTN #2: In this case, the two main ΔV_{th} levels are located always above the RTN #1 level. For instance, at recovery time of 25 s, RTN #1 stays at level L1, thus the RTN #2 switches between levels L1 and L3. On the contrary, at the recovery time of 30 s, the RTN #1 has dropt to level L0 and the RTN #2 switches between L0 and L2.

Once the ΔV_{th} levels are identified by the w-TLP, the method assigns, to each sample in the ΔV_{th} trace, the closest ΔV_{th} level obtained with the w-TLP. This step quantifies the ΔV_{th} recovery trace, removing the background noise and leaving only ΔV_{th} levels associated to slow and fast defects capture/emissions. From the clean traces, η and τ_{e} related to the BTI aging component in the trace can be obtained, for the modeling of the BTI phenomenon [52]. It must be emphasized that the complete process can be automatized, to evaluate the large number of traces needed for accurate statistical analysis.

5 Conclusions

In this chapter, some advanced RTN characterization techniques have been presented. On the one hand, some of the solutions proposed to overcome the main experimental limitations linked to the characterization of RTN have been revised. First, since a statistical characterization of the phenomenon is needed, an array-based chip has been proven as an area-efficient solution, which can partially contribute to the reduction of the testing time. Second, a new experimental setup has been presented, which provides a higher time resolution than standard equipment, so that faster trapping/detrapping events can be captured. To analyze the obtained data, the weighted time lag plot (w-TLP) method has been presented, being an efficient and robust method to evaluate the defect-related parameters, even when the background noise is large. The potentiality of the method has been demonstrated through several examples: RTN analysis in RRAM and CMOS devices, the evaluation of the statistical distributions of the drain current in MOSFETs in the presence of RTN and the identification of slow emissions and RTN transients during BTI testing.

Acknowledgments This work has been supported in part by the TEC2013-45638-C3-1-R and TEC2016-75151-C3-1-R Projects (funded by the Spanish AEI and ERDF).

References

1. M.J. Kirton, M.J. Uren, Noise in solid-state microstructures: a new perspective on individual defects, interface states and low-frequency (1/f) noise. Adv. Phys. **38**(4), 367–468 (1989)
2. K.R. Farmer, Discrete conductance fluctuations and related phenomena in metal-oxide-silicon device structures, in *Proc. Insulating Films on Semiconductors*, 1991, pp. 1–18
3. E. Simoen, C. Claeys, Random telegraph signals in silicon-on-insulator metal-oxide-semiconductor transistors. J. Appl. Phys. **75**(7), 3647–3653 (1994)
4. K.S. Ralls, W.J. Skocpol, L.D. Jackel, R.E. Howard, L.A. Fetter, R.W. Epworth, D.M. Tennant, Discrete resistance switching in submicrometer silicon inversion layers: individual interface traps and (1/f?) noise. Phys. Rev. Lett. **52**(3), 228–231 (1984)
5. R. Huang, X.B. Jiang, S.F. Guo, P.P. Ren, P. Hao, Z.Q. Yu, Z. Zhang, Y.Y. Wang, R.S. Wang, Variability-and reliability-aware design for 16/14nm and beyond technology, in *2017 IEEE International Electron Devices Meeting (IEDM)*, pp. 12.4.1–12.4.4
6. S.S. Chung, Recent advances of RTN technique towards the understanding of the gate dielectric reliability in trigate FinFETs, in *2016 IEEE 23rd International Symposium on the Physical and Failure Analysis of Integrated Circuits (IPFA)*, 2016, pp. 33–37
7. C.G. Theodorou, E.G. Ioannidis, S. Haendler, N. Planes, E. Josse, C.A. Dimitriadis, G. Ghibaudo, New LFN and RTN analysis methodology in 28 and 14nm FD-SOI MOSFETs, in *2015 IEEE International Reliability Physics Symposium*, pp. XT.1.1–XT.1.6

8. C. Marquez, N. Rodriguez, F. Gamiz, A. Ohata, Electrical characterization of random telegraph noise in Back-biased ultrathin silicon-on-insulator MOSFETs, in *Eurosoi-Ulis*, 2016
9. C. Claeys, M.G.C. de Andrade, Z. Chai, W. Fang, B. Govoreanu, B. Kaczer, W. Zhang, E. Simoen, Random telegraph signal noise in advanced high performance and memory devices, in *31st Symposium on Microelectronics Technology and Devices (SBMicro)*, 2016
10. Y. Zhang, X. Jiang, J. Wang, S. Guo, Y. Fang, R. Wang, M. Luo, R. Huzng, Impacts of random telegraph noise (RTN) on the energy-delay tradeoffs of logic circuits, in *2016 China Semiconductor Technology International Conference (CSTIC)*, pp.1–4
11. K. Ito, T. Matsumoto, S. Nishizawa, H. Sunagawa, K. Kobayashi H. Onodera, The impact of RTN on performance fluctuation in CMOS logic circuits, in *2011 International Reliability Physics Symposium*, pp. CR.5.1–CR.5.4
12. D.S. Yaney, C.Y. Lu, R.A. Kohler, M.J. Kelly, J.T. Nelson, A meta-stable leakage phenomenon in DRAM charge storage—variable hold time, in *Tech. Dig. IEDM*, 1987, pp. 336–339
13. P.J. Restle, J.W. Park, B.F. Floyd, DRAM variable retention time, in *Tech. Dig. IEDM*, 1992, pp. 807–810
14. H. Qiu et al., Impact of random telegraph noise on write stability in silicon-on-thin-BOX (SOTB) SRAM cells at low supply voltage in sub-0.4V regime, in *VLSI Technology*, 2015
15. A. Goda, C. Miccoli, C. Monzio Compagnoni, Time dependent threshold-voltage fluctuations in NAND flash memories: from basic physics to impact on array operation, in *Tech. Dig. IEDM*, 2015
16. B. Zimmer, O. Thomas, S. Oon Toh, T. Vincent, K. Asanovi'c, B. Nikoli'c, Joint impact of random variations and RTN on dynamic writeability in 28nm bulk and FDSOI SRAM, in *44th European Solid State Device Research Conference (ESSDERC)*, 2014
17. K.C. Akyel, L. Ciampolini, O. Thomas, D. Turgis, G. Ghibaudo, UTBB FD-SOI front- and back-gate coupling aware random telegraph signal impact analysis on a 6T SRAM, in *SOI-3D-Subthreshold Microelectronics Technology Unified Conference (S3S)*, 2014
18. J.P. Noel et al., Multi-VT UTBB FDSOI device architectures for low-power CMOS circuit. IEEE Trans. Electron Devices **58**(8), 2473–2482 (2011)
19. H. Akinaga, H. Shima, Resistive random access memory (RRAM) based on metal oxides. Proc. IEEE **98**(12), 2237–2251 (2010)
20. G. Burr, B. Kurdi, J. Scott, C. Lam, K. Gopalakrishnan, R. Shenoy, Overview of candidate device technologies for storage-class memory. IBM J. Res. Dev. **52**(4/5), 449–464 (2008)
21. J.-K. Lee, J.-W. Lee, J. Park, S.-W. Chung, J.S. Roh, S.-J. Hong, I. Cho, H.-I. Kwon, J.-H. Lee, Extraction of trap location and energy from random telegraph noise in amorphous TiOx resistance random access memories. Appl. Phys. Lett. **98**, 143502-1–143502-3 (2011)
22. F. Puglisi, Random telegraph noise analysis as a tool to link physical device features to electrical reliability in nanoscale devices, in *IIRW*, 2016
23. A.S.M. Shamsur Rouf, Z. Çelik-Butler, *Oxide Trap-Induced RTS in MOSFETs*, Chapter 17 of this book, 2019
24. C. Claeys et al., Random telegraph signal noise in advanced high performance and memory devices, in *Symposium on Microelectronics Technology and Device*, 2016, pp. 1–6
25. A. Yonezawa, A. Teramoto, T. Obara, R. Kuroda, S. Sugawa, T. Ohmi, The study of time constant analysis in random telegraph noise at the subthreshold voltage region, in *2011 IEEE International Conference on IC Design & Technology*, pp. 1–4
26. V. Velayudhan, J. Martin-Martinez, R. Rodriguez, M. Porti, M. Nafria, X. Aymerich, C. Medina, F. Gamiz, TCAD simulation of interface traps related variability in bulk decananometer MOSFETS, in *2014 5th European Workshop on CMOS Variability (VARI)*, pp. 1–6
27. B. Kaczer, S. M. Amoroso, R. Hussin, A. Asenov, J. Franco, P. Weckx, Ph.J. Roussel, G. Rzepa, T. Grasser, N. Horiguchi, On the distribution of the FET threshold voltage shifts due to individual charged gate oxide defects, in *2016 IEEE International Integrated Reliability Workshop (IIRW)*, pp. 18–20
28. F. Adamu-Lema, C. Compagnoni, O. Badami, V. Georgiev, A. Asenov, *RTN and Its Intrinsic Interaction with Statistical Variability Sources in Advanced Nanoscale Devices: A Simulation Study*, Chapter 13 of this book, 2019

29. E. Simoen, C. Claeys, Random telegraph signal: a local probe for single point defect studies in solid-state devices. Mater. Sci. Eng. B **91–92**, 136–143 (2002)
30. J. Franco et al., Impact of single charged gate oxide defects on the performance and scaling of nanoscaled FETs, in *Proc. IEEE Int. Rel. Phys. Symp. (IRPS)*, Apr. 2012, pp. 5A.4.1–5A.4.6
31. M. Toledano-Luque, B. Kaczer, J. Franco, Ph.J. Roussel, M. Bina, T. Grasser, M. Cho, P. Weckx, G. Groeseneken, Degradation of time dependent variability due to interface state generation, in *2013 Symposium on VLSI Technology*, pp. T190–T191
32. N. Ayala, J. Martin-Martinez, R. Rodriguez, M. Nafria, X. Aymerich, Unified characterization of RTN and BTI for circuit performance and variability simulation, in *2012 Proceedings of the European Solid-State Device Research Conference (ESSDERC)*, pp. 266–269
33. M. Toledano-Luque, B. Kaczer, E. Simoen, R. Degraeve, J. Franco, Ph.J. Roussel, T. Grasser, G. Groeseneken, Correlation of single trapping and detrapping effects in drain and gate currents of nanoscaled nFETs and pFETs, in *2012 IEEE International Reliability Physics Symposium (IRPS)*, pp. XT.5.1–XT.5.6
34. T. Grasser, K. Rott, H. Reisinger, M. Waltl, J. Franco, B. Kaczer, A unified perspective of RTN and BTI, in *2014 IEEE International Reliability Physics Symposium*, pp. 4A.5.1–4A.5.7
35. J. Martin-Martinez, B. Kaczer, M. Toledano-Luque, R. Rodriguez, M. Nafria, X. Aymerich, G. Groeseneken, Probabilistic defect occupancy model for NBTI, in *2011 International Reliability Physics Symposium*, pp. XT.4.1–XT.4.6
36. T. Nagumo, K. Takeuchi, S. Yowogawa et al., New analysis methods for comprehensive understanding of random telegraph noise, in *Tech. Dig. IEDM*, Dec. 2009, pp. 32.1.1–32.1.4
37. C. Monzio Compagnoni, R. Gusmeroli, A.S. Spinelli, et al., Statistical model for random telegraph noise in flash memories. IEEE Trans. Electron Devices **55**(1), 388–395 (2008)
38. Y. Yuzhelevski, M. Yuzhelevski, G. Jung, Random telegraph noise analysis in time domain. Rev. Sci. Instrum. **71**(4), 1681–1688 (2000)
39. J. Martin-Martinez, J. Diaz, R. Rodriguez, M. Nafria, X. Aymerich, New weighted time lag method for the analysis of random telegraph signals. IEEE Electron Device Lett. **35**(4), 479–481 (2014)
40. K. Abe, A. Teramoto, S. Shigetoshim et al., Understanding of traps causing random telegraph noise based on experimentally extracted time constants and amplitude, in *Proc. IRPS*, 2011, pp. 4A.4.1–4A.4.6
41. C. Marquez, O. Huerta, A.I. Tec-Chim, F. Guarin, E.A. Gutierrez-D, F. Gamiz, *Systematic Characterization of Random Telegraph Noise and Its Dependence with Magnetic Fields in MOSFET Devices*, Chapter 4 of this book, 2019
42. https://www.tek.com/sites/default/files/media/media/resources/4225-PMUDataSht.pdf
43. https://www.keysight.com/en/pd-1443698-pn-B1500A-A30re/waveform-generator-fast-measurement-unit-wgfmu-module-for-the-b1500a?cc=ES&lc=eng
44. N. Raghavan, R. Degraeve, A. Fantini, L. Goux, S. Strangio, B. Govoreanu, D.J. Wouters, G. Groeseneken, M. Jurczak, Microscopic origin of random telegraph noise fluctuations in aggressively scaled RRAM and its impact on read disturb variability, in *IEEE International, Reliability Physics Symposium (IRPS)*, 2013, pp. 5E.3.1–5E.3.7
45. F.M. Puglisi, P. Pavan, L. Larcher, A. Padovani, Analysis of RTN and cycling variability in HfO2 RRAM devices in LRS, in *European Solid State Device Research Conference (ESSDERC)*, 2014, pp. 246–249
46. D. Ielmini, F. Nardi, C. Cagli, Resistance-dependent amplitude of random telegraph-signal noise in resistive switching memories. Appl. Phys. Lett. **96**(5), 053503_1–053503_3 (2010)
47. H.-S.P. Wong, H.-Y. Lee, S. Yu, Y.-S. Chen, Y. Wu, P.-S. Chen, B. Lee, F.T. Chen, M.-J. Tsai, Metal-oxide RRAM. Proc. IEEE **100**(6), 1951–1970 (2012)
48. C.S. Chen, L. Li, Q. Lim, H.H. Teh, N.F.B. Omar, C.L. Ler, J.T. Watt, A compact array for characterizing 32k transistors in wafer scribe lanes, in *2014 IEEE Conference on Microelectronic Test Structures*, pp. 227–232
49. P. Weckx, B. Kaczer, C. Chen, P. Raghavan, D. Linten, A. Mocuta, Relaxation of time-dependent NBTI variability and separation from RTN, in *Proc. IRPS*, 2017

50. P. Weckx, B. Kaczer, C. Chen, J. Franco,E. Bury, K. Chanda, J. Watt, Ph.J. Roussel, F. Catthoor, G. Groeseneken, Characterization of time-dependent variability using 32k transistor arrays in an advanced HK/MG technology, in *Proc. IRPS*, 2015
51. A. Whitcombe, S. Taylor, M. Denham, V. Milovanovic, B. Nikoli, On-chip I-V variability and random telegraph noise characterization in 28 nm CMOS, in *46th European Solid-State Device Research Conference (ESSDERC)*, 2016, pp. 248–251
52. J. Diaz-Fortuny, J. Martin-Martinez, R. Rodriguez, M. Nafria, R. Castro-Lopez, E. Roca, F.V. Fernandez, E. Barajas, X. Aragones, D. Mateo, A transistor array chip for the statistical characterization of process variability, RTN and BTI/CHC aging, in *14th International Conference on Synthesis, Modeling, Analysis and Simulation Methods and Applications to Circuit Design (SMACD)*, 2017
53. J. Diaz-Fortuny, J. Martin-Martinez, R. Rodriguez, M. Nafria, R. Castro-Lopez, E. Roca, F.V. Fernandez, TARS: a toolbox for statistical reliability modeling of CMOS devices, in *14th International Conference on Synthesis, Modeling, Analysis and Simulation Methods and Applications to Circuit Design (SMACD)*, 2017
54. J. Martin-Martinez, J. Diaz, R. Rodriguez, M. Nafria, X. Aymerich, E. Roca, F.V. Fernandez, A. Rubio, Characterization of random telegraph noise and its impact on reliability of SRAM sense amplifiers, in *2014 5th European Workshop on CMOS Variability (VARI)*
55. J. Diaz-Fortuny, J. Martin-Martinez, R. Rodriguez, M. Nafria, R. Castro-Lopez, E. Roca, F.V. Fernandez, A noise and RTN-removal smart method for parameters extraction of CMOS aging compact models, in *2018 Joint International EUROSOI Workshop and International Conference on Ultimate Integration on Silicon (EUROSOI-ULIS)*
56. B. Kaczer et al., Ubiquitous relaxation in BTI stressing-new evaluation and insights, in *IEEE Int. Reliab. Phys. Symp. Proc.*, 2008, pp. 20–27

An Overview on Statistical Modeling of Random Telegraph Noise in the Frequency Domain

Thiago H. Both, Maurício Banaszeski da Silva, Gilson I. Wirth, Hans P. Tuinhout, Adrie Zegers-van Duijnhoven, Jeroen A. Croon, and Andries J. Scholten

1 Introduction

The well-known benefits of MOS transistor scaling are lower power consumption, increased speed, and high integration density. However, area scaling has also introduced some downsides, such as increased parametric mismatch variation and, discussed in this chapter, increased low-frequency noise (LFN). It has being reported extensively in the literature that the power spectral density (PSD) of the low-frequency noise is inversely proportional to the device area. Therefore, in noise sensitive applications, such as analog and radio-frequency applications, the LFN potentially is a major limiter to device scaling.

On top of that, it is also discussed in the literature that the LFN variability in highly scaled devices is extremely high, where the LFN of devices with a few hundred nanometers of length and width can vary by many orders of magnitude [1], as exemplified in Fig. 1a, based on our own measurements. This high variability poses an additional challenge for using such scaled devices in noise sensitive circuits. Therefore, a statistical model for the LFN must be made available to circuit designers, making it possible to evaluate whether the LFN of a device could be a limiter in circuit performance and yield.

T. H. Both (✉)
Universidade Federal de Pelotas (UFPel), Pelotas, Brazil
e-mail: thboth@ufpel.edu.br

M. Banaszeski da Silva
Universidade Federal de Santa Maria (UFSM), Santa Maria, Brazil

G. I. Wirth
Universidade Federal do Rio Grande do Sul (UFRGS), Porto Alegre, Brazil

H. P. Tuinhout · A. Zegers-van Duijnhoven · J. A. Croon · A. J. Scholten
NXP Semiconductors, Eindhoven, The Netherlands

T. Grasser (ed.), *Noise in Nanoscale Semiconductor Devices*,
https://doi.org/10.1007/978-3-030-37500-3_15

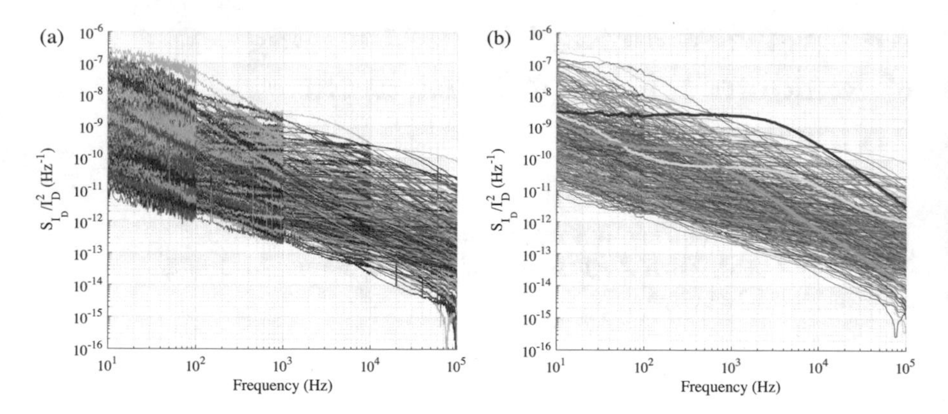

Fig. 1 (**a**) LF noise PSD of 320 nMOS device with $W \times L = 0.3 \times 0.04\,\mu\text{m}^2$. (**b**) Highlight of the noise PSD of 3 devices, demonstrating typical Lorentzian-like spectra

The major contribution to MOSFET low-frequency is attributed to oxide defects that capture and emit carriers from and to the device channel. The switching behavior of each individual defect results in a Lorentzian spectrum in the frequency domain [2–8]. In large-area devices, the summation of a large number of defects gives the noise PSD the characteristic $1/f$-like behavior, commonly referred to as flicker noise or $1/f$ noise. In contrast, in very small-area devices, the individual contribution of each defect is usually visible, and the PSDs have the distinct Lorentzian behaviors associated with random telegraph noise (RTN), as shown in Fig. 1b. From this explanation, $1/f$ noise and RTN are one and the same effect. In small-area devices, this effect manifests itself as RTN, whereas in large-area devices it manifests itself as $1/f$-like noise. This issue is also discussed in detail in Chapter 18 of this book [9]. The large variability of the LFN of small-area devices is due to the stochastic nature of the parameters of these Lorentzian spectra.

The next sections of this chapter present a review of the works [10–14] in which the authors of this chapter introduced a novel statistical LFN model, based on RTN. This model proves suited to describe LFN of very large devices (with an almost perfect $1/f$ PSD) as well as very small devices (which have just a few defects).

2 Dielectric Defects as the Origin of the LFN

The novel RTN-based model for LFN is derived by considering that the low-frequency noise is the result of defects near the dielectric-silicon interface of MOS devices [4, 15]. The atomistic modeling of dielectric defects that may lead to RTN is discussed in detail in Chapter 19 of this book [16]. These defects, or traps, are known to capture and release carriers from and into the channel. The capture and emission of carriers by a single trap makes the drain current vary between two fixed

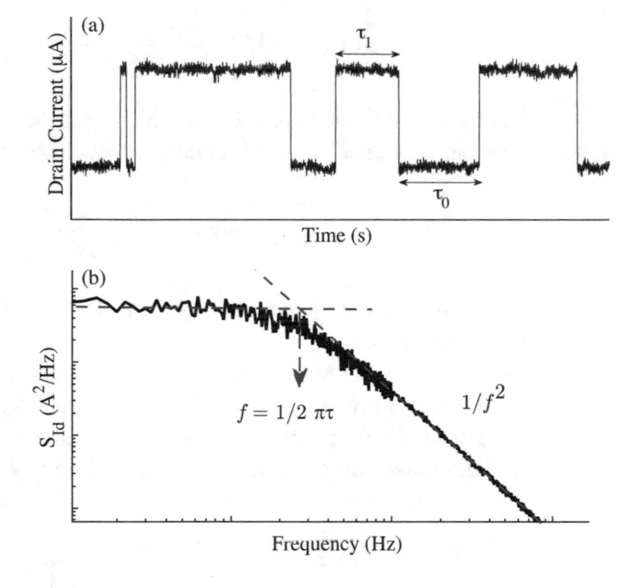

Fig. 2 (**a**) Illustration of the RTN given by the impact of a single defect in the oxide; τ_0 and τ_1 are the times to emit and capture a carrier, respectively. (**b**) Power spectrum density of the RTN given by a single defect

levels with stochastic low- and high-level times, referred to as *random telegraph signal*, illustrated in Fig. 2a.

The PSD, $S(f)$, of the random telegraph noise, caused by a single defect, is calculated as [17]

$$S(f) = \frac{4\Delta I^2}{(\overline{\tau}_0 + \overline{\tau}_1)\left[\left(\frac{1}{\overline{\tau}_0} + \frac{1}{\overline{\tau}_1}\right)^2 + (2\pi f)^2\right]}. \tag{1}$$

This is known as a Lorentzian function, where $\overline{\tau}_0$ and $\overline{\tau}_1$ are the average emission and capture times, respectively, and ΔI is the amplitude of the drain current fluctuation. In the log-log domain the Lorentzian function consists of a plateau and a region with a $1/f^2$ dependence, as shown in Fig. 2b.

Equation (1) can be rewritten by substituting

$$\beta = \frac{\overline{\tau}_0}{\overline{\tau}_1} \tag{2}$$

and

$$\frac{1}{\tau} = \frac{1}{\overline{\tau}_0} + \frac{1}{\overline{\tau}_1}, \tag{3}$$

resulting in

$$S(f) = 4\Delta I^2 \frac{\beta}{(1+\beta)^2} \frac{\tau}{1+(2\pi f \tau)^2}, \quad (4)$$

where β and τ are the ratio between emission and capture times, and the characteristic time constant, respectively. From detailed balance [4],

$$\beta = \frac{\overline{\tau}_0}{\overline{\tau}_1} = g \exp\left(\frac{E_F - E_T}{kT}\right), \quad (5)$$

where g, E_F, and E_T are the degeneracy (usually equal to one), Fermi energy, and trap energy, respectively.

By treating the low-frequency noise as a consequence of the capture and emission of channel carriers from multiple oxide defects (or traps),the drain current noise PSD can be derived from the discrete summation of the contributions of each individual defect (assuming that they are uncorrelated), resulting in

$$S_{I_d}(f) = 4\sum_{i=1}^{N_{tot}} \Delta I^2 \frac{\beta}{(1+\beta)^2} \frac{\tau}{1+(2\pi f \tau)^2}, \quad (6)$$

where N_{tot} and τ are the total number of traps in the devices, and the characteristic time constant of the traps, respectively.

When the characteristic time (τ) of the defects in a device is log-uniformly distributed, or in other words, when the corner frequencies of the Lorentzians spectra are uniformly distributed on a logarithmic scale, the device noise will have a $1/f$ behavior as demonstrated in Fig. 3.

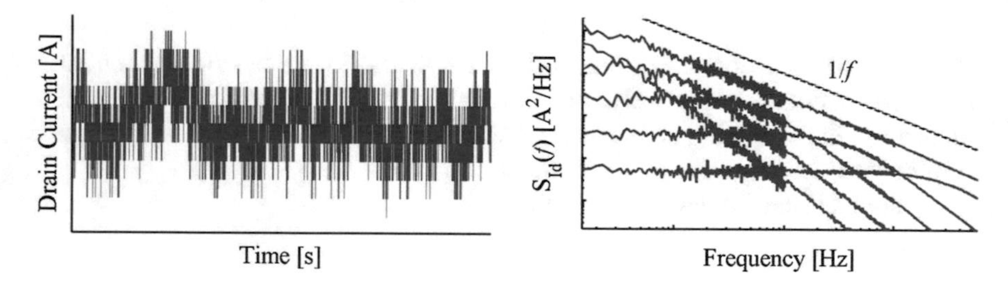

Fig. 3 Illustration of the noise given by the summation of five traps with log-uniformly distributed time constants in time domain (left) and frequency domain (right) obtained through Monte Carlo simulation. The blue spectra are associated with five individual traps with log-uniformly spread-out time constants and the red curve is the sum of the five spectra

3 Statistical Model for the Low-Frequency Noise

Based on the works of [10, 14, 18], the LFN at a given frequency is commonly found to be approximately log-normally distributed. By definition, a log-normal distribution is normally distributed on a log-scale, i.e., if X is a log-normally distributed random variable, then Y = ln (X) is normally distributed.

The quantile-quantile plots (QQ-plot) of several LFN measurements from different populations are displayed in Fig. 4. The natural logarithm of the noise quantity (sorted data) is displayed on the y-axis. On the x-axis standard normal quantiles are given. If the resulting plot is a straight line, the data are distributed normally. Though small deviations are observed, it is clear that the measurements follow a straight line over several orders of magnitude, indicating a normal behavior, meaning the S_{I_d} of both n and pMOS transistors are well-described by a log-normal distribution.

The statistical RTN-based model proposed by [10, 11] to describe the statistics of the LFN provides an analytical expression to calculate the noise statistics. The RTN-based model is fundamentally similar to number fluctuation models, as both assume capturing and emission of carriers by traps, and the LFN PSD is a result of the summation of Lorentzians. In comparison to the McWhorter [2] and Hung [19, 20] models, however, the statistical RTN-based model employs discrete random quantities, which allow for a statistical description of the LFN for a device population.

Assuming that the LF noise PSD of a device is given by the summation of uncorrelated RTNs, described by (6), the expected value and the variance of the noise can be calculated, considering that the properties of each defect—current deviation (ΔI_d), characteristic time (τ) and energy (E_T) –and that the number of defects in each transistor (N_{tot}) are random variables, under the following assumptions:

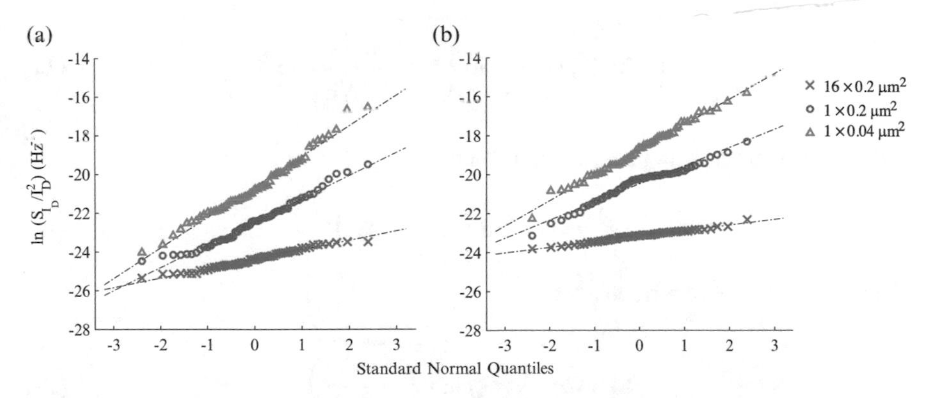

Fig. 4 Q-Q plot of the measured LFN PSD at 20 Hz for 60 DUT populations (40-nm node) of (**a**) nMOS and (**b**) pMOS devices. Bias conditions are $V_{gs} = 0.9$ V and $V_{ds} = 50$ mV

(i) the number of traps in a device is a Poisson distributed random variable;
(ii) traps are assumed uniformly distributed along the device channel;
(iii) τ is a log-uniformly distributed random variable;
(iv) $\beta/(1+\beta)^2$ and $\beta^2/(1+\beta)^4$ are approximated by delta functions of energy multiplied by kT and $kT/6$, respectively;
(v) ΔI_d, N_{tot} and τ are independent.

The expected value of the low-frequency noise related to RTN is given by [10, 11]

$$\mathrm{E}\left[S_{I_d}(f)\right] = \frac{kTI_d^2}{WL^2 f}\int_0^L \mathrm{E}\left[\Delta\widetilde{I_d}^2 | \mathrm{X}_T = x\right]\frac{N_{tr}(E_{Fn})}{\gamma}dx, \tag{7}$$

and the variance by

$$\mathrm{Var}\left[S_{I_d}(f)\right] = \frac{kTI_d^4}{3\pi^2 f^2 W^3 L^4}\int_0^L \mathrm{E}\left[\Delta\widetilde{I_d}^4 | \mathrm{X}_T = x\right]\frac{N_{tr}(E_{Fn})}{\gamma}dx, \tag{8}$$

where k, T, I_d, W, L, and f are the Boltzmann constant, temperature, drain current, transistor width, transistor length, and frequency. The trap position along the channel, X_T, is assumed to be a uniformly distributed random variable; $\Delta\widetilde{I_d}$ is the normalized current deviation, given by

$$\Delta\widetilde{I_d} = \frac{WL\Delta I_d}{I_d}, \tag{9}$$

and $N_{tr}(E_{fn})/\gamma$ is the trap density in the area, energy and in the log-domain of time constants (cm^{-2}eV^{-1} per neper) at a given quasi-Fermi level (E_{Fn}).

Based on the statistics of the log-normal distribution, the variability of the low-frequency noise is

$$\sigma\left[\ln(S_{I_d})\right] = \sqrt{\ln\left(1 + \frac{\mathrm{Var}\left[S_{I_d}\right]}{\mathrm{E}\left[S_{I_d}\right]^2}\right)}. \tag{10}$$

Additionally, by defining a parameter K as

$$K = WL\mathrm{Var}\left[S_{I_d}\right]/\mathrm{E}\left[S_{I_d}\right]^2 \tag{11}$$

the variability can be written as

$$\sigma\left[\ln(S_{I_d})\right] = \sqrt{\ln\left(1 + \frac{K}{WL}\right)}, \tag{12}$$

In this definition, K is the quantity, in units of $length^2$, that can be used to compare the variability behavior among different technologies, geometries and bias conditions.

3.1 Trap Impact on the MOSFET Drain Current

The variability of the trap impact on the current, ΔI_d, is associated with several factors, such as (random) dopant positioning, oxide charges, and position of the trap inside the oxide. The literature indicates that the resulting distribution of ΔI_d can be assumed to be exponential [21–24]. Based on this, we assume that for a given trap position along the channel the current deviation can be approximated by an exponential distribution.

From the exponential distribution the second and fourth raw moments used to calculate the expected value, in (7), and the variance, in (8), are calculated as

$$\mathrm{E}\left[\Delta\tilde{I}_d^2|X_T\right] = 2E\left[\Delta\tilde{I}_d|X_T\right]^2 \tag{13}$$

$$\mathrm{E}\left[\Delta\tilde{I}_d^4|X_T\right] = 24E\left[\Delta\tilde{I}_d\,|X_T\right]^4. \tag{14}$$

To accommodate devices with a doping gradient(e.g., halo-implanted devices), it is furthermore assumed that the expected value of the current deviation dependent on the random position of the trap is given by

$$\mathrm{E}\left[\Delta\tilde{I}_d|X_T\right] = \delta\tilde{I}_d(x_t), \tag{15}$$

with $\delta\tilde{I}_d(x_t) = WL \times \delta I_d(x_t)/I_d$, where $\delta I_d(x_t)$ is the current deviation calculated for defects at the device interface, using the number-mobility fluctuation model [19, 20] and a modified Klaassen–Prins equation [25, 26]. Therefore,

$$\delta\tilde{I}_d(x_t) = WL\frac{\delta I_d(x_t)}{I_d} = L\left(\frac{f(x_t)}{\int_0^L f(x_t)dx}\right)\left(\frac{R(x_t)}{N_{inv}(x_t)} + \alpha\mu(x_t)\right), \tag{16}$$

with

$$f(x) = \exp\left[-\int_0^x \frac{1}{g}\frac{\partial g}{\partial x}dx\right], \tag{17}$$

where g is the channel conductance, N_{inv} the carrier density, α the scattering coefficient, and $R = C_{\mathrm{inv}}/(C_{\mathrm{inv}} + C_{\mathrm{ox}} + C_{\mathrm{dep}})$, with C_{inv} the inversion capacitance, C_{ox} the oxide capacitance and C_{dep} the depletion capacitance.

3.2 Statistical Model for the Low-Frequency Noise in MOSFETs

Using the assumptions above the expected value, the variance and K are calculated as

$$\mathrm{E}\left[S_{I_d}(f)\right] = 2\frac{kT}{f}\frac{I_d^2}{WL^2}\int_0^L \delta\tilde{I}_d^2(x)\frac{N_{tr}(E_{Fn})}{\gamma}dx, \tag{18}$$

$$\mathrm{Var}\left[S_{I_d}(f)\right] = 24\frac{kT}{3\pi^2 f^2}\frac{I_d^4}{W^3L^4}\int_0^L \delta\tilde{I}_d^4(x)\frac{N_{tr}(E_{Fn})}{\gamma}dx, \tag{19}$$

$$K = 6\frac{L}{3\pi^2 kT}\frac{\int_0^L \delta I_d^4(x)N_{tr}(E_{Fn})/\gamma dx}{\left(\int_0^L \delta I_d^2(x)N_{tr}(E_{Fn})/\gamma dx\right)^2}. \tag{20}$$

From (20), K relates the variability of the LFN to the trap density and to the distribution of the current deviation for different trap positions along the channel.

Using log-normal statistics the average of the logarithm of the noise is given by

$$\mu\left[\ln\left(S_{I_d}(f)\right)\right] = \ln\left(\mathrm{E}\left[S_{I_d}(f)\right]\right) - \frac{1}{2}\sigma\left[\ln\left(S_{I_d}\right)\right]^2 \tag{21}$$

and corner values of the noise distribution can be estimated through

$$S_{I_d}(f)_{|n_c-\sigma} = e^{\mu\left[\ln\left(S_{I_d}(f)\right)\right]+n_C\sigma\left[\ln\left(S_{I_d}\right)\right]}, \tag{22}$$

where n_C accounts for the number of standard deviations on a log-scale; e.g., $n_C = 3$ provides the three-sigma corner [12]. Equation (22) can be rewritten in the form of

$$S_{I_d}(f)_{|n_c-\sigma} = \frac{\exp\left(n_C\sqrt{\ln\left(1+\frac{K}{WL}\right)}\right)}{\sqrt{1+\frac{K}{WL}}}\mathrm{E}\left[S_{I_d}(f)\right], \tag{23}$$

in which becomes clear that K is a key quantity for calculating the low-frequency noise statistics.

3.3 Area Dependence Under Simplified Conditions

When the MOS device is under strong inversion, with high gate (V_{gs}) and low drain bias applied (V_{ds}), the device channel can be assumed to be uniformly inverted from source to drain. Under this condition, the current deviation (ΔI_d) and the trap density (N_{tr}/γ) can be considered independent of the position of the defect along the channel, and the equations that describe the statistics of the LFN, in (18), (19) and (20), can be simplified as

$$\mathrm{E}\left[S_{I_d}(f)\right] = 2\frac{kT}{f}\frac{I_d^2}{WL}\left(\frac{R}{N_{inv}}+\alpha\mu\right)^2\frac{N_{tr}(E_{Fn})}{\gamma}, \tag{24}$$

$$\mathrm{Var}\left[S_{I_d}(f)\right] = 24\frac{kT}{3\pi^2 f^2}\frac{I_d^4}{(WL)^3}\left(\frac{R}{N_{inv}}+\alpha\mu\right)^4\frac{N_{tr}(E_{Fn})}{\gamma}, \tag{25}$$

and

$$K = \frac{2}{\pi^2 kT}\frac{1}{N_{tr}/\gamma}, \tag{26}$$

respectively.

Under these simplifications the expected value depends inversely on the area, the variance depends inversely on third power of the area, while K is area independent. Figures 5 and 6 demonstrate the results obtained when the derived expected value, variance and K equations are calculated from experimental data of n-channel devices under approximately uniformly inverted channel.

Figure 7 shows the area scaling of the variability of the noise calculated using (12) compared to the experimental results. From the figure, the statistical LFN model and the experimental results show a strong dependence of the variability on device area. Such a strong dependence is in accordance with many other works in the literature [27–30]. Moreover, the statistical LFN model and the measurements

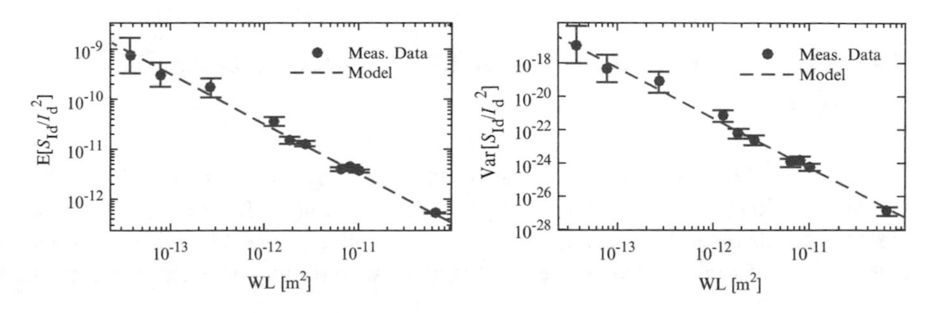

Fig. 5 Area scaling of the expected value (left) and variance (right) of the LF noise, for n-channel devices in a 140-nm technology. Bias conditions are $V_{ds} = 0.1$ V and $V_{gs} = 1.4$ V

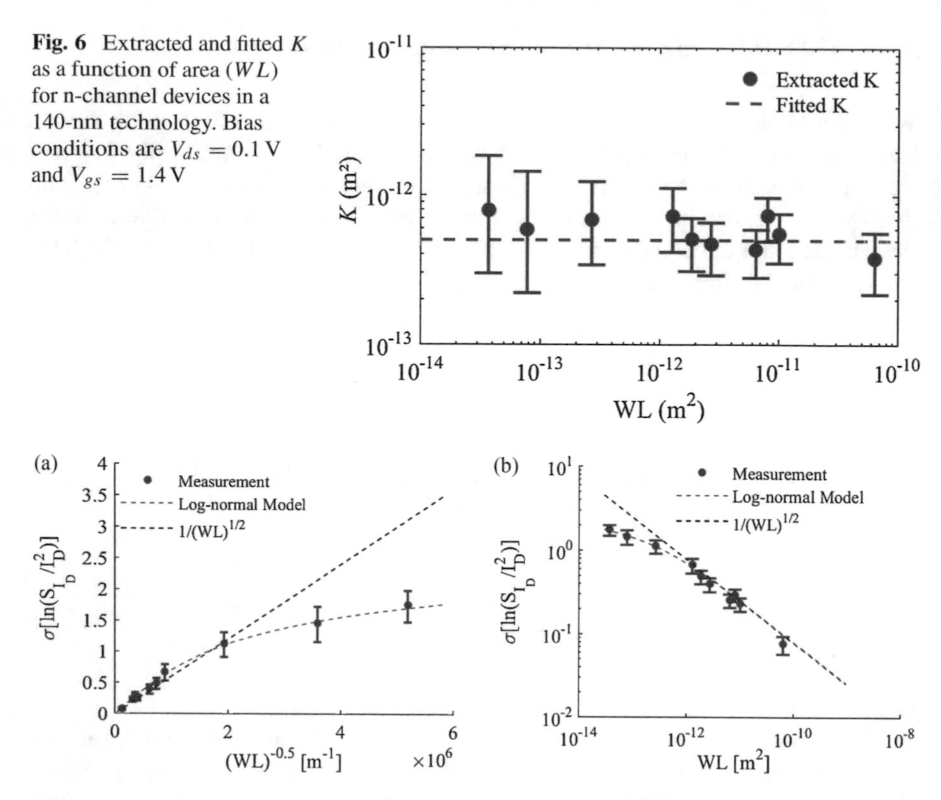

Fig. 6 Extracted and fitted K as a function of area (WL) for n-channel devices in a 140-nm technology. Bias conditions are $V_{ds} = 0.1$ V and $V_{gs} = 1.4$ V

Fig. 7 Standard deviation of $\ln(S_{I_d}/I_d^2)$ as a function of (**a**) $1/\sqrt{WL}$ and (**b**) WL for n-channel devices in 140-nm technology. Bias conditions are $V_{ds} = 0.1$ V and $V_{gs} = 1.4$ V

in Fig. 7 shows a different area dependence than the $1/\sqrt{WL}$ behavior that we know from, e.g., current and capacitance mismatch [31]. This is particularly relevant because this $1/\sqrt{\text{area}}$ behavior is quite frequently, but erroneously, used in circuit simulation parameter decks.

3.4 Short-Versus Long-Channel Devices and the Drain Bias Effect

The section above analyzed the low-frequency noise under a simplified bias condition, which results in a uniformly inverted channel. To analyze the noise statistics for different bias conditions, let us first divide the MOS devices into two categories, uniformly doped devices and devices with doping gradients around the source and drain diffusions (a.k.a. halo-implanted devices). Halo-implanted devices will be analyzed in the next section. For a device without a doping gradient along the channel, the average current deviation for a trap at the interface and at a position x_t along the channel is calculated using number-mobility correlation and the Klaassen–Prins equation [25], and the normalized current deviation is given by

$$\delta \tilde{I}_d(x_t) = \left(\frac{R(x_t)}{N_{inv}(x_t)} + \alpha \mu(x_t) \right), \tag{27}$$

From this equation, it can be seen that the effect of a large drain bias is to reduce the carrier density close to the drain, and as a consequence, an increase in the current deviation for traps located at this region. Therefore, switching traps closer to the drain will have a larger impact on the current fluctuations than those located closer to the source. This difference between the current deviation caused by traps closer to the device source and drain increases the variability of the LFN, as shown by the K calculation in (20), for clarity repeated below

$$K = 6\frac{L}{3\pi^2 kT} \frac{\int_0^L \delta I_d^4(x) N_{tr}(E_F)/\gamma dx}{\left(\int_0^L \delta I_d^2(x) N_{tr}(E_F)/\gamma dx\right)^2}.$$

Due to velocity saturation, the carrier density at the drain side of short-channel devices is much higher than that of long-channel devices when a large drain bias is applied. Thus, for short-channel devices the difference between the current deviation for traps located closer to the source and drain is much smaller than in long-channel devices. Therefore, the increase in the LFN variability in short-channel devices, when the device is operated under saturation, is much smaller, or even negligible, compared to longer channel devices, as shown by the K values extracted from experimental data in Fig. 8.

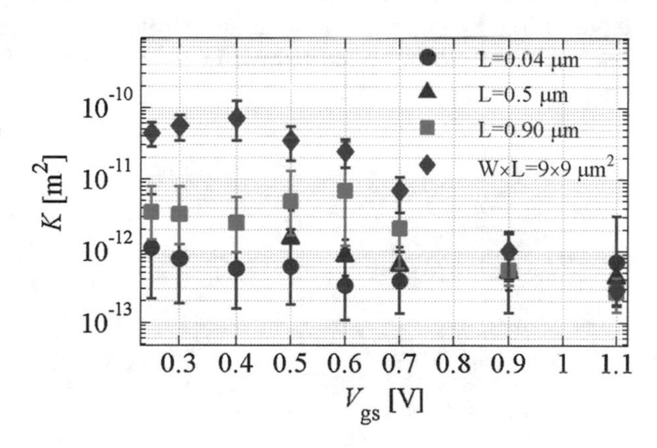

Fig. 8 Extracted K of 68 thick-oxide nMOS devices, without halo implants, in a 65-nm technology, with V_{ds} = 0.5 V

3.5 Short-Versus Long-Channel Devices with Halo Implants

To analyze the variability of halo-implanted devices, the current deviation caused by a trap is calculated again using a modified Klaassen–Prins equation to account for the doping gradient along the channel. The average current deviation, normalized by the area, on drain current caused by a trap at the device interface is given by (16)

$$\delta \tilde{I}_d(x_t) = WL\frac{\delta I_d(x_t)}{I_d} = L\left(\frac{f(x_t)}{\int_0^L f(x_t)dx}\right)\left(\frac{R(x_t)}{N_{inv}(x_t)} + \alpha\mu(x_t)\right), \tag{28}$$

with

$$f(x) = \exp\left[-\int_0^x \frac{1}{g}\frac{\partial g}{\partial x}dx\right]. \tag{29}$$

The effect of the term $f(x)$ is to amplify the current deviation produced by traps located in the halo-implanted regions. This amplification is more pronounced the larger the difference between the doping concentration in the channel and in the halo-implanted region is. Additionally, the trap impact on the current is further enhanced if the device is operated in weak inversion [11]. In long-channel devices, this effect makes very small regions (halo-implanted regions) of the channel to induce very large current deviations. As the current deviation in these small regions increases, adding large Lorentzians in the LFN spectrum, the LFN variability in these devices increase as well. Therefore, this effect creates a strong gate bias dependence on the LFN variability, increasing it the closer the devices get into weak inversion. In short-channel devices the halo-implanted regions are overlapped or represent a much larger portion of the channel and the increase in the variability is much smaller or even negligible. Moreover, the effect described in Sect. 3.4 is amplified by the presence of halo implants, as a large drain bias drives the halo region closer to the drain into weak inversion. Figure 9 shows the effect of the halo implantation on the K value, extracted from experimental data, for different channel length and bias conditions.

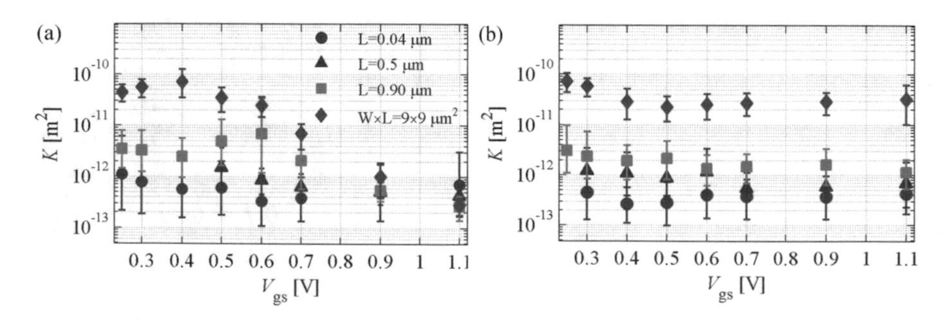

Fig. 9 Extracted K of 63 thin-oxide n-channel devices, with halo implants, in 40-nm technology. (**a**) $V_{ds} = 0.05\,\text{V}$, (**b**) $V_{ds} = 0.55\,\text{V}$

3.6 The Statistical LFN Model Versus Experimental Data

Figures 10 and 11 show the comparison of the statistical LFN model, represented by (18), (19), (20), and (12), with experimental results for two different device geometries, and for various bias conditions. The values from the statistical model were calculated using TCAD simulations to extract the current deviation, using (16), for traps at different positions along the device interface. Then these values were used in the model equations. The simulation was performed using the TAURUS MEDICI tool, from Synopsys®. The TCAD deck was calibrated to reproduce the I–V characteristics of the experimental devices. Therefore, the only parameters left for calculating the LFN statistics is the trap density N_{tr}/γ and the scattering coefficient α.

Figures 12 and 13 show the same comparison done for pMOS devices. Differently from nMOS devices, pMOS devices require a trap density that is nonuniform in energy, shown in Fig. 14. This explains the different behavior between the $1/f$

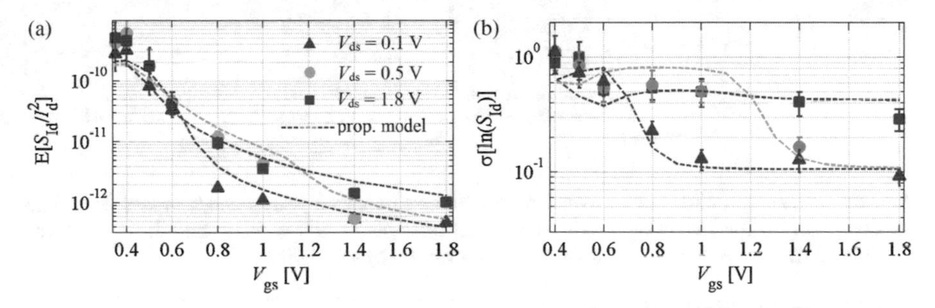

Fig. 10 Model prediction and experimental data of expected values and standard deviations versus gate bias for nMOS devices, with halo implants, in a 140-nm technology, $W = 8\,\mu\text{m}$ and $L = 8\,\mu\text{m}$. Calculated using $N_{tr}/\gamma = 1.1 \times 10^9\,\text{cm}^{-2}\text{eV}^{-1}$ per neper and $\alpha = 0.8 \times 10^{-5}$ V s. (**a**) Expected value, (**b**) variability

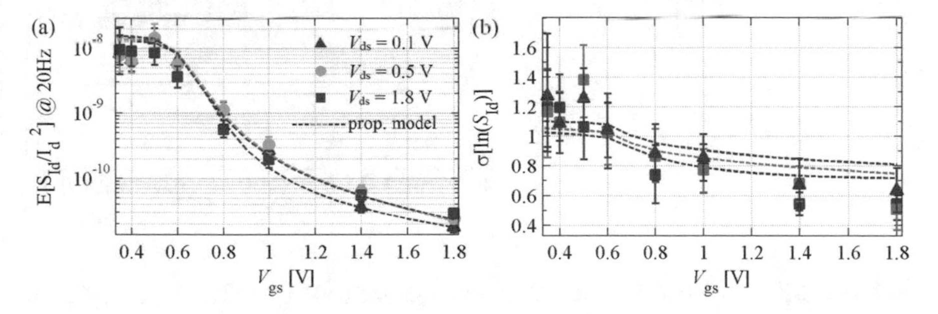

Fig. 11 Model prediction and experimental data of expected values and standard deviations versus gate bias for nMOS devices, with halo implants, in a 140-nm technology, $W = 8\,\mu\text{m}$ and $L = 0.14\,\mu\text{m}$. Calculated using $N_{tr}/\gamma = 1.3 \times 10^9\,\text{cm}^{-2}\text{eV}^{-1}$ per neper and $\alpha = 0.8 \times 10^{-5}$ V s. (**a**) Expected value, (**b**) variability

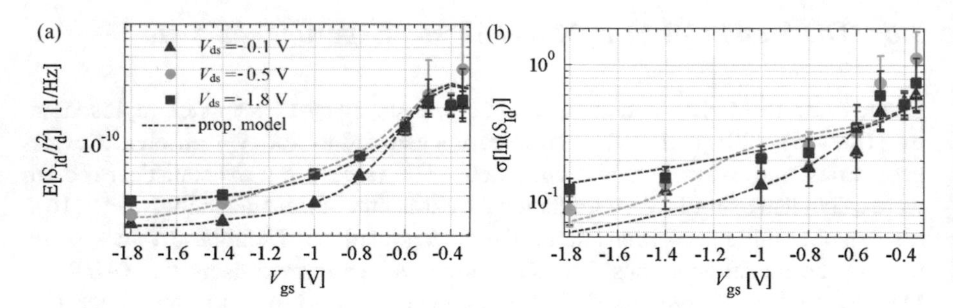

Fig. 12 Model prediction versus experimental data of expected values and standard deviations versus gate bias for pMOS devices in a 140-nm technology, $W = 8\,\mu$m and $L = 1\,\mu$m. Calculated using $\alpha = 1 \times 10^{-5}$ V s. (**a**) Expected value, (**b**) variability

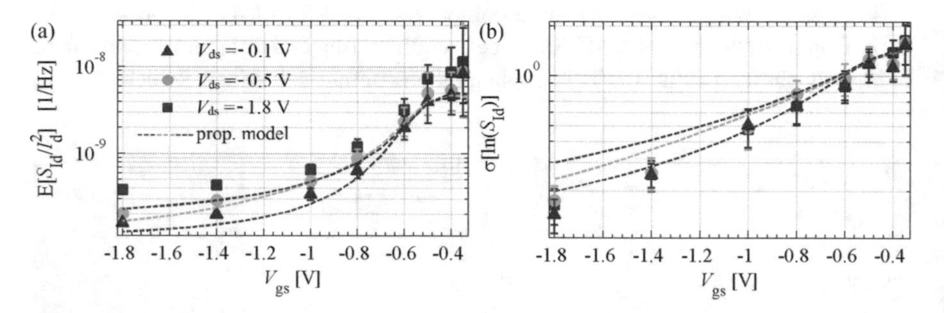

Fig. 13 Model prediction versus experimental data of expected values and standard deviations versus gate bias for pMOS devices in a 140-nm technology, $W = 8\,\mu$m and $L = 0.14\,\mu$m. Calculated using $\alpha = 1 \times 10^{-5}$ V s. (**a**) Expected value, (**b**) variability

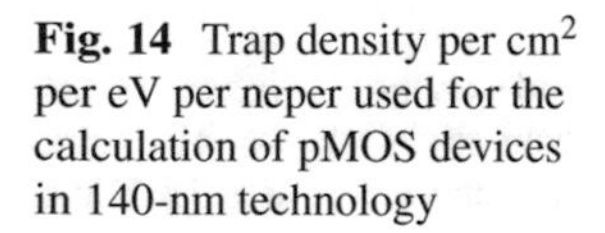

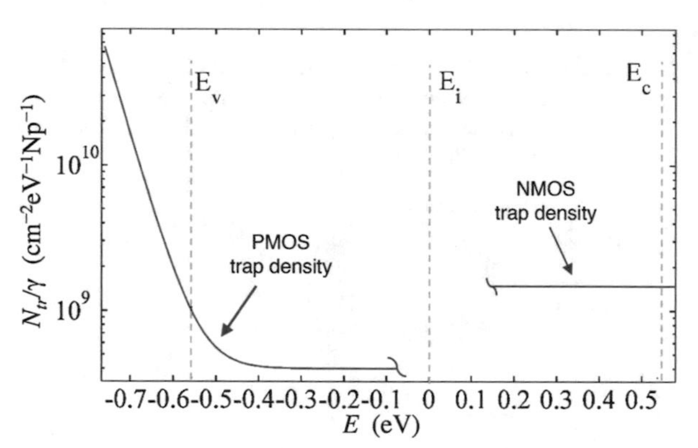

Fig. 14 Trap density per cm^2 per eV per neper used for the calculation of pMOS devices in 140-nm technology

noise of nMOS and pMOS devices that is well-described in literature [5, 32–35], but with divergence in explaining the mechanism behind this difference. In the context of the statistics of the LFN of pMOS, the trap distribution shown in Fig. 14 indicates that decreasing $|V_{gs}|$ effectively reduces the trap density. This effect counteracts that of δI_d with V_{gs}, which increases with decreasing $|V_{gs}|$. Therefore, due to these opposing effects, the dependence of E[$S_{Id}(f)$] with V_{gs} is reduced. In turn, the trap density reduction with decreasing $|V_{gs}|$ increases the variability, as K is inversely proportional to N_{tr}/γ. This effect explains the smaller dependence of E[$S_{Id}(f)$] and the larger dependence of the variability with the gate bias for pMOS compared to nMOS devices. The increased trap density close to the edges of the Si bandgap—leading to a defect band or the tails of normal distributions, resembling a U-shape trap distribution — is in agreement with observations in the context of NBTI in SiGe pFETs with a high-k stack [32], as well as with RTN noise measurements [36, 37].

4 Correlation Coefficient Analysis

We proposed a novel noise analysis technique for the interpretation and modeling of LFN in MOSFETs [13, 14]. The technique consists of estimating the correlation between the noise power density of a given population at a given condition (such as bias, temperature, frequency), with the PSDs of the same population at a different condition — e.g., $S_{I_d}(f)$ and $S_{I_d}(f + \Delta f)$.

This analysis proves extremely sensitive to the physical mechanisms assumed for the interpretation and modeling of the LFN in MOSFETs. Using this methodology it is possible to draw valuable conclusions about the frequency dependence of the fundamental noise sources underlying the $1/f$ noise in MOSFETs, as well as the origin of the trapping mechanism.

Under the assumption that the LFN PSD follows a log-normal distribution (Fig. 4), the correlation coefficient from experimental data is calculated using the maximum likelihood estimators of the log-normal distribution. The correlation coefficient between two log-normally distributed random variables, X and Y, is given by [14]

$$\mathrm{R}\left[\mathrm{X},\mathrm{Y}\right]=\frac{e^{\rho\sigma_x\sigma_y}-1}{\sqrt{\left(e^{\sigma_x^2}-1\right)\left(e^{\sigma_y^2}-1\right)}}, \qquad (30)$$

where σ_x and σ_y are the standard deviations of ln(X) and ln(Y), respectively (remember that ln(X) and ln(Y) are normally distributed), and ρ is the correlation coefficient between ln(X) and ln(Y).

4.1 Frequency Autocorrelation

The frequency autocorrelation analysis, originally presented in [13, 14], provides valuable information on the frequency dependence of the fundamental noise sources underlying the $1/f$ noise. An analytical expression for the frequency autocorrelation was derived using the RTS-framework in [13, 14], resulting in

$$\mathrm{R}\left[S(f_1), S(f_2)\right] = 2\frac{\ln\left(\frac{f_1}{f_2}\right)}{\left(f_1^2 - f_2^2\right)} f_1 f_2, \tag{31}$$

where f_1 and f_2 are the frequencies at which the PSDs of the population are compared, and R the correlation coefficient.

It is important to notice that the resulting shape of R-function only depends on the spectra of the fundamental noise sources and their superposition. If, for instance, the observed $1/f$ noise would be the sum of microscopic $1/f$ noise sources with a perfect slope of -1, R would be equal to 1 for all Δf. Therefore, this correlation coefficient forms a signature of the frequency dependence of the fundamental noise sources underlying the $1/f$ noise.

Equation (31) can be written in terms of a variable $\Delta f'$, which is unitless, if $f_1 = f_{\text{ref}}$ and $f_2 = f_{\text{ref}}\,\Delta f'$. In this case, (31) simplifies to

$$\mathrm{R}\left[S(f_1), S(f_2)\right] = 2\frac{\ln\left(\Delta f'\right)}{\left(1 - \Delta f'^2\right)} \Delta f'. \tag{32}$$

This demonstrates the fact that the shape of the correlation coefficient curve is independent of the reference frequency, f_{ref}, when analyzed in log-scale. Figure 15 depicts the correlation coefficient as a function of $\Delta f'$.

Figure 15 illustrates that for small frequency increments (small $\Delta f'$), devices that have higher noise levels at f_{ref} are more likely to have higher noise levels at $f_{\text{ref}}\Delta f'$. Conversely, for large frequency increments (large $\Delta f'$), as the correlation coefficient approaches zero, it is not possible to predict which devices have higher noise levels at $f_{\text{ref}}\Delta f'$ based on their levels at f_{ref} (i.e., they are statistically uncorrelated).

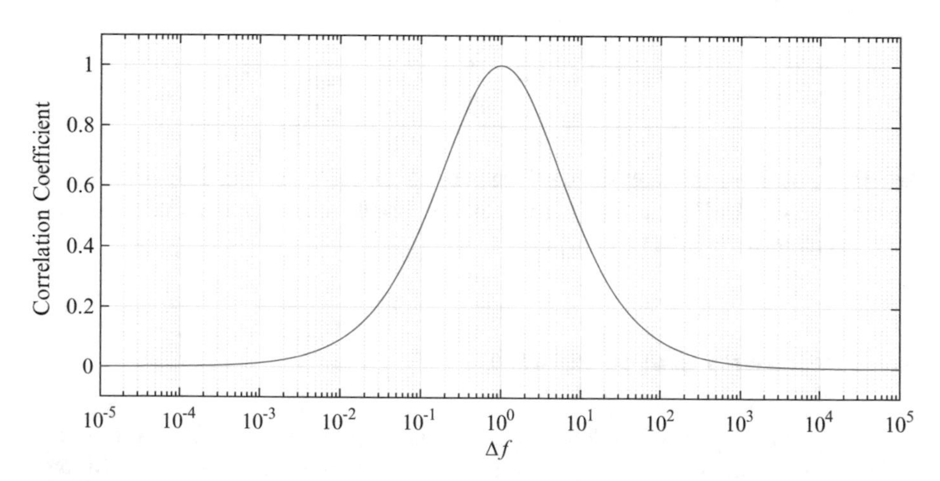

Fig. 15 Correlation coefficient as a function of $\Delta f'$, calculated using (32)

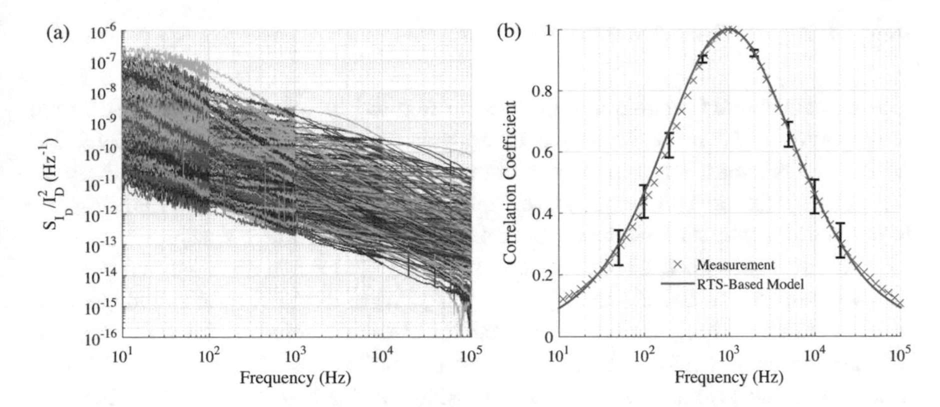

Fig. 16 (**a**) LFN spectra of a 320-DUT small-area NFET population (40-nm technology); and (**b**) extracted (blue markers) versus theoretically calculated (red line) correlation coefficients. Transistor dimensions are $W = 0.3\,\mu$m and $L = 0.04\,\mu$m; bias conditions are $V_{gs} = 1.1$ V and $V_{ds} = 0.1$ V

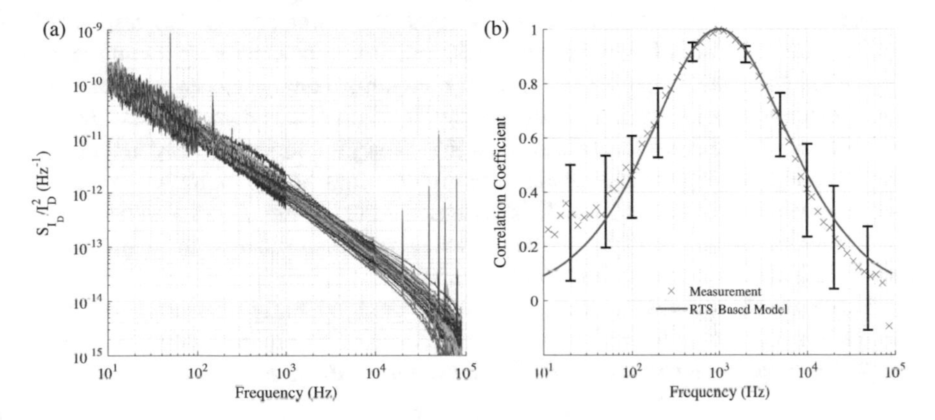

Fig. 17 (**a**) LFN spectra of a 60-DUT large-area PFET population (40-nm technology); and (**b**) extracted (blue markers) versus theoretically calculated (red line) correlation coefficients. Transistor dimensions are $W = 16\,\mu$m and $L = 0.2\,\mu$m; bias conditions are $V_{gs} = 1.1$ V and $V_{ds} = 0.05$ V

Experimental results presented in Figs. 16 and 17 for a 40-nm technology nodes indicate that the correlation coefficient as a function of frequency for both N and PMOS transistors follows the predicted behavior. Moreover, this behavior is observed regardless of device area [13, 14]. This is a strong indication that the observed $1/f$ noise for these MOS devices is indeed primarily composed of a superposition of Lorentzian spectra, arising from multiple superimposed Random Telegraph Signals (RTS).

4.2 Temperature Autocorrelation

The autocorrelation technique can also be applied to measurements at different temperatures. $1/f$ noise is known to be weakly temperature dependent [2, 34] despite RTN being strongly temperature dependent [4, 38]. The sensitivity of the autocorrelation technique proves useful to draw conclusions regarding the temperature dependence of the characteristic time constant of the traps.

Analogously to Sect. 4.1, the correlation coefficient R can be computed in terms of temperature, i.e., $S_{I_d}(T)$ and $S_{I_d}(T + \Delta T)$. Were the LFN spectra independent of temperature, strong correlation would be observed between the noise PSD at different measured temperatures. This would be the case if elastic tunneling is assumed as the primary capture and emission mechanism. On the other hand, if the LFN spectra are strongly dependent on temperature, a de-correlation would be observed between the noise PSD at different measured temperatures. This is the case if thermal activation is assumed as the primary capture and emission mechanism.

The de-correlation of measured PSD observed in Fig. 18 with increasing ΔT indicates that, at the reference frequency (100 Hz), different traps are responsible for the LFN PSD at different temperatures, which is consistent with a temperature dependent mechanism. Moreover, through a combined analysis between frequency and temperature autocorrelation [14], shown in Fig. 19, it is demonstrated that the characteristic time constants of the traps are strongly temperature dependent, i.e., traps become faster with increasing temperature, causing the entire noise PSD to shift towards higher frequencies. This is demonstrated in Fig. 19, in which the peak of correlation indeed shifts in frequency when the noise PSDs of the same device population are compared at different temperatures.

These results provide a strong case for predominantly thermally activated trapping process even for $1/f$-like spectra. This is in accordance with the findings of [3, 39]. This should be accounted for in compact LFN models. It should be noted

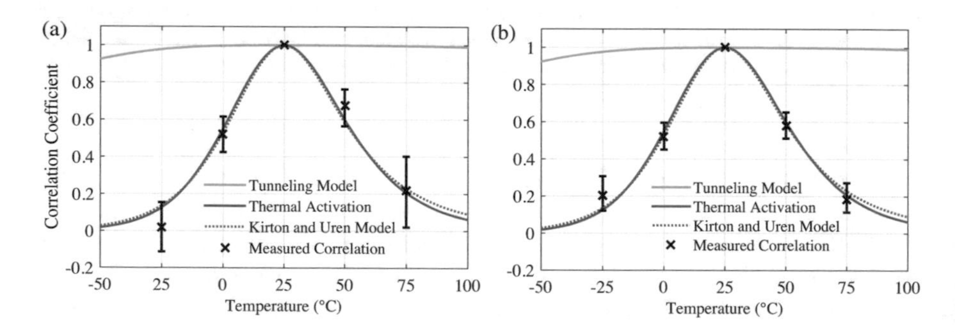

Fig. 18 Correlation coefficient as a function of temperature ($T_{ref} = 25$ °C) at 100 Hz for (**a**) two independent 80-DUT NFET populations ($1 \times 0.04\,\mu m^2$) and (**b**) two independent 78-DUT NFET populations ($1 \times 1\,\mu m^2$). The first population of each geometry was measured at −25, 0, and 25 °C; whereas the second was measured at 25, 50, and 75 °C. All populations were measured on the same wafer

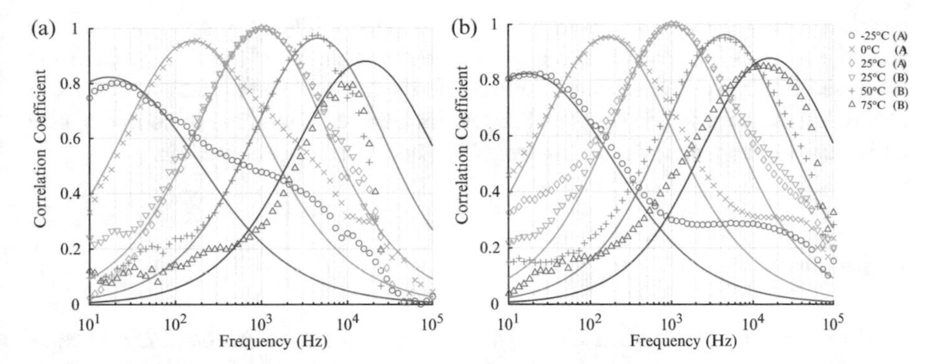

Fig. 19 Correlation coefficient w.r.t. the reference temperature of $T_{ref} = 25$ °C and to the reference frequency of $f_{\text{ref}} = 1$ kHz. Extracted from measured data (symbols) from two 78-DUT NFET, 40-nm technology, populations (A and B) from the same wafer. Device geometry is $W = 1$ μm and $L = 1$ μm; bias conditions were (**a**) $V_{gs} = 1.1$ V, $V_{gs} = 0.1$ V and (**b**) $V_{gs} = 0.7$ V, $V_{gs} = 0.5$ V. Solid lines indicate theoretical prediction

that even though the frequency-temperature autocorrelation follows the previously discussed bell shape, it does not fully reach 1 when $|\Delta T| > 0$, which indicates that the stochastic process underlying the LFN contains at least two independent stochastic components as is, e.g., the case for the Kirton and Uren model [4].

5 Summary and Conclusion

The statistical methods presented in this chapter are applicable to n-channel and p-channel devices from different mixed-signal CMOS technology nodes, being well suited for both large and small-area devices, and for both $1/f$-like and Lorentzian dominated spectra. Using the log-normal nature of noise distribution, we also explained why variability of RTN ($\sigma[\ln(S_{I_d})]$) does not follow a $1/\sqrt{\text{area}}$ dependence.

The noise variability ($\sigma[\ln(S_{I_d})]$) of long-channel devices is shown to be a strong function of drain and gate biases, while the variability of short-channel devices depends weakly on bias. The effect of the halo-implanted regions in the LF noise statistics, which can considerably increase the noise variability of long-channel devices under weak inversion and saturation, is also understood in the context of the presented statistical model. The introduction of K allows the comparison between the noise variability among different device technologies, geometries and biases. Under a given bias condition, K is technology specific and can be used to compare the dielectric quality of different technologies.

The correlation coefficient analysis as a function of frequency reaffirms that the $1/f$ noise in MOSFETs is composed by a summation of Lorentzians. The temperature dependence of the LFN evaluated using the autocorrelation technique provides a strong case for predominantly thermally activated trapping process, in accordance with the temperature dependence observed by [3] and [39]. Both the discussed RTN-based variability analysis and the autocorrelation techniques prove highly suitable to assess the physical and mathematical correctness of contemporary and future implementations of low-frequency noise in compact models.

References

1. G. Ghibaudo, O. Roux-dit-Buisson, Low frequency fluctuations in scaled down silicon CMOS devices status and trends, in *ESSDERC '94: 24th European Solid State Device Research Conference* (1994), pp. 693–700
2. A. McWhorter, 1/f Noise and Germanium Surface Properties, (1955)
3. P. Dutta, P.M. Horn, Low-frequency fluctuations in solids: $1/f$ noise. Rev. Mod. Phys. **53**, 497–516 (1981)
4. M. Kirton, M. Uren, Noise in solid-state microstructures: a new perspective on individual defects, interface states and low-frequency $(1/f)$ noise. Adv. Phys. **38**(4), 367–468 (1989)
5. E. Simoen, C. Claeys, On the flicker noise in submicron silicon MOSFETs. Solid-State Electron. **43**(5), 865–882 (1999)
6. A.J. Scholten, L.F. Tiemeijer, R. van Langevelde, R.J. Havens, A.T.A. Zegers-van Duijnhoven, V.C. Venezia, Noise modeling for RF CMOS circuit simulation. IEEE Trans. Electron Devices **50**, 618–632 (2003)
7. G.I. Wirth, J. Koh, R. da Silva, R. Thewes, R. Brederlow, Modeling of statistical low-frequency noise of deep-submicrometer MOSFETs, IEEE Trans. Electron Devices **52**, 1576–1588 (2005)
8. G.I. Wirth, R. da Silva, R. Brederlow, Statistical model for the circuit bandwidth dependence of low-frequency noise in deep-submicrometer MOSFETs. IEEE Trans. Electron Devices **54**, 340–345 (2007)
9. A.S.M.S. Rouf, Z. Çelik Butler, Oxide trap-induced RTS in MOSFETs, in *Noise in Nanoscale Semiconductor Devices*, ed. by T. Grasser (Springer, Cham, 2019). https://doi.org/10.1007/978-3-030-37500-3_17
10. M. Banaszeski da Silva, H. Tuinhout, A.T.A. Zegers-van Duijnhoven, G.I. Wirth, A. Scholten, A physics-based RTN variability model for MOSFETs, in *2014 IEEE International Electron Devices Meeting* (2014), pp. 35.2.1–35.2.4
11. M. Banaszeski da Silva, H.P. Tuinhout, A.T.A. Zegers-van Duijnhoven, G.I. Wirth, A.J. Scholten, A physics-based statistical RTN Model for the low frequency noise in MOSFETs. IEEE Trans. Electron Devices **63**, 3683–3692 (2016)
12. M. Banaszeski da Silva, H.P. Tuinhout, A.T.A. Zegers-van Duijnhoven, G.I. Wirth, A.J. Scholten, A compact model for the statistics of the low-frequency noise of MOSFETs with laterally uniform doping, IEEE Trans. Electron Devices **64**, 3331–3336 (2017)
13. T.H. Both, J.A. Croon, M.B. da Silva, H.P. Tuinhout, A.T.A. Zegers-van Duijnhoven, A.J. Scholten, G.I. Wirth, A variability-based analysis technique revealing physical mechanisms of MOSFET low-frequency noise, in *2017 International Conference of Microelectronic Test Structures (ICMTS)* (2017), pp. 1–5
14. T.H. Both, J.A. Croon, M.B. da Silva, H.P. Tuinhout, A.J. Scholten, A.T.A. Zegers-van Duijnhoven, G.I. Wirth, Autocorrelation analysis as a technique to study physical mechanisms of MOSFET low-frequency noise.IEEE Trans. Electron Devices **64**, 2919–2926 (2017)
15. T. Grasser, K. Rott, H. Reisinger, M. Waltl, J. Franco, B. Kaczer, A unified perspective of RTN and BTI, in *2014 IEEE International Reliability Physics Symposium* (2014), pp. 4A.5.1–4A.5.7

16. Y.W.M. Dominic Waldhoer, A.-M.B. El-Sayed, T. Grasser, Atomistic modeling of oxide defects, in Noise in Nanoscale Semiconductor Devices, ed. by T. Grasser (Springer, Cham, 2019). https://doi.org/10.1007/978-3-030-37500-3_18
17. S. Machlup, Noise in semiconductors: spectrum of a two-parameter random signal. J. Appl. Phys. **25**(3), 341–343 (1954)
18. B. Yu, X. Li, J. Yonemura, Z. Wu, J.S. Goo, C. Thuruthiyil, A. Icel, Modeling local variation of low-frequency noise in MOSFETs via sum of lognormal random variables, in *Proceedings of the IEEE 2012 Custom Integrated Circuits Conference* (2012), pp. 1–4
19. K.K. Hung, P.K. Ko, C. Hu, Y.C. Cheng, A physics-based MOSFET noise model for circuit simulators. IEEE Trans. Electron Devices **37**, 1323–1333 (1990)
20. K.K. Hung, P.K. Ko, C. Hu, Y.C. Cheng, A unified model for the flicker noise in metal-oxide-semiconductor field-effect transistors. IEEE Trans. Electron Devices **37**, 654–665 (1990)
21. H.H. Mueller, M. Schulz, Random telegraph signal: an atomic probe of the local current in field-effect transistors. J. Appl. Phys. **83**(3), 1734–1741 (1998)
22. A. Asenov, R. Balasubramaniam, A.R. Brown, J.H. Davies, RTS amplitudes in decananometer MOSFETs: 3-D simulation study. IEEE Trans. Electron Devices **50**, 839–845 (2003)
23. M.F. Bukhori, S. Roy, A. Asenov, Simulation of statistical aspects of charge trapping and related degradation in bulk MOSFETs in the presence of random discrete dopants. IEEE Trans. Electron Devices **57**, 795–803 (2010)
24. B. Kaczer, P.J. Roussel, T. Grasser, G. Groeseneken, Statistics of multiple trapped charges in the gate oxide of deeply scaled MOSFET devices—application to NBTI. IEEE Electron Device Lett. **31**, 411–413 (2010)
25. F. Klaassen, J. Prins, Thermal noise of MOS transistors. Philips Res. Rep. **22**, 505–514 (1967)
26. A.S. Roy, C.C. Enz, J. Sallese, Noise modeling in lateral nonuniform MOSFET. IEEE Trans. Electron Devices **54**, 1994–2001 (2007)
27. D. Lopez, S. Haendler, C. Leyris, G. Bidal, G. Ghibaudo, Low-frequency noise investigation and noise variability analysis in high-*k*/metal gate 32-nm CMOS transistors. IEEE Trans. Electron Devices **58**, 2310–2316 (2011)
28. E.G. Ioannidis, S. Haendler, A. Bajolet, T. Pahron, N. Planes, F. Arnaud, R.A. Bianchi, M. Haond, D. Golanski, J. Rosa, C. Fenouillet-Beranger, P. Perreau, C.A. Dimitriadis, G. Ghibaudo, Low frequency noise variability in high-k/metal gate stack 28nm bulk and FD-SOI CMOS transistors, in *2011 International Electron Devices Meeting* (2011), pp. 18.6.1–18.6.4
29. E.G. Ioannidis, S. Haendler, A. Bajolet, J. Rosa, J. Manceau, C.A. Dimitriadis, G. Ghibaudo, Evolution of low frequency noise and noise variability through CMOS bulk technology nodes, in *2013 22nd International Conference on Noise and Fluctuations (ICNF)* (2013), pp. 1–4
30. E. Ioannidis, S. Haendler, C. Theodorou, S. Lasserre, C. Dimitriadis, G. Ghibaudo, Evolution of low frequency noise and noise variability through CMOS bulk technology nodes from 0.5 μm down to 20 nm. Solid-State Electron. **95**, 28–31 (2014)
31. M.J.M. Pelgrom, A.C.J. Duinmaijer, A.P.G. Welbers, Matching properties of MOS transistors. IEEE J. Solid-State Circuits **24**, 1433–1439 (1989)
32. J.H. Scofield, N. Borland, D.M. Fleetwood, Reconciliation of different gate-voltage dependencies of 1/f noise in n-MOS and p-MOS transistors. IEEE Trans. Electron Devices **41**, 1946–1952 (1994)
33. L.K.J. Vandamme, X. Li, D. Rigaud, 1/f noise in MOS devices, mobility or number fluctuations? IEEE Trans. Electron Devices **41**, 1936–1945 (1994)
34. J. Chang, A.A. Abidi, C.R. Viswanathan, Flicker noise in CMOS transistors from subthreshold to strong inversion at various temperatures. IEEE Trans. Electron Devices **41**, 1965–1971 (1994)
35. L.K.J. Vandamme, F.N. Hooge, What do we certainly know about 1/f noise in MOSTs. IEEE Trans. Electron Devices **55**, 3070–3085 (2008)

36. A. van der Wel, E. Klumperink, E. Hoekstra, B. Nauta, Relating random telegraph signal noise in metal-oxide-semiconductor transistors to interface trap energy distribution. Appl. Phys. Lett. **87**(18), 183507 (2005)
37. M. Nour, Z. Çelik Butler, A. Sonnet, F. Hou, S. Tang, G. Mathur, A stand-alone, physics-based, measurement-driven model and simulation tool for random telegraph signals originating from experimentally identified MOS gate-oxide defects. IEEE Trans. Electron Devices **63**, 1428–1436 (2016)
38. T. Grasser, H. Reisinger, W. Goes, T. Aichinger, P. Hehenberger, P.J. Wagner, M. Nelhiebel, J. Franco, B. Kaczer, Switching oxide traps as the missing link between negative bias temperature instability and random telegraph noise, in *2009 IEEE International Electron Devices Meeting (IEDM)* (2009), pp. 1–4
39. C. Surya, T.Y. Hsiang, A thermal activation model for $1/f^y$ noise in Si-MOSFETs. Solid-State Electron. **31**(5), 959–964 (1988)

Defect-Based Compact Modeling of Random Telegraph Noise

Pieter Weckx, Ben Kaczer, Marko Simicic, Bertrand Parvais, and Dimitri Linten

1 Introduction

Recent advances in characterizing and modeling degradation as time-dependent variability have led to a paradigm shift in reliability data analysis and simulation of deeply scaled CMOS circuits [2, 12, 17, 23, 27]. Besides the already established time-zero variability, this additional stochastic device degradation can further impact yield and performance of sensitive digital and analog circuitry. Stochastic device degradation due to individual oxide defects like random telegraph noise (RTN) and bias temperature instability (BTI) is of particular interest. Both NBTI and PBTI have been shown to follow the compound exponential-Poisson (EP) distribution on a wide variety of commercial and research grade technologies [2, 8, 12, 23, 33]. This universality of the observations made by several research groups indicates that the EP distribution is independent of the characterization method applied. In this chapter, we will further elaborate on the origins of the Poisson distributed number of defects and the memoryless behavior of the BTI distribution with respect to time. Moreover, RTN can be described by the same set of statistical equations. RTN distinguishes itself as being a process in steady-state, while BTI is considered a process that is out of steady-state. This universal model makes it possible to capture the complete workload dependence of combined RTN and BTI degradation in a manner that is useful for circuit simulation. For this, a model is derived to calculate the activity and contributions of individual oxide defects, stochastically distributed in each transistor.

P. Weckx (✉) · B. Kaczer · M. Simicic · B. Parvais · D. Linten
IMEC, Leuven, Belgium
e-mail: pieter.weckx@imec.be

T. Grasser (ed.), *Noise in Nanoscale Semiconductor Devices*,
https://doi.org/10.1007/978-3-030-37500-3_16

Various methodologies based on modeling the activity of defects exist in the state of the art (SotA). Initial implementations were designed for RTN under transient SPICE (simulation program with integrated circuit emphasis) simulations. Here, an equivalent electrical model for each trap is used to model the defect occupancy probability [28, 37, 38]. Other approaches calculate the trap activity outside the electrical simulation [1] to predict the degradation behavior. The first approach allows to simulate the degradation "on the fly" during a transient simulation, which naturally couples the degradation to the circuit behavior but is restricted to an electrical equivalent circuit of the defects. Moreover, they do not allow fully coupled simulations, i.e., degradation being correlated to the actual circuit behavior during runtime. An enhancement referred to as the atomistic approach of calculating RTN or BTI degradation circumvents these issues and was first introduced in [13]. In this implementation a circuit netlist uses annotated devices, i.e., devices that each have a certain number of associated defects, which degrade the device performance on the fly. This naturally couples the degradation to the circuit behavior and moreover is intrinsically fully workload dependent. This model is a custom modification of the BSIM compact model and relies on external scripting tools to annotate the netlist. Consequently, it is not very practical for use in industry standard electronic design automation (EDA) tools, and was, moreover, significantly increasing simulation run times. The ultimate compact model needs to have a set of equations to be implemented, capable of handling both the workload dependence and stochastic nature of the defects. Finally, these equations need to be incorporated in a compact model solvable at runtime.

2 Defect-Centric Modeling for BTI

This Section will discuss the defect-centric modeling of BTI variability. As RTN and BTI are closely related [12], RTN can be described by the same set of statistical equations. BTI but distinguishes itself as being a process in equilibrium, and BTI is considered a process that is out of equilibrium.

2.1 BTI as Time-Dependent Variability

Initial studies on NBTI induced variability were performed in the early 2000s [7] due to concerns for device mismatch in analog circuits. The ever decreasing dimensions of CMOS transistors also created a growing issue for SRAM cell yield and stability. Hence, more BTI measurement and analysis were performed on single minimum sized devices [10, 26, 27]. First observations about the total ΔV_{TH} distribution were the increase in variance for different amounts of mean degradation and the gate area related scaling of the variance.

In the following, we will divide the empirical distribution models into normal distribution models and non-normal distribution models. The use of, e.g., normal statistics originates from the practical requirements of circuit simulation, allowing for easy evaluation and extrapolation towards higher quantiles. Although further investigation of the total ΔV_{TH} for large sample sizes revealed a deviation from normal statistics, highlighting the need for a more rigorous statistical examination.

Statistical Moments and Normal Approximation

A first order modeling approach for the observed total ΔV_{TH} assumes it to be normally distributed by fitting the mean (μ) and variance (σ^2). This approach had been used in [17] as a benchmark for comparing the BTI induced ΔV_{TH} variability against the mismatch based time-zero V_{TH} variability. This pragmatic way of using normal statistics reduces the problem to a purely empirical form and does not provide any insight into the underlying physics.

However, an attempt was made earlier [26] to explain the observed increase in variance for increased mean shifts of ΔV_{TH} from a statistical point of view. Using the charge sheet approximation, one can directly relate the total number of charged defects N to the total ΔV_{TH} by

$$\Delta V_{\mathrm{TH}} \propto \frac{q}{Area} \frac{t_{\mathrm{INV}}}{\epsilon_{SiO_2}} N = \eta_0 N. \tag{1}$$

It is commonly assumed that the random fluctuation in the number of charged defects follows a Poisson distribution. This underlying Poisson process provides a correlation between the variance and mean of the total ΔV_{TH} distribution, observed in [26], since for a Poisson process

$$\sigma^2(N) = \mu(N). \tag{2}$$

However, this only considers the variation originating from *the number of defects*. There is also a contribution from the *defect impact* due to the spatial distribution of defects and their interaction with the random dopants of the channel as shown by TCAD simulation [3]. To account for this added source of variability, i.e., over-dispersion compared to a pure Poisson process, an empirical parameter K was added to the equation [26]. The mean and variance of the total ΔV_{TH} is then given by

$$\mu(\Delta V_{\mathrm{TH}}) = \eta_0 \mu(N) \tag{3}$$

$$\mu(\Delta V_{\mathrm{TH}}) = \sigma^2(\Delta V_{\mathrm{TH}}) = K \eta_0 \mu(\Delta V_{\mathrm{TH}}), \tag{4}$$

such that the variance of the total ΔV_{TH} can be written as

$$\sigma^2(\Delta V_{\mathrm{TH}}) = K\eta_0\mu(\Delta V_{\mathrm{TH}}), \tag{5}$$

where K was shown to be about 2–3 through simulation [3] and measurements [10, 26]. The work of [26] provided the first formulation for the variance of the total ΔV_{TH} as a function of the mean.

Compound Poisson Distribution

A more rigorous, statistical examination was performed in [27]. Here the underlying stochastic process is linked to the creation and repassivation of interface defects consistent with the, at that time recognized, reaction diffusion process for NBTI. This resulted in treating the total ΔV_{TH} as the combination of "created" C and "destructed" D interface defects, each uncorrelated Poisson distributed variates, such that

$$\mu(C - D) = \mu(C) - \mu(D) \tag{6}$$

$$\sigma^2(C - D) = \sigma^2(C) + \sigma^2(D). \tag{7}$$

This provided the needed over-dispersion compared to a normal Poisson process since

$$\frac{\sigma^2(C - D)}{\mu(C - D)} = \frac{\mu(C) + \mu(D)}{\mu(C) - \mu(D)} > 1. \tag{8}$$

The distribution of two independent Poisson distributed random variates is also referred to as a Skellam distribution [29]. The total ΔV_{TH} is then given by

$$\Delta V_{\mathrm{TH}} = \sum_{i=1}^{N_c(t)} S_i - \sum_{i=N_c(t)+1}^{N_c(t)+N_d(t)} S_i, \tag{9}$$

where N_c are the created and N_d the destructed interface defects. Each defect has a unique, distributed impact on the V_{TH} given by S_i such that

$$\mu(S) = \eta_0. \tag{10}$$

The first moment of the total ΔV_{TH} reads

$$\mu(\Delta V_{\mathrm{TH}}) = \eta_0(\mu(C) - \mu(D)). \tag{11}$$

The second moment, using the Blackwell–Girshick equation [5],[1] reads

[1] $\sigma^2(\Delta V_{\mathrm{TH}}) = \mu(N)\left(\mu(S_i)^2 + \sigma^2(S_i)\right)$.

$$\sigma^2(\Delta V_{\mathrm{TH}}) = \eta_0^2 \left(\mu(C) + \mu(D)\right) \left(1 + \sigma^2(S/\eta_0)\right) \tag{12}$$

and hence

$$\sigma^2(\Delta V_{\mathrm{TH}}) = \frac{\mu(C) + \mu(D)}{\mu(C) - \mu(D)} \left(1 + \sigma^2(S/\eta_0)\right) \eta_0 \mu(\Delta V_{\mathrm{TH}}), \tag{13}$$

which provides a formulation for the over-dispersion factor K of Eq. (5). In [27] an approximation is provided for large values of $N = C - D$ where the compound Skellam distribution is approximated by a compound Poisson distribution, i.e.,

$$\Delta V_{\mathrm{TH}} \approx \sum_{i=1}^{N_c(t)} S_i, \tag{14}$$

which, under the proposition of the central limit theorem,[2] can be shown to result in the following cumulative distribution function:

$$F(\Delta V_{\mathrm{TH}}) \approx \sum_{n=0}^{N} \frac{\mathrm{e}^{-N} N^n}{n!} \Phi\left(\frac{\Delta V_{\mathrm{TH}}/\eta_0 - n}{\sqrt{n(K-1)}}\right), \tag{15}$$

where $\Phi()$ is the standard normal CDF and K the over-dispersion factor, which needs to be larger than 1.

Origin of NBTI Variability

The statistical models described in sections "Statistical Moments and Normal Approximation" and "Compound Poisson Distribution" will be classified as empirical or semi-empirical, since some of their parameters used have no physical meaning and are solely used as fitting parameters. In contrast to that, Kaczer et al. [12] showed that, by combining the individual defect properties, it is possible to completely describe the total ΔV_{TH} by physically meaningful parameters. Here, a well-defined distribution for the single defect impact was established, leading to a closed-form solution for the ΔV_{TH} distribution without the use of empirical parameters.

According to [12], the total ΔV_{TH} is given by the sum of the contribution of a number of defects N,

$$\Delta V_{\mathrm{TH}} = \sum_{i=0}^{N(t)} S_i, \tag{16}$$

[2]The central limit theorem states the arithmetic mean of a sufficiently large number of independent random samples will be approximately normally distributed.

with the number of defects N Poisson distributed around the mean N_T and where each defect has an well-defined exponentially distributed impact S_i with a mean value of η. Equation (16) is thus a compound Poisson-exponential distribution of which an exact analytical formulation can be obtained. In a first step, the distribution of ΔV_{TH} for a fixed number of defects n is considered. This will be the distribution of the sum of n identically and independently distributed exponential random variables. This distribution is the so-called Erlang distribution [6] with probability density function given by

$$f_n(\Delta V_{\mathrm{TH}}) = \frac{\Delta V_{\mathrm{TH}}{}^{n-1} \mathrm{e}^{-\frac{\Delta V_{\mathrm{TH}}}{\eta}}}{(n-1)!\eta^n} \tag{17}$$

and cumulative distribution function by

$$F_n(\Delta V_{\mathrm{TH}}) = \frac{\gamma\left(n, \frac{\Delta V_{\mathrm{TH}}}{\eta}\right)}{(n-1)!} \tag{18}$$

with $\gamma()$ the incomplete gamma function and parameter η the mean defect impact. The CDF is plotted in Fig. 1 for various numbers of defects n.

Since the number of defects is, however, Poisson distributed, the total ΔV_{TH} cumulative distribution function is calculated as a Poisson weighted sum over all possible defect combinations as

$$F_N(\Delta V_{\mathrm{TH}}) = \sum_{n=0}^{\infty} \frac{\mathrm{e}^{-N_T} N_T^n}{n!} F_n(\Delta V_{\mathrm{TH}}, \eta) \tag{19}$$

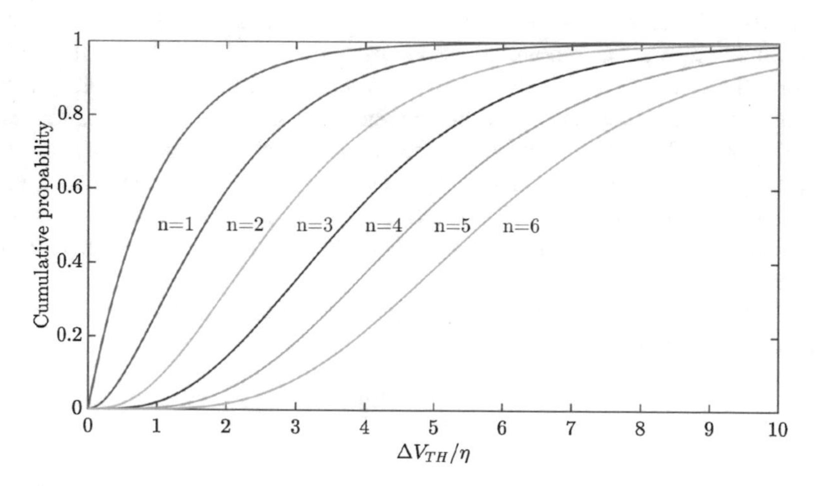

Fig. 1 The Erlang cumulative distribution function of Eq. (18) plotted for different numbers of n defects

i.e.,

$$F_N(\Delta V_{\mathrm{TH}}) = \sum_{n=0}^{\infty} \frac{\mathrm{e}^{-N_T} N_T^n}{n!} \frac{\gamma\left(n, \frac{\Delta V_{\mathrm{TH}}}{\eta}\right)}{(n-1)!}, \tag{20}$$

which gives a corresponding probability density function

$$f_N(\Delta V_{\mathrm{TH}}) = \mathrm{e}^{-N_T}\left[\delta(\Delta V_{\mathrm{TH}}) + N_T \frac{\mathrm{e}^{-\frac{\Delta V_{\mathrm{TH}}}{\eta}}}{\eta}\, {}_0\mathscr{F}_1\left(2; N_T \frac{\Delta V_{\mathrm{TH}}}{\eta}\right)\right], \tag{21}$$

where ${}_0\mathscr{F}_1(2; x)$ is a confluent hypergeometric limit function [22]. In Fig. 2 an illustration is given on how the defect-centric cumulative distribution function of Eq. (20) is the sum of Poisson weighted Erlang cumulative distributions. The first two moments[3] of the distribution are

$$\mu(\Delta V_{\mathrm{TH}}) = N_T \eta \tag{22}$$

and

$$\sigma^2(\Delta V_{\mathrm{TH}}) = 2 N_T \eta^2, \tag{23}$$

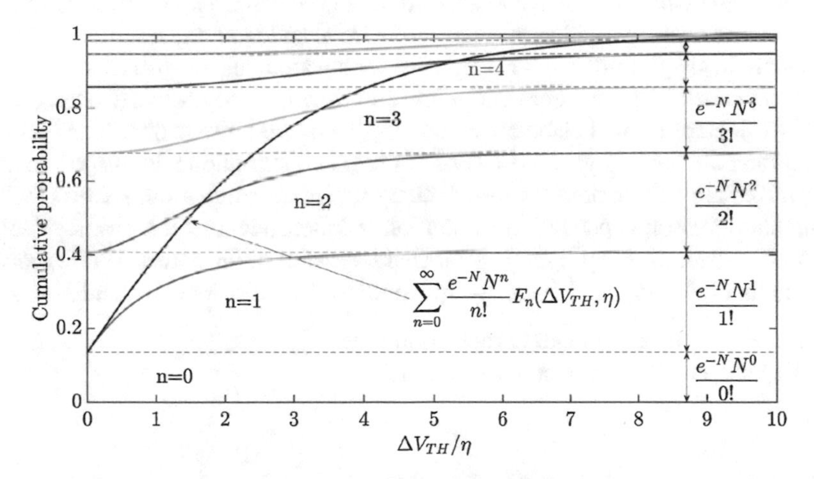

Fig. 2 Illustration on how the defect-centric cumulative distribution function of Eq. (20) (black) is the sum of Poisson scaled Erlang cumulative distributions with different values of n. These Erlang contributions are illustrated for $n = 0..6$ with an offset corresponding to their Poisson cumulative probability

[3]Using Blackwell–Girshick's equation.

such that the variance of the total ΔV_{TH} can be written as

$$\sigma^2(\Delta V_{TH}) = 2\eta\mu(\Delta V_{TH}), \tag{24}$$

giving a theoretically derived constant for the over-dispersion parameter K (Eq. (5)) to **equal exactly 2**. The compound Poisson-exponential distribution is thus completely described by *only two physically meaningful parameters*, the average number of charged defects N and the average defect impact η on the V_{TH}. Moreover, it has to be noted that the parameter η does not necessarily equal η_0 as used in Eq. (5) that is calculated assuming a charge sheet approximation for the defect impact. This defect-centric approach is by now well established within the reliability community because of its high predictive power in the extreme tail regions of NBTI distributions. It has been verified experimentally, across a wide variety of commercial and research grade technologies, from planar architectures to FinFET, in bulk and SOI, from SiO_2 and SiON gate oxides to High-k metal gate (HK-MG) and for different channel material like silicon and III–V materials [2, 8, 14–16, 23, 30, 33].

2.2 *Time Invariance of the Compound Poisson-Exponential Distribution*

As discussed in the previous section, the compound Poisson-exponential distribution has been observed on a wide variety of commercial and research grade technologies. Although the distribution has been reported for a wide range of stress and relaxation times, its dependence on stress and relaxation time has only recently been investigated [21]. Here we will elaborate on the origins of the Poisson distributed number defects and the "memoryless" behavior of the BTI distribution with respect to time. The universality of the observations made by several research groups shows that the compound Poisson-exponential distribution is independent of the characterization method applied. This highlights several characteristics of the defect-centric process, namely that the compound Poisson-exponential ΔV_{TH} distribution should be:

- independent of the amount of relaxation time
- independent of the amount of stress time.

Origin of Poisson Distributed Defects After Relaxation

When measuring the ΔV_{TH} there is always some relaxation involved and

$$\Delta V_{TH} = \Delta V_{TH,\,\text{stress}} - \Delta V_{TH,\,\text{relax}}. \tag{25}$$

Therefore we should write the defect-centric equation Eq. (16) as

$$\Delta V_{\mathrm{TH}} = \sum_{i=0}^{N_c} S_i - \sum_{j=0}^{N_d} S_i, \tag{26}$$

where N_c and N_d are the number of charged and discharged defects. Equation (26) is similar to Eq. (9), proposed by Rauch et al. [27], however, with two major differences. Firstly, at that time, the exact distribution of S_i was unknown and a normal one was used for simplicity. We now understand the distribution to be approximately exponential. Secondly, Rauch assumed N_c and N_d to be uncorrelated. However, causality dictates that N_d is a dependent random variable on N_c since no more traps can discharge then have been charged initially. Thus the process N_d is conditionally dependent on the process N_c since

$$N_d \leq N_c. \tag{27}$$

In general, during the relaxation process we can state that

$$N_{t+\Delta t} \leq N_t. \tag{28}$$

Now let N_{r1} and N_{r2} be the average number of remaining traps, charged at time t_{r1} and t_{r2} due to a stress and relaxation process (Fig. 3) such that

$$N_{r1} = N_c - N_{d1} \tag{29}$$

$$N_{r2} = N_c - N_{d2}, \tag{30}$$

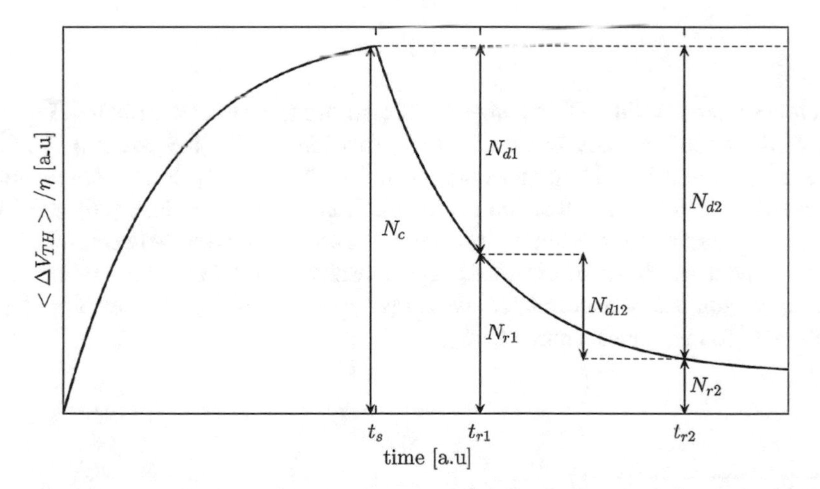

Fig. 3 Illustration of the stress and relaxation of BTI. At the end of the stress phase a number of defects N_c will be charged. During the relaxation phase a number of defects N_d will discharge, resulting in a remainder of charged defects $N_r = N_c - N_d$

which we can rewrite as

$$N_{r1} = N_{r2} + N_{d12}, \tag{31}$$

where $N_{d12} = N_{d2} - N_{d1}$, the fraction of defects discharged between t_{r1} and t_{r2}. Since the ΔV_{TH} distribution is observed to be compound Poisson-exponential at time t_{r1}, Eq. (26) can be written as

$$\Delta V_{\mathrm{TH},t_{r1}} = \sum_{i=0}^{N_{r1}} S_i, \tag{32}$$

where N_{r1} is observed to be Poisson distributed by measurement. Moreover, since the same can be said about N_{r2}, and by using equality (31), Raikov's theorem[4] then states that N_{r2} and N_{d12} must be independent and Poisson distributed themselves. It can then readily be shown by mathematical induction that the same holds for any combination of remaining N_R and discharged defects N_D and we can make the following statements:

- The ΔV_{TH} is compound Poisson-exponential distributed, independent of time
- The ΔV_{TH} is independent of the previously discharged $\Delta V_{\mathrm{TH},relax}$

Combining Eqs. (32), (26), and (31) we can write the BTI defect-centric process in one master set of equations as

$$\sum_{i=0}^{N_c} S_i = \sum_{i=0}^{N_r} S_i + \sum_{j=0}^{N_d} S_i \tag{33}$$

$$N_c = N_r + N_d \tag{34}$$

$$N_d \perp N_r, \tag{35}$$

where each term in Eq. (33) is compound Poisson-exponential distributed. Figure 4 shows the experimentally observed ΔV_{TH} distribution for the charged (C), discharged (D), and remaining (R) components of NBTI [33]. Each component is observed to follow the defect-centric distribution as given by Eqs. (26) and (32). Additional confirmation is found by examining the correlation between the D and R component as shown in Fig. 5. As hypothesized in Eq. (33), these components should be completely uncorrelated, which is supported by the very low correlation coefficient found in measurement.

[4]Raikov's theorem, named after Dmitry Raikov, states that if the sum of two independent non-negative random variables X and Y has a Poisson distribution, then both X and Y themselves must follow a Poisson distribution [24].

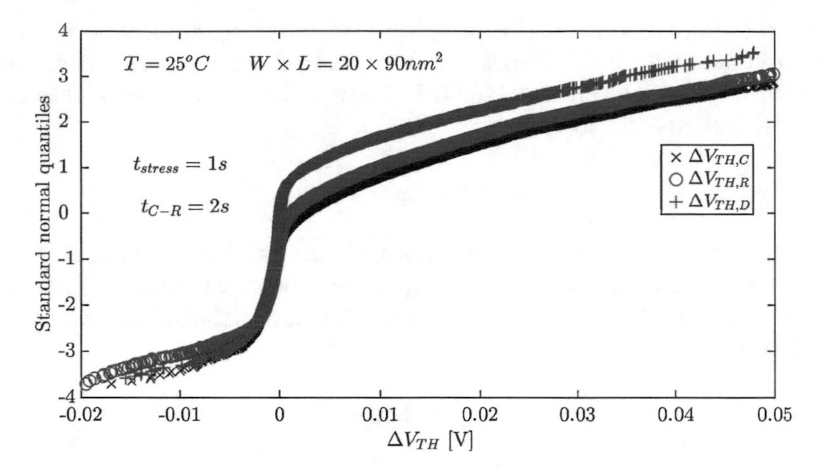

Fig. 4 The ΔV_{TH} distribution for the C, D, and R components of NBTI. Each component is observed to follow the defect-centric distribution as given by Eqs. (26) and (32). Observe the opposite shifts (negative values) that originate due to RTN which will be discussed in Sect. 3

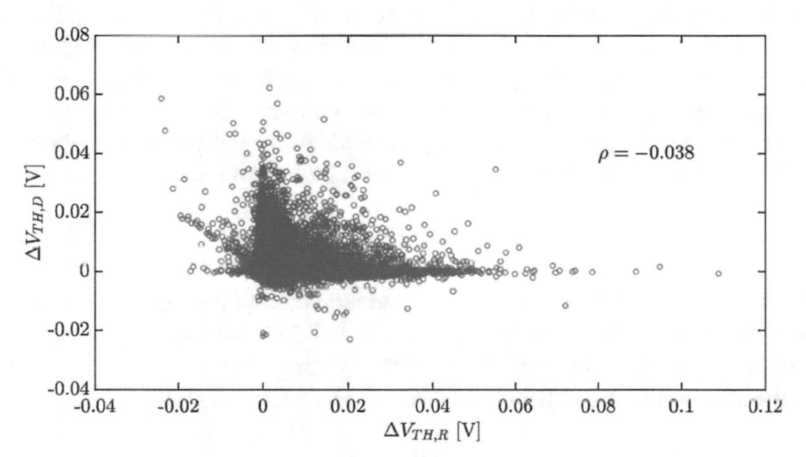

Fig. 5 Scatter plot of the ΔV_{TH} D and R components for visual inspection of the correlation. The calculated correlation value $\rho = -0.038$ indicates an insignificant correlation, further justifying our statements made in Eq. (33)

Origin of Poisson Distributed Defects During Stress

In order to fully understand the origin of the compound Poisson-exponential distribution and its relation to time, we answer the "existential question" of "why the number of charged defects is Poisson distributed?" There are only two ways of generating Poisson distributed variates:

- there is a pre-existing Poisson distributed defect population of which a fraction gets charged independent on the underlying Poisson distribution
- there is a "memoryless" defect generating mechanism

These two mechanisms can also be referred to as "charging of defects already present in the oxide due to fabrication" and "charged defects created under stress conditions" and will be discussed in the following. The total number of defects can be a combination of the two as

$$N = N_{x,fab} + N_{gen}, \tag{36}$$

where x denotes the fraction of as-fabricated defects charged (total number of fabricated defect equals N_{fab}) and N_{gen} is the number of generated defects. If all defects are pre-existing, i.e., as-fabricated, it can be assumed they are Poisson distributed as

$$N_{fab} \sim \text{Poisson}. \tag{37}$$

This is a valid assumption since the Poisson distribution is an approximation of the binomial distribution that describes the probability of finding a given number of defects within a fixed volume if that probability is small and the probability is independent in space.[5] If the process of the defects getting charged is governed by a Markov process, each defect has an associated stochastic capture time τ_c, which is independent from the amount of defects. The process of charging the defects is thus a random sampling of the original population. With more stress time, more samples are taken but are always random and independent. Therefore, at any given stress time the number of charged defects will always be Poisson distributed, i.e.,

$$N_{x,fab} \sim \text{Poisson} \tag{38}$$

Furthermore, this is independent from the kinetic charging process taking place as long as that process is sampling the original Poisson distribution randomly. This moreover implies that the fraction of defects getting charged N_{c12} is independent from the previous already charged number of defects N_{c1} such that

$$N_{c2} = N_{c1} + N_{c12} \tag{39}$$

$$N_{c1} \perp N_{c12}, \tag{40}$$

which is a similar expression as the one found for the relaxation in Eq. (33). Now we focus on the number of generated defects. Since the number of observed defects is Poisson distributed, the number of generated charged defects should also be Poisson distributed by Eqs. (36) and (38). The generation process of defects should therefore result in a Poisson distributed number of defects at any fixed point in time. The only known process for creating Poisson distributed numbers is that of a memoryless process where the inter-arrival times are exponentially distributed.

[5] Also referred to as the *"Poisson limit theorem"* or *"law of rare events."*

These exponentially distributed inter-arrival times originate from the property that the probability of a defect being created is independent of time. This is a process in which events occur continuously and independently at a constant average rate. For every $t > 0$ the number of defects created in the time interval $[0, t]$ follows the Poisson distribution with mean λt, if and only if the sequence of inter-arrival times are independent and identically distributed exponential random variables having mean $1/\lambda$. This can be a valid generating process but has implications on the mean and variance of the total distribution. Firstly the mean number of generated defects will increase linear in time and secondly so does the variance

$$\mu(N) = \lambda t \sigma^2(N) = \lambda t \tag{41}$$

such that

$$\mu(\Delta V_{\mathrm{TH}}) = R\eta t, \tag{42}$$

where $R = \lambda$ is the defect generating rate in s^{-1} and η the mean defect impact. The time dependence of NBTI and PBTI however has been shown to approximately follow a power-law with respect to time

$$\mu(\Delta V_{\mathrm{TH}}) \approx At^n \tag{43}$$

with n the time exponent having a value of ~0.15 – 0.20 and ~0.25 for NBTI and PBTI, respectively. Consequently, if the generating process is occurring it will generate defects according to a power-law with a time exponent equal to 1. This will be in stark contrast to the observed power-law slopes for both NBTI and PBTI, as illustrated in Fig. 6. No evidence in literature was found showing such behavior nor was it observed in our own conducted experiments. *This leads to the assumption that BTI is caused by the charging of Poisson distributed pre-existing defects or at least it is the equivalent process that involves Poisson distributed precursors.*

3 Defect-Centric RTN Modeling

Random telegraph noise (RTN), arguably caused by the same defect-centric mechanism as BTI [12], will manifest itself as an opposite V_{TH} shift which can also be observed when measuring BTI (i.e., as a positive NBTI or negative PBTI shift) as shown in Fig. 7. This “opposite” tail has independently been observed in various works [2, 23, 27]. Thorough investigation performed in [33] shows that it contains relevant information about the distribution of defect properties.

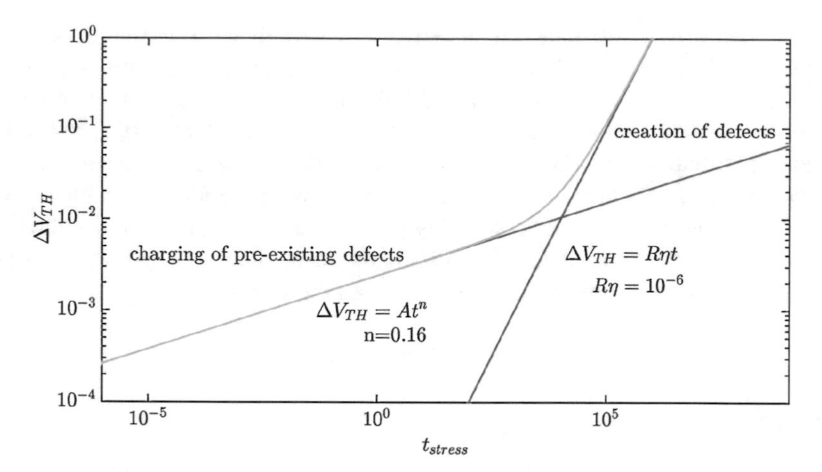

Fig. 6 Illustration on how a defect generating mechanism would be observed in the time dependence of the mean ΔV_{TH}. A quick onset would be observed due to the linear increase of defects with time, i.e., a power-law slope equal to 1. The magnitude of the defect generating rate R (chosen arbitrary here) will shift the curve with respect to time

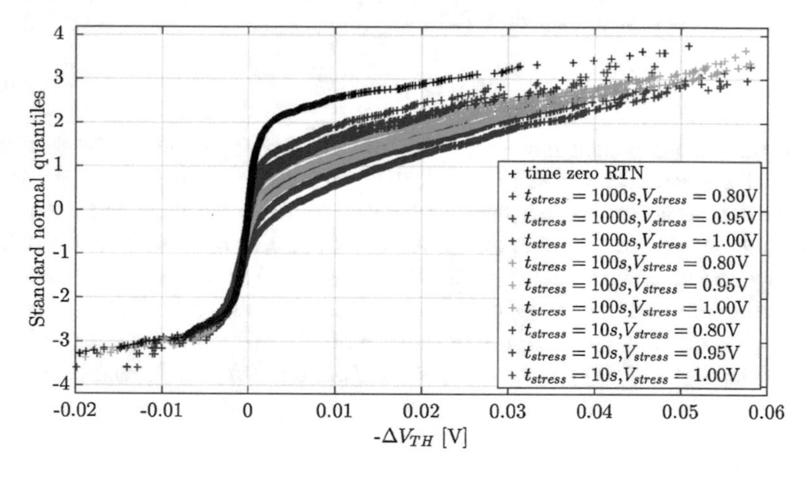

Fig. 7 NBTI time-dependent variability measured for different voltages and stress times. Time-zero noise distribution is added for comparison. Opposite shifts are observed originating from RTN where the number of RTN traps in the tail seems independent of the applied stress

3.1 Time Constants and Markov Processes

The BTI impact can be modeled, in a simplified way, based on two-state trap kinetics. In this approximation, a defect can either be occupied or not. The defect occupied results in a threshold voltage shift ΔV_{TH}. The rate at which a defect can capture or emit a carrier is determined by time constants, dependent on the gate voltage. The kinetics of the oxide traps can, to a first approximation, be modeled as a two-state continuous time Markov chain, referring to the memoryless property of the stochastic process. The oxide defect switches between an occupied and an unoccupied state (see Fig. 8) where the time it stays in a state is exponentially distributed as

$$\Delta t_c \sim \exp(\tau_c) \tag{44}$$

and

$$\Delta t_e \sim \exp(\tau_e), \tag{45}$$

where the capture time τ_c and the emission time τ_e is the average time it takes to respectively capture or emit a carrier in the defect, or to change to a charged or uncharged state, as illustrated in Fig. 9.

A Markov chain process is characterized by first order kinetics

$$\frac{\mathrm{d}P_c}{\mathrm{d}t} = k_e(1 - P_c) - k_c P_c, \tag{46}$$

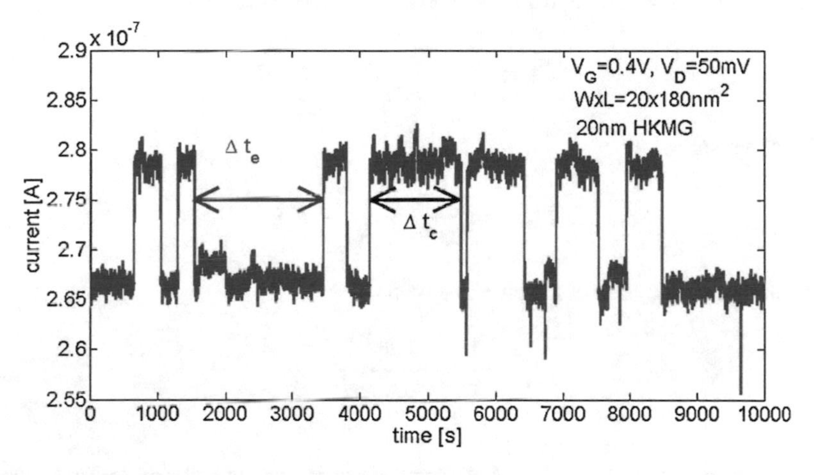

Fig. 8 Typical RTN current trace showing discrete current fluctuations associated with the charging and discharging of a single oxide defect

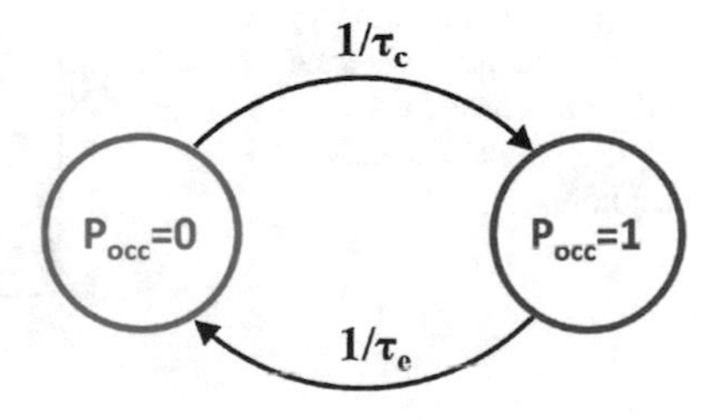

Fig. 9 A simple two-state Markov chain depicting the transition rates

where k_e and k_c are the transition rates for changing from an occupied to an unoccupied state and vice versa. The first order kinetics can also be written in terms of the capture and emission constants τ_c ans τ_e, which are the inverse of the transition rates, as

$$\frac{\mathrm{d}P_c}{\mathrm{d}t} = \frac{1 - P_c}{\tau_c} - \frac{P_c}{\tau_e}. \tag{47}$$

The unconditional probability for being in the occupied (P_1) or unoccupied state (P_0) reads

$$P_1 = \frac{\tau_e}{\tau_e + \tau_c} \tag{48}$$

and

$$P_0 = \frac{\tau_c}{\tau_e + \tau_c} \tag{49}$$

The conditional probability for capturing a carrier when the occupancy is zero and emitting a carrier when the occupancy equals 1 reads (Fig. 10)

$$P_{0\to 1} = \frac{\tau_e}{\tau_e + \tau_c}\left(1 - \mathrm{e}^{-\frac{t}{\tau}}\right) \tag{50}$$

and vice versa,

$$P_{1\to 0} = \frac{\tau_c}{\tau_e + \tau_c}\left(1 - \mathrm{e}^{-\frac{t}{\tau}}\right) \tag{51}$$

where the effective time constant τ is given by

$$\frac{1}{\tau} = \frac{1}{\tau_e} + \frac{1}{\tau_c}. \tag{52}$$

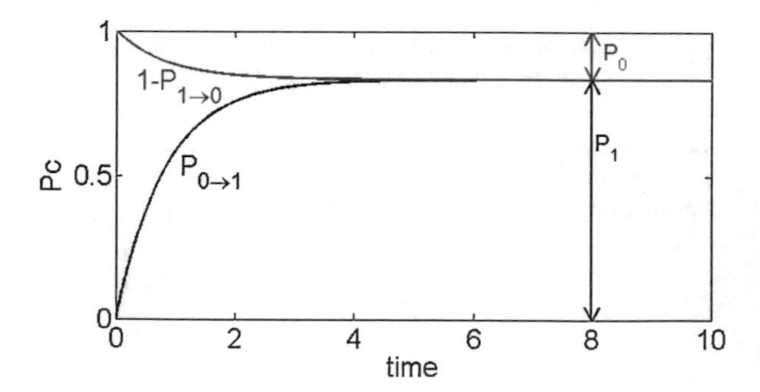

Fig. 10 The probability of the defect being occupied when the initial state is occupied (red) or unoccupied (black) changes over time with time constant $(\frac{1}{\tau_e} + \frac{1}{\tau_c})^{-1}$ to a steady-state condition

A more general solution for the occupancy probability reads

$$P_c(t + \Delta t) = \frac{\tau_e}{\tau_e + \tau_c}\left(P_c(t) - \mathrm{e}^{-\frac{\Delta t}{\tau}}\right). \tag{53}$$

3.2 *Unconditional RTN Events*

We will first focus on the unconditional case for observing RTN events. This means that when we observe RTN at two different points in time, t_1 and t_2, the observations are assumed to be uncorrelated. Then the unconditional probability for observing an occupied trap (occ = 1) or an unoccupied trap (occ = 0) at any point in time reads:

$$P_{\mathrm{occ}=1} = \frac{\tau_e}{\tau_e + \tau_c}, \tag{54}$$

$$P_{\mathrm{occ}=0} = \frac{\tau_c}{\tau_e + \tau_c}. \tag{55}$$

The probability of having a positive or negative shift is then equal and given by:

$$P_{+shift,\Delta t} = P_{0\rightarrow 1,\Delta t} = P_{\mathrm{occ}=0,t_1} P_{\mathrm{occ}=1,t_2} \tag{56}$$

$$P_{+shift,\Delta t} = \frac{\tau_e \tau_c}{(\tau_e + \tau_e)^2} \tag{57}$$

and

$$P_{-shift,\Delta t} = P_{1\rightarrow 0,\Delta t} = P_{\mathrm{occ}=1,t_1} P_{\mathrm{occ}=0,t_2} \tag{58}$$

$$P_{-shift,\Delta t} = \frac{\tau_e \tau_c}{(\tau_e + \tau_e)^2}, \tag{59}$$

where Δt is the time difference $t_2 - t_1$.

Zero values can also be observed when the state of the defect is unchanged between both measurement times

$$P_{0shift} = \frac{\tau_e^2}{(\tau_e + \tau_e)^2} + \frac{\tau_c^2}{(\tau_e + \tau_e)^2} \tag{60}$$

such that we can write the equality

$$P_{+shift} + P_{-shift} + P_{0shift} = 1. \tag{61}$$

The ΔV_{TH} for measured RTN is then given by

$$\Delta V_{\mathrm{TH},\Delta t} = \Delta V_{\mathrm{TH},t_2} - \Delta V_{\mathrm{TH},t_1}, \tag{62}$$

where $\Delta V_{\mathrm{TH},t_1}$ and $\Delta V_{\mathrm{TH},t_2}$ are the values measured at t_1 and t_2, respectively. Since the occupation of defect states is described by a Markov process (a process without memory), the two ΔV_{TH} samples at t_0 and t are independently drawn values from the same distribution, described by exponential-Poisson statistics such that

$$\Delta V_{\mathrm{TH},\Delta t} = \sum_{i=0}^{N_c(\Delta t)} S_i - \sum_{j=0}^{N_d(\Delta t)} \Delta S_i, \tag{63}$$

where N_c and N_d are, respectively, the number of defects charging and discharging between time $t_2 - t_1 = \Delta t$. For RTN, following Eqs. (57) and (59), we can state that the number of traps charging and discharging is equal, $N_c = N_d = N$, and that the impacts S_i and S_j are both exponentially distributed around mean η. Therefore, the probability density function for the RTN can then be obtained by means of convoluting two defect-centric distributions as

$$f_{\mathrm{RTN}}(\Delta V_{\mathrm{TH}}) = f_{N,\eta}(-\Delta V_{\mathrm{TH}}) * f_{N,\eta}(\Delta V_{\mathrm{TH}}). \tag{64}$$

If, however, the number of switching traps between t and t_0 is small, such that the positive or negative RTN values are almost entirely caused by a single switching trap, both tails can be described by an exponential-Poisson density function. The ΔV_{TH} probability density function can then be approximated by

$$f_{\mathrm{RTN}}(\Delta V_{\mathrm{TH}}) = \quad f_{N,\eta}(-\Delta V_{\mathrm{TH}}) \quad \Delta V_{\mathrm{TH}} < 0 \tag{65}$$

$$f_{\mathrm{RTN}}(\Delta V_{\mathrm{TH}}) = 1 - 2\mathrm{e}^{-N}(\Delta V_{\mathrm{TH}}) \; \Delta V_{\mathrm{TH}} = 0 \tag{66}$$

$$f_{\mathrm{RTN}}(\Delta V_{\mathrm{TH}}) = \quad f_{N,\eta}(\Delta V_{\mathrm{TH}}) \quad \Delta V_{\mathrm{TH}} > 0, \tag{67}$$

where $1 - 2\mathrm{e}^{-N}(\Delta V_{\mathrm{TH}})$ accounts for the zero values observed for no traps changing their state between t and t_0. Figure 11 illustrates how the different fractions of positive and negative ΔV_{TH} combine to form a total ΔV_{TH} RTN distribution.

Using the properties of compound Poisson processes we can write the variance for the $\Delta V_{\mathrm{TH,RTN}}$ of Eq. 63 as

$$\sigma^2(\Delta V_{\mathrm{TH}}) = 2\eta^2(N_c) + 2\eta^2(N_d) = 4\eta^2 N. \tag{68}$$

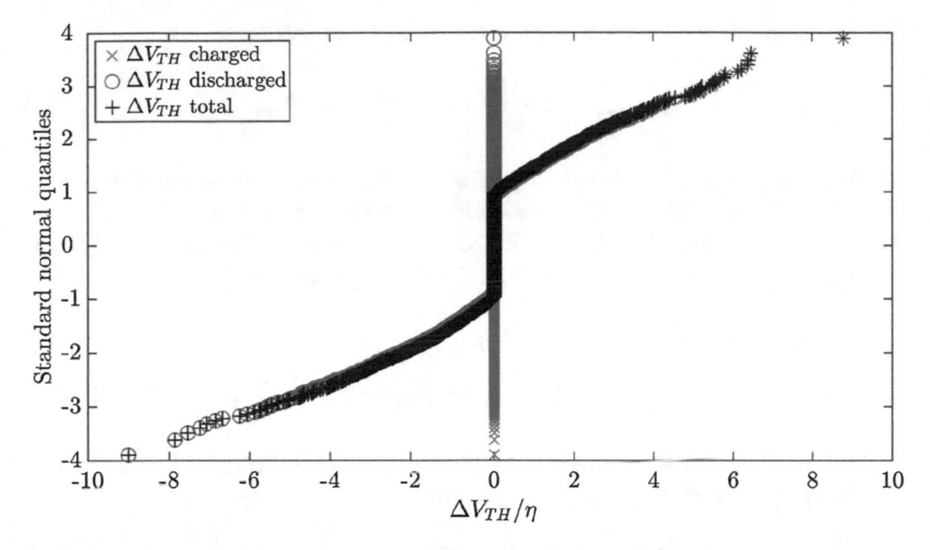

Fig. 11 A Monte Carlo simulation of the RTN ΔV_{TH} distribution. The RTN ΔV_{TH} distribution (black) is a convolution of two equal exponential-Poisson distributions having opposite sign (blue and red). When the number of switching traps is sufficiently small the tails of the combined RTN distribution can be approximated by an exponential-Poisson probability density function

3.3 *Conditional RTN Events*

In reality, when measuring RTN at two different points in time, t_1 and t_2, the observations are not uncorrelated as discussed in the previous subsection. When a defect is charged or discharged at t_1, the probability of that defect getting discharged or charged depends on its emission and capture time constant, respectively. The conditional probability for capturing a carrier when the occupancy is zero and emitting a carrier when the occupancy equals 1 reads

$$P_{\text{capture, occ=0}} = \frac{\tau_e}{\tau_e + \tau_c}\left(1 - \mathrm{e}^{-\frac{\Delta t}{\tau}}\right) \tag{69}$$

$$P_{\text{emission, occ=1}} = 1 - \frac{\tau_e}{\tau_e + \tau_c}\left(1 - \mathrm{e}^{-\frac{\Delta t}{\tau}}\right) - \mathrm{e}^{-\frac{\Delta t}{\tau}}, \tag{70}$$

where τ_e and τ_c are the defect time constants. After some rewriting, the probability for observing an RTN event between 2 consecutive measurement points with time Δt apart equals

$$P_{\text{RTN}} = P_{0\to1} + P_{1\to0} = 2\frac{\tau_e\tau_c}{(\tau_e + \tau_c)^2}\left(1 - \mathrm{e}^{-\frac{\Delta t}{\tau}}\right). \tag{71}$$

The probability increases for larger Δt and saturates to a value

$$P_{\mathrm{RTN},\Delta t=\infty} = 2\frac{\tau_e \tau_c}{(\tau_e + \tau_c)^2} = 2P_{\mathrm{occ}=1}P_{\mathrm{occ}=0}, \tag{72}$$

which corresponds to the probability of observing an RTN event unconditionally given by Eqs. (57) and (59). Equation 63 still holds for conditional RTN events but there is a correlation of N_c and N_d with Δt. This will be reflected in the variance as well which we can express mathematically using a covariance as

$$\sigma^2\left(\Delta V_{\mathrm{TH},t_2-t_1}\right) = \sigma^2\left(\Delta V_{\mathrm{TH},t_1} - \Delta V_{\mathrm{TH},t_2}\right) \tag{73}$$

$$\sigma^2\left(\Delta V_{\mathrm{TH},t_2-t_1}\right) = \sigma^2\left(\Delta V_{\mathrm{TH},t_1}\right) + \sigma^2\left(\Delta V_{\mathrm{TH},t_2}\right) - 2\mathrm{cov}\left(\Delta V_{\mathrm{TH},t_1}, \Delta V_{\mathrm{TH},t_2}\right). \tag{74}$$

Since the RTN is a wide-sense stationary process,[6] $\sigma^2(\Delta V_{\mathrm{TH},t_1}) = \sigma^2(\Delta V_{\mathrm{TH},t_2})$ and we can rewrite as

$$\sigma^2(\Delta V_{\mathrm{TH},\Delta t}) = (1 - R(\Delta t))2\sigma^2(\Delta V_{\mathrm{TH}}), \tag{75}$$

where $R(\Delta t)$ is the autocorrelation function given as

$$R(\Delta t) = \frac{\mathrm{cov}\left(\Delta V_{\mathrm{TH},t_1}, \Delta V_{\mathrm{TH},t_2}\right)}{\sigma^2(\Delta V_{\mathrm{TH}})} \tag{76}$$

and $\sigma^2(\Delta V_{\mathrm{TH}})$ is the unconditional RTN variance given by Eq. (68). The probability of an RTN event can be expressed in terms of widely distributed time constants such that

$$\sigma^2(\Delta V_{\mathrm{TH},\Delta t}) = 2\eta^2 \int_0^\infty \int_0^\infty 2\frac{\tau_e \tau_c}{(\tau_e + \tau_c)^2}\left(1 - e^{-\frac{\Delta t}{\tau}}\right) f(\tau_c, \tau_e)d\tau_c d\tau_e, \tag{77}$$

where $f(\tau_c, \tau_e)$ is the bivariate probability density function for τ_c and τ_e. Analyzing Eq. (75) we can state that the autocorrelation has a value of 1 for $t = 0$ and converges to 0 for $t = \infty$ which is also evident from Eq. (72). This means that the bigger the time difference between two measurement points the higher the variance of RTN will become and will eventually saturate to a fixed value, as illustrated in Fig. 12.

Figure 13 shows the number of switching RTN traps as a function of increasing time difference Δt for different stress cases [35]. The number of switching traps increases logarithmically with time as expected from observing widely distributed log-normal capture and emission times in a limited measurement window. This increase in number of switching traps and consequently the variance is explained by the effect of the autocorrelation for wide-sense stationary processes.

[6] A strict wide-sense stationary process is a stochastic process whose joint probability distribution does not change when shifted in time.

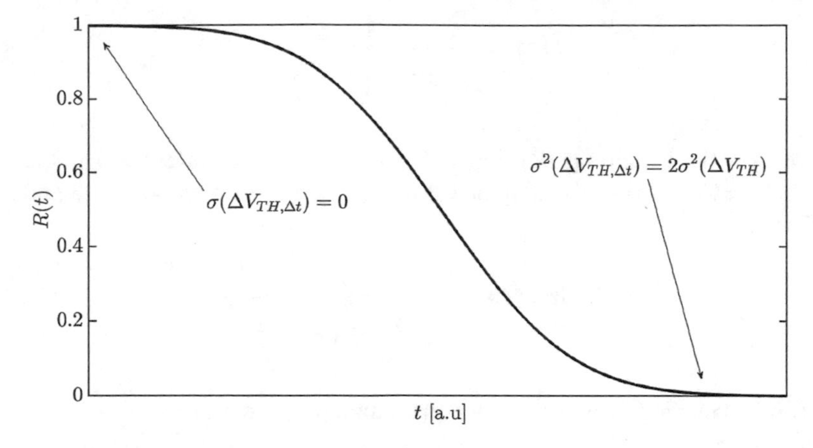

Fig. 12 Change in the autocorrelation as a function of time difference between the two measurement points. For long time differences, higher variance of RTN will become apparent which will eventually saturate to zero

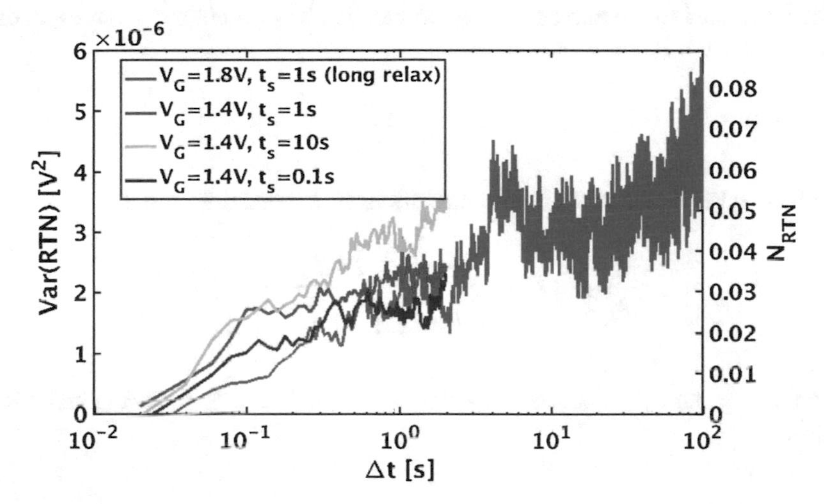

Fig. 13 Extraction of the RTN average number of traps as a function of increasing time difference Δt for different stress cases [35]

3.4 *Noise Modeling*

Single defect RTN behavior has been shown to exhibit a so-called Lorentzian power spectral density (PSD) distribution [19]. This Lorentzian, also known as Cauchy distribution, has as probability density function of the form of

$$f(x) = \frac{1}{\pi\gamma\left(1 + \left(\frac{x-x_0}{\gamma}\right)^2\right)}, \tag{78}$$

where x_0 is the location parameter, and γ is the scale parameter. A single defect creating a δV_{TH} during charging and discharging has been shown to reproduce a PSD of

$$S_{V_{\mathrm{TH}}}(f) = \frac{2A\delta V_{\mathrm{TH}}}{\pi f_c\left(1 + \left(\frac{f}{f_c}\right)^2\right)}, \tag{79}$$

Here, A is a loading factor scaling the maximum power density

$$A = \frac{\tau_e \tau_c}{(\tau_e + \tau_c)^2} \tag{80}$$

which is related to the trap occupancy probability. The parameter γ corresponds to the so-called corner frequency f_c where

$$f_c = \frac{1}{2\pi}\frac{\tau_e + \tau_c}{\tau_e \tau_c}. \tag{81}$$

The total ΔV_{TH} is the sum of the contributions S_i of each defect in the oxide

$$S_{V_{\mathrm{TH}}}(f) = \sum_{i=1}^{n} \frac{2A_i S_i}{\pi f_{c,i}\left(1 + \left(\frac{f}{f_{c,i}}\right)^2\right)}, \tag{82}$$

where A_i, S_i, and $f_{c,i}$ depend on the trap's time constants $\tau_{e,i}$ and $\tau_{c,i}$ and impact S_i [4].

4 Defect-Based Compact Modeling for RTN and BTI Variability

Workload dependent BTI and RTN degradation can be modeled in various ways depending on the implementation method used for circuit simulation. The various implementation methods are listed below

- **External**: the BTI is calculated using an external program/tool using a predetermined workload. The BTI degradation is then annotated to the netlist by changing the V_{TH} parameters of the transistors [1, 9, 18, 32].

- **RC equivalent circuit**: the BTI charging and discharging is mimicked by an RC equivalent circuit which is connected to the gate of the transistor. Transistors can degrade during simulation [28, 37].
- **BSIM compact model**: the BTI model is incorporated into the BSIM code [13].
- **Verilog-A wrapper**: the BTI model is incorporated into a Verilog-A wrapper [34].

BTI can be simulated using an external simulator and is added to the circuit simulation as annotation. Calculating the BTI degradation externally has the advantage of being fast and efficient when using dedicated mathematical software. Since the workload is needed for calculating the BTI degradation, upfront abstraction of that workload will need to be provided to the external simulator. The external simulation approach is typically used in a pre-stress and post-stress simulation approach. This approach is capable of simulating static BTI and RTN stochastic degradation but cannot do a fully coupled simulation as the circuit simulator and BTI degradation calculation are separated.

Other candidates are the direct incorporation into BSIM and using a Verilog-A wrapper. Direct incorporation into, e.g. the BSIM core model is, however, undesired since it would be restricted to that BSIM model. The reliability compact model should be compatible with any compact device model. Moreover, the compact modeling approach must meet all must-have requirements for proper incorporation in the standard API for commercial EDA tools, listed in [25]. Therefore it is argued that the Verilog-A wrapper will be the most suited approach to incorporate reliability degradation into circuit simulation. In the following we will discuss how the statistical and kinetic equations governing BTI and RTN behavior are translated into such a Verilog-A based model wrapper.

4.1 Combined Dynamic BTI and RTN Emulation

During a dynamic simulation transistor parameters can change as a result of the applied workloads. This results in a fully coupled simulation where workloads impact degradation, and degradation in turn impacts the workloads. This simulation, first proposed by [13], is a complete emulation of the BTI and RTN degradation mechanism. A circuit netlist is annotated with defect traps for each transistor. Each defect trap can respond to the applied workload and be either occupied or unoccupied. There are generally two approaches to determine the stochastic charge state over time. In a first approach the occupancy probability is calculated and evaluated at each time step. If the defect is in the charged state, the probability for emission equals

$$P_e(t + \Delta t) = P_{e,\infty}\left(1 - \mathrm{e}^{-\frac{\Delta t}{\tau(t)}}\right). \tag{83}$$

If the defect is in the uncharged state, the probability for charging equals

$$P_c(t + \Delta t) = P_{c,\infty}\left(1 - \mathrm{e}^{-\frac{\Delta t}{\tau(t)}}\right), \tag{84}$$

where

$$P_{e,\infty} = \frac{\tau_c(t)}{\tau_e(t) + \tau_c(t)} \tag{85}$$

$$P_{c,\infty} = \frac{\tau_e(t)}{\tau_e(t) + \tau_c(t)} \tag{86}$$

are the steady-state probabilities for the emission and capture event, respectively. At each time step, the capture or emission probability is compared to a uniform random number between 0 and 1 to determine if the charging or discharging event happens.

The second approach is to calculate an "event time" for the charging and discharging of the defect. At each change of the occupancy state, the time to the next state change is calculated by drawing an exponential random variable from the time constants as

$$\Delta t_c \sim \exp(\tau_c) \tag{87}$$

$$\Delta t_e \sim \exp(\tau_e) \tag{88}$$

This latter approach requires less computational resources compared to the former approach because the calculation is not performed at every time step. However, the last approach is only accurate if the time constants are assumed to be constant till the next event, which cannot be guaranteed. In fact we are predicting the future by calculating the next event without accurately knowing the future workload.

4.2 A Verilog-A Reliability Compact Model

The Verilog-A modeling language for analog behavior perfectly suits the needs for BTI and RTN implementation in the lowest level of SPICE simulation. It allows to program analog circuit components with highly non-linear behavior and support many mathematical constructs like distributions and a large set of basic functions. Figure 14 shows the hierarchical structure of the device compact model. A Verilog-A module wraps around the core BSIM compact model and allows to alter the currents and voltages seen by the core device. Implementing a V_{TH} shift can be as straightforward as adding a variable voltage source on the gate of the core device to comprehend all BTI and RTN defect related reactions. The high level data flow for the Verilog-A module is given in Fig. 15. In the following, we will focus on the BTI induced ΔV_{TH}.

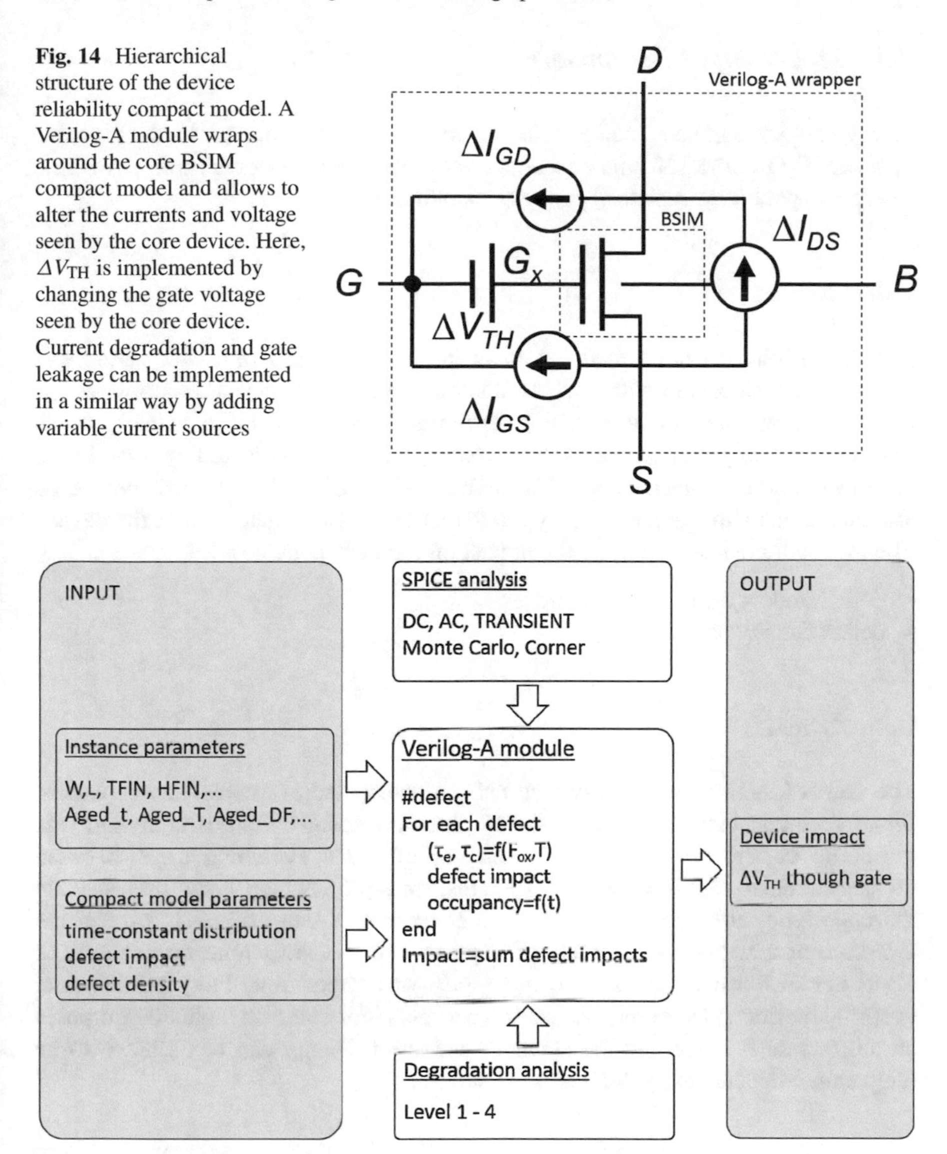

Fig. 14 Hierarchical structure of the device reliability compact model. A Verilog-A module wraps around the core BSIM compact model and allows to alter the currents and voltage seen by the core device. Here, ΔV_{TH} is implemented by changing the gate voltage seen by the core device. Current degradation and gate leakage can be implemented in a similar way by adding variable current sources

Fig. 15 Data flow showing the input and output parameters of the Verilog-A module. Input parameters consist of instance parameters, unique to each device such as device dimensions and the aged parameters, and compact model parameters describing the BTI and RTN. The SPICE analysis combined with the degradation level analysis defines the internal algorithms. Output parameters impact the behavior of the device through, ΔV_{TH} in our case

4.3 Model Input Parameters

The Verilog-A compact model relies on many input parameters to describe the complete BTI and RTN time-dependent variability. We can break this down into a defect impact (electrostatics) part and a kinetics part.

Defect Impact

The defect impact can be modeled as an impact on critical transistor parameters like V_{TH}, sub-threshold slope SS, or transconductance g_m to name a few. ΔV_{TH} has been shown to be exponentially distributed around the mean value η which scales reciprocally with gate area. Each defect is thus characterized by a stochastic ΔV_{TH} which is calculated using η. The number of defects will be Poisson distributed around a mean value determined by the defect density and gate area of the device. The input parameters governing the impact on the device are

- defect impact distribution parameter η
- defect density parameter n_d

Defect Kinetics

The defect kinetics can be described using capture and emission time constants. These time constants are distributed and also dependent on temperature and gate bias. The current most advanced modeling of oxide switching traps is based on non-radiative-multiphonon (NMP) transitions [36], which takes into account the underlying atomic organization. The NMP model can accurately predict the temperature and bias dependency of capture and emission time constants. The NMP model however relies on many input parameters describing the adiabatic potential barrier and structural relaxation energies. For compact modeling purposes an abstraction is made and the apparent activation energy can be assumed to be temperature dependent by Arrhenius' equation as

$$\tau \approx A\mathrm{e}^{-\frac{E_a}{k_B T}}, \tag{89}$$

where A is a constant and E_a the activation energy. Furthermore the activation energy can be assumed to be linearly dependent on the oxide electric field such that

$$\tau \approx \tau_0 \mathrm{e}^{-\frac{E_{a0}+\alpha F_{ox}}{k_B T}}, \tag{90}$$

where α is a linear dependence coefficient modeling the gate bias dependence of the time constants (through oxide field F_{ox}). The oxide field will be approximated as

$$F_{ox} \approx \frac{V_{OV}}{t_{\text{inv}}} \approx \frac{V_G - V_{\text{TH},0}}{t_{\text{inv}}}. \tag{91}$$

The coefficient α is a distributed parameter that lumps and approximates the NMP model behavior. From a compact modeling point of view we can furthermore represent the time constant by a generic function, dependent on gate bias and temperature. By rearranging Eq. (90) we can introduce additional coefficients which capture both the temperature and the gate bias dependence as

$$\tau \approx \mathrm{e}^{a+bF_{ox}+c\frac{1}{T}+d\frac{F_{ox}}{T}}. \tag{92}$$

Compared to Eq. (90), coefficient b is added to capture any additional voltage dependence. The logarithm of the time constant can thus modeled as a polynomial surface

$$\log(\tau_e) \approx a_e + b_e x + c_e y + d_e xy \tag{93}$$

$$\log(\tau_c) \approx a_c + b_c x + c_c y + d_c xy, \tag{94}$$

where $x = F_{ox}$ and $y = T^{-1}$. Each parameter a, b, c, and d is assumed to be normally distributed with cross-correlations. If we define a parameter vector $\boldsymbol{X}$ as

$$\boldsymbol{X} = \left[a_e\ b_e\ c_e\ d_e\ a_c\ b_c\ c_c\ d_c\right]^T, \tag{95}$$

where the entries are random variables, each with finite variance, then the covariance matrix $\boldsymbol{\Sigma}$ is the matrix whose (i, j) entry is the covariance

$$\boldsymbol{\Sigma}_{i,j} = \text{Cov}(\boldsymbol{X}_i, \boldsymbol{X}_j). \tag{96}$$

We also define $\boldsymbol{M}$ whose (i) entry is the expectation value

$$\boldsymbol{M}_i = \text{E}(\boldsymbol{X}_i). \tag{97}$$

These two matrices $\boldsymbol{\Sigma}$ and $\boldsymbol{M}$ fully define the multivariate normally distributed parameter space describing the temperature (T) and field (F_{ox}) dependence of the time constants τ_e and τ_c together with all cross-correlations. Using the mean and covariance matrices it is possible to generate correlated stochastic parameters of $\boldsymbol{X}$. Since Verilog-A does not support multivariate random number generation, we will generate the parameters using a decomposition of $\boldsymbol{\Sigma}$ defined as

$$\boldsymbol{\Sigma} = \boldsymbol{L}\boldsymbol{L}^{\boldsymbol{T}}, \tag{98}$$

where $\boldsymbol{L}$ is a lower triangular matrix with real and positive diagonal entries, and $\boldsymbol{L}^{\boldsymbol{T}}$ denotes the conjugate transpose of $\boldsymbol{L}$. The random parameter vector can then be calculated as

$$\boldsymbol{X} = \boldsymbol{M} + \boldsymbol{L}\boldsymbol{R}, \tag{99}$$

where $\boldsymbol{R}$ is a vector whose components are independent standard normal variates. The entries of $\boldsymbol{R}$ can be readily generated in Verilog-A.

With this compact modeling approach the 4-state NMP transitions can be approximated.

4.4 Initialization

Each simulation is initialized by attributing a given number of defects to each device which is Poisson distributed around the mean value $n_d WL$, where n_d is the defect density. Each defect is then characterized by its capture and emission time constants combined with gate bias and temperature dependencies. After the creation of defects and their parameters, the next step is to initialize the occupancy state of each defect. The occupancy probability at time-zero is given by the steady-state condition as

$$P_{\text{occ}} = \left.\frac{\tau_e}{\tau_e + \tau_c}\right|_{F_{ox,0}}, \tag{100}$$

where the field $F_{ox,0}$ equals the field in the oxide at time-zero. We can also incorporate past stress which is given by the aged parameters.

$$P_{\text{occ}} = (P_H - P_L)\frac{1 - \mathrm{e}^{-\frac{DF}{f\tau_c^*}}}{1 - \mathrm{e}^{-\frac{1}{f}\left(\frac{DF}{\tau_c^*} + \frac{1-DF}{\tau_e^*}\right)}}\left(1 - \mathrm{e}^{-t\left(\frac{DF}{\tau_c^*} + \frac{1-DF}{\tau_e^*}\right)}\right) + P_L, \tag{101}$$

where

$$\begin{aligned} P_H &= \left.\frac{\tau_e}{\tau_c + \tau_e}\right|_{F_{ox}=F_H} \\ P_L &= \left.\frac{\tau_e}{\tau_c + \tau_e}\right|_{F_{ox}=F_L} \end{aligned} \tag{102}$$

which are the steady-state occupancy probabilities corresponding to the aged higher (H) and lower voltage (L). The time constants

$$\begin{aligned} \tau_c^* &= \left.\left(\frac{1}{\tau_c} + \frac{1}{\tau_e}\right)^{-1}\right|_{F_{ox}=F_H} \\ \tau_e^* &= \left.\left(\frac{1}{\tau_c} + \frac{1}{\tau_e}\right)^{-1}\right|_{F_{ox}=F_L} \end{aligned} \tag{103}$$

are equivalent time constants corresponding to the aged higher (H) and lower voltage (L). After calculating the defect occupancy probability P_{occ}, it is compared against a uniform random number (between 0 and 1) to evaluate the occupancy state. Finally, the total ΔV_{TH} is calculated by summing all defect impacts multiplied by their occupancy state.

A stream of consecutive workloads can also be calculated as

$$P_{\mathrm{occ}[i]} = (P_H[i] - P_L[i]) \frac{1 - \exp - \frac{DF[i]}{f \tau_c[i]^*}}{1 - \exp^{-\frac{1}{f[i]}\left(\frac{DF[i]}{\tau_c[i]^*} + \frac{1-DF[i]}{\tau_e[i]^*}\right)}} \left(1 \exp -t \left(\frac{DF[i]}{\tau_c[i]^*} + \frac{1-DF[i]}{\tau_e[i]^*}\right)\right)$$
$$+ P_L[i] + (P_{\mathrm{occ}}[i-1] - P_L[i]) \exp -t[i] \left(\frac{DF[i]}{\tau_c[i]^*} + \frac{1-DF[i]}{\tau_e[i]^*}\right) \tag{104}$$

such that the initial degraded state is calculated from one workload phase to another. The aging parameters duty cycle DF, frequency f, high gate bias V_H, low gate bias V_L, and time t can therefore be single scalars or vectors of equal size, where each entry corresponds to one workload.

4.5 Simulation

Figure 16 shows the simulation flow chart the Verilog-A compact model follows. If the high level flag RELMOD is used a reliability incorporated simulation is invoked. Each simulation is then initialized by attributing a Poisson distributed number of defects to each device in the stochastic (Monte Carlo) mode. For a corner based simulation (e.g., mean behavior) a sufficient large sample size (e.g., 10 k) is used to describe the mean behavior (see Fig. 17).

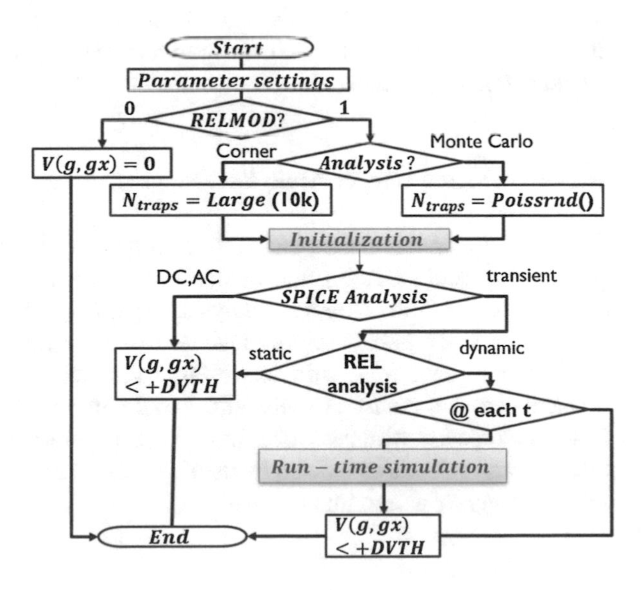

Fig. 16 Simulation flow chart

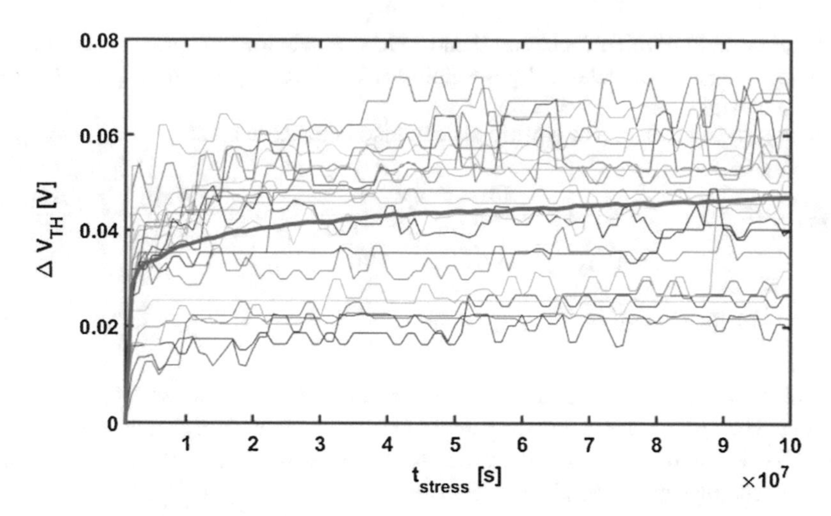

Fig. 17 Single device ΔV_{TH} results for both mean and Monte-Carlo based TRAN simulation under nominal operation conditions

AC, DC, and Transient Analysis

Only in the case of the SPICE analysis being transient (tran) can the defect occupancy state be updated at runtime. In all other analyses the defect occupancy states will be unchanged from those calculated in the initialization. Depending on the defect occupancy state, the probability for emission P_e or capture P_c is calculated and evaluated for each time step using Eqs. (83) and (84). The change of defect occupancy state is evaluated by comparing P_c or P_e against a uniform random number (between 0 and 1). After each state change, P_c or P_e is reset accordingly. In DC and AC analyses, the defect occupancy states will be unchanged from those calculated in their initialization.

Monte Carlo and Corner Analysis

By stochastically calculating the contributions of all individual defects, distributed in each transistor, Monte Carlo analysis comes naturally to the BTI simulation. In order to perform a corner analysis, e.g., a typical corner representing the mean degradation behavior, we make use of the same stochastic algorithms but simulate only once for a sufficiently large number of defects. The defect impact is calculated as a random exponential variate around the mean value of η, which scales reciprocally with gate area. In order to accommodate corner simulation, the defects mean value is calculated by using the mean number of defects N_{avg} and the defect density n_d as and input variable

$$\eta = \frac{n_d}{N_{avg}} \eta_0, \tag{105}$$

such that we have eliminated the gate area dependency. If the defect number is set to a high value, induced by a corner simulation (e.g., 10,000 defects), the defect impact will scale appropriately without having to artificially alter gate areas.

4.6 Measurements and Calibration

In order to calibrate the compact model a dedicated extraction procedure was developed in [34] based on eMSM measurements [11] techniques combined with precise Id-Vg and BTI measurements [33]. The compact model assumes a distribution attributed to recoverable BTI and RTN and another one for permanent BTI. In order to aid the fitting convergence, no emission is assumed for the permanent BTI. Consequently, a minimum set of the model requires 16 parameters to describe the defects kinetics (Fig. 18). Furthermore, 2 defect densities for permanent and recoverable BTI, n_R and n_P, are used to model the absolute impact of the degradation. A single η value is used for both the permanent and recoverable part as shown in [35].

The 16 kinetic model parameters and 2 defect density parameters can be extracted by dedicated fitting of eMSM data for different gate bias and temperature conditions as shown in Fig. 19. Of these parameters it was found that the distribution and moreover the weak correlation (as also found in [20]) between capture and emission coupling factors α_e and α_e are crucial for capturing the gate bias dependency. The

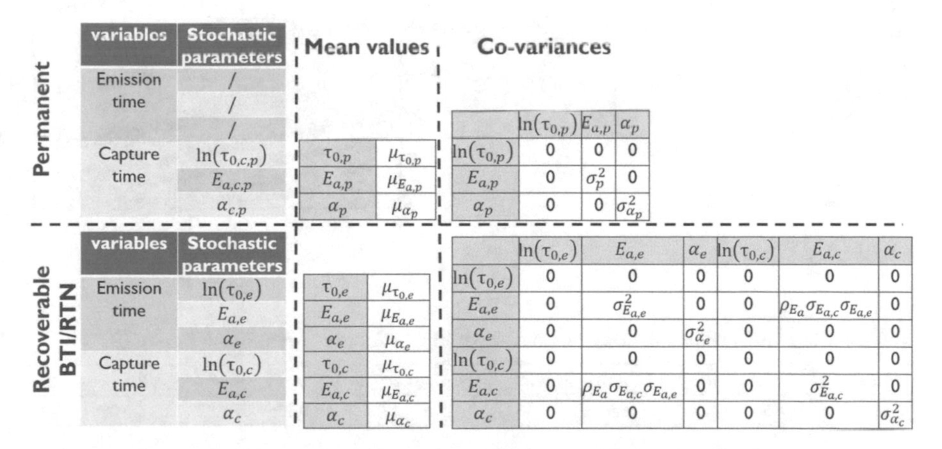

Fig. 18 Physical model parameters making up the variables (τ_e and τ_c) for describing the recoverable BTI/RTN and permanent BTI. The physical model parameters are assumed to be normally distributed having a mean value, standard deviation, and correlation to other parameters given by the covariance matrix

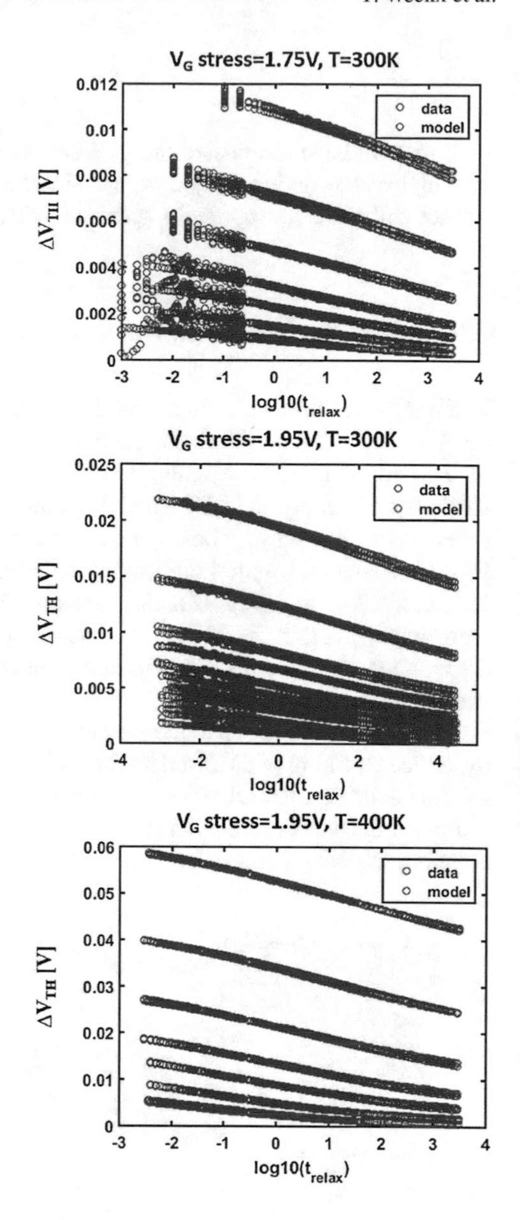

Fig. 19 Time constant distribution model fitted on eMSM data sets for different gate biases and temperatures using a single set of parameters [34]

Table 1 Recoverable kinetic model parameters of Fig. 18 for the fitting shown in Fig. 19

$\mu(\ln(\tau_{0,eR}))$	$\mu(E_{a,eR})$	$\mu(\alpha_{eR})$	$\mu(\ln(\tau_{0,cR}))$	$\mu(E_{a,cR})$	$\mu(\alpha_{cR})$	$\sigma^2(E_{a,eR})$	$\sigma^2(\alpha_{eR})$	$\sigma^2(E_{a,cR})$	$\sigma^2(\alpha_{cR})$	ρ_{E_a}
6.56	0.965	0.102	−18.9	1.46	0.173	1.05	0.176	0.338	0.0981	0.12

Table 2 Permanent kinetic model parameters of Fig. 18 for the fitting shown in Fig. 19

$\mu(\ln(\tau_{0,cP}))$	$\mu(E_{a,cP})$	$\mu(\alpha_{cP})$	$\sigma^2(E_{a,cP})$	$\sigma^2(\alpha_{cP})$
−4.78	1.95	0.495	0.00961	0.0685

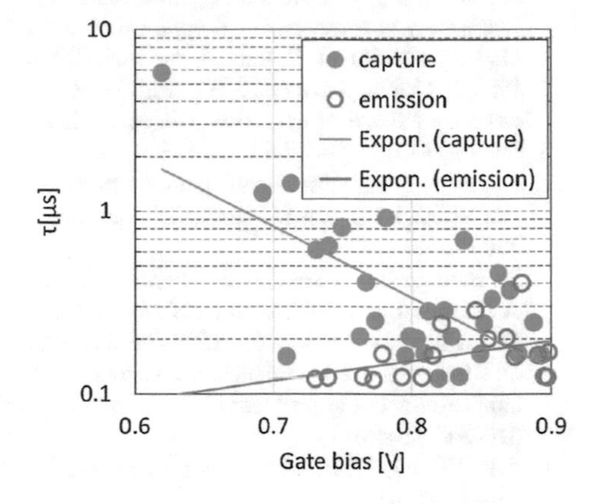

Fig. 20 Gate bias dependency of τ_e and τ_e is fitted with an exponential to extract the coupling factors of α_e and α_e [34]

parameters values can be found in Tables 1 and 2. In addition to eMSM, RTN trace measurements using precise Id-Vg [31] were performed to directly extract the distribution of α_e and α_e verifying our initial fitting of the eMSM data (Fig. 20). Finally, two-point BTI measurements [33] were used to accurately extract η from the exponential-Poisson distribution completing the parameter set of the compact model.

5 Conclusion

A comprehensive overview on state-of-the-art BTI distribution models has been given, of which the compound exponential-Poisson model is considered the leading in the field. The observed opposite ΔV_{TH} tail in BTI measurement data was shown to be originating from RTN. A complete physics-based defect-centric model is used to describe combined BTI and RTN. The analytical formulation describing the statistics of the defect characteristics, and the resulting total ΔV_{TH} will serve as a basis for reliability data analysis and simulations. A Verilog-A based wrapper was described that allows to incorporate all BTI and RTN related electrostatics and kinetics in standard EDA tools as a "black box" without any custom simulation flow. It can therefore be used in either a manual configuration for academic purposes or be integrated as is into standard EDA tools or industry standard simulation flows.

References

1. K.V. Aadithya, A. Demir, S. Venugopalan, J. Roychowdhury, SAMURAI: an accurate method for modelling and simulating non-stationary random telegraph noise in SRAMs, in *2011 Design, Automation Test in Europe Conference Exhibition (DATE)* (2011), pp. 1–6
2. D. Angot, V. Huard, L. Rahhal, A. Cros, X. Federspiel, A. Bajolet, Y. Carminati, M. Saliva, E. Pion, F. Cacho, A. Bravaix, BTI variability fundamental understandings and impact on digital logic by the use of extensive dataset, in *2013 IEEE International Electron Devices Meeting (IEDM)* (2013), pp. 15.4.1–15.4.4
3. A. Asenov, Random dopant induced threshold voltage lowering and fluctuations in sub-0.1 μm MOSFET's: a 3-D "atomistic" simulation study. IEEE Trans. Electron Devices **45**(12), 2505–2513 (1998)
4. M. Banaszeski da Silva, H.P. Tuinhout, A. Zegers-van Duijnhoven, G.I. Wirth, A.J. Scholten, A Physics-Based Statistical RTN Model for the Low Frequency Noise in MOSFETs. IEEE Trans. Electron Devices **63**, 3683–3692 (2016)
5. D. Blackwell, M.A. Girshick, On functions of sequences of independent chance vectors with applications to the problem of the "Random Walk" in k dimensions. Ann. Math. Stat. **17**(3), 310–317 (1946)
6. A.K. Erlang, Telefon-Ventetider. Et Stykke Sandsynlighedsregning. Matematisk tidsskrift. B, 25–42 (1920)
   E. Brockmeyer, H.L. Halstrm, A. Jensen, A.K. Erlang, The life and works of A.K. Erlang. Trans. Dan. Acad. Tech. Sci. **2** (1948)
7. Y. Chen, J. Zhou, S. Tedja, F. Hui, A. Oates, Stress-induced MOSFET mismatch for analog circuits, in *2001 IEEE International of Integrated Reliability Workshop Final Report* (2001), pp. 41–43
8. J. Franco, B. Kaczer, N. Waldron, J. Roussel, A. Alian, M. Pourghaderi, Z. Ji, T. Grasser, T. Kauerauf, S. Sioncke, N. Collaert, A. Thean, G. Groeseneken, RTN and PBTI-induced time-dependent variability of replacement metal-gate high-k InGaAs FinFETs, in *2014 IEEE International of Electron Devices Meeting (IEDM)* (2014), pp. 20.2.1–20.2.4
9. K.U. Giering, C. Sohrmann, G. Rzepa, L. Hei, T. Grasser, R. Jancke, NBTI modeling in analog circuits and its application to long-term aging simulations, in *2014 IEEE International of Integrated Reliability Workshop Final Report (IIRW)* (2014), pp. 29–34
10. V. Huard, C. Parthasarathy, C. Guerin, T. Valentin, E. Pion, M. Mammasse, N. Planes, L. Camus, NBTI degradation: from transistor to SRAM arrays, in *2008 IEEE International of Reliability Physics Symposium (IRPS 2008)* (2008), pp. 289–300
11. B. Kaczer, T. Grasser, P. Roussel, J. Martin-Martinez, R. O'Connor, B. O'Sullivan, G. Groeseneken, Ubiquitous relaxation in BTI stressing: new evaluation and insights, in *2008 IEEE International of Reliability Physics Symposium (IRPS 2008)* (2008), pp. 20–27
12. B. Kaczer, T. Grasser, P. Roussel, J. Franco, R. Degraeve, L.-A. Ragnarsson, E. Simoen, G. Groeseneken, H. Reisinger, Origin of NBTI variability in deeply scaled pFETs, in *2010 IEEE International of Reliability Physics Symposium (IRPS)*, pp. 26–32 (2010)
13. B. Kaczer, S. Mahato, V.V. de Almeida Camargo, M. Toledano-Luque, P.J. Roussel, T. Grasser, F. Catthoor, P. Dobrovolny, P. Zuber, G. Wirth, G. Groeseneken, Atomistic approach to variability of bias-temperature instability in circuit simulations, in *2011 IEEE International of Reliability Physics Symposium (IRPS)* (2011), pp. XT.3.1–XT.3.5
14. B. Kaczer, C. Chen, P. Weckx, P. Roussel, M. Toledano-Luque, J. Franco, M. Cho, J. Watt, K. Chanda, G. Groeseneken, T. Grasser, Maximizing reliable performance of advanced CMOS circuits: a case study, in *2014 IEEE International of Reliability Physics Symposium* (2014), pp. 2D.4.1–2D.4.6

15. B. Kaczer, J. Franco, M. Cho, T. Grasser, P. Roussel, S. Tyaginov, M. Bina, Y. Wimmer, L. Procel, L. Trojman, F. Crupi, G. Pitner, V. Putcha, P. Weckx, E. Bury, Z. Ji, A. De Keersgieter, T. Chiarella, N. Horiguchi, G. Groeseneken, A. Thean, Origins and implications of increased channel hot carrier variability in nFinFETs, in *2015 IEEE International of Reliability Physics Symposium (IRPS)* (2015), pp. 3B.5.1–3B.5.6
16. B. Kaczer, J. Franco, P. Roussel, G. Groeseneken, T. Chiarella, N. Horiguchi, T. Grasser, Extraction of the random component of time-dependent variability using matched pairs. IEEE Electron Device Lett. **36**(4), 300–302 (2015)
17. A. Kerber, T. Nigam, Challenges in the characterization and modeling of BTI induced variability in metal gate/High-k CMOS technologies, in *2013 IEEE International of Reliability Physics Symposium (IRPS)* (2013), pp. 2D.4.1–2D.4.6
18. H. Kufluoglu, V. Reddy, A. Marshall, J. Krick, T. Ragheb, C. Cirba, A. Krishnan, C. Chancellor, An extensive and improved circuit simulation methodology for NBTI recovery, in *2010 IEEE International Reliability Physics Symposium (IRPS)* (2010), pp. 670–675
19. C. Leyris, S. Pilorget, M. Marin, M. Minondo, H. Jaouen, Random telegraph signal noise SPICE modeling for circuit simulators, in *2007 IEEE International European Solid State Device Research Conference (ESSDERC)* (2007), pp. 187–190
20. H. Miki, N. Tega, M. Yamaoka, D.J. Frank, A. Bansal, M. Kobayashi, K. Cheng, C.P. D'Emic, Z. Ren, S. Wu, J.B. Yau, Y. Zhu, M.A. Guillorn, D.G. Park, W. Haensch, E. Leobandung, K. Torii, Statistical measurement of random telegraph noise and its impact in scaled-down high-k/metal-gate MOSFETs, in *2012 IEEE International Electron Devices Meeting (IEDM)*, (2012), pp. 19.1.1–19.1.4
21. D. Nouguier, G. Ghibaudo, X. Federspiel, M. Rafik, D. Roy, Characterization and modeling of NBTI permanent and recoverable components variability, in *2013 IEEE International Reliability Physics Symposium (IRPS)* (2016), pp. XT-08-1–XT-08-6
22. F. Tricomi, Sulle funzioni ipergeometriche confluenti. Annali di Matematica Pura ed Applicata. **26**(1), 141–175 (1947). https://doi.org/10.1007/BF02415375
23. C. Prasad, M. Agostinelli, J. Hicks, S. Ramey, C. Auth, K. Mistry, S. Natarajan, P. Packan, I. Post, S. Bodapati, M. Giles, S. Gupta, S. Mudanai, K. Kuhn, Bias temperature instability variation on SiON/Poly, HK/MG and trigate architectures, in *2014 IEEE International Reliability Physics Symposium* (2014), pp. 6A.5.1–6A.5.7
24. D. Raikov, On the decomposition of Poisson laws. Proc. USSR Acad. Sci. **14**, 9–11 (1937)
25. A. Ramadan, Compact model council's standard circuit simulator interface for reliability modeling, in *2013 IEEE International Reliability Physics Symposium (IRPS)* (2013), pp. 2A.5.1–2A.5.6
26. S. Rauch, The statistics of NBTI-induced v_t and β mismatch shifts in pMOSFETs. IEEE Trans. Device Mater. Reliab. **2**(4), 89–93 (2002)
27. S. Rauch, Review and reexamination of reliability effects related to NBTI-induced statistical variations. IEEE Trans. Device Mater. Reliab.**7**(4), 524–530 (2007)
28. H. Reisinger, T. Grasser, K. Ermisch, H. Nielen, W. Gustin, C. Schlnder, Understanding and modeling AC BTI, in *2011 IEEE International Reliability Physics Symposium (IRPS)*, pp. 6A.1.1–6A.1.8 (2011)
29. J.G. Skellam, The frequency distribution of the difference between two Poisson variates belonging to different populations. J. R. Stat. Soc. **109**(3), 296–296 (1946)
30. M. Toledano-Luque, B. Kaczer, J. Franco, P. Roussel, M. Bina, T. Grasser, M. Cho, P. Weckx, G. Groeseneken, Degradation of time dependent variability due to interface state generation, in *2013 Symposium on VLSI Technology (VLSIT)* (2013), pp. T190–T191
31. M. Toledano-Luque, R. Degraeve, P.J. Roussel, V. Luong, B. Tang, J.G. Lisoni, C. Tan, A. Arreghini, G. Van den bosch, G. Groeseneken, J. Van Houdt, Statistical spectroscopy of switching traps in deeply scaled vertical poly-Si channel for 3D memories, in *2013 International Electron Devices Meeting (IEDM)* (2013), pp. 21.3.1–21.3.4
32. W. Wang, S. Yang, S. Bhardwaj, S. Vrudhula, F. Liu, Y. Cao, The impact of NBTI effect on combinational circuit: modeling, simulation, and analysis. IEEE Trans. Very Large Scale Integr. VLSI Syst. **18**(2) , 173–183 (2010)

33. P. Weckx, B. Kaczer, C. Chen, J. Franco, E. Bury, K. Chanda, J. Watt, P. Roussel, F. Catthoor, G. Groeseneken, Characterization of time-dependent variability using 32 k transistor arrays in an advanced HK/MG technology, In *2015 IEEE International Reliability Physics Symposium (IRPS)* (2015), pp. 3B.1.1–3B.1.6
34. P. Weckx, M. Simicic, K. Nomoto, M. Ono, B. Parvais, B. Kaczer, P. Raghavan, D. Linten, K. Sawada, H. Ammo, S. Yamakawa, A. Spessot, D. Verkest, A. Mocuta, Defect-based compact modeling for RTN and BTI variability, in *2015 IEEE International Reliability Physics Symposium (IRPS)* (2017), pp. CR-7.1–CR-7.6
35. P. Weckx, B. Kaczer, C. Chen, P. Raghavan, D. Linten, A. Mocuta, Relaxation of time-dependent NBTI variability and separation from RTN, in *2015 IEEE International Reliability Physics Symposium (IRPS)* (2017), pp. XT-9.1–XT-9.5
36. F. Schanovsky, W. Goes, T. Grasser, *Bias Temperature Instability for Devices and Circuits* (Springer, Berlin, 2014), pp. 409–446. ch. Advanced Modeling of Oxide Defects
37. Y. Ye, C.C. Wang, Y. Cao, Simulation of random telegraph noise with 2-stage equivalent circuit, in *2010 IEEE/ACM International Conference on Computer-Aided Design (ICCAD)* (2010), pp. 709–713
38. K. Zhao, J.H. Stathis, B.P. Linder, E. Cartier, A. Kerber, PBTI under dynamic stress: from a single defect point of view, in *2011 IEEE International Reliability Physics Symposium (IRPS)* (2011), pp. 4A.3.1–4A.3.9

Oxide Trap-Induced RTS in MOSFETs

A. S. M. Shamsur Rouf and Zeynep Çelik-Butler

1 Introduction

When an inversion layer carrier gets trapped, it changes both the number and mobility of the inversion layer carriers. These changes result in a discrete switching in the channel resistance, commonly known as random telegraph signals. With the miniaturization of MOSFETs, RTS has become a prominent issue. As the device size gets smaller, the probability of observing switching in the output due to a singular defect becomes greater. Hence, RTS measurements are usually taken in devices of gate area $<1\ \mu m^2$.

RTS was first observed by Kandiah and coworkers in JFETs in 1978 [1]. Later, Ralls et al. studied fluctuations due to individual defects at cryogenic temperatures in 1984 [2]. Since then, extensive research has been done on RTS in electronic devices. RTS has been studied to characterize electron traps in nMOSFETs [3–9], as well as hole defects in pMOSFETs [10–14]. In the last few years, RTS has also been observed and studied in scaled SRAMs and DRAMs [15, 16], RRAMs [17], and flash memories [18].

Activation of a single trap residing near the oxide/semiconductor interface results in a two-level signal (Fig. 1), known as simple RTS. Complex RTS may also be observed due to multiple traps. Complex RTS may be observed when two dependent traps become active together (Fig. 2), or switching due to one fast trap gets modulated by an RTS due to a slower trap (known as envelope switching) (Fig. 3). The switching pattern between different RTS levels, and the number of levels indicate the number of traps active [19]. Unlike the RTS reported in [20, 21], and [22], these signals have different separation magnitudes between different

A. S. M. Shamsur Rouf · Z. Çelik-Butler (✉)
Electrical Engineering Department, University of Texas at Arlington, Arlington, TX, USA
e-mail: zbutler@uta.edu

T. Grasser (ed.), *Noise in Nanoscale Semiconductor Devices*,
https://doi.org/10.1007/978-3-030-37500-3_17

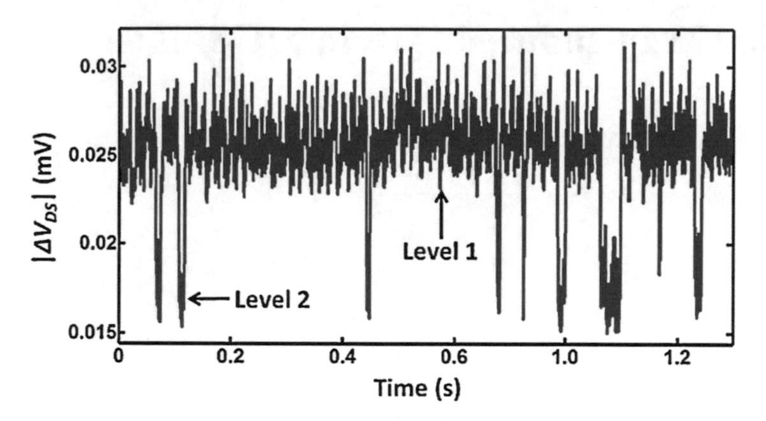

Fig. 1 A sample of two-level RTS measured at the MOSFET drain–source voltage [14]

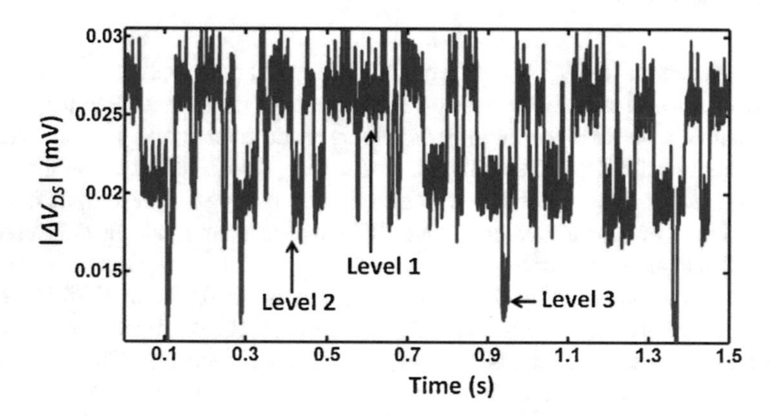

Fig. 2 A three-level RTS. All transitions take place between levels 1 ↔ 3, and 2 ↔ 3, which means that occupancy of one trap depends on that of the other [14]

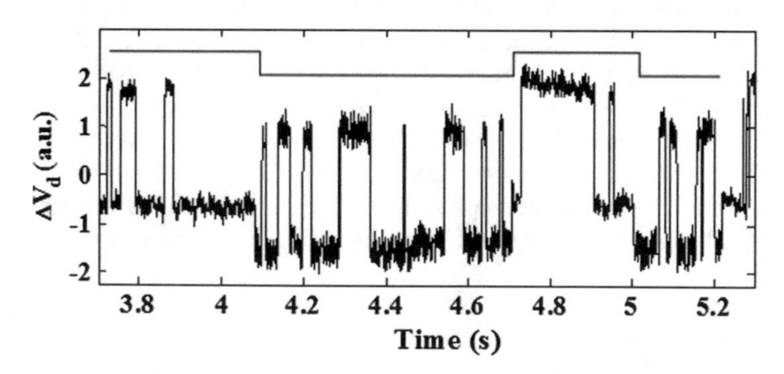

Fig. 3 A complex RTS showing envelope transition. Switching due to a fast trap is modulated by switching due to a slow trap as shown at the top [8]

RTS levels, and permanent multilevel traces. This indicates that these signals are observed because of multiple defects, not due to metastable states of a single defect. Analyses methods of such RTS traces are explained in the latter part of this chapter.

One advantage of RTS over other trap characterization techniques such as time-dependent defect spectroscopy (TDDS) or charge pumping (CP) is that it can detect the charge state of the trap before and after switching. In case of electron traps, traps that are neutral before capturing an electron and negatively charged upon capturing an electron are repulsive center traps, whereas traps that are positively charged before trapping an electron and neutral after capturing an electron are attractive center. On the other hand, for hole traps, traps with neutral charge state before hole capture and positive charge state after hole capture are repulsive centers, whereas traps with negative charge state before hole capture and neutral charge state after hole capture are attractive center traps. Although both attractive and repulsive center traps are observed in nMOSFETs and pMOSFETs, the location of these traps in any device is totally random.

Several oxide traps have been reported so far to be responsible for degradation in MOSFETs. Oxygen vacancy defects are known to be responsible for RTS in nMOSFETs [23]. E' centers can capture and emit holes [24, 25]. Hydrogen-related defects have been linked with stress-induced leakage current (SILC) [26] and negative temperature bias instability (NBTI) [27]. E'_{δ} and E'_{γ} centers are considered to be good candidates for 1/*f* noise in MOSFETs [28]. To minimize the RTS in the scaled devices to a certain extent, it is essential to study and understand the physical nature of the traps responsible for RTS in MOSFETs. Hence, RTS in MOSFETs need to be studied in detail.

2 RTS Theory

Random capture and emission of channel carriers by gate oxide defects through phonon-assisted tunneling are known to cause RTS [29]. The average capture time for a carrier is given by [29]

$$\overline{\tau}_{\mathrm{c}} = \frac{1}{n\overline{v}_{\mathrm{th}}\sigma} \tag{1}$$

where n is the uniform inversion layer electron density per unit volume, $\overline{v}_{\mathrm{th}}$ is the average channel electron thermal velocity, and σ is the trap capture cross section. To get captured, a channel carrier needs to overcome a certain energy barrier known as capture activation energy (ΔE_{B}). Trap capture cross section can be written in terms of temperature and capture activation energy as [29]

$$\sigma = \sigma_0 \mathrm{e}^{-(\Delta E_{\mathrm{B}}/k_{\mathrm{B}}T)} \tag{2}$$

where σ_0 is the capture cross-sectional pre-factor, T is the absolute temperature, and k_{B} is the Boltzmann's constant. The principle of detailed balance can be used to get the expression for average emission time ($\overline{\tau}_{\mathrm{e}}$) [29]

$$\overline{\tau}_{\mathrm{e}} = \frac{\mathrm{e}^{[(\Delta E_{\mathrm{B}}+\Delta E_{\mathrm{CT}})/k_{\mathrm{B}}T]}}{g\sigma_0 N_{\mathrm{C}}(8k_{\mathrm{B}}T/\pi m^*)^{1/2}} \tag{3}$$

where $g = 1$ is the degeneracy factor for traps, N_{C} is the effective density of states for electrons, and m^* is the electron effective mass. σ_0 can be extracted from the Arrhenius plots of $\overline{\tau}_{\mathrm{c}}$ and $\overline{\tau}_{\mathrm{e}}$. The σ_0 values obtained from these two plots are independent of each other. However, it has been observed that the two σ_0 values obtained from experimentally measured $\overline{\tau}_{\mathrm{c}}$ and $\overline{\tau}_{\mathrm{e}}$ using Eqs. (1) and (2) differ from each other by two orders of magnitude [30]. To resolve this discrepancy, an approach used by Kirton and Uren in [29] is followed. The excess energy after the trapping and detrapping process is considered as Gibbs free energy. Hence, ΔE_{CT} is split into trap binding enthalpy (ΔH) and change in entropy (ΔS). Therefore, $\Delta E_{\mathrm{CT}} = \Delta H - T\Delta S$. Consequentially, the σ_0 found from the Arrhenius plot of $\overline{\tau}_{\mathrm{e}}$ is not the capture cross-sectional pre-factor, but it is in fact σ_0 multiplied by $\Delta S/k_{\mathrm{B}}$. Therefore, Eq. (3) can be modified to

$$\overline{\tau}_{\mathrm{e}} = \frac{1}{g(8k_{\mathrm{B}}T/\pi m^*)^{1/2} N_{\mathrm{C}}\sigma_0 \mathrm{e}^{\Delta S/k_{\mathrm{B}}}} \mathrm{e}^{[(\Delta E_{\mathrm{B}}+\Delta H)/k_{\mathrm{B}}T]} \tag{4}$$

Upon getting captured, the trapped electron relaxes to the trap energy level, emitting several phonons, which corresponds to relaxation energy (E_{R}). The energy band diagram and configuration coordinate diagram indicating the capture and emission process and the trap energy parameters are shown in Fig. 4a, b.

The ratio of $\overline{\tau}_{\mathrm{c}}$ to $\overline{\tau}_{\mathrm{e}}$ can be expressed as [29, 30]

$$\ln\left(\frac{\overline{\tau}_{\mathrm{c}}}{\overline{\tau}_{\mathrm{e}}}\right) = -\frac{1}{k_{\mathrm{B}}T}\left[\left(E_{\mathrm{C_{ox}}} - E_{\mathrm{T}}\right) - \left(E_{\mathrm{C}} - E_{\mathrm{F}}\right) - \varphi_0 + q\psi_{\mathrm{S}} + \frac{qx_{\mathrm{T}}}{T_{\mathrm{ox}}}\left(V_{\mathrm{GS}} - V_{\mathrm{FB}} - \psi_{\mathrm{S}}\right)\right] \tag{5}$$

where E_{T} is the trap energy level, q is the electronic charge, $E_{\mathrm{C_{ox}}}$ the oxide conduction band edge, E_{C} the bulk silicon conduction band edge, E_{F} the Fermi energy level in silicon, T_{ox} the oxide thickness, x_{T} the oxide trap location measured from the Si–SiO_2 interface, ψ_{S} the amount of band bending at the Si–SiO_2 interface, V_{GS} the applied gate to source voltage, V_{FB} the flat band voltage, and $\varphi_0 = 3.1$ eV is the difference between the conduction band energies of SiO_2 and Si at the Si–SiO_2 interface. Equation (5) can be differentiated with respect to V_{GS} to obtain x_{T}

$$x_{\mathrm{T}} = \frac{T_{\mathrm{ox}}\left[(k_{\mathrm{B}}T/q)\left\{d\ln\left(\overline{\tau}_{\mathrm{c}}/\overline{\tau}_{\mathrm{e}}\right)/dV_{\mathrm{GS}}\right\} + (d\psi_{\mathrm{s}}/dV_{\mathrm{GS}})\right]}{(d\psi_{\mathrm{s}}/dV_{\mathrm{GS}}) - 1} \tag{6}$$

Then, $(E_{\mathrm{C_{ox}}} - E_{\mathrm{T}})$ can be calculated from Eq. (5). Using the parabolic approximation of the conduction band in the coordinate diagram in Fig. 4, a relation is established between E_{R} and ΔE_{B} [31]

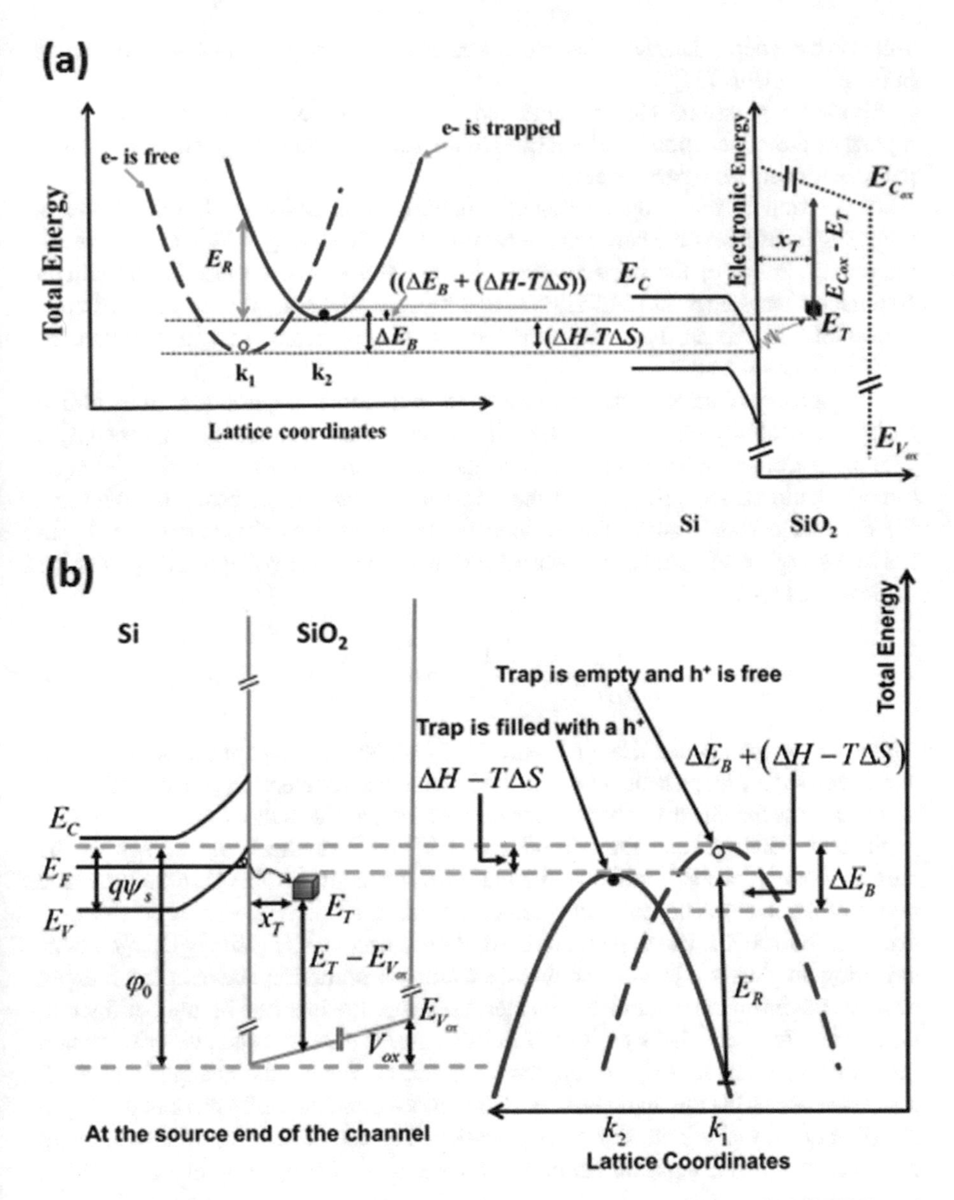

Fig. 4 Configuration coordinate diagram with all the trap parameters of (**a**) an nMOSFET [7] and (**b**) a pMOSFET [13]

$$\Delta E_{\mathrm{B}} = \frac{(E_{\mathrm{R}} + \Delta E_{\mathrm{CT}})^2}{4E_{\mathrm{R}}} \tag{7}$$

This makes $\overline{\tau}_{\mathrm{c}}$ strongly dependent on the applied bias, which was not a part of the original model proposed by Kirton [29], where ΔE_{B} was assumed to be constant. $\overline{\tau}_{\mathrm{e}}$

will also have dependence on bias. However, this dependence is not always observed in the experiments [22].

Similar expressions can be obtained to analyze hole traps in pMOSFETs replacing the conduction band energies with valence band energies, and electron parameters with hole parameters.

In addition to the phonon-assisted tunneling probability as described above, tunneling is also possible between the two states shown in Fig. 4 [32, 33]. However, the energy levels of the traps responsible for RTS are very close to the silicon conduction band edge in nMOSFETs, and close to the silicon valence band edge in pMOSFETs. Hence, the tunneling is too fast to be detected in the timescales used in RTS measurements.

Another important RTS parameter is RTS magnitude. According to the Unified Noise and Mobility Fluctuations (UNMF) model, when a carrier gets trapped by a gate oxide defect, in addition to the change in the inversion layer carrier number, remote Coulomb scattering due to the interaction between the charged defect and the channel carriers causes fluctuations in the channel carrier mobility [34]. In the linear region of operation, the fluctuation in the drain voltage, ΔV_{DS}, can be expressed as [34]

$$\frac{\Delta V_{DS}}{V_{DS}} = -\left(\frac{1}{N} \pm \alpha\mu\right)\frac{1}{WL_{\mathrm{eff}}} \tag{8}$$

where V_{DS} is the applied drain to source voltage, W and L_{eff} are the device width and effective length, respectively, N is the inversion layer electron concentration per unit area, μ is the effective channel carrier mobility, and α is the screened scattering coefficient. The first and the second terms in the parenthesis are known as the number fluctuations and mobility fluctuations, respectively. The sign between the two fluctuation terms depends on whether the trap is attractive (−) or repulsive (+). For an nMOSFET, if the trap is repulsive, then it becomes negatively charged after capturing an electron. Hence, the remote Coulomb scattering between the charged trap and the inversion layer electrons increases with the increase in number fluctuations. Therefore, positive sign is used in this case. If the trap is an attractive center, then it becomes neutral upon trapping an electron. Hence, the change in channel resistance due to the fluctuations in electron numbers and mobility work in the opposite directions, making the sign to be negative. The use of positive or negative sign can be established using a similar concept in the case of a hole trap in a pMOSFET.

3 Charge Quantization Effects on RTS

Another matter of concern is the charge quantization effects in the inversion layer of MOSFETs. With the scaling of device dimensions and the oxide thickness and higher substrate doping concentration, the strong transverse electric field in the inversion layer causes quantization in the carrier energy levels [35, 36].

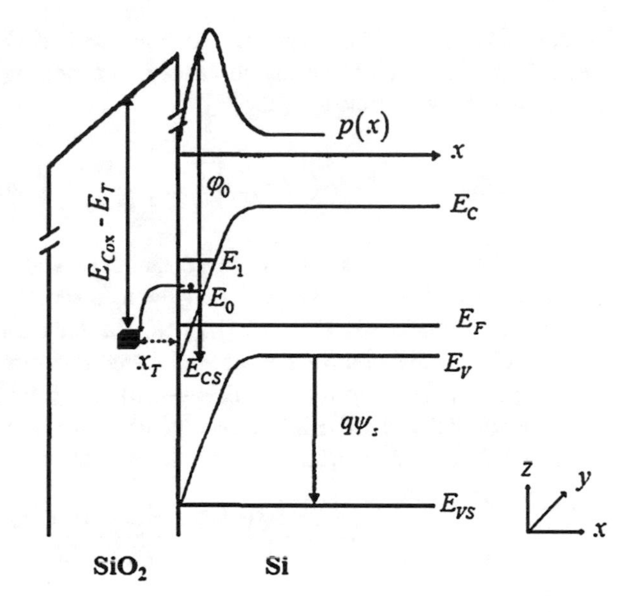

Fig. 5 Band diagram for an nMOSFET showing the probability function $p(x)$ into the substrate [9] (Used with permission from [9] for reuse and edit)

The quantization effects may lead to an increase in the threshold voltage and gate oxide capacitance. In addition, these effects will change the capture and emission energies as well as the probability function into the substrate, $p(x)$ in the inversion layer (Fig. 5). In a 3-D treatment of RTS, the average emission time is given by Shockley–Read–Hall statistics

$$\overline{\tau}_{\mathrm{e}} = \frac{1}{\overline{e}} = \frac{\overline{\tau}_{\mathrm{c}}}{g} \exp\left[-\frac{E_{\mathrm{T}} - E_{\mathrm{F}}}{k_{\mathrm{B}}T}\right] = \frac{\exp\left[\left(E_{\mathrm{F}} - E_{\mathrm{T}}\right)/k_{\mathrm{B}}T\right]}{\sigma \overline{v}_{\mathrm{th}} n} \tag{9}$$

where $\overline{e}$ is the emission coefficient. If an electric field is strong enough to create a triangular potential well at the Si–SiO_2 interface (Fig. 5), and its width into the substrate is smaller than the carrier wavelength, then according to the Schrodinger wave equation, the conduction band energies are split into discrete energy levels, known as electric sub-bands. Taking the quantization effects into consideration, the mean capture and emission times can be expressed as [9]

$$\overline{\tau}_{\mathrm{c}} = \frac{1}{\overline{c} n_{2\mathrm{D}} \int_{0}^{\overline{x}} \frac{p(x)}{\overline{x}} dx} = \frac{1}{\sigma\,(2\mathrm{D})\, \overline{v}_{\mathrm{th}} n_{2\mathrm{D}} \int_{0}^{\overline{x}} \frac{p(x)}{\overline{x}} dx} \tag{10}$$

$$\overline{\tau}_{\mathrm{e}} = \frac{1}{\overline{e}_{\mathrm{n}}} = \frac{\exp\left[\left(E_{\mathrm{F}} - E_{\mathrm{T}}\right)/k_{\mathrm{B}}T\right]}{\sigma\,(2\mathrm{D})\, \overline{v}_{\mathrm{th}} n_{2\mathrm{D}} \int_{0}^{\overline{x}} \frac{p(x)}{\overline{x}} dx} \tag{11}$$

where σ(2D) is the 2-D capture cross section, n_{2D} is the 2-D electron concentration, and $\overline{x}$ is the inversion layer charge centroid location. $p(x)$ can be found using the Stern–Howard wave function [35, 37]

$$p(x) = \frac{b^3}{2}x^2 \exp\left(-bx\right), \quad b = \left[\frac{12qm_1}{\hbar^2 \varepsilon_{Si}\varepsilon_0}\left(Q_B + \frac{11}{32}Q_{inv}\right)\right]^{1/3} \tag{12}$$

Here, $\hbar$ is the reduced Planck's constant, ε_0 and ε_{Si} are the permittivity of free space and dielectric constant of silicon, respectively, m_l is the longitudinal electron effective mass, Q_B and Q_{inv} are the bulk and inversion layer charge, respectively. If most of the electrons are located at the ground energy level E_0, then [35] $n_{2D} = N_C(2D) \exp\left[(E_F - (E_{CS} + E_0))/k_BT\right]$, where E_{CS} is the conduction band energy at the Si–SiO$_2$ interface, $N_C(2D)$ is the 2-D effective density of states. $N_C(2D) = 2k_BTm_t/\hbar^2\pi$, where m_t is the transverse electron effective mass. Hence,

$$\overline{\tau}_e = \frac{\exp\left[\left(E_{CS} - E_T + E_0\right)/k_BT\right]}{\sigma\left(2D\right)\overline{v}_{th}N'_C\left(2D\right)} \tag{13}$$

where

$$N'_C\left(2D\right) = N_C\left(2D\right)\int_0^{\overline{x}} \frac{p(x)}{\overline{x}}dx \approx 2k_BTm_tb/5\hbar^2\pi \tag{14}$$

Therefore,

$$\overline{\tau}_e = \frac{\exp\left[\left(E_{CS} - E_T + E_0\right)/k_BT\right]}{\sigma\left(2D\right)\overline{v}_{th}\left(2k_BTm_tb/5\hbar^2\pi\right)} \tag{15}$$

To get captured, the electrons will need the extra energy $\Delta E_0 = E_0 - E_{CS}$. ΔE_0 can be found using the triangular potential well assumption [37]

$$\Delta E_0 = \left(\frac{\hbar^2}{2m_l}\right)^{1/3}\left[\frac{9\pi q}{8}E_S\right]^{2/3} \tag{16}$$

where E_S is the surface electric field.

In addition to the capture and emission times, the quantization effect also modifies the RTS magnitude. In Eq. (8), it is assumed that $\mu^{-1} = \mu_n^{-1} + \mu_t^{-1} = \mu_n^{-1} + \alpha N_t$, where μ_t and μ_n are the mobility limited by oxide charge scattering and other mechanisms, respectively, and N_t is the number of occupied traps per unit area. n_{2D} and $p(x)$ can be varied by changing source to substrate voltage (V_{SB}). N can be obtained by integrating the inversion layer carrier concentration n_{2D} and charge profile $p(x)$.

$$\frac{1}{N}=\frac{1}{\int n_{2D}p(x)dx}=\left\{\frac{2k_BTm_t}{\pi\hbar^2}\exp\left[-\left(E_{CS}+\Delta E_0-E_F\right)/k_BT\right]\times\int_0^{\overline{x}}p(x)dx\right\}^{-1} \tag{17}$$

For $\overline{x}=3/b$, $\int_0^{\overline{x}}p(x)dx\approx 0.58$.

$$\frac{1}{N}=\left\{\frac{2k_BTm_t}{\pi\hbar^2}\exp\left[-\left(E_{CS}+\Delta E_0-E_F\right)/k_BT\right]\times 0.58\right\}^{-1} \tag{18}$$

Using Surya and Hsiang's 2-D surface mobility fluctuation model [38], μ_t can be expressed as

$$\mu_t^{-1}=\frac{m^*q^3}{8\hbar\pi\varepsilon_{av}^2E_p}\int dx\int dE\times\int_0^{\pi/2}\frac{\exp\left(-4kx\sin\phi\right)\sin^2\phi}{\left(\sin\phi+\frac{c}{2k}\right)^2}d\phi N_t\left(E,x\right) \tag{19}$$

where $\varepsilon_{av}=\left(\varepsilon_{Si}+\varepsilon_{SiO_2}\right)/2$, ε_{SiO_2} is the dielectric constant of SiO_2, ϕ is the half of the angle between the 2-D plane wave states k and k', E_p is the energy where $Ef(E)g(E)$ peaks, where E is the electron energy, $f(E)$ is the Fermi–Dirac distribution function, $g(E)$ is the density of states, c is the screening constant written as [38]

$$c=\frac{2q^2d_vm^*}{4\hbar^2\pi\varepsilon_{Si}}\left\{1-\exp\left[-\left(\frac{\hbar^2\pi N}{k_BTd_vm^*}\right)\right]\right\} \tag{20}$$

where d_v is the degeneracy of electrons. If most of the electrons are assumed to be occupying the lowest energy of conduction band, then $k=0.8(2\pi/a_{Si})$, where a_{Si} is the silicon lattice constant. Equation (20) can be simplified if a single trap is considered. Hence, $N_t(E,x)=N_t\delta(E-E_T)\times\delta(x-x_T)$, where energy at x_T is equal to E_T and x_T the trap location from the Si–SiO_2 interface. If the trap is considered to be at the Si–SiO_2 interface, i.e., $x_T=0$, and there is no screening ($b=0$), then,

$$\mu_t^{-1}=\frac{m^*q^3}{8\hbar\pi\varepsilon_{av}^2E_p}\frac{\pi}{2}N_t \tag{21}$$

Hence,

$$\alpha=\frac{m^*q^3}{16\hbar\pi\varepsilon_{av}^2E_p}\pi \tag{22}$$

Considering screening,

$$\mu_{\mathrm{t}}^{-1} = \frac{m^* q^3}{8\hbar\pi\varepsilon_{\mathrm{av}}^2 E_{\mathrm{p}}} \int_0^{\pi/2} \frac{\exp(-4kx\sin\phi)\sin^2\phi}{\left(\sin\phi + \frac{c}{2k}\right)^2} d\phi N_{\mathrm{t}} \tag{23}$$

Using the fitting parameter x_{T}, α can be calculated from the measured RTS magnitude using the mobility expression. Using the 2-D model, RTS measurements have been done to extract x_{T} and α [9]. This model has also been proved to be valid for capture and emission times in deep micron MOSFETs [9].

4 Experimental Setup and Measurement Techniques

RTS measurements were taken on differently sized nMOSFETs and pMOSFETs to investigate the electron and hole trap properties. To verify the device functionality, a semiconductor parameter analyzer (SPA) was used. The plots of drain current (I_{D}) with respect to gate voltage (V_{G}) and drain voltage (V_{D}) were observed for this purpose. The threshold voltage, V_{T} was extracted from the x-axis intercept of the tangent drawn on $\sqrt{I_{\mathrm{D}}}$ at the point where $d\sqrt{I_{\mathrm{D}}}/dV_{\mathrm{G}}$ is minimum.

After verifying the device functionality and extracting the DC characteristics, RTS measurements were taken inside a shielded room so that the effects of the external noise sources were minimized (Fig. 6). The gate and drain of the devices were biased using a custom battery-operated biasing circuitry. All DC-operating equipment such as the biasing circuitry and the preamplifier were kept inside the shielded room, and all AC-operating equipment such as oscilloscope, SPA, and impedance analyzer were kept outside the shielded room. The output was connected to an oscilloscope through a low-noise preamplifier. During the RTS measurements, to eliminate the effect of the 60 Hz frequency components coming from the power line, the substrate, source, and back-gate terminals of the device were shorted to the battery virtual ground and were connected to the earth ground.

If RTS was observed on the oscilloscope, time domain traces were recorded at that temperature at 6–10 points by varying the gate voltage while keeping the drain voltage constant. For each bias point, depending on the sampling frequency, the total duration of the trace was varied to make sure that at least 500 transitions were recorded. After finishing the measurements at room temperature, the temperature was dropped. To lower the temperature, a passive continuous flow, open cycle cryogenic system was used. An auto-tuning temperature controller was used to observe the change in temperature and stabilize it at the desired value. With the lowering of temperature, thermally activated process slow down. Hence, the capture and emission times are increased resulting in, fewer transitions. If the largest oscilloscope time span is not sufficient to record 500 transitions at once, multiple RTS traces are recorded and stitched together to obtain enough switching events needed for RTS analyses.

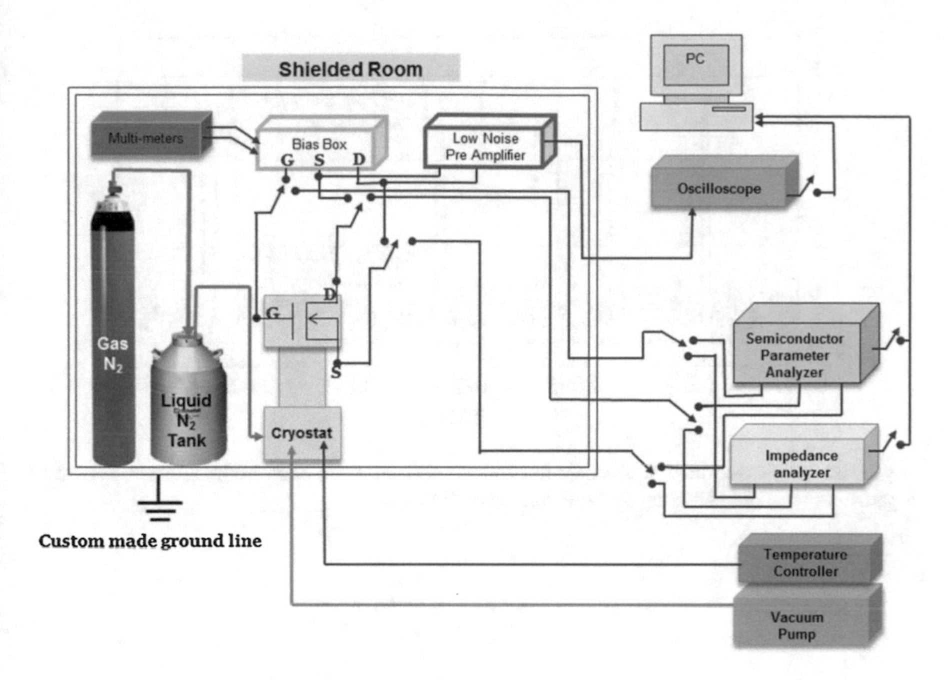

Fig. 6 Variable temperature RTS measurement setup

A precision impedance analyzer was used to obtain the *C–V* characteristics. Split *C–V* measurements were performed and at each temperature RTS traces were recorded. Using the procedure mentioned in [13], the *C–V* characteristics were used to calculate the equivalent oxide thickness (T_{ox}) using $C_{ox} = \varepsilon_{SiO_2}/T_{ox}$ and the inversion layer carrier concentration per unit area (*N*) using $N = Q_{inv}/q$. Here, C_{ox} is the measured oxide capacitance per unit area.

For observing the effects of hot carrier stress on RTS, the devices were stressed using the SPA. All stressing was done at room temperature so that the carriers get enough energy to cause degradation in the oxide. The RTS measurements were taken immediately after stressing the devices so that the devices do not relax back to the prestress condition.

5 RTS Analyses

A sample two-level RTS signal is shown in Fig. 7. The limits of each of the RTS levels were defined so that each signal value is assigned to a level (Fig. 7). A MATLAB code was developed to extract the average time spent at each level. The probability of switching from one RTS level to another follows Poisson's statistics. Hence, the average time spent at each RTS level shows an exponential distribution [13] (Fig. 8). The average time spent at each RTS level is calculated using

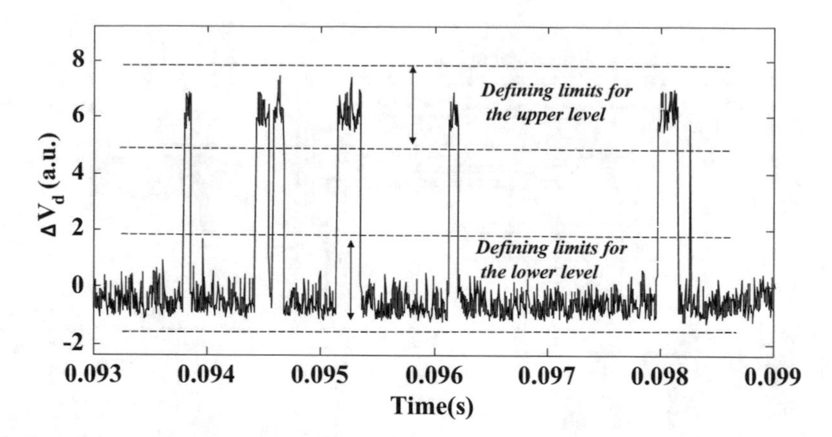

Fig. 7 Sample two-level RTS observed due to a single trap. Each point in the RTS is assigned to a particular level defined using an upper and lower limit [8]

$$\overline{\tau} = \sum_{m=1}^{M} t_m \left| F_m \right| / \sum_{m=1}^{M} \left| F_m \right| \tag{24}$$

where t_m is the time span for each bin m, M is the number of bins in total, and F_m is the number of switching occurrences that have taken place within the time span t_m. From the increase or decrease of the average time of each RTS level, it can be determined if the trap is attractive or repulsive. In a MOSFET, $\overline{\tau}_c$ indicates the average time for which the trap is empty, and $\overline{\tau}_e$ indicates the average time for which the trap is full. For nMOSFETs where the channel carrier is electrons, the trap occupancy function can be written as $f_t = \overline{\tau}_e / \left(\overline{\tau}_c + \overline{\tau}_e\right)$. Therefore, the ratio of mean capture to emission time can be written as $\overline{\tau}_c / \overline{\tau}_e = \left(1 - f_t\right) / f_t$, where f_t is the probability that the trap is filled with an electron [29]. With the increase of V_{GS}, the probability that the trap will be filled with an electron increases. Hence, the $\overline{\tau}_c / \overline{\tau}_e$ value decreases with the increase of V_{GS}.

For repulsive electron-trapping center defects, when an electron gets captured, it increases the channel resistance. Hence, the upper level of ΔV_{DS} corresponds to the state where the trap is filled with an electron, whereas for an attractive center trap, capture of an electron decreases the channel resistance. Therefore, the upper level of ΔV_{DS} corresponds to the state when the trap is empty. Therefore, for an RTS in an nMOSFET, if the mean time spent at the lower level ($= \overline{\tau}_c$) decreases with the increase of V_{GS}, then the trap is a repulsive center. If the mean time spent at the upper level ($= \overline{\tau}_c$) decreases with the increase in V_{GS}, then the trap is an attractive center.

On the other hand, in pMOSFETs, where the channel carriers are holes, the ratio of average capture to emission time can be expressed as $\overline{\tau}_c / \overline{\tau}_e = f_t / \left(1 - f_t\right) = \left(1 - f_{th}\right) / f_{th}$, where f_{th} is the probability that the trap will be empty of an electron, i.e., filled with a hole. With the increase of $|V_{GS}|$, the probability that the trap will

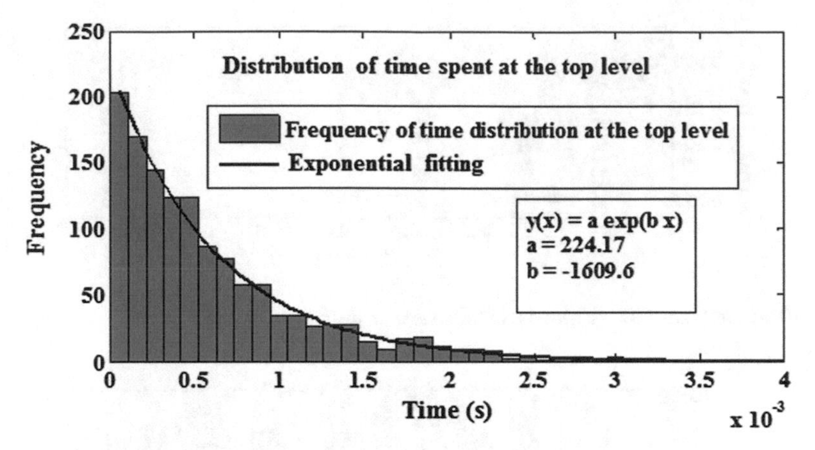

Fig. 8 Exponential distribution of time spent at any RTS level [8]

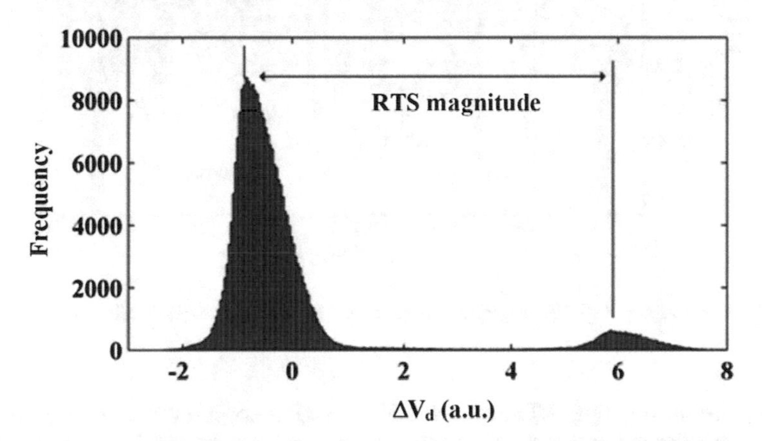

Fig. 9 RTS amplitude extraction for a two-level RTS. RTS amplitude is calculated form the difference between the two peaks from the corresponding Gaussians. Similar way is used to extract the RTS magnitude of signals with more than two levels [8]

be filled with a hole decreases, resulting in a decrease in $\overline{\tau}_c/\overline{\tau}_e$. For repulsive hole-trapping traps, capture of a hole increases the channel resistance, whereas trapping a hole decreases the channel resistance. Hence, for the repulsive centers, the upper level of $|\Delta V_{DS}|$ corresponds to the state when the trap is filled with a hole, whereas for the attractive center traps, the upper level of $|\Delta V_{DS}|$ corresponds to the state when the trap is empty. Therefore, for hole traps, if the mean time spent at the lower level of $|\Delta V_{DS}|$ $(= \overline{\tau}_c)$ decreases with the increase of $|V_{GS}|$, the trap is a repulsive center. If the average time spent at the upper level of $|\Delta V_{DS}|$ $(= \overline{\tau}_c)$ shows a decrease with the increase of $|V_{GS}|$, the trap is an attractive center.

The RTS magnitude is calculated from the difference between the peaks of the corresponding Gaussian distribution in the RTS amplitude histogram (Fig. 9).

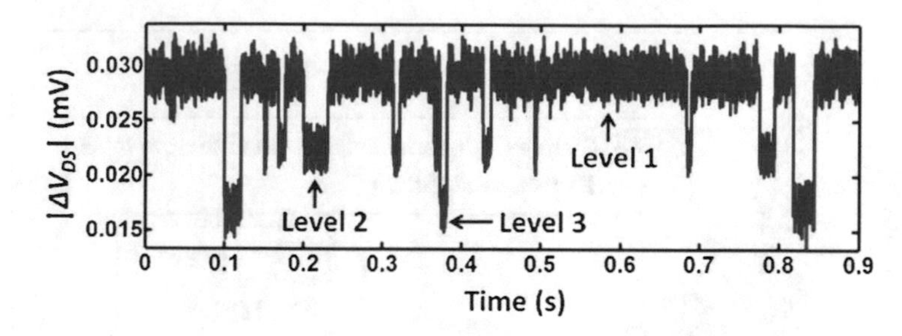

Fig. 10 A three-level RTS where the transitions take place only between levels 1 ↔ 2, and 1 ↔ 3 [14]

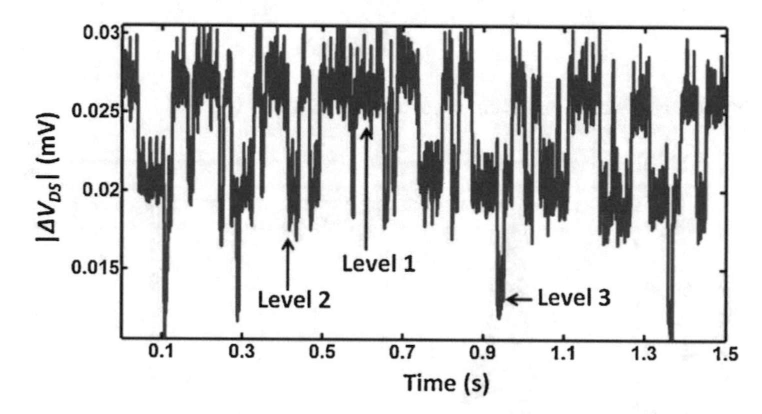

Fig. 11 A three-level RTS where the transitions take place only between levels 1 ↔ 2, and 2 ↔ 3 [14]

In addition to simple RTS, complex RTS is also observed when more than one trap becomes active at the same time. Two different three-level RTS traces are shown in Figs. 10 and 11. In Fig. 10, all switching events take place from level 1. On the other hand, in Fig. 11, all the transitions take place from level 2. Since both traces depict attractive trapping centers, in the first case, traps are independent, whereas, in the latter case, occupancy of one trap depends on that of the second. Capture and emission time extraction of these types of signals have been described at the later parts of this chapter.

Another type of complex four-level RTS is shown in Fig. 12, where two traps are active together. The switching due to a fast trap is modulated by the switching of another slow trap (envelope switching). Complex signals of this pattern are analyzed by separating the RTS due to the individual traps. To separate the fast RTS, levels 1 and 3, and 2 and 4 are merged together (Fig. 13) [8]. The capture and emission times of the merged signal are extracted using the same concept as a simple two-level signal. The capture and emission times for the slow trap are extracted by calculating the time spent at the lower envelope and the upper envelope (Fig. 14) [8]. The RTS magnitude due to each trap in the complex RTS signal is calculated from the difference of the corresponding Gaussian peaks in the amplitude histogram.

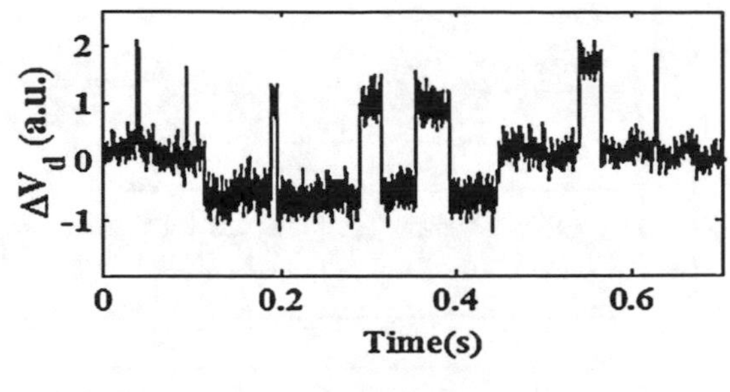

Fig. 12 Example of envelope transition in RTS. Switching due to a fast trap is shifted by a slow trap [8]

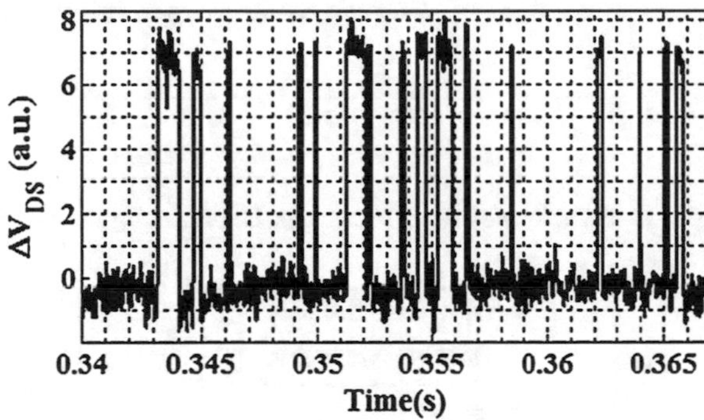

Fig. 13 RTS after merging together the switching levels due to the fast trap from Fig. 12 [8]

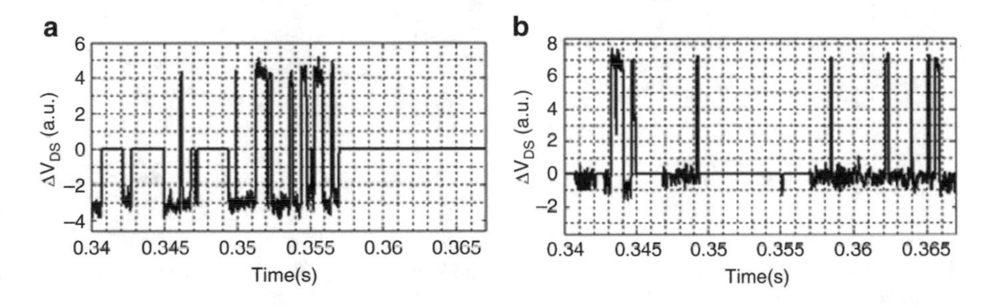

Fig. 14 (**a**) Separation of the lower transition and (**b**) upper transitions due to the slow trap [8]

6 RTS in nMOSFETs

Several trap species have been reported for oxide degradation in MOSFETs. Oxygen vacancy defects are known to cause RTS in MOSFETs [23]. However, due to variation in local bonding, different oxygen vacancies have different structures exhibiting different relaxation energies [39]. The trap energy parameters such as ΔE_B, E_R, ΔH, and ΔS extracted through RTS measurements can be used to find the structural configuration of the defect responsible.

To study the gate oxide defects, RTS measurements were carried out in the linear region on several nMOSFETs with area 0.36–0.55 μm^2. All nMOSFETs reported here showed simple two-level RTS at all temperatures. A sample RTS is shown in Fig. 15. From the dependence of mean time spent at the lower level and upper level

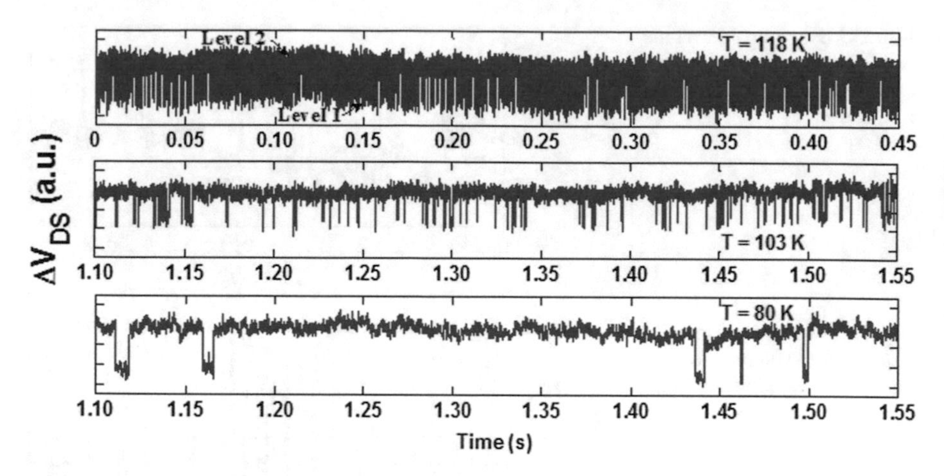

Fig. 15 Sample two-level RTS in an nMOSFET at $V_{GS} = 1.10$ V, $V_{DS} = 0.30$ V [7]

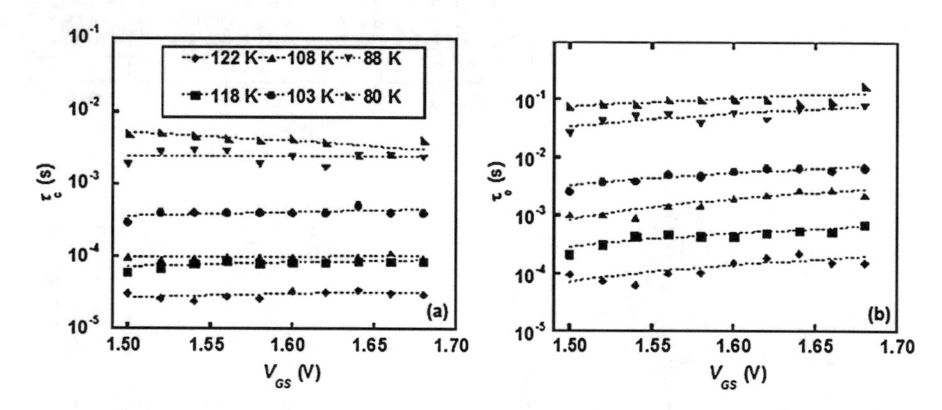

Fig. 16 Plot of (**a**) capture times and (**b**) emission times as a function of V_{GS} at different temperatures. $V_{DS} = 0.25$ V [42]

on V_{GS}, all traps were found to be repulsive center traps. The average capture and emission times are shown as a function of V_{GS} in Fig. 16.

The calculated value of $x_T \approx 1.5$ nm. This is higher than the Si–O bond length of 0.15–0.17 nm [40], which rules out the Pb centers (trivalent silicon defects at the Si–SiO_2 interface) from the possible defect candidates.

The capture cross section of the defects was calculated using Eq. (1), and they were found to be much lower than the capture cross sections found using other techniques [23, 39, 41]. The σ_0 and ΔE_B values were calculated from the Arrhenius plot of σ (Fig. 17), and the ΔH and ΔS values were extracted from the Arrhenius plot of normalized $\overline{\tau}_e$ (Fig. 17) as discussed before. The $\Delta S/k_B$ values were 6.00–9.17. The $(\Delta H - T\Delta S)$ values varied from −41 to 25 meV at room temperature,

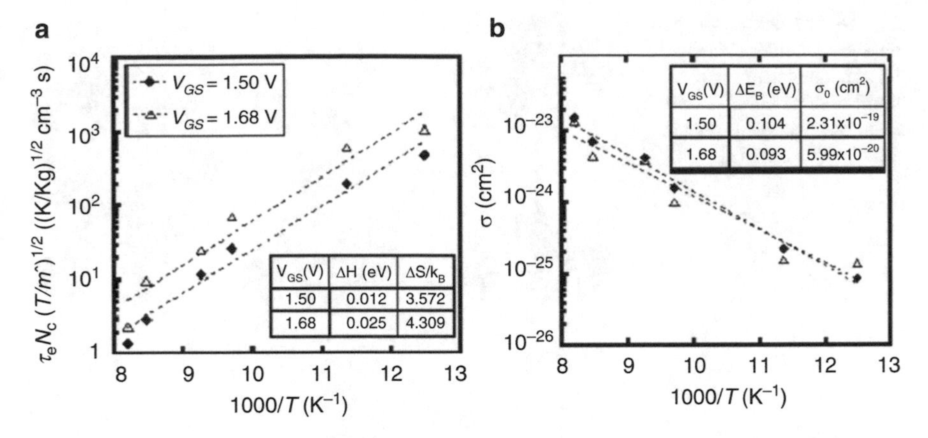

Fig. 17 Arrhenius plot of (**a**) normalized emission time and (**b**) capture cross section [42]

Table 1 Summary of trap parameters of all the studied defects [42]

Trap species	E_R (eV)	ΔE_B (eV)	Transition pattern	$E_{C_{ox}} - E_T$ (eV)
V^0 ODC II	0.8–2.1 [39] Capturing an e−	0.1–0.5 [23]		0/−3.3 [43]
Nitrogen defect		0.9 [44]		2.5–2.87 [45–48]
Hydrogen bridge with Si atoms	1.9–3.1 [49] Capturing an e−	1.5 [50]	+/0	3.9–4.1
			0/+	8.1–8.7
			0/−	4.6–5.8
			−/0	7.8–8.1 [51]
E'_γ	0.3–0.4 [52, 53] Capturing a h+	0.1–0.4 [52, 53]	+/0	4.9–5.1 [51]
			0/+	6.6–6.9 [51]
			0/−	–
			−/0	–
E'_δ	0.6–1.6 [54] Capturing a h+	1.2 [55]	+/0	5.4–7.3 [51]
			0/+	8.5–8.7 [51]
Fluorine		0.5–1.0 [44]	0/+	1.0 [44]

and the $\left(E_{C_{ox}} - E_T\right)$ values were found to be around 3 eV. This indicated that the defect energy level is very close to the silicon conduction band edge.

The ΔE_B values were found to be 0.1–0.35 eV, implying a low potential barrier for capturing electrons. The E_R values were calculated solving Eq. (7), and found to be 0.60–1.63 eV. Summary of all the studied defects and their parameters are shown in Table 1. Comparing with the already reported relaxation energies of different structural defects in SiO_2, the neutral oxygen deficiency center (V^0 ODC II) was found to be the most suitable defect (Fig. 18) [23, 39, 40].

An RTS simulation tool (RTSSIM) was developed in the MATLAB platform that is based on the first principles data to accurately reconstruct the RTS data

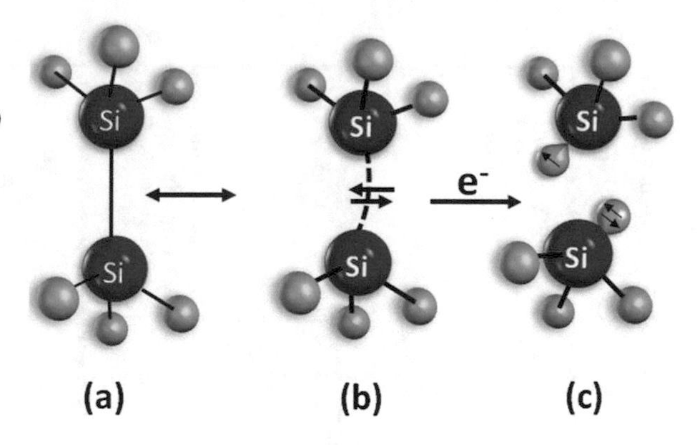

Fig. 18 (**a**) Neutral oxygen deficiency center (V^0 ODC II) at relaxed state, (**b**) unrelaxed V^0 ODC II, and (**c**) negatively charged oxygen deficiency center [7]

[56]. Although other RTS simulation tools have been reported [15, 57–65], they all seem to possess certain limitations. Even though the number of traps becoming active at the same time, as well as trap locations in the oxide and along the channel significantly affect the resultant RTS, most of the RTS tools assume the traps to be residing at the Si–SiO_2 interface [58, 61, 63, 64], or uniformly distributed in the oxide [57, 59]. As discussed before, the RTS amplitude consists of fluctuations in both the channel carrier number and the correlated mobility. Apart from a few reports [58, 59, 62], mostly only number fluctuations have been taken into consideration in these studies. In addition, either experimentally found trap energies or U-shaped energy distributions along the silicon bandgap has been considered in these tools. To the best of our knowledge, no RTS tool has used the physical trap parameters in simulation. Lastly, the RTS magnitude depends largely on whether the trap is repulsive or attractive. None of the RTS simulation tools takes the charge state of the trap into consideration.

In RTSSIM, an experimentally verified exponentially decreasing trap distribution from the Si–SiO_2 interface has been assumed [66]. Both number and mobility fluctuation terms are considered in the simulation. To obtain accurate RTS traces, variable temperature RTS measurements were done to extract the trap characteristics. The experimentally obtained trap parameters as well as other parameters found in the literature were provided as input for the simulation. Both attractive and repulsive nature of traps is considered in RTSSIM. The number of active traps in a device is calculated using the trap density and device area. It is also important to note that there was no limitation on the number of active traps. RTSSIM is capable to make the transition from RTS to flicker noise.

At the beginning, RTSSIM asks the user to provide trap and device parameters as input as shown in Table 2. The user is asked to provide the V_{GS}, V_{DS}, and temperature ranges for simulation. Based on the user-provided bias and temperature ranges, the equilibrium electron (n_0) and hole (p_0) concentration are calculated. For low temperatures, carrier freeze-out is taken into account. Nonuniform doping concentration is assumed along the channel, which allows halo doping for the channel implants.

Table 2 User-provided and RTSSIM calculated parameters [56]

User input		Calculated by RTSSIM	
Device	Trap	Device	Trap
Channel L and W (μm)	ΔE_B (eV)	p_0 (cm^{-3})	$N_{tA_{Active}}$
			$N_{tD_{Active}}$
			$N_{tAA_{Active}}$
Halo-implanted region length, L_{Halo} (μm)	x_T or $\overline{x}_T$ (nm)	n_0 (cm^{-3})	ΔH (eV)
Doping N_A or N_D (cm^{-3})	$E_{C_{ox}} - E_T$ (eV)	n (cm^{-3})	$\Delta S/k_B$
Halo doping N_{Ahalo} or N_{Dhalo} (cm^{-3})	α_0 (V s) and α_1 (V s)	ψ_S (eV)	$E_T - E_F$ (eV)
V_{GS} range (V)		C_{ox} (F/cm^2)	ΔE_{CT} (eV)
V_{DS} range (V)		N (cm^{-2})	σ_0 (cm^2)
T range (77–350) (K)		μ (cm^2/V s)	E_R (eV)
Threshold voltage V_T (V)		T_{ox} (nm)	$\overline{\tau}_c$&$\overline{\tau}_e$ (s)
g_D vs. V_{GS} (A/V)		L_{eff} (μm)	ΔV_{DS} (V)
C–V curve or T_{ox} (nm)			ΔE_B (eV)[a]
Trap density, N_{tA}, N_{tAA}, and/or N_{tD} (cm^{-2})			x_T (nm)[a]
			$E_{C_{ox}} - E_T$ (eV)[a]
			y_T (μm)

[a]For input parameter verification only

Three options are given to the user to provide input gate-oxide capacitance: (1) C–V characteristics to calculate the gate-channel capacitance (C_{GC}), (2) specific values of C_{GC} at accumulation, $V_{GS} = 0$, and inversion; and (3) oxide thickness T_{ox}. Inversion layer carrier concentration (N) can be calculated using either of the first two cases [67]. For the third case. N can be extracted through $N = C_{ox}(V_{GS} - V_T)/q$. ψ_S is calculated solving Poisson's equation [68]

$$N = \frac{\sqrt{2}\varepsilon_{Si}k_BT}{q^2L_D} \times \left(\sqrt{\left(u_S + e^{-u_S} - 1\right) + \left(\frac{n_0}{p_0}\right)\left(e^{u_S} - u_S - 1\right)} - \sqrt{u_S}\right) \quad (25)$$

where $u_S = q\psi_S/k_BT$ and $L_D = (\varepsilon_{Si}k_BT/q^2p_0)^{1/2}$ is the Debye length.

After calculating the basic device and trap parameters, the simulation begins. The primary parameters of an RTS trace are $\overline{\tau}_c$, $\overline{\tau}_e$, and ΔV_{DS}. ΔV_{DS} is calculated using the user-provided input parameters α_0 and α_1. $\overline{\tau}_c$ is calculated for each trap using Eq. (1). n is calculated through $n = n_0e^{q\psi_S/k_BT}$. σ is found through σ_0 and ΔE_B using [69]

$$\sigma_0 = \sqrt{\frac{E_1}{E_R}}\left(\frac{\pi^2}{2}\right)\left(\frac{\hbar^2}{2m^*k_BT}\right) \quad (26)$$

where E_1 is considered equal to 0.06 eV [70].

The user also needs to provide the trap location, x_T, or the average trap location, $\overline{x}_T$. If $\overline{x}_T$ is provided as input, then x_T is selected using an exponentially decreasing probability from the Si–SiO_2 interface [66, 71]. The probability is calculated using the random generator function expand ($\overline{x}_T$) function. However, x_T is limited to T_{ox}. The trap location along the channel, y_T, is selected randomly. $\ln(\overline{\tau}_c/\overline{\tau}_e)$ is calculated using Eq. (5) from the provided $\left(E_{C_{ox}} - E_T\right)$ and x_T. $(\Delta H - T\Delta S)$ is computed using the modified form of Eq. (4) [56]

$$(\Delta H - T\Delta S) = (k_B T)\left[\ln\left(\frac{g(8k_B T/\pi m^*)^{1/2} N_C}{n\overline{v}_{th}}\frac{\overline{\tau}_e}{\overline{\tau}_c}\right)\right] \tag{27}$$

ΔE_{CT} and E_R are found from the calculated $(\Delta H - T\Delta S)$ and provided ΔE_B using Eq. (7). Then $\overline{\tau}_e$ is extracted from Eq. (4). μ is found from the channel conductance g_D and N using $\mu = g_D L/WqN$ [72]. α in Eq. (8) can be modeled as [34] $\alpha = \alpha_0 + \alpha_1 \ln(NWL_{eff})$, where α_0 and α_1 are fitting parameters. L_{eff} is found using the gate to channel capacitance in the accumulation and inversion regions using $L_{eff} = L\left(1 - C_{GC_{acc}}/C_{GC_{inv}}\right)$ [67].

Most of the articles published so far have reported RTS due to single repulsive center trap [7, 8, 30], attractive center trap, or two repulsive center traps whose occupancy depends on each other. Based on the already published literature, RTSSIM was developed to construct RTS due to three types of defects: (a) repulsive center or acceptor (A) [7, 8, 30]; (b) two dependent repulsive centers or acceptors (AA) [8, 73–77]; and (c) attractive center or donor (D) [2, 4, 8]. The user has the option to choose any single option, or any one of the nine permutations with two options at a time. If the user provides Null for any one set in the options, RTSSIM will proceed with the other set of trap type options. If Null is given as input both the sets of trap type options, then the simulation terminates.

RTSSIM takes care of the transition from a single trap to multiple traps. The number of active traps is calculated using the provided areal trap density. The number of active traps in a device has been reported to follow Poisson's distribution [19, 78, 79], and the trap density is known to follow a U-shaped distribution along the silicon bandgap [66, 80]. The total number of active traps ($N_{tA_{active}}$, $N_{tAA_{active}}$, and $N_{tD_{active}}$) is calculated using these distributions.

The RTS variability is taken into account in RTSSIM using the variation in trap number, energy, and location that changes with each simulation run for the same user provided parameters. To statistically analyze the average noise and the variability, the same simulation needs to be run multiple times, or add a loop in the code in the program to run the program several times.

After calculating all the trap and RTS parameters at each bias point and temperature, the RTS trace is reconstructed. All traps are assumed to be empty at the beginning. For a two-level switching between levels S1 and S2 due to a repulsive center trap, the probability that the trap will capture an electron from the channel is given by

$$P_{12}\,(1 \rightarrow 2) = dt/\overline{\tau}_1 = dt/\overline{\tau}_{\rm c} \tag{28}$$

The probability that the electron is emitted back to the channel can be written as [29]

$$P_{21}\,(2 \rightarrow 1) = dt/\overline{\tau}_2 = dt/\overline{\tau}_{\rm e} \tag{29}$$

Therefore, as a function of time, the probability that the trap will remain occupied is

$$P_{12}(t) = \overline{\tau}_2/\left(\overline{\tau}_1 + \overline{\tau}_2\right)\left[\exp - \left(1/\overline{\tau}_1 + 1/\overline{\tau}_2\right)t - 1\right] \tag{30}$$

The probability that the electron will remain trapped is [29, 80]

$$P_{22}(t) = \overline{\tau}_2/\left(\overline{\tau}_1 + \overline{\tau}_2\right)\exp - \left(1/\overline{\tau}_1 + 1/\overline{\tau}_2\right)t + \overline{\tau}_1/\left(\overline{\tau}_1 + \overline{\tau}_2\right) \tag{31}$$

The time domain RTS trace is constructed for each active trap using Poisson's statistics. A new random number is generated replacing the old one, and new sequence of sample is created for each time interval. The new random number is generated using the MATLAB ceiling function ceil($-$log(rand)$\overline{\tau}_{1/2}F_{\rm S}$), where rand is a MATLAB function used for generating a uniformly distributed random number, and $F_{\rm S}$ is the sampling frequency, given by $F_{\rm S} = 3\,(1/\overline{\tau}_1 + \overline{\tau}_2)$ [81]. In case of two active traps whose electron occupancy depends on each other, the probability of switching from one level to the others is computed by finding the highest transition probability.

The power spectral density (PSD) ($S_{V_{\rm DS}}$) for each generated RTS trace is calculated using Welch's PSD [29, 82]. For 20 or more traps active, RTSSIM automatically switches from RTS to 1/f noise mode. The total PSD for 20 or more active traps is computed by summing the Lorentzian spectra [57]:

$$S_{V_{\rm DS}}(f) = \sum_{i=1}^{N_{\rm t_{active}}} k_i/\left[1 + \left(f/f_{0i}\right)^2\right] \tag{32}$$

where f_{0i} and k_i are the corner frequency and PSD magnitude due to ith trap, $f_0 = 1/\left(2\pi\overline{\tau}\right)$, $1/\overline{\tau} = 1/\overline{\tau}_{\rm c} + 1/\overline{\tau}_{\rm e}$, and $k_i = 4\overline{\tau}^2\Delta V_{\rm DS}^2/\left(\overline{\tau}_{\rm c} + \overline{\tau}_{\rm e}\right)$. If the particular values are not known, then they can be taken from the literature [83, 84].

To verify the RTS model described above, RTS measurements were done at variable temperatures (88–300 K) using the setup and techniques described before. In total, 22 devices were measured, from which RTS data of three nMOSFETs are shown here, referred as TA, TB, and TC, whose device areas are $W \times L = (1 \times 0.55)$, (1×0.50), and (0.60×0.60) μm^2, respectively. $T_{\rm ox}$ for TA, TB, and TC are 7.5, 12.9, and 7.5 nm, respectively. Two-level RTS was found in each of the devices. The bias conditions for the three devices were: (1) For TA,

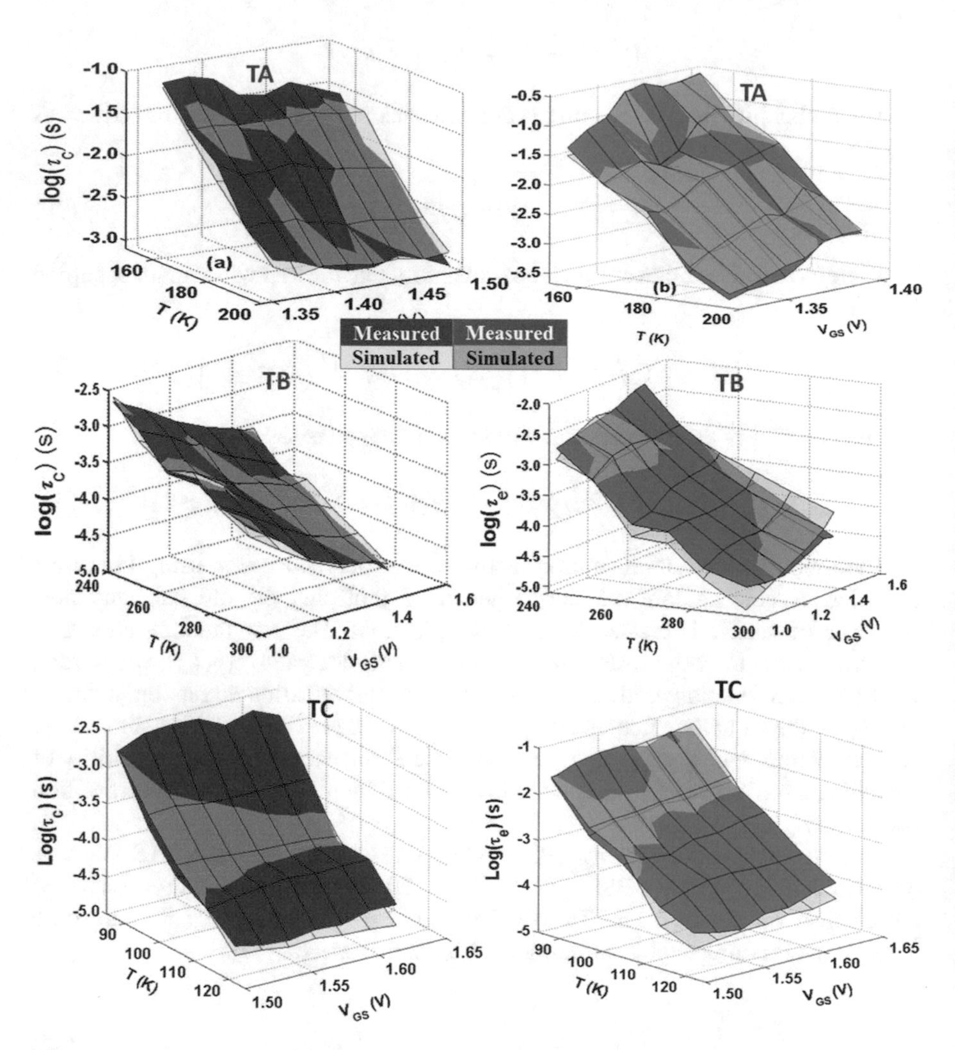

Fig. 19 (**a**) Measured and (**b**) simulated capture and emission times as a function of gate voltage and temperature for TA, TB, and TC. V_{DS} = 0.20, 0.30, and 0.25 V for TA, TB, and TC, respectively [56]

V_{GS} = 1.36–1.5 V, V_{DS} = 0.2 V, and T = 154–195 K; (2) For TB, V_{GS} = 1.00–1.45 V, V_{DS} = 0.3 V, and T = 217–300 K; (3) For TC, V_{GS} = 1.50–1.68 V, V_{DS} = 0.25 V, T = 88–122 K. The capture and emission times (Fig. 19), and the RTS magnitude (Fig. 21) were extracted using the procedure discussed before. Using the RTS analysis technique mentioned before, traps in all three devices were identified as repulsive centers. ΔE_{B} and σ_0 were extracted from an Arrhenius plot of σ found from the experimentally obtained $\overline{\tau}_{\mathrm{c}}$. ΔH and ΔS were extracted from the Arrhenius plot of normalized $\overline{\tau}_{\mathrm{e}}$ through (4). The trap energy parameters for TA are shown in Table 3. x_{T} was calculated using (6). The normalized x_{T} are shown in Table 4. E_{R}

Table 3 Experimentally measured and simulated ΔE_B, ΔH, and $\Delta S/k_B$ for device TA

	Measured			Simulated		
V_{GS} (V)	ΔE_B (eV)	ΔH (eV)	$\Delta S/k_B$	ΔE_B (eV)	ΔH (eV)	$\Delta S/k_B$
1.36	0.236	0.060	6.780	0.222	0.065	7.226
1.40	0.250	0.073	6.869	0.223	0.075	7.233
1.44	0.232	0.104	8.108	0.221	0.082	7.041
1.48	0.230	0.076	5.579	0.221	0.096	7.297
1.50	0.216	0.080	5.538	0.217	0.098	7.050

Input ΔE_B is 0.234 eV, $V_{DS} = 0.20$ V [56]

Table 4 Experimentally measured and simulated x_T/T_{ox} for device TA

T (K)	Measured x_T/T_{ox}	Simulated x_T/T_{ox}
195	0.354	0.342
185	0.351	0.335
174	0.335	0.295
163	0.312	0.280
154	0.297	0.293

Average x_T/T_{ox} provided to RTSSIM is 0.310 [56]

Table 5 Input parameters provided by the used and the simulation conditions for the three devices [56]

Parameter	TA	TB	TC
V_{GS} (V)	1.36–1.50	1.00–1.45	1.50–1.62
V_{DS} (V)	0.20	0.30	0.25
T (K)	154–195	242–295	88–122
ΔE_B (eV)	0.234	0.308	0.240
$\left(E_{C_{ox}} - E_T\right)$ (eV)	2.87	3.00	3.10
x_T/T_{ox}	0.310	0.134	0.061
α_0 (V s)	7.79×10^{-13} (195 K) 1.57×10^{-12} (163 K) 1.77×10^{-12} (154 K)	5.08×10^{-13} (295 K) 5.85×10^{-13} (283 K)	1.14×10^{-12} (122 K)
α_1 (V s)	-7.48×10^{-14} (195 K) -1.55×10^{-13} (163 K) -1.74×10^{-13} (154 K)	-5.81×10^{-14} (295 K) -6.69×10^{-14} (283 K)	-1.12×10^{-13} (122 K)

was found solving (7). The average E_R values for TA, TB, and TC were found to be 1.11, 1.25, and 0.60 eV, respectively. As mentioned before, the obtained ΔE_B and E_R values showed resemblance to those of an unrelaxed neutral oxygen deficiency center (V^0 ODC II) [7].

In order to verify RTSSIM, the simulation was run 22 times using the same bias and temperature conditions, and experimentally found ΔE_B, x_T, $\left(E_{C_{ox}} - E_T\right)$, α_0, and α_1. The simulated $\overline{\tau}_c$, $\overline{\tau}_e$, E_R, $\left(E_{C_{ox}} - E_T\right)$, ΔE_B, α_0, and x_T showed a standard deviation of 3%, 4%, 1%, 0.1%, 15%, 1%, and 2%, respectively (Table 5).

From the measured simulated values shown in Table 3, the mean variation in ΔE_B, ΔH, and $\Delta S/k_B$ for the device TA is found to be 3%, 6%, and 8%, respectively.

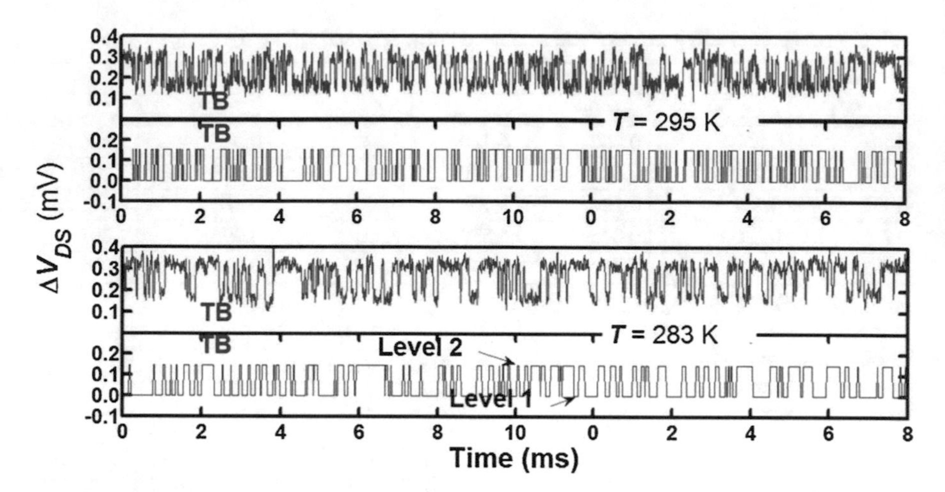

Fig. 20 Measured and reconstructed two-level RTS for device TB at 295 and 283 K at $V_{GS} = 1.10$ V, $V_{DS} = 0.30$ V [56]

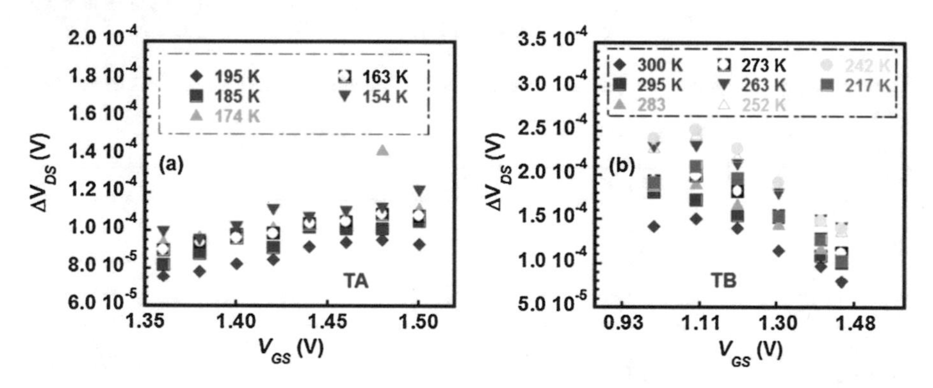

Fig. 21 RTS magnitude of (**a**) TA and (**b**) TB. $V_{DS} = 0.20$ and 0.30 for TA and TB, respectively [56]

The mean variation in the measured and simulated x_T was found to be around 6% (Table 4). The average variation in the simulated and measured values of ΔE_B, ΔH, $\Delta S/k_B$, and x_T for TB was 1%, 9%, 11%, and 11%, respectively.

An example of the experimentally observed and reconstructed RTS traces is shown in Fig. 20. The RTS amplitudes found through the experimentally observed RTS traces are shown in Fig. 21. The corner frequency and PSD (Fig. 22) of the measured and generated RTS traces showed a variation of around 18% because of the combined effects of variations in $\overline{\tau}_c$ and $\overline{\tau}_e$.

The accuracy of reconstructing a three-level RTS in RTSSIM was also checked by comparing the experimental and simulated RTS and trap parameters. Figure 23a, b shows the measured and generated three-level RTS because of two

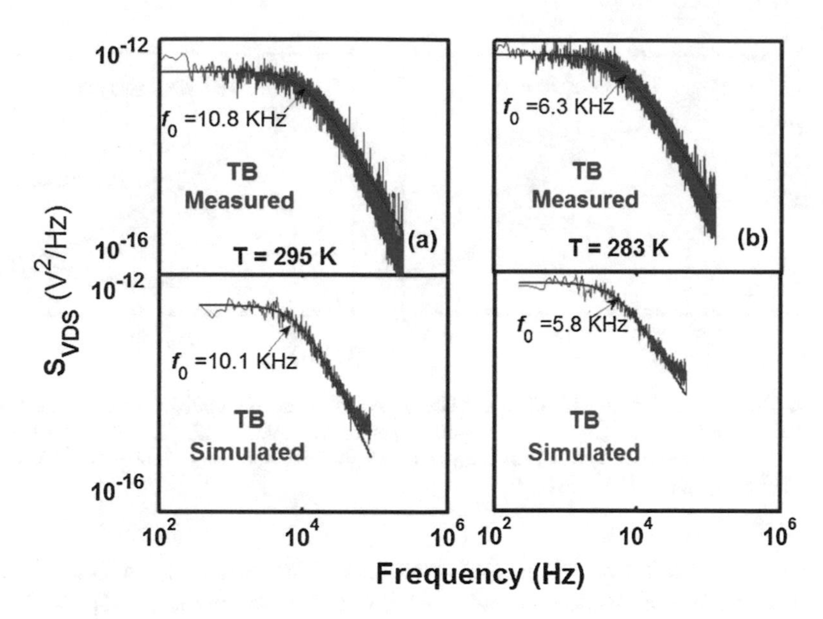

Fig. 22 Measured and simulated PSD of TB at (**a**) 295 K and (**b**) 283 K. $V_{GS} = 1.10$ V, $V_{DS} = 0.30$ V [56]

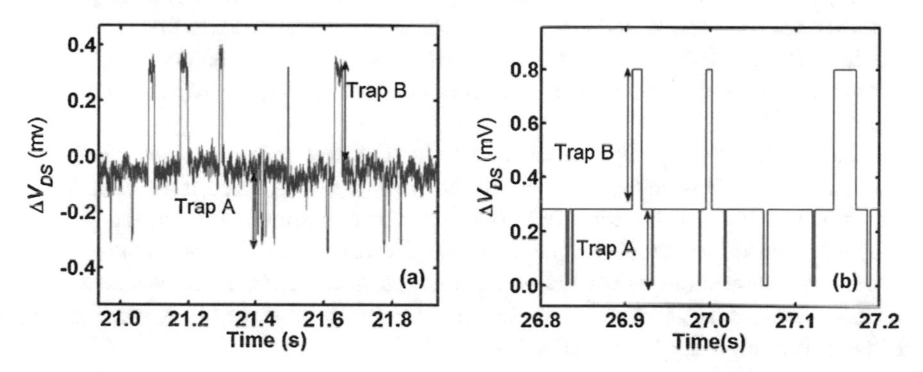

Fig. 23 (**a**) Measured and (**b**) simulated three-level RTS with two dependent repulsive centers active. Device $W \times L = (0.6 \times 0.6)\ \mu m^2$. Measured $\overline{\tau}_{cA} = 2.5$ ms, $\overline{\tau}_{eA} = 74$ ms, $\overline{\tau}_{cB} = 240$ ms, $\overline{\tau}_{eB} = 12$ ms. Simulated $\overline{\tau}_{cA} = 1.6$ ms, $\overline{\tau}_{eA} = 52$ ms, $\overline{\tau}_{cB} = 337$ ms, $\overline{\tau}_{eB} = 11$ ms [56]

active traps whose electron occupancy depends on each other. The figures show that RTSSIM is capable of successfully reconstructing three-level RTS as well.

The transition from RTS to 1/f noise in RTSSIM was verified by checking the scalability with the UNMF model [34]

$$S_{V_{\mathrm{DS}}} = \frac{k_{\mathrm{B}} T V_{\mathrm{DS}}^2}{\lambda f W L_{\mathrm{eff}}} \left(\frac{1}{N} \pm \alpha \mu \right) N_{\mathrm{T}} \left(E_{\mathrm{F}} \right) \tag{33}$$

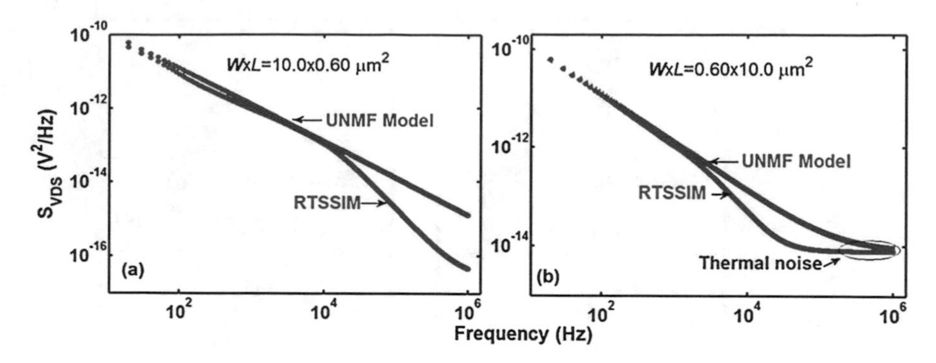

Fig. 24 Simulated PSD using RTSSIM and UNMF for (**a**) 35 acting repulsive centers for a device $W \times L = (10 \times 0.6)\ \mu\text{m}^2$, (**b**) 39 acting repulsive centers for a device $W \times L = (0.6 \times 10)\ \mu\text{m}^2$. $V_{GS} = 1.52$ V, $V_{DS} = 0.25$ V, $T = 295$ K. $\alpha_0 = 4.56 \times 10^{-13}$ V s, $\alpha_1 = -3.02 \times 10^{-14}$ V s, $N_{tA} = 6 \times 10^8\ \text{cm}^{-2}$, $\overline{x}_T = 1$ nm [56]

where λ is the wave attenuation coefficient for electrons, which can be found through the Wentzel–Kramers–Brillouin approximation [34] as $\lambda = 2/\hbar\left(2qm^*_{ox}\varphi_0\right)^{1/2}$, where m^*_{ox} is the effective mass of electrons in the oxide, and $N_T(E_F)$ is the trap density per unit volume per unit energy. Assuming that traps only within $\pm 3k_BT$ are active, and that the traps are uniformly distributed up to $2\overline{x}_T$ from the Si–SiO_2 interface with an average distance of $\overline{x}_T$, N_T can be computed using the areal trap density and average trap distance through $N_T = N_t/\left(3k_BT\left(2\overline{x}_T\right)\right)$ ($\text{cm}^{-3}\ \text{eV}^{-1}$).

Thermal noise (S_{V_R}) is also calculated in RTSSIM using $S_{V_R} = 4k_BT/g_D$. The thermal noise is added to the total PSD due to all activated traps (Fig. 24b). As shown in Fig. 24, the total PSD obtained through simulation and the UNMF model agree well for a wide frequency range. However, in RTSSIM, the individual Lorentzians with finite corner frequencies of the active traps are added, while the UNMF assumes 1/*f* form. That is why a deviation is observed between the two PSDs at very high frequencies (Fig. 24).

7 RTS in pMOSFETs

In order to study the hole trap properties, extensive amount of RTS measurements was done on pMOSFETs. Even though RTS has been used as a trap characterization tool for the last few decades, RTS in pMOSFETs has been severely understudied compared to that in nMOSFETs. Most of the published articles on RTS in pMOSFETs confined their studies to physical trap location from the Si–SiO_2 interface and along the channel, capture cross section, and capture activation energies [10–12, 85]. There have been articles about the effect of substrate bias on the RTS parameters as well [86, 87]. Effects of drain voltage on the capture and emission time [88], and lateral trap position [89] have also been examined. The physical nature of the hole traps responsible for negative bias temperature instability (NBTI) has also been investigated using TDDS measurements and density functional theory (DFT) [26, 90–93]. Although from the examined defect properties it has been claimed that similar defects are responsible for NBTI and RTS in pMOSFETs [26, 90, 92], our results suggest otherwise. Investigation of the hole trap properties would need substantial amount of RTS measurements as have been done in case of nMOSFETs.

Variable temperature (165–295 K) RTS measurements were taken on 15 pMOSFETs in total, among which four devices, named according to their dimensions have been reported here: (1) PMOS500 of $W \times L = (1 \times 0.5)\ \mu m^2$, (2) PMOS550 of $W \times L = (1 \times 0.55)\ \mu m^2$, (3) PMOS600 of $W \times L = (1 \times 0.6)\ \mu m^2$, and (4) PMOS700 of $W \times L = (1 \times 0.7)\ \mu m^2$. The RTS measurements were taken on these devices using the same setup and technique as described in the earlier portion of this chapter. T_{ox} of these devices was 12.1 nm.

Two-level RTS was observed in all four devices at each bias point and temperature (Fig. 25). The bias condition and temperature ranges of the four devices are:

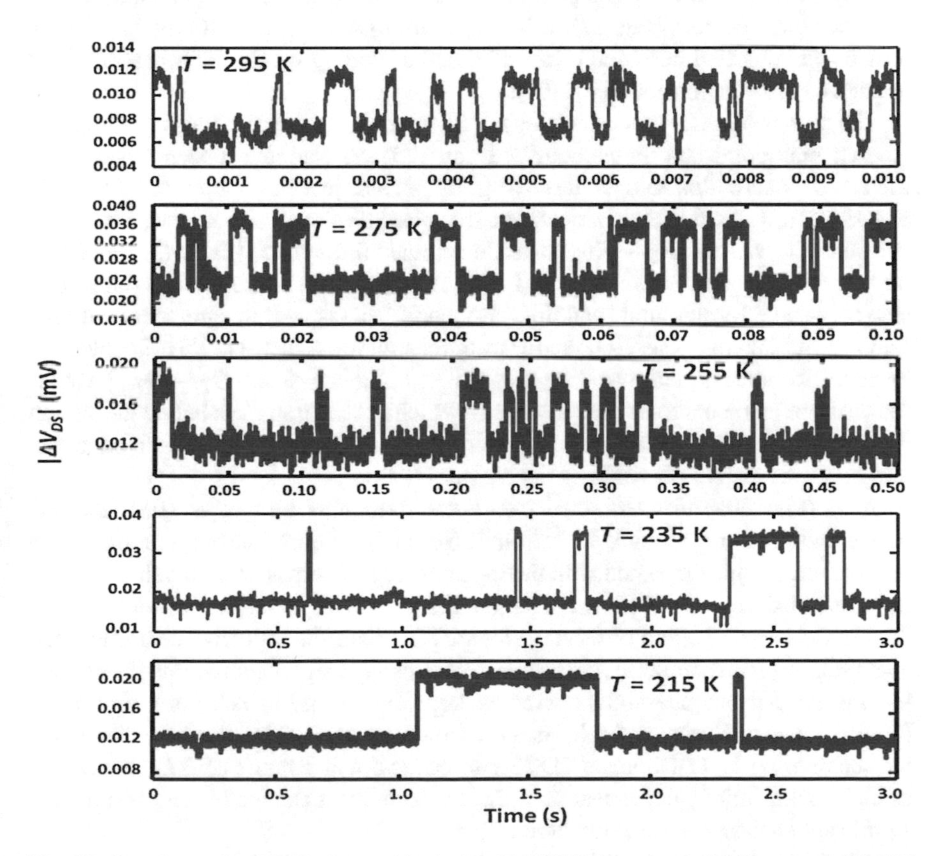

Fig. 25 Sample two-level RTS observed in device PMOS550. $V_{GS} = -1.4$ V, $V_{DS} = -0.4$ V [13]

(1) For PMOS500, $V_{GS} = -2.15$ to -2.50 V, $V_{DS} = -0.5$ V, $T = 175$–265 K; (2) For PMOS550, $V_{GS} = -1.50$ to -2.30 V, $V_{DS} = -0.4$ V, $T = 215$–295 K; (3) For PMOS600, $V_{GS} = -1.65$ to -1.90 V, $V_{DS} = -0.5$ V, $T = 215$–295 K; and (4) For PMOS700, $V_{GS} = -1.80$ to -2.30 V, $V_{DS} = -0.5$ V, $T = 245$–265 K.

The mean times spent at both the lower and upper level were calculated using Poisson's statistics as discussed before. From the pattern of the average time spent at each level as a function of V_{GS}, the lower level of $|\Delta V_{DS}|$ for PMOS500 belong to $\overline{\tau}_c$, while the upper $|\Delta V_{DS}|$ level corresponded to $\overline{\tau}_c$ for the other three devices. Therefore, according to the theory explained earlier, PMOS500 was identified to have a repulsive center trap (denoted by 0/+), whereas PMOS550, PMOS600, and PMOS700 had attractive centers (denoted by −/0). The capture and emission times of all four devices as a function of V_{GS} and temperature are shown in Fig. 26. According to the UNMF model, the number and mobility fluctuations are added in (8) for a repulsive center, while an attractive center will have these fluctuation terms subtracted. Hence, the normalized RTS magnitude of a repulsive center is higher than that of an attractive center. That is why the normalized RTS magnitude of PMOS500 was found to be around one order of magnitude higher than that of the other three devices (Fig. 27). It is important to note that in MOSFETs of large dimensions reported here, the effect of discrete doping on the RTS magnitude is negligible in strong inversion [64, 94].

The trap position x_T was calculated for all four traps using (6). ψ_S was computed at each bias point and temperature solving (25). x_T varied between 0.4–1.4 nm for all traps across all temperatures (Fig. 28). This, just like the electron traps in nMOSFETs, rules the Pb centers from the possible defect candidates. Using the calculated x_T values, $\left(E_T - E_{V_{ox}}\right)$ was determined through (5) (Table 6). There was no correlation found between x_T and RTS mean capture and emission times, and also between RTS amplitude and time constants. This is well in agreement with the findings of Nagumo et al. [95], and also agrees with the fact that RTS is governed by multiphonon-assisted tunneling, not direct tunneling. The $\left(E_T - E_{V_{ox}}\right)$ values for the traps came out to be around 4.7 eV, which put the traps just below the silicon valence band edge. As the Fermi level is around the silicon valence band edge, the trap becomes easily accessible by the channel holes.

ΔE_B (Fig. 30a) was extracted using the Arrhenius plot of σ (Fig. 29a). σ varied between 10^{-27} and 10^{-23} cm^2 for all the traps across the measured temperatures, which matched with the reported capture cross-section values of hole traps responsible for RTS [96]. The larger ΔE_B value of the attractive centers (Fig. 30a) is the reason behind observing slower RTS than the repulsive center and the nMOSFETs [7]. Although ΔE_B is supposed to change significantly with the change in V_{GS}, the narrow gate voltage window kept the change in ΔE_B in a tiny range [7, 29]. This small range of ΔE_B is a drawback of using RTS as a trap diagnostic tool comparing to TDDS since TDDS can cover a wider range of ΔE_B. However, as $\overline{\tau}_c$ is exponentially dependent on ΔE_B, even slightest change in ΔE_B results in a significant change in $\overline{\tau}_c$ as a function of V_{GS}.

ΔH (Fig. 30b) and ΔS (Fig. 30c) values were evaluated from the Arrhenius plot of normalized mean emission time constant (Fig. 29b). The positive values

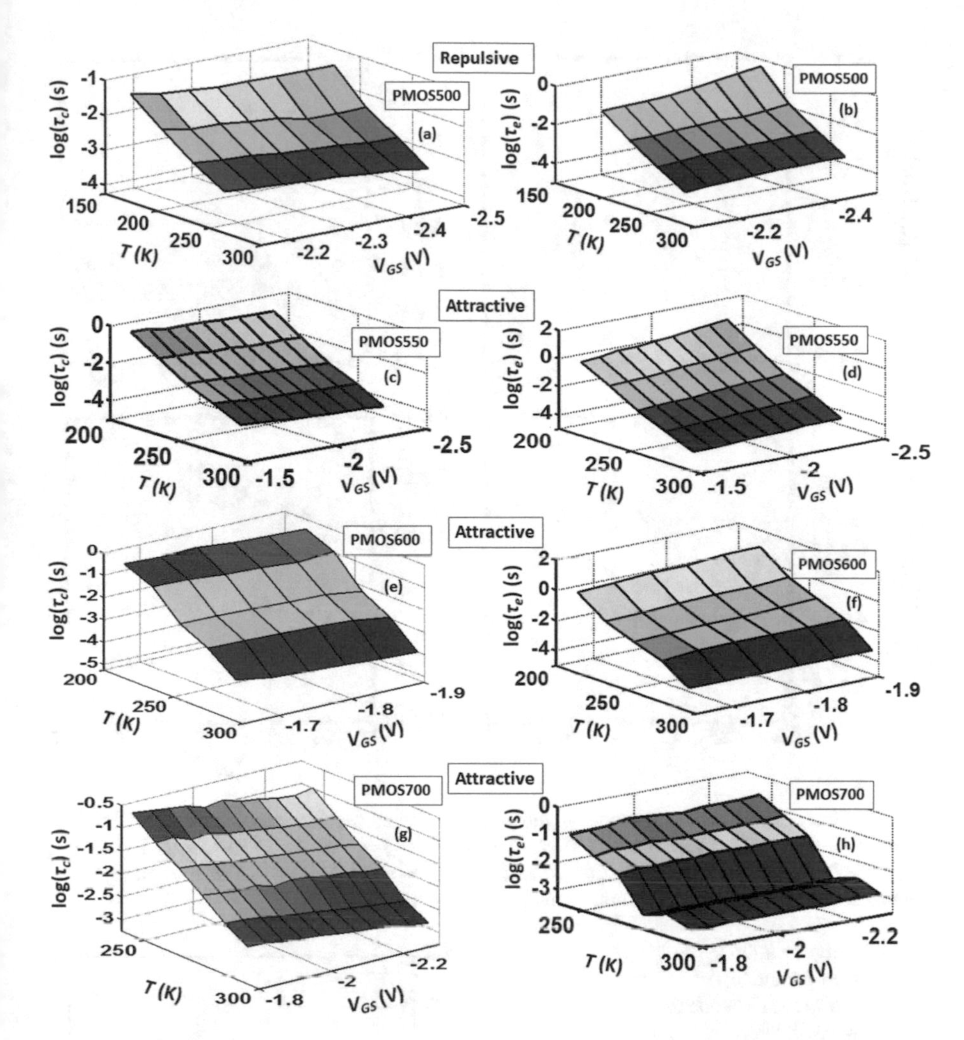

Fig. 26 Measured average hole capture times for (**a**) PMOS500, (**c**) PMOS550, (**e**) PMOS600, and (**g**) PMOS700. Average hole emission times for (**b**) PMOS500, (**d**) PMOS550, (**f**) PMOS600, and (**h**) PMOS700. For PMOS500, PMOS600, and PMOS700 $V_{\mathrm{DS}} = -0.5$ V. For PMOS550, $V_{\mathrm{DS}} = -0.4$ V [13]

of ΔH indicated an endothermic process, whereas negative ΔH values inferred that the process was exothermic. The mean ($\Delta H - T\Delta S$) values varied from -53.61 to 8.14 meV, which once again proved that the traps resided just below the silicon valence band edge. E_{R} values were calculated following the same method as nMOSFETs, using (7) (Fig. 30d).

$E_{\mathrm{T}} - E_{\mathrm{V_{ox}}} \approx 4.7$ eV eliminates two certain types of E' centers from the possible hole defects in our devices. They are: (1) twofold coordinated silicon (II–Si) neutral oxygen vacancy centers and (2) a pair of under coordinated oxygen atom and over-

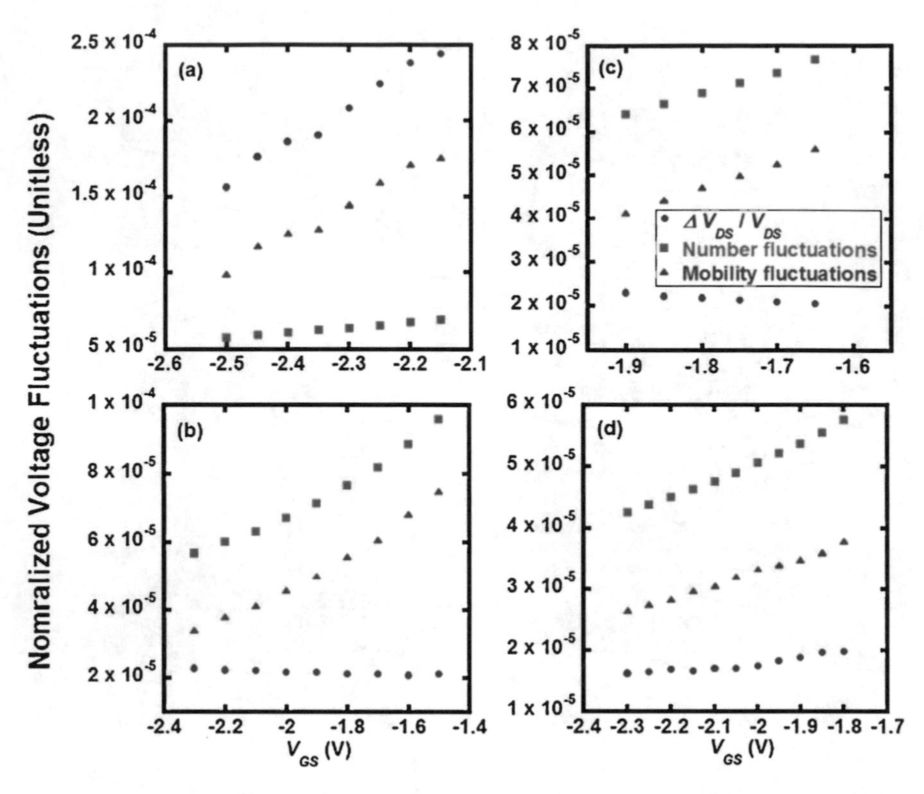

Fig. 27 Normalized RTS magnitude, number, and mobility fluctuations for (**a**) PMOS500 at $T = 265$ K, $V_{DS} = -0.5$ V, (**b**) PMOS550 at $T = 295$ K, $V_{DS} = -0.4$ V, (**c**) PMOS600 $T = 295$ K, $V_{DS} = -0.5$ V, and (**d**) PMOS700 at $T = 295$ K, $V_{DS} = -0.5$ V. PMOS500, being a repulsive center, showed higher RTS magnitude than the other three defects [13]

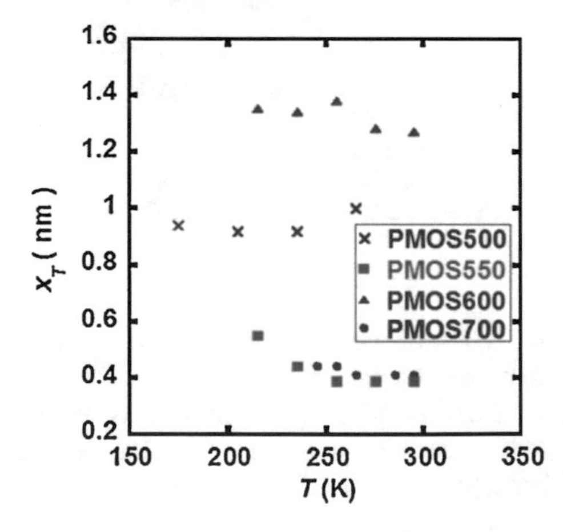

Fig. 28 Trap location at different temperatures. As expected, trap location does not change with temperature. The variation is due to the error in computation, which gets larger with decreasing temperature and the resultant decrease in number of RTS transitions. Also, it does not indicate whether the trap is attractive or repulsive [13]

Table 6 Summary of the energy parameters of all the studied defects

Trap species	Trap type	Capture activation energy (ΔE_B) (eV)	Relaxation energy (E_R) (eV)	$E_T - E_{V_{ox}}$ (eV)	$E_T - E_V$ (eV)
Attractive centers here [13]	–/0	0.47–0.55	1.91–2.28	4.69–4.81	–0.04 to +0.02
D-III-Si	–/0		0.5–2.3 [25]	4.8–5.6 [25]	
H bridge defect	–/0		2.5 [28], 1.71 [98]	4–6.5 [28], 3.8–5.2 [51]	
III-Si/V-Si	–/0		1.7–2.1 [25]	5.0–5.6 [25]	
III-O/V-Si	–/0		~1.7 [25]	5.1–6.3 [25]	
NOV	–/0		0.5–1 [25]	~7.1 [25]	
II-Si	–/0		~1.2 [25]	~6.9 [25]	
Repulsive center here [13]	0/+	0.18–0.20	0.81–0.89	4.68 – 4.73	–0.07 to –0.04
III-O/III-Si	0/+		0.3–0.4 [4], 0.5–1.8 [25]	3.9-5.3 [25]	~0 [51][a]
H bridge defect	0/+	0.1–1.0 [26]	1.5–2.0 [26][b], 2.0–2.5 [28], 2.22 [98]	4–6.5 [28], 4.3– 4.9 [51]	–1.0 to +0.5 [26], [92]
Hydroxyl E' center	0/+	0.2–1.0 [26], ~0.4 [92]	~2.0 [26][c]		–1 to +1 [26], [92]
E'_δ center	0/+			2.9–3.1 [51]	
NOV	0/+	1.5–2.5 [26]	~1.0 [25], ~ 3.0 [26]	1.6–2.4 [25]	–1.3 to –3.7 [26], [92]
III-Si/V-Si	0/+		~1.0 [25]	2.1–2.9 [25]	
III-O/V-Si	0/+		0.7–1.1 [25]	2.2–2.5 [25]	

Blue highlights: agreement with the attractive trapping centers reported our manuscript
Yellow highlights: agreement with the repulsive center reported in our manuscript
Gray highlights: disagreement with the values reported in our manuscript
[a]Referred as the "puckered configuration of E'_γ" in [51]
[b]Calculated from Fig. 13 of [26]
[c]Calculated from Fig. 16 of [26]

coordinated silicon atom (III–O/V–Si). Since these traps have a shallow energy level in the SiO_2 bandgap [25], or in other words, close to $E_{V_{ox}}$ or $E_{C_{ox}}$, they can be eliminated from consideration. $E_T - E_{V_{ox}}$ values of hydrogen bridge defects and hydroxyl E' centers match with that of the pMOSFETs reported here [26, 28, 51, 92, 97]. However, higher capture activation energy, and structural relaxation implies that neither hydrogen-related defects [26, 28, 51, 92, 97], nor hydroxyl E' center [26, 92] is responsible for the RTS reported here.

A particular type of E' center, a pair of under-coordinated silicon defects (D–III–Si) has been reported to exhibit relaxation energy of 1–2 eV, calculated through generalized gradient approximation of density functional theory (DFT-GGA) [25]. These defects also stay at around 3.8–5.6 eV higher than the $E_{V_{ox}}$ in the SiO_2 bandgap. These results match with the E_R shown by the aforementioned attractive centers. There is a doubly occupied sp^3 in one of the defect pairs, one of which gets neutralized upon capturing a hole, increasing the O–Si–O bond angle to ~107° from

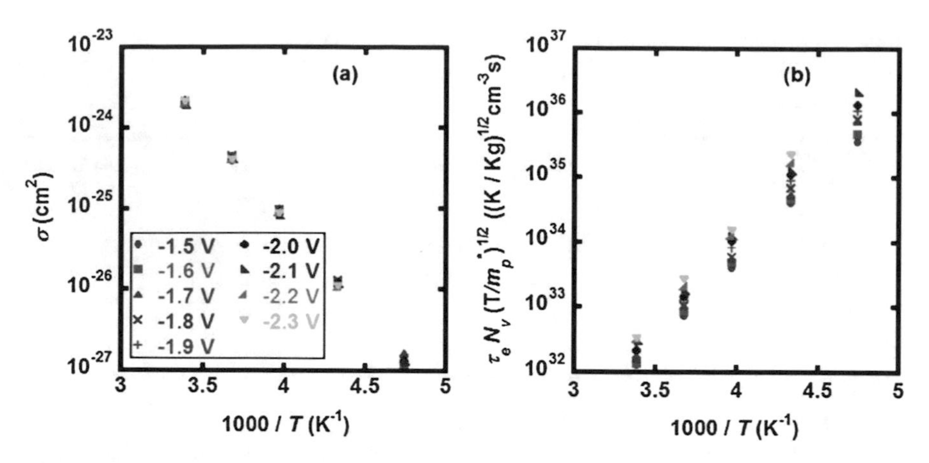

Fig. 29 Arrhenius plot of (**a**) trap capture cross section and (**b**) normalized mean emission time for PMOS550. $V_{\mathrm{DS}} = -0.4$ V [13]

~103° (Fig. 31a) [25]. Hence, this defect behaves like an attractive center.

$$\equiv \mathrm{Si}^{\cdot} + \equiv \mathrm{Si}{:}^{-} + \mathrm{h}^{+} \leftrightarrow \equiv \mathrm{Si}^{\cdot} + \equiv \mathrm{Si}^{\cdot} \tag{34}$$

where $\equiv$ is three silicon–oxygen bonds, $^{\cdot}$ is a single unpaired electron, and : is paired electrons in a single sp^3 orbital.

Another type of E' center, a puckered/back-projected oxygen vacancy center [51, 98] (or a pair of under-coordinated silicon atom and over-coordinated oxygen atom, as mentioned in [25]) showed $E_{\mathrm{T}} - E_{\mathrm{V_{ox}}}$ values of 3.8–5.2 eV, and E_{R} values of ~0.8–1.5 eV [25]. The parameters were calculated using DFT-GGA, and showed an excellent match with the $E_{\mathrm{T}} - E_{\mathrm{V_{ox}}}$ and E_{R} of the repulsive center, PMOS500. Upon capturing a hole, this defect becomes positively charged from neutral charge state, which is a similar switching pattern to a repulsive center. Trapping a hole neutralizes one of the two dangling bond electrons (Fig. 31b), increasing the O–Si–O bond angle to ~108–109° from ~103° [25].

$$^{+}\equiv \mathrm{O} + \equiv \mathrm{Si}{:}^{-} + \mathrm{h}^{+} \leftrightarrow {}^{+}\equiv \mathrm{O} + \equiv \mathrm{Si}^{\cdot} \tag{35}$$

A summary of all the possible defect candidates is shown in Table 6. It is important to mention that the SiO_2 bandgap correction was taken into consideration [13] while finding the $E_{\mathrm{T}} - E_{\mathrm{V_{ox}}}$ values from [25]. Although Kimmel et al. [39] reported the puckered silicon defect to be less stable than the neutral dimer silicon vacancy configuration, and argued that the puckered configuration would go back to the neutral structure very quickly, this puckered/back-projected silicon vacancy is the only defect that showed matching energy parameters with the repulsive enter, PMOS500.

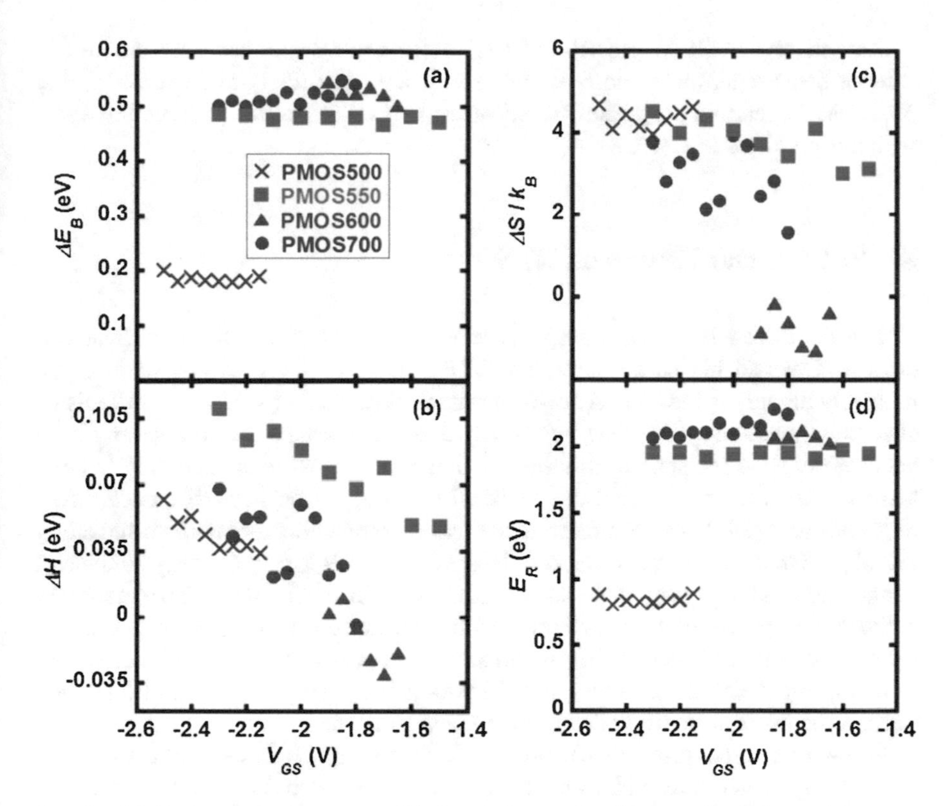

Fig. 30 (**a**) Capture activation energy, (**b**) change in enthalpy, (**c**) normalized change in entropy, and (**d**) relaxation energy. $V_{DS} = -0.5$ V for PMOS500, PMOS600, and PMOS700; for PMOS550, $V_{DS} = -0.4$ V. In (**d**), $T = 295$ K for PMOS550, PMOS600, and PMOS700; $T = 265$ K for PMOS500 [13]

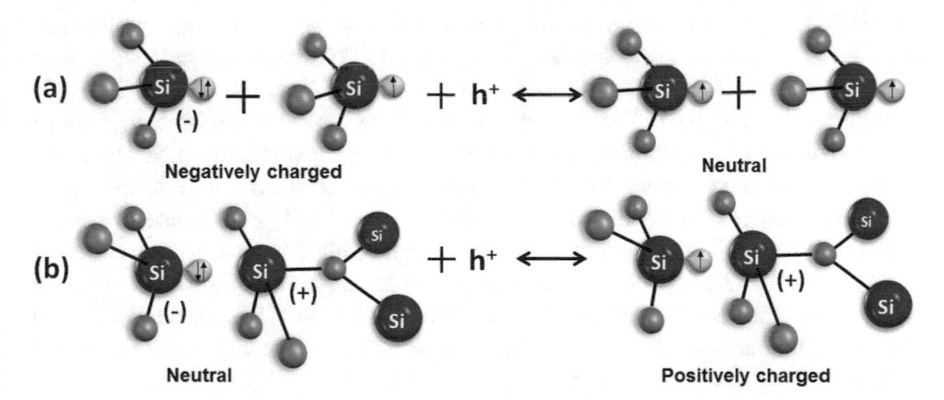

Fig. 31 (**a**) A pair of dissociated threefold silicon atoms (D–III–Si) and (**b**) puckered/back-projected silicon vacancy defect. The bond lengths and angles are not to scale [13]

As both the D–III–Si and III–O/III–Si defects are positioned near the silicon valence band edge, these two types of defects are most likely to be active during pMOSFET operation [25]. The RTS measurements discussed in this section agree with this prediction.

8 Hot-Carrier Effects on RTS

Hot-carrier effects have been a major concern in terms of reliability in ultra-scaled devices. Channel hot carrier stressing (CHC) at $V_{GS} = V_{DS}$ is known to cause relatively higher degradation in device parameters than other stressing mechanisms at room temperature [99]. New traps created at the interface due to stressing have been responsible for shift in threshold voltage and transconductance [100, 101]. Most of the articles published on electrical stressing so far have discussed either time constants [89, 102], position from the oxide–semiconductor interface and along the SiO_2 bandgap for stress-induced defects [103, 104], or modeling of channel degradation and RTS magnitude due to stressing [105, 106]. Oxide trap generation, activation, or passivation mechanisms due to stressing are still not clear. In order to characterize the oxide defects in both prestress and post-stressed conditions, as well as to observe the effects of stress on RTS parameters, massive amount of RTS data needs to be taken at carefully determined stress intervals.

To investigate the properties of process-induced traps, RTS measurements were taken on 5 V rated pMOSFETs of different sizes (≤ 1 μm^2) at room temperature using the same experimental setup and techniques as discussed before. T_{ox} of all the devices was 12.1 nm. Like the previous RTS measurements, all measurements here were done in the linear region of operation as well. After finishing the RTS measurements, CHC was applied to the devices at $V_{GS} = V_{DS} = -6.5$ V using the SPA. RTS measurements after a particular step of stressing were commenced as soon as the stressing was finished. Stressing and RTS measurements were done on eight pMOSFETs in total. Results of one pMOSFET of $W \times L = (1 \times 0.7)$ μm^2 are presented here. The RTS measurements were done within a V_{GS} range of −1.6 to −3.6 V. While measuring the RTS due to a particular trap, the V_{GS} was varied keeping the V_{DS} constant. Since different traps became active at different gate and drain voltages, therefore to catch all the active traps, V_{DS} was varied between −0.2 and −0.6 V. Cumulative stress of 3000 s was applied to the device at different stress intervals. The summary of the stress intervals, number of RTS levels, and the responsible traps is provided in Table 7. While taking the RTS measurements, the sampling frequency and the total time span of the oscilloscope were varied to record at least 500 switching events as usual.

A three-level RTS was observed in the device in prestress condition, where all the transitions took place between levels 1 $\leftrightarrow$ 2 and 1 $\leftrightarrow$ 3 (Fig. 32a). Three levels in an RTS can be attributed to two traps [19], denoted as traps A and B here (Table 7). A new three-level RTS was observed upon 200 s of stressing, where the transitions took place between levels 1 $\leftrightarrow$ 2, and 2 $\leftrightarrow$ 3 (Fig. 33b). This new pattern in

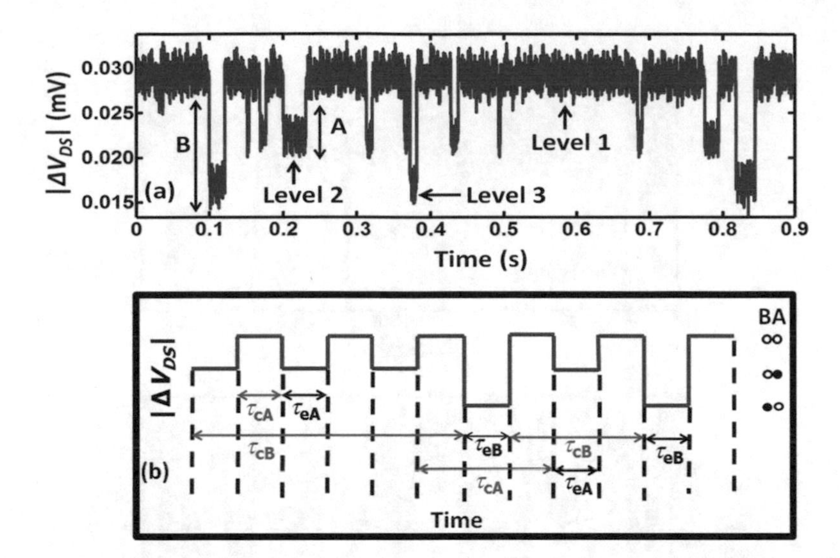

Fig. 32 (**a**) Example of RTS observed at $V_{GS} = -2.4$ V, $V_{DS} = -0.4$ V when traps A and B are active. Switching between levels 1 $\leftrightarrow$ 2 took place due to trap A, while trap B switched between levels 1 $\leftrightarrow$ 3; (**b**) Calculation of capture and emission times. Since both A and B are attractive centers, the upper level of the RTS magnitude indicates the capture time, and the time spent at the bottom level is the emission time. The two circles at the right indicate whether the traps are empty or full. An empty circle represents the state when the trap is empty, while a filled circle represents that the trap has captured a hole [14]

Table 7 Summary of active traps and RTS levels at different stress intervals [14]

Cumulative stress time (s)	Number of RTS levels	Traps active
0, 5, 10, 40, 70, 100	3	A, B
200	2, 3	A, C
300	2	A
400	2, 3	A, C
500	3	A, B
600	2	A
700	3	A, B
800	2, 3, 4	A, B, D
900, 1000, 1200	2, 3	A, C, D
1500	3	A, B
2000	2, 3	E, F
2500	2	G, H
3000	No RTS	N/A

switching was observed because of disappearance of trap B, and introduction of a new trap C (Table 7). Only trap A was active at 300, and 600 s of stressing, showing a two-level RTS (Fig. 33a). A new trap D was observed after 800 s of stressing (Fig. 34a), which later added an additional level to the original RTS observed due to

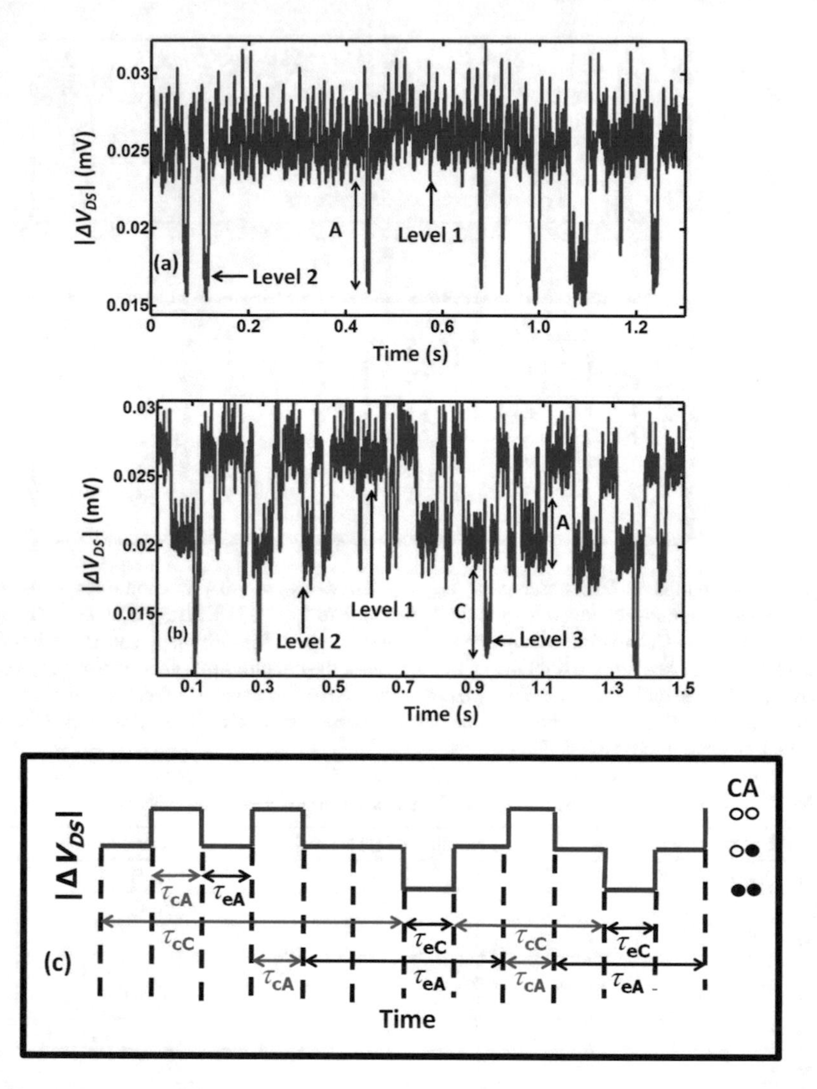

Fig. 33 (**a**) Portion of RTS observed at $V_{GS} = -2.3$ V, $V_{DS} = -0.4$ V after 200 s of stressing with only trap A is active. The time spent at level 1 is the capture time, while level 2 corresponds to the emission time; (**b**) RTS observed at $V_{GS} = -2.9$ V, $V_{DS} = -0.4$ V, 200 s of stressing, when both traps A and C are active. (**c**) Calculation of capture and emission times for A and C. Instead of $1 \leftrightarrow 3$, the transitions to level 3 occur from level 2. Therefore, C is filled with a hole only if A is filled with a hole [14]

traps A and B (Fig. 34b). After 2000 s of stressing, none of traps A, B, C, or D was observed. Rather, two new traps E and F were observed, resulting in a three-level RTS (Fig. 35a, b). Both E and F disappeared upon further stressing at 2500 s, and two new traps G and H were observed at two different bias ranges, each causing a two-level RTS (Fig. 36a, b). After further stressing, all the aforementioned traps

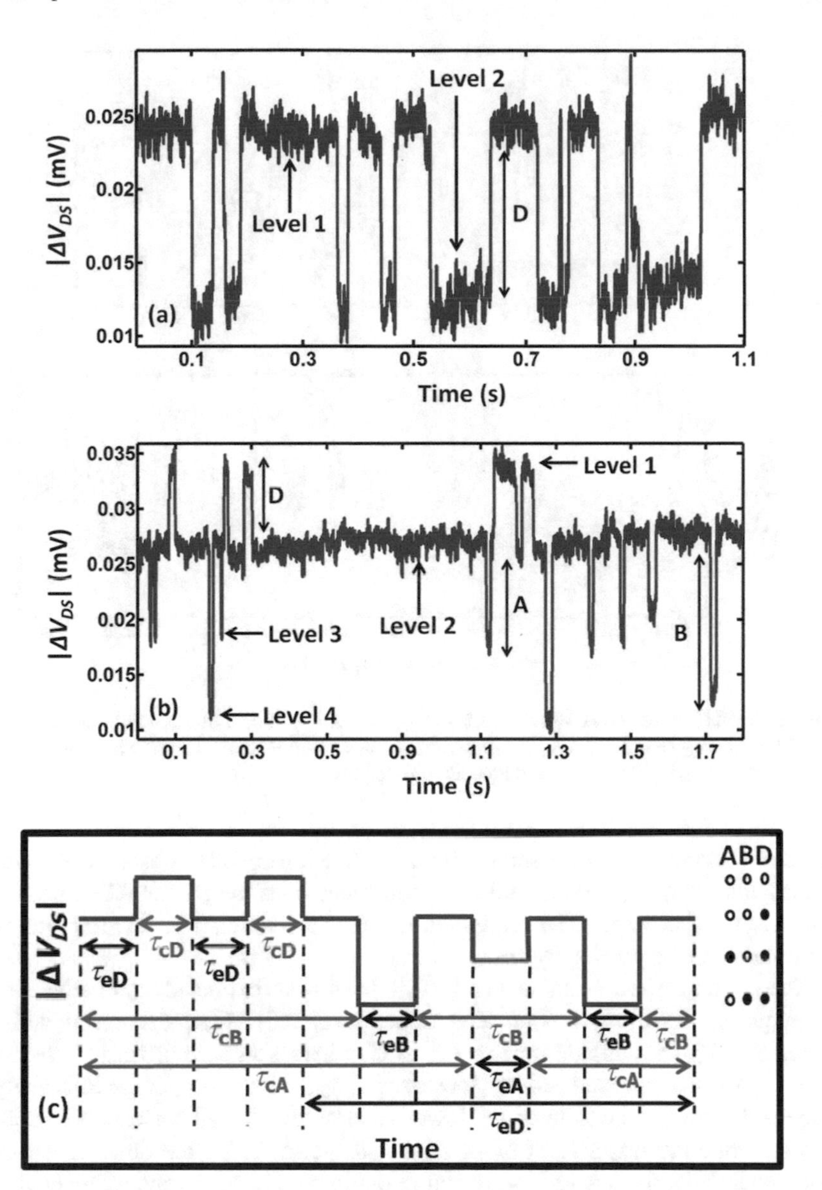

Fig. 34 (**a**) Sample RTS trace at $V_{GS} = -1.6$ V, $V_{DS} = -0.6$ V after 800 s of stressing with only trap D is active. (**b**) RTS observed at $V_{GS} = -2.3$ V, $V_{DS} = -0.6$ V after 800 s of stressing when traps A, B, and D are active resulting in a four-level complex RTS. (**c**) Calculation of capture and emission times. As all three traps are attractive centers, the upper level of the RTS magnitude belongs to the capture time, and the lower level belongs to the emission time [14]

disappeared, and no RTS was observed (Table 7). It is apparent form Table 7 that traps B and C showed volatile behavior with stress, getting activated and deactivated randomly. Similar observations have also been reported by Grasser et al. [107].

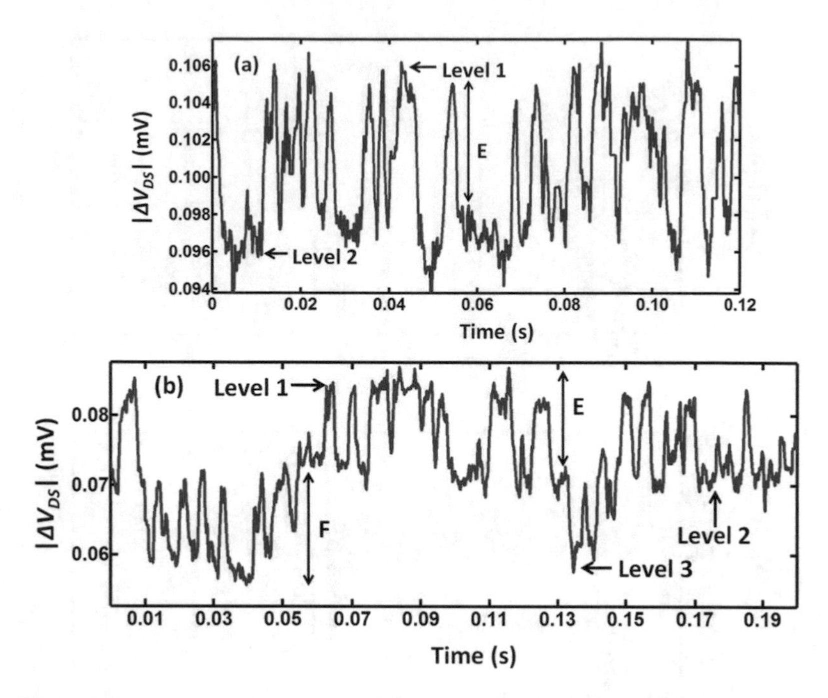

Fig. 35 (**a**) RTS observed at $V_{GS} = -2.1$ V, $V_{DS} = -0.2$ V, 2000 s of stressing, due to only trap E; (**b**) RTS at $V_{GS} = -2.3$ V, $V_{DS} = -0.3$ V with traps E and F active. Capture and emission time were calculated using the same technique as Fig. 2c [14]

Whenever a new RTS was observed, it was different from other observed RTS traces in switching pattern, capture and emission times, bias ranges, or RTS magnitude. This ensured that a new RTS originated from the activation of a new trap, not from the instability of a previously observed trap.

The average time spent at each RTS level was extracted using the similar technique as described before. The upper level of $|\Delta V_{DS}|$ decreased with the increase of V_{GS} magnitude for traps A to G, whereas in case of trap H, the lower level of $|\Delta V_{DS}|$ decreased with the increase in $|V_{GS}|$. Hence, as per the RTS analyses discussed earlier in the chapter, the lower level of the $|\Delta V_{DS}|$ corresponds to $\overline{\tau}_c$ for trap H, while the upper level of $|\Delta V_{DS}|$ belongs to $\overline{\tau}_c$ for the other seven traps. Consequentially, traps A to G were found out to be attractive centers, whereas trap H was found to be a repulsive center.

As the trap properties rely largely on the capture and emission times, it is very important to extract the capture and emission times of each trap in a complex RTS. Capture and emission times extraction procedure of simple RTS traces has already been explained before. For the three-level RTS signal observed due to traps A and B (Fig. 32a), as shown in Fig. 32b, level 1 belongs to $\overline{\tau}_c$ for both A and B, level 2 belongs to $\overline{\tau}_c$ for B, and $\overline{\tau}_e$ for A, and level 3 belongs to $\overline{\tau}_c$ for A, and $\overline{\tau}_e$ for B. For the three-level RTS due to A and C, level 1 corresponds to $\overline{\tau}_c$ for both A and C, level 2 corresponds to $\overline{\tau}_c$ for C, and $\overline{\tau}_e$ for A, and level 3 corresponds to $\overline{\tau}_e$ for both

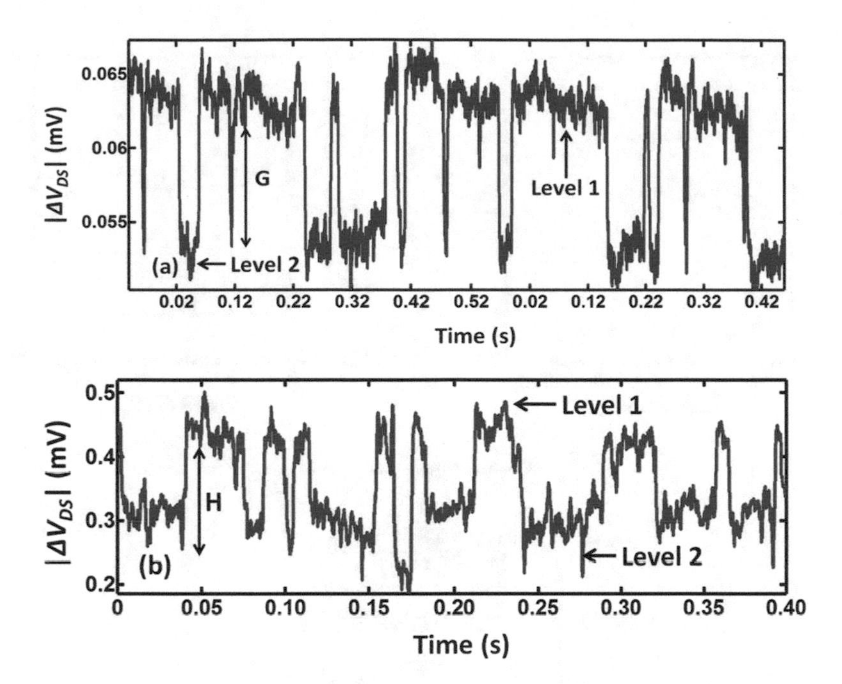

Fig. 36 (**a**) Portion of two-level RTS trace at $V_{GS} = -1.7$ V, $V_{DS} = -0.2$ V, 2500 s of stress with only trap G active; (**b**) RTS at $V_{GS} = -2.1$ V, $V_{DS} = -0.6$ V, 2500 s of stress, with only trap H active. As H is a repulsive center, in (**b**), level 1 corresponds to emission time, while level 2 corresponds to capture time [14]

A and C (Fig. 33c). In case of the four-level complex RTS due to traps A, B, and D, level 1 indicates $\overline{\tau}_c$ for A, B, and D, level 2 indicates $\overline{\tau}_c$ for A and B, and $\overline{\tau}_e$ for D, level 3 indicates $\overline{\tau}_c$ for B, and $\overline{\tau}_e$ for A and D, and finally level 4 indicates $\overline{\tau}_c$ for A, and $\overline{\tau}_e$ for B and D (Fig. 34c). $\overline{\tau}_c$ and $\overline{\tau}_e$ for traps E and F were extracted using the same technique as traps A and C. Traps G and H resulted in two-level RTS each, and hence, extraction of capture and emission time constants for these two traps was straight forward. Like trap A (Fig. 37a), time constants of none of the traps showed any particular dependence on stress time.

As mentioned before, the RTS magnitude due to each trap was extracted from the difference in the peaks of the corresponding Gaussians. For each trap, the same V_{GS} was applied at all the stress intervals to ensure a similar bias condition through all stress times. However, the threshold voltage might shift with stress, and hence, the number of channel holes will not be the same at the same V_{GS} at each stress. To remedy this problem, while calculating the inversion layer hole concentration, the change in threshold voltage (ΔV_T) was taken into account, and the channel hole concentration was determined at the corresponding $V_{GS} - |\Delta V_T|$. Unlike traps A to G, the number and mobility fluctuation terms in (8) are added for trap H. Therefore, as expected in case of a repulsive center, trap H exhibited much higher RTS magnitude than the other seven, attractive, traps (Fig. 38a–h). However, neither

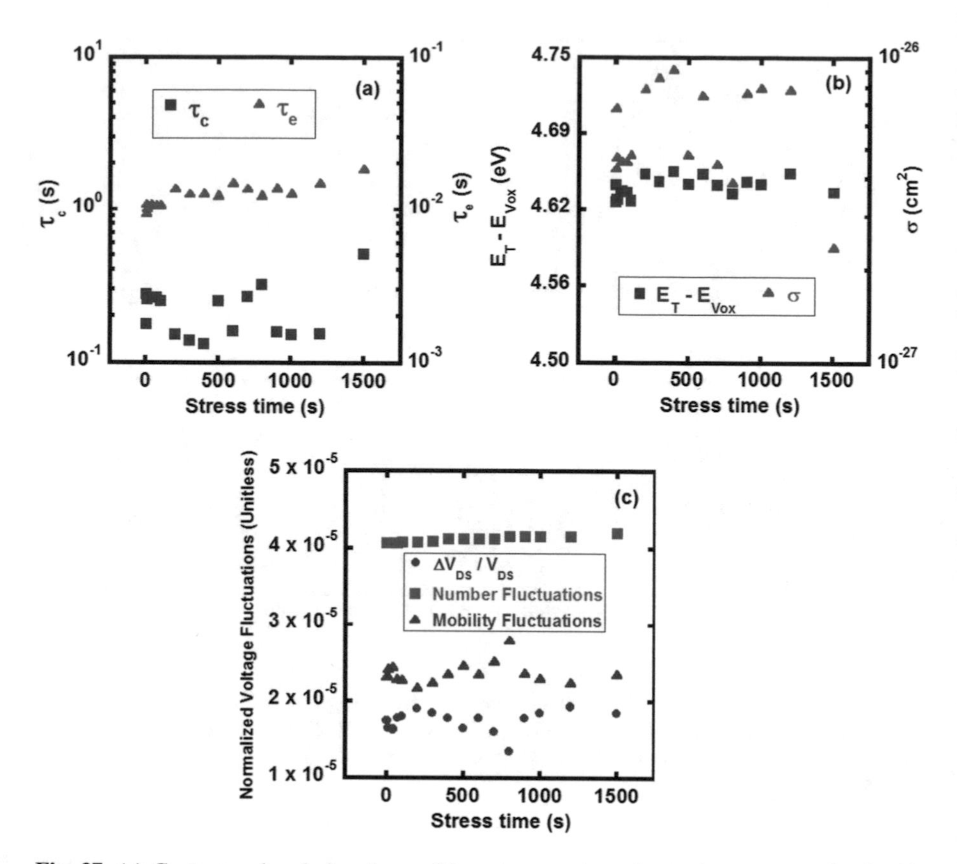

Fig. 37 (**a**) Capture and emission times; (**b**) capture cross section and trap energy level with respect to oxide valence band edge; (**c**) normalized RTS magnitude, number, and mobility fluctuations of trap A at different stress intervals at $V_{GS} = -2.4$ V, $V_{DS} = -0.4$ V. Just like trap A, no particular dependence of the trap parameters on stress time was observed for the other traps as well [14]

RTS magnitude, nor number fluctuations, or mobility fluctuations due to any trap exhibited a particular pattern with respect to varying V_{GS} (shown for trap A in Fig. 37c).

x_T was calculated for each trap using Eq. (6), and $E_T - E_{V_{ox}}$ was found using (5). x_T for all traps varied between 0.60 and 1.65 nm (Fig. 39), while $E_T - E_{V_{ox}}$ varied from 4.60 to 4.78 eV (Table 8). These values once again suggest that the traps are not Pb centers [40], and the defects responsible for RTS (both process-induced and stress-induced) stay just below the silicon valence band edge in the energy band diagram, and hence become easily accessible by the channel holes (Fig. 4b). σ was calculated through (1). Since both σ and $E_T - E_{V_{ox}}$ values are extracted from $\overline{\tau}_c$ and $\overline{\tau}_e$, the random behavior of the time constants led to no particular pattern in σ and $E_T - E_{V_{ox}}$ as a function of stress time (Fig. 37b). Therefore, application of stress had no effect on the RTS or trap parameters. Rather it affected only the generation

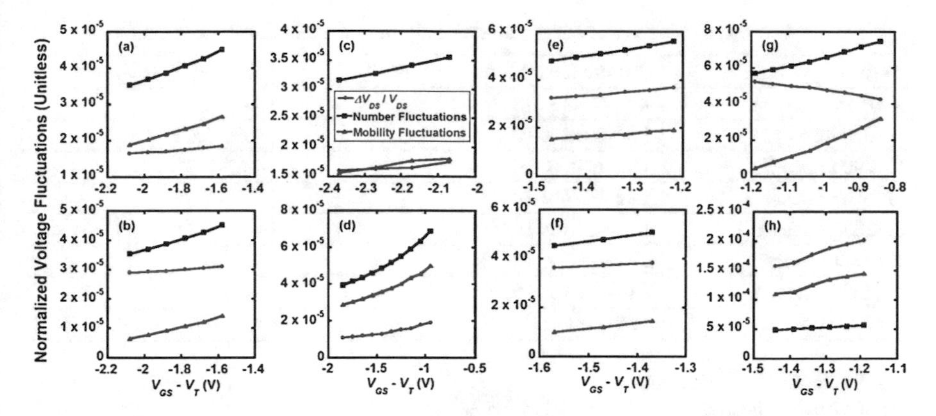

Fig. 38 RTS magnitude normalized to drain voltage, number fluctuations, and mobility fluctuations for (**a**) trap A at $V_{DS} = -0.4$ V, 0 s stress; (**b**) trap B at $V_{DS} = -0.4$ V, 0 s stress; (**c**) trap C at $V_{DS} = -0.4$ V, 200 s stress; (**d**) trap D at $V_{DS} = -0.6$ V, 800 s stress; (**e**) trap E at $V_{DS} = -0.2$ V, 2000 s stress; (**f**) trap F at $V_{DS} = -0.3$ V, 2000 s stress; (**g**) trap G at $V_{DS} = -0.2$ V, 2500 s stress; (**h**) trap H at $V_{DS} = -0.6$ V, 2500 s stress. As H is a repulsive center, unlike the other seven traps, the number and mobility fluctuations are added resulting in larger RTS magnitude [14]

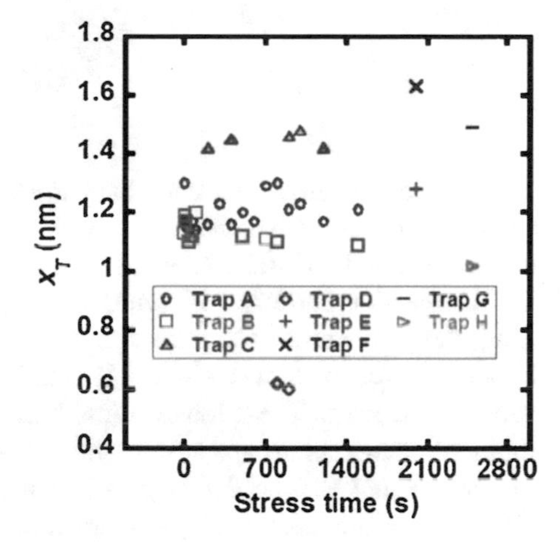

Fig. 39 Trap location of the traps at different stress intervals. The trap distances indicate that none of the traps is Pb center [14]

and passivation of the defects. The σ and $E_T - E_{V_{ox}}$ values were found to be in a similar range as the hole traps reported in the previous section, the attractive centers being a pair of dissociated silicon defects (D–III–Si), and the repulsive center a puckered/back-projected silicon vacancy defect.

When an electric field is applied, a polarization is introduced in the lattice molecules, leading to distortion in the lattice [108]. Hence the oxide molecules will start to feel a local electric field that is much larger than the applied electric field. For the aforementioned stressing conditions, the activation energy to break a Si–Si bond in a neutral silicon vacancy decreases to 0.77 eV from 1.15 eV [108], making this breakdown process possible to take place at room temperature. The structure formed after the Si–Si bond breakage (Fig. 40) is similar to the structure of the D–III–Si

Table 8 Capture cross sections and energy levels of all traps at a particular stress condition [14]

Trap	Capture cross section (σ) (cm^2)	$E_T - E_{V_{ox}}$ (eV)
A (0 s stress)	$4.92–7.92 \times 10^{-27}$	4.61–4.68
B (0 s stress)	$2.61–5.88 \times 10^{-27}$	4.61–4.68
C (200 s stress)	$4.83–6.35 \times 10^{-27}$	4.67–4.71
D (800 s stress)	$3.42–6.37 \times 10^{-26}$	4.74–4.85
E (2000 s stress)	$2.12–4.58 \times 10^{-25}$	4.66–4.71
F (2000 s stress)	$2.49–4.43 \times 10^{-25}$	4.58–4.63
G (2500 s stress)	$7.11–8.19 \times 10^{-26}$	4.64–4.67
H (2500 s stress)	$5.65–5.81 \times 10^{-26}$	4.72–4.74

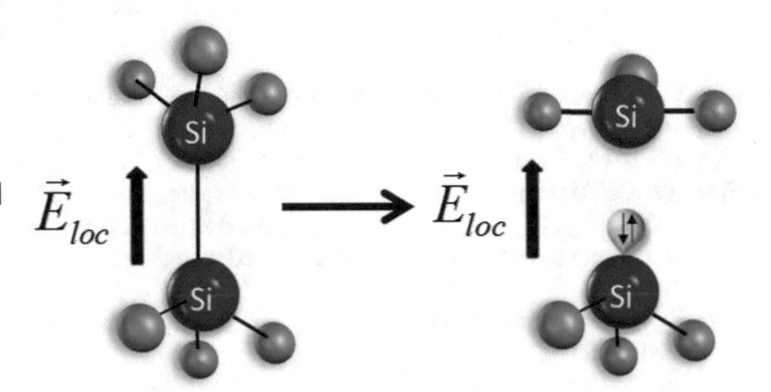

Fig. 40 Si–Si bond breakage mechanism in a neutral oxygen vacancy center. The resultant structure has two electrons in the sp^3 orbital, and can act as a hole trap [14]

defect reported earlier that contains paired electrons in an sp^3 orbital. Considering the similarity in the structure, σ and $E_T - E_{V_{ox}}$ values, the stress-induced defects are believed to be D–III–Si defects.

The reaction barrier of E' centers for hydrogen species is around 0.4 eV [109, 110], and the barrier for releasing a hydrogen molecule (H_2) by reacting with another hydrogen atom is ~0.5 eV [111]. Therefore, both reactions can take place at room temperature. These low reaction barriers of E' centers with hydrogen indicate that hydrogen capture and release might be the reason behind the volatile nature of traps B and C. The defect gets passivated by reacting with a hydrogen atom (H) or molecule (H_2), and gets reactivated when it reacts with another hydrogen atom, releasing H_2 (Fig. 41a, b). Reactivation of an already passivated trap might be another reason behind the volatility of traps B and C.

$$\mathrm{Si}^{\cdot} + \mathrm{H}/\mathrm{H}_2 \rightarrow \mathrm{Si{-}H} + \mathrm{H} \quad (\text{for } \mathrm{H}_2) \tag{36}$$

$$\mathrm{Si{-}H} + \mathrm{H} \rightarrow \mathrm{Si}^{\cdot} + \mathrm{H}_2 \tag{37}$$

where Si – H represents silicon–hydrogen bond.

Here, the source of hydrogen remains as the primary matter of concern. There have been numerous reports where plenty of hydrogen has been mentioned to remain in bound form at the Si–SiO_2 interface [112, 113], or captured by the oxide defects

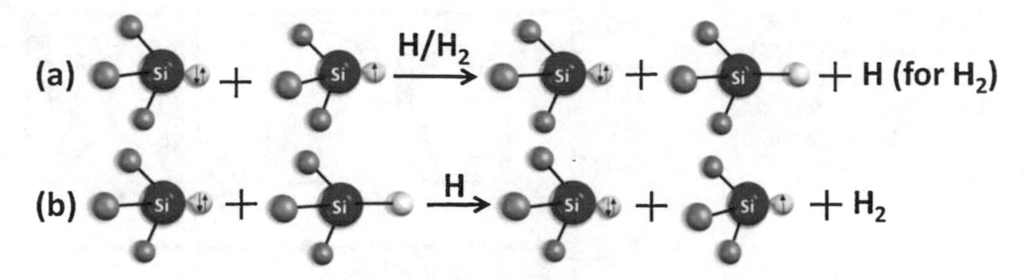

Fig. 41 (**a**) Passivation of the D–III–Si defect center via reaction with hydrogen resulting in a Si–H bond; (**b**) reactivation of the D–III–Si defect center through reaction of the passivated structure with another hydrogen atom, releasing H_2 [14]

[114]. Application of CHC generates new interface traps releasing the hydrogen species [101, 115]. These hydrogen species travel towards the gate, and get captured by the oxide defects. Even though according to the H/H_2 reaction–diffusion (RD) model, the stress-released hydrogen species are supposed to move out towards the bulk from the Si–SiO_2 interface [116], in our case, application of negative potential drifts the stress released hydrogen towards the gate [117].

To investigate the effects of stress on the trap energy parameters, RTS measurements were conducted at variable temperatures (255–295 K) using the same setup and technique described before. Stress was applied at suitable intervals at room temperature on 5 V rated devices at $V_{GS} = V_{DS} = -6.5$ V. T_{ox} of all the devices was 12.1 nm. The RTS measurements were taken as soon as the stressing was done. As mentioned earlier, the RTS traces were recorded in time domain in an oscilloscope, of which sampling frequency and time span were varied to record at least 500 transitions between the levels. Measurements were taken on 10 devices in total. Results of four devices are presented here: (1) SP41 of $W \times L = (1 \times 1)$ μm^2, in which a process-induced trap P41 was observed, which disappeared after 50 s of stress, and two stress-induced traps, $S41_{50A}$ and $S41_{50B}$ were observed at two different bias ranges. (2) SP221 of $W \times L = (1 \times 0.5)$ μm^2, in which a process-induced trap, P221 was observed, which disappeared after stress, (3) SP222 of $W \times L = (1 \times 0.5)$ μm^2, in which a process-induced trap, P222 was observed, which disappeared after stress. After 200 s of stress, a stress-induced trap $S222_{200}$ was observed, and (4) SP61 of $W \times L = (1 \times 0.7)$ μm^2, in which a process-induced trap, P61 was observed, which got deactivated after stress. Two stress-induced traps, $S61_{50}$ and $S61_{100}$ were observed after 50 and 100 s of stress, respectively. Hence, four process-induced traps, and five stress-induced traps were observed, all exhibiting a two-level RTS (Fig. 42). Depending on the active range of a particular trap, V_{GS} was varied from -1.25 to -2.7 V, V_{DS} was varied between -0.05 V, and $-$ 0.5 V, and temperature was varied between 255 K and 295 K. The fact that the process-induced traps disappeared with stressing once again confirms that stressing did not have any effect on the RTS or trap parameters. Instead, stressing impacts only the oxide trap generation and passivation mechanism.

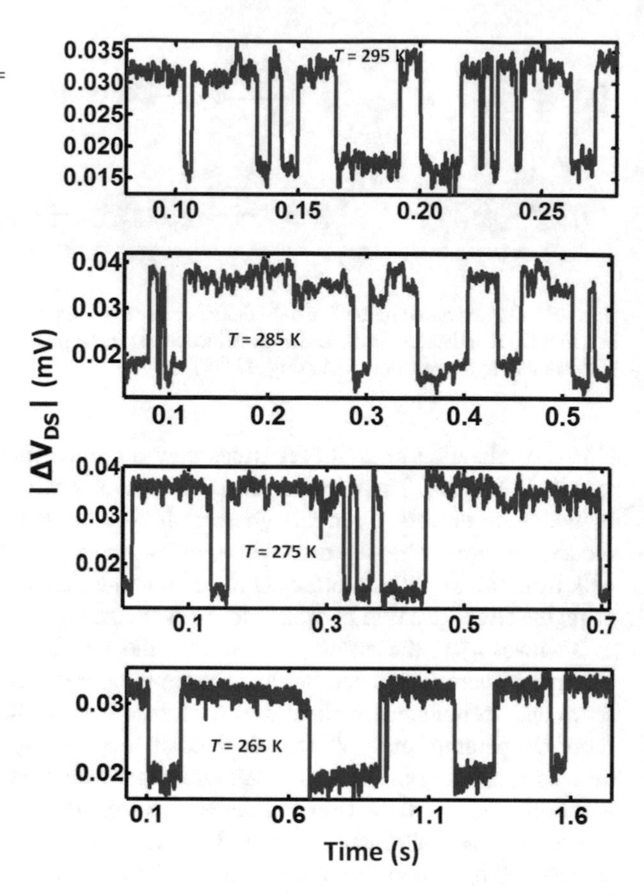

Fig. 42 Two-level RTS observed due to P41 at $V_{GS} = -2.0$ V, $V_{DS} = -0.5$ V. Similar two-level RTS was observed due to the other traps

As discussed earlier, the RTS analyses started with extracting the average time spent at each RTS level using the previously mentioned method. From the behavior of the mean times at the RTS upper and lower levels as a function of V_{GS} magnitude, the upper level was found to correspond to $\overline{\tau}_c$ for all the traps. Hence, all of the traps were identified as attractive centers. Their RTS magnitudes calculated through the corresponding Gaussian peaks also justified this since RTS magnitudes (Fig. 43) of all the traps were within a similar range as that of the attractive centers reported earlier in the chapter.

x_T was calculated using (6), σ was calculated through (1), and $E_T - E_{V_{ox}}$ was calculated using (5). Once again, the x_T values (Fig. 44) discarded the Pb centers from the possible defects [40], and the $E_T - E_{V_{ox}}$ values (Table 9) placed the defects just below the silicon valence band edge. The σ values (Table 9) were in the similar range as the other hole traps discussed earlier.

ΔE_B (Fig. 45a) was calculated using the Arrhenius plot of σ, and ΔH (Fig. 45b) and ΔS (Fig. 45c) were extracted from the Arrhenius plot of normalized $\overline{\tau}_e$. E_R (Fig. 45d) was calculated solving (7). The ΔE_B, E_R, and $E_T - E_{V_{ox}}$ values of the stress-induced traps match with those of the attractive center hole defects studied

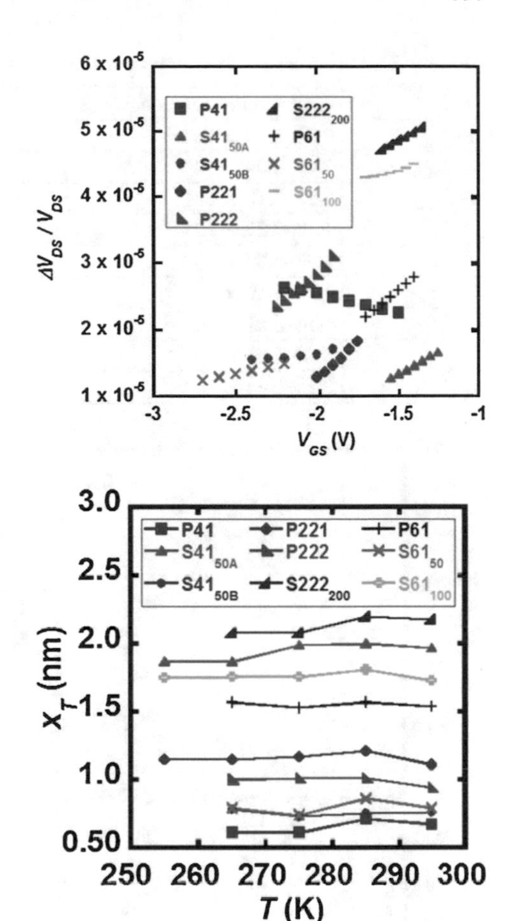

Fig. 43 Normalized RTS magnitude of all the traps. The RTS magnitudes of all the traps are in the similar range as those of the attractive centers discussed earlier

Fig. 44 Trap locations of all the traps at different temperatures. As discussed before, the trap locations indicate that these defects are not Pb centers

earlier in the chapter. This confirms the claim made earlier in this section that the stress-induced attractive centers are in fact dissociated silicon defects (D–III–Si).

9 Comparison of RTS with Other Trap Characterization Techniques

Comparing with other trap characterization techniques, RTS has certain advantages as well as limitations. Charge pumping is such a technique being used to characterize hole traps. The hole capture cross sections found using RTS [96] exhibited much lower values than that found using charge-pumping technique [118–120]. This is because of the difference in time constants of the defects these two techniques are probing into. Charge-pumping probes into the oxide–semiconductor interface traps, whereas RTS probes into the oxide defects responsible for output noise.

Table 9 Capture cross sections and energy levels of all nine traps

Trap	Capture cross-section (σ) (cm^2)	$E_T - E_{V_{ox}}$ (eV)	E_R (eV)
P41	2.34–3.90 × 10^{-26}	4.68–4.75	2.28–2.35
$S41_{50A}$	7.47–8.94 × 10^{-26}	4.56–4.63	1.66–1.97
$S41_{50B}$	2.88–5.10 × 10^{-26}	4.71–4.77	2.11–2.21
P221	1.19–2.27 × 10^{-25}	4.68–4.72	2.10–2.26
P222	1.70–2.54 × 10^{-26}	4.70–4.74	2.16–2.26
$S222_{200}$	1.00–3.59 × 10^{-25}	4.56–4.63	1.61–1.71
P61	4.76–8.81 × 10^{-26}	4.61–4.67	1.98–2.08
$S61_{50}$	3.50–5.01 × 10^{-26}	4.73–4.78	1.82–2.05
$S61_{100}$	2.26–6.44 × 10^{-25}	4.62–4.68	1.46–1.63

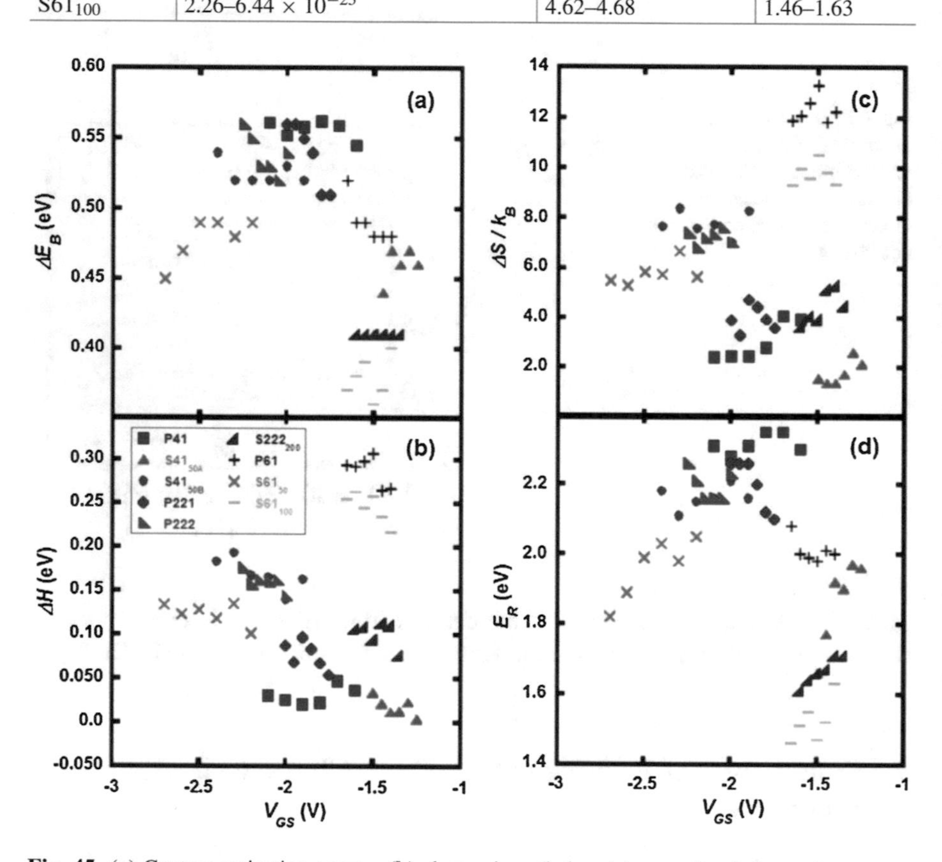

Fig. 45 (**a**) Capture activation energy, (**b**) change in enthalpy, (**c**) normalized change in entropy, and (**d**) relaxation energy. $V_{DS} = -0.05$ V for P222; $V_{DS} = -0.1$ V for $S41_{50A}$, P61, $S61_{100}$; $V_{DS} = -0.15$ V for $S222_{200}$; $V_{DS} = -0.2$ V for $S61_{50}$; $V_{DS} = -0.3$ V for $S41_{50B}$; $V_{DS} = -0.5$ V for P41, P221; In (**d**), $T = 295$ K

RTS probes into different traps than the time-dependent defect spectroscopy (TDDS) technique as well [26, 27, 92]. RTS can detect traps with switching states, whereas TDDS detects traps that undergo high relaxation after charging, causing gate leakage. RTS can detect traps with ratio of time constants ~0.1–10, whereas TDDS can characterize traps of much wider range of time constants.

A major advantage of using RTS over the other techniques is that RTS can detect the charge state of the defects before and after the capture of a carrier, which has a large impact on the noise magnitude as shown earlier. However, the energy window of the traps detected by RTS is limited to $\pm 3k_BT$ of the Fermi energy level [42], while the other two techniques are able to diagnose traps of much wider energy range.

10 Discussions

The most common defects observed in amorphous SiO_2 are due to a missing oxygen or additional hydrogen atom. Among all these defects, the E' centers are capable of capturing both electrons and holes, as they have been observed in both nMOSFETs and pMOSFETs.

The capture and emission of electrons and holes are due to similar mechanisms except the fact that in nMOSFETs, emission of an electron results in a free electron in the conduction band, whereas in case of pMOSFETs, emission of a hole results in the capture of a bonded electron from the valence band.

Surprisingly, the capture activation energy and relaxation energy values of the repulsive centers in both nMOSFETs and pMOSFETs showed similar values. This indicates that the puckered/back-projected silicon vacancy defect is able to capture an electron from the valence band and emit it to the conduction band, which means that this defect can exhibit multiple charge states. This agrees with the observations reported by Cobden et al. [96]. However, the physical structure of the defect was not identified back then.

The enthalpy and change in entropy values of the hole defects (both attractive and repulsive) were found to follow the same trend line. The fact that only the under-coordinated silicon atom undergoes structural change in both the defects might be the reason behind this. However, the trap PMOS550 that shows RTS at a lower drain voltage than the other traps showed higher entropy values. This can be explained if that defect is assumed to be near the drain side of the channel, meaning that a higher change in energy will be needed for emitting a hole to the valence band.

Trap volatility has also been reported by Grasser et al. [107] using TDDS experiments, where the volatile behavior of the traps were attributed to the capture and release of hydrogen by the hydroxyl E' centers. However, the reaction barrier for hydroxyl E' centers with hydrogen varies between 1.25 to 2.45 eV [93]. These reactions are highly unlikely to take place at room temperature. The experiments reported in [107] were done at an elevated temperature (125 °C), which supports the fact that TDDS and RTS indeed probe into two different types of traps [121].

11 Conclusion

RTS analysis is an effective technique to obtain information about both electron and hole oxide traps in MOSFETs. In scaled devices, a strong transverse field and higher substrate doping can result into discrete energy levels of the charge carriers. In that case, to get back to the channel, the carrier needs to obtain some extra energy. While analyzing the capture and emission times, carrier mobility, and screened scattering coefficients, these effects need to be taken into account. Unlike other trap characterization techniques, RTS can detect if the responsible defect is an attractive center, or a repulsive center, which significantly affects the output noise magnitude. The capture and emission of charge carriers into the trap is a thermally activated process. Analyzing RTS at different temperatures allows obtaining the trap energy parameters such as capture activation energy, relaxation energy, change in enthalpy and entropy that are involved with the carrier trapping and detrapping process. These parameters can then be utilized to study the structural behavior of the defects. In nMOSFETs, only repulsive center defects were found, and the responsible trap was found to be a neutral oxygen deficiency center (V^0 ODC II), whereas in pMOSFETs, both attractive and repulsive center traps were observed. The attractive center was a pair of threefold dissociated silicon atoms (D–III–Si), while the repulsive center was found to be a puckered/back-projected silicon vacancy defect. An RTS simulation tool RTSSIM was developed which can successfully reconstruct the RTS signal using the provided trap parameters for a single attractive center defect, single repulsive center trap, and two dependent repulsive center traps. Hot carrier effects are also a growing concern in the ultra-scaled devices. Channel hot carrier stress was found to generate new traps, and cause additional RTS in the output signal. The stress-induced traps were found to have the same structure as the process-induced traps. Application of stress did not seem to affect any of the RTS or trap parameters. Instead, it had impact only on the trap generation and passivation mechanism, which was later verified in variable temperature measurements. Trap volatility was observed as a result of application of CHC. Hydrogen species were suggested to be responsible behind such behavior of the traps. Traps get neutralized after capturing a hydrogen atom, and get reactivated after reacting with another hydrogen releasing H_2. Information thus obtained about the physical structure of the oxide-trapping centers causing RTS is helpful in developing processing techniques to minimize their number and impact on MOSFET performance.

Acknowledgments We would like to thank Texas Instruments for supplying some of the devices used in this study.

References

1. K. Kandiah, F.B. Whiting, Low frequency noise in junction field effect transistors. Solid State Electron. **21**(8), 1079–1088 (1978). https://doi.org/10.1016/0038-1101(78)90188-0

2. K.S. Ralls, W.J. Skocpol, L.D. Jackel, R.E. Howard, L.A. Fetter, R.W. Epworth, D.M. Tennant, Discrete resistance switching in submicrometer silicon inversion layers: individual interface traps and low-frequency *(1/f?)* noise. Phys. Rev. Lett. **52**(3), 228–231 (1984). https://doi.org/10.1103/PhysRevLett.52.228
3. N. Sghaier, L. Militaru, M. Trabelsi, N. Yacoubi, A. Souifi, Analysis of slow traps centres in submicron MOSFETs by random telegraph signal technique. Microelectron. J. **38**(4–5), 610–614 (2007). https://doi.org/10.1016/j.mejo.2007.02.003
4. Z. Shi, J.P. Miéville (Mieville), M. Dutoit, Random telegraph signals in deep submicron n-MOSFETs. IEEE Trans. Electron Devices **41**(7), 1161–1168 (1994). https://doi.org/10.1109/16.293343
5. N.V. Amarasinghe, Z. Çelik-Butler, A. Zlotnicka, F. Wang, Model for random telegraph signals in sub-micron MOSFETS. Solid State Electron. **47**(9), 1443–1449 (2003). https://doi.org/10.1016/S0038-1101(03)00100-X
6. J.P. Campbell, J. Qin, K.P. Cheung, L.C. Yu, J.S. Suehle, A. Oates, K. Sheng, Random telegraph noise in highly scaled nMOSFETs, in *IRPS*, 2009, pp. 382–388. https://doi.org/10.1109/IRPS.2009.5173283
7. M. Nour, Z. Çelik-Butler, A. Sonnet, F.C. Hou, S. Tang, Random telegraph signals originating from unrelaxed neutral oxygen vacancy centres in SiO_2. Electron. Lett. **51**(20), 1610–1611 (2015). https://doi.org/10.1049/el.2015.2074
8. M.I. Mahmud, *Investigation of Degradation in Advanced Analog MOS Technologies* (Univ. of Texas Arlington, Arlington, TX, 2013)
9. Z. Celik Butler, F. Wang, Effect of quantization on random telegraph signals observed in deep-submicron MOSFETs. Microelectron. Reliab. **40**(11), 1823–1831 (2000). https://doi.org/10.1016/S0026-2714(00)00083-4
10. M. Schulz, A. Karmann, Individual, attractive defect centers in the SiO_2-Si interface of μm-sized MOSFETs. Appl. Phys. A **52**(2), 104–111 (1991). https://doi.org/10.1007/BF00323724
11. J.H. Scofield, N. Borland, D.M. Fleetwood, Random telegraph signals in small gate area p-MOS transistors. AIP Conf. Proc. **285**, 386–389 (1993). https://doi.org/10.1063/1.44673
12. J.H. Scofield, N. Borland, D.M. Fleetwood, Temperature-independent switching rates for a random telegraph signal in a silicon metal-oxide-semiconductor field-effect transistor at low temperatures. Appl. Phys. Lett. **76**(22), 3248–3250 (2000). https://doi.org/10.1063/1.126596
13. A.S.M.S. Rouf, Z. Çelik-Butler, F.C. Hou, S. Tang, G. Mathur, Two types of E′ centers as gate oxide defects responsible for hole trapping and random telegraph signals in pMOSFETs. IEEE Trans. Electron Devices **65**(10), 4527–4534 (2018). https://doi.org/10.1109/TED.2018.2866229
14. A.S.M.S. Rouf, Z. Çelik-Butler, F.C. Hou, S. Tang, G. Mathur, Channel hot carrier induced volatile oxide traps responsible for random telegraph signals in a submicron pMOSFET, Solid State Electron. 164, 107745 (2020). https://doi.org/10.1016/j.sse.2019.107745.
15. K.V. Aadithya, A. Demir, S. Venugopalan, J. Roychowdhury, Accurate prediction of random telegraph noise effects in SRAMs and DRAMs. IEEE Trans. Comput. Aided Design Integr. Circuits Syst. **32**(1), 73–86 (2013). https://doi.org/10.1109/TCAD.2012.2212897
16. J. Martin-Martinez, R. Rodriguez, M. Nafria, G. Torrens, S.A. Bota, J. Segura, F. Moll, A. Rubio, Statistical characterization and modeling of random telegraph noise effects in 65nm SRAMs cells, in *14th International Conference on Synthesis Modeling Analysis and Simulation Methods and Applications to Circuit Design (SMACD)*, 2017, pp. 1–4. https://doi.org/10.1109/SMACD.2017.7981610
17. D. Veksler, G. Bersuker, L. Vandelli, A. Padovani, L. Larcher, A. Muraviev, B. Chakrabarti, E. Vogel, D.C. Gilmer, P.D. Kirsch, Random telegraph noise (RTN) in scaled RRAM devices, in *Proc. IEEE Int. Rel. Phys. Symp. (IRPS)*, 2013, pp. MY.10.1–MY.10.4. https://doi.org/10.1109/IRPS.2013.6532101
18. H. Kurata, K. Otsuga, A. Kotabe, S. Kajiyama, T. Osabe, Y. Sasago, S. Narumi, K. Tokami, S. Kamohara, O. Tsuchiya, Random telegraph signal in flash memory: its impact on scaling of multilevel flash memory beyond the 90-nm node. IEEE J. Solid State Circuits **42**(6), 1362–1369 (2007). https://doi.org/10.1109/JSSC.2007.897158

19. S. Realov, K.L. Shepard, Analysis of random telegraph noise in 45-nm CMOS using on-chip characterization system. IEEE Trans. Electron Devices **60**(5), 1716–1722 (2013). https://doi.org/10.1109/TED.2013.2254118
20. M.J. Uren, M.J. Kirton, S. Collins, Anomalous telegraph noise in small-area silicon metal-oxide-semiconductor field-effect transistors. Phys. Rev. B **37**(14), 8346–8350 (1988). https://doi.org/10.1103/PhysRevB.37.8346
21. T. Grasser, H. Reisinger, P.J. Wagner, F. Schanovsky, W. Goes, B. Kaczer, The time dependent defect spectroscopy (TDDS) for the characterization of the bias temperature instability, in *IRPS*, 2010, pp. 2A.2.1–2A.2.10. https://doi.org/10.1109/IRPS.2010.5488859
22. T. Grasser, Stochastic charge trapping in oxides: from random telegraph noise to bias temperature instabilities. Microelectron. Reliab. **52**(1), 39–70 (2012). https://doi.org/10.1016/j.microrel.2011.09.002
23. A.L. Shluger, K.P. McKenna, Models of oxygen vacancy defects involved in degradation of gate dielectrics, in *IRPS*, 2013, pp. 5A.1.1–5A.1.9. https://doi.org/10.1109/IRPS.2013.6532018
24. N.L. Anderson, R.P. Vedula, P.A. Schultz, R.M. Van Ginhoven, A. Strachan, First-principles investigation of low energy E′ center precursors in amorphous silica. Phys. Rev. Lett. **106**(20), 2064021–2064024 (2011). https://doi.org/10.1103/PhysRevLett.106.206402
25. N.L. Anderson, R.P. Vedula, P.A. Schultz, R.M. van Ginhoven, A. Strachan, Defect level distributions and atomic relaxations induced by charge trapping in amorphous silica. Appl. Phys. Lett. **100**(17), 172908-1–172908-4 (2012). https://doi.org/10.1063/1.4707340
26. T. Grasser, W. Goes, Y. Wimmer, F. Schanovsky, G. Rzepa, M. Waltl, K. Rott, H. Reisinger, V.V. Afanas'ev, A. Stesmans, A.M. El-Sayed, A.L. Shluger, On the microscopic structure of hole traps in pMOSFETs, in *IEDM Tech. Dig.*, 2014, pp. 530–533. https://doi.org/10.1109/IEDM.2014.7047093
27. T. Grasser, K. Rott, H. Reisinger, M. Waltl, P. Wagner, F. Schanovsky, W. Goes, G. Pobegen, B. Kaczer, Hydrogen-related volatile defects as the possible cause for the recoverable component of NBTI, in *IEDM Tech. Dig.*, 2013, pp. 409–412. https://doi.org/10.1109/IEDM.2013.6724637
28. D.M. Fleetwood, H.D. Xiong, Z.Y. Lu, C.J. Nicklaw, J.A. Felix, R.D. Schrimpf, S.T. Pantelides, Unified model of hole trapping 1/f noise and thermally stimulated current in MOS devices. IEEE Trans. Nucl. Sci. **49**(6), 2674–2683 (2002). https://doi.org/10.1109/TNS.2002.805407
29. M.J. Kirton, M.J. Uren, Noise in solid-state microstructures: a new perspective on individual defects, interface states and low frequency (1/f) noise. Adv. Phys. **38**(4), 367–468 (1989). https://doi.org/10.1080/00018738900101122
30. N.V. Amarasinghe, Z. Çelik-Butler, A. Keshavarz, Extraction of oxide trap properties using temperature dependence of random telegraph signals in submicrometer MOSFETs. J. Appl. Phys. **89**(10), 5526–5532 (2001). https://doi.org/10.1063/1.1367404
31. D. Veksler, G. Bersuker, H. Park, C. Young, K.Y. Lim, W. Taylor, S. Lee, H. Shin, The critical role of the defect structural relaxation for interpretation of noise measurements in MOSFETs, in *Proc. IEEE IIRW*, 2009, pp. 102–105. https://doi.org/10.1109/IRWS.2009.5383021
32. L. Vandelli, A. Padovani, L. Larcher, R.G. Southwick III, W.B. Knowlton, G. Bersuker, A physical model of the temperature dependence of the current through SiO_2 /HfO_2 stacks. IEEE Trans. Electron Devices **58**(9), 2878–2887 (2011). https://doi.org/10.1109/TED.2011.2158825
33. A. Padovani, D.Z. Gao, A.L. Shluger, L. Larcher, A microscopic mechanisms of dielectric breakdown in SiO_2 films: an insight from multi-scale modeling. J. Appl. Phys. **121**(15), 155101–155110 (2017).https://doi.org/10.1063/1.4979915
34. K.K. Hung, P.K. Ko, C. Hu, Y.C. Cheng, A unified model for the flicker noise in metal-oxide-semiconductor field-effect transistors. IEEE Trans. Electron Devices **37**(3), 654–665 (1990). https://doi.org/10.1109/16.47770
35. R.R. Siergiej, M.H. White, N.S. Saks, Theory and measurement of quantization effects on Si-SiO_2 interface trap modeling. Solid State Electron. **35**(6), 843–854 (1992). https://doi.org/10.1016/0038-1101(92)90287-M

36. M.J. Van Dort, P.H. Woerlee, A.J. Walker, C.A. Juffermans, H. Lifka, Influence of high substrate doping levels on the threshold voltage and the mobility of deep-submicrometer MOSFET's. IEEE Trans. Electron Devices **39**(4), 932–938 (1992). https://doi.org/10.1109/16.127485.
37. F. Stern, W.E. Howard, Properties of semiconductor surface inversion layers in the electric quantum limit. Phys. Rev. **163**(3), 816–835 (1967). https://doi.org/10.1103/PhysRev.163.816
38. C. Surya, T.Y. Hsiang, Surface mobility fluctuations in metal-oxide-semiconductor field-effect transistors. Phys. Rev. B **35**(12), 6343–6347 (1987). https://doi.org/10.1103/PhysRevB.35.6343
39. A.V. Kimmel, P.V. Sushko, A.L. Shluger, G. Bersuker, Positive and negative oxygen vacancies in amorphous silica. ECST **19**(2), 3–17 (2009). https://doi.org/10.1149/1.3122083
40. R. Salh, Defect related luminescence in silicon dioxide network: a review, in *Crystalline Silicon: Properties and Uses*, ed. by S. Basu, (InTech, Rijeka, 2011), p. 137. https://doi.org/10.5772/22607
41. D. Vuillaume, A. Bravaix, Charging and discharging properties of electron traps created by hot carrier injections in gate oxide of n-channel metal oxide semiconductor field effect transistor. J. Appl. Phys. **73**(5), 2559–2563 (1993). https://doi.org/10.1063/1.353065
42. M. Nour, *Measurements, Modeling, and Simulation of Semiconductor/Gate Dielectric Defects Using Random Telegraph Signals* (Univ. of Texas Arlington, Arlington, TX, 2015)
43. P.V. Sushko, S. Mukhopadhyay, A.M. Stoneham, A.L. Shluger, Oxygen vacancies in amorphous silica: structure and distribution of properties. Microelectron. Eng. **80**, 292–295 (2005). https://doi.org/10.1016/j.mee.2005.04.083
44. V.J. Kappor, W.D. Brown, *Silicon Nitride and Silicon Dioxide Thin Insulator Films* (The Electrochemical Society Inc., Pennington, NJ, 1997)
45. R. Perera, A. Ikeda, R. Hattori, Y. Kuroki, Trap assisted leakage current conduction in thin silicon oxynitride films grown by rapid thermal oxidation combined microwave excited plasma nitridation combined microwave excited plasma nitridation. Microelectron. Eng. **65**(4), 357–370 (2003). https://doi.org/10.1016/S0167-9317(02)01025-0
46. S. Suzuki, D.K. Schroder, Y. Hayashi, Carrier conduction in ultrathin nitrided oxide films. J. Appl. Phys. **60**(10), 3616–3621 (1986). https://doi.org/10.1063/1.337568
47. S. Fleischer, P.T. Lai, Y.C. Cheng, A new method for extracting the trap energy in insulators. J. Appl. Phys. **73**(7), 3348–3351 (1993). https://doi.org/10.1063/1.352934
48. X.R. Cheng, Y.C. Cheng, B.Y. Liu, Nitridation-enhanced conductivity behavior and current transport mechanism in thin thermally nitrided SiO_2. J. Appl. Phys. **63**(3), 797–802 (1988). https://doi.org/10.1063/1.340072
49. A. Alkauskas, A. Pasquarello, Alignment of hydrogen–related defect levels at the Si–SiO_2 interface. Phys. B Condens. Matter **401–402**, 546–549 (2007). https://doi.org/10.1016/j.physb.2007.09.018
50. S.N. Rashkeev, D.M. Fleetwood, R.D. Schrimpf, S.T. Pantelides, Dual behavior of H+ at Si–SiO_2 interfaces: mobility versus trapping. Appl. Phys. Lett. **81**(10), 1839–1841 (2002). https://doi.org/10.1063/1.1504879
51. W. Goes, M. Karner, V. Sverdlov, T. Grasser, Charging and discharging of oxide defects in reliability issues. IEEE Trans. Device Mater. Reliab. **8**(3), 491–500 (2008). https://doi.org/10.1109/TDMR.2008.2005247
52. A.C. Pineda, S.P. Karna, Electronic structure theory of radiation–induced defects in Si/SiO_2, in *HPCERC*, 2000, pp. 1–39
53. M. Boero, A. Pasquarello, J. Sarnthein, R. Car, Structure and hyperfine parameters of E'_1 centers in α-quartz and in vitreous SiO_2. Phys. Rev. Lett. **78**(5), 887–890 (1997). https://doi.org/10.1103/PhysRevLett.78.887
54. N.L. Anderson, Structural, thermodynamic, electronic, and magnetic, characterization of point defect in amorphous silica. PhD dissertation, School of Materials Engineering, Purdue University, West Lafayette, IN, 2012
55. Z.Y. Lu, C.J. Nicklaw, D.M. Fleetwood, R.D. Schrimpf, S.T. Pantelides, Structure, properties, and dynamics of oxygen vacancies in amorphous SiO_2. Phys. Rev. Lett. **89**(28), 285505-1–285505-4 (2002). https://doi.org/10.1103/PhysRevLett.89.285505

56. M. Nour, Z. Çelik-Butler, A. Sonnet, F.C. Hou, S. Tang, G. Mathur, A stand-alone, physics-based, measurement-driven model and simulation tool for random telegraph signals originating from experimentally identified MOS gate-oxide defects. IEEE Trans. Electron Devices **63**(4), 1428–1436 (2016). https://doi.org/10.1109/TED.2016.2528218
57. N.H. Hamid, A.F. Murray, S. Roy, Time-domain modeling of low frequency noise in deep-submicrometer MOSFET. IEEE Trans. Circuits Syst. I, Reg. Papers **55**(1), 245–257 (2008). https://doi.org/10.1109/TCSI.2007.910543
58. T.B. Tang, A.F. Murray, S. Roy, Methodology of statistical RTS noise analysis with charge-carrier trapping models. IEEE Trans. Circuits Syst. I, Reg. Papers **57**(5), 1062–1070 (2010). https://doi.org/10.1109/TCSI.2010.2043988
59. M.N. Higashi, H. Sasaki, H.S. Momose, T. Ohguro, Y. Mitani, T. Ishihara, K. Matsuzawa, Unified transient and frequency domain noise simulation for random telegraph noise and flicker noise using a physics-based model. IEEE Trans. Electron Devices **61**(12), 4197–4203 (2014). https://doi.org/10.1109/TED.2014.2365015
60. A.P. Van der Wel, E.A.M. Klumperink, L.K.J. Vandamme, B. Nauta, Modeling random telegraph noise under switched bias conditions using cyclostationary RTS noise. IEEE Trans. Electron Devices **50**(5), 1378–1384 (2003). https://doi.org/10.1109/TED.2003.813247
61. G. Wirth, D. Vasileska, N. Ashraf, L. Brusamarello, R.D. Guistina, P. Srinivasan, Compact modeling and simulation of random telegraph noise under non-stationary conditions in the presence of random dopants. Microelectron. Reliab. **52**(12), 2955–2961 (2012). https://doi.org/10.1016/j.microrel.2012.07.011
62. G.I. Wirth, J. Koh, R. Da Silva, R. Thewes, R. Brederlow, Modeling of statistical low-frequency noise of deep-submicrometer MOSFETs. IEEE Trans. Electron Devices **52**(7), 1576–1588 (2005). https://doi.org/10.1109/TED.2005.850955
63. A. Ghetti, M. Bonanomi, C.M. Compagnoni, A.S. Spinelli, A.L. Lacaita, A. Visconti, Physical modeling of single-trap RTS statistical distribution in flash memories, in *Proc. IEEE Int. Rel. Phys. Symp.*, 2008, pp. 610–615. https://doi.org/10.1109/RELPHY.2008.4558954
64. A. Asenov, R. Balasubramaniam, A.R. Brown, J.H. Davies, RTS amplitudes in decananometer MOSFETs: 3-D simulation study. IEEE Trans. Electron Devices **50**(3), 839–845 (2003). https://doi.org/10.1109/TED.2003.811418
65. H. Awano, H. Tsutsui, H. Ochi, T. Sato, Bayesian estimation of multi-trap RTN parameters using Markov chain Monte Carlo method. IEICE Trans. Fundam. Electron. Commun. Comput. Sci. **E95-A**(12), 2272–2283 (2012). https://doi.org/10.1587/transfun.E95.A.2272
66. Z. Çelik-Butler, T.Y. Hsiang, Spectral dependence of full-size image $1/f\gamma$ noise on gate bias in *n*-MOSFETs. Solid-State Electron. **30**(4), 419–423 (1987). https://doi.org/10.1016/j.sse.2017.06.003
67. D.K. Schroder, *Semiconductor Material and Device Characterization*, 3rd edn. (Wiley, Hoboken, NJ, 2006)
68. B.G. Streetman, S.K. Banerjee, Field-effect transistors, in *Solid State Electronic Devices*, 6th edn., (Prentice-Hall, Upper Saddle River, NJ, 2006)
69. Y. Son, T. Kang, S. Park, H. Shin, A simple model for capture and emission time constants of random telegraph signal noise. IEEE Trans. Nanotechnol. **10**(6), 1352–1356 (2011). https://doi.org/10.1109/TNANO.2011.2142401
70. C.H. Henry, D.V. Lang, Nonradiative capture and recombination by multiphonon emission in GaAs and GaP. Phys. Rev. B **15**(2), 989–1016 (1977). https://doi.org/10.1103/PhysRevB.15.989
71. D. Bauza, Y. Maneglia, In-depth exploration of Si–SiO_2 interface traps in MOS transistors using the charge pumping technique. IEEE Trans. Electron Devices **44**(12), 2262–2266 (1997). https://doi.org/10.1109/16.644648
72. B. Van Zeghbroeck, Principles of Electronic Devices [Online], 2011. Available: http://ecee.colorado.edu/~bart/book/
73. T. Nagumo, K. Takeuchi, S. Yokogawa, K. Imai, Y. Hayashi, New analysis methods for comprehensive understanding of random telegraph noise, in *Proc. IEEE IEDM*, 2009, pp. 1–4. https://doi.org/10.1109/IEDM.2009.5424230

74. K.R. Farmer, C.T. Rogers, R.A. Buhrman, Localized-state interactions in metal-oxide-semiconductor tunnel diodes. Phys. Rev. Lett. **58**(21), 2255–2258 (1987). https://doi.org/10.1103/PhysRevLett.58.2255
75. M. Nour, M.I. Mahmud, Z. Celik-Butler, D. Basu, S. Tang, F.C. Hou, R. Wise, Variability of random telegraph noise in analog MOS transistors, in *Proc. 22nd ICNF*, 2013, pp. 1–4. https://doi.org/10.1109/ICNF.2013.6578978
76. K. Kandiah, M.O. Deighton, F.B. Whiting, A physical model for random telegraph signal currents in semiconductor devices. J. Appl. Phys. **66**(2), 937–948 (1989). https://doi.org/10.1063/1.343523
77. T. Obara, A. Teramoto, A. Yonezawa, R. Kuroda, S. Sugawa, T. Ohmi, Analyzing correlation between multiple traps in RTN characteristics, in *Proc. IEEE Int. Rel. Phys. Symp.*, 2014, pp. 4A.6.1–4A.6.7. https://doi.org/10.1109/IRPS.2014.6860644
78. R. Da Silva, L.C. Lamb, G.I. Wirth, Collective Poisson process with periodic rates: applications in physics from micro-to nanodevices. Philos. Trans. R. Soc. A Math. Phys. Eng. Sci. **369**(1935), 307–321 (2011). https://doi.org/10.1098/rsta.2010.0258
79. K. Takeuchi, T. Nagumo, S. Yokogawa, K. Imai, Y. Hayashi, Single-charge-based modeling of transistor characteristics fluctuations based on statistical measurement of RTN amplitude, in *Proc. Symp. VLSI Technol.*, 2009, pp. 54–55
80. R. Da Silva, G.I. Wirth, L. Brusamarello, A novel and precise time domain description of MOSFET low frequency noise due to random telegraph signals. Int. J. Mod. Phys. B **24**(30), 5885–5894 (2010). https://doi.org/10.1142/S0217979210057535
81. S. Miller, D. Childers, *Probability and Random Processes*, 2nd edn. (Academic, Waltham, MA, 2012)
82. P.D. Welch, The use of fast Fourier transform for the estimation of power spectra: a method based on time averaging over short, modified periodograms. IEEE Trans. Audio Electroacoust. **15**(2), 70–73 (1967). https://doi.org/10.1109/TAU.1967.1161901
83. N.V. Amarasinghe, Random telegraph signals in submicron MOSFETs. PhD dissertation, Dept. Elect. Eng., SMU, Dallas, TX, 2001
84. A. Godoy, F. Gámiz, A. Palma, J.A. Jiménez-Tejada, J. Banqueri, J.A. López-Villanueva, Influence of mobility fluctuations on random telegraph signal amplitude in *n*-channel metal–oxide–semiconductor field-effect transistors. J. Appl. Phys. **82**(9), 4621–4628 (1997). https://doi.org/10.1063/1.366200
85. E. Simoen, C. Claeys, Random telegraph signal: a local probe for single point defect studies in solid-state devices. Mater. Sci. Eng. B **91–92**, 136–143 (2002). https://doi.org/10.1016/S0921-5107(01)00963-1
86. E. Simoen, C. Claeys, Substrate bias effect on the capture kinetics of random telegraph signals in submicron p-channel silicon metal-oxide-semiconductor transistors. Appl. Phys. Lett. **66**(5), 598–600 (1995). https://doi.org/10.1063/1.114025
87. E. Simoen, C. Claeys, Substrate bias effect on the random telegraph signal parameters in submicrometer silicon p-metal-oxide-semiconductor transistors. J. Appl. Phys. **77**(2), 910–914 (1995). https://doi.org/10.1063/1.359018
88. Y. Illarinov, M. Bina, S. Tyaginov, K. Rott, B. Kaczer, H. Reisinger, T. Grasser, Extraction of the lateral position of border traps in nanoscale MOSFETs. IEEE Trans. Electron Devices **62**(9), 2730–2737 (2015). https://doi.org/10.1109/TED.2015.2454433
89. B. Ullman, M. Jech, K. Puschkarsky, G.A. Rott, M. Waltl, Y. Illarinov, H. Reisinger, T. Grasser, Impact of mixed negative Bias temperature instability and hot carrier stress on MOSFET characteristics—part I: experimental. IEEE Trans. Electron Devices **66**(1), 232–240 (2019). https://doi.org/10.1109/TED.2018.2873419
90. T. Grasser, K. Rott, H. Reisinger, M. Waltl, J. Franco, B. Kaczer, Unified perspective of RTN and BTI, in *IRPS*, 2014, pp. 4A.5.1–4A.5.7. https://doi.org/10.1109/IRPS.2014.6860643
91. W. Goes, Y. Wimmer, A.M. El-Sayed, G. Rzepa, M. Jech, A.L. Shluger, T. Grasser, Identification of oxide defects in semiconductor devices: a systematic approach linking DFT to rate equations and experimental evidence. Microelectron. Reliab. **87**, 286–320 (2018). https://doi.org/10.1016/j.microrel.2017.12.021

92. Y. Wimmer, A.M. El-Sayed, W. Gös, T. Grasser, A.L. Shluger, Role of hydrogen in volatile behaviour of defects in SiO_2-based electronic devices. Proc. R. Soc. A **472**(2190), 20160009-1–20160009-23 (2016). https://doi.org/10.1098/rspa.2016.0009
93. A.M. El-Sayed, M.B. Watkins, T. Grasser, V.V. Afanas'ev, A.L. Shluger, Hydrogen-induced rupture of strained Si—O bonds in amorphous silicon dioxide. Phys. Rev. Lett. **114**(11), 1155031–1155035 (2015). https://doi.org/10.1103/PhysRevLett.114.115503
94. A. Asenov, R. Balasubramaniam, A.R. Brown, J.H. Davies, S. Saini, Random telegraph signal amplitudes in sub 100 nm (decanano) MOSFETs: a 3D 'Atomistic' simulation study, in *Int. Electron Dev. Meeting, Tech. Dig. IEDM*, 2000 (Cat. No. 00CH37138), pp. 12.3.1–12.3.4. https://doi.org/10.1109/IEDM.2000.904311
95. T. Nagumo, K. Takeuchi, T. Hase, Y. Hayashi, Statistical characterization of trap position, energy, amplitude and time constants by RTN measurement of multiple individual traps, in *IEDM Tech. Dig.*, 2010, pp. 628–631. https://doi.org/10.1109/IEDM.2010.5703437
96. D.H. Cobden, M.J. Uren, M.J. Kirton, Entropy measurements on slow Si/SiO_2 interface states. Appl. Phys. Lett. **56**(13), 1245–1247 (1990). https://doi.org/10.1063/1.102527
97. P.E. Blochl, J.H. Stathis, Hydrogen electrochemistry and stress induced leakage current in silica. Phys. Rev. Lett. **83**(2), 372–375 (1999). https://doi.org/10.1103/PhysRevLett.83.372
98. C.J. Nicklaw, Z.Y. Lu, D.M. Fleetwood, R.D. Schrimf, S.T. Pantelides, The structure, properties, and dynamics of oxygen vacancies in amorphous SiO_2. IEEE Trans. Nucl. Sci. **49**(6), 2667–2673 (2002). https://doi.org/10.1109/TNS.2002.805408
99. G. La Rosa, F. Guarin, S. Rauch, A. Acovic, J. Lukaitis, E. Crabbe, NBTI-channel hot carrier effects in PMOSFETs in advanced CMOS technologies, in *Proc. Int. Rel. Phys. Symp.*, 1997, pp. 282–286. https://doi.org/10.1109/RELPHY.1997.584274
100. Z. Chen, K. Hess, J. Lee, J.W. Lyding, E. Rosenbaum, I. Kizilyalli, S. Chetlur, R. Huang, On the mechanism for interface trap generation in MOS transistors due to channel hot carrier stressing. IEEE Electron Device Lett. **21**(1), 24–26 (2000). https://doi.org/10.1109/55.817441
101. S. Tyaginov, Physics-based modelling of hot carrier degradation, in *Hot carrier degradation in semiconductor devices*, ed. by T. Grasser, (Springer International Publishing, New York, 2015), p. 106
102. M. Jech, B. Ullman, G. Rzepa, S. Tyaginov, A. Grill, M. Waltl, D. Jabs, C. Jungemann, T. Grasser, Impact of mixed negative Bias temperature instability and hot carrier stress on MOSFET characteristics—part II: theory. IEEE Trans. Electron Devices **66**(1), 241–248 (2019). https://doi.org/10.1109/TED.2018.2873421
103. P. Fang, K.K. Hung, P.K. Ko, C. Hu, Hot-electron-induced traps studied through the random telegraph noise. IEEE Electron Device Lett. **12**(6), 273–275 (1991). https://doi.org/10.1109/55.82058
104. D. Kang, J. Kim, D. Lee, B.-G. Park, J.D. Lee, H. Shin, Extraction of vertical lateral locations and energies of hot-electrons-induced traps through the random telegraph noise. Jpn. J. Appl. Phys. **48**(4S), 04C034–1–04C034–4 (2009). https://doi.org/10.1143/JJAP.48.04C034
105. E. Simoen, B. Dierickx, C. Claeys, Hot-carrier degradation of the random telegraph signal amplitude in submicrometer Si MOSTs. Appl. Phys. A **57**(3), 283–289 (1993). https://doi.org/10.1007/BF00332604
106. E. Simoen, C. Claeys, Hot-carrier stress effects on the amplitude of random telegraph signals in small area Si p-MOSFETS. Microelectron. Reliab. **37**(7), 1015–1019 (1997). https://doi.org/10.1016/S0026-2714(96)00263-6
107. T. Grasser, M. Waltl, W. Goes, A. El-Sayed, A. Shluger, B. Kaczer, On the volatility of oxide defects: activation deactivation and transformation, in *Proc. Int. Rel. Phys. Symp. (IRPS)*, 2015, pp. 5A.3.1–5A.3.8.https://doi.org/10.1109/IRPS.2015.7112739
108. J.W. McPherson, H.C. Mogul, Underlying physics of the thermochemical E model in describing low-field time-dependent dielectric breakdown in SiO_2 films. J. Appl. Phys. **84**(3), 1513–1523 (1998). https://doi.org/10.1063/1.368217
109. M. Vitiello, N. Lopez, F. Illas, G. Pacchioni, H_2 cracking at SiO_2 defect centers. J. Phys. Chem. A **104**(20), 4674–4684 (2000). https://doi.org/10.1021/jp993214f

110. B.R. Tuttle, D.R. Hughart, R.D. Schrimpf, D.M. Fleetwood, S.T. Pantelides, Defect interactions of H_2 in SiO_2: implications for ELDRS and latent interface trap buildup. IEEE Trans. Nucl. Sci. **57**(6), 3046–3053 (2010). https://doi.org/10.1109/TNS.2010.2086076.
111. G. Lucovsky, H. Yang, Z. Jing, J.L. Whitten, Hydrogen atom participation in metastable defect formation at Si-SiO_2 interfaces. Appl. Surf. Sci. **117–118**, 192–197 (1997). https://doi.org/10.1016/S0169-4332(97)80077-3
112. D.L. Griscom, Diffusion of radiolytic molecular hydrogen as a mechanism for the post-irradiation buildup of interface states in SiO_2-on-Si structures. J. Appl. Phys. **58**(7), 2524–2533 (1985). https://doi.org/10.1063/1.335931
113. E.H. Poindexter, Chemical reactions of hydrogenous species in the Si/SiO_2 system. J. Noncryst. Solids **187**, 257–263 (1995). https://doi.org/10.1016/0022-3093(95)00146-8
114. D. Fink, J. Krauser, D. Nagengast, T.A. Murphy, J. Erxmeier, L. Palmetshofer, D. Bräunig, A. Weidinger, Hydrogen implantation and diffusion in silicon and silicon dioxide. Appl. Phys. A **61**(4), 381–388 (1995). https://doi.org/10.1007/BF01540112
115. D.J. DiMaria, Defect generation in field-effect transistors under channel-hot-electron stress. J. Appl. Phys. **87**(12), 8707–8715 (2000). https://doi.org/10.1063/1.373600
116. S. Mahapatra, N. Goel, S. Desai, S. Gupta, B. Jose, S. Mukhopadhyay, K. Joshi, A. Jain, A.E. Islam, M.A. Alam, A comparative study of different physics-based NBTI models. IEEE Trans. Electron Devices **60**(3), 901–916 (2013). https://doi.org/10.1109/TED.2013.2238237
117. F.B. McLean, A framework for understanding radiation-induced interface states in SiO_2 MOS structures. IEEE Trans. Nucl. Sci. **27**(6), 1651–1657 (1980). https://doi.org/10.1109/TNS.1980.4331084
118. N.S. Saks, M.G. Ancona, Determination of interface trap capture cross sections using three-level charge pumping. IEEE Electron Device Lett. **ED-11**(8), 339–341 (1990). https://doi.org/10.1109/55.57927
119. N.S. Saks, Measurement of single interface trap capture cross sections with charge pumping. Appl. Phys. Lett. **70**(25), 3380–3382 (1997). https://doi.org/10.1063/1.119177
120. L. Militaru, A. Souifi, Study of a single dangling bond at the SiO_2/Si interface in deep submicron metal-oxide-semiconductor transistors. Appl. Phys. Lett. **83**(12), 2456–2458 (2003). https://doi.org/10.1063/1.1608493
121. H. Reisinger, The time-dependent defect spectroscopy, in *Bias Temperature Instability for Devices and Circuits*, ed. by T. Grasser, (Springer International Publishing, New York, 2014), p. 75. https://doi.org/10.1007/978-1-4614-7909-3

Atomistic Modeling of Oxide Defects

Dominic Waldhoer, Al-Moatasem Bellah El-Sayed, Yannick Wimmer, Michael Waltl, and Tibor Grasser

1 Introduction

Random telegraph noise (RTN) refers to a stochastic discrete signal observed in the drain current I_D at a constant gate voltage V_G. This noise signal is believed to be caused by oxide defects, which can exchange charges with the device substrate or gate [1]. Charges trapped at oxide defects alter the device electrostatics, leading to a shift of the threshold voltage, denoted as ΔV_{th}. Since each discrete step in a random telegraph signal (RTS) stems from a single charge trapping/emission event of a specific defect, the analysis of these signals allows the experimental extraction of single-defect parameters. Typically, the impact of a single defect is only resolvable in small-area devices due to their small oxide capacitance. In contrast, large-area devices show a smooth ΔV_{th} degradation during operation at elevated temperatures and high gate voltages. This phenomenon is commonly known as bias temperature instability (BTI) [2] and is a serious reliability concern in modern MOSFET devices. It has been demonstrated in recent works that BTI and RTN are both linked to oxide defects and can therefore be modeled within the same framework [3]. Oxide defects can be either classified as hole traps or electron traps. In devices with a-SiO_2 dielectric, usually negative BTI (NBTI) caused by hole traps is the dominant degradation mechanism. In contrast, the increasing use of high-κ materials like HfO_2 also leads to significant positive BTI (PBTI) degradation, which is associated with electron traps in these materials [4].

D. Waldhoer (✉) · M. Waltl
Christian Doppler Laboratory for Single-Defect Spectroscopy in Semiconductor Devices, Institute for Microelectronics, Vienna, Austria
e-mail: waldhoer@iue.tuwien.ac.at; waltl@iue.tuwien.ac.at

A.-M. B. El-Sayed · Y. Wimmer · T. Grasser
Institute for Microelectronics, Vienna, Austria
e-mail: elsayed@iue.tuwien.ac.at; wimmer@iue.tuwien.ac.at; grasser@iue.tuwien.ac.at

T. Grasser (ed.), *Noise in Nanoscale Semiconductor Devices*,
https://doi.org/10.1007/978-3-030-37500-3_18

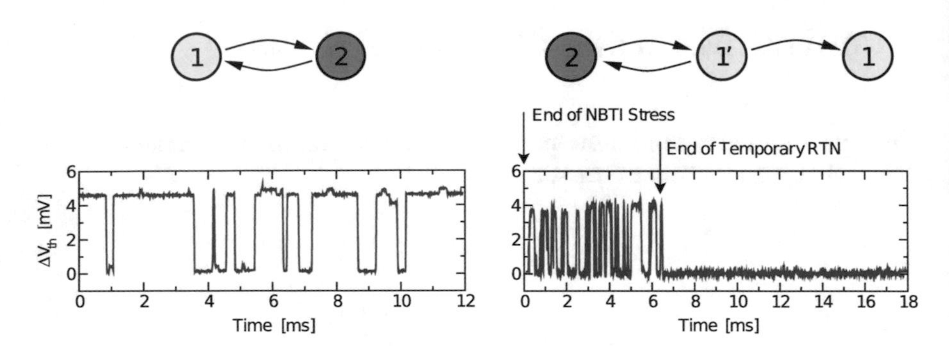

Fig. 1 RTN signals measured on a small-area pMOS. **Left:** A continuous RTN signal, which is produced by an oxide defect with two different states. **Right:** After BTI stress, often a so-called temporary RTN signal can be observed. In this case, the noise signal disappears after some time, indicating an additional electrically inactive defect state. Modified from [6]

Examples of frequently observed random telegraph signals are depicted in Fig. 1. The left figure shows a RTN signal produced by a two-state defect continuously switching between its two different charge states. The RTN signal of a fundamentally different defect is shown in Fig. 1 (right). In this case, the signal disappears after a certain time. This finding indicates a defect with at least three different states. The frequent occurrence of such signals in measurements lead to the development of the widely successful 4-state nonradiative multiphonon (NMP) model for the description of RTN and BTI [5, 6]. A brief review of this model is given in Sect. 4.

The various parameters of the 4-state NMP model can be fitted to data obtained from electrical defect characterization measurements. Techniques like RTS analysis [7] and time-dependent defect spectroscopy (TDDS) [8] allow the extraction of transition barriers and trap levels for individual defects. On the other hand the impact of a whole defect ensemble on device degradation can be monitored by applying measure-stress-measure (MSM) sequences [9]. A detailed description of experimental parameter extraction is given in Chapter 7 of this book [10]. Although these techniques allow for the electrical characterization of oxide defects, they can only provide limited information about the underlying physical and chemical nature of the defects. Using clean semiconductor-oxide samples instead of fully processed devices, additional insights into the atomistic structure of defects can be gained by electron spin resonance (ESR) spectroscopy. This technique is sensitive to defects with unpaired electrons. Using this method, various types of paramagnetic E′ centers were discovered in SiO_2 [11] as well as P_b centers at the Si/SiO_2 interface [12]. ESR measurements also confirmed the existence of hydrogenated E′ centers in amorphous SiO_2 [13]. Further experimental data also shows that hydrogen-related defects seem to play an important role for the RTN and NBTI phenomena in devices with SiO_2 gate dielectric [14–16].

Although all these experimental methods provide some information about the atomistic nature of oxide defects, a more detailed knowledge is desired in order to improve the device reliability and to justify and enhance existing reliability models.

In this context atomistic modeling provides a valuable tool to explore defects and their behavior theoretically. Over the last decades density functional theory (DFT) has become arguably the most popular ab-initio method in computational chemistry. Due to the ever increasing computational power in combination with the development of highly efficient algorithms, it has become feasible to simulate a collection of thousands of atoms quantum mechanically with DFT [17]. Even though the simulation of an entire device from first principles is still out of reach, it is possible to calculate defect parameters in bulk materials. In this chapter we will summarize the state-of-the-art methodology for point defect calculations using DFT. Furthermore we will review recent DFT results with an emphasis on hole traps in a-SiO_2 and electron traps in a-HfO_2, and compare them to experimental findings. It will be demonstrated that hydrogen-related defects like the hydrogen bridge (HB) or the hydroxyl-E$'$ center (H-E$'$) can act as hole traps in a-SiO_2 in agreement with NBTI and RTN measurements. PBTI in high-κ gate stacks on the other hand is likely caused by intrinsic electron traps in a-HfO_2.

2 DFT in a Nutshell

Atomistic modeling of materials has become an important tool in many fields of research. This approach allows the theoretical study of material properties like internal geometry, electronic structures, energy levels, or optical and vibrational absorption spectra just to name a few. The various methods of computational chemistry can be roughly separated in two groups, namely empirical and first principle (ab-initio) methods. Empirical approaches mostly assume an analytical force field between atoms and treat the resulting atomic movements entirely within classical mechanics. These methods are useful to determine geometric structures of molecules or solids. Empirical methods allow the study of very large molecules containing millions of atoms and are therefore very popular in biochemistry. However, these methods cannot give any information about the electronic structure of matter. For this task, ab-initio methods, which approximately solve the Schrödinger equation of the system, are needed. The study of defects, especially the estimation of their energy levels, therefore relies on ab-initio simulations. These approaches are computationally very expensive and scale considerably with the number of particles in the system, making them only usable for small molecules with just a handful of atoms. Among ab-initio methods, density functional theory (DFT) is arguably the most commonly used approach, due to its comparably low computational costs. DFT enables the quantum mechanical simulation of several thousand atoms on currently available computer hardware and is therefore very well suited for the study of defects in oxides. In the following we will give a brief introduction to general DFT. The modeling of amorphous structures and defects with DFT will be discussed later on.

2.1 Many-Body Schrödinger Equation

All non-relativistic properties of a system can, in principle, be determined by solving its time-independent many-body Schrödinger equation

$$\hat{H}\Psi(\tilde{\boldsymbol{r}}, \tilde{\boldsymbol{R}}) = E\Psi(\tilde{\boldsymbol{r}}, \tilde{\boldsymbol{R}}). \tag{1}$$

Here $\hat{H}$ represents the standard many-body Hamiltonian given by

$$\hat{H} = \hat{T}_n + \hat{T}_e + \hat{V} + \hat{U}\,, \tag{2}$$

which accounts for the kinetic energies of nuclei $(\hat{T}_n)$ and electrons $(\hat{T}_e)$, as well as the Coulomb interactions between nuclei and electrons $(\hat{V})$ and between electrons themselves $(\hat{U})$. The system's wavefunction Ψ is a function of all electronic and ionic coordinates, denoted by $\tilde{\boldsymbol{r}}$ and $\tilde{\boldsymbol{R}}$, respectively. A first common step in order to solve Eq. 1 is to assume a wavefunction, which is separable into an electronic wavefunction ψ and a vibrational wavefunction η

$$\Psi(\tilde{\boldsymbol{r}}, \tilde{\boldsymbol{R}}) = \psi(\tilde{\boldsymbol{r}}; \tilde{\boldsymbol{R}})\eta(\tilde{\boldsymbol{R}}). \tag{3}$$

Electrons have a much lower mass than nuclei and can, therefore, be assumed to adapt to changes in the ionic coordinates $\tilde{\boldsymbol{R}}$ almost instantly. Applying this fact to the separation Ansatz Eq. 3 yields an electronic Schrödinger equation

$$(\hat{T}_e + \hat{V} + \hat{U})\psi(\tilde{\boldsymbol{r}}; \tilde{\boldsymbol{R}}) = V(\tilde{\boldsymbol{R}})\psi(\tilde{\boldsymbol{r}}; \tilde{\boldsymbol{R}}) \tag{4}$$

and a vibrational Schrödinger equation

$$(\hat{T}_n + V(\tilde{\boldsymbol{R}}))\eta(\tilde{\boldsymbol{R}}) = E\eta(\tilde{\boldsymbol{R}}). \tag{5}$$

This approach is commonly known as the Born–Oppenheimer approximation [18]. Within this approximation the nuclei are assumed to move like classical particles in a potential $V(\tilde{\boldsymbol{R}})$ generated by the electrons. This potential is referred to as potential energy surface (PES) and plays a central role in the description of defects within NMP theory as will be outlined in Sect. 4.

2.2 Hohenberg–Kohn Theorems

Ab-initio methods aim at solving the electronic Schrödinger equation (Eq. 4). This equation is exceptionally hard to solve due to the electron–electron interaction operator $\hat{U}$, which prohibits a separation in single-electron wavefunctions. In fact, even a numerical solution of the electronic equation without further physical approximations is unfeasible even for small systems. One can approximate $\hat{U}$ by the interaction of each individual electron with the mean field caused by all electrons. This so-called Hartree–Fock (HF) method leads to Schrödinger equations of single electrons moving in an effective potential. These equations can be solved self-consistently for the single-electron wavefunctions. Although the HF method can produce a rough estimate of the electronic structure, it completely neglects electron correlation effects and is generally not accurate enough for practical calculations. The so-called post-HF methods like Møller–Plesset perturbation theory or the coupled cluster approach can mitigate these drawbacks at the price of greatly increased computational demands. Further details on HF and other wavefunction based methods can be found in [19].

When only the electronic ground state ψ_0 is of interest, a fundamentally different approach for solving Eq. 4 is offered by two important theorems proven by Hohenberg and Kohn [20]. These theorems establish a one-to-one mapping between the ground state electron density $n_0(\boldsymbol{r})$ and the ground state electronic wavefunction ψ_0, and can be stated as [21]:

1. The ground state wavefunction is a unique functional $\psi_0 = \psi[n_0]$ of the ground state electron density. As a consequence, every ground state observable $\hat{O}$ can be defined as a density functional as well

$$O[n_0] = \langle \psi[n_0] | \hat{O} | \psi[n_0] \rangle .$$

2. The energy functional $E[n]$ obtains its global minimum for the true ground state electron density n_0.

These theorems allow the reformulation of the many-body Schrödinger equation in terms of a variational problem with respect to the electron density $n(\boldsymbol{r})$. Therefore, instead of searching explicitly for the $3N$ dimensional wavefunction ψ_0, knowledge about the 3 dimensional electron density $n(\boldsymbol{r})$ is sufficient to derive all other ground state properties.

2.3 Kohn–Sham Equations

Although the Hohenberg–Kohn theorems provide a vast simplification of the problem at hand, they only prove the existence of a kinetic energy functional $T[n]$ and an electron interaction functional $U[n]$. However, these functionals are unknown for interacting electrons. In order to apply DFT to real systems, most often the method of Kohn–Sham (KS) orbitals [22] is employed. In this approach the electron density is assumed to be constructed by a set of non-interacting single-electron wavefunctions

$$n(\boldsymbol{r}) = \sum_{i=1}^{N_{\mathrm{el}}} |\phi_i(\boldsymbol{r})|^2 \,. \tag{6}$$

For such a system of non-interacting electrons, the resulting kinetic energy is simply given by

$$T_{\mathrm{s}}[n] = -\frac{1}{2}\sum_{i=1}^{N_{\mathrm{el}}} \langle \phi_i | \nabla^2 | \phi_i \rangle \,. \tag{7}$$

In classical electrodynamics the interaction functional can be expressed by the so-called Hartree-energy

$$U_{\mathrm{H}}[n] = \frac{1}{2}\iint \frac{n(\boldsymbol{r})\, n(\boldsymbol{r}')}{|\boldsymbol{r}-\boldsymbol{r}'|} \mathrm{d}^3\boldsymbol{r}\,\mathrm{d}^3\boldsymbol{r}'. \tag{8}$$

Although T_{s} and U_{H} do not account for the quantum mechanical exchange and correlation effects, they represent the largest contributions to the true functionals T and U. All the unknown quantum mechanical many-body effects are collected in a so-called exchange-correlation functional $E_{\mathrm{XC}}[n]$. The functional of the total energy can, therefore, be written as

$$E[n] = T_{\mathrm{s}}[n] + U_{\mathrm{H}}[n] + V[n] + E_{\mathrm{XC}}[n] \,. \tag{9}$$

Using the representation of the electron density by KS orbitals and applying the Euler–Lagrange formalism to Eq. 9 yields a system of eigenvalue problems for the individual orbitals, the so-called Kohn–Sham equations

$$\left(-\frac{1}{2}\nabla^2 + v_{\mathrm{eff}}(\boldsymbol{r})\right)\phi_i(\boldsymbol{r}) = \epsilon_i \phi_i(\boldsymbol{r}) \,. \tag{10}$$

These equations describe individual non-interacting electrons in an effective potential $v_{\mathrm{eff}}(\boldsymbol{r})$, which is given by the functional derivative

$$v_{\mathrm{eff}}(\boldsymbol{r}) = \frac{\delta}{\delta n}(V + U_{\mathrm{H}} + E_{\mathrm{XC}}) = v_{\mathrm{ext}}(\boldsymbol{r}) + \int \frac{n(\boldsymbol{r}')}{|\boldsymbol{r}-\boldsymbol{r}'|}\mathrm{d}^3\boldsymbol{r}' + \frac{\delta E_{\mathrm{XC}}}{\delta n} \,. \tag{11}$$

Similar to the Hartree–Fock method, these equations can be solved iteratively with an initial guess for either the electron density n or the KS orbitals ϕ_i. When these self-consistent field iterations converge, the system's total energy is given by [21]

$$E^{\mathrm{tot}} = \sum_{i}^{N_{\mathrm{el}}} \epsilon_i - U_{\mathrm{H}}[n_0] + E_{\mathrm{XC}}[n_0] - \int \left.\frac{\delta E_{\mathrm{XC}}}{\delta n}\right|_{n_0} n_0(\boldsymbol{r})\mathrm{d}^3\boldsymbol{r} \,. \tag{12}$$

It should be noted that the KS orbitals for themselves do not have a strict physical meaning, they just represent a fictitious system of non-interacting electrons, which produce the same electron density as the real system. However, it is a common practice to interpret the KS eigenenergies ϵ_i as electronic density of states and derive bandgaps with them. Although the KS system is not intended to represent any quantity beside the electron density, this approach works very well. A rigorous proof which justifies this practice is to the best of our knowledge still missing.

2.4 *Exchange-Correlation Functionals and Pseudopotentials*

As already mentioned, the exact exchange-correlation functional E_{XC} is unknown and is probably so complex that it would be of no use in practical DFT calculations. To overcome this, several different approximations for E_{XC} have been proposed. These approximations are often fitted to well-known experimental data in order to give accurate values for specific physical quantities of interest. Strictly speaking, DFT is not entirely an ab-initio method, but rather bridges the gap between purely empirical classical force-field methods and accurate, but extremely expensive wavefunction based ab-initio approaches. Therefore an important part of DFT simulations is the selection of a XC functional, which is well suited for the system at hand or at least to chemically similar systems.

The first and simplest XC approximation assumes a homogenous electron gas and was proposed by Kohn and Hohenberg in their seminal work [20]. In this so-called local density approximation (LDA) the XC functional is determined only by the local electron density and can be expressed as

$$E_{\mathrm{XC}}^{\mathrm{LDA}} = \int n(\boldsymbol{r})\, \epsilon_{\mathrm{XC}}(n)\, \mathrm{d}^3\boldsymbol{r}. \tag{13}$$

The function ϵ_{XC} is very well known from the Thomas–Fermi model [23] and quantum mechanical Monte Carlo simulations [24]. Although the LDA can give surprisingly good results in molecules, it generally overestimates binding energies and is not suitable for solids [25]. Furthermore, this approximation naturally fails for strongly localized charges with rapid variations of the electron density, which is highly relevant for the study of charged defects.

The main problem with LDA is its local dependency on the electron density, whereas the exact functional has to be a non-local property. To overcome this, the so-called semi-local or generalized gradient approximations (GGA) were introduced. These functionals include the density gradient and are of the form

$$E_{\mathrm{XC}}^{\mathrm{GGA}} = \int n(\boldsymbol{r})\, \epsilon_{\mathrm{XC}}^{\mathrm{GGA}}(n, \nabla n)\, \mathrm{d}^3\boldsymbol{r}. \tag{14}$$

Many different GGA functionals were proposed in the literature, among the most popular ones are the PBE [26] and PW91 [27] functionals. These functionals often provide better results than LDA when applied to suitable systems [28].

LDA and GGA can give reasonable results for geometries and activation energies; however, they consistently underestimate the electronic bandgap. This phenomenon is known as the infamous DFT-bandgap problem. For the study of charge transfer in oxide defects, the position of the charge trapping level with respect to the band edges is of utmost importance. Therefore these methods are inadequate for the simulation of defects with DFT. In order to solve the bandgap-problem, the so-called hybrid functionals have been developed. They are based on GGA functionals, but mix in a portion of the exact two-electron non-local exchange energy from the HF method [29]

$$E_{\mathrm{X}}^{\mathrm{HF}}[n] = -\sum_{i>j} \iint \frac{\phi_i^*(\boldsymbol{r})\,\phi_j^*(\boldsymbol{r}')\,\phi_j(\boldsymbol{r})\,\phi_i(\boldsymbol{r}')}{|\boldsymbol{r}-\boldsymbol{r}'|}\mathrm{d}^3\boldsymbol{r}\,\mathrm{d}^3\boldsymbol{r}'. \quad (15)$$

Using hybrid functionals allows an accurate prediction of the electronic bandgap, commonly within 10% of the experimental value [30]. Most DFT studies on defects therefore use hybrid functionals in order to predict trap levels accurately. The most frequently used hybrid functionals are HSE [31], PBE0 [32], and B3LYP [33]. Most results reviewed in this chapter were obtained with a variant of the PBE0 functional. It should be kept in mind that hybrid functionals are far more computationally expensive than simple LDA or GGA schemes. Due to the four-center integral in Eq. 15 the computational costs scale with $\mathcal{O}(N^4)$ and therefore become a limiting factor in the feasible structure size. The detrimental scaling can be mitigated by using certain approximation schemes like the Auxiliary Density Matrix Method (ADMM) [34], which can calculate the HF exchange accurately on a reduced basis set.

The computational cost of any DFT calculation obviously depends on the number of electrons in the system. Furthermore, the wavefunctions of the inner electrons are rapidly oscillating in space due to their high energy. A quantum mechanical description of all electrons in the system would therefore need a very large basis set in order to accurately treat electrons in the vicinity of the core. The use of such large basis sets requires a lot of computational effort and is often even unfeasible for large systems. In order to keep computational costs low, the so-called frozen-core approximation is frequently employed. In this approximation only valence electrons are treated within DFT, whereas the effect of the core electrons is reduced to a screened Coulomb potential for the atomic core, the so-called pseudopotential. This approach is justified, since core electrons hardly ever engage in chemical bonding. The pseudopotential is created by optimizating parameters of an analytical potential against the results of an all-electron calculation for each chemical element of interest. It should be noted that this procedure already employs a certain XC functional for the all-electron calculations; therefore, pseudopotentials are linked to the functional being used for its creation.

3 Modeling of Amorphous Structures

For the study of oxide defects a suitable model for the host oxide is needed. In essence, there are two different approaches for modeling the oxide surrounding the defect. From a chemical perspective, a small cluster of the host material inside a closed simulation cell can be used to study defects. In order to increase the accuracy of this rather crude model, the isolated cluster can be embedded in a potential, which mimics the impact of the surrounding host material. This is known as the embedded cluster approximation (ECA) [35]. A more physical approach uses a model containing a few hundred atoms of the host material and enforces periodic boundary conditions on the system in order to mimic an extended bulk material [36]. Both of these methods were improved over the years and can provide accurate descriptions of point defects. However, since all results presented in this work are obtained with periodic models of amorphous materials, in the following we will only focus on this method. A description of defect modeling within the embedded cluster method can be found in [37].

3.1 Melt-Quench Technique

Periodic boundary conditions are widely used in solid-state physics to describe crystals and their band structures. Theoretical modeling of defects in crystalline materials is a rich field, which provides valuable insights into many phenomena like doping, diffusion processes, or the performance of electronic devices. In fact, the first theoretical studies on oxide defects were conducted in crystals like α-quartz [38]. However, deposited gate oxides are generally amorphous. In the context of device reliability the modeling of oxide defects therefore requires knowledge about the distribution of defect parameters within the amorphous gate oxide. Furthermore it has been demonstrated that some defects occurring in a-SiO_2 like the hydroxyl-E′ center are not stable in α-quartz [39] and therefore cannot be studied in crystalline models. To obtain a more complete picture, it is thus required to also study defects in amorphous model structures.

The most popular approach to create amorphous model structures simulates the melting of a crystal within a periodic cell, followed by an equilibration of the melt at high temperatures and a rapid cooling afterwards. This physically motivated technique is called melt-quench method and is illustrated in Fig. 2. Hereby the melting and cooling of the material is simulated using molecular dynamics (MD) [36]. In MD simulations the temporal evolution of a system is studied within the framework of classical mechanics, where particles move in an interatomic force field according to Newton's laws of motion. However, the study of the system behavior at varying temperatures or pressures within MD requires the ability to control these thermodynamic quantities. This is usually achieved by extending the equations of motion with terms which enforce, e.g., the

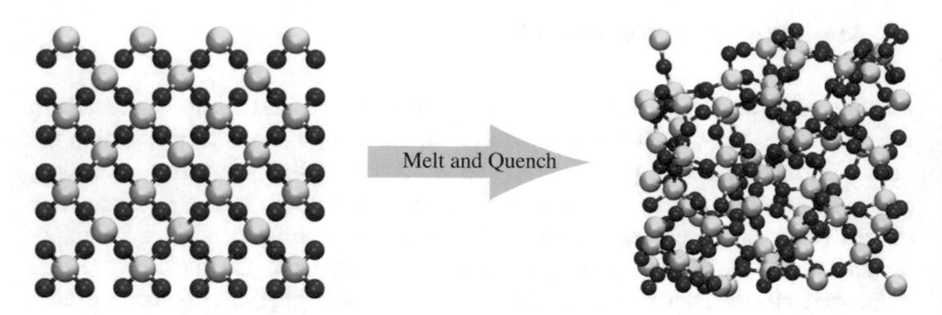

Fig. 2 Initial crystalline SiO_2 and final a-SiO_2 structure of a melt-quench run. The yellow and red spheres indicate Si and O atoms, respectively. The initial configuration is a $2 \times 2 \times 2$ supercell of β-cristobalite, a high-temperature crystalline polymorph of SiO_2, containing 216 atoms. This crystal is an ideal starting point, since it has a cubic lattice and its density of 2.33 g/cm^3 is close to the experimental value of 2.2 g/cm^3 for a-SiO_2. The crystal was melted at 5000 K for 1 ns and afterwards cooled with a rate of 6 K/ps. Note that the volume of the simulation cell is allowed to change during this process [36]

temperature (thermostat) or the pressure (barostat). More information regarding MD and different thermostat and barostat models can be found in [40]. The simulation of melt and quench processes typically has to cover a time span of several nanoseconds to allow the structure to properly equilibrate. Due to the large amount of energy and atomic force evaluations needed, DFT is rarely used during these steps. Instead, the MD simulation is performed with classical interatomic force fields. The geometry and cell parameters of the resulting amorphous structures are then further optimized with DFT to reduce internal stress and to obtain a proper electronic structure. The used force field must be parameterized carefully and must be able to describe chemical reactions of the involved chemical elements in order to produce physically meaningful structures. For example, a-SiO_2 structures can be created with the ReaxFF [41] force field, a suitable force field for a-HfO_2 is presented in [42].

It should be mentioned at this point that often highly unphysical parameters are needed for a melt-quench run. For example, the creation of amorphous SiO_2 structures requires temperatures of up to 5000 K. The high temperature can be ascribed to the use of empirical force fields, the absence of nucleation centers as well as the confinement of the system inside the simulation cell. Also an extremely high cooling rate of up to 6 K/ps is used to significantly reduce the number of needed simulation timesteps [36].

3.2 Structural Verification

The internal structure of amorphous materials strongly depends on the involved manufacturing processes as well as preparation conditions like temperature and pressure. For example, the structure can be partially crystallized due to annealing.

This ambiguity makes modeling of amorphous materials for a specific application particularly hard. For that reason, the structures obtained from the melt-quench technique should be compared to available experimental data of the material system in question. Basic structural tests can include comparisons to the experimental density and electronic structure. Furthermore, the spatial behavior of the band edges across the structures should be checked with an analysis of the local density of states. This is necessary to detect spurious electric fields, which lead to artificial band bending. More details on structural verification can be found in [36]. In the following we briefly discuss two powerful structural model tests from experimental data.

Particularly useful information about the internal structure of amorphous solids can be gained from the structure factor $S(Q)$ obtained from X-ray or neutron diffraction measurements. For isotropic materials the structure factor can be written as [43]

$$S(Q) = \frac{1}{N}\sum_{i,k} f_i f_k \frac{\sin(Q\,|\boldsymbol{R}_i - \boldsymbol{R}_k|)}{Q\,|\boldsymbol{R}_i - \boldsymbol{R}_k|}. \tag{16}$$

Here, f_i and f_k are the atomic form factors, describing the scattering amplitudes of the involved atoms. Q is the amplitude of the scattering vector and R_i,R_k are the atomic positions. The structure factor can be obtained from experiment but is also easily accessible from the amorphous model structures. A comparison of $S(Q)$ between experiment and atomistic model structures for a-SiO_2 is given in Fig. 3. As can be seen, these models are in good agreement with experimental data, indicating the validity of the melt-quench method and the used force fields.

Another way to test model structures is offered by infrared absorption and Raman scattering experiments. In these measurements incoming photons from an IR laser couple inelastically to vibrational states of the target material. The resulting spectra are therefore linked to the vibrational density of states (VDOS) in the

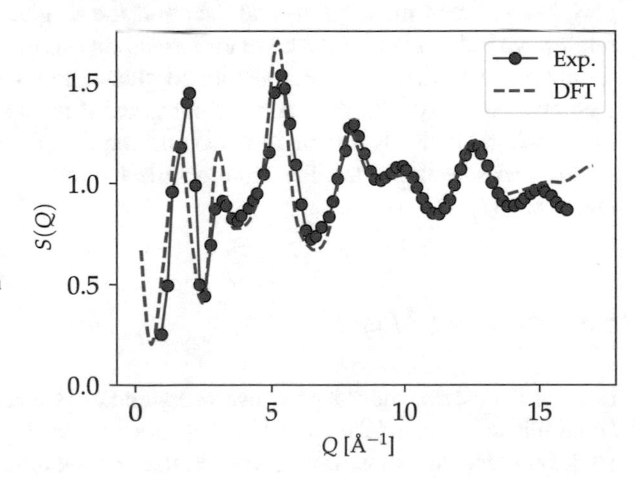

Fig. 3 Comparison of structure factors between experiments [44] and DFT optimized a-SiO_2 models obtained from the melt-quench procedure using the ReaxFF force field. The model structures show good agreement with the experimental data. The match of signal peak positions for large Q values indicate a proper long-range order of the used models. Modified from [36]

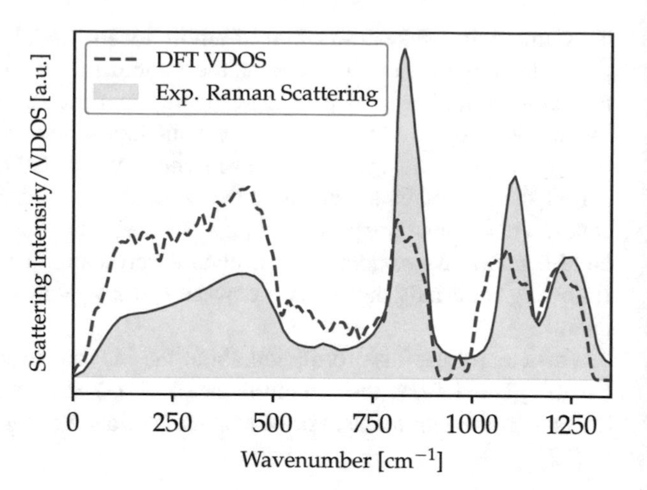

Fig. 4 Comparison between experimental Raman scattering intensities [46] and the vibrational density of states (VDOS) in an a-SiO_2 model obtained from DFT calculations. As can be seen, the positions of the signal peaks match well between DFT and experiment

target. Since the VDOS of a structure can be predicted with DFT calculations, such measurements can be used to compare model structures to real systems. Within the harmonic approximation the phonon modes of the structure can be derived from the force constant matrix, which is defined by

$$\Phi_{jk} = \frac{1}{\sqrt{M_j M_k}} \frac{\partial^2 E^{\text{tot}}}{\partial R_j \partial R_k}\bigg|_{\tilde{\boldsymbol{R}}=\tilde{\boldsymbol{R}}_0} = \frac{1}{\sqrt{M_j M_k}} \frac{\partial F_k}{\partial R_j}\bigg|_{\tilde{\boldsymbol{R}}=\tilde{\boldsymbol{R}}_0} . \tag{17}$$

Here M_j is the mass of the j-th ion. The force constant matrix essentially describes the curvature of the potential energy surface $E^{\text{tot}}(\tilde{\boldsymbol{R}})$ at the minimum configuration $\tilde{\boldsymbol{R}}_0$. By definition this matrix is positive-definite and its eigenvalues ω_j^2 are the squared classical vibrational frequencies of the structure [45]. Figure 4 shows the resulting phonon density of states alongside experimental Raman scattering intensities for the used a-SiO_2 models. There is good agreement on the peak positions between the experimental data and the results obtained from DFT. This finding is another piece of evidence for the validity of the used structural models. It should be noted that within this simplified classical model only peak positions are predicted correctly. The simulation of complete Raman and IR absorption spectra from ab-initio methods is more involved and requires the consideration of quantum mechanical selection rules. For more details we refer the interested reader to the literature [47].

3.3 *Interface Models*

In order to describe the charge transfer between oxide defects and device substrate, an atomistic model of the oxide/channel interface as depicted in Fig. 5 is needed. Such interface structures can be created using a modification of the melt-quench

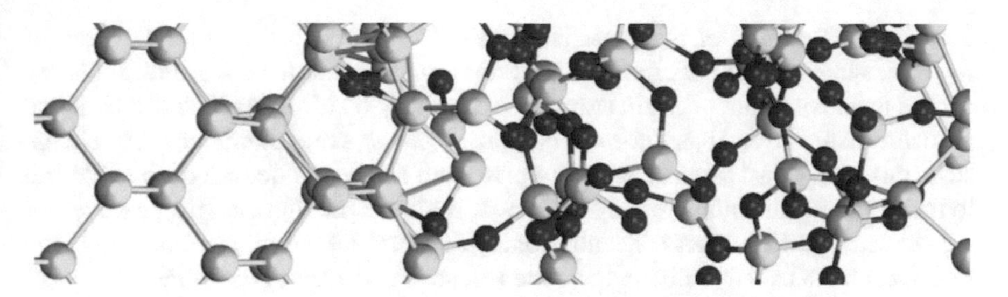

Fig. 5 Atomistic model for an interface between crystalline Si (left) and a-SiO_2 (right) [36]

method, as is outlined in [36]. However, obtaining an accurate interface model is far more difficult than simple bulk oxide models due to the limited experimental knowledge of the transition region between oxide and substrate. When analyzing charged defects in those structures with DFT another problem arises. Simply adding electrons or holes to the system does not guarantee a charge localization at the oxide defect. Since DFT determines the electronic ground state by minimizing the energy functional, additional charges may simply be delocalized in the substrate valence or conduction band if energetically favorable. Although charges can be forced into certain regions of the simulation cell with constrained DFT (CDFT) [48], such calculations are very expensive and thus rarely found in literature. In fact, all results presented in this work are obtained from bulk oxide models. In these bulk models the charge transfer between oxide defect and the substrate valence or conduction band is not accessible within DFT. This has some implications on the modeling of charge transitions as will be outlined in Sect. 4.3.

3.4 Formation Energies and Thermodynamic Trap Levels

The valence and conduction band of the substrate act as charge reservoirs for the oxide defects. Since the substrate is not included in the DFT models, the energy of charge carriers in these reservoirs must be included explicitly in the DFT energies of the different defect charge states. The standard approach to account for the carrier energy in the reservoir is to compare the so-called formation energies of the different defect states

$$E_{\mathrm{Q}}^{\mathrm{f}}(\tilde{\boldsymbol{R}}) = E_{\mathrm{Q}}^{\mathrm{tot}}(\tilde{\boldsymbol{R}}) - E_{\mathrm{bulk}}^{\mathrm{tot}} - \sum_i \mu_i n_i + qE_{\mathrm{F}} + E^{\mathrm{cor}}. \tag{18}$$

Here $E_{\mathrm{Q}}^{\mathrm{f}}$ denotes the formation energy of the defect state with a net charge

$$Q \in \{\ldots, -1, 0, +1, \ldots\},$$

the total DFT energy of this state is termed $E_{\mathrm{Q}}^{\mathrm{tot}}$ and $E_{\mathrm{bulk}}^{\mathrm{tot}}$ is the total energy of the defect-free neutral bulk oxide. The terms $\mu_i n_i$ account for the chemical energy needed to add or remove certain atoms to create the defect from a pristine bulk. Since all defects studied in this work are composed of the same atoms in every charge state, this term is just a constant energy offset and can be omitted. Contributions due to the chemical potential are only important if total formation energies are needed, e.g., to estimate the defect concentration. E_{F} denotes the Fermi level of the charge reservoir and is usually referenced to the valence band edge of the oxide

$$E_{\mathrm{F}} = E_{\mathrm{VBM}} + \epsilon_{\mathrm{F}}. \tag{19}$$

E_{VBM} is the energy of the valence band maximum of the oxide. It can be approximated by the energy of the highest occupied KS-orbital obtained by an electronic density of states analysis of the bulk system [49].

When dealing with charged defects in a periodic cell, the electrostatic potential within the cell would be infinite due to the slowly decaying Coulomb contributions of all other repeated cells. To avoid this problem, a net charge within the periodic cell is compensated by a fictional background charge of opposite sign to keep the cell potential finite. Although this approach leads to a bounded electrostatic potential, it introduces an unwanted interaction between the defect charge and the neutralizing background charge. Furthermore, there is also a spurious electrostatic interaction between the charged defect and its periodic images. Especially for small simulation cells this interaction must be balanced out by a charge correction term E^{cor}. Several methods for this charge correction have been proposed, among the most popular schemes are the Lany–Zunger [50] and the Freysoldt–Neugebauer–Van de Walle [51] corrections. An excellent review of the different charge correction schemes for defects in bulk materials can be found in [52].

The defect formation energy is an important quantity to evaluate the thermodynamical stability of the different defect states. As can be seen in Fig. 6, a certain

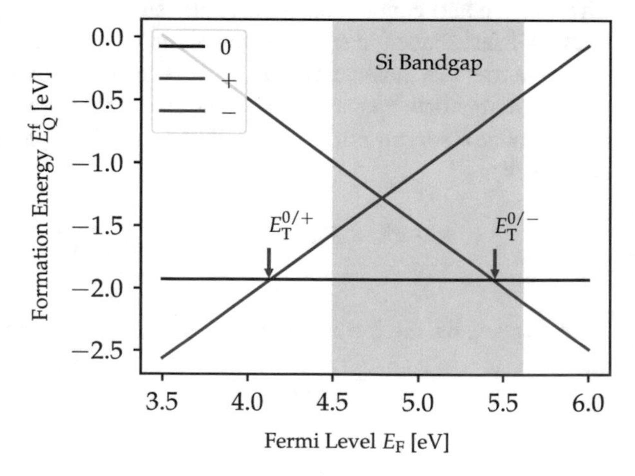

Fig. 6 Formation energies of a hydrogen bridge defect in a-SiO_2 in different charge states. The Fermi level is given with respect to the SiO_2 valence band. The shaded region indicates the bandgap of the Si substrate. The charge trapping levels are given by the Fermi levels at which two formation energies cross. These points are easily detectable in experiments. Modified from [53]

charge state is only stable in a certain range of the Fermi level. The boundary between two such regions is called the charge trapping level (CTL). When the Fermi level matches the charge trapping level, the corresponding defect states are equally stable, which leads to a 50% occupancy in equilibrium and therefore results in maximal power of the RTN signal. Thus the CTL is easily accessible in experiments by varying the gate bias voltage [6] and is an important parameter for comparing defect candidates in DFT to experimental data. In the case of defects in SiO_2 and HfO_2 such comparisons are given in Sect. 5.4 and Sect. 6.3, respectively.

4 The Four-State NMP Model

The discovery of degradation mechanisms like BTI and RTN sparked the development of several models to describe these phenomena. Earlier models were based on a classical reaction-diffusion process involving atomic hydrogen [54, 55]. These models were widely accepted in industrial applications. However, new experimental findings obtained from techniques like TDDS are incompatible with these models [56]. Modern modeling attempts focus on reaction-limited processes involving state changes of defects in the oxide or at the interface. Early oxide defect models were derived from the famous Shockley–Reed–Hall (SRH) model [57, 58], which was originally derived for defects in the device channel. The SRH model was extended in order to account for the tunneling of charges to the oxide [59] and transition barriers for the experimentally observed temperature-activated charge transfer [1]. Although these adapted models must be considered as milestones in reliability physics, they were lacking a solid physical foundation. In this section, we briefly recap the fundamentals of the 4-state nonradiative multiphonon model [6, 60–62], which is capable of accurately describing BTI and RTN in a unified framework. This model treats the charge transfer reactions between defects and device substrate within the nonradiative multiphonon theory (NMP) [63, 64]. An in-depth discussion on the 4-state NMP model can be found in [65].

4.1 Experimental Evidence for the Four-State NMP Model

Electrical defect characterizing methods like the temporal analysis of RTN signals or TDDS can be used to extract statistics of the time constants for charge transfers in the devices. Furthermore, the temperature and bias dependency of these time constants can be measured. A detailed description of experimental parameter extraction is given in Chapter 7 of this book [10]. The following key experimental findings lead to the development of the 4-state NMP model [65, 66]:

- As shown in Fig. 7 the characteristic time constants for charge capture and emission, denoted by τ_c and τ_e, respectively, are very sensitive to temperature.

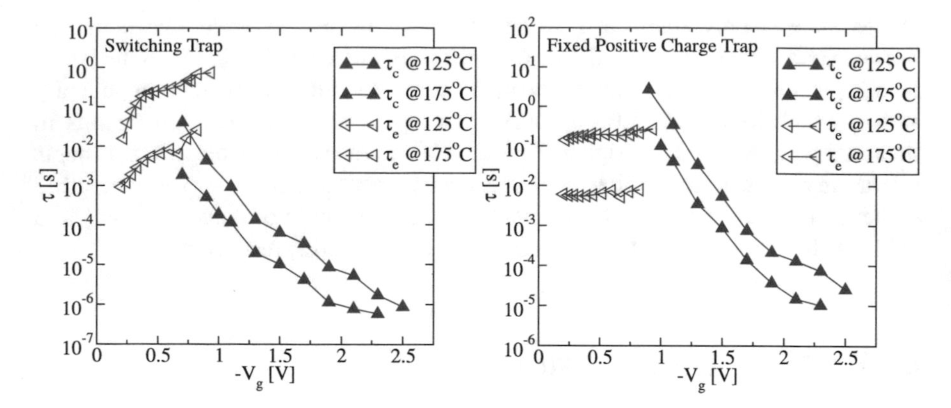

Fig. 7 Experimental capture and emission time constants of two different defects obtained through TDDS measurements. τ_c and τ_e strongly decrease with higher temperatures. **Left:** For a switching trap, both charge capture and emission depend exponentially on the applied gate bias. **Right:** In contrast, τ_e is nearly bias-independent for a so-called fixed charge trap. Adapted from [67]

This indicates a thermally activated process being involved in charge transfer reactions between the oxide and the substrate. The mean time constants can be expressed as

$$\tau_i \propto \exp\left(\frac{E_{A,i}}{k_B T}\right) \tag{20}$$

according to Arrhenius's law with an activation energy $E_{A,i}$. This finding rules out a SRH-like mechanism for charge trapping in oxide defects. However, charge transfer based on NMP theory is consistent with these results.

- The charge capture times depend exponentially on the applied gate bias V_G, which can be easily explained for a NMP process as will be demonstrated in Sect. 4.3. Furthermore, defects can be categorized in fixed traps and switching traps depending on the bias-sensitivity of the emission times, see Fig. 7 (left) and (right), respectively. In the case of fixed traps, the emission time is nearly independent of the bias. This indicates the existence of metastable states, which allow different competing reaction paths.
- It was demonstrated that a high-frequency bias significantly increases the resulting capture times. This finding can also be explained with additional defect states, which limit the overall transition rates.

Most of these experimental findings were obtained for hole traps responsible for NBTI in SiO_2 based devices. However, recent results suggest a similar behavior for electron traps in high-κ materials like HfO_2. Therefore, it can be assumed that the 4-state model is also applicable to electron traps in these materials as well [68, 69].

4.2 State Diagram

Within the 4-state model defects are treated as Markov chains with two stable states (1, 2) and two metastable states ($1'$, $2'$). Using Markov chains for the description implicitly assumes the defect to be memory-less. This means, the defect behavior is solely determined by its current state but not the entire history of the defect. This assumption is crucial for modeling and is justified by the extremely fast relaxation into the new configuration after a transition [70]. The resulting state diagram is depicted in Fig. 8. The states 1 and $1'$ represent electrically neutral defect configurations, whereas 2 and $2'$ are charged states with either an electron or hole trapped at the defect site. In this model, a shift in the threshold voltage can only be observed when the defect changes its charge state, i.e., during the transitions $1 \Leftrightarrow 2'$ and $2 \Leftrightarrow 1'$. The thermal transitions $1 \Leftrightarrow 1'$ and $2 \Leftrightarrow 2'$ account for possible structural relaxation of the defect without charge transfer. Such transitions are not directly detectable with electrical measurements, since they do not change the device electrostatics. However, the experimentally observed decorrelation of τ_c and τ_e as well as the occurrence of fixed traps are clear evidence for their existence [71].

4.3 Transition Rates

The abstract description in the 4-state model with Markov chains is completely agnostic towards the microscopic nature of the defect. However, the parameters within the model allow the calculation of the macroscopic defect behavior, in partic-

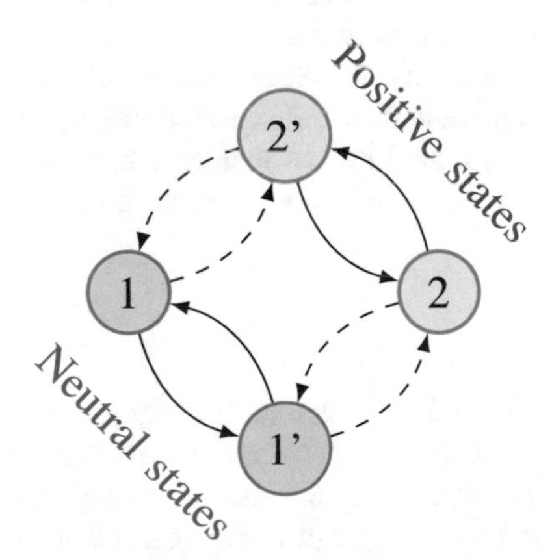

Fig. 8 The 4-State model for oxide defects. The defect is assumed to exist in a stable and metastable configuration both in the neutral and charged state. Here, the metastable states are indicated with a prime. Thermal transitions are represented by solid lines between states with the same charge, whereas NMP charge transitions are drawn as dashed lines between states with different charges

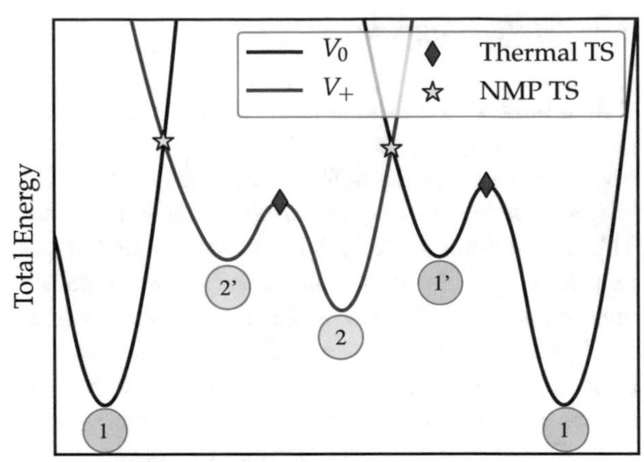

Fig. 9 Schematic energy profiles between the different states of the 4-state model. In this picture the states are given by the minima of the energy profiles. The transition rate for thermal transitions is determined by the point with highest energy along the reaction path. In the classical limit of NMP theory charge transitions occur at the intersection points of the surfaces. Modified from [72]

ular τ_c and τ_e as a function of the gate voltage V_G, from a solid physical foundation and are therefore also accessible through experiments. The model parameters can be also extracted from DFT calculations, which allows the identification of suitable defect candidates to describe the measurement data.

In the 4-state model all physics is condensed into the transition rates k_{ij} from an initial state i to a final state j. Since these transitions are temperature-activated, a certain energy barrier E_{ij} has to be overcome. The barriers in the 4-state model can be depicted with potential energy profiles as shown in Fig. 9. In these plots the electronic potential energy surfaces $V_i(\tilde{\boldsymbol{R}})$ are evaluated along a path between the initial and final state, resulting in energy profiles $V_i(q)$ as a function of the so-called configuration coordinate q. Using this picture, the defect states lie in minimum configurations of the potential energy surfaces. In the case of the thermal transitions $1 \Leftrightarrow 1'$ and $2 \Leftrightarrow 2'$ the transition barrier is determined by the classical transition state (TS), whereas the transition state of the NMP transitions is in a classical picture given by the intersection point between the corresponding potential energy surfaces. In the following we will discuss how the resulting barriers are linked to transition rates and how they can be determined from DFT simulations.

Thermal Transitions

In thermal transitions there is no charge transfer between the defect and the device substrate. Thus the defect is described by the same potential energy surface throughout the transition. Assuming a classical movement of the nuclei on the energy surface, the mean thermal transition rates can be determined within classical transition state theory [73, 74] and are given by

$$\tau_{ij}^{-1} = k_{ij} = \nu_0 \exp\left(-\frac{E_{ij}}{k_B T}\right). \tag{21}$$

Here E_{ij} is the thermal barrier, which is determined by the transition state along the reaction path. ν_0 is known as the attempt frequency and is related to the vibrational frequency within the harmonic approximation. In most systems, a value of $\nu_0 = k_B T / h \approx 10^{13}$ Hz can be assumed for the attempt frequency [65].

In order to calculate the transition barrier E_{ij} for a specific defect in DFT, finding the TS between the two states on the high-dimensional potential energy surface is required. Finding a transition state is more involved than the determination of minimum states, which can be easily found with various optimization schemes [75]. The standard approach for finding transition states in DFT relies on the nudged elastic band (NEB) method [76]. This method uses a band of intermediate configurations, which connects the two states of interest. The total energy of the whole band is then minimized, driving the intermediate configurations towards the minimum energy path between the states. The TS is then given by the configuration with highest energy along this path.

Nonradiative Multiphonon Transitions

Contrary to thermal transitions, NMP transitions describe the transfer of charges. This involves a change of the electronic state and thus a crossing between potential energy surfaces as shown in Fig. 10. In this case, the transition is governed by Fermi's Golden Rule in combination with the Franck-Condon principle [78, 79]. The resulting transition rate can be expressed as [65]

$$k_{ij} = A_{ij} f_{ij} , \tag{22}$$

where A_{ij} is the electronic matrix element and f_{ij} is the so-called lineshape function.

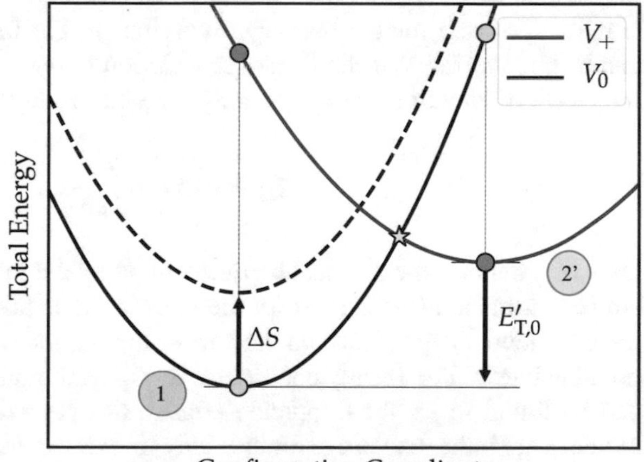

Fig. 10 NMP transition within the harmonic approximation. Here the energy profiles are approximated by parabolas. The small circles indicate the DFT energies, which are used to calculate the harmonic approximation. An applied gate bias shifts the energy profiles relative to each other by ΔS. This leads to bias dependent intersection points (IP) and transition barriers. Adapted from [77]

The electronic matrix element, which describes the coupling between the defect states and the device substrate, is given by

$$A_{ij} = \frac{2\pi}{\hbar} \left| \langle \psi_i | \hat{H}' | \psi_j \rangle \right|^2 \tag{23}$$

with the perturbation operator $\hat{H}'$. A_{ij} is not obtainable from bulk oxide DFT simulations since information about the wavefunctions of the charge reservoirs in the substrate valence and conduction band is not available. Instead, A_{ij} is frequently approximated by a semi-classical approach

$$A_{ij} \approx v_{\mathrm{th}} \sigma \vartheta. \tag{24}$$

Here $v_{\mathrm{th}} = \sqrt{8k_{\mathrm{B}}T/\pi m^*}$ denotes the thermal velocity of the charge carriers in the substrate, σ is an effective capture cross section, which is fitted to experimental data (e.g., $\sigma = 3.0 \times 10^{-14}\,\mathrm{cm}^2$ [80]). ϑ is a tunneling factor given by

$$\vartheta = \exp\left(-\frac{x_{\mathrm{t}}}{x_0}\right), \tag{25}$$

where x_0 is an effective tunneling length from a WKB approximation [81] and x_{t} is the distance of the defect from the interface.

The lineshape function f_{ij} accounts for the vibrational overlaps between the two states. Assuming a thermal equilibrium in the initial state, the lineshape function can be written as

$$f_{ij} = \underset{\alpha}{\mathrm{ave}} \left(\langle \eta_{i\alpha} | \eta_{j\beta} \rangle \, \delta \left(E_{i\alpha} - E_{i\beta} \right) \right). \tag{26}$$

Here ave stands for the thermal average within a canonical ensemble. However, implementations of the NMP model in device simulators typically use a much simpler approach for the lineshape function. In the limit of high temperatures it can be shown [62] that the lineshape function is only determined by the classical intersection point of the corresponding potential energy surfaces

$$f_{ij} \approx \exp\left(-\frac{E_{ij}}{k_{\mathrm{B}}T}\right). \tag{27}$$

Here, E_{ij} denotes the classical barrier defined by the intersection point. However, a similar problem arises here as for the thermal transitions discussed earlier. Due to the exponential dependence on the barrier height, the reaction path with the lowest possible barrier dominantly contributes to f_{ij}. Although this minimum energy path can be found in DFT by applying constraint optimization algorithms [82], such methods are rather expensive for the study of oxide defects, especially when dealing with an ensemble of defects as is the case in amorphous oxides. Instead, the potential energy surfaces are approximated by parabolas as indicated in Fig. 10. Within this approximation the energy profiles are solely determined by a fit to the energies of four DFT single-point calculations. The resulting parabolas are then used to calculate the intersection point and the classical lineshape function f_{ij}.

One important aspect, which has to be considered, is that the formation energies of the defect are influenced by the applied gate bias. On the one hand this influence is introduced by the change of the Fermi level within the charge reservoirs. On the other hand the difference in electrostatic potential between defect and device channel results in an additional energy shift. This effect is accounted for by a field-dependent trap level. Assuming a charge-free dielectric, a simple linear model for the trap can be used

$$E_\mathrm{T} = E_{\mathrm{T},0} + \Delta S \quad \text{with} \quad \Delta S = QFx_\mathrm{t}. \tag{28}$$

Here, $E_{\mathrm{T},0}$ is the trap level at flatband conditions and F is the electric field inside the oxide. Furthermore, oxide defects can potentially exchange charges with many states within the semiconductor valence or conduction band. The total transition rate is therefore given by a summation over all band states. For example in the case of hole traps this results in the transition rates

$$k_{12'} = \int_{-\infty}^{\infty} D(E) f_p(E) A_{12'}(E, E_\mathrm{T}) f_{12'}(E, E_\mathrm{T}) \mathrm{d}E \tag{29a}$$

$$k_{2'1} = \int_{-\infty}^{\infty} D(E) f_n(E) A_{2'1}(E, E_\mathrm{T}) f_{2'1}(E, E_\mathrm{T}) \mathrm{d}E. \tag{29b}$$

Here D is the substrate's density of states, f_n and f_p are the occupation probabilities for electrons and holes, respectively. Note that the matrix element A_{ij} as well as the lineshape function f_{ij} are expressed as a function of the trap level E_T and the energy E of the charge reservoir state. Applied to a pMOS device and given the additional assumption that defects mostly interact with the Si valence band edge, Eq. 29 leads to a simple analytical expressions for the transitions rates:

$$k_{12'} = p v_\mathrm{th} \sigma \vartheta \exp\left(-\frac{E_{12'}}{k_\mathrm{B}T}\right) \tag{30a}$$

$$k_{2'1} = p v_\mathrm{th} \sigma \vartheta \exp\left(-\frac{E_{2'1} + E_{\mathrm{V,Si}} - E_\mathrm{F}}{k_\mathrm{B}T}\right). \tag{30b}$$

The resulting NMP transition rates reflect the experimental findings. Firstly, charge transitions depend exponentially on the gate bias, which changes the barriers $E_{12'}$ and $E_{2'1}$. Furthermore, the NMP rates explain the observed thermal activation. From Eq. 29 or 30 one can extract activation energies for defect candidates with DFT, which also allows a comparison to experimental data.

In our derivations we mainly focused on hole traps and their interaction with the substrate's valence band; however, similar equations for transition rates can also be derived for electron traps. The full set of rate equations alongside an in-depth discussion of the NMP model can be found in [65].

5 Defects in Amorphous SiO_2

SiO_2 is the native oxide of the Si substrate and has therefore been used as gate dielectric since the early days of microelectronics. Although SiO_2 is increasingly replaced by high-κ materials like HfO_2 in order to improve the electrostatic channel control, the study of defects in a-SiO_2 is still highly relevant for reliability concerns. Defects in SiO_2 also play a role in modern high-κ gate stack structures since a thin SiO_2 layer always forms at the Si interface during oxide deposition. Defects in this layer can potentially trap charges rather easily due to the small distance to the channel and therefore contribute considerably to BTI and RTN in high-k devices. Modern devices also use nitrided SiO_2 (SiON) in order to decrease the gate leakage current. As was shown experimentally in [83], defects in SiON behave similarly to the ones in pure SiO_2; therefore, it is sufficient to model defects in the much simpler SiO_2 system. The Si/SiO_2 material system is well understood both experimentally and theoretically; a plethora of data is therefore available as a reference for theoretical defect studies using DFT. Furthermore, defect densities can be kept low in the Si oxidation process, which allows an experimental characterization of individual defects in small-area devices using techniques like TDDS. In the following, the key findings of recent ab-initio studies on defects in a-SiO_2 are summarized. All results presented here were obtained from periodic random network models containing 216 atoms, as described in Sect. 3.

5.1 Oxygen Vacancies

Presumably the most studied defect in silica is the oxygen vacancy [84–86]. It is formed when a two-coordinated O atom is missing from the a-SiO_2 network. This defect forms naturally during Si oxidation and the resulting defect concentration depends on processing parameters such as temperature and oxygen partial pressure [87]. Depending on the local defect environment, the oxygen vacancy can exist in various stable and metastable states with different charges [88]; it thus supports the aforementioned 4-state defect model, see Fig. 11. In the primary neutral configuration, the two Si atoms forming the vacancy bind together, which leads to long ranging distortions in the flexible a-SiO_2 network surrounding the defect.

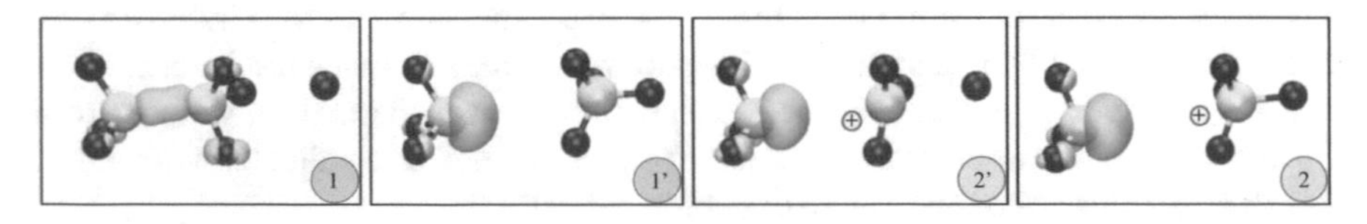

Fig. 11 Oxygen vacancy in a-SiO_2 with its different configurations mapped to the states in the 4-state model. The turquoise bubbles indicate the HOMO electron distribution [72]

Since all electrons in this configuration are paired, the state 1 of the OV does not carry a net spin and thus cannot be detected by spin-resonance measurements. If the vacancy traps a hole (State $2'$), e.g., one of the electrons in the Si–Si bond is removed, the bond can break and the Si atoms move apart from each other, resulting in a Si dangling bond. In other cases, hole trapping only weakens the Si–Si bond instead of breaking it, the resulting spin density is then shared between the two Si atoms, which is known as the dimer configuration [88, 89]. In both cases, there is an unpaired electron, which is detectable in ESR measurements. It is assumed that these two configurations correspond to the experimentally observed E' center defect in a-SiO_2 [90]. In addition, the Si atom without a dangling bond can move through the plane spanned by its O neighbors. Depending on the local environment, the backprojected Si atom can be stabilized by a nearby O (see Fig. 11 state $1'$ and 2). This process is termed puckering and is associated with a thermally activated transition without charge transfer, e.g., the transitions $1 \Leftrightarrow 1'$ and $2 \Leftrightarrow 2'$ in the 4-state model. The positive state 2 is believed to be the frequently measured E'_γ center [90].

Although the OV defect is in principle compatible with the 4-state model, multiple theoretical studies [72, 91, 92] clearly show that the hole trap level of this defect is too far below the silicon valence band edge ($E_T - E_{V,Si} \approx -3.5\,eV$). In order for the OV to trap a hole, thick oxides and large electric fields would be needed. However, in a modern pMOS device with a thin oxide layer the OV remains neutral under all relevant bias conditions and cannot contribute to NBTI or RTN through to hole trapping.

5.2 Hydrogen-Induced Defects

Although some models suggested oxygen vacancies as the cause of NBTI [93], the focus has recently shifted towards hydrogen-related oxide traps [14, 94]. During the forming gas anneal process, hydrogen is used to passivate Si dangling bonds at the Si/SiO_2 interface. This crucial step is necessary to reduce the high density of interface traps, which would otherwise occur and significantly degrade the device characteristics. However, experimental [95] as well as theoretical [96, 97] studies indicate that hydrogen easily diffuses in SiO_2 with activation energies as low as 0.2 eV. Furthermore it was shown that there is a connection between hydrogen concentration during anneal and the resulting density of electrically active oxide traps [14, 94]. DFT studies indicate that H can form promising candidates for active defects like the hydrogen bridge or the hydroxyl-E$'$ center, which will be discussed in the following.

Hydrogen Bridge

Although the oxygen vacancy itself stays neutral under typical bias conditions, it acts as a precursor site for another related defect, the hydrogen bridge (HB) [98]. The HB is formed by the reaction of atomic hydrogen with an OV. DFT studies in α-quartz conducted by Blöchl et al. showed that the HB defect might be responsible for stress-induced leakage current (SILC) in gate oxides [99]. The formation of a HB defect from an interstitial H atom and an OV is nearly barrierless, whereas the reverse reaction of H release from the HB has a thermal barrier of 2.8 eV on average [100]. These reaction kinetics in combination with the abundance of vacancies and H suggest that a significant amount of HB defects are formed.

The most stable neutral configuration of the HB defect is shown as state 1 in Fig. 12. Here the atomic H binds to only one of the Si atoms forming the vacancy, leaving the other one undercoordinated. The resulting ESR from this dangling bond was first identified by Nelson and Weeks [101]. The Si–H bond length is rather constant at 1.47 Å, whereas the weak interaction with the second Si atom is strongly dependent on the local environment of the defect, which results in a large spread of the corresponding Si–H distances [72, 100]. The HB can capture a hole resulting in a configuration depicted in Fig. 12 state 2′. Here the remaining 2 electrons are shared between the two Si and the H atom, resulting in a nearly symmetrical configuration. Similar to the aforementioned OV in a-SiO_2, the HB can undergo a puckering transition in both charge states, where one Si atom moves through the plane spanned by its O neighbors.

Hydroxyl-E′ Center

In amorphous SiO_2 the Si–O bond lengths and angles do not have well defined values, but are widely distributed. It was shown in [100] that sites with distorted, but otherwise intact, Si–O bonds can interact with atomic hydrogen resulting in new defects, which do not exist in crystalline SiO_2. One such defect, the so-called hydroxyl-E′ center (H-E′), arises from the interaction of H with strained Si–O bonds, which are significantly longer than the equilibrium bond length of 1.61 Å in α-quartz [39]. In this case, H is able to bind to the oxygen atom and break one of the two Si–O bonds, resulting in a hydroxyl group facing a Si dangling bond as depicted in Fig. 13 (state 1). Such a configuration is more stable than the interstitial H by 0.8 eV on average. Due to the significant amount of strained bonds in a-SiO_2 and its

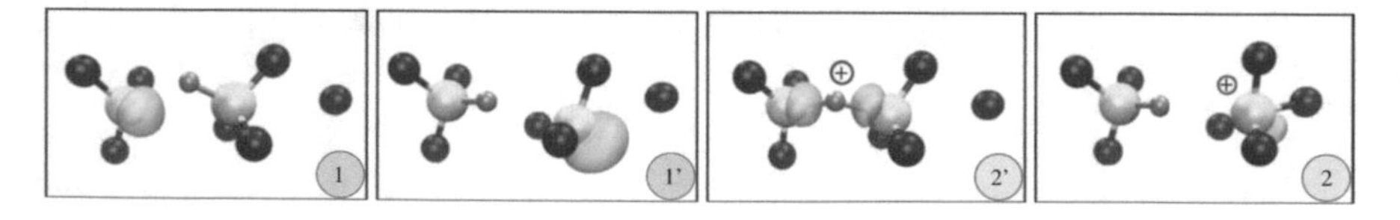

Fig. 12 Different configurations of the hydrogen bridge defect within the 4-state model [72]

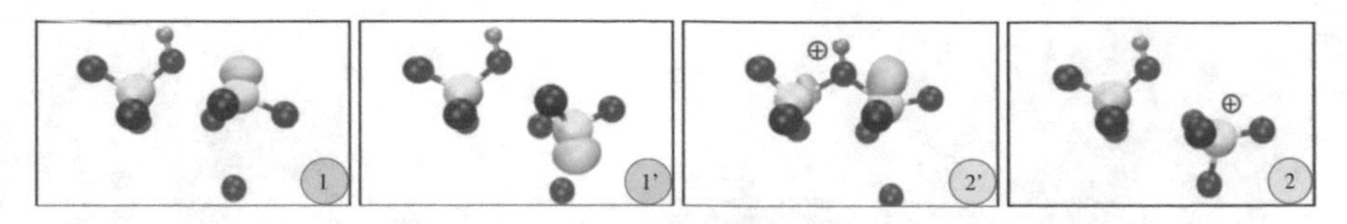

Fig. 13 Different configurations of the hydroxyl-E′ center within the 4-state model [72]

thermodynamical stability, the H-E′ center is expected to be formed in significant quantities, much larger than any other defect. By trapping a hole, the broken Si–O bond is restored and the positive H ion is bonded to the partially negative O atom by Coulomb interaction. The hydroxyl-E′ center also supports a 4-state defect model via a similar puckering transition as those of the OV and HB. Due to its thermodynamical trap levels (see Sect. 3.4) near the Si valence band, this defect is a promising candidate for explaining NBTI and RTN in pMOS devices [72].

In addition to the H-E′ center, another hydrogen-related defect in a-SiO_2 has been discovered recently [100]. H can donate its electron to a Si atom with a shortened bond (<1.6 Å) and binds electrostatically to the neighboring O atom. This results in a defect termed $[SiO_4/H]^0$ center where the Si–O–Si angle is stretched and an additional electron is localized at one of the Si atoms. Naturally, such a defect is formed more easily when H encounters a site with an already stretched Si–O–Si angle. It was shown that this defect is only slightly more stable (0.1–0.2 eV) than an interstitial H atom in a-SiO_2. Due to its lower thermodynamic stability and lower barriers for the H atom to become interstitial again compared to the H-E′ center, it was concluded that the interaction of H with the a-SiO_2 network primarily leads to H-E′ centers instead of $[SiO_4/H]^0$ centers.

5.3 Charge Trapping at Intrinsic Sites

Most atomistic studies on BTI and RTN are focused on charge trapping at defective sites of the host material. However, certain measured defect bands could not be linked to any known defects. One example of such an unidentified band are deep electron traps in SiO_2 with a trap level of 2.8 eV below the a-SiO_2 conduction band [102, 103], which are suspected to play a major role for PBTI in SiC based power devices. The observed high trap density of up to 5.0×10^{19} cm^{-3} in this defect band is not consistent with previous assumptions [104] that oxygen-deficient defects like the E′ center are responsible for these electron traps.

Based on these findings it was suggested that electron trapping does not only occur at defects but can happen spontaneously at certain sites in a-SiO_2. A recent theoretical study [105] found that the LUMO of the used a-SiO_2 models is partially localized at particularly long Si–O bonds, as depicted in Fig. 14 (left). Upon electron capture, this electronic state is filled and causes a pronounced structural distortion, which leads to the additional electron being localized at one particular Si atom

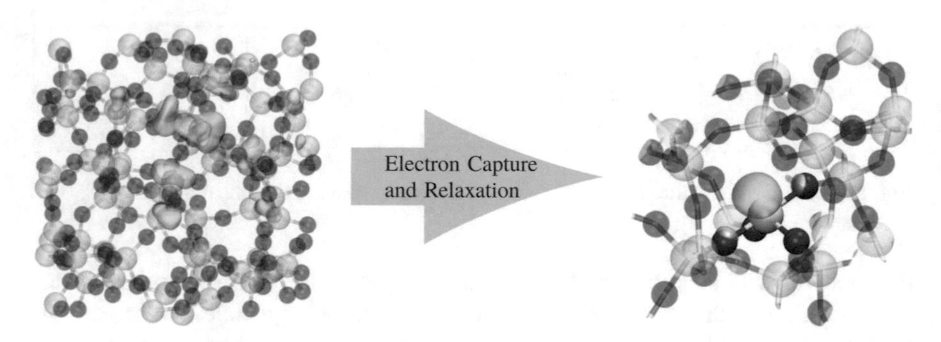

Fig. 14 Intrinsic electron trapping in a-SiO_2. **Left:** The LUMO is partially localized at certain sites in the structure. **Right:** Upon electron injection this orbital is filled and collapses onto a single Si atom. Reproduced from [105]

as shown in Fig. 14 (right). This structural relaxation is accompanied by a large energy gain resulting in a deep electron trap level on average 3.17 eV below the SiO_2 conduction band. It was further shown that the additional electron always spontaneously localizes at sites with O–Si–O angles exceeding 132°. Based on molecular dynamic models, the concentration of such sites was estimated to be around 4×10^{19} cm^{-3}, which is in excellent agreement with the experimentally observed trap density at this position in the bandgap.

For the sake of completeness it should be mentioned that intrinsic hole trapping was also observed in a-SiO_2 [106]. However, spectroscopic studies [107] together with later DFT investigations [108] suggest a hole trap level deep below the Si valence band. It can thus be assumed that intrinsic hole trapping does not contribute to NBTI or RTN in Si/SiO_2 based devices.

5.4 Comparison to Experimental Data

Due to the technological importance of SiO_2 in microelectronics and optics, defects in this material have been studied extensively both in experiment and theory. Here we compare the results from various DFT investigations for the aforementioned defect candidates with experimentally observed defect bands in devices with an a-SiO_2 oxide. In this discussion we will focus on comparisons to easily accessible quantities like thermodynamic trap levels and activation energies for charge transfer from electrical measurements. Detailed discussions on other parameters within the 4-state model like thermal barriers or relative stabilities can be found in [66, 72, 92].

Trap Level Distributions

Efforts were made to reduce the modeling complexity within the 4-state model, while at the same time keeping most of its predictive capabilities and capture the essence of underlying defect physics. This resulted in an effective 2-state model implemented in a Compact-Physics framework (Comphy [83]). Based on electrical MSM sequences defect bands for NBTI and PBTI degradation were extracted for multiple commercial device architectures within this framework. A comparison between DFT calculations and the resulting experimental trap levels for the 28 nm foundry planar technology is shown in Fig. 15. In the case of NBTI degradation it can be seen that the hydrogen-related defects HB and H-E′ act as hole traps close to the Si valence band in good agreement with the experimental evidence. Although the OV defect can capture holes as discussed in Sect. 5.1, the resulting trap level is far too deep to explain the observed NBTI degradation. Under typical bias conditions this defect type is mostly neutral and thus does not interfere with the device electrostatics. The same holds true for intrinsic hole traps. Note that in order to be most consistent with the 2-state model used in Comphy, in the comparison to DFT calculations only OV, HB and H-E′ defects were considered which also showed an effective 2-state character [66].

PBTI degradation in Si/SiO_2-based devices is very weak and difficult to measure accurately. Thus the extracted experimental electron trap levels have some uncertainty and are drawn as dashed lines. Figure 15 shows that intrinsic electron traps described in Sect. 5.3 may be responsible for this weak observed trap band. The hydrogen-related defect candidates discussed in Sect. 5.2 are known to be

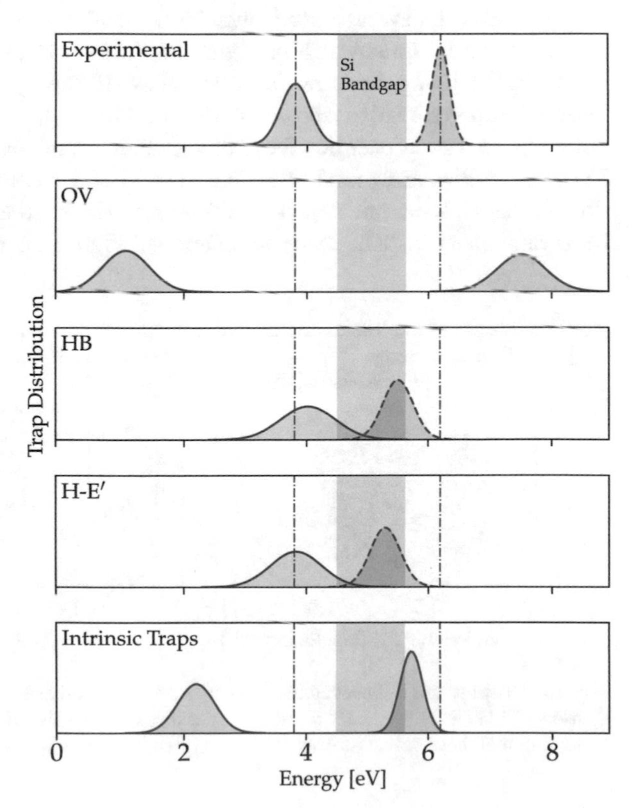

Fig. 15 Schematic trap level distributions from experiments and DFT predictions for various defect candidates in a-SiO_2. The shaded area marks the position of the Si bandgap. Red and blue lines indicate the distribution of hole and electron traps, respectively. The dashed lines show distributions from insufficient data sets, which may need further investigation. Sources for hole and electron trapping, respectively: Experimental [83], OV [66, 109], H-E′ [66, 100], HB [66, 100], intrinsic traps [105, 108]. Note that in the case of intrinsic traps the relaxed defect orbital energy was plotted, since the thermodynamic trap level was not calculated in these studies. However, these values give an estimate of the corresponding trap levels

able of capturing an electron [100], the resulting trap level are also lying near the experimental range. However, since these studies focused primarily on hole trapping mechanisms, only little information on electron trapping is available for these defects. Although PBTI is rather unimportant in Si/SiO_2 devices, the extracted defect band may play an important role in SiC/SiO_2 device degradation due to the different band alignments. A detailed theoretical study on electron traps in SiO_2 is thus desired in future works to better understand the degradation mechanisms in these devices.

Finally it should be emphasized that calculations of defect trap levels from DFT depend on the used XC functionals and are strongly affected by the resulting bandgap. The given distributions should therefore be considered as *qualitative* guidelines for the identification of possible defect candidates rather than a quantitative analysis. In order to get more accurate estimates for charge trap levels, higher levels of theory like the GW approximation [110] could be used to calculate correction terms for the studied defects.

Defect Activation Energies

Although the position of the charge trap level is an important indicator to judge whether or not a given defect candidate can explain the experimental data, it only provides the average occupancy in thermal equilibrium at a given gate bias. However, there is no information about the time scales on which a charge transition would occur. Using the transition rates of NMP theory (Eq. 29) together with the harmonic approximation shown in Fig. 10 allows the extraction of capture and emission activation energies for a given defect candidate from DFT simulations. These quantities are useful since they are also accessible from experiments and do not depend on the defect position relative to the interface, which would be meaningless in bulk oxide simulations. Figure 16 shows the distribution of

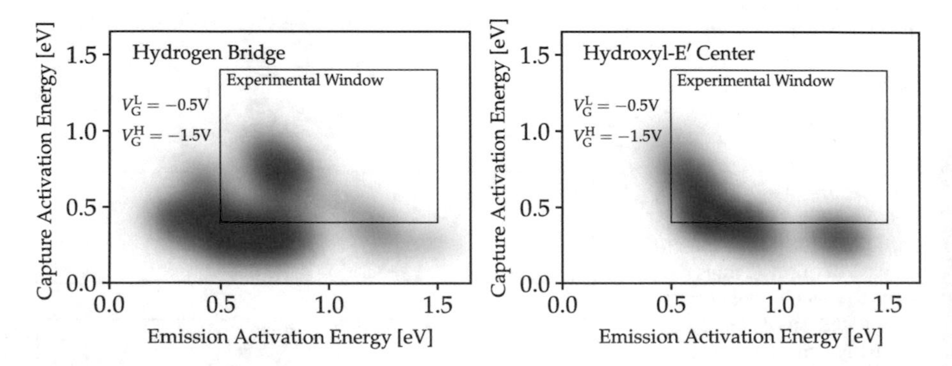

Fig. 16 Distributions of hole capture and emission activation energies for the hydrogen bridge and hydroxyl-E′ center. Significant portions of the distributions lie within the experimental window, indicating that these defects are active under typical bias conditions. Reproduced from [92]

activation energies for the hydrogen bridge and the hydroxyl-E′ center alongside the experimental window in which RTN and BTI can be observed. As can be seen, a considerable amount of these defect candidates lie within this window, indicating that they are visible during RTS or TDDS measurements [92]. These findings suggest that hydrogen-related defects are likely candidates for hole trapping in SiO_2. Although activation energies for intrinsic electron traps are not available, based on their relaxation energy of 0.8 eV one might speculate that these sites could act as fast electron traps. However, this needs to be investigated in future works.

Defect Volatility

Defects in a-SiO_2 are known to frequently disappear during TDDS measurements and stay inactive for a certain amount of time [111]. Such a defect behavior is termed volatility. Assuming a first-order thermally activated process as the underlying cause of volatility, the corresponding activation energies can be estimated by an Arrhenius law. Given an inactive timespan of at least 1200 s at room temperature during experiments results in an activation energy of around $E_A^{vol} \approx 1.0$ eV. Furthermore measurements on large-area devices with varying H content suggest a reaction involving hydrogen as the root cause of volatility [112]. It is therefore worth taking a closer look at the kinetics of hydrogen, especially the repassivation of the HB and HE center defects. In the following the findings of our previous works [72, 113] regarding the role of hydrogen in defect volatility are summarized.

In the case of the hydrogen bridge, volatility is expected to occur when the H is released and an oxygen vacancy is formed. Since the Si–H bond is very strong, the thermal barrier for the repassivation of HB defects is rather high. Nudged elastic band calculations for the dissociation of the Si–H bond in the positive charge state at multiple defect sites showed that there is a minimal barrier of 2.6 eV for this reaction. Similar findings also hold true in the neutral charge state. Such high barriers clearly indicate that HB defects cannot account for the observed volatility. On the other hand, dissociation of a neutral H E′ center has a reaction barrier of 1.66 eV on average, with a standard deviation of 0.37 eV. Although the barrier is smaller than for the HB defect, this process still does not provide a satisfactory explanation for defect volatility. However, it has been demonstrated, that the H^+ ion of the positive H-E′ center can dissociate with barriers sometimes even lower than 1.0 eV. In this case, the H^+ ion could hop to a nearby O site, forming a new state termed 0^+ as shown in Fig. 17. It was found that the states $2'$ and 0^+ are energetically similar, leading to comparable barriers for the forward and backward reaction. However, since in a-SiO_2 all O sites differ from each other, the new state 0^+ may not support the formation of a new electrically active H-E′ center, which leads to the disappearance of the defect during measurements. Since volatility is frequently observed in experiments, these theoretical findings suggest that the H-E′ center might be a better defect candidate to explain the measured trap behavior compared to the HB.

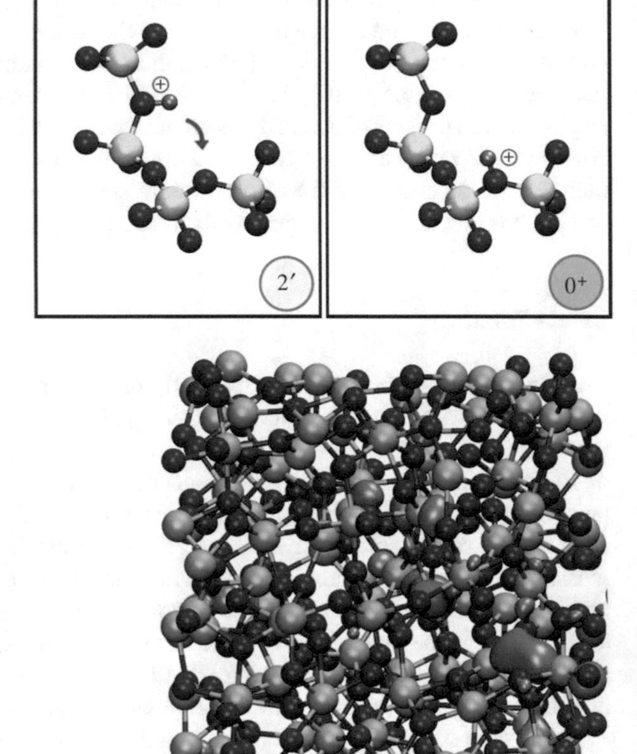

Fig. 17 Possible mechanism behind defect volatility. The H^+ ion of the positive hydroxyl-E′ center can hop to nearby O sites with barriers as low as 1.0 eV. This is possible since the ion is only bound via Coulomb attraction and therefore no chemical bond breaking is required during this process [72]

Fig. 18 Periodic a-HfO_2 model with 324 atoms. Most O atoms within the structure are threefold-coordinated, indicating a strong ionic binding component, which is absent in a-SiO_2. The majority of Hf atoms are 6-fold coordinated; however, a significant portion is also 5- or 7-coordinated. The orange bubble represents the partially localized LUMO, showing precursor sites for intrinsic charge trapping. Reproduced from [114]

6 Defects in Amorphous HfO_2

Although SiO_2 was the primary dielectric in microelectronics for decades, further downscaling of devices required the use of novel gate dielectrics. For example, the decreasing size of transistors also demands a reduction of the effective oxide thickness in order to keep the electrostatic control over the channel. However, such thin oxides have an increased leakage current, which drastically increases the power consumption and requirements for heat dissipation. To overcome this scaling issue, a thick layer of a high-κ material like hafnia (HfO_2) is used to decrease the leakage current while keeping control over the channel. HfO_2 can exist in a crystalline cubic phase and a metastable amorphous phase. HfO_2, together with other metal oxides like ZrO_2 and Al_2O_3, is a non-glass forming oxide. This implies that, contrary to SiO_2, the coordination numbers and oxidation states vary throughout the structure for the same kind of atoms. Due to the higher degree of disorder compared to a-SiO_2, see Fig. 18, it is suspected that intrinsic charge trapping at under- or overcoordinated sites plays a fundamental role for BTI and RTN in these materials [114]. Non-glass forming oxide films on substrates tend to be less stable than comparable SiO_2 films and can undergo structural transformations during processing steps. For example, HfO_2 films are known to partially crystallize during annealing [115]. Due to this additional structural uncertainty, atomistic modeling of non-glass forming metal oxides is particularly challenging. Devices with high-κ dielectrics like HfO_2 or other amorphous metal oxides are significantly affected by PBTI degradation, indicating that electron trapping is more pronounced in these materials than hole trapping. Although the density of hole traps can be rather high in HfO_2, the energetic position relative to the substrate valence band edge render them mostly inactive in common gate stack structures like $HfO_2/SiO_2/Si$ [83].

6.1 Oxygen Vacancies

Similar to SiO_2, oxygen vacancies can naturally form in a-HfO_2 due to local oxygen deficiency during deposition. Oxygen vacancies in a-HfO_2 can either form at two-, three-, or four-coordinated O sites. They are known to exist in neutral, positive, and negative states. Due to the ionic bonding character in HfO_2, oxygen vacancies behave like F-centers in ionic crystals, i.e., additional charges are strongly localized within the vacancy [116]. Since oxygen vacancies can also exist in negative charge states it was assumed that they could be responsible for the high density of electron traps observed in high-κ devices. However, recent theoretical studies [117] showed that the corresponding trap level $E_{\mathrm{T}}^{0/-}$ lies closely below the HfO_2 conduction band and is therefore too shallow to explain the experimental findings based on optical [117] and electrical [83] measurements. As shown in Fig. 19, the positively charged vacancy is thermodynamically stable for Fermi levels below $E_{\mathrm{F}} < 4.3\,\mathrm{eV}$. This results in a hole trap level $E_{\mathrm{T}}^{0/+}$ above the Si conduction band edge. The majority of these vacancies will therefore form fixed positive oxide charges.

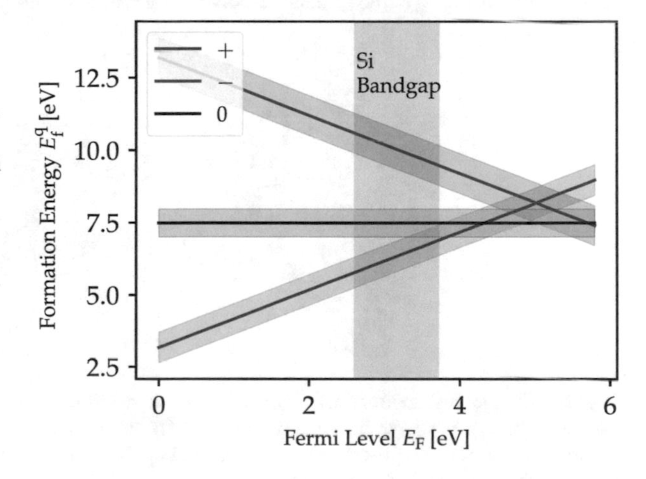

Fig. 19 Formation energies for the oxygen vacancy in a-HfO_2. The spreading of formation energies due to the amorphous host is indicated by the shaded strips. Adapted from [117]

Beside their possible role as charge traps, oxygen vacancies are also investigated as a potential cause of resistive switching in novel hafnia-based RRAM devices. It has been demonstrated recently that electron injection in HfO_2 can create Frenkel-pairs consisting of oxygen vacancies and interstitial O^{-2} ions. Furthermore, preexisting oxygen vacancies act as electron traps and can facilitate the formation of these Frenkel-pairs nearby, resulting in a stable divacancy [118]. This process suggests a possible aggregation of oxygen vacancies, which leads to a significant reduction of resistivity due to formation of a conductive filament.

6.2 Intrinsic Charge Trapping

Numerous studies [83, 117] clearly show that electron trapping characteristics in HfO_2 have a pronounced dependence on the chemical details of film deposition as well as the parameters during annealing. These findings suggest electron traps that are sensitive to the phase of the oxide film, e.g., amorphous or crystalline. A possible explanation for these electron traps was proposed in [119, 120]. Based on their theoretical investigations the authors proposed that electron trapping could take place at intrinsic sites similar to the process already discussed for a-SiO_2. In the case of a-HfO_2, undercoordinated Hf-ions or elongated Hf–O bonds serve as precursor sites. Trapped electrons at such sites will be highly localized at two or three Hf atoms as shown in Fig. 20 (left). Additionally, the d-orbitals of the involved Hf atoms also carry substantial parts of the excess electron as indicated by the ring-shaped charge distributions. The extra electron leads to a distortion, where the Hf cations are pulled towards the trapped charge distribution, whereas the O anions are pushed outwards. These electron traps are therefore called polarons due to similarities with the identically named self-trapped charges in crystals. Contrary to crystalline HfO_2, the formation of polarons in a-HfO_2 is accompanied by a large relaxation energy of 0.8 eV on average, which makes electron polarons stable even at room temperature.

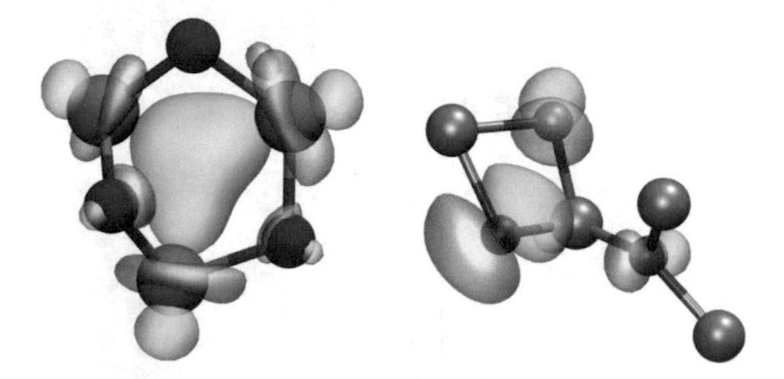

Fig. 20 Charge distributions of intrinsic traps in a-HfO_2. **Left:** In an electron polaron the additional charge is shared between two or three Hf atoms. The ring-shaped distributions indicate a participation of the Hf d-orbitals. **Right:** A hole polaron localizes on several O atoms near the trapping site. Reproduced from [114]

Similarly, holes have been demonstrated to localize at certain O atoms, forming a so-called hole polaron as shown in Fig. 20 (right) with a relaxation energy of 0.7 eV [119].

6.3 *Trap Level Distributions*

Similar to Sect. 5.4 we now compare the resulting trap levels of the discussed defects with the fitted model parameters from experiments using the Comphy framework. As shown in Fig. 21, the experimentally observed hole trap level in a-HfO_2 lies within the Si bandgap. This explains the rather weak NBTI response from hafnia in ordinary HfO_2/SiO_2/Si stacks. However, the response of this defect band is much larger when a few additional layers of Al_2O_3 are deposited on the stack due to different energetic alignment with respect to the gate Fermi level [83]. Electrical characterizations on HfO_2/SiO_2/Si stacks show a large hole trap density in the HfO_2 region [83], suggesting HfO_2 related defects as its origin. In contrast, other investigations found that the measured hole trap density is nearly unaffected by the thickness of the deposited HfO_2 layer, indicating a dominant role of defect states at the HfO_2/SiO_2 interface for the NBTI phenomenon [118, 121]. As can be seen in Fig. 21, DFT calculations suggest that the hole trap band in a-HfO_2 consists of intrinsic hole polarons, whereas the oxygen vacancy clearly is not a suitable candidate due to its very high trap level. This finding might also give an explanation for the experimental discrepancies regarding hole trapping in hafnia. Hole polarons require an amorphous oxide in order to be stable. During annealing steps the HfO_2 layer tends to partially crystallize, leading to different experimental estimates for the trap location depending on annealing conditions.

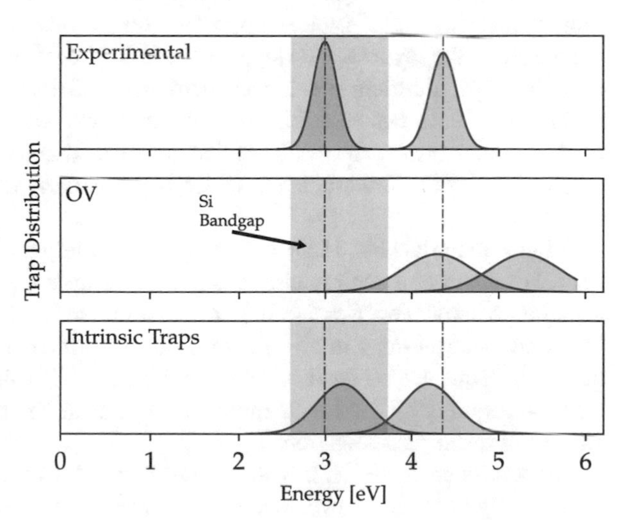

Fig. 21 Schematic trap level distributions from experiments and DFT predictions for defect candidates in a-HfO_2, similar to Fig. 15. Sources: Experimental [83], OV [117], intrinsic traps [119]. Note that in the case of intrinsic traps the relaxed defect orbital energy was plotted, since the thermodynamic trap level was not calculated in these studies. However, these values give an estimate of the corresponding trap levels

A similar picture can be drawn for the measured electron trap band. Here the predicted electron trap level for the oxygen vacancy is very close to the HfO_2 conduction band and is thus too shallow to account for the experimental data. However, DFT predictions for electron polarons match the measured trap levels perfectly. This finding is also supported by time-dependent DFT studies on the photo-absorption spectra of electron polarons, which fits nicely to optical measurements in a-HfO_2 samples [114]. In summary, all these findings lead to the conclusion that intrinsic electron and hole polarons play a fundamental role in the BTI degradation mechanism of hafnia-based devices.

7 Summary and Outlook

In this chapter we have investigated the microscopic structure of possible defect candidates responsible for RTN and BTI in SiO_2 and HfO_2, two of the most important gate oxide materials in microelectronics. Using molecular dynamics together with the melt-quench procedure allows for the realistic atomic modeling of amorphous gate oxides. Density functional theory can then be used to create defects inside these model structures and study their behavior from first principles. In combination with the 4-state NMP model, important defect parameters like the thermodynamic trap level or charge transition barriers and activation energies can be extracted from DFT. These parameters are also accessible from experimental measurement techniques like the analysis of RTN signals or time-dependent defect spectroscopy. This allows for the identification of relevant defects and might lead to strategies for improving device reliability.

Contrary to previous assumptions, we showed that RTN and NBTI in devices with SiO_2 gate oxide cannot be explained by oxygen vacancies since they stay mostly inactive under typical operating conditions. However, hydrogen-related defects like the hydrogen bridge and the hydroxyl-E$'$ center are likely acting as hole traps. Furthermore it was demonstrated that strained structures within amorphous silica act as intrinsic electron traps. Although there is a need for further investigations, the available data suggests that these intrinsic traps might be responsible for PBTI degradation, which is particularly important for power devices based on SiC.

High-κ materials like HfO_2 are non-glass forming metal oxides. Due to the high degree of disorder in their amorphous phase, intrinsic charge trapping seems to play a major role for device reliability in these materials. It was shown that intrinsic hole and electron traps in HfO_2 match the experimental charge trap levels nearly perfectly. However, so far it is unknown if these intrinsic traps possess metastable states as required by the 4-state model. Also, the charge trapping dynamics of these sites need further investigation.

All results presented in this work are obtained from bulk oxide models. This is a frequently used and convenient model simplification. However, many defects are expected to be near the interface, where the assumption of bulk-like oxide properties is likely breaking down. For example, the oxide near the interface is usually strained and might not be stoichiometric. In order to study the impact of these effects on defects, theoretical studies on interface models should be conducted in the future.

References

1. M. Kirton, M. Uren, Noise in solid-state microstructures: a new perspective on individual defects, interface states and low-frequency (1/f) noise. Adv. Phys. **38**(4), 367–468 (1989)
2. Y. Miura, Y. Matukura, Investigation of silicon-silicon dioxide interface using MOS structure. Jpn. J. Appl. Phys. **5**(2), 180 (1966)
3. T. Grasser, K. Rott, H. Reisinger, M. Waltl, J. Franco, B. Kaczer, A unified perspective of RTN and BTI, in *2014 IEEE International Reliability Physics Symposium* (2014), pp. 4A.5.1–4A.5.7
4. M. Chudzik, B. Doris, R. Mo, J. Sleight, E. Cartier, C. Dewan, D. Park, H. Bu, W. Natzle, W. Yan, C. Ouyang, K. Henson, D. Boyd, S. Callegari, R. Carter, D. Casarotto, M. Gribelyuk, M. Hargrove, W. He, Y. Kim, B. Linder, N. Moumen, V.K. Paruchuri, J. Stathis, M. Steen, A. Vayshenker, X. Wang, S. Zafar, T. Ando, R. Iijima, M. Takayanagi, V. Narayanan, R. Wise, Y. Zhang, R. Divakaruni, M. Khare, T.C. Chen, High-performance high- amp metal gates for 45nm CMOS and beyond with gate-first processing, in *2007 IEEE Symposium on VLSI Technology* (2007), pp. 194–195
5. F. Schanovsky, O. Baumgartner, V. Sverdlov, T. Grasser, J. Comput. Electron. **11**(3), 218–224 (2012)
6. T. Grasser, Stochastic charge trapping in oxides: from random telegraph noise to bias temperature instabilities. Microelectron. Reliab. **52**(1), 39–70 (2012)
7. H. Miki, N. Tega, M. Yamaoka, D.J. Frank, A. Bansal, M. Kobayashi, K. Cheng, C.P. D'Emic, Z. Ren, S. Wu, J.B. Yau, Y. Zhu, M.A. Guillorn, D.G. Park, W. Haensch, E. Leobandung, K. Torii, Statistical measurement of random telegraph noise and its impact in scaled-down high-k metal-gate MOSFETs, in *2012 International Electron Devices Meeting* (2012), pp. 19.1.1–19.1.4
8. T. Grasser, H. Reisinger, P.-J. Wagner, B. Kaczer, Time-dependent defect spectroscopy for characterization of border traps in metal-oxide-semiconductor transistors. Phys. Rev. B **82**, 245318 (2010)
9. T. Grasser, P. Wagner, P. Hehenberger, W. Gos, B. Kaczer, A rigorous study of measurement techniques for negative bias temperature instability, in *2007 IEEE International Integrated Reliability Workshop Final Report* (2007), pp. 6–11
10. B. Stampfer, A. Grill, M. Waltl, Advanced electrical characterization of single oxide defects utilizing noise signals, in *Noise in Nanoscale Semiconductor Devices*, ed. by T. Grasser (Springer, Cham, 2019). https://doi.org/10.1007/978-3-030-37500-3_7
11. P. Lenahan, Dominating defects in the MOS system: P_b and E' centers, in *Defects in Microelectronic Materials and Devices*, ed. by D. Fleetwood, R. Schrimpf, S. Pantelides (Taylor and Francis/CRC Press, Boca Raton, 2008)
12. A. Stesmans, V.V. Afanas'ev, Electron spin resonance features of interface defects in thermal (100) Si/SiO_2. J. Appl. Phys. **83**(5), 2449–2457 (1998)
13. J.F. Conley, P.M. Lenahan, Room temperature reactions involving silicon dangling bond centers and molecular hydrogen in amorphous SiO_2 thin films on silicon, Appl. Phys. Lett. **62**(1), 40–42 (1993)
14. A.T. Krishnan, S. Chakravarthi, P. Nicollian, V. Reddy, S. Krishnan, Negative bias temperature instability mechanism: the role of molecular hydrogen. Appl. Phys. Lett. **88**(15), 153518 (2006)

15. T. Grasser, B. Kaczer, W. Goes, An energy-level perspective of bias temperature instability, in *2008 IEEE International Reliability Physics Symposium* (2008), pp. 28–38
16. T. Aichinger, S. Puchner, M. Nelhiebel, T. Grasser, H. Hutter, Impact of hydrogen on recoverable and permanent damage following negative bias temperature stress, in *2010 IEEE International Reliability Physics Symposium* (2010), pp. 1063–1068
17. S. Mohr, L.E. Ratcliff, L. Genovese, D. Caliste, P. Boulanger, S. Goedecker, T. Deutsch, Accurate and efficient linear scaling DFT calculations with universal applicability. Phys. Chem. Chem. Phys. **17**, 31360–31370 (2015)
18. M. Born, R. Oppenheimer, Zur quantentheorie der molekeln. Ann. Phys. **389**(20), 457–484 (1927)
19. A. Szabo, *Modern Quantum Chemistry: Introduction to Advanced Electronic Structure Theory* (McGraw-Hill, New York, 1989)
20. P. Hohenberg, W. Kohn, Inhomogeneous electron gas. Phys. Rev. **136**, B864–B871 (1964)
21. E. Engel, *Density Functional Theory: An Advanced Course* (Springer, Berlin, 2011)
22. W. Kohn, L.J. Sham, Self-consistent equations including exchange and correlation effects. Phys. Rev. **140**, A1133–A1138 (1965)
23. P.A.M. Dirac, Note on exchange phenomena in the Thomas atom. Math. Proc. Cambridge Philos. Soc. **26**(3), 376–385 (1930)
24. D.M. Ceperley, B.J. Alder, Ground state of the electron gas by a stochastic method. Phys. Rev. Lett. **45**, 566–569 (1980)
25. P.P. Rushton, Towards a non-local density functional description of exchange and correlation. PhD Thesis, University of Durham, 2002
26. J.P. Perdew, K. Burke, M. Ernzerhof, Generalized gradient approximation made simple [Phys. Rev. Lett. 77, 3865 (1996)]. Phys. Rev. Lett. **78**, 1396–1396 (1997)
27. J.P. Perdew, J.A. Chevary, S.H. Vosko, K.A. Jackson, M.R. Pederson, D.J. Singh, C. Fiolhais, Atoms, molecules, solids, and surfaces: applications of the generalized gradient approximation for exchange and correlation. Phys. Rev. B **46**, 6671–6687 (1992)
28. D.C. Langreth, M.J. Mehl, Beyond the local-density approximation in calculations of ground-state electronic properties. Phys. Rev. B **28**, 1809–1834 (1983)
29. A.D. Becke, A new mixing of Hartree-Fock and local density-functional theories. J. Chem. Phys. **98**(2), 1372–1377 (1993)
30. A.J. Garza, G.E. Scuseria, Predicting band gaps with hybrid density functionals. J. Phys. Chem. Lett. **7**, 08 (2016)
31. J. Heyd, G.E. Scuseria, M. Ernzerhof, Hybrid functionals based on a screened coulomb potential. J. Chem. Phys. **118**(18), 8207–8215 (2003)
32. J.P. Perdew, M. Ernzerhof, K. Burke, Rationale for mixing exact exchange with density functional approximations. J. Chem. Phys. **105**(22), 9982–9985 (1996)
33. A.D. Becke, Density functional thermochemistry. III. The role of exact exchange. J. Chem. Phys. **98**(7), 5648–5652 (1993)
34. M. Guidon, J. Hutter, J. VandeVondele, Auxiliary density matrix methods for Hartree-Fock exchange calculations. J. Chem. Theory Comput. **6**(8), 2348–2364 (2010). PMID: 26613491
35. V.B. Sulimov, P.V. Sushko, A.H. Edwards, A.L. Shluger, A.M. Stoneham, Asymmetry and long-range character of lattice deformation by neutral oxygen vacancy in α-quartz. Phys. Rev. B **66**, 024108 (2002)
36. A.-M. El-Sayed, Atomistic modelling of charge trapping defects in silicon dioxide. PhD thesis, University College London, 2015
37. D. Muñoz Ramo, J.L. Gavartin, A.L. Shluger, G. Bersuker, Spectroscopic properties of oxygen vacancies in monoclinic HfO_2 calculated with periodic and embedded cluster density functional theory. Phys. Rev. B **75**, 205336 (2007)
38. P.E. Blöchl, J.H. Stathis, Hydrogen electrochemistry and stress-induced leakage current in silica. Phys. Rev. Lett. **83**, 372–375 (1999)
39. A.-M. El-Sayed, M.B. Watkins, T. Grasser, V.V. Afanas'ev, A.L. Shluger, Hydrogen-induced rupture of strained Si–O bonds in amorphous silicon dioxide. Phys. Rev. Lett. **114**, 115503 (2015)

40. T. Schlick, *Molecular Modeling and Simulation: An Interdisciplinary Guide* (Springer Science+Business Media, LLC, New York 2010)
41. T.P. Senftle, S. Hong, M.M. Islam, S.B. Kylasa, Y. Zheng, Y.K. Shin, C. Junkermeier, R. Engel-Herbert, M.J. Janik, H.M. Aktulga, T. Verstraelen, A. Grama, A.C.T. van Duin, The ReaxFF reactive force-field: development, applications and future directions. npj Comput. Mater. **2**, 15011 EP (2016). Review Article
42. G. Broglia, G. Or, L. Larcher, M. Montorsi, Molecular dynamics simulation of amorphous HfO_2 for resistive RAM applications. Modell. Simul. Mater. Sci. Eng. **22**, 065006 (2014)
43. S.R. Elliott, *Physics of Amorphous Materials* (Longman, New York, 1984)
44. D. Price, J. Carpenter, Scattering function of vitreous silica. J. Non Cryst. Solids **92**(1), 153–174 (1987)
45. R. Martin, *Electronic Structure: Basic Theory and Practical Methods* (Cambridge University Press, Cambridge, 2008)
46. P. McMillan, B. Piriou, R. Couty, A Raman study of pressure-densified vitreous silica. J. Chem. Phys. **81**(10), 4234–4236 (1984)
47. D.R. Fredkin, A. Komornicki, S.R. White, K.R. Wilson, Ab initio infrared and Raman spectra. J. Chem. Phys. **78**(12), 7077–7092 (1983)
48. B. Kaduk, T. Kowalczyk, T. Van Voorhis, Constrained density functional theory. Chem. Rev. **112**(1), 321–370 (2012). PMID: 22077560
49. C.G. Van de Walle, J. Neugebauer, First-principles calculations for defects and impurities: applications to III-Nitrides. J. Appl. Phys. **95**(8), 3851–3879 (2004)
50. S. Lany, A. Zunger, Assessment of correction methods for the band-gap problem and for finite-size effects in supercell defect calculations: case studies for ZnO and GaAs. Phys. Rev. B **78**, 235104 (2008)
51. C. Freysoldt, J. Neugebauer, C.G. Van de Walle, Fully ab initio finite-size corrections for charged-defect supercell calculations. Phys. Rev. Lett. **102**, 016402 (2009)
52. H.-P. Komsa, T.T. Rantala, A. Pasquarello, Finite-size supercell correction schemes for charged defect calculations. Phys. Rev. B **86**, 045112 (2012)
53. A.-M. El-Sayed, Y. Wimmer, W. Goes, T. Grasser, V.V. Afanas'ev, A.L. Shluger, Theoretical models of hydrogen-induced defects in amorphous silicon dioxide. Phys. Rev. B **92**, 014107 (2015)
54. K.O. Jeppson, C.M. Svensson, Negative bias stress of MOS devices at high electric fields and degradation of MNOS devices. J. Appl. Phys. **48**(5), 2004–2014 (1977)
55. M. Houssa, M. Aoulaiche, S. De Gendt, G. Groeseneken, M.M. Heyns, A. Stesmans, Reaction-dispersive proton transport model for negative bias temperature instabilities. Appl. Phys. Lett. **86**(9), 093506 (2005)
56. F. Schanovsky, T. Grasser, On the microscopic limit of the RD model, in *Bias Temperature Instability for Devices and Circuits*, ed. by T. Grasser (Springer, New York, 2014)
57. R.N. Hall, Electron-hole recombination in germanium. Phys. Rev. **87**, 387–387 (1952)
58. W. Shockley, W.T. Read, Statistics of the recombinations of holes and electrons. Phys. Rev. **87**, 835–842 (1952)
59. A. McWorther, 1/f noise and related surface effects in germanium. Sc.D Thesis, Massachusetts Institute of Technology, 1955
60. F. Schanovsky, W. Gös, T. Grasser, An advanced description of oxide traps in MOS transistors and its relation to DFT. J. Comput. Electron. **9**, 135–140 (2010)
61. F. Schanovsky, W. Gös, T. Grasser, Multiphonon hole trapping from first principles. J. Vac. Sci. Technol. B **29**(1), 01A201 (2011)
62. F. Schanovsky, O. Baumgartner, V. Sverdlov, T. Grasser, A multi scale modeling approach to non-radiative multi phonon transitions at oxide defects in MOS structures. J. Comput. Electron. **11**, 218–224 (2012)
63. K. Huang, A. Rhys, Theory of light absorption and non-radiative transitions in F-centres. Proc. Roy. Soc. Lond. A: Math. Phys. Eng. Sci. **204**(1078), 406–423 (1950)
64. C.H. Henry, D.V. Lang, Nonradiative capture and recombination by multiphonon emission in GaAs and GaP. Phys. Rev. B **15**, 989–1016 (1977)

65. W. Goes, F. Schanovsky, T. Grasser, Advanced modeling of oxide defects, in *Bias Temperature Instability for Devices and Circuits*, ed. by T. Grasser (Springer, New York, 2014)
66. W. Goes, Y. Wimmer, A.-M. El-Sayed, G. Rzepa, M. Jech, A. Shluger, T. Grasser, Identification of oxide defects in semiconductor devices: a systematic approach linking DFT to rate equations and experimental evidence. Microelectron. Reliab. **87**, 286–320 (2018)
67. T. Grasser, H. Reisinger, P. Wagner, B. Kaczer, F. Schanovsky, W. Goes, The time dependent defect spectroscopy (TDDS) for the characterization of the bias temperature instability, in *Proceedings of the International Reliability Physics Symposium* (2010), pp. 16–25
68. M. Waltl, G. Rzepa, A. Grill, W. Gös, J. Franco, B. Kaczer, L. Witters, J. Mitard, N. Horiguchi, T. Grasser, Superior NBTI in high-k SiGe transistors–part I: experimental. IEEE Trans. Electron Devices **64**(5), 2092–2098 (2017)
69. M. Waltl, G. Rzepa, A. Grill, W. Gös, J. Franco, B. Kaczer, L. Witters, J. Mitard, N. Horiguchi, T. Grasser, Superior NBTI in high-k SiGe transistors–part II: theory. IEEE Trans. Electron Devices **64**(5), 2099–2105 (2017)
70. F. Schanovsky, Atomistic modeling in the context of the bias temperature instability. Ph.D Thesis, TU Wien, 2013
71. W. Goes, Hole trapping and the negative bias temperature instability. Ph.D Thesis, TU Wien, 2011
72. Y. Wimmer, A.-M. El-Sayed, W. Gös, T. Grasser, A.L. Shluger, Role of hydrogen in volatile behaviour of defects in SiO_2-based electronic devices. Proc. Roy. Soc. A: Math. Phys. Eng. Sci. **472**(2190), 20160009 (2016)
73. H. Eyring, The activated complex in chemical reactions. J. Chem. Phys. **3**(2), 107–115 (1935)
74. M.G. Evans, M. Polanyi, Some applications of the transition state method to the calculation of reaction velocities, especially in solution. Trans. Faraday Soc. **31**, 875–894 (1935)
75. R. Fletcher, *Practical Methods of Optimization* (Wiley, Chichester, 1987)
76. H. Jónsson, G. Mills, K.W. Jacobsen, Nudged elastic band method for finding minimum energy paths of transitions, in *Classical and Quantum Dynamics in Condensed Phase Simulations* (World scientific, Singapore, 1998), pp. 385–404
77. Y. Wimmer, Hydrogen related defects in amorphous SiO_2 and the negative bias temperature instability. PhD Thesis, TU Vienna, E360, 2017
78. J. Franck, E.G. Dymond, Elementary processes of photochemical reactions. Trans. Faraday Soc. **21**, 536–542 (1926)
79. E. Condon, A theory of intensity distribution in band systems. Phys. Rev. **28**, 1182–1201 (1926)
80. J.F. Conley, P.M. Lenahan, W.F. McArthur, Preliminary investigation of the kinetics of postoxidation rapid thermal anneal induced hole-trap-precursor formation in microelectronic SiO_2 films. Appl. Phys. Lett. **73**(15), 2188–2190 (1998)
81. T. Tewksbury, Relaxation effects in MOS devices due to tunnel exchange with near-interface oxide traps. Ph.D Thesis, Massachusetts Institute of Technology, 1992
82. T. Chachiyo, J.H. Rodriguez, A direct method for locating minimum-energy crossing points (MECPs) in spin-forbidden transitions and nonadiabatic reactions. J. Chem. Phys. **123**(9), 094711 (2005)
83. G. Rzepa, J. Franco, B. O'Sullivan, A. Subirats, M. Simicic, G. Hellings, P. Weckx, M. Jech, T. Knobloch, M. Waltl, P.J. Roussel, D. Linten, B. Kaczer, T. Grasser, Comphy—a compact-physics framework for unified modeling of BTI. Microelectron. Reliab. **85**, 49–65 (2018)
84. K.L. Yip, W.B. Fowler, Electronic structure of $E_1{}'$ centers in SiO_2. Phys. Rev. B **11**, 2327–2338 (1975)
85. E.P. O'Reilly, J. Robertson, Theory of defects in vitreous silicon dioxide. Phys. Rev. B **27**, 3780–3795 (1983)
86. J.K. Rudra, W.B. Fowler, F.J. Feigl, Model for the E_2' center in alpha quartz. Phys. Rev. Lett. **55**, 2614–2617 (1985)
87. M.K. Schurman, M. Tomozawa, Equilibrium oxygen vacancy concentrations and oxidant diffusion in germania, silica, and germania-silica glasses. J. Non-Cryst. Solids **202**(1), 93–106 (1996)

88. S. Mukhopadhyay, P.V. Sushko, A.M. Stoneham, A.L. Shluger, Modeling of the structure and properties of oxygen vacancies in amorphous silica. Phys. Rev. B **70**, 195203 (2004)
89. N. Richard, L. Martin-SaMOS, G. Roma, Y. Limoge, J.-P. Crocombette, First principle study of neutral and charged self-defects in amorphous SiO_2. J. Non-Cryst. Solids **351**(21), 1825–1829 (2005)
90. S. Pantelides, Z.-Y. Lu, C. Nicklaw, T. Bakos, S. Rashkeev, D. Fleetwood, R. Schrimpf, The E' center and oxygen vacancies in SiO_2. J. Non-Cryst. Solids **354**(2), 217–223 (2008). Physics of Non-Cryst. Solids 11.
91. F. Schanovsky, O. Baumgartner, W. Gös, T. Grasser, A detailed evaluation of model defects as candidates for the bias temperature instability, in *Proceedings of the International Conference on Simulation of Semiconductor Processes and Devices (SISPAD)* (2013), pp. 1–4
92. T. Grasser, W. Gös, Y. Wimmer, F. Schanovsky, G. Rzepa, M. Waltl, K. Rott, H. Reisinger, V.V. Afanas'Ev, A. Stesmans, A.-M. El-Sayed, A.L. Shluger, On the microscopic structure of hole traps in pMOSFETs, in *2014 IEEE International Electron Devices Meeting* (2014), pp. 530–533. Talk: International Electron Devices Meeting (IEDM), San Francisco (15 Dec 2014–17 Dec 2014)
93. P.M. Lenahan, J.P. Campbell, A.T. Krishnan, S. Krishnan, A model for NBTI in nitrided oxide MOSFETs which does not involve hydrogen or diffusion. IEEE Trans. Device Mater. Reliab. **11**, 219–226 (2011)
94. T. Aichinger, S. Puchner, M. Nelhiebel, T. Grasser, H. Hutter, Impact of hydrogen on recoverable and permanent damage following negative bias temperature stress, in *2010 IEEE International Reliability Physics Symposium* (2010), pp. 1063–1068
95. K. Kajihara, L. Skuja, M. Hirano, H. Hosono, Diffusion and reactions of hydrogen in F_2-laser-irradiated SiO_2 glass. Phys. Rev. Lett. **89**, 135507 (2002)
96. J. Godet, A. Pasquarello, Proton diffusion mechanism in amorphous SiO_2. Phys. Rev. Lett. **97**, 155901 (2006)
97. S.A. Sheikholeslam, H. Manzano, C. Grecu, A. Ivanov, Reduced hydrogen diffusion in strained amorphous SiO2: understanding ageing in MOSFET devices. J. Mater. Chem. C **4**, 8104–8110 (2016)
98. J. Isoya, J.A. Weil, L.E. Halliburton, EPR and ab initio SCF-MO studies of the Si-H-Si system in the E'_4 center of α-quartz. J. Chem. Phys. **74**(10), 5436–5448 (1981)
99. P.E. Blöchl, First-principles calculations of defects in oxygen-deficient silica exposed to hydrogen. Phys. Rev. B **62**, 6158–6179 (2000)
100. A.-M. El-Sayed, Y. Wimmer, W. Goes, T. Grasser, V.V. Afanas'ev, A.L. Shluger, Theoretical models of hydrogen-induced defects in amorphous silicon dioxide. Phys. Rev. B **92**, 014107 (2015)
101. C.M. Nelson, R.A. Weeks, Trapped electrons in irradiated quartz and silica: I, optical absorption. J. Am. Ceram. Soc. **43**(8), 396–399 (1960)
102. V.V. Afanas'ev, A. Stesmans, Interfacial defects in SiO_2 revealed by photon stimulated tunneling of electrons. Phys. Rev. Lett. **78**, 2437–2440 (1997)
103. N.S. Saks, A.K. Agarwal, Hall mobility and free electron density at the SiC/SiO_2 interface in 4H-SiC. Appl. Phys. Lett. **77**(20), 3281–3283 (2000)
104. A.A. Karpushin, A.N. Sorokin, V.A. Gritsenko, Si–Si bond as a deep trap for electrons and holes in silicon nitride. JETP Lett. **103**, 171–174 (2016)
105. A.-M. El-Sayed, M.B. Watkins, V.V. Afanas'ev, A.L. Shluger, Nature of intrinsic and extrinsic electron trapping in SiO_2. Phys. Rev. B **89**, 125201 (2014)
106. D.L. Griscom, Self-trapped holes in amorphous silicon dioxide. Phys. Rev. B **40**, 4224–4227 (1989)
107. Y. Sasajima, K. Tanimura, Optical transitions of self-trapped holes in amorphous SiO_2. Phys. Rev. B **68**, 014204 (2003)
108. A. Kimmel, P. Sushko, A. Shluger, Structure and spectroscopic properties of trapped holes in silica. J. Non-Cryst. Solids **353**(5), 599–604 (2007). SiO2, Advanced Dielectrics and Related Devices 6

109. D.Z. Gao, J. Strand, M.S. Munde, A.L. Shluger, Mechanisms of oxygen vacancy aggregation in SiO_2 and HfO_2. Front. Phys. **7**, 43 (2019)
110. F. Aryasetiawan, O. Gunnarsson, The GW method. Rep. Prog. Phys. **61**, 237–312 (1998)
111. T. Grasser, K. Rott, H. Reisinger, M. Waltl, P.-J. Wagner, F. Schanovsky, W. Gös, G. Pobegen, B. Kaczer, Hydrogen-related volatile defects as the possible cause for the recoverable component of NBTI, in *2013 International Electron Devices Meeting (IEDM) Technical Digest* (2013), pp. 409–412. Talk: International Electron Devices Meeting (IEDM), Washington (09 Dec 2013–11 Dec 2013)
112. T. Aichinger, M. Nelhiebel, S. Einspieler, T. Grasser, In situ poly heater-a reliable tool for performing fast and defined temperature switches on chip. IEEE Trans. Device Mater. Reliab. **10**, 3–8 (2010)
113. Y. Wimmer, W. Gös, A.-M. El-Sayed, A. L. Shluger, T. Grasser, A Density-functional study of defect volatility in amorphous silicon dioxide, in *Proceedings of the International Conference on Simulation of Semiconductor Processes and Devices (SISPAD)* (2015), pp. 44–47. Talk: International Conference on Simulation of Semiconductor Processes and Devices (SISPAD), Washington (09 Sep 2015–11 Sep 2015)
114. J. Strand, M. Kaviani, D. Gao, A.-M. El-Sayed, V.V. Afanas'ev, A.L. Shluger, Intrinsic charge trapping in amorphous oxide films: status and challenges. J. Phys.: Condens. Matter **30**, 233001 (2018)
115. F. Bohra, B. Jiang, J.-M. Zuo, Textured crystallization of ultrathin hafnium oxide films on silicon substrate. Appl. Phys. Lett. **90**(16), 161917 (2007)
116. A. Shluger, *Defects in Oxides in Electronic Devices* (Springer International Publishing, Cham, 2018), pp. 1–22
117. F. Cerbu, O. Madia, D.V. Andreev, S. Fadida, M. Eizenberg, L. Breuil, J.G. Lisoni, J.A. Kittl, J. Strand, A.L. Shluger, V.V. Afanas'ev, M. Houssa, A. Stesmans, Intrinsic electron traps in atomic-layer deposited HfO_2 insulators. Appl. Phys. Lett. **108**(22), 222901 (2016)
118. S.R. Bradley, G. Bersuker, A.L. Shluger, Modelling of oxygen vacancy aggregates in monoclinic HfO_2: can they contribute to conductive filament formation? J. Phys. Condens. Matter **27**, 415401 (2015)
119. M. Kaviani, J. Strand, V.V. Afanas'ev, A.L. Shluger, Deep electron and hole polarons and bipolarons in amorphous oxide. Phys. Rev. B **94**, 020103 (2016)
120. J. Strand, M. Kaviani, V.V. Afanas'ev, J.G. Lisoni, A.L. Shluger, Intrinsic electron trapping in amorphous oxide. Nanotechnology **29**, 125703 (2018)
121. V.V. Afanas'ev, A. Stesmans, Stable trapping of electrons and holes in deposited insulating oxides: Al_2O_3, ZrO_2, and HfO_2. J. Appl. Phys. **95**(5), 2518–2526 (2004)

The Langevin–Boltzmann Equation for Noise Calculation

Christoph Jungemann

1 Introduction

In the semi-classical framework the Langevin–Boltzmann equation (LBE) is the fundamental equation for particle transport and noise [1, 2]. The LBE combines band structures and scattering processes calculated by quantum mechanical methods with Newton's classical equations of motion and transport is modeled on a microscopic level [3, 4]. The charge carriers are accelerated by the electric field described by Newton's laws. Since scattering is assumed to be instantaneous in the framework of the LBE, the particles do not move during scattering events and only the velocity changes abruptly. This scattering leads not only to, for example, finite drift velocities but also to noise. Since the electron carries a charge, it induces a displacement current, which is related to the velocity of the particle and thus changes abruptly, when the particle's velocity changes abruptly [5]. The displacement-current fluctuations lead directly to fluctuations of the terminal currents in the quasi-stationary approximation [6–8]. Thus, the scattering processes, which determine the transport, are also the source of the noise and in the framework of the LBE both effects are treated consistently and no further information is needed to calculate the noise. This is a huge advantage of the LBE compared to less CPU-intensive models (e.g., drift-diffusion (DD) or hydrodynamic (HD) models, compact models), which require explicit noise sources beyond the transport model and therefore additional information.

C. Jungemann (✉)
Chair of Electromagnetic Theory, RWTH Aachen University, Aachen, Germany
e-mail: cj@ithe.rwth-aachen.de

T. Grasser (ed.), *Noise in Nanoscale Semiconductor Devices*,
https://doi.org/10.1007/978-3-030-37500-3_19

In the past two different methods have been established to solve the LBE. The by far most often used approach is the Monte Carlo method [9, 10], which directly simulates the microscopic transport processes described above. The stochastic Monte Carlo method is well understood and easily implemented. Since the method simulates particle transport in the time domain, the power spectral densities of fluctuations are obtained by recording correlation functions, which are then Fourier transformed into the frequency domain. The minimum frequency that is accessible is inversely proportional to the duration of the simulation. Thus, investigation of slow processes and rare events is difficult [11]. The second approach is based on a deterministic solution of the LBE by numerical means [12, 13]. This method has various disadvantages. It is much more difficult to implement and requires sophisticated methods in order to obtain numerical stability and is very memory intensive [14]. On the other hand, noise is calculated directly in the frequency domain based on small-signal analysis. The full frequency range from zero to THz is accessible and rare events cause no problems. In addition, inclusion of the Pauli principle in the scattering integral is straightforward [1, 15]. Since the fundamental noise sources and the Green's functions are directly accessible, it is possible to analyze the origin of the terminal current noise within the device.

Some theoretical aspects of the LBE are discussed in this chapter, where the numerical details will be left out, which are well documented in [14]. Instead we will describe the general approach, its most important features and its relation to modeling of noise by balance equations derived from the LBE. In the next section the LBE and Poisson equation (PE) for the quasi-stationary potential are introduced. The fundamental equations for the calculation of noise in the framework of the small-signal approximation, the noise sources, and the transfer functions are presented. The transfer function for the terminal current is discussed in detail. Next, the admittance parameters based on the LBE and PE are analyzed and it is shown that the admittance parameters are reciprocal for equilibrium. The transfer function for the terminal current is derived for equilibrium and it is shown that the terminal current noise satisfies the Nyquist theorem. Explicit analytical expressions for some important quantities are derived for a simple resistor, which give some insight into the general concepts and are useful for debugging purposes. In addition, some properties of the approach are demonstrated by numerical solutions. In the next section, the relation between the LBE and balance equations based on the macroscopic relaxation time approximation is investigated with emphasis on the noise sources. The macroscopic relaxation time approximation of the scattering integral destroys the consistency of transport and noise modeling (e.g., violation of fluctuation–dissipation theorems at equilibrium [16]). By modification of the Langevin forces it is ensured that the noise sources reproduce the power spectral densities (PSD) of the variables of the balance equations under homogeneous bulk conditions similar to the derivation of the transport parameters [17]. Some analytical results for a homogeneous resistor are given and some numerical results.

2 The Langevin–Boltzmann Equation

Discussion of the LBE is limited to the stationary case of electrons in a 3D k-space of a homogeneous semiconductor material with a single conduction band and the Pauli principle is neglected for the sake of clarity and brevity.[1] The LBE for this simple system is given by

$$\left\{\frac{\partial}{\partial t} - \frac{e}{\hbar}\boldsymbol{E}(\boldsymbol{r},t)\,\nabla_{\boldsymbol{k}} + \boldsymbol{v}(\boldsymbol{k})\,\nabla_{\boldsymbol{r}}\right\} f(\boldsymbol{r},\boldsymbol{k},t)$$
$$= \frac{\Omega}{(2\pi)^3}\int W(\boldsymbol{k}|\boldsymbol{k}')(\boldsymbol{r}) f(\boldsymbol{r},\boldsymbol{k}',t) - W(\boldsymbol{k}'|\boldsymbol{k})(\boldsymbol{r}) f(\boldsymbol{r},\boldsymbol{k},t)\mathrm{d}^3k' + \xi(\boldsymbol{r},\boldsymbol{k},t)\,, \tag{1}$$

where the distribution function $f(\boldsymbol{r},\boldsymbol{k},t)$ is a stochastic variable due to the Langevin force $\xi(\boldsymbol{r},\boldsymbol{k},t)$, which is assumed to have a Gaussian probability density function and to be delta-correlated in time and real space [1]. $\boldsymbol{v}(\boldsymbol{k})$ is the group velocity depending on the wave vector $\boldsymbol{k}$ and $W(\boldsymbol{k}|\boldsymbol{k}')(\boldsymbol{r})$ the transition rate of an instantaneous scattering event from state $\boldsymbol{k}'$ to $\boldsymbol{k}$ at position $\boldsymbol{r}$ in real space. $\hbar$ is the reduced Planck constant, e the positive elementary charge, and Ω the system volume. $\boldsymbol{E} = -\nabla_{\boldsymbol{r}}\varphi$ is the electric field strength and φ the quasi-stationary potential, which is the result of the PE

$$\nabla_{\boldsymbol{r}}\,[\kappa(\boldsymbol{r})\nabla_{\boldsymbol{r}}\varphi(\boldsymbol{r},t)] = -e\,(N_{\mathrm{D}}(\boldsymbol{r}) - n(\boldsymbol{r},t))\,, \tag{2}$$

where κ is the permittivity, N_{D} the donor concentration, and n the electron density

$$n(\boldsymbol{r},t) = \frac{1}{(2\pi)^3}\int f(\boldsymbol{r},\boldsymbol{k},t)\mathrm{d}^3k\,. \tag{3}$$

In the electron density spin degeneracy is not explicitly accounted for by a factor of two. Instead, it is assumed that the distribution function contains both spin directions. For a more detailed discussion see [15].

The distribution function is split into a noiseless stationary part f_0 and a fluctuation δf

$$f(\boldsymbol{r},\boldsymbol{k},t) = f_0(\boldsymbol{r},\boldsymbol{k}) + \delta f(\boldsymbol{r},\boldsymbol{k},t)\,. \tag{4}$$

The fluctuations are assumed to be so small that the system's response at a given bias point is linear. In addition, the linearized system is time invariant (LTI) because of the assumption that the underlying system is stationary. Due to the above definition

[1]For the case of a quasi-2D electron gas the reader is referred to [18]. Inclusion of the Pauli principle, generation/recombination processes, and multiple bands can be found in [13–15].

and the linear system behavior the expected value of the fluctuation vanishes. The time-independent part of the distribution function must satisfy the stationary noiseless BE

$$\left\{-\frac{e}{\hbar}\boldsymbol{E}_0(\boldsymbol{r})\,\nabla_{\boldsymbol{k}}+\boldsymbol{v}(\boldsymbol{k})\,\nabla_{\boldsymbol{r}}\right\}f_0(\boldsymbol{r},\boldsymbol{k})$$
$$=\frac{\Omega}{(2\pi)^3}\int W(\boldsymbol{k}|\boldsymbol{k}')(\boldsymbol{r})f_0(\boldsymbol{r},\boldsymbol{k}')-W(\boldsymbol{k}'|\boldsymbol{k})(\boldsymbol{r})f_0(\boldsymbol{r},\boldsymbol{k})\mathrm{d}^3k' \quad (5)$$

and PE

$$\nabla_{\boldsymbol{r}}\left[\kappa(\boldsymbol{r})\nabla_{\boldsymbol{r}}\varphi_0(\boldsymbol{r})\right]=-e\left(N_\mathrm{D}(\boldsymbol{r})-\frac{1}{(2\pi)^3}\int f_0(\boldsymbol{r},\boldsymbol{k})\mathrm{d}^3k\right). \quad (6)$$

Since Gaussian noise filtered by an LTI system remains Gaussian, the correlation function of the fluctuations is sufficient to calculate the noise transfer through the LTI system [19]. The correlation function of the Langevin force can be calculated with the noiseless solution of the stationary BE [20]

$$c_{\xi\xi}(\boldsymbol{r},\boldsymbol{k},t;\boldsymbol{r}',\boldsymbol{k}',t')$$
$$=\Omega\left[\left(\int W(\boldsymbol{k}''|\boldsymbol{k})(\boldsymbol{r})f_0(\boldsymbol{r},\boldsymbol{k})+W(\boldsymbol{k}|\boldsymbol{k}'')(\boldsymbol{r})f_0(\boldsymbol{r},\boldsymbol{k}'')\mathrm{d}^3k''\right)\delta(\boldsymbol{k}-\boldsymbol{k}')\right.$$
$$\left.-W(\boldsymbol{k}|\boldsymbol{k}')(\boldsymbol{r})f_0(\boldsymbol{r},\boldsymbol{k}')-W(\boldsymbol{k}'|\boldsymbol{k})(\boldsymbol{r})f_0(\boldsymbol{r},\boldsymbol{k})\right]\delta(\boldsymbol{r}-\boldsymbol{r}')\delta(t-t'). \quad (7)$$

Due to the instantaneous nature of the scattering events the correlation function is local in time and real space. The distribution function still represents both spin directions, although they are degenerate. For sufficiently fast decaying correlations the Wiener–Khinchin theorem allows to convert the correlation function from the time domain into the frequency domain by Fourier transform yielding the PSD, where a one-sided PSD is considered in this work, resulting in an additional factor of two [19]

$$S_{\xi\xi}(\boldsymbol{r};\boldsymbol{k};\boldsymbol{k}';\omega)\delta(\boldsymbol{r}-\boldsymbol{r}')$$
$$=2\Omega\left[\left(\int W(\boldsymbol{k}''|\boldsymbol{k})(\boldsymbol{r})f_0(\boldsymbol{r},\boldsymbol{k})+W(\boldsymbol{k}|\boldsymbol{k}'')(\boldsymbol{r})f_0(\boldsymbol{r},\boldsymbol{k}'')\mathrm{d}^3k''\right)\delta(\boldsymbol{k}-\boldsymbol{k}')\right.$$
$$\left.-W(\boldsymbol{k}|\boldsymbol{k}')(\boldsymbol{r})f_0(\boldsymbol{r},\boldsymbol{k}')-W(\boldsymbol{k}'|\boldsymbol{k})(\boldsymbol{r})f_0(\boldsymbol{r},\boldsymbol{k})\right]\delta(\boldsymbol{r}-\boldsymbol{r}'). \quad (8)$$

ω is the angular frequency. The PSD is white due to the delta correlation in time [1].

In the simple LBE (1) only the product of the electric field and the gradient in the k-space of the distribution function is nonlinear resulting in a linear system that contains the linearized LBE and PE

$$\begin{aligned}
&\frac{\partial \delta f(\boldsymbol{r},\boldsymbol{k},t)}{\partial t} + \frac{e}{\hbar}\nabla_{\boldsymbol{r}}\varphi_0(\boldsymbol{r})\,\nabla_{\boldsymbol{k}}\delta f(\boldsymbol{r},\boldsymbol{k},t) \\
&+\frac{e}{\hbar}\nabla_{\boldsymbol{r}}\delta\varphi(\boldsymbol{r},t)\,\nabla_{\boldsymbol{k}} f_0(\boldsymbol{r},\boldsymbol{k}) + \boldsymbol{v}(\boldsymbol{k})\,\nabla_{\boldsymbol{r}}\delta f(\boldsymbol{r},\boldsymbol{k},t) \\
&-\frac{\Omega}{(2\pi)^3}\int W(\boldsymbol{k}|\boldsymbol{k}')(\boldsymbol{r})\delta f(\boldsymbol{r},\boldsymbol{k}',t) - W(\boldsymbol{k}'|\boldsymbol{k})(\boldsymbol{r})\delta f(\boldsymbol{r},\boldsymbol{k},t)\mathrm{d}^3k' = \xi(\boldsymbol{r},\boldsymbol{k},t)\,, \\
&\quad\nabla_{\boldsymbol{r}}\left[\kappa(\boldsymbol{r})\nabla_{\boldsymbol{r}}\delta\varphi(\boldsymbol{r},t)\right] - \frac{e}{(2\pi)^3}\int \delta f(\boldsymbol{r},\boldsymbol{k},t)\mathrm{d}^3k = 0\,.
\end{aligned} \tag{9}$$

The variables are the fluctuations of the distribution function and the quasi-stationary potential. In order to solve the linear system, Green's functions are introduced for the distribution function $\delta f(\boldsymbol{r},\boldsymbol{k},t) \to G_f(\boldsymbol{r},\boldsymbol{k},t;\boldsymbol{r}',\boldsymbol{k}',t')$ and the potential $\delta\varphi(\boldsymbol{r},t) \to G_\varphi(\boldsymbol{r},t;\boldsymbol{r}',\boldsymbol{k}',t')$, which are the solutions of the linear system for a unit source, which generates a single electron at time t' in the state $(\boldsymbol{r}',\boldsymbol{k}')$ with either spin up or down, where a factor of $(2\pi)^3$ cancels in the definition of the Green's functions and the correlation functions

$$\begin{aligned}
&\frac{\partial G_f(\boldsymbol{r},\boldsymbol{k},t;\boldsymbol{r}',\boldsymbol{k}',t')}{\partial t} \\
&+\frac{e}{\hbar}\nabla_{\boldsymbol{r}}\varphi_0(\boldsymbol{r})\,\nabla_{\boldsymbol{k}} G_f(\boldsymbol{r},\boldsymbol{k},t;\boldsymbol{r}',\boldsymbol{k}',t') \\
&+\frac{e}{\hbar}\nabla_{\boldsymbol{r}} G_\varphi(\boldsymbol{r},t;\boldsymbol{r}',\boldsymbol{k}',t')\,\nabla_{\boldsymbol{k}} f_0(\boldsymbol{r},\boldsymbol{k}) \\
&+\boldsymbol{v}(\boldsymbol{k})\,\nabla_{\boldsymbol{r}} G_f(\boldsymbol{r},\boldsymbol{k},t;\boldsymbol{r}',\boldsymbol{k}',t') \\
&-\frac{\Omega}{(2\pi)^3}\int W(\boldsymbol{k}|\boldsymbol{k}'')(\boldsymbol{r})G_f(\boldsymbol{r},\boldsymbol{k}'',t;\boldsymbol{r}',\boldsymbol{k}',t') \\
&-W(\boldsymbol{k}''|\boldsymbol{k})(\boldsymbol{r})G_f(\boldsymbol{r},\boldsymbol{k},t;\boldsymbol{r}',\boldsymbol{k}',t')\mathrm{d}^3k'' = \delta(\boldsymbol{r}-\boldsymbol{r}')\delta(\boldsymbol{k}-\boldsymbol{k}')\delta(t-t')\,, \\
&\quad\nabla_{\boldsymbol{r}}\left[\kappa(\boldsymbol{r})\nabla_{\boldsymbol{r}} G_\varphi(\boldsymbol{r},t;\boldsymbol{r}',\boldsymbol{k}',t')\right] - \frac{e}{(2\pi)^3}\int G_f(\boldsymbol{r},\boldsymbol{k},t;\boldsymbol{r}',\boldsymbol{k}',t')\mathrm{d}^3k = 0\,.
\end{aligned} \tag{10}$$

Since the linear system does not distinguish between the two spin directions, the spin index does not explicitly occur. The Green's functions vanish for $t' > t$, because the system is causal. In addition, the Green's functions depend only on $t - t'$, because the system is time invariant. Fourier transformation w.r.t. $t - t'$ yields the transfer functions (Green's function in the frequency domain)

$$\begin{aligned}
&\mathrm{i}\omega G_f(\boldsymbol{r},\boldsymbol{k};\boldsymbol{r}',\boldsymbol{k}';\omega) \\
&+\frac{e}{\hbar}\nabla_{\boldsymbol{r}}\varphi_0(\boldsymbol{r})\,\nabla_{\boldsymbol{k}} G_f(\boldsymbol{r},\boldsymbol{k};\boldsymbol{r}',\boldsymbol{k}';\omega) \\
&+\frac{e}{\hbar}\nabla_{\boldsymbol{r}} G_\varphi(\boldsymbol{r};\boldsymbol{r}',\boldsymbol{k}';\omega)\,\nabla_{\boldsymbol{k}} f_0(\boldsymbol{r},\boldsymbol{k}) \\
&+\boldsymbol{v}(\boldsymbol{k})\,\nabla_{\boldsymbol{r}} G_f(\boldsymbol{r},\boldsymbol{k};\boldsymbol{r}',\boldsymbol{k}';\omega)
\end{aligned}$$

$$
\begin{aligned}
&-\frac{\Omega}{(2\pi)^3}\int W(\boldsymbol{k}|\boldsymbol{k}'')(\boldsymbol{r})G_f(\boldsymbol{r},\boldsymbol{k}'';\boldsymbol{r}',\boldsymbol{k}';\omega)\\
&-W(\boldsymbol{k}''|\boldsymbol{k})(\boldsymbol{r})G_f(\boldsymbol{r},\boldsymbol{k};\boldsymbol{r}',\boldsymbol{k}';\omega)\mathrm{d}^3k''=\delta(\boldsymbol{r}-\boldsymbol{r}')\delta(\boldsymbol{k}-\boldsymbol{k}')\,,
\end{aligned}
$$

$$
\begin{aligned}
&\nabla_{\boldsymbol{r}}\left[\kappa(\boldsymbol{r})\nabla_{\boldsymbol{r}}G_\varphi(\boldsymbol{r};\boldsymbol{r}',\boldsymbol{k}';\omega)\right]\\
&\quad-\frac{e}{(2\pi)^3}\int G_f(\boldsymbol{r},\boldsymbol{k};\boldsymbol{r}',\boldsymbol{k}';\omega)\mathrm{d}^3k=0\,. \qquad (11)
\end{aligned}
$$

With the transfer functions the PSDs can be calculated. In the case of the distribution function the cross-PSD, which is nonlocal in real and k-space, is given by the Wiener-Lee theorem [19], which states that the PSD at the output of an LTI system is given by the PSD at the input times the transfer function and times the complex conjugate of the transfer function

$$
\begin{aligned}
&S_{ff}(\boldsymbol{r},\boldsymbol{k};\boldsymbol{r}',\boldsymbol{k}';\omega)\\
&=\int\int\int G_f(\boldsymbol{r},\boldsymbol{k};\boldsymbol{r}_0,\boldsymbol{k}_1;\omega)S_{\xi\xi}(\boldsymbol{r}_0;\boldsymbol{k}_1;\boldsymbol{k}_2;\omega)G_f^\star(\boldsymbol{r}',\boldsymbol{k}';\boldsymbol{r}_0,\boldsymbol{k}_2;\omega)\mathrm{d}^3k_1\mathrm{d}^3k_2\mathrm{d}^3r_0\\
&=2\Omega\int\int\int\left[G_f(\boldsymbol{r},\boldsymbol{k};\boldsymbol{r}_0,\boldsymbol{k}_1;\omega)-G_f(\boldsymbol{r},\boldsymbol{k};\boldsymbol{r}_0,\boldsymbol{k}_2;\omega)\right]W(\boldsymbol{k}_1|\boldsymbol{k}_2)(\boldsymbol{r}_0)f_0(\boldsymbol{r}_0,\boldsymbol{k}_2)\\
&\qquad\left[G_f(\boldsymbol{r}',\boldsymbol{k}';\boldsymbol{r}_0,\boldsymbol{k}_1;\omega)-G_f(\boldsymbol{r}',\boldsymbol{k}';\boldsymbol{r}_0,\boldsymbol{k}_2;\omega)\right]^\star\mathrm{d}^3k_1\mathrm{d}^3k_2\mathrm{d}^3r_0\,. \qquad (12)
\end{aligned}
$$

This formula has a very simple interpretation. The difference of the two transfer functions in the square brackets describes the generation of an electron in the state $(\boldsymbol{r}_0,\boldsymbol{k}_1)$ (positive transfer function) and the annihilation of an electron in the state $(\boldsymbol{r}_0,\boldsymbol{k}_2)$ (negative transfer function). Thus, the term in the first square brackets is the transfer function for the response of the distribution function at $(\boldsymbol{r},\boldsymbol{k})$ to a scattering event from state $(\boldsymbol{r}_0,\boldsymbol{k}_2)$ into $(\boldsymbol{r}_0,\boldsymbol{k}_1)$ and no charge is created or destroyed. The term $W(\boldsymbol{k}_1|\boldsymbol{k}_2)(\boldsymbol{r}_0)f_0(\boldsymbol{r}_0,\boldsymbol{k}_2)$ is the rate of this transition, which is a Poisson process and its white PSD is given by twice its rate (similar to shot noise). Thus, the integrand (cross-PSD) describes the filtering of a scattering event by the transport equations (LTI system), where the cross-PSD is given by the product of the PSD for the scattering event with the transfer function from scattering to the response of the distribution function at $(\boldsymbol{r},\boldsymbol{k})$ and the complex conjugate of the other transfer function for $(\boldsymbol{r}',\boldsymbol{k}')$. Since the individual scattering events are independent, the total cross-PSD is obtained by integration over the real space and the two k-space arguments involved in the scattering process.

This approach is easily extended to more complex cases. If the Pauli principle is to be included in the calculation, the transition rate in Eq. (12) has to be multiplied by $[1-f_0(\boldsymbol{r}_0,\boldsymbol{k}_1)]$, where in this factor the distribution function stands only for one spin direction [1, 15]. In addition, the Pauli principle has to be included in the LBE and the equations for the calculation of the transfer functions become somewhat more complex. In the case of electron–electron scattering, noise can be calculated by considering the transfer functions of the two involved electrons for their initial and final states resulting in four transfer functions in the square brackets and by replacing the transition rate by the one for electron–electron scattering [1]. In the case that phonon transport is included in the calculations, the emission and absorption of a phonon during a scattering process can be included by adding the corresponding transfer function to the square brackets. Since the phonon is either generated or annihilated, the square brackets contain only three transfer functions (two for the electron and one for the phonon) [21].

In this manner the PSD of various quantities can be calculated. As the most important example the case of terminal current noise is discussed next. With the particle current density $\boldsymbol{j}$

$$\boldsymbol{j}(\boldsymbol{r},t) = \frac{1}{(2\pi)^3}\int \boldsymbol{v}(\boldsymbol{k}) f(\boldsymbol{r},\boldsymbol{k},t)\mathrm{d}^3k \tag{13}$$

the current of the kth terminal is given for fixed terminal voltages by the extended Ramo–Shockley theorem [8]

$$I_k(t) = e\int \boldsymbol{j}(\boldsymbol{r},t)\,\nabla_{\boldsymbol{r}} h_k(\boldsymbol{r})\mathrm{d}^3r = \frac{e}{(2\pi)^3}\int\int \nabla_{\boldsymbol{r}} h_k(\boldsymbol{r})\,\boldsymbol{v}(\boldsymbol{k}) f(\boldsymbol{r},\boldsymbol{k},t)\mathrm{d}^3k\mathrm{d}^3r\,, \tag{14}$$

where $h_k(\boldsymbol{r})$ is the basis function for the kth contact. It is the fundamental solution of the PE for zero charge

$$\nabla_{\boldsymbol{r}}\;[\kappa(\boldsymbol{r})\nabla_{\boldsymbol{r}} h_k(\boldsymbol{r})] = 0\,, \tag{15}$$

where a normalized potential of one is applied to the kth contact and zero to all others. Calculation of the cross-PSD of the kth and k'th terminal currents is straightforward

$$S_{I_k I_{k'}}(\omega) = \frac{e^2}{(2\pi)^6}\int\int\int\int \nabla_{\boldsymbol{r}} h_k(\boldsymbol{r})\,\boldsymbol{v}(\boldsymbol{k}) S_{ff}(\boldsymbol{r},\boldsymbol{k};\boldsymbol{r}',\boldsymbol{k}';\omega)\boldsymbol{v}(\boldsymbol{k}')\,\nabla_{\boldsymbol{r}'} h_{k'}(\boldsymbol{r}')\,\mathrm{d}^3k\mathrm{d}^3k'\mathrm{d}^3r\mathrm{d}^3r'\,. \tag{16}$$

This calculation is extremely CPU intensive, because it includes in total three integrals over the real space and four over k-space considering also the integrals in Eq. (12) and the transfer functions depend on two sets of arguments. The number of integrals can be reduced by introducing a transfer function for the terminal current

$$G_{I_k}(\boldsymbol{r}',\boldsymbol{k}';\omega) = \frac{e}{(2\pi)^3}\int\int \nabla_{\boldsymbol{r}} h_k(\boldsymbol{r})\,\boldsymbol{v}(\boldsymbol{k}) G_f(\boldsymbol{r},\boldsymbol{k};\boldsymbol{r}',\boldsymbol{k}';\omega)\mathrm{d}^3k\mathrm{d}^3r\,. \tag{17}$$

The transfer function of the kth terminal current depends only on the second set of arguments in Eq. (11) and it can be calculated directly as the solution of the adjoint version of Eq. (11) without having to calculate the Green's function G_f. With the adjoint method the calculation of this transfer function does not require more CPU time than solving the linear system (11) once [22]. This transfer function is the ratio of the kth terminal current and a current injected into the device at position $\boldsymbol{r}'$ with wave vector $\boldsymbol{k}'$ and frequency ω and thus a generalization of Shockley's impedance field [23]. With the transfer functions of the terminal currents the PSD is obtained

$$S_{I_k I_{k'}}(\omega) = 2\Omega \int\int\int \left[G_{I_k}(\boldsymbol{r}_0, \boldsymbol{k}_1; \omega) - G_{I_k}(\boldsymbol{r}_0, \boldsymbol{k}_2; \omega)\right] W(\boldsymbol{k}_1|\boldsymbol{k}_2)(\boldsymbol{r}_0) f_0(\boldsymbol{r}_0, \boldsymbol{k}_2) \left[G_{I_{k'}}(\boldsymbol{r}_0, \boldsymbol{k}_1; \omega) - G_{I_{k'}}(\boldsymbol{r}_0, \boldsymbol{k}_2; \omega)\right]^\star \mathrm{d}^3k_1 \mathrm{d}^3k_2 \mathrm{d}^3r_0 \,. \quad (18)$$

$\boldsymbol{r}_0$ is the position in real space of the scattering event. With

$$K_{I_k I_{k'}}(\boldsymbol{r}_0, \omega) = 2\Omega \int\int \left[G_{I_k}(\boldsymbol{r}_0, \boldsymbol{k}_1; \omega) - G_{I_k}(\boldsymbol{r}_0, \boldsymbol{k}_2; \omega)\right] W(\boldsymbol{k}_1|\boldsymbol{k}_2)(\boldsymbol{r}_0) f_0(\boldsymbol{r}_0, \boldsymbol{k}_2) \left[G_{I_{k'}}(\boldsymbol{r}_0, \boldsymbol{k}_1; \omega) - G_{I_{k'}}(\boldsymbol{r}_0, \boldsymbol{k}_2; \omega)\right]^\star \mathrm{d}^3k_1 \mathrm{d}^3k_2 \,, \quad (19)$$

the local contribution to the terminal current noise can be determined, which is proportional to the local rate of scattering events and the product of the transfer functions [13]. Thus, frequent scattering alone is no indicator for a strong contribution to terminal current noise, because the difference of the transfer functions in the square brackets can be small at the same time [24, 25].

2.1 Reciprocity

At equilibrium the linear response of an electron device should be reciprocal for a vanishing magnetic field [1, 3]. In order to show this, the distribution function is split into two factors

$$f(\boldsymbol{r}, \boldsymbol{k}, t) = g(\boldsymbol{r}, \boldsymbol{k}, t) \underbrace{\exp\left(\frac{e\varphi(\boldsymbol{r}, t) - \varepsilon(\boldsymbol{k}) - \varepsilon_{\mathrm{cb}}}{k_{\mathrm{B}} T_0}\right)}_{=h(\boldsymbol{r}, \boldsymbol{k}, t)} \,. \quad (20)$$

For equilibrium the second factor on the right-hand side is the solution of the stationary BE (Boltzmann distribution function), where $\varepsilon(\boldsymbol{k})$ is the band energy relative to the conduction band edge $\varepsilon_{\mathrm{cb}}$. The band edge is chosen in such a way that the correct particle density is obtained at equilibrium including spin degeneracy. Thus, at equilibrium $g(\boldsymbol{r}, \boldsymbol{k}, t) = 1$ holds. Since the potential depends in general on time, the exponential factor modifies the time derivative in the BE and the noiseless BE (Eq. (1) for zero Langevin force) for $g(\boldsymbol{r}, \boldsymbol{k}, t)$ is given by

$$\left\{\frac{\partial}{\partial t} + \frac{1}{V_\mathrm{T}}\frac{\partial \varphi}{\partial t} - \frac{e}{\hbar}\boldsymbol{E}(\boldsymbol{r},t)\,\nabla_{\boldsymbol{k}} + \boldsymbol{v}(\boldsymbol{k})\cdot\nabla_{\boldsymbol{r}}\right\} g(\boldsymbol{r},\boldsymbol{k},t)$$
$$= -\frac{\Omega}{(2\pi)^3}\int W(\boldsymbol{k}'|\boldsymbol{k})(\boldsymbol{r})\left[g(\boldsymbol{r},\boldsymbol{k},t) - g(\boldsymbol{r},\boldsymbol{k}',t)\right]\mathrm{d}^3k' \,, \quad (21)$$

where the principle of detailed balance [3] $W(\boldsymbol{k}|\boldsymbol{k}')(\boldsymbol{r})\exp\left(\frac{\varepsilon(\boldsymbol{k})-\varepsilon(\boldsymbol{k}')}{k_\mathrm{B}T_0}\right) = W(\boldsymbol{k}'|\boldsymbol{k})(\boldsymbol{r})$ has been used to symmetrize the scattering integral. $V_\mathrm{T} = k_\mathrm{B}T_0/e$ is the thermal voltage. With the rearranged scattering integral

$$W\{g\} = \frac{\Omega}{(2\pi)^3}\int W(\boldsymbol{k}'|\boldsymbol{k})(\boldsymbol{r})[g(\boldsymbol{r},\boldsymbol{k},t) - g(\boldsymbol{r},\boldsymbol{k}',t)]\mathrm{d}^3k' \,, \quad (22)$$

the streaming operator

$$L\{g\} = \left\{-\frac{e}{\hbar}\boldsymbol{E}(\boldsymbol{r},t)\,\nabla_{\boldsymbol{k}} + \boldsymbol{v}(\boldsymbol{k})\,\nabla_{\boldsymbol{r}}\right\} g(\boldsymbol{r},\boldsymbol{k},t) \quad (23)$$

and splitting of g into an even part, which yields densities, and an odd part, which yields fluxes, [26, 27]

$$g^\mathrm{e}(\boldsymbol{r},\boldsymbol{k},t) = \frac{g(\boldsymbol{r},\boldsymbol{k},t) + g(\boldsymbol{r},-\boldsymbol{k},t)}{2}$$
$$g^\mathrm{o}(\boldsymbol{r},\boldsymbol{k},t) = \frac{g(\boldsymbol{r},\boldsymbol{k},t) - g(\boldsymbol{r},-\boldsymbol{k},t)}{2} \quad (24)$$

we get two coupled BEs

$$\frac{\partial g^\mathrm{e}}{\partial t} + \frac{g^\mathrm{e}}{V_\mathrm{T}}\frac{\partial \varphi}{\partial t} + L\{g^\mathrm{o}\} + W\{g^\mathrm{e}\} = 0$$

$$\frac{\partial g^\mathrm{o}}{\partial t} + \frac{g^\mathrm{o}}{V_\mathrm{T}}\frac{\partial \varphi}{\partial t} + L\{g^\mathrm{e}\} + W\{g^\mathrm{o}\} = 0 \,, \quad (25)$$

where it has been assumed that $W(\boldsymbol{k}'|\boldsymbol{k}) = W(-\boldsymbol{k}'|-\boldsymbol{k})$ holds and the scattering integral takes the same form for the even and odd equation, which is, for example, the case for transition rates that depend only on the scattering angle. The streaming operator shows the usual odd/even coupling due to the derivative w.r.t. the wave vector and the velocity which are both odd in $\boldsymbol{k}$.

To solve the BEs boundary conditions are required. If particles can enter the device only through contacts, the particle flux (proportional to the odd part g^o) must be zero through all surfaces which are not part of contacts. On the contacts connected to the semiconductor we use Dirichlet boundary conditions for the densities (even part g^e), which forces the even distribution function towards a Boltzmann distribution

$$g^{\mathrm{e}}(\boldsymbol{r},\boldsymbol{k},t)=\exp\left(-\frac{V_k}{V_{\mathrm{T}}}\right)\qquad \text{for}\quad \boldsymbol{r}\in\partial D_k \tag{26}$$

and the distribution function is given by its equilibrium value. V_k is the bias applied to the kth contact and ∂D_k its domain in real space. The potential must satisfy a similar Dirichlet boundary condition on contacts and the electric flux density homogeneous Neumann boundary conditions on the other parts of the surface.

Linearization of the BEs and PE around a stationary bias point yields an LTI system, which is best solved in the frequency domain

$$\mathrm{i}\omega\left(g^{\mathrm{e}}+\frac{g_0^{\mathrm{e}}}{V_{\mathrm{T}}}\varphi\right)+L_0\{g^{\mathrm{o}}\}+W\{g^{\mathrm{e}}\}=-\frac{e}{\hbar}\nabla_{\boldsymbol{r}}\varphi\,\nabla_{\boldsymbol{k}}g_0^{\mathrm{o}}$$

$$\mathrm{i}\omega\left(g^{\mathrm{o}}+\frac{g_0^{\mathrm{o}}}{V_{\mathrm{T}}}\varphi\right)+L_0\{g^{\mathrm{e}}\}+W\{g^{\mathrm{o}}\}=-\frac{e}{\hbar}\nabla_{\boldsymbol{r}}\varphi\,\nabla_{\boldsymbol{k}}g_0^{\mathrm{e}}$$

$$\nabla_{\boldsymbol{r}}\,[\kappa\nabla_{\boldsymbol{r}}\varphi]-\frac{e}{(2\pi)^3}\int\left(g^{\mathrm{e}}+g_0^{\mathrm{e}}\frac{\varphi}{V_{\mathrm{T}}}\right)h_0\mathrm{d}^3k=0 \tag{27}$$

with the small-signal quantities $g^{\mathrm{e}}(\boldsymbol{r},\boldsymbol{k},\omega)$, $g^{\mathrm{o}}(\boldsymbol{r},\boldsymbol{k},\omega)$, $\varphi(\boldsymbol{r},\omega)$, and $L_0\{\}=-e/\hbar\boldsymbol{E}_0\,\nabla_{\boldsymbol{k}}+\boldsymbol{v}\,\nabla_{\boldsymbol{r}}$. The small-signal electron density in the PE contains a factor $(g^{\mathrm{e}}+g_0^{\mathrm{e}}\varphi/V_{\mathrm{T}})$, which is the same as the term in the brackets of the time derivative in the BE for g^{e}. As will be shown below, this symmetry is necessary to obtain reciprocity at equilibrium. If the small-signal BEs and PE are discretized, this symmetry must be conserved, otherwise reciprocity is lost [18].

The above system of equations is linear in all three small-signal quantities, and these can be expressed as superpositions of fundamental solutions for the individual contacts

$$\begin{aligned}
g^{\mathrm{e}}(\boldsymbol{r},\boldsymbol{k},\omega)&=\sum_{i=1}^{N_{\mathrm{C}}}\eta_i^{\mathrm{e}}(\boldsymbol{r},\boldsymbol{k},\omega)V_i\ ,\\
g^{\mathrm{o}}(\boldsymbol{r},\boldsymbol{k},\omega)&=\sum_{i=1}^{N_{\mathrm{C}}}\eta_i^{\mathrm{o}}(\boldsymbol{r},\boldsymbol{k},\omega)V_i\ ,\\
\varphi(\boldsymbol{r},\omega)&=\sum_{i=1}^{N_{\mathrm{C}}}\tau_i(\boldsymbol{r},\omega)V_i\ .
\end{aligned} \tag{28}$$

V_i is the small-signal bias applied to the ith contact and N_{C} the number of contacts of the device. The fundamental solutions must satisfy the linear equations

$$\mathrm{i}\omega\left(\eta_i^{\mathrm{e}} + g_0^{\mathrm{e}}\frac{\tau_i}{V_{\mathrm{T}}}\right) + L_{\mathrm{O}}\{\eta_i^{\mathrm{o}}\} + W\{\eta_i^{\mathrm{e}}\} = -\frac{e}{\hbar}\nabla_{\boldsymbol{r}}\tau_i\,\nabla_{\boldsymbol{k}}g_0^{\mathrm{o}}\,,$$

$$\mathrm{i}\omega\left(\eta_i^{\mathrm{o}} + g_0^{\mathrm{o}}\frac{\tau_i}{V_{\mathrm{T}}}\right) + L_{\mathrm{O}}\{\eta_i^{\mathrm{e}}\} + W\{\eta_i^{\mathrm{o}}\} = -\frac{e}{\hbar}\nabla_{\boldsymbol{r}}\tau_i\,\nabla_{\boldsymbol{k}}g_0^{\mathrm{e}}\,,$$

$$\nabla_{\boldsymbol{r}}\,[\kappa\nabla_{\boldsymbol{r}}\tau_i] - \frac{e}{(2\pi)^3}\int\left(\eta_i^{\mathrm{e}} + g_0^{\mathrm{e}}\frac{\tau_i}{V_{\mathrm{T}}}\right)h_{\mathrm{O}}\mathrm{d}^3k = 0\,, \tag{29}$$

the Neumann boundary conditions for the fluxes, and normalized Dirichlet ones for densities and potentials on all contacts

$$\begin{aligned}\eta_i^{\mathrm{e}}(\boldsymbol{r},\boldsymbol{k},\omega) &= -\frac{\delta_{i,k}}{V_{\mathrm{T}}}\exp\left(-\frac{V_{k,0}}{V_{\mathrm{T}}}\right)\\ \tau_i(\boldsymbol{r},\omega) &= \delta_{i,k}\end{aligned} \tag{30}$$

for $\boldsymbol{r} \in \partial D_k$ and $k = 1, \ldots, N_{\mathrm{C}}$. With the admittance parameter

$$Y_{ki} = \int_{\partial D_k}\left[\frac{e}{(2\pi)^3}\int \boldsymbol{v}\left(\eta_i^{\mathrm{o}} + g_0^{\mathrm{o}}\frac{\tau_i}{V_{\mathrm{T}}}\right)h_{\mathrm{O}}\mathrm{d}^3k + \mathrm{i}\omega\kappa\nabla_{\boldsymbol{r}}\tau_i\right]\mathrm{d}^2\boldsymbol{r}\,, \tag{31}$$

where the infinitesimal surface element $\mathrm{d}^2\boldsymbol{r}$ has an outward orientation, the small-signal current of the kth terminal is given by

$$I_k = \sum_{i=1}^{N_{\mathrm{C}}} Y_{ki}V_i\,. \tag{32}$$

At equilibrium the odd part of the stationary distribution function vanishes ($g_0^{\mathrm{o}} = 0$) and the even part is given by $g_0^{\mathrm{e}} = 1$ due to $V_{k,0} = 0$ for all k. Thus, the gradients on the RHS of the BEs in Eq. (29) are zero. The admittance is given for equilibrium by

$$\begin{aligned}Y_{ki}^{\mathrm{eq}} &= \frac{e}{(2\pi)^3}\int_{\partial D_k}\int \boldsymbol{v}\eta_i^{\mathrm{o}}h_{\mathrm{O}}\mathrm{d}^3k\,\mathrm{d}^2\boldsymbol{r} \quad + \mathrm{i}\omega\int_{\partial D_k}\kappa\nabla_{\boldsymbol{r}}\tau_i\,\mathrm{d}^2\boldsymbol{r}\\ &= -\frac{eV_{\mathrm{T}}}{(2\pi)^3}\oint_{\partial D}\int \eta_k^{\mathrm{e}}\boldsymbol{v}\eta_i^{\mathrm{o}}h_{\mathrm{O}}\mathrm{d}^3k\,\mathrm{d}^2\boldsymbol{r} + \mathrm{i}\omega\oint_{\partial D}\tau_k\kappa\nabla_{\boldsymbol{r}}\tau_i\,\mathrm{d}^2\boldsymbol{r}\,.\end{aligned} \tag{33}$$

∂D is the total surface of the device. Extension of the surface integral to the total surface is possible, because τ_k and $-V_{\mathrm{T}}\eta_k^{\mathrm{e}}$ are one on the kth contact and zero on all other contacts. The surface parts outside the contacts do not contribute to the surface integral because of the homogeneous Neumann boundary conditions for the fluxes. The integral over the total surface can be converted with the divergence theorem into an integral over the device domain D

$$
\begin{aligned}
Y_{ki}^{\mathrm{eq}} &= -\frac{eV_{\mathrm{T}}}{(2\pi)^3}\int_D\int \nabla_{\boldsymbol{r}}\left[\eta_k^{\mathrm{e}}\boldsymbol{v}\eta_i^{\mathrm{o}}h_0\right]\mathrm{d}^3k\mathrm{d}^3r + \mathrm{i}\omega\int_D \nabla_{\boldsymbol{r}}\left[\tau_k\kappa\nabla_{\boldsymbol{r}}\tau_i\right]\mathrm{d}^3r \\
&= -\frac{eV_{\mathrm{T}}}{(2\pi)^3}\int_D\int L_0\left\{\eta_k^{\mathrm{e}}\eta_i^{\mathrm{o}}h_0\right\}\mathrm{d}^3k\mathrm{d}^3r - \frac{eV_{\mathrm{T}}}{(2\pi)^3}\int_D\int \frac{e}{\hbar}\boldsymbol{E}_0\nabla_{\boldsymbol{k}}\left[\eta_k^{\mathrm{e}}\eta_i^{\mathrm{o}}h_0\right]\mathrm{d}^3k\mathrm{d}^3r \\
&\quad +\mathrm{i}\omega\int_D \kappa\nabla_{\boldsymbol{r}}\tau_k\,\nabla_{\boldsymbol{r}}\tau_i\mathrm{d}^3r + \mathrm{i}\omega\int_D \tau_k\nabla_{\boldsymbol{r}}\left[\kappa\nabla_{\boldsymbol{r}}\tau_i\right]\mathrm{d}^3r\,. \qquad (34)
\end{aligned}
$$

The integral

$$
\int \nabla_{\boldsymbol{k}}\left[\eta_k^{\mathrm{e}}\eta_i^{\mathrm{o}}h_0\boldsymbol{E}_0\right]\mathrm{d}^3k = \oint \eta_k^{\mathrm{e}}\eta_i^{\mathrm{o}}h_0\boldsymbol{E}_0\,\mathrm{d}^2\boldsymbol{k} = 0 \qquad (35)
$$

vanishes, because on the surface of the k-space the energy is infinitely large and the exponential function in h_0 goes to zero. With Eq. (29) we get

$$
\begin{aligned}
Y_{ki}^{\mathrm{eq}} &= -\frac{eV_{\mathrm{T}}}{(2\pi)^3}\int_D\int \left(L_0\left\{\eta_k^{\mathrm{e}}\right\}\eta_i^{\mathrm{o}} + \eta_k^{\mathrm{e}}L_0\left\{\eta_i^{\mathrm{o}}\right\}\right)h_0\mathrm{d}^3k\mathrm{d}^3r + \mathrm{i}\omega\int_D \kappa\nabla_{\boldsymbol{r}}\tau_k\,\nabla_{\boldsymbol{r}}\tau_i\mathrm{d}^3r \\
&\quad +\mathrm{i}\omega\frac{e}{(2\pi)^3}\int_D\int \tau_k\left(\eta_i^{\mathrm{e}} + \frac{\tau_i}{V_{\mathrm{T}}}\right)h_0\mathrm{d}^3k\mathrm{d}^3r \\
&= \frac{eV_{\mathrm{T}}}{(2\pi)^3}\int_D\int \left(W\{\eta_k^{\mathrm{o}}\}\eta_i^{\mathrm{o}} + \eta_k^{\mathrm{e}}W\{\eta_i^{\mathrm{e}}\}\right)h_0\mathrm{d}^3k\mathrm{d}^3r + \mathrm{i}\omega\int_D \kappa\nabla_{\boldsymbol{r}}\tau_k\,\nabla_{\boldsymbol{r}}\tau_i\mathrm{d}^3r \\
&\quad +\mathrm{i}\omega\frac{eV_{\mathrm{T}}}{(2\pi)^3}\int_D\int \left[\eta_k^{\mathrm{o}}\eta_i^{\mathrm{o}} + \left(\eta_k^{\mathrm{e}} + \frac{\tau_k}{V_{\mathrm{T}}}\right)\left(\eta_i^{\mathrm{e}} + \frac{\tau_i}{V_{\mathrm{T}}}\right)\right]h_0\mathrm{d}^3k\mathrm{d}^3r\,. \qquad (36)
\end{aligned}
$$

With

$$
\begin{aligned}
&\int\int W(\boldsymbol{k}'|\boldsymbol{k})(\boldsymbol{r})\eta_k^{\mathrm{o}}(\boldsymbol{r},\boldsymbol{k}',t)\eta_i^{\mathrm{o}}(\boldsymbol{r},\boldsymbol{k},t)\exp\left(\frac{-\varepsilon(\boldsymbol{k})}{k_{\mathrm{B}}T_0}\right)\mathrm{d}^3k'\mathrm{d}^3k \\
&= \int\int W(\boldsymbol{k}|\boldsymbol{k}')(\boldsymbol{r})\eta_k^{\mathrm{o}}(\boldsymbol{r},\boldsymbol{k},t)\eta_i^{\mathrm{o}}(\boldsymbol{r},\boldsymbol{k}',t)\exp\left(\frac{-\varepsilon(\boldsymbol{k}')}{k_{\mathrm{B}}T_0}\right)\mathrm{d}^3k'\mathrm{d}^3k \\
&= \int\int W(\boldsymbol{k}|\boldsymbol{k}')(\boldsymbol{r})\exp\left(\frac{\varepsilon(\boldsymbol{k})-\varepsilon(\boldsymbol{k}')}{k_{\mathrm{B}}T_0}\right)\eta_k^{\mathrm{o}}(\boldsymbol{r},\boldsymbol{k},t)\eta_i^{\mathrm{o}}(\boldsymbol{r},\boldsymbol{k}',t)\exp\left(\frac{-\varepsilon(\boldsymbol{k})}{k_{\mathrm{B}}T_0}\right)\mathrm{d}^3k'\mathrm{d}^3k \\
&= \int\int W(\boldsymbol{k}'|\boldsymbol{k})(\boldsymbol{r})\eta_i^{\mathrm{o}}(\boldsymbol{r},\boldsymbol{k}',t)\eta_k^{\mathrm{o}}(\boldsymbol{r},\boldsymbol{k},t)\exp\left(\frac{-\varepsilon(\boldsymbol{k})}{k_{\mathrm{B}}T_0}\right)\mathrm{d}^3k'\mathrm{d}^3k \qquad (37)
\end{aligned}
$$

we obtain

$$
\int W\{\eta_k^{\mathrm{o}}\}\eta_i^{\mathrm{o}}h_0\mathrm{d}^3k = \int W\{\eta_i^{\mathrm{o}}\}\eta_k^{\mathrm{o}}h_0\mathrm{d}^3k \qquad (38)
$$

and similarly

$$\int W\{\eta_k^{\mathrm{e}}\}\eta_i^{\mathrm{e}} h_0 \mathrm{d}^3 k = \int W\{\eta_i^{\mathrm{e}}\}\eta_k^{\mathrm{e}} h_0 \mathrm{d}^3 k \,. \tag{39}$$

Thus, reciprocity results

$$Y_{ki}^{\mathrm{eq}} = Y_{ik}^{\mathrm{eq}} \,. \tag{40}$$

This relation holds for arbitrary frequencies and all indices i, k. It must also hold in the discrete case and is one of the fundamental checks for a small-signal simulation.

Furthermore, the device must be passive at equilibrium and the real part of the admittance matrix should be positive semidefinite for nonzero inelastic scattering. Since the fundamental solutions are real-valued on the contacts, the admittance can be also calculated by multiplication with the complex conjugate of the fundamental solutions

$$Y_{ki}^{\mathrm{eq}} = -\frac{eV_{\mathrm{T}}}{(2\pi)^3} \oint_{\partial D} \int \left(\eta_k^{\mathrm{e}}\right)^{\star} \boldsymbol{v} \eta_i^{\mathrm{o}} h_0 \mathrm{d}^3 k \, \mathrm{d}^2 \boldsymbol{r} + \mathrm{i}\omega \oint_{\partial D} \tau_k^{\star} \kappa \boldsymbol{\nabla}_{\boldsymbol{r}} \tau_i \, \mathrm{d}^2 \boldsymbol{r} \,. \tag{41}$$

The sum of the admittance matrix and its Hermitian yields twice its real part due to reciprocity

$$\begin{aligned}
2\Re\left\{Y_{ki}^{\mathrm{eq}}\right\} &= Y_{ki}^{\mathrm{eq}} + \left(Y_{ik}^{\mathrm{eq}}\right)^{\star} \\
&= -\frac{eV_{\mathrm{T}}}{(2\pi)^3} \oint_{\partial D} \int \left(\eta_k^{\mathrm{e}}\right)^{\star} \boldsymbol{v} \eta_i^{\mathrm{o}} h_0 \mathrm{d}^3 k \, \mathrm{d}^2 \boldsymbol{r} + \mathrm{i}\omega \oint_{\partial D} \tau_k^{\star} \kappa \boldsymbol{\nabla}_{\boldsymbol{r}} \tau_i \, \mathrm{d}^2 \boldsymbol{r} \\
&\quad -\frac{eV_{\mathrm{T}}}{(2\pi)^3} \oint_{\partial D} \int \eta_i^{\mathrm{e}} \boldsymbol{v} \left(\eta_k^{\mathrm{o}}\right)^{\star} h_0 \mathrm{d}^3 k \, \mathrm{d}^2 \boldsymbol{r} - \mathrm{i}\omega \oint_{\partial D} \tau_i \kappa \boldsymbol{\nabla}_{\boldsymbol{r}} \tau_k^{\star} \, \mathrm{d}^2 \boldsymbol{r} \\
&= \frac{eV_{\mathrm{T}}}{(2\pi)^3} \int_D \int \left(\left(W\{\eta_k^{\mathrm{o}}\}\right)^{\star} \eta_i^{\mathrm{o}} + \left(\eta_k^{\mathrm{e}}\right)^{\star} W\{\eta_i^{\mathrm{e}}\} \right) h_0 \mathrm{d}^3 k \mathrm{d}^3 r \\
&\quad + \mathrm{i}\omega \int_D \kappa \boldsymbol{\nabla}_{\boldsymbol{r}} \tau_k^{\star} \, \boldsymbol{\nabla}_{\boldsymbol{r}} \tau_i \mathrm{d}^3 r \\
&\quad + \mathrm{i}\omega \frac{eV_{\mathrm{T}}}{(2\pi)^3} \int_D \int \left[-\left(\eta_k^{\mathrm{o}}\right)^{\star} \eta_i^{\mathrm{o}} + \left(\eta_k^{\mathrm{e}} + \frac{\tau_k}{V_{\mathrm{T}}}\right)^{\star} \left(\eta_i^{\mathrm{e}} + \frac{\tau_i}{V_{\mathrm{T}}}\right) \right] h_0 \mathrm{d}^3 k \mathrm{d}^3 r \\
&\quad + \frac{eV_{\mathrm{T}}}{(2\pi)^3} \int_D \int \left(W\{\eta_i^{\mathrm{o}}\} \left(\eta_k^{\mathrm{o}}\right)^{\star} + \eta_i^{\mathrm{e}} \left(W\{\eta_k^{\mathrm{e}}\}\right)^{\star} \right) h_0 \mathrm{d}^3 k \mathrm{d}^3 r \\
&\quad - \mathrm{i}\omega \int_D \kappa \boldsymbol{\nabla}_{\boldsymbol{r}} \tau_i \, \boldsymbol{\nabla}_{\boldsymbol{r}} \tau_k^{\star} \mathrm{d}^3 r \\
&\quad - \mathrm{i}\omega \frac{eV_{\mathrm{T}}}{(2\pi)^3} \int_D \int \left[-\eta_i^{\mathrm{o}} \left(\eta_k^{\mathrm{o}}\right)^{\star} + \left(\eta_i^{\mathrm{e}} + \frac{\tau_i}{V_{\mathrm{T}}}\right) \left(\eta_k^{\mathrm{e}} + \frac{\tau_k}{V_{\mathrm{T}}}\right)^{\star} \right] h_0 \mathrm{d}^3 k \mathrm{d}^3 r \\
&= \frac{2eV_{\mathrm{T}}}{(2\pi)^3} \int_D \int \left(\left(W\{\eta_k^{\mathrm{o}}\}\right)^{\star} \eta_i^{\mathrm{o}} + \left(\eta_k^{\mathrm{e}}\right)^{\star} W\{\eta_i^{\mathrm{e}}\} \right) h_0 \mathrm{d}^3 k \mathrm{d}^3 r
\end{aligned}$$

$$
\begin{aligned}
&= \frac{2eV_\mathrm{T}}{(2\pi)^3}\int_D\int \left(\left(\eta_k^\mathrm{o}\right)^\star W\{\eta_i^\mathrm{o}\} + \left(\eta_k^\mathrm{e}\right)^\star W\{\eta_i^\mathrm{e}\}\right) h_0 \mathrm{d}^3k\mathrm{d}^3r \\
&= \frac{2eV_\mathrm{T}}{(2\pi)^3}\int_D\int \left[\eta_k^\mathrm{e}+\eta_k^\mathrm{o}\right]^\star W\{[\eta_i^\mathrm{e}+\eta_i^\mathrm{o}]\} h_0 \mathrm{d}^3k\mathrm{d}^3r \\
&= \frac{2eV_\mathrm{T}}{(2\pi)^3}\int_D\int \left[\eta_k^\mathrm{e}-\eta_k^\mathrm{o}\right]^\star W\{[\eta_i^\mathrm{e}-\eta_i^\mathrm{o}]\} h_0 \mathrm{d}^3k\mathrm{d}^3r \,.
\end{aligned} \tag{42}
$$

The last two equalities hold because the scattering integral $W\{\}$ is an even function of $\boldsymbol{k}$ for η_i^e and an odd one for η_i^o, and the integral over the additional terms vanishes because the corresponding integrand is in both cases an odd function of $\boldsymbol{k}$.

With

$$
\eta_k(\boldsymbol{r},\boldsymbol{k},\omega) = \eta_k^\mathrm{e}(\boldsymbol{r},\boldsymbol{k},\omega) + \eta_k^\mathrm{o}(\boldsymbol{r},\boldsymbol{k},\omega) \tag{43}
$$

the second last row of (42) is given by

$$
\Re\left\{Y_{ki}^\mathrm{eq}\right\} = \frac{eV_\mathrm{T}}{(2\pi)^3}\int_D\int \eta_k^\star W\{\eta_i\} h_0 \mathrm{d}^3k\mathrm{d}^3r \,. \tag{44}
$$

First we will show that the main diagonal elements of this matrix are positive for $\omega > 0$. In this case the integral

$$
\begin{aligned}
0 &< \int\int W(\boldsymbol{k}'|\boldsymbol{k})|\eta_k(\boldsymbol{k}) - \eta_k(\boldsymbol{k}')|^2 h_0(\boldsymbol{k})\mathrm{d}^3k'\mathrm{d}^3k \\
&= \int\int W(\boldsymbol{k}'|\boldsymbol{k})\left[|\eta_k(\boldsymbol{k})|^2 + |\eta_k(\boldsymbol{k}')|^2 - 2\Re\left\{\eta_k^\star(\boldsymbol{k})\eta_k(\boldsymbol{k}')\right\}\right] h_0(\boldsymbol{k})\mathrm{d}^3k'\mathrm{d}^3k
\end{aligned} \tag{45}
$$

is positive, where only the important arguments are shown. Due to the detailed balance at equilibrium, we get in analogy to (38)

$$
\int\int W(\boldsymbol{k}'|\boldsymbol{k})|\eta_k(\boldsymbol{k})|^2 h_0(\boldsymbol{k})\mathrm{d}^3k'\mathrm{d}^3k = \int\int W(\boldsymbol{k}'|\boldsymbol{k})|\eta_k(\boldsymbol{k}')|^2 h_0(\boldsymbol{k})\mathrm{d}^3k'\mathrm{d}^3k \tag{46}
$$

and

$$
\int\int W(\boldsymbol{k}'|\boldsymbol{k})\eta_k^\star(\boldsymbol{k})\eta_k(\boldsymbol{k}')h_0(\boldsymbol{k})\mathrm{d}^3k'\mathrm{d}^3k = \int\int W(\boldsymbol{k}'|\boldsymbol{k})\eta_k^\star(\boldsymbol{k}')\eta_k(\boldsymbol{k})h_0(\boldsymbol{k})\mathrm{d}^3k'\mathrm{d}^3k \,. \tag{47}
$$

This yields

$$
\int\int W(\boldsymbol{k}'|\boldsymbol{k})\left[|\eta_k(\boldsymbol{k})|^2 - \eta_k^\star(\boldsymbol{k})\eta_k(\boldsymbol{k}')\right] h_0(\boldsymbol{k})\mathrm{d}^3k'\mathrm{d}^3k = \int \eta_k^\star W\left\{\eta_k\right\} h_0\mathrm{d}^3k > 0 \tag{48}
$$

and

$$\Re\left\{Y_{kk}^{\mathrm{eq}}\right\} > 0\,. \tag{49}$$

The sum of the fundamental solutions over all contacts is constant for g

$$\sum_{k=1}^{N_{\mathrm{C}}} \eta_k = -\frac{1}{V_{\mathrm{T}}} \tag{50}$$

and φ

$$\sum_{k=1}^{N_{\mathrm{C}}} \tau_k = 1\,. \tag{51}$$

This corresponds to applying to all contacts the same small-signal bias, which shifts the potential by this value and does not change the distribution function f. Since this solution satisfies the small-signal LBE and PE together with all boundary conditions, it is a solution of the boundary value problem. For a sum over a row of the matrix $\Re\left\{Y_{ki}^{\mathrm{eq}}\right\}$ we obtain with this result

$$\sum_{i=1}^{N_{\mathrm{C}}} \Re\left\{Y_{ki}^{\mathrm{eq}}\right\} = \frac{eV_{\mathrm{T}}}{(2\pi)^3}\int_D\int \eta_k^{\star} W\left\{\sum_{i=1}^{N_{\mathrm{C}}} \eta_i\right\} h_0 \mathrm{d}^3k\mathrm{d}^3r = 0 \tag{52}$$

because the scattering integral vanishes for a constant argument. This result is consistent with the fact that the sum over all terminal currents must be zero (Kirchhoff's current law). In addition, the off-diagonal elements are nonpositive

$$\Re\left\{Y_{kk}^{\mathrm{eq}}\right\} = -\sum_{i=1, i\neq k}^{N_{\mathrm{C}}} \Re\left\{Y_{ki}^{\mathrm{eq}}\right\} = \sum_{i=1, i\neq k}^{N_{\mathrm{C}}} \left|\Re\left\{Y_{ki}^{\mathrm{eq}}\right\}\right| \tag{53}$$

and with the Gershgorin theorem it follows that all eigenvalues are nonnegative [28]. Therefore, the matrix is positive semidefinite and the device passive at equilibrium. This is another important check for the small-signal parameters.

2.2 Nyquist Theorem

At equilibrium fluctuation–dissipation theorems must hold and thus the Nyquist theorem [1]. With Eq. (20) the transfer function of the distribution function is given by

$$G_f(\boldsymbol{r},\boldsymbol{k};\boldsymbol{r}',\boldsymbol{k}';\omega) = \Bigg[G_{g^e}(\boldsymbol{r},\boldsymbol{k};\boldsymbol{r}',\boldsymbol{k}';\omega) + G_{g^o}(\boldsymbol{r},\boldsymbol{k};\boldsymbol{r}',\boldsymbol{k}';\omega) + \left(g_0^e(\boldsymbol{r},\boldsymbol{k}) + g_0^o(\boldsymbol{r},\boldsymbol{k})\right)\frac{G_\varphi(\boldsymbol{r};\boldsymbol{r}',\boldsymbol{k}';\omega)}{V_T}\Bigg] h_0 \,. \quad (54)$$

At equilibrium these transfer functions have to satisfy

$$\left[\mathrm{i}\omega\left(G_{g^e} + \frac{G_\varphi}{V_T}\right) + L_0\{G_{g^o}\} + W\{G_{g^e}\}\right] h_0 = \frac{\delta(\boldsymbol{k}-\boldsymbol{k}') + \delta(\boldsymbol{k}+\boldsymbol{k}')}{2}\delta(\boldsymbol{r}-\boldsymbol{r}')$$

$$\left[\mathrm{i}\omega G_{g^o} + L_0\{G_{g^e}\} + W\{G_{g^o}\}\right] h_0 = \frac{\delta(\boldsymbol{k}-\boldsymbol{k}') - \delta(\boldsymbol{k}+\boldsymbol{k}')}{2}\delta(\boldsymbol{r}-\boldsymbol{r}')$$

$$\boldsymbol{\nabla}_{\boldsymbol{r}}\left[\kappa\boldsymbol{\nabla}_{\boldsymbol{r}} G_\varphi\right] - \frac{e}{(2\pi)^3}\int\left(G_{g^e} + \frac{G_\varphi}{V_T}\right) h_0 \mathrm{d}^3k = 0 \,. \quad (55)$$

The even part G_{g^e} must satisfy Dirichlet boundary conditions on the contacts and on the other parts of the surface homogeneous Neumann boundary conditions must hold for the fluxes w.r.t. the first argument in real space.

The transfer function for the terminal current is given at equilibrium by

$$\begin{aligned} G_{I_k}(\boldsymbol{r}',\boldsymbol{k}';\omega) &= \frac{e}{(2\pi)^3}\int_{\partial D_k}\int \boldsymbol{v}G_{g^o}h_0\mathrm{d}^3k\,\mathrm{d}^2\boldsymbol{r} + \mathrm{i}\omega\int_{\partial D_k}\kappa\boldsymbol{\nabla}_{\boldsymbol{r}}G_\varphi\,\mathrm{d}^2\boldsymbol{r} \\ &= -\frac{eV_T}{(2\pi)^3}\oint_{\partial D}\int \eta_k^e\boldsymbol{v}G_{g^o}h_0\mathrm{d}^3k\,\mathrm{d}^2\boldsymbol{r} + \mathrm{i}\omega\oint_{\partial D}\tau_k\kappa\boldsymbol{\nabla}_{\boldsymbol{r}}G_\varphi\,\mathrm{d}^2\boldsymbol{r} \,, \end{aligned} \quad (56)$$

where the fluxes again have to satisfy the homogeneous Neumann boundary condition on the part of the surface that does not belong to the contacts. Similar to Eq. (34) we obtain

$$\begin{aligned} G_{I_k}(\boldsymbol{r}',\boldsymbol{k}';\omega) &= -\frac{eV_T}{(2\pi)^3}\int_D\int \boldsymbol{\nabla}_{\boldsymbol{r}}\left[\eta_k^e\boldsymbol{v}G_{g^o}h_0\right]\mathrm{d}^3k\mathrm{d}^3r + \mathrm{i}\omega\int_D\boldsymbol{\nabla}_{\boldsymbol{r}}\left[\tau_k\kappa\boldsymbol{\nabla}_{\boldsymbol{r}}G_\varphi\right]\mathrm{d}^3r \\ &= -\frac{eV_T}{(2\pi)^3}\int_D\int L_0\left\{\eta_k^e G_{g^o}h_0\right\}\mathrm{d}^3k\mathrm{d}^3r \\ &\quad +\mathrm{i}\omega\int_D\kappa\boldsymbol{\nabla}_{\boldsymbol{r}}\tau_k\,\boldsymbol{\nabla}_{\boldsymbol{r}}G_\varphi\mathrm{d}^3r + \mathrm{i}\omega\int_D\tau_k\boldsymbol{\nabla}_{\boldsymbol{r}}\left[\kappa\boldsymbol{\nabla}_{\boldsymbol{r}}G_\varphi\right]\mathrm{d}^3r \end{aligned} \quad (57)$$

and

$$\begin{aligned} G_{I_k}(\boldsymbol{r}',\boldsymbol{k}';\omega) &= -\frac{eV_T}{(2\pi)^3}\int_D\int\left(L_0\left\{\eta_k^e\right\}G_{g^o} + \eta_k^e L_0\left\{G_{g^o}\right\}\right)h_0\mathrm{d}^3k\mathrm{d}^3r \\ &\quad +\mathrm{i}\omega\int_D\kappa\boldsymbol{\nabla}_{\boldsymbol{r}}\tau_k\,\boldsymbol{\nabla}_{\boldsymbol{r}}G_\varphi\mathrm{d}^3r \end{aligned}$$

$$
\begin{aligned}
&+\mathrm{i}\omega\frac{e}{(2\pi)^3}\int_D\int \tau_k\left(G_{g^{\mathrm{e}}}+\frac{G_\varphi}{V_{\mathrm{T}}}\right)h_0\mathrm{d}^3k\mathrm{d}^3r\\
=\;&\frac{eV_{\mathrm{T}}}{(2\pi)^3}\int_D\int\left[\mathrm{i}\omega\eta_k^{\mathrm{o}}+W\{\eta_k^{\mathrm{o}}\}\right]G_{g^{\mathrm{o}}}h_0\mathrm{d}^3k\mathrm{d}^3r\\
&+\frac{eV_{\mathrm{T}}}{(2\pi)^3}\int_D\int\eta_k^{\mathrm{e}}\left[\mathrm{i}\omega\left\{G_{g^{\mathrm{e}}}+\frac{G_\varphi}{V_{\mathrm{T}}}\right\}+W\{G_{g^{\mathrm{e}}}\}\right]h_0\mathrm{d}^3k\mathrm{d}^3r\\
&+\mathrm{i}\omega\int_D\kappa\nabla_{\boldsymbol{r}}\tau_k\,\nabla_{\boldsymbol{r}}G_\varphi\mathrm{d}^3r\\
&+\mathrm{i}\omega\frac{eV_{\mathrm{T}}}{(2\pi)^3}\int_D\int\frac{\tau_k}{V_{\mathrm{T}}}\left(G_{g^{\mathrm{e}}}+\frac{G_\varphi}{V_{\mathrm{T}}}\right)h_0\mathrm{d}^3k\mathrm{d}^3r\\
&-\frac{eV_{\mathrm{T}}}{(2\pi)^3}\eta_k^{\mathrm{e}}(\boldsymbol{r}',\boldsymbol{k}',\omega)\\
=\;&\frac{eV_{\mathrm{T}}}{(2\pi)^3}\int_D\int\left(W\{\eta_k^{\mathrm{o}}\}G_{g^{\mathrm{o}}}+\eta_k^{\mathrm{e}}W\{G_{g^{\mathrm{e}}}\}\right)h_0\mathrm{d}^3k\mathrm{d}^3r\\
&+\mathrm{i}\omega\int_D\kappa\nabla_{\boldsymbol{r}}\tau_k\,\nabla_{\boldsymbol{r}}G_\varphi\mathrm{d}^3r\\
&+\mathrm{i}\omega\frac{eV_{\mathrm{T}}}{(2\pi)^3}\int_D\int\left[\eta_k^{\mathrm{o}}G_{g^{\mathrm{o}}}+\left(\eta_k^{\mathrm{e}}+\frac{\tau_k}{V_{\mathrm{T}}}\right)\left(G_{g^{\mathrm{e}}}+\frac{G_\varphi}{V_{\mathrm{T}}}\right)\right]h_0\mathrm{d}^3k\mathrm{d}^3r\\
&-\frac{eV_{\mathrm{T}}}{(2\pi)^3}\eta_k^{\mathrm{e}}(\boldsymbol{r}',\boldsymbol{k}',\omega)\,. \qquad (58)
\end{aligned}
$$

The last term is due to the delta-functions in the BE for the even part (55). With

$$
\int W\{\eta_k^{\mathrm{o}}\}G_{g^{\mathrm{o}}}h_0\mathrm{d}^3k=\int W\{G_{g^{\mathrm{o}}}\}\eta_k^{\mathrm{o}}h_0\mathrm{d}^3k \qquad (59)
$$

and

$$
\int W\{G_{g^{\mathrm{e}}}\}\eta_k^{\mathrm{e}}h_0\mathrm{d}^3k=\int W\{\eta_k^{\mathrm{e}}\}G_{g^{\mathrm{e}}}h_0\mathrm{d}^3k \qquad (60)
$$

we obtain

$$
\begin{aligned}
&G_{I_k}(\boldsymbol{r}',\boldsymbol{k}';\omega)+\frac{eV_{\mathrm{T}}}{(2\pi)^3}\eta_k^{\mathrm{e}}(\boldsymbol{r}',\boldsymbol{k}',\omega)\\
&\quad=\frac{eV_{\mathrm{T}}}{(2\pi)^3}\int_D\int\left(W\{G_{g^{\mathrm{o}}}\}\eta_k^{\mathrm{o}}+W\{\eta_k^{\mathrm{e}}\}G_{g^{\mathrm{e}}}\right)h_0\mathrm{d}^3k\mathrm{d}^3r\\
&\quad+\mathrm{i}\omega\frac{eV_{\mathrm{T}}}{(2\pi)^3}\int_D\int\left[\eta_k^{\mathrm{o}}G_{g^{\mathrm{o}}}+\left(\eta_k^{\mathrm{e}}+\frac{\tau_k}{V_{\mathrm{T}}}\right)\left(G_{g^{\mathrm{e}}}+\frac{G_\varphi}{V_{\mathrm{T}}}\right)\right]h_0\mathrm{d}^3k\mathrm{d}^3r\\
&\quad+\mathrm{i}\omega\int_D\kappa\nabla_{\boldsymbol{r}}\tau_k\,\nabla_{\boldsymbol{r}}G_\varphi\mathrm{d}^3r\,. \qquad (61)
\end{aligned}
$$

Due to the Dirichlet boundary conditions we have $G_{g^{\mathrm{e}}}(\boldsymbol{r},\boldsymbol{k};\boldsymbol{r}',\boldsymbol{k}';\omega)=0$ and $G_\varphi(\boldsymbol{r};\boldsymbol{r}',\boldsymbol{k}';\omega)=0$ for $\boldsymbol{r}\in\delta D_k$ for all contacts. Therefore, the following integral vanishes

$$
\begin{aligned}
0 &= -\frac{eV_\mathrm{T}}{(2\pi)^3}\oint_{\partial D}\int G_{g^\mathrm{e}}\boldsymbol{v}\eta_k^\mathrm{o}h_0\mathrm{d}^3k\,\mathrm{d}^2\boldsymbol{r} + \mathrm{i}\omega\oint_{\partial D} G_\varphi\kappa\nabla_{\boldsymbol{r}}\tau_k\,\mathrm{d}^2\boldsymbol{r}\\
&= -\frac{eV_\mathrm{T}}{(2\pi)^3}\int_D\int\nabla_{\boldsymbol{r}}\left[G_{g^\mathrm{e}}\boldsymbol{v}\eta_k^\mathrm{o}h_0\right]\mathrm{d}^3k\mathrm{d}^3r + \mathrm{i}\omega\int_D\nabla_{\boldsymbol{r}}\left[G_\varphi\kappa\nabla_{\boldsymbol{r}}\tau_k\right]\mathrm{d}^3r\\
&= -\frac{eV_\mathrm{T}}{(2\pi)^3}\int_D\int L_0\left\{\eta_k^\mathrm{o}G_{g^\mathrm{e}}h_0\right\}\mathrm{d}^3k\mathrm{d}^3r\\
&\quad+\mathrm{i}\omega\int_D\kappa\nabla_{\boldsymbol{r}}\tau_k\,\nabla_{\boldsymbol{r}}G_\varphi\mathrm{d}^3r + \mathrm{i}\omega\int_D G_\varphi\nabla_{\boldsymbol{r}}\left[\kappa\nabla_{\boldsymbol{r}}\tau_k\right]\mathrm{d}^3r\\
&= -\frac{eV_\mathrm{T}}{(2\pi)^3}\int_D\int\left(L_0\left\{\eta_k^\mathrm{o}\right\}G_{g^\mathrm{e}} + \eta_k^\mathrm{o}L_0\left\{G_{g^\mathrm{e}}\right\}\right)h_0\mathrm{d}^3k\mathrm{d}^3r\\
&\quad+\mathrm{i}\omega\frac{eV_\mathrm{T}}{(2\pi)^3}\int_D\int\frac{G_\varphi}{V_\mathrm{T}}\left(\eta_k^\mathrm{e} + \frac{\tau_k}{V_\mathrm{T}}\right)h_0\mathrm{d}^3k\mathrm{d}^3r\\
&\quad+\mathrm{i}\omega\int_D\kappa\nabla_{\boldsymbol{r}}\tau_k\,\nabla_{\boldsymbol{r}}G_\varphi\mathrm{d}^3r\\
&= \frac{eV_\mathrm{T}}{(2\pi)^3}\int_D\int\left(W\{\eta_k^\mathrm{e}\}G_{g^\mathrm{e}} + \eta_k^\mathrm{o}W\{G_{g^\mathrm{o}}\}\right)h_0\mathrm{d}^3k\mathrm{d}^3r\\
&\quad+\mathrm{i}\omega\frac{eV_\mathrm{T}}{(2\pi)^3}\int_D\int\left[\eta_k^\mathrm{o}G_{g^\mathrm{o}} + \left(G_{g^\mathrm{e}} + \frac{G_\varphi}{V_\mathrm{T}}\right)\left(\eta_k^\mathrm{e} + \frac{\tau_k}{V_\mathrm{T}}\right)\right]h_0\mathrm{d}^3k\mathrm{d}^3r\\
&\quad+\mathrm{i}\omega\int_D\kappa\nabla_{\boldsymbol{r}}\tau_k\,\nabla_{\boldsymbol{r}}G_\varphi\mathrm{d}^3r - \frac{eV_\mathrm{T}}{(2\pi)^3}\eta_k^\mathrm{o}(\boldsymbol{r}',\boldsymbol{k}',\omega)
\end{aligned}
\tag{62}
$$

and we finally obtain at equilibrium

$$
G_{I_k}(\boldsymbol{r}',\boldsymbol{k}';\omega) = -\frac{eV_\mathrm{T}}{(2\pi)^3}\left[\eta_k^\mathrm{e}(\boldsymbol{r}',\boldsymbol{k}',\omega) - \eta_k^\mathrm{o}(\boldsymbol{r}',\boldsymbol{k}',\omega)\right] = -\frac{eV_\mathrm{T}}{(2\pi)^3}\eta_k(\boldsymbol{r}',-\boldsymbol{k}',\omega)\,. \tag{63}
$$

This relation is another important check for small-signal simulations, since it equates two rather different solutions.

With Eqs. (18) and (42) we can calculate the noise at equilibrium

$$
\begin{aligned}
S_{I_kI_i}(\omega) &= 2\Omega\int\int\int\left[G_{I_k}(\boldsymbol{r},\boldsymbol{k};\omega) - G_{I_k}(\boldsymbol{r},\boldsymbol{k}';\omega)\right]W(\boldsymbol{k}'|\boldsymbol{k})(\boldsymbol{r})h_0\\
&\qquad\left[G_{I_i}(\boldsymbol{r},\boldsymbol{k};\omega) - G_{I_i}(\boldsymbol{r},\boldsymbol{k}';\omega)\right]^\star\mathrm{d}^3k'\mathrm{d}^3k\mathrm{d}^3r\\
&= 4(2\pi)^3\int\int G_{I_k}W\{G_{I_i}^\star\}h_0\mathrm{d}^3k\mathrm{d}^3r\\
&= \frac{4e^2V_\mathrm{T}^2}{(2\pi)^3}\int_D\int\left[\eta_k^\mathrm{e} - \eta_k^\mathrm{o}\right]W\{\left[\eta_i^\mathrm{e} - \eta_i^\mathrm{o}\right]^\star\}h_0\mathrm{d}^3k\mathrm{d}^3r\\
&= 4k_\mathrm{B}T_0\Re\{Y_{ki}\}
\end{aligned}
\tag{64}
$$

and the Nyquist theorem results.

2.3 Analytical Solutions for a Homogeneous Resistor

In Fig. 1 a 1D resistor with a constant donor concentration of N_D is shown. The applied stationary bias is zero and no current flows in the device. The stationary electron density is $n_0 \approx N_\mathrm{D}$ assuming that $N_\mathrm{D} \gg n_\mathrm{i}$, where n_i is the intrinsic density of silicon, and the hole density is negligible $p_0 \ll n_\mathrm{i}$. The stationary potential is constant in the device and the stationary distribution function is given by

$$f_0(\boldsymbol{r}, \boldsymbol{k}) = \exp\left(\frac{e\varphi_0 - \varepsilon(\boldsymbol{k}) - \varepsilon_\mathrm{cb}}{k_\mathrm{B} T_0}\right) \tag{65}$$

and

$$g_0^\mathrm{e} = 1\,, \qquad g_0^\mathrm{o} = 0\,. \tag{66}$$

If a small-signal bias V under the sinusoidal steady-state condition is applied to the contact on the RHS, a small-signal potential results

$$\varphi(\boldsymbol{r}, \omega) = \frac{x}{L} V\,. \tag{67}$$

In this special case of a homogeneous 1D device with Dirichlet boundary conditions at both ends plasma oscillations do not occur and the potential does not depend on frequency. The even part of the first term of the small-signal distribution function is given by

$$g^\mathrm{e}(\boldsymbol{r}, \boldsymbol{k}, \omega) = -\frac{\varphi(\boldsymbol{r}, \omega)}{V_\mathrm{T}} = -\frac{V}{V_\mathrm{T}} \frac{x}{L} \tag{68}$$

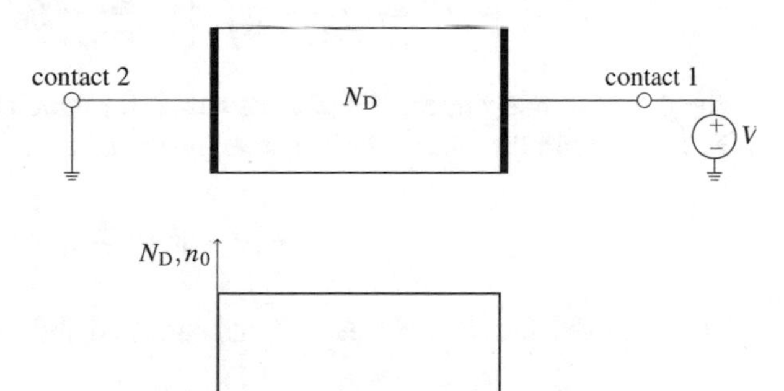

Fig. 1 Donor concentration and stationary electron density in a simple 1D silicon resistor

and the small-signal electron density is zero. The odd part evaluates to

$$g^{\mathrm{o}}(\boldsymbol{r},\boldsymbol{k},\omega) = -\frac{\tau_{\mathrm{mic}}(\varepsilon(\boldsymbol{k}))}{1+\mathrm{i}\omega\tau_{\mathrm{mic}}(\varepsilon(\boldsymbol{k}))}\boldsymbol{v}(\boldsymbol{k})\,\frac{\boldsymbol{E}}{V_{\mathrm{T}}} = \frac{\tau_{\mathrm{mic}}(\varepsilon(\boldsymbol{k}))v_x(\boldsymbol{k})}{1+\mathrm{i}\omega\tau_{\mathrm{mic}}(\varepsilon(\boldsymbol{k}))}\frac{V}{V_{\mathrm{T}}L}\,, \tag{69}$$

where it has been assumed that the microscopic relaxation time, which is the solution of the following integral equation

$$\boldsymbol{v} = W\{\tau_{\mathrm{mic}}\boldsymbol{v}\}\,, \tag{70}$$

depends on the wave vector only via the energy, which is the case for a large class of scattering processes [29, 30]. These quantities satisfy the system of equations (27) and the boundary conditions. They are therefore the solution of the problem. The corresponding fundamental solutions for the contact on the RHS are

$$\eta_1^{\mathrm{e}} = -\frac{1}{V_{\mathrm{T}}}\frac{x}{L}\,, \qquad \eta_1^{\mathrm{o}} = \frac{\tau_{\mathrm{mic}}(\varepsilon(\boldsymbol{k}))v_x(\boldsymbol{k})}{1+\mathrm{i}\omega\tau_{\mathrm{mic}}(\varepsilon(\boldsymbol{k}))}\frac{1}{V_{\mathrm{T}}L}\,, \tag{71}$$

and

$$\tau_1 = \frac{x}{L}\,. \tag{72}$$

The self-admittance of the one-port evaluates with (36) and the cross section A of the device to

$$Y_{11} = \left[\frac{e}{(2\pi)^3 V_{\mathrm{T}}}\int \frac{\tau_{\mathrm{mic}}v_x^2}{1+\mathrm{i}\omega\tau_{\mathrm{mic}}} f_0 \mathrm{d}^3k + \mathrm{i}\omega\kappa\right]\frac{A}{L}\,, \tag{73}$$

where

$$D = \frac{1}{n_0}\frac{1}{(2\pi)^3}\int \frac{\tau_{\mathrm{mic}}v_x^2}{1+\mathrm{i}\omega\tau_{\mathrm{mic}}} f_0 \mathrm{d}^3k \tag{74}$$

is the position-independent diffusion constant, for which the Einstein relation $D = \mu V_{\mathrm{T}}$ holds. With the conductivity $\sigma = en_0\mu$ the admittance is given by

$$Y_{11} = [\sigma + \mathrm{i}\omega\kappa]\,\frac{A}{L}\,. \tag{75}$$

The Transfer function of contact 1 can be calculated with (63)

$$G_{I_1} = \frac{e}{(2\pi)^3}\left[\frac{x}{L} + \frac{\tau_{\mathrm{mic}}v_x}{1+\mathrm{i}\omega\tau_{\mathrm{mic}}}\frac{1}{L}\right]\,. \tag{76}$$

In the expression for the PSD of the current fluctuations (18), the even part of the fundamental solution cancels, because it does not depend on the wave vector. The odd part is position-independent and yields

$$S_{I_1,I_1} = \frac{4e^2}{(2\pi)^3} \int \frac{\tau_{\mathrm{mic}} v_x^2}{1+\omega^2\tau_{\mathrm{mic}}^2} f_0 \mathrm{d}^3k \frac{A}{L} = 4k_{\mathrm{B}} T_0 \Re\{Y_{11}\}\,, \tag{77}$$

the Nyquist theorem.

2.4 Numerical Solutions for an N^+NN^+ Resistor

A simple 1D $\mathrm{N^+NN^+}$ silicon resistor is simulated by numerical means as described in [13] and generation/recombination processes are neglected. For an applied bias of 1 V the minimum electron density increases compared to equilibrium (Fig. 2). The device shows the usual punch-through behavior of this kind of structures. The minimum of the potential, which is also the top of the energy barrier for electrons, moves from the middle of the device to the left (Fig. 3). In addition, the barrier is lowered and a current flows. The electrons gain a lot of energy from the electric field when they move from left to right (Fig. 4). The local contribution to the terminal current noise (19) is shown in Fig. 5 at a low frequency of 1 GHz. In the case of equilibrium (zero bias) the contribution is roughly inversely proportional to the conductivity and thus the electron density. The strongest contribution comes from the middle of the device, where the energy barrier is located. If a bias is applied,

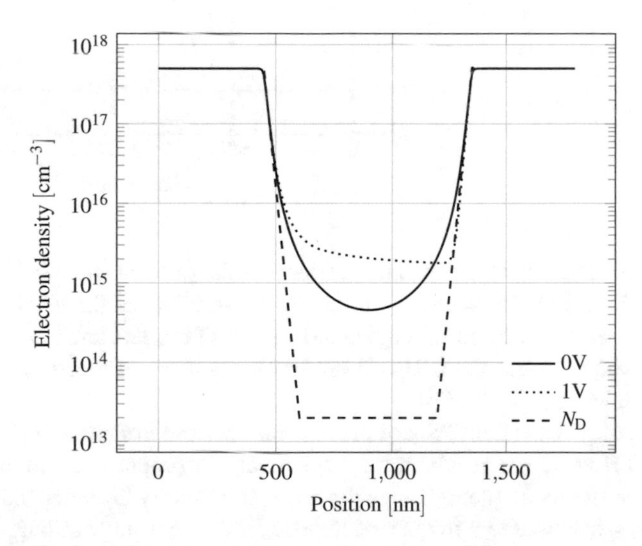

Fig. 2 Electron density at 0 (solid) and 1 V (dotted) DC bias and donor concentration (dashed) in a simple 1D silicon $\mathrm{N^+NN^+}$ resistor at room temperature

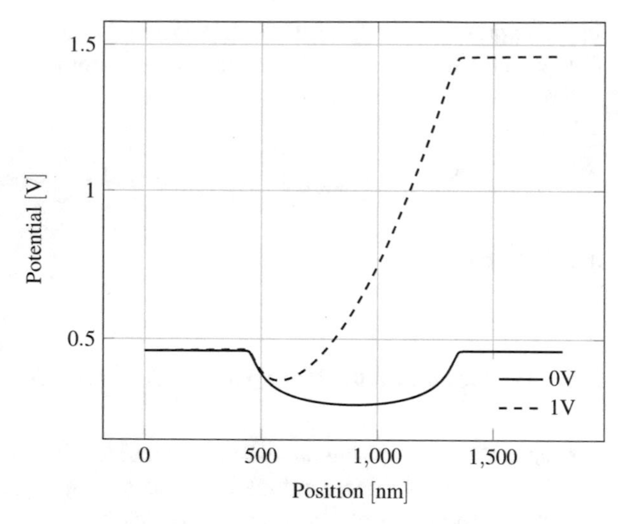

Fig. 3 Potential at 0 (solid) and 1 V (dashed) DC bias in the simple 1D silicon N^+NN^+ resistor at room temperature

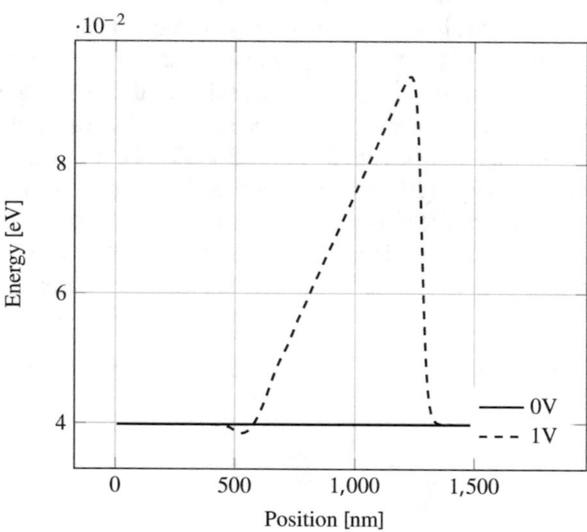

Fig. 4 Energy at 0 (solid) and 1 V (dashed) DC bias in a simple 1D silicon N^+NN^+ resistor at room temperature

the situation changes very much. The barrier and the peak of the local contribution move to the left. The hot electrons on the right-hand side of the barrier contribute little to the total noise, because they cannot move to the left back over the barrier due to the strong electric field. This is similar to the case of electrons in a bipolar transistor or a MOSFET [13, 24].

The frequency-dependent PSD of the terminal current is shown in Fig. 6. It has a peak in the THz range after which it decays inversely proportional to the square of the frequency. In this frequency range the noise shows only little dependence on the applied bias, whereas at low frequencies the dependence is quite strong. In addition, thermal noise is shown for the case of equilibrium and the Nyquist theorem (64) is satisfied with excellent accuracy.

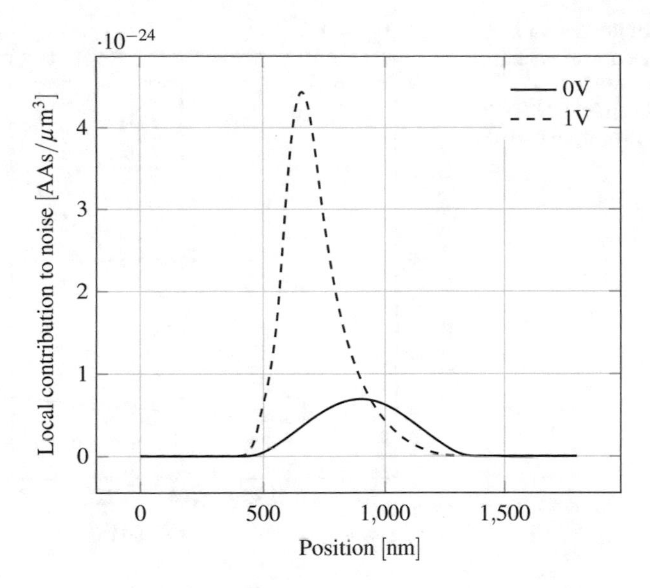

Fig. 5 Local contribution to the terminal current noise at 0 (solid) and 1 V (dashed) DC bias in a simple 1D silicon N^+NN^+ resistor at room temperature and 1 GHz

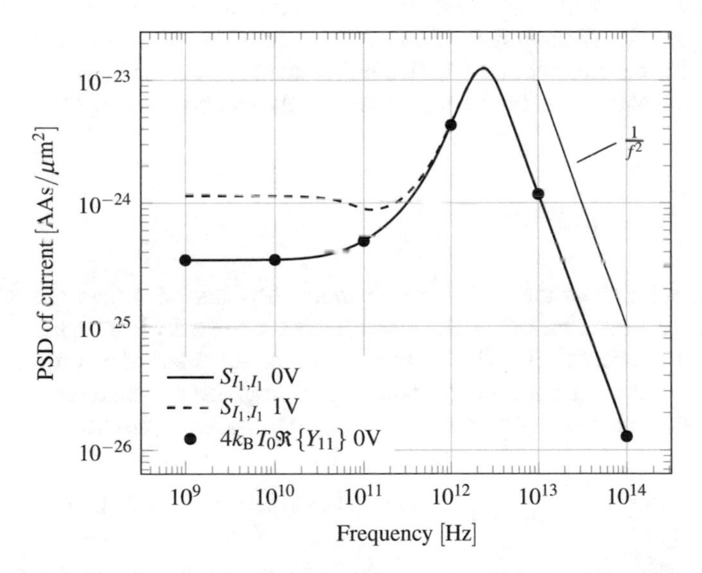

Fig. 6 PSD of the current at 0 (solid) and 1 V (dashed) DC bias and the corresponding Johnson–Nyquist noise (solid circles) in a simple 1D silicon N^+NN^+ resistor at room temperature as a function of frequency

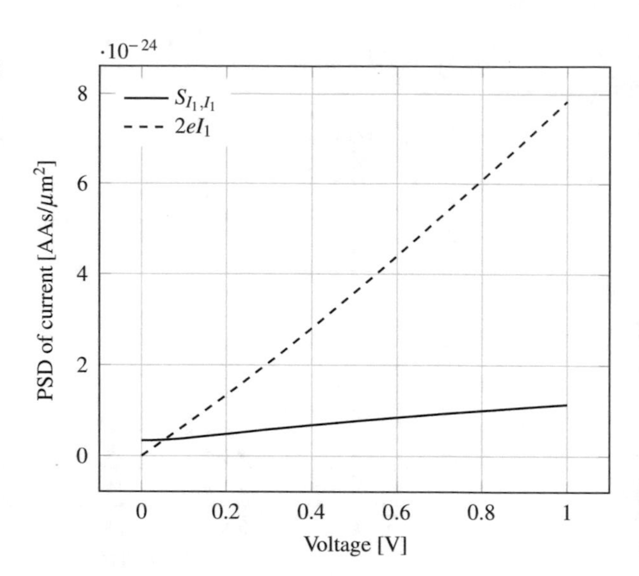

Fig. 7 PSD of the current at 1 GHz as a function of the DC bias and shot noise in a simple 1D silicon N^+NN^+ resistor at room temperature

The PSD of the current at 1 GHz is shown as a function of the bias in Fig. 7. At high bias the PSD is much smaller than shot noise $2eI_1$. This is due to the negative feedback via the potential, which leads to a correlation of the electrons injected over the barrier and suppression of the shot noise. If the feedback via the potential is turned off, the electrons are injected independently over the barrier and the noise becomes full shot noise. Further numerical results can be found in [13, 14].

3 Balance Equations

Since the solution of the LBE is much more CPU intensive than the solution of balance equations (e.g., DD or HD model), these are usually used to calculate noise on the device level [23, 31–33]. Balance equations are obtained by projection of the LBE onto a set of microscopic quantities $X(\boldsymbol{k})$ (e.g., group velocity), which yields corresponding macroscopic quantities $x(\boldsymbol{r}, t)$ (e.g., particle current density) [34–36]

$$x(\boldsymbol{r}, t) = \frac{1}{(2\pi)^3} \int X(\boldsymbol{k}) f(\boldsymbol{r}, \boldsymbol{k}, t) \mathrm{d}^3 k = n(\boldsymbol{r}, t) \langle X \rangle (\boldsymbol{r}, t) \,, \tag{78}$$

where the distribution function still includes both spin directions, which are again assumed to be degenerate. With the flux density

$$\boldsymbol{j}_X = n \langle X \boldsymbol{v} \rangle \,, \tag{79}$$

and the projected Langevin force

$$\xi_{\dot{X}} = \frac{1}{(2\pi)^3} \int X(\boldsymbol{k}) \xi(\boldsymbol{r}, \boldsymbol{k}, t) \mathrm{d}^3 k \tag{80}$$

projection of the LBE (1) yields the balance equation

$$\frac{\partial x}{\partial t} + \frac{e}{\hbar} n \boldsymbol{E} \langle \nabla_{\boldsymbol{k}} X \rangle + \nabla_{\boldsymbol{r}} \boldsymbol{j}_X + n \langle W\{X\} \rangle = \xi_{\dot{X}} \tag{81}$$

Linearization of this formula results in the small-signal balance equation

$$\begin{aligned}\frac{\partial \delta x}{\partial t} + \frac{e}{\hbar} \delta n \boldsymbol{E}_0 \langle \nabla_{\boldsymbol{k}} X \rangle_0 + \frac{e}{\hbar} n_0 \delta \boldsymbol{E} \langle \nabla_{\boldsymbol{k}} X \rangle_0 + \frac{e}{\hbar} n_0 \boldsymbol{E}_0 \, \delta \langle \nabla_{\boldsymbol{k}} X \rangle \\ + \nabla_{\boldsymbol{r}} \delta \boldsymbol{j}_X + \delta n \langle W\{X\} \rangle_0 + n_0 \delta \langle W\{X\} \rangle = \xi_{\dot{X}} \,. \end{aligned} \tag{82}$$

This approach is sometimes called acceleration-fluctuation scheme, because in the case of the velocity as the microscopic quantity the corresponding projected Langevin force describes an acceleration [37]. The cross-PSD of the projected Langevin forces of two microscopic quantities X, Y can be calculated with Eq. (8)

$$\begin{aligned} S_{\xi_{\dot{X}} \xi_{\dot{Y}}} \delta(\boldsymbol{r} - \boldsymbol{r}') = \frac{2\Omega}{(2\pi)^6} \int \int \left[X(\boldsymbol{k}) - X(\boldsymbol{k}') \right] W(\boldsymbol{k}|\boldsymbol{k}')(\boldsymbol{r}) f_0(\boldsymbol{r}, \boldsymbol{k}') \\ \left[Y(\boldsymbol{k}) - Y(\boldsymbol{k}') \right]^\star \mathrm{d}^3 k \mathrm{d}^3 k' \delta(\boldsymbol{r} - \boldsymbol{r}') \,. \end{aligned} \tag{83}$$

In [38] this PSD is evaluated by Monte Carlo simulation with a scattering-based estimator, whereas in this work a deterministic solver for the LBE is used [13]. The PSD can be expressed with Eq. (22) by

$$\begin{aligned} S_{\xi_{\dot{X}} \xi_{\dot{Y}}} &= 2 n_0 \langle X W\{Y^\star\} + Y^\star W\{X\} \rangle_0 \\ &\quad + \frac{2\Omega}{(2\pi)^6} \int X(\boldsymbol{k}) Y^\star(\boldsymbol{k}) \int W(\boldsymbol{k}|\boldsymbol{k}')(\boldsymbol{r}) f_0(\boldsymbol{r}, \boldsymbol{k}') \\ &\quad - W(\boldsymbol{k}'|\boldsymbol{k})(\boldsymbol{r}) f_0(\boldsymbol{r}, \boldsymbol{k}) \mathrm{d}^3 k' \mathrm{d}^3 k \\ &= 2 n_0 \langle X W\{Y^\star\} + Y^\star W\{X\} \rangle_0 \\ &\quad + 2 \nabla_{\boldsymbol{r}} \left(n_0 \langle X Y^\star \boldsymbol{v} \rangle_0 \right) + \frac{2e}{\hbar} n_0 \langle \nabla_{\boldsymbol{k}} (X Y^\star) \rangle_0 \, \boldsymbol{E}_0(\boldsymbol{r}) \,. \end{aligned} \tag{84}$$

The integral over $\boldsymbol{k}'$ is the scattering integral of the stationary BE (5), which is replaced by the streaming term.

In order to obtain a closed system of equations, the scattering integral has to be simplified. The usual approach is to use the macroscopic relaxation time approximation

$$\langle W\{X\} \rangle \approx \frac{\langle X \rangle - \langle X \rangle^{\mathrm{eq}}}{\tau_X} \,, \tag{85}$$

where $\langle X \rangle^{\text{eq}}$ is evaluated for an equilibrium distribution function [39]. An alternative approach is based on the microscopic relaxation time (70). In this case the scattering integral can be solved exactly by projecting onto $X = \tau_{\text{mic}}\boldsymbol{v}$

$$\boldsymbol{j} = n\langle W\{\tau_{\text{mic}}\boldsymbol{v}\}\rangle \,. \tag{86}$$

For the sake of brevity and clarity the case of the DD model for isotropic parabolic bands is discussed next. Assuming a particle number conserving scattering integral, projection of the LBE onto $X = 1$ yields the continuity equation with $\xi_1 = 0$

$$\frac{\partial \delta n}{\partial t} + \nabla_{\boldsymbol{r}}\, \delta\boldsymbol{j} = 0\,. \tag{87}$$

With $X = \boldsymbol{v}$ the balance equation for the current density is obtained

$$\frac{\partial \delta\boldsymbol{j}}{\partial t} + \delta n \frac{e}{m}\boldsymbol{E}_0 + n_0 \frac{e}{m}\delta\boldsymbol{E} + \frac{k_{\text{B}}T_0}{m}\nabla_{\boldsymbol{r}}\delta n + \frac{\delta\boldsymbol{j}}{\tau_v} = \boldsymbol{\xi}_{\dot{\boldsymbol{v}}}\,. \tag{88}$$

m is the constant isotropic mass of the parabolic band structure and the usual approximation for the DD model is used in the diffusion term ($\langle \boldsymbol{v}^2\rangle/3 \approx k_{\text{B}}T_0/m$), which leads to the Einstein relation, where T_0 is the constant ambient temperature [40]. With the isotropic mobility $\mu = e\tau_v/m$ and diffusion constant $D = \mu V_{\text{T}}$, the constitutive equation of the DD model results

$$\delta\boldsymbol{j} + \tau_v \frac{\partial \delta\boldsymbol{j}}{\partial t} + \mu\left(\delta n \boldsymbol{E}_0 + n_0 \delta\boldsymbol{E} + V_{\text{T}}\nabla_{\boldsymbol{r}}\delta n\right) = \tau_v \boldsymbol{\xi}_{\dot{\boldsymbol{v}}}\,. \tag{89}$$

The macroscopic relaxation time τ_v and thus the mobility μ are assumed to be constants in this case for the sake of simplicity. This approximation has no consequences for the following discussion of the local noise source and extension to a macroscopic relaxation time that depends on the particle gas state is straightforward.

The terminal current noise can be calculated again based on the linearized balance equations and PE

$$\frac{\partial \delta n}{\partial t} + \nabla_{\boldsymbol{r}}\, \delta\boldsymbol{j} = \xi_1$$

$$\delta\boldsymbol{j} + \tau_v \frac{\partial \delta\boldsymbol{j}}{\partial t} - \mu\left(\delta n \nabla_{\boldsymbol{r}}\varphi_0 + n_0 \nabla_{\boldsymbol{r}}\delta\varphi - V_{\text{T}}\nabla_{\boldsymbol{r}}\delta n\right) = \tau_v \boldsymbol{\xi}_{\dot{\boldsymbol{v}}}$$

$$\nabla_{\boldsymbol{r}}\left[\kappa \nabla_{\boldsymbol{r}}\delta\varphi\right] - e\delta n = 0\,. \tag{90}$$

This linear system of equations contains four scalar balance equations for the microscopic variables $1, v_x, v_y, v_z$ and four Langevin forces $\xi_1, \xi_{\dot{v}_x}, \xi_{\dot{v}_y}, \xi_{\dot{v}_z}$. Since all these balance equations contain Langevin forces, for each individual Langevin force a set of transfer functions $G_{\alpha}^{\beta}(\boldsymbol{r}, \boldsymbol{r}', \omega)$ has to be introduced in the frequency domain, where the indices α, β run over all microscopic variables with $\beta = 1, v_x, v_y, v_z$

$$
\begin{aligned}
\mathrm{i}\omega G_1^{\beta} + \frac{\partial G_{v_x}^{\beta}}{\partial x} + \frac{\partial G_{v_y}^{\beta}}{\partial y} + \frac{\partial G_{v_z}^{\beta}}{\partial z} &= \delta_{1,\beta}\delta(\boldsymbol{r} - \boldsymbol{r}') \\
(1 + \mathrm{i}\omega\tau_v)G_{v_x}^{\beta} - \mu\left(G_1^{\beta}\frac{\partial \varphi_0}{\partial x} + n_0\frac{\partial G_{\varphi}^{\beta}}{\partial x} - V_{\mathrm{T}}\frac{\partial G_1^{\beta}}{\partial x}\right) &= \delta_{v_x,\beta}\delta(\boldsymbol{r} - \boldsymbol{r}') \\
(1 + \mathrm{i}\omega\tau_v)G_{v_y}^{\beta} - \mu\left(G_1^{\beta}\frac{\partial \varphi_0}{\partial y} + n_0\frac{\partial G_{\varphi}^{\beta}}{\partial y} - V_{\mathrm{T}}\frac{\partial G_1^{\beta}}{\partial y}\right) &= \delta_{v_y,\beta}\delta(\boldsymbol{r} - \boldsymbol{r}') \\
(1 + \mathrm{i}\omega\tau_v)G_{v_z}^{\beta} - \mu\left(G_1^{\beta}\frac{\partial \varphi_0}{\partial z} + n_0\frac{\partial G_{\varphi}^{\beta}}{\partial z} - V_{\mathrm{T}}\frac{\partial G_1^{\beta}}{\partial z}\right) &= \delta_{v_z,\beta}\delta(\boldsymbol{r} - \boldsymbol{r}') \\
\nabla_{\boldsymbol{r}}\left[\kappa\nabla_{\boldsymbol{r}} G_{\varphi}^{\beta}\right] - eG_1^{\beta} &= 0\,. \qquad (91)
\end{aligned}
$$

In the case of the DD model four sets of transfer functions are obtained. We can define the transfer functions for the terminal current by

$$
\begin{aligned}
G_{I_k}^{\beta}(\boldsymbol{r}', \omega) = \int \Big[& e\left(G_{v_x}^{\beta}(\boldsymbol{r}, \boldsymbol{r}', \omega)\boldsymbol{e}_x + G_{v_y}^{\beta}(\boldsymbol{r}, \boldsymbol{r}', \omega)\boldsymbol{e}_y + G_{v_z}^{\beta}(\boldsymbol{r}, \boldsymbol{r}', \omega)\boldsymbol{e}_z\right) \\
& + \mathrm{i}\omega\kappa\nabla_{\boldsymbol{r}} G_{\varphi}^{\beta}(\boldsymbol{r}, \boldsymbol{r}', \omega)\Big]\, \mathrm{d}^2\boldsymbol{r}\,. \qquad (92)
\end{aligned}
$$

The PSD of the terminal current is obtained with the Wiener-Lee theorem

$$
\begin{aligned}
S_{I_k I_{k'}}(\omega) = \int & \begin{pmatrix} G_{I_k}^{1}(\boldsymbol{r}, \omega) \\ G_{I_k}^{v_x}(\boldsymbol{r}, \omega) \\ G_{I_k}^{v_y}(\boldsymbol{r}, \omega) \\ G_{I_k}^{v_z}(\boldsymbol{r}, \omega) \end{pmatrix}^{\mathrm{T}}
\begin{pmatrix}
S_{\xi_{\mathrm{i}}\xi_{\mathrm{i}}}(\boldsymbol{r}) & \tau_v S_{\xi_{\mathrm{i}}\xi_{v_x}}(\boldsymbol{r}) & \tau_v S_{\xi_{\mathrm{i}}\xi_{v_y}}(\boldsymbol{r}) & \tau_v S_{\xi_{\mathrm{i}}\xi_{v_z}}(\boldsymbol{r}) \\
\tau_v S_{\xi_{v_x}\xi_{\mathrm{i}}}(\boldsymbol{r}) & \tau_v^2 S_{\xi_{v_x}\xi_{v_x}}(\boldsymbol{r}) & \tau_v^2 S_{\xi_{v_x}\xi_{v_y}}(\boldsymbol{r}) & \tau_v^2 S_{\xi_{v_x}\xi_{v_z}}(\boldsymbol{r}) \\
\tau_v S_{\xi_{v_y}\xi_{\mathrm{i}}}(\boldsymbol{r}) & \tau_v^2 S_{\xi_{v_y}\xi_{v_x}}(\boldsymbol{r}) & \tau_v^2 S_{\xi_{v_y}\xi_{v_y}}(\boldsymbol{r}) & \tau_v^2 S_{\xi_{v_y}\xi_{v_z}}(\boldsymbol{r}) \\
\tau_v S_{\xi_{v_z}\xi_{\mathrm{i}}}(\boldsymbol{r}) & \tau_v^2 S_{\xi_{v_z}\xi_{v_x}}(\boldsymbol{r}) & \tau_v^2 S_{\xi_{v_z}\xi_{v_y}}(\boldsymbol{r}) & \tau_v^2 S_{\xi_{v_z}\xi_{v_z}}(\boldsymbol{r})
\end{pmatrix} \\
& \begin{pmatrix} G_{I_{k'}}^{1}(\boldsymbol{r}, \omega) \\ G_{I_{k'}}^{v_x}(\boldsymbol{r}, \omega) \\ G_{I_{k'}}^{v_y}(\boldsymbol{r}, \omega) \\ G_{I_{k'}}^{v_z}(\boldsymbol{r}, \omega) \end{pmatrix}^{\star} \mathrm{d}r^3\,. \qquad (93)
\end{aligned}
$$

This general approach can be easily extended to a larger number of balance equations (e.g., HD model) [17]. With the assumption that there are no generation/recombination processes ($\xi_{\mathrm{i}} = 0$), the equation reduces to

$$S_{I_k I_{k'}}(\omega) = \int \tau_v^2(\boldsymbol{r}) \begin{pmatrix} G_{I_k}^{v_x}(\boldsymbol{r},\omega) \\ G_{I_k}^{v_y}(\boldsymbol{r},\omega) \\ G_{I_k}^{v_z}(\boldsymbol{r},\omega) \end{pmatrix}^{\mathrm{T}} \begin{pmatrix} S_{\xi_{\dot{v_x}}\xi_{\dot{v_x}}}(\boldsymbol{r}) & S_{\xi_{\dot{v_x}}\xi_{\dot{v_y}}}(\boldsymbol{r}) & S_{\xi_{\dot{v_x}}\xi_{\dot{v_z}}}(\boldsymbol{r}) \\ S_{\xi_{\dot{v_y}}\xi_{\dot{v_x}}}(\boldsymbol{r}) & S_{\xi_{\dot{v_y}}\xi_{\dot{v_y}}}(\boldsymbol{r}) & S_{\xi_{\dot{v_y}}\xi_{\dot{v_z}}}(\boldsymbol{r}) \\ S_{\xi_{\dot{v_z}}\xi_{\dot{v_x}}}(\boldsymbol{r}) & S_{\xi_{\dot{v_z}}\xi_{\dot{v_y}}}(\boldsymbol{r}) & S_{\xi_{\dot{v_z}}\xi_{\dot{v_z}}}(\boldsymbol{r}) \end{pmatrix} \begin{pmatrix} G_{I_{k'}}^{v_x}(\boldsymbol{r},\omega) \\ G_{I_{k'}}^{v_y}(\boldsymbol{r},\omega) \\ G_{I_{k'}}^{v_z}(\boldsymbol{r},\omega) \end{pmatrix}^{\star} \mathrm{d}r^3$$

$$= \int \tau_v^2(\boldsymbol{r}) \boldsymbol{G}_{I_k}^{\mathrm{T}}(\boldsymbol{r},\omega) \hat{S}_{\boldsymbol{\xi}_{\dot{v}}\boldsymbol{\xi}_{\dot{v}}}(\boldsymbol{r}) \boldsymbol{G}_{I_{k'}}^{\star}(\boldsymbol{r},\omega) \mathrm{d}r^3 \,. \tag{94}$$

The matrix-valued PSD of the local noise source for the velocity is given with Eq. (83) by[2]

$$\hat{S}_{\boldsymbol{\xi}_{\dot{v}}\boldsymbol{\xi}_{\dot{v}}}(\boldsymbol{r}) = \frac{2\Omega}{(2\pi)^6} \int\int \left[\boldsymbol{v}(\boldsymbol{k}) - \boldsymbol{v}(\boldsymbol{k}')\right] W(\boldsymbol{k}|\boldsymbol{k}')(\boldsymbol{r}) f_0(\boldsymbol{r},\boldsymbol{k}') \left[\boldsymbol{v}(\boldsymbol{k}) - \boldsymbol{v}(\boldsymbol{k}')\right]^{\mathrm{T}} \mathrm{d}^3k \mathrm{d}^3k' \,. \tag{95}$$

Due to the macroscopic relaxation time approximation this definition of the local noise source leads to problems as will be shown next.

Similar to the transport parameters of the balance equations (μ in the case of the DD model, macroscopic relaxation times in the case of the HD model [39–41]) the noise sources are evaluated under homogeneous bulk conditions for a constant electric field [17, 38]. Since the problems with the macroscopic relaxation time approximation already occur in the case of a homogeneous system, the following discussion is restricted to a homogeneous one, for which the derivatives w.r.t. position in Eq. (91) vanish, and the electron density and electric field are constants ($G_1^\beta = 0$, $G_\varphi^\beta = 0$) and only Green's functions for the velocity occur

$$\begin{aligned} (1 + \mathrm{i}\omega\tau_v) G_{v_x}^\beta &= \delta_{v_x,\beta} \\ (1 + \mathrm{i}\omega\tau_v) G_{v_y}^\beta &= \delta_{v_y,\beta} \\ (1 + \mathrm{i}\omega\tau_v) G_{v_z}^\beta &= \delta_{v_z,\beta} \,. \end{aligned} \tag{96}$$

Due to the assumption of a homogeneous system, the dependence on the position in the real space vanishes completely. This yields a square matrix of Green's functions for the velocity ($\hat{I}$ is the identity matrix)

[2]The PSD is white (frequency-independent) because the microscopic quantities do not depend on the frequency and the Langevin force of the LBE is delta-correlated [1].

$$\hat{G} = \begin{pmatrix} G_{v_x}^{v_x} & G_{v_x}^{v_y} & G_{v_x}^{v_z} \\ G_{v_y}^{v_x} & G_{v_y}^{v_y} & G_{v_y}^{v_z} \\ G_{v_z}^{v_x} & G_{v_z}^{v_y} & G_{v_z}^{v_z} \end{pmatrix} = \frac{1}{1 + \mathrm{i}\omega\tau_v}\hat{I} \,, \tag{97}$$

which in the case of the DD model has a very simple structure. Even if the mobility were not assumed to be a constant, this equation would not change, because in the DD model the mobility is under homogeneous bulk conditions a function of the electric field, which does not fluctuate. With the Wiener-Lee theorem the PSD of the velocity is obtained

$$\hat{S}_{\boldsymbol{v}\boldsymbol{v}} = \tau_v^2 \hat{G}\hat{S}_{\xi_{\dot{v}}\xi_{\dot{v}}}\hat{G}^\dagger = \frac{\tau_v^2}{1 + \omega^2\tau_v^2}\hat{S}_{\xi_{\dot{v}}\xi_{\dot{v}}} \,. \tag{98}$$

The superscript † indicates the complex conjugate of the transpose. In the general case of balance equations with a vector of microscopic variables $\boldsymbol{X}$ (e.g., for the HD model $v_x, v_y, v_z, \varepsilon, \varepsilon v_x, \varepsilon v_y, \varepsilon v_z$, where ε is the energy) the general relationship is obtained

$$\hat{S}_{\boldsymbol{X}\boldsymbol{X}} = \hat{G}\hat{\tau}\hat{S}_{\xi_{\dot{X}}\xi_{\dot{X}}}\hat{\tau}\hat{G}^\dagger \tag{99}$$

with the diagonal matrix of the relaxation times $[\hat{\tau}]_{i,j} = \tau_{X_i}\delta_{i,j}$. The matrix of the Green's functions has in the general case also off-diagonal elements, because the transport parameters (e.g., mobility) depend on the macroscopic variables (energy in the case of the HD model). The PSD $\hat{S}_{\boldsymbol{X}\boldsymbol{X}}$ can be calculated with Eq. (12) adapted to the homogeneous case.

Under equilibrium conditions the scattering integral of the stationary BE (5) in Eq. (84) vanishes due to the principle of detailed balance [3]. For $\boldsymbol{X} = \boldsymbol{Y} = \boldsymbol{v}$ the PSD of the Langevin forces is therefore given without approximations for a single particle in a homogeneous bulk system by

$$\hat{S}^{\mathrm{eq}}_{\xi_{\dot{v}}\xi_{\dot{v}}} = 4\langle W\{\boldsymbol{v}\}\boldsymbol{v}^{\mathrm{T}}\}\rangle_0^{\mathrm{eq}} = 4\left\langle \frac{\boldsymbol{v}\boldsymbol{v}^{\mathrm{T}}}{\tau_{\mathrm{mic}}} \right\rangle_0^{\mathrm{eq}} , \tag{100}$$

where the microscopic relaxation time depends on the wave vector only via the energy. Under equilibrium conditions the PSD is isotropic and it can be replaced by a scalar ($\boldsymbol{v}\boldsymbol{v}^{\mathrm{T}} \rightarrow \boldsymbol{v}^2/3$)

$$\hat{S}^{\mathrm{eq}}_{\xi_{\dot{v}}\xi_{\dot{v}}} = \frac{4}{3}\left\langle \frac{\boldsymbol{v}^2}{\tau_{\mathrm{mic}}} \right\rangle_0^{\mathrm{eq}} \hat{I} = S^{\mathrm{eq}}_{\xi_{\dot{v}}\xi_{\dot{v}}}\hat{I} \,. \tag{101}$$

Furthermore, under equilibrium conditions fluctuation–dissipation relations must hold [1]. The corresponding fluctuation–dissipation theorem for the velocity at zero frequency

$$\tau_v^2 S_{\xi_{\dot{v}}\xi_{\dot{v}}}^{\mathrm{eq}} = S_{vv} = 4D^{\mathrm{eq}} \tag{102}$$

relates the PSD of the velocity fluctuations to the diffusion constant[3]

$$D^{\mathrm{eq}} = \frac{\tau_v}{3}\left\langle \boldsymbol{v}^2 \right\rangle_0^{\mathrm{eq}} = \frac{k_{\mathrm{B}} T_0 \tau_v}{m} = \frac{1}{3}\left\langle \tau_{\mathrm{mic}} \boldsymbol{v}^2 \right\rangle_0^{\mathrm{eq}}, \tag{103}$$

which can be calculated exactly with the microscopic relaxation time [16, 29]. Since at equilibrium the Einstein relation ($D = \mu V_{\mathrm{T}}$) holds, this definition of the macroscopic relaxation time is consistent with the definition based on the mobility ($\mu = e\tau_v/m$). Therefore, the macroscopic relaxation time is given at equilibrium by

$$\tau_v = \frac{\left\langle \tau_{\mathrm{mic}} \boldsymbol{v}^2 \right\rangle_0^{\mathrm{eq}}}{\left\langle \boldsymbol{v}^2 \right\rangle_0^{\mathrm{eq}}}. \tag{104}$$

The corresponding fluctuation–dissipation theorem on the other hand requires

$$\tau_v = \frac{\left\langle \boldsymbol{v}^2 \right\rangle_0^{\mathrm{eq}}}{\left\langle \frac{\boldsymbol{v}^2}{\tau_{\mathrm{mic}}} \right\rangle_0^{\mathrm{eq}}}. \tag{105}$$

The two definitions are only consistent, if $\tau_{\mathrm{mic}} = \tau_v$, in which case the macroscopic relaxation time approximation holds also on the level of the LBE. Since this is only the case for a very simple scattering integral, the macroscopic relaxation time approximation in general violates the fluctuation–dissipation theorem and fails already in the case of equilibrium (a more detailed discussion can be found in [16]). In Fig. 8 the ratio of the two definitions of the macroscopic velocity relaxation time

$$r_v = \frac{\left\langle \tau_{\mathrm{mic}} \boldsymbol{v}^2 \right\rangle_0^{\mathrm{eq}} \left\langle \frac{\boldsymbol{v}^2}{\tau_{\mathrm{mic}}} \right\rangle_0^{\mathrm{eq}}}{\left(\left\langle \boldsymbol{v}^2 \right\rangle_0^{\mathrm{eq}}\right)^2} \tag{106}$$

is shown as a function of the donor concentration for a homogeneous silicon system at room temperature for 0 Hz and equilibrium. The term on the LHS of (102), which is based on the exact Langevin forces and the macroscopic relaxation time approximation, clearly fails to reproduce the PSD of the velocity.

In order to solve this problem, the noise source (84) of the balance equations has to be defined in a different manner. In the case of the DD model the usual approach is to express the PSD in the device by the fluctuation–dissipation theorem together with the Einstein relation [23, 31, 32]

[3] At equilibrium this definition of the diffusion constant is also consistent with the definition based on the autocorrelation function of the velocity [16, 42].

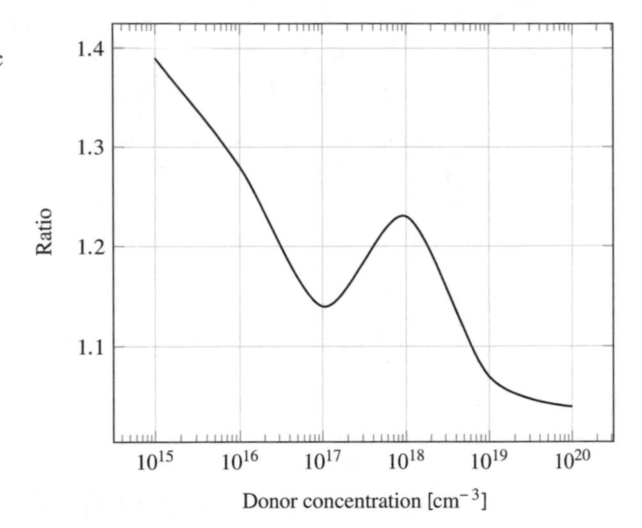

Fig. 8 Ratio of the two definitions of the macroscopic velocity relaxation time for a silicon bulk system at room temperature for zero frequency and equilibrium

$$\tau_v^2 S_{\xi_{\dot{v}}\xi_{\dot{v}}} \approx 4n V_{\mathrm{T}} \mu \,. \tag{107}$$

The advantage of this formulation is that at equilibrium it yields the correct result by definition and it does not require any new information beyond the mobility for the case of nonequilibrium. On the other hand, under nonequilibrium the expression is no longer correct and the error can be quite large [43, 44]. To overcome this problem, in [16, 17] the concept of modified Langevin forces was introduced based on Eq. (99) for a homogeneous system

$$\hat{\tau} \hat{S}'_{\xi_{\dot{X}}\xi_{\dot{X}}} \hat{\tau} = \hat{G}^{-1} \hat{S}_{XX} \left(\hat{G}^{\dagger}\right)^{-1} , \tag{108}$$

where $\hat{S}'_{\xi_{\dot{X}}\xi_{\dot{X}}}$ is the PSD of the modified Langevin forces. Thus, instead of calculating $\hat{S}_{\xi_{\dot{X}}\xi_{\dot{X}}}$ the PSD $\hat{S}_{XX}$ is calculated, which is subsequently transformed into the PSD of the modified Langevin forces $\hat{S}'_{\xi_{\dot{X}}\xi_{\dot{X}}}$. This definition ensures that the PSDs of all microscopic quantities are reproduced exactly under homogeneous bulk conditions as it is the case for the corresponding expectations (transport parameters, e.g., mobility). In addition, this approach also corrects all errors introduced during the formulation of the balance equations, which otherwise might lead to a violation of the fluctuation–dissipation theorem.

In the case of the DD model the transformation matrix is given at zero frequency by the identity matrix and the local noise source is given by the PSD of the velocity [43]. In Fig. 9 the PSD of the velocity is shown together with the usual approximation based on the Einstein relation. At equilibrium the correct result is obtained, whereas far from equilibrium the Einstein relation fails and underestimates noise.

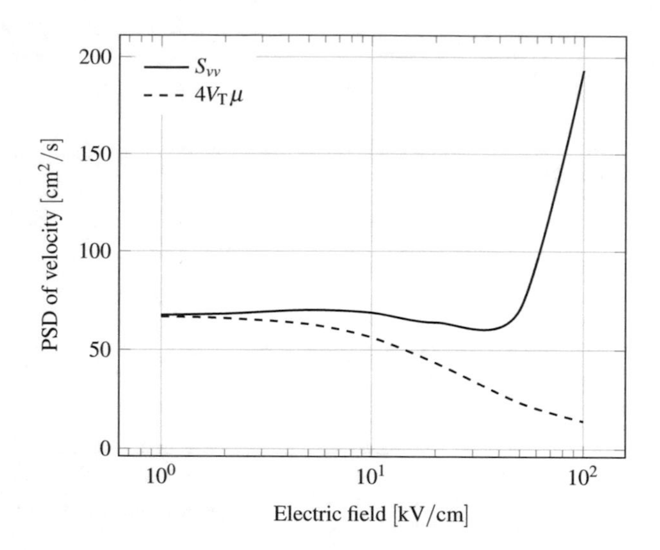

Fig. 9 PSD of the velocity parallel to the electric field evaluated by the LBE and the Einstein relation for a silicon bulk system with a donor concentration of $10^{17}/\mathrm{cm}^3$ at room temperature for zero frequency

3.1 Analytical Solutions for a Homogeneous Resistor

Again the linear response at equilibrium of the homogeneous 1D resistor shown in Fig. 1 is investigated. The small-signal DD model for this homogeneous device is given for the sinusoidal steady-state with $J_x = -ej_x$, $n_0 = N_{\mathrm{D}}$ by

$$\frac{\partial J_x}{\partial x} - \mathrm{i}\omega e n = 0$$
$$J_x = e\frac{\mu}{1+\mathrm{i}\omega\tau_v}\left(-N_{\mathrm{D}}\frac{\partial \varphi}{\partial x} + V_{\mathrm{T}}\frac{\partial n}{\partial x}\right)$$
$$\frac{\partial^2 \varphi}{\partial x^2} = \frac{e}{\kappa}n \,. \tag{109}$$

The boundary conditions for the small-signal electron density are

$$n(x=0) = n(x=L) = 0 \tag{110}$$

and for the potential

$$\varphi(x=0) = 0\,, \qquad \varphi(x=L) = V\,. \tag{111}$$

The solution is very similar to the case of the BE

$$\begin{aligned} \varphi(x) &= \frac{x}{L}V \\ n(x) &= 0 \\ J_x(x) &= -\frac{eN_{\mathrm{D}}\mu}{1+\mathrm{i}\omega\tau_v}\frac{V}{L} \end{aligned} \tag{112}$$

and

$$Y_{11} = \left(\frac{eN_{\mathrm{D}}\mu}{1+\mathrm{i}\omega\tau_v} + \mathrm{i}\omega\kappa\right)\frac{A}{L}\,. \tag{113}$$

If the microscopic relaxation time τ_{mic} in (73) is replaced by the macroscopic τ_v, the BE and DD model yield the same solutions for this device.

The Green's functions are the solutions of the corresponding 1D small-signal DD model

$$\frac{\partial G_{J_x}(x,x')}{\partial x} - \mathrm{i}\omega e G_n(x,x') = \frac{1}{A}\delta(x-x') \tag{114}$$

$$G_{J_x}(x,x') = e\frac{\mu}{1+\mathrm{i}\omega\tau_v}\left(-N_{\mathrm{D}}\frac{\partial G_\varphi(x,x')}{\partial x} + V_{\mathrm{T}}\frac{\partial G_n(x,x')}{\partial x}\right) \tag{115}$$

$$\frac{\partial^2 G_\varphi(x,x')}{\partial x^2} = \frac{e}{\kappa}G_n(x,x') \tag{116}$$

with the boundary conditions

$$G_n(x=0,x') = G_n(x=L,x') = 0 \tag{117}$$

and

$$G_\varphi(x=0,x') = G_\varphi(x=L,x') = 0\,. \tag{118}$$

The solutions are

$$\begin{aligned} G_n(x,x') &= -\frac{p}{bA}\left(\frac{\sinh(px)\sinh(p(L-x'))}{\sinh(pL)} + \Theta(x-x')\sinh(p(x'-x))\right) \\ G_\varphi(x,x') &= \frac{e}{\kappa bA}\left(\frac{L-x'}{L}x - \Theta(x-x')(x-x')\right) + \frac{e}{\kappa p^2}G_n^n(x,x') \end{aligned} \tag{119}$$

with

$$b = \frac{\mu}{1+\mathrm{i}\omega\tau_v}N_{\mathrm{D}}\frac{e^2}{\kappa} + \mathrm{i}\omega e \tag{120}$$

$$p^2 = \frac{b}{e\frac{\mu}{1+\mathrm{i}\omega\tau_v}V_{\mathrm{T}}}\,. \tag{121}$$

With these the transfer functions of the current for terminal 1 and 2 can be calculated

$$G_{I_1}(x') = -\frac{x'}{L}, \qquad G_{I_2}(x') = -\frac{L - x'}{L}. \tag{122}$$

The sum of both transfer functions is -1, which is exactly the unit current injected at position x' due to the delta function in the continuity equation.

These Green's functions are calculated for a source in the continuity equation (114), whereas for the calculation of noise we need a source in the constitutive equation. In the case of the DD model the Green's functions for the source in the continuity equation can be easily transformed into the other ones by taking the gradient [23]. The terminal current noise is then given with Eqs. (98), (102) by

$$S_{I_1,I_1} = 4e^2 A \int_0^L \left| \frac{\partial G_{I_1}}{\partial x'} \right|^2 \frac{N_{\mathrm{D}} \mu V_{\mathrm{T}}}{1 + \omega^2 \tau_v^2} \mathrm{d}x' = 4k_{\mathrm{B}} T_0 \Re\{Y_{11}\} \tag{123}$$

and again for equilibrium thermal noise is obtained.

3.2 Numerical Solutions for an N^+NN^+ Resistor

In Fig. 10 the electron density is shown for the N^+NN^+ resistor and the results of the BE and DD are quite similar, where the device size was chosen such that the DD approximation holds.

The terminal current noise calculated by the DD model is also similar to the one by the BE for this device (Fig. 11). The small difference at zero bias is due to the differences in the transport models. The DD model yields in the case of built-in fields a higher conductance than the BE [45]. Further numerical results can be found in [24, 44].

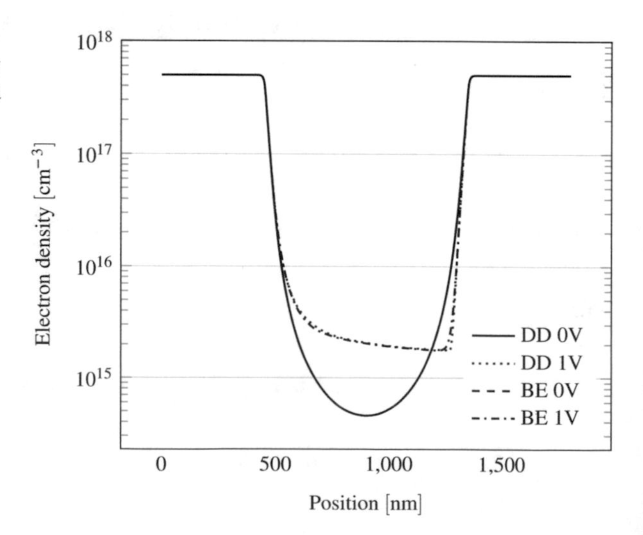

Fig. 10 Electron density at 0 and 1 V DC bias calculated by the DD model and the BE in a simple 1D silicon N^+NN^+ resistor at room temperature

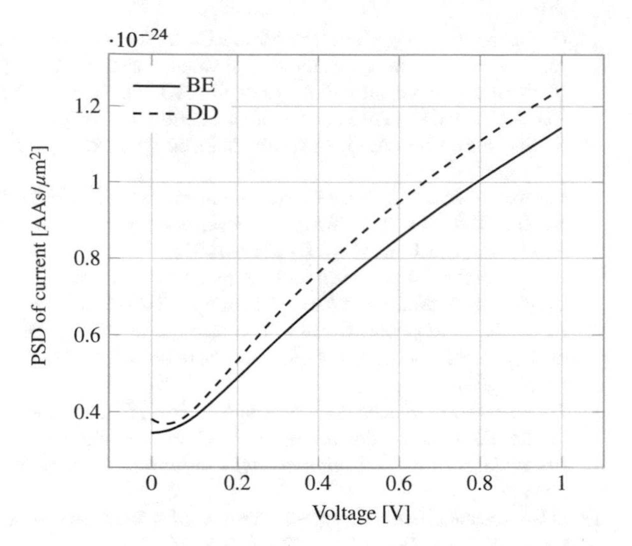

Fig. 11 PSD of the current at 1 GHz as a function of the DC bias calculated by the BE and DD model with modified Langevin forces in a simple 1D silicon N^+NN^+ resistor at room temperature

4 Conclusions

In this chapter it has been shown how to apply the LBE to noise calculations in semiconductor devices. The fundamental concepts were introduced in the framework of semi-classical transport theory and equations for the PSDs of the most important quantities (e.g., terminal currents) based on Green's functions and local noise sources were derived. This approach gives more insight into the origin of noise in devices than the Monte Carlo method, which was demonstrated by a numerical example. In order to facilitate the debugging process during the development of a deterministic solver for the LBE, a few analytical examples were given and important relations were derived, which must hold at equilibrium. In addition, noise modeling in the framework of balance equations was discussed and balance equations including noise were derived. It was shown that the macroscopic relaxation time approximation leads to serious errors in the modeling of noise and how this problem can be circumvented by introduction of the modified Langevin forces, which re-established the consistency of transport and noise modeling on the level of balance equations. This was demonstrated by comparison of noise results obtained by the LBE and DD model for an N^+NN^+ structure. Furthermore, some analytical results were presented for debugging purposes.

References

1. S. Kogan, *Electronic Noise and Fluctuations in Solids* (Cambridge University Press, Cambridge/New York/Melbourne, 1996)
2. S.V. Gantsevich, V.L. Gurevich, R. Katilius, Theory of fluctuations in nonequilibrium electron gas. Nuovo Cimento **2**, 5 (1979)
3. O. Madelung, *Introduction to Solid State Theory* (Springer, Berlin, 1978)

4. C. Jacoboni, L. Reggiani, The Monte Carlo method for the solution of charge transport in semiconductors with applications to covalent materials. Rev. Mod. Phys. **55**, 645–705 (1983). [Online]. Available: http://link.aps.org/doi/10.1103/RevModPhys.55.645
5. J.D. Jackson, *Classical Electrodynamics*, 2nd edn. (Wiley, New York, 1975)
6. W. Shockley, Currents to conductors induced by a moving point charge. J. Appl. Phys. **9**, 635–636 (1938)
7. S. Ramo, Currents induced by electron motion. Proc. IRE **27**, 584–585 (1939)
8. H. Kim, H.S. Min, T.W. Tang, Y.J. Park, An extended proof of the Ramo-Shockley theorem. Solid–State Electron. **34**, 1251–1253 (1991)
9. C. Moglestue, Monte-Carlo particle modelling of noise in semiconductors, in *International Conference on Noise and Fluctuations*, pp. 23–25 (1983)
10. L. Varani, L. Reggiani, T. Kuhn, T. González, D. Pardo, Microscopic simulation of electronic noise in semiconductor materials and devices. IEEE Trans. Electron Devices **41**(11), 1916–1925 (1994)
11. T. Gonzalez, J. Mateos, M.J. Martin-Martinez, S. Perez, R. Rengel, B.G. Vasallo, D. Pardo, Monte Carlo simulation of noise in electronic devices: limitations and perspectives, in *Proceedings of the 3rd International Conference on Unsolved Problems of Noise*, pp. 496–503 (2003)
12. C.E. Korman, I.D. Mayergoyz, Semiconductor noise in the framework of semiclassical transport. Phys. Rev. B **54**, 17620–17627 (1996)
13. C. Jungemann, A deterministic approach to RF noise in silicon devices based on the Langevin Boltzmann equation. IEEE Trans. Electron Devices **54**(5), 1185–1192 (2007)
14. S.-M. Hong, A.T. Pham, C. Jungemann, in *Deterministic Solvers for the Boltzmann Transport Equation.* Computational Microelectronics, ed. by S. Selberherr (Springer, Wien/New York, 2011)
15. C. Jungemann, A deterministic solver for the Langevin Boltzmann equation including the Pauli principle, in *SPIE: Fluctuations and Noise*, vol. 6600, pp. 660007–1–660007–12 (2007)
16. C. Jungemann, B. Meinerzhagen, *Hierarchical Device Simulation: The Monte-Carlo Perspective.* Computational Microelectronics (Springer, Wien/New York, 2003)
17. C. Jungemann, B. Neinhüs, B. Meinerzhagen, Hierarchical 2–D DD and HD noise simulations of Si and SiGe devices: Part I—Theory. IEEE Trans. Electron Devices **49**(7), 1250–1257 (2002)
18. D. Ruić, C. Jungemann, Numerical aspects of noise simulation in MOSFETs by a Langevin-Boltzmann solver. J. Comput. Electron. **14**(1), 21–36 (2015). [Online]. Available: http://dx.doi.org/10.1007/s10825-014-0642-4
19. A. Papoulis, *Probability, Random Variables, and Stochastic Processes*, 3rd edn. (McGraw-Hill, New York, 1991)
20. S.M. Kogan, Equations for the correlation functions using a generalized Keldysh technique. Phys. Rev. A **44**, 8072–8082 (1991)
21. M. Ramonas, C. Jungemann, A deterministic approach to noise in a non-equilibrium electron-phonon system based on the Boltzmann equation. J. Comput. Electron. **14**(1), 43–50 (2015). [Online]. Available: http://dx.doi.org/10.1007/s10825-014-0627-3
22. F.H. Branin, Network sensitivity and noise analysis simplified. IEEE Trans. Circuit Theory **20**, 285–288 (1973)
23. W. Shockley, J.A. Copeland, R.P. James, The impedance field method of noise calculation in active semiconductor devices, in *Quantum Theory of Atoms, Molecules and Solid State*, ed. by P.O. Lowdin (Academic, New York, 1966), pp. 537–563
24. C. Jungemann, B. Neinhüs, C.D. Nguyen, A.J. Scholten, L.F. Tiemeijer, B. Meinerzhagen, Numerical modeling of RF noise in scaled MOS devices. Solid-State Electron. **50**, 10–17 (2006)
25. C. Jungemann, B. Meinerzhagen, Do hot electrons cause excess noise? Solid-State Electron. **50**, 674–679 (2006)

26. C. Jungemann, A.-T. Pham, B. Meinerzhagen, C. Ringhofer, M. Bollhöfer, Stable discretization of the Boltzmann equation based on spherical harmonics, box integration, and a maximum entropy dissipation principle. J. Appl. Phys. **100**(2), 024502–1–13 (2006)
27. C. Ringhofer, Numerical methods for the semiconductor Boltzmann equation based on spherical harmonics expansions and entropy discretizations. Transp. Theory Stat. Phys. **31**(4–6), 431–452 (2002)
28. J. Stoer, R. Bulirsch, *Einführung in die Numerische Mathematik*, vol. 2, 2nd edn. (Springer, Berlin/Heidelberg/New York, 1978)
29. W. Brauer, H.W. Streitwolf, *Theoretische Grundlagen der Halbleiterphysik*, 2nd edn. (Vieweg, Braunschweig, 1977)
30. E.M. Lifshitz, L.P. Pitaevskii, *Physical Kinetics*. Courses of Theoretical Physics, vol. 10 (Butterworth-Heinemann, Oxford, 1981)
31. A. van der Ziel, *Noise in Solid State Devices and Circuits* (Wiley, Canada, 1986)
32. F. Bonani, G. Ghione, *Noise in Semiconductor Devices, Modeling and Simulation*. Advanced Microelectronics (Springer, Berlin/Heidelberg/New York, 2001)
33. P. Shiktorov, E. Starikov, V. Gruzinskis, T. Gonzalez, J. Mateos, D. Pardo, L. Reggiani, L. Varani, J.C. Vaissiere, Transfer-field methods for electronic noise in submicron semiconductor structures. Riv. Nuovo Cimento **24**(9), 1–72 (2001)
34. R. Stratton, Diffusion of hot and cold electrons in semiconductor barriers. Phys. Rev. **126**, 2002–2013 (1962)
35. N.G. van Kampen, *Stochastic Process in Physics and Chemistry* (North-Holland Publishing, Amsterdam, 1981)
36. K. Bløtekjær, Transport equations for electrons in two-valley semiconductors. IEEE Trans. Electron Devices **17**(1), 38–47 (1970)
37. P. Shiktorov, E. Starikov, V. Gruzinskis, Acceleration fluctuation scheme for diffusion noise sources within a generalized impedance field method. Phys. Rev. B **57**, 11866–11869 (1998)
38. P. Shiktorov, E. Starikov, V. Gruzinskis, T. Gonzalez, J. Mateos, D. Pardo, L. Reggiani, L. Varani, J.C. Vaissere, Langevin forces and generalized transfer fields for noise modeling in deep submicron devices. IEEE Trans. Electron Devices **47**(10), 1992–1998 (2000)
39. R. Thoma, A. Emunds, B. Meinerzhagen, H.J. Peifer, W.L. Engl, Hydrodynamic equations for semiconductors with nonparabolic band structures. IEEE Trans. Electron Devices **38**(6), 1343–1352 (1991)
40. S. Selberherr, *Analysis and Simulation of Semiconductor Devices* (Springer, Wien, 1984)
41. T. Grasser, R. Kosik, C. Jungemann, H. Kosina, S. Selberherr, Nonparabolic macroscopic transport models for device simulation based on bulk Monte Carlo data. J. Appl. Phys. **97**, 093710–1–12 (2005)
42. H.S. Min, D. Ahn, Langevin noise sources for the Boltzmann transport equations with the relaxation-time approximation in nondegenerate semiconductors. J. Appl. Phys. **58**, 2262–2265 (1985)
43. J.-P. Nougier, Fluctuations and noise of hot carriers in semiconductor materials and devices. IEEE Trans. Electron Devices **41**(11), 2034–2049 (1994)
44. C. Jungemann, B. Neinhüs, S. Decker, B. Meinerzhagen, Hierarchical 2–D DD and HD noise simulations of Si and SiGe devices: part II—Results. IEEE Trans. Electron Devices **49**(7), 1258–1264 (2002)
45. C. Jungemann, T. Grasser, B. Neinhüs, B. Meinerzhagen, Failure of moments-based transport models in nanoscale devices near equilibrium. IEEE Trans. Electron Devices **52**(11), 2404–2408 (2005)

Benchmark Tests for MOSFET Thermal Noise Models

Andries J. Scholten, G. D. J. Smit, R. M. T. Pijper, and L. F. Tiemeijer

1 Introduction

In today's semiconductor industry, many traditional integrated device manufacturers (IDMs) are moving away from chip manufacturing, and transforming into fabless companies that use foundry services for manufacturing their ICs. This is especially true in the field of advanced CMOS technologies. In these companies-under-transformation, the work of the modeling engineer is changing: instead of building models from scratch themselves, most companies choose to use the modeling packages that are delivered by the foundries. As a consequence, the work of the modeling engineer is changing from model creation to model verification.

The amount of model verification needed depends on the application field. If the foundry technology is used to make ICs with analog/RF blocks, it is an absolute must to verify analog/RF model properties, in order to prevent unnecessary and expensive design cycles.

In this chapter, we focus on one particular aspect of analog/RF model verification, namely the verification of RF noise modeling. There are two reasons to be skeptical about RF noise models. First, measurement of noise, and RF noise in particular, is a difficult and specialist topic. One should not take for granted that every company has the required expertise to carry out this task successfully. A second reason to check RF noise models is that the most popular compact MOSFET models are BSIM4 [1] and BSIMBULK [2], which are not particularly strong and certainly not predictive when it comes to RF noise. As a consequence, the model may be correct at one bias condition (e.g., where the device was measured) but may still fail at other bias conditions of interest.

A. J. Scholten (✉) · G. D. J. Smit · R. M. T. Pijper · L. F. Tiemeijer
NXP Semiconductors, Eindhoven, The Netherlands
e-mail: andries.scholten@nxp.com

T. Grasser (ed.), *Noise in Nanoscale Semiconductor Devices*,
https://doi.org/10.1007/978-3-030-37500-3_20

In this chapter, we demonstrate a method to address the validity of thermal noise models, without having access to experimental data. The method is based on a set of benchmark tests/sanity checks that we have developed over the years, and have been published partly in previous work [3–5]. This chapter is organized as follows: First, in Sect. 2, we will give a general introduction into MOSFET thermal noise modeling. Then, in Sect. 3, we will describe the measurements that we use to illustrate the benchmark tests that will be introduced in Sect. 4. In that section, we will explain the benchmark tests one by one, and, where possible, illustrate them using experimental data from a 40-nm CMOS technology. Finally, in Sect. 5, we will apply the tests to a 28-nm high-k metal-gate technology, for which we only have a model card, and no experimental data. The outcome of these tests can be used to give feedback to the foundry that provided this technology, to improve their model.

2 MOSFET Thermal Noise Models

2.1 Fundamentals

The modeling of thermal noise of long-channel MOSFET devices is well established through the works of Klaassen and Prins [6] and van der Ziel [7]. The basic idea behind their approach is to divide the MOSFET channel into small segments of length Δx. Each channel segment is supposed to have equilibrium thermal noise, given by the Nyquist law

$$S_{I,\Delta x} = 4\, k_{\mathrm{B}}\, T\, \frac{g(x)}{\Delta x}\ , \tag{1}$$

where $g(x)$ is the local channel conductivity as it appears in the drain current equation

$$I_{\mathrm{D}} = g(x)\, \frac{\mathrm{d}V}{\mathrm{d}x}\ . \tag{2}$$

To proceed, one has to calculate the effect of the local noise source, given by (1), on the current in the drain terminal. In other words, one has to calculate the *transfer* of the local noise to the drain terminal. Next, assuming each channel segment fluctuates independently, the total noise can be obtained by integration of the product of local noise and transfer function over the channel. This leads to

$$S_{I_{\mathrm{D}}} = \frac{4\, k_{\mathrm{B}}\, T}{I_{\mathrm{D}}\, L^2} \int_{V_{\mathrm{S}}}^{V_{\mathrm{D}}} g(V)^2\, \mathrm{d}V\ , \tag{3}$$

a result that is known as the Klaassen–Prins equation [6].

A refinement of (3), coined "improved Klaassen–Prins equations," was put forward by Paasschens et al. [4], who took into account the effect of velocity saturation. They retained the same local noise source as given in (1), but accounted for the effect of velocity saturation in the transfer to drain and gate terminals. Please refer to [4] for the details.

2.2 *The Excess-Noise Controversy*

The thermal noise of short-channel MOSFETs in saturation has led to quite some controversy in the literature. These discussions were triggered by early reports of Abidi [8] (down to 0.7 μm channel length) and Jindal [9] (down to 0.25 μm channel length), claiming up to an order-of-magnitude noise enhancement w.r.t. long-channel theory, and relating those observations to hot-electron effects. In the late nineties, circuit designers started to develop RF CMOS applications, and, based on these early excess-noise reports, feared poor RF noise performance. This fear of strong excess noise triggered new studies on MOSFET thermal noise, such as [10, 11]. Also these new studies reported strong excess noise.

In a series of publications [3, 12, 13], however, we have shown that this so-called excess noise, contrary to what was believed at that time, is largely absent in commercial CMOS technologies down to about 100 nm. Most likely, the strong enhancement factors found in the earlier reports are due to a combination of (1) large parasitic resistances, such as gate resistance, (2) the noise associated with strong impact ionization current, and (3) measurement errors. Our modeling approach, based on the improved Klaassen–Prins equations [4], and taking into account channel-length modulation [14] and parasitic resistance, was shown to give very reasonable noise predictions without invoking microscopic excess noise.

2.3 *Excess Noise in Sub-100-nm CMOS Technology*

In the previous section, we saw that excess thermal noise, in CMOS technologies $\geq$100 nm, is largely absent, contradicting the claims made in early reports. For sub-100-nm CMOS technologies, however, the situation changed considerably [15, 16]. The modeling approach, as outlined above, was no longer able to predict the measured RF noise data. The long-feared "excess noise" now started to become reality. However, it should be noted that the largest enhancement that was found experimentally (for a 40-nm n-channel MOSFET) was only about 2, which is still much less than what was suggested (for much longer channels) in the earlier papers [8–11]; see Fig. 1.

In [15, 16], our previous modeling approach was extended with a field-dependent noise enhancement factor (to account for non-equilibrium effects) that is applied to the local noise source. That approach, combined with careful gate resistance model-

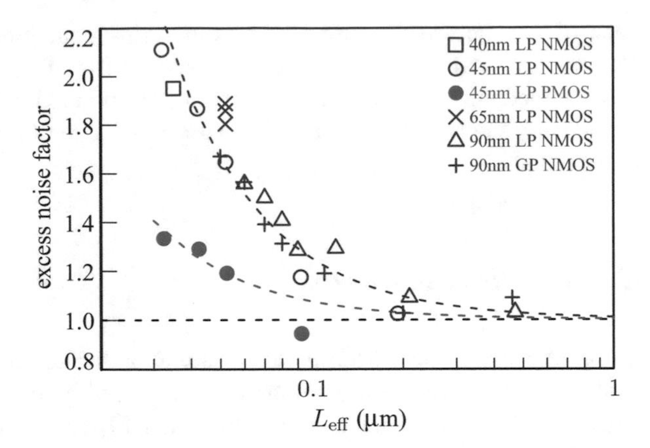

Fig. 1 Excess noise factor versus effective channel length for four technologies [15, 16]. The contribution of gate resistance has been removed from the data. The data points were taken at $f = 10$ GHz, V_{DS} at the supply voltage, and V_{GS} around maximum g_{m}. The colored dashed lines are a guide to the eye. The black dotted line is pure Nyquist noise. The inset shows an overview of all measured devices (LP = low power, GP = general purpose). For the exact meaning of "excess noise factor" please refer to [15, 16]

ing, was shown to describe the measured noise accurately for CMOS technologies down to the 40-nm node.

3 Measurement Details and Definitions

In this section, we describe the RF-noise measurement system and the MOSFET test structures that are used to demonstrate the validity of the benchmark tests. Then, we will introduce some dimensionless quantities that are derived from these measurements. Those quantities will be useful in formulating the benchmark tests in Sect. 4.

3.1 Measurement System

We performed noise measurements on a commercially available 40-nm CMOS technology. The devices used have been laid out in a common-source ground-signal-ground configuration. In total, we have measured six different NMOS transistors, and three different PMOS transistors, see Table 1. The device set consists of thin- and thick-oxide devices, with different V_{th} flavors (labeled "VT1" and "VT2"), different number of fingers N_{f}, as well as different finger width W_{f} and length L_{f}.

Table 1 Overview of measured devices

	Type	ID	Oxide	N_f	W_f (μm)	L_f (nm)
#1	NMOS	VT2 20 × 5/0.04	Thin	20	5	40
#2	NMOS	VT1 60 × 2/0.04	Thin	60	2	40
#3	NMOS	VT1 20 × 1/0.04	Thin	20	1	40
#4	NMOS	VT1 20 × 1/0.054	Thin	20	1	54
#5	NMOS	VT1 20 × 1/0.1	Thin	20	1	100
#6	NMOS	GO2 20 × 5/0.27	Thick	20	5	270
#7	PMOS	VT2 20 × 5/0.04	Thin	20	5	40
#8	PMOS	VT1 60 × 2/0.04	Thin	60	2	40
#9	PMOS	GO2 20 × 5/0.27	Thick	20	5	270

All devices have single-sided gate connection

On these nine devices, on-wafer RF-noise measurements were taken in the 1–50 GHz range. The measurements were performed as a function of frequency and bias condition. At each frequency, a constellation of 15 source states was used to extract the noise parameters NF_{min}, R_{N}, and Γ_{opt} [17]. These noise parameters were subsequently converted to noise spectral densities $S_{\mathrm{I_D}}$, $S_{\mathrm{I_G}}$, and cross-spectral density $S_{I_{\mathrm{G}},I_{\mathrm{D}}}$. The measurements were de-embedded down to the device level, using open, short, and load dummies.

3.2 *The White-Noise Gamma Factor*

Noise is usually represented in terms of noise power spectral densities, e.g., $S_{I_{\mathrm{D}}}$ and $S_{I_{\mathrm{G}}}$. For our benchmark tests, however, it is more convenient to make use of dimensionless quantities. For drain current noise, we will make use of the so-called white-noise gamma factor γ. This quantity is defined as

$$\gamma \equiv \frac{S_{I_{\mathrm{D}}}}{4\,k_{\mathrm{B}}\,T\,g_{\mathrm{DS0}}}\,. \tag{4}$$

Here, $S_{I_{\mathrm{D}}}$ is the noise power spectral density of the drain current, measured/simulated at a certain bias condition ($V_{\mathrm{GS}}, V_{\mathrm{DS}}$) and frequency f. The quantity g_{DS0} is the drain conductance $\mathrm{Re}(Y_{\mathrm{DD}})$, taken at the same frequency f and gate-source voltage V_{GS} as $S_{I_{\mathrm{D}}}$, but at a drain-source voltage of $V_{\mathrm{DS}} = 0$ V. Although the definition of γ, derived from quantities at different bias conditions, may seem a bit awkward, it turns out to be very useful in practice.

Some care should be taken with an alternative, g_{m}-based, γ that is regularly found, in particular in lecture notes from universities, e.g., [18]. We will denote it here by γ' for clarity. It is given by:

$$\gamma' \equiv \frac{S_{I_\mathrm{D}}}{4\,k_\mathrm{B}\,T\,g_\mathrm{m}} \,. \tag{5}$$

This alternative γ' (sometimes referred to as α [6], Γ [19], or ϵ [20]) is not "wrong." However, it may lead to confusion because it shares the property with the g_DS0-based γ that it is $2/3$ for a long-channel MOSFET in saturation (as we will see in Sect. 4.5). However, it attains very different values in other regimes of operation. In particular at $V_\mathrm{DS} = 0$ V, the g_m-based γ' is not useful because it becomes infinite. In this chapter, we will therefore stick to γ as defined by (4).

3.3 The Fano Factor

Another convenient dimensionless quantity is the so-called Fano factor [21]. It measures how close the noise is to shot noise $2 \cdot q \cdot I$. For the drain current noise, we define F as

$$F \equiv \frac{S_{I_\mathrm{D}}}{2q\,I_\mathrm{D}} \,. \tag{6}$$

As we will see, a MOSFET in weak inversion exhibits full shot noise, corresponding to $F = 1$.

3.4 The β Factor

To quantify the induced gate noise, we will make use of the quantity β that was introduced by van der Ziel [7], in his formula for induced gate noise in saturation. Here, we will generalize its usage and define

$$\beta \equiv \frac{5\,g_\mathrm{DS0}\,S_{I_\mathrm{G}}}{4\,k_\mathrm{B}\,T\,(\omega\,C_\mathrm{GC})^2} \,. \tag{7}$$

4 Overview of Thermal Noise Benchmark Tests

In this section, we will collect a series of noise benchmark tests, and illustrate them with the help of actual measurements on a 40-nm CMOS technology. The tests are summarized are Table 2.

Table 2 Overview of benchmark tests for thermal noise

	Bias	Channel length	Quantity	Test	Remark
#1	$V_{DS} = 0$ V	All	S_{I_D}	$\gamma = 1$	
#2	$V_{DS} = 0$ V	All	S_{I_G}	$\beta = 5/12$	
#3	$V_{DS} = 0$ V	All	c	$c = 0$	In the limit $f \downarrow 0$ Hz
#4	Weak inversion	All	S_{I_D}	$F = 1$	Disregard S_{I_G} contributions from gate to drain
#5	Saturation	Long	S_{I_D}	$\gamma = 2/3$	
#6	Saturation	Long	S_{I_G}	$\beta = 4/3$	
#7	Saturation	Long	c	$c = 0.4j$	
#8	Saturation	Short	S_{I_D}	γ enhancement in line with Fig. 1	Switch off gate resistance in model
#9	Saturation	All	S_{I_D}	$\gamma_{D,NMOS} \geq \gamma_{D,PMOS}$	Switch off gate resistance in model
#10	Saturation	All	S_{I_D}	Different V_{th} flavors should nearly coincide	When plotted against I_D

4.1 Benchmark Test #1: Drain Current Noise at $V_{DS} = 0$ V

The Nyquist law for a passive multiport device reads [22]:

$$S_I = 4\,k_B\,T\,\mathrm{Re}\,(Y) \ , \tag{8}$$

where S_I is the matrix of noise current spectral densities, which, for a MOSFET in standard two-port configuration, is

$$S_I = \begin{pmatrix} S_{I_G} & S_{I_G,I_D} \\ S_{I_D,I_G} & S_{I_D} \end{pmatrix} \ , \tag{9}$$

and Y is the corresponding admittance matrix. Applying (8) to the MOSFET drain current noise immediately yields

$$S_{I_D} = 4\,k_B\,T\,\mathrm{Re}\,(Y_{DD}) \ , \tag{10}$$

or, in terms of γ,

$$\gamma = 1 \ . \tag{11}$$

This equation is very fundamental and holds for any MOSFET, as long as $V_{DS} =$ 0 V. This is confirmed by Fig. 2, where we plot measured values of γ as a function of V_{DS} for a variety of NMOS and PMOS devices. For all devices, it is observed

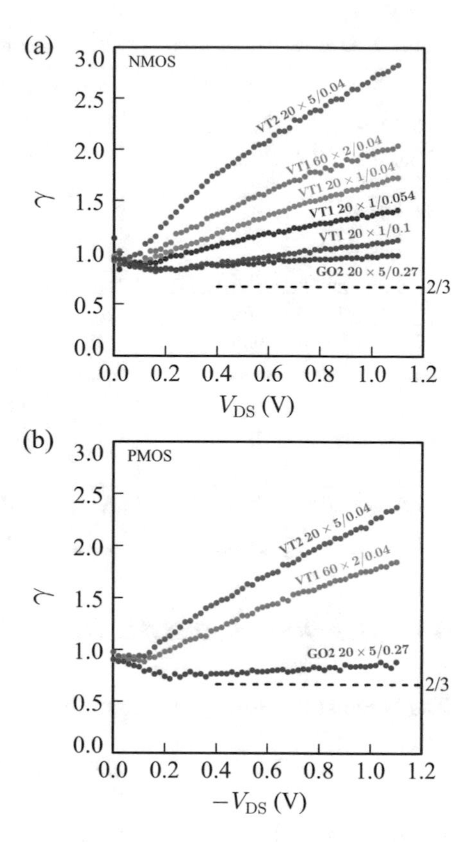

Fig. 2 Measured white-noise gamma factors, defined by (4), as a function of V_{DS} for (**a**) NMOS, and (**b**) PMOS transistors. For all measurements, the frequency is $f = 30$ GHz and the temperature is 21 °C. $V_{GS} = 0.8$ V for NMOS transistors and $V_{GS} = -0.8$ V for PMOS transistors. Various transistor flavors and geometries are represented by different colors. The annotation of the curves indicates the measured device, and corresponds to the column "ID" in Table 1. The dashed line corresponds to the theoretical long-channel limit in saturation, i.e., $\gamma = 2/3$

that γ approaches the value of 1 indeed (on average, the measured γ at $V_{DS} = 0$ V is a modest ∼5% low, which can be attributed to imperfections in the measurement system used).

The $\gamma = 1$ check can be applied as a sanity check for a compact drain current noise model. For instance, this sanity check was applied in the development of the PSP thermal noise model [4, 5]. Later in this chapter, in Sect. 5, we will apply it in the assessment of the RF noise model for a 28-nm CMOS technology.

4.2 Benchmark Test #2: Gate Current Noise at $V_{DS} = 0$ V

Equation (8) is not restricted to drain current noise. It can also be applied to the MOSFET gate current noise at $V_{DS} = 0$ V, leading to

$$S_{I_G} = 4\,k_B\,T\,\mathrm{Re}\,(Y_{GG})\ . \tag{12}$$

Unfortunately, it is not straightforward to apply (12) as a compact-model sanity check, because compact MOSFET models almost always make use of the so-called quasi-static approximation [23]. As a consequence, $\mathrm{Re}\,(Y_{\mathrm{GG}})$ will be zero for the core MOSFET model without parasitic resistances.

To circumvent this, we use the approximation of (12) that was published in [4],

$$S_{I_{\mathrm{G}}} = 4\,k_{\mathrm{B}}\,T\;(\omega\,C_{GC})^2\;\frac{1}{12\,g_{\mathrm{DS0}}}\;, \tag{13}$$

where C_{GC} is the channel-to-gate capacitance, and $1/(12 \cdot g_{\mathrm{DS0}})$ is the MOSFET input resistance at $V_{\mathrm{DS}} = 0$ V. In other words, (13) is just the Nyquist noise of the MOSFET input resistance, connected in series with the channel-to-gate capacitance. Note that this equation is also found in [24], but with an incorrect explanation in terms of non-equilibrium noise.

In terms of the quantity β, introduced in Sect. 3.4, (13) reduces to

$$\beta = \frac{5}{12}\;, \tag{14}$$

which is a simple test that can be carried out on a MOSFET model with induced gate noise.

4.3 Benchmark Test #3: Correlation Coefficient at $V_{\mathrm{DS}} = 0$ V

Finally, we can use (8) to predict the correlation coefficient between gate and drain current noise at $V_{\mathrm{DS}} = 0$ V. From (8), it immediately follows that

$$S_{\mathrm{I_G},I_{\mathrm{D}}} = 4\,k_{\mathrm{B}}\,T\,\mathrm{Re}\,(Y_{\mathrm{GD}})\;, \tag{15}$$

from which we can calculate the correlation coefficient

$$c = \frac{S_{\mathrm{I_G},I_{\mathrm{D}}}}{\sqrt{S_{\mathrm{I_G}}\,S_{\mathrm{I_D}}}}\;. \tag{16}$$

It immediately follows that the imaginary part of the correlation coefficient should vanish at $V_{\mathrm{DS}} = 0$ V. Using device symmetry, one can further derive that for the real part of the correlation coefficient

$$\lim_{f\downarrow 0}\,(\mathrm{Re}\,(c)) = 0\;. \tag{17}$$

Consequently, the complex correlation coefficient c is expected to become zero at $V_{DS} = 0$ V. This serves as another simple sanity check for a physics-based thermal noise model.

4.4 Benchmark Test #4: Drain Current Noise in Weak Inversion

In the weak-inversion condition, charge carriers are injected into the channel over the source barrier, and travel independently to the drain. Under these conditions (and V_{DS} much larger than the thermal voltage ϕ_T), the noise at the drain terminal is expected to be "shot noise"

$$S_{I_D} = 2q\, I_D\ . \tag{18}$$

The terms "shot noise" and "thermal noise" are both terms referring to macroscopic noise, i.e., noise emanating from the device terminals. They are due to the same microscopic phenomenon called "diffusion noise." Therefore, the Klaassen–Prins equation, (3) also reduces to shot noise in weak inversion [3, 25].

For the specific case of long channels, the result (18) can be expressed in terms of the white-noise gamma factor as $\gamma = 1/2$ [5]. However, (18) holds for *all* channel lengths. Therefore we prefer to express (18) in terms of the Fano factor, introduced in Sect. 3.3. The weak-inversion test (18) then simply reads

$$F = 1\ , \tag{19}$$

and should be satisfied for all channel lengths.

The validity of (18) is demonstrated experimentally using two examples. The first example, shown in Fig. 3a, clearly shows that the measured drain current noise spectral density converges to $2qI_D$ at low values of V_{GS}. The second example, shown in Fig. 3b, is more complicated. Going to lower V_{GS} we first observe a region where the noise is close to $2qI_D$. At even lower V_{GS}, however, the measured noise levels off. This plateau is related to the relatively large finger width of this device, $W_f = 5\ \mu$m. This leads to a relatively large effective gate resistance, the thermal noise of which runs to the drain because of capacitive coupling

$$S_{I_D} = 4\, k_B\, T\, R_G\ (\omega\, C_{DG})^2\ . \tag{20}$$

This additional noise contribution causes the plateau in S_{I_D}, observed at the lowest V_{GS} values. Therefore, when applying the Fano-factor test, (19), to a compact model, it is better to switch off the gate resistance first.

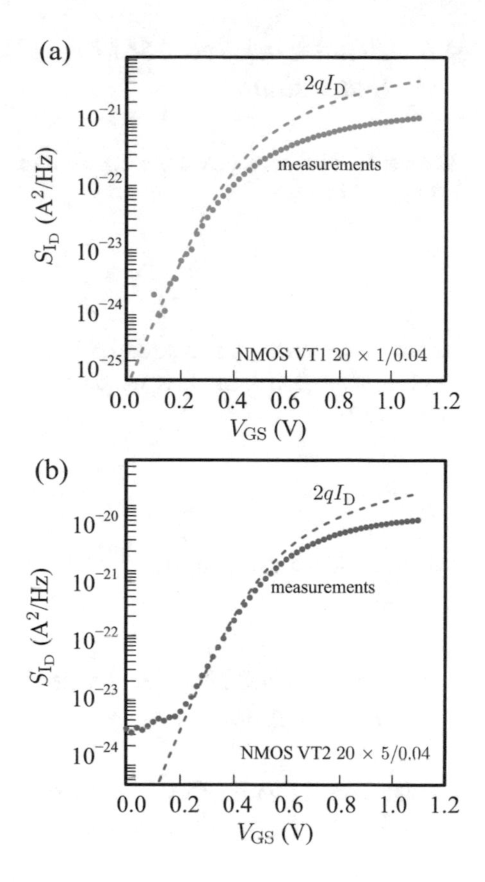

Fig. 3 Drain current noise power spectral density as a function of V_{GS} for (**a**) the NMOS device VT1 20 × 1/0.04, and (**b**) the NMOS device VT2 20 × 5/0.04. For both devices, $V_{DS} = 1.1$ V, $f = 10$ GHz, and $T = 21$ °C. Markers are measured values; dashed lines represent full shot noise $2qI_D$

4.5 *Benchmark Test #5: Long-Channel Drain Current Noise in Saturation*

As explained in Sect. 2.1, the thermal noise of long-channel MOSFETs in saturation is well established [6, 7]. In the saturation region, the drain current thermal noise result can be expressed in terms of terms of γ as

$$\gamma = \frac{2}{3} . \tag{21}$$

A good noise model is expected to approach $\gamma = 2/3$ for sufficiently long channel lengths.

In our set of measurements, Fig. 2, we do not have transistors that actually reach $\gamma = 2/3$. Still, we can see that γ decreases when going to longer channel lengths, and already reaches about 1 for the 270-nm thick-oxide devices. The remaining deviation from the long-channel limit of 2/3 is due to the fact that our transistors are not long enough. As a consequence, there is some noise enhancement due to (1) channel-length modulation [14] and (2) parasitic resistances.

4.6 Benchmark Test #6: Long-Channel Gate Current Noise in Saturation

Van der Ziel [7] also gave an expression for the induced gate noise of a long-channel MOSFET. It reads

$$S_{I_G} = \beta\, 4\, k_B\, T\, \frac{\omega^2\, C_{CG}^2}{5\, g_{D0}} , \tag{22}$$

where the dimensionless quantity β is the same that was introduced in Sect. 3.4, and has the value of $4/3$. In other words, for a long-channel MOSFET in saturation it is expected that

$$\beta = \frac{4}{3} , \tag{23}$$

which is, indeed, close to what we have measured in [3].

4.7 Benchmark Test #7: Long-Channel Correlation Coefficient in Saturation

For the correlation coefficient between gate and drain current noise of a long-channel MOSFET, van der Ziel found [7]

$$c = 0.395\, j . \tag{24}$$

A good noise model is expected to approach $c = 0.395 \cdot j$ for sufficiently long channel lengths. Unfortunately, actual measurements of the correlation coefficient are often too "noisy" to clearly reveal this value.

4.8 Benchmark Test #8: Short-Channel Drain Current Noise in Saturation

As discussed in Sect. 2.3, sub-100-nm MOSFETs show a considerably enhanced thermal noise due to non-equilibrium effects. On top of that, the thermal noise of the effective gate resistance R_G may further enhance the experimentally observed S_{I_D}, by the amount

$$\Delta S_{I_D} = 4\, k\, T\, R_G\, |Y_{DG}|^2 . \tag{25}$$

and thus enhance γ by the amount of

$$\Delta\gamma = \frac{R_G \, |Y_{DG}|^2}{g_{DS0}} . \tag{26}$$

These effects make testing of a short-channel compact noise model in saturation less straightforward then the previously discussed $V_{DS} = 0$ V and weak-inversion regions (Sects. 4.1–4.4). Still, it is possible to define a qualitative sanity check by demanding that γ is in line with the expected intrinsic γ enhancement, as given by the trend lines in Fig. 1. When performing this test, the gate resistance should be switched off, because, as expressed by (26), the thermal noise of the gate resistance adds to the drain current noise and may therefore obscure the intrinsic thermal noise of the MOSFET.

4.9 Benchmark Test #9: PMOS vs. NMOS

From Fig. 1, it is clear that the intrinsic noise enhancement of a PMOS transistor is smaller than that of the corresponding NMOS transistor. If the NMOS and PMOS transistors have similar gate resistance, this statement remains true for the additional γ enhancement due to gate resistance, as given by (26). Indeed, as demonstrated in Fig. 4, this is what is observed experimentally, when we compare NMOS and PMOS transistors with exactly the same layout. A compact noise model is expected to reproduce this behavior. Some care should be taken with modern CMOS technologies where the hole mobility is almost equal to the electron mobility, as a consequence of stress engineering; in this case, γ is expected to be similar for NMOS and PMOS.

4.10 Benchmark Test #10: Comparing V_{th} Flavors

Today's advanced CMOS technologies offer a large range of transistors with different V_{th} flavors, which are fabricated by varying the channel or pocket doping concentrations. Based on our understanding of the physics of thermal noise, these doping variations should only have a minor impact, and these different V_{th} flavors are expected to show similar thermal noise characteristics, if the threshold voltage difference is accounted for.

Our experiments confirm this expectation. To make a fair comparison between devices with different V_{th}, we will plot the measured drain current noise spectral density as a function of drain current rather than V_{GS}. Because the devices used for this investigation have different geometries, we divide both the drain current itself and its current noise spectral density by their total width. The result, for a VT1 and VT2 device, is shown in Fig. 5a. It is observed that (1) both for NMOS

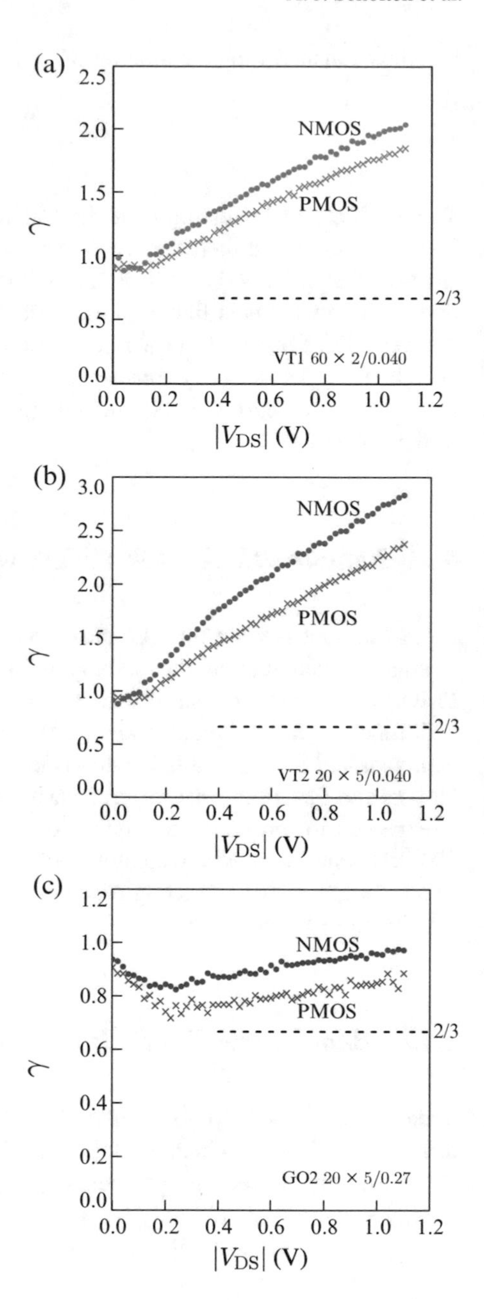

Fig. 4 Direct comparison of the white-noise gamma factor between NMOS and PMOS devices. The comparison is given for (**a**) the VT1 $60 \times 2/0.04$ device, (**b**) the VT2 $20 \times 5/0.04$ device, and (**c**) the GO2 $20 \times 5/0.27$ device

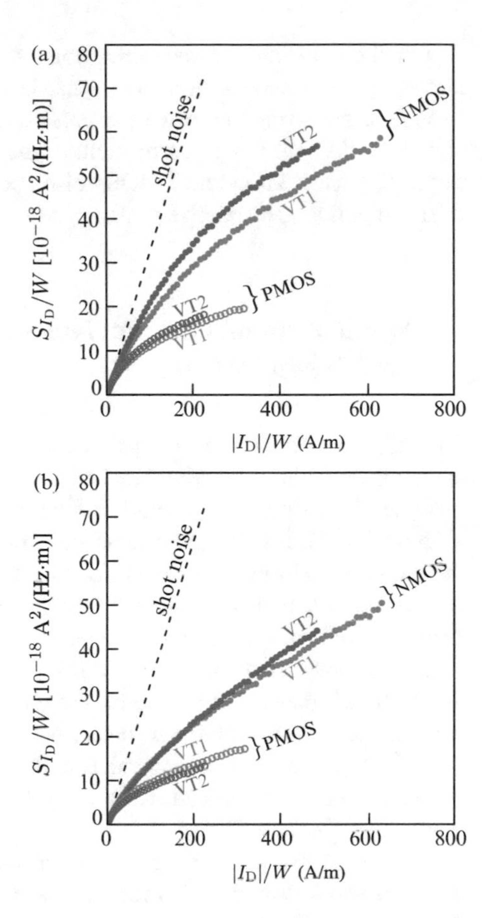

Fig. 5 (**a**) Drain current noise power spectral density as a function of $|I_D|$, both normalized w.r.t. total device width $W = W_f \cdot N_f$, for the VT1 60 × 2/0.04 device, and the VT1 × 5/0.04 device, at a frequency of $f = 30$ GHz, and a temperature of $T = 21$ °C. Filled and open circles represent NMOS ($V_{DS} = 1.1$ V) and PMOS ($V_{DS} = -1.1$ V) data, respectively. The dashed line represents full shot noise. (**b**) Same as plot (**a**), but now the estimated contributions of gate resistance noise have been subtracted

and PMOS, the two different V_{th} flavors have very comparable noise behavior, and (2) at the same current density, short-channel PMOS devices have significantly less noise than their NMOS counterparts (cf. Sect. 4.9).

Unfortunately, this experiment suffers from the imperfection that the layouts of the two devices are not the same: the VT2 device has a larger finger width and, hence, a larger effective gate resistance. We can correct for this, using

$$S_{I_D,\,\mathrm{corr}} = S_{I_D} - 4\,k_B\,T\,R_G\,|Y_{DG}|^2\,, \tag{27}$$

and estimating R_G from the measured value of $\mathrm{Re}(1/Y_{GG})$. The result is shown in Fig. 5b. It is seen that VT1 and VT2 devices overlap even better now. Note that the remaining (small) differences depend sensitively on the exact value of the extracted gate resistances. Therefore, this plot cannot be used to say anything about the magnitude and sign of the remaining, small, difference between VT1 and VT2 devices. The purpose of this plot is merely to show that the intrinsic noise spectral density of these different V_{th}-flavors is very close, both in the NMOS and PMOS case.

Finally, note that the measured noise spectral densities converge to shot noise at low I_{DS}, corresponding to the weak-inversion regime (see Sect. 4.4). In strong inversion, however, the measured noise spectral densities remain well below full shot noise (which would correspond to the ultimate non-equilibrium case). At the highest $|V_{GS}|$, NMOS and PMOS noise spectral densities are about 30% and 20% of full shot noise, respectively.

5 Application of Thermal Noise Benchmark Tests to a 28-nm CMOS Model Card

We will now apply the benchmark tests of Table 2 to the model card of a commercially available high-k metal-gate 28-nm CMOS technology. The technology has as many as six different V_{th}-flavors, which we will denote as VT1, VT2, VT3, VT4, VT5, and VT6. It is not our intention to promote or criticize a particular foundry; therefore, we do not disclose the foundry that is providing the technology at hand; we just remark that the results presented here are quite typical for what we have seen in model cards of multiple foundries.

The foundry model card uses BSIM4 [1], version 4.5, with `tnoimod=1`, i.e., the "holistic" thermal noise model. As will be explained in Sect. 5.1, this model is expected to give induced gate noise, based on its topology. However, it does not do so in any practical implementation that we have tested. Therefore, we restrict ourselves to the six drain current noise related tests in Table 2, and skip benchmark tests #2, #3, #6, and #7.

Because we want to focus on the thermal noise model, we have first disabled flicker noise contributions in the simulation model. As a consequence, the simulated drain current noise is white over a large frequency range, and we can simply take the simulations at $f = 1$ MHz. The models at hand are the so-called baseline models, as opposed to "RF" models, and therefore do not have gate resistance, nor the noise associated with that, that would obscure the results of the benchmark tests. The models do have external source and drain resistances, however, giving rise to an S_{I_G} contribution that is proportional to f^2, and similar deviations from white noise in S_{I_D} at very high frequencies. This does not affect the results, however, because of the 1 MHz frequency that was chosen.

5.1 *The BSIM4 Thermal Noise Model*

Before we carry out the actual benchmark tests, we will first explain the background of the BSIM4 "holistic" noise model [1]. As shown in Fig. 6, it has two uncorrelated noise sources, which are denoted by S_V and S_I. Evidently, these two noise sources both lead to noise in the drain current. On top of that, however, the noise source S_V

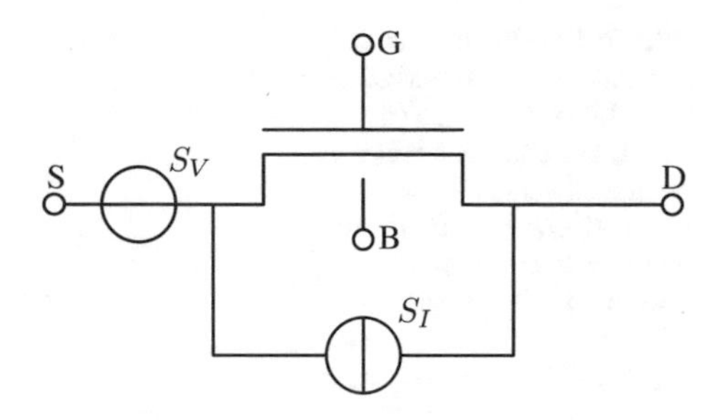

Fig. 6 Topology of the BSIM4 "holistic" thermal noise model [1]

is also intended to give channel thermal noise in the gate current, a phenomenon known as induced gate noise. Unfortunately, however, in all circuit simulators that we have tested, there is no sign of actual existence of induced gate noise when using `tnoimod=1`. Therefore, we will restrict our discussions to the noise emanating from the drain terminal.

The noise source S_I is given by [1]

$$S_I = 4k_\mathrm{B}T\,\frac{V_\mathrm{ds,\,eff}}{I_\mathrm{ds}}\,\left[g_\mathrm{DS} + \beta_\mathrm{noi}\,(g_\mathrm{m} + g_\mathrm{mb})\right]^2 - S_V\,(g_\mathrm{m} + g_\mathrm{DS} + g_\mathrm{mb})^2\;. \quad (28)$$

Because S_V and S_I are not correlated, the drain current noise power spectral density is given by

$$S_{I_\mathrm{D}} = 4k_\mathrm{B}T\,\frac{V_\mathrm{ds,\,eff}}{I_\mathrm{ds}}\,\left[g_\mathrm{DS} + \beta_\mathrm{noi}\,(g_\mathrm{m} + g_\mathrm{mb})\right]^2\;, \quad (29)$$

where the parameter β_noi is a function of bias and channel length:

$$\beta_\mathrm{noi} = \mathbf{RNOIA}\left[1 + \mathbf{TNOIA}\,L_\mathrm{eff}\left(\frac{V_\mathrm{gsteff}}{E_\mathrm{sat}\,L_\mathrm{eff}}\right)^2\right]\;. \quad (30)$$

For the meaning of the symbols, please refer to the BSIM4 manual [1]. The combined bias and channel-length dependence of (30) can be used to model short-channel excess noise (see Sect. 2.3). It should be noted that neither (29), nor (30) is based on physics. The results, obtained with this model, depend strongly on the value of the adjustable parameters **RNOIA** and **TNOIA**.

5.2 *Benchmark Test #1: the $V_\mathrm{DS} = 0$ V Condition*

In this section, we look at the test results for the $V_\mathrm{DS} = 0$ V condition, where we expect the Nyquist law to apply. Although this is not the most frequently used bias condition, it is of practical importance in RF circuits like the passive mixer [26].

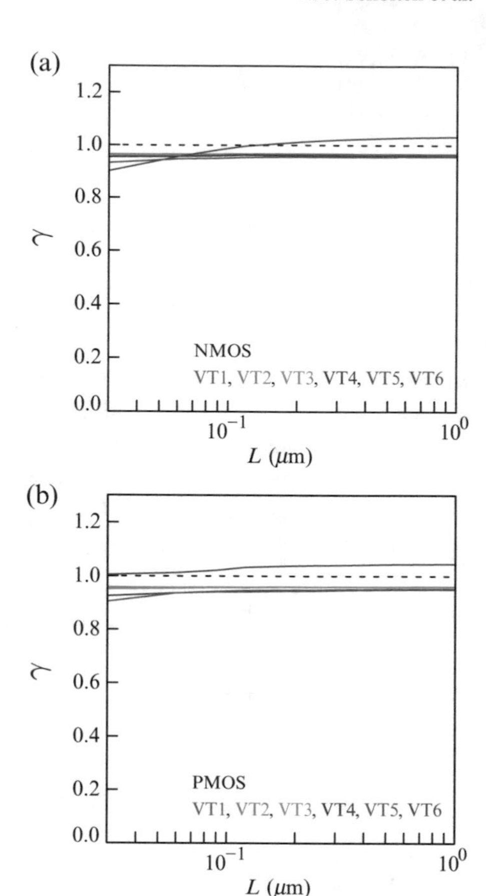

Fig. 7 Benchmark test #1 carried out for (**a**) NMOS and (**b**) PMOS devices. The white-noise gamma factor γ is plotted versus drawn channel length L. Different colors reflect different threshold voltage variants VT1, VT2, VT3, VT4, VT5, and VT6. The drain-source voltage is $V_{DS} = 0$ V. The gate-source voltage is $V_{GS} = 0.7$ V for NMOS and $V_{GS} = -0.7$ V for PMOS. The dashed lines indicate the expected behavior $\gamma = 1$

In Fig. 7, the results of test #1 are displayed for NMOS and PMOS devices of all the six V_{th}-flavors. We observe that, in all cases, γ is very close to 1, as demanded by test # 1. The small deviations that can be observed are always less than 10% and are not expected to result in any problems in practice.

These results can be understood when inspecting (29). At $V_{DS} = 0$ V, both g_m and g_{mb} are zero, so that (29) becomes

$$S_{I_D} = 4k_B T \, \frac{V_{ds,\,eff}}{I_{ds}} \, {g_{DS}}^2 \,. \tag{31}$$

It is reasonable to assume that

$$\lim_{V_{DS}\to 0} \frac{V_{ds,\,eff}}{I_{ds}} = \frac{1}{g_{DS}} \,, \tag{32}$$

so that the Nyquist law $S_{I_D} = 4k_B T \, g_{DS}$ is recovered. Therefore, benchmark test # 1 is, indeed, expected to be passed when using this model, independent of the values of the parameters **RNOIA** and **TNOIA**. It is not clear what causes the small discrepancies from this ideal behavior, as observed in Fig. 7.

5.3 Benchmark Test #4: The Weak-Inversion Region

Now we turn to the weak-inversion region, i.e., test #4 in Table 2. This region is of importance in low-power RF design. The results for the Fano factor F, which should approach 1 in this region, are shown as a function of V_{GS} in Fig. 8. We see that none of the devices passes this test. Apart from the lack of absolute accuracy, the simulations show unexpected scatter between various V_{th}-flavors. For instance, the VT6 flavor remains a factor of about 3 below the expected level, whereas the VT5 device shows a factor of about 3 above it (both for NMOS and PMOS). The potential danger is that a circuit designer would choose a particular V_{th}-flavor based on these—evidently wrong—model predictions.

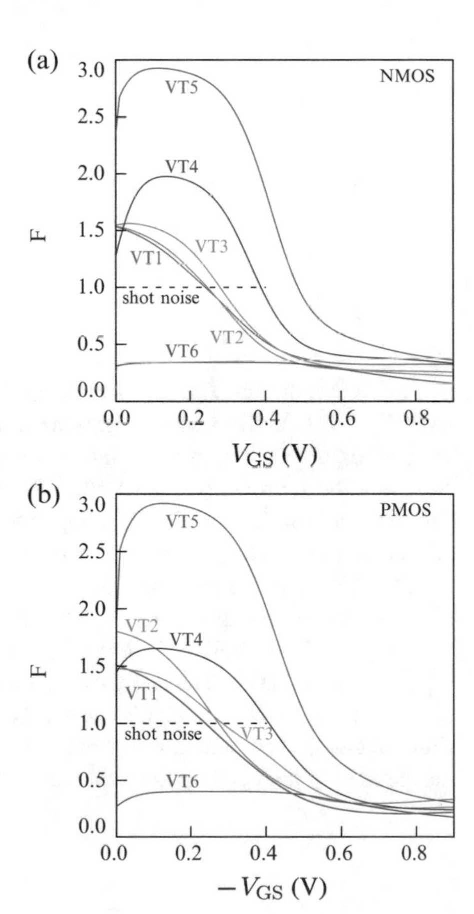

Fig. 8 Benchmark test #4 carried out for (**a**) NMOS and (**b**) PMOS devices. The Fano factor F is plotted versus V_{GS}. The drain-source voltage is $V_{DS} = 0.9$ V for NMOS and $V_{DS} = -0.9$ V for PMOS. Different colors reflect different threshold voltage variants, as in Fig. 7. The dashed lines indicate $F = 1$, which is expected to be reached in weak inversion

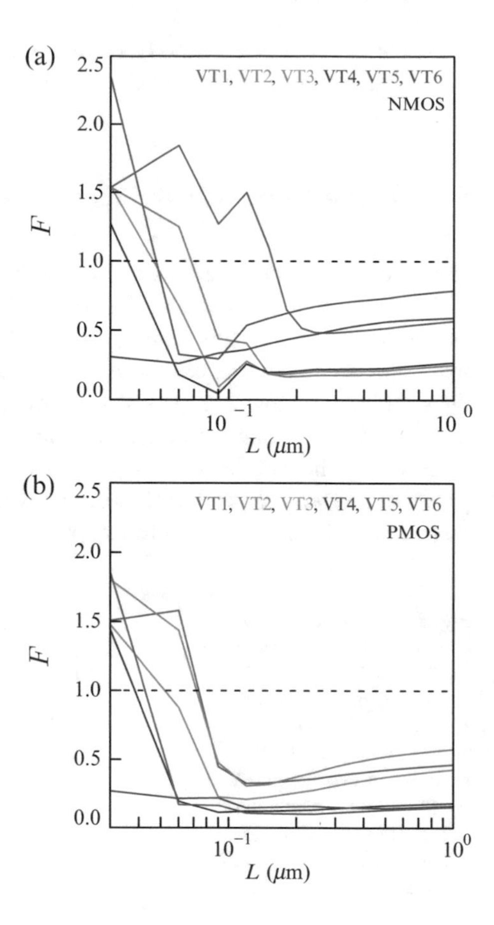

Fig. 9 Benchmark test #4 carried out for (**a**) NMOS and (**b**) PMOS devices. The Fano factor F is plotted versus drawn channel length L, for $V_{GS} = 0$ V. The drain-source voltage is $V_{DS} = 0.9$ V for NMOS and $V_{DS} = -0.9$ V for PMOS. Different colors reflect different threshold voltage variants, as in Fig. 7. The dashed lines indicate $F = 1$, i.e., the expected behavior for all channel lengths

Figure 9 plots the simulated Fano factor as a function of channel length, at a fixed $V_{GS} = 0$ V. We observe some unexpected channel-length dependence of the Fano factor F, which, in turn, depends on the V_{th}-flavor and on the channel type. Note that the sharp edges for some of the devices (especially the VT1 NMOS) are due to the use of the so-called binning. This is an artificial way to capture geometry dependence of model parameters by stitching various geometry ranges together.

It should be noted that the deviation from $F = 1$ in weak inversion cannot be remedied by adjusting the noise parameters, at least not without sacrificing other bias regions. The "holistic" thermal noise model in BSIM4 has two parameters; **RNOIA** and **TNOIA**. The former is needed to get proper long-channel behavior in saturation; see Sect. 5.4. The latter can be used to address noise enhancement for short channels. The behavior in weak inversion then simply follows, and cannot be independently adjusted by the model engineer.

5.4 Benchmark Tests #5, #8, #9, and #10: The Saturation Region

Now we turn to the most important region, the saturation region. This is the typical bias region for, e.g., LNA design. In Fig. 10, we plot the white-noise gamma factor γ as a function of channel length, for various V_{th}-flavors, and for both NMOS and PMOS.

All these devices are expected to show $\gamma \approx 2/3$ for long channel length (test #5). Some of them actually do, but others (VT1 and VT5 NMOS) are a bit too pessimistic, and yet others (VT2, VT3, and VT4 PMOS) are too optimistic.

It should be noted that the successful passing of this benchmark test #5 is not guaranteed by the BSIM4 model equations themselves. It depends on the value of the model parameter β_{noi}, which should be close to its default value of $\sqrt{3}/3 = 0.577$. This can be understood as follows: neglecting g_{DS} as well as g_{mb}, the $S_{I_{\mathrm{D}}}$ Eq. (29) becomes

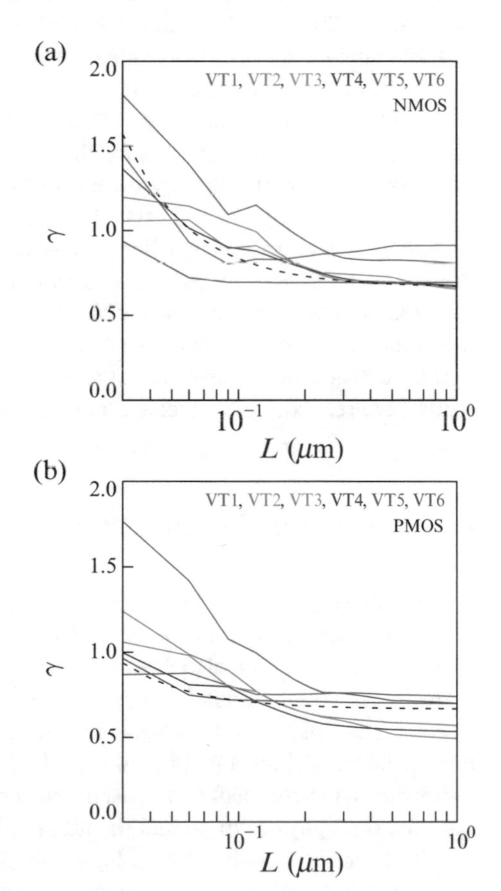

Fig. 10 Benchmark tests #5 and #8 carried out for (**a**) NMOS and (**b**) PMOS devices. The white-noise gamma factor γ in saturation is plotted versus drawn channel length L. The bias conditions are $V_{\mathrm{GS}} = 0.7$ V and $V_{\mathrm{DS}} = 0.9$ V for NMOS, and $V_{\mathrm{GS}} = -0.7$ V and $V_{\mathrm{DS}} = -0.9$ V for PMOS. Different colors reflect different threshold voltage variants, as in Fig. 7. The dashed line is the expected behavior, starting at $\gamma = 2/3$ for long channels, and rising towards short channels according to the trend lines in Fig. 1

$$S_{I_\mathrm{D}} = 4k_\mathrm{B}T\,\frac{V_\mathrm{ds,eff}}{I_\mathrm{ds}}\,\beta_\mathrm{noi}^2\,g_\mathrm{m}^2\,. \tag{33}$$

Using simple long-channel square-law theory, one can derive that $V_\mathrm{ds,eff}/I_\mathrm{ds} = 2/g_\mathrm{D0}$ and $g_\mathrm{m} = g_\mathrm{D0}$. Now, using $\beta_\mathrm{noi} = \sqrt{3}/3$ we get

$$S_{I_\mathrm{D}} = 4k_\mathrm{B}T\,g_\mathrm{D0}\,\frac{2}{3} \tag{34}$$

which corresponds to the well-known Van der Ziel long-channel result and thus benchmark test #5.

Towards shorter channels, γ is expected to rise according to the trend lines derived in Fig. 1 (test #8). These trend lines, multiplied by the long-channel value of 2/3, are displayed as the dashed lines in Fig. 10. It is seen that some devices (e.g., the VT4 NMOS and VT6 PMOS) exhibit close to the expected behavior, whereas others (e.g., the VT5 NMOS and VT5 PMOS) are too pessimistic, and yet another (the VT6 NMOS) is too optimistic. Again, it is not just the absolute accuracy of the noise model that is problematic. The most significant problem is the inconsistency: Fig. 10 could make a circuit designer believe that the VT6 NMOS device is a better low-noise device than the other flavors.

Now let us look at benchmark test #9 ($\gamma_\mathrm{D,NMOS} \geq \gamma_\mathrm{D,PMOS}$), with the help from Fig. 1. In Fig. 10, it is seen that, for the VT4 flavor, this test is passed; both NMOS and PMOS follow the expected trend lines (dashed lines) for short channels. For others, however, e.g., the VT5 flavor, this test is not passed.

Finally, we carry out benchmark test #10; see Fig. 11. Here, the drain current noise spectral density is plotted as a function of drain current. Just like in Fig. 5, both quantities are normalized w.r.t. device width. Clearly, the six V_th-flavors do not coincide, neither for NMOS nor for PMOS, as they are expected to. At a fixed current density, the difference between V_th-flavors may vary up to a factor of 2, which is unrealistic. We conclude that benchmark test #10 is not passed.

6 Summary and Conclusion

In this chapter, we have given an overview of RF noise benchmark tests. These tests are of importance in the development of MOSFET/FinFET compact-model equations. But they are also useful in the assessment of actual model cards, in case there are no experimental data available. We have demonstrated this using a model card for a 28-nm high-k metal-gate technology. Overall, the simulated results, using this model card, show a number of points where the model can be improved. This is partly due to the compact model used: the "holistic" noise model in BSIM4 does not have the right physics to predict the noise behavior correctly in all bias regions and for all geometries, neither does it have enough parameters to empirically correct for that. Another part, however, is due to the way that the model is used by the foundry

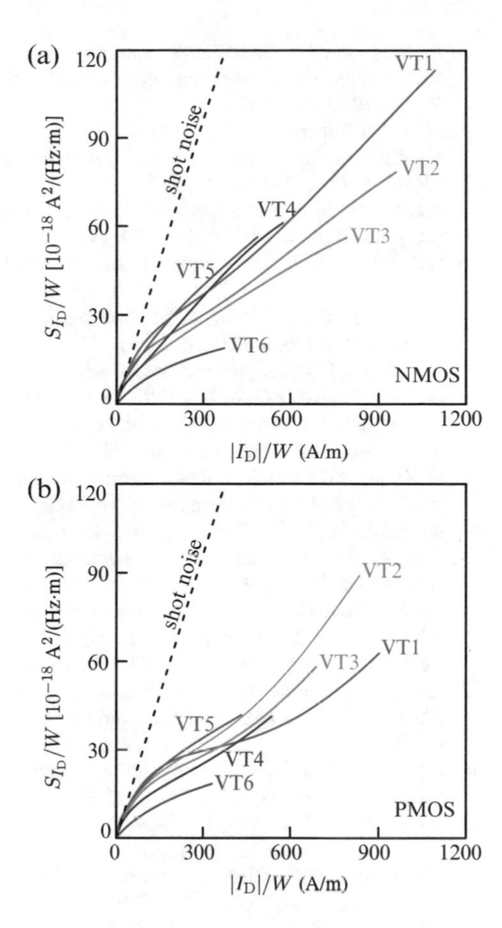

Fig. 11 Benchmark tests #10 carried out for (**a**) NMOS and (**b**) PMOS devices. The drain current noise spectral density is plotted versus drain current, where both quantities are normalized w.r.t. total channel width. For NMOS, $V_{DS} = 0.9$ V, and V_{GS} is swept from 0 V to 0.9 V. For PMOS, $V_{DS} = -0.9$ V, and V_{GS} is swept from 0 V to -0.9 V. Different colors reflect different threshold voltage variants, as shown in the graphs. They are expected to overlap, but do not do that in the simulation model. For reference, the dashed line represents full shot noise

in question: clearly, more attention needs to be payed to guarantee consistency between models for different V_{th} flavors and channel types. Even if the model would not be very accurate in an absolute sense, more consistency would prevent circuit designers from making the wrong choices, either in device or bias condition, in their designs.

References

1. N. Paydavosi, T.H. Morshed, D.D. Lu, W. Yang, M.V. Dunga, X. Xi, J. He, W. Liu, K.M. Cao, X. Jin, J.J. Ou, M. Chan, A.M. Niknejad, C. Hu, BSIM4v4.8.0 MOSFET Model - User's Manual. [Online]. Available: http://bsim.berkeley.edu/models/bsim4/
2. H. Agarwal, C. Gupta, H.-L. Chang, S. Khandelwal, J.P. Duarte, Y.S. Chauhan, S. Salahuddin, C. Hu, BSIM-BULK106.2.0 MOSFET Compact Model - Technical Manual. [Online]. Available: http://bsim.berkeley.edu/models/bsimbulk/

3. A.J. Scholten, L.F. Tiemeijer, R. van Langevelde, R.J. Havens, A.T.A. Zegers-van Duijnhoven, V.C. Venezia, Noise modeling for RF CMOS circuit simulation. IEEE Trans. Electron Devices **50**(3), 618–632 (2003)
4. J.C.J. Paasschens, A.J. Scholten, R. van Langevelde, Generalizations of the Klaassen–Prins equation for calculating the noise of semiconductor devices. IEEE Trans. Electron Devices **52**(11), 2643–2472 (2005)
5. A.J. Scholten, R. van Langevelde, L.F. Tiemeijer, D.B.M. Klaassen, Compact modeling of noise in CMOS, in *IEEE Custom Integrated Circuits Conf. (CICC)*, pp. 711–716 (2006)
6. F.M. Klaassen, J. Prins, Thermal noise of MOS transistors. Philips Res. Rep. **22**, 505–514 (1967)
7. A. van der Ziel, *Noise in Solid State Devices and Circuits* (Wiley, New York, Chichester, Brisbane, Toronto, Singapore, 1986)
8. A.A. Abidi, High-frequency noise measurements on FET's with small dimensions. IEEE Trans. Electron Devices **ED-33**(11), 1801–1805 (1986)
9. R.P. Jindal, Hot-electron effects on channel thermal noise in fine-line nMOS field-effect transistors. IEEE Trans. Electron Devices **ED-33**(9), 1395–1397 (1986)
10. P. Klein, An analytical thermal noise model of deep submicron MOSFET's for circuit simulation with emphasis on the BSIM3v3 SPICE model, in *Proc. 28th Eur. Solid-State Device Research Conf. (ESSDERC)*, pp. 460–463 (1998)
11. P. Klein, An analytical thermal noise model of deep submicron MOSFET's. IEEE Electron Device Lett. **20**(8), 399–401 (1999)
12. A.J. Scholten, H.J. Tromp, L.F. Tiemeijer, R. van Langevelde, R.J. Havens, P.W.H. de Vreede, R.F.M. Roes, P.H. Woerlee, A.H. Montree, D.B.M. Klaassen, Accurate thermal noise model for deep-submicron CMOS, in *IEDM Tech. Dig.*, pp. 155–158 (1999)
13. R. van Langevelde, J.C.J. Paasschens, A.J. Scholten, R.J. Havens, L.F. Tiemeijer, D.B.M. Klaassen, New compact model for induced gate current noise, in *IEDM Tech. Dig.*, pp. 867–870 (2003)
14. C.H. Chen, M.J. Deen, Channel noise modeling of deep submicron MOSFETs. IEEE Trans. Electron Devices **49**(8), 1484–1487 (2002)
15. G.D.J. Smit, A.J. Scholten, R.M.T. Pijper, R. van Langevelde, L.F. Tiemeijer, D.B.M. Klaassen, Experimental demonstration and modeling of excess RF noise in sub-100-nm CMOS technologies. IEEE Electron Device Lett. **31**(8), 884–886 (2010)
16. G.D.J. Smit, A.J. Scholten, R.M.T. Pijper, L.F. Tiemeijer, R. van der Toorn, D.B.M. Klaassen, RF-noise modeling in advanced CMOS technologies. IEEE Trans. Electron Devices **61**(2), 245–254 (2014)
17. G. Gonzalez, *Microwave Transistor Amplifiers* (Prentice-Hall, Upper Saddle River, NJ, 1996)
18. A. Konczakowska, B.M. Wilamowski, Noise in Semiconductor Devices. [Online]. Available: http://www.eng.auburn.edu/~wilambm/pap/2011/K10147_C011.pdf
19. F. Svelto, Noise analysis of submicron PMOS in PWELL. Nucl. Phys. B (Proc. Suppl.) **61B**, 539–544 (1998)
20. S. Tedja, J. Van der Spiegel, H.H. Williams, Analytical and experimental studies of thermal noise in MOSFET's. IEEE Trans. Electron Devices **41**(311), 2069–2075 (1994)
21. C. Beenakker, C. Schönenberger, Quantum shot noise. Phys. Today **56**, 37–42 (2003)
22. R.Q. Twiss, Nyquist's and Thevenin's theorems generalized for nonreciprocal linear networks. J. Appl. Phys. **26**(5), 599–602 (1955)
23. Y.P. Tsividis, *Operation and Modeling of the MOS Transistor* (McGraw-Hill, New York, 1987)
24. R.P. Jindal, Effect of induced gate noise at zero drain bias in field-effect transistors. IEEE Trans. Electron Devices **52**(3), 432–434 (2005)
25. G. Reimbold, P. Gentil, White noise of MOS transistors operating in weak inversion. IEEE Trans. Electron Devices **29**(11), 1722–1725 (1982)
26. S. Chehrazi, R. Bagheri, A.A. Abidi, Noise in passive FET mixers: a simple physical model, in *IEEE Custom Integrated Circuits Conference (CICC)*, pp. 375–378 (2004)

Index

T. Grasser (ed.), *Noise in Nanoscale Semiconductor Devices*,
https://doi.org/10.1007/978-3-030-37500-3

D

G

M

O

P

Q

R

S

U

V

Printed in the United States
by Baker & Taylor Publisher Services